Complete Practice of NCERT for
JEE • 11 & 12

MASTER THE NCERT MATHEMATICS-I

Thousands of MCQs Based on Each & Every Topic of Latest NCERT

Class-XI

BL SHARMA

Co-authored by
Naveen Chandra Joshi
Alokmani Tripathi
Priyanka Sharma

ARIHANT PRAKASHAN (Series), MEERUT

Arihant Prakashan (Series), Meerut

Administrative & Production Offices

Regd. Office

'Ramchhaya' 4577/15, Agarwal Road, Darya Ganj, New Delhi -110002
Tele: 011- 47630600, 43518550

Head Office

Kalindi, TP Nagar, Meerut (UP) - 250002
Tel: 0121-7156203, 7156204

Sales & Support Offices

Agra, Ahmedabad, Bengaluru, Bareilly, Chennai, Delhi, Guwahati, Hyderabad, Jaipur, Jhansi, Kolkata, Lucknow, Nagpur & Pune.

ISBN 978-93-24197-02-3

PRICE ₹435.00

PO No : TXT-XX-XXXXXXX-X-XX

Published by Arihant Publications (India) Ltd.

For further information about the books published by Arihant, log on to www.arihantbooks.com or e-mail at info@arihantbooks.com

Follow us on   

PREFACE

In the present dynamic scenario of competitions, NCERT Books are gaining importance day by day. On an average about 30% questions asked in various competitions are from NCERT. Similarly with the incorporation of Board weightage in competitions the necessity of dual synchronous preparation *i.e.,* to study NCERT for School Exams and for Competitive Exams, is increasing day by day. **Arihant's Master the NCERT Series** is framed with the philosophy of such synchronous preparation in mind.

The book consists of 16 chapters in each of which Questions are framed in Two Categories; First category has questions divided topically while Second category has Miscellaneous Questions from the chapter as a whole. The questions in first category i.e., Topical Questions have been prepared in the same order as in NCERT.

Some special points associated with the distribution of the questions are

- The Questions of First Category have been divided into Two Levels;
  Level I Questions are directly in patch up with the theory of NCERT especially in flow. The solutions of these problems are framed in such a way to give the maximum exposure of NCERT theory.
- Thus, in Level I questions of the book we have tried to connect the student with the theory of NCERT.
- Questions of Level II are compiled to provide the insight of competitions regarding the topic concerned. This set contains elevated levelwise questions i.e., more difficult questions as compared to level I.
- The questions of Second Category i.e., Miscellaneous, are the toughest one and are arranged in random manner to provide tough testing to a student's learning.

Hence, while level one question's are related to learning, the questions of rest of the levels are framed for exposure and practice.

The variety in types of questions framed will be helpful in analysis of self-performance, and exposures to face tough problems of competitions. **Previous Years' Engineering Entrance Questions** have also been incorporated at both the levels so that the students get the exposure of type of questions asked in various competitions on the same topic.

We are thankful to Mr. Katar Singh , Mr. Arvind Agarwal and Mr. Mukesh Kumar for their valuable support in the making of this book.

Huge efforts have been made from our side to make this book error free. Any suggestions for further improvement and pointing out mistakes and misprints in the book are welcome.

AUTHORS

CONTENTS

1

Sets

Sets and Their Representations

Level I

1. The theory of sets was developed by
(a) Einstein (b) Georg Cantor
(c) Aryabhatt (d) Albert Girard

2. A crowd of people is a
(a) collection (b) well-defined collection
(c) not a collection (d) None of these

3. The number which has only two factors one and itself, is called
(a) natural number (b) prime number
(c) integer (d) rational number

4. Which of the following is/are well-defined collections?
(i) The vowels in the English alphabet.
(ii) The collection of books.
(iii) The rivers of India.
(iv) The collection of most talented batsmen of India.
(a) (i) and (ii) (b) Only (i) (c) (i) and (iii) (d) (i) and (iv)

5. Match each of the following sets in Column I with the symbols in Column II and choose the correct option from the codes given below.

Column I	Column II
A. The set of natural numbers.	1. Z
B. The set of integers.	2. R
C. The set of rational numbers.	3. Z^+
D. The set of real numbers.	4. N
E. The set of positive integers.	5. Q

Codes

	A	B	C	D	E
(a)	4	1	2	2	3
(b)	4	1	5	2	3
(c)	5	4	3	1	2
(d)	5	2	3	4	1

6. The set of positive rational numbers is represented by ...X... . Here, X refers to
(a) R (b) R^+
(c) Q^+ (d) Q^-

7. A well-defined collection of objects is called
(a) set (b) function
(c) relation (d) None of these

8. Statement I Objects, elements and members of a set are not synonymous terms.
Statement II Sets are usually denoted by capital letters A, B, C, X, Y, Z, etc.
(a) Statement I is true
(b) Statement II is true
(c) Both are true
(d) Both are false

9. Statement I The elements of a set are represented by small letters a, b, c, x, y, z, etc.
Statement II If a is an element of A, then we can write $a \notin A$.
(a) Statement I is true (b) Statement II is true
(c) Both are true (d) Both are false

10. If b is not an element of A, then we write
(a) $b \in A$ (b) $A \in b$
(c) $b \notin A$ (d) $A \notin b$

11. The number of methods to represent a set are
(a) 2 (b) 3
(c) 4 (d) 5

12. In ...X... form, the elements are being separated by commas and are enclosed within braces { }. Here, X refers to
(a) roster (b) set-builder
(c) roster and set-builder (d) None of these

13. The set of natural numbers which divide 42 is
(a) {1, 2, 3} (b) {1, 2, 3, 6, 7, 14, 21}
(c) {1, 2, 3, 6, 7, 14, 21, 42} (d) {42}

14. The set of all vowels in the English alphabet in roster form is
(a) $\{a, i, e, o, u\}$ (b) $\{a, e, o, u, v\}$
(c) $\{a, e, i, h, v\}$ (d) $\{e, i, o, u, v\}$

15. The set of odd natural numbers is represented by ...X... . Here, X refers to
(a) {1, 3, 5} (b) {1, 3, 5, 7, 9}
(c) {2, 3, 5, 7, 9, ...} (d) {1, 3, 5, 7, 9, ...}

16. The set of all letters of the word 'SCHOOL' is represented by
I. {S, C, H, O, O, L}
II. {S, C, H, O, L}
III. {C, H, L, O, S}
IV. {S, C, H, L}

The correct code is
(a) I and II (b) I, II and III
(c) II and III (d) I, II, III and IV

17. In ...A... form, all the elements of a set possess a single common property which is not possessed by any element outside the set. Here, A refers to
(a) roster (b) set-builder
(c) roster and set-builder (d) None of these

18. The sets represented in set-builder form are
$A = \{x : x$ is a natural number which divides 42$\}$
$B = \{a, e, i, o, u\}$
$C = \{x : x$ is an odd natural number$\}$

Choose the correct option.
(a) A and B (b) A and C
(c) B and C (d) A, B and C

19. The solution set of the equation $x^2 + x - 2 = 0$ in roster form is ...X... . Here, X refers to
(a) {−1, 2} (b) {1, −2}
(c) {−1, 1} (d) {2, −2}

20. The set $\{x : x$ is a positive integer and $x^2 < 40\}$ can be written in roster form as
(a) {1, 3, 5, 7} (b) {1, 2, 3, 4, 5}
(c) {1, 2, 3, 4, 5, 6} (d) {2, 4, 6}

21. The set $A = \{1, 4, 9, 16, 25, \dots\}$ can be written in set-builder form as
(a) $\{x : x$ is the square of an integer$\}$
(b) $\{x : x$ is the square of a natural number$\}$
(c) $\{x : x = n^2$, where $n \in R\}$
(d) None of the above

22. The set $\left\{\frac{1}{2}, \frac{2}{3}, \frac{3}{4}, \frac{4}{5}, \frac{5}{6}, \frac{6}{7}\right\}$ in the set-builder form is
(a) $\{x : x = \frac{n}{n+1}$, where $n \in N$ and $1 < n < 6\}$
(b) $\{x : x = \frac{n}{n+1}$, where $n \in N$ and $1 \le n < 6\}$
(c) $\{x : x = \frac{n}{n+1}$,, where $n \in N$ and $1 \le n \le 6\}$
(d) None of the above

23. Match each of the sets in Column I described in the roster form with the same set in Column II described in the set-builder form and choose the correct option from the codes given below.

Column I	Column II
A. {P, R, I, N, C, A, L}	1. $\{x : x$ is a positive integer and is a divisor of 18$\}$
B. {0}	2. $\{x : x$ is an integer and $x^2 - 9 = 0\}$
C. {1, 2, 3, 6, 9, 18}	3. $\{x : x$ is an integer and $x + 1 = 1\}$
D. {3, −3}	4. $\{x : x$ is a letter of the word 'PRINCIPAL'$\}$

Codes

	A	B	C	D
(a)	1	2	3	4
(b)	4	3	2	1
(c)	4	3	1	2
(d)	3	1	2	4

24. Which of the following are the sets?
I. The collection of all months of a year beginning with the letter J.
II. The collection of ten most talented writers of India.
III. A team of eleven best cricket batsmen of the world.
IV. The collection of all boys in your class.
(a) I and II
(b) I and III
(c) I and IV
(d) I, II and III

25. Which of the following are the sets?

I. A collection of novels written by the writer Munshi Premchand.
II. The collection of all even integers.
III. The collection of questions in this chapter.
IV. A collection of most dangerous animals of the world.

(a) I, II, III and IV (b) I, II and III
(c) I and II (d) II, III and IV

26. Statement I "The collection of all natural numbers less than 100" is a set.

Statement II A set is a well-defined collection of the distinct objects.

(a) Statement I is true; Statement II is true; Statement II is a correct explanation for Statement I
(b) Statement I is true; Statement II is true; Statement II is not a correct explanation for Statement I
(c) Statements I is true; Statement II is false
(d) Statements I is false; Statement II is true

27. Let $A = \{1, 2, 3, 4, 5, 6\}$. Then,

I. $5 \in A$ II. $2 \in A$
III. $8 \in A$ IV. $0 \notin A$
V. $4 \notin A$ VI. $10 \in A$

The correct option is

(a) I, II and IV (b) I, II, III and VI
(c) I, IV and V (d) I, II and V

28. The set $A = \{x : x$ is an integer and $-3 < x < 7\}$ in roster form is ...X... . Here, X refers to

(a) $\{-3, -2, -1, 0, 1, 2, 3, 4, 5, 6, 7\}$
(b) $\{-2, -1, 0, 1, 2, 3, 4, 5, 6\}$
(c) $\{0, 1, 2, 3, 4, 5, 6\}$
(d) $\{1, 2, 3, 4, 5, 6\}$

29. The set $B = \{x : x$ is a natural number less than $6\}$ in roster form is ...Y... . Here, Y refers to

(a) $\{0, 1, 2, 3, 4, 5\}$ (b) $\{1, 2, 3, 4, 5\}$
(c) $\{0, 1, 2, 3, 4, 5, 6\}$ (d) $\{1, 2, 3, 4, 5, 6\}$

30. The set $C = \{x : x$ is a two-digit natural number such that the sum of its digits is $8\}$ in roster form is

(a) $\{17, 26, 35, 44, 53, 62, 71, 80\}$
(b) $\{17, 26, 35, 44, 53, 62, 80\}$
(c) $\{08, 17, 26, 35, 44, 53, 62, 80\}$
(d) $\{17, 26, 35, 53, 62, 80\}$

31. Statement I The set $D = \{x : x$ is a prime number which is a divisor of $60\}$ in roster form is $\{1, 2, 3, 4, 5\}$.

Statement II The set E = the set of all letters in the word 'TRIGONOMETRY', in the roster form is {T, R, I, G, O, N, M, E, Y}.

(a) Statement I is true (b) Statement II is true
(c) Both are true (d) Both are false

32. The set of all letters in the word 'BETTER' in roster form is

(a) {B, E, T, R} (b) {B, T, R}
(c) {B, E, R} (d) {B, R}

33. The set $\{3, 6, 9, 12, ...\}$ in the set-builder form is

(a) $\{x : x$ is a multiple of $6\}$
(b) $\{x : x = 3n$, where $n \in N\}$
(c) $\{x : x = 3n$, where $n \in N$ and $1 < x < 4\}$
(d) $\{x : x = 3n$, where $n \in N$ and $1 \le n \le 4\}$

34. The set $\{2, 4, 8, 16, 32\}$ in the set-builder form is

(a) $\{x : x$ is a multiple of $2\}$
(b) $\{x : x = 2^n$, where $n \in N\}$
(c) $\{x : x = 2^n$, where $n \in N$ and $1 \le n \le 5\}$
(d) $\{x : x = 2n$, where $n \in N$ and $1 \le n \le 5\}$

35. Statement I The set $\{5, 25, 125, 625\}$ in the set-builder form is $\{x : x = 5^n, n \in N\}$.

Statement II The set $\{2, 4, 6, ...\}$ in the set-builder form is $\{x : x = 2^n, n \in N\}$.

(a) Statement I is true (b) Statement II is true
(c) Both are true (d) Both are false

36. Assertion (A) The set $\{1, 4, 9, ... 100\}$ in the set-builder form is $\{x : x = n^2$, where $n \in N$ and $1 \le n \le 10\}$.

Reason (R) In roster form, the order in which the elements are listed is, immaterial.

(a) Both A and R are correct; R is the correct explanation of A
(b) Both A and R are correct; R is not the correct explanation of A
(c) A is correct; R is incorrect
(d) R is correct; A is incorrect

37. The set $A = \{x : x$ is an odd natural number$\}$ in the roster form is ...X... . Here, X refers to

(a) $\{1, 3, 5, 7\}$ (b) $\{1, 3, 5, 7, ...\}$
(c) $\{3, 5, 7, ...\}$ (d) None of these

38. The elements of the set $\{x : x$ is an integer, $-\frac{1}{2} < x < \frac{9}{2}\}$ can be represented as ...Y... . Here, Y refers to

(a) $\{-1, 0, 1, 2, 3, 4\}$ (b) $\{0, 1, 2, 3, 4\}$
(c) $\{0, 1, 2, 3, 4, 5\}$ (d) $\{1, 2, 3, 4\}$

39. The elements of the set $\{x : x$ is an integer, $x^2 \le 4\}$ can be represented as ...Z... . Here, Z refers to

(a) $\{-2, 2\}$ (b) $\{-1, 0, 1\}$
(c) $\{-2, -1, 0, 1, 2\}$ (d) $\{0, 1, 2\}$

40. The element of the set $\{x : x$ is a letter of the word 'LOYAL'$\}$ can be represented as

(a) {L, O, Y, A, L} (b) {L, O, Y, A}
(c) {L, Y} (d) None of these

41. **Statement I** The set $\{x : x$ is a month of a year not having 31 days$\}$ in roster form is {February, April, June, September, November}.

Statement II The set $F = \{x : x$ is a consonant in the English alphabet which precedes $k\}$ in roster form is $F = \{b, c, d, f, g, h, j\}$.

(a) Statement I is true (b) Statement II is true
(c) Both are true (d) Both are false

42. Match each of the sets in Column I in the roster form with the same set in Column II described in set-builder form and choose the correct option from the codes given below.

Column I		Column II	
A.	{1, 2, 3, 6}	1.	$\{x : x$ is a prime number and a divisor of 6$\}$
B.	{2, 3}	2.	$\{x : x$ is an odd natural number less than 10$\}$
C.	{M, A, T, H, E, I, C, S}	3.	$\{x : x$ is a natural number and divisor of 6$\}$
D.	{1, 3, 5, 7, 9}	4.	$\{x : x$ is a letter of the word 'MATHEMATICS'$\}$

Codes

	A	B	C	D		A	B	C	D
(a)	3	1	4	2	(b)	1	1	4	2
(c)	3	3	4	2	(d)	2	3	4	1

Level II

43. The set $\{x : x^2 = x, x \in N\}$ can be expressed in roster form as

(a) {0, 1} (b) {1} (c) {0} (d) { }

44. The set $\left\{x : \dfrac{x-1}{x+3} = 4, x \in R\right\}$ in roster form is

(a) $\left\{\dfrac{-11}{3}\right\}$ (b) $\left\{\dfrac{-12}{3}\right\}$
(c) $\left\{\dfrac{-13}{3}\right\}$ (d) $\left\{\dfrac{-14}{3}\right\}$

45. The set $\{x : x^4 - 5x^2 + 6 = 0, x \in R\}$ in roster form is ...X... . Here, X refers to

(a) $\{\sqrt{2}, \sqrt{3}\}$ (b) $\{-\sqrt{3}, -\sqrt{2}\}$
(c) $\{-\sqrt{3}, -\sqrt{2}, \sqrt{2}, \sqrt{3}\}$ (d) $\{-\sqrt{2}, \sqrt{2}\}$

46. Which of the following are true?

I. $35 \in \{x : x$ has exactly four positive factors$\}$
II. $128 \in \{x :$ the sum of all the positive factors of x is $2x\}$
III. $2 \notin \{x : x^4 - 5x^3 + 2x^2 - 112x + 6 = 0\}$
IV. $5 \notin \{x : x$ is a prime factor of 180$\}$

(a) I and II (b) I and III
(c) I, II and IIII (d) II, III and IV

47. The set $\{x :$ where $x \in N$ and $1 < x \leq 2\}$ can be written as ...A... . Here, A refers to

(a) {1, 2} (b) {1} (c) {2} (d) { }

Directions (Q. Nos. 48-50) *If* $Y = \{1, 2, 3, \ldots, 10\}$ *and a represents any element of Y.*

48. The set containing all the elements which satisfy $a \in Y$ but $a^2 \notin Y$ is

(a) {1, 2, 3} (b) {3, 4, 5, 6}
(c) {4, 5, 6, 7, 8, 9, 10} (d) {5, 6, 7, 8, 9, 10}

49. The set containing all the elements satisfy $a + 2 = 6$, $a \in Y$ is

(a) {2} (b) {3}
(c) {4} (d) {5}

50. The set containing all the elements satisfy a is less than 5 and $a \in Y$ is

(a) {1, 2, 3, 4} (b) {1, 2, 3, 4, 5}
(c) {5, 6, 7, 8, 9, 10} (d) {6, 7, 8, 9, 10}

51. If $Y = \{x : x$ is a positive factor of the number $2^{p-1}(2^p - 1)$, where $2^p - 1$ is a prime number$\}$, then Y can be represented in roster form as [NCERT Exemplar]

(a) {2}
(b) {1, 2}
(c) $\{1, 2, 2^2, \ldots, 2^{p-1}\}$
(d) $\{1, 2, 2^2, 2^3, \ldots, 2^{p-1}, (2^p - 1)\}$

52. The set $\{x : x$ is a positive integer less than 6 and $3^x - 1$ is an even number$\}$ in roster form is

(a) {1, 2, 3, 4, 5} (b) {1, 2, 3, 4, 5, 6}
(c) {2, 4, 6} (d) {1, 3, 5}

53. Which of the following is/are true?

I. $17 \notin \{x : x$ has exactly two positive factors$\}$
II. $7745 \in \{x : x$ is a multiple of 36$\}$

(a) Only I is true (b) Only II is true
(c) Both are true (d) Both are false

54. **Assertion** (A) The set $\{x : x \in N$ and $2^x + 3^x$ is an odd natural number$\}$ in roster form is $\{0, 1, 2, 3, \ldots\}$.

Reason (R) The sum of an even number and an odd number is always odd.

(a) Both A and R are correct; R is the correct explanation of A
(b) Both A and R are correct; R is not the correct explanation of A
(c) A is correct; R is incorrect
(d) R is correct; A is incorrect

55. The set $\{x : x = e^n, n \in N\}$ in roster form is

(a) $\{1, e, e^2, e^3, \ldots\}$ (b) $\{e, e^2, e^3, \ldots\}$
(c) R (d) None of these

The Empty Set

Level I

56. If $A = \{x : x$ is a student of class XI presently studying in a school$\}$, then the number of elements in set A are

(a) finite (b) infinite
(c) does not exist (d) None of these

57. If $B = \{x : x$ is a student presently studying in both classes X and XI$\}$. Then, the number of elements in set B are

(a) finite (b) infinite
(c) zero (d) None of these

58. A set which does not contain any element, is called

(a) empty or null or void set (b) finite set
(c) infinite set (d) power set

59. The empty set is represented by

I. ϕ II. $\{\phi\}$
III. $\{\ \}$ IV. $\{\{\ \}\}$

(a) I and II (b) I and III (c) II and III (d) I and IV

60. **Statement I** The set $\{x : x$ is a student of class XI studying in secondary school$\}$ is not empty set.

Statement II The set $\{x : x$ is a student presently studying in both classes X and XI$\}$ is an empty set.

(a) Statement I is true (d) Statement II is true
(c) Both are true (d) Both are false

61. Which of the following is/are true?

I. If $A = \{x : 1 < x < 2,\ x \in N\}$, then A is the empty set.

II. If $B = \{x : x^2 - 2 = 0$ and x is a rational number$\}$, then B is the empty set.

(a) Only I is true (b) Only II is true
(c) Both I and II are true (d) None of these

62. **Statement I** The set $A = \{x : x$ is an even prime number greater than $2\}$ is the empty set.

Statement II The set $B = \{x : x^2 = 4,\ x$ is odd$\}$ is not an empty set.

(a) Statement I is true (b) Statement II is true
(c) Both are true (d) Both are false

63. Which of the following are the examples of null set?

I. Set of odd natural numbers divisible by 2.

II. Set of even prime numbers.

(a) Only I (b) Only II
(c) Both I and II (d) None of these

64. Identify the null set from the following

I. $\{x : x$ is a natural number, $x < 5$ and $x > 7\}$

II. $\{y : y$ is a point common to any two parallel lines$\}$

(a) Only I (b) Only II
(c) Both I and II (d) None of these

Level II

65. If a set is denoted as $A = \phi$, then number of elements in A is

(a) 0 (b) 1 (c) 2 (d) 3

66. Which of the following is/are empty set?

$A = \{x : x^2 + x + 1 = 0$ and $x \in Z\}$

$B = \{x : x$ is a prime number and $19 < x < 23\}$

(a) A (b) B
(c) Both A and B (d) None of these

67. The set $A = \{x : x^4 - x^3 - x^2 = 0$ and $x \in N\}$ represent a ...X... . Here, X refers to

(a) null set (b) singleton set
(c) infinite set (d) None of these

68. **Assertion** (A) The set $A = \{x : x$ is a natural number and $5 < x < 3\}$ is the empty set.

Reason (R) An empty set contains no elements.

(a) Both A and R are correct; R is the correct explanation of A
(b) Both A and R are correct; R is not the correct explanation of A
(c) A is correct; R is incorrect
(d) R is correct; A is incorrect

69. **Statement I** The set $\{x : x$ is a real number and $x^2 - 1 = 0\}$ is the empty set.

Statement II The set $A = \{x : x \in R,\ x^2 = 16$ and $2x = 6\}$ is an empty set.

(a) Statement I is true (b) Statement II is true
(c) Both are true (d) Both are false

70. Consider the sets

$A = \{x : x$ is a real number and $x^2 + 1 = 0\}$
$B = \{x : x$ is a real number and $x^2 - 16 = 0\}$
$C = \{x : x$ is a real number and $x^2 = x + 2\}$

Which of the following is/are null set?

(a) A (b) A and B
(c) A and C (d) A, B and C

Finite and Infinite Sets

Level I

71. Let $A = \{1, 2, 3, 4, 5\}$. Then, the number of elements in A are

(a) 1 (b) 2 (c) 3 (d) 5

72. By number of elements of a set S, we mean the

(a) number of same elements in S
(b) number of distinct elements in S
(c) total elements in S
(d) None of the above

73. The number of distinct elements of a set S is represented by ...X... . Here, X refers to

(a) S (b) $n(S)$ (c) $\{S\}$ (d) $n\{S\}$

74. Let S be any set and $n(S)$ is a natural number. Then, the set S is ...Y... . Here, Y refers to

(a) empty set (b) infinite set
(c) non-empty finite set (d) None of these

75. Statement I The set $A = \{a, b, c, d, e, g\}$ is finite set.

Statement II The set $B = \{$men living presently in different parts of the world$\}$ is finite set.

(a) Statement I is true (b) Statement II is true
(c) Both are true (d) Both are false

76. Which of the following sets are finite?

W = Set of days of the week
S = Set of the solutions of the equation $x^2 - 16 = 0$
G = Set of points on a line

(a) W and S (b) W and G
(c) S and G (d) W, S and G

77. Which of the following sets is/are infinite?

I. $\{x : x \in N \text{ and } (x-1)(x-2) = 0\}$
II. $\{x : x \in N \text{ and } x \text{ is prime}\}$
III. $\{x : x \in N \text{ and } x^2 = 4\}$

(a) Only I (b) Only II (c) I and II (d) II and III

78. Which of the following are true?

I. $\{x : x \in N \text{ and } 2x - 1 = 0\}$ is finite.
II. $\{x : x \in N \text{ and } x \text{ is odd}\}$ is finite.
III. $\{x : x \in Z \text{ and } x \text{ is positive}\}$ is infinite.

(a) I and II (b) I and III
(c) II and III (d) I, II and III

79. The set of months of a year is ...X... set. Here, X refers to

(a) empty (b) finite (c) infinite (d) singleton

80. The set $\{1, 2, 3, ...\}$ is ...Y... set. Here, Y refers to

(a) null (b) finite (c) infinite (d) singleton

81. Assertion (A) The set $\{1, 2, 3, ..., 99, 100\}$ is finite.

Reason (R) A set consists of a definite number of elements is called finite.

(a) Both A and R are correct; R is the correct explanation of A
(b) Both A and R are correct; R is not the correct explanation of A
(c) A is correct; R is incorrect
(d) R is correct; A is incorrect

82. Statement I The set of positive integers greater than 100 is infinite.

Statement II The set of prime numbers less than 99 is finite.

(a) Statement I is true (b) Statement II is true
(c) Both are true (d) Both are false

83. Select the infinite set from the following.

I. The set of lines which are parallel to the X-axis.
II. The set of numbers which are multiples of 5.
III. The set of letters in the English alphabet.

(a) I and II (b) II and III
(c) I and III (d) None of these

84. State which of the following is/are true?

I. The set of animals living on the Earth is finite.
II. The set of circles passing through the origin (0, 0) is infinite.

(a) Only I (b) Only II
(c) I and II (d) None of these

Level II

85. The set $\{x : x^4 = 1 \text{ and } x \in R\}$ is ...X... set. Here, X refers to

(a) empty
(b) finite
(c) infinite
(d) singleton

86. The set $\{x : x \in R \text{ and } 0 < x < 1\}$ is ...Y... set. Here, Y refers to

(a) empty
(b) infinite
(c) finite
(d) None of the above

87. **Statement I** The set of concentric circles in a plane is infinite.

Statement II The set $\{x : x^2 - 3 = 0$ and x is rational$\}$ is finite.

(a) Statement I is true (b) Statement II is true
(c) Both are true (d) Both are false

Directions (Q. Nos. 89-89) *Let* $A = \{a_n : n \in N, a_{n+1} = 3a_n$ *and* $a_1 = 1\}$.

88. The set A can be represent in roster form as

(a) $\{1, 2, 3, 4, ...\}$ (b) $\{1, 2^2, 3^2, 4^2, ...\}$
(c) $\{1, 3, 3^2, 3^3, ...\}$ (d) $\{1, 3, 3, 3, ...\}$

89. The set A is ...X... set. Here, X refers to

(a) empty (b) finite (c) infinite (d) singleton

Directions (Q. Nos. 90-92) *Let* $A = \{3, 4, 5, 6, 7\}$. *If n represents any member of A.*

90. The set containing all numbers represented by $n + 1$ is

(a) $\{2, 3, 4, 5, 6\}$ (b) $\{3, 4, 5, 6, 7\}$
(c) $\{4, 5, 6, 7, 8\}$ (d) $\{5, 6, 7, 8, 9\}$

91. The set containing all numbers represented by $n^2 - 1$ is

(a) $\{8, 15, 24, 35, 48\}$ (b) $\{10, 17, 26, 37, 50\}$
(c) $\{9, 16, 25, 36, 49\}$ (d) $\{3, 4, 5, 6, 7\}$

92. The set A is ...X... set. Here, X refers to

(a) empty (b) finite
(c) infinite (d) universal set

93. **Assertion** (A) The set $X = \{9^n - 8n - 1 : n \in N\}$ is a finite set.

Reason (R) The set containing definite number of elements is finite set.

(a) Both A and R are correct; R is the correct explanation of A
(b) Both A and R are correct; R is not the correct explanation of A
(c) A is correct; R is incorrect
(d) R is correct; A is incorrect

94. Let $S = \{x : x$ is a positive multiple of 3 less than $100\}$ and $P = \{x : x$ is a prime number less than $20\}$. Then, $n(S) + n(P)$ is equal to [NCERT Exemplar]

(a) 34 (b) 41
(c) 33 (d) 30

Equal Sets

Level I

95. Two sets A and B are said to be equal, if they have exactly the ...X... elements. Here, X refers to

(a) same (b) different (c) equal (d) unequal

96. If two sets A and B are equal, then

(a) $A \in B$ (b) $A = B$ (c) $A \neq B$ (d) $A \notin B$

97. If $A \neq B$, then we can say two sets A and B are ...Y... . Here, Y refers to

(a) equal (b) unequal
(c) disjoint (d) None of these

98. Let $A = \{1, 2, 3, 4\}$ and $B = \{3, 1, 4, 2\}$. Then,

(a) $A \in B$ (b) $A = B$
(c) $A \neq B$ (d) $A \notin B$

99. **Statement I** If A = set of prime numbers less than 6 and B = set of prime factors of 30, then $A = B$.

Statement II If $P = \{1, 2, 3\}$ and $Q = \{2, 2, 1, 3, 3\}$, then P and Q are not equal.

(a) Statement I is true (b) Statement II is true
(c) Both are true (d) Both are false

100. Choose the pair of equal sets from the following

$A = \{0\}$, $B = \{x : x > 15$ and $x < 5\}$,
$C = \{x : x - 5 = 0\}$, $D = \{x : x^2 = 25\}$,
$E = \{x : x$ is an integral positive root of the equation $x^2 - 2x - 15 = 0\}$

(a) A and B (b) C and D (c) C and E (d) B and C

101. If A = the set of letters in 'ALLOY' and B = the set of letters in 'LOYAL', then A and B are ...X... . Here, X refers to

(a) equal (b) unequal
(c) disjoint (d) None of these

102. Let $A = \{x : x \in Z$ and $x^2 \leq 4\}$ and $B = \{x : x \in R$ and $x^2 - 3x + 2 = 0\}$. Then,

(a) $A = B$ (b) $A \neq B$ (c) $A \in B$ (d) $A \notin B$

103. Which of the following are true?

I. $\{a, b, c, d\} = \{d, c, b, a\}$
II. $\{4, 8, 12, 16\} = \{8, 4, 16, 18\}$
III. $\{2, 4, 6, 8, 10\} = \{x : x$ is positive even integer and $x \leq 10\}$

IV. $\{x : x \text{ is a multiple of } 10\} = \{10, 15, 20, 25, 30, ...\}$
V. $\{4, 8, 12, 14\} = \{2, 4, 8, 12\}$

(a) I and II are true (b) II and III are true
(c) I and III are true (d) II, IV and V are true

104. Which of the following is/are correct?

I. $\{x : x$ is letter in the word FOLLOW$\} = \{y : y$ is letter in the word WOLF$\}$
II. $\{x : x \text{ is a solution of } x^2 + 5x + 6 = 0\} = \{2, 3\}$
III. $\{1, 2, 3, 4\} = \{3, 1, 4, 2\}$
IV. $\{1, -1\} = \{-1, 1\}$
V. $\{O, G\} = \{0, 1\}$

(a) I and II are correct (b) I, III and IV are correct
(c) Only V is correct (d) All are correct

Level II

105. If $A = \{1, 2, 3\}$, $B = \{x \in R : x^2 - 2x + 1 = 0\}$, $C = \{1, 2, 2, 3\}$ and $D = \{x \in R : x^3 - 6x^2 + 11x - 6 = 0\}$, then the equal sets are

(a) A and B (b) A and C
(c) A, B and C (d) A, C and D

106. **Assertion** (A) If $A = \{0, 1, 2\}$ and $B = \{x : x \in R, 0 \le x \le 2\}$, then $A = B$.

Reason (R) If two sets are equal, then they have exactly the same elements.

(a) Both A and R are correct; R is the correct explanation of A
(b) Both A and R are correct; R is not the correct explanation of A
(c) A is correct; R is incorrect
(d) R is correct; A is incorrect

107. **Statement I** Let $X = \{1, 2, 3, 4, 5\}$. If n represents any member of X. Then, $A = \{n : 2n \notin X\}$ and $B = \{n : n \text{ is greater than } 4\}$ are not equal.

Statement II The set $C = \{x : x \in Z, -4 < x \le 6\}$ and the set $D = \{x : x \in N, -4 < x \le 6\}$ are equal.

(a) Only Statement I is true (b) Only Statement II is true
(c) Both are true (d) Both are false

108. Match each of the sets in Column I equal with the sets in Column II and choose the correct option from the codes given below.

Column I	Column II
A. $\{x : x$ is a letter in the word 'FINITE'$\}$	1. $\{x : x \in N \text{ and } x^2 + 25 = 0\}$
B. $\{x : x \in N, x \text{ is odd}, x < 5\}$	2. $\{x : x \in N \text{ and } x^2 + x - 6 = 0\}$
C. $\{x : x$ is an even prime number$\}$	3. $\{x : x$ is a letter in the word 'INFINITE'$\}$
D. $\{x \in R : x > x\}$	4. $\{1, 1, 3\}$

Codes

	A	B	C	D		A	B	C	D
(a)	3	4	2	1	(b)	1	2	3	4
(c)	2	1	4	3	(d)	4	3	1	2

Subset

Level I

109. Consider

X = Set of all students in your school.
Y = Set of all students in your class.

Then, which of the following is true?

(a) Every element of Y is also an element of X
(b) Every element of X is also an element of Y
(c) Every element of Y is not an element of X
(d) Every element of X is not an element of Y

110. If every element of A is also an element of B, then

(a) A is the subset of B
(b) B is the subset of A
(c) A is not the subset of B
(d) None of these

111. Match the following statements in Column I with their symbolic forms in Column II and choose the correct option from the codes given below.

Column I	Column II
A. A is a subset of B.	1. if and only if
B. If $A \subset B$ and $B \subset A$, then	2. $A \subset B$
C. A is not a subset of B.	3. $A = B$
D. If $a \in A \Rightarrow a \in B$, then	4. $A \not\subset B$
E. The symbol "$\Leftrightarrow$" means	

Codes

	A	B	C	D	E		A	B	C	D	E
(a)	4	3	1	2	3	(b)	2	3	4	2	1
(c)	1	2	3	4	3	(d)	4	3	2	1	4

112. Which of the followinig is/are true?

I. Every set A is a subset of itself.
II. Empty set is a subset of every set.

(a) Only I is true (b) Only II is true
(c) Both I and II are true (d) None of these

113. The set of rational number Q, ...X... the set of real number R. Here, X refers to

(a) $\in$ (b) $\subset$
(c) $\not\subset$ (d) $\notin$

114. Let $A = \{1, 3, 5\}$ and $B = \{x : x$ is an odd natural number less than 6$\}$.

Then, which of the following are true?

I. $A \subset B$ II. $B \subset A$
III. $A = B$ IV. $A \not\subset B$

(a) I and II are true (b) I and III are true
(c) I, II and III are true (d) I, II and IV are true

115. **Assertion** (A) Let $A = \{1, 2, 3\}$ and $B = \{1, 2, 3, 4\}$. Then, $A \subset B$.

Reason (R) If every element of X is also an element of Y, then X is a subset of Y.

(a) Both A and R are correct; R is the correct explanation of A
(b) Both A and R are correct; R is not the correct explanation of A
(c) A is correct; R is incorrect
(d) R is correct; A is incorrect

116. Match each of the subset in Column I with a set in Column II and choose the correct option from the codes given below.

Column I	Column II
A. $\{x : x$ is a student of class XI of your school$\}$	1. The set of real numbers
B. $\{x : x$ is a circle with radius 1$\}$	2. $\{x : x$ is a student of your school$\}$
C. $\{x : x$ is an even natural number$\}$	3. $\{x : x$ is a circle in any plane$\}$
D. $\{x : x \in R, 0 < x < 7\}$	

Codes

	A	B	C	D		A	B	C	D
(a)	1	2	3	2	(b)	3	1	2	3
(c)	2	3	1	1	(d)	2	1	3	2

117. If $A \subset B$ and $A \neq B$, then

(a) A is called a proper subset of B
(b) A is called a super set of B
(c) A is not a subset of B
(d) B is a subset of A

118. Which of the following is/are correct?

I. $\{2, 3, 4\} \not\subset \{1, 2, 3, 4, 5\}$
II. $\{a, b, c\} \not\subset \{b, c, d\}$
III. $\{x : x$ is an equilateral triangle in a plane$\} \subset \{x : x$ is a triangle in the same plane$\}$
IV. $\{x : x$ is an even natural number$\} \not\subset \{x : x$ is an integer$\}$

(a) I and III are correct (b) II and III are correct
(c) Only IV is correct (d) All are correct

119. A set has only one element, is called

(a) singleton set (b) empty set
(c) infinite set (d) None of these

Directions (Q. Nos. 120-122) *Consider the sets* $\phi, A = \{1, 3\}, B = \{1, 5, 9\}$ *and* $C = \{1, 3, 5, 7, 9\}$.

120. ϕ ...X... B. Here, X refers to

(a) $\in$ (b) $\notin$ (c) $\subset$ (d) $\not\subset$

121. A ...Y... B. Here, Y refers to

(a) $\in$ (b) $\notin$
(c) $\subset$ (d) $\not\subset$

122. A ...Z... C, and B...K...C. Here, Z and K refer to respectively

(a) $\in, \in$ b) $\notin, \notin$ (c) $\subset, \subset$ (d) $\not\subset, \not\subset$

123. Let A, B, C be three sets. If $A \in B$ and $B \subset C$, then

(a) $A \subset C$ (b) $A \not\subset C$
(c) $A \in C$ (d) $A \notin C$

124. Which of the following are true?

I. The set of natural numbers N is the subset of the set of real numbers R.
II. The set of natural numbers N is the subset of the set of integers Z.
III. The set of rational numbers Q is the subset of the set of integers Z.

(a) I and II are true (b) II and III are true
(c) I and III are true (d) All are true

125. Let N = Set of natural numbers.

Z = Set of integers.
Q = Set of rational numbers.
R = Set of real numbers.
T = Set of irrational numbers.

Then,

(a) $Z \subset N \subset Q \subset R$ (b) $N \subset Z \subset Q \subset R, T \subset R$
(c) $R \subset Q \subset Z \subset N$ (d) $R \subset Q \subset N \subset Z, N \subset T$

126. The set of real numbers $\{x : a < x < b\}$ is called

(a) open interval (b) closed interval
(c) semi-open interval (d) semi-closed interval

127. Match the following sets in Column I with the intervals in Column II and choose the correct option from the codes given below.

Column I	Column II
A. $\{x : x \in R, a < x < b\}$	1. $(a, b]$
B. $\{x \in R : a \le x \le b\}$	2. $[a, b)$
C. The set of real numbers x such that $a \le x < b$	3. (a, b)
D. $\{x : x \in R$ and $a < x \le b\}$	4. $[a, b]$

Codes

	A	B	C	D
(a)	4	1	2	3
(b)	2	3	4	1
(c)	1	2	3	4
(d)	3	4	2	1

128. If $A = (-3, 5)$ and $B = [-7, 9]$, then

(a) $A = B$ (b) $A \subset B$
(c) $B \subset A$ (d) None of these

129. The set of non-negative real numbers is represented by ...X... . Here, X refers to

(a) $(0, \infty)$ (b) $[0, \infty]$ (c) $[0, \infty)$ (d) $(0, \infty]$

130. The set of negative real numbers is denoted by ...Y... . Here, Y refers to

(a) $(-\infty, 0)$ (b) $[-\infty, 0]$
(c) $(-\infty, 0]$ (d) $[-\infty, 0)$

131. The interval represented by

(a) (a, b) (b) $[a, b]$
(c) $[a, b)$ (d) $(a, b]$

132. The interval $[a, b)$ is represented on the number line as

(a)  (b) 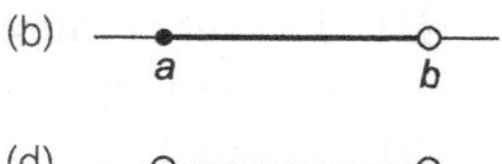

(c) (d)

133. The length of interval (a, b) is

(a) a (b) $a - b$
(c) $b - a$ (d) b

134. Statement I The interval $\{x : x \in R, -4 < x \le 6\}$ is represented by $(-4, 6]$.

Statement II The interval $\{x : x \in R, -12 < x < -10\}$ is represented by $[-12, -10]$.

(a) Statement I is true
(b) Statement II is true
(c) Both are true
(d) Both are false

135. Which of the following are true?

I. $\{a, b\} \not\subset \{b, c, a\}$
II. $\{a, e\} \subset \{x : x$ is a vowel in the English alphabet$\}$
III. $\{1, 2, 3\} \subset \{1, 3, 5\}$
IV. $\{a\} \subset \{a, b, c\}$
V. $\{a\} \in \{a, b, c\}$
VI. $\{x : x$ is an even natural number less than $6\} \subset \{x : x$ is a natural number which divides $36\}$.

(a) I, II, IV, V (b) II, IV, VI
(c) I, III, V (d) II, IV, V, VI

Directions (Q. Nos. 136-137) *Consider* $A = \{1, 2, \{3, 4\}, 5\}$

136. Which of the following are true?

I. $\{3, 4\} \subset A$ II. $\{\{3, 4\}\} \subset A$
III. $1 \subset A$ IV. $\{1, 2, 3\} \subset A$
V. $1 \in A$ VI. $\{1, 2, 5\} \subset A$

(a) II, V, VI (b) I, V, IV, VI (c) I, IV, V (d) II, IV, V

137. Which of the following are incorrect?

I. $\{3, 4\} \in A$ II. $\{1, 2, 5\} \in A$
III. $\phi \in A$ IV. $\phi \subset A$
V. $\{\phi\} \subset A$

(a) I, II, V (b) II, III, V
(c) II, III, IV (d) I, IV, V

138. Match the following sets given in Column I with the subsets in Column II and choose the correct option from the codes given below.

Column I	Column II
A. $\{a\}$	1. ϕ
B. $\{a, b\}$	2. $\phi, \{a\}$
C. $\{1, 2, 3\}$	3. $\phi, \{a\}, \{b\}, \{a, b\}$
D. ϕ	4. $\{\ \}, \{1\}, \{2\}, \{3\}, \{1, 2\}, \{2, 3\}, \{1, 3\}, \{1, 2, 3\}$

Codes

	A	B	C	D		A	B	C	D
(a)	4	1	2	3	(b)	1	2	3	4
(c)	2	3	4	1	(d)	3	4	1	2

139. Match the following intervals in Column I with the sets in Column II and choose the correct option from the codes given below.

Column I	Column II
A. $(-3, 0)$	1. $\{x : x \in R, 6 < x \le 12\}$
B. $[6, 12]$	2. $\{x : x \in R, -23 \le x < 5\}$
C. $(6, 12]$	3. $\{x : x \in R, -3 < x < 0\}$
D. $[-23, 5)$	4. $\{x : x \in R, 6 \le x \le 12\}$

Codes

	A	B	C	D		A	B	C	D
(a)	1	2	3	4	(b)	3	4	1	2
(c)	4	1	2	3	(d)	2	3	4	1

Level II

140. If $A = \{1, 2, 3\}$, then the number of proper subsets is

(a) 8 (b) 7 (c) 6 (d) 5

141. Which set is the subset of all given sets?

(a) $\{1, 2, 3, 4, \ldots\}$ (b) $\{1\}$
(c) $\{0\}$ (d) $\{\ \}$

142. Which of the following is true? [HOTS]

(a) $a \in \{\{a\}, b\}$ (b) $\{b, c\} \subset \{a, \{b, c\}\}$
(c) $\{a, b\} \subset \{a, \{b, c\}\}$ (d) None of these

143. Match the following subsets in Column I with their super sets in Column II and choose the correct option from the codes given below. [HOTS]

Column I	Column II
A. $\{2, \phi\}$	1. $\{\phi, \{\phi\}, \{1, \phi\}\}$
B. $\{\phi, \{1, \phi\}\}$	2. $\{\phi, 1, 4, 5\}$
C. $\{\{4, 5\}\}$	3. $\{\phi, \{1\}, 2\}$
D. $\{4, 5\}$	4. $\{1, 2, \{4, 5\}\}$
E. $\{1, \{4, 5\}\}$	
F. $\{\phi, \{1\}\}$	

Codes

	A	B	C	D	E	F
(a)	1	2	4	3	2	1
(b)	4	3	2	1	4	3
(c)	3	1	4	2	4	3
(d)	1	2	3	4	3	2

144. The number of non-empty subsets of the set $\{1, 2, 3, 4\}$ is [AMU 1998]

(a) 15 (b) 14 (c) 16 (d) 17

145. **Assertion** (A) The subsets of the set $\{1, \{2\}\}$ are $\{\ \}$, $\{1\}$, $\{\{2\}\}$ and $\{1, \{2\}\}$.

Reason (R) The total number of proper subsets of a set containing n elements is $2^n - 1$.

(a) Both A and R are correct; R is the correct explanation of A
(b) Both A and R are correct; R is not the correct explanation of A
(c) A is correct; R is incorrect
(d) R is correct; A is incorrect

146. Two finite sets have m and n elements respectively. The total number of subsets of first set is 56 more than the total number of subsets of the second set. The value of m and n respectively are [NCERT Exemplar]

(a) 7, 6 (b) 5, 1 (c) 6, 3 (d) 8, 7

147. Let P be the set of prime numbers and $S = \{t : 2^t - 1 \text{ is a prime}\}$. Then,

(a) $S \subset P$ (b) $P \subset S$
(c) $S = P$ (d) None of these

148. If $A \subset B$, $B \subset C$ and $C \subset A$, then

(a) $A = B$
(b) $B = C$
(c) $A = C$
(d) None of the above

149. **Statement I** If $A = \{x : x \in R, x \geq 4\}$ and $B = \{x : x \in R, x > 5\}$, then $B \subset A$.

Statement II If $X = \{8^n - 7n - 1 : n \in N\}$ and $Y = \{49(n-1) : n \in N\}$, then $X \subset Y$.

(a) Statement I is true
(b) Statement II is true
(c) Both are true
(d) Both are false

150. Match the following intervals in Column I with the sets in the set-builder form in Column II and choose the correct option from the codes given below.

Column I	Column II
A. $(-5, 0)$	1. $\{x : x \in R \text{ and } -20 \leq x < 3\}$
B. $[3, 4]$	2. $\{x : x \in R \text{ and } -5 < x \leq 11\}$
C. $(-5, 11]$	3. $\{x : x \in R \text{ and } -5 < x < 0\}$
D. $[-20, 3)$	4. $\{x : x \in R \text{ and } 3 \leq x \leq 4\}$

Codes

	A	B	C	D
(a)	2	1	4	3
(b)	3	4	2	1
(c)	1	2	3	4
(d)	4	3	2	1

Power Set

Level I

151. Power set of a set A is the set of all the ...X... of the set A. Here, X refers to

(a) sets
(b) subsets
(c) proper subsets
(d) None of the above

152. The power set of a set A is denoted by

(a) P (b) $P\{A\}$
(c) $P(A)$ (d) (A)

153. **Assertion** (A) The power set of the set $\{1, 2\}$ is the set $\{\phi, \{1\}, \{2\}, \{1,2\}\}$.

Reason (R) The power set is set of all subsets of the set.

(a) Both A and R are correct; R is the correct explanation of A
(b) Both A and R are correct; R is not the correct explanation of A
(c) A is correct; R is incorrect
(d) R is correct; A is incorrect

154. The number of elements of the power set of a set containing m elements is

(a) 2^{m-1} (b) m (c) m^2 (d) 2^m

155. If $A = \phi$, then number of elements in $P(A)$ is ...X... . Here, X refers to

(a) 0 (b) 1 (c) 2 (d) 3

Level II

156. If $A = \{a, \{b\}\}$, then $P(A)$ is

(a) $\{\phi, \{a\}, \{b\}, \{a, b\}\}$ (b) $\{\phi, \{a\}, \{\{b\}\}, \{a, b\}\}$
(c) $\{\phi, \{a\}, \{\{b\}\}, \{a, \{b\}\}\}$ (d) None of these

157. The number of elements in $P\,[P(P(\phi))]$ is

(a) 2
(b) 3
(c) 4
(d) 5

158. If $A = \{1, 2, 3\}$ and $B = \{2, 4\}$, then the set of all sets X satisfying $X \subset A$ and $X \not\subset B$ is

(a) $\{\{1\}, \{3\}\}$
(b) $\{\{1\}, \{3\}, \{1, 2\}, \{1, 3\}, \{2, 3\}\}$
(c) $\{\{1\}, \{3\}, \{1, 2\}, \{1, 3\}, \{2, 3\}, \{1, 2, 3\}\}$
(d) None of the above

Universal Set

Level I

159. While studying the system of numbers, we are interested in the set of natural numbers and its subsets such as the set of all prime numbers, and so forth. This basic set is called

(a) power set (b) subset
(c) universal set (d) None of these

160. The universal set is usually denoted by

(a) A (b) U
(c) Y (d) None of these

161. The universal set for the set of all rational numbers is ...X... . Here, X refers to

(a) N (b) Z (c) Q^+ (d) R

162. In human population studies, the universal set is the set of ...A... . Here, A refers to

(a) all people in India (b) all people in a locality
(c) all people in the world (d) None of these

163. Which of the following is/are the universal set(s) for the set of isosceles triangles?

I. Set of right angled triangles.
II. Set of scalene triangles.
III. Set of all triangles in a plane.

(a) I (b) III
(c) II and III (d) None of these

164. Given the sets $A = \{1, 3, 5\}$, $B = \{2, 4, 6\}$ and $C = \{0, 2, 4, 6, 8\}$. Then, which of the following may be considered as universal set(s) for all the three sets A, B and C?

I. $\{0, 1, 2, 3, 4, 5, 6\}$
II. ϕ
III. $\{0, 1, 2, 3, 4, 5, 6, 7, 8, 9, 10\}$
IV. $\{1, 2, 3, 4, 5, 6, 7, 8\}$

(a) I (b) III (c) I, III (d) III, IV

Level II

165. The universal set for the sets $A = \{x : x^2 - 5x + 6 = 0\}$ and $B = \{x : x^2 - 3x + 2 = 0\}$ having least number of elements is

(a) $\{1, 2\}$ (b) $\{1, 3\}$
(c) $\{1, 2, 3\}$ (d) $\{0, 1, 2, 3\}$

166. Assertion (A) Set of English alphabets is the universal set for the set of vowels in English alphabets.

Reason (R) The set of vowels is the subset of set of consonants in the English alphabets.

(a) Both A and R are correct; R is the correct explanation of A
(b) Both A and R are correct; R is not the correct explanation of A
(c) A is correct; R is incorrect
(d) R is correct; A is incorrect

167. Statement I Let X be any set. Then, $P(X)$ be the universal set of X.

Statement II If $A \subset B$, then B can be considered universal set of A.

(a) Statement I is true
(b) Statement II is true
(c) Both are true
(d) Both are false

Venn Diagrams

Level I

168. Most of the relationships between sets can be represented by means of diagrams which are known as

(a) rectangles (b) circles
(c) Venn diagrams (d) triangles

169. Venn diagrams are named after the English Logician

(a) John Venn (1800-1883) (b) John Venn (1834-1880)
(c) John Venn (1880-1934) (d) John Venn (1834-1883)

170. Venn diagram consists of

(a) rectangles and closed curves
(b) rectangles and open curves
(c) triangles and straight lines
(d) None of the above

171. The universal set is represented by usually a ...X... and its subsets by ...Y.... Here, X and Y refer to

(a) circle, rectangles (b) rectangle, circles
(c) circle, circles (d) rectangle, lines

Directions (Q. Nos. 172-173) *Consider the Venn diagram*

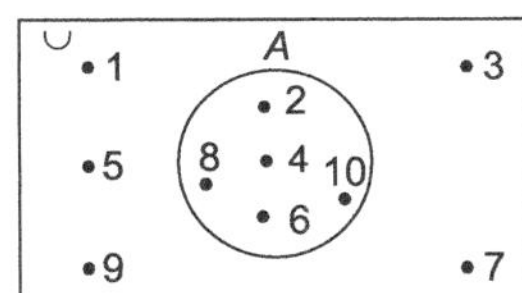

172. In the Venn diagram, the universal set U is

(a) {1, 3, 5, 7, 9} (b) {2, 4, 6, 8, 10}
(c) {1, 2, 3, 4, 5, 6, 7, 8, 9, 10} (d) None of these

173. The subset A of the universal set U is

(a) {1, 5, 9} (b) {3, 7}
(c) {2, 4, 6, 8, 10} (d) {1, 3, 5, 7, 9}

174. If $U = \{1, 2, 3, 4, \ldots, 10\}$ is the universal set of A, B and $A = \{2, 4, 6, 8, 10\}$, $B = \{4, 6\}$ are subsets of U, then given sets can be represented by Venn diagram as

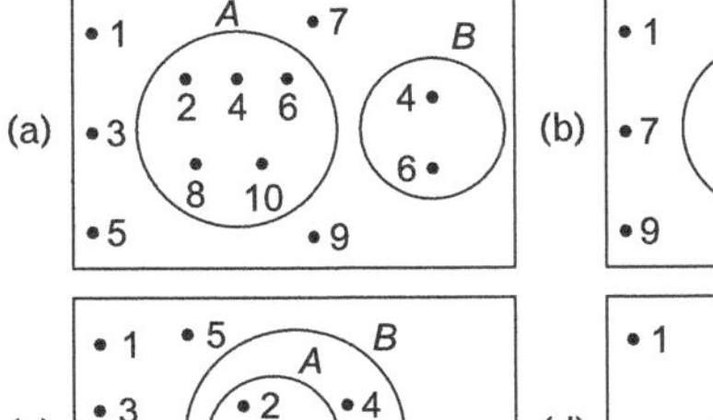

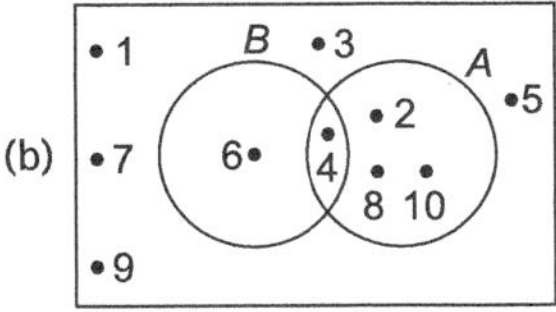

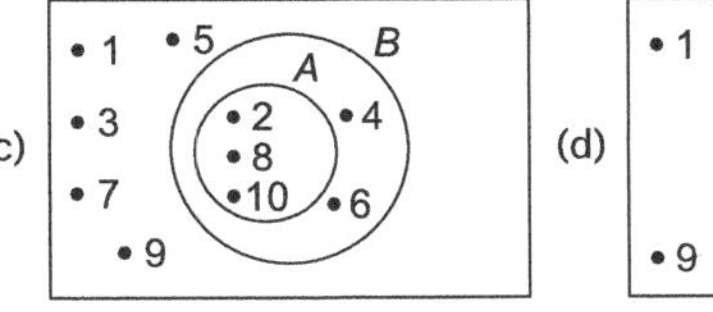

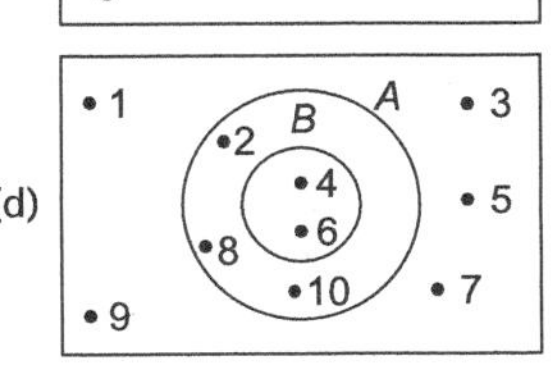

Level II

175. **Statement I** If $A \subset B$ for any two sets A and B.

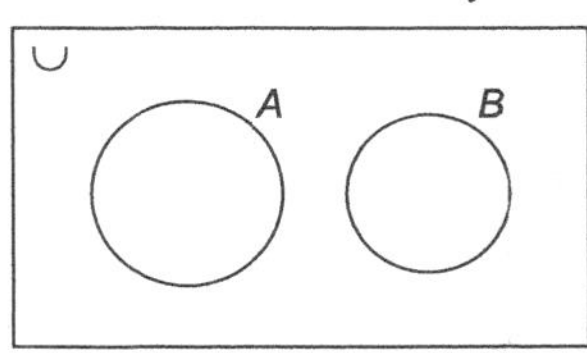

Then, above Venn diagram represents correct relationship between A and B.

Statement II If $A \subset B$, then all elements of A is also in B.

(a) Statement I is true; Statement II is true; Statement II is a correct explanation for Statement I
(b) Statement I is true; Statement II is true; Statement II is not a correct explanation for Statement I
(c) Statement I is true; Statement II is false
(d) Statement I is false; Statement II is true

Directions (Q. Nos. 176-177) *Let E is the set of students studying English in a school, M is the set of students studying Mathematics in the same school, U is the set of all students in that school.*

176. The Venn diagram, which represents all the students who study Mathematics, English, but some students who study English do not study Mathematics is [NCERT Exemplar]

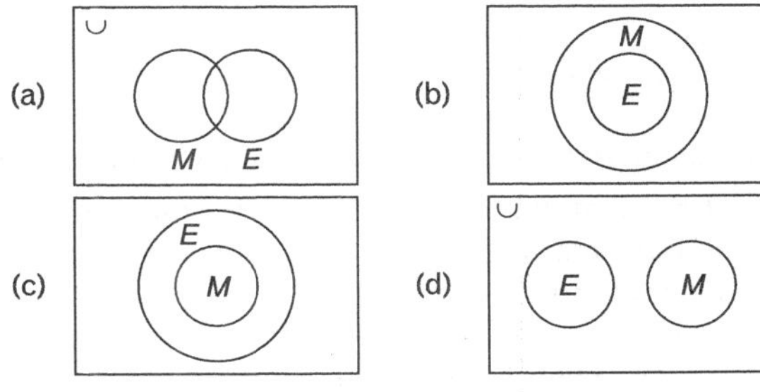

177. Some of the students study Mathematics but do not study English, some study English but do not study Mathematics, and some study both. It can be represented by Venn diagram as [NCERT Exemplar]

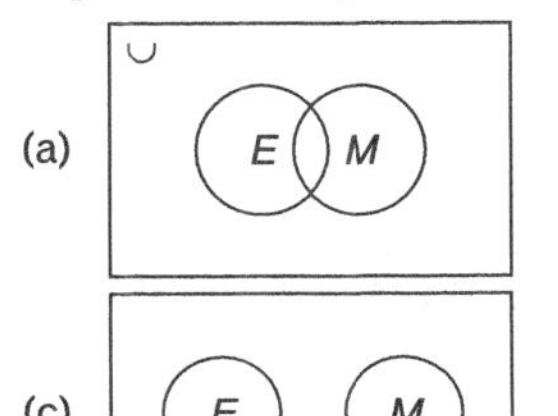

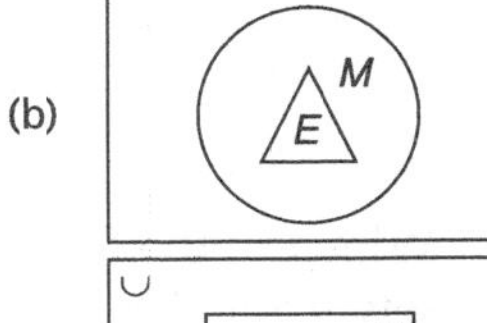

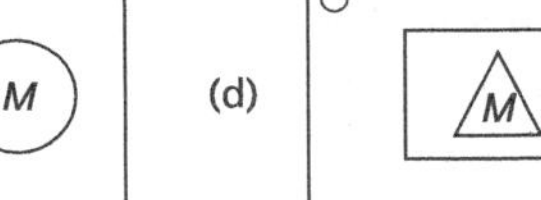

Operations on Sets

Union

Level I

178. Let A and B be any two sets. The union of A and B is the set
(a) containing the elements of A
(b) containing the elements of B
(c) containing the elements of A and B both
(d) None of the above

179. Statement I In the union of two sets A and B, the common elements being taken only once.

Statement II The symbol '$\cap$' is used to denote the union.
(a) Statement I is true
(b) Statement II is true
(c) Both are true
(d) Both are false

180. If $A = \{2, 4, 6, 8\}$ and $B = \{6, 8, 10, 12\}$, then $A \cup B$ is
(a) $\{2, 4, 6, 8\}$ (b) $\{6, 8, 10, 12\}$
(c) $\{6, 8\}$ (d) $\{2, 4, 6, 8, 10, 12\}$

181. Let $A = \{a, e, i, o, u\}$ and $B = \{a, i, u\}$. Then, $A \cup B$ is ...X... . Here, X refers to
(a) A (b) B
(c) A and B (d) None of these

182. Assertion (A) Let X = {Ram, Geeta, Akbar} be the set of students of class XI, who are in school hockey team. Let Y = {Geeta, David, Ashok} be set of students from class XI who are in the school football team. Then, $X \cup Y$ is the set {Ram, Geeta, Akbar, David, Ashok}.

Reason (R) The union of two sets is the set which consists some of the elements of both the sets.
(a) Both A and R are correct; R is the correct explanation of A
(b) Both A and R are correct; R is not the correct explanation of A
(c) A is correct; R is incorrect
(d) R is correct; A is incorrect

183. Which of the following represent the union of two sets A and B?
(a) $\{x : x \in A\}$ (b) $\{x : x \in A \text{ and } x \in B\}$
(c) $\{x : x \in B\}$ (d) $\{x : x \in A \text{ or } x \in B\}$

184. Which of the following represent the union of two sets A and B?

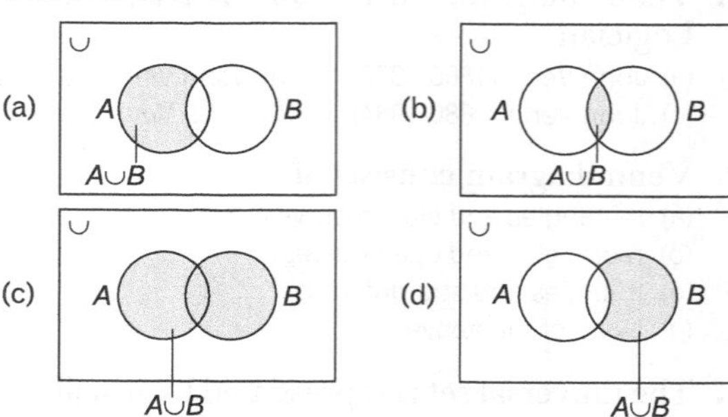

185. Match the following terms in Column I with terms in Column II and choose the correct option from the codes given below.

	Column I		Column II
A.	$A \cup B$	1.	A
B.	$(A \cup B) \cup C$	2.	U
C.	$A \cup \phi$	3.	$B \cup A$
D.	$A \cup A$	4.	$A \cup (B \cup C)$
E.	$U \cup A$		

Codes

	A	B	C	D	E		A	B	C	D	E
(a)	4	2	3	4	1	(b)	2	1	3	4	3
(c)	3	4	1	1	2	(d)	1	2	3	4	4

186. Match the sets in Column I with their union sets in Column II and choose the correct option from the codes given below.

Column I	Column II
A. $X = \{1, 3, 5\}$ $Y = \{1, 2, 3\}$	1. $\{1, 2, 3\}$
B. $A = \{x : x$ is a natural number multiple of 3$\}$ $B = \{x : x$ is a natural number less than 6$\}$	2. $\{1, 2, 3, 5\}$
C. $A = \{x : x$ is a natural number and $1 < x \le 6\}$ $B = \{x : x$ is natural number and $6 < x < 10\}$	3. $\{1, 2, 3, 4, 5, 6, 9, \ldots\}$
D. $A = \{1, 2, 3\}, B = \phi$	4. $\{2, 3, 4, \ldots, 9\}$
E. $A = \{a, e, i, o, u\}, B = \{a, o, e\}$	5. $\{a, e, i, o, u\}$

Codes

	A	B	C	D	E
(a)	4	3	1	2	5
(b)	3	4	5	1	2
(c)	2	3	4	1	5
(d)	1	2	3	4	5

187. **Statement I** Let $A = \{a, b\}$ and $B = \{a, b, c\}$. Then, $A \not\subset B$.

Statement II If $A \subset B$, then $A \cup B = B$.

(a) Statement I is true
(b) Statement II is true
(c) Both are true
(d) Both are false

Directions (Q. Nos. 188-191) *Consider* $A = \{1, 2, 3, 4\}$, $B = \{3, 4, 5, 6\}$, $C = \{5, 6, 7, 8\}$ *and* $D = \{7, 8, 9, 10\}$

188. The sets $A \cup B$ and $B \cup C$ are respectively

(a) {1, 2, 3, 4}, {1, 2, 3}
(b) {3, 4, 5, 6}, {2, 3}
(c) {1, 2, 3, 4, 5, 6}, {3, 4, 5, 6, 7, 8}
(d) {1, 2, 5, 6}, {5, 6, 7, 8}

189. The sets $A \cup B \cup C$ is

(a) {1, 2, 3, 4, 5, 6} (b) {3, 4, 5, 6, 7, 8}
(c) {1, 2, 7, 8} (d) {1, 2, 3, 4, 5, 6, 7, 8}

190. The sets $B \cup D$ and $B \cup C \cup D$ are

(a) same (b) different
(c) either same or different (d) None of these

191. The sets $A \cup C$ ...K... $A \cup B \cup D$. Here, K refers to

(a) $\subset$ (b) $\not\subset$ (c) $\in$ (d) $\notin$

Intersection

192. The intersection of two sets A and B is the set which consists of

(a) elements of A
(b) elements of B
(c) elements of A and B
(d) common elements of A and B

193. **Statement I** The symbol '$\cap$' is used to denote the intersection of sets.

Statement II The intersection of two sets A and B can be written symbolically as $A \cap B = \{x : x \in A$ or $x \in B\}$.

(a) Statement I is true (b) Statement II is true
(c) Both are true (d) Both are false

194. Let $A = \{2, 4, 6, 8\}$ and $B = \{6, 8, 10, 12\}$. Then, $A \cap B$ is

(a) {2, 4} (b) {6, 8}
(c) {10, 12} (d) {2, 4, 6, 8, 10, 12}

195. Let X = {Ram, Geeta, Akbar} be the set of students of class XI, who are in school hockey team and Y = {Geeta, David, Ashok} be the set of students from class XI, who are in the school football team. Then, $X \cap Y$ is

(a) {Ram, Geeta} (b) {Ram}
(c) {Geeta} (d) None of these

196. Let $A = \{1, 2, 3, 4, 5, 6, 7, 8, 9, 10\}$ and $B = \{2, 3, 5, 7\}$. Then, which of following is true?

(a) $A \cap B = A$ (b) $A \cap B = B$
(c) $A \cap B \not\subset B$ (d) None of these

197. The shaded portion represented in the Venn diagram is

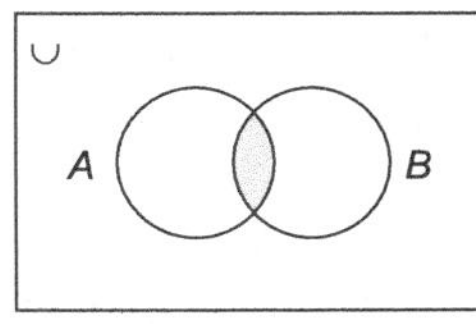

(a) $A \cup B$ (b) $A \cap B$ (c) $A - B$ (d) $B - A$

198. Two sets are said to be ...X..., if $A \cap B$ = ...Y.... Here, X and Y refer to

(a) disjoint, A (b) disjoint, B (c) disjoint, ϕ (d) same, ϕ

199. Let $A = \{2, 4, 6, 8\}$ and $B = \{1, 3, 5, 7\}$. Then, A and B are ...X... . Here, X refers to

(a) equal
(b) disjoint
(c) either equal or disjoint
(d) None of the above

200. Two disjoint sets A and B can be represented by means of Venn diagram as

(a) 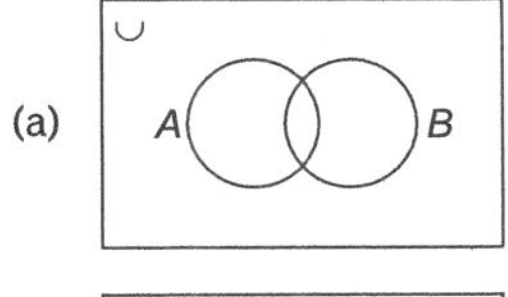

(b) 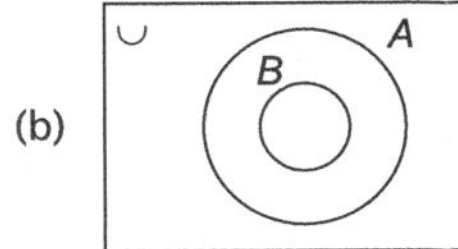

(c) 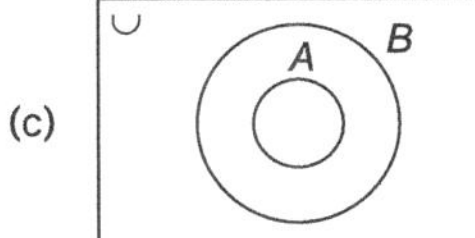

(d) 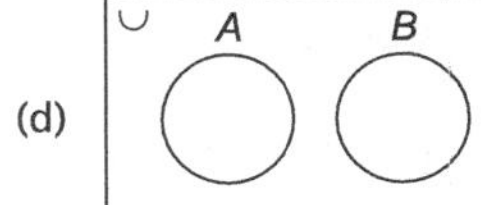

201. Match the following sets in Column I with the equal sets in Column II and choose the correct option from the codes given below.

Column I	Column II
A. $A \cap B$	1. $(A \cap B) \cup (A \cap C)$
B. $(A \cap B) \cap C$	2. A
C. $\phi \cap A$	3. $A \cap (B \cap C)$
D. $U \cap A$	4. $B \cap A$
E. $A \cap A$	5. ϕ
F. $A \cap (B \cup C)$	

Codes

	A	B	C	D	E	F
(a)	5	1	4	3	1	2
(b)	3	4	2	1	5	4
(c)	4	3	5	2	2	1
(d)	1	2	3	4	5	2

202. If we draw the Venn diagram of $A \cap (B \cup C)$ and $(A \cap B) \cup (A \cap C)$, we find that the set $A \cap (B \cup C)$ and $(A \cap B) \cup (A \cap C)$ are

(a) equal
(b) disjoint
(c) either equal or disjoint
(d) None of these

203. Which of the following pairs of sets are disjoint?

I. $\{1, 2, 3, 4\}$ and $\{x : x$ is a natural number and $4 \leq x \leq 6\}$
II. $\{a, e, i, o, u\}$ and $\{c, d, e, f\}$
III. $\{x : x$ is an even integer$\}$ and $\{x : x$ is an odd integer$\}$

(a) I and II
(b) II
(c) III
(d) II and III

204. Let $A = \{x : x$ is natural number$\}$
$B = \{x : x$ is an even natural number$\}$
$C = \{x : x$ is an odd natural number$\}$
$D = \{x : x$ is a prime number$\}$

Then, match the following terms in Column I with the terms of Column II and choose the correct option from the codes given below.

Column I	Column II
A. $A \cap B$	1. ϕ
B. $A \cap C$	2. $\{2\}$
C. $A \cap D$	3. B
D. $B \cap C$	4. C
E. $B \cap D$	5. $\{x : x$ is an odd prime number$\}$
F. $C \cap D$	6. D

Codes

	A	B	C	D	E	F
(a)	3	4	6	1	2	5
(b)	1	2	3	4	5	6
(c)	4	5	6	3	2	1
(d)	2	1	4	5	6	3

Directions (Q. Nos. 205-207) *Let* $A = \{3, 5, 7, 9, 11\}$, $B = \{7, 9, 11, 13\}$, $C = \{11, 13, 15\}$ *and* $D = \{15, 17\}$.

205. The set $A \cap C \cap D$ is

(a) $\{11\}$
(b) $\{15\}$
(c) $\{13\}$
(d) ϕ

206. Which of the following is/are true?

I. $A \cap B = B \cap C$
II. $A \cap C = B \cap D$
III. $A \cap (B \cup C) = A \cap (B \cup D)$
IV. $A \cap D = (A \cap B) \cap (B \cup C)$

(a) I and III
(b) III and IV
(c) IV
(d) None of the above

207. The set $(A \cup D) \cap (B \cup C)$ is

(a) $\{7, 9, 11\}$
(b) $\{7, 9, 11, 13\}$
(c) $\{7, 9, 11, 15\}$
(d) $\{7, 11, 13, 15\}$

208. Match the intersection of sets of Column I with their respective values in Column II and choose the correct option from the codes given below.

Column I		Column II
A. $X = \{1, 3, 5\}$	$Y = \{1, 2, 3\}$	1. ϕ
B. $A = [a, e, i, o, u\}$	$B = \{a, b, c\}$	2. $\{3\}$
C. $A = \{x : x$ is a natural number and multiple of 3$\}$ $B = \{x : x$ is a natural number less than 6$\}$		3. $\{a\}$
D. $A = \{x : x$ is a natural number and $1 < x \leq 6\}$ $B = \{x : x$ is a natural number and $6 < x < 10\}$		4. $\{1, 3\}$
E. $A = \{1, 2, 3\}, B = \phi$		

Codes

	A	B	C	D	E
(a)	2	3	4	2	1
(b)	4	3	2	1	1
(c)	1	2	3	4	2
(d)	3	1	4	2	3

209. Which of the following are disjoint sets?

I. $\{2, 3, 4, 5\}$ and $\{3, 6\}$
II. $\{a, e, i, o, u\}$ and $\{a, b, c, d\}$
III. $\{2, 6, 10, 14\}$ and $\{3, 7, 11, 15\}$
IV. $\{2, 6, 10\}$ and $\{3, 7, 11\}$

(a) I and II
(b) II and III
(c) III and IV
(d) II and IV

Difference *of* Sets

210. $A - B$ is the set of elements which belongs to ...X... but not to ...Y... . Here, X and Y refer to

(a) A and B (b) B and B (c) A and A (d) B and A

211. Let $A = \{1, 2, 3, 4, 5, 6\}$ and $B = \{2, 4, 6, 8\}$. Then,

(a) $A - B = B - A$ (b) $A - B \neq B - A$
(c) $A - B \subset B - A$ (d) $B - A \subset A - B$

212. Let $V = \{a, e, i, o, u\}, V - B = \{e, o\}$ and $B - V = \{k\}$. Then, the set B is [HOTS]

(a) $\{a, i, u\}$ (b) $\{a, e, k, u\}$ (c) $\{a, i, k, u\}$ (d) $\{a, e, i, k, u\}$

213. Which of the following represent $A - B$?

(a) $\{x : x \in A$ and $x \in B\}$ (b) $\{x : x \in A$ and $x \notin B\}$
(c) $\{x : x \in A$ or $x \in B\}$ (d) $\{x : x \in A$ or $x \notin B\}$

214. Statement I The set $A - B$ can be represented by Venn diagram as

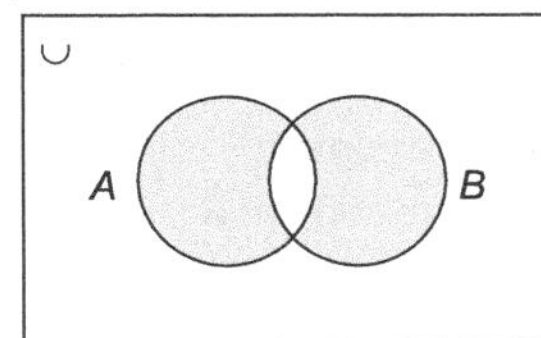

Statement II The sets $A - B$, $A \cap B$ and $B - A$ are mutually disjoint sets.

(a) Statement I is true
(b) Statement II is true
(c) Both are true
(d) Both are false

215. Assertion (A) If R is the set of real numbers and Q is the set of rational numbers, then $R - Q$ is the set of integers.

Reason (R) The set of real numbers consists of set of rational numbers and set of irrational numbers.

(a) Both A and R are correct; R is the correct explanation of A
(b) Both A and R are correct; R is not the correct explanation of A
(c) A is correct; R is incorrect
(d) R is correct; A is incorrect

216. If $X = \{a, b, c, d\}$ and $Y = \{f, b, d, g\}$, then match the following terms in Column I with terms of Column II and choose the correct option from the codes given below.

Column I		Column II	
A.	$X - Y$	1.	$\{b, d\}$
B.	$Y - X$	2.	$\{a, c\}$
C.	$X \cap Y$	3.	$\{a, b, c, d, f, g\}$
D.	$X \cup Y$	4.	$\{f, g\}$

Codes

	A	B	C	D
(a)	2	4	1	3
(b)	1	2	3	4
(c)	4	3	2	1
(d)	3	1	4	2

Directions (Q. Nos. 217-220)

Let $A = \{3, 6, 9, 12, 15, 18, 21\}$,
$B = \{4, 8, 12, 16, 20\}$,
$C = \{2, 4, 6, 8, 10, 12, 14, 16\}$
and $D = \{5, 10, 15, 20\}$.

217. The set $A - B$ is ...K... . Here, K refers to

(a) $\{3, 6, 9, 15, 21\}$ (b) $\{4, 8, 16, 20\}$
(c) $\{3, 6, 9, 15, 18, 21\}$ (d) $\{3, 9, 15, 21\}$

218. Which of the following is true?

(a) $A - C = A - D$ (b) $A - D \subset B - A$
(c) $C - A = D - A$ (d) $B - C \neq B - D$

219. The set $(C - B) \cap (D - B)$ is

(a) $\{15\}$ (b) $\{20\}$
(c) $\{10\}$ (d) ϕ

220. The set $(C - D) \cup (D - C)$ is ...X... . Here, X refers to

(a) $\{2, 4, 5, 6, 8, 10, 12, 14, 15, 16, 20\}$
(b) $\{2, 4, 5, 6, 8, 12, 14, 15, 16, 20\}$
(c) $\{2, 4, 5, 6, 8, 10, 12, 14, 16, 20\}$
(d) $\{2, 4, 5, 6, 8, 10, 12, 14, 15, 16\}$

Level II

221. If $A = \{1, 2, 3, 4, 5\}$, $B = \{2, 4, 6\}$ and $C = \{3, 4, 6\}$, then $(A \cup B) \cap C$ is [OJEE 2004]

(a) $\{3, 4, 6\}$ (b) $\{1, 2, 3\}$
(c) $\{1, 4, 3\}$ (d) None of these

222. If the sets A and B are defined as

$A = \{(x, y) : y = \frac{1}{x}, 0 \neq x \in R\}$ and

$B = \{(x, y) : y = -x, x \in R\}$, then

(a) $A \cap B = A$ (b) $A \cap B = B$
(c) $A \cap B = \phi$ (d) None of these

223. Which of the following are correct?

I. $A - B = A - (A \cap B)$
II. $A = (A \cap B) \cup (A - B)$
III. $A - (B \cup C) = (A - B) \cup (A - C)$

(a) I, II (b) II, III
(c) I, II, III (d) None of these

224. **Assertion** (A) If A, B and C are three non-empty sets, then $(A \cap B) \cap (B \cap C) \cap (C \cap A)$ is equal to $A \cap B \cap C$.

Reason (R) If A, B and C are three non-empty sets, then $(A \cup B) \cup (A \cup C)$ is equal to $A \cup B \cup C$.

(a) Both A and R are correct; R is the correct explanation of A
(b) Both A and R are correct; R is not the correct explanation of A
(c) A is correct; R is incorrect
(d) R is correct; A is incorrect

225. If $aN = \{ax : x \in N\}$, then $3N \cap 7N$ is equal to

(a) $3N$ (b) $7N$
(c) N (d) $21N$

226. If A and B are non-empty subsets of a set, then $(A - B) \cup (B - A)$ equals to

(a) $(A \cap B) \cup (A \cup B)$ (b) $(A \cup B) - (A - B)$
(c) $(A \cup B) - (A \cap B)$ (d) $(A \cup B) - B$

227. If $A = \{1, 2, 3, 4, 5\}$, $B = \{x : x^2 - 5x + 6 = 0\}$ and $C = \{x : x^2 = 4 \text{ or } x = 1\}$, then

(a) $A \cap (B \cap C) = \phi$
(b) $A \cap (B \cap C) \neq \phi$
(c) Both (a) and (b)
(d) None of the above

228. Match the following sets in Column I equal with the sets in Column II and choose the correct option from the codes given below.

	Column I		Column II
A.	$(A \cup B) - A$	1.	$A \cup B$
B.	$A - (B \cup C)$	2.	$B - A$
C.	$A \cup (B - A)$	3.	$(A - B) \cup (A \cap C)$
D.	$A - (B - C)$	4.	$(A - B) \cap (A - C)$

Codes

	A	B	C	D
(a)	4	1	2	3
(b)	3	4	2	1
(c)	2	4	1	3
(d)	1	2	3	4

229. **Statement I** Let $X = \{x : x$ is a positive integer < 8 and x is a multiple of 3 or 5$\}$ and $Y = \{x : x$ is even number $\leq 7\}$. Then, $X \cap Y = \phi$.

Statement II Consider $A = \{1, 2\}$ and $B = \{2, 3\}$. Then, $P(A \cup B) \not\subset P(A) \cup P(B)$.

(a) Statement I is true
(b) Statement II is true
(c) Both are true
(d) Both are false

230. The smallest set A such that $A \cup \{1, 2\} = \{1, 2, 3, 5, 9\}$ is

(a) $\{2, 3, 5\}$ (b) $\{3, 5, 9\}$
(c) $\{1, 2, 5, 9\}$ (d) None of these

231. The shaded region in the given figure is

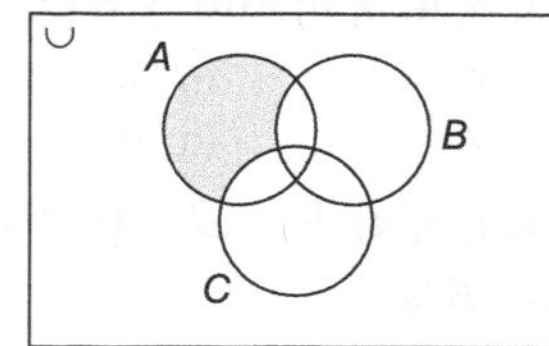

(a) $A \cap (B \cup C)$
(b) $A \cup (B \cap C)$
(c) $A \cap (B - C)$
(d) $A - (B \cup C)$

232. **Assertion** (A) Three sets A, B, C are such that $A = B \cap C$ and $B = C \cap A$, then $A = B$.

Reason (R) If $A = \{x, y\}$, then $A \cap P(A) = A$.

(a) Both A and R are correct; R is the correct explanation of A
(b) Both A and R are correct; R is not the correct explanation of A
(c) A is correct; R is incorrect
(d) R is correct; A is incorrect

233. If $A = \{(x, y) : x^2 + y^2 = 25\}$ and $B = \{(x, y) : x^2 + 9y^2 = 144\}$, then $A \cap B$ contains

(a) one point
(b) two points
(c) three points
(d) four points

234. If A_n is the set of first n prime numbers, then $\bigcup_{n=2}^{10} A_n$ is equal to

(a) $\{2, 3, 5, 7, 11, 13, 17, 19\}$
(b) $\{2, 3, 5, 7, 11, 13, 17, 19, 23, 29\}$
(c) $\{3, 5\}$
(d) $\{2, 3\}$

235. If A, B, C are three non-empty sets such that any two of them are disjoint, then $(A \cup B \cup C) \cap (A \cap B \cap C)$ is equal to

(a) A (b) B
(c) C (d) ϕ

236. Let A, B, C are three non-empty sets. If $A \subset B$ and $B \subset C$, then which of the following is true?

(a) $B - A = C - B$
(b) $A \cap B \cap C = B$
(c) $A \cup B = B \cap C$
(d) $A \cup B \cup C = A$

Complement of a Set

Level I

237. Let U be the set of all prime numbers and $A = \{x : x \in U$ and x is not a divisor of $42\}$.

Then, the set of those elements of U which do not belong to A, is

(a) $\{2, 3\}$ (b) $\{2, 3, 5\}$ (c) $\{2, 3, 7\}$ (d) $\{3, 5, 7\}$

238. Statement I Let U be the universal set and A be the subset of U. Then, complement of A is the set of element of A.

Statement II The complement of a set A can be represented by A'.

(a) Statement I is true (b) Statement II is true
(c) Both are true (d) Both are false

239. Let $U = \{1, 2, 3, 4, \ldots, 10\}$ and $A = \{1, 3, 5, 7, 9\}$. Then, the set A' is ...X... . Here, X refers to

(a) $\{2, 4, 6, 8, 10\}$ (b) $\{1, 3, 5, 7, 9\}$
(c) $\{1, 2, 3, 4, 5\}$ (d) $\{5, 6, 7, 8, 9, 10\}$

240. Let U be universal set of all the students of class XI of a coeducational school and A be the set of all girls in class XI. Then, A' is the set of

(a) all boys in the school (b) all boys in class XI
(c) all girls in the school (d) all girls in class XI

241. Which of the following is/are true?

I. If A is a subset of the universal set U, then its complement A' is also a subset of U.

II. If $U = \{1, 2, 3, \ldots, 10\}$ and $A = \{1, 3, 5, 7, 9\}$, then $(A')' = A$.

(a) Only I is true (b) Only II is true
(c) Both I and II are true (d) None of these

Directions (Q. Nos. 242-244) *Let* $U = \{1, 2, 3, 4, 5, 6\}$, $A = \{2, 3\}$ *and* $B = \{3, 4, 5\}$.

242. A' is the set ...X... . Here, X refers to

(a) $\{4, 5\}$ (b) $\{1, 4, 5\}$
(c) $\{1, 4, 5, 6\}$ (d) $\{4, 5, 6\}$

243. B' is the set ...Y... . Here, Y refers to

(a) $\{2, 3\}$ (b) $\{1, 2, 6\}$
(c) $\{1, 2\}$ (d) $\{1, 2, 3, 6\}$

244. Which of the following is true?

(a) $(A \cup B)' = A' \cup B'$
(b) $(A \cup B)' = A' \cap B'$
(c) $(A \cap B)' = A' \cap B'$
(d) None of the above

245. The complement of the union of two sets is the intersection of their complements and the complement of the intersection of two sets is the union of their complement. These are called ...X... laws. Here, X refers to

(a) idempotent (b) commutative
(c) De-Morgan's (d) associative

246. In the Venn diagram, the shaded portion represents

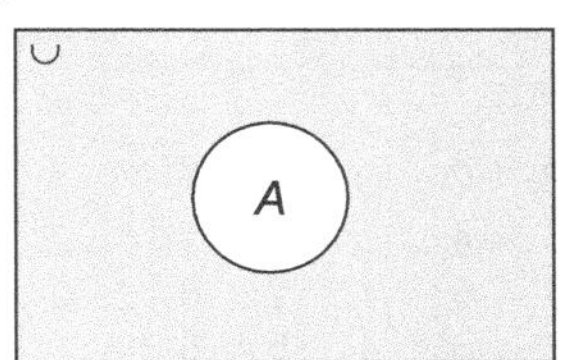

(a) complement of set A (b) universal set
(c) set A (d) None of these

247. Match the following sets in Column I equal with the sets in Column II and choose the correct option from the codes given below.

	Column I		Column II
A.	$A \cup A'$	1.	$A' \cap B'$
B.	$A \cap A'$	2.	$A' \cup B'$
C.	$(A \cup B)'$	3.	U
D.	$(A \cap B)'$	4.	ϕ
E.	ϕ'	5.	A
F.	U'		
G.	$(A')'$		

Codes

	A	B	C	D	E	F	G
(a)	1	2	3	4	5	3	2
(b)	3	4	1	2	3	4	5
(c)	4	3	2	1	4	5	3
(d)	5	4	3	2	1	4	1

Directions (Q. Nos. 248-249) *Let* $U = \{1, 2, 3, 4, 5, 6, 7, 8, 9\}$, $A = \{1, 2, 3, 4\}$, $B = \{2, 4, 6, 8\}$ *and* $C = \{3, 4, 5, 6\}$.

248. The sets A' and B' are ...X... and ...Y..., respectively. Here, X and Y refer to

(a) $\{3, 4, 5, 6\}$ and $\{1, 3, 5, 7, 9\}$
(b) $\{5, 6, 7, 8, 9\}$ and $\{1, 3, 4, 5, 6\}$
(c) $\{5, 6, 7, 8, 9\}$ and $\{1, 3, 5, 7, 9\}$
(d) $\{4, 5, 6, 7, 8, 9\}$ and $\{3, 5, 7, 9\}$

249. Which of the following is true?

(a) $(A \cup C)' = (A \cup B)'$ (b) $(A')' = B$
(c) $(B - C)' = B' - C'$ (d) None of these

250. Taking the set of natural numbers as the universal set, match the complement set of following sets in Column I with the sets in Column II and choose the correct option from the codes given below.

Column I	Column II
A. $\{x : x$ is an even natural number$\}$	1. $\{x : x$ is not divisible by 15$\}$
B. $\{x : x$ is a prime number$\}$	2. $\{x : x$ is not a perfect square$\}$
C. $\{x : x$ is a multiple of 3$\}$	3. $\{x : x$ is an odd natural number$\}$
D. $\{x : x$ is a natural number divisible by 3 and 5$\}$	4. $\{x : x$ is not a prime number$\}$
E. $\{x : x$ is a perfect square$\}$	5. $\{x : x$ is not a multiple of 3$\}$
F. $\{x : x$ is a perfect cube$\}$	6. $\{x : x$ is not a perfect cube$\}$

Codes

	A	B	C	D	E	F
(a)	6	5	4	3	2	1
(b)	3	4	5	1	2	6
(c)	1	2	3	4	5	6
(d)	2	3	4	5	6	1

251. If the set of natural numbers is the universal set, $A = \{x : x + 5 = 8$ and $x \in U\}$, $B = \{x : 2x + 5 = 9$ and $x \in U\}$, $C = \{x : x \geq 7$ and $x \in U\}$ and $D = \{x : x \in N$ and $2x + 1 > 10\}$, then which of the following are true?

I. $A' = \{1, 2, 3, 4, \ldots\}$
II. $B' = \{2, 3, 4, 5, 6, \ldots\}$
III. $C' = \{7, 8, 9, 10, \ldots\}$
IV. $D' = \{5, 6, 7, 8, \ldots\}$

(a) I and II (b) II and III
(c) III and IV (d) None of these

252. If $U = \{1, 2, 3, 4, 5, 6, 7, 8, 9\}$, $A = \{2, 4, 6, 8\}$ and $B = \{2, 3, 5, 7\}$, then $(A \cup B)' = \ldots X\ldots$ and $(A \cap B)' = \ldots Y\ldots$. Here, X and Y refer to

(a) A' and $A' \cup B'$ (b) $A' \cap B'$ and B'
(c) $A' \cap B'$ and $A' \cup B'$ (d) A' and B'

253. **Statement I** The Venn diagram of $(A \cup B)'$ and $A' \cap B'$ are same.

Statement II The Venn diagram of $(A \cap B)'$ and $A' \cup B'$ are different.

(a) Statement I is true (b) Statement II is true
(c) Both are true (d) Both are false

254. Let U be the set of all triangles in a plane. If A is the set of all triangles with atleast one angle different from 60°, then A' is the

(a) set of right angled triangle
(b) set of equilateral triangle
(c) set of isosceles triangle
(d) None of the above

255. The sets $\phi' \cap A = \ldots X\ldots$ and $U' \cap A = \ldots Y\ldots$. Here, X and Y refer to

(a) A and ϕ (b) ϕ and A
(c) U and ϕ (d) A and U

256. The sets $A \cup A' = \ldots K\ldots$ and $A \cap A' = \ldots P\ldots$. Here, K and P refer to respectively

(a) U and ϕ (b) ϕ and U (c) A and A' (d) A' and U

Level II

257. If $A = \{x : x = 2x, x \in N\}$ and $B = \{x : x = 2x - 1, x \in N\}$, then $(A \cap B)'$ is

(a) A (b) B (c) ϕ (d) U

258. Let A and B be two sets. Then, $(A \cup B)' \cup (A' \cap B)$ is equal to

(a) A' (b) B'
(c) A (d) None of these

259. **Assertion** (A) For any two sets A and B, $A - B \subset B'$.

Reason (R) If A be any set, then $A \cap A' = \phi$.

(a) Both A and R are correct; R is the correct explanation of A
(b) Both A and R are correct; R is not the correct explanation of A
(c) A is correct; R is incorrect
(d) R is correct; A is incorrect

260. **Statement I** For any two sets A and B, $A - (A - B) = A' \cap B'$.

Statement II Let $U = \{1, 2, 3, 4, \ldots, 10\}$, $A = \{1, 5\}$ and $B = \{6, 7, 8\}$. Then, $(A \cap B')$ is A.

(a) Statement I is true (b) Statement II is true
(c) Both are true (d) Both are false

261. Let U be the universal set and $A \cup B \cup C = U$. Then, $\{(A - B) \cup (B - C) \cup (C - A)\}'$ is equal to

(a) $A \cup B \cup C$ (b) $A \cup (B \cap C)$
(c) $A \cap B \cap C$ (d) $A \cap (B \cup C)$

262. The set $(A \cap B')' \cup (B \cap C)$ is equal to $\ldots X\ldots$. Here, X refers to [NCERT Exemplar]

(a) $A' \cup B \cup C$ (b) $A' \cup B$ (c) $A' \cup C'$ (d) $A' \cap B$

263. If $U = \{1, 2, 3, 4, 5, 6, 7, 8, 9, 10\}$, $A = \{1, 2, 3, 5\}$, $B = \{2, 4, 6, 7\}$ and $C = \{2, 3, 4, 8\}$, then which of the following is true?

(a) $(B \cup C)' = \{1, 5, 9, 10\}$
(b) $(C - A)' = \{1, 2, 3, 5, 6, 7, 9, 10\}$
(c) Both (a) and (b)
(d) None of the above

264. The value of $(A \cup B \cup C) \cap (A \cap B' \cap C')' \cap C'$ is

(a) $B \cap C'$ (b) $B' \cap C'$ (c) $B \cap C$ (d) $A \cap B \cap C$

Practical Problems on Union and Intersection of Two Sets

Level I

265. Let A and B be finite sets. If $A \cap B = \phi$, then $n(A \cup B)$ is ...X... . Here, X refers to

(a) $n(A)$ (b) $n(B)$
(c) $n(A) + n(B)$ (d) $n(A) + n(B) - n(A \cap B)$

266. If A and B are finite sets, then $n(A \cup B)$ is equal to

(a) $n(A) + n(B)$ (b) $n(A) + n(B) - n(A \cap B)$
(c) $n(A) + n(B) - n(A - B)$ (d) $n(A) + n(B) + n(B - A)$

267. **Statement I** If A, B and C are finite sets, then $n(A \cup B \cup C) = n(A) + n(B) + n(C) - n(A \cap B) - n(B \cap C) - n(A \cap C) + n(A \cap B \cap C)$.

Statement II If A, B and C are mutually pairwise disjoint, then $n(A \cup B \cup C) = n(A) + n(B) + n(C) - n(A \cap B) - n(B \cap C) - n(A \cap C)$.

(a) Statement I is true
(b) Statement II is true
(c) Both are true
(d) Both are false

268. If X and Y are two sets such that $X \cup Y$ has 50 elements, X has 28 elements and Y has 32 elements, then number of elements $X \cap Y$ have

(a) 10 (b) 46 (c) 110 (d) 54

269. In a school, there are 20 teachers who teach Mathematics or Physics. Of these, 12 teach Mathematics and 4 teach both Physics and Mathematics. Then, number of teachers who teach Physics, is

(a) 28 (b) 12 (c) 4 (d) 36

270. In a class of 35 students, 24 like to play cricket and 16 like to play football. Also, each student likes to play atleast one of the two games. The number of students who like to play both cricket and football, is

(a) 27 (b) 43 (c) 5 (d) 75

271. In a survey of 400 students in a school, 100 were listed as taking apple juice, 150 as taking orange juice and 75 were listed as taking both apple as well as orange juice.

Then, which of the following is/are true?

I. 150 students were taking atleast one juice.
II. 225 students were taking neither apple juice nor orange juice.

(a) Only I is true (b) Only II is true
(c) Both I and II are true (d) None of these

Directions (Q. Nos. 272-274) *There are 200 individuals with a skin disorder of these, 120 had been exposed to the chemical C_1, 50 to chemical C_2 and 30 to both the chemicals C_1 and C_2.*

272. The number of individuals exposed to chemical C_1 but not chemical C_2 is

(a) 90 (b) 80 (c) 150 (d) 120

273. The number of individuals exposed to chemical C_2 but not chemical C_1 is

(a) 80 (b) 20 (c) 50 (d) 10

274. The number of individuals exposed to chemical C_1 or chemical C_2 is

(a) 100 (b) 120 (c) 140 (d) 50

275. Match the following terms in Column I with terms of Column II and choose the correct option from the codes given below.

	Column I		Column II
A.	If X and Y are two sets such that $n(X) = 17$, $n(Y) = 23$ and $n(X \cup Y) = 38$, then $n(X \cap Y)$ is	1.	30
B.	In a group of 400 people, 250 can speak Hindi and 200 can speak English. The number of people who can speak both Hindi and English, is	2.	2
C.	If S and T are two sets such that S has 21 elements, T has 32 elements and $S \cap T$ has 11 elements. The number of elements $S \cup T$ have, is	3.	50
D.	If X and Y are two sets such that X has 40 elements, $X \cup Y$ has 60 elements and $X \cap Y$ has 10 elements, then number of elements in Y is	4.	42

Codes

	A	B	C	D
(a)	3	4	2	1
(b)	2	3	4	1
(c)	3	1	4	2
(d)	2	1	4	3

276. If X and Y are two sets such that $X \cup Y$ has 18 elements, X has 8 elements and Y has 15 elements, then $n(X \cap Y)$ is equal to

(a) 3 (b) 5
(c) 4 (d) 7

277. In a group of 70 people, 37 like coffee, 52 like tea and each person likes atleast one of the two drinks. The number of people who like both coffee and tea, is ...X... . Here, X refers to

(a) 89 (b) 55
(c) 85 (d) 19

278. In a group of 65 people, 40 like cricket, 10 like both cricket and tennis.

Then, which of the following is/are true?

I. The number of people who like tennis only and not cricket, is 25.

II. The number of people who like tennis, is 30.

(a) Only I is true
(b) Only II is true
(c) Both I and II are true
(d) None of the above

279. In a committee, 50 people speak French, 20 speak Spanish and 10 speak both Spanish and French. The number of people who speak atleast one of these two languages, is

(a) 40 (b) 60
(c) 20 (d) 80

Level II

280. Let A and B be two sets such that $n(A)=0.16$, $n(B)=0.14$ and $n(A \cup B)=0.25$. Then, $n(A \cap B)$ is equal to

(a) 0.3 (b) 0.5 (c) 0.05 (d) None of these

281. In a city, 20% of the population travels by car, 50% travels by bus and 10% travels by both car and bus. Then, person travelling by car or bus, is

(a) 80% (b) 40%
(c) 60% (d) 70%

282. Assertion (A) If $n(A)=3$, $n(B)=6$ and $A \subset B$, then the number of elements in $A \cup B$ is 9.

Reason (R) If A and B are disjoint, then $n(A \cup B)$ is $n(A)+n(B)$.

(a) Both A and R are correct; R is the correct explanation of A
(b) Both A and R are correct; R is not the correct explanation of A
(c) A is correct; R is incorrect
(d) R is correct; A is incorrect

283. A survey shows that 63% of the Americans like cheese whereas 76% like apple. If x% of the Americans like both cheese and apple, then

(a) $x = 39$ (b) $x = 63$
(c) $39 \le x \le 63$ (d) None of these

284. From 50 students taking examinations in Mathematics, Physics and Chemistry, each of the students has passed in atleast one of the subject, 37 passed Mathematics, 24 Physics and 43 Chemistry. Atmost 19 passed Mathematics and Physics, atmost 29 Mathematics and Chemistry and atmost 20 Physics and Chemistry. Then, the largest numbers that could have passed all three examinations, are [NCERT Exemplar]

(a) 12 (b) 14 (c) 15 (d) 16

285. Statement I If $n(U)=20$, $n(A)=12$, $n(B)=9$ and $n(A \cap B)=4$, where U is the universal set and A and B are subsets of U, then $n(A \cup B)'=3$.

Statement II If $n(A)=11$, $n(B)=5$ and $n(C)=4$ for three disjoint sets A, B and C, then $n(A \cup B \cup C)$ is 21.

(a) Statement I is true (b) Statement II is true
(c) Both are true (d) Both are false

286. Each set X_r contains 5 elements and each set Y_r contains 2 elements and $\bigcup_{r=1}^{20} X_r = S = \bigcup_{r=1}^{n} Y_r$. If each element of S belongs to exactly 10 of the X_r's and to exactly 4 of the Y_r's, then n is equal to

(a) 10 (b) 20 (c) 100 (d) 50

287. In a certain town, 25% families own a phone and 15% own a car, 65% families own neither a phone nor a car, 2000 families own both a car and a phone.

Consider the following statements in this regard

I. 10% families own both a car and a phone.

II. 35% families own either a car or a phone.

III. 40000 families live in the town.

Which of the following are correct?

(a) I and II are correct
(b) I and III are correct
(c) II and III are correct
(d) All are correct

288. In a class of 175 students, the following data shows the number of students opting one or more subjects.

Mathematics 100; Physics 70; Chemistry 40; Mathematics and Physics 30; Mathematics and Chemistry 28; Physics and Chemistry 23; Mathematics, Physics and Chemistry 18.

Then, number of students opted Mathematics alone is

(a) 35 (b) 48 (c) 60 (d) 20

Miscellaneous Exercise

289. **Statement I** The set of letters needed to spell 'CATARACT' and the set of letters needed to spell 'TRACT' are equal.

Statement II All the subsets of the set $\{-1, 0, 1\}$ are ϕ, $\{-1\}$, $\{0\}$, $\{1\}$, $\{-1, 0\}$, $\{-1, 1\}$ and $\{0, 1\}$.

(a) Statement I is true (b) Statement II is true
(c) Both are true (d) Both are false

290. If $A \cup B = A \cap B$, then which of the following is true?

(a) $A \subset B$ (b) $B \subset A$
(c) $A = B$ (d) All of these

291. **Assertion** (A) For any sets A and B, $P(A \cap B) = P(A) \cap P(B)$.

Reason (R) Two sets A and B are equal, if $A \subset B$ and $B \subset A$.

(a) Both A and R are correct; R is the correct explanation of A
(b) Both A and R are correct; R is not the correct explanation of A
(c) A is correct; R is incorrect
(d) R is correct; A is incorrect

292. A market research group conducted a survey of 1000 consumers and reported that 720 consumers like product A and 450 consumers like the product B, then least number that must have like both products, is

(a) 730 (b) 170
(c) 180 (d) 1270

293. A college awarded 38 medals in football, 15 in basketball and 20 in cricket. If these medals went to a total of 58 men and only three men got medals in all the three sports, then number of medals received in exactly two of the three sports is

(a) 18 (b) 15
(c) 9 (d) 24

294. Let $A = \{x : x \in R \text{ and satisfy } x^2 - 8x + 12 = 0\}$
$B = \{2, 4, 6\}$, $C = \{2, 4, 6, 8, \ldots\}$ and $D = \{6\}$.
Then, which of following are true?

I. $A \subset B$ II. $A \subset C$
III. $B \subset C$ IV. $C \subset D$

(a) I, II (b) I, II and III
(c) II, III and IV (d) I, II, III and IV

295. Let A, B and C be the sets such that $A \cup B = A \cup C$ and $A \cap B = A \cap C$. Then,

(a) $A = B$ (b) $B = C$
(c) $A = C$ (d) None of these

296. Determine which of the following are true?

I. If $x \in A$ and $A \in B$, then $x \in B$.
II. If $A \subset B$ and $B \in C$, then $A \in C$.
III. If $A \subset B$ and $B \subset C$, then $A \subset C$.
IV. If $A \not\subset B$ and $B \not\subset C$, then $A \not\subset C$.
V. If $x \in A$ and $A \not\subset B$, then $x \in B$.
VI. If $A \subset B$ and $x \notin B$, then $x \notin A$.

(a) I, IV and VI (b) III, VI
(c) II, III and V (d) All of these

297. **Statement I** The four conditions $A \subset B$, $A - B = \phi$, $A \cup B = B$ and $A \cap B = A$ are equivalent.

Statement II If $A \subset B$, then $C - B \not\subset C - A$.

(a) Statement I is true (b) Statement II is true
(c) Both are true (d) Both are false

298. If $P(A) = P(B)$, then

(a) $A \subset B$ (b) $B \subset A$
(c) $A = B$ (d) All of these

299. Which of the following is/are true?

I. Let A and B are two sets. Then, $P(A) \cup P(B) = P(A \cup B)$.
II. If $A \cap B = A \cap C$, then $B = C$.
III. Let A and B be sets. If $A \cap X = B \cap X = \phi$ and $A \cup X = B \cup X$ for some set X, then $A = B$.

(a) Only I is true (b) I and II are true
(c) III is true (d) All are true

300. Match the following sets in Column I equal with the sets in Column II and choose the correct option from the codes given below.

Column I	Column II
A. $(A \cap B) \cup (A - B)$	1. B
B. $A \cup (B - A)$	2. A
C. $B \cup (A \cap B)$	3. $(A \cup B)$
D. $A \cap (A \cup B)$	

Codes

	A	B	C	D
(a)	1	2	3	1
(b)	2	3	1	2
(c)	3	2	1	2
(d)	2	1	3	3

301. If $A \cap B$, $B \cap C$ and $A \cap C$ are non-empty sets and $A \cap B \cap C = \phi$, then A, B and C are respectively

(a) $\{1, 3, 5\}$, $\{5, 7, 9\}$ and $\{3, 5, 7\}$
(b) $\{1, 2, 3, 4\}$, $\{4, 5, 6\}$ and $\{2, 4, 6, 8\}$
(c) $\{1, 3, 5\}$, $\{1, 2, 4\}$ and $\{3, 4, 5\}$
(d) $\{2, 4, 6\}$, $\{1, 3, 5\}$ and $\{3, 4, 5\}$

302. In a survey of 600 students in a school, 150 students were found to be taking tea and 225 taking coffee, 100 were taking both tea and coffee. Then, the number of students were taking neither tea nor coffee, is

(a) 225 (b) 350
(c) 275 (d) 325

303. In a group of students, 100 students know Hindi, 50 know English and 25 know both. Each of the students knows either Hindi or English. The number of students in the group is

(a) 50
(b) 125
(c) 75
(d) 175

Directions (Q. Nos. 304-305) *In a survey of 60 people, it was found that 25 people read newspaper H, 26 read newspaper T, 26 read newspaper I, 9 read H and I, 11 read H and T, 8 read T and I, 3 read all three newspapers.*

304. The number of people who read atleast one of the newspaper, is

(a) 60 (b) 46
(c) 52 (d) 58

305. The number of people who read exactly one newspaper, is

(a) 30
(b) 18
(c) 24
(d) 22

306. In a survey, it was found that 21 people like product A, 26 like product B and 29 like product C. If 14 people like products A and B, 12 people like products C and A, 14 people like products B and C and 8 like all the three products, then number of people who liked product C only, is

(a) 7
(b) 15
(c) 11
(d) 9

Answer *with* Explanations

1. *(b)* The theory of sets was developed by German Mathematician Georg Cantor.

2. *(a)* A crowd of people is a collection of objects of a particular kind.

3. *(b)* A number which has only two factors one and itself, is called a prime number.

4. *(c)* In (i) and (iii), we can definitely decide whether a given particular object belongs to a given collection or not. For example, we can say that the river Nile does not belong to the collection of rivers of India. On the other hand, the river Ganga belongs to this collection.

Again, the collection of most talented batsmen of India and the collection of books is not well-defined, because the criterion for determining most talented batsman and collection of particular kind of books may vary from person- to-person.

5. *(b)* N : The set of natural numbers
Z : The set of all integers
Q : The set of rational numbers
R : The set of real numbers
Z^+ : The set of positive integers

6. *(c)* Q^+ : The set of positive rational numbers.

7. *(a)* A well-defined collection of objects is called set.

8. *(b)* Some points related to sets are

I. Objects, elements and members of a set are synonymous terms.
II. Sets are usually denoted by capital letters *A, B, C, X, Y, Z*, etc.

9. *(a)* The elements of a set are represented by small letters *a, b, c, x, y, z*, etc.

If *a* is an element of a set *A*, then we can say that '*a* belongs to *A*'. The Greek symbol '$\in$' (epsilon) is used to denote the phrase 'belongs to'. Thus, we write $a \in A$.

10. *(c)* If '*b*' is not an element of a set *A*, then we write $b \notin A$ and read '*b* does not belong to *A*'.

11. *(a)* There are two methods of representing a set

I. Roster or tabular form
II. Set-builder form

12. *(a)* In roster form, all the elements of a set are listed, the elements are being separated by commas and are enclosed within braces { }. For example, the set of all even positive integers less than 7 is described in roster form as {2, 4, 6}.

In roster form, the order in which the elements are listed, is immaterial. Thus, the above can also be represented as {2, 6, 4}.

13. *(c)* Factors of 42 are 1, 2, 3, 6, 7, 14, 21, 42.

The set of natural numbers which divide 42, is represented by {1, 2, 3, 6, 7, 14, 21, 42}.

14. *(a)* The vowels in English alphabet are *a, e, i, o, u*.

The set of vowels in roster form can be represented by $\{a, e, i, o, u\}$.

15. *(d)* The set of odd natural numbers is represented by {1, 3, 5, ...}. The dots tell us that the list of odd numbers continue indefinitely.

16. *(c)* While writing the set in roster form, an element is not generally repeated, *i.e.*, all elements are taken as distinct. The set of letters forming the word 'SCHOOL' is {S, C, H, O, L} or {H, O, L, C, S}. Here, the order of listing elements has no relevance. We can also express it as {S, C, H, O, L}.

17. *(b)* In set-builder form, all the elements of a set possess a single common property which is not possessed by any element outside the set. For example, in the set $\{a, e, i, o, u\}$, all the elements possess a common property, namely, each of them is a vowel in the English alphabet and no other letter possess this property. Denoting this set by *V*, we write

$$V = \{x : x \text{ is a vowel in English alphabet}\}$$

It may be observed that we describe the element of the set by using a symbol *x* (any other symbol like the letters *y, z*, etc., could be used) which is followed by a colon " : ". After the sign of colon, we write the characteristic property possessed by the elements of the set and then enclose the whole description within braces. The above description of the set *V* is read as "the set of all *x* such that *x* is a vowel of the English alphabet".

In this description, the braces stand for "the set of all", the colon stands for "such that".

18. *(b)* In set-builder form, all the elements of a set possess a single common property which is not possessed by any element outside the set. Clearly, *A* and *C* represent the set in set-builder form.

19. *(b)* The given equation can be written as

$$x^2 + 2x - x - 2 = 0$$

or $$(x - 1)(x + 2) = 0$$

i.e., $$x = 1, -2$$

$\therefore$ The solution set of the given equation can be written in roster form as $\{1, -2\}$.

20. *(c)* The required numbers are 1, 2, 3, 4, 5, 6. So, the given set in the roster form is {1, 2, 3, 4, 5, 6}.

21. *(b)* We may write the set A as

$A = \{x : x \text{ is the square of a natural number}\}$

Alternatively, we can write

$A = \{x : x = n^2, \text{where } n \in N\}$.

22. *(c)* We see that each member in the given set has the numerator one less than the denominator. Also, the numerator begins from 1 and do not exceed 6. Hence, in the set-builder form, the given set is $\{x : x = \frac{n}{n+1}$, where $n \in N$ and $1 \le n \le 6\}$.

23. *(c)* Since in (4), there are 9 letters in the word 'PRINCIPAL' and two letters P and I are repeated, so (A) matches (4). Similarly, (B) matches (3) as $x + 1 = 1 \Rightarrow x = 0$. Also, 1, 2, 3, 6, 9, 18 are all divisors of 18 and so (C) matches (1). Finally, $x^2 - 9 = 0 \Rightarrow x = 3, -3$ and so (D) matches (2).

24. *(c)* The collection of all months of a year beginning with the letter J and the collection of all boys in your class are well-defined. But the collection of ten most talented writers of India and a team of eleven best cricket batsmen of the world may vary from person-to-person, so these are not well-defined. Hence, I and IV represent the sets.

25. *(b)* The collection of most dangerous animals of the world may vary from animal-to-animal, so it is not well-defined collection. Therefore, it is not a set. Hence, (i), (ii) and (iii) are the sets as all of them are well-defined collection.

26. *(a)* "The collection of all natural numbers less than 100", is a well-defined collection. So, it is a set.

27. *(a)* 2, 4 and 5 are the elements of set A. So, we write $2 \in A$, $5 \in A$ and $4 \in A$. 0, 8 and 10 are not the elements of set A. So, we write $0 \notin A$, $8 \notin A$ and $10 \notin A$.

28. *(b)* The required numbers are –2, –1, 0, 1, 2, 3, 4, 5, 6. So, the given set in roster form is $\{-2, -1, 0, 1, 2, 3, 4, 5, 6\}$.

29. *(b)* The required numbers are 1, 2, 3, 4, 5. So, the given set in roster form is {1, 2, 3, 4, 5}.

30. *(a)* The two-digit natural numbers, sum of whose digits is 8, are 17, 26, 35, 44, 53, 62, 71, 80. So, the given set in the roster form is {17, 26, 35, 44, 53, 62, 71, 80}.

31. *(b)* We can write $60 = 2 \times 2 \times 3 \times 5$

$\therefore$ Prime factors of 60 are 2, 3 and 5.

Hence, the set D in roster form is {2, 3, 5}.

There are 12 letters in the word 'TRIGONOMETRY' out of which three letters T, R and O are repeated.

Hence, set E in the roster form is {T, R, I, G, O, N, M, E, Y}.

32. *(a)* There are 6 letters in the word 'BETTER' and two letters E and T are repeated. So, the given set can be represented in roster form as {B, E, T, R}.

33. *(d)* We see that each member in the given set is a multiple of 3 and $1 \le n \le 4$. Hence, the set-builder form of the given set is $\{x : x = 3n, \text{where } n \in N \text{ and } 1 \le n \le 4\}$.

34. *(c)* The given set in the set-builder form is

$\{x : x = 2^n, \text{ where } x \in N \text{ and } 1 \le n \le 5\}$

or $\{x : x = 2^n, \text{ where } x \in N \text{ and } 1 \le n < 6\}$.

35. *(d)* The set {5, 25, 125, 625} in set-builder form is $\{x : x = 5^n, \text{ where } x \in N \text{ and } 1 \le n \le 4\}$.

The set {2, 4, 6, ...} in the set-builder form is $\{x : x = 2n$, where $x \in N\}$ or $\{x : x$ is an even natural number$\}$.

36. *(b)* We see that each member in the given set is the square of a natural number. Hence, the given set in set-builder form is $\{x : x = n^2, \text{ where } n \in N \text{ and } 1 \le n \le 10\}$. Therefore, A is true. Also, R is true but it is not a correct explanation of A.

37. *(b)* The given set can be represent in roster form as {1, 3, 5, 7, ...}.

38. *(b)* The required numbers are 0, 1, 2, 3, 4. So, the given set in roster form is {0, 1, 2, 3, 4}.

39. *(c)* The required numbers are –2, –1, 0, 1, 2. So, the given set can be represented as $\{-2, -1, 0, 1, 2\}$.

40. *(b)* There are 5 letters in the word 'LOYAL' and L is repeated So, it can be represented as {L, O, Y, A}.

41. *(c)* The months not containing 31 days are February, April, June, September, November. So, the given set in roster form is {February, April, June, September, November}.

The given set can be represented in roster form as $F = \{b, c, d, e, f, g, h, j\}$. Hence, both the statements are true.

42. *(a)* Factors of 6 are 1, 2, 3, 6, so (A) matches (3) and (B) matches (1). Clearly, (C) matches (4). Also, 1, 3, 5, 7, 9 are odd natural numbers and so (D) matches (2).

43. *(b)* The natural number whose square is the number itself is 1. So, the given set in the roster form is {1}.

44. *(c)* Since, $\frac{x-1}{x+3} = 4$

$$\Rightarrow \quad x - 1 = 4x + 12$$

$$\Rightarrow \quad x = -\frac{13}{3}$$

So, the given set in roster form is $\left\{-\frac{13}{3}\right\}$.

45. *(c)* $x^4 - 5x^2 + 6 = 0$

$$\Rightarrow \quad x^4 - 3x^2 - 2x^2 + 6 = 0$$

$$\Rightarrow \quad (x^2 - 3)(x^2 - 2) = 0$$

$$\Rightarrow \quad x = \sqrt{2}, -\sqrt{2}, \sqrt{3}, -\sqrt{3}$$

So, the given set in roster form is $\{-\sqrt{3}, -\sqrt{2}, \sqrt{2}, \sqrt{3}\}$.

46. *(b)* Factors of 35 are 1, 5, 7 and 35.

So, I is true.

Factors of 128 are 1, 2, 4, 8, 16, 32, 64 and 128.

Sum of factors of 128 $= 1 + 2 + 4 + 8 + 16 + 32 + 64 + 128$
$= 255 \neq 2\,(128)$

So, II is false.

Let $f(x) = x^4 - 5x^3 + 2x^2 - 112x + 6$

$f(2) = (2)^4 - 5(2)^3 + 2(2)^2 - 112(2) + 6$
$= 16 - 40 + 8 - 224 + 6$
$= -234$

$\therefore$ 2 is not the solution of $x^4 - 5x^3 + 2x^2 - 112x + 6 = 0$.

So, III is true.

Prime factors of 180 and 2, 3, 5.

$\therefore\ 5 \in \{x : x \text{ is a prime factor of } 180\}$

So, IV is false.

47. *(c)* The required number is 2. So, the given set in roster form is {2}.

48. *(c)* Y contains the square of 1, 2 and 3. So, the given set is {4, 5, 6, 7, 8, 9, 10}.

49. *(c)* $\because a + 2 = 6$

$\Rightarrow \quad a = 4$

Hence, the given set is {4}.

50. *(a)* The required numbers are 1, 2, 3, 4. So, the given set is {1, 2, 3, 4}.

51. *(d)* The factors of $2^{p-1}(2^p - 1)$, where $2^p - 1$ is a prime number, are $1, 2, 2^2, 2^3, \ldots, 2^{p-1}$ and $(2^p - 1)$. So, the given set in roster form is $\{1, 2, 2^2, 2^3, \ldots, 2^{p-1}, (2^p - 1)\}$.

52. *(a)* Since, $3^x - 1$ is an even number for all $x \in Z^+$. So, the given set in roster form is {1, 2, 3, 4, 5}.

53. *(d)* The factors of 17 are 1 and 17.

$\therefore$ I is false.

Since, 7745 is not a multiple of 36.

$\therefore$ II is false.

54. *(d)* 2^x is an even number $\forall\ x \in N$.

3^x is an odd number $\forall\ x \in N$.

$2^x + 3^x$ is an odd natural number (as sum of even and odd numbers is an odd number)

$\therefore$ The given set in roster form is {5, 13, 35, ...}.

Hence, A is false but R is true.

55. *(b)* The set $\{x : x = e^n, x \in N\}$ in roster form is $\{e^1, e^2, e^3, \ldots\}$.

56. *(a)* We can count the number of students presently studying in class XI in the school. Thus, the set A contains a finite number of elements.

57. *(c)* A student cannot study simultaneously in both classes X and XI. Thus, the set B contains no element at all.

58. *(a)* A set which does not contain any element, is called empty set or null set or void set.

59. *(b)* The empty set is denoted by the symbol ϕ or { }.

60. *(c)* It is not possible, there is no student in class XI in a secondary school.

$\therefore$ Statement I is true.

It is also not possible that a student presently studying in both the classes X and XI. So, the set in Statement II is empty set. Hence, it is true.

61. *(c)* There is no natural number lies between 1 and 2. So, A is an empty set.

The equation $x^2 - 2 = 0$ is not satisfied by any rational value of x. So, B is the empty set.

62. *(a)* 2 is the only even prime number. So, A is the empty set.

The equation $x^2 = 4$ is not satisfied by any odd value of x. So, B is the empty set.

63. *(a)* There is no odd number which is divisible by 2.

2 is the only prime number which is even.

64. *(c)* There is no natural number which is simultaneously greater than 7 and less than 5.

There is no point common to any two parallel lines.

65. *(a)* The empty set contains no element.

66. *(c)* The equation $x^2 + x + 1 = 0$ is not satisfied by any integral value of x. So, A is the empty set.

20, 21 and 22 lie between 19 and 23.

Since, all numbers 20, 21 and 22 have more than two factors.

Hence, there is no prime number between 19 and 23.

67. *(a)* Given equation can be written as

$$x^2(x^2 - x - 1) = 0$$

$$\Rightarrow \quad x = 0, \frac{1 \pm \sqrt{5}}{2}$$

$\therefore$ There is no natural value of x which satisfies $x^4 - x^3 - x^2 = 0$.

68. *(a)* There is no natural number which is simultaneously greater than 5 and less than 3.

So, A is true. R is also true and a correct explanation of A.

69. *(b)* The set of real numbers which satisfy $x^2 - 1 = 0$, is $\{-1, 1\}$.

So, Statement I is false.

Given, $x^2 = 16$ and $2x = 6$

$x = 4, -4$ and $x = 3$

$\therefore$ There is no real x which simultaneously satisfied $x^2 = 16$ and $2x = 6$.

So, Statement II is true.

70. *(a)* Since, $x^2 + 1 = 0 \Rightarrow x^2 = -1$

There is no real number whose square is negative.

$\therefore$ A is empty set.

Now, $x^2 - 16 = 0$

$\Rightarrow x^2 = 16 \Rightarrow x = -4, 4$

So, B is not the empty set.

Again, $x^2 = x + 2$

$\Rightarrow x^2 - x - 2 = 0$

$\Rightarrow (x-2)(x+1) = 0 \Rightarrow x = -1, 2$

So, C is not the empty set.

71. *(d)* We observe that A contains 5 elements.

72. *(b)* By number of elements of a set S, we mean the number of distinct elements in S.

73. *(b)* The number of elements of a set S is represented by $n(S)$.

74. *(c)* If $n(S)$ is a natural number, then S is a non-empty finite set.

75. *(c)* We know that, a set which is empty or consists of a definite number of elements, is called finite, otherwise the set is called infinite. Since, set A contains finite number of elements. So, it is a finite set. We do not know the number of elements in B, but it is some natural number. So, B is also finite.

76. *(a)* W contains finite number of elements as there are seven days in a week. So, it is finite.

The equation $x^2 - 16 = 0$ has two solutions. So, S is finite.

There are infinite points on a line. So, G is infinite.

Note *All infinite sets cannot be described in the roster form.*

77. *(b)* Given, $(x-1)(x-2) = 0$

$\Rightarrow x = 1, 2 \in N$

$\therefore$ The set $\{x : x \in N \text{ and } (x-1)(x-2) = 0\}$ is finite.

There are infinite natural prime numbers. So, $\{x : x \in N$ and x is prime$\}$ is infinite.

2 is the only natural number which satisfies $x^2 = 4$. So, $\{x : x \in N \text{ and } x^2 = 4\}$ is finite.

78. *(b)* There is no natural number which satisfies $2x - 1 = 0$. So, $\{x : x \in N \text{ and } 2x - 1 = 0\}$ is an empty set, which is finite set.

There are infinite odd natural numbers. So, $\{x : x \in N$ and x is odd$\}$ is infinite.

There are infinite positive integers. So, $\{x : x \in Z$ and x is positive$\}$ is infinite.

79. *(b)* A year contains 12 months. So, the set of months of a year is finite.

80. *(c)* The set $\{1, 2, 3, \ldots\}$ contains infinite number of elements. So, it is infinite.

81. *(a)* The given set contains 100 elements. So, it is finite. Hence, A is true. R is also true and a correct explanation for A.

82. *(c)* There are infinite positive integer greater than 100. So, the set of positive integers greater than 100 is infinite.

There are 25 prime numbers less than 99.

So, the set of prime numbers less than 99 is finite.

83. *(a)* There are infinite lines parallel to X-axis. So, the set of lines parallel to X-axis is infinite.

There are infinite numbers which are multiple of 5. So, the set of numbers, which are multiple of 5, is infinite.

There are 26 letters in the English alphabet. So, the set of letters in the English alphabet is finite.

84. *(c)* We do not know the number of animals living on the Earth, but it is some natural number. So, the set of animals living on the Earth is finite. There are infinite circles passing through the origin (0, 0). So, the set of circles passing through the origin (0, 0) is infinite.

85. *(b)* Given equation is $x^4 = 1$

$\Rightarrow x^4 - 1 = 0$

$\Rightarrow (x^2 - 1)(x^2 + 1) = 0$

$\Rightarrow (x-1)(x+1) = 0 \quad (\because x^2 \neq -1, x \in R)$

$\Rightarrow x = -1, 1$

Hence, the given set is $\{-1, 1\}$.

86. *(b)* There are infinite real numbers lie between 0 and 1. So, the given set is infinite.

87. *(c)* There are infinite concentric circles in a plane. So, the given set is infinite.

Now, $x^2 - 3 = 0$

or $x^2 = 3$

or $x = \pm\sqrt{3}$

Thus, there is no rational number satisfied $x^2 - 3 = 0$. So, given set is null set.

88. *(c)* Given, $a_{n+1} = 3a_n$ and $a_1 = 1$

$$a_2 = 3a_1 = 3$$
$$a_3 = 3a_2 = 3^2$$
$$a_4 = 3a_3 = 3^3$$

So, the given set A in roster form is $\{1, 3, 3^2, 3^3, \dots\}$.

89. *(c)* Clearly, the set A is infinite.

90. *(c)* The given set is $\{(3+1), (4+1), (5+1), (6+1), (7+1)\}$ *i.e.*, $\{4, 5, 6, 7, 8\}$.

91. *(a)* The given set is $\{(3^2-1), (4^2-1), (5^2-1), (6^2-1), (7^2-1)\}$ *i.e.*, $\{8, 15, 24, 35, 48\}$.

92. *(b)* Set A contains 5 elements. So, it is finite.

93. *(d)* Since, the set of natural numbers is infinite. Then, the set $\{9^n - 8n - 1 : n \in N\}$ is also infinite.

Hence, A is incorrect but R is correct.

94. *(b)* $S = \{3, 6, 9, \dots, 99\}$

$$n(S) = 33$$
$$P = \{2, 3, 5, 7, 11, 13, 17, 19\}$$
$$n(P) = 8$$
$$n(S) + n(P) = 33 + 8 = 41$$

95. *(a)* If every element of A is also an element of B and if every element of B is also an element of A, then the sets A and B are equal. Clearly, the two sets have exactly the same elements.

96. *(b)* If two sets A and B are equal, then we write $A = B$.

97. *(b)* Two sets are said to be unequal, if they are not equal and we write $A \neq B$.

98. *(b)* Since, both sets have same elements. So, $A = B$.

99. *(a)* $A = \{2, 3, 5\}$

Factors of 30 are 1, 2, 3, 5, 6, 10, 15 and 30.

So, $B = \{2, 3, 5\}$

$\therefore$ $A = B$

Hence, Statement I is true.

We know that, a set does not change, if one or more elements of the set are repeated.

Hence, P and Q are equal.

100. *(c)* Since, $0 \in A$ and 0 does not belong to any of the sets B, C, D and E, it follows that $A \neq B, A \neq C, A \neq D, A \neq E$.

Since, $B = \phi$ but none of the other sets are empty. Therefore, $B \neq C, B \neq D$ and $B \neq E$. Also, $C = \{5\}$ but $-5 \in D$, hence $C \neq D$.

Since, $E = \{5\}, C = E$. Further, $D = \{-5, 5\}$ and $E = \{5\}$, we find that $D \neq E$. Thus, the only one pair of equal sets is C and E.

101. *(a)* $A = \{\text{A, L, O, Y}\}$, $B = \{\text{L, O, Y, A}\}$

Thus, A and B are equal.

102. *(b)* $A = \{-2, -1, 0, 1, 2\}$

Given equation is $x^2 - 3x + 2 = 0$

$\Rightarrow$ $(x-1)(x-2) = 0$

$\Rightarrow$ $x = 1, 2$

$\therefore$ $B = \{1, 2\}$

Since, $0 \in A$ and $0 \notin B$, A and B are not equal sets.

103. *(c)* I. Clearly, $\{a, b, c, d\} = \{d, c, b, a\}$

II. $12 \in \{4, 8, 12, 16\}$ but $12 \notin \{8, 4, 16, 18\}$

$\therefore$ $\{4, 8, 12, 16\} \neq \{8, 4, 16, 18\}$

III. $\{x : x$ is positive even integer and $x \leq 10\}$
$= \{2, 4, 6, 8, 10\}$

IV. $\{x : x$ is a multiple of $10\} = \{10, 20, 30, 40, \dots\}$

$\therefore$ $\{x : x$ is a multiple of $10\} \neq \{10, 15, 20, 25, \dots\}$

V. $14 \in \{4, 8, 12, 14\}$ but $14 \notin \{2, 4, 8, 12\}$

$\therefore$ $\{4, 8, 12, 14\} \neq \{2, 4, 8, 12\}$

104. *(b)* I. In L.H.S. there are 6 letters in the word 'FOLLOW' and two letters O and l are repeated, therefore L.H.S. $= \{F, L, O, W\}$ and in R.H.S. there are four different letters, therefore R.H.S. $= \{W, O, L, F\}$. Also, we know that two sets are equal if the number of elements and elements are same (irrespective of arrangements). Thus, L.H.S. = R.H.S.

II. Solution of $x^2 + 5x + 6 = 0$ is given by

$$x^2 + 2x + 3x + 6 = 0$$
$$x(x+2) + 3(x+2) = 0$$
$$(x+2)(x+3) = 0$$

$\Rightarrow$ $x = -2, -3$

$\therefore$ The solution set is $\{-2, -3\}$. Thus LHS $\neq$ RHS.

III. $\{1, 2, 3, 4,\} = \{3, 1, 4, 2\}$, as elements and number of elements are same.

IV. $\{1, -1\} = \{-1, 1\}$, as elements and number of elements are same.

V. $\{O, G\} \neq \{0, 1\}$, as elements in the sets are not same.

105. *(d)* $A = \{1, 2, 3\}$

$$B = \{x \in R : x^2 - 2x + 1 = 0\} = \{1\}$$
$$C = \{1, 2, 2, 3\} = \{1, 2, 3\}$$
$$D = \{x \in R : x^3 - 6x^2 + 11x - 6 = 0\} = \{1, 2, 3\}$$

$\therefore$ A, C and D are equal.

106. *(d)* $A = \{0, 1, 2\}$

$$B = \{x \in R : 0 \leq x \leq 2\}$$

There are infinite real numbers between 0 and 2. So, A and B are not equal.

Hence, A is false but R is true.

107. *(a)* $A = \{3, 4, 5\}$

$$B = \{5\}$$

So, A and B are not equal.

$$C = \{-3, -2, -1, 0, 1, 2, 3, 4, 5, 6\}$$
$$D = \{1, 2, 3, 4, 5, 6\}$$

So, C and D are not equal.

108. *(a)* In (A) on the left and (3) on the right have 5 different letters F, I, N, T, E. So, (A) matches (3). (B) represents the set $\{1, 3\}$. So, (B) matches (4). 2 is the only even prime number and $x^2 + x - 6 = 0 \Rightarrow (x+3)(x-2) = 0 \Rightarrow x = 2, -3$. So, (2) can be written as $\{2\}$. Therefore, (C) matches (2) and so, (D) matches (1).

109. *(a)* We note that every element of Y is also an element of X, as if a student is in your class, then he is also in your school.

110. *(a)* A set A is said to be a subset of a set B, if every element of A is also an element of B.

111. *(b)* If A is a subset of B, we write $A \subset B$ and if A is not a subset of B, then we write $A \not\subset B$.

In other words, $A \subset B$ if $a \in A$, then $a \in B$.

Now, if $A \subset B \Rightarrow$ Every element of A is in B and

$B \subset A \Rightarrow$ Every element of B is in A, then

we can say A and B are the same set, so that we have $A \subset B$ and $B \subset A \Leftrightarrow A = B$

where "$\Leftrightarrow$" is a symbol for two way implications, and usually read as if and only if (briefly written as "iff").

112. *(c)* From the definition of subset, it follows that every set is a subset of itself. Since, the empty set ϕ has no element, we agree to say that ϕ is a subset of every set.

113. *(b)* The set Q of rational numbers is a subset of the set R of real numbers, and we write $Q \subset R$.

114. *(c)* $A = \{1, 3, 5\}$

$B = \{x : x$ is an odd natural number less than $6\}$
$= \{1, 3, 5\}$

Since, every element of A is in B, so $A \subset B$.

Every element of B is in A, so $B \subset A$.

Then, $A = B$.

115. *(a)* Since, every element of A is in B, so $A \subset B$.

$\therefore$ A is true. Also, R is true and a correct explanation of A.

116. *(c)* Since, a student of class XI of your school is also a student of your school. So, (A) matches (2). Clearly, (B) matches (3). Since, set of natural numbers is a subset of real numbers. So, (C) matches (1). (D) represents the set of real numbers lie between 0 and 7. So, (D) matches (1).

117. *(a)* If $A \subset B$ and $A \neq B$, then A is called a proper subset of B and B is called a super set of A.

118. *(b)* It is clear that $\{2, 3, 4\} \subset \{1, 2, 3, 4, 5\}$

$\{a, b, c\} \not\subset \{b, c, d\}$

$\{x = x$ is an equilateral triangle in a plane$\}$

$\subset \{x = x$ is a triangle in the same plane$\}$

and $\{x = x$ is an even natural number$\}$

$\subset \{x = x$ is an integer$\}$

Hence, II and III are correct.

119. *(a)* If a set A has only one element, we call it a singleton set. Thus, $\{a\}$ is a singleton set.

120. *(c)* Since, null set is the subset of every set.

$\therefore \quad \phi \subset B$

121. *(d)* $3 \in A$ but $3 \notin B$. So, $A \not\subset B$.

122. *(c)* Since, every element of A is in C. So, $A \subset C$.

$B \subset C$, as each element of B is also an element of C.

123. *(b)* Let $A = \{1\}$, $B = \{\{1\}, 2\}$ and $C = \{\{1\}, 2, 3\}$.

Here, $A \in B$ as $A = \{1\}$ and $B \subset C$ but $A \not\subset C$ as $1 \in A$ but $1 \notin C$.

124. *(a)* Every natural number is an integer and every integer is a rational number. Every rational number is a real number.

125. *(b)* Refer solution (132)

$$T = \{x : x \in R \text{ and } x \in Q\}$$

126. *(a)* Let $a, b \in R$ and $a < b$. Then, the set of real numbers $\{x : a < x < b\}$ is called an open interval. And a, b do not belong to this interval.

127. *(d)* The open interval $a < x < b$ is represented by (a, b) or $]a, b[$. The interval $a \le x \le b$ contain end points also is called closed interval and is denoted by $[a, b]$. The interval $a \le x < b$ closed at the end a and open at the end b, *i.e.*, $[a, b)$. Similarly, the interval $a < x \le b$ is represented by $(a, b]$.

128. *(b)* The interval $A = (-3, 5)$ can be written as $\{x : -3 < x < 5\}$.

The interval $B = [-7, 9]$ can be written as $\{x : -7 \le x \le 9\}$.

Hence, $\quad A \subset B$

129. *(c)* The set of non-negative real numbers is denoted by $[0, \infty)$.

130. *(a)* The set of negative real numbers is denoted by $(-\infty, 0)$.

131. *(b)* The interval in the figure is $[a, b]$.

132. *(b)* (filled dot at a, open dot at b) represents $[a, b)$.

(open dot at a, filled dot at b) represents $(a, b]$.

(open dot at a, open dot at b) represents (a, b).

(filled dot at a, filled dot at b) represents $[a, b]$.

133. *(c)* The length of the interval (a, b) is $b - a$.

Note *The number $(b - a)$ is called the length of any intervals (a, b), $[a, b]$, $[a, b)$ or $(a, b]$.*

134. *(a)* The interval $\{x : x \in R, -4 < x \le 6\}$ is represented by $(-4, 6]$.

The interval $\{x : x \in R, -12 < x < -10\}$ is represented by $(-12, -10)$.

135. *(b)* I. $\{a, b\} \subset \{b, c, a\}$

II. Since, a and e are vowels, then $\{a, e\} \subset \{x : x$ is a vowel in the English alphabet$\}$
$\therefore$ II is true.

III. $\because \{1, 2, 3\} \not\subset \{1, 3, 5\}$
$\therefore$ III is false.

IV. $\because \{a\} \subset \{a, b, c\}$
$\therefore$ IV is true.

V. $\because \{a\}$ represent a subset of $\{a, b, c\}$
$\therefore \{a\} \notin \{a, b, c\}$
$\therefore$ V is false.

VI. $\because$ The given set can be written as
$\{2, 4\} \subset \{1, 2, 3, 4, 6, 9, 12, 18, 36\}$
$\therefore$ VI is true.

136. *(a)* $\because$ $\{3, 4\}$ is an element of A.

$\therefore \quad \{3, 4\} \in A$

$\Rightarrow \quad \{\{3, 4\}\} \subset A$

1 is also an element of A, so $1 \in A$

$1, 2, 5 \in A$

$\Rightarrow \quad \{1, 2, 5\} \subset A$

$\because \quad 3 \notin A \Rightarrow \{1, 2, 3\} \not\subset A$

137. *(b)* Refer solution (144)

$\because \phi$ is the subset of every set.

$\therefore \quad \phi \subset A$

138. *(c)* The number of subsets of a set containing n elements are 2^n.

ϕ is the subset of every set. So, ϕ is the subset of $\{a, b\}$. We see that $\{a\}$ and $\{b\}$ are the subsets of $\{a, b\}$. Every set is a subset of itself. So, $\{a, b\}$ is a subset of $\{a, b\}$.

Hence, subsets of $\{a, b\}$ are ϕ, $\{a\}$, $\{b\}$ and $\{a, b\}$.

Similarly, we can find the subsets of others.

139. *(b)* The interval $(-3, 0)$ is open at both the ends. So, it can be represent in set-builder form as $\{x : x \in R, -3 < x < 0\}$.

The interval $[6, 12]$ is closed at both end. So, it can be represent as $\{x : x \in R, 6 \le x \le 12\}$.

Similarly, we can represent other intervals.

140. *(b)* $n(A) = 3$

Then, the number of proper subsets $= 2^{n(A)} - 1$

$= 2^3 - 1 = 8 - 1 = 7$

141. *(d)* We know that, null set is the subset of all sets.

142. *(d)* a is not an element of $\{\{a\}, b\}$

$\therefore \quad a \notin \{\{a\}, b\}$

$\{b, c\}$ is the element of $\{a, \{b, c\}\}$

$\therefore \quad \{b, c\} \in \{a, \{b, c\}\}$

$b \in \{a, b\}$ but $b \notin \{a, \{b, c\}\}$

$\therefore \quad \{a, b\} \not\subset \{a, \{b, c\}\}$

143. *(c)* A. $2 \in \{\phi, \{1\}, 2\}$ and $\phi \in \{\phi, \{1\}, 2\}$

$\therefore \quad \{2, \phi\} \subset \{\phi, \{1\}, 2\}$

B. $\phi, \{1, \phi\} \in \{\phi, \{\phi\}, \{1, \phi\}\}$
$\therefore \{\phi, \{1, \phi\}\} \subset \{\phi, \{\phi\}, \{1, \phi\}\}$

C. $\{4, 5\} \in \{1, 2, \{4, 5\}\}$
$\therefore \{\{4, 5\}\} \subset \{1, 2, \{4, 5\}\}$

D. $4, 5 \in \{\phi, 1, 4, 5\}$
$\therefore \{4, 5\} \subset \{\phi, 1, 4, 5\}$

E. $1, \{4, 5\} \in \{1, 2, \{4, 5\}\}$
$\therefore \{1, \{4, 5\}\} \subset \{1, 2, \{4, 5\}\}$

F. $\phi, \{1\} \in \{\phi, \{1\}, 2\}$
$\therefore \{\phi, \{1\}\} \subset \{\phi, \{1\}, 2\}$

144. *(a)* Total number of subsets of $\{1, 2, 3, 4\} = 2^4 = 16$

But ϕ is the subset of every set.

$\therefore$ Number of non-empty subsets $= 16 - 1 = 15$

145. *(b)* 1 and $\{2\}$ are the element of $\{1, \{2\}\}$

So, the subsets of the set $\{1, \{2\}\}$ are ϕ, $\{1\}$, $\{\{2\}\}$ and $\{1, \{2\}\}$.

Hence, A is true.

We know, total number of proper subsets of a set containing n elements is $2^n - 1$.

Hence, R is true. But R is not the correct explanation of A.

146. *(c)* Let A and B are such sets, then

$n(A) = m, n(B) = n$

Total number of subsets of $A = 2^m$

Total number of subsets of $B = 2^n$

So, $\quad 2^m - 2^n = 56$

$\Rightarrow \quad 2^n(2^{m-n} - 1) = 56 = 8 \times 7$

$\Rightarrow \quad 2^n(2^{m-n} - 1) = 2^3 \times 7$

$\Rightarrow \quad n = 3, 2^{m-n} - 1 = 7$

$\Rightarrow \quad 2^{m-3} = 2^3$

On compairing powers we get,

$\Rightarrow \quad m = 6$

147. *(a)* Let $x \notin P$

$\Rightarrow x$ is a composite number.

Let us now assume that $x \in S$

$\Rightarrow \quad 2^x - 1 = m$ (where m is a prime number)

$\Rightarrow \quad 2^x = m + 1$

which is not true for all composite numbers, say for $x = 4$ because $2^4 = 16$ which cannot be equal to the sum of any prime number m and 1.

Thus, we arrive at a contradiction

$\Rightarrow \quad x \notin S$

Thus, when $x \notin P$, we arrive at $x \notin S$.

So, $\quad S \subset P$

148. *(c)* Since, $A \subset B$ and $B \subset C$

$\Rightarrow \quad A \subset C \quad$ ...(i)

Given, $\quad C \subset A \quad$ ...(ii)

From Eqs. (i) and (ii), we get

$A = C$

149. *(c)* The set A can be written as $[4, \infty)$ and set B can be written as $(5, \infty)$.

Clearly, $\quad B \subset A$

$\therefore$ Statement I is true.

Let, $\quad x_n = 8^n - 7n - 1 = (1+7)^n - 7n - 1$

$= {}^nC_2\, 7^2 + {}^nC_3\, 7^3 + \ldots + {}^nC_n\, 7^n$

$= 49({}^nC_2 + {}^nC_3 \cdot 7 + \ldots\ {}^nC_n\, 7^{n-2})$ for $n \ge 2$

and for $\quad n = 1, x_n = 0$

Thus, X contains all positive integral multipliers of the form $49k_n$, where $k_n = ({}^nC_2 + {}^nC_3 \cdot 7 \ldots {}^nC_n \cdot 7^{n-2})$.

But, Y contains all positive integral multiples of 49 including 0. Thus, $X \subset Y$.

$\therefore$ Statement II is true.

150. *(b)* Refer solution (147).

151. *(b)* The collection of all subsets of a set A is called the power set of A.

152. *(c)* The power set is denoted by $P(A)$.

153. *(a)* The subsets of the sets $\{1, 2\}$ are ϕ, $\{1\}$, $\{2\}$ and $\{1, 2\}$.

So, $\quad P(A) = \{\phi, \{1\}, \{2\}, \{1, 2\}\}$

$\therefore$ A is true. Also, R is true and R is the correct explanation of A.

154. *(d)* Let A is a set with $n(A) = m$

$\because$ Number of elements of power set of a given set is equal to number of subsets of the set.

$\therefore \quad n[P(A)] = 2^m$

155. *(b)* If $A = \phi$, then $n(A) = 0$

$\therefore \quad n[P(A)] = 2^0 = 1$

156. *(c)* $A = \{a, \{b\}\}$

The subsets of A are ϕ, $\{a\}$, $\{\{b\}\}$ and $\{a, \{b\}\}$.

$\therefore \quad P(A) = \{\phi, \{a\}, \{\{b\}\}, \{a, \{b\}\}\}$

157. *(c)* $n[P(\phi)] = 2^0 = 1 \quad [\because n(\phi) = 0]$

$n[P(P(\phi))] = 2^1 = 2$

$n[P\{P(P(\phi)\}] = 2^2 = 4$

158. *(c)* We have, $X \subset A$ and $X \not\subset B$

X is a subset of A but X is not a subset of B.

$\Rightarrow \quad \{1\}, \{3\}, \{1, 2\}, \{1, 3\}, \{2, 3\}, \{1, 2, 3\}$

Hence, the set of all sets X is

$\{\{1\}, \{3\}, \{1, 2\}, \{1, 3\}, \{2, 3\}, \{1, 2, 3\}\}$.

159. *(c)* In set theory, universal set is the set that contains all sets under consideration. So, in given problem, the basic set is universal set.

160. *(b)* The universal set is usually denoted by U.

161. *(d)* The set of rational number Q is the subset of real number R. Therefore, the universal set for Q is R.

162. *(c)* In human population studies, the universal set is the set of all people in the world.

163. *(b)* From all the three sets, set of all triangles in a plane is the universal set for set of isosceles triangle.

164. *(b)* The universal set must contain the elements 0, 1, 2, 3, 4, 5, 6 and 8.

165. *(c)* $A = \{x : x^2 - 5x + 6 = 0\} = \{2, 3\}$

$B = \{x : x^2 - 3x + 2 = 0\} = \{1, 2\}$

Hence, the universal set for A and B with least elements is $\{1, 2, 3\}$.

166. *(c)* Since, the set of vowels is the subset of the set of English alphabets. So, the set of English alphabets is the universal set for set of vowels in English alphabets.

Hence, A is true.

We know set of vowels is not the subset of the set of consonants.

Hence, R is false.

167. *(b)* Let $X = \{1, 2\}$

$\therefore \quad P(X) = \{\phi, \{1\}, \{2\}, \{1, 2\}\}$

$\because \quad 1 \in X$ but $1 \notin P(X)$

$\therefore \quad X \not\subset P(X)$

Hence, $P(X)$ is not the universal set for X.

By the definition of universal set, Statement II is true.

168. *(c)* Most of the relationships between sets can be represented by Venn diagrams.

169. *(d)* Venn diagrams are named after the English Logician, John Venn (1834-1883).

170. *(a)* Venn diagrams consist of rectangles and closed curves usually circles.

171. *(b)* The universal set is represented usually by a rectangle and its subsets by circles.

172. *(c)* The universal set contains all the elements in Venn diagram.

$\therefore \quad U = \{1, 2, 3, 4, \ldots, 10\}$

173. *(c)* The set $A = \{2, 4, 6, 8, 10\}$

$\left(\because \text{ the element in the circle are the elements of the set representing by circle}\right)$

174. *(d)* $U = \{1, 2, 3, 4, \ldots, 10\}$

$$A = \{2, 4, 6, 8, 10\}$$
$$B = \{4, 6\}$$

$\because$ All the elements of B are also in A.

$\therefore \quad B \subset A$

$\Rightarrow$ Set B lies inside A in the Venn diagram

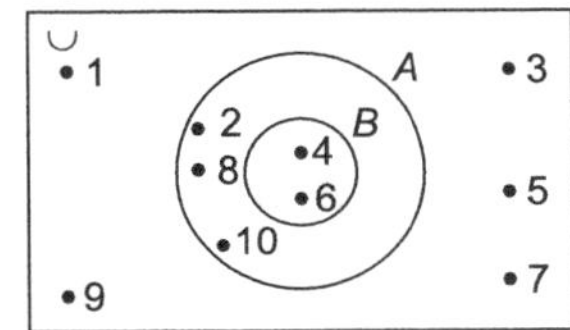

175. *(d)* If $A \subset B$, then every element of A is in B. Hence, this can be represented by Venn diagram as

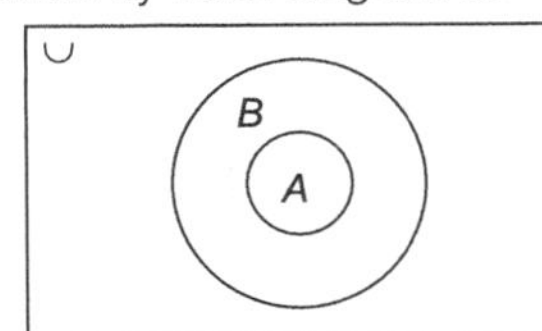

So, Statement I is false and Statement II is true.

176. *(c)* Since, all the students who study Mathematics, study English and some students who study English, do not study Mathematics. Then,

$$M \subset E$$

Thus, Venn diagram is

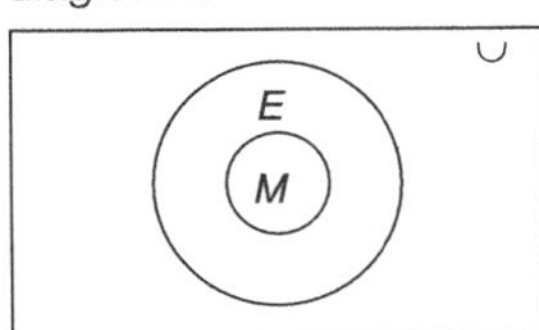

177. *(a)* Since, there are some students who study both English and Mathematics, some English only and some Mathematics only.

Thus, Venn diagram is

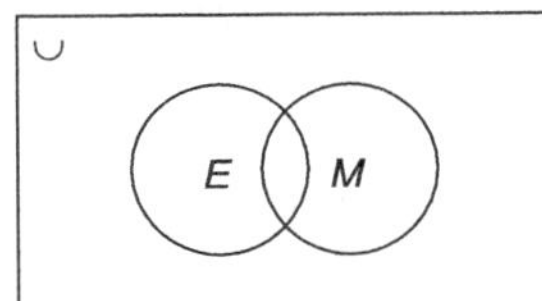

178. *(c)* The union of A and B is the set which consists of all the elements of A and all the elements of B. Common elements taken only once.

179. *(a)* Let A and B be two sets. Symbolically, the union of A and B write as $A \cup B$ and the common elements of A and B being taken only once.

180. *(d)* $A = \{2, 4, 6, 8\}$, $B = \{6, 8, 10, 12\}$

The common elements of A and B are 6 and 8 which have been taken only once.

$\therefore \quad A \cup B = \{2, 4, 6, 8, 10, 12\}$

181. *(a)* $A = \{a, e, i, o, u\}$

$B = \{a, i, u\}$

$$A \cup B = \{a, e, i, o, u\} = A$$

Note *We observe that* $B \subset A$.
Hence, if $B \subset A$, *then* $A \cup B = A$.

182. *(c)* $X = \{\text{Ram, Geeta, Akbar}\}$

$Y = \{\text{Geeta, David, Akbar}\}$

$\therefore X \cup Y = \{\text{Ram, Geeta, Akbar, David, Ashok}\}$

Hence, A is correct.

The union of two sets A and B is the set C which consists of all those elements which are either in A or in B (including those which are in both).

Hence, R is incorrect.

183. *(d)* In symbols, we write the union of two sets A and B as

$$A \cup B = \{x : x \in A \text{ or } x \in B\}$$

184. *(c)* The union of two sets A and B can be represented by a Venn diagram as

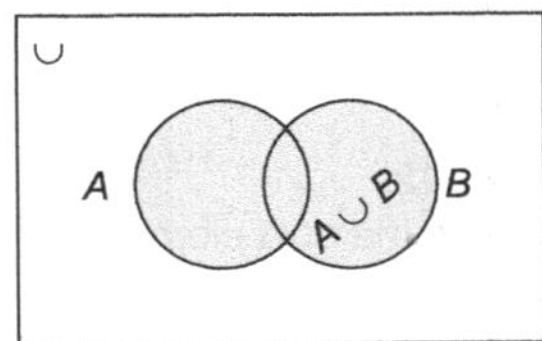

185. *(c)* Here are some properties of the operation of union.

A. $A \cup B = B \cup A$ (commutative law)
B. $(A \cup B) \cup C = A \cup (B \cup C)$ (associative law)
C. $A \cup \phi = A$ (law of identity, ϕ is the identity of U)
D. $A \cup A = A$ (idempotent law)
E. $U \cup A = U$ (law of U)

186. *(c)* A. $X = \{1, 3, 5\}$

$$Y = \{1, 2, 3\}$$
$$X \cup Y = \{1, 2, 3, 5\}$$

$\therefore$ (A) matches (3).

B. $A = \{3, 6, 9, 12, \ldots\}$

$$B = \{1, 2, 3, 4, 5\}$$
$$A \cup B = \{1, 2, 3, 4, 5, 6, 9, 12, \ldots\}$$

$\therefore$ (B) matches (3).

C. $A = \{2, 3, 4, 5, 6\}$

$B = \{7, 8, 9\}$

$A \cup B = \{2, 3, 4, 5, 6, 7, 8, 9\}$

$\therefore$ (C) matches (4).

So, (D) matches (1) and (E) matches (5).

187. *(b)* $A = \{a, b\}, B = \{a, b, c\}$

Since, all the elements of A are in B. So,

$A \subset B$

Hence, Statement I is false.

$\because$ $A \subset B$

$\Rightarrow$ $A \cup B = B$

Therefore, Statement II is true.

188. *(c)* $A \cup B = \{1, 2, 3, 4, 5, 6\}$

$B \cup C = \{3, 4, 5, 6, 7, 8\}$

189. *(d)* $A \cup B = \{1, 2, 3, 4, 5, 6\}$

and $C = \{5, 6, 7, 8\}$

$\therefore$ $A \cup B \cup C = \{1, 2, 3, 4, 5, 6, 7, 8\}$

190. *(a)* $B \cup D = \{3, 4, 5, 6, 7, 8, 9, 10\}$

$B \cup C \cup D = \{3, 4, 5, 6, 7, 8, 9, 10\}$

191. *(a)* $A \cup C = \{1, 2, 3, 4, 5, 6, 7, 8\}$

$A \cup B \cup D = \{1, 2, 3, 4, 5, 6, 7, 8, 9, 10\}$

$\Rightarrow$ $A \cup C \subset A \cup B \cup D$

192. *(d)* The intersection of two sets A and B is the set of all elements which are common to both A and B.

193. *(a)* The intersection of two sets A and B is the set of those elements which belong to both A and B. Symbolically, we write

$A \cap B = \{x : x \in A \text{ and } x \in B\}$

194. *(b)* $A = \{2, 4, 6, 8\}$

$B = \{6, 8, 10, 12\}$

Since, 6 and 8 are the only elements which are common to both A and B. Then,

$A \cap B = \{6, 8\}$

195. *(c)* Here, $X = \{\text{Ram, Geeta, Akbar}\}$

and $Y = \{\text{Geeta, David, Ashok}\}$

Then, $X \cap Y = \{\text{Geeta}\}$

196. *(b)* $A = \{1, 2, 3, 4, \dots, 10\}$

$B = \{2, 3, 5, 7\}$

$\therefore$ $A \cap B = \{2, 3, 5, 7\} = B$

197. *(b)* The intersection of two sets A and B is set of all those elements which belong to both A and B *i.e.*, $A \cap B$.

The shaded portion in the figure given in the question indicates the intersection of A and B *i.e.*, $A \cap B$.

198. *(c)* If A and B are two sets such that $A \cap B = \phi$, then A and B are called disjoint sets.

199. *(b)* $A = \{2, 4, 6, 8\}$

$B = \{1, 3, 5, 7\}$

$A \cap B = \phi$

Hence, A and B are disjoint.

200. *(d)* The disjoint sets can be represented by means of Venn diagram as

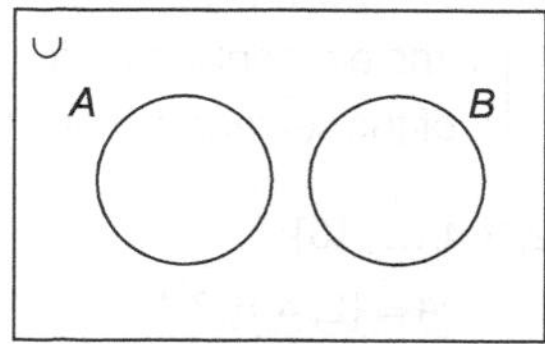

201. *(c)* Some properties of operation of intersection are as follows

A. $A \cap B = B \cap A$ (commutative law)

B. $(A \cap B) \cap C = A \cap (B \cap C)$ (associative law)

C. $\phi \cap A = \phi$ (law of ϕ)

D. $U \cap A = A$ (law of U)

E. $A \cap A = A$ (idempotent law)

F. $A \cap (B \cup C) = (A \cap B) \cup (A \cap C)$ (disributive law)

202. *(a)*

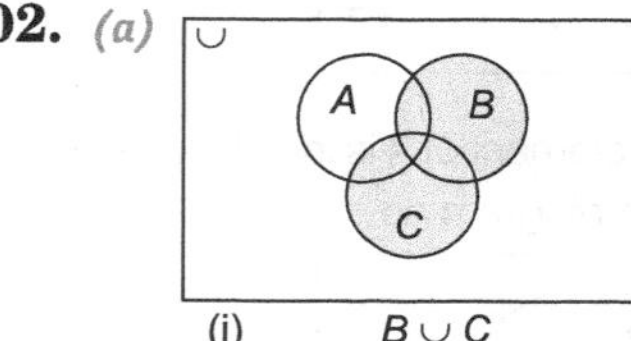

(i) $B \cup C$

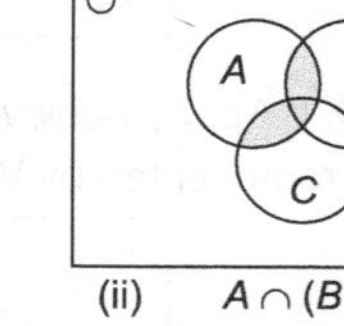

(ii) $A \cap (B \cup C)$

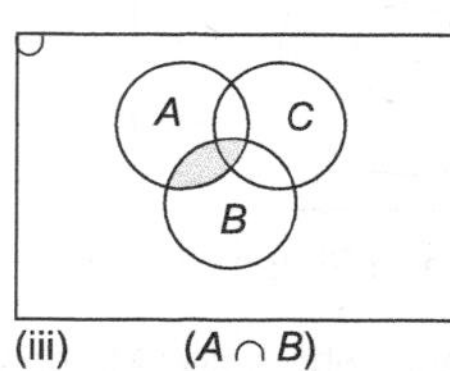

(iii) $(A \cap B)$

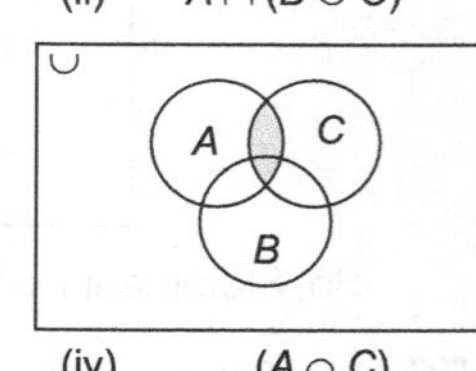

(iv) $(A \cap C)$

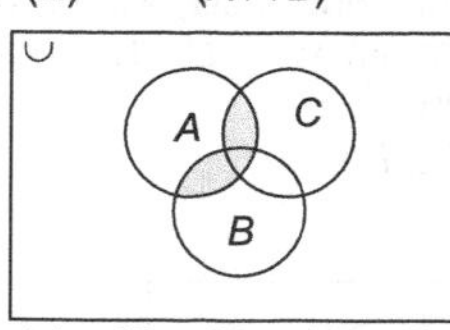

(v) $(A \cap B) \cup (A \cap C)$

From the above diagrams, we observe that

$A \cap (B \cup C) = (A \cap B) \cup (A \cap C)$

203. *(c)* I. Let, $A = \{1, 2, 3, 4\}$

$B = \{x : x \text{ is a natural number and } 4 \le x \le 6\}$

$= \{4, 5, 6\}$

$A \cap B = \{4\}$

Hence, A and B are not disjoint.

II. Let, $P = \{a, e, i, o, u\}$

$Q = \{c, d, e, f\}$

$P \cap Q = \{e\}$

Hence, P and Q are not disjoint.

III. Let, $X = \{x : x \text{ is an even integer}\}$
$= \{\dots -6, -4, -2, 2, 4, 6, \dots\}$
and $Y = \{x : x \text{ is an odd integer}\}$
$= \{\dots -5, -3, -1, 1, 3, 5, \dots\}$
$\therefore \quad X \cap Y = \phi$
Hence, X and Y are disjoint.

204. *(a)* $A = \{1, 2, 3, 4, \dots\}$
$B = \{2, 4, 6, 8, \dots\}$
$C = \{1, 3, 5, 7, \dots\}$
$D = \{x : x \text{ is a prime number}\}$
$= \{2, 3, 5, 7, 11, 13, \dots\}$
$\therefore$ A. $A \cap B = \{2, 4, 6, 8, \dots\} = B$
B. $A \cap C = \{1, 3, 5, 7, \dots\} = C$
C. $A \cap D = \{2, 3, 5, 7, \dots\} = D$
D. $B \cap C = \phi$
E. $B \cap D = \{2\}$
F. $C \cap D = \{3, 5, 7, 11, 13, \dots\}$

205. *(d)* $A = \{3, 5, 7, 9, 11\}, C = \{11, 13, 15\}$
and $D = \{15, 17\}$
$C \cap D = \{15\}$
$A \cap C \cap D = \phi$

206. *(d)* I. $A \cap B = \{7, 9, 11\}, B \cap C = \{11, 13\}$
$\therefore \quad A \cap B \neq B \cap C$
II. $A \cap C = \{11\}, B \cap D = \{\ \}$ or ϕ
$\therefore \quad A \cap C \neq B \cap D$
III. $B \cup C = \{3, 5, 7, 9, 11, 13\}$
$A \cap (B \cup C) = \{3, 5, 7, 9, 11\}$
$B \cup D = \{7, 9, 11, 13, 15, 17\}$
$A \cap (B \cup D) = \{7, 9, 11\}$
$\therefore \quad A \cap (B \cup C) \neq A \cap (B \cup D)$
IV. $A \cap D = \phi$
$A \cup B = \{3, 5, 7, 9, 11, 13\}$
$B \cup C = \{7, 9, 11, 13, 15\}$
$(A \cup B) \cap (B \cup C) = \{7, 9, 11, 13\}$
$\therefore \quad A \cap D \neq (A \cup B) \cap (B \cup C)$

207. *(c)* $A \cup D = \{3, 5, 7, 9, 11, 15, 17\}$
$B \cup C = \{7, 9, 11, 13, 15\}$
$(A \cup D) \cap (B \cup C) = \{7, 9, 11, 15\}$

208. *(b)* We know that, the intersection of two or more sets is the set of common elements.
A. $X \cap Y = \{1, 3, 5\} \cap \{1, 2, 3\} = \{1, 3\}$
B. $A \cap B = \{a, e, i, o, u\} \cap \{a, b, c\} = \{a\}$
C. $A \cap B = \{3, 6, 9, \dots\} \cap \{1, 2, 3, 4, 5\} = \{3\}$
D. $A \cap B = \{2, 3, 4, 5, 6\} \cap \{7, 8, 9\} = \phi$
E. $A \cap B = \{1, 2, 3\} \cap \phi = \phi$

209. *(c)* I. Let $A = \{2, 3, 4, 5\}$ and $B = \{3, 6\}$
$\therefore \quad A \cap B = \{3\}$
Hence, A and B are not disjoint.
II. Let $P = \{a, e, i, o, u\}$
and $Q = \{a, b, c, d\}$
$\therefore \quad P \cap Q = \{a\}$
Hence, P and Q are not disjoint.
III. Let $X = \{2, 6, 10, 14\}$
and $Y = \{3, 7, 11, 15\}$
$\therefore \quad X \cap Y = \{\ \}$ or ϕ
Hence, X and Y are disjoint.
IV. Let $E = \{2, 6, 10\}$
and $F = \{3, 7, 11\}$
$\therefore \quad E \cap F = \phi$ or $\{\ \}$
Hence, E and F are disjoint.

210. *(a)* The difference of sets A and B in this order is the set of elements which belong to A but not to B. Symbolically, we write $A - B$ and read as "A minus B".

211. *(b)* $A = \{1, 2, 3, 4, 5, 6\}, B = \{2, 4, 6, 8\}$
Since, the elements 1, 3, 5 belong to A but not to B.
$\therefore \quad A - B = \{1, 3, 5\}$
$B - A = \{8\}$
Since, the element 8 belongs to B and not to A.

212. *(c)* $V = \{a, e, i, o, u\}$
$V - B = \{e, o\}$
i.e., e and o are the elements belong to V but not to B.
$B - V = \{k\}$
i.e., k is the element belongs to B but not to V.
$\therefore \quad B = \{a, i, u, k\}$

213. *(b)* Using the set-builder form, we can write the definition of difference as
$$A - B = \{x : x \in A \text{ and } x \notin B\}$$

214. *(b)* In Venn diagram, difference of two sets A and B can be represented as

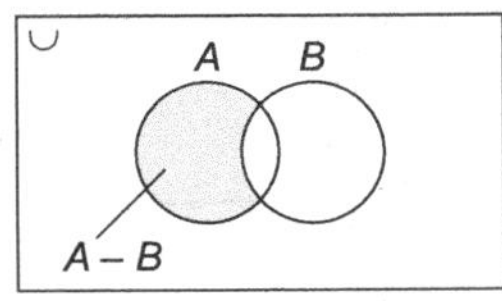

Hence, Statement I is false
Now,

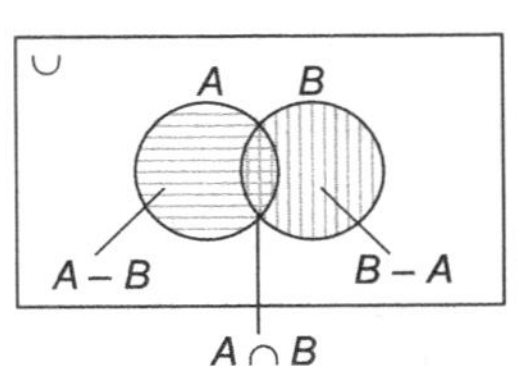

It is clear from the diagram $A - B$, $A \cap B$ and $B - A$ are mutually sets.
Hence, Statement II is true.

215. *(d)* Since, set of real numbers consists of all rational and irrational numbers. Therefore, $R - Q$ is the set of irrational numbers.

Hence, A is incorrect and R is correct.

216. *(a)* $X = \{a, b, c, d\}$ and $Y = \{f, b, d, g\}$

$\therefore \quad X - Y = \{a, c\}$

$Y - X = \{f, g\}$

$X \cap Y = \{b, d\}$

$X \cup Y = \{a, b, c, d, f, g\}$

217. *(c)* $A - B = \{3, 6, 9, 15, 18, 21\}$

218. *(d)* $A - C = \{3, 9, 15, 18, 21\}$

$A - D = \{3, 6, 9, 12, 18, 21\}$

$\therefore \quad A - C \neq A - D$

$B - A = \{4, 8, 16, 20\}$

$\therefore \quad A - D \not\subset B - A$ (as $3 \in A - D$ but $3 \notin B - A$)

$C - A = \{2, 4, 8, 10, 14, 16\}$

$D - A = \{5, 10, 20\}$

$\therefore \quad C - A \neq D - A$

$B - C = \{20\}$

$B - D = \{4, 8, 12, 16\}$

$\therefore \quad B - C \neq B - D$

219. *(c)* $C - B = \{2, 6, 10, 14\}$

$D - B = \{5, 10, 15\}$

$\therefore \quad (C - B) \cap (D - B) = \{10\}$

220. *(b)* $C - D = \{2, 4, 6, 8, 12, 14, 16\}$

$D - C = \{5, 15, 20\}$

$\therefore \quad (C - D) \cup (D - C)$

$= \{2, 4, 5, 6, 8, 12, 14, 15, 16, 20\}$

221. *(a)* $A \cup B = \{1, 2, 3, 4, 5, 6\}$

$(A \cup B) \cap C = \{3, 4, 6\}$

222. *(c)* The graph of A and B is represented by

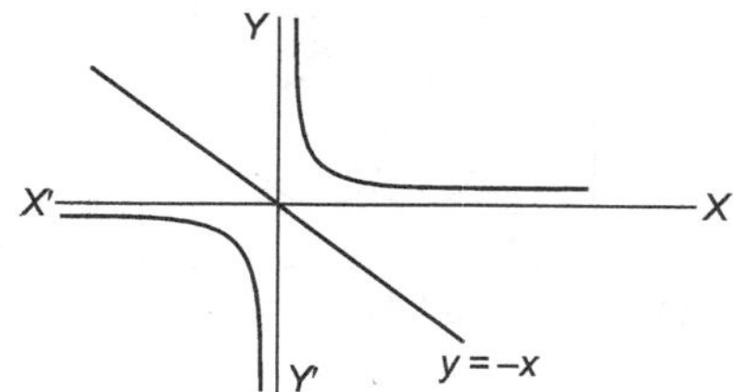

We see that two graphs do not intersect each other. Therefore,

$$A \cap B = \phi$$

223. *(a)* I.

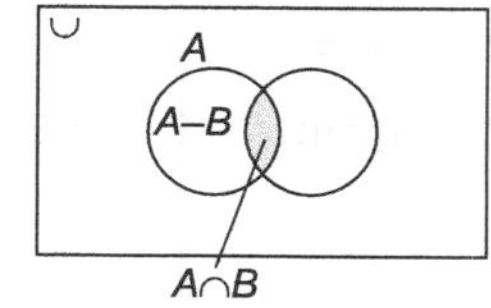

It is clear from the Venn diagram

$$A - B = A - (A \cap B)$$

II. Also, it is clear from above diagram

$$A = (A \cap B) \cup (A - B)$$

III.

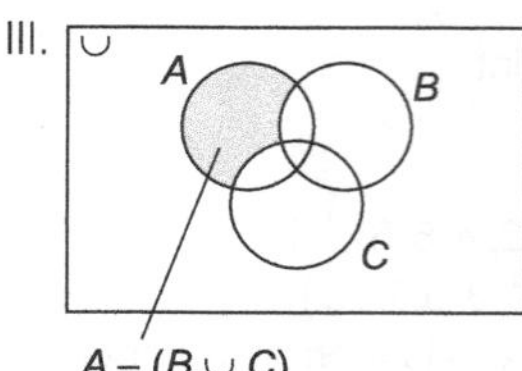

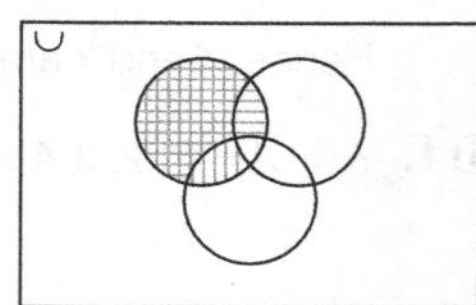

≡ (horizontal) lines : $A - C$
III (vertical) lines : $A - B$

It is clear from the diagrams

$$A - (B \cup C) = (A - B) \cap (A - C)$$

224. *(b)* If A, B and C are three non-empty sets.

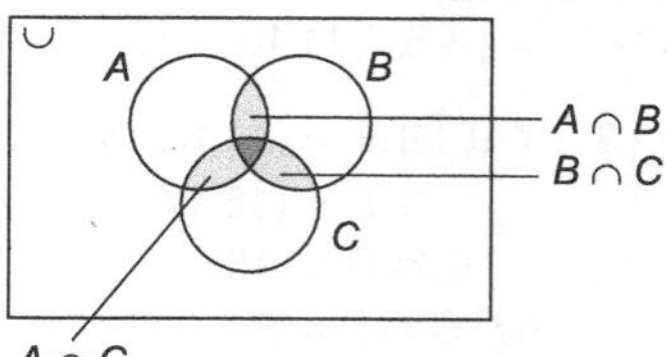

Clearly, $(A \cap B) \cap (B \cap C) \cap (A \cap C) = A \cap B \cap C$

Again,

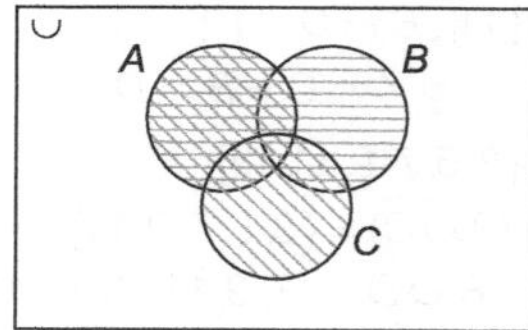

≡ (horizontal) lines : $A \cup B$
\\\ (slanted) lines : $A \cup C$

Clearly, $(A \cup B) \cup (A \cup C) = A \cup B \cup C$

Hence, both A and R are correct but R is not a correct explanation of A.

225. *(d)* $aN = \{ax : x \in N\}$, $3N = \{3x : x \in N\}$

$= \{x : x$ is a positive multiple of $3\}$

$7N = \{7x : x \in N\}$

$= \{x : x$ is a positive multiple of $7\}$

$3N \cap 7N = \{x : x$ is a positive multiple of $21\}$

$= \{21x : x \in N\} = 21N$

226. *(c)*

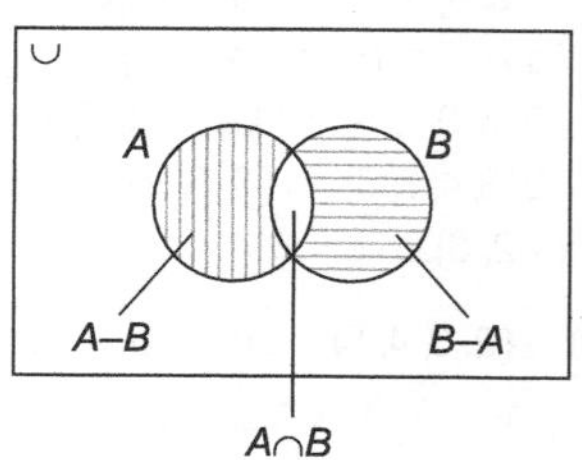

Clearly, $(A \cup B) - (A \cap B) = (A - B) \cup (B - A)$

227. *(b)* $A = \{1, 2, 3, 4, 5\}$

$B = \{x : x^2 - 5x + 6 = 0\} = \{2, 3\}$

$C = \{x : x^2 = 4 \text{ or } x = 1\} = \{-2, 1, 2\}$

$\therefore \quad A \cap (B \cap C) = 2 \neq \phi$

228. *(c)*

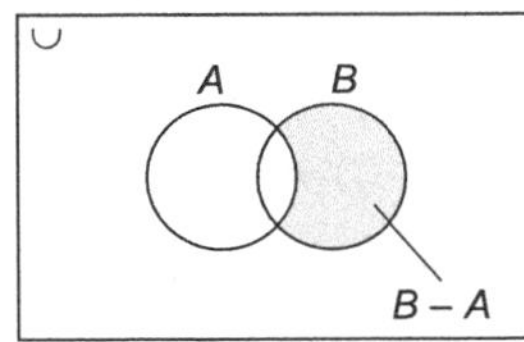

Clearly, $(A \cup B) - A = B - A$

So, (A) matches (2).

and $\quad A \cup (B - A) = A \cup B$

So, (C) matches (1).

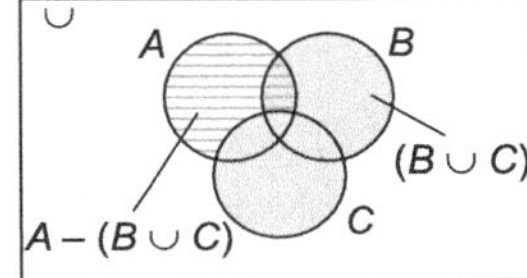

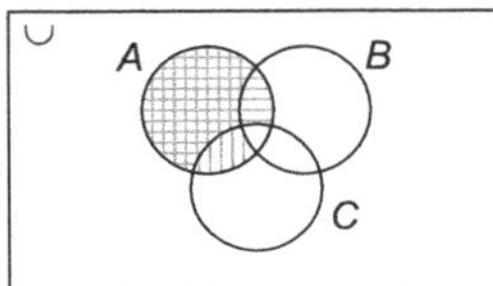

III represents $A - B$,
≡ represents $A - C$,
represents $(A - B) \cap (A - C)$

It is clear from above diagram

$A - (B \cup C) = (A - B) \cap (A - C)$

$\therefore$ (B) matches (4). Hence, (D) matches (3).

$A - (B - C) = (A - B) \cup (A \cap C)$

229. *(b)* $X = \{x : x \text{ is a positive integer} < 8 \text{ and } x \text{ is a multiple of } 3 \text{ or } 5\}$

$= \{3, 5, 6\}$

$Y = \{x : x \text{ is an even number } \leq 7\}$

$= \{2, 4, 6\}$

$X \cap Y = \{6\} \neq \phi$

Now, $\quad A = \{1, 2\}, B = \{2, 3\}$

$P(A) = \{\phi, \{1\}, \{2\}, \{1, 2\}\}$

$P(B) = \{\phi, \{2\}, \{3\}, \{2, 3\}\}$

$P(A) \cup P(B) = \{\phi, \{1\}, \{2\}, \{3\}, \{1, 2\}, \{2, 3\}\}$

$A \cup B = \{1, 2, 3\}$

$P(A \cup B) = \{\phi, \{1\}, \{2\}, \{3\}, \{1, 2\}, \{2, 3\}, \{1, 3\}, \{1, 2, 3\}\}$

$\{1, 2, 3\} \in P(A \cup B)$

But $\quad \{1, 2, 3\} \notin P(A) \cup P(B)$

$\therefore \quad P(A \cup B) \not\subset P(A) \cup P(B)$

230. *(b)* The smallest set $A = \{3, 5, 9\}$, such that

$A \cup \{1, 2\} = \{1, 2, 3, 5, 9\}$

231. *(d)* The shaded region in the figure is $A - (B \cup C)$.

232. *(c)* We have, $\quad A = B \cap C$

$\Rightarrow \quad A = (C \cap A) \cap C \quad (\because B = C \cap A)$

$\Rightarrow \quad A = C \cap A \Rightarrow A = B$

Now, $\quad A = \{x, y\}$

$P(A) = \{\phi, \{x\}, \{y\}, \{x, y\}\}$

$A \cap P(A) = \phi$

233. *(d)* Clearly, A is the set of points on the circle $x^2 + y^2 = 25$ and B is the set of points on the ellipse $x^2 + 9y^2 = 144$. These two intersect at four points. Hence, $A \cap B$ contains four points.

234. *(b)* If A_n is the set of first n prime numbers, then $\bigcup_{n=2}^{10} A_n$ is the set of first 10 prime numbers *i.e.*, {2, 3, 5, 7, 11, 13, 17, 19, 23, 29}.

235. *(d)* If A, B and C are three non-empty sets such that any two of them are disjoint, then

$A \cap B \cap C = \phi$

$\therefore \quad (A \cup B \cup C) \cap (A \cap B \cap C)$

$= (A \cup B \cup C) \cap \phi = \phi$

236. *(c)* If $A \subset B$ and $B \subset C$, then these sets is represented in Venn diagram as

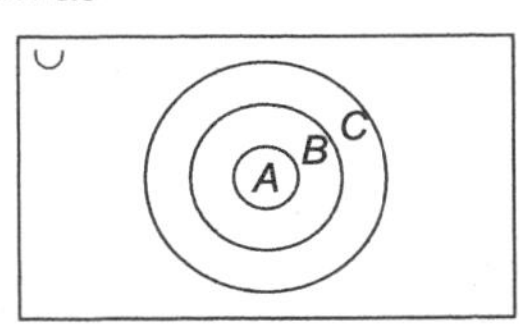

Clearly, $\quad A \cup B = B$

and $\quad B \cap C = B$

Hence, $\quad A \cup B = B \cap C$

237. *(c)* U = Set of prime numbers

$A = \{x : x \in U \text{ and } x \text{ is not a divisor of } 42\}$

$2 \in U$ and 2 is a divisor of 42

$\Rightarrow \quad 2 \notin A$

$3 \in U$ and 3 is a divisor of 42

$\Rightarrow \quad 3 \notin A$

$7 \in U$ and 7 is a divisor of 42

$\Rightarrow \quad 7 \notin A$

$\therefore$ Set of the elements which do not belong to A is {2, 3, 7}.

238. *(b)* Let U be the universal set and A is a subset of U. Then, the complement of A is the set of all elements of U which are not the elements of A. Symbolically, we write A' to denote the complement of A with respect to U. Thus,

$A' = \{x : x \in U \text{ and } x \notin A\}$

Obviously, $\quad A' = U - A$

239. *(a)* $A' = U - A$

= Set of those elements of U which are not in A

$= \{2, 4, 6, 8, 10\}$

240. (b) U = Set of all the students of class XI

A = Set of all girls in class XI

$A' = U - A$

= Set of all boys in class XI

241. (c) If A is a subset of the universal set U, then its complement A' is also a subset of U.

We have, $A' = \{2, 4, 6, 8, 10\}$

Hence, $(A')' = \{x : x \in U \text{ and } x \notin A'\}$

$= \{1, 3, 5, 7, 9\} = A$

It is clear from the definition of the complement that for any subset of the universal set U, we have

$$(A')' = A$$

242. (c) $A' = U - A = \{1, 4, 5, 6\}$

243. (b) $B' = U - B = \{1, 2, 6\}$

244. (b) Clearly, $A' = \{1, 4, 5, 6\}, B' = \{1, 2, 6\}$.

Hence, $A' \cap B' = \{1, 6\}$.

Also, $A \cup B = \{2, 3, 4, 5\}$, so that $(A \cup B)' = \{1, 6\}$

$(A \cup B)' = \{1, 6\} = A' \cap B'$

It can be shown that the above result is true in general. If A and B are any two subsets of the universal set U, then

$$(A \cup B)' = A' \cap B'$$

Similarly, $(A \cap B)' = A' \cup B'$

245. (c) The complement of the union of two sets is the intersection of their complements and the complement of the intersection of two sets is the union of their complements. These are called De-Morgan's laws. These are named after the Mathematician De-Morgan.

246. (a) In the given figure, the shaded portion represents complement of set A.

247. (b) By properties of complement of a set,

A. $A \cup A' = U$

B. $A \cap A' = \phi$

By De-Morgan's laws

C. $(A \cup B)' = A' \cap B'$

D. $(A \cap B)' = A' \cup B'$

By laws of empty set and universal set

E. $\phi' = U$ and

F. $U' = \phi$

By law of double complementation

G. $(A')' = A$

248. (c) $A' = U - A = \{5, 6, 7, 8, 9\}$

$B' = U - B = \{1, 3, 5, 7, 9\}$

249. (d) $A \cup C = \{1, 2, 3, 4, 5, 6\}$

$(A \cup C)' = U - (A \cup C) = \{7, 8, 9\}$

$A \cup B = \{1, 2, 3, 4, 6, 8\}$

$(A \cup B)' = U - (A \cup B) = \{5, 7, 9\}$

$\therefore (A \cup C)' \neq (A \cup B)'$

We know, $(A')' = A$

$B - C = \{2, 8\}$

$(B - C)' = U - (B - C)$

$= \{1, 3, 4, 5, 6, 7, 9\}$

$B' - C' = (U - B) - (U - C)$

$= \{1, 3, 5, 7, 9\} - \{1, 2, 7, 8, 9\}$

$= \{3, 5\}$

$\therefore \quad (B - C)' \neq B' - C'$

250. (b) U = Set of natural numbers

Let $A = \{x : x \text{ is an even natural number}\}$

$A' = U - A$

$= \{x : x \text{ is an odd natural number}\}$

$B = \{x : x \text{ is a prime number}\}$

$B' = U - B$

$= \{x : x \text{ is not a prime number}\}$

$C = \{x : x \text{ is a multiple of } 3\}$

$C' = U - C$

$= \{x : x \text{ is not a multiple of } 3\}$

$D = \{x : x \text{ is a natural number divisible by 3 and 5}\}$

$= \{x : x \text{ is a natural number divisible by } 15\}$

$D' = U - D$

$= \{x : x \text{ is a natural number not divisible by } 15\}$

$E = \{x : x \text{ is a perfect square}\}$

$E' = U - E$

$= \{x : x \text{ is not a perfect square}\}$

$F = \{x : x \text{ is a perfect cube}\}$

$F' = U - F = \{x : x \text{ is not a perfect cube}\}$

251. (d) U = Set of natural numbers

$A = \{x : x + 5 = 8\}$

$= \{3\}$

$A' = U - A$

$= \{1, 2, 4, 5, 6, \dots\}$

$B = \{x : 2x + 5 = 9\}$

$= \{2\}$

$B' = U - B$

$= \{1, 3, 4, 5, 6, \dots\}$

$C = \{x : x \geq 7\}$

$= \{7, 8, 9, 10, \dots\}$

$C' = U - C$

$= \{1, 2, 3, 4, 5, 6\}$

$D = \{x : x \in N \text{ and } 2x + 1 = 10\} = \phi$

$D' = U - D = U$

252. (c) $A' = U - A = \{1, 3, 5, 7, 9\}$

$B' = U - B = \{1, 4, 6, 8, 9\}$

$A \cup B = \{2, 3, 5, 6, 7, 8\}$

$(A \cup B)' = U - (A \cup B) = \{1, 9\}$

$A' \cap B' = \{1, 9\}$

$\therefore \quad (A \cup B)' = A' \cap B'$

$$A \cap B = \{2\}$$
$$(A \cap B)' = U - (A \cap B)$$
$$= \{1, 3, 4, 5, 6, 7, 8, 9\}$$
$$A' \cup B' = \{1, 3, 4, 5, 6, 7, 8, 9\}$$
$$\therefore \quad (A \cap B)' = A' \cup B'$$

253. *(c)*

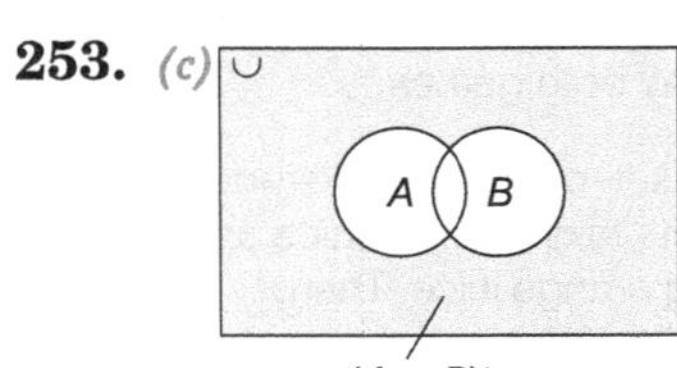

$(A \cup B)'$

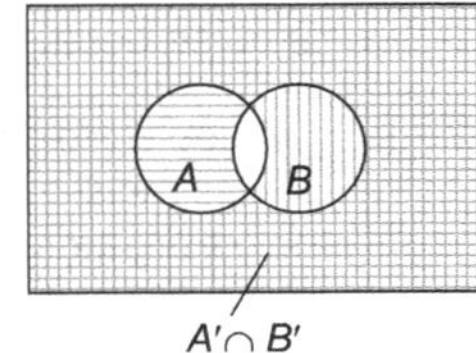

$A' \cap B'$

Clearly, $(A \cup B)'$ and $A' \cap B'$ are same.

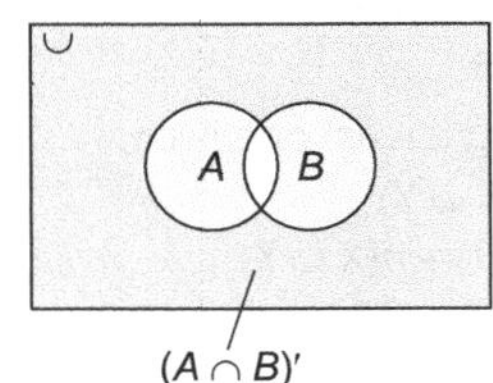

$(A \cap B)'$

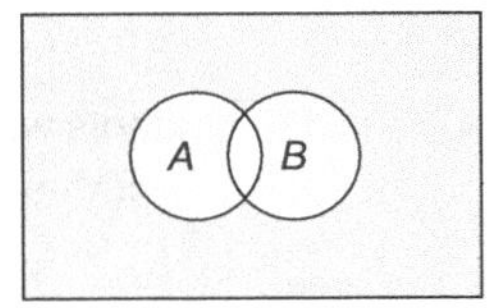

The shaded region is $A' \cup B'$

Cleary, $(A \cap B)'$ and $A' \cup B'$ are same.

254. *(b)* $A =$ Set of all triangles atleast one angle different from 60°.

$A' =$ Set of all triangles with all angles is of 60°.

255. *(a)* $\phi' \cap A = U \cap A = A$

$U' \cap A = \phi \cap A = \phi$

256. *(a)* $A \cup A' = A \cup (U - A))$

$A \cup A' = U$

and $A \cap A' = A \cap (U - A) = \phi$

257. *(d)* $A = \{2, 4, 6, 8, \dots\}$

$$B = \{1, 3, 5, 7, \dots\}$$
$$A \cap B = \phi$$
$$(A \cap B)' = U$$

258. *(a)* $(A \cup B)' \cup (A' \cap B)$

$= (A' \cap B') \cup (A' \cap B)$

[by De-Morgan's law, $(A \cup B)' = A' \cap B'$]

$= A' \cap (B' \cup B)$ (by distribution of $\cup$ over $\cap$)

$= A' \cap U = A'$

259. *(b)* Let $x \in A - B$

$\Rightarrow \quad x \in A$ and $x \notin B$

$\Rightarrow \quad x \in A$ and $x \in B'$

$\Rightarrow \quad x \in B'$

$\therefore \quad A - B \subset B'$

It is true $A \cap A' = \phi$ (by complement laws)

Hence, both A and R are correct but R is not a correct explanation of A.

260. *(b)* We know, $A - (A - B) = A \cap B$ (refer Venn diagram)

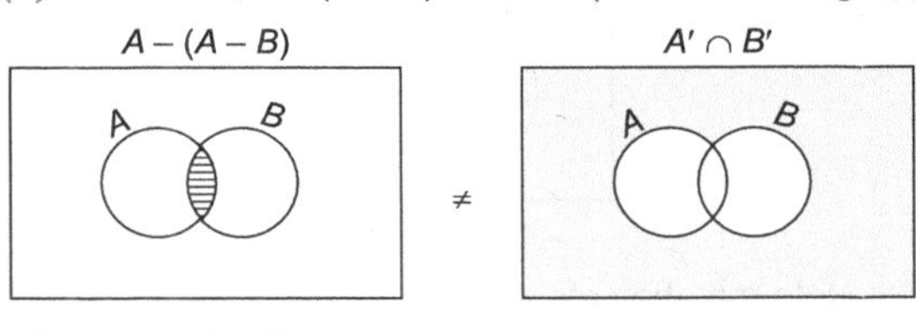

$\therefore$ Statement I is false.

Now, $B' = \{1, 2, 3, 4, 5, 9, 10\}$

$A \cap B' = \{1, 5\} = A$

$\therefore$ Statement II is true.

261. *(c)* It is clear from the Venn diagram,

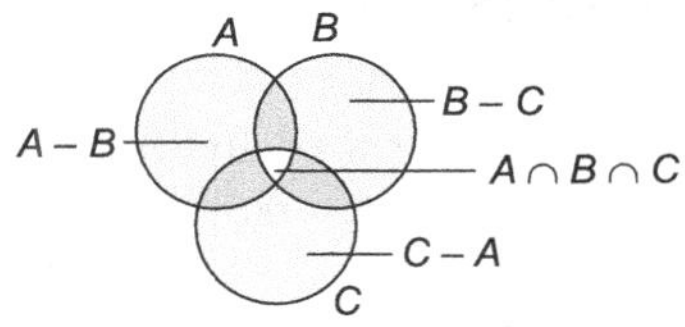

$(A - B) \cup (B - C) \cup (C - A)$ represents shaded region.

$\therefore [(A - B) \cup (B - C) \cup (C - A)]'$ represents unshaded region, which is $A \cap B \cap C$.

262. *(b)* $(A \cap B')' \cup (B \cap C)$

$= [A' \cup (B')'] \cup (B \cap C)$ (by De-Morgan's law)

$= (A' \cup B) \cup (B \cap C)$ $[\because (B')' = B]$

$= [(A' \cup B) \cup B] \cap [(A' \cup B) \cup C]$ (by distributive law)

$= (A' \cup B) \cap [(A' \cup B) \cup C]$

$= A' \cup B$

263. *(c)* $B \cup C = \{2, 3, 4, 6, 7, 8\}$

$$(B \cup C)' = U - (B \cup C) = \{1, 5, 9, 10\}$$
$$C - A = \{4, 8\}$$
$$(C - A)' = \{1, 2, 3, 5, 6, 7, 9, 10\}$$

264. *(a)* $(A \cap B' \cap C')' = A' \cup (B')' \cup (C')'$

$= A' \cup B \cup C$

$\therefore (A \cup B \cup C) \cap (A \cap B' \cap C')'$

$= (A \cup B \cup C) \cap (A' \cup B \cup C)$

$= (A \cap A') \cup (B \cup C)$ (by distributive law)

$= \phi \cup (B \cup C) = B \cup C$

Hence, $(A \cup B \cup C) \cap (A \cap B' \cap C')' \cup C'$

$= (B \cup C) \cap C'$

$= (B \cap C') \cup (C \cap C')$ (by distributive law)

$= B \cap C' \cup \phi = B \cap C'$

265. *(c)* If $A \cap B = \phi$, then A and B can be represented by Venn diagram as

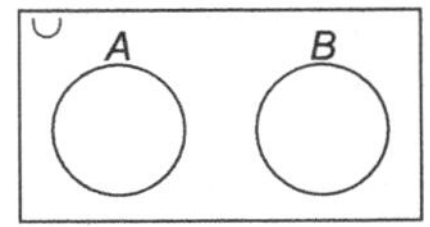

Clearly, from the diagram,

$$n(A \cup B) = n(A) + n(B)$$

266. *(b)*

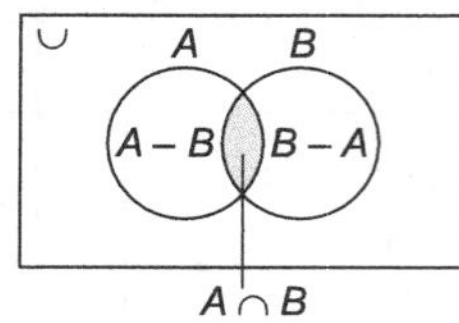

Note that the sets $A - B$, $A \cap B$ and $B - A$ are disjoint and their union is $A \cup B$. Therefore,

$$n(A \cup B) = n(A - B) + n(A \cap B) + n(B - A)$$
$$= n(A - B) + n(A \cap B) + n(B - A) + n(A \cap B) - n(A \cap B)$$
$$= n(A) + n(B) - n(A \cap B)$$

267. *(a)* If A, B and C are finite sets, then

$$n(A \cup B \cup C) = n(A) + n(B \cup C) - n[A \cap (B \cup C)]$$
$$[\because n(A \cup B) = n(A) + n(B) - n(A \cap B)]$$
$$= n(A) + n(B) + n(C) - n(B \cap C) - n[A \cap (B \cup C)] \quad \text{...(i)}$$

Since, $A \cap (B \cup C) = (A \cap B) \cup (A \cap C)$, we get

$$n[A \cap (B \cup C] = n(A \cap B) + n(A \cap C) - n[(A \cap B) \cap (A \cap C)]$$
$$= n(A \cap B) + n(A \cap C) - n(A \cap B \cap C)$$

Therefore, $n(A \cup B \cup C) = n(A) + n(B) + n(C) - n(A \cap B) - n(B \cap C) - n(A \cap C) + n(A \cap B \cap C)$

Now, if A, B and C are mutually pairwise disjoint, then

$$A \cap B = \phi = B \cap C = A \cap C = A \cap B \cap C$$

$\therefore \quad n(A \cup B \cup C) = n(A) + n(B) + n(C)$

268. *(a)* Given that,

$$n(X \cup Y) = 50, n(X) = 28, n(Y) = 32, n(X \cap Y) = ?$$

By the formula,

$n(X \cup Y) = n(X) + n(Y) - n(X \cap Y)$, we find that

$$n(X \cap Y) = n(X) + n(Y) - n(X \cup Y)$$
$$= 28 + 32 - 50 = 10$$

269. *(b)* Let M denote the set of teachers who teach Mathematics and P denote the set of teachers who teach Physics. In the statement of the problem, the word 'or' gives us a clue of union and the word 'and' gives us a clue of intersection. We have,

$$n(M \cup P) = 20, n(M) = 12$$

and $\quad n(M \cap P) = 4$

We wish to determine $n(P)$.

Using the result

$n(M \cup P) = n(M) + n(P) - n(M \cap P)$, we obtain

$$20 = 12 + n(P) - 4$$

Thus, $\quad n(P) = 12$

Hence, 12 teachers teach Physics.

270. *(c)* Let X be the set of students who like to play cricket and Y be the set of students who like to play football. Then, $X \cup Y$ is the set of students who like to play atleast one game, and $X \cap Y$ is the set of students who like to play both games.

Given, $n(X) = 24, n(Y) = 16, n(X \cup Y) = 35, n(X \cap Y) = ?$

Using the formula $n(X \cup Y) = n(X) + n(Y) - n(X \cap Y)$, we get

$$35 = 24 + 16 - n(X \cap Y)$$

Thus, $\quad n(X \cap Y) = 5$

i.e., 5 students like to play both games.

271. *(b)* Let U denote the set of surveyed students and X denote the set of students taking apple juice and Y denote the set of students taking orange juice. Then,

$n(U) = 400, n(X) = 100, n(Y) = 150$

and $\quad n(X \cap Y) = 75$

$$n(X \cup Y) = n(X) + n(Y) - n(X \cap Y)$$
$$= 100 + 150 - 75$$
$$= 175$$

$\therefore$ 175 students were taking atleast one juice.

$$n(X' \cap Y') = n(X \cup Y)'$$
$$= n(U) - n(X \cup Y)$$
$$= 400 - 175$$
$$= 225$$

Hence, 225 students were taking neither apple juice nor orange juice.

272. *(a)* Let U denote the universal set consisting of individuals suffering from the skin disorder, A denote the set of individuals exposed to the chemical C_1 and B denote the set of individuals exposed to the chemical C_2.

Here, $n(U) = 200, n(A) = 120, n(B) = 50$ and $n(A \cap B) = 30$

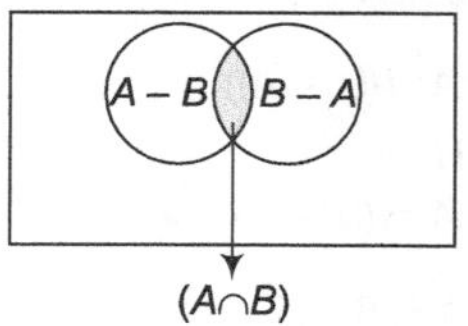

From the Venn diagram, we have

$$A = (A - B) \cup (A \cap B)$$
$$n(A) = n(A - B) + n(A \cap B)$$
(since, $A - B$ and $A \cap B$ are disjoint)

or $\quad n(A - B) = n(A) - n(A \cap B)$
$$= 120 - 30$$
$$= 90$$

Hence, the number of individuals exposed to chemical C_1 but not to chemical C_2 is 90.

273. *(b)* From the above diagram, we have

$$B = (B - A) \cup (A \cap B)$$

and so, $\quad n(B) = n(B - A) + n(A \cap B)$
(since, $B - A$ and $A \cap B$ are disjoint)

or $\quad n(B - A) = n(B) - n(A \cap B)$
$$= 50 - 30 = 20$$

Thus, the number of individuals exposed to chemical C_2 but not to chemical C_1 is 20.

274. *(c)* $n(A \cup B) = n(A) + n(B) - n(A \cap B)$

$$= 120 + 50 - 30 = 140$$

Thus, the number of individuals exposed either to chemical C_1 or to chemical C_2 is 140.

275. *(b)* A. $n(X \cup Y) = n(X) + n(Y) - n(X \cap Y)$

$$38 = 17 + 23 - n(X \cap Y)$$
$$n(X \cap Y) = 2$$

B. Let H be set of people who speak Hindi and E be set of people who speak English.

Given, $n(H \cup E) = 400, n(H) = 250$

and $n(E) = 200$

$\therefore$ $n(H \cup E) = n(H) + n(E) - n(H \cap E)$

$$400 = 250 + 200 - n(H \cap E)$$

$\therefore$ $n(H \cap E) = 50$

Thus, 50 people speak both Hindi and English.

C. Given, $n(S) = 21, n(T) = 32$

and $n(S \cap T) = 11$

$\therefore$ $n(S \cup T) = n(S) + n(T) - n(S \cap T)$

$$= 21 + 32 - 11 = 42$$

D. Given, $n(X) = 40, n(X \cup Y) = 60$

and $n(X \cap Y) = 10$

$\therefore$ $n(X \cup Y) = n(X) + n(Y) - n(X \cap Y)$

$$60 = 40 + n(Y) - 10$$

$\therefore$ $n(Y) = 30$

276. *(b)* Given, $n(X \cup Y) = 18$

$$n(X) = 8, n(Y) = 15$$

Using the identity $n(X \cup Y) = n(X) + n(Y) - n(X \cap Y)$, we have

$$18 = 8 + 15 - n(X \cap Y)$$

$\Rightarrow$ $n(X \cap Y) = 23 - 18 = 5$

277. *(d)* Let A be the set who like coffee and B be the set who like tea.

Then, $n(A \cup B) = 70, n(A) = 37$ and $n(B) = 52$

$$n(A \cup B) = n(A) + n(B) - n(A \cap B)$$
$$70 = 37 + 52 - n(A \cap B)$$
$$n(A \cap B) = 19$$

Thus, 19 people like both coffee and tea.

278. *(a)* Let X denote the set who like cricket and Y denote the set who like tennis.

Given, $n(X \cup Y) = 65, n(X) = 40$

and $n(X \cap Y) = 10$

$\therefore$ $n(X \cup Y) = n(x) + n(Y) - n(X \cap Y)$

$$65 = 40 + n(Y) - 10$$
$$n(Y) = 35$$
$$n(Y - X) = n(Y) - n(X \cap Y)$$
$$= 35 - 10 = 25$$

Thus, 25 people like tennis only and 35 people like tennis.

279. *(b)* Let X denote the set of people who speak French and Y denote the set of people who speak Spanish.

Given, $n(X) = 50, n(Y) = 20$

and $n(X \cap Y) = 10$

$\therefore$ $n(X \cup Y) = n(X) + n(Y) - n(X \cap Y)$

$$= 50 + 20 - 10 = 60$$

Thus, 60 people speak atleast one of these two languages.

280. *(c)* Since, $n(A \cup B) = n(A) + n(B) - n(A \cap B)$

$$0.25 = 0.16 + 0.14 - n(A \cap B)$$
$$n(A \cap B) = 0.30 - 0.25$$
$$= 0.05$$

281. *(c)* Let X be the set of people travels by car and Y be the set of people travels by bus.

Then, $n(X) = 20\%$

$$n(Y) = 50\%$$
$$n(X \cap Y) = 10\%$$

$\therefore$ $n(X \cup Y) = n(X) + n(Y) - n(X \cap Y)$

$$= 20 + 50 - 10$$
$$= 60\%$$

282. *(d)* $A \subset B \Rightarrow n(A \cup B) = n(B) = 6$

$\therefore$ A is incorrect.

If A and B are disjoint, then $n(A \cap B) = 0$.

$$n(A \cup B) = n(A) + n(B) - n(A \cap B)$$
$$= n(A) + n(B)$$

$\therefore$ R is correct.

283. *(c)* Given, $n(C) = 63, n(A) = 76$ and $n(C \cap A) = x$

We know that,

$$n(C \cup A) = n(C) + n(A) - n(C \cap A)$$
$$100 = 63 + 76 - x$$
$$x = 39$$

and $n(C \cap A) \le n(C)$

$\Rightarrow$ $x \le 63$

$\therefore$ $39 \le x \le 63$

284. *(b)* Let M be the set of students passing in Mathematics, P be the set of students passing in Physics and C be the set of students passing in Chemistry.

Now, $n(M \cup P \cup C) = 50, n(M) = 37, n(P) = 24, n(C) = 43$

$n(M \cap P) \le 19, n(M \cap C) \le 29, n(P \cap C) \le 20$ (given)

$$n(M \cup P \cup C) = n(M) + n(P) + n(C) - n(M \cap P)$$
$$- n(M \cap C) - n(P \cap C) + n(M \cap P \cap C) \le 50$$

$\Rightarrow$ $37 + 24 + 43 - 19 - 29 - 20 + n(M \cap P \cap C) \le 50$

$\Rightarrow$ $n(M \cap P \cap C) \le 50 - 36$

$\Rightarrow$ $n(M \cap P \cap C) \le 14$

Thus, the largest possible number that could have passed all the three examinations, is 14.

285. (a) $n(A \cup B) = n(A) + n(B) - n(A \cap B)$

$$= 12 + 9 - 4 = 17$$

$$n(A \cup B)' = n(U) - n(A \cup B)$$

$$= 20 - 17 = 3$$

$\therefore$ Statement I is true.

If A, B and C are disjoint sets, then

$$n(A \cup B \cup C) = n(A) + n(B) + n(C)$$

$$= 11 + 5 + 4 = 20$$

$\therefore$ Statement II is false.

286. (b) Since, $n(X_r) = 5$, $\bigcup_{r=1}^{20} X_r = S$, we get $n(S) = 100$.

But each element of S belongs to exactly 10 of the X_r's.

So, $\frac{100}{10} = 10$ are the number of distinct elements in S.

Also, each element of S belong to exactly 4 of the Y_r's and each Y_r contains 2 elements. If S has n numbers of Y_r in it,

then $\frac{2n}{4} = 10$

which gives $n = 20$

287. (c) Let there be x families in the town. Let P and C denote the set of families using phone and car, respectively. Then,

$$n(P) = \frac{25x}{100} = \frac{x}{4},$$

$$n(C) = \frac{15x}{100} = \frac{3x}{20}$$

and $$n(\bar{P} \cap \bar{C}) = \frac{65x}{100} = \frac{13x}{20}$$

Also, $$n(P \cap C) = 2000$$

Now, $$n(\bar{P} \cap \bar{C}) = \frac{13x}{20}$$

$$\Rightarrow \quad n(\overline{P \cup C}) = \frac{13x}{20}$$

$$\Rightarrow \quad n(U) - n(P \cup C) = \frac{13x}{20}$$

$$\Rightarrow \quad x - \{n(P) + n(C) - n(P \cap C)\} = \frac{13x}{20}$$

$$\Rightarrow \quad x - \left(\frac{x}{4} + \frac{3x}{20} - 2000\right) = \frac{13x}{20}$$

$$\Rightarrow \quad \frac{x}{20} = 2000$$

$$\Rightarrow \quad x = 40000$$

Thus, Statement III is correct

10% of the total families in the town is 4000 and it is given that 2000 families own both a car and a phone. So, Statement I is not correct.

Now, $$n(P \cup C) = n(P) + n(C) - n(P \cap C)$$

$$\Rightarrow \quad n(P \cup C) = \frac{x}{4} + \frac{3x}{20} - 2000$$

$$\Rightarrow \quad n(P \cup C) = 10000 + 6000 - 2000 = 14000$$

$$\Rightarrow \quad n(P \cup C) = 35\% \text{ of } 40000$$

Thus, Statement II is correct.

288. (c) Given, $n(M) = 100, n(P) = 70, n(C) = 40$

$$n(M \cap P) = 30, n(M \cap C) = 28$$

$$n(P \cap C) = 23$$

and $n(M \cap P \cap C) = 18$

$$\therefore \quad n(M \cap P' \cap C') = n\,[M \cap (P \cup C)']$$

$$= n(M) - n\,[M \cap (P \cup C)]$$

$$[\because n(A \cap B') = n(A) - n(B)]$$

$$= n(M) - n\,[(M \cap P) \cup (M \cap C)]$$

$$= n(M) - [n(M \cap P) + n(M \cap C) - n(M \cap P \cap C)]$$

$$= 100 - [30 + 28 - 18]$$

$$= 100 - 40$$

$$= 60$$

289. (a) Let X be the set of letters in word 'CATARACT'. Then,

$$X = \{C, A, T, R\}$$

Let Y be the set of letters in word 'TRACT'. Then,

$$Y = \{T, R, A, C, T\} = \{T, R, A, C\}$$

Since, every element in X is in Y and every element in Y is in X. It follows that $X = Y$.

Hence, Statement I is true.

Let $A = \{-1, 0, 1\}$. The subset of A having no element is the empty set ϕ. The subsets of A having one element are $\{-1\}$, $\{0\}$ and $\{1\}$. The subsets of A having two elements are $\{-1, 0\}$, $\{-1, 1\}$ and $\{0, 1\}$. The subset of A having three elements of A is A itself. So, all the subsets of A are $\phi, \{-1\}$, $\{0\}$, $\{1\}$, $\{-1, 0\}$, $\{-1, 1\}$, $\{0, 1\}$ and $\{-1, 0, 1\}$.

Hence, Statement II is false.

290. (d) Let $a \in A$. Then, $a \in A \cup B$. Since, $A \cup B = A \cap B$, $a \in A \cap B$. So, $a \in B$.

Therefore, $A \subset B$. Similarly, if $b \in B$, then $b \in A \cup B$.

Since, $A \cup B = A \cap B, b \in A \cap B$.

So, $b \in A$. Therefore, $B \subset A$. Thus, $A = B$.

291. (a) Let $X \in P(A \cap B)$. Then, $X \subset A \cap B$. So, $X \subset A$ and $X \subset B$. Therefore, $X \in P(A)$ and $X \in P(B)$ which implies $X \in P(A) \cap P(B)$. This gives $P(A \cap B) \subset P(A) \cap P(B)$. Let, $Y \in P(A) \cap P(B)$. Then, $Y \in P(A)$ and $Y \in P(B)$. So, $Y \subset A$ and $Y \subset B$. Therefore, $Y \subset A \cap B$, which implies $Y \in P(A \cap B)$. This gives $P(A) \cap P(B) \subset P(A \cap B)$.

Hence, $P(A \cap B) = P(A) \cap P(B)$.

So, A is true. Also, R is true and a correct explanation of A.

292. (b) Let U be the set of consumers questioned, S be the set of consumers who liked the product A and T be the set of consumers who liked the product B.

Given that, $n(U) = 1000, n(S) = 720, n(T) = 450$

So, $$n(S \cup T) = n(S) + n(T) - n(S \cap T)$$

$$= 720 + 450 - n(S \cap T)$$

$$= 1170 - n(S \cap T)$$

Therefore, $n(S \cup T)$ is maximum when $n(S \cap T)$ is least. But $S \cup T \subset U$ implies $n(S \cup T) \le n(U) = 1000$. So, maximum values of $n(S \cup T)$ is 1000. Thus, the least value of $n(S \cap T)$ is 170. Hence, the least number of consumers who liked both products, is 170.

293. *(c)* Let F, B and C denote the set of men who received medals in football, basketball and cricket, respectively.

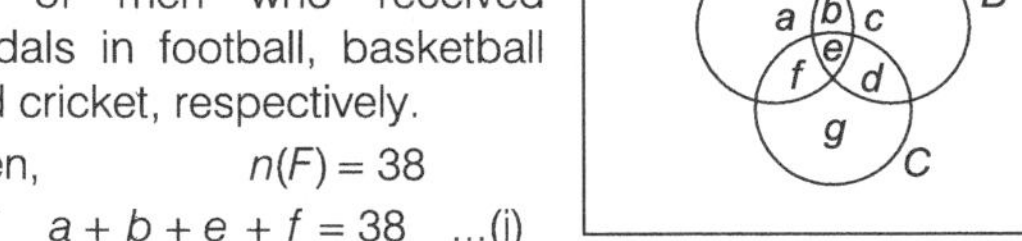

Then, $n(F) = 38$

$\Rightarrow \quad a + b + e + f = 38 \quad$ ...(i)

$n(B) = 15$

$\Rightarrow \quad b + c + e + d = 15 \quad$ ...(ii)

$n(C) = 20$

$\Rightarrow \quad d + e + f + g = 20 \quad$ ...(iii)

$n(F \cup B \cup C) = 58$

$\Rightarrow \quad a + b + c + d + e + f + g = 58 \quad$ ...(iv)

$n(F \cap B \cap C) = 3$

$\Rightarrow \quad e = 3 \quad$ ...(v)

From Eqs. (ii) and (iv), we have

$a + 15 + f + g = 58$

$\Rightarrow \quad a + f + g = 43 \quad$ ...(vi)

Subtracting Eq. (vi) from Eq. (i), we have

$b + e - g = -5 \quad$ ...(vii)

Adding Eqs. (iii) and (vii), we have

$d + e + f + b + e = 15$

$\Rightarrow \quad d + f + b = 9$

Hence, number of people who got medals in exactly two of the three sports, is 9.

Alternate Method

Let F, B and C denote the set of men who received medals in football, basketball and cricket, respectively.

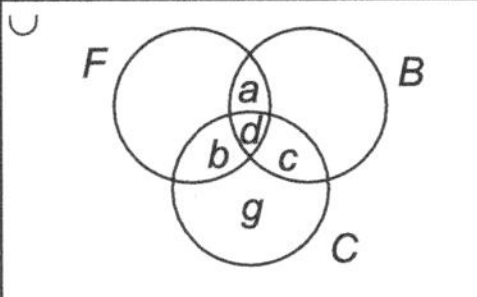

Then,

$n(F) = 38, n(B) = 15, n(C) = 20$

$n(F \cup B \cup C) = 58$ and $n(F \cap B \cap C) = 3$.

Therefore, $n(F \cup B \cup C) = n(F) + n(B) + n(C) - n(F \cap B)$
$- n(F \cap C) - n(B \cap C) + n(F \cap B \cap C)$,

gives $n(F \cap B) + n(F \cap C) + n(B \cap C) = 18$

Consider the Venn diagram as given in figure.

Here, a denotes the number of men who got medals in football and basketball only, b denotes the number of men who got medals in football and cricket only, c denotes the number of men who got medals in basketball and cricket only and d denotes the number of men who got medals in all the three.

Thus, $d = n(F \cap B \cap C) = 3$ and $a + d + b + d + c + d = 18$

Therefore, $a + b + c = 9$, which is the number of people who got medals in exactly two of the three sports.

294. *(b)* $A = \{x : x \in R \text{ and } x^2 - 8x + 12 = 0\}$

$= \{2, 6\}$

$B = \{2, 4, 6\}, C = \{2, 4, 6, 8, \dots\}$

and $\quad D = \{6\}$

Clearly, $A \subset B, A \subset C, B \subset C$

But $C \not\subset D$ as $4 \in C$ but $4 \notin D$

295. *(b)* Given, $A \cup B = A \cup C$

$B \cap (A \cup B) = B \cap (A \cup C)$

$(B \cap A) \cup (B \cap B) = (B \cap A) \cup (B \cap C)$

$(B \cap A) \cup B = (A \cap B) \cup (B \cap C)$

$B = (A \cap B) \cup (B \cap C) \quad$ ...(i)

Again, $\quad A \cup B = A \cup C$

$C \cap (A \cup B) = C \cap (A \cup C)$

$(C \cap A) \cup (C \cap B) = (C \cap A) \cup (C \cap C)$

$(A \cap B) \cup B \cap C) = (C \cap A) \cup C \quad (\because A \cap B = A \cap C)$

$(A \cap B) \cup (B \cap C) = C \quad$ ...(ii)

From Eqs. (i) and (ii), we get

$B = C$

296. *(b)* I. If $x \in A$ and $A \in B \Rightarrow x \notin B$

As $\quad A = \{1, 2\}, B = \{\{1, 2\}, 3\}$

Then, $\quad 1 \in A$ but $1 \notin B$

II. If $A \subset B$ and $B \in C \Rightarrow A \notin C$

As $A = \{1\}, B = \{1, 2\}$ and $C = \{\{1, 2\}, 3\}$

III. If $A \subset B$ and $B \subset C$

$\Rightarrow$ Every element of A is in B and every element of B is in C.

$\Rightarrow$ Every element of A is also in C.

$\Rightarrow \quad A \subset C$

IV. If $A \not\subset B$ and $B \not\subset C$, then A may be subset of C.

As $A = \{1, 2\}, B = \{3\}, C = \{1, 2, 4, 5\}$

V. If $x \in A$ and $A \not\subset B \Rightarrow x \notin B$

As $A = \{1, 2\}, B = \{3, 4, 5\}$

Here, $1 \in A$ and $A \not\subset B$ but $1 \notin B$

VI. If $A \subset B$ and $x \notin B$, then $x \notin A$

As $A \subset B$

$\Rightarrow$ Every element of A is in B.

i.e., there does not exist any element of A which does not belong to B.

297. *(a)* Consider the Venn diagram

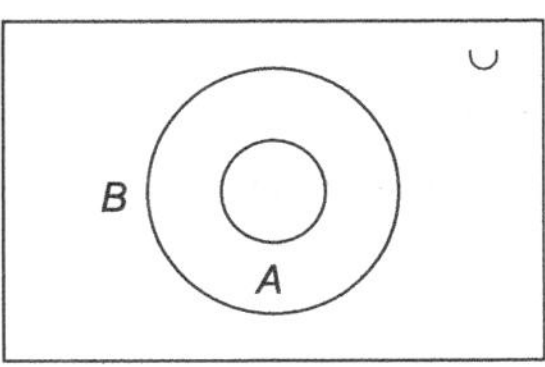

Here, $A \subset B$, $A - B = \phi$, $A \cup B = B$ and $A \cap B = A$

Hence, these four conditions are equivalent.

So, Statement I is true.

Now, let $x \in C - B$

$\Rightarrow \quad x \in C$ and $x \notin B$

$\Rightarrow \quad x \in C$ and $x \notin A \quad (\because A \subset B)$

$\Rightarrow \quad x \in C - A$

$\therefore \quad C - B \subset C - A$

Hence, Statement II is not true.

298. *(d)* Let $x \in A$

$\Rightarrow$ There exists $X \in P(A)$ such that $x \in X$

$\Rightarrow \quad X \in P(B) \quad [\because P(A) = P(B)]$

$\Rightarrow \quad x \in B$

$\Rightarrow \quad A \subset B \quad$ ...(i)

Similarly, we can prove $B \subset A$ ...(ii)

From Eqs. (i) and (ii), we get

$$A = B$$

299. *(c)* I. Consider $A = \{1, 2\}, B = \{2, 3\}$

$$P(A) = \{\phi, \{1\}, \{2\}, \{1, 2\}\}$$
$$P(B) = \{\phi, \{2\}, \{3\}, \{2, 3\}\}$$
$$P(A) \cup P(B) = \{\phi, \{1\}, \{2\}, \{3\}, \{1, 2\}, \{2, 3\}\}$$
$$A \cup B = \{1, 2, 3\}$$
$$P(A \cup B) = \{\phi, \{1\}, \{2\}, \{3\}, \{1, 2\}, \{2, 3\}, \{1, 3\}, \{1, 2, 3\}\}$$
$$\therefore \quad P(A) \cup P(B) \neq P(A \cup B)$$

II. Consider $A = \{1, 2\}, B = \{2, 3\}, C = \{2, 4\}$

Here, $A \cap B = \{2\}$ and $A \cap C = \{2\}$

But $B \neq C$

III. Given, $A \cup X = B \cup X$

$\Rightarrow \quad A \cap (A \cup X) = A \cap (B \cup X)$

$\Rightarrow (A \cap A) \cup (A \cap X) = (A \cap B) \cup (A \cap X)$

$\Rightarrow \quad A \cup \phi = (A \cap B) \cup \phi$

$(\because A \cap X = B \cap X = \phi)$

$\Rightarrow \quad A = A \cap B \quad$ ...(i)

Again, $A \cup X = B \cup X$

$B \cap (A \cup X) = B \cap (B \cup X)$

$(B \cap A) \cup (B \cap X) = (B \cap B) \cup (B \cap X)$

$(B \cap A) \cup \phi = B \cup \phi \quad (\because A \cap X = B \cap X = \phi)$

$A \cap B = B \quad$ ...(ii)

From Eqs. (i) and (ii), we get

$$A = B$$

300. *(b)* A. $(A \cap B) \cup (A - B) = (A \cup (A - B)) \cap B \cup (A - B)$

$= A \cap (A \cup B) = A$

B. $A \cup (B - A) = A \cup B$

C. $B \cup (A \cap B) = (B \cup A) \cap (B \cup B)$

$= (B \cup A) \cap B = B$

D. $A \cap (A \cup B) = A$

301. *(c)* In (c), $A \cap B = \{1\}, B \cap C = \{4\}$,

$A \cap C = \{3, 5\}$ and $A \cap B \cap C = \phi$

302. *(d)* Let U be the universal set, A be the set of those students who were taking tea and B be the set of those students who were taking coffee.

Then, $\quad n(U) = 600, n(A) = 150, n(B) = 225$

and $\quad n(A \cap B) = 100$

$\therefore \quad n(A \cup B) = n(A) + n(B) - n(A \cap B)$

$= 150 + 225 - 100 = 275$

$n(A' \cap B') = n(A \cup B)'$

$= n(U) - n(A \cup B)$

$= 600 - 275 = 325$

303. *(b)* Let H be the set of those students who know Hindi and E be the set of those students who know English.

Then, $n(H) = 100, n(E) = 50$ and $n(H \cap E) = 25$

$\therefore \quad n(H \cup E) = n(H) + n(E) - n(H \cap E)$

$= 100 + 50 - 25 = 125$

Hence, the number of students in the group is 125.

304. *(c)* Let H be the set of those people who read newspaper H, T be the set of those people who read newspaper T and I be the set of those people who read newspaper I.

Then, $\quad n(H) = 25, n(T) = 26, n(I) = 26$

$n(H \cap I) = 9, n(H \cap T) = 11, n(T \cap I) = 8$

and $\quad n(H \cap T \cap I) = 3$

$\therefore \quad n(H \cup T \cup I) = n(H) + n(T) + n(I) - n(H \cap I)$

$- n(H \cap T) - n(T \cap I) + n(H \cap T \cap I)$

$= 25 + 26 + 26 - 9 - 11 - 8 + 3$

$= 80 - 28 = 52$

305. *(a)* From above solution,

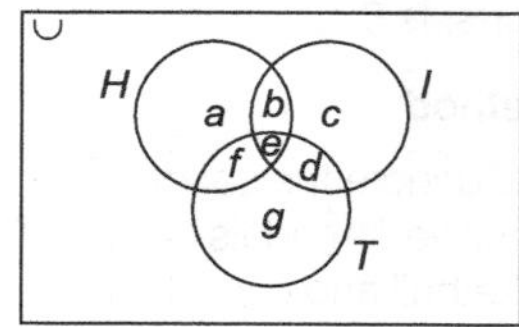

$n(H \cup T \cup I) = 52$

$a + b + c + d + e + f + g = 52 \quad$ ...(i)

$n(H \cap I) = 9$

$\Rightarrow \quad b + e = 9 \quad$ ...(ii)

$n(H \cap T) = 11$

$\Rightarrow \quad f + e = 11 \quad$ ...(iii)

$n(T \cap I) = 8$

$\Rightarrow \quad e + d = 8 \quad$ ...(iv)

$n(H \cap T \cap I) = 3$

$\Rightarrow \quad e = 3$

Adding Eqs. (ii), (iii) and (iv) and then subtracting from Eq. (i), we get

$$a + c + g - 2e = 24$$
$$a + c + g - 6 = 24$$
$$a + c + g = 30$$

$\therefore$ Number of people who read exactly one newspaper, is 30.

306. *(c)* Let A be the set who like product A, B be the set who like product B and C be the set who like product C.

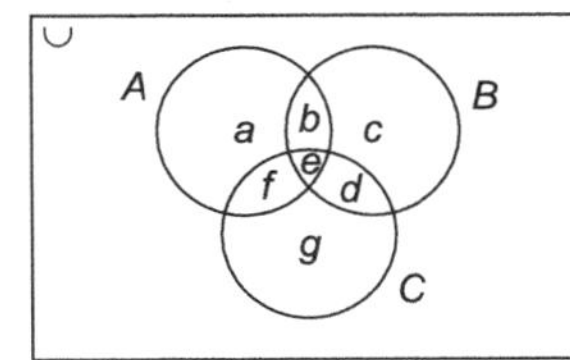

Then, $n(A)=21, n(B)=26, n(C)=29$

$n(A\cap B)=14, n(B\cap C)=14$

$n(A\cap C)=12, n(A\cap B\cap C)=8$

$$a+b+e+f=21 \quad \text{...(i)}$$
$$b+c+d+e=26 \quad \text{...(ii)}$$
$$d+e+f+g=29 \quad \text{...(iii)}$$
$$b+e=14 \quad \text{...(iv)}$$
$$e+d=14 \quad \text{...(v)}$$
$$f+e=12 \quad \text{...(vi)}$$
$$e=8 \quad \text{...(vii)}$$

Substituting the values of e in Eqs. (iv), (v) and (vi), we get

$$b=6, d=6 \quad \text{and} \quad f=4$$

Substituting the values of d, e and f in Eq. (iii), we get

$$6+8+4+g=29$$
$$\therefore \quad g=11$$

Number of people who like product C only, is 11.

Alternate Method

$$n(C\cap A'\cap B')=n(C)-n(C\cap A)-n(C\cap B) + n(A\cap B\cap C)$$
$$=29-12-14+8$$
$$=37-26$$
$$=11$$

2

Relations *and* Functions

Cartesian Products of Sets

Level I

1. Which of the following is/are the relation(s)?
(a) A number m is less than number n
(b) Line l is parallel to line m
(c) Set A is a subset of B
(d) All of the above

2. Let $A = \{\text{red, blue}\}$ and $B = \{b, c, s\}$, where, b, c and s represent a particular bag, coat and shirt, respectively. Then, number of pairs of coloured objects can be made from these two sets, is
(a) 2 (b) 3 (c) 5 (d) 6

3. Which of the following is an ordered pair?
(a) (p, q), $p \in P$ and $q \in Q$ (b) $[p, q]$, $p \in P$ and $q \in Q$
(c) $\{p, q\}$, $p \in P$ and $q \in Q$ (d) All of these

4. Consider the following statements

Statement I The Cartesian product of two non-empty sets P and Q is denoted as $P \times Q$ and $P \times Q = \{(p, q) : p \in P, q \in Q\}$.

Statement II If $A = \{\text{red, blue}\}$ and $B = \{b, c, s\}$, then $A \times B = \{(\text{red}, b), (\text{red}, c), (\text{red}, s), (\text{blue}, b), (\text{blue}, c), (\text{blue}, s)\}$.

Choose the correct option.
(a) Statement I is true
(b) Statement II is true
(c) Both are true
(d) Both are false

5. The Cartesian product of two sets P and Q *i.e.*, $P \times Q = \phi$, if
(a) either P or Q is the null set
(b) neither P nor Q is null set
(c) Both (a) and (b)
(d) None of the above

6. If $A = \{DL, MP, KA\}$ and $B = \{01, 02, 03\}$, then set of all ordered pairs from A and B in that order represents P and number of elements in ...P... is ...Q... . Here, P and Q refer to
(a) $(A \cup B)$; 6 (b) $(A \cap B)$; 3 (c) $A \times B$; 9 (d) $A \times B$; 6

7. If $A = \{a_1, a_2\}$ and $B = \{b_1, b_2, b_3, b_4\}$, then $A \times B$ is equal to
(a) $\{(a_1, b_1), (a_2, b_2)\}$
(b) $\{(a_1, b_1), (a_2, b_2), (a_3, b_3), (a_4, b_4)\}$
(c) $\{(a_1, b_1), (a_1, b_2), (a_1, b_3), (a_1, b_4)\}$
(d) $\{(a_1, b_1), (a_1, b_2), (a_1, b_3), (a_1, b_4), (a_2, b_1), (a_2, b_2), (a_2, b_3), (a_2, b_4)\}$

8. **Assertion** (A) If $(x + 1, y - 2) = (3, 1)$, then $x = 2$ and $y = 3$.

Reason (R) Two ordered pairs are equal, if their corresponding elements are equal.
(a) Both A and R are correct; R is the correct explanation of A
(b) Both A and R are correct; R is not the correct explanation of A
(c) A is correct; R is incorrect
(d) R is correct; A is incorrect

9. If A and B are two non-empty sets and $n(A) = p$ and $n(B) = q$, then $n(A \times B)$ is equal to

(a) $p + q$ (b) $p - q$
(c) pq (d) None of these

10. If A and B are non-empty sets and either A or B is infinite, then $A \times B$ is ...X... . Here, X refers to

(a) infinite (b) finite
(c) empty (d) None of these

11. Let A be any non empty set. Then, $A \times A \times A = \{(a, b, c) : a, b, c \in A\}$. Here, (a, b, c) is called an

(a) ordered pair (b) ordered triplet
(c) Both (a) and (b) (d) None of these

12. If $P = \{a, b, c\}$ and $Q = \{r\}$, then

(a) $P \times Q = Q \times P$ (b) $P \times Q \neq Q \times P$
(c) $P \times Q \subset Q \times P$ (d) None of these

13. Let $A = \{1, 2, 3\}$, $B = \{3, 4\}$ and $C = \{4, 5, 6\}$. Then, match the following in Column I with the sets of ordered pairs in Column II and choose the correct option from the codes given below.

Column I	Column II
A. $A \times (B \cap C)$	1. $\{(1, 4), (2, 4), (3, 4)\}$
B. $(A \times B) \cap (A \times C)$	2. $\{(1, 3), (1, 4), (1, 5), (1, 6), (2, 3), (2, 4), (2, 5), (2, 6), (3, 3), (3, 4), (3, 5), (3, 6)\}$
C. $A \times (B \cup C)$	3. $\{(3, 4)\}$
D. $(A \times B) \cup (A \times C)$	
E. $(A \cap B) \times (B \cap C)$	

Codes

	A	B	C	D	E		A	B	C	D	E
(a)	1	2	1	2	3	(b)	2	2	1	1	3
(c)	1	1	2	2	3	(d)	2	1	2	3	2

14. If $P = \{1, 2\}$, then $P \times P \times P = \{(1, 1, 1), (1, 1, 2), (1, 2, 1), (2, 1, 2), (2, 2, 1), (2, 2, 2)\}$. The ordered triplet missing in $P \times P \times P$ are

(a) (1, 2); (2, 1) (b) (1, 2, 1); (2, 1, 1)
(c) (1, 2, 2); (2, 1, 2) (d) (1, 2, 2); (2, 1, 1)

15. If R is the set of real numbers, then Cartesian product $R \times R \times R$ represents

(a) coordinates of all points in two dimensional space
(b) coordinates of all points in three dimensional space
(c) Both (a) and (b)
(d) None of the above

16. If $A \times B = \{(p, q), (p, r), (m, q), (m, r)\}$, then A and B are ...X..., respectively. Here, X refers to

(a) $\{q, r\}$ and $\{p, m\}$ (b) $\{p, m\}$ and $\{q, r\}$
(c) $\{p, q\}$ and $\{m, r\}$ (d) $\{q, m\}$ and $\{p, r\}$

17. Consider the following statements

Statement I If $\left(\frac{x}{3} + 1, y - \frac{2}{3}\right) = \left(\frac{5}{3}, \frac{1}{3}\right)$, then the values of x and y are 2 and 1, respectively.

Statement II If the set A has 3 elements and the set $B = \{3, 4, 5\}$, then the number of elements in $(A \times B)$ is 6.

Choose the correct option.

(a) Statement I is true
(b) Statement II is true
(c) Both are true
(d) Both are false

18. If $G = \{7, 8\}$ and $H = \{5, 4, 2\}$, then $G \times H$ ...A... $H \times G$. Here, A refers to

(a) = (b) $\neq$ (c) $\in$ (d) $\subset$

19. Which of the following is/are true?

I. If $P = \{m, n\}$ and $Q = \{n, m\}$, then $P \times Q = \{(m, n), (n, m)\}$.
II. If A and B are non-empty sets, then $A \times B$ is a non-empty set of ordered pairs (x, y), such that $x \in A$ and $y \in B$.
III. If $A = \{1, 2\}$ and $B = \{3, 4\}$, then $A \times (B \cap \phi) = \phi$

(a) I and II are true (b) II and III are true
(c) I and III are true (d) All are true

20. If $A = \{-1, 1\}$, then $A \times A \times A$ is equal to

(a) $\{(1, 1), (-1, -1)\}$
(b) $\{(1, 1, 1), (-1, -1, -1)\}$
(c) $\{(1, 1, 1), (1, 1, -1), (1, -1, 1), (-1, 1, 1), (-1, 1, -1), (-1, -1, 1)\}$
(d) $\{(1, 1, 1), (1, 1, -1), (1, -1, 1), (1, -1, -1), (-1, 1, 1), (-1, 1, -1), (-1, -1, 1), (-1, -1, -1)\}$

21. If $A \times B = \{(a, x), (a, y), (b, x), (b, y)\}$, then sets A and B are ...X... and ...Y... , respectively. Here, X and Y refer to

(a) $\{a, x\}$; $\{b, y\}$ (b) $\{a, b\}$; $\{x, y\}$
(c) $\{x, y\}$; $\{a, b\}$ (d) $\{a, y\}$; $\{b, x\}$

22. Let $A = \{1, 2\}$, $B = \{1, 2, 3, 4\}$, $C = \{5, 6\}$ and $D = \{5, 6, 7, 8\}$. Then, which of the following is/are true?

I. $A \times B = A \times C$
II. $(A \times C) \subset (B \times D)$
III. $A \times (B \cap C) = (A \times B) \cap (A \times C)$

(a) I and II are true (b) Only II is true
(c) Only III is true (d) II and III are true

23. Let $A = \{1, 2\}$, $B = \{3, 4\}$. Then, number of subsets of $A \times B$ is

(a) 4 (b) 8
(c) 18 (d) 16

24. **Assertion** (A) If $(x, 1)$, $(y, 2)$ and $(z, 1)$ are in $A \times B$ and $n(A) = 3$, $n(B) = 2$, then $A = \{x, y, z\}$ and $B = \{1, 2\}$.

Reason (R) If $n(A) = 3$ and $n(B) = 2$, then $n(A \times B) = 6$.

(a) Both A and R are correct; R is the correct explanation of A
(b) Both A and R are correct; R is not the correct explanation of A
(c) A is correct; R is incorrect
(d) R is correct; A is incorrect

25. The Cartesian product $A \times A$ has 9 elements among which are found $(-1, 0)$ and $(0, 1)$. Then, the remaining elements of $A \times A$ are

(a) $\{(0, 0), (0, 1), (0, -1), (-1, 1), (-1, 0), (1, 0), (1, -1)\}$
(b) $\{(-1, 1), (-1, 0), (0, 0), (0, -1), (0, 1), (1, 0), (1, -1)\}$
(c) $\{(0, -1), (1, 0), (1, -1)\}$
(d) None of the above

Level II

26. If $A = \{1, 2, 3, 4\}$ and $B = \{5, 7, 9\}$, then

(a) $A \times B = B \times A$
(b) $n(A \times B) = n(B \times A)$
(c) Both (a) and (b)
(d) None of the above

27. If $(1, 3)$, $(2, 5)$ and $(3, 3)$ are three elements of $A \times B$ and total number of elements in $A \times B$ is 6, then remaining elements of $A \times B$ are

(a) $(1, 5), (2, 3), (3, 5)$ (b) $(5, 1), (3, 2), (5, 3)$
(c) $(1, 5), (2, 3), (5, 3)$ (d) None of these

28. Consider the following statements

Statement I If $(4x + 3, y) = (3x + 5, -2)$, then $x = 2$ and $y = -2$.

Statement II If $A = \{-1, 3, 4\}$, then $A \times A$ is $\{(-1, -1), (-1, 3), (-1, 4), (3, -1), (4, -1), (3, 4)\}$.

Choose the correct option.

(a) Statement I is true (b) Statement II is true
(c) Both are true (d) Both are false

29. If A, B and C are three sets, then [HOTS]

(a) $A \times (B \cap C) = (A \times B) \cap (A \times C)$
(b) $A \times (B' \cup C')' = (A \times B) \cap (A \times C)$
(c) Both (a) and (b)
(d) None of the above

30. Let A and B are two sets having 3 elements in common. If $n(A) = 5$ and $n(B) = 4$, then $n[(A \times B) \cap (B \times A)]$ is equal to [J & K CET 2008]

(a) 20 (b) 25
(c) 16 (d) 9

31. If $A = \{1, 2\}$ and $B = \{3, 4\}$, then $A \times (B \cap \phi)$ is equal to

(a) ϕ (b) A (c) B (d) $A \times B$

32. Consider the following statements

Statement I If $A \subseteq B$ and $C \subseteq D$, then $A \times C \subseteq B \times D$.

Statement II If $A = \{2, 3\}$, $B = \{4, 5\}$ and $C = \{5, 6\}$, then $A \times (B \cup C) = A \times (B \cap C)$.

Choose the correct option.

(a) Statement I is true (b) Statement II is true
(c) Both are true (d) Both are false

33. If $A = \{2, 4, 6, 9\}$ and $B = \{4, 6, 18, 27, 54\}$, $a \in A$ and $b \in B$, then the set of ordered pairs such that a is factor of b and $a < b$, is

(a) $\{(2, 4), (2, 6), (2, 18), (2, 54)\}$
(b) $\{(6, 18), (6, 54), (9, 18), (9, 27), (9, 54)\}$
(c) $\{(2, 4), (2, 6), (2, 18), (2, 54), (6, 18), (6, 54), (9, 18), (9, 27), (9, 54)\}$
(d) $\{(2, 4), (2, 6), (2, 18), (2, 54), (6, 18), (6, 54), (4, 4), (6, 6), (9, 18), (9, 27), (9, 54), (27, 54)\}$

34. If $A = \{a, b, c\}$, $B = \{b, c, d\}$ and $C = \{a, d, c\}$, then $(A - B) \times (B \cap C)$ is equal to [Jamia Millia Islamia 2008]

(a) $\{(a, c), (a, d)\}$ (b) $\{(a, b), (c, d)\}$
(c) $\{(c, a), (d, a)\}$ (d) $\{(a, c), (a, d), (b, d)\}$

35. Let $Y = \{1, 2, 3, 4, 5\}$, $A = \{1, 2\}$, $B = \{3, 4, 5\}$ and ϕ denotes null set. If $(A \times B)$ denotes Cartesian product of the sets A and B, then $(Y \times A) \cap (Y \times B)$ is [OJEE 2003]

(a) Y (b) A (c) B (d) ϕ

36. **Assertion** (A) Let A and B are non-empty sets such that $A \subseteq B$. Then, $A \times C \subseteq B \times C$.

Reason (R) For any two sets A and B $A \times B = B \times A$.

(a) Both A and R are correct; R is the correct explanation of A
(b) Both A and R are correct; R is not the correct explanation of A
(c) A is correct; R is incorrect
(d) R is correct; A is incorrect

37. Consider the following statements

Statement I If $n(U) = 10$, $n(A) = 4$ and $n(B) = 5$, then $n(A \times B') = 20$.

Statement II If the ordered pairs $(x, -1)$ and $(5, y)$ belong to the set $\{(a, b) : b = 2a - 3\}$, then $x = 2$ and $y = 1$.

Choose the correct option.

(a) Statement I is true (b) Statement II is true
(c) Both are true (d) Both are false

Relations

Level I

38. Let $P = \{a, b, c\}$, $Q = \{\text{Ali, Bhanu, Binoy, Chandra, Divya}\}$ and $R = \{(x, y) : x$ is the first letter of the name $y, x \in P, y \in Q\}$. Then,

(a) $R \subset P \times Q$
(b) A visual representation of R can be an arrow diagram
(c) Both (a) and (b)
(d) None of the above

39. A relation from A to B is a ...K... of the Cartesian product $A \times B$, where A and B are non-empty sets. Here, K refers to

(a) subset (b) super set
(c) universal set (d) None of these

40. Consider the following statements

Statement I If R is a relation from A to B, then domain of R is the set A.

Statement II The set of all second elements in a relation R from a set A to a set B is called codomain of R.

Choose the correct option.

(a) Statement I is true (b) Statement II is true
(c) Both are true (d) Both are false

41. A relation is represented by

(a) Roster method
(b) Set-builder method
(c) Both (a) and (b)
(d) None of the above

42. Let $A = \{1, 2, 3, 4, 5, 6\}$ and a relation R from A to A is defined by $R = \{(x, y) : y = x + 1\}$. Then, which of the following is/are true?

I. This relation cannot depict by an arrow diagram.
II. Domain of R is $\{1, 2, 3, 4, 5\}$.
III. Range = Codomain

(a) Only I is true (b) Only II is true
(c) Only III is true (d) II and III are true

Directions (Q. Nos. 43-46) *The figure shows a relation R between the sets P and Q.*

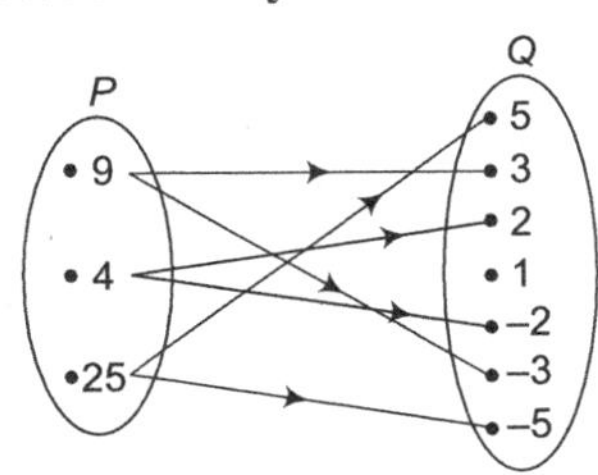

43. The relation R in Set-builder form is

(a) $\{(x, y) : x \in P, y \in Q\}$
(b) $\{(x, y) : x \in Q, y \in P\}$
(c) $\{(x, y) : x$ is the square of $y, x \in P, y \in Q\}$
(d) $\{(x, y) : y$ is the square of $x, x \in P, y \in Q\}$

44. The relation R in Roster form is

(a) $\{(9, 3), (4, 2), (25, 5)\}$
(b) $\{(9, -3), (4, -2), (25, -5)\}$
(c) $\{(9, -3), (9, 3), (4, -2), (4, 2), (25, -5), (25, 5)\}$
(d) None of the above

45. The domain of the relation R is

(a) $\{4, 9, 25\}$ (b) $\{2, 3, 5\}$
(c) $\{-2, -3, -5\}$ (d) $\{-4, -9, -25\}$

46. The range of the relation R is

(a) $\{2, 3, 4\}$ (b) $\{-2, -3, -5\}$
(c) $\{-5, -3, -2, 2, 3, 5\}$ (d) $\{-5, -3, -2, 1, 2, 3, 5\}$

47. Assertion (A) Let $A = \{1, 2\}$ and $B = \{3, 4\}$. Then, number of relations from A to B is 16.

Reason (R) If $n(A) = p$ and $n(B) = q$, then number of relations is 2^{pq}.

(a) Both A and R are correct; R is the correct explanation of A
(b) Both A and R are correct; R is not the correct explanation of A
(c) A is correct; R is incorrect
(d) R is correct; A is incorrect

48. Let $A = \{1, 2, 3, 4, \ldots, 14\}$. A relation R from A to A is defined by $R = \{(x, y) : 3x - y = 0$, where $x, y \in A\}$. Then, match the terms of Column I with the terms of Column II and choose the correct option from the codes given below.

	Column I		Column II
A.	In Roster form, the relation R is	1.	$\{1, 2, 3, 4\}$
B.	The domain of R is	2.	$\{3, 6, 9, 12\}$
C.	The range of R is	3.	$\{1, 2, 3, 4, \ldots, 14\}$
D.	The codomain of R is	4.	$\{(1, 3), (2, 6), (3, 9), (4, 12)\}$

Codes

	A	B	C	D
(a)	4	3	2	1
(b)	4	1	2	3
(c)	4	2	1	3
(d)	4	3	1	2

49. A relation R on the set N of natural numbers is defined by $R = \{(x, y) : y = x + 5$, where x is a natural number less than 4; $x, y \in N\}$.

Then, which of the following is/are true?

I. In Roster form, the relation R is $\{(1, 6), (2, 7), (3, 8)\}$.

II. The domain of R is $\{6, 7, 8\}$.

III. The range of R is $\{1, 2, 3\}$.

(a) Only I is true
(b) Only II is true
(c) I and II are true
(d) All are true

50. Let $A = \{1, 2, 3, 5\}$ and $B = \{4, 6, 9\}$ and a relation R from A to B is defined by $R = \{(x, y)$: the difference between x and y is odd; $x \in A$ and $y \in B\}$. Then Roster form of R is ...K... . Here, K refers to

(a) $\{(1,4), (1, 6), (1, 9)\}$
(b) $\{(1, 4), (1, 6), (2, 9), (3, 4), (3, 6), (5, 9)\}$
(c) $\{(1, 4), (1, 6), (2, 9), (3, 4), (3, 6), (5, 4), (5, 6)\}$
(d) None of the above

51. The figure shows a relationship between the sets P and Q. Then, which of the following is/are true?

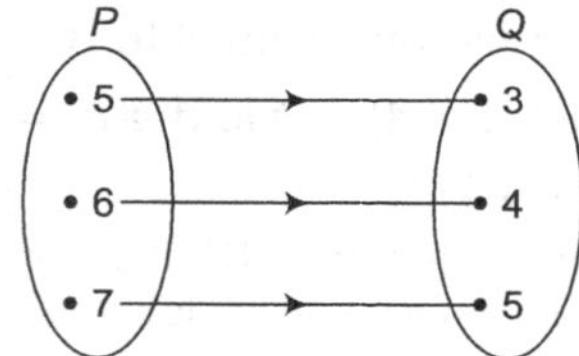

I. The relation in Set-builder form is $\{(x, y) : x - y = 2, x \in P$ and $y \in Q\}$.

II. The relation in Roster form is $\{(5, 3), (6, 4), (7, 5)\}$.

III. The domain of this relation is $\{3, 4, 5\}$.

IV. The range of this relation is $\{5, 6, 7\}$.

(a) I and II are true
(b) III and IV are true
(c) II, III and IV are true
(d) All are true

52. Let $A = \{1, 2, 3, 4, 6\}$. If R is the relation on A defined by $\{(a, b) : a, b \in A, b$ is exactly divisible by $a\}$.

Assertion (A) The relation R in Roster form is $\{(6, 3), (6, 2), (4, 2)\}$.

Reason (R) The domain and range of R is $\{1, 2, 3, 4, 6\}$,

(a) Both A and R are correct; R is the correct explanation of A
(b) Both A and R are correct; R is not the correct explanation of A
(c) A is correct; R is incorrect
(d) R is correct; A is incorrect

53. Let R be a relation defined by

$R = \{(x, x + 5) : x \in \{0, 1, 2, 3, 4, 5\}\}$

Then, consider the following statements

Statement I The domain of R is $\{0, 1, 2, 3, 4, 5\}$.

Statement II The range of R is $\{0, 1, 2, 3, 4, 5\}$.

Choose the correct option.

(a) Statement I is true
(b) Statement II is true
(c) Both are true
(d) Both are false

54. The relation $R = \{(x, x^3) : x$ is a prime number less than 10$\}$ in Roster form is

(a) $\{(2, 8), (3, 27), (5, 125), (7, 343), (9, 729)\}$
(b) $\{(3, 27), (5, 125), (7, 343)\}$
(c) $\{(2, 8), (3, 27), (5, 125), (7, 343)\}$
(d) $\{(1, 1), (2, 8), (3, 27), (5, 125), (7, 343)\}$

55. Let $A = \{x, y, z\}$ and $B = \{1, 2\}$. Then, the number of relations from A to B is

(a) 32 (b) 64 (c) 128 (d) 8

56. Let R be the relation on Z defined by $R = \{(a, b) : a, b \in Z, a - b$ is an integer$\}$. Then

(a) domain of R is $\{2, 3, 4, 5, \ldots\}$
(b) range of R is Z
(c) Both (a) and (b)
(d) None of the above

Level II

57. Let R be a relation from Q to Q defined by $R = \{(a, b) : a, b \in Q$ and $a - b \in Z\}$. Then, which of the following are true?

I. $(a, a) \in R$ for all $a \in Q$

II. $(a, b) \in R \Rightarrow (b, a) \in R$

III. (a, b) and $(b, c) \in R \Rightarrow (a, c) \in R$

(a) I and II are true
(b) II and III are true
(c) I and III are true
(d) All are ture

58. The domain and range of the relation R given by $R = \{(x, y) : y = x + \frac{6}{x}$; where $x, y \in N$ and $x < 6\}$ are ...X... and ...Y... . Here, X and Y refer to [NCERT Exemplar]

(a) $\{1, 2, 3\}$ and $\{3, 5\}$
(b) $\{1, 2, 3\}$ and $\{5, 7\}$
(c) $\{5, 7\}$ and $\{1, 2, 3\}$
(d) $\{3, 5\}$ and $\{1, 2, 3\}$

59. Let $A = \{x, y, z\}$ and $B = \{a, b, c, d\}$. Then, which one of the following is not a relation from A to B? [Kerala CEE 2010]

(a) $\{(x, a), (x, c)\}$
(b) $\{(y, c), (y, d)\}$
(c) $\{(z, a), (z, d)\}$
(d) $\{(z, b), (y, b), (a, d)\}$
(e) $\{(x, c)\}$

60. **Assertion** (A) Let $A=\{1, 2, 3, 5\}, B=\{4, 6, 9\}$ and $R=\{(x, y) : |x-y|$ is odd, $x \in A$, $y \in B\}$. Then, domain of R is $\{1, 2, 3, 5\}$.

Reason (R) $|x|$ is always positive $\forall x \in R$.

(a) Both A and R are correct; R is the correct explanation of A
(b) Both A and R are correct; R is not the correct explanation of A
(c) A is correct; R is incorrect
(d) R is correct; A is incorrect

61. Consider the following statements

Statement I The domain of the relation, $A=\{(x, y): x, y \in Z, xy=4\}$ is $\{-2, -1, 0, 1, 2\}$.

Statement II The range of the relation, $R_1=\{(a, b): a, b \in N$ and $2a+b=10\}$ is $\{2, 4, 6, 8\}$.

Choose the correct option.

(a) Statement I is true (b) Statement II is true
(c) Both are true (d) Both are false

62. R is a relation on N given by $R=\{(x, y): 4x+3y=20\}$.

Which of the following belongs to R? [KCET 2008]

(a) $(-4, 12)$ (b) $(5, 0)$
(c) $(3, 4)$ (d) $(2, 4)$

63. Let $A=\{1, 2, 3, 4\}$, $B=\{1, 4, 9, 16, 25\}$ and R be a relation defined from A to B as $R=\{(x, y): x \in A, y \in B$ and $y=x^2\}$. Then, range and codomain of R are ...X... and ...Y... . Here, X and Y refer to

(a) $\{1, 2, 3, 4\}$ and $\{1, 4, 9, 16, 25\}$
(b) $\{1, 2, 3, 4\}$ and $\{1, 4, 9, 16\}$
(c) $\{1, 4, 9, 16\}$ and $\{1, 4, 9, 16, 25\}$
(d) $\{1, 4, 9, 16, 25\}$ and $\{1, 4, 9, 16, 25\}$

64. Let $A=\{1, 2, 3\}$ and $B=\{-1, 0, 1\}$. Then, which of the following is not the relation from A to B?

(a) $R=\{(1, -1), (2, 0), (3, 1)\}$
(b) $R=\{(1, -1), (1, 0), (2, 0), (2, 1)\}$
(c) $R=\{(2, 0), (2, 1), (1, -1), (3, 1), (3, 0)\}$
(d) $R=\{(1, -1), (-1, 1), (2, 0), (0, 2), (3, 1), (1, 3)\}$

65. Let $A=\{3, 5\}$, $B=\{7, 11\}$ and $R=\{(a, b): a \in A, b \in B, a+b$ is odd$\}$. Then, R in Roster form, is

(a) $\{(3, 7), (5, 11)\}$ (b) $\{(3, 11), (5, 7)\}$
(c) $\{\ \}$ (d) None of these

66. **Assertion** (A) The domain of the relation $R=\{(x+2, x+4) : x \in N, x< 8\}$ is $\{3, 4, 5, 6, 7, 8, 9\}$.

Reason (R) The range of the relation $R=\{(x+2, x+4): x \in N, x< 8\}$ is $\{1, 2, 3, 4, 5, 6, 7\}$.

(a) Both A and R are correct; R is the correct explanation of A
(b) Both A and R are correct; R is not the correct explanation of A
(c) A is correct; R is incorrect
(d) R is correct; A is incorrect

67. Consider the following statements

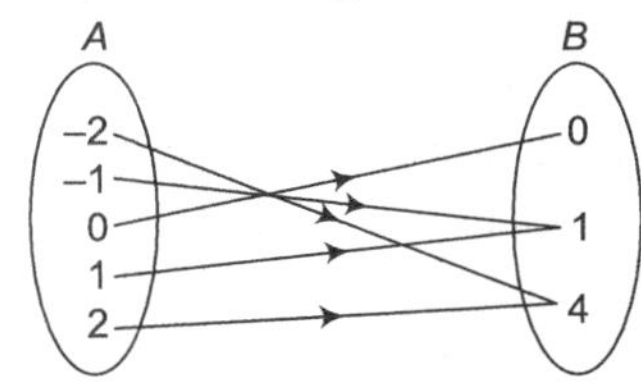

Statement I The figure shows a relationship between the sets A and B. Then, the relation in Set-builder form is $\{(x, y): y=x^2, x, y \in N$ and $-2 \le x \le 2\}$.

Statement II The above Relation in Roster form is $\{(-1, 1), (-2, 4), (0, 0), (1, 1), (2, 4)\}$.

Choose the correct option.

(a) Statement I is true (b) Statement II is true
(c) Both are true (d) Both are false

68. If $A=\{1, 2\}$ and $B=\{3, 4, 5\}$. Then match the terms of Column I with the terms of Column II and choose the correct option from the codes given below.

	Column I		Column II
A.	Number of relations from A to A	1.	512
B.	Number of relations from B to B	2.	64
C.	Number of relations from A to B	3.	16
D.	Number of relations from B to A		

Codes

	A	B	C	D
(a)	3	1	2	2
(b)	3	2	1	1
(c)	1	2	3	1
(d)	2	2	1	3

69. The relation R defined on the set of natural numbers as $\{(a, b): a$ differs from b by $3\}$ is given by

(a) $\{(4, 1), (5, 2), (6, 3), \ldots\}$ (b) $\{(1, 3), (2, 6), (3, 9), \ldots\}$
(c) $\{(1, 4), (2, 5), (3, 6), \ldots\}$ (d) None of these

70. If R is a relation on a finite set A having $(n-1)$ elements, then the number of relations on A is

(a) 2^{n-1} (b) $2^{(n-1)^2}$
(c) $(n-1)^2$ (d) $(n-1)^{n-1}$

71. If $R=\{(x, y): x, y \in Z, x^2+y^2 \le 4\}$ is a relation, then domain of R is

(a) $\{0, 1, 2\}$ (b) $\{0, -1, -2\}$
(c) $\{-2, -1, 0, 1, 2\}$ (d) None of these

Functions

Level I

72. Let A and B are non-empty sets. Then, which of the following are true?

I. Every relation is a function.
II. Every function is a relation.
III. If f is a function from A to B and $(a, b) \in f$, then b is called the image of a under f.

(a) I and II are true (b) II and III are true
(c) I and III are true (d) All are true

73. Consider the following statements

Statement I If $A = \{1, 2, 3, 4, 5, 6\}$ and R is a relation defined by $\{R = (x, y) : y = x + 1\}$, then R is a function from A to A.

Statement II

If $P = \{9, 4, 25\}$, $Q = \{-5, -3, -2, 1, 2, 3, 5\}$ and R is relation from P to Q such that

$R = \{(9, 3), (9, -3), (4, 2), (4, -2), (25, 5), (25, -5)\}$, then R is not function from P to Q.

Choose the correct option.

(a) Statement I is true
(b) Statement II is true
(c) Both are true
(d) Both are false

74. Let N be the set of natural numbers and the relation R be defined such that $\{R = (x, y) : y = 2x, x, y \in N\}$. Then,

(a) R is a function
(b) R is not a function
(c) domain range and codomain is N
(d) None of the above

75. There are three relations R_1, R_2 and R_3 such that

$R_1 = \{(2, 1), (3, 1)(4, 2)\}$,
$R_2 = \{(2, 2), (2, 4), (3, 3), (4, 4)\}$

and $R_3 = \{(1, 2), (2, 3), (3, 4), (4, 5), (5, 6), (6, 7)\}$. Then,

(a) R_1 and R_2 are functions (b) R_2 and R_3 are functions
(c) R_1 and R_3 are functions (d) Only R_1 is a function

76. If the domain of a function is either R (real numbers) or a subset of R, then it is called a ...K.... Here, K refers to

(a) real function
(b) real valued function
(c) real function or real valued function
(d) None of the above

77. Let N be the set of natural numbers and $f : N \to N$ is a real valued function defined by $f(x) = 2x + 1$. Then, match the terms of Column I with the terms of Column II and choose the correct option from the codes given below.

	Column I		Column II
A.	$f(1)$	1.	15
B.	$f(2)$	2.	9
C.	$f(4)$	3.	5
D.	$f(5)$	4.	3
E.	$f(7)$	5.	11

Codes

	A	B	C	D	E		A	B	C	D	E
(a)	3	1	5	2	4	(b)	4	3	1	5	2
(c)	4	3	2	5	1	(d)	3	4	1	5	2

78. The graph of an identity function on R is

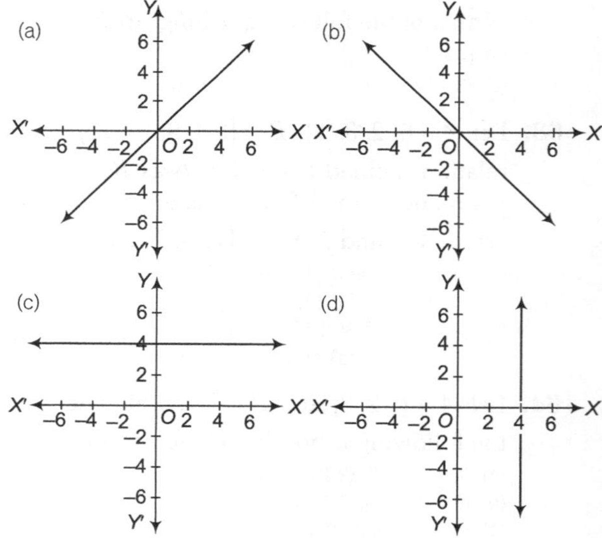

79.

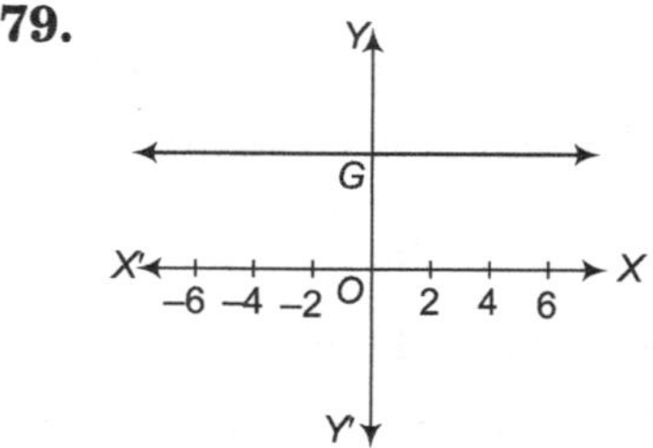

If G represents the name of the function in above graph, then G is a/an

(a) identity function (b) constant function
(c) modulus function (d) None of these

80. Which of the following is/are the polynomial functions?

I. $f(x) = x^3 - x^2 + 2 \ \forall \ x \in R$

II. $f(x) = x^4 + \sqrt{2}x \ \forall x \in R$

III. $f(x) = x^{2/3} + 2x \ \forall \ x \in R$

(a) I, II (b) II, III (c) I, II, III (d) Only I

81. The graph of the function $f: R \to R$ and $f(x) = x^2$, $x \in R$, is

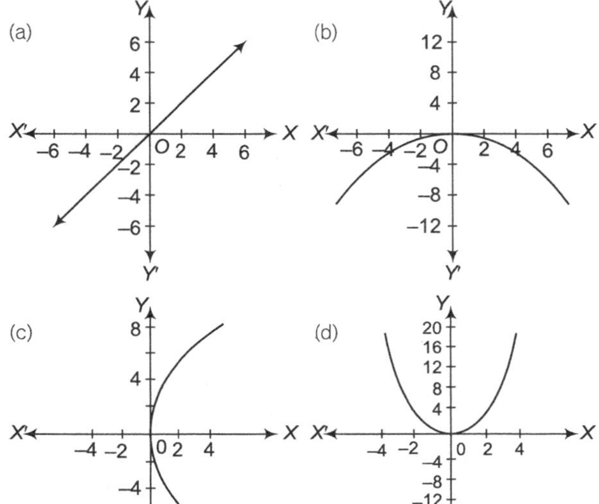

82. The graph of the function $f: R \to R$ defined by $f(x) = x^3$, $x \in R$, is

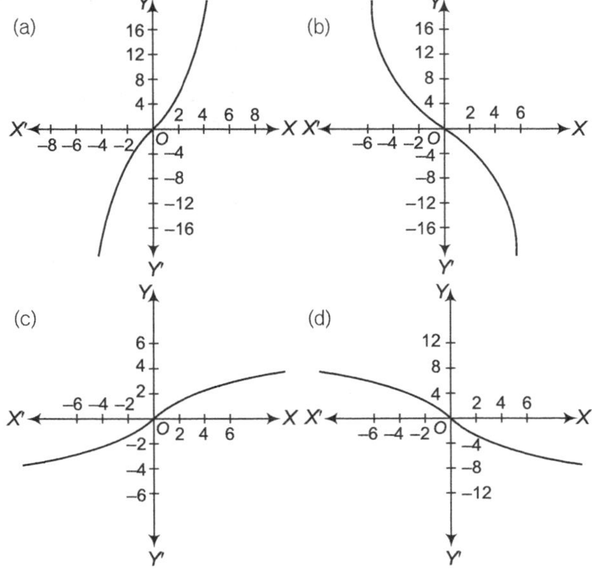

83. Which of the following represent rational functions, if $f(x)$ and $g(x)$ are polynomial functions?

(a) $f(x) \cdot g(x)$

(b) $\frac{f(x)}{g(x)}, g(x) \neq 0$

(c) $\frac{f(x)}{g(x)}$

(d) None of the above

84. The graph of real value of function $f: R - \{0\} \to R$ defined by $f(x) = \frac{1}{x}$, $x \in R - \{0\}$, is

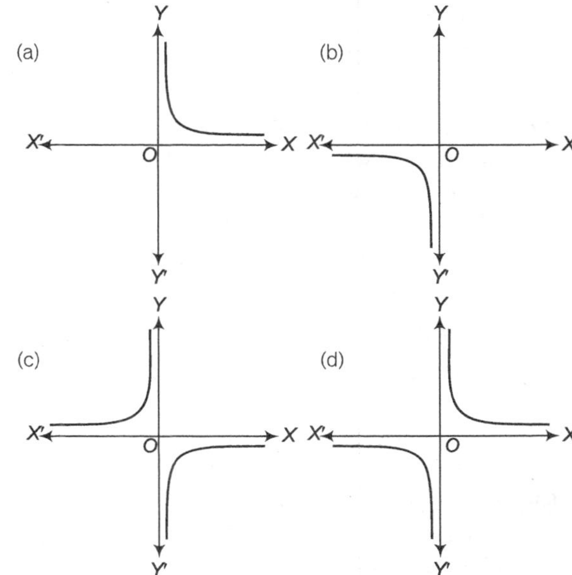

85. Consider the following statements

Statement I The graph of modulus function is

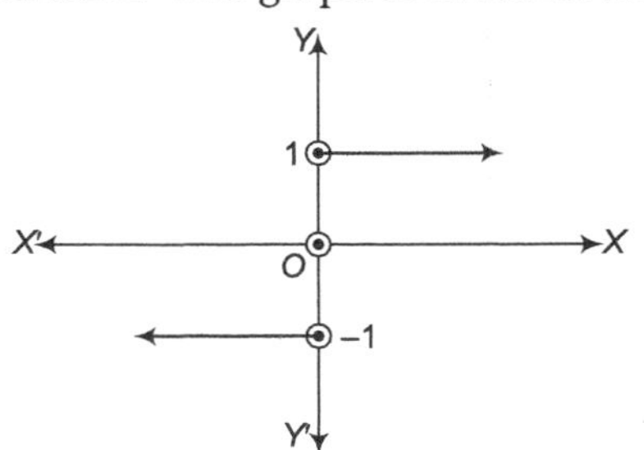

Statement II The Signum function is $f: R \to R$ defined by $f(x) = [x]$.

Choose the correct option.

(a) Statement I is true

(b) Statement II is true

(c) Both are true

(d) Both are false

86. 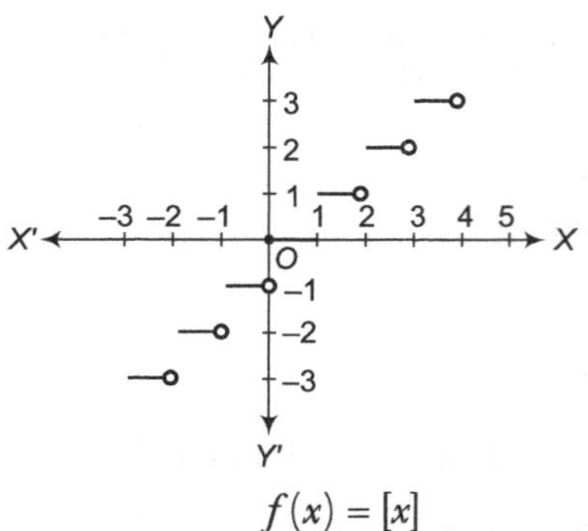

$f(x) = [x]$

Which of the following options identify the above graph?

(a) Modulus function (b) Greatest integer function
(c) Signum function (d) None of these

87. Let $f : X \to R$ and $g : X \to R$ be any two real functions, where $X \subset R$. Then, which of the following is/are true?

I. $(f + g)(x) = f(x) + g(x)\ \forall\ x \in X$
II. $(f - g)(x) = f(x - x)\ \forall\ x \in X$
III. $(\alpha f)(x) = f(\propto x)\ \forall\ x \in X$
IV. $(f\,g)(x) = f(x)\ g(x)\ \forall\ x \in X$
V. $\left(\dfrac{f}{g}\right)(x) = \left(\dfrac{f(x)}{g(x)}\right)\ \forall\ x \in X, g(x) \neq 0$

(a) I, III and IV are true (b) I, II, III and V are true
(c) I and III are true (d) I, IV and V are true

88. Let $f(x) = x^2$ and $g(x) = 2x + 1$ be two real functions. Then, match the functions given in Column I with the expressions in Column II and choose the correct option from the codes given below.

	Column I		Column II
A.	$(f + g)(x)$	1.	$x^2 - 2x - 1$
B.	$(f - g)(x)$	2.	$x^2 + 2x + 1$
C.	$(fg)(x)$	3.	$\dfrac{x^2}{2x+1}, x \neq -\dfrac{1}{2}$
D.	$\left(\dfrac{f}{g}\right)(x)$	4.	$2x^3 + x^2$

Codes

	A	B	C	D		A	B	C	D
(a)	2	1	4	3	(b)	4	1	2	3
(c)	2	1	3	4	(d)	2	4	1	3

89. Let $f(x) = \sqrt{x}$ and $g(x) = x$ be two functions defined over the set of non-negative real numbers. Then, which of the following is/are true?

I. $(f + g)(x) = \sqrt{x}\,(\sqrt{x} + 1)$
II. $(f - g)(x) = \sqrt{x}\,(\sqrt{x} - 1)$
III. $(fg)(x) = x^{\frac{2}{3}}$
IV. $\left(\dfrac{f}{g}\right)(x) = \dfrac{1}{\sqrt{x}}, x \neq 0$

(a) I and II are true (b) I, III and IV are true
(c) I and IV are true (d) None of these

90. Which of the following relations are functions?

$R_1 = \{(2, 1), (5, 1), (8, 1), (11, 1), (14, 1), (17, 1)\}$
$R_2 = \{(2, 1), (4, 2), (6, 3), (8, 4), (10, 5), (12, 6), (14, 7)\}$
$R_3 = \{(1, 3), (1, 5), (2, 5)\}$

(a) R_1 and R_2 (b) R_2 and R_3
(c) R_1, R_2 and R_3 (d) None of these

91. Consider the following statements

Statement I The domain of function $f(x) = -|x|$ is R.

Statement II The range of function $f(x) = -|x|$ is R.

Choose the correct option.

(a) Statement I is true (b) Statement II is true
(c) Both are true (d) Both are false

92. The domain and range of the real function $\sqrt{9 - x^2}$ are ...X... and ...Y... respectively. Here, X and Y refer to

(a) R_1 and (0, 3) (b) [−3, 3] and [−3, 3]
(c) [−3, 3] and [0, 3] (d) [0, 3] and [0, 3]

93. A function f is defined by $f(x) = 2x - 5$, then

(a) $f(0) = f(7)$ (b) $f(-3) = -11$
(c) Both (a) and (b) (d) None of the above

94. The function t which maps temperature in degree Celsius into temperature in degree Fahrenheit, is defined by $t(C) = \dfrac{9C}{5} + 32$. Then, match the terms of Column I with the terms of Column II and choose the correct option from the codes given below.

	Column I		Column II
A.	$t(0)$	1.	$\dfrac{412}{5}$
B.	$t(28)$	2.	100
C.	$t(-10))$	3.	32
D.	The value of C when $t(C) = 212$	4.	14

Codes

	A	B	C	D
(a)	3	1	4	2
(b)	2	1	4	3
(c)	3	4	1	2
(d)	3	4	2	1

95. Consider the following statements

Statement I The range of the function $f(x) = 2 - 3x, x \in R, x > 0$ is R.

Statement II The range of the function $f(x) = x^2 + 2$ is R.

Coose the correct option.

(a) Statement I is true (b) Statement II is true
(c) Both are true (d) Both are false

96. The range of the function $f(x) = x$, x is a real number ...X... . Here, X refers to

(a) N (b) Z
(c) $[0, \infty]$ (d) R

97. Let R be the set of real numbers. A real function $f : R \to R$ is defined by $f(x) = x + 10$. Then, the graph of this function is

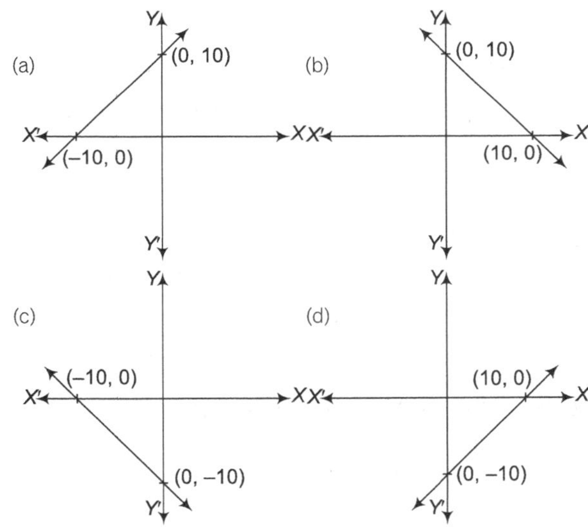

98. The function f defined by $f(x) = mx + c$, where m and c are constants, $x \in R$, is called ...X... . Here, X refers to

(a) logarithmic function (b) exponential function
(c) linear function (d) quadratic function

99. The relation f and g are defined by

$$f(x) = \begin{cases} x^2, & 0 \le x \le 3 \\ 3x, & 3 \le x \le 10 \end{cases}$$

and $g(x) = \begin{cases} x^2, & 0 \le x \le 2 \\ 3x, & 2 \le x \le 10 \end{cases}$, then

(a) f is a function
(b) g is a function
(c) Both (a) and (b)
(d) None of the above

Level II

100. Which of the following relations are functions?

$R_1 = \{(1, 1), (2, 2), (3, 3), (4, 4), (4, 5)\}$
$R_2 = \{(1, 1), (2, 3), (3, 5), (4, 7)\}$
$R_3 = \{(1, 1), (2, 4), (3, 9), (4, 16), (5, 25)\}$

(a) R_1 and R_2 (b) R_2 and R_3 (c) R_1 and R_3 (d) R_1, R_2 and R_3

101. If $A = \{1, 2, 3, 4\}$, then which of the following is/are functions in the given set A?

(a) $R_1 = \{(x, y) : x, y \in A \text{ and } y = x + 1\}$
(b) $R_2 = \{(x, y) : x, y \in A \text{ and } x + y = 5\}$
(c) Both (a) and (b)
(d) None of the above

102. Consider the following statements

Statement I The following arrow diagram represents a function.

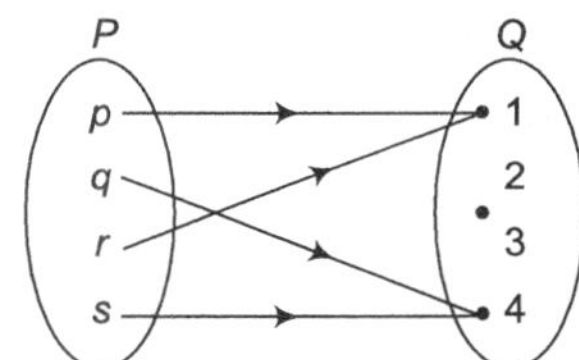

Statement II Let $f : R - \{3\} \to R$ be defined by $f(x) = \dfrac{x^2 - 9}{x - 3}$ and $g : R \to R$ be defined by $g(x) = x + 3$. Then, $f = g$.

Choose the correct option.

(a) Statement I is true (b) Statement II is true
(c) Both are true (d) Both are false

103. Let $f : R \to R$ be such that $f(x) = 2^x$. Then,

(a) Range of f is R (b) $\{x : f(x) = 1\} = \{0, 1\}$
(c) $f(x + y) = f(x) \cdot f(y)$ (d) None of these

104. The domain for which the functions $f(x) = 2x^2 - 1$ and $g(x) = 1 - 3x$ is equal *i.e.*, $f(x) = g(x)$, is

(a) $\{0, 2\}$ (b) $\left\{\frac{1}{2}, -2\right\}$ (c) $\left\{-\frac{1}{2}, 2\right\}$ (d) $\left\{\frac{1}{2}, 2\right\}$

105. Assertion (A) The domain of the function $f(x) = \dfrac{x}{x^2 + 3x + 2}$ is $R - \{-1, -2\}$.

Reason (R) If f is a function from A to B, then domain of f is B.

(a) Both A and R are correct; R is the correct explanation of A
(b) Both A and R are correct; R is not the correct explanation of A
(c) A is correct; R is incorrect
(d) R is correct; A is incorrect

106. The domain of the real function $f(x) = \dfrac{1}{\sqrt{4 - x^2}}$ is [J & K CET 2009]

(a) the set of all real numbers
(b) the set of all positive real numbers
(c) $(-2, 2)$
(d) $[-2, 2]$

107. If $f(x) = x^3 - \dfrac{1}{x^3}$, then $f(x) + f\left(\dfrac{1}{x}\right)$ is equal to [NCERT Exemplar]

(a) $2x^3$ (b) $\dfrac{2}{x^3}$ (c) 0 (d) 1

108. Consider the following statements

Statement I

If $f(x) = x + \dfrac{1}{x}$, then $[f(x)]^3 = f(x^3) + 3f\left(\dfrac{1}{x}\right)$.

Statement II If $f(x) = (x - a)^2 (x - b)^2$, then $f(a + b)$ is 0.

Choose the correct option.

(a) Statement I is true (b) Statement II is true
(c) Both are true (d) Both are false

109. If $f(x) = \dfrac{x+1}{x-1}$, then the value of $f\{f(x)\}$ is

(a) x (b) $-x$
(c) 0 (d) 1

110. The domain of the function f, defined by $f(x) = \dfrac{1}{\sqrt{x - |x|}}$ is [NCERT Exemplar]

(a) R (b) R^+
(c) R^- (d) None of these

111. Let $f(x) = 2x + 5$ and $g(x) = x^2 + x$. Then, match the functions given in Column I with the expressions in Column II and choose the correct option from the codes given below.

Column I	Column II
A. $(f+g)x$	1. $2x^3 + 7x^2 + 5x$
B. $(f-g)x$	2. $x^2 + 3x + 5$
C. $(fg)(x)$	3. $\dfrac{2x+5}{x^2+x}$, $x \neq 0, -1$
D. $\left(\dfrac{f}{g}\right)(x)$	4. $5 + x - x^2$

Codes

	A	B	C	D
(a)	2	4	1	3
(b)	4	1	2	3
(c)	2	1	4	3
(d)	1	4	2	3

112. The range of the function $f(x) = \dfrac{x}{1+x^2}$ is [DCE 2009]

(a) $(-\infty, \infty)$ (b) $[-1, 1]$ (c) $\left[-\dfrac{1}{2}, \dfrac{1}{2}\right]$ (d) $[\sqrt{-2}, \sqrt{2}]$

113. If a function $f : R \to R$ be defined by

$$f(x) = \begin{cases} 3x - 2, & x < 0 \\ 1, & x = 0 \\ 4x + 1, & x > 0 \end{cases}$$

Then, the values of $f(-1)$, $f(0)$ and $f(1)$ are ...X..., ...Y... and ...Z..., respectively. Here, X, Y and Z refer to

(a) -5, 1 and 5 (b) -5, -2 and 1
(c) -3, 1 and 5 (d) 1, -2 and 5

114. If $f(x) = ax + b$, where a and b are integers, $f(-1) = -5$ and $f(3) = 3$, then a and b are equal to [NCERT Exemplar]

(a) $a = -3, b = -1$ (b) $a = 2, b = -3$
(c) $a = 0, b = 2$ (d) $a = 2, b = 3$

115. **Assertion** (A) Let f and g be two real functions given by

$f = \{(0, 1), (2, 0), (3, -4), (4, 2), (5, 1)\}$ and
$g = \{(1, 0), (2, 2), (3, -1), (4, 4), (5, 3)\}$.

Then, domain of $f \cdot g$ is given by $\{2, 3, 4, 5\}$.

Reason (R) Let f and g be two real functions. Then, $(f \cdot g)(x) = f\{g(x)\}$.

(a) Both A and R are correct; R is the correct explanation of A
(b) Both A and R are correct; R is not the correct explanation of A
(c) A is correct; R is incorrect
(d) R is correct; A is incorrect

Miscellaneous Exercise

116. Let $f = \{(1, 1), (2, 3), (0, -1), (-1, -3)\}$ be a linear function from Z to Z. Then, $f(x)$ is

(a) $x - 1$ (b) $2x - 1$
(c) $2x + 1$ (d) $x + 1$

117. Consider the following statements

Statement I The domain of the function $f(x) = \dfrac{x^2 + 3x + 5}{x^2 - 5x + 4}$ is $R - \{1, 4\}$.

Statement II The graph of the function defined by $f(x)=\begin{cases}1-x, & x<0\\ 1\ , & x=0\\ x+1, & x>0\end{cases}$ is

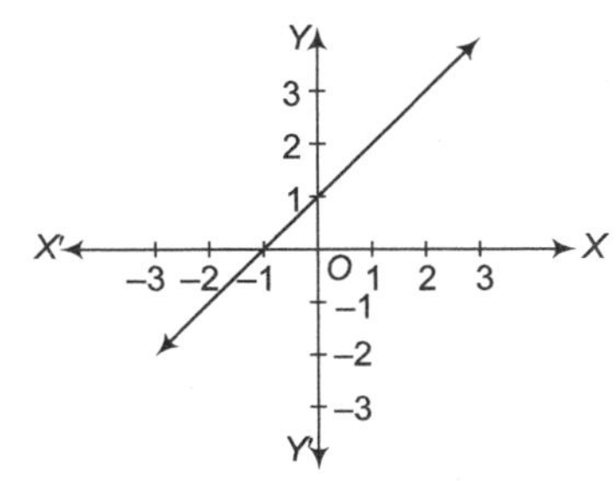

Choose the correct option.
(a) Statement I is true
(b) Statement II is true
(c) Both are true
(d) Both are false

118. If $f(x)=x^2$, then the value of $\dfrac{f(1\cdot1)-f(1)}{(1\cdot1-1)}$ is ...A.... Here, A refers to
(a) 0.1 (b) 2.1
(c) 0 (d) 21

119. The domain of the function $f(x)=\dfrac{x^2+2x+1}{x^2-8x+12}$ is
(a) $R-\{-1\}$
(b) $R-\{-2,-6\}$
(c) $R-\{2,6\}$
(d) $R-\{2,10\}$

120. **Assertion** (A) The domain of the real function f defined by $f(x)=\sqrt{x-1}$ is $R-\{1\}$.
Reason (R) The range of the function defined by $f(x)=\sqrt{x-1}$ is $[0,\infty)$.
(a) Both A and R are correct; R is the correct explanation of A
(b) Both A and R are correct; R is not the correct explanation of A
(c) A is correct; R is incorrect
(d) R is correct; A is incorrect

121. Consider the following statements
Statement I The domain of real function f defined by $f(x)=|x-1|$, is R.
Statement II The range of real function f defined by $f(x)=|x-1|$ is R.
Choose the correct option.
(a) Statement I is true
(b) Statement II is true
(c) Both are true
(d) Both are false

122. Let $f=\left\{\left(x,\dfrac{x^2}{1+x^2}\right): x\in R\right\}$ be a function from R to R. Then, the range of f is
(a) (0, 1) (b) $R-\{0,1\}$ (c) [0, 1) (d) [0, 1]

123. Let $f,\ g: R\to R$ be defined, respectively by $f(x)=x+1$ and $g(x)=2x-3$. Then,
(a) $(f+g)(x)=3x-4$
(b) $(f-g)(x)=x+4$
(c) $\left(\dfrac{f}{g}\right)(x)=\dfrac{x+1}{2x-3}$, provided $x\neq\dfrac{3}{2}$
(d) All of the above

124. Let $f=\{(1,1),(2,3),(0,-1),(-1,-3)\}$ be a function from Z to Z defined by $f(x)=ax+b$ for some integers a and b. Then, the values of a and b are
(a) $a=2, b=-1$ (b) $a=-1, b=2$
(c) $a=-1, b=-1$ (d) $a=-2, b=1$

125. Let R be a relation from N to N defined by $R=\{(a,b): a, b\in N \text{ and } a=b^2\}$. Then, which of the following is/are true?
I. $(a,a)\in R$ for all $a\in N$.
II. $(a,b)\in R$ implies $(b,a)\in R$.
III. $(a,b)\in R, (b,c)\in R$ implies $(a,c)\in R$.
(a) I and II are true (b) II and III are true
(c) All are true (d) None of these

126. Let $A=\{1,2,3,4\}$, $B=\{1,5,9,11,15,16\}$ and $f=\{(1,5),(2,9),(3,1),(4,5),(2,11)\}$. Then,
(a) f is a relation from A to B (b) f is a function from A to B
(c) Both (a) and (b) (d) None of these

127. Consider the following statements
Statement I Let f be the subset of $Z\times Z$ defined by $f=\{(ab, a+b): a, b\in Z\}$. Then, f is a function from Z to Z.
Statement II Let $A=\{9,10,11,12,13\}$ and let $f: A\to N$ be defined by
$f(x)$ = The highest prime factor of n.
Then, the range of f is $\{2,3,5\}$.
Choose the correct option.
(a) Statement I is true
(b) Statement II is true
(c) Both are true
(d) Both are false

128. If $n(A)=10$ and $n(B)=5$, then
(a) $n(A\times B)=50$
(b) number of relations from A to B is 2^{50}
(c) $n(A\times B)=n(B\times A)$
(d) All of the above

129. Consider the following statements

Statement I The ordered pair 5, 2 belongs to the relation $R = \{(x, y) : y = x - 5, x, y \in Z\}$.

Statement II If $(x - 2, y + 5) = \left(-2, \frac{1}{3}\right)$, then $x = 4$ and $y = -\frac{14}{3}$.

Choose the correct option. [NCERT Exemplar]

(a) Statement I is true (b) Statement II is true
(c) Both are true (d) Both are false

130. If $A = \{2, 3, 4, 5\}$ and $B = \{3, 6, 7, 10\}$. R is a relation defined by $R = \{(a, b) : a$ is relatively prime to b, $a \in A$ and $b \in B\}$, then domain of R is

(a) $\{2, 3, 5\}$ (b) $\{3, 5\}$ (c) $\{2, 3, 4\}$ (d) $\{2, 3, 4, 5\}$

131. If $A = \{1, 2, 4\}$, $B = \{2, 4, 5\}$ and $C = \{2, 5\}$, then $(A - B) \times (B - C)$ is equal to

(a) $\{(1, 2), (1, 5), (2, 5)\}$ (b) $\{(1,4)\}$
(c) $(1, 4)$ (d) None of these

132. Let $n(A) = m$ and $n(B) = n$. Then, the total number of non-empty relations that can be defined from A to B, is [NCERT Exemplar]

(a) m^n (b) $n^m - 1$ (c) 2^{mn} (d) $2^{mn} - 1$

133. If S is a relation on R defined by

$S = \{(a, b) \Leftrightarrow 1 + ab > 0, a, b \in R\}$, then [HOTS]

(a) $(a, a) \in S \ \forall a \in R$
(b) $(a, b) \in S$ and $(b, c) \in S \Rightarrow (a, c) \in S \ \forall a, b$ and $c \in R$
(c) Both (a) and (b)
(d) None of the above

134. Consider the following statements

Statement I The domain of the relation $R = \{(a, b) : a \in N, a < 5, b = 4\}$ is $\{1, 2, 3, 4\}$.

Statement II The range of the relation $S = \{(a, b) : b = |a - 1|, a \in Z$ and $|a| \le 3\}$ is $\{1, 2, 3, 4\}$.

Choose the correct option.

(a) Statement I is true (b) Statement II is true
(c) Both are true (d) Both are false

135. If S is set with 10 elements and $A = \{(x, y) : x, y \in S, x \ne y\}$, then number of elements in A is

(a) 100 (b) 90 (c) 50 (d) 45

136. If $n(A) = 4$, $n(B) = 3$ and $n(A \times B \times C) = 24$, then $n(C)$ is equal to

(a) 288 (b) 1 (c) 12 (d) 2

137. If $f(x) = \log_e\left(\frac{1 + x}{1 - x}\right)$, then $f\left(\frac{2x}{1 + x^2}\right)$ is equal to

(a) $[f(x)]^2$ (b) $[f(x)]^3$ (c) $2f(x)$ (d) $3f(x)$

138. Consider the following statements

Statement I If $f : R \to R$ and $g : R \to R$ are defined by $f(x) = 2x + 3$ and $g(x) = x^2 + 7$, then the values of x such that $g\{f(x)\} = 8$ are -1 and 2.

Statement II If $f : R \to R$ be given by $f(x) = \frac{4^x}{4^x + 2}$ for all $x \in R$, then $f(x) + f(1 - x) = 1$.

Choose the correct option.

(a) Statement I is true (b) Statement II is true
(c) Both are true (d) Both are false

139. If A, B and C are three sets, then [HOTS]

(a) $A \times (B \cup C) = (A \times B) \cup (A \times C)$
(b) $A \times (B - C) = (A \times B) - (A \times C)$
(c) Both (a) and (b)
(d) None of the above

140. The domain of relation

$R = \{(x, y) : x^2 + y^2 = 16, x, y \in Z\}$ is

(a) $\{0, 1, 2, 3, 4\}$
(b) $\{-4, -3, -2, -1\}$
(c) $\{-4, -3, -2, -1, 0, 1, 2, 3, 4\}$
(d) None of the above

141. Match the terms of Column I with the terms of Column II and choose the correct option from the codes given below.

	Column I		Column II
A.	If $P = \{x : x < 3, x \in N\}$ and $Q = \{x : x \le 2, x \in N\}$, then $(P \cup Q) \times (P \cap Q)$ is equal to	1.	$\{(0, 8), (8, 0), (0, -8), (-8, 0)\}$
B.	If $A = \{x : x \in W, x < 2\}$, $B = \{x : x \in N, 1 < x < 5\}$ and $C = \{3, 5\}$, where W is the set of whole numbers, then $A \times (B \cap C)$ is equal to	2.	$\{(-2, 3), (-1, 5), (0, 7), (1, 9), (2, 11)\}$
C.	If $R = \{(x, y) : x$ and y are integers and $x^2 + y^2 = 64\}$ is a relation, then R in Roster form is	3.	$\{(0, 3), (1, 3)\}$
D.	If $R = \{(x, y) : y = 2x + 7, x \in Z$ and $-2 \le x \le 2\}$ is a relation, then R is	4.	$\{(1, 1), (1, 2), (2, 1), (2, 2)\}$

Codes

	A	B	C	D
(a)	4	3	1	2
(b)	3	4	1	2
(c)	4	3	2	1
(d)	3	1	4	2

Answer *with* Explanations

1. *(d)* In all three we observe that the two objects link with a certain property. So, all of these are relation.

2. *(d)* The pairs of coloured objects are (red, b), (red, c), (red, s), (blue, b), (blue, c), (blue, s)

Thus, we get 6 distinct objects.

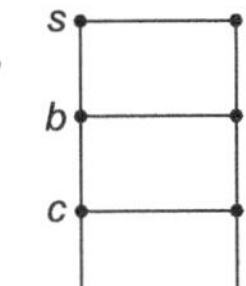

3. *(a)* We know that, an ordered pair of elements taken from any two sets P and Q is a pair of elements written in small brackets and grouped together in a particular order, *i.e.*, (p, q), $p \in P$ and $q \in Q$.

4. *(c)* P and Q are two non-empty sets. The Cartesian product $P \times Q$ is the set of all ordered pairs of elements from P and Q, *i.e.*, $P \times Q = \{(p, q) : p \in P \text{ and } q \in Q\}$

Now, $A = \{\text{red, blue}\}$, $B = \{b, c, s\}$

$A \times B =$ Set of all ordered pairs

$= \{(\text{red}, b), (\text{red}, c), (\text{red}, s), (\text{blue}, b), (\text{blue}, c), (\text{blue}, s)\}$

5. *(a)* If either P or Q is the null set, then $P \times Q$ will be an empty set, *i.e.*, $P \times Q = \phi$

6. *(c)* $A = \{DL, MP, KA\}$ and $B = \{01, 02, 03\}$

The set of all ordered pairs from A and B is represented by $A \times B$.

$$A \times B = \{(DL, 01), (DL, 02), (DL, 03), (MP, 01), (MP, 02), (MP, 03), (KA, 01), (KA, 02), (KA, 03)\}$$

$\therefore$ Number of element in $(A \times B)$ is 9.

7. *(d)* If $A = \{a_1, a_2\}$, $B = \{b_1, b_2, b_3, b_4\}$, then.

$$A \times B = \{(a_1, b_1), (a_1, b_2), (a_1, b_3), (a_1, b_4), (a_2, b_1), (a_2, b_2), (a_2, b_3), (a_2, b_4)\}$$

8. *(a)* Two ordered pairs are equal, if and only if the corresponding first elements are equal and the second elements are also equal.

Given, $(x + 1, y - 2) = (3, 1)$.

Then, by the definition

$$x + 1 = 3 \quad \text{and} \quad y - 2 = 1$$

$$\Rightarrow \quad x = 2 \quad \text{and} \quad y = 3$$

9. *(c)* If there are p elements in A and q elements in B, then there will be pq elements in $A \times B$, *i.e.*, if $n(A) = p$ and $n(B) = q$, then $n(A \times B) = pq$

10. *(a)* If A and B are non-empty sets and either A or B is an infinite set, then $A \times B$ is also infinite.

11. *(b)* Let A be a non-empty set, then $A \times A \times A = \{(a, b, c) : a, b, c \in A\}$

Here, (a, b, c) is called a ordered triplet.

12. *(b)* Given, $P = \{a, b, c\}$ and $Q = \{r\}$

$$P \times Q = \{(a, r), (b, r), (c, r)\}$$
$$Q \times P = \{(r, a), (r, b), (r, c)\}$$

Since, by the the definition of equality of ordered pairs, the pair (a, r) is not equal to the pair (r, a), we conclude that

$$P \times Q \neq Q \times P$$

However, the number of elements in each set will be the same.

13. *(c)* Given, $A = \{1, 2, 3\}$, $B = \{3, 4\}$ and $C = \{4, 5, 6\}$

A. $B \cap C = \{4\}$

$\therefore A \times (B \cap C) = \{(1, 4), (2, 4), (3, 4)\}$

B. $A \times B = \{(1, 3), (1, 4), (2, 3), (2, 4), (3, 3), (3, 4)\}$

$A \times C = \{(1, 4), (1, 5), (1, 6), (2, 4), (2, 5), (2, 6), (3, 4), (3, 5), (3, 6)\}$

$\therefore (A \times B) \cap (A \times C) = \{(1, 4), (2, 4), (3, 4)\}$

C. $B \cup C = \{3, 4, 5, 6\}$

$\therefore A \times (B \cup C) = \{(1, 3), (1, 4), (1, 5), (1, 6), (2, 3), (2, 4), (2, 5), (2, 6), (3, 3), (3, 4), (3, 5), (3, 6)\}$

D. $(A \times B) \cup (A \times C) = \{(1, 3), (1, 4), (1, 5), (1, 6), (2, 3), (2, 4), (2, 5), (2, 6), (3, 3), (3, 4), (3, 5), (3, 6)\}$

E. $A \cap B = \{3\}$, $B \cap C = \{4\}$

$\therefore (A \cap B) \times (B \cap C) = \{3, 4\}$

14. *(d)* Given, $P = \{1, 2\}$

$$P \times P = \{1, 2\} \times \{1, 2\} = \{(1, 1), (1, 2), (2, 1), (2, 2)\}$$

$\therefore \quad P \times P \times P = \{(1, 1, 1), (1, 1, 2), (1, 2, 1), (1, 2, 2), (2, 1, 1), (2, 1, 2), (2, 2, 1), (2, 2, 2)\}$

$\therefore$ Missing ordered triplet are (1, 2, 2) and (2, 1, 1).

15. *(b)* The Cartesian product $R \times R$ represents the set $R \times R = \{(x, y) : x, y \in R\}$, which represents the coordinate of all the point in two dimensional space and the Cartesian product $R \times R \times R$ represents the set $R \times R \times R = \{(x, y, z) : x, y, z \in R\}$, which represents the coordinates of all the points in three dimensional space.

16. *(b)* $A =$ Set of first elements $= \{p, m\}$

$B =$ Set of second elements $= \{q, r\}$

17. *(a)* I. By the definition of equality of two ordered pair

$$\frac{x}{3} + 1 = \frac{5}{3} \quad \text{and} \quad y - \frac{2}{3} = \frac{1}{3}$$

$$\Rightarrow \quad \frac{x}{3} = \frac{5}{3} - 1 \quad \text{and} \quad y = \frac{1}{3} + \frac{2}{3}$$

$$\Rightarrow \quad x = 2 \quad \text{and} \quad y = 1$$

II. $n(A) = 3$ and $n(B) = 3$

$$n(A \times B) = n(A) \times n(B) = 3 \times 3 = 9$$

18. *(b)* Given, $G = \{7, 8\}$ and $H = \{5, 4, 2\}$

$$G \times H = \{(7, 5), (7, 4), (7, 2), (8, 5), (8, 4), (8, 2)\}$$
$$H \times G = \{(5, 7), (5, 8), (4, 7), (4, 8), (2, 7), (2, 8)\}$$

We know that, $(7, 5) \neq (5, 7)$, similarly others.

$\therefore \quad G \times H \neq H \times G$

19. *(b)* I. $P = \{m, n\}$ and $Q = \{n, m\}$

$$P \times Q = \{(m, n), (m, m), (n, n), (n, m)\}$$

II. True (refer Q. 4)

III. $A = \{1, 2\}, B = \{3, 4\}$

$$B \cap \phi = \phi$$
$\therefore \quad A \times (B \cap \phi) = A \times \phi = \phi$

20. *(d)* Given, $A = \{-1, 1\}$

$$A \times A = \{-1, 1\} \times \{-1, 1\}$$
$$= \{(-1, -1), (-1, 1), (1, -1), (1, 1)\}$$
$$A \times A \times A = \{(-1, -1, -1), (-1, -1, 1), (-1, 1, -1), (-1, 1, 1), (1, -1, -1), (1, -1, 1), (1, 1, -1), (1, 1, 1)\}$$

21. *(b)* A = Set of first elements $= \{a, b\}$

B = Set of second elements $= \{x, y\}$

22. *(d)* I. $A \times B = \{(1, 1), (1, 2), (1, 3), (1, 4), (2, 1), (2, 2), (2, 3), (2, 4)\}$

$$A \times C = \{(1, 5), (1, 6), (2, 5), (2, 6)\}$$

$\therefore A \times B \neq A \times C \quad [\because (1, 1) \in A \times B \text{ but } (1, 1) \notin A \times C]$

Hence, I is not true.

II. $\because A \subset B$ and $C \subset D$

$\therefore \quad A \times C \subset B \times D$

Therefore, II is true.

III. $B \cap C = \{\ \}$ or ϕ

$$A \times (B \cap C) = A \times \phi = \phi$$
$$(A \times B) \cap (A \times C) = \phi$$

So, $\quad A \times (B \cap C) = (A \times B) \cap (A \times C)$

Hence, III is true.

23. *(d)* $n(A) = 2$ and $n(B) = 2$

$$n(A \times B) = n(A) \times n(B)$$
$$= 2 \times 2 = 4$$

$\therefore$ Number of sub set of $A \times B = 2^{n(A \times B)}$

$$= 2^4 = 16$$

24. *(b)* A = Set of first elements $= \{x, y, z\}$

B = Set of second elements $= \{1, 2\}$

$\therefore$ A is correct.

Now, $\quad n(A) = 3, n(B) = 2$

$$n(A \times B) = n(A) \times n(B)$$
$$= 3 \times 2 = 6$$

$\therefore$ R is correct but R is not a correct explanation of A.

25. *(d)* If $A \times A$ has 9 elements, then

$$n(A \times A) = 9$$
$\Rightarrow \quad n(A) \times n(A) = 9$

$\Rightarrow \quad \{n(A)\}^2 = 9$

$\Rightarrow \quad n(A) = 3$

$\therefore A$ contains 3 elements.

Since, $(-1, 0), (0, 1) \in A \times A$

$\therefore \quad A = \{-1, 0, 1\}$

$$A \times A = \{(-1, -1), (-1, 0), (-1, 1), (0, -1), (0, 0), (0, 1), (1, -1), (1, 0), (1, 1)\}$$

Hence, the remaining elements of $A \times A$ are $(-1, -1), (-1, 1), (0, -1), (0, 0), (1, -1), (1, 0), (1, 1)$.

26. *(b)* Given, $A = \{1, 2, 3, 4\}$ and $B = \{5, 7, 9\}$

$$A \times B = \{1, 5), (1, 7), (1, 9), (2, 5), (2, 7), (2, 9), (3, 5), (3, 7), (3, 9), (4, 5), (4, 7), (4, 9)\}$$
$$B \times A = \{(5, 1), (5, 2), (5, 3), (5, 4), (7, 1), (7, 2), (7, 3), (7, 4), (9, 1), (9, 2), (9, 3), (9, 4)\}$$

By the definition of ordered pair $(5, 1) \neq (1, 5)$

$\therefore \quad A \times B \neq B \times A$

$$n(A \times B) = 12 = n(B \times A)$$

27. *(a)* A = Set of first elements of the ordered pairs of $A \times B$

$= \{1, 2, 3\}$

B = Set of second element of the ordered pairs of $A \times B$

$= \{3, 5\}$

$$A \times B = \{(1, 3), (1, 5), (2, 3), (2, 5), (3, 3), (3, 5)\}$$

So, the remaining elements of $A \times B$ are

$$(1, 5), (2, 3), (3, 5)$$

28. *(a)* I. Given, $(4x + 3, y) = (3x + 5, -2)$

Two ordered pairs are equal when their corresponding elements are equal.

$$4x + 3 = 3x + 5 \quad \text{and} \quad y = -2$$
$$4x - 3x = 5 - 3$$
$$x = 2$$

II. Now, $A = \{-1, 3, 4\}$

$\therefore A \times A = \{(-1, -1), (-1, 3), (-1, 4), (3, -1), (3, 3), (3, 4), (4, -1), (4, 3), (4, 4)\}$

29. *(c)* Let $(a, b) \in A \times (B \cap C)$

$\Rightarrow \quad a \in A$ and $b \in (B \cap C)$

$\Rightarrow \quad a \in A$ and $(b \in B$ and $b \in C)$

$\Rightarrow \quad (a \in A$ and $b \in B)$ and $(a \in A$ and $b \in C)$

$\Rightarrow \quad (a, b) \in A \times B$ and $(a, b) \in (A \times C)$

$$(a, b) \in (A \times B) \cap (A \times C)$$

$\Rightarrow \quad A \times (B \cap C) \subset (A \times B) \cap (A \times C) \quad \ldots \text{(i)}$

Again, let $\quad (x, y) \in (A \times B) \cap (A \times C)$

$$(x, y) \in A \times B \quad \text{and} \quad (x, y) \in A \times C$$

$\Rightarrow \quad (x \in A \text{ and } y \in B) \quad \text{and} \quad (x \in A \text{ and } y \in C)$

$\Rightarrow \quad x \in A \quad \text{and} \quad (y \in B \text{ and } y \in C)$

$\Rightarrow \quad x \in A \text{ and } y \in (B \cap C)$

$\Rightarrow \quad (x, y) \in A \times (B \cap C)$

$\Rightarrow \quad (A \times B) \cap (A \times C) \subset A \times (B \cap C) \quad \ldots(\text{ii})$

From Eqs. (i) and (ii), we get

$$A \times (B \cap C) = (A \times B) \cap (A \times C) \quad \ldots(\text{iii})$$

Now, $A \times (B' \cup C')'$

$= A \times [(B')' \cap (C')']$ (by De-Morgan's law)

$= A \times (B \cap C)$ $[\because (A')' = A]$

$= (A \times B) \cap (A \times C)$ [by Eq. (iii)]

30. *(d)* Given, $n(A) = 5, n(B) = 4, n(A \cap B) = 3$

$n[(A \times B) \cap (B \times A)]$

$= n[(A \cap B) \times (B \cap A)]$

$= n(A \cap B) \times n(B \cap A) \quad [\because n(A \cap B) = n(B \cap A)]$

$= 3 \times 3 = 9$

31. *(a)* Given, $A = \{1, 2\}, B = \{3, 4\}$

$$(B \cap \phi) = \phi$$

$\therefore \quad A \times (B \cap \phi) = A \times \phi = \phi$

32. *(a)* I. Let $(a, b) \in A \times C$

$\Rightarrow \quad a \in A \quad \text{and} \quad b \in C$

$\Rightarrow \quad a \in B \quad \text{and} \quad b \in D \quad (\because A \subseteq B : a \in A \Rightarrow a \in B)$

$\Rightarrow \quad (a, b) \in B \times D$

$\therefore \quad A \times C \subseteq B \times D$

II. Now, $A = \{2, 3\}, B = \{4, 5\}, C = \{5, 6\}$

$B \cup C = \{4, 5, 6\}$

$\therefore \quad A \times (B \cup C) = \{(2, 4), (2, 5), (2, 6), (3, 4), (3, 5), (3, 6)\}$

$B \cap C = \{5\}$

$A \times (B \cap C) = \{(2, 5), (3, 5)\}$

$\therefore \quad A \times (B \cup C) \neq A \times (B \cap C)$

33. *(c)* The required set of ordered pairs is

$\{(2, 4), (2, 6), (2, 18), (2, 54), (6, 18), (6, 54), (9, 18), (9, 27), (9, 54)\}$

34. *(a)* $A = \{a, b, c\}, B = \{b, c, d\}$ and $C = \{a, d, c\}$

$$A - B = \{a\}, B \cap C = \{c, d\}$$

$\therefore \quad (A - B) \times (B \cap C) = \{(a, c), (a, d)\}$

35. *(d)* $(Y \times A) \cap (Y \times B)$

$= Y \times (A \cap B)$

$= Y \times \phi = \phi \quad [\because A = \{1, 2\}, B = \{3, 4, 5\}, A \cap B = \phi]$

36. *(c)* Let $(a, b) \in A \times C$

$\Rightarrow \quad a \in A \quad \text{and} \quad b \in C \quad (\because A \subseteq B, a \in A \Rightarrow a \in B)$

$\Rightarrow \quad a \in B \quad \text{and} \quad b \in C$

$\Rightarrow \quad (a, b) \in (B \times C)$

$\Rightarrow \quad A \times C \subseteq B \times C$

Hence, A is correct.

For any two sets, we know that in general

$$A \times B \neq B \times A$$

$\therefore$ R is incorrect.

37. *(a)* I. Given, $n(U) = 10, n(A) = 4, n(B) = 5$

$n(B') = n(U) - n(B)$

$= 10 - 5 = 5$

$n(A \times B') = n(A) \times n(B')$

$= 4 \times 5 = 20$

II. Now, $(x, -1) \in \{(a, b) : b = 2a - 3\}$

$\Rightarrow \quad -1 = 2x - 3$

$\Rightarrow \quad x = 1$

Again, $(5, y) \in \{(a, b) : b = 2a - 3\}$

$y = 2(5) - 3 = 7$

38. *(c)* Given, $P = \{a, b, c\}$

and $Q = \{\text{Ali, Bhanu, Binoy, Chandra, Divya}\}$

$P \times Q = \{(a, \text{Ali}), (a, \text{Bhanu}), (a, \text{Binoy}) \ldots\ldots (c, \text{Divya})\}$

$R = \{(a, \text{Ali}), (b, \text{Bhanu}), (b, \text{Binoy}), (c, \text{Chandra})\}$

We observe that, every element of R is in $P \times Q$.

$\therefore \quad R \subset P \times Q$

A visual representation of this relation R is shown as

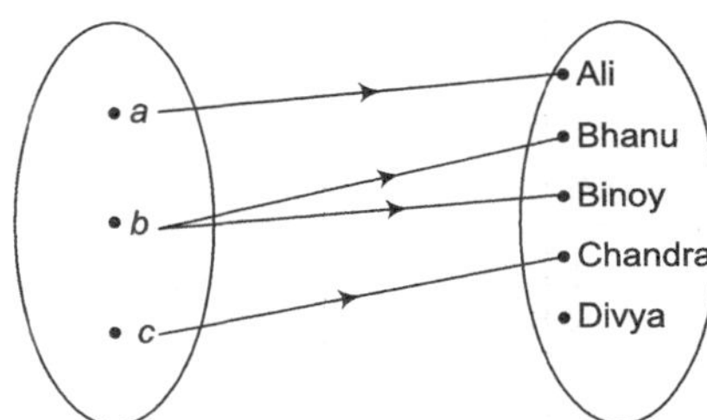

Given diagram is called an arrow diagram.

39. *(a)* A relation R from a non-empty set A to a non-empty set B is a subset of the Cartesian product $A \times B$. The subset is derived by describing a relationship between the first element and the second element of the ordered pairs in $A \times B$. The second element is called the image of the first element.

40. *(d)* The set of first elements of the ordered pairs in a relation R from a set A to a set B is called the domain of the relation $R \Rightarrow$ Domain $< A$.

The set of all second elements in a relation R from a set A to a set B is called the range of the relation R. The whole set B is called the codomain of the relation R.

Note that range $\subseteq$ codomain.

41. *(c)* A relation may be represented algebraically either by the Roster method or by the Set-builder method.

An arrow diagram is a visual representation of a relation.

42. *(b)* I. By the definition of the relation,

$R = \{(1, 2), (2, 3), (3, 4), (4, 5), (5, 6)\}$.

The corresponding arrow diagram is shown in figure

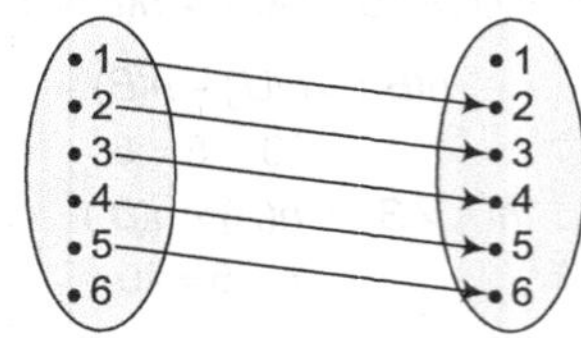

II. We can see that the domain $= \{1, 2, 3, 4, 5\}$

III. Similarly, the range $= \{2, 3, 4, 5, 6\}$

and the codomain $= \{1, 2, 3, 4, 5, 6\}$

43. *(c)* It is obvious that the relation R is 'x is the square of y'.

In Set-builder form, $R = \{(x, y): x$ is the square of $y, x \in P, y \in Q\}$

44. *(c)* In Roster form, $R = \{(9, 3), (9, -3), (4, 2), (4, -2), (25, 5), (25, -5)\}$

45. *(a)* Domain of R = Set of first elements of ordered pairs in R

$$= \{4, 9, 25\}$$

46. *(c)* Range of R = Set of second elements of ordered pairs in R

$$= \{-5, 5, -3, 3, -2, 2\}$$

47. *(a)* The total number of relation that can be defined from a set A to a set B is the number of possible subset of $A \times B$. If $n(A) = p$ and $n(B) = q$, then $n(A \times B) = pq$ and the total number of relation is 2^{pq}.

Given, $A = \{1, 2\}$ and $B = \{3, 4\}$

$\therefore$ $A \times B = \{(1, 3), (1, 4), (2, 3), (2, 4)\}$

Since, $n(A \times B) = 4$, the number of subsets of $A \times B$ is 2^4. Therefore, the number of relation from A to B will be $2^4 = 16$.

Note *A relation R from A to A is also stated as a relation on A.*

48. *(b)* We have, $3x - y = 0 \Rightarrow y = 3x$

For,
$$x = 1, y = 3 \in A$$
$$x = 2, y = 6 \in A$$
$$x = 3, y = 9 \in A$$
$$x = 4, y = 12 \in A$$
$$x = 5, y = 15 \notin A$$

A. In Roster form,

$$R = \{(1, 3), (2, 6), (3, 9), (4, 12)\}$$

B. Domain of R = Set of first element of ordered pairs in R

$$= \{1, 2, 3, 4\}$$

C. Range of R = Set of second element of ordered pairs in R

$$= \{3, 6, 9, 12\}$$

D. Codomain of R is the set A.

49. *(a)* We have, $y = x + 5, x < 4, x, y \in N$

For,
$$x = 1, y = 6$$
$$x = 2, y = 7$$
$$x = 3, y = 8$$

$\therefore$ In Roster form $R = \{(1, 6), (2, 7), (3, 8)\}$

Domain of $R = \{1, 2, 3\}$

Range of $R = \{6, 7, 8\}$

50. *(c)* Difference between x and y is odd, if one of them is odd.

$\therefore$ In Roster form $R = \{(1, 4), (1, 6), (2, 9), (3, 4), (3, 6), (5, 4), (5, 6)\}$

51. *(a)* Clearly, the difference between the elements in P and Q is 2. So, in Set-builder form,

Relation $R = \{(x, y): x - y = 2, x \in P$ and $y \in Q\}$

In Roster form, $R = \{(5, 3), (6, 4), (7, 5)\}$

Domain of $R = \{5, 6, 7\}$

Range of $R = \{3, 4, 5\}$

52. *(d)* In Roster form $R = \{(1, 1), (1, 2), (1, 3), (1, 4), (1, 6), (2, 4), (2, 6), (2, 2), (4, 4), (6, 6), (3, 3), (3, 6)\}$

Domain of R = Set of first element of ordered pairs in $R = \{1, 2, 3, 4, 6\}$

Range of $R = \{1, 2, 3, 4, 6\}$

53. *(a)* The given relation in Roster form is

$$R = \{(0, 5), (1, 6), (2, 7), (3, 8), (4, 9), (5, 10)\}$$

Domain of $R = \{0, 1, 2, 3, 4, 5\}$

Range of $R = \{5, 6, 7, 8, 9, 10\}$

54. *(c)* Prime numbers less than 10 are 2, 3, 5, 7.

Hence, relation R in Roster form is

$$R = \{(2, 8), (3, 27), (5, 125), (7, 343)\}$$

55. *(b)* $n(A) = 3, n(B) = 2$

$\Rightarrow$ $n(A \times B) = 6$

Total number of relation from A to $B = 2^{n(A \times B)}$

$$= 2^6 = 64$$

56. *(d)* The difference of two integers is also an integer.

$\therefore$ Domain of $R = Z$

Range of $R = Z$

57. *(d)* Given, relation $R = \{(a, b): a, b \in Q$ and $a - b \in Z\}$

I. Since, $a - a = 0 \in Z$, it follows that $(a, a) \in R$.

II. $(a, b) \in R$ implies that $a - b \in Z$. So, $b - a \in Z$. Therefore, $(b, a) \in R$.

III. (a, b) and $(b, c) \in R$ implies that $a - b \in Z, b - c \in Z$.
So, $a - c = (a - b) + (b - c) \in Z$.
Therefore, $(a, c) \in R$.

58. *(b)* Given, $R = \{(x, y): y = x + \frac{6}{x}$, where $x, y \in N$ and $x < 6\}$

For,
$$x = 1, y = 1 + \frac{6}{1} = 7$$
$$x = 2, y = 2 + \frac{6}{2} = 5$$

$$x = 3, y = 3 + \frac{6}{3} = 5$$
$$x = 4, y = 4 + \frac{6}{4} = \frac{11}{2} \notin N$$
$$x = 5, y = 5 + \frac{6}{5} = \frac{31}{5} \notin N$$

$\therefore$ The given relation in Roster form is

$$R = \{(1, 7), (2, 5), (3, 5)\}$$

Hence, domain of $R = \{1, 2, 3\}$

Range of $R = \{7, 5\}$

59. *(d)* In option (d), $a \notin A$

$\therefore$ It is not a relation.

60. *(b)* Given, $R = \{(x, y) : |x - y| \text{ is odd}, x \in A, y \in B\}$

The relation R in Roster form is

$R = \{(1, 4), (1, 6), (2, 9), (3, 4), (3, 6), (5, 4), (5, 6)\}$

$\therefore$ Domain of $R = \{1, 2, 3, 5\}$

So, A is true.

It is also true $|x|$ is always positive. But R is not the correct explanation of A.

61. *(b)* I. Given, $R = \{(x, y) : x, y \in Z, xy = 4\}$

$\because$ $xy = 4 \Rightarrow y = \frac{4}{x} \in Z$

$\Rightarrow$ 4 is exactly divisible by x.

$\Rightarrow$ The value of x are $-2, -1, 1, 2, 4, -4$

$\therefore$ R in Roster form is,

$R = \{(-2, -2), (-1, -4), (1, 4), (2, 2), (4, 1), (-4, -1)\}$

Hence, domain of $R = \{-2, -1, 1, 2, 4, -4\}$

II. Now, $R_1 = \{(a, b) : a, b \in N \text{ and } 2a + b = 10\}$

$\because$ $2a + b = 10$

$\Rightarrow$ $b = 10 - 2a$

For, $a = 1, b = 10 - 2(1) = 8$

$a = 2, b = 10 - 2(2) = 6$

$a = 3, b = 10 - 2(3) = 4$

$a = 4, b = 10 - 2(4) = 2$

$a = 5, b = 10 - 2(5) = 0 \notin N$

$\therefore$ R_1 in Roster form is $R_1 = \{(1, 8), (2, 6), (3, 4), (4, 2)\}$

Hence, range of $R_1 = \{2, 4, 6, 8\}$

62. *(d)* Since $(-4, 0) \notin N$ and R is a relation on N, so, $(-4, 12), (5, 0) \notin R$

Now, $4(3) + 3(4) = 20 \Rightarrow 24 = 20$, which is not true

$\therefore$ $(3, 4) \notin R$.

Again, $4(2) + 3(4) = 20$

$\Rightarrow$ $20 = 20$, which is true

$\therefore$ $(2, 4) \in R$

63. *(c)* In Roster form, relation R is

$$R = \{(1, 1), (2, 4), (3, 9), (4, 16)\}$$

$\therefore$ Range of $R = \{1, 4, 9, 16\}$

Codomain of $R = B = \{1, 4, 9, 16, 25\}$

64. *(d)* $\because -1, 0, -3, \notin A$

$\therefore$ $R = \{(1, -1), (-1, 1), (2, 0), (0, 2), (3, 1), (1, 3)\}$ is not a relation from A to B.

65. *(c)* $\because 3 + 7 = 10, 3 + 11 = 14, 5 + 7 = 12, 5 + 11 = 16.$

$\therefore$ There is no element that belongs to R.

Hence, $R = \{\ \}$

66. *(c)* The given relation in Roster form is,

$R = \{(3, 5), (4, 6), (5, 7), (6, 8), (7, 9), (8, 10), (9, 11)\}$

$\therefore$ Domain of $R = \{3, 4, 5, 6, 7, 8, 9\}$

So, A is true.

Range of $R = \{5, 6, 7, 8, 9, 10, 11\}$

So, R is false.

67. *(b)* The relation shown in figure is represented in Roster form as

$$R = \{(-2, 4), (-1, 1), (0, 0), (1, 1), (2, 4)\}$$

We observe that, second element of each ordered pair is the square of first element.

$\therefore$ In Set-builder form,

$$R = \{(x, y) : y = x^2, x, y \in Z$$

and $-2 \le x \le 2\}$ $(\because 0 \notin N)$

68. *(a)* We have, $A = \{1, 2\}, B = \{3, 4, 5\}$

Here, $n(A) = 2, n(B) = 3$

A. $n(A \times A) = n(A) \times n(A) = 2 \times 2 = 4$

$\therefore$ Number of relation from A to $A = 2^{n(A \times A)} = 2^4 = 16$

B. $n(B \times B) = n(B) \times n(B) = 3 \times 3 = 9$

$\therefore$ Number of relation from B to $B = 2^{n(B \times B)} = 2^9 = 512$

C. $n(A \times B) = n(A) \times n(B) = 2 \times 3 = 6$

$\therefore$ Number of relation from A to $B = 2^{n(A \times B)} = 2^6 = 64$

D. $n(B \times A) = n(B) \times n(A) = 3 \times 2 = 6$

$\therefore$ Number of relation from B to $A = 2^{n(B \times A)} = 2^6 = 64$

69. *(c)* Given, $R = \{(a, b) : a \text{ differs from } b \text{ by } 3\}$

$\because$ $b - a = 3$

$\Rightarrow$ $b = a + 3$

For, $a = 1, b = 4$

$a = 2, b = 5$

$a = 3, b = 6$

... ...

... ...

$\therefore$ $R = \{(1, 4), (2, 5), (3, 6) \ldots\ldots\}$

70. *(b)* We have, set A having $(n - 1)$ elements

i.e., $n(A) = n - 1$

$n(A \times A) = n(A) \times n(A)$

$= (n - 1)(n - 1) = (n - 1)^2$

$\therefore$ Number of relation on $A = 2^{n(A \times A)} = 2^{(n-1)^2}$

71. *(c)* Given, $R = \{(x, y): x, y \in Z, x^2 + y^2 \le 4\}$

In Roster form,

$R = \{(-2, 0), (-1, -1), (-1, 1), (-1, 0), (0, -2), (0, -1), (0, 0), (0, 1), (0, 2), (1, 1), (1, 0), (1, -1), (2, 0)\}$

Hence, domain of $R = \{-2, -1, 0, 1, 2\}$

Graphically

$x^2 + y^2 = a^2$ represent circle here. Equation of circle is $x^2 + y^2 \le a^2$. Hence, we can also take interior integral points.

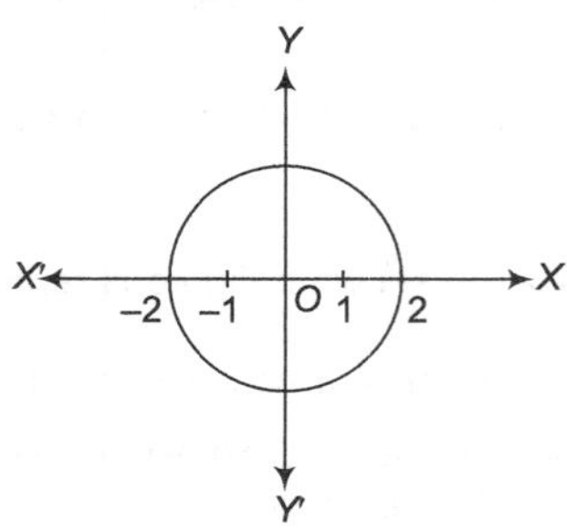

72. *(b)* A special type of relation is called function. There are many terms such as 'map' or 'mapping' used to denote the function.

A relation f from a set A to a set B is said to be a function, if every element of set A has one and only one image in set B.

In other words, a function f is a relation from a non-empty set A to a non-empty set B such that the domain of f is A and no two distinct ordered pairs in f have the same first element.

If f is a function from A to B and $(a, b) \in f$, then $f(a) = b$, where b is called the image of a under f and a is called the pre-image of b under f.

The function f from A to B is denoted by $f : A \to B$.

73. *(b)* I. Given,

$A = \{1, 2, 3, 4, 5, 6\}$, R is a relation on A defined by $R = \{(x, y) : y = x + 1\}$

$R = \{(1, 2), (2, 3), (3, 4), (4, 5), (5, 6)\}$

R is not a function because 6 has no image.

II. Now, $P = \{9, 4, 25\}$

and $Q = \{-5, -3, -2, 1, 2, 3, 5\}$

$\because \quad R = \{(9, 3), (9, -3), (4, 2), (4, -2), (25, 5), (25, -5)\}$

R is not a function because 9, 4, 25 has two different images in Q.

74. *(a)* $R = \{(1, 2), (2, 4), (3, 6), (4, 8)\ldots\ldots\}$

Since, every natural number N has one and only one image, this relation R is a function.

The domain of R is the set of natural number *i.e.*, N. The codomain is also N, and the range is the set of even natural numbers.

75. *(c)* Since 2, 3, 4 are the elements of domain of R_1 having their unique images, this relation R_1 is a function.

Since, the same first element 2 corresponds to two different images 2 and 4, this relation R_2 is not a function.

Since, every element has one and only one image, this relation R_3 is a function.

76. *(a)* A function which has either R or one of its subsets as its range is called a real valued function. Further, if its domain is also either R or a subset of R, it is called a real function.

77. *(c)* $f(x) = 2x + 1$

For,
$x = 1, \ f(1) = 2(1) + 1 = 3$
$x = 2, \ f(2) = 2(2) + 1 = 5$
$x = 4, \ f(4) = 2(4) + 1 = 9$
$x = 5, \ f(5) = 2(5) + 1 = 11$
$x = 7, \ f(7) = 2(7) + 1 = 15$

78. *(a)* Let R be the set of real numbers. Define the real valued function $f : R \to R$ by $y = f(x) = x$ for each $x \in R$. Such a function is called the identity function. Here, the domain and range of f are R. The graph is a straight line as shown in figure given below. It passes through the origin.

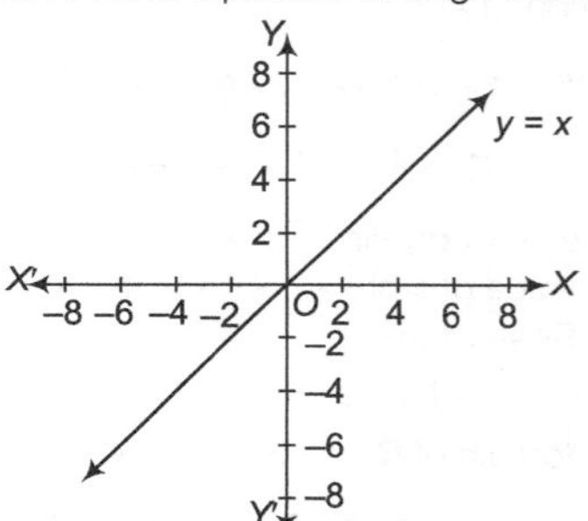

79. *(b)* **Constant function** Define the function $f : R \to R$ by $y = f(x) = c$, $x \in R$, where c is a constant. Here, domain of f is R and its range is $\{c\}$.

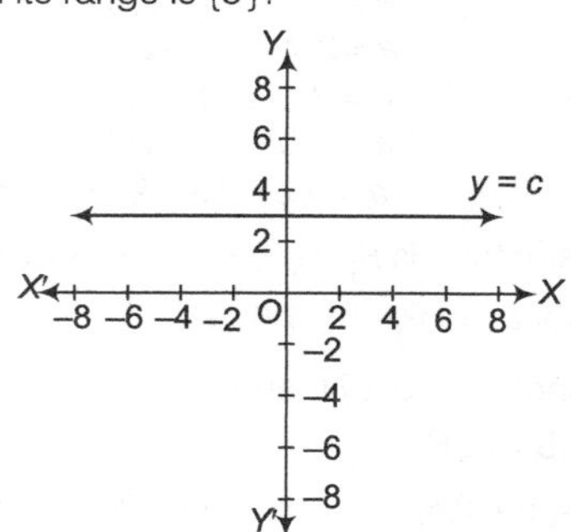

e.g. The graph of $f(x) = 3$, is a line passing through (0, 3) and parallel to X-axis.

80. *(a)* **Polynomial function** A function $f : R \to R$ is said to be polynomial function, if for each x in R, $y = f(x) = a_0 + a_1 x + a_2 x^2 + \cdots + a_n x^n$, where n is a non-negative integer and $a_0, a_1, a_2, \ldots, a_n \in R$.

The function $f(x) = x^3 - x^2 + 2$ and $f(x) = x^4 + \sqrt{2}x$ are polynomial functions whereas $f(x) = x^{2/3} + 2x$ is not a function because in first term power of x is not an integer.

81. *(d)* First construct the table given below

x	−4	−3	−2	−1	0	1	2	3	4
$y = f(x) = x^2$	16	9	4	1	0	1	4	9	16

Domain of $f = \{x : x \in R\}$. Range of $f = \{x^2 : x \in R\}$. The graph of f is given by,

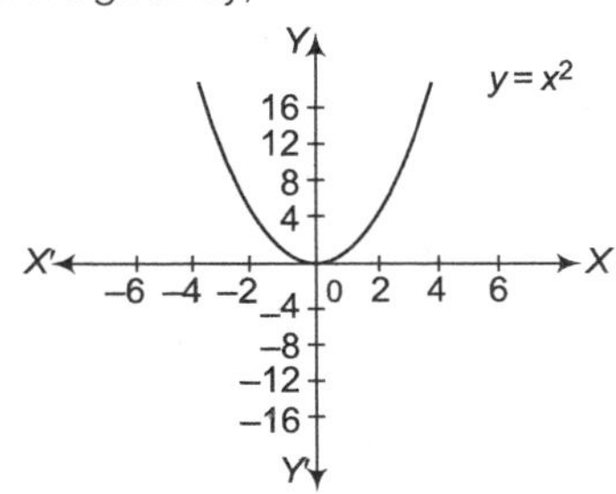

82. *(a)* We have, $f(0) = 0$, $f(1) = 1$, $f(-1) = -1$, $f(2) = 8$, $f(-2) = -8$, $f(3) = 27$; $f(-3) = -27$ etc.

Therefore, $f = \{(x, x^3) : x \in R\}$

The graph of f is given by,

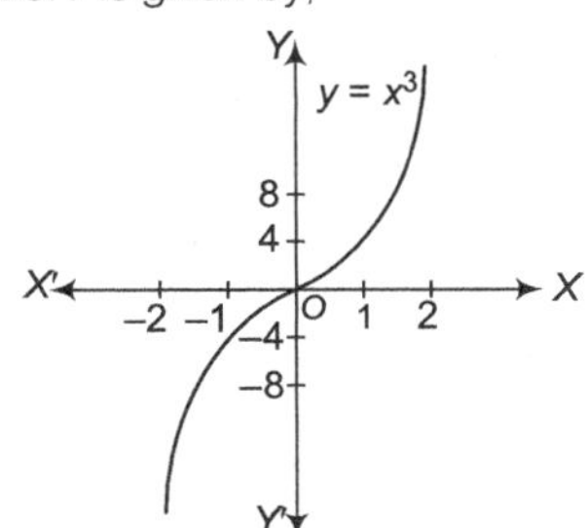

83. *(b)* Rational functions are functions of the type $\frac{f(x)}{g(x)}$, where $f(x)$ and $g(x)$ are polynomial functions of x defined in a domain, where $g(x) \neq 0$.

84. *(d)* First construct the table given below

x	−2	−1.5	−1	−0.5	0.25	0.5	1	1.5	2
$y = \frac{1}{x}$	−0.5	−0.67	−1	−2	4	2	1	0.67	0.5

The domain is all real numbers except 0 and its range is also all real numbers except 0. The graph of f is given by,

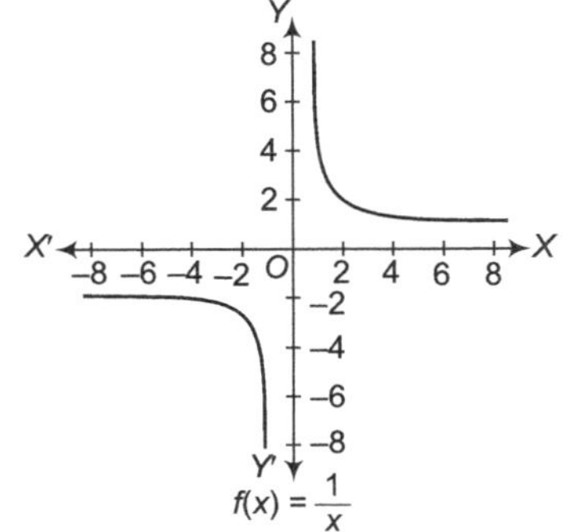

$f(x) = \frac{1}{x}$

85. *(d)* I. **The modulus function** The function $f : R \to R$ defined by $f(x) = |x|$ for each $x \in R$ is called modulus function. For each non-negative value of x, $f(x)$ is equal to x. But for negative values of x, the value of $f(x)$ is the negative of the value of x, *i.e.*,

$$f(x) = \begin{cases} x, & x \geq 0 \\ -x, & x < 0 \end{cases}$$

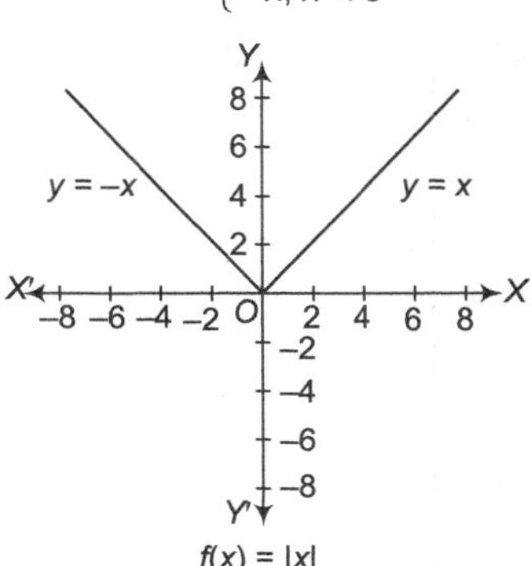

$f(x) = |x|$

The graph of the modulus function is given in above figure.

II. **Signum function** The function $f : R \to R$ defined by

$$f(x) = \begin{cases} 1, & \text{if } x > 0 \\ 0, & \text{if } x = 0 \\ -1, & \text{if } x < 0 \end{cases}$$

is called the signum function. The domain of the signum function is R and the range is the set $\{-1, 0, 1\}$. The graph of the signum function is given by the figure.

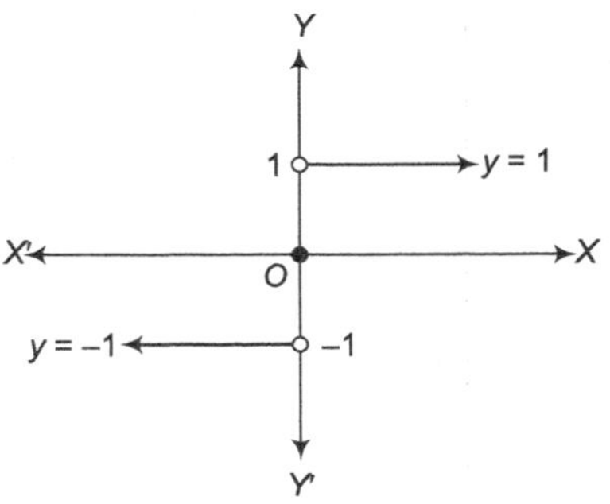

$f(x) = \frac{|x|}{x}$, $x \neq 0$ and 0 for $x = 0$

86. *(b)* **Greatest integer function** The function $f : R \to R$ defined by $f(x) = [x]$, $x \in R$ assumes the value of the greatest integer, less than or equal to x. Such a function is called the greatest integer function.

From the definition of $[x]$, we can see that

$[x] = -1$ for $-1 \leq x < 0$
$[x] = 0$ for $0 \leq x < 1$
$[x] = 1$ for $1 \leq x < 2$
$[x] = 2$ for $2 \leq x < 3$ and so on.

The graph of the function is given in the question.

87. *(d)* **Addition of two real functions** Let $f: X \to R$ and $g: X \to R$ be any two real functions, where $X \subset R$. Then, we define $(f+g): X \to R$ by $(f+g)(x) = f(x) + g(x)$, for all $x \in X$.

Subtraction of a real function from another Let $f: X \to R$ and $g: X \to R$ be any two real functions, where $X \subset R$. Then, we define $(f-g): X \to R$ by $(f-g)(x) = f(x) - g(x)$, for all $x \in X$.

Multiplication by a scalar Let $f: X \to R$ be a real valued function and α be a scalar. Here by scalar, we mean a real number. Then, the product αf is a function from X to R defined by $(\alpha f)(x) = \alpha f(x)$, $x \in X$.

Multiplication of two real functions The product (or multiplication) of two real functions $f: X \to R$ and $g: X \to R$ is a function $fg: X \to R$ defined by $(fg)(x) = f(x)\,g(x)$, for all $x \in X$.

This is also called pointwise multiplication.

Quotient of two real functions Let f and g be two real functions defined from $X \to R$, where $X \subset R$. The quotient of f by g denoted by $\frac{f}{g}$ is a function defined by $\left(\frac{f}{g}\right)(x) = \frac{f(x)}{g(x)}$, provided $g(x) \neq 0$, $x \in X$.

88. *(a)* Since, Domain of f = Domain of g

We have, $(f+g)(x) = x^2 + 2x + 1$,

$$(f-g)(x) = x^2 - 2x - 1,$$

$$(fg)(x) = x^2(2x+1) = 2x^3 + x^2,$$

$$\left(\frac{f}{g}\right)(x) = \frac{x^2}{2x+1}, x \neq -\frac{1}{2}$$

89. *(c)* Domain of $f = R^+$, Domain of $g = R$

Domain of $f \cap$ Domain of $g = R^+ \cap R = R^+$

$\therefore$ Domain of $f+g$, $f-g$, fg are R^+ and domain of $\frac{f}{g}$ is $R^+ - \{g(x) \neq 0\}$.

We have, $(f+g)(x) = \sqrt{x} + x$, $(f-g)(x) = \sqrt{x} - x$,

$$(fg)\,x = \sqrt{x}(x) = x^{\frac{3}{2}}$$

and $$\left(\frac{f}{g}\right)(x) = \frac{\sqrt{x}}{x} = x^{-\frac{1}{2}}, x \neq 0$$

90. *(a)* In R_1 and R_2, every element has unique image. So, R_1 and R_2 are functions. But R_3 has two images 3 and 5 for element 1. So, R_3 is not a function.

91. *(a)* We have, $f(x) = -|x|$

$\because$ $f(x)$ is defined for all R

$\therefore$ Domain of $f = R$

We know that,

$$-\infty < x < \infty$$

$\Rightarrow$ $0 \leq |x| < \infty$

$\Rightarrow$ $-\infty < -|x| \leq 0$

$\Rightarrow$ $-\infty < f(x) \leq 0$

$\therefore$ Range of $f = (-\infty, 0)$

92. *(c)* Let $f(x) = \sqrt{9 - x^2}$

$f(x)$ is defined, if $9 - x^2 \geq 0$

$\Rightarrow$ $x^2 - 9 \leq 0$

$\Rightarrow$ $(x+3)(x-3) \leq 0$

$\Rightarrow$ $-3 \leq x \leq 3$

$[\because (x-a)(x-b) \leq 0, a \leq x \leq b, a < b]$

$\Rightarrow$ $x \in [-3, 3]$

$\therefore$ Domain of $f = [-3, 3]$

Now, let $y = f(x)$, then

$$y = \sqrt{9 - x^2}$$

$$y^2 = 9 - x^2 \quad \text{(on squaring)}$$

$$x = \pm\sqrt{9 - y^2}$$

x assumes real values, if $9 - y^2 \geq 0$

$\Rightarrow$ $y^2 - 9 \leq 0$

$\Rightarrow$ $-3 \leq y \leq 3$

$\Rightarrow$ $y \in [-3, 3]$

But $f(x) = y \geq 0$.

$\therefore$ Range of $f = [0, 3]$

93. *(b)* Given, $f(x) = 2x - 5$

Put $x = 0$, $f(0) = 2(0) - 5 = -5$

$x = 7$, $f(7) = 2(7) - 5 = 9$

$\therefore$ $f(0) \neq f(7)$

$x = -3$, $f(-3) = 2(-3) - 5$
$= -6 - 5 = -11$

94. *(a)* Given, $t(C) = \frac{9C}{5} + 32$

A. Put $C = 0$, $t(0) = \frac{9(0)}{5} + 32 = 32$

B. $C = 28$, $t(28) = \frac{9(28)}{5} + 32 = \frac{252 + 160}{5}$
$= \frac{412}{5}$

C. $C = -10$, $t(-10)$
$= \frac{9(-10)}{5} + 32 = \frac{-90 + 160}{5}$
$= \frac{70}{5} = 14$

D. Now, $t(C) = 212$

$\Rightarrow$ $\frac{9C}{5} + 32 = 212$

$\Rightarrow$ $\frac{9C}{5} = 180$

$\Rightarrow$ $C = 100$

95. *(d)* I. We have, $f(x) = 2 - 3x$, $x \in R$, $x > 0$

Let $f(x) = y$, then

$$y = 2 - 3x$$

$\Rightarrow$ $3x = 2 - y$

$\Rightarrow$ $x = \frac{2 - y}{3}$

$\because \quad x > 0$

$\Rightarrow \quad \dfrac{2-y}{3} > 0$

$\Rightarrow \quad 2 - y > 0$

$\Rightarrow \quad 2 > y$

$\therefore \quad y < 2$

Hence, range of $f = (-\infty, 2)$

$\therefore$ Statement I is false.

II. Now, $f(x) = x^2 + 2$

Let $y = f(x)$, then

$y = x^2 + 2$

$\Rightarrow \quad x = \sqrt{y-2}$

x assumes real values, if $y - 2 \geq 0$

$\Rightarrow \quad y \geq 2$

$\Rightarrow \quad y \in [2, \infty]$

$\therefore$ Range of $f = [2, \infty]$

96. *(d)* Given, $f(x) = x$, x is a real number.

$f(x)$ is defined for all $x \in R$

$\therefore \quad f(x) \in R$

Hence, range of f is R.

97. *(a)* We have,

$f(x) = x + 10$

$f(0) = 10, f(1) = 11, f(2) = 12 \ldots\ldots f(10) = 20$ etc., and

$f(-1) = 9, f(-2) = 8 \ldots f(-10) = 0 \ldots$

Therefore, the graph of the given function is

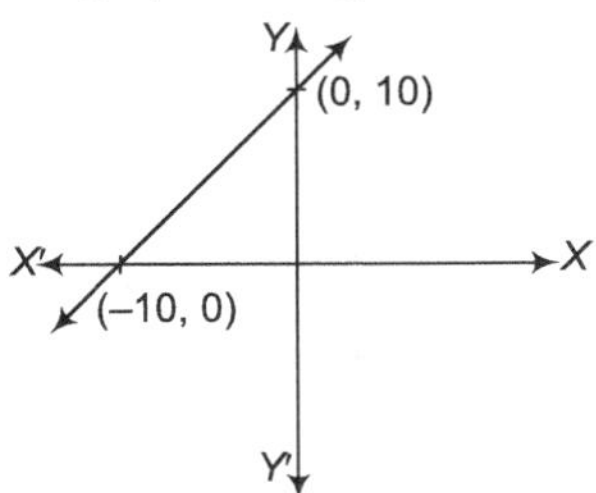

98. *(c)* The function f defined by $f(x) = mx + c$, $x \in R$, is called linear function, where m and c are constants.

99. *(a)* We have, $f(x) = \begin{cases} x^2, & 0 \leq x \leq 3 \\ 3x, & 3 \leq x \leq 10 \end{cases}$

Since, every element has unique image by f. So, f is a function.

Now, $\quad g(x) = \begin{cases} x^2, & 0 \leq x \leq 2 \\ 3x, & 2 \leq x \leq 10 \end{cases}$

Since, $\quad g(2) = 2^2 = 4$

and $\quad g(2) = 3(2) = 6.$

2 has two images. So, g is not a function.

100. *(b)* In R_1, 4 has two images. So, R_1 is not a function. In R_2 and R_3 every element has unique image. So, R_2 and R_3 are functions.

101. *(b)* Consider $R_1 = \{(x, y) : x, y \in A \text{ and } y = x + 1\}$

$= \{(1, 2)\ (2, 3), (3, 4)\}$

Since, 4 has no image. So, R_1 is not a function.

Now, $\quad R_2 = \{(x, y) : x, y \in A \text{ and } x + y = 5\}$

$= \{(1, 4), (2, 3), (3, 2), (4, 1)\}$

Since, every element of A has its image in A.

So, R_2 is a function in A.

102. *(a)* I. In arrow diagram, every element of P has its unique image in Q. So, it represent a function.

II. Domain of $f = R - \{3\}$.

Domain of $g = R$

$\because \quad D_f \neq D_g$

We know that, two functions are equal when their domain and range are equal and same element in their domain produce same image.

$\therefore \quad f \neq g$

103. *(c)* Since, 2^x is positive for every $x \in R$. So, $f(x) = 2^x$ is a positive real number for every $x \in R$.

$\therefore \quad$ Range of $f = R^+$

Now, $\quad f(x) = 1$

$\Rightarrow \quad 2^x = 1 \Rightarrow 2^x = 2^0 \Rightarrow x = 0$

$\therefore \quad \{x : f(x) = 1\} = \{0\}$

We have, $\quad f(x + y) = 2^{x+y}$

$= 2^x \cdot 2^y = f(x) \cdot f(y)$

$\therefore \quad f(x + y) = f(x) \cdot f(y)$ for all $x, y \in R$.

104. *(b)* For $f(x) = g(x)$

$\Rightarrow \quad 2x^2 - 1 = 1 - 3x$

$\Rightarrow \quad 2x^2 + 3x - 2 = 0$

$\Rightarrow \quad 2x^2 + 4x - x - 2 = 0$

$\Rightarrow \quad 2x(x + 2) - 1(x + 2) = 0$

$\Rightarrow \quad (x + 2)(2x - 1) = 0$

$\Rightarrow \quad x = -2, \dfrac{1}{2}$

$\therefore$ The domain for which the function $f(x) = g(x)$ is $\left\{-2, \dfrac{1}{2}\right\}$.

105. *(c)* f is a rational function of the form $\dfrac{g(x)}{h(x)}$, where $g(x) = x$ and $h(x) = x^2 + 3x + 2$, $h(x) \neq 0$.

Above function is defined when $h(x) \neq 0$

$\because \quad h(x) \neq 0$

$\Rightarrow \quad x^2 + 3x + 2 \neq 0$

$\Rightarrow \quad (x + 1)(x + 2) \neq 0$

$\Rightarrow \quad x \neq -1, -2$

Hence, the domain of the given function is

$R - \{-1, -2\}$.

So, A is correct.

If f is a function from A to B, then domain of $f = A$

So, R is incorrect.

106. *(c)* $f(x)$ assumes real values, if

$$4 - x^2 > 0$$
$$\Rightarrow \quad x^2 - 4 < 0$$
$$\Rightarrow \quad (x+2)(x-2) < 0$$
$$\Rightarrow \quad x \in (-2, 2)$$
$$\therefore \quad \text{Domain of } f = (-2, 2)$$

107. *(c)* Given, $f(x) = x^3 - \dfrac{1}{x^3}$

$$f\left(\frac{1}{x}\right) = \frac{1}{x^3} - x^3$$
$$\therefore \quad f(x) + f\left(\frac{1}{x}\right) = x^3 - \frac{1}{x^3} + \frac{1}{x^3} - x^3 = 0$$

108. *(a)* I. Given, $f(x) = x + \dfrac{1}{x}$

$$f(x^3) = x^3 + \frac{1}{x^3}$$
$$[f(x)]^3 = \left(x + \frac{1}{x}\right)^3$$
$$= x^3 + \frac{1}{x^3} + 3\left(x + \frac{1}{x}\right)$$
$$= f(x^3) + 3f(x)$$
$$= f(x^3) + 3f\left(\frac{1}{x}\right) \quad \left[\because f\left(\frac{1}{x}\right) = \frac{1}{x} + x = f(x)\right]$$

$\therefore$ Statement I is true.

II. Now, we have, $f(x) = (x-a)^2 (x-b)^2$

$$f(a+b) = (a+b-a)^2 (a+b-b)^2 = b^2a^2$$

$\therefore$ Statement II is false.

109. *(a)* We have, $f(x) = \dfrac{x+1}{x-1}$

$$f\{f(x)\} = f\left(\frac{x+1}{x-1}\right)$$
$$= \frac{\frac{x+1}{x-1} + 1}{\frac{x+1}{x-1} - 1} \quad \left[\text{replace } x \text{ by } \frac{x+1}{x-1} \text{ in } f(x)\right]$$
$$= \frac{x+1+x-1}{x+1-x+1}$$
$$= \frac{2x}{2} = x$$

110. *(d)* We have, $f(x) = \dfrac{1}{\sqrt{x - |x|}}$

where, $x - |x| = x - x = 0$, if $x \geq 0$

$x - (-x) = 2x$, if $x < 0$

Thus, $\dfrac{1}{\sqrt{x-|x|}}$ is not defined for $x \in R$.

Hence, f is not defined for any $x \in R$.

111. *(a)* Domain of $f = R$

Domain of $g = R$

Domain of $f \cap$ Domain of $g = R$

A. $f + g : R \to R$ is given by

$$(f+g)(x) = f(x) + g(x)$$
$$= 2x + 5 + x^2 + x$$
$$= x^2 + 3x + 5$$

B. $f - g : R \to R$ is defined as

$$(f-g)(x) = f(x) - g(x)$$
$$= 2x + 5 - x^2 - x$$
$$= 5 + x - x^2$$

C. $(fg)(x) = f(x) \cdot g(x)$

$$= (2x+5)(x^2+x)$$
$$= 2x^3 + 2x^2 + 5x^2 + 5x$$
$$= 2x^3 + 7x^2 + 5x$$

D. $g(x) = 0$

$$\therefore \quad x^2 + x = 0$$
$$\Rightarrow \quad x(x+1) = 0$$
$$\Rightarrow \quad x = 0, -1$$

$$\text{Domain of}\left(\frac{f}{g}\right) = \text{Domain of } f \cap \text{Domain of } g - \{0, -1\}$$
$$= R - \{0, -1\}$$

Thus, $\dfrac{f}{g} : R - \{0, -1\} \to R$ is given by

$$\left(\frac{f}{g}\right)(x) = \frac{f(x)}{g(x)}$$
$$= \frac{2x+5}{x^2+x}$$

112. *(c)* We have, $f(x) = \dfrac{x}{1+x^2}$

Let, $f(x) = y$

$$\therefore \quad y = \frac{x}{1+x^2}$$
$$\Rightarrow \quad x^2y + y = x$$
$$\Rightarrow \quad yx^2 - x + y = 0$$

x assumes real values, if $\Delta \geq 0$

$$(-1)^2 - 4(y)(y) \geq 0$$
$$\Rightarrow \quad 1 - 4y^2 \geq 0$$
$$\Rightarrow \quad 4y^2 - 1 \leq 0$$
$$\Rightarrow \quad (2y+1)(2y-1) \leq 0$$
$$\Rightarrow \quad y \in \left[-\frac{1}{2}, \frac{1}{2}\right]$$
$$\therefore \quad \text{Range of } f = \left[-\frac{1}{2}, \frac{1}{2}\right]$$

113. *(a)* We have,

$$f(x) = \begin{cases} 3x - 2, & x < 0 \\ 1, & x = 0 \\ 4x + 1, & x > 0 \end{cases}$$

$$\therefore \quad f(-1) = 3(-1) - 2 = -5$$
$$f(0) = 1$$
$$f(1) = 4(1) + 1$$
$$= 5$$

114. *(b)* We have,

$$f(x) = ax + b$$
$$f(-1) = a(-1) + b$$
$$-5 = -a + b \quad \ldots(\text{i})$$
$$f(3) = a(3) + b$$
$$3 = 3a + b \quad \ldots(\text{ii})$$

Subtracting Eqs. (i) from (ii), we get

$$3 - (-5) = (3a + b) - (-a + b)$$
$$3 + 5 = 3a + b + a - b$$
$$8 = 4a$$
$$a = 2$$

From Eq. (i), we get

$$-5 = -2 + b$$
$$\therefore \quad b = -3$$

115. *(c)* Domain of $f = \{0, 2, 3, 4, 5\}$

Domain of $g = \{1, 2, 3, 4, 5\}$

Domain of $f \cdot g$ = Domain of $f \cap$ Domain of $g = \{2, 3, 4, 5\}$

Hence, A is true.

If f and g be two real functions.

Then, $(f \cdot g)(x) = f(x) \cdot g(x)$

Hence, R is false.

116. *(b)* Since, $f(x)$ is a linear function, $f(x) = mx + c$. Also, since $(1, 1), (0, -1) \in R$, $f(1) = m + c = 1$ and $f(0) = c = -1$. This gives $m = 2$ and $f(x) = 2x - 1$.

117. *(a)* I. Rational function $\dfrac{f(x)}{g(x)}$ is defined when $g(x) \neq 0$, where $f(x)$ and $g(x)$ are polynomial function. Since, $x^2 - 5x + 4 = (x - 4)(x - 1)$, function f is polynomial, defined for all real numbers except at $x = 4$ and $x = 1$. Hence, the domain of f is $R - \{1, 4\}$.

Hence, Statement I is true.

II. Now, $f(x) = 1 - x$, $x < 0$, this gives

$$f(-4) = 1 - (-4) = 5,$$
$$f(-3) = 1 - (-3) = 4,$$
$$f(-2) = 1 - (-2) = 3$$
$$f(-1) = 1 - (-1) = 2; \text{etc,}$$

and $f(1) = 2, f(2) = 3, f(3) = 4$

$f(4) = 5$ and so on for $f(x) = x + 1$, $x > 0$.

Thus, the graph of f is as shown in figure.

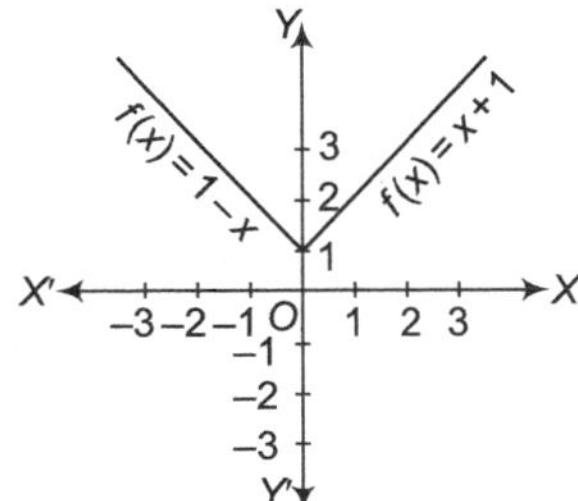

Hence, Statement II is false.

118. *(b)* We have, $f(x) = x^2$

$$\frac{f(1.1) - f(1)}{1.1 - 1} = \frac{(1.1)^2 - 1^2}{1.1 - 1}$$
$$= \frac{1.21 - 1}{0.1} = \frac{0.21}{0.1} = 2.1$$

119. *(c)* $f(x)$ is defined, if $x^2 - 8x + 12 \neq 0$

or $\quad x^2 - 6x - 2x + 12 \neq 0$

$\Rightarrow \quad x(x - 6) - 2(x - 6) \neq 0$

$\Rightarrow \quad (x - 6)(x - 2) \neq 0$

$\Rightarrow \quad x \neq 6, 2$

$\therefore \quad$ Domain of $f = R - \{2, 6\}$

120. *(d)* We have, $f(x) = \sqrt{x - 1}$

$f(x)$ is defined, if $x - 1 \geq 0$

i.e., $\quad x \geq 1$

$\therefore \quad$ Domain of $f = [1, \infty)$

Hence, A is incorrect.

Let $\quad f(x) = y$

Then, $\quad y = \sqrt{x - 1}$

$\Rightarrow \quad y^2 = x - 1$

$\Rightarrow \quad x = y^2 + 1$

Since, $y \geq 0$ and $x \in [1, \infty) \Rightarrow$ Range of $f = [0, \infty)$.

121. *(a)* We have, $f(x) = |x - 1|$

$f(x)$ is defined for all $x \in R$

$\therefore \quad$ Domain for $f = R$

$$-\infty < x < \infty$$
$$-\infty < x - 1 < \infty$$
$$0 \leq |x - 1| < \infty$$
$$\because \quad 0 \leq f(x) < \infty$$

Since, $f(x)$ will take only positive values.

$\therefore \quad$ Range of $f = [0, \infty)$

122. *(c)* We have, $f(x) = \dfrac{x^2}{1 + x^2}$

Let $f(x) = y$, then

$$y = \frac{x^2}{1 + x^2}$$
$$x^2 y + y = x^2$$
$$y = x^2(1 - y)$$
$$x = \sqrt{\frac{y}{1 - y}}$$

$\therefore \quad x \in R$

So, $\quad \dfrac{y}{1 - y} \geq 0$

$\Rightarrow \quad \dfrac{y}{y - 1} \leq 0$

$\Rightarrow \quad 0 \leq y < 1$

$\Rightarrow \quad y \in [0, 1)$

Hence, range of $f = [0, 1)$

123. *(c)* We have, $f(x)=x+1, g(x)=2x-3$

$$(f+g)(x)=f(x)+g(x) = x+1+2x-3=3x-2$$

$$(f-g)(x)=f(x)-g(x) = x+1-2x+3=-x+4$$

$$\left(\frac{f}{g}\right)(x)=\frac{f(x)}{g(x)}, \text{ provided } g(x)\neq 0$$

$$=\frac{x+1}{2x-3}, \text{ provided } 2x-3\neq 0$$

i.e., $x\neq\frac{3}{2}$

124. *(a)* $\therefore f(1)=1, f(2)=3, f(0)=-1$

$$f(0)=-1$$

$\Rightarrow \quad a(0)+b=-1 \Rightarrow b=-1$

$f(1)=1 \Rightarrow a(1)+b=1$

$\Rightarrow \quad a+b=1$

$\Rightarrow \quad a-1=1 \Rightarrow a=2$

125. *(d)* $\therefore 2\neq 2^2 \Rightarrow (2,2)\notin R$

Hence, $(a,a)\notin R$ for all $a\in N$

Let $(a,b)\in R$ for all $a,b\in N$

$\Rightarrow \quad a=b^2$ (by definition)

$\Rightarrow \quad b\neq a^2$

$\Rightarrow \quad (b,a)\notin R$ for all $a,b\in N$

Let $(a,b)\in R$ and $(b,c)\in R$ for all $a,b,c\in N$

$\Rightarrow \quad a=b^2$ and $b=c^2$

$\Rightarrow \quad a=c^4$

$\Rightarrow \quad (a,c)\notin R$, for all $a,b,c\in N$

126. *(a)* Since, first elements of the ordered pairs in f belongs to A and second elements of the ordered pairs belongs to B. So, f is a relation from A to B.

Now, 2 has two different images 9 and 11.

So, f is not a function.

127. *(d)* I. We have, $f(ab)=a+b\ \forall\ a,b\in Z$

$$f(0\times 2)=0+2$$
$$f(0)=2$$
$$f(0\times 5)=0+5$$
$$f(0)=5$$

$\because$ 0 has more than one image.

$\therefore f$ is not a function from Z to Z.

II. Now, $A=\{9,10,11,12,13\}$

$f: A\to N, f(x)=$ The highest prime factor of x

$\therefore \quad f=\{(9,3),(10,5),(11,11),(12,3),(13,13)\}$

$\therefore$ Range of $f=\{3,5,11,13\}$

128. *(d)* $n(A)=10, n(B)=5$

$$n(A\times B)=n(A)\times n(B)=10\times 5=50$$
$$n(B\times A)=n(B)\times n(A)=5\times 10=50$$

$\therefore \quad n(A\times B)=n(B\times A)$

Total number of relation from A to $B=2^{n(A\times B)}=2^{50}$

129. *(d)* I. We have, $y=x-5$

$$2=5-5, 2=0$$

which is not true

$\therefore \quad (5,2)\notin R$

II. Now, $(x-2, y+5)=\left(-2,\frac{1}{3}\right)$

By the definition of equality of ordered pair,

$$x-2=-2 \text{ and } y+5=\frac{1}{3}$$

$\Rightarrow \quad x=0$ and $y=\frac{1}{3}-5=\frac{-14}{3}$

130. *(d)* In Roster form relation R is,

$R=\{(2,3),(2,7),(3,7),(3,10),(4,3),(4,7),(5,3),(5,6),(5,7)\}$

$\therefore \quad$ Domain of $R=\{2,3,4,5\}$

131. *(b)* $A-B=\{1\}, B-C=\{4\}$

$\therefore \quad (A-B)\times(B-C)=\{(1,4)\}$

132. *(d)* $n(A)=m, n(B)=n$

$$n(A\times B)=n(A)\times n(B) = mn$$

Total number of subsets of $(A\times B)=2^{mn}$

Total number of non-empty relation $=2^{mn}-1$

133. *(a)* Let $a\in R$, then

$$1+a.a=1+a^2>0$$

$\Rightarrow \quad (a,a)\in S\ \forall\ a\in R$

Clearly $\left(\frac{1}{2},-1\right)\in S$ and $(-1,-5)\in S$,

as $1+\left(\frac{1}{2}\right)(-1)=\frac{1}{2}>0$ and $1+(-1)(-5)=6>0$

But $\quad 1+\left(\frac{1}{2}\right)(-5)=1-\frac{5}{2}$

$$=-\frac{3}{2} \ngtr 0$$

So, $\quad \left(\frac{1}{2},-5\right)\notin S$

134. *(a)* I. In Roster form,

$R=\{(1,4),(2,4),(3,4),(4,4)\}$

$\therefore$ Domain of $R=\{1,2,3,4\}$

II. Here, $\quad |a|\leq 3$

$\Rightarrow \quad -3\leq a\leq 3$

Hence, $\quad a=-3,-2,-1,0,1,2,3$

In Roster form,

$S=\{(-3,4),(-2,3),(-1,2),(0,1),(1,0),(2,1),(3,2)\}$

$\therefore$ Range of $S=\{0,1,2,3,4\}$

135. *(b)* $n(S) = 10$

A = Set of ordered pairs on S such that both elements of the ordered pairs are different

Number of ordered pairs on $S = n(S \times S)$
$$= n(S) \times n(S)$$
$$= 10 \times 10 = 100$$

Number of ordered pairs on S such that both elements are same = 10

$\therefore$ $n(A)$ = Number of ordered pairs on S such that both elements are different = 100 − 10 = 90

136. *(d)* $n(A) = 4, n(B) = 3$
$$n(A \times B \times C) = 24$$
$$n(A) \times n(B) \times n(C) = 24$$
$$4 \times 3 \times n(C) = 24$$
$$n(C) = 2$$

137. *(c)* We have, $f(x) = \log_e\left(\frac{1+x}{1-x}\right)$
$$f\left(\frac{2x}{1+x^2}\right) = \log_e\left(\frac{1+\frac{2x}{1+x^2}}{1-\frac{2x}{1+x^2}}\right)$$
$$= \log_e\left(\frac{1+x^2+2x}{1+x^2-2x}\right)$$
$$= \log_e\left(\frac{1+x}{1-x}\right)^2$$
$$= 2\log_e\left(\frac{1+x}{1-x}\right) \quad (\because \log a^b = b\log a)$$
$$= 2f(x)$$

138. *(b)* I. We have, $f(x) = 2x + 3, g(x) = x^2 + 7$
$$g[f(x)] = 8$$
$\Rightarrow$ $g(2x+3) = 8$
$\Rightarrow$ $(2x+3)^2 + 7 = 8$
$\Rightarrow$ $(2x+3)^2 = 1$
$\Rightarrow$ $2x + 3 = \pm 1,$
$2x + 3 = -1$
or $2x + 3 = 1$, then
$\Rightarrow$ $x = -1, x = -2$

Hence, Statement I is false.

II. Now, $f(x) = \dfrac{4^x}{4^x + 2}$
$$f(1-x) = \frac{4^{1-x}}{4^{1-x}+2}$$
$$\therefore \quad f(x) + f(1-x) = \frac{4^x}{4^x+2} + \frac{4^{1-x}}{4^{1-x}+2}$$
$$= \frac{4^x}{4^x+2} + \frac{4/4^x}{\frac{4}{4^x}+2}$$
$$= \frac{4^x}{4^x+2} + \frac{4}{4+2\cdot 4^x}$$
$$= \frac{4^x}{4^x+2} + \frac{2}{4^x+2}$$
$$= \frac{4^x+2}{4^x+2} = 1$$

Hence, Statement II is true.

139. *(c)* Let $(a, b) \in A \times (B \cup C)$

$\Rightarrow$ $a \in A$ and $b \in (B \cup C)$
$\Rightarrow$ $a \in A$ and $(b \in B$ or $b \in C)$
$\Rightarrow$ $(a \in A$ and $b \in B)$ or $(a \in A$ and $b \in C)$
$\Rightarrow$ $(a, b) \in (A \times B)$ or $(a, b) \in (A \times C)$
$\Rightarrow$ $(a, b) \in (A \times B) \cup (A \times C)$
$\Rightarrow$ $(a, b) \in (A \times B) \cup (A \times C)$
$\Rightarrow$ $A \times (B \cup C) \subseteq (A \times B) \cup (A \times C)$ ...(i)

Similarly, we can show
$(A \times B) \cup (A \times C) \subseteq A \times (B \cup C)$ ...(ii)

By Eqs. (i) and (ii), we get
$$A \times (B \cup C) = (A \times B) \cup (A \times C)$$

Now, let $(x, y) \in A \times (B - C)$
$\Rightarrow$ $x \in A$ and $y \in (B - C)$
$\Rightarrow$ $x \in A$ and $(y \in B$ and $y \notin C)$
$(x \in A$ and $y \in B)$ and $(x \in A$ and $y \notin C)$
$\Rightarrow$ $(x, y) \in A \times B$ and $(x, y) \notin A \times C$
$(x, y) \in (A \times B) - (A \times C)$
$\Rightarrow$ $A \times (B - C) \subseteq (A \times B) - (A \times C)$ ...(iii)

Similarly, we can show
$(A \times B) - (A \times C) \subseteq A \times (B - C)$ ...(iv)

From Eqs. (iii) and (iv), we get
$$A \times (B - C) = (A \times B) - (A \times C)$$

140. *(d)* We have, $(x, y) \in R$, if $x^2 + y^2 = 16$

i.e., $y = \pm\sqrt{16 - x^2}$

For, $x = 0, y = \pm 4$
$x = \pm 4, y = 0$

We observe that no other values of $x, y \in Z$, which satisfy $x^2 + y^2 = 16$.
$$R = \{(0, 4), (0, -4), (4, 0)(-4, 0)\}$$
$\therefore$ Domain of $R = \{0, 4, -4\}$

141. *(a)* A. $P = \{1, 2\}, Q = \{1, 2\}$
$$P \cup Q = \{1, 2\} \; P \cap Q = \{1, 2\}$$
$$\therefore (P \cup Q) \times (P \cap Q) = \{(1, 1), (1, 2), (2, 1), (2, 2)\}$$

B. $A = \{0, 1\}, B = \{2, 3, 4\}, C = \{3, 5\}$
$$B \cap C = \{3\}$$
$$A \times (B \cap C) = \{(0, 3), (1, 3)\}$$

C. Refer solution 140
$$R = \{(0, 8), (0, -8), (8, 0), (-8, 0)\}$$

D. $R = \{(-2, 3), (-1, 5), (0, 7), (1, 9)(2, 11)\}$

3

Trigonometric Functions

Angles

Level I

1. The measure of rotation of a given ray about its initial point is called

(a) angular momentum (b) circle
(c) angle (d) None of these

2. Consider the figure given below

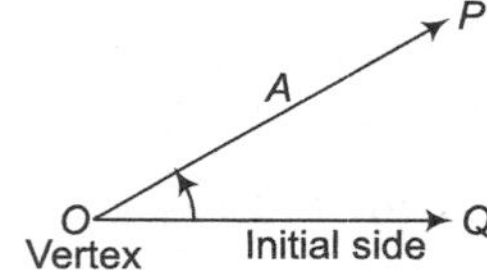

Here, A refers to

(a) intermediate side (b) terminal side
(c) positive side (d) original side

3. Consider the figure given below

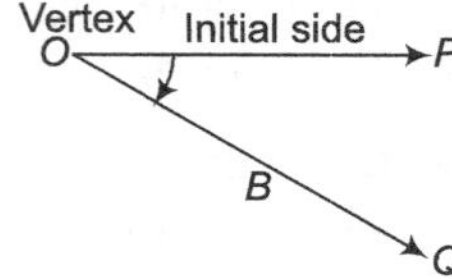

Here, B refers to

(a) terminal side
(b) original side
(c) positive side
(d) negative side

4. The point of rotation is called

(a) angle (b) vertex (c) origin (d) axis

5. Which among the following is/are correct?

(a) The angle is called negative, if the rotation is clockwise
(b) The angle is called positive, if the rotation is anti-clockwise
(c) The amount of rotation performed to get the terminal side from the initial side is called the measure of an angle
(d) All the above are correct

6. The definition of an angle suggests a unit for

(a) rotating angles (b) measuring angles
(c) angular momentum (d) None of these

7. In the figure shown below, what can you say about the angle formed by the position of the initial side?

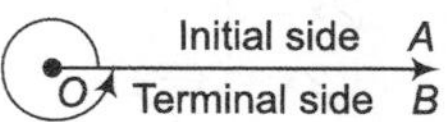

(a) One complete revolution (b) Zero revolution
(c) Linear angle (d) Right angle

8. If a rapidly spinning wheel is making an angle of say 15 revolutions per second, then this is an example of

(a) small angle (b) obtuse angle
(c) large angle (d) None of these

9. The most commonly used units for measurement of angles are

(a) degree measure and radian measure
(b) grade measure and radian measure
(c) grade measure and degree measure
(d) None of the above

10. If a rotation from the initial side to terminal side is ...A... of a revolution, the angle is said to have a measure of one degree, written as ...B... Here, A and B respectively refer to

(a) $\left(\frac{1}{180}\right)^{th}$ and 2° (b) $\left(\frac{1}{360}\right)^{th}$ and 1°

(c) $\left(\frac{1}{90}\right)^{th}$ and 1° (d) $\left(\frac{1}{360}\right)^{th}$ and 360°

11. Which among the following statement is/are true?

I. 1 degree (or 1°) is divided into 59 s.
II. 1 min (or 1′) is divided into 160 s.
III. 1 degree (or 1°) is divided into 600 min.
IV. 1 min (or 1′) is divided into 60 s.

(a) I is true (b) II is true (c) III is true (d) IV is true

12. 1 min (or 1′) is equivalent to ...A... of a degree and 1 s (or 1″) is equivalent to ...B... of a minute. Here, A and B respectively refer to

(a) $\left(\frac{1}{60}\right)^{th}$ and $\left(\frac{1}{180}\right)^{th}$ (b) $\left(\frac{1}{60}\right)^{th}$ and $\left(\frac{1}{60}\right)^{th}$

(c) 60th and 180th (d) $\left(\frac{1}{60}\right)^{th}$ and $\left(\frac{5}{60}\right)^{th}$

13. Which of the following figures are not showing its measures properly?

I.
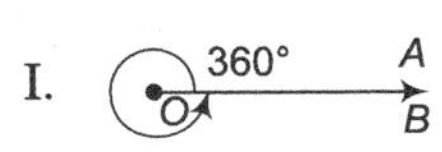

II.
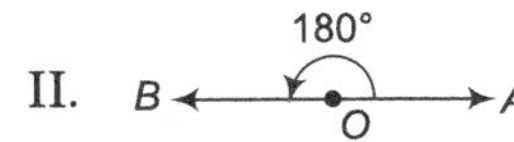

III.
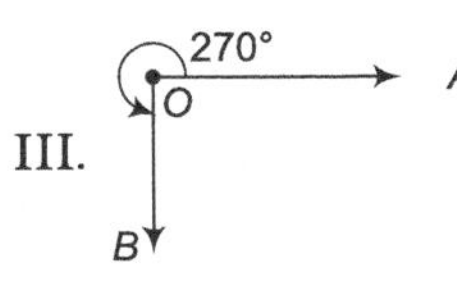

IV.
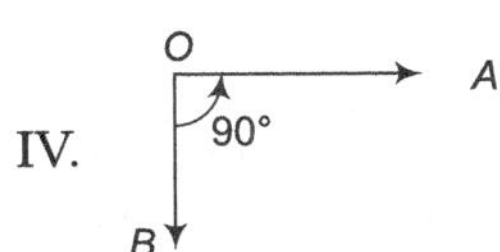

V.
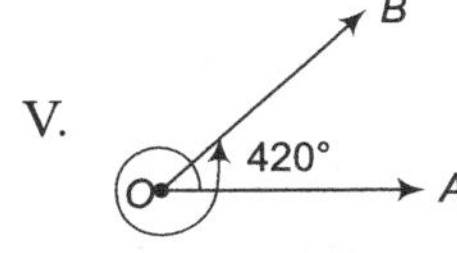

VI.
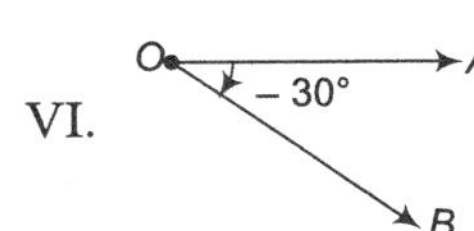

VII.
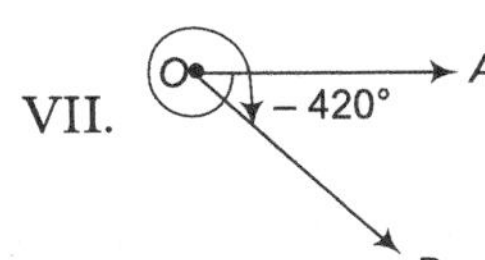

VIII.
B O A
180°

(a) I, II, III and IV (b) V, VI, VII and VIII
(c) III and IV (d) IV and VIII

14. Which among the following is/are true for a unit circle ?

I. In a unit circle, radius of circle is 1 unit.
II. In a unit circle, circumference of circle, is 1 unit.
III. Area of unit circle is π unit^2.
IV. Circumference of unit circle is 2π unit.

(a) Only I is true (b) I and II
(c) I, III and IV (d) III and IV

15. Angle subtended at the centre by an arc of length 1 unit in a unit circle is said to have a measure of

(a) 1 degree (b) 1 grade
(c) 1 radian (d) 1 arc

16. Which among the following figure is not a correct measure of radian ?

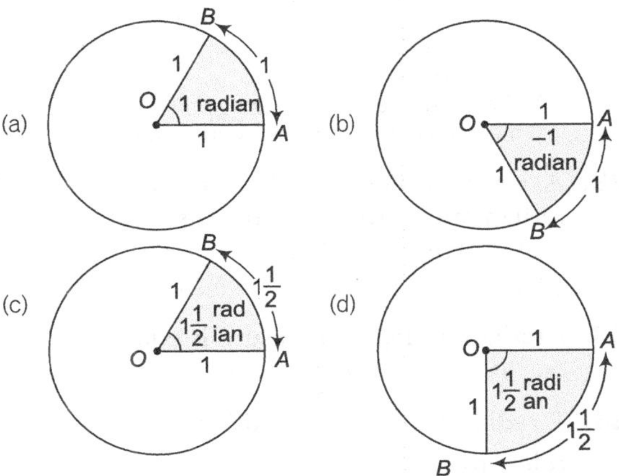

17. One complete revolution of the initial side subtends an angle of

(a) π radian (or π^c) (b) 2π radian (or $2\pi^c$)

(c) $\frac{\pi}{2}$ radian $\left(\text{or } \frac{\pi^c}{2}\right)$ (d) $2n\pi$ radian (or $2n\pi^c$)

18. Equal arcs of a circle subtend ...A... at the centre. Here, A refers to

(a) equal angle (b) obtuse angle
(c) acute angle (d) right angle

19. In a circle of radius r, an arc of length r will subtend an angle of

(a) r radian (b) 1 radian
c) π radian (d) 2π radian

20. In a circle of radius r, an arc of length l subtends an angle θ radian at the centre, then θ is equal to

(a) $\frac{l}{r}$ (b) lr

(c) $\frac{r}{l}$ (d) None of these

21. Consider the unit circle with centre O given below

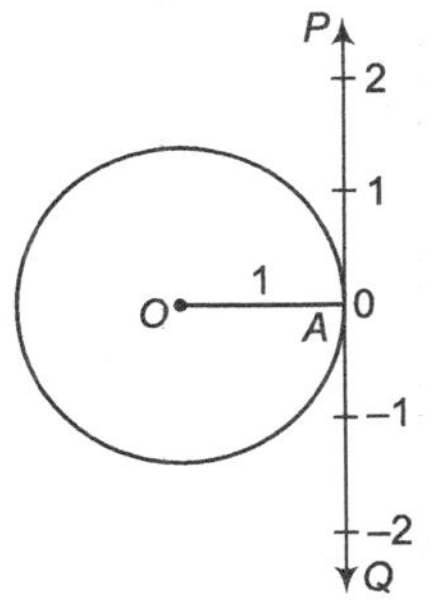

Here, QAP is tangent to the circle at A. Also, the point A is at real number zero. AP and AQ are representing positive real numbers and negative real numbers respectively on this line. On the basis of above information, choose the statement which is incorrect?

(a) AP is in the anti-clockwise direction along the circle
(b) AQ is in the clockwise direction along the circle
(c) Every real number on line PQ will correspond to a radian measure and conversely
(d) The length an arc of the circle will give the degree measure of the angle subtended by the arc at the centre

22. A circle subtends at the centre an angle whose radian measure is 2π and its degree measure is ...A.... Here, A refers to

(a) 360° (b) 180° (c) 90° (d) 270°

23. π radian in degree measure is equal to

(a) 18° (b) 180° (c) 200° (d) 360°

24. 1 radian= ...A... = 57°16′ approximately. Here, A refers to

(a) $\frac{180°}{\pi}$ (b) $\frac{360°}{\pi}$
(c) $\frac{90°}{\pi}$ (d) None of these

25. 1 degree = ...B... = 0.01746 radian (approx). Here, B refers to

(a) $\frac{\pi}{180}$ degree (b) $\frac{\pi}{180}$ radian (c) $\frac{\pi}{90}$ radian (d) $\frac{\pi}{1800}$ radian

Directions (Q. Nos. 26-27) *Match the terms of Column I with the terms of Column II and choose the correct option from the codes given below.*

26.

Column I (Radian Measure)		Column II (Degree Measure)	
A.	$\frac{\pi}{6}$	1.	45°
B.	$\frac{\pi}{4}$	2.	90°
C.	$\frac{\pi}{3}$	3.	30°
D.	$\frac{\pi}{2}$	4.	60°

Codes

	A	B	C	D		A	B	C	D
(a)	3	1	4	2	(b)	3	2	4	1
(c)	2	4	1	3	(d)	1	2	3	4

27.

Column I (Radian Measure)		Column II (Degree Measure)	
A.	π	1.	225°
B.	2π	2.	270°
C.	$3\pi/2$	3.	360°
D.	$5\pi/4$	4.	180°

Codes

	A	B	C	D		A	B	C	D
(a)	4	3	1	2	(b)	4	3	2	1
(c)	1	2	3	4	(d)	2	3	4	1

28. State which among the following is incorrect ?

I. Angle $\theta°$ means the angle whose degree measure is θ.
II. Angle β means the angle whose degree measure is β.
III. Radian measure $= \frac{\pi}{180} \times$ Degree measure
IV. Degree measure $= \frac{180}{\pi} \times$ Radian measure

(a) I is incorrect (b) II is incorrect
(c) III is incorrect (d) IV is incorrect

29. Radian measure of 40°20′ is equal to

(a) $\frac{120\pi}{504}$ radian (b) $\frac{121\pi}{540}$ radian
(c) $\frac{121\pi}{3}$ radian (d) None of these

30. Approximate degree measure of 6 radians is equal to

(a) 343° 38′ (b) 343° 38′11″ (c) 243° 38′11″
(d) 143° 38′11″

31. The radius of circle in which a central angle of 60° intercepts an arc of length 37.4 cm is $\left(\text{use } \pi = \frac{22}{7}\right)$.

(a) 17.7 cm (b) 37.5 cm (c) 35.7 cm (d) 18.35 cm

32. The minute hand of a watch is 1.5 cm long. The distance travelled by the minute hand in 40 minutes is equal to

(a) 3.28 cm (b) 4.28 cm
(c) 5.28 cm (d) 6.28 cm

33. If the arcs of the same lengths in two circles subtend angles 65° and 110° at the centre, then the ratio of their radii is equal to

(a) 22 : 13 (b) 11 : 13
(c) 10 : 51 (d) None of these

34. Match the terms of Column I with the terms of Column II and choose the correct option from the codes given below.

Column I (Degree Measure)		Column II (Radian Measure)	
A.	25°	1.	$\frac{26\pi}{9}$
B.	– 47° 30′	2.	$4\pi/3$
C	240°	3.	$\frac{-19\pi}{72}$
D	520°	4.	$\frac{5\pi}{36}$

Codes

	A	B	C	D		A	B	C	D
(a)	4	1	2	3	(b)	4	3	2	1
(c)	1	3	2	4	(d)	1	4	3	2

35. Match the terms of Column I with the terms of Column II and choose the correct option from the codes given below $\left[\text{use } \pi = \frac{22}{7}\right]$.

	Column I (Radian Measure)		Column II (Degree Measure)
A.	$\frac{11}{16}$	1.	300°
B.	-4	2.	210°
C.	$5\pi/3$	3.	39° 22′ 30″
D.	$7\pi/6$	4.	− 229° 5′ 27″

Codes

	A	B	C	D		A	B	C	D
(a)	3	4	2	1	(b)	1	4	2	3
(c)	3	4	1	2	(d)	2	4	1	3

36. A wheel makes 360 revolutions in 1 min. It turns in 1s by

(a) 6π radian (b) 12π radian
(c) 18π radian (d) 24π radian

37. The degree measure of the angle subtended at the centre of a circle of radius 100 cm by an arc of length 22 cm is $\left[\text{use } \pi = \frac{22}{7}\right]$

(a) 12°36′ (b) 22°36′ (c) 10°30′ (d) 12°30′

38. In a circle of diameter 40 cm, the length of a chord is 20 cm, then the length of minor arc of the chord is

(a) $\frac{20\pi}{7}$ cm (b) $\frac{\pi}{3}$ cm (c) $\frac{10\pi}{3}$ cm (d) $\frac{20\pi}{3}$ cm

39. If in two circles, arcs of the same length subtend angles 60° and 75° at the centre, then the ratio of their radii is

(a) $\frac{5}{3}$ (b) $\frac{2}{5}$ (c) $\frac{5}{4}$ (d) $\frac{4}{3}$

40. Which among the following is/are true?

I. If a pendulum of length 75 cm, describes an arc of length 10 cm, then it swings by $\frac{2}{15}$ radian.

II. If a pendulum of length 75 cm describes an arc of length 15 cm, then it swings by $\frac{1}{5}$ radian.

III. If a pendulum of length 75 cm describes an arc of length 21 cm, then it swings by $\frac{7}{25}$ radian.

(a) I and II are true (b) Only II is true
(c) Only III is true (d) All are true

Level II

41. Which among the following statement is incorrect ?

I. The angle between two consecutive digits in a clock is 30°.

II. The minute hand rotates through an angle of 6° in a minute.

III. The hour hand rotates through an angle of 30° in one hour *i.e.*, $(1/2)°$ in a minute.

IV. The second hand rotates through an angle of 180° in a minute.

(a) I is incorrect (b) II is incorrect
(c) III is incorrect (d) IV is incorrect

42. Which among the following statement is/are true?

I. The unit of measurement is degree in sexagesimal system.

II. The unit of measurement is radian in circular system.

III. The circumference of a circle always bears a constant ratio denoted as π to its diameter.

IV. All the above are correct.

(a) I is true (b) II is true (c) III is true (d) IV is true

43. Consider the following statements

Statement I The degree measure corresponding to (-2) radian is $-114°$ 19 min.

Statement II The degree measure of a given radian measure $= \frac{180}{\pi} \times$ Radian measure.

Choose the correct option.

(a) Statement I is true; Statement II is true; Statement II is a correct explanation for Statement I
(b) Statement I is true; Statement II is true; Statement II is not a correct explanation for Statement I
(c) Statement I is true; Statement II is false
(d) Statement I is false; Statement II is true

44. Match the terms of Column I with the terms of Column II and choose the correct option from the codes given below.

	Column I (Degree Measure)		Column II (Radian Measure)
A.	37° 30′	1.	$\left(\frac{17\pi}{9}\right)^c$
B.	5° 37′ 30″	2.	$\left(\frac{121\pi}{540}\right)^c$
C.	40° 20′	3.	$(5\pi/24)^c$
D.	340°	4.	$(\pi/32)^c$

Codes

	A	B	C	D		A	B	C	D
(a)	3	4	2	1	(b)	3	2	4	1
(c)	3	1	4	2	(d)	2	4	1	3

45. Assertion (A) Radian measure of 504° is $\frac{14\pi}{5}$ radian.

Reason (R) Radian measure of a given degree measure $= \frac{\pi}{180} \times$ Degree measure.

(a) Both A and R are correct; R is the correct explanation of A
(b) Both A and R are correct; R is not the correct explanation of A
(c) A is correct; R is incorrect
(d) R is correct; A is incorrect

46. Match the terms of Column I with the terms of Column II and choose the correct option from the codes given below .

	Column I (Radian Measure)		Column II (Degree Measure)
A.	$\left(\frac{4\pi}{15}\right)^c$	1.	14° 19′ 6″
B.	$\left(\frac{1}{4}\right)^c$	2.	48°
C.	$\left(\frac{\pi}{8}\right)^c$	3.	22° 30′
D.	$\left(\frac{11}{16}\right)^c$	4.	39° 22′ 30″

Codes

	A	B	C	D		A	B	C	D
(a)	1	2	3	4	(b)	2	1	4	3
(c)	2	1	3	4	(d)	4	1	2	3

47. If the angles of a triangle are in A P and the number of degrees in the least is to the number of radians in the greatest is 60 : π, then the angles in radian are [HOTS]

(a) $\frac{\pi}{8}, \frac{\pi}{4}, \frac{\pi}{2}$ (b) $\frac{\pi}{3}, \frac{\pi}{6}, \frac{\pi}{5}$ (c) $\frac{\pi}{6}, \frac{\pi}{3}, \frac{\pi}{2}$ (d) $\frac{2\pi}{9}, \frac{\pi}{3}, \frac{4\pi}{9}$

48. Assertion (A) The angular measurement of a regular decagon in radian is $\frac{4\pi}{5}$.

Reason (R) The angle of n sided regular polygon is equal to $(2n - 4) \times 90°$

(a) Both A and R are correct; R is the correct explanation of A
(b) Both A and R are correct; R is not the correct explanation of A
(c) A is correct; R is incorrect
(d) R is correct; A is incorrect

49. If a circular wire of radius 3 cm cut and bent, so as to lie along the circumference of a hoop whose radius is 48 cm, then the angle in degrees which is subtended at the centre of hoop is [NCERT Exempler]

(a) $\left(23\frac{1}{2}\right)^\circ$ (b) $\left(22\frac{1}{2}\right)^\circ$ (c) $\left(24\frac{1}{2}\right)^\circ$ (d) 25°

50. If the distance of moon from the earth is 360000 km and its diameter subtends an angle of 31′ at the eye of the observer, then the diameter of the moon is [HOTS]

(a) 1247.62 km (b) 2247.62 km
(c) 3247.62 km (d) None of these

Trigonometric Functions

Level I

Directions (Q. Nos. 51-53) *Consider the information given below*

Let $P\ (a, b)$ be any point on the unit circle given below, which has its centre at origin O. It is given that $\angle AOP = x$ radian.

Now, answer the questions based on the figure given below.

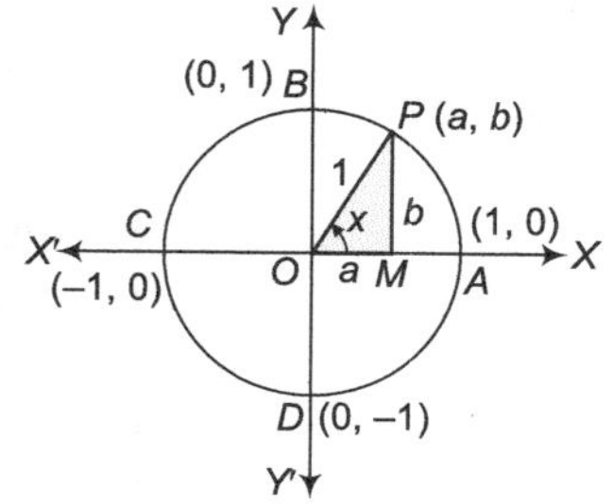

51. Which among the following statements can be concluded from the above figure.

I. For every point on the unit circle, we have $\cos^2 x + \sin^2 x = 1$

II. $\cos x = \frac{a}{1}$, so $\cos 0° = \frac{1}{1} = 1$

III. $\sin x = \frac{b}{1}$, so $\sin\frac{3\pi}{2} = \frac{-1}{1} = -1$

IV. $\sin \pi = 0$

Choose the correct options

(a) Only I
(b) Only I and II
(c) Only III
(d) All of the above

52. If the value of $\sin x = 0$, then the value of x is equal to

(a) $(n+1)\pi, n \in z$ (b) $n\pi, n \in z$
(c) $2n\pi, n \in z$ (d) $(2n+1)\pi, n \in z$

53. If the value of $\cos x = 0$, then the value of x is equal to

(a) $(2n+1)\pi, n \in z$ (b) $2n\pi, n \in z$
(c) $(2n+1)\frac{\pi}{2}, n \in z$ (d) None of these

54. Match the terms of Column I with the terms of Column II and choose the correct option from the codes given below.

Column I	Column II
A. cosec x	1. $\frac{1}{\cos x}, x \neq (2n+1)\frac{\pi}{2}, n \in Z$
B. sec x	2. $\frac{1}{\sin x}, x \neq n\pi, n \in Z$
C. tan x	3. $\frac{\cos x}{\sin x}, x \neq n\pi, n \in Z$
D. cot x	4. $\frac{\sin x}{\cos x}, x \neq (2n+1)\frac{\pi}{2}, n \in Z$

Codes

	A	B	C	D
(a)	2	1	4	3
(b)	2	4	1	3
(c)	1	2	3	4
(d)	2	1	3	4

55. Which among the following statements is true?

I. $\sin^2 x - \cos^2 x = 1$
II. $1 - \tan^2 x = \sec^2 x$
III. $1 - \cot^2 x = \text{cosec}^2 x$
IV. $\sin^2 x = 1 - \cos^2 x$

(a) I is true (b) II is true (c) III is true (d) IV is true

56. Consider the information given below

The values of trigonometric functions for some angles are shown below and for some angles, the values are missing and marked as A, B, C and D.

	0	$\frac{\pi}{6}$	$\frac{\pi}{4}$	$\frac{\pi}{3}$	$\frac{\pi}{2}$	π	$\frac{3\pi}{2}$	2π
sin	0	$\frac{1}{2}$	$\frac{1}{\sqrt{2}}$	A	1	0	−1	0
cos	1	B	$\frac{1}{\sqrt{2}}$	$\frac{1}{2}$	0	−1	0	1
tan	0	C	1	D	not defined	0	not defined	0

The values of A, B, C and D are respectively.

(a) $\frac{\sqrt{3}}{2}, \frac{\sqrt{3}}{2}, \sqrt{3}, \frac{1}{\sqrt{3}}$ (b) $\frac{\sqrt{3}}{2}, \frac{1}{\sqrt{3}}, \sqrt{3}, \frac{\sqrt{3}}{2}$
(c) $\frac{\sqrt{3}}{2}, \frac{\sqrt{3}}{2}, \frac{1}{\sqrt{3}}, \sqrt{3}$ (d) $\sqrt{3}, \frac{1}{\sqrt{3}}, \frac{\sqrt{3}}{2}, \frac{\sqrt{3}}{2}$

57. Consider the figure as shown and it is shown that $P(a, b)$ be a point on the unit circle with centre at O.

Now, which among the following can be concluded from the shown figure.

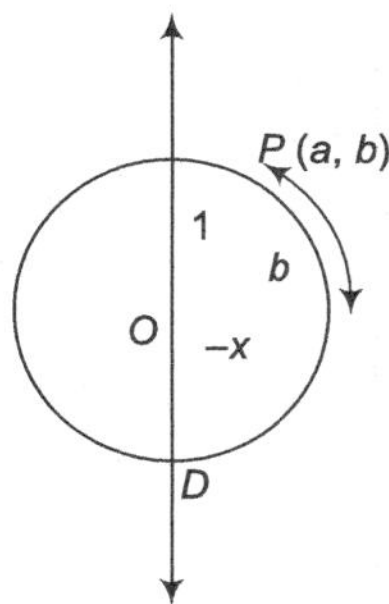

I. $\cos x = a$
II. $\cos(-x) = a$
III. $\sin x = b$
IV. $\sin(-x) = -b \Rightarrow \sin(-x) = -\sin x$

Choose the correct option.

(a) Only I is true (b) Only II is true
(c) I, II and III are true (d) All are true

58. Consider the information given below

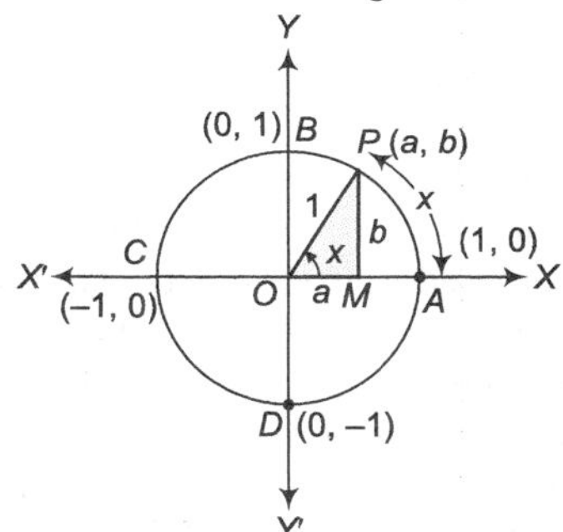

A point $P(a, b)$ is point on the unit circle as shown above such that its centre lies on origin.

Based on the above information, choose which among the following is incorrect?

(a) $\sin x$ is positive for $0 < x < \pi$
(b) $\sin x$ is negative for $\pi < x < 2\pi$
(c) $\cos x$ is positive for $0 < x < \frac{\pi}{2}$ and $\frac{3\pi}{2} < x < 2\pi$
(d) $\cos x$ is positive for $\frac{\pi}{2} < x < \frac{3\pi}{2}$

59. Consider the statements given below

I. $\sin x$ is positive in first and second quadrants.
II. cosec x is negative in third and fourth quadrants.
III. $\tan x$ and $\cot x$ are negative in second and fourth quadrant.
IV. $\cos x$ and $\sec x$ is positive in first and fourth quadrants.

Choose the correct option.

(a) All are correct
(b) Only I and IV are correct
(c) None is correct
(d) Only III and IV are correct

60. Consider the statements given below

I. Range of sine and cosine functions are $[-1, 1]$
II. Domain of tangent function is same as that of cotangent function.
III. Range of secant and cosecant function are same.
IV. Domain of tangent function is same as secant function.

Choose which among the following is correct?

(a) Only I, III and IV are correct
(b) Only II, III and IV are correct
(c) Only III and II are correct
(d) None of the above

61. Choose the correct statement among the following.

I. $\sin x$ increases from 0 to 1 in first quadrant and decreases from 0 to -1 in third quadrant.
II. $\cos x$ increases from -1 to 0 in third quadrant and decreases from 1 to 0 in first quadrant.
III. $\tan x$ increases from 0 to ∞ in third quadrant and increases from $-\infty$ to 0 in second quadrant.

(a) I is correct
(b) II is correct
(c) III is correct
(d) All are correct

62. Choose the statement which is incorrect among the following.

I. $\cot x$ decreases from 0 to $-\infty$ in first quadrant and increases from 0 to ∞ in third quadrant.
II. $\sec x$ increases from $-\infty$ to -1 in second quadrant and decreases from ∞ to 1 in fourth quadrant.
III. $\operatorname{cosec} x$ increases from 1 to ∞ in second quadrant and decreases from -1 to $-\infty$ in fourth quadrant.

(a) I is incorrect
(b) II is incorrect
(c) III is incorrect
(d) IV is incorrect

63. Consider the statements given below

I. Domain of $\sin x$ is R.
II. Range of $\sin x$ is $[-1, 1]$.
III. The value of $\sin x$ repeats after an interval of 2π.
IV. $\sin x$ decreases in second and third quadrant.

Choose the correct option.

(a) Only I, II, and III are correct
(b) Only II and III are correct
(c) Only III and IV are correct
(d) All are correct

64. Choose correct statements from the following.

I. Domain of $\cos x$ is R.
II. Range of $\cos x$ is $[-1, 1]$.
III. The value of $\cos x$ repeats after an interval of 2π.

(a) I is correct
(b) II is correct
(c) III is correct
(d) All are correct

65. Consider the information given below

I. Domain of $\tan x$ is
$$R^+ - \left\{(2n+1)\frac{\pi}{2} : n \in Z\right\}$$
II. Range of $\tan x$ is R.
III. Graph of $\tan x$ repeats after an interval of π *i.e.*, $\tan(\pi + x) = \tan x$

Choose which among the following is correct.

(a) I is correct
(b) II is correct
(c) II and III are correct
(d) All are correct

66. Which among the following is/are true?

I. The values of $\operatorname{cosec} x$ repeat after an interval of 2π.
II. The values of $\sec x$ repeat after an interval of 2π.
III. The values of $\cot x$ repeat after an interval of π.

(a) I is true
(b) II is true
(c) III is true
(d) All are true

67. If $\cos x = -\frac{3}{5}$ and x lies in third quadrant, then which among the following is/are correct?

I. $\sin x = -\frac{4}{5}$
II. $\operatorname{cosec} x = \frac{-5}{4}$
III. $\sec x = \frac{-5}{3}$
IV. $\cot x = \frac{4}{3}$
V. $\tan x = \frac{3}{4}$

(a) Only I and II are correct
(b) Only IV and V are correct
(c) Only I, II and III are correct
(d) All are correct

68. If $\cot x = \frac{-5}{12}$ and x lies in second quadrant, then choose the correct option.

(a) $\sec x = \pm\frac{13}{5}$
(b) $\tan x = \frac{12}{5}$
(c) $\sin x = -\frac{13}{12}$ and $\operatorname{cosec} x = \frac{12}{13}$
(d) $\cos x = \frac{-5}{13}$

69. The value of $\sin\frac{31\pi}{3}$ is

(a) $\frac{\sqrt{3}}{2}$
(b) $-\frac{\sqrt{3}}{2}$
(c) $-\frac{1}{\sqrt{2}}$
(d) $\frac{1}{\sqrt{2}}$

70. The value of $\cos(-1710°)$ is ...A.... Here, A refers to

(a) -1
(b) 1
(c) $\frac{1}{2}$
(d) None of these

71. If $\cos x = \frac{-1}{2}$ and x lies in third quadrant, then

(a) $\tan x = -\sqrt{3}$ and $\cot x = \frac{-1}{\sqrt{3}}$

(b) $\sec x = 2$ and $\text{cosec } x = \frac{2}{\sqrt{3}}$

(c) $\sin x = \frac{-\sqrt{3}}{2}$

(d) All are correct

72. If $\sin x = \frac{3}{5}$ and x lies in second quadrant, then

(a) $\cos x = -\frac{4}{5}$

(b) $\tan x = \frac{4}{3}$ and $\cot x = \frac{3}{4}$

(c) $\sec x = \frac{5}{4}$ and $\text{cosec } x = \frac{-5}{3}$

(d) All are correct

73. The value of $\sin 765°$ is equal to

(a) $-\frac{1}{\sqrt{2}}$ (b) $-\frac{1}{2}$ (c) $\frac{1}{2}$ (d) $\frac{1}{\sqrt{2}}$

74. The value of $\text{cosec}(-1410)°$ is equal to

(a) 1 (b) $\frac{1}{2}$

(c) 2 (d) None of these

75. Consider the information given below

$\tan x = -\frac{5}{12}$, where x lies in second quadrant.

Statement I The values of $\cot x$, $\sin x$, $\cos x$, $\text{cosec } x$, $\sec x$ are $\frac{-12}{5}, \frac{5}{13}, \frac{-12}{13}, \frac{13}{5}, \frac{-13}{12}$, respectively.

Statement II In second quadrant, the value of $\sin x$ and $\text{cosec } x$ are positive.

Choose the correct option.

(a) Statement I is true; Statement II is true; Statement II is a correct explanation for Statement I

(b) Statement I is true; Statement II is true; Statement II is not a correct explanation for Statement I

(c) Statement I is true; Statement II is false

(d) Statement I is false; Statement II is true

76. The value of $\tan \frac{19\pi}{3}$ is

(a) $\frac{1}{\sqrt{3}}$ (b) $\sqrt{3}$ (c) $-\sqrt{3}$ (d) $-\frac{1}{\sqrt{3}}$

77. If $\cot x = \frac{3}{4}$ and x lies in third quadrant, then

(a) $\tan x = -\frac{4}{3}$ and $\sec x = \pm\frac{5}{3}$

(b) $\sec x = \frac{-5}{3}$ and $\sin x = -\frac{4}{5}$

(c) $\text{cosec } x = \frac{5}{4}$ and $\cos x = \frac{5}{3}$

(d) Both (b) and (c) are correct

78. The value of $\sin\left(\frac{-11\pi}{3}\right)$ is

(a) $\frac{2}{\sqrt{3}}$ (b) $\frac{-\sqrt{3}}{2}$

(c) $-\frac{1}{2}$ (d) None of these

79. The value of $\cot\left(\frac{-15\pi}{4}\right)$ is

(a) $\frac{-1}{\sqrt{3}}$ (b) 1

(c) $\sqrt{3}$ (d) $-\sqrt{3}$

80. Consider the information given below

$\sec x = \frac{13}{5}$ and x lies in fourth quadrant.

Statement I The values of $\cos x$, $\sin x$, $\text{cosec } x$, $\tan x$, $\cot x$ are $\frac{5}{13}, \frac{-12}{13}, \frac{-13}{12}, \frac{-12}{5}, \frac{-5}{12}$, respectively.

Statement II In fourth quadrant the values of $\cos x$ and $\sec x$ are positive.

Choose the correct option.

(a) Statement I is true; Statement II is true; Statement II is a correct explanation for Statement I

(b) Statement I is true; Statement II is true; Statement II is not a correct explanation for Statement I

(c) Statement I is true; Statement II is false

(d) Statement I is false; Statement II is true

Level II

81. If $\frac{\cos A}{3} = \frac{\cos B}{4} = \frac{1}{5}, -\frac{\pi}{2} < A < 0, -\frac{\pi}{2} < B < 0,$ then value of $2\sin A + 4\sin B$ is

(a) 4 (b) −2

(c) −4 (d) 0

82. The value of $\left(1+\cos\frac{\pi}{6}\right)\left(1+\cos\frac{\pi}{3}\right)\left(1+\cos\frac{2\pi}{3}\right)\left(1+\cos\frac{7\pi}{6}\right)$ is

(a) $\frac{3}{16}$ (b) $\frac{3}{8}$

(c) $\frac{3}{4}$ (d) $\frac{1}{2}$

83. ΔABC is right angled at C, then $\tan A + \tan B$ is equal to

(a) $\frac{b^2}{ac}$ (b) $a+b$ (c) $\frac{a^2}{bc}$ (d) $\frac{c^2}{ab}$

84. If $\tan\theta = \frac{1}{\sqrt{7}}$, then $\left(\frac{\text{cosec}^2\theta - \sec^2\theta}{\text{cosec}^2\theta + \sec^2\theta}\right)$ is equal to

(a) $\frac{1}{2}$ (b) $\frac{3}{4}$ (c) $\frac{5}{4}$ (d) (2)

85. If $x = h + a\sec\theta$ and $y = k + b\,\text{cosec}\,\theta$. Then,

(a) $\frac{a^2}{(x+h)^2} - \frac{b^2}{(y+k)^2} = 1$ (b) $\frac{a^2}{(x-h)^2} + \frac{b^2}{(y-k)^2} = 1$

(c) $\frac{(x-h)^2}{a^2} + \frac{(y-k)^2}{b^2} = 1$ (d) $\frac{(x-h)^2}{a^2} - \frac{(y-k)^2}{b^2} = 1$

86. If $\tan\theta = \frac{-4}{3}$, then $\sin\theta$ is [NCERT Exemplar]

(a) $\frac{-4}{5}$ but not $\frac{4}{5}$ (b) $\frac{-4}{5}$ or $\frac{4}{5}$

(c) $\frac{4}{5}$ but not $-\frac{4}{5}$ (d) None of these

87. Consider the following statements

Statement I If $A = \cos^2\theta + \sin^4\theta$ for all values of θ, then $\frac{3}{4} \le A \le 1$.

Statement II $\sin^2\theta + \cos^4\theta = 1$.

Choose the correct option.

(a) Statement I is true; Statement II is true; Statement II is a correct explanation for Statement I
(b) Statement I is true; Statement II is true; Statement II is not a correct explanation for Statement I
(c) Statement I is true; Statement II is false
(d) Statement I is false; Statement II is true

88. If θ lies in second quadrant, then $\sqrt{\frac{1-\sin\theta}{1+\sin\theta}} + \sqrt{\frac{1+\sin\theta}{1-\sin\theta}}$ is equal to [HOTS]

(a) $2\cos\theta$ (b) $2\sec\theta$ (c) $-2\cos\theta$ (d) $-2\sec\theta$

89. Match the terms of Column I with the terms of Column II and choose the correct option from the codes given below.

	Column I		Column II
A.	$\sin\frac{25\pi}{3}$	1.	$-\sqrt{3}$
B.	$\cos\frac{41\pi}{4}$	2.	$\frac{\sqrt{3}}{2}$
C.	$\tan\left(\frac{-16\pi}{3}\right)$	3.	1
D.	$\cot\frac{29\pi}{4}$	4.	$\frac{1}{\sqrt{2}}$

Codes

	A	B	C	D
(a)	2	4	1	3
(b)	2	1	4	3
(c)	3	1	4	2
(d)	3	4	1	2

90. Consider the following statements

Statement I $\text{cosec}\left(\frac{-33\pi}{4}\right) = -\sqrt{2}$

Statement II $\text{cosec}\,(2n\pi + \theta) = \text{cosec}\theta$

Choose the correct option.

(a) Statement I is true; Statement II is true; Statement II is a correct explanation for Statement I
(b) Statement I is true; Statement II is true; Statement II is not a correct explanation for Statement I
(c) Statement I is true; Statement II is false
(d) Statement I is false, Statement II is true

91. Consider the following statements

Statement I $\sec 6\pi = 1$

Statement II $\sec(2n\pi + \theta) = \sec\theta$

Choose the correct option.

(a) Statement I is true; Statement II is true; Statement II is a correct explanation for Statement I
(b) Statement I is true; Statement II is true; Statement II is not a correct explanation for Statement I
(c) Statement I is true; Statement II is false
(d) Statement I is false; Statement II is true

92. If $\sin x = \frac{-2\sqrt{6}}{5}$ and x lies in III quadrant, then the value of $\cot x$ is

(a) $\frac{3}{\sqrt{6}}$ (b) $-\frac{3}{\sqrt{6}}$ (c) $\frac{1}{2\sqrt{6}}$ (d) $-\frac{1}{2\sqrt{6}}$

93. If $\sec x = -2$ and $\pi < x < \frac{3\pi}{2}$, then $\sin x$ is equal to

(a) $-\frac{\sqrt{3}}{2}$ (b) $\frac{\sqrt{3}}{2}$ (c) $-\frac{1}{2}$ (d) $\frac{1}{2}$

94. If $\text{cosec}\,x = -\frac{2}{\sqrt{3}}$ and x lies in quadrant IV, then the value of $\tan x$ is

(a) $\sqrt{3}$ (b) $-\sqrt{3}$ (c) $\frac{1}{\sqrt{3}}$ (d) $-\frac{1}{\sqrt{3}}$

95. If $\cos\theta = \frac{-3}{5}$ and $\pi < \theta < \frac{3\pi}{2}$, then the value of $\left(\frac{\text{cosec}\,\theta + \cot\theta}{\sec\theta - \tan\theta}\right)$ is equal to

(a) $\frac{1}{3}$ (b) $\frac{2}{3}$

(c) $\frac{13}{2}$ (d) None of these

96. If θ lies in second quadrant and $\sin\theta = \frac{3}{5}$, then which of the following is true ?

(a) $\sec\theta = -\frac{5}{4}$ (b) $\cot\theta = -\frac{4}{3}$

(c) $2\sec\theta - 3\cot\theta = \frac{3}{2}$ (d) All the above are true

Trigonometric Functions of Sum and Difference of Two Angles

Level I

97. The value of $\cos(x + y)$ is equal to

(a) $\cos x \cdot \cos y + \sin x \cdot \sin y$
(b) $\cos x \cdot \sin y + \sin x \cdot \cos y$
(c) $\cos x \cdot \cos y - \sin x \cdot \sin y$
(d) None of the above

98. Which among the following is correct identity?
(a) $\cos(x - y) = \cos x \cdot \cos y + \sin x \cdot \sin y$
(b) $\cos(x - y) = \sin x \cdot \cos y + \cos x \cdot \sin y$
(c) $\sin(x - y) = \cos x \cdot \cos y + \sin x \cdot \sin y$
(d) All are correct

99. Match the terms of Column I with the terms of Column II and select the correct options from the codes given below.

Column I	Column II
A. $\cos\left(\frac{\pi}{2} - x\right)$	1. $-\sin x$
B. $\sin\left(\frac{\pi}{2} - x\right)$	2. $\cos x$
C. $\cos\left(\frac{\pi}{2} + x\right)$	3. $\sin x$
D. $\sin\left(\frac{\pi}{2} + x\right)$	

Codes

	A	B	C	D
(a)	1	2	3	3
(b)	3	1	2	3
(c)	3	2	1	2
(d)	1	2	3	3

100. The value of $\sin(x + y)$ is equal to

(a) $\sin x \cdot \cos y + \cos x$
(b) $\sin x \cdot \cos y + \cos x \cdot \sin y$
(c) $\sin x \cdot \cos y - \cos x$
(d) $\sin x \cdot \cos y - \cos x \cdot \sin y$

101. Match the terms of Column I with the terms of Column II and select the correct options which is suitable for given codes.

Column I	Column II
A. $\cos(\pi - x)$	1. $-\cos x$
B. $\sin(\pi - x)$	2. $-\sin x$
C. $\sin(\pi + x)$	3. $\cos x$
D. $\cos(\pi + x)$	4. $\sin x$
E. $\cos(2\pi - \pi)$	
F. $\sin(2\pi - x)$	

Codes

	A	B	C	D	E	F
(a)	1	4	2	1	3	2
(b)	1	2	4	1	3	1
(c)	2	4	1	2	3	2
(d)	1	2	2	4	3	1

102. The value of $\tan(x + y)$, if none of the angles x, y and $(x + y)$ is an odd multiple of $\frac{\pi}{2}$ is

(a) $\frac{\tan x - \tan y}{1 - \tan x \cdot \tan y}$ (b) $\frac{\tan x + \tan y}{1 - \tan x \cdot \tan y}$
(c) $\frac{\tan x + \tan y}{1 + \tan x \cdot \tan y}$ (d) $\frac{1 + \tan x \cdot \tan y}{\tan x + \tan y}$

103. Which among the following is correct?
(a) $\tan(x - y) = \frac{\tan x + \tan y}{1 - \tan x \cdot \tan y}$
(b) $\tan(x + y) = \frac{\tan x - \tan y}{1 + \tan x \cdot \tan y}$
(c) $\tan(x - y) = \frac{\tan x - \tan y}{1 + \tan x \cdot \tan y}$
(d) None of the above

104. If none of the angles x, y and $(x + y)$ is a multiple of π, then
(a) $\cot(x + y) = \frac{\cot x \cdot \cot y - 1}{\cot y + \cot x}$
(b) $\cot(x - y) = \frac{\cot x \cdot \cot y + 1}{\cot y - \cot x}$
(c) (a) and (b) are true
(d) (a) and (b) are not true

105. If $x \neq n\pi + \frac{\pi}{2}$, then the value of $\cos 2x$ is equal to

I. $\cos^2 x - \sin^2 x$ II. $2\cos^2 x - 1$
III. $1 - 2\sin^2 x$ IV. $\frac{1 - \tan^2 x}{1 + \tan^2 x}$

(a) Only I, II and III are true
(b) Only II and III are true
(c) Only II and IV are true
(d) All are true

106. The value of $\sin 2x$ is equal to $\left(\text{where, } x \neq n\pi + \frac{\pi}{2}\right)$

(a) $2\sin x \cdot \cos x$
(b) $\frac{2\tan x}{1 + \tan^2 x}$
(c) Both (a) and (b) are true
(d) None of the above

107. Match the terms of Column I with the terms of Column II and choose the correct option from the codes given below.

Column I	Column II
A. $\tan 2x$	1. $\dfrac{3\tan x - \tan^3 x}{1 - 3\tan^2 x}$
B. $\sin 3x$	2. $4\cos^3 x - 3\cos x$
C. $\cos 3x$	3. $3\sin x - 4\sin^3 x$
D. $\tan 3x$	4. $\dfrac{2\tan x}{1 - \tan^2 x}$

Codes

	A	B	C	D		A	B	C	D
(a)	4	3	1	2	(b)	4	3	2	1
(c)	1	2	3	4	(d)	1	3	2	4

108. Match the terms of Column I with the terms of Column II and choose the correct option from the codes given below.

Column I	Column II
A. $\cos x + \cos y$	1. $-2\sin\dfrac{x+y}{2}\cdot\sin\dfrac{x-y}{2}$
B. $\cos x - \cos y$	2. $2\cos\dfrac{x+y}{2}\cdot\cos\dfrac{x-y}{2}$
C. $\sin x + \sin y$	3. $2\cos\dfrac{x+y}{2}\cdot\sin\dfrac{x-y}{2}$
D. $\sin x - \sin y$	4. $2\sin\dfrac{x+y}{2}\cdot\cos\dfrac{x-y}{2}$

Codes

	A	B	C	D		A	B	C	D
(a)	2	1	4	3	(b)	1	2	4	3
(c)	1	2	3	4	(d)	2	1	3	4

109. Consider the statements given below

I. $2\cos x \cdot \cos y = \cos(x+y) - \cos(x-y)$

II. $-2\sin x \cdot \sin y = \cos(x+y) - \cos(x-y)$

III. $2\sin x \cdot \cos y = \sin(x+y) - \sin(x-y)$

IV. $2\cos x \cdot \sin y = \sin(x+y) + \sin(x-y)$

Choose the correct statements.

(a) I is correct (b) II is correct
(c) Both I and II are correct (d) III is correct

110. The value of

$3\sin\frac{\pi}{6}\sec\frac{\pi}{3} - 4\sin\frac{5\pi}{6}\cot\frac{\pi}{4}$ is equal to

(a) 1 (b) 2 (c) -1 (d) -2

111. The value of $\sin 15°$ is

(a) $\frac{\sqrt{3}+1}{2\sqrt{2}}$ (b) $\frac{\sqrt{3}-1}{2\sqrt{2}}$ (c) $\frac{\sqrt{3}}{\sqrt{2}}$ (d) $\frac{\sqrt{3}}{2\sqrt{2}}$

112. The value of $\tan\frac{13\pi}{12}$ is

(a) 2 (b) $2+\sqrt{3}$
(c) $2-\sqrt{3}$ (d) $\sqrt{3}$

113. $\frac{\sin(x+y)}{\sin(x-y)}$ is equal to ...A... . Here, A refers to

(a) $\tan x \cdot \tan y$ (b) $\frac{\tan x + \tan y}{\tan x - \tan y}$
(c) $\tan(x+y)$ (d) $\cot(x-y)$

114. $\tan 3x - \tan 2x - \tan x$ is equal to

(a) $\tan 3x \cdot \tan 2x$ (b) $\tan 3x \cdot \tan 2x \cdot \tan x$
(c) $\tan 2x \cdot \tan x$ (d) None of these

115. The value of $\cos\left(\frac{\pi}{4}+x\right)+\cos\left(\frac{\pi}{4}-x\right)$ is equal to

(a) $\sqrt{2}\cos x$ (b) $\sqrt{3}\cos x$
(c) $\cos x$ (d) $\cos 2x$

116. The value of $\frac{\cos 7x + \cos 5x}{\sin 7x - \sin 5x}$ is equal to

(a) $\cot x$ (b) $2\tan x$
(c) $\tan x$ (d) $2\cot x$

117. $\frac{\sin 5x - 2\sin 3x + \sin x}{\cos 5x - \cos x}$ is equal to

(a) $\cot x$ (b) $2\tan x$
(c) $2\cot x$ (d) $\tan x$

118. The value of

$\sin^2\frac{\pi}{6} + \cos^2\frac{\pi}{3} - \tan^2\frac{\pi}{4}$ is equal to ...A... . Here, A refers to

(a) $\frac{1}{2}$ (b) $-\frac{1}{2}$ (c) $\frac{3}{2}$ (d) $-\frac{3}{2}$

119. $2\sin^2\frac{\pi}{6} + \operatorname{cosec}^2\frac{7\pi}{6}\cdot\cos^2\frac{\pi}{3}$ is equal to

(a) $\frac{3}{2}$ (b) $-\frac{3}{2}$ (c) $\frac{1}{2}$ (d) $-\frac{1}{2}$

120. $\cot^2\frac{\pi}{6} + \operatorname{cosec}\frac{5\pi}{6} + 3\tan^2\frac{\pi}{6}$ is equal to

(a) 1 (b) 3
(c) 5 (d) None of these

121. $2\sin^2\frac{3\pi}{4} + 2\cos^2\frac{\pi}{4} + 2\sec^2\frac{\pi}{3}$ is equal to

(a) 4 (b) 6
(c) 8 (d) 10

122. Consider the statements given below

I. The value of $\sin 75°$ is $\frac{\sqrt{3}+1}{2\sqrt{2}}$.

II. The value of $\tan 15°$ is $2+\sqrt{3}$.

III. $\tan 15° = \tan(45° - 30°) = \frac{\tan 45° - \tan 30°}{1 + \tan 45° \cdot \tan 30°}$

Choose the correct option.

(a) I is correct (b) II is correct
(c) I and III are correct (d) II and III are correct

123. $\cos\left(\frac{\pi}{4}-x\right)\cdot\cos\left(\frac{\pi}{4}-y\right)-\sin\left(\frac{\pi}{4}-x\right)\cdot\sin\left(\frac{\pi}{4}-y\right)$ is equal to

(a) $\sin(x-y)$ (b) $\sin(x+y)$
(c) $\cos(x+y)$ (d) $\cos(x-y)$

124. The value of $\dfrac{\tan\left(\frac{\pi}{4}+x\right)}{\tan\left(\frac{\pi}{4}-x\right)}$ is equal to

(a) $(1+\tan x)^2$ (b) $(1-\tan x)^2$
(c) $\left[\frac{(1+\tan x)}{(1-\tan x)}\right]^2$ (d) None of these

125. Consider the following statements

Statement I $\dfrac{\cos(\pi+x)\cdot\cos(-x)}{\sin(\pi-x)\cdot\cos\left(\frac{\pi}{2}+x\right)}=\cot^2 x$

Statement II $\cos(\pi+\theta)=-\cos\theta$ and $\cos(-\theta)=\cos\theta$.

Also, $\sin(\pi-\theta)=\sin\theta$ and $\sin(-\theta)=-\sin\theta$.

Choose the correct option.

(a) Statement I is true; Statement II is true; Statement II is a correct explanation for Statement I
(b) Statement I is true; Statement II is true; Statement II is not correct explanation for Statement I
(c) Statement I is true; Statement II is false
(d) Statement I is false; Statement II is true

126. The value of

$$\cos\left(\frac{3\pi}{2}+x\right)\cdot\cos(2\pi+x)\left[\cot\left(\frac{3\pi}{2}-x\right)+\cot(2\pi+x)\right]$$ is equal to

(a) 0 (b) 1
(c) 2 (d) −1

127. The value of $\sin(n+1)x\cdot\sin(n+2)x+\cos(n+1)x\cos(n+2)x$ is equal to

(a) $\sin x$ (b) $\cos x$
(c) $\cot x$ (d) $\tan x$

128. $\cos\left(\frac{3\pi}{4}+x\right)-\cos\left(\frac{3\pi}{4}-x\right)$ is equal to

(a) $\sqrt{2}\sin x$ (b) $-2\sin x$
(c) $-\sqrt{2}\sin x$ (d) None of these

129. $\sin^2 6x-\sin^2 4x$ is equal to ...A.... Here, A refers to

(a) $\sin 2x\sin 10x$
(b) $\sin 2x$
(c) $\sin 10x$
(d) $\cos 2x\cdot\cos 10x$

130. $\cos^2 2x-\cos^2 6x$ is equal to

(a) $\sin 9x\cdot\sin 8x$ (b) $\sin 8x\cdot\sin 4x$
(c) $\cos 4x\cdot\cos 8x$ (d) None of these

131. $\sin 2x+2\sin 4x+\sin 6x$ is equal to

(a) $4\cos^2 x\cdot\sin 2x$ (b) $4\cos^2 x\cdot\sin 4x$
(c) $2\cos^2 x\cdot\sin 4x$ (d) None of these

132. Consider the following statements

Statement I $\cot 4x(\sin 5x+\sin 3x)=\cot x(\sin 5x-\sin 3x)$

Statement II $\sin 5x+\sin 3x=2\sin 4x\cdot\cos x$

Choose the correct option.

(a) Statement I is true; Statement II is true; Statement II is a correct explanation for Statement I
(b) Statement I is true; Statement II is true; Statement II is not a correct explanation for Statement I
(c) Statement I is true; Statement II is false
(d) Statement I is false; Statement II is true

133. Match the terms of Column I with the terms of Column II and choose the correct option from the codes given below.

Column I	Column II
A. $\dfrac{\cos 9x-\cos 5x}{\sin 17x-\sin 3x}$	1. $\tan 4x$
B. $\dfrac{\sin 5x+\sin 3x}{\cos 5x+\cos 3x}$	2. $-\dfrac{\sin 2x}{\cos 10x}$
C. $\cos 4x$	3. $\tan 2x$
D. $\dfrac{\sin x+\sin 3x}{\cos x+\cos 3x}$	4. $1-8\sin^2 x\cdot\cos^2 x$

Codes

	A	B	C	D
(a)	1	2	3	4
(b)	2	1	3	4
(c)	2	1	4	3
(d)	1	3	2	4

134. $\dfrac{\sin x-\sin y}{\cos x+\cos y}$ is equal to

(a) $\tan\frac{x+y}{2}$
(b) $\sin\frac{x+y}{2}$
(c) $\tan\frac{x-y}{2}$
(d) $\sin\frac{x-y}{2}$

135. $\dfrac{\sin x-\sin 3x}{\sin^2 x-\cos^2 x}$ is equal to

(a) $\sin 2x$
(b) $\cos 2x$
(c) $\tan 2x$
(d) None of the above

Level II

136. Match the terms of Column I with the terms of Column II and choose the correct option from the codes given below.

Column I	Column II	
A. $\sin \frac{A}{2}$	1. $\pm\sqrt{\frac{1-\cos A}{1+\cos A}}$	$+$, if $\frac{A}{2}$ lies in I or II quadrant; $-$, if $\frac{A}{2}$ lies in III or IV quadrant
B. $\cos \frac{A}{2}$	2. $\pm\sqrt{\frac{1-\cos A}{2}}$	$+$, if $\frac{A}{2}$ lies in I or IV quadrant; $-$, if $\frac{A}{2}$ lies in II or III quadrant
C. $\tan \frac{A}{2}$	3. $\pm\sqrt{\frac{1+\cos A}{2}}$	$+$, if $\frac{A}{2}$ lies in I or III quadrant; $-$, if $\frac{A}{2}$ lies in II or IV quadrant

Codes

	A	B	C		A	B	C
(a)	2	3	1	(b)	2	1	3
(c)	1	2	3	(d)	3	1	2

137. Match the terms of Column I with the terms of Column II and choose the correct option from the codes given below.

Column I	Column II
A. $\sin 18°$	1. $\frac{\sqrt{10-2\sqrt{5}}}{4}$
B. $\cos 18°$	2. $\sqrt{\frac{10+2\sqrt{5}}{4}}$
C. $\cos 36°$	3. $\frac{\sqrt{5}-1}{4}$
D. $\sin 36°$	4. $\frac{\sqrt{5}+1}{4}$

Codes

	A	B	C	D		A	B	C	D
(a)	1	2	3	4	(b)	2	3	4	1
(c)	3	4	2	1	(d)	3	2	4	1

138. The value of $\frac{\cos 4x + \cos 3x + \cos 2x}{\sin 4x + \sin 3x + \sin 2x}$ is equal to

(a) $\tan 3x$ (b) $\cot 3x$
(c) $\sin 3x$ (d) None of these

139. $\cot x \cdot \cot 2x - \cot 2x \cdot \cot 3x - \cot 3x \cdot \cot x$ is equal to

(a) -1 (b) 0 (c) 1 (d) $\frac{1}{2}$

140. Consider the following statements

Statement I The value of $\sqrt{3}$ cosec $20° - \sec 20°$ is equal to 4.

Statement II $\sin(x-y) = \sin x \cdot \cos y - \cos x \cdot \sin y$

Choose the correct option.

(a) Statement I is true; Statement II is true; Statement II is a correct explanation for Statement I
(b) Statement I is true; Statement II is true; Statement II is not a correct explanation for Statement I
(c) Statement I is true; Statement II is false
(d) Statement I is false; Statement II is true

141. The value of $\tan 4x$ is equal to

(a) $\frac{4\tan x\,(1-\tan^2 x)}{1+6\tan^2 x-\tan^4 x}$ (b) $\frac{4\tan x\,(1-\tan^2 x)}{1-6\tan^2 x+\tan^4 x}$
(c) $\frac{\tan x\,(1-\tan^2 x)}{1+6\tan^2 x-\tan^4 x}$ (d) None of these

142. The value of $\cos 6x$ is equal to

(a) $\cos^6 x - 48\cos^3 x + 18\cos^2 x - 1$
(b) $3\cos^2 x - 4\cos^3 x + 18\cos^2 x$
(c) $32\cos^6 x - 48\cos^4 x + 18\cos^2 x - 1$
(d) None of the above

143. Consider the following statements

Statement I $\tan 9° - \tan 27° - \tan 63° + \tan 81° = 4$

Statement II $\tan 9° + \cot 9° = \frac{2}{\sin 18°}$

and $\cot 27° + \tan 27° = \frac{2}{\cos 36°}$

Choose the correct option. [NCERT Exemplar]

(a) Statement I is true; Statement II is true; Statement II is a correct explanation for Statement I
(b) Statement I is true; Statement II is true; Statement II is not a correct explanation for Statement I
(c) Statement I is true; Statement II is false
(d) Statement I is false; Statement II is true

144. Which among the following is/are true statements? [NCERT Exemplar]

I. $\frac{\sec 8\theta - 1}{\sec 4\theta - 1} = \frac{\tan 8\theta}{\tan 2\theta}$ II. $1 - \cos 8\theta = 2\sin^2 4\theta$

III. $\sin 8\theta = 2\sin 4\theta \cdot \cos 4\theta$

(a) I is true (b) II is true (c) III is true (d) All are true

145. Consider the following statements

Statement I $\left(1+\cos\frac{\pi}{8}\right)\left(1+\cos\frac{3\pi}{8}\right)\left(1+\cos\frac{5\pi}{8}\right)\left(1+\cos\frac{7\pi}{8}\right) = \frac{1}{8}$

Statement II $\sin^2\theta = \frac{1}{2}(1-\cos 2\theta)$

Choose the correct option.

(a) Statement I is true; Statement II is true; Statement II is a correct explanation for Statement I
(b) Statement I is true; Statement II is true; Statement II is not a correct explanation for Statement I
(c) Statement I is true; Statement II is false
(d) Statement I is false; Statement II is true

146. The greatest value of $\sin x \cdot \cos x$ is

(a) 1 (b) 2 (c) $\sqrt{2}$ (d) $\frac{1}{2}$

147. The value of $\sin 20° \cdot \sin 40° \cdot \sin 60° \cdot \sin 80°$ is

(a) $\frac{-3}{16}$ (b) $\frac{5}{16}$ (c) $\frac{3}{16}$ (d) $\frac{1}{16}$

148. Consider the following statements

Statement I If $3\tan(\theta - 15°) = \tan(\theta + 15°)$

$$0° < \theta < 90°, \text{ then } \theta = \frac{\pi}{4}$$

Statement II $\tan 3\theta = \dfrac{3\tan\theta - \tan^3\theta}{1 - 3\tan^2\theta}$

Choose the correct option.

(a) Statement I is true; Statement II is true; Statement II is a correct explanation for Statement I
(b) Statement I is true; Statement II is true; Statement II is not a correct explanation for Statement I
(c) Statement I is true; Statement II is false
(d) Statement I is false; Statement II is true

149. Match the terms of Column I with the terms of Column II and choose the correct option from the codes given below. [NCERT Exemplar]

Column I	Column II
A. $\dfrac{1-\cos x}{\sin x}$	1. $\cot^2 \dfrac{x}{2}$
B. $\dfrac{1+\cos x}{1-\cos x}$	2. $\cot \dfrac{x}{2}$
C. $\dfrac{1+\cos x}{\sin x}$	3. $\lvert \cos x + \sin x \rvert$
D. $\sqrt{1+\sin 2x}$	4. $\tan \dfrac{x}{2}$

Codes

	A	B	C	D
(a)	1	2	4	3
(b)	4	1	2	3
(c)	4	2	1	3
(d)	1	2	3	4

150. The value of $\cos\theta \cdot \cos\frac{\theta}{2} - \cos 3\theta \cdot \cos\frac{9\theta}{2}$ is equal to [HOTS]

(a) $\sin 4\theta . \sin\frac{7\theta}{2}$
(b) $\sin 8\theta$
(c) $\sin 7\theta + \sin 8\theta$
(d) $\sin 7\theta \cdot \sin 8\theta$

Trigonometric Equations

Level I

151. Match the terms of Column I with the terms of Column II and choose the correct option from the codes given below.

Column I	Column II
A. Equations involving trigonometric functions of a variable are called	1. Principal solutions
B. The values of $\sin x$ and $\cos x$ repeat after an interval of	2. General solutions
C. The values of $\tan x$ repeat after an interval of	3. π
D. The solutions of a trigonometric equation for which $0 \le x < 2\pi$ are called	4. Trigonometric equations
E. The expression involving integer 'n' which gives all solutions of a trigonometric equation is called the	5. 2π

Codes

	A	B	C	D	E		A	B	C	D	E
(a)	1	3	4	2	5	(b)	4	5	3	1	2
(c)	4	3	5	1	2	(d)	2	3	5	1	4

152. The principal solutions of the equation $\sin x = \frac{\sqrt{3}}{2}$ are

(a) $\frac{13\pi}{6}$ and $\frac{11\pi}{6}$ (b) $\frac{5\pi}{4}$ and $\frac{11\pi}{4}$
(c) $\frac{5\pi}{6}$ and $\frac{11\pi}{6}$ (d) None of these

153. The equation $\tan x = \frac{-1}{\sqrt{3}}$ have its principal solutions as

(a) $\frac{\pi}{3}$ and $\frac{2\pi}{3}$ (b) $\frac{2\pi}{3}$ and $\frac{\pi}{4}$
(c) $\frac{5\pi}{6}$ and $\frac{11\pi}{6}$ (d) $\frac{\pi}{4}$ and $\frac{\pi}{6}$

154. For any real numbers x and y, $\sin x = \sin y$ implies
(a) $x = n\pi + (-1)^n Z$, where $n \in Z$
(b) $x = n\pi + y$, where $n \in Z$
(c) $x = n\pi + (-1)^n y$, where $n \in Z$
(d) None of the above

155. Consider the following statements
Statement I For any real numbers x and y, $\cos x = \cos y$ implies $x = 2n\pi \pm y$, where $n \in Z$.
Statement II $\sin \frac{x+y}{2} = 0$ or $\sin \frac{x-y}{2} = 0$
Choose the correct option.
(a) Statement I is true; Statement II is true; Statement II is a correct explanation for Statement I
(b) Statement I is true; Statement II is true; Statement II is not a correct explanation for Statement I
(c) Statement I is true; Statement II is false
(d) Statement I is false; Statement II is true

156. If x and y are not odd multiples of $\frac{\pi}{2}$, then $\tan x = \tan y$ implies
(a) $x = n\pi + y$, where $n \in Z$ (b) $x = n\pi - y$, where $n \in Z$
(c) $x = n\pi \pm y$, where $n \in Z$ (d) None of these

157. The solution of $\sin x = -\frac{\sqrt{3}}{2}$ is
(a) $x = n\pi + (-1)^n \frac{4\pi}{3}$, where $n \in Z$
(b) $x = n\pi + (-1)^n \frac{2\pi}{3}$, where $n \in Z$
(c) $x = n\pi + (-1)^n \frac{3\pi}{3}$, where $n \in Z$
(d) None of the above

158. If $\cos x = \frac{1}{2}$, then
(a) $x = 2n\pi + \frac{\pi}{3}$, where $n \in Z$ (b) $x = 2n\pi - \frac{\pi}{3}$, where $n \in Z$
(c) $x = 2n\pi \pm \frac{\pi}{3}$, where $n \in Z$ (d) None of these

159. Consider the following statements
Statement I If $\tan 2x = -\cot\left(x + \frac{\pi}{3}\right)$, then $x = n\pi + \frac{5\pi}{6}, n \in Z$.
Statement II $\tan x = \tan y \Rightarrow x = n\pi + y$, where $n \in Z$.
Choose the correct option.
(a) Statement I is true; Statement II is true; Statement II is a correct explanation for Statement I
(b) Statement I is true; Statement II is true; Statement II is not a correct explanation for Statement I
(c) Statement I is true; Statement II is false
(d) Statement I is false; Statement II is true

160. Consider the following statements
Statement I If $\sin 2x - \sin 4x + \sin 6x = 0$, then $x = \frac{n\pi}{4}$ or $x = n\pi \pm \frac{\pi}{6}$, where $n \in Z$.
Statement II $\sin x = \sin y \Rightarrow x = n\pi + (-1)^n y$, $n \in Z$ and $\cos x = \cos y \Rightarrow x = 2n\pi \pm y, n \in Z$.
Choose the correct option.
(a) Statement I is true; Statement II is true; Statement II is a correct explanation for Statement I
(b) Statement I is true; Statement II is true; Statement II is not a correct explanation for Statement I
(c) Statement I is true; Statement II is false
(d) Statement I is false; Statement II is true

161. The solution of $2\cos^2 x + 3\sin x = 0$ is
(a) $x = 2n\pi \pm \frac{2\pi}{3}, n \in Z$ (b) $x = 2n\pi \pm \frac{7\pi}{6}, n \in Z$
(c) $x = n\pi + (-1)^n \frac{7\pi}{6}, n \in Z$ (d) $x = n\pi + (-1)^n \frac{2\pi}{3}, n \in Z$

162. Consider the information given below.
$$\tan x = \sqrt{3}$$
Based on the above information choose the correct statement among the following.
(a) The principal value is $x = \frac{2\pi}{3}$
(b) General solution is $x = n\pi + \frac{2\pi}{3}, n \in Z$
(c) The principal solution is $x = \pi/3$
(d) General solution is $x = n\pi + \frac{4\pi}{13}, n \in Z$

163. Consider the following statements
Statement I If $\sec x = 2$, then principal value is $x = \frac{\pi}{3}$.
Statement II General solution of $\sec x = 2$ is $x = 2n\pi \pm \frac{\pi}{3}, n \in Z$.
Choose the correct option.
(a) Statement I is true; Statement II is true; Statement II is a correct explanation for Statement I
(b) Statement I is true; Statement II is true; Statement II is not a correct explanation for Statement I
(c) Statement I is true; Statement II is false
(d) Statement I is false; Statement II is true

164. If $\cot x = -\sqrt{3}$, then
(a) Its principal value is $\frac{5\pi}{6}$
(b) Its general solution is $n\pi + \frac{5\pi}{6}, n \in Z$
(c) Only (a) is true
(d) Both (a) and (b) are true

165. General solution of cosec $x = -2$ is

(a) $x = 2n\pi + (-1)^n \frac{7\pi}{6}, n \in Z$

(b) $x = 2n\pi \pm \frac{7\pi}{6}, n \in Z$

(c) $x = n\pi + (-1)^n \frac{7\pi}{6}, n \in Z$

(d) $x = n\pi \pm (-1)^n \frac{7\pi}{6}, n \in Z$

166. General solution of $\cos 4x = \cos 2x$ is

(a) $x = \frac{n\pi}{3}$ or $x = n\pi, n \in Z$

(b) $x = n\pi$ or $x = \frac{3n\pi}{4}, n \in Z$

(c) $x = 2n\pi$ or $x = n\pi + (-1)^n \frac{\pi}{3}, n \in Z$

(d) None of the above

167. If $\cos 3x + \cos x - \cos 2x = 0$, then

(a) $x = (2n+1)^3 \frac{\pi}{4}, n \in Z$

(b) $x = 2n\pi + \frac{3\pi}{2}, n \in Z$

(c) $x = 2n\pi \pm \frac{4\pi}{3}, n \in Z$

(d) $x = 2n\pi \pm \frac{\pi}{3}, n \in Z$

168. If $\sin 2x + \cos x = 0$, then which among the following is/are true?

I. $\cos x = 0$

II. $\sin x = -\frac{1}{2}$

III. $x = (2n+1)\frac{\pi}{2}, n \in Z$

IV. $x = n\pi + (-1)^n \frac{7\pi}{6}, n \in Z$

(a) I is true (b) I and II are true

(c) I, II and III are true (d) All are true

169. If $\sec^2 2x = 1 - \tan 2x$, then

(a) $x = \frac{n\pi}{2} + \frac{3\pi}{4}, n \in Z$ (b) $x = \frac{n\pi}{3}, n \in Z$

(c) $x = \frac{n\pi}{2} + \frac{3\pi}{8}, n \in Z$ (d) None of these

Level II

170. If $\sin x + \sin 3x + \sin 5x = 0$, then

(a) $x = \frac{n\pi}{3}$

(b) $x = n\pi + \frac{\pi}{3}$

(c) $x = n\pi - \frac{\pi}{3}$

(d) All are correct

171. If $\sec x \cdot \cos 5x + 1 = 0$, where $0 < x \le \frac{\pi}{2}$, then the value of x is/are

(a) $\frac{\pi}{6}$ (b) $\frac{\pi}{4}$

(c) $\frac{\pi}{2}$ (d) Both (a) and (b)

172. Which among the following is/are correct statement/ statements?

I. The general value of θ satisfying the equations $\sin^2\theta = \sin^2\alpha$, $\cos^2\theta = \cos^2\alpha$ and $\tan^2\theta = \tan^2\alpha$ is given by $\theta = n\pi \pm \alpha$.

II. The general value of θ satisfying equations $\sin\theta = \sin\alpha$ and $\cos\theta = \cos\alpha$ simultaneously is given by $\theta = 2n\pi \pm \alpha_1$ $n \in Z$.

(a) I is correct

(b) II is correct

(c) Both I and II are correct

(d) Both I and II are incorrect

173. If $a\cos\theta + b\sin\theta = c$, then which among the following is/are correct?

I. $\cos(\theta - \alpha) = \frac{c}{r}$, where $r = \sqrt{a^2 + b^2}$ and $\tan\alpha = \frac{b}{a}$.

II. Maximum and minimum values of the expression a $a\cos\theta + b\sin\theta$ are $\sqrt{a^2+b^2}$ and $-\sqrt{a^2+b^2}$ respectively, where a and b are constants.

(a) I is correct (c) II is correct

(c) Both I and II are correct (d) Both I and II are incorrect

174. The value of θ for $\sin\theta - \sin 3\theta + \sin 5\theta = 0$, is [NCERT Exemplar]

(a) $\frac{n\pi}{3}, n \in Z$ (b) $\frac{2n\pi}{5}, n \in Z$

(c) $\frac{3n\pi}{2}, n \in Z$ (d) None of these

175. The number of solutions possible for the equation

$$2\tan^2 x + \sec^2 x = 2 \text{ for } 0 \le x \le 2\pi.$$

(a) 1 (b) 2

(c) 3 (d) 4

176. If $\sin\theta$ and $\cos\theta$ are the roots of the equation

$$ax^2 - bx + c = 0,$$

then a, b and c satisfy the relation

(a) $a^2 + b^2 + 2ac = 0$

(b) $a^2 - b^2 + 2ac = 0$

(c) $a^2 + c^2 + 2ab = 0$

(d) $a^2 - b^2 - 2ac = 0$

177. Consider the following statements

Statement I The general value of θ satisfying the equation $2\sin^2\theta - 3\sin\theta - 2 = 0$ is

$$n\pi + (-1)^{n+1}\frac{\pi}{6}.$$

Statement II $\sin\theta = \sin\alpha \Rightarrow \theta = n\pi + (-1)^n\alpha$

Choose the correct option.

(a) Statement I is true; Statement II is true; Statement II is a correct explanation for Statement I
(b) Statement I is true; Statement II is true; Statement II is not a correct explanation for Statement I
(c) Statement I is true; Statement II is false
(d) Statement I is false; Statement II is true

178. Solution of the equation $3\tan(\theta - 15) = \tan(\theta + 15)$ is

(a) $\theta = \frac{n\pi}{2} + (-1)^n\frac{\pi}{4}$
(b) $\theta = n\pi + (-1)^n\frac{\pi}{3}$
(c) $\theta = n\pi - \frac{\pi}{3}$
(d) $\theta = n\pi - \frac{\pi}{4}$

179. If $\sin 6\theta + \sin 4\theta + \sin 2\theta = 0$, then the general value of θ is [WB JEE 2010]

(a) $\frac{n\pi}{4}, n\pi \pm \frac{\pi}{3}$
(b) $\frac{n\pi}{4}, n\pi \pm \frac{\pi}{6}$
(c) $\frac{n\pi}{4}, 2n\pi \pm \frac{\pi}{3}$
(d) $\frac{n\pi}{4}, 2n\pi \pm \frac{\pi}{6}$

180. The number of points of intersection of $2y = 1$ and $y = \sin x$, in $-2\pi \le x \le 2\pi$ is [VITEEE 2010; HOTS]

(a) 1 (b) 2
(c) 3 (d) 4

181. If $5\cos 2\theta + 2\cos^2\frac{\theta}{2} + 1 = 0$, when $(0 < \theta < \pi)$, then the values of θ are [WB JEE 2009]

(a) $\frac{\pi}{3} \pm \pi$ (b) $\frac{\pi}{3}, \cos^{-1}\left(\frac{3}{5}\right)$
(c) $\cos^{-1}\left(\frac{3}{5}\right) \pm \pi$ (d) $\frac{\pi}{3}, \pi - \cos^{-1}\left(\frac{3}{5}\right)$

182. The value of x in $\left(0, \frac{\pi}{2}\right)$ satisfying the equation $\sin x \cos x = \frac{1}{4}$ is [Kerala CEE 2010]

(a) $\frac{\pi}{6}$ (b) $\frac{\pi}{3}$ (c) $\frac{\pi}{8}$ (d) $\frac{\pi}{4}$
(e) $\frac{\pi}{12}$

183. The number of solutions of the equation $\sin x \cdot \cos 3x = \sin 3x \cdot \cos 5x$ in $\left[0, \frac{\pi}{2}\right]$ is

(a) 6 (b) 5 (c) 4 (d) 3

184. The number of solutions of $\cos 2\theta = \sin\theta$ in $(0, 2\pi)$ is [Kerala CEE 2010]

(a) 1
(b) 2
(c) 3
(d) 4
(e) 0

185. The maximum and minimum values of $6\sin x\cos x + 4\cos 2x$ are respectively [HOTS]

(a) 5 and –5
(b) 3 and –3
(c) 4 and –4
(d) None of these

186. The maximum value of $\cos^2\left(\frac{\pi}{3} - x\right) - \cos^2\left(\frac{\pi}{3} + x\right)$ is [DCE 2008]

(a) $-\frac{\sqrt{3}}{2}$
(b) $\frac{1}{2}$
(c) $\frac{\sqrt{3}}{2}$
(d) $\frac{3}{2}$

187. The maximum value of $5\cos\theta + 3\cos\left(\theta + \frac{\pi}{3}\right) + 3$ is [UP SEE 2008]

(a) 5
(b) 11
(c) 10
(d) –1

188. Consider the following statements

Statement I The maximum value of

$$f(x) = \sin x\,(1 + \cos x) \text{ is } \frac{3\sqrt{3}}{4}.$$

Statement II $f(x)$ is maximum at $x = \frac{\pi}{6}$.

Choose the correct option.

(a) Statement I is true; Statement II is true; Statement II is a correct explanation for Statement I
(b) Statement I is true; Statement II is true; Statement II is not a correct explanation for Statement I
(c) Statement I is true; Statement II is false
(d) Statement I is false; Statement II is true

Simple Applications of Sine and Cosine Formulae

Level I

189. Consider the figure given below

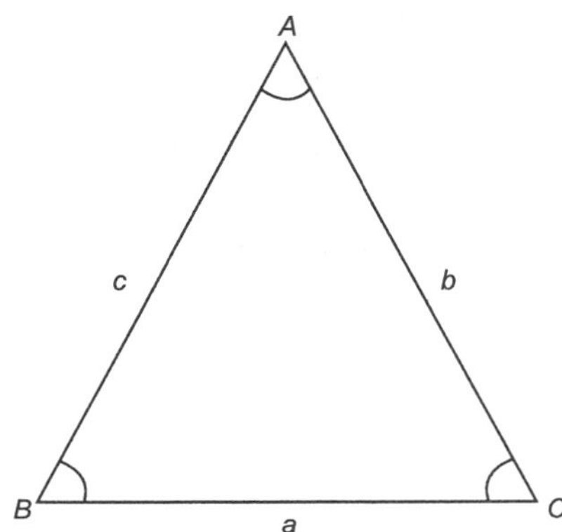

Based on the above figure, which among the following statements is true?

I. $\frac{\sin A}{a} = \frac{\sin B}{b}$

II. $\frac{\sin B}{b} = \frac{\sin C}{c}$

III. $\frac{\sin A}{a} = \frac{\sin B}{b} = \frac{\sin C}{c}$

(a) I is true (b) II is true
(c) III is true (d) All are true

190. Let A, B and C be angles of a triangle and a, b and c be lengths of sides opposite to angles A, B and C respectively, then which of the following is/are correct?

I. $a^2 = b^2 + c^2 - 2bc \cos A$

II. $b^2 = c^2 + a^2 - 2ca \cos B$

III. $c^2 = a^2 + b^2 - 2ab \cos C$

(a) I is correct (b) II is correct
(c) III is correct (d) All are correct

191. Which among the following is/are called Napier's Analog in a ΔABC?

(a) $\tan \frac{B-C}{2} = \frac{b-c}{b+c} \cot \frac{A}{2}$

(b) $\tan \frac{C-A}{2} = \frac{c-a}{c+a} \cot \frac{B}{2}$

(c) $\tan \frac{A-B}{2} = \frac{a-b}{a+b} \cot \frac{C}{2}$

(d) All of the above

192. Consider the following statements

Statement I In any triangle ABC, the value of $a \sin (B-C) + b \sin (C-A) + c \sin (A-B)$ is 0.

Statement II $\frac{\sin A}{a} = \frac{\sin B}{b} = \frac{\sin C}{c}$

Choose the correct option.

(a) Statement I is true; Statement II is true; Statement II is a correct explanation for Statement I
(b) Statement I is true; Statement II is true; Statement II is not a correct explanation for Statement I
(c) Statement I is true; Statement II is false
(d) Statement I is false; Statement II is true

193. The angle of elevation of the top point P of the vertical tower PQ of height h from a point A is 45° and from a point B, the angle of elevation is 60°, where B is a point at a distance d from the point A measured along the line AB which makes an angle 30° with AQ, then the value of d is

(a) $h(\sqrt{3}+1)$ (b) $h(\sqrt{3}-1)$ (c) h (d) $\frac{h(\sqrt{3}+1)}{2}$

194. A lamp post is situated at the middle point M of the side AC of a triangular plot ABC with $BC = 7$ m, $CA = 8$ m and $AB = 9$ m. Lamp post subtends an angle 15° at the point B. The height of the lamp post is

(a) $h = 7$ m (b) $h = (7 + \sqrt{3})$m
(c) $h = (2 - \sqrt{3})$ m (d) None of these

Directions (Q. Nos. 195-197)

Consider the information given below

In $\Delta ABC, a = 18, b = 24,$ and $c = 30$. Answer the following questions.

195. The value of $\cos A$ is equal to

(a) $\frac{4}{5}$ (b) $\frac{3}{5}$ (c) 0 (d) 1

196. The value of $\cos B$ is equal to

(a) $\frac{3}{5}$ (b) $\frac{4}{5}$ (c) 1 (d) 0

197. The value of $\cos C$ is equal to

(a) $\frac{1}{2}$ (b) $\frac{3}{2}$
(c) $\frac{4}{5}$ (d) None of these

198. If in a triangle ABC, it is given that $a = 18$, $b = 24$, and $c = 30$, then which of following is/are correct?

I. The value of $\sin A$ is $\frac{3}{5}$.

II. The value of $\sin B$ is $\frac{4}{5}$.

III. The value of $\sin C$ is 1.

(a) I is correct (b) II is correct
(c) III is correct (d) All are correct

199. In any ΔABC, match the terms of Column I with the terms of Column II and choose the correct option from the codes given below.

Column I	Column II
A. $\frac{a+b}{c}$	1. $\sin\left(\frac{A-B}{2}\right)/\cos\frac{C}{2}$
B. $\frac{a-b}{c}$	2. $\cos\frac{A-B}{2}/\sin\frac{C}{2}$
C. $\sin\frac{B-C}{2}$	3. $\frac{b-c}{a}\cdot\cos\frac{A}{2}$

Codes

	A	B	C		A	B	C
(a)	2	1	3	(b)	1	2	3
(c)	2	3	1	(d)	3	1	2

200. $a(b\cos C - c\cos B)$ is equal to

(a) $b^2 + c^2$ (b) $b^2 - c^2$
(c) $c^2 - b^2$ (d) None of these

201. $(b+c)\cos\left(\frac{B+C}{2}\right)$ is equal to

(a) $a\cos\left(\frac{B-C}{2}\right)$
(b) $b\cos\left(\frac{B-C}{2}\right)$
(c) $b\cos\left(\frac{C-A}{2}\right)$
(d) None of these

202. Consider the following statements

Statement I $\frac{\cos A}{a} + \frac{\cos B}{b} + \frac{\cos C}{c} = \frac{a^2+b^2+c^2}{2abc}$

Statement II $\cos A = \frac{b^2+c^2+a^2}{2abc}$

Choose the correct option.

(a) Statement I is true; Statement II is true; Statement II is a correct explanation for Statement I
(b) Statement I is true; Statement II is true; Statement II is not a correct explanation for Statement I
(c) Statement I is true; Statement II is false
(d) Statement I is false; Statement II is true

203. The value of $\frac{b^2-c^2}{a^2}\sin 2A + \frac{c^2-a^2}{b^2}\sin 2B + \frac{a^2-b^2}{c^2}\sin 2C$ is equal to

(a) 0 (b) 1 (c) 2 (d) −1

204. Consider the following statements

Statement I $a(\cos C - \cos B)$ is equal to $2(b-c)\cos^2\frac{A}{2}$.

Statement II $a\cos A + b\cos B + c\cos C = 2a\sin B\cdot\sin C$

Choose the correct option.

(a) Statement I is true; Statement II is true; Statement II is a correct explanation for Statement I
(b) Statement I is true; Statement II is true; Statement II is not a correct explanation for Statement I
(c) Statement I is true; Statement II is false
(d) Statement I is false; Statement II is true

205. In any ΔABC, $\frac{(b^2-c^2)}{a^2}$ is equal to

(a) $\frac{\sin(B+C)}{\sin(B-C)}$ (b) $\frac{\sin(B-C)}{\sin(B+C)}$
(c) $\sin(B+C)$ (d) $\sin(B-C)$

206. $(b^2-c^2)\cot A + (c^2-a^2)\cot B + (a^2-b^2)\cot C$ is equal to

(a) +1 (b) 2
(c) 3 (d) None of these

207. A tree stands vertically on a hill side which makes an angle of 15° with the horizontal. From a point on the ground 35 m down the hill from the base of the tree, the angle of elevation of the top of the tree is 60°. The height of the tree is

(a) $31\sqrt{2}$ m (b) $33\sqrt{2}$ m
(c) $35\sqrt{2}$ m (d) $34\sqrt{2}$ m

208. Two ships leave a port at the same time. One goes 24 km/h in the direction N 45° E and other travels 32 km/h in the direction S 75°E. The distance between the ships at the end of 3 h is approximately

(a) 81.4 km (b) 82 km
(c) 85 km (d) 86.4 km

209. Two trees, A and B are on the same side of a river. From a point C in the river the distance of the trees A and B is 250 m and 300 m, respectively. If the angle C is 45°. Then, the distance between the trees (use $\sqrt{2} = 1.44$) is approximately

(a) 210 m (b) 214 m
(c) 215.5 m (d) 218 m

Level II

210. In a ΔABC, if $\angle A = 60°$ and $a = 5$, $b = 4$, then c is a root of the equation

(a) $c^2 - 5c + 9 = 0$ (b) $c^2 + 4c + 9 = 0$

(c) $c^2 - 4c - 9 = 0$ (d) None of these

211. In a ΔABC, $2ac \sin \dfrac{A - B + C}{2}$ is equal to

(a) $a^2 + b^2 - c^2$

(b) $a^2 + c^2 + b^2$

(c) $b^2 + c^2 - a^2$

(d) $a^2 + c^2 - b^2$

212. If angles A, B and C are in AP, then $\dfrac{a + c}{b}$ is equal to

(a) $2\cos \dfrac{A - C}{2}$ (b) $2\sin \dfrac{A + C}{2}$

(c) $2\cos \dfrac{A + C}{2}$ (d) $2\tan \dfrac{A + B}{2}$

213. Angles of a triangle are in the ratio 4 : 1 : 1. The ratio between its greatest side and perimeter is [UP SEE 2008]

(a) $\dfrac{3}{2 + \sqrt{3}}$ (b) $\dfrac{1}{2 + \sqrt{3}}$

(c) $\dfrac{\sqrt{3}}{\sqrt{3} + 2}$ (d) $\dfrac{2}{2 + \sqrt{3}}$

214. The sides BC, CA and AB of a ΔABC are of lengths a, b, and c respectively. If D is the mid-point of BC and AD is perpendicular to AC, then the value of $\cos A \cos C$ is [DCE 2009]

(a) $\dfrac{3(a^2 - c^2)}{2ac}$ (b) $\dfrac{3(a^2 - c^2)}{3bc}$

(c) $\dfrac{(a^2 - c^2)}{3ac}$ (d) $\dfrac{2(c^2 - a^2)}{3ac}$

215. If one side of a triangle is double the other and the angles opposite to these sides differ by $60°$, then the triangle is [KCET 2009]

(a) obtuse angled (b) acute angled

(c) isosceles (d) right angled

216. In ΔABC, $a = 2$, $b = 3$ and $\sin A = \dfrac{2}{3}$, then B is equal to

(a) 30° (b) 60° (c) 90° (d) 120°

217. In a ΔABC, if $\sin A \sin B = \dfrac{ab}{c^2}$, then the triangle is

(a) equilateral

(b) isosceles

(c) right angled

(d) obtuse angled

218. The sides of a triangle are respectively 7 cm, $4\sqrt{3}$ cm and $\sqrt{13}$ cm, then the smallest angle of the triangle is

(a) $\dfrac{\pi}{6}$ b) $\dfrac{\pi}{3}$ (c) $\dfrac{\pi}{4}$ (d) $\dfrac{\pi}{5}$

219. Let ABC be a triangle such that $\angle ACB = \dfrac{\pi}{6}$ and let a, b and c denote the lengths of the sides opposite to A, B and C respectively. The value(s) of x for which $a = x^2 + x + 1$, $b = x^2 - 1$ and $c = 2x + 1$ is/are [IIT JEE 2010]

(a) $-(2 + \sqrt{3})$ (b) $1 + \sqrt{3}$

(c) $2 + \sqrt{3}$ (d) $4\sqrt{3}$

220. If $a = 2\sqrt{2}$, $b = 6$, $A = 45°$, then

(a) no triangle is possible

(b) one triangle is possible

(c) two triangles are possible

(d) either no triangle or two triangles are possible

Miscellaneous Exercise

221. Radian measure of 25° is

(a) $\left(\dfrac{3\pi}{36}\right)^c$ (b) $\left(\dfrac{5\pi}{36}\right)^c$

(c) $\left(\dfrac{8\pi}{35}\right)^c$ (d) None of these

222. Degree measure of 11 radian is

(a) 630°

(b) 330°

(c) 460°

(d) None of the above

223. Match the terms of Column I with the terms of Column II and choose the correct option from the codes given below.

	Column I		Column II
A.	1 radian is equal to	1.	0.01746 radian
B.	1° is equal to	2.	57° 16′ (approx)
C.	3° 45′ is equal to	3.	$\dfrac{9\pi}{32}$ radian
D.	50° 37′ 30″ is equal to	4.	$\dfrac{\pi}{48}$ radian

Codes

	A	B	C	D
(a)	1	4	3	2
(b)	2	4	1	3
(c)	2	1	4	3
(d)	3	1	4	2

224. If $\sin x = \frac{3}{5}$ and $\cos y = \frac{-12}{13}$, where x and y both lie in second quadrant, then the value of $\sin(x+y)$ is equal to

(a) $-\frac{65}{56}$ (b) $-\frac{56}{65}$
(c) $-\frac{31}{56}$ (d) $-\frac{56}{31}$

225. $\cos 2x \cdot \cos\frac{x}{2} - \cos 3x \cdot \cos\frac{9x}{2}$ is equal to

(a) $-\sin 5x \cdot \sin\frac{5x}{2}$ (b) $\sin 5x \cdot \sin\frac{5x}{2}$
(c) $\sin 3x \cdot \sin\frac{3x}{2}$ (d) None of these

226. The value of $\tan\frac{\pi}{8}$ is equal to

(a) $1-\sqrt{2}$ (b) $\sqrt{2}-1$
(c) $\frac{\sqrt{2}}{3}$ (d) $\sqrt{3}$

227. If $\tan x = \frac{3}{4}$ and $\pi < x < \frac{3\pi}{2}$, then which of the following is/are correct?

(a) $\sin\frac{x}{2} = \frac{3}{\sqrt{10}}$ (b) $\cos\frac{x}{2} = \frac{1}{\sqrt{10}}$
(c) $\tan\frac{x}{2} = 3$ (d) All are correct

228. The value of $\cos^2 x + \cos^2\left(x+\frac{\pi}{3}\right) + \cos^2\left(x-\frac{\pi}{3}\right)$ is equal to

(a) $\frac{1}{2}$ (b) 2
(c) 3 (d) None of the above

229. Consider the following statements

Statement I $2\cos\frac{\pi}{3}\cos\frac{9\pi}{13} + \cos\frac{3\pi}{13} + \cos\frac{5\pi}{13} = 0$

Statement II $\cos\frac{10\pi}{13} + \cos\frac{3\pi}{13} = 0$

and $\cos\frac{8\pi}{13} + \cos\frac{5\pi}{13} = 0$

Choose the correct option.

(a) Statement I is true; Statement II is true; Statement II is a correct explanation for Statement I
(b) Statement I is true; Statement II is true; Statement II is not a correct explanation for Statement I
(c) Statement I is true; Statement II is false
(d) Statement I is false; Statement II is true

230. Consider the following statements

Statement I $(\cos x + \cos y)^2 + (\sin x - \sin y)^2 = 4\cos^2\frac{x+y}{2}$

Statement II $(\sin 3x + \sin x)\sin x + (\cos 3x - \cos x)\cdot\cos x = 0$

Choose the correct option.

(a) Statement I is true
(b) Statement II is true
(c) Both the Statements are true
(d) Both the Statements are false

231. Consider the following statements

Statement I $(\cos x - \cos y)^2 + (\sin x - \sin y)^2 = 4\sin^2\left(\frac{x-y}{2}\right)$

Statement II $\sin^2 x + \cos^2 x = 1$

Choose the correct option.

(a) Statement I is true; Statement II is true; Statement II is a correct explanation for Statement I
(b) Statement I is true; Statement II is true; Statement II is not a correct explanation for Statement I
(c) Statement I is true; Statement II is false
(d) Statement I is false; Statement II is true

232. $\sin x + \sin 3x + \sin 5x + \sin 7x$ is equal to [HOTS]

(a) $4\cos x \cdot \cos 2x \sin 3x$ (b) $4\cos 2x \cdot \cos x \cdot \sin x$
(c) $4\cos x \cdot \cos 2x \cdot \sin 4x$ (d) None of the above

233. $\frac{(\sin 7x + \sin 5x) + (\sin 9x + 3x)}{(\cos 7x + \cos 5x) + (\cos 9x + \cos 3x)}$ is equal to

(a) $\tan 3x$ (b) $\cot 3x$ (c) $\tan 6x$ (d) $\cot 6x$

234. $\sin 3x + \sin 2x - \sin x$ is equal to

(a) $4\sin x \cdot \cos\frac{3x}{2} \cdot \cos\frac{x}{2}$ (b) $2\sin x \cdot \cos\frac{3x}{2} \cdot \cos\frac{x}{2}$
(c) $3\sin x \cdot \cos\frac{3x}{2} \cdot \cos\frac{x}{2}$ (d) None of these

235. If $\tan x = -\frac{4}{3}$ and x is in quadrant II, then which among the following is/are correct?

I. $\sin\frac{x}{2} = \frac{2\sqrt{5}}{5}$

II. $\cos\frac{x}{2} = \frac{\sqrt{5}}{5}$

III. $\tan\frac{x}{2} = \frac{2}{5}$

(a) I is correct (b) II is correct
(c) I and II are correct (d) All are correct

236. If $\cos x = -\frac{1}{3}$ and x is in III quadrant, then which among the following is/are correct ?

I. $\cos\frac{x}{2} = -\frac{\sqrt{3}}{3}$ II. $\sin\frac{x}{2} = \frac{\sqrt{6}}{3}$

III. $\tan\frac{x}{2} = -\sqrt{2}$

(a) I is correct (b) II is correct
(c) III is correct (d) All are correct

237. If $\sin x = \frac{1}{4}$ and x is in quadrant II, then

I. $\sin\frac{x}{2} = \frac{\sqrt{15}}{2}$ II. $\cos\frac{x}{2} = \frac{\sqrt{8-2\sqrt{15}}}{4}$

III. $\tan\frac{x}{2} = 4 \pm \sqrt{15}$

(a) I is correct (b) II is correct
(c) III is correct (d) All are correct

238. Consider the following statements

Statement I If α and β are the solutions of the equation $a\tan\theta + b\sec\theta = c$, then $\tan(\alpha+\beta) = \frac{2ac}{a^2 - c^2}$.

Statement II $\tan(A+B) = \frac{\tan A + \tan B}{1 - \tan A \tan B}$

Choose the correct option.

(a) Statement I is true; Statement II is true; Statement II is a correct explanation for Statement I
(b) Statement I is true; Statement II is true; Statement II is not a correct explanation for Statement I
(c) Statement I is true; Statement II is false
(d) Statement I is false; Statement II is true

239. $2\sin^2\beta + 4\cos(\alpha+\beta)\cdot\sin\alpha\cdot\sin\beta + \cos 2(\alpha+\beta)$ is equal to [HOTS]

(a) $\sin 2\alpha$ (b) $\cos 2\alpha$
(c) $\tan 2\alpha$ (d) $\cot 2\alpha$

240. If angle θ is divided into two parts such that the tangent of one part is K times the tangent to other, and ϕ is their difference, then $\sin\theta$ is equal to [HOTS]

(a) $\frac{K+1}{K-1}\sin\frac{\theta}{2}$ (b) $\frac{K+1}{K-1}\sin\frac{\phi}{2}$
(c) $\frac{K+1}{K-1}\sin\phi$ (d) $\frac{K-1}{K+1}\sin\phi$

241. If $\sqrt{3}\cos\theta + \sin\theta = \sqrt{2}$, then θ is equal to

(a) $2m\pi + \frac{5\pi}{12}, m \in Z$
(b) $2m\pi - \frac{\pi}{12}, m \in Z$
(c) Both (a) and (b) are correct
(d) Both (a) and (b) are incorrect

242. Consider the following statements

Statement I The inequality

$$2^{\sin\theta} + 2^{\cos\theta} \geq 2^{1-\frac{1}{\sqrt{2}}}$$

holds true for all real values of θ.

Statement II Arithmetic mean of two numbers is greater than or equal to their geometric mean.

Choose the correct option. [HOTS]

(a) Statement I is true; Statement II is true; Statement II is a correct explanation for Statement I
(b) Statement I is true; Statement II is true; Statement II is not a correct explanation for Statement I
(c) Statement I is true; Statement II is false
(d) Statement I is false; Statement II is true

243. The value of $\cos\frac{\pi}{5}\cos\frac{2\pi}{5}\cos\frac{4\pi}{5}\cos\frac{8\pi}{5}$ is equal to

(a) $\frac{1}{16}$ (b) 0
(c) $\frac{-1}{8}$ (d) $\frac{-1}{16}$

244. If $m\sin\theta = n\sin(\theta + 2\alpha)$, then $\tan(\theta+\alpha)\cdot\cot\alpha$ is equal to [NCERT Exemplar]

(a) $\frac{m+n}{m-n}$ (b) $\frac{m-n}{m+n}$
(c) $\frac{m+n}{mn}$ (d) $\frac{m-n}{mn}$

245. If $\tan(A+B) = p$, $\tan(A-B) = q$, then $\tan 2A$ is equal to [NCERT Exemplar]

(a) $\frac{p-q}{1+pq}$ (b) $\frac{p+q}{1-2pq}$
(c) $\frac{p+q}{1-pq}$ (d) None of these

246. If $\frac{\sin(x+y)}{\sin(x-y)} = \frac{a+b}{a-b}$, then $\frac{\tan x}{\tan y}$ is equal to

(a) $\frac{b}{a}$ (b) ab
(c) $\frac{a}{b}$ (d) $(ab)^2$

247. If $\tan\theta = \frac{\sin\alpha - \cos\alpha}{\sin\alpha + \cos\alpha}$, then which among the following is true?

I. $\sin\alpha + \cos\alpha = \sqrt{2}$
II. $\sin\alpha + \cos\alpha = \sqrt{2}\sec\theta$
III. $\sin\alpha + \cos\alpha = \sqrt{2}\cos\theta$
IV. $\sin\alpha + \cos\alpha = \sqrt{2}\sin\theta$

(a) I is true (b) II is true
(c) III is true (d) IV is true

Answer *with* Explanations

1. *(c)* An angle is the measure of rotation of a given ray about its initial point.

2. *(b)* Here, A refers to terminal side. Also, it shows rotation in anti-clockwise direction.

3. *(a)* Here, B refers to terminal side. Also, it shows rotation in clockwise direction.

4. *(b)* The point of rotation is said to be vertex.

5. *(d)* In anti-clockwise rotation, the angle is said to be positive.

In clockwise rotation, the angle is said to be negative.

The measure of an angle is the amount of rotation performed to get the terminal side from the initial side.

So, all the statements are correct.

6. *(b)* Definition of an angle suggest a unit for measuring angles.

7. *(a)* In shown figure, the position of the initial side is indicating one complete revolution *i.e.*, 360° or 2π is the measure of this one revolution.

8. *(c)* The wheel is rotating by 15 revolutions per second. So, it is clear to say that this is an example of large angle.

9. *(a)* The most commonly used units for measurement of angles are degree measure and radian measure.

10. *(b)* A refers to $\left(\frac{1}{360}\right)^{th}$ and B refers to 1°.

11. *(d)* In degree measure, 1 degree is divided into 60 min and 1 min is divided into 60 s, so only option (d) is correct.

12. *(b)* One sixtieth of a degree is called a minute written as 1′, and one sixtieth of a minute is called a second, written as 1″.

Thus, $1° = 60'$

$1' = 60''$

13. *(d)* We know that, if the direction of rotation is anti-clockwise, the angle is said to be positive and if the direction of rotation is clockwise, then the angle is negative. So, here figures (i), (ii), (iii), (v), (vi), (vii) are having its measure correctly.

In figure (iv), the rotation of the angle is in clockwise direction, so its measure is – 90° and In figure (viii), the rotation of the angle is in clockwise direction so its measure is – 180°.

So, figures (i), (ii), (iii), (v), (vi), (vii) are correct and figures (iv) and (viii) are incorrect.

14. *(c)* In a unit circle,

We have, $r = 1$ unit

$\therefore$ Circumference of circle $= 2\pi r = 2\pi \times 1$

$= 2\pi$ unit

and Area of circle $= \pi r^2 = \pi \times 1^2$

$= \pi$ unit^2

$\therefore$ Only (i), (iii) and (iv) are true.

15. *(c)* Angle subtended at the centre by an arc of length 1 unit in a unit circle is said to have a measure of 1 radian.

16. *(d)* In the figure (i) to (iv), OA is the initial side and OB is the terminal side. The figures show the angles whose measures are 1 radian, -1 radian, $1\frac{1}{2}$ radian and $-1\frac{1}{2}$ radian.

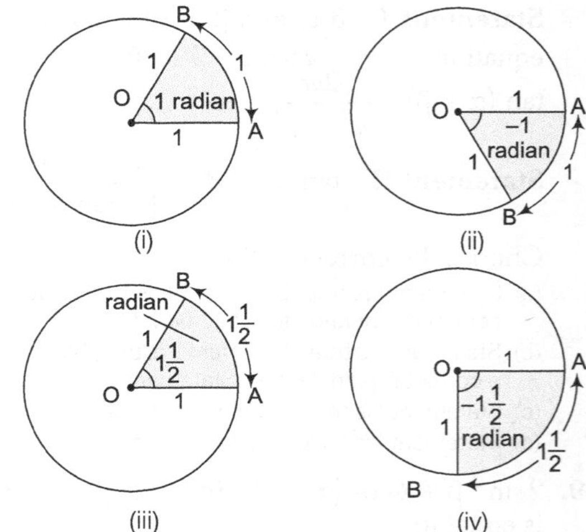

In figure (iv), the rotation is in clockwise direction, so its measurement should be $-1\frac{1}{2}$ radian.

17. *(b)* One complete revolution of the initial side subtends an angle of 360° *i.e.*, 2π radian (or $2\pi^c$).

18. *(a)* Equal arcs of a circle subtends equal angle at the centre.

19. *(b)* In a circle of radius r, an arc of length r will subtend an angle whose measure is 1 radian.

20. *(a)* An arc of length l will subtend an angle whose measure is $\frac{l}{r}$ radian. Thus, if in a circle of radius r, an arc of length l subtends an angle θ radian at the centre, we have,

$$\theta = \frac{l}{r} \quad \text{or} \quad l = r\theta \quad (\text{where } \theta \text{ is in radian})$$

21. *(d)* This problem is based on the relation between radian and real numbers which is given below.

Consider the unit circle with centre O. Let A be any point on the circle. Consider, OA as initial side of an angle. Then, the length of an arc of the circle will give the radian measure of the angle which the arc will subtend at the centre of the circle. Consider the line PAQ which is tangent to the circle at A.

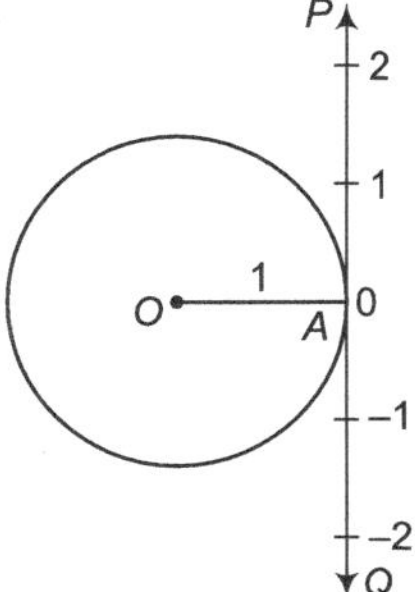

Let the point A represents the real number zero, AP represents positive real number and AQ represents negative real numbers. If we rope the line AP in the anti-clockwise direction along the circle and AQ in the clockwise direction, then every real number will correspond to a radian measure and conversely. Thus, radian measures and real numbers can be considered as one and the same. So, we can conclude that the length of an arc of the circle gives the radian measure of the angle subtended by the arc at the centre.

22. *(a)* $\because 2\pi$ radian $= 360°$

23. *(b)* π radian $= 180°$

24. *(a)* $\because$ 1 radian $= \dfrac{180°}{\pi} = 57°16'$ (approx)

$\therefore$ A refers to $\dfrac{180°}{\pi}$.

25. *(b)* $\because$ 1 degree $= \dfrac{\pi}{180}$ radian $= 0.01746$ radian (approx.)

(Here, we use $\pi = 22/7$ as its approx. value)

26. *(a)*

	Column I (Radian Measure)		Column II (Degree Measure)
A.	$\frac{\pi}{6}$	3.	30°
B.	$\frac{\pi}{4}$	1.	45°
C.	$\pi/3$	4.	60°
D.	$\pi/2$	2.	90°

27. *(b)*

	Column I (Radian Measure)		Column II (Degree Measure)
A.	π	4.	180°
B.	2π	3.	360°
C.	$3\pi/2$	2.	270°
D.	$5\pi/4$	1.	225°

28. *(b)* Since, angles are measured either in degrees or in radians, we follow the convention that whenever we write angle $\theta°$, we mean the angle whose degree measure is θ and whenever we write angle β, we mean the angle whose radian measure is β.

So, when an angle is expressed in radian, the word 'radian' is frequently omitted. Also, we know that

$$\text{Radian measure} = \frac{\pi}{180} \times \text{Degree measure}$$

$$\text{Degree measure} = \frac{180}{\pi} \times \text{Radian measure}$$

Hence, in the given statements only option (b) is incorrect.

29. *(b)* We know that, $180° = \pi$ radian

Hence, $40°20' = 40\frac{1}{3}$ degree

$= \dfrac{\pi}{180} \times \dfrac{121}{3}$ radian $= \dfrac{121\pi}{540}$ radian.

Therefore, $40°20' = \dfrac{121\pi}{540}$ radian.

30. *(b)* We know that, π radian $= 180°$

Hence, 6 radians $= \dfrac{180}{\pi} \times 6$ degree

$= \dfrac{1080 \times 7}{22}$ degree

$= 343\dfrac{7}{11}$ degree

$= 343° + \dfrac{7 \times 60}{11}$ min (as $1° = 60'$)

$= 343° + 38' + \dfrac{2}{11}$ min (as $1' = 60''$)

$= 343° + 38' + 10.9''$

$= 343°38'11''$ approximately.

Hence, 6 radians $= 343°\ 38'\ 11''$ approximately.

31. *(c)* Here, $l = 37.4$ cm and $\theta = 60° = \dfrac{60\pi}{180}$ radian $= \dfrac{\pi}{3}$

Hence, by $r = \dfrac{l}{\theta}$, we have

$$r = \frac{37.4 \times 3}{\pi} = \frac{37.4 \times 3 \times 7}{22} = 35.7 \text{ cm}$$

32. *(d)* In 60 min, the minute hand of watch completes one revolution. Therefore, in 40 min the minute hand turns through $\frac{2}{3}$ of a revolution. Therefore, $\theta = \frac{2}{3} \times 360°$ or $\frac{4\pi}{3}$ radian. Hence, the required distance travelled is given by

$l = r\,\theta = 1.5 \times \dfrac{4\pi}{3}$ cm $= 2\pi$ cm $= 2 \times 3.14$ cm $= 6.28$ cm.

33. *(a)* Let r_1 and r_2 be the radii of the two circles. Given that,

$$\theta_1 = 65° = \frac{\pi}{180} \times 65 = \frac{13\pi}{36} \text{ radian}$$

and $\theta_2 = 110° = \dfrac{\pi}{180} \times 110 = \dfrac{22\pi}{36}$ radian

Let l be the length of each of the arc. Then, $l = r_1\theta_1 = r_2\theta_2$, which gives

$$\frac{13\pi}{36} \times r_1 = \frac{22\pi}{36} \times r_2, \text{ i.e., } \frac{r_1}{r_2} = \frac{22}{13}$$

Hence, $r_1 : r_2 = 22 : 13$.

34. *(b)* We know that,

Radian measure $= \frac{\pi}{180} \times$ Degree measure

A. Radian measure of $25° = \frac{\pi}{180} \times 25° = \frac{5\pi}{36}$

B. We know that, $30' = \left(\frac{1}{2}\right)^{\circ}$ $\quad (\because 60' = 1°)$

$\therefore -47°30' = -\left(47\frac{1}{2}\right)^{\circ} = \left(\frac{-95}{2}\right)^{\circ}$

$\therefore$ Radian measure of $(-47°30') = \frac{\pi}{180} \times \left(\frac{-95}{2}\right)$

$= \frac{-19\pi}{72}$

C. Radian measure of $240° = \frac{\pi}{180} \times 240 = \frac{4\pi}{3}$

D. Radian measure of $520° = \frac{\pi}{180} \times 520 = \frac{26\pi}{9}$

So, A $\to$ 4, B $\to$ 3, C $\to$ 2, D $\to$ 1

$\therefore$ Correct answer is (b).

35. *(c)* We know that,

Degree measure $= \frac{180}{\pi} \times$ Radian measure

A. Degree measure of $\frac{11}{16} = \left(\frac{180}{\pi} \times \frac{11}{16}\right)^{\circ}$

$= \left(\frac{180}{22} \times \frac{11}{16} \times 7\right)^{\circ}$ $\quad \left(\because \pi = \frac{22}{7}\right)$

$= \left(\frac{90 \times 7}{16}\right)^{\circ} = \left(\frac{315}{8}\right)^{\circ}$

$= \left(39\frac{3}{8}\right)^{\circ} = 39° \left(\frac{3}{8} \times 60\right)'$ $\quad (\because 1° = 60')$

$= 39° \left(22\frac{1}{2}\right)'$

$= 39°22' \left(\frac{1}{2} \times 60\right)''$ $\quad (\because 1' = 60'')$

$= 39°22'30''$

B. Degree measure of $-4 = \left(\frac{180}{\pi} \times -4\right)^{\circ}$

$= \left(\frac{180}{22} \times -4 \times 7\right)^{\circ}$ $\quad \left(\because \pi = \frac{22}{7}\right)$

$= \left(\frac{90 \times (-28)}{11}\right)^{\circ}$

$= -\left(\frac{2520}{11}\right)^{\circ} = -\left(229\frac{1}{11}\right)^{\circ}$

$= -229° \left(\frac{1}{11} \times 60'\right)$ $\quad (\because 1° = 60')$

$= -229° \left(5\frac{5}{11}\right)'$

$= -229°\, 5' \left(\frac{5}{11} \times 60\right)''$ $\quad (\because 1' = 60'')$

$= -229°\, 5'\, 27.3''$

$\approx -229°\, 5'\, 27''$ (approx)

C. Degree measure of $\frac{5\pi}{3} = \left(\frac{180}{\pi} \times \frac{5\pi}{3}\right)^{\circ} = 300°$

D. Degree measure of $\frac{7\pi}{6} = \left(\frac{180}{\pi} \times \frac{7\pi}{6}\right)^{\circ} = 210°$

Thus, we see that correct option is (c).

$\Rightarrow$ A $\leftrightarrow$ 3, B $\leftrightarrow$ 4, C $\leftrightarrow$ 1, D $\leftrightarrow$ 2

36. *(b)* In one minute, wheel turns by $(360 \times 2\pi)$ radian

$(\because$ 1 revolution $= 2\pi$ radian$)$

$\therefore$ In 1 s, wheel turns by $\left(\frac{360 \times 2\pi}{60}\right)$ radian.

or, In 1 s, wheel turns by 12π radian.

37. *(a)* Here, radius $(r) = 100$ cm and length of arc $(l) = 22$ cm

$\because$ Angle subtended at the centre $= \frac{\text{arc } (l)}{\text{radius } (r)}$

$= \frac{22}{100}$ radian

$= \left(\frac{22}{100} \times \frac{180}{\pi}\right)^{\circ}$

$\left(\because \text{Degree measure} = \frac{180}{\pi} \times \text{Radian measure}\right)$

$= \frac{22 \times 18 \times 7}{10 \times 22} = 12° \left(\frac{6}{10} \times 60°\right)'$

$= 12°\, 36'$

38. *(d)* Since, diameter of the circle = 40 cm

$\therefore$ Radius of the circle $(r) = 20$ cm

and length of the chord $(l) = 20$ cm

From the shown figure, it is clear that $\angle AOB = 60° = \frac{\pi}{3}$ radian

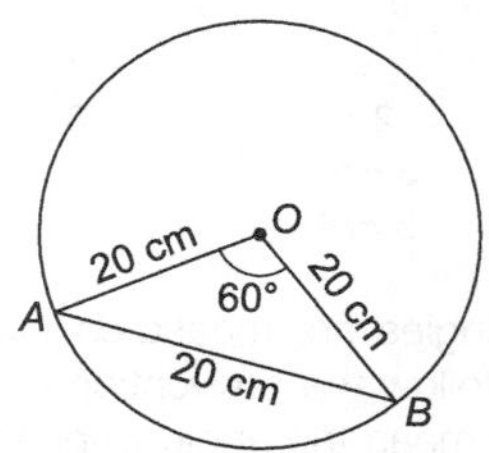

$\therefore$ Length of minor arc of the chord (l) = radius of circle (r) $\times$ angle subtended in radian by the minor arc (θ)

$= 20 \times \frac{\pi}{3} = \frac{20\pi}{3}$ cm

39. *(c)* Since, length of arcs are equal,

$\therefore \quad l_1 = l_2 = l$ (say)

Let the length of radius of two circles be r_1 and r_2, respectively.

Also, given that $\theta_1 = 60° \times \frac{\pi}{180} = \frac{\pi}{3}$ radian

and $\theta_2 = 75° \times \frac{\pi}{180} = \frac{5\pi}{12}$ radian.

We know that,

$$l = r\theta \Rightarrow r = \frac{l}{\theta}$$

$\therefore \quad r_1 = \frac{l_1}{\theta_1} = \frac{l}{\frac{\pi}{3}} = \frac{3l}{\pi}$

and $r_2 = \frac{l_2}{\theta_2} = \frac{l}{\frac{5\pi}{12}} = \frac{12l}{5\pi}$

So, $\frac{r_1}{r_2} = \frac{3l/\pi}{12l/5\pi} = \frac{15}{12} = \frac{5}{4}$

40. *(d)* We know that, θ (in radian) $= \frac{l}{r}$

(i) $\theta = \frac{10}{75} = \frac{2}{15}$ rad (ii) $\theta = \frac{15}{75} = \frac{1}{5}$ rad (iii) $\theta = \frac{21}{75} = \frac{7}{25}$ rad

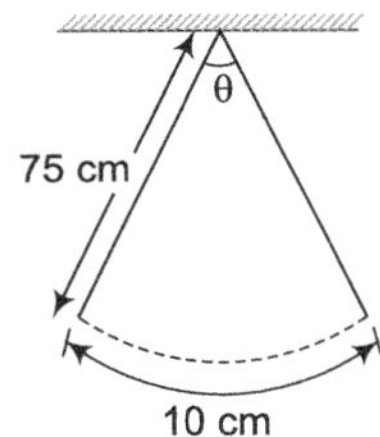

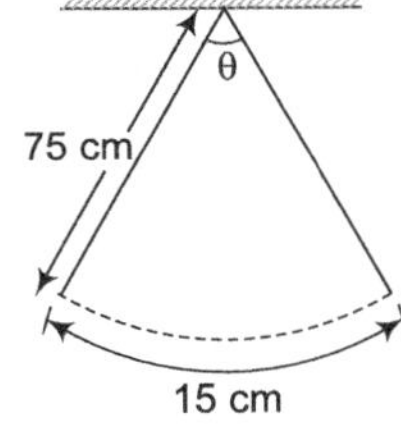

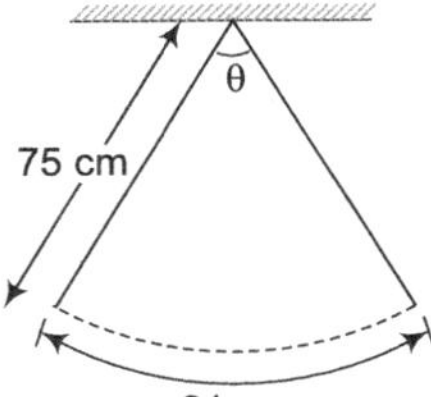

Hence, option (d) is correct.

41. *(d)* The number of digits in a clock is 12.

So, the angle between two consecutive digits in a clock is $= \frac{360°}{12} = 30°$ or $\frac{\pi}{6}$ radian.

The minute hand rotates through 360° in 60 min.

So, in 1 min it rotates through $\frac{360°}{60} = 6°$.

The hour hand rotates through 360° in 12 h. So, in one hour it rotates through $\frac{360°}{12} = 30°$ and in one minute it rotates through $\frac{30°}{60} = \left(\frac{1}{2}\right)^\circ$. The second hand rotates through 360° in a minute.

Thus, we see that option (d) is incorrect.

42. *(d)* Usually we follow two types of conventions for measuring angles *i.e.*, (i) Sexagesimal system (ii) Circular system.

In sexagesimal system, the unit of measurement is degree.

In circular system of measurement, the unit of measurement is radian.

The circumference of a circle always bears a constant ratio to its diameter. This constant ratio is a number denoted by π which is taken approximately as $\frac{22}{7}$ for all practical purpose.

Hence, option (d) is correct.

43. *(d)* Statement II is true.

$\therefore$ Degree measure of (– 2) radian

$$= \frac{180}{\pi} \times -2$$

$$= \frac{180}{22} \times -2 \times 7 \quad \left(\because \pi = \frac{22}{7}\right)$$

$$= \left(-\frac{1260}{11}\right)^\circ = -\left(114\frac{6}{11}\right)^\circ$$

$$= -114° \left(\frac{6}{11} \times 60\right)'$$

$$= -114° \, 32' \left(\frac{8}{11} \times 60\right)''$$

$$= -114° \, 32' \, 43.6''$$

$$= -114° \, 32' \, 44'' \text{ (approx.)}$$

So, Statement I is false.

So, correct option is (d).

44. *(a)* A. $37°30' = \left(37\frac{1}{2}\right)^\circ = \left(\frac{75°}{2}\right)^\circ$

$$= \left(\frac{75}{2} \times \frac{\pi}{180}\right)^c = \left(\frac{5\pi}{24}\right)^c$$

B. $37'30'' = \left(37\frac{30}{60}\right)' = \left(37\frac{1}{2}\right)' = \left(\frac{75}{2}\right)'$

$$= \left(\frac{75}{2} \times \frac{1}{60}\right)^\circ = \left(\frac{5}{8}\right)^\circ$$

So, $5° \, 37' \, 30'' = \left(5\frac{5}{8}\right)^\circ = \left(\frac{45}{8}\right)^\circ$

$$= \left(\frac{45}{8} \times \frac{\pi}{180}\right)^c = \left(\frac{\pi}{32}\right)^c$$

C. $\because 20' = \left(\frac{20}{60}\right)^\circ = \frac{1}{3}^\circ$

$\therefore \quad 40°20' = \left(40\frac{1}{3}\right)^\circ = \left(\frac{121}{3}\right)^\circ$

$$= \left(\frac{121}{3} \times \frac{\pi}{180}\right)^c = \left(\frac{121\pi}{540}\right)^c$$

D. $340° = \left(340 \times \frac{\pi}{180}\right)^c = \left(\frac{17\pi}{9}\right)^c$

$\therefore$ A $\leftrightarrow$ 3, B $\leftrightarrow$ 4, C $\leftrightarrow$ 2, D $\leftrightarrow$ 1

45. *(a)* $504° = 504 \times \frac{\pi}{180} = \frac{56\pi}{20}$

$= \left(\frac{14\pi}{5}\right)^c$

Thus, both A and R are correct and R is the correct explanation of A.

46. *(c)* A. $\left(\frac{4\pi}{15}\right)^c = \left(\frac{4\pi}{15} \times \frac{180}{\pi}\right)^\circ = 48°$

B. $\left(\frac{1}{4}\right)^c = \left(\frac{1}{4} \times \frac{180}{\pi}\right)^\circ = \left(\frac{1}{4} \times \frac{180}{22} \times 7\right)^\circ$

$= \left(\frac{315}{22}\right)^\circ = \left(14\frac{7}{22}\right)^\circ$

$= 14° \left(\frac{7}{22} \times 60\right)' = 14° \left(\frac{210}{11}\right)'$

$= 14° \left(19\frac{1}{11}\right)' = 14° 19' \left(\frac{1}{11} \times 60\right)''$

$= 14° 19' 6''$ (approx.)

C. $\left(\frac{\pi}{8}\right)^c = \left(\frac{\pi}{8} \times \frac{180}{\pi}\right)^\circ = \left(22\frac{1}{2}\right)^\circ$

$= 22° \left(\frac{1}{2} \times 60\right)'$

$= 22° 30'$

D. $\left(\frac{11}{16}\right)^c = \left(\frac{11}{16} \times \frac{180}{\pi}\right)^\circ$

$= \left(\frac{11}{16} \times \frac{180}{22} \times 7\right)^\circ$

$= \left(\frac{45 \times 7}{8}\right)^\circ = \left(\frac{315}{8}\right)^\circ = \left(39\frac{3}{8}\right)^\circ$

$= 39° \left(\frac{3}{8} \times 60\right)' = 39° \left(\frac{45}{2}\right)'$

$= 39° 22' \left(\frac{1}{2} \times 60\right)''$

$= 39° 22' 30''$

Hence, $A \leftrightarrow 2, B \leftrightarrow 1, C \leftrightarrow 3, D \leftrightarrow 4.$

47. *(c)* Let the angles of a triangle be

$(a - d)°, a°$ and $(a + d)°$.

So, $a - d + a + a + d = 180°$

$\Rightarrow a = 60°$

$\therefore$ Angles in degree are $(60 - d)°, 60°, (60 + d)°$

$\Rightarrow (60 - d)°$ is the least angle and $(60 + d)°$ is the greatest angle.

Now, the greatest angle in radian $= \left[(60 + d) \times \frac{\pi}{180}\right]^c$

Thus, we get

$$\frac{(60 - d)}{(60 + d)\frac{\pi}{180}} = \frac{60}{\pi}$$

$\Rightarrow 60\pi - d\pi = 60 \times 60 \times \frac{\pi}{180} + 60 \times d \times \frac{\pi}{180}$

$\Rightarrow 60\pi - d\pi = 20\pi + \frac{d\pi}{3}$

$\Rightarrow d\left(\frac{\pi}{3} + \pi\right) = 40\pi$

$\Rightarrow d = 40\pi \times \frac{3}{4\pi} = 30°$

Hence, the angles are $(60 - 30)°, 60°, (60 + 30)°$

i.e., $30°, 60°, 90°$ or $\frac{\pi^c}{6}, \frac{\pi^c}{3}, \frac{\pi^c}{2}$.

48. *(c)* We know that, in a regular polygon of n side, the angle is equal to $\left(\frac{2n - 4}{n}\right) \times$ right angle.

$\therefore$ The angular measurement of a regular decagon in radian is equal to $\left(\frac{2 \times 10 - 4}{10}\right) \times \frac{\pi}{2} = \left(\frac{4\pi}{5}\right)^c$

So, A is correct but R is incorrect.

49. *(b)* The angle θ, in radian, subtended by the arc at the centre of the circle is given by

$$\theta = \frac{\text{arc}}{\text{radius}} = \frac{\text{circumference of circular wire}}{\text{radius of circular hoop}}$$

$$\theta = \frac{2\pi \times 3}{48} = \frac{\pi}{8} \text{ radian}$$

$$= \left(\frac{\pi}{8} \times \frac{180}{\pi}\right)^\circ = \left(22\frac{1}{2}\right)^\circ$$

50. *(c)* Let AB be the diameter of the moon and let O be the eye of the observer. Since, the distance between the earth and the moon is quite large, so we take diameter AB as arc AB. Let d be the diameter of the moon. Then, d = arc AB.

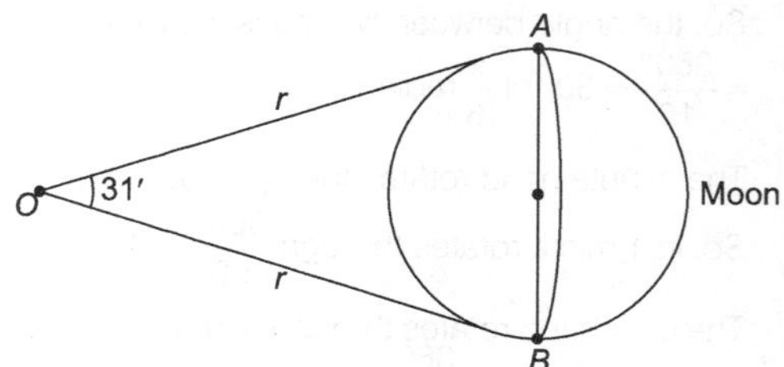

Now, $\theta = 31' = \left(\frac{31}{60}\right)^\circ = \left(\frac{31}{60} \times \frac{\pi}{180}\right)^c$

and $r = 360000$ km

$\therefore \theta = \frac{\text{arc}}{\text{radius}} = \frac{\text{diameter of the moon}}{\text{moon's distance from the earth}}$

$\Rightarrow \frac{31}{60} \times \frac{\pi}{180} = \frac{d}{360000}$

$$\Rightarrow \quad d = \left(\frac{31}{60} \times \frac{\pi}{180} \times 360000\right) \text{km}$$

$$\Rightarrow \quad d = 3247.62 \text{ km}$$

Hence, the diameter of the moon is 3247.62 km.

Solutions (Q. Nos. 51-53)

51. *(d)* Consider a unit circle with centre at origin of the coordinate axes. Let $P(a, b)$ be any point on the circle with $\angle AOP = x$ radian, *i.e.*, length of arc $AP = x$ (see figure)

We define $\cos x = a$ and $\sin x = b$. Since, ΔOMP is a right angled triangle, we have

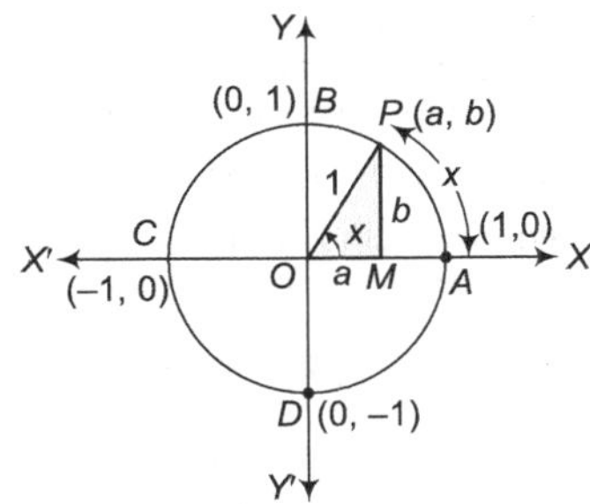

$$OM^2 + MP^2 = OP^2 \quad \text{or} \quad a^2 + b^2 = 1$$

Thus, for every point on the unit circle, we have

$$a^2 + b^2 = 1 \quad \text{or} \quad \cos^2 x + \sin^2 x = 1$$

Since, one complete revolution subtends an angle of 2π radian at the centre of the circle $\angle AOB = \frac{\pi}{2}$, $\angle AOC = \pi$ and $\angle AOD = \frac{3\pi}{2}$. All angles which are integral multiples of $\frac{\pi}{2}$ are called quadrantal angles. The coordinates of the points A, B, C and D are respectively $(1, 0), (-1, 0)$ and $(0, -1)$. Therefore, for quadrantal angles, we have

$$\cos 0° = 1, \sin 0° = 0,$$
$$\cos \frac{\pi}{2} = 0, \ \sin \frac{\pi}{2} = 1$$
$$\cos \pi = -1, \ \sin \pi = 0$$
$$\cos \frac{3\pi}{2} = 0, \ \sin \frac{3\pi}{2} = -1$$
$$\cos 2\pi = 1, \ \sin 2\pi = 0$$

Hence, all the statements are true.

52. *(b)* Now, if we take one complete revolution from the point P, we again come back to same point P. Thus, we also observe that if x increases (or decreases) by any integral multiple of 2π, the values of sine and cosine functions do not change. Thus,

$$\sin(2n\pi + x) = \sin x, n \in Z, \cos(2n\pi + x) = \cos x, n \in Z$$

Further, $\sin x = 0$, if $x = 0, \pm 2\pi, \pm 3\pi, \ldots$, *i.e.*, when x is an integral multiple of π and $\cos x = 0$, if $x = \pm \frac{\pi}{2}, \pm \frac{3\pi}{2}, \pm \frac{3\pi}{2}, \ldots$ *i.e.*, $\cos x$ vanishes when x is an odd multiple of $\frac{\pi}{2}$. Thus,

$\sin x = 0 \Rightarrow x = n\pi$, where n is any integer.

53. *(c)*

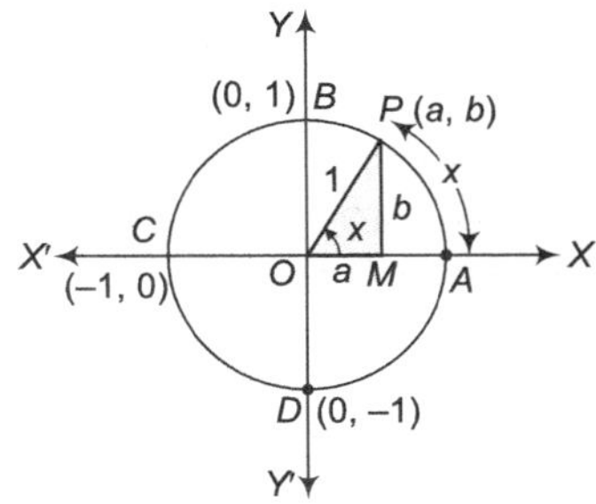

$\cos x = 0$ implies $x = (2n + 1)\frac{\pi}{2}$, where n is any integer.

54. *(a)* We can define other trigonometric functions in terms of sine and cosine functions.

$$\text{cosec } x = \frac{1}{\sin x}, x \neq n\pi, \text{ where } n \text{ is any integer.}$$

$$\sec x = \frac{1}{\cos x}, x \neq (2n+1)\frac{\pi}{2}, \text{ where } n \text{ is any integer.}$$

$$\tan x = \frac{\sin x}{\cos x}, x \neq (2n+1)\frac{\pi}{2}, \text{ where } n \text{ is any integer.}$$

$$\cot x = \frac{\cos x}{\sin x}, x \neq n\pi, \text{ where } n \text{ is any integer.}$$

55. *(d)* Refering to the solution of question 61, we have shown that for all real x,

$$\sin^2 x + \cos^2 x = 1 \quad \ldots(i)$$

It follows that

$$1 + \tan^2 x = \sec^2 x \quad \ldots(ii)$$

[on dividing Eq. (i) with $\cos^2 x$]

and $$1 + \cot^2 x = \text{cosec}^2 x \quad \ldots(iii)$$

[on dividing Eq. (i) with $\sin^2 x$]

Hence, only option (d) is correct.

56. *(c)* In earlier classes, we have discussed the values of trigonometric ratios for 0°, 30°, 45°, 60° and 90°. The values of trigonometric functions for these angles are same as that of trigonometric ratios studied in earlier classes. Thus, we have the following table.

	0	$\frac{\pi}{6}$	$\frac{\pi}{4}$	$\frac{\pi}{3}$	$\frac{\pi}{2}$	π	$\frac{3\pi}{2}$	2π
sin	0	$\frac{1}{2}$	$\frac{1}{\sqrt{2}}$	$\frac{\sqrt{3}}{2}$	1	0	-1	0
cos	1	$\frac{\sqrt{3}}{2}$	$\frac{1}{\sqrt{2}}$	$\frac{1}{2}$	0	-1	0	1
tan	0	$\frac{1}{\sqrt{3}}$	1	$\sqrt{3}$	not defined	0	not defined	0

Hence, option (c) is correct.

Note *The values of cosec x, sec x and cot x are the reciprocal of the values of sin x, cos x and tan x, respectively.*

57. *(d)* Given, $P(a, b)$ be a point on the unit circle with centre at the origin such that $\angle AOP = x$. If $\angle AOQ = -x$, then the coordinates of the point Q will be $(a, -b)$. Therefore,

$$\cos(-x) = \cos x = a$$

and $$\sin(-x) = -\sin x = -b$$

Hence, all the conclusions are true.

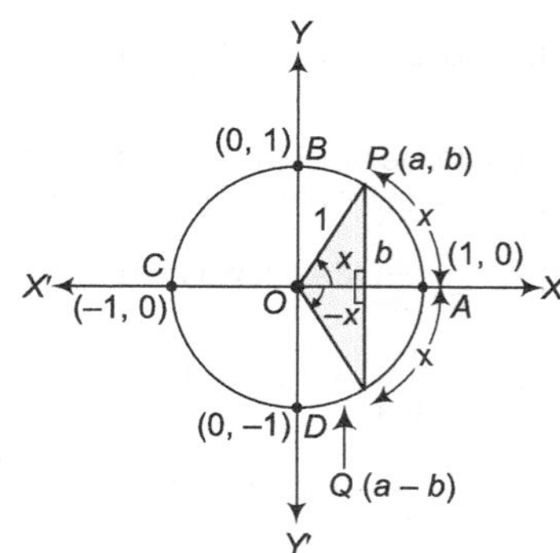

58. *(d)* Since, for every point $P(a, b)$ on the unit circle, $-1 \le a \le 1$ and $-1 \le b \le 1$, we have $-1 \le \cos x \le 1$ and $-1 \le \sin x \le 1$ for all x. We have learnt in previous classes that in the first quadrant $\left(0 < x < \frac{\pi}{2}\right)$ a and b both are positive, in the second quadrant $\left(\frac{\pi}{2} < x < \pi\right)$ a is negative and b is positive, in the third quadrant $\left(\pi < x < \frac{3\pi}{2}\right)$ a and b are both negative and in the fourth quadrant $\left(\frac{3\pi}{2} < x < 2\pi\right)$ a is positive and b is negative. Therefore, $\sin x$ is positive for $0 < x < \pi$ and negative for $\pi < x < 2\pi$. Similarly, $\cos x$ is positive for $0 < x < \frac{\pi}{2}$, negative for $\frac{\pi}{2} < x < \frac{3\pi}{2}$ and also positive for $\frac{3\pi}{2} < x < 2\pi$.

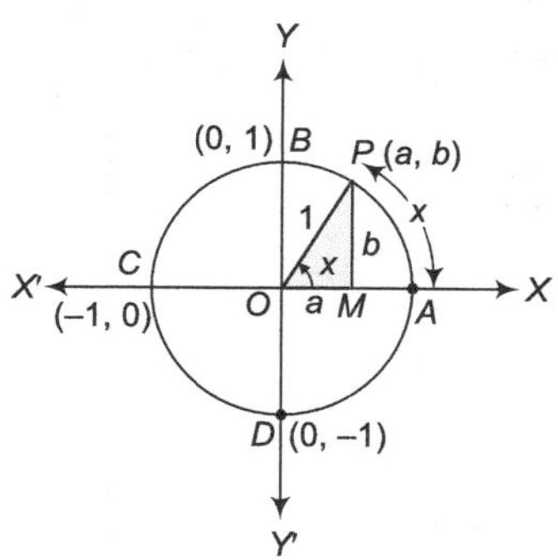

So, $\cos x$ is negative for $\pi/2 < x < \frac{3\pi}{2}$.

59. *(a)* The signs of trigonometric functions in different quadrants are shown below.

	I	II	IV	IV
sin x	+	+	–	–
cos x	+	–	–	+
tan x	+	–	+	–
cosec x	+	+	–	–
sec x	+	–	–	+
cot x	+	–	+	–

According to the above table, option (a) is correct.

60. *(a)* From the definition of sine and cosine functions, we observe that they are defined for all real numbers, further we observe that for each real number x,

$$-1 \le \sin x \le 1 \text{ and } -1 \le \cos x \le 1 \text{ and } y = \cos x$$

The domain of $y = \sin x$ and $y = \cos x$ is the set of all real numbers and range is the interval $[-1, 1]$ *i.e.*, $-1 \le y \le 1$.

Since, $\operatorname{cosec} x = \frac{1}{\sin x}$, the domain of $y = \operatorname{cosec} x$ is the set $\{x : x \in R \text{ and } x \ne n\pi, n \in Z\}$ and range is the set $\{y : y \in R, y \ge 1 \text{ or } y \le -1\}$. Similarly, the domain of $y = \sec x$ is the set $\{x : x \in R \text{ and } x \ne (2n+1)\frac{\pi}{2}, n \in Z\}$ and range is the set $\{y : y \in R, y \le -1 \text{ or } y \ge 1\}$. The domain of $y = \tan x$ is the set $\{x : x \in R \text{ and } x \ne (2n+1)\frac{\pi}{2}, n \in Z\}$ and range is the set of all real numbers. The domain of $y = \cot x$ is the set $\{x : x \in R \text{ and } x \ne n\pi, n \in Z\}$ and the range is the set of all real numbers.

Thus, we get the following table for domain and range of trigonometric functions.

Functions	Domain	Range
sine	R	$[-1, 1]$
cosine	R	$[-1, 1]$
tan	$R - \{(2n+1)\frac{\pi}{2} ; n \in Z\}$	R
cot	$R - \{n\pi : n \in Z\}$	R
sec	$R - \{(2n+1)\frac{\pi}{2} : n \in Z\}$	$R - (-1, 1)$
cosec	$R - \{n\pi : n \in Z\}$	$R - (-1, 1)$

Solutions (Q. Nos. 61-62)

From the figure it is clear that in the first quadrant, as x increases from 0 to $\frac{\pi}{2}$, $\sin x$ increases from 0 to 1.

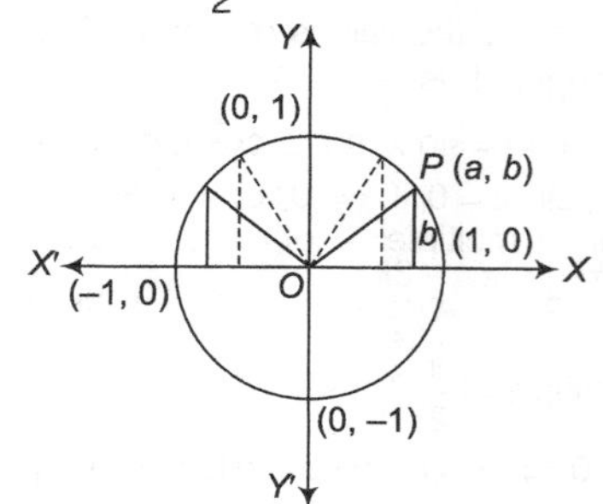

In the second quadrant, as x increases from $\frac{\pi}{2}$ to π, sin x decreases from 1 to 0. In the third quadrant, as x increases from π to $\frac{3\pi}{2}$, sin x decreases from 0 to – 1and finally, in the fourth quadrant, sin x increases from – 1to 0 as x increases from $\frac{3\pi}{2}$ to 2π. Similarly, we get the behaviour of other trigonometric functions and get the following table.

	I quadrant	II quadrant	III quadrant	IV quadrant
sin	increases from 0 to 1	decreases from 1 to 0	decreases from 0 to –1	increases from –1 to 0
cos	decreases from 1 to 0	decreases from 0 to –1	increases from –1 to 0	increases from 0 to 1
tan	increases from 0 to ∞	increase from –∞ to 0	increases from 0 to ∞	increases from –∞ to 0
cot	decreases from ∞ to 0	decreases from 0 to –∞	decreases from ∞ to 0	decreases from 0 to –∞
sec	increases from 1 to ∞	increases from –∞ to –1	decreases from –1 to –∞	decreases from ∞ to 1
cosec	decreases from ∞ to 1	increases from 1 to ∞	increases from –∞ to –1	decreases from –1 to –∞

Remark In the above table, the statement tan x increases from 0 to – ∞ (infinity) for $0 < x < \frac{\pi}{2}$ simply means that tan x increases as x increases for $0 < x < \frac{\pi}{2}$ and assumes arbitrarily large positive values as x approaches to $\frac{\pi}{2}$. Similarly, to say that cosec x decreases from – 1 to – ∞ (minus infinity) in the fourth quadrant means that cosec x decreases for $x \in \left(\frac{3\pi}{2}, 2\pi\right)$ and assumes arbitrarily large negative values as x approaches to 2π. The symbols ∞ and – ∞ simply specify certain types of behaviour of functions and variables.

61. *(d)* Thus, we see that all the given statements are true.

62. *(a)* Only option (a) is incorrect.

63. *(d)* We know that domain of sin x is R and range of sin x is $[-1, 1]$. Also, we have seen that the value of sin x repeats after an interval of 2π. Using this knowledge we get the graph of sin x as given below.

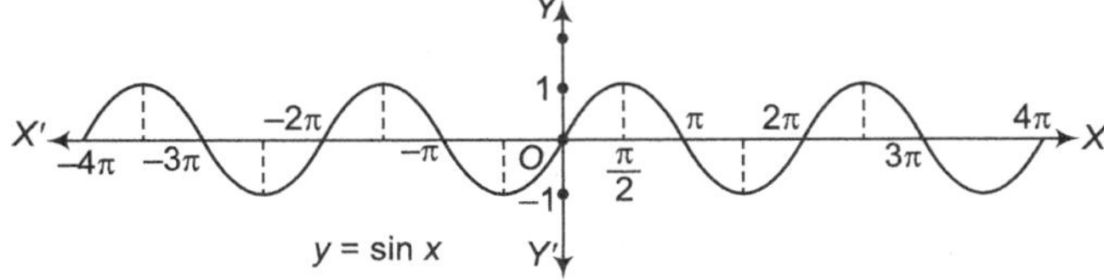

Thus, It is clear to say that sin x decreases in second and third quadrant.

64. *(d)* We know that the domain of cos x is real number.

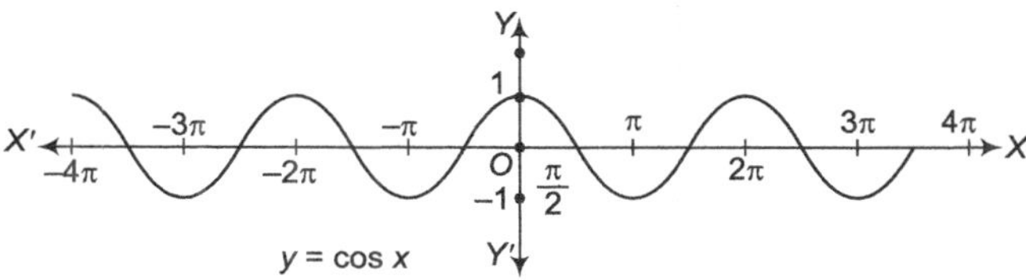

Range of trigonometric function cos x is $[-1, 1]$. Also, using behaviour of cos x, we get the graph of cos x as give of cos x repeats after an interval of 2π *i.e.*, $\cos(2x\pi + x) = \cos x, n \in Z$.

65. *(c)* We have, seen that domain and range of tan x is $R - \left\{(2n+1)\frac{\pi}{2} : n \in Z\right\}$ and R respectively. Also, the following the behaviour of trigonometric function tan x, we get the graph as shown below

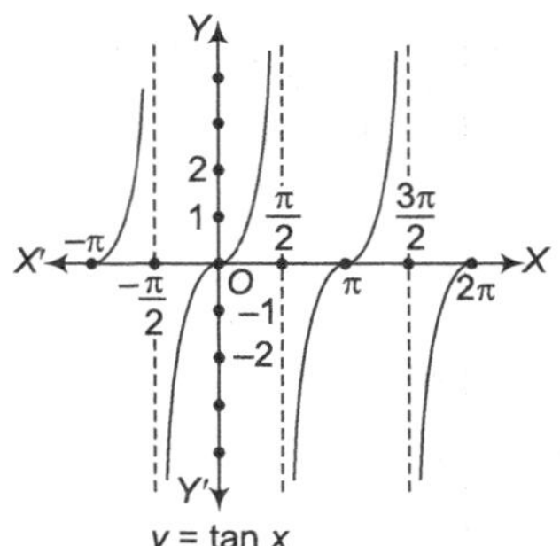

Now, it can be understood that the graph of tan x repeats after an interval of π *i.e.*, $\tan(\pi + x) = \tan x$.

66. *(d)* Using behaviour of trigonometric functions we can draw the graphs of $y = \cot x$, $y = \text{cosec } x$ and $y = \sec x$ as shown below.

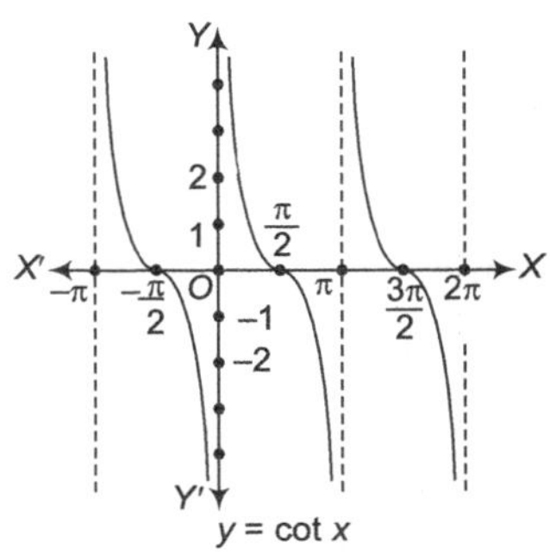

So, we see that the values of cot x repeat after an interval of π.

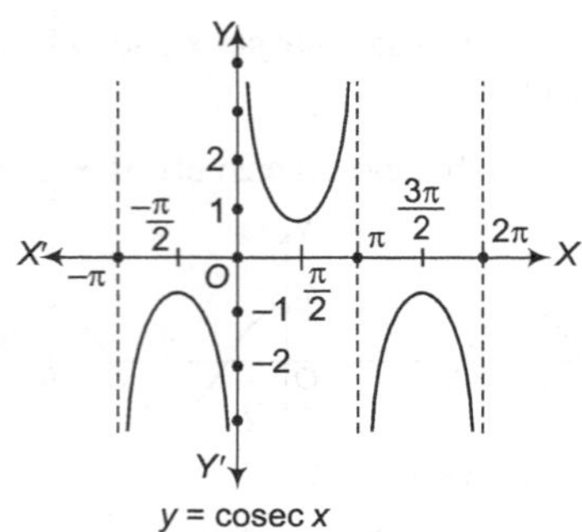

$y = \text{cosec } x$

Also, we can see that the values of cosec x repeat after an interval of 2π by using above graph of $y = \text{cosec } x$. Similarly, we can say that the values of sec x repeat after an interval of 2π by using the graph of $y = \sec x$ as shown below.

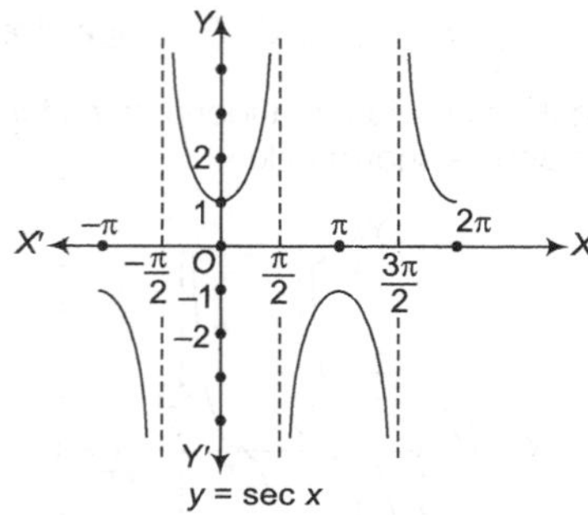

$y = \sec x$

Hence, it is concluded that all the given statements are true.

67. *(c)* Since, $\cos x = -\frac{3}{5}$, we have $\sec x = \frac{-5}{3}$

Now, $\sin^2 x + \cos^2 x = 1$, *i.e.*, $\sin^2 x = 1 - \cos^2 x$

$$\Rightarrow \quad \sin^2 x = 1 - \frac{9}{25} = \frac{16}{25}$$

Hence, $\sin x = \pm \frac{4}{5}$

Since, x lies in third quadrant, $\sin x$ is negative.

Therefore,

$$\sin x = -\frac{4}{5}$$

which also gives

$$\text{cosec } x = -\frac{5}{4}$$

Further, we have

$$\tan x = \frac{\sin x}{\cos x} = \frac{4}{3}$$

and $\cot x = \frac{\cos x}{\sin x} = \frac{3}{4}$

68. *(d)* Since, $\cot x = \frac{-5}{12}$, we have $\tan x = -\frac{12}{5}$

Now, $\sec^2 x = 1 + \tan^2 x$

$$= 1 + \frac{144}{25} = \frac{169}{25}$$

Hence, $\sec x = \pm \frac{13}{5}$

Since, x lies in second quadrant, $\sec x$ will be negative.

Therefore, $\sec x = -\frac{13}{5}$,

which also gives $\cos x = -\frac{5}{13}$

Further, we have

$$\sin x = \tan x \cos x = \left(-\frac{12}{5}\right) \times \left(-\frac{5}{13}\right) = \frac{12}{13}$$

and $\text{cosec } x = \frac{1}{\sin x} = \frac{13}{12}$

69. *(a)* We know that the values of $\sin x$ repeats after an interval of 2π. Therefore,

$$\sin \frac{31\pi}{3} = \sin\left(10\pi + \frac{\pi}{3}\right) = \sin \frac{\pi}{3} = \frac{\sqrt{3}}{2}$$

70. *(d)* We know that the values of $\cos x$ repeats after an interval of 2π or 360°.

Therefore,

$$\cos(-1710°) = \cos(-1710° + 5 \times 360°)$$
$$= \cos(-1710° + 1800°) = \cos 90° = 0$$

71. *(c)* $\cos x = -\frac{1}{2}$

Given that x lies in third quadrant.

i.e., $\pi < x < \frac{3\pi}{2}$

We have, $\sin^2 x + \cos^2 x = 1 \Rightarrow \sin^2 x = 1 - \cos^2 x$

$$\Rightarrow \quad \sin^2 x = 1 - \left(-\frac{1}{2}\right)^2$$

$$\Rightarrow \quad \sin^2 x = 1 - \frac{1}{4} = \frac{3}{4}$$

$$\Rightarrow \quad \sin x = \pm \frac{\sqrt{3}}{2}$$

$\because$ In third quadrant $\sin x$ is negative, so we will leave its positive value.

i.e., $\sin x = -\frac{\sqrt{3}}{2}$

Now, $\tan x = \frac{\sin x}{\cos x} = \frac{-\frac{\sqrt{3}}{2}}{-\frac{1}{2}} = \sqrt{3}$,

$$\cot x = \frac{1}{\tan x} = \frac{1}{\sqrt{3}}$$

and $\sec x = \frac{1}{\cos x} = -2$,

$$\text{cosec } x = \frac{1}{\sin x} = -\frac{2}{\sqrt{3}}$$

72. *(a)* $\sin x = \frac{3}{5}$

Given that x lies in second quadrant.

i.e., $\frac{\pi}{2} < x < \pi$

$\because \quad \sin^2 x + \cos^2 x = 1$

$\Rightarrow \quad \cos^2 x = 1 - \sin^2 x = 1 - \left(\frac{3}{5}\right)^2$

$= 1 - \frac{9}{25} = \frac{25-9}{25} = \frac{16}{25}$

$\Rightarrow \quad \cos x = \pm \frac{4}{5}$

$\because$ In second quadrant, $\cos x$ is negative, so we will leave its positive value.

i.e., $\cos x = -\frac{4}{5}$

$\Rightarrow \quad \tan x = \frac{\sin x}{\cos x}$

$= \frac{\frac{3}{5}}{-\frac{4}{5}} = -\frac{3}{4},$

$\Rightarrow \quad \cot x = \frac{1}{\tan x} = -\frac{4}{3},$

$\Rightarrow \quad \sec x = \frac{1}{\cos x} = -\frac{5}{4}$

and $\quad \operatorname{cosec} x = \frac{1}{\sin x} = \frac{5}{3}$

73. *(d)* $\because \sin 765° = \sin (360 \times 2 + 45)°$

$= \sin 45° = \frac{1}{\sqrt{2}} \quad \{\because \sin (2n\pi + \theta) = \sin \theta\}$

74. *(c)* As, we know that

$\operatorname{cosec}(-\theta) = -\operatorname{cosec}\theta$

$\therefore \quad \operatorname{cosec}(-1410°) = -\operatorname{cosec}(360 \times 4 - 30)°$

$= -(-\operatorname{cosec} 30°)$

$= \operatorname{cosec} 30°$

$[\because \operatorname{cosec}(2n\pi - \theta) = -\operatorname{cosec}\theta]$

$= 2$

75. *(b)* In second quadrant, $\frac{\pi}{2} < x < \pi$

Here, $\quad \tan x = -\frac{5}{12}$

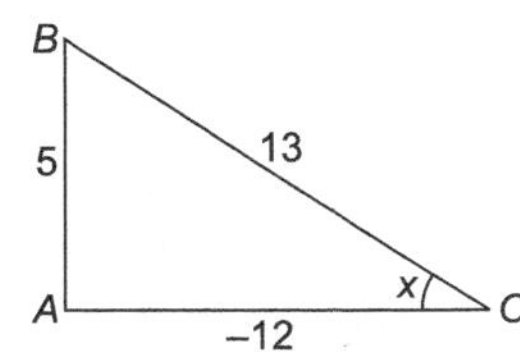

$\therefore \quad \cot x = \frac{1}{\tan x} = -\frac{12}{5}$

($\because$ in second quadrant $\sin x$ and $\operatorname{cosec} x$ are positive)

$\therefore \quad \sin x = \frac{5}{13}, \operatorname{cosec} x = \frac{13}{5}, \cos x = \frac{-12}{13}, \sec x = \frac{-13}{12}$

Alternate Method

$\tan x = -\frac{5}{12}$

Given that, x lies in second quadrant, *i.e.,* $\frac{\pi}{2} < x < \pi$

$\cot x = \frac{1}{\tan x} = -\frac{12}{5}$

Now, $\sec^2 x = 1 + \tan^2 x = 1 + \left(-\frac{5}{12}\right)^2$

$= 1 + \frac{25}{144} = \frac{169}{144}$

$\Rightarrow \quad \sec^2 x = \left(\frac{13}{12}\right)^2$

$\Rightarrow \quad \sec x = \pm \frac{13}{12}$

$\because$ x lies in second quadrant, so we take negative sign.

$\therefore \quad \sec x = -\frac{13}{12}$

$\Rightarrow \quad \cos x = \frac{1}{\sec x} = \frac{-12}{13}$

Now, $\quad \tan x = \frac{\sin x}{\cos x}$

$\Rightarrow \quad -\frac{5}{12} = \frac{\sin x}{-\frac{12}{13}}$

$\Rightarrow \quad \sin x = \frac{5}{13}$

and $\quad \operatorname{cosec} x = \frac{1}{\sin x}$

$= \frac{13}{5}$

Hence, option (b) is correct.

76. *(b)* $\tan \frac{19\pi}{3} = \tan\left(6\pi + \frac{\pi}{3}\right)$

$= \tan\left(2\pi \times 3 + \frac{\pi}{3}\right)$

$= \tan \frac{\pi}{3} \quad [\because \tan (2n\pi + \theta) = \tan \theta]$

$= \sqrt{3}$

77. *(b)* $\cot x = \frac{3}{4}$

Given that x lies in third quadrant.

i.e., $\pi < x < \frac{3\pi}{2}$

$\Rightarrow \quad \tan x = \frac{1}{\cot x} = \frac{4}{3}$

$\because \quad \sec^2 x = \tan^2 x + 1 = \frac{16}{9} + 1$

$\Rightarrow \quad \sec^2 x = \frac{25}{9}$

$\Rightarrow \quad \sec x = \pm \frac{5}{3}$

$\because$ x lies in third quadrant, so we take only its negative value.

i.e., $\quad \sec x = -\frac{5}{3}$

$\Rightarrow \quad \cos x = -\frac{3}{5}$

$\because \quad \tan x = \frac{\sin x}{\cos x}$

$\Rightarrow \quad \frac{4}{3} = \frac{\sin x}{-\frac{3}{5}}$

$\Rightarrow \quad \sin x = -\frac{4}{5}$

and $\quad \text{cosec } x = -\frac{5}{4}$

78. *(d)* $\sin\left(\frac{-11\pi}{3}\right) = -\sin\left(\frac{11\pi}{3}\right)$ $[\because \sin(-\theta) = \sin\theta]$

$= -\sin\left(4\pi - \frac{\pi}{3}\right) = -\sin\left(2\pi \times 2 - \frac{\pi}{3}\right)$

$= -\left(-\sin\frac{\pi}{3}\right) = \sin\left(\frac{\pi}{3}\right) = \frac{\sqrt{3}}{2}$

$[\because \sin(2n\pi - \theta) = -\sin\theta]$

79. *(b)* $\cot\left(-\frac{15\pi}{4}\right) = -\cot\left(\frac{15\pi}{4}\right)$ $[\because \cot(-\theta) = -\cot\theta]$

$= -\cot\left(4\pi - \frac{\pi}{4}\right) = -\cot\left(2\pi \times 2 - \frac{\pi}{4}\right)$

$= -\left(-\cot\frac{\pi}{4}\right) = \cot\frac{\pi}{4} = 1$

$[\because \cot(2n\pi - \theta) = -\cot\theta]$

80. *(b)* $\sec x = \frac{13}{5}$

Given that x lies in fourth quadrant.

i.e., $\quad \frac{3\pi}{2} < x < 2\pi$

$\Rightarrow \quad \cos x = \frac{5}{13}$

$\because \quad \sin^2 x + \cos^2 x = 1$

$\Rightarrow \quad \sin^2 x = 1 - \cos^2 x$

$= 1 - \left(\frac{5}{13}\right)^2$

$= 1 - \frac{25}{169}$

$= \frac{169 - 25}{169} = \frac{144}{169} = \left(\frac{12}{13}\right)^2$

$\Rightarrow \quad \sin x = \pm \frac{12}{13}$

Since, x lies in fourth quadrant, so we take only negative sign.

i.e., $\quad \sin x = -\frac{12}{13}$

$\tan x = \frac{\sin x}{\cos x} = \frac{-\left(\frac{12}{13}\right)}{\left(\frac{5}{13}\right)} = -\frac{12}{5}$

$\Rightarrow \text{cosec } x = \frac{1}{\sin x} = \frac{-13}{12}, \cot x = \frac{1}{\tan x} = -\frac{5}{12}$

81. *(c)* $\cos A = \frac{3}{5}, \cos B = \frac{4}{5}$

$\because \quad \angle A$ and $\angle B$ lies on the 4th quadrant.

$\therefore \quad \sin A = -\sqrt{1 - \frac{9}{25}}, \sin B = -\sqrt{1 - \frac{16}{25}}$

$\Rightarrow \quad \sin A = -\frac{4}{5}, \sin B = -\frac{3}{5}$

$\therefore 2\sin A + 4\sin B = 2\left(-\frac{4}{5}\right) + 4\left(-\frac{3}{5}\right)$

$= -\frac{8}{5} - \frac{12}{5} = -\frac{20}{5} = -4$

82. *(a)* $\left(1 + \cos\frac{\pi}{6}\right)\left(1 + \cos\frac{\pi}{3}\right)\left(1 + \cos\frac{2\pi}{3}\right)\left(1 + \cos\frac{7\pi}{6}\right)$

$= \left(1 + \frac{\sqrt{3}}{2}\right)\left(1 + \frac{1}{2}\right)\left(1 - \frac{1}{2}\right)\left(1 - \frac{\sqrt{3}}{2}\right)$

$= \left(1 - \frac{3}{4}\right)\left(1 - \frac{1}{4}\right)$

$= \frac{1}{4} \times \frac{3}{4} = \frac{3}{16}$

83. *(d)* $\tan A = \frac{a}{b}, \tan B = \frac{b}{a}$

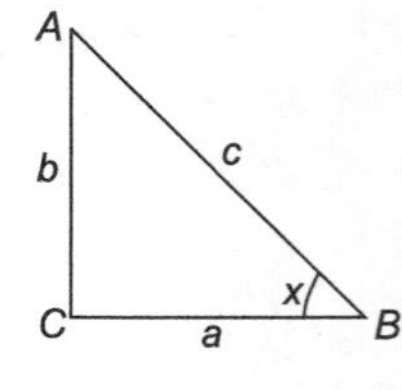

$\therefore \quad \tan A + \tan B = \frac{a^2 + b^2}{ab}$

Since, $\quad a^2 + b^2 = c^2$

$\therefore \quad \tan A + \tan B = \frac{c^2}{ab}$

84. *(b)* Given, $\tan\theta = \frac{1}{\sqrt{7}} \Rightarrow \cot\theta = \sqrt{7}$

Now, $\frac{(\text{cosec}^2\theta - \sec^2\theta)}{(\text{cosec}^2\theta + \sec^2\theta)}$

$= \frac{(1 + \cot^2\theta - 1 - \tan^2\theta)}{(1 + \cot^2\theta + 1 + \tan^2\theta)}$

$= \frac{\cot^2\theta - \tan^2\theta}{2 + \cot^2\theta + \tan^2\theta}$

$$= \frac{(\sqrt{7})^2 - \left(\frac{1}{\sqrt{7}}\right)^2}{2 + (\sqrt{7})^2 + \left(\frac{1}{\sqrt{7}}\right)^2}$$

$$= \frac{49-1}{7} \times \frac{7}{63+1}$$

$$= \frac{48}{64} = \frac{3}{4}$$

85. *(b)* Given equations can be rewritten as

$$\cos\theta = \frac{a}{x-h}, \sin\theta = \frac{b}{y-k}$$

Now, $\frac{a^2}{(x-h)^2} + \frac{b^2}{(y-k)^2} = 1$ $\quad (\because \cos^2\theta + \sin^2\theta = 1)$

86. *(b)* Since, $\tan\theta = -\frac{4}{3}$ is negative, θ lies either in second quadrant or in fourth quardant. Thus, $\sin\theta = \frac{4}{5}$ if θ lies in the second quadrant or $\sin\theta = -\frac{4}{5}$, if θ lies in the fourth quadrant.

87. *(c)* We have, $A = \cos^2\theta + \sin^4\theta$

$$= \cos^2\theta + \sin^2\theta\sin^2\theta \le \cos^2\theta + \sin^2\theta$$

Therefore, $A \le 1$

Also, $A = \cos^2\theta + \sin^4\theta = (1-\sin^2\theta) + \sin^4\theta$

$$= \left(\sin^2\theta - \frac{1}{2}\right)^2 + \left(1 - \frac{1}{4}\right)$$

$$= \left(\sin^2\theta - \frac{1}{2}\right)^2 + \frac{3}{4} \ge \frac{3}{4}$$

Hence, $\frac{3}{4} \le A \le 1.$

88. *(d)* We have,

$$\sqrt{\frac{1-\sin\theta}{1+\sin\theta}} + \sqrt{\frac{1+\sin\theta}{1-\sin\theta}} = \frac{1-\sin\theta}{\sqrt{1-\sin^2\theta}} + \frac{1+\sin\theta}{\sqrt{1-\sin^2\theta}}$$

$$= \frac{2}{\sqrt{\cos^2\theta}} = \frac{2}{|\cos\theta|}$$

$(\because \sqrt{\alpha^2} = |\alpha|$ for every real number $\alpha)$

Given that, θ lies in the second quadrant, so $|\cos\theta| = -\cos\theta$ [since, $\cos\theta < 0$].

Hence, the required value of the expression is $\frac{2}{-\cos\theta} = -2\sec\theta.$

89. *(a)* A. $\sin\frac{25\pi}{3} = \sin\left(8\pi + \frac{\pi}{3}\right) = \sin\frac{\pi}{3} = \frac{\sqrt{3}}{2}$

$[\because \sin(2n\pi + \theta) = \sin\theta]$

B. $\cos\frac{41\pi}{4} = \cos\left(10\pi + \frac{\pi}{4}\right) = \cos\frac{\pi}{4} = \frac{1}{\sqrt{2}}$

$[\because \cos(2n\pi + \theta) = \cos\theta]$

C. $\tan\left(\frac{-16\pi}{3}\right) = -\tan\frac{16\pi}{3}$ $\quad [\because \tan(-\theta) = -\tan\theta]$

$$= -\tan\left(5\pi + \frac{\pi}{3}\right) = -\tan\frac{\pi}{3} = -\sqrt{3}$$

$[\because \tan(n\pi + \theta) = \tan\theta]$

D. $\cot\frac{29\pi}{4} = \cot\left(7\pi + \frac{\pi}{4}\right) = \cot\frac{\pi}{4} = 1$

$[\because \cot(n\pi + \theta) = \cot\theta]$

$\therefore A \leftrightarrow 2, B \leftrightarrow 4, C \leftrightarrow 1, D \leftrightarrow 3$

90. *(a)* $\operatorname{cosec}\frac{-33\pi}{4} = -\operatorname{cosec}\frac{33\pi}{4}$

$$= -\operatorname{cosec}\left(8\pi + \frac{\pi}{4}\right)$$

$$= -\operatorname{cosec}\frac{\pi}{4}$$

$= -\sqrt{2}$ $\quad [\because \operatorname{cosec}(2n\pi + \theta) = \operatorname{cosec}\theta]$

91. *(a)* $\sec 6\pi = \sec(6\pi + 0)$

$= \sec 0 = 1$ $\quad [\because \sec(2n\pi + \theta) = \sec\theta]$

92. *(c)* $\because \cos^2 x = (1 - \sin^2 x) = \left(1 - \frac{24}{25}\right) = \frac{1}{25}$

$(\because \sin x$ and $\cos x$ are negative in III quadrant)

$\Rightarrow \quad \cos x = -\frac{1}{5}$

$\therefore \quad \cot x = \frac{\cos x}{\sin x} = \frac{-1/5}{-2\sqrt{6}/5} = \frac{1}{2\sqrt{6}}$

93. *(a)* $\because \pi < x < \frac{3\pi}{2}$

So, $\sin x$ is negative,

$\therefore \quad \sec x = -2 \Rightarrow \cos x = -\frac{1}{2}$

$\therefore \quad \sin^2 x = 1 - \cos^2 x = 1 - \frac{1}{4} = \frac{3}{4}$

$\Rightarrow \quad \sin x = \frac{-\sqrt{3}}{2}$

94. *(b)* Since, in fourth quadrant only $\sec x$ and $\cos x$ is positive.

$\therefore \quad \operatorname{cosec} x = \frac{-2}{\sqrt{3}} \Rightarrow \sin x = \frac{-\sqrt{3}}{2}$

So, $\cos^2 x = 1 - \sin^2 x = 1 - \frac{3}{4} = \frac{1}{4}$

$\Rightarrow \quad \cos x = \frac{1}{2}$ (+ ve)

$\therefore \quad \tan x = \frac{\sin x}{\cos x} = \frac{-\sqrt{3}/2}{1/2}$

$$= -\sqrt{3}$$

95. *(d)* In III quadrant, only $\tan\theta$ and $\cot\theta$ are positive.

$$\sin^2\theta = (1 - \cos^2\theta) = \left(1 - \frac{9}{25}\right) = \frac{16}{25}$$

$\Rightarrow \quad \sin\theta = \pm\sqrt{\frac{16}{25}} = \pm\frac{4}{5} \Rightarrow \sin\theta = \frac{-4}{5}$

$\therefore \quad \tan\theta = \left(\frac{-4}{5}\times\frac{5}{-3}\right) = \frac{4}{3}$

and $\quad \cot\theta = \frac{3}{4} \Rightarrow \text{cosec}\,\theta = \frac{-5}{4}$

and $\quad \sec\theta = \frac{-5}{3}$

$$\therefore \quad \frac{(\text{cosec}\,\theta + \cot\theta)}{(\sec\theta - \tan\theta)} = \frac{\left(\frac{-5}{4}+\frac{3}{4}\right)}{\left(\frac{-5}{3}-\frac{4}{3}\right)} = \frac{\left(\frac{-2}{4}\right)}{\left(\frac{-9}{3}\right)} = \frac{-2}{4}\times\frac{3}{-9} = \frac{1}{6}$$

96. *(d)* In II quadrant, only $\sin\theta$ and $\text{cosec}\,\theta$ are positive.

$\therefore \quad \cos^2\theta = (1-\sin^2\theta) = \left(1-\frac{9}{25}\right) = \frac{16}{25}$

$\Rightarrow \quad \cos\theta = -\sqrt{\frac{16}{25}} = \frac{-4}{5}$

$\therefore \quad \sec\theta = \frac{-5}{4}$

and $\quad \cot\theta = \frac{\cos\theta}{\sin\theta} = \frac{-4}{5}\times\frac{5}{3} = \frac{-4}{3}$

$$\therefore \quad (2\sec\theta - 3\cot\theta) = \left\{2\times\frac{(-5)}{4} - 3\times\frac{(-4)}{3}\right\} = \left(\frac{-5}{2}+4\right) = \frac{3}{2}$$

97. *(c)* Consider the unit circle with centre at the origin. Let x be the angle P_4OP_1 and y be the angle P_1OP_2. Then, $(x+y)$ is the angle P_4OP_2. Also, let $(-y)$ be the angle P_4OP_3. Therefore, P_1, P_2, P_3 and P_4 will have the coordinates $P_1(\cos x, \sin x)$, $P_2[\cos(x+y), \sin(x+y)]$, $P_3[\cos(-y), \sin(-y)]$ and $P_4(1, 0)$.

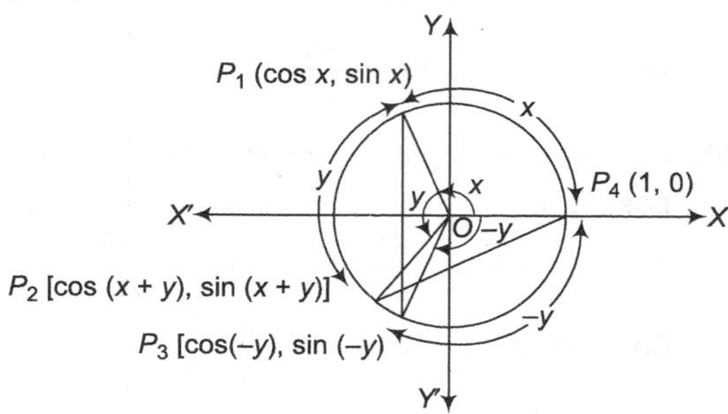

Consider the triangles P_1OP_3 and P_2OP_4. They are congruent. Therefore, P_1P_3 and P_2P_4 are equal. By using distance formula, we get

$$P_1P_3^2 = [\cos x - \cos(-y)]^2 + [\sin x - \sin(-y)]^2$$
$$= (\cos x - \cos y)^2 + (\sin x + \sin y)^2$$
$$= \cos^2 x + \cos^2 y - 2\cos x\cos y + \sin^2 x + \sin^2 y + 2\sin x\sin y$$
$$= 2 - 2(\cos x\cos y - \sin x\sin y)$$

Also, $P_2P_4^2 = [1-\cos(x+y)]^2 + [0-\sin(x+y)]^2$
$$= 1 - 2\cos(x+y) + \cos^2(x+y) + \sin^2(x+y)$$
$$= 2 - 2\cos(x+y)$$

Since, $P_1P_3 = P_2P_4$, we have $P_1P_3^2 = P_2P_4^2$

Therefore, $2-2(\cos x\cos y - \sin x\sin y) = 2 - 2\cos(x+y)$

Hence, $\cos(x+y) = \cos x\cos y - \sin x\sin y$

98. *(a)* Referring to previous question, replacing y by $(-y)$ in the identity $\cos(x+y) = \cos x\cdot\cos y - \sin x\cdot\sin y$, we get

$\cos[(x+(-y)] = \cos x\cdot\cos(-y) - \sin x\cdot\sin(-y)$

$\Rightarrow \quad \cos(x-y) = \cos x\cdot\cos y + \sin x\cdot\sin y$

99. *(c)* Replacing x by $\frac{\pi}{2}$ and y by x in above identity (as used in previous question), we get

A. $\cos\left(\frac{\pi}{2}-x\right) = \cos\frac{\pi}{2}\cdot\cos x + \sin\frac{\pi}{2}\cdot\sin x$

$\Rightarrow \quad \cos\left(\frac{\pi}{2}-x\right) = \sin x \quad \left(\because \sin\frac{\pi}{2} = 1 \text{ and } \cos\frac{\pi}{2} = 0\right)$

Thus, we have $\sin x = \cos\left(\frac{\pi}{2}-x\right)$

B. $\sin\left(\frac{\pi}{2}-x\right) = \cos\left[\frac{\pi}{2}-\left(\frac{\pi}{2}-x\right)\right]$ $\left(\text{on replacing } x \text{ by } \frac{\pi}{2}-x\right)$

$\therefore \quad \sin\left(\frac{\pi}{2}-x\right) = \cos x$

C. $\cos\left(\frac{\pi}{2}+x\right) = -\sin x$

D. $\sin\left(\frac{\pi}{2}+x\right) = \cos x$

Hence, option (c) is correct.

A ↔ 3, B ↔ 2, C ↔ 1, D ↔ 2

Note *The angles $\frac{n\pi}{2}\pm\theta$ are called allied or related angles and $\theta \pm n\times 360°$ are called coterminal angles. For general deduction, we have the following rules. The value of any trigonometric function for $\left(\frac{n\pi}{2}\pm\theta\right)$ is numerically equal to*

(a) the value of the same function, if n is an even integer with algebraic sign of the function as per the quadrant in which angles lie.

(b) corresponding cofunction of θ, if n is an odd integer with algebraic sign of the function for the quadrant in which it lies. Here, sine and cosine; tan and cot; sec and cosec are cofunctions of each other.

100. *(b)* We know that,

$$\sin(x+y)=\cos\left[\frac{\pi}{2}-(x+y)\right]=\cos\left[\left(\frac{\pi}{2}-x\right)-y\right]$$

$$=\cos\left(\frac{\pi}{2}-x\right)\cdot\cos y+\sin\left(\frac{\pi}{2}-x\right)\cdot\sin y$$

$\Rightarrow$ $\sin(x+y)=\sin x\cdot\cos y+\cos x\cdot\sin y$

In the above identity by replacing y by $(-y)$, we get,

$$\sin(x-y)=\sin x\cdot\cos y-\cos x\cdot\sin y$$

101. *(a)* By taking suitable values of x and y in the identities we get the following results

$$\cos(\pi-x)=-\cos x\,;\ \sin(\pi-x)=\sin x$$
$$\cos(\pi+x)=-\cos x\,;\ \sin(\pi+x)=-\sin x$$
$$\cos(2\pi-x)=\cos x\,;\ \sin(2\pi-x)=-\sin x$$

Note *Similar results for* $\tan x$, $\cot x$, $\sec x$ *and* $\operatorname{cosec} x$ *can be obtained from the results of* $\sin x$ *and* $\cos x$.

102. *(b)* Since, none of the x, y and $(x+y)$ is an odd multiple of $\frac{\pi}{2}$, it follows that $\cos x$, $\cos y$ and $\cos(x+y)$ are non-zero.
Now,

$$\tan(x+y)=\frac{\sin(x+y)}{\cos(x+y)}=\frac{\sin x\cos y+\cos x\sin y}{\cos x\cos y-\sin x\sin y}$$

On dividing numerator and denominator by $\cos x\cos y$, we have

$$\tan(x+y)=\frac{\dfrac{\sin x\cos y}{\cos x\cos y}+\dfrac{\cos x\sin y}{\cos x\cos y}}{\dfrac{\cos x\cos y}{\cos x\cos y}-\dfrac{\sin x\sin y}{\cos x\cos y}}=\frac{\tan x+\tan y}{1-\tan x\tan y}$$

103. *(c)* Referring to the previous question,
If we replace y by $(-y)$ in the identity, we get

$$\tan(x-y)=\tan[x+(-y)]$$
$$=\frac{\tan x+\tan(-y)}{1-\tan x\tan(-y)}=\frac{\tan x-\tan y}{1+\tan x\tan y}$$

$\therefore$ $$\tan(x-y)=\frac{\tan x-\tan y}{1+\tan x\cdot\tan y}$$

104. *(c)* Since, none of the x, y and $(x+y)$ is multiple of π, we find that $\sin x$, $\sin y$ and $\sin(x+y)$ are non-zero. Now,

$$\cot(x+y)=\frac{\cos(x+y)}{\sin(x+y)}=\frac{\cos x\cos y-\sin x\sin y}{\sin x\cos y+\cos x\sin y}$$

On dividing numerator and denominator by $\sin x\sin y$, we have

$$\cot(x+y)=\frac{\cot x\cot y-1}{\cot y+\cot x}$$

On replacing y by $(-y)$ in above identity, we get

$$\cot(x-y)=\frac{\cot x\cdot\cot y+1}{\cot y-\cot x}$$

105. *(d)* We know that,

$$\cos(x+y)=\cos x\cos y-\sin x\sin y$$

On replacing y by x, we get

$$\cos 2x=\cos^2 x-\sin^2 x$$
$$=\cos^2 x-(1-\cos^2 x)$$
$$=2\cos^2 x-1$$

Again, $\cos 2x=\cos^2 x-\sin^2 x$
$$=1-\sin^2 x-\sin^2 x=1-2\sin^2 x$$

We have, $\cos 2x=\cos^2 x-\sin^2 x$
$$=\frac{\cos^2 x-\sin^2 x}{\cos^2+\sin^2 x}$$

On dividing numerator and denominator by $\cos^2 x$, we get

$$\cos 2x=\frac{1-\tan^2 x}{1+\tan^2 x},\ x\neq n\pi+\frac{\pi}{2},$$ where n is an integer.

106. *(c)* We have, $\sin(x+y)=\sin x\cos y+\cos x\sin y$

On replacing y by x, we get

$$\sin 2x=2\sin x\cos x.$$

Again, $$\sin 2x=\frac{2\sin x\cos x}{\cos^2 x+\sin^2 x}$$

On dividing each term by $\cos^2 x$, we get

$$\sin 2x=\frac{2\tan x}{1+\tan^2 x}$$

107. *(b)* A. We know that,

$$\tan(x+y)=\frac{\tan x+\tan y}{1-\tan x\tan y}$$

On replacing y by x, we get

$$\tan 2x=\frac{2\tan x}{1-\tan^2 x}$$

B. We have,

$$\sin 3x=\sin(2x+x)$$
$$=\sin 2x\cos x+\cos 2x\sin x$$
$$=2\sin x\cos x\cos x+(1-2\sin^2 x)\sin x$$
$$=2\sin x(1-\sin^2 x)+\sin x-2\sin^3 x$$
$$=2\sin x-2\sin^3 x+\sin x-2\sin^3 x$$
$$=3\sin x-4\sin^3 x$$

C. We have,

$$\cos 3x=\cos(2x+x)$$
$$=\cos 2x\cos x-\sin 2x\sin x$$
$$=(2\cos^2 x-1)\cos x-2\sin x\cos x\sin x$$
$$(2\cos^2 x-1)\cos x-2\cos x(1-\cos^2 x)$$
$$=2\cos^3 x-\cos x-2\cos x+2\cos^3 x$$
$$=4\cos^3 x-3\cos x$$

D. We have, $\tan 3x = \tan (2x + x)$

$$= \frac{\tan 2x + \tan x}{1 - \tan 2x \tan x}$$

$$= \frac{\frac{2\tan x}{1-\tan^2 x} + \tan x}{1 - \frac{2\tan x \cdot \tan x}{1-\tan^2 x}}$$

$$= \frac{2\tan x + \tan x - \tan^3 x}{1 - \tan^2 x - 2\tan^2 x}$$

$$= \frac{3\tan x - \tan^3 x}{1 - 3\tan^2 x}$$

$\therefore$ A $\leftrightarrow$ 4, B $\leftrightarrow$ 3, C $\leftrightarrow$ 2, D $\leftrightarrow$ 1

108. *(a)* We know that,

$$\cos (x + y) = \cos x \cos y - \sin x \sin y \quad \text{...(i)}$$

and $$\cos (x - y) = \cos x \cos y + \sin x \sin y \quad \text{...(ii)}$$

On adding and subtracting Eqs. (i) and (ii), we get

$$\cos (x + y) + \cos (x - y) = 2\cos x \cos y \quad \text{...(iii)}$$

and $$\cos (x + y) - \cos (x - y) = -2\sin x \sin y \quad \text{...(iv)}$$

Further $$\sin (x + y) = \sin x \cos y + \cos x \sin y \quad \text{...(v)}$$

and $$\sin (x - y) = \sin x \cos y - \cos x \sin y \quad \text{...(vi)}$$

On adding and subtracting Eqs. (v) and (vi), we get

$$\sin (x + y) + \sin (x - y) = 2\sin x \cos y \quad \text{...(vii)}$$

and $$\sin (x + y) - \sin (x - y) = 2\cos x \sin y \quad \text{...(viii)}$$

Let $x + y = \theta$ and $x - y = \phi$. Therefore,

$$x = \left(\frac{\theta + \phi}{2}\right) \text{ and } y = \left(\frac{\theta - \phi}{2}\right)$$

Substituting the values of x and y in Eqs. (iii), (iv), (vii) and (viii), we get

$$\cos \theta + \cos \phi = 2\cos\left(\frac{\theta + \phi}{2}\right)\cos\left(\frac{\theta - \phi}{2}\right)$$

$$\cos \theta - \cos \phi = -2\sin\left(\frac{\theta + \phi}{2}\right)\sin\left(\frac{\theta - \phi}{2}\right)$$

$$\sin \theta + \sin \phi = 2\sin\left(\frac{\theta + \phi}{2}\right)\cos\left(\frac{\theta - \phi}{2}\right)$$

$$\sin \theta - \sin \phi = 2\cos\left(\frac{\theta + \phi}{2}\right)\sin\left(\frac{\theta - \phi}{2}\right)$$

Since, θ and ϕ can take any real values, we can replace θ by x and ϕ by y.

Thus, we get

$$\cos x + \cos y = 2\cos\frac{x + y}{2}\cos\frac{x - y}{2};$$

$$\cos x - \cos y = -2\sin\frac{x + y}{2}\sin\frac{x - y}{2}$$

$$\sin x + \sin y = 2\sin\frac{x + y}{2}\cos\frac{x - y}{2};$$

$$\sin x - \sin y = 2\cos\frac{x + y}{2}\sin\frac{x - y}{2}$$

Hence, option (a) is correct.

$\therefore$ A $\leftrightarrow$ 2, B $\leftrightarrow$ 1, C $\leftrightarrow$ 4, D $\leftrightarrow$ 3

109. *(b)* As a part of identities from above, we can also show that

I. $2\cos x \cos y = \cos (x + y) + \cos (x - y)$
II. $-2\sin x \sin y = \cos (x + y) - \cos (x - y)$
III. $2\sin x \cos y = \sin (x + y) + \sin (x - y)$
IV. $2\cos x \sin y = \sin (x + y) - \sin (x - y)$

Hence, option (b) is correct.

110. *(a)* We have, $3\sin\frac{\pi}{6}\sec\frac{\pi}{3} - 4\sin\frac{5\pi}{6}\cot\frac{\pi}{4}$

$$= 3 \times \frac{1}{2} \times 2 - 4\sin\left(\pi - \frac{\pi}{6}\right) \times 1 = 3 - 4\sin\frac{\pi}{6}$$

$$= 3 - 4 \times \frac{1}{2} = 1$$

111. *(b)* We have,

$$\sin 15^\circ = \sin (45^\circ - 30^\circ)$$

$$= \sin 45^\circ \cos 30^\circ - \cos 45^\circ \sin 30^\circ$$

$$= \frac{1}{\sqrt{2}} \times \frac{\sqrt{3}}{2} - \frac{1}{\sqrt{2}} \times \frac{1}{2} = \frac{\sqrt{3} - 1}{2\sqrt{2}}$$

112. *(c)* We have,

$$\tan\frac{13\pi}{12} = \tan\left(\pi + \frac{\pi}{12}\right) = \tan\frac{\pi}{12} = \tan\left(\frac{\pi}{4} - \frac{\pi}{6}\right)$$

$$= \frac{\tan\frac{\pi}{4} - \tan\frac{\pi}{6}}{1 + \tan\frac{\pi}{4}\tan\frac{\pi}{6}} = \frac{1 - \frac{1}{\sqrt{3}}}{1 + \frac{1}{\sqrt{3}}} = \frac{\sqrt{3} - 1}{\sqrt{3} + 1} = 2 - \sqrt{3}$$

113. *(b)* We have, $\dfrac{\sin (x + y)}{\sin (x - y)} = \dfrac{\sin x \cos y + \cos x \sin y}{\sin x \cos y - \cos x \sin y}$

On dividing the numerator and denominator by $\cos x \cos y$, we get

$$\frac{\sin (x + y)}{\sin (x - y)} = \frac{\tan x + \tan y}{\tan x - \tan y}$$

114. *(b)* We know that, $3x = 2x + x$

Therefore, $\tan 3x = \tan (2x + x)$

$\Rightarrow$ $$\tan 3x = \frac{\tan 2x + \tan x}{1 - \tan 2x \tan x}$$

$\Rightarrow$ $\tan 3x - \tan 3x \tan 2x \tan x = \tan 2x + \tan x$

$\Rightarrow$ $\tan 3x - \tan 2x - \tan x = \tan 3x \tan 2x \tan x$

$\Rightarrow$ $\tan 3x \tan 2x \tan x = \tan 3x - \tan 2x - \tan x$

115. *(a)* As we know, $\cos x + \cos y = 2\cos\frac{x + y}{2} \cdot \cos\frac{x - y}{2}$,

So, by using this identity, we get

$$\cos\left(\frac{\pi}{4} + x\right) + \cos\left(\frac{\pi}{4} - x\right)$$

$$= 2\cos\left(\frac{\frac{\pi}{4}+x+\frac{\pi}{4}-x}{2}\right)\cos\left(\frac{\frac{\pi}{4}+x-\left(\frac{\pi}{4}-x\right)}{2}\right)$$

$$= 2\cos\frac{\pi}{4}\cos x = 2\times\frac{1}{\sqrt{2}}\cos x = \sqrt{2}\cos x$$

116. *(a)* We know that,

$$\cos x + \cos y = 2\cos\frac{x+y}{2}\cdot\cos\frac{x-y}{2}$$

$$\sin x - \sin y = 2\cos\frac{x+y}{2}\cdot\sin\frac{x-y}{2}$$

Thus, we have $\frac{\cos 7x + \cos 5x}{\sin 7x - \sin 5x}$

$$= \frac{2\cos\frac{7x+5x}{2}\cos\frac{7x-5x}{2}}{2\cos\frac{7x+5x}{2}\sin\frac{7x-5x}{2}} = \frac{\cos x}{\sin x} = \cot x$$

117. *(d)* We have,

$$\frac{\sin 5x - 2\sin 3x + \sin x}{\cos 5x - \cos x} = \frac{\sin 5x + \sin x - 2\sin 3x}{\cos 5x - \cos x}$$

$$= \frac{2\sin 3x\cos 2x - 2\sin 3x}{-2\sin 3x\sin 2x} = -\frac{\sin 3x(\cos 2x - 1)}{\sin 3x\sin 2x}$$

$$= \frac{1-\cos 2x}{\sin 2x} = \frac{2\sin^2 x}{2\sin x\cos x} = \tan x$$

118. *(b)* $\sin^2\frac{\pi}{6} + \cos^2\frac{\pi}{3} - \tan^2\frac{\pi}{4}$

$$= \left(\frac{1}{2}\right)^2 + \left(\frac{1}{2}\right)^2 - (1)^2 = \frac{1}{4} + \frac{1}{4} - 1 = \frac{1}{2} - 1 = -\frac{1}{2}$$

119. *(a)* $2\sin^2\frac{\pi}{6} + \operatorname{cosec}^2\frac{7\pi}{6}\cos^2\frac{\pi}{3}$

$$= 2\times\left(\frac{1}{2}\right)^2 + \operatorname{cosec}^2\left(\pi+\frac{\pi}{6}\right)\cos^2\frac{\pi}{3}$$

$$= 2\times\frac{1}{4} + \operatorname{cosec}^2\frac{\pi}{6}\cos^2\frac{\pi}{3}$$

$[\because \operatorname{cosec}(\pi+\theta) = -\operatorname{cosec}\theta]$

$$= \frac{1}{2} + (2)^2\times\left(\frac{1}{2}\right)^2 = \frac{1}{2} + 1 = \frac{3}{2}$$

120. *(d)* $\cot^2\frac{\pi}{6} + \operatorname{cosec}\frac{5\pi}{6} + 3\tan^2\frac{\pi}{6}$

$$= (\sqrt{3})^2 + \operatorname{cosec}\left(\pi - \frac{\pi}{6}\right) + 3\left(\frac{1}{\sqrt{3}}\right)^2$$

$$= 3 + \operatorname{cosec}\frac{\pi}{6} + 3\times\frac{1}{3} \quad [\because \operatorname{cosec}(\pi-\theta) = \operatorname{cosec}\theta]$$

$$= 3 + 2 + 1 = 6$$

121. *(d)* $2\sin^2\frac{3\pi}{4} + 2\cos^2\frac{\pi}{4} + 2\sec^2\frac{\pi}{3}$

$$= 2\sin^2\left(\pi - \frac{\pi}{4}\right) + 2\cos^2\frac{\pi}{4} + 2\sec^2\frac{\pi}{3}$$

$$= 2\sin^2\frac{\pi}{4} + 2\times\left(\frac{1}{\sqrt{2}}\right)^2 + 2(2)^2 \quad [\because \sin(\pi-\theta) = \sin\theta]$$

$$= 2\times\left(\frac{1}{\sqrt{2}}\right)^2 + 2\times\frac{1}{2} + 2\times 4$$

$$= 1 + 1 + 8 = 10$$

122. *(c)* (i) $\sin 75° = \sin(45° + 30°)$

$$\sin(45° + 30°) = \sin 45°\cos 30° + \cos 45°\sin 30°$$

$[\because \sin(A+B) = \sin A\cos B + \cos A\sin B]$

$$= \frac{1}{\sqrt{2}}\times\frac{\sqrt{3}}{2} + \frac{1}{\sqrt{2}}\times\frac{1}{2} = \frac{\sqrt{3}+1}{2\sqrt{2}}$$

(ii) $\tan 15° = \tan(45° - 30°)$

$$= \frac{\tan 45° - \tan 30°}{1 + \tan 45°\tan 30°}\left[\because \tan(A-B) = \frac{\tan A - \tan B}{1+\tan A\tan B}\right]$$

$$= \frac{1-\frac{1}{\sqrt{3}}}{1+\frac{1}{\sqrt{3}}} = \frac{\sqrt{3}-1}{\sqrt{3}+1}$$

$$= \frac{\sqrt{3}-1}{\sqrt{3}+1}\times\frac{\sqrt{3}-1}{\sqrt{3}-1} = \frac{(\sqrt{3}-1)^2}{3-1}$$

$$= \frac{3+1-2\sqrt{3}}{2} = \frac{4-2\sqrt{3}}{2}$$

$$= 2 - \sqrt{3}$$

(iii) Clearly, it is true.

Hence, only (i) and (iii) are true.

123. *(b)* $\cos\left(\frac{\pi}{4}-x\right)\cos\left(\frac{\pi}{4}-y\right) - \sin\left(\frac{\pi}{4}-x\right)\sin\left(\frac{\pi}{4}-y\right)$

Let $\frac{\pi}{4} - x = A$ and $\frac{\pi}{4} - y = B$

Then, $\cos A\cos B - \sin A\sin B = \cos(A+B)$

$$= \cos\left(\frac{\pi}{4} - x + \frac{\pi}{4} - y\right)$$

$$= \cos\left(\frac{\pi}{2} - x - y\right)$$

$$= \cos\left[\frac{\pi}{2} - (x+y)\right] = \sin(x+y)$$

124. *(c)* $\frac{\tan\left(\frac{\pi}{4}+x\right)}{\tan\left(\frac{\pi}{4}-x\right)} = \frac{\tan\frac{\pi}{4} + \tan x}{1 - \tan\frac{\pi}{4}\tan x}\times\frac{1+\tan\frac{\pi}{4}\tan x}{\tan\frac{\pi}{4} - \tan x}$

$$\left[\because \tan(A+B) = \frac{\tan A + \tan B}{1-\tan A\tan B} \text{ and } \tan(A-B) = \frac{\tan A - \tan B}{1+\tan A\tan B}\right]$$

$$= \frac{1+\tan x}{1-\tan x}\times\frac{1+\tan x}{1-\tan x} = \left(\frac{1+\tan x}{1-\tan x}\right)^2$$

125. *(a)* I. $\dfrac{\cos(\pi + x)\cos(-x)}{\sin(\pi - x)\cos\left(\dfrac{\pi}{2} + x\right)} = \dfrac{(-\cos x)(\cos x)}{(\sin x)\ (-\sin x)}$

$$\left[\begin{array}{r} \because \cos(\pi + \theta) = -\cos\theta \\ \cos(-\theta) = \cos\theta \\ \sin(\pi - \theta) = \sin\theta \\ \sin(-\theta) = -\sin\theta \end{array}\right]$$

$$= \frac{\cos^2 x}{\sin^2 x} = \cot^2 x$$

So, both the statements are true and Statement II is the correct explanation of Statement I.

126. *(b)* $\cos\left(\dfrac{3\pi}{2} + x\right)\cos(2\pi + x)\left[\cot\left(\dfrac{3\pi}{2} - x\right) + \cot(2\pi + x)\right]$

$$= (\sin x)(\cos x)[\tan x + \cot x]$$

$$\left[\begin{array}{l} \because \cot\left(\dfrac{3\pi}{2} - \theta\right) = \tan\theta \\ \text{and} \cos\left(\dfrac{3\pi}{2} + \theta\right) = \sin\theta \end{array}\right]$$

$$= \sin x \cos x\left[\frac{\sin x}{\cos x} + \frac{\cos x}{\sin x}\right]$$

$$= \sin x \cos x\left[\frac{\sin^2 x + \cos^2 x}{\cos x \sin x}\right] = 1$$

127. *(b)* $\sin(n + 1)x \sin(n + 2)x + \cos(n + 1)x \cos(n + 2)x$

$$= \cos[(n + 1)x - (n + 2)x]$$

$$[\because \cos(A - B) = \cos A \cos B + \sin A \sin B]$$

$$= \cos[(n + 1)x - (n + 2)x]$$

$$= \cos[nx + x - nx - 2x] = \cos x$$

128. *(c)* $\cos\left(\dfrac{3\pi}{4} + x\right) - \cos\left(\dfrac{3\pi}{4} - x\right)$

$$= -2\sin\frac{\frac{3\pi}{4} + x + \frac{3\pi}{4} - x}{2}\sin\frac{\frac{3\pi}{4} + x - \frac{3\pi}{4} + x}{2}$$

$$= -2\sin\frac{3\pi}{4}\sin x$$

$$= -2\sin\left(\pi - \frac{\pi}{4}\right)\sin x$$

$$= -2\sin\frac{\pi}{4}\sin x = -2 \times \frac{1}{\sqrt{2}}\sin x = -\sqrt{2}\sin x$$

129. *(a)* $\sin^2 6x - \sin^2 4x = \sin(6x + 4x)\sin(6x - 4x)$

$$= \sin 10x \sin 2x$$

$$[\because \sin^2 A - \sin^2 B = \sin(A + B)\sin(A - B)]$$

130. *(b)* $\cos^2 2x - \cos^2 6x$

$$= (1 - \sin^2 2x) - (1 - \sin^2 6x) \quad (\because \sin^2 A + \cos^2 A = 1)$$

$$= \sin^2 6x - \sin^2 2x$$

$$= \sin(6x + 2x)\sin(6x - 2x)$$

$$[\because \sin^2 A - \sin^2 B = \sin(A + B)\sin(A - B)]$$

$$= \sin 8x \sin 4x$$

131. *(b)* $\sin 2x + 2\sin 4x + \sin 6x$

$$= 2\sin\frac{6x + 2x}{2}\cos\frac{6x - 2x}{2} + 2\sin 4x$$

$$\left(\because \sin A + \sin B = 2\sin\frac{A + B}{2}\cos\frac{A - B}{2}\right)$$

$$= 2\sin 4x \cos 2x + 2\sin 4x = 2\sin 4x(\cos 2x + 1)$$

$$= 2\sin 4x(2\cos^2 x - 1 + 1) = 2\sin 4x(2\cos^2 x)$$

$$= 4\cos^2 x \sin 4x$$

132. *(b)* I. $\cot 4x(\sin 5x + \sin 3x)$

$$= \cot 4x \cdot 2\sin\frac{5x + 3x}{2}\cos\frac{5x - 3x}{2}$$

$$\left(\because \sin C + \sin D = 2\sin\frac{C + D}{2}\cos\frac{C - D}{2}\right)$$

$$= \frac{\cos 4x}{\sin 4x} \times 2\sin 4x \cos x$$

$$= 2\cos 4x \cos x$$

Now, $\cot x(\sin 5x - \sin 3x)$

$$= \cot x \cdot 2\cos\frac{5x + 3x}{2}\sin\frac{5x - 3x}{2}$$

$$\left(\because \sin C - \sin D = 2\cos\frac{C + D}{2}\sin\frac{C - D}{2}\right)$$

$$= \frac{\cos x}{\sin x} \times 2\cos 4x \sin x = 2\cos 4x \cos x$$

$\therefore$ LHS = RHS

Hence, Statement I is true.

Also, Statement II is also true but not a correct explanation of Statement I.

133. *(c)* A. $\dfrac{\cos 9x - \cos 5x}{\sin 17x - \sin 3x} = \dfrac{-2\sin\dfrac{9x + 5x}{2}\sin\dfrac{9x - 5x}{2}}{2\cos\dfrac{17x + 3x}{2}\sin\dfrac{17x - 3x}{2}}$

(by formulae)

$$= \frac{-\sin 7x \sin 2x}{\cos 10x \sin 7x} = -\frac{\sin 2x}{\cos 10x}$$

B. $\dfrac{\sin 5x + \sin 3x}{\cos 5x + \cos 3x} = \dfrac{2\sin\dfrac{5x + 3x}{2}\cos\dfrac{5x - 3x}{2}}{2\cos\dfrac{5x + 3x}{2}\cos\dfrac{5x - 3x}{2}}$

$$= \frac{\sin 4x \cos x}{\cos 4x \cos x} = \tan 4x$$

C $\cos 4x = 1 - 2\sin^2 2x = 1 - 2(\sin 2x)^2$

$$= 1 - 2(2\sin x \cdot \cos x)^2$$

$$= 1 - 8\sin^2 x \cdot \cos^2 x$$

D. $\dfrac{\sin x + \sin 3x}{\cos x + \cos 3x} = \dfrac{\sin 3x + \sin x}{\cos 3x + \cos x}$

$$= \frac{2\sin\dfrac{3x + x}{2}\cos\dfrac{3x - x}{2}}{2\cos\dfrac{3x + x}{2}\cos\dfrac{3x - x}{2}}$$

$$= \frac{\sin 2x \cos x}{\cos 2x \cos x} = \tan 2x$$

$\therefore$ A $\leftrightarrow$ 2, B $\leftrightarrow$ 1, C $\leftrightarrow$ 4, D $\leftrightarrow$ 3

134. *(c)* $$\frac{\sin x - \sin y}{\cos x + \cos y} = \frac{2\cos\frac{x+y}{2}\cdot\sin\frac{x-y}{2}}{2\cos\frac{x+y}{2}\cdot\cos\frac{x-y}{2}}$$

$$= \tan\frac{x-y}{2}$$

135. *(d)* $$\frac{\sin x - \sin 3x}{\sin^2 x - \cos^2 x} = \frac{\sin 3x - \sin x}{\cos^2 x - \sin^2 x}$$

$$= \frac{2\cos\frac{3x+x}{2}\cdot\sin\frac{3x-x}{2}}{\cos 2x}$$

$$= \frac{2\cos 2x\cdot\sin x}{\cos 2x}$$

$$= 2\sin x$$

136. *(a)*

A. $$\sin\frac{A}{2} = \pm\sqrt{\frac{1-\cos A}{2}}\begin{cases} +, \text{ if } \frac{A}{2} \text{ lies in I or II quadrant} \\ -, \text{ if } \frac{A}{2} \text{ lies in III or IV quadrant}\end{cases}$$

B. $$\cos\frac{A}{2} = \pm\sqrt{\frac{1+\cos A}{2}}\begin{cases} +, \text{ if } \frac{A}{2} \text{ lies I or IV quadrant} \\ -, \text{ if } \frac{A}{2} \text{ lies in II or III quadrant}\end{cases}$$

C. $$\tan\frac{A}{2} = \pm\sqrt{\frac{1-\cos A}{1+\cos A}}\begin{cases} +, \text{ if } \frac{A}{2} \text{ lies in I or III quadrant} \\ -, \text{ if } \frac{A}{2} \text{ lies in II or IV quadrant}\end{cases}$$

$\therefore$ A $\leftrightarrow$ 2, B $\leftrightarrow$ 3, C $\leftrightarrow$ 1

137. *(d)* A. Let $\theta = 18°$.

$$\Rightarrow \quad 5\theta = 90°$$

$$\Rightarrow \quad 2\theta + 3\theta = 90°$$

Then, $2\theta = 90° - 3\theta$

Therefore, $\sin 2\theta = \sin(90° - 3\theta) = \cos 3\theta$

$$\Rightarrow \quad \sin 2\theta = 4\cos^3\theta - 3\cos\theta$$

Since, $\cos\theta \neq 0$, we get

$$2\sin\theta = 4\cos^2\theta - 3 = 1 - 4\sin^2\theta$$

$$\Rightarrow 4\sin^2\theta + 2\sin\theta - 1 = 0$$

Hence, $$\sin\theta = \frac{-2 \pm \sqrt{4+16}}{8}$$

$$= \frac{-1 \pm \sqrt{5}}{4}$$

Since, $\theta = 18°$, $\sin\theta > 0$, therefore

$$\sin 18° = \frac{\sqrt{5}-1}{4}$$

B. $$\cos 18° = \sqrt{1 - \sin^2 18°}$$

$$= \sqrt{1 - \frac{6-2\sqrt{5}}{16}} = \sqrt{\frac{10+2\sqrt{5}}{4}}$$

Now, we can easily find $\cos 36°$ and $\sin 36°$ as follows.

C. $$\cos 36° = 1 - 2\sin^2 18°$$

$$= 1 - \frac{6-2\sqrt{5}}{8} = \frac{2+2\sqrt{5}}{8} = \frac{\sqrt{5}+1}{4}$$

Hence, $$\cos 36° = \frac{\sqrt{5}+1}{4}$$

D. $$\sin 36° = \sqrt{1 - \cos^2 36°}$$

$$= \sqrt{1 - \frac{6+2\sqrt{5}}{16}}$$

$$= \sqrt{\frac{10-2\sqrt{5}}{4}}$$

$\therefore$ A $\leftrightarrow$ 3, B $\leftrightarrow$ 2, C $\leftrightarrow$ 4, D $\leftrightarrow$ 1

138. *(b)* $$\frac{\cos 4x + \cos 3x + \cos 2x}{\sin 4x + \sin 3x + \sin 2x}$$

$$= \frac{(\cos 4x + \cos 2x) + \cos 3x}{(\sin 4x + \sin 2x) + \sin 3x}$$

$$= \frac{2\cos\frac{4x+2x}{2}\cos\frac{4x-2x}{2} + \cos 3x}{2\sin\frac{4x+2x}{2}\cos\frac{4x-2x}{2} + \sin 3x}$$

(by formulae)

$$= \frac{2\cos 3x\cos x + \cos 3x}{2\sin 3x\cos x + \sin 3x}$$

$$= \frac{\cos 3x\,(2\cos x + 1)}{\sin 3x\,(2\cos x + 1)} = \cot 3x$$

139. *(c)* We know that, $\cot 3x = \cot(2x + x)$

$$\Rightarrow \quad \cot 3x = \frac{\cot 2x\cdot\cot x - 1}{\cot 2x + \cot x}$$

$\Rightarrow \cot 3x\cdot\cot 2x + \cot 3x\cdot\cot x = \cot 2x\cdot\cot x - 1$

$\Rightarrow \cot 3x\cdot\cot 2x + \cot 3x\cdot\cot x - \cot 2x\cdot\cot x = -1$

$\Rightarrow \cot x\cdot\cot 2x - \cot 2x\cdot\cot 3x - \cot 3x\cdot\cot x = 1$

140. *(a)* $$\sqrt{3}\operatorname{cosec} 20° - \sec 20° = \frac{\sqrt{3}}{\sin 20°} - \frac{1}{\cos 20°}$$

$$= \frac{\sqrt{3}\cos 20° - \sin 20°}{\sin 20°\cos 20°} = 4\left(\frac{\frac{\sqrt{3}}{2}\cos 20° - \frac{1}{2}\sin 20°}{2\sin 20°\cos 20°}\right)$$

$$= 4\left(\frac{\sin 60°\cos 20° - \cos 60°\sin 20°}{\sin 40°}\right)$$

$[\because \sin x\cdot\cos y - \cos x\cdot\sin y = \sin(x-y)]$

$$= 4\left(\frac{\sin(60° - 20°)}{\sin 40°}\right) = 4$$

141. *(b)* $\tan 4x = \tan 2(2x) = \dfrac{2\tan 2x}{1-\tan^2 2x} = \dfrac{2\dfrac{2\tan x}{1-\tan^2 x}}{1-\left(\dfrac{2\tan x}{1-\tan^2 x}\right)^2}$

$$= \frac{4\tan x}{1-\tan^2 x} \times \frac{(1-\tan^2 x)^2}{(1-\tan^2 x)^2 - 4\tan^2 x}$$

$$= \frac{4\tan x\,(1-\tan^2 x)}{1+\tan^4 x - 2\tan^2 x - 4\tan^2 x}$$

$$= \frac{4\tan x\,(1-\tan^2 x)}{1-6\tan^2 x + \tan^4 x}$$

142. *(c)* $\cos 6x = \cos 2(3x) = 2\cos^2 3x - 1 = 2(\cos 3x)^2 - 1$

$(\because \cos 2\theta = 2\cos^2\theta - 1)$

$= 2(4\cos^3 x - 3\cos x)^2 - 1$

$(\because \cos 3\theta = 4\cos^3\theta - 3\cos\theta)$

$= 2(16\cos^6 x + 9\cos^2 x - 24\cos^4 x) - 1$

$= 32\cos^6 x - 48\cos^4 x + 18\cos^2 x - 1$

143. *(a)* We have $\tan 9° - \tan 27° - \tan 63° + \tan 81°$

$= \tan 9° + \tan 81° - \tan 27° - \tan 63°$

$= \tan 9° + \tan(90° - 9°) - \tan 27° - \tan(90° - 27°)$

$= \tan 9° + \cot 9° - (\tan 27° + \cot 27°)$ ...(i)

Also, $\tan 9° + \cot 9° = \dfrac{1}{\sin 9° \cos 9°} = \dfrac{2}{\sin 18°}$ ...(ii)

$\left(\because \dfrac{\sin x}{\cos x} + \dfrac{\cos x}{\sin x} = \dfrac{1}{\sin x \cdot \cos x}\right)$

Similarly, $\tan 27° + \cot 27°$

$= \dfrac{1}{\sin 27° \cos 27°} = \dfrac{2}{\sin 54°} = \dfrac{2}{\cos 36°}$ ...(iii)

Using Eqs. (ii) and (iii) in Eq. (i), we get

$\tan 9° - \tan 27° - \tan 63° + \tan 81°$

$$= \frac{2}{\sin 18°} - \frac{2}{\cos 36°}$$

$$= \frac{2\times 4}{\sqrt{5}-1} - \frac{2\times 4}{\sqrt{5}+1} = 4$$

144. *(d)* We have, $\dfrac{\sec 8\theta - 1}{\sec 4\theta - 1} = \dfrac{(1-\cos 8\theta)\cos 4\theta}{\cos 8\theta\,(1-\cos 4\theta)}$

$\left(\because \sec 8\theta - 1 = \dfrac{1-\cos 8\theta}{\cos 8\theta}\right)$

$= \dfrac{2\sin^2 4\theta \cos 4\theta}{\cos 8\theta\,(2\sin^2 2\theta)}$ $(\because 1-\cos 2\theta = 2\sin^2\theta)$

$$= \frac{\sin 4\theta\,(2\sin 4\theta \cos 4\theta)}{2\cos 8\theta \sin^2 2\theta}$$

$= \dfrac{\sin 4\theta \sin 8\theta}{2\cos 8\theta \sin^2 2\theta}$ $(\because \sin 2\theta = 2\sin\theta\cdot\cos\theta)$

$$= \frac{2\sin 2\theta \cos 2\theta \sin 8\theta}{2\cos 8\theta \sin^2 2\theta} = \frac{\tan 8\theta}{\tan 2\theta}$$

145. *(a)* $\left(1+\cos\dfrac{\pi}{8}\right)\left(1+\cos\dfrac{3\pi}{8}\right)\left(1+\cos\dfrac{5\pi}{8}\right)\left(1+\cos\dfrac{7\pi}{8}\right)$

$= \left(1+\cos\dfrac{\pi}{8}\right)\left(1+\cos\dfrac{3\pi}{8}\right)\left(1+\cos\left(\pi-\dfrac{3\pi}{8}\right)\right)\left(1+\cos\left(\pi-\dfrac{\pi}{8}\right)\right)$

$= \left(1-\cos^2\dfrac{\pi}{8}\right)\left(1-\cos^2\dfrac{3\pi}{8}\right)$

$\left[\because 1+\cos\left(\pi-\dfrac{3\pi}{8}\right) = 1-\cos\dfrac{3\pi}{8}\right.$ and $\left. 1+\cos\left(\pi-\dfrac{\pi}{8}\right) = 1-\cos\dfrac{\pi}{8}\right]$

$= \sin^2\dfrac{\pi}{8}\sin^2\dfrac{3\pi}{8}$

$= \dfrac{1}{4}\left(1-\cos\dfrac{\pi}{4}\right)\left(1-\cos\dfrac{3\pi}{4}\right)$ $\left[\because \sin^2\theta = \dfrac{1}{2}(1-\cos 2\theta)\right]$

$= \dfrac{1}{4}\left(1-\cos\dfrac{\pi}{4}\right)\left(1+\cos\dfrac{\pi}{4}\right)$

$= \dfrac{1}{4}\left(1-\cos^2\dfrac{\pi}{4}\right) = \dfrac{1}{4}\left(1-\dfrac{1}{2}\right) = \dfrac{1}{8}$

$[\because (a-b)(a+b) = a^2 - b^2]$

146. *(d)* $\sin x\cdot\cos x = \dfrac{1}{2}\sin 2x \le \dfrac{1}{2}$ $(\because |\sin 2x| \le 1)$

147. *(c)* $\sin 20° \sin 40° \sin 60° \sin 80°$

$= \dfrac{\sqrt{3}}{2}\sin 20° \sin(60° - 20°)\sin(60° + 20°)$

$\left(\because \sin 60° = \dfrac{\sqrt{3}}{2}\right)$

$= \dfrac{\sqrt{3}}{2}\sin 20° [\sin^2 60° - \sin^2 20°]$

$[\because \sin(A+B)\cdot\sin(A-B) = \sin^2 A - \sin^2 B]$

$= \dfrac{\sqrt{3}}{2}\sin 20°\left[\dfrac{3}{4} - \sin^2 20°\right]$

$= \dfrac{\sqrt{3}}{2}\times\dfrac{1}{4}[3\sin 20° - 4\sin^3 20°]$

$= \dfrac{\sqrt{3}}{2}\times\dfrac{1}{4}(\sin 60°)$ $(\because \sin 3A = 3\sin A - 4\sin^3 A)$

$= \dfrac{\sqrt{3}}{2}\times\dfrac{1}{4}\times\dfrac{\sqrt{3}}{2} = \dfrac{3}{16}$

148. *(b)* Given that, $3\tan(\theta - 15°) = \tan(\theta + 15°)$ which can be rewritten as

$$\frac{\tan(\theta + 15°)}{\tan(\theta - 15°)} = \frac{3}{1}$$

Applying componendo and dividendo, we get

$$\frac{\tan(\theta + 15°) + \tan(\theta - 15°)}{\tan(\theta + 15°) - \tan(\theta - 15°)} = 2$$

$$\frac{\sin(\theta+15°)\cos(\theta-15°)+\sin(\theta-15°)\cos(\theta+15°)}{\sin(\theta+15°)\cos(\theta-15°)-\sin(\theta-15°)\cos(\theta-15°)}=2$$

$$\Rightarrow \quad \frac{\sin 2\theta}{\sin 30°}=2 \text{ i.e., } \sin 2\theta = 1$$

$$\Rightarrow \quad 2\theta=\frac{\pi}{2}$$

$$\therefore \quad \theta=\frac{\pi}{4}$$

149. *(b)* A. $\frac{1-\cos x}{\sin x}=\frac{2\sin^2\frac{x}{2}}{2\sin\frac{x}{2}\cos\frac{x}{2}}=\frac{\sin\frac{x}{2}}{\cos\frac{x}{2}}=\tan\frac{x}{2}$

B. $\frac{1+\cos x}{1-\cos x}=\frac{2\cos^2\frac{x}{2}}{2\sin^2\frac{x}{2}}=\cot^2\frac{x}{2}$

C. $\frac{1+\cos x}{\sin x}=\frac{2\cos^2\frac{x}{2}}{2\sin\frac{x}{2}\cos\frac{x}{2}}=\cot\frac{x}{2}$

D. $\sqrt{1+\sin 2x}=\sqrt{\sin^2 x+\cos^2 x+2\sin x\cos x}$

$=\sqrt{(\sin x+\cos x)^2}=|(\sin x+\cos x)|$

$\therefore A \leftrightarrow 4, B \leftrightarrow 1, C \leftrightarrow 2, D \leftrightarrow 3$

150. *(a)* We have, $\cos\theta\cdot\cos\frac{\theta}{2}-\cos 3\theta\cdot\cos\frac{9\theta}{2}$

$=\frac{1}{2}\left[2\cos\theta\cdot\cos\frac{\theta}{2}-2\cos 3\theta\cdot\cos\frac{9\theta}{2}\right]$

$=\frac{1}{2}\left[\cos\left(\theta+\frac{\theta}{2}\right)+\cos\left(\theta-\frac{\theta}{2}\right)-\left\{\cos\left(3\theta+\frac{9\theta}{2}\right)+\cos\left(3\theta-\frac{9\theta}{2}\right)\right\}\right]$

$=\frac{1}{2}\left[\cos\frac{3\theta}{2}+\cos\frac{\theta}{2}-\left\{\cos\frac{15\theta}{2}+\cos\left(-\frac{3\theta}{2}\right)\right\}\right]$

$=\frac{1}{2}\left[\cos\frac{3\theta}{2}+\cos\frac{\theta}{2}-\cos\frac{15\theta}{2}-\cos\frac{3\theta}{2}\right]$

$=\frac{1}{2}\left[2\sin\frac{\theta+15\theta}{4}\cdot\sin\frac{15\theta-\theta}{4}\right]$

$=\sin 4\theta\cdot\sin\frac{7\theta}{2}$

151. *(b)* Equations involving trigonometric functions of a variable are called trigonometric equations. We have, already learnt that the values of $\sin x$ and $\cos x$ repeat after an interval of 2π and the values of $\tan x$ repeat after an interval of π. The solutions of a trigonometric equation for which $0 \le x < 2\pi$ are called principal solutions. The expression involving integer 'n' which gives all solutions of a trigonometric equation is called the general solution. We shall use 'Z' to denote the set of integers.

$A \leftrightarrow 4, B \leftrightarrow 5, C \leftrightarrow 3, D \leftrightarrow 1, E \leftrightarrow 2$

152. *(d)* We know that, $\sin\frac{\pi}{3}=\frac{\sqrt{3}}{2}$ and $\sin\frac{2\pi}{3}=\sin\left(\pi-\frac{\pi}{3}\right)$

$=\sin\frac{\pi}{3}=\frac{\sqrt{3}}{2}$

$\therefore$ Principal solutions are $x=\frac{\pi}{3}$ and $\frac{2\pi}{3}$

153. *(c)* We know that, $\tan\frac{\pi}{6}=\frac{1}{\sqrt{3}}$.

$\tan\left(\pi-\frac{\pi}{6}\right)=-\tan\frac{\pi}{6}=-\frac{1}{\sqrt{3}}$

and $\tan\left(2\pi-\frac{\pi}{6}\right)=-\tan\frac{\pi}{6}=-\frac{1}{\sqrt{3}}$

Thus, $\tan\frac{5\pi}{6}=\tan\frac{11\pi}{6}=-\frac{1}{\sqrt{3}}$

Therefore, principal solutions are $\frac{5\pi}{6}$ and $\frac{11\pi}{6}$.

154. *(c)* If $\sin x=\sin y$, then

$\sin x-\sin y=0$

$\Rightarrow \quad 2\cos\frac{x+y}{2}\sin\frac{x-y}{2}=0$

which gives $\cos\frac{x+y}{2}=0 \Rightarrow \sin\frac{x-y}{2}=0$

Therefore, $\frac{x+y}{2}=(2n+1)\frac{\pi}{2}$

$\Rightarrow \quad \frac{x-y}{2}=n\pi,$ where $n\in Z$

i.e., $x=(2n+1)\pi-y$

$\Rightarrow \quad x=2n\pi+y$, where $n\in Z$

Hence, $x=(2n+1)\pi+(-1)^{2n+1}y$

$\Rightarrow \quad x=2n\pi+(-1)^{2n}y$, where $n\in Z$.

Combining these two results, we get $x=n\pi+(-1)^n y$, where $n\in Z$.

155. *(a)* If $\cos x=\cos y$, then

$\cos x-\cos y=0$ i.e., $-2\sin\frac{x+y}{2}\sin\frac{x-y}{2}=0$

Thus, $\sin\frac{x+y}{2}=0$

$\Rightarrow \quad \sin\frac{x-y}{2}=0$

Therefore, $\frac{x+y}{2}=n\pi$

$\Rightarrow \quad \frac{x-y}{2}=n\pi$, where $n\in Z$

i.e., $x=2n\pi-y \Rightarrow x=2n\pi+y$, where $n\in Z$

Hence, $x=2n\pi\pm y$, where $n\in Z$

So, both the statements are true and Statement II is the correct explanation of Statement I.

156. *(a)* If $\tan x = \tan y$, then $\tan x - \tan y = 0$

or $\dfrac{\sin x \cos y - \cos x \sin y}{\cos x \cos y} = 0$

which gives $\sin(x - y) = 0$

Therefore, $x - y = n\pi$, *i.e.*, $x = n\pi + y$, where $n \in Z$

157. *(a)* We have, $\sin x = -\dfrac{\sqrt{3}}{2} = -\sin\dfrac{\pi}{3} = \sin\left(\pi + \dfrac{\pi}{3}\right)$

$= \sin\dfrac{4\pi}{3}$

Hence, $\sin x = \sin\dfrac{4\pi}{3}$, which gives

$x = n\pi + (-1)^n\dfrac{4\pi}{3}$, where $n \in Z$.

Note *$\frac{4\pi}{3}$ is one such value of x for which $\sin x = -\frac{\sqrt{3}}{2}$. One may take any other value of x for which $\sin x = -\frac{\sqrt{3}}{2}$. The solutions obtained will be the same although these may apparently look different.*

158. *(c)* We have, $\cos x = \dfrac{1}{2} = \cos\dfrac{\pi}{3}$

Therefore $x = 2n\pi \pm \dfrac{\pi}{3}$, where $n \in Z$.

159. *(a)* We have, $\tan 2x = -\cot\left(x + \dfrac{\pi}{3}\right) = \tan\left(\dfrac{\pi}{2} + x + \dfrac{\pi}{3}\right)$

$\Rightarrow \tan 2x = \tan\left(x + \dfrac{5\pi}{6}\right)$

Therefore, $2x = n\pi + \left(x + \dfrac{5\pi}{6}\right)$, where $n \in Z$

$(\because \tan x = \tan y \Rightarrow x = n\pi + y, n \in Z)$

$\Rightarrow x = n\pi + \dfrac{5\pi}{6}$, where $n \in Z$.

160. *(a)* The equation can be written as

$\sin 6x + \sin 2x - \sin 4x = 0$

$\Rightarrow 2\sin 4x\cos 2x - \sin 4x = 0$

i.e., $\sin 4x(2\cos 2x - 1) = 0$

Therefore, $\sin 4x = 0$ or $\cos 2x = \dfrac{1}{2}$

i.e., $\sin 4x = 0$ or $\cos 2x = \cos\dfrac{\pi}{3}$

Hence, $4x = n\pi$ or $2x = 2n\pi \pm \dfrac{\pi}{3}$, where $n \in Z$

$\left[\begin{array}{l}\because \sin x = \sin y \Rightarrow x = n\pi + (-1)^n y, n \in Z \\ \text{and } \cos x = \cos y \Rightarrow x = 2n\pi \pm y, n \in Z\end{array}\right]$

i.e., $x = \dfrac{n\pi}{4} \Rightarrow x = n\pi \pm \dfrac{\pi}{6}$ where $n \in Z$

161. *(c)* The equation can be written as

$2(1 - \sin^2 x) + 3\sin x = 0$

$\Rightarrow 2\sin^2 x - 3\sin x - 2 = 0$

$\Rightarrow (2\sin x + 1)(\sin x - 2) = 0$

Hence, $\sin x = -\dfrac{1}{2}$ or $\sin x = 2$

But $\sin x = 2$ is not possible.

Therefore, $\sin x = -\dfrac{1}{2} = \sin\dfrac{7\pi}{6}$

Hence, the solution is given by

$x = n\pi + (-1)^n\dfrac{7\pi}{6}$, where $n \in Z$

162. *(c)* $\tan x = \sqrt{3}$

$\tan x = \tan\dfrac{\pi}{3}$

$\Rightarrow \tan\left(\pi + \dfrac{\pi}{3}\right) = \tan\dfrac{\pi}{3}$ or $\tan\dfrac{4\pi}{3}$

$\Rightarrow x = \dfrac{\pi}{3}$ or $\dfrac{4\pi}{3}$

Here, principal value is $x = \dfrac{\pi}{3}$

We know that, if $\tan x = \tan\alpha$, then

General solution is $x = n\pi + \alpha$

$\therefore x = n\pi + \dfrac{\pi}{3}$, where $n \in Z$

163. *(b)* $\sec x = 2$

$\Rightarrow \cos x = \dfrac{1}{2}$

$\Rightarrow \cos x = \cos\dfrac{\pi}{3} \Rightarrow \cos\left(2\pi - \dfrac{\pi}{3}\right)$

$x = \dfrac{\pi}{3} \Rightarrow \dfrac{5\pi}{3}$

Here, principal value is $x = \dfrac{\pi}{3}$, we know that if $\cos x = \cos\alpha$ then general solution, $x = 2n\pi \pm \alpha$

$\therefore x = 2n\pi \pm \dfrac{\pi}{3}$, $(\because n \in Z)$

164. *(d)* $\cot x = -\sqrt{3}$

$\Rightarrow \tan x = -\dfrac{1}{\sqrt{3}}$

$\Rightarrow \tan x = -\tan\dfrac{\pi}{6}$

$\Rightarrow \tan x = \tan\left(\pi - \dfrac{\pi}{6}\right)$ or $\tan\left(2\pi - \dfrac{\pi}{6}\right)$

[$\because$ $\tan x$ is negative in 2nd and 4th quadrant and $\tan(\pi - \theta) = -\tan\theta$ and $\tan(2\pi - \theta) = -\tan\theta$]

$\tan x = \tan\dfrac{5\pi}{6}$ or $\tan\dfrac{11\pi}{6}$

$\Rightarrow x = \dfrac{5\pi}{6}$ or $\dfrac{11\pi}{6}$

Here, principal value is $x = \dfrac{5\pi}{6}$

We know that, if $\tan x = \tan\alpha$, then

General solution, $x = n\pi + \alpha$

$x = n\pi + \dfrac{5\pi}{6}$ $(\because n \in Z)$

165. *(c)* $\text{cosec } x = -2$

$\Rightarrow \quad \sin x = -\frac{1}{2}$

$\Rightarrow \quad \sin x = -\sin \frac{\pi}{6}$

$\Rightarrow \quad \sin x = \sin\left(\pi + \frac{\pi}{6}\right) \text{ or } \sin\left(2\pi - \frac{\pi}{6}\right)$

[$\because$ $\sin x$ is negative in 3rd and 4th quadrant, or $\sin(\pi + \theta) = \sin(2\pi - \theta) = -\sin\theta$]

$\sin x = \sin \frac{7\pi}{6} \text{ or } \sin \frac{11\pi}{6}$

Here, principal value is

$x = \frac{7\pi}{6}$

General solution, $\quad x = n\pi + (-1)^n \alpha$

$\Rightarrow \quad x = n\pi + (-1)^n \frac{7\pi}{6} \quad (\because n \in Z)$

166. *(a)* Given, $\cos 4x = \cos 2x$

$\Rightarrow \quad \cos 4x - \cos 2x = 0$

$\Rightarrow \quad -2\sin\frac{4x + 2x}{2} \sin\frac{4x - 2x}{2} = 0$

$\Rightarrow \quad \sin 3x \sin x = 0$

$\Rightarrow \quad \sin 3x = 0 \text{ or } \sin x = 0$

$\Rightarrow \quad x = \frac{n\pi}{3} \text{ or } x = n\pi \quad (\because n \in Z)$

167. *(d)* $\cos 3x + \cos x - \cos 2x = 0$

$\Rightarrow \quad 2\cos\frac{3x + x}{2}\cos\frac{3x - x}{2} - \cos 2x = 0$

$\left(\because \cos C + \cos D = 2\cos\frac{C + D}{2}\cos\frac{C - D}{2}\right)$

$\Rightarrow \quad 2\cos 2x \cos x - \cos 2x = 0$

$\Rightarrow \quad \cos 2x = 0 \quad \Rightarrow \quad 2\cos x - 1 = 0$

$\Rightarrow \quad 2x = (2n + 1)\frac{\pi}{2} \quad \Rightarrow \quad \cos x = \frac{1}{2}$

$\Rightarrow \quad x = (2n + 1)\frac{\pi}{4} \quad \Rightarrow \quad \cos x = \cos\frac{\pi}{3} \quad (\because n \in Z)$

or $\quad x = 2n\pi \pm \frac{\pi}{3}$

168. *(d)* $\sin 2x + \cos x = 0$

$\Rightarrow \quad 2\sin x \cos x + \cos x = 0$

$(\because \sin 2x = 2\sin x \cos x)$

$\Rightarrow \quad \cos x(2\sin x + 1) = 0$

$\Rightarrow \quad \cos x = 0 \text{ or } \sin x = -\frac{1}{2}$

When $\cos x = 0$,

Then, $\quad x = (2n + 1)\frac{\pi}{2}$

When $\sin x = -\frac{1}{2}$,

Then, $\quad \sin x = -\sin\frac{\pi}{6}$

$\sin x = \sin\left(\pi + \frac{\pi}{6}\right) \quad [\because \sin(\pi + \theta) = -\sin\theta]$

$\sin x = \sin\frac{7\pi}{6}$

$\Rightarrow \quad x = n\pi + (-1)^n \frac{7\pi}{6} \quad (n \in Z)$

169. *(c)* $\sec^2 2x = 1 - \tan 2x$

$1 + \tan^2 2x = 1 - \tan 2x \quad (\because \sec^2 x - \tan^2 x = 1)$

$\Rightarrow \quad \tan^2 2x + \tan 2x = 0$

$\Rightarrow \quad \tan 2x(\tan 2x + 1) = 0$

$\Rightarrow \quad \tan 2x = 0$

or $\quad \tan 2x = -1$

When $\quad \tan 2x = 0$

Then, $\quad 2x = n\pi$

$\Rightarrow \quad x = \frac{n\pi}{2}$

When, $\quad \tan 2x = -1$

Then, $\quad \tan 2x = -\tan\frac{\pi}{4}$

$\tan 2x = \tan\left(\pi - \frac{\pi}{4}\right)$

$= \tan\frac{3\pi}{4} \quad [\because \tan(\pi - \theta) = -\tan\theta]$

$\therefore \quad 2x = n\pi + \frac{3\pi}{4}$

$\Rightarrow \quad x = \frac{n\pi}{2} + \frac{3\pi}{8}$

170. *(d)* $\sin x + \sin 3x + \sin 5x = 0$

$\Rightarrow \quad (\sin 5x + \sin x) + \sin 3x = 0$

$\Rightarrow \quad 2\sin\frac{5x + x}{2}\cos\frac{5x - x}{2} + \sin 3x = 0$

$\left(\because \sin C + \sin D = 2\sin\frac{C + D}{2}\cos\frac{C - D}{2}\right)$

$\Rightarrow \quad 2\sin 3x \cos 2x + \sin 3x = 0$

$\Rightarrow \quad \sin 3x(2\cos 2x + 1) = 0$

Either $\quad \sin 3x = 0$

or $\quad 2\cos 2x + 1 = 0$

When $\sin 3x = 0$,

Then, $\quad 3x = n\pi$

$\Rightarrow \quad x = \frac{n\pi}{3}$

When $\quad 2\cos 2x + 1 = 0$

$\Rightarrow \quad \cos 2x = -\frac{1}{2}$

$\Rightarrow \quad \cos 2x = -\cos\frac{\pi}{3}$

$\Rightarrow \quad \cos 2x = \cos\left(\pi - \frac{\pi}{3}\right)$

$[\because \cos(\pi - \theta) = -\cos\theta]$

$\Rightarrow \quad \cos 2x = \cos \frac{2\pi}{3}$

$\Rightarrow \quad 2x = 2n\pi \pm \frac{2\pi}{3}$

$\Rightarrow \quad x = n\pi \pm \frac{\pi}{3}, n \in Z$

171. *(d)* Given, $\sec x \cos 5x + 1 = 0$

$\Rightarrow \quad \frac{1}{\cos x} \cdot \cos 5x + \frac{1}{1} = 0$

$\Rightarrow \quad \frac{\cos 5x + \cos x}{\cos x} = 0$

$\Rightarrow \quad \cos 5x + \cos x = 0$

$\Rightarrow 2 \cos \frac{5x + x}{2} \cos \frac{5x - x}{2} = 0$

$\left(\because \cos C + \cos D = 2 \cos \frac{C + D}{2} \cos \frac{C - D}{2}\right)$

$\Rightarrow \quad 2 \cos 3x \cos 2x = 0$

$\Rightarrow \quad \cos 3x = 0 \quad \text{or} \quad \cos 2x = 0$

$\Rightarrow \quad 3x = (2n + 1)\frac{\pi}{2} \quad \text{or} \quad 2x = (2n + 1)\frac{\pi}{2}$

$\Rightarrow \quad x = (2n + 1)\frac{\pi}{6} \quad \text{or} \quad x = (2n + 1)\frac{\pi}{4}$

If $n = 0$, then $\quad x = \frac{\pi}{6} \quad \text{or} \quad x = \frac{\pi}{4}$

If $n = 1$, then $\quad x = 3 \times \frac{\pi}{6} = \frac{\pi}{2}$

or $\quad x = \frac{3\pi}{4} > \frac{\pi}{2}$

Therefore, the values of x are $\frac{\pi}{6}, \frac{\pi}{4}$.

Note *$\cos x = 0 \Rightarrow x = (2n + 1)\frac{\pi}{2}$. But $x \neq (2n \pm 1)\frac{\pi}{2}$*

172. *(a)* (i) The general value of θ satisfying any of the equations $\sin^2 \theta = \sin^2 \theta, \cos^2 \theta = \cos^2 \alpha$ and $\tan^2 \theta = \tan^2 \alpha$ is given by $\theta = n\pi \pm \alpha$.

(ii) The general value of θ satisfying equations $\sin \theta = \sin \alpha$ simultaneously is given by $\theta = 2n\pi + \alpha, n \in Z$.

So, Statement I is correct and Statement II is incorrect.

173. *(c)* To find the solution of the equation of the form $a \cos \theta + b \sin \theta = c$, we put

$a = r \cos \alpha$ and $b = r \sin \alpha$, so that $r^2 = a^2 + b^2$ and $\tan \alpha = \frac{b}{a}$.

Thus, we find $a \cos \theta + b \sin \theta = c$ changed into the form

$r (\cos \theta \cos \alpha + \sin \theta \sin \alpha) = c$

or $r \cos (\theta - \alpha) = c$ and hence $\cos (\theta - \alpha) = \frac{c}{r}$. This gives the solution of the given equation.

Maximum and minimum values of the expression $a \cos \theta + b \sin \theta$ are $\sqrt{a^2 + b^2}$ and $-\sqrt{a^2 - b^2}$ respectively, where a and b are constants.

174. *(a)* We have $\sin \theta - \sin 3\theta + \sin 5\theta = 0$

$\Rightarrow \quad (\sin \theta + \sin 5\theta) - \sin 3\theta = 0$

$\Rightarrow \quad 2 \sin 3\theta \cos 2\theta - \sin 3\theta = 0$

$\Rightarrow \quad \sin 3\theta (2 \cos 2\theta - 1) = 0$

$\Rightarrow \quad \sin 3\theta = 0$

or $\quad \cos 2\theta = \frac{1}{2}$

When, $\quad \sin 3\theta = 0$

then $\quad 3\theta = n\pi \quad \text{or} \quad \theta = \frac{n\pi}{3}$

When, $\quad \cos 2\theta = \frac{1}{2} = \cos \frac{\pi}{3}$

then $\quad 2\theta = 2n\pi \pm \frac{\pi}{3}$

$\Rightarrow \quad \theta = n\pi \pm \frac{\pi}{6}$

which gives $\quad \theta = (6n + 1)\frac{\pi}{6}$

or $\quad \theta = (6n - 1)\frac{\pi}{6}$

$\Rightarrow \quad \theta = \frac{n\pi}{3} \quad \text{or} \quad n\pi \pm \pi/6$

175. *(d)* Here, $2 \tan^2 x + \sec^2 x = 2$

$\Rightarrow \quad 2 \tan^2 x + 1 + \tan^2 x = 2$

which gives $\tan x = \pm \frac{1}{\sqrt{3}}$

If we take $\tan x = \frac{1}{\sqrt{3}}$, then $x = \frac{\pi}{6}$ or $\frac{7\pi}{6}$

Again, if we take $\tan x = \frac{-1}{\sqrt{3}}$, then $x = \frac{5\pi}{6}$ or $\frac{11\pi}{6}$

Therefore, the possible solutions of above equations are

$x = \frac{\pi}{6}, \frac{5\pi}{6}, \frac{7\pi}{6}$ and $\frac{11\pi}{6}$ where $0 \le x \le 2\pi$

176. *(b)* Given that, $\sin \theta$ and $\cos \theta$ are the roots of the equation $ax^2 - bx + c = 0$, so $\sin \theta + \cos \theta = \frac{b}{a}$ and $\sin \theta \cos \theta = \frac{c}{a}$.

Using the identity $(\sin \theta + \cos \theta)^2 = \sin^2 \theta + \cos^2 \theta + 2 \sin \theta \cos \theta$, we have

$$\frac{b^2}{a^2} = 1 + \frac{2c}{a}$$

or $\quad a^2 - b^2 + 2ac = 0$

177. *(a)* $2 \sin^2 - 3 \sin \theta - 2 = 0$

$\Rightarrow \quad (2 \sin \theta + 1)(\sin \theta - 2) = 0$

$\Rightarrow \quad \sin\theta = -\frac{1}{2} \qquad (\because \sin\theta \neq 2)$

$\Rightarrow \quad \theta = n\pi + (-1)^n\left(-\frac{\pi}{6}\right)$

$\Rightarrow \quad \theta = n\pi + (-1)^n\left(-\frac{\pi}{6}\right)$

$\Rightarrow \quad \theta = n\pi + (-1)^{n+1}\frac{\pi}{6}$

178. *(a)* Given, $3\tan(\theta - 15) = \tan(\theta + 15)$

$$\frac{\tan A}{\tan B} = \frac{3}{1},$$

where $A = \theta + 15°$, $B = \theta - 15°$

On applying componendo and dividendo, we get

$\Rightarrow \quad \frac{\tan A + \tan B}{\tan A - \tan B} = \frac{3+1}{3-1}$

$\Rightarrow \quad \frac{\sin(A+B)}{\sin(A-B)} = 2$

$\Rightarrow \quad \sin 2\theta = 2\sin 30°$

$\Rightarrow \quad \sin 2\theta = 2\cdot\frac{1}{2} = 1 = \sin\frac{\pi}{2}$

$\Rightarrow \quad 2\theta = n\pi + (-1)^n\frac{\pi}{2}$

$\Rightarrow \quad \theta = \frac{n\pi}{2} + (-1)^n\frac{\pi}{4}$

179. *(a)* $\sin 6\theta + \sin 4\theta + \sin 2\theta = 0$

$\Rightarrow \quad (\sin 6\theta + \sin 2\theta) + \sin 4\theta = 0$

$\Rightarrow \quad 2\sin 4\theta\cos 2\theta + \sin 4\theta = 0$

$\Rightarrow \quad \sin 4\theta(2\cos 2\theta + 1) = 0$

$\therefore$ Either $\sin 4\theta = 0$ or $\cos 2\theta = -\frac{1}{2}$

When $\quad \sin 4\theta = 0$

$\Rightarrow \quad 4\theta = n\pi$

$\Rightarrow \quad \theta = \frac{n\pi}{4}$

and when $\quad \cos 2\theta = -\frac{1}{2} = \cos\frac{2\pi}{3}$

$\Rightarrow \quad 2\theta = 2n\pi \pm \frac{2\pi}{3}$

$\Rightarrow \quad \theta = n\pi \pm \frac{\pi}{3}$

180. *(d)* $\sin x = \frac{1}{2} \Rightarrow \sin x = \sin\frac{\pi}{6}$

$\Rightarrow \quad x = n\pi + (-1)^n\frac{\pi}{6}$

For $\quad -2\pi \le x \le 2\pi$

$$x = \frac{\pi}{6}, \frac{5\pi}{6}, -\frac{7\pi}{6}, -\frac{11\pi}{6}$$

$\therefore$ Number of points of intersection of two given curves are 4.

Alternate Method

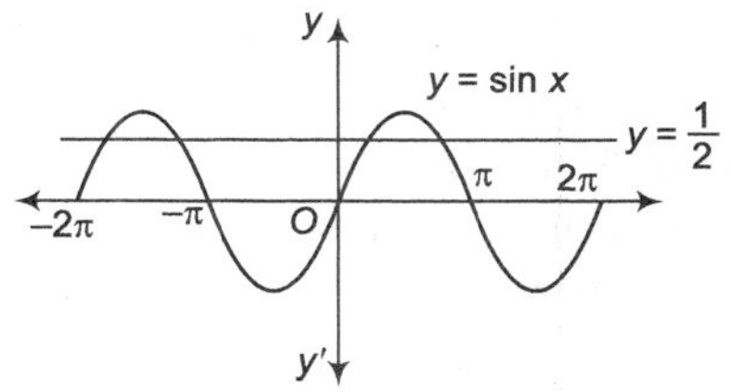

The number of points of intersection are 4.

181. *(d)* Given equation is

$$5\cos 2\theta + 2\cos^2\frac{\theta}{2} + 1 = 0$$

$\Rightarrow \quad 5(2\cos^2\theta - 1) + 1 + \cos\theta + 1 = 0$

$\Rightarrow \quad 10\cos^2\theta + \cos\theta - 3 = 0$

$\Rightarrow \quad (2\cos\theta - 1)(5\cos\theta + 3) = 0$

$\Rightarrow \quad \cos\theta = \frac{1}{2}$

or $\quad \cos\theta = -\frac{3}{5}$

$\Rightarrow \quad \theta = \frac{\pi}{3}$ or $\theta = \pi - \cos^{-1}\left(\frac{3}{5}\right)$

182. *(e)* $2\sin x\cos x = \frac{1}{2}$

$\Rightarrow \quad \sin 2x = \frac{1}{2} = \sin\frac{\pi}{6}$

$\Rightarrow \quad 2x = n\pi + (-1)^n\frac{\pi}{6}$

$\Rightarrow \quad x = \frac{n\pi}{2} + (-1)^n\frac{\pi}{12}$

For $\quad x \in \left(0, \frac{\pi}{2}\right)$

$\therefore \quad x = \frac{\pi}{12} \qquad (\because n = 0)$

183. *(b)* Given equation is $\sin x.\cos 3x = \sin 3x.\cos 5x$

$\Rightarrow \quad 2\sin x\cos 3x - 2\sin 3x\cos 5x = 0$

$\Rightarrow \quad \sin(3x + x) - \sin(3x - x) - \sin(3x + 5x) + \sin(5x - 3x) = 0$

$\Rightarrow \quad \sin 4x - \sin 2x - \sin 8x + \sin 2x = 0$

$\quad \sin 4x - \sin 8x = 0$

$\Rightarrow \quad 2\cos\left(\frac{4x + 8x}{2}\right)\sin\left(\frac{4x - 8x}{2}\right) = 0$

$\Rightarrow \quad -2\cos 6x\sin 2x = 0$

$\Rightarrow \quad \cos 6x = 0$ or $\sin 2x = 0$

$\Rightarrow \quad 6x = (2n+1)\frac{\pi}{2}$ or $x = \frac{n\pi}{2}$

$\Rightarrow \quad x = (2n+1)\frac{\pi}{12}$ or $x = \frac{n\pi}{2}$

$\Rightarrow \quad x = 0, \frac{\pi}{2}, \frac{\pi}{12}, \frac{3\pi}{12}, \frac{5\pi}{12} \in \left[0, \frac{\pi}{2}\right]$

$\therefore$ Number of solutions is 5.

184. *(c)* $\cos 2\theta = \sin\theta$

$$\Rightarrow \quad 1 - 2\sin^2\theta = \sin\theta$$
$$\Rightarrow \quad 2\sin^2\theta + \sin\theta - 1 = 0$$
$$\Rightarrow \quad 2\sin^2\theta + 2\sin\theta - \sin\theta - 1 = 0$$
$$\Rightarrow 2\sin\theta(\sin\theta + 1) - 1(\sin\theta + 1) = 0$$
$$\Rightarrow \quad (\sin\theta + 1)(2\sin\theta - 1) = 0$$
$$\Rightarrow \quad \sin\theta = -1, \sin\theta = \frac{1}{2}$$
$$\Rightarrow \quad \sin\theta = \sin\frac{3\pi}{2}, \sin\theta = \sin\frac{\pi}{6}$$
$$\Rightarrow \quad \theta = n\pi + (-1)^n\frac{3\pi}{2}$$
or
$$\theta = m\pi + (-1)^m\frac{\pi}{6}$$
For
$$\theta \in (0, 2\pi)$$
$$\theta = \frac{3\pi}{2}, \frac{\pi}{6}, \frac{5\pi}{6}$$

185. *(a)* We have, $6\sin x \cdot \cos x + 4\cos 2x$

$$= 3\sin 2x + 4\cos 2x$$

which is of the form $a\sin\theta + b\cos\theta$

$$\therefore \quad -\sqrt{3^2 + 4^2} \le 3\sin 2x + 4\cos 2x \le \sqrt{3^2 + 4^2} \quad \left(\because -\sqrt{a^2 + b^2} \le a\sin\theta + b\cos\theta \le \sqrt{a^2 + b^2}\right)$$

$$\Rightarrow \quad -5 \le 3\sin 2x + 4\cos 2x \le 5$$

Thus, maximum and minimum values of $6\sin x \cdot \cos x + 4\cos 2x$ are 5 and -5 respectively.

186. *(c)* $\cos^2\left(\frac{\pi}{3} - x\right) - \cos^2\left(\frac{\pi}{3} + x\right)$

$$= \sin\left(\frac{\pi}{3} - x + \frac{\pi}{3} + x\right)\sin\left(\frac{\pi}{3} + x - \frac{\pi}{3} + x\right)$$
$$= \sin\frac{2\pi}{3}\sin 2x = \frac{\sqrt{3}}{2}\sin 2x$$

(since, maximum value of $\sin 2x$ is 1)

Therefore, its maximum value is $\frac{\sqrt{3}}{2}$.

187. *(c)* $5\cos\theta + 3\cos\left(\theta + \frac{\pi}{3}\right) + 3$

$$= 5\cos\theta + 3\left(\cos\theta\cos\frac{\pi}{3} - \sin\theta\sin\frac{\pi}{3}\right) + 3$$
$$= 5\cos\theta + \frac{3}{2}\cos\theta - \frac{3\sqrt{3}}{2}\sin\theta + 3$$
$$= \frac{13}{2}\cos\theta - \frac{3\sqrt{3}}{2}\sin\theta + 3$$
$$\therefore \text{ Maximum value} = 3 + \sqrt{\left(\frac{13}{2}\right)^2 + \left(-\frac{3\sqrt{3}}{2}\right)^2}$$
$$= 3\sqrt{\frac{196}{4}} = 3 + 7 = 10$$

188. *(c)* Given, $f(x) = \sin x(1 + \cos x)$

It is maximum at $x = \frac{\pi}{3}$

$$\therefore \quad f\left(\frac{\pi}{3}\right) = \sin\frac{\pi}{3}\left(1 + \cos\frac{\pi}{3}\right)$$
$$= \frac{\sqrt{3}}{2}\left(1 + \frac{1}{2}\right) = \frac{3\sqrt{3}}{4}$$

189. *(d)* Here, ABC be a triangle. By angle A, we mean the angle between the sides AB an AC which lies between 0° and 180°. The angles B and C are similarly defined. The sides AB, BC and CA opposite to the vertices C, A and B will be denoted by c, a and b respectively

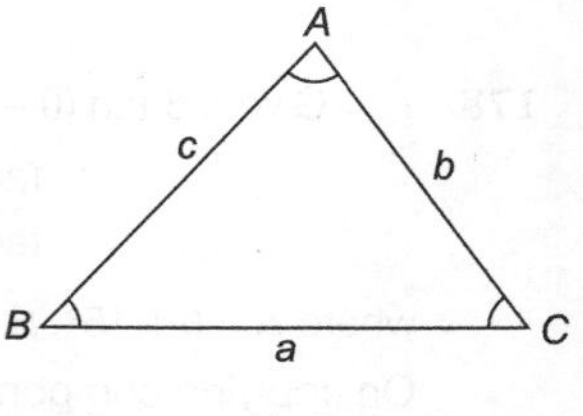

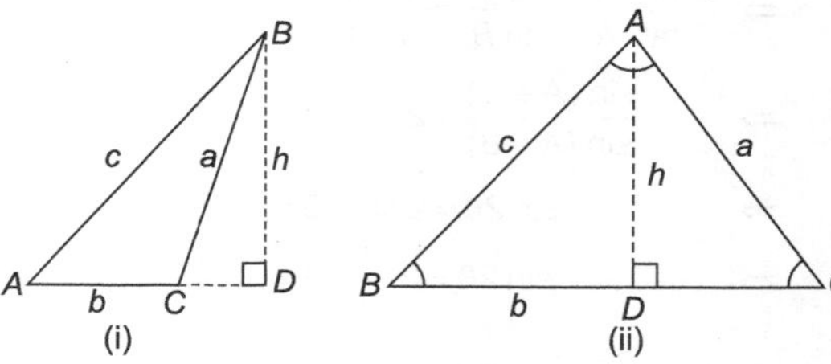

Let ABC be either of the triangles as shown in Fig, (i) and (ii)

The altitude h is drawn from the vertex B to meet the side AC in point D [in (i) AC is produced of meet the altitude in D]. From the right angled ΔABD in figure we have

$$\sin A = \frac{h}{c}, \text{ i.e., } h = c\sin A \quad \ldots\text{(i)}$$

and $$\sin(180° - C) = \frac{h}{a} \Rightarrow h = a\sin C, \quad \ldots\text{(ii)}$$

From Eqs. (i) and (ii), we get

$$c\sin A = a\sin C, \text{ i.e., } \frac{\sin A}{a} = \frac{\sin C}{c} \quad \ldots\text{(iii)}$$

Similarly, we can prove that

$$\frac{\sin A}{a} = \frac{\sin B}{b} \quad \ldots\text{(iv)}$$

From Eqs. (iii) and (iv), we get

$$\frac{\sin A}{a} = \frac{\sin B}{b} = \frac{\sin C}{c}$$

For ΔABC in figures (ii), Eqs. (iii) and (iv) follow similarly.

The above formulae states that, in any triangle, sides are proportional to the sines of the opposite angles. This formula is also known as sine formula.

190. *(d)* Let ABC be triangle as given in Fig. (i) and (ii)

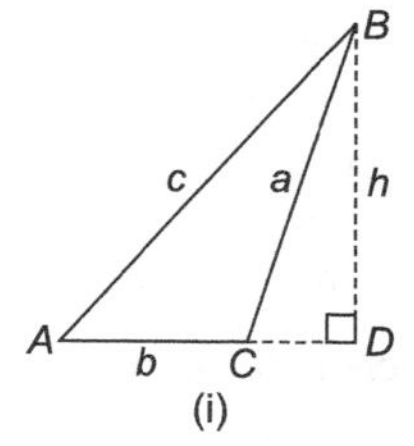

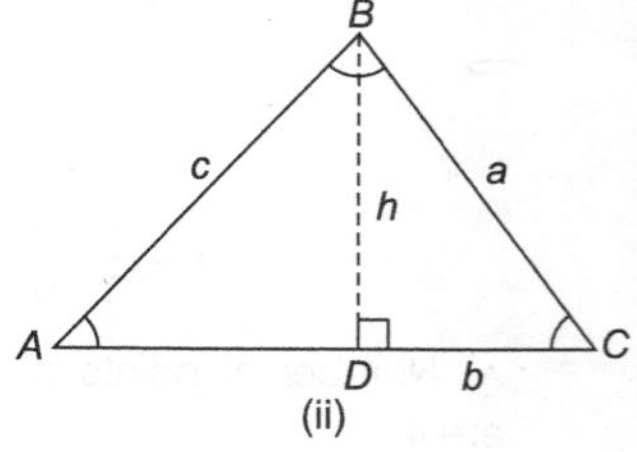

Refering to Fig. (ii), we have

$$BC^2 = BD^2 + DC^2 = BD^2 + (AC - AD)^2$$
$$= BD^2 + AD^2 + AC^2 - 2AC \cdot AD$$
$$= AB^2 + AC^2 - 2AC\ AB \cos A$$

or $a^2 = b^2 + c^2 - 2bc \cos A$

Similarly, we can obtain

$$b^2 = c^2 + a^2 - 2\,ca \cos B$$

and $c^2 = a^2 + b^2 - 2ab \cos C$

Same equations can be obtained for Fig. (i), where C is obtuse. A convenient form of the cosine formulae, when angles are to be found are as follows:

$$\cos A = \frac{b^2 + c^2 - a^2}{2bc}$$
$$\cos B = \frac{c^2 + a^2 - b^2}{2ac}$$
$$\cos C = \frac{a^2 + b^2 - c^2}{2ab}$$

The above formulae are called the cosine formulae.

191. *(d)* By sine rule, we have

$$\frac{a}{\sin A} = \frac{b}{\sin B} = \frac{c}{\sin C} = k \text{ (say)}$$

Therefore, $\frac{b-c}{b+c} = \frac{k (\sin B - \sin C)}{k (\sin B + \sin C)}$

$$= \frac{2 \cos \frac{B+C}{2} \sin \frac{B-C}{2}}{2 \sin \frac{B+C}{2} \cos \frac{B-C}{2}}$$
$$= \cot \frac{(B+2)}{2} \tan \frac{(B-C)}{2}$$
$$= \cot \left(\frac{\pi}{2} - \frac{A}{2}\right) \tan \left(\frac{B-C}{2}\right)$$
$$= \frac{\tan \frac{B-C}{2}}{\cot \frac{A}{2}}$$

Therefore, $\tan \frac{B-C}{2} = \frac{b-c}{b+c} \cot \frac{A}{2}$

Similarly, $\tan \left(\frac{A-B}{2}\right) = \frac{a-b}{a-c} \cot \frac{C}{2}$

and $\tan \left(\frac{C-A}{2}\right) = \frac{c-a}{c+a} \cot \frac{B}{2}$

These results are well known as Napier's Analogies.

192. *(a)* Consider

$$a \sin (B - C) = a [\sin B \cos C - \cos B \sin C] \quad \ldots\text{(i)}$$

Now, $\frac{\sin A}{a} = \frac{\sin B}{b} = \frac{\sin C}{c} = K$ (say)

Therefore, $\sin A = aK, \sin B = bK, \sin C = cK$

Substituting the values of $\sin B$ and $\sin C$ in Eq. (i) and using cosine formulae, we get

$$a \sin (B - C) = a \left[bK \left(\frac{a^2 + b^2 - c^2}{2\,ab}\right) - cK \left(\frac{c^2 + a^2 - b^2}{2\,ac}\right)\right]$$
$$= \frac{K}{2} (a^2 + b^2 - c^2 - c^2 - a^2 + b^2) = K (b^2 - c^2)$$

Similarly, $b \sin (C - A) = K (c^2 - a^2)$

and $c \sin (A - B) = K (a^2 - b^2)$

Hence, $a \sin (B - C) + b \sin (C - A) + c \sin(A - B)$

$$= K (b^2 - c^2 + c^2 - a^2 + a^2 - b^2)$$
$$= 0$$

193. *(b)* From the figure, we have

$$\angle PAQ = 45°, \angle BAQ = 30°, \angle PBH = 60°$$

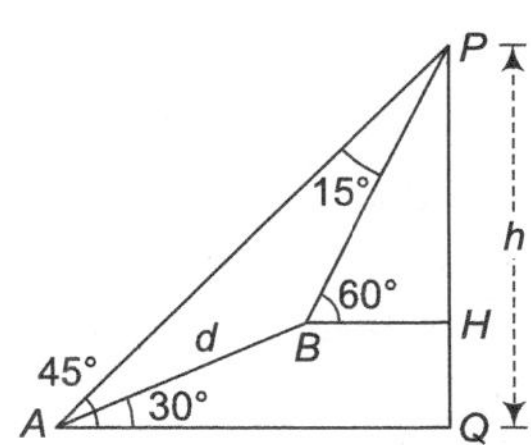

Clearly, $\angle APQ = 45°, \angle BPH = 30°$, giving $\angle APB = 15°$

Again, $\angle PAB = 15°$

$\Rightarrow$ $\angle ABP = 150°$

For $\Delta\ APQ$, we have

$$AP^2 = h^2 + h^2 = 2h^2$$
$$(\because \angle APQ = \angle PAQ = 45°, \therefore AQ = PQ)$$

$\Rightarrow$ $AP = \sqrt{2}h$

Applying sine formulae in ΔABP, we get

$$\frac{AB}{\sin 15°} = \frac{AP}{\sin 150°}$$

$\Rightarrow$ $\frac{d}{\sin 15°} = \frac{\sqrt{2}h}{\sin 150°}$

i.e., $d = \frac{\sqrt{2}h \sin 15°}{\sin 30°} = h (\sqrt{3} - 1)$

194. *(d)* From the figure, we have $AB = 9\,m = c, BC = 7\,m = a$ and $AC = 8\,m = b$.

M is the mid-point of the side AC at which lamp post MP of height h (say) is located. Again, it is given that lamp post subtends an angle θ (say) at B which is 15°.

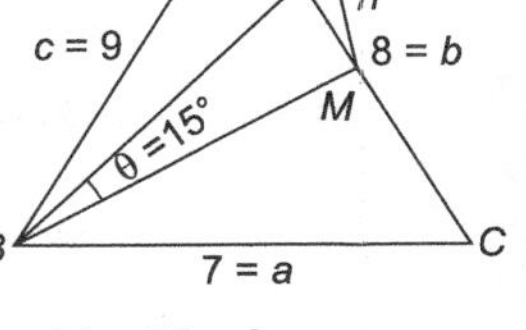

Applying cosine formulae in ΔABC, we have

$$\cos C = \frac{a^2 + b^2 - c^2}{2ab} = \frac{49 + 64 - 81}{2 \times 7 \times 8} = \frac{2}{7} \quad \ldots\text{(i)}$$

Similarly, using cosine formulae in ΔBMC, we get

$$BM^2 = BC^2 + CM^2 - 2\,BC \times CM \cos C.$$

Here, $CM = \frac{1}{2} CA = 4$, since M is the mid-point of AC.

Therefore, using Eq. (i), we get

$$BM^2 = 49 + 16 - 2 \times 7 \times 4 \times \frac{2}{7}$$

$$\therefore \quad BM = 7$$

Thus, from ΔBMP right angled at M, we have

$$\tan \theta = \frac{PM}{BM} = \frac{h}{7}$$

$$\Rightarrow \quad \frac{h}{7} = \tan(15°) = 2 - \sqrt{3}$$

$$\Rightarrow \quad h = 7(2 - \sqrt{3}) \text{ m}$$

195. *(a)* $a = 18, b = 24$ and $c = 30$

$$\therefore \quad \cos A = \frac{b^2 + c^2 - a^2}{2bc} = \frac{24^2 + 30^2 - 18^2}{2 \times 24 \times 30}$$

$$= \frac{576 + 900 - 324}{1440} = \frac{1152}{1440} = \frac{128}{160}$$

$$= \frac{8}{10} = \frac{4}{5}$$

196. *(a)* $\cos B = \frac{c^2 + a^2 - b^2}{2ac}$

$$= \frac{900 + 324 - 576}{2 \times 18 \times 30} = \frac{648}{1080} = \frac{72}{120} = \frac{3}{5}$$

197. *(d)* $\cos C = \frac{a^2 + b^2 - c^2}{2ab}$

$$= \frac{324 + 576 - 900}{2 \times 18 \times 24} = 0$$

198. *(d)* Since, $a = 18, b = 24$ and $c = 30$,

and $\quad 30^2 = 24^2 + 18^2 \quad (\because c^2 = a^2 + b^2)$

So, we see that $\angle C = 90°$ and by sine formulae

$$\frac{a}{\sin A} = \frac{b}{\sin B} = \frac{c}{\sin c}$$

$$\Rightarrow \quad \frac{\sin A}{a} = \frac{\sin B}{b} = \frac{\sin C}{c} = K \quad \text{(say)}$$

$$\Rightarrow \quad \frac{\sin A}{18} = \frac{\sin B}{24} = \frac{\sin C}{30} = K$$

$\therefore \quad \sin A = 18K, \sin B = 24K$

and $\quad \sin C = 30K$.

But $\quad \angle C = 90° \Rightarrow \sin C = \sin 90° = 1$

$$\Rightarrow \quad 30K = 1 \Rightarrow K = \frac{1}{30}$$

$$\therefore \quad \sin A = 18K = 18 \times \frac{1}{30} = \frac{3}{5}$$

$$\sin B = 24K = 24 \times \frac{1}{30} = \frac{4}{5}$$

and $\quad \sin C = 30K = 30 \times \frac{1}{30} = 1$

$$\therefore \quad \sin A = \frac{3}{5}, \sin B = \frac{4}{5} \text{ and } \sin C = 1$$

199. *(a)* A. $\frac{a+b}{c} = \frac{K \sin A + K \sin B}{K \sin C}$

$$\left[\because \frac{a}{\sin A} = \frac{b}{\sin B} = \frac{c}{\sin C} = K \text{ (say)}\right]$$

$$= \frac{2 \sin \frac{A+B}{2} \cdot \cos \frac{A-B}{2}}{2 \sin \frac{C}{2} \cdot \cos \frac{C}{2}}$$

Since, in a ΔABC, $A + B + C = 180°$

$$\Rightarrow \quad \frac{A}{2} + \frac{B}{2} + \frac{C}{2} = 90°$$

$$\Rightarrow \quad \frac{A+B}{2} = 90 - \frac{C}{2}$$

$$\therefore \quad \frac{a+b}{c} = \frac{2 \sin\left(90 - \frac{C}{2}\right) \cdot \cos \frac{A-B}{2}}{2 \sin \frac{C}{2} \cdot \cos \frac{C}{2}}$$

$$= \frac{\cos \frac{C}{2} \cdot \cos \frac{A-B}{2}}{\sin \frac{C}{2} \cdot \cos \frac{C}{2}}$$

$$= \frac{\cos \frac{A-B}{2}}{\sin \frac{C}{2}}$$

B. Similarly, $\frac{a-b}{c} = \frac{K(\sin A - \sin B)}{K \sin C}$

(using sine formulae)

$$= \frac{2 \cos \frac{A+B}{2} \cdot \sin \frac{A-B}{2}}{2 \sin \frac{C}{2} \cdot \cos \frac{C}{2}}$$

$$= \frac{\cos\left(90 - \frac{C}{2}\right) \cdot \sin \frac{A-B}{2}}{\sin \frac{C}{2} \cdot \cos \frac{C}{2}}$$

$$= \frac{\sin \frac{C}{2} \cdot \sin \frac{A-B}{2}}{\sin \frac{C}{2} \cdot \cos \frac{C}{2}}$$

$$= \frac{\sin \frac{A-B}{2}}{\cos \frac{C}{2}}$$

C. $\sin \frac{B-C}{2} = \frac{\sin \frac{B-C}{2} \times \cos A/2}{\cos A/2}$

$$\because \quad = \frac{\sin \frac{B-C}{2} \times \cos A/2}{\sin\left(\frac{\pi}{2} - \frac{A}{2}\right)}$$

$$= 2 \frac{\sin \frac{B-C}{2} \cdot \cos A/2 \cdot \cos \frac{B+C}{2}}{2 \cdot \sin \frac{B+C}{2} \cdot \cos \frac{B+C}{2}}$$

$$= \frac{\left[\sin\left(\frac{B-C}{2}+\frac{B+C}{2}\right)+\sin\left(\frac{B-C}{2}-\frac{B+C}{2}\right)\right]\cdot\cos\frac{A}{2}}{\sin(B+C)}$$

$$= \frac{\sin B+\sin(-C)}{\sin A}\cos\frac{A}{2}$$

$$= \frac{\sin B-\sin C}{\sin A}\cos\frac{A}{2}$$

$$= \frac{b-c}{a}\cos\frac{A}{2}$$

$\therefore \quad A \leftrightarrow 2, B \leftrightarrow 1, C \leftrightarrow 3$

200. *(b)* $a(b\cos C - c\cos B)$

$$= a\left[b\left(\frac{a^2+b^2-c^2}{2ab}\right)-c\left(\frac{a^2+c^2-b^2}{2ac}\right)\right]$$

$$= \frac{a^2+b^2-c^2-a^2-c^2+b^2}{2}$$

$$= b^2-c^2$$

201. *(a)* $(b+c)\cos\left(\frac{B+C}{2}\right)$

$$= K(\sin B+\sin C)\cos\frac{B+C}{2}$$

$$= K\,2\sin\frac{B+C}{2}\cos\frac{B-C}{2}\cos\frac{B+C}{2}$$

$$= K\sin(B+C)\cos\frac{B-C}{2}$$

$$= K\sin A\cos\frac{B-C}{2}$$

$$= a\cos\frac{B-C}{2}$$

202. *(c)* I. $\frac{\cos A}{a}+\frac{\cos B}{b}+\frac{\cos C}{c}$

$$= \frac{b^2+c^2-a^2}{2bc\cdot a}+\frac{c^2+a^2-b^2}{2ac\cdot b}+\frac{a^2+b^2-c^2}{2ab\cdot c}$$

$$= \frac{b^2+c^2-a^2+c^2+a^2-b^2+a^2+b^2-c^2}{2abc}$$

$$= \frac{a^2+b^2+c^2}{2abc}$$

II. We know that, $\cos A = \frac{b^2+c^2-a^2}{2bc}$

$\therefore$ Only Statement I is true.

203. *(a)* By the sine rule, we have

$$\frac{a}{\sin A}=\frac{b}{\sin B}=\frac{c}{\sin C}=K \text{ (say)}$$

$\Rightarrow \quad a = K\sin A, b = K\sin B$

and $\quad c = K\sin C.$

Applying sine rule and cosine formula, we have

$$\frac{(b^2-c^2)}{a^2}\sin 2A = \frac{(b^2-c^2)}{a^2}\cdot(2\sin A\cos A)$$

$$= \frac{(b^2-c^2)}{a^2}\left(\frac{2a}{K}\right)\cdot\frac{(b^2+c^2-a^2)}{2bc}$$

$$\left[\because \sin A = \frac{a}{K} \text{ and } \cos A = \frac{(b^2+c^2-a^2)}{2bc}\right]$$

$$= \frac{1}{(Kabc)}\cdot(b^2-c^2)(b^2+c^2-a^2) \quad \ldots\text{(i)}$$

Similarly, $\frac{(c^2-a^2)}{b^2}\sin 2B = \frac{1}{(Kabc)}\cdot(c^2-a^2)(c^2+a^2-b^2)\ldots\text{(ii)}$

and, $\frac{(a^2-b^2)}{c^2}\sin 2C = \frac{1}{(Kabc)}\cdot(a^2-b^2)(a^2+b^2-c^2)\ldots\text{(iii)}$

From Eqs. (i), (ii) and (iii), we get

$$\frac{(b^2-c^2)}{a^2}\sin 2A+\frac{(c^2-a^2)}{b^2}\sin 2B+\frac{(a^2-b^2)}{c^2}\sin 2C$$

$$= \frac{1}{(Kabc)}\cdot[(b^2-c^2)(b^2+c^2-a^2)+(c^2-a^2)(c^2+a^2-b^2)+(a^2-b^2)(a^2+b^2-c^2)]$$

$$= \frac{1}{(Kabc)}\times 0 = 0$$

$$\therefore \frac{(b^2-c^2)}{a^2}\sin 2A+\frac{(c^2-a^2)}{b^2}\sin 2B+\frac{(a^2-b^2)}{c^2}\sin 2C = 0$$

204. *(b)* I. Applying cosine formula, we have

$a(\cos C - \cos B)$

$$= a\left[\frac{(a^2+b^2-c^2)}{2ab}-\frac{(a^2+c^2-b^2)}{2ac}\right]$$

$$= \frac{(a^2c+b^2c-c^3-a^2b-bc^2+b^3)}{2bc}$$

$$= \frac{(b^3-c^3)+(b^2c-bc^2)-(a^2b-a^2c)}{2bc}$$

$$= \frac{(b^3-c^3)+bc(b-c)-a^2(b-c)}{2bc}$$

$$= (b-c)\frac{[(b^2+c^2+bc)+(bc-a^2)]}{2bc}$$

$$= (b-c)\cdot\left[\frac{(b^2+c^2-a^2)}{2bc}+\frac{2bc}{2bc}\right]$$

$$= (b-c)\left[\frac{(b^2+c^2-a^2)}{2bc}+1\right] = (b-c)(1+\cos A)$$

$$= 2(b-c)\cos^2\frac{A}{2}$$

$$\therefore a(\cos C-\cos B) = 2(b-c)\cos^2\frac{A}{2}$$

II. Applying sine rule, we have

$$\frac{a}{\sin A}=\frac{b}{\sin B}=\frac{c}{\sin C}=K \text{ (say)}$$

$a = K\sin A, b = K\sin B$ and $c = K\sin C$

Now, $a\cos A + b\cos B + c\cos C$

$$= K\sin A\cos A + K\sin B\cos B + K\sin C\cos C$$

$$= \frac{1}{2} K (\sin 2A + \sin 2B + \sin 2C)$$
$$= K [\sin (A + B) \cos (A - B) + \sin C \cos C]$$
$$= K [\sin (\pi - C) \cos (A - B) + \sin C \cos C]$$
$$= K \sin C [\cos (A - B) + \cos C]$$
$$= K \sin C [\cos (A - B) + \cos \{\pi - (A + B)\}]$$
$$= K \sin C [\cos (A - B) - \cos (A + B)]$$
$$= K \sin C \times 2 \sin A \sin B$$
$$= 2 (K \sin A) \sin B \sin C$$
$$= 2a \sin B \sin C$$

Hence, $a \cos A + b \cos B + c \cos C = 2a \sin B \sin C$

205. *(b)* $\because \dfrac{(b^2 - c^2)}{a^2} = \dfrac{K^2 \sin^2 B - K^2 \sin^2 C}{K^2 \sin^2 A}$

$$= \frac{(\sin^2 B - \sin^2 C)}{\sin^2 A} = \frac{\sin (B + C) \cdot \sin (B - C)}{\sin^2 (B + C)}$$

$$\left\{\begin{array}{l} \because \ (A + B + C) = \pi \Rightarrow A = \pi - (B + C) \\ \therefore \ \sin A = \sin [\pi - (B + C)] = \sin (B + C) \end{array}\right\}$$

$$= \frac{\sin (B - C)}{\sin (B + C)}$$

Hence, $\dfrac{\sin (B - C)}{\sin (B + C)} = \dfrac{(b^2 - c^2)}{a^2}$

206. *(d)* By the sine rule, we have

$$\frac{a}{\sin A} = \frac{b}{\sin B} = \frac{c}{\sin C}$$

$$\Rightarrow \quad \frac{\sin A}{a} = \frac{\sin B}{b} = \frac{\sin C}{c} = K \text{ (say)}$$

$\Rightarrow \sin A = Ka, \sin B = Kb$ and $\sin C = Kc$

$$\therefore \quad (b^2 - c^2) \cot A = \frac{(b^2 - c^2) \cdot \cos A}{\sin A}$$
$$= \frac{(b^2 - c^2)}{Ka} \cdot \frac{(b^2 + c^2 - a^2)}{2bc}$$
$$= \frac{1}{(2Kabc)} \cdot (b^2 + c^2)(b^2 + c^2 - a^2)$$
$$= \frac{(b^4 - c^4 - a^2b^2 + a^2c^2)}{2Kabc}$$

Similarly, $(c^2 - a^2) \cot B = \dfrac{(c^4 - a^4 - b^2c^2 + a^2b^2)}{2Kabc}$

and $\quad (a^2 - b^2) \cot C = \dfrac{(a^4 - b^4 - a^2c^2 + b^2c^2)}{2Kabc}$

$(b^2 - c^2) \cot A + (c^2 - a^2) \cot B + (a^2 - b^2) \cot C$

$$= \frac{1}{2Kabc} [(b^4 - c^4 - a^2b^2 + a^2c^2)$$
$$+ (c^4 - a^4 - b^2c^2 + a^2b^2 + (a^4 - b^4 - a^2c^2 + b^2c^2)]$$
$$= \frac{1}{2Kabc} \times 0 = 0$$

Hence, $(b^2 - c^2) \cot A + (c^2 - a^2) \cot B$
$+ (a^2 - b^2) \cot C = 0$

207. *(c)* Let PAQ be the hill, AB be the tree and PCD be the horizontal. Let P be the point of observation.

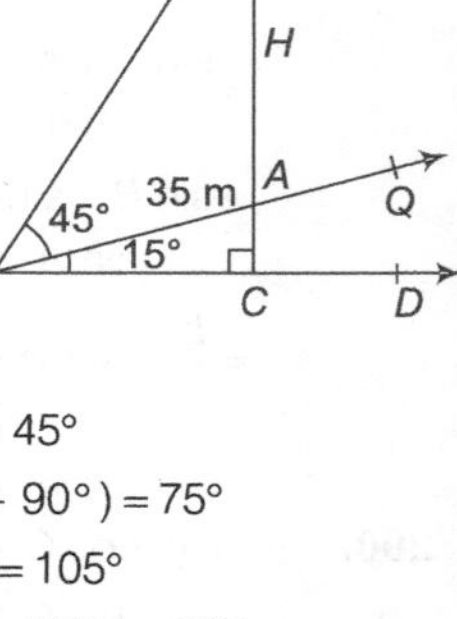

Produce BA to meet at C. Let $AB = H$ m.

Then, $\angle DPA = 15°$, $PA = 35$ m

$\angle CPB = 60°$

and $\angle PCA = 90°$.

$\therefore \quad \angle APB = (60° - 15°) = 45°$

In ΔPAC, $\quad \angle PAC = 180° - (15° + 90°) = 75°$

$\therefore \quad \angle PAB = (180° - 75°) = 105°$

and $\quad \angle PBA = 180° - (45° + 105°) = 30°$

Applying sine rule on ΔPAB, we get

$$\frac{PA}{\sin \angle PBA} = \frac{AB}{\sin \angle APB} \Rightarrow \frac{35}{\sin 30°} = \frac{H}{\sin 45°}$$

$$\therefore \quad 35 \times 2 = H \times \sqrt{2} \Rightarrow H = 35\sqrt{2} \text{ m}$$

Hence, the height of the tree is $35\sqrt{2}$ m.

208. *(d)* Let A and B be the positions of the two ships at the end of 3 h.

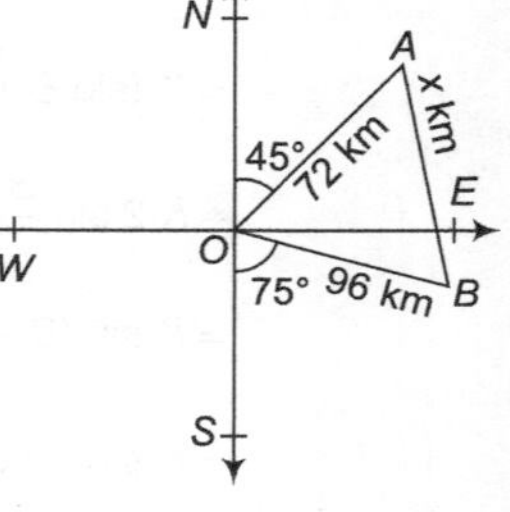

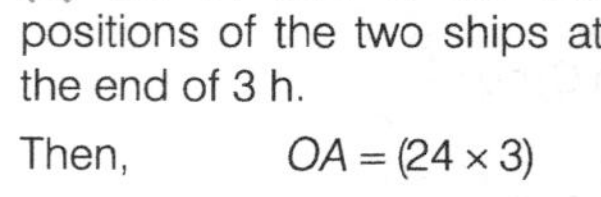

Then, $\quad OA = (24 \times 3)$
$= 72$ km

And, $\quad OB = (32 \times 3)$
$= 96$ km

Let $\quad AB = x$ km

We have, $\angle NOA = 45°$

and $\quad \angle SOB = 75°$

$\therefore \quad \angle AOB = 180° - (45° + 75°) = 60°$

Using cosine formula on ΔAOB, we get

$$\cos 60° = \frac{(72)^2 + (96)^2 - x^2}{2 \times 72 \times 96}$$

$$\Rightarrow \quad (72)^2 + (96)^2 - x^2 = \frac{1}{2} \times 2 \times 72 \times 96$$
$$\Rightarrow \quad 5184 + 9216 - x^2 = 6912$$
$$\Rightarrow \quad x^2 = 14400 - 6912 = 7488$$
$$\Rightarrow \quad x = \sqrt{7488} = 86.53 \text{ km}$$

Hence, the distance between the ships at the end of 3 h is 86.53 km $\approx$ 86.4 km (approx).

209. *(a)* Let A and B be the trees and C be a point in the river. Then, $CA = 250$ m, $CB = 300$ m and $\angle ACB = 45°$

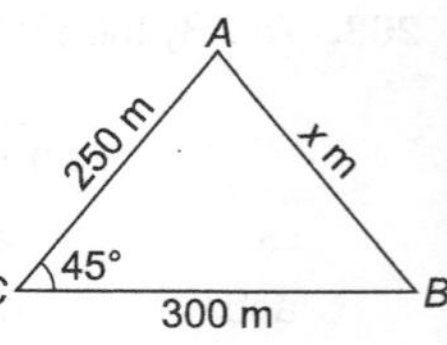

Let $AB = x$ m

Applying cosine formula on ΔACB, we get,

$$\cos C = \frac{a^2 + b^2 - c^2}{2ab}$$

$$\Rightarrow \quad \cos 45° = \frac{(300)^2 + (250)^2 - x^2}{2 \times 300 \times 250}$$

$$\Rightarrow \quad \frac{1}{\sqrt{2}} = \frac{90000 + 62500 - x^2}{150000}$$

$$\Rightarrow \quad \frac{152500 - x^2}{150000} = \frac{1}{\sqrt{2}} \times \frac{\sqrt{2}}{\sqrt{2}} = \frac{\sqrt{2}}{2}$$

$$\Rightarrow \quad 152500 - x^2 = 75000 \times \sqrt{2} = 75000 \times 1.44 = 108000$$

$$\Rightarrow \quad x^2 = 152500 - 108000 = 44500$$

$$\Rightarrow \quad x = \sqrt{44500} = 210.95 \text{ m}$$

$$\approx 210 \text{ m (approx.)}$$

210. *(c)* Given, $\angle A = 60°, a = 5, b = 4$

$$\therefore \quad \cos A = \frac{b^2 + c^2 - a^2}{2bc}$$

$$\Rightarrow \quad \cos 60° = \frac{1}{2} = \frac{16 + c^2 - 25}{8c}$$

$$\Rightarrow \quad 4c = c^2 - 9$$

$$\Rightarrow \quad c^2 - 4c - 9 = 0$$

211. *(d)* $2ac \sin\left(\frac{A - B + C}{2}\right)$

$$= 2\,ac \sin\left(\frac{180° - 2B}{2}\right)$$

$$= 2\,ac \sin (90° - B) = 2ac \cos B = a^2 + c^2 - b^2$$

212. *(a)* $A + B + C = \pi$

$$2B = A + C = \pi - B \quad (\because A, B \text{ and } C \text{ are in AP})$$

$$\Rightarrow \quad 3B = \pi \Rightarrow B = \frac{\pi}{3}$$

$$\therefore \quad \frac{a + c}{b} = \frac{\sin A + \sin C}{\sin B}$$

$$= \frac{2 \sin \frac{A + C}{2} \cos \frac{A - C}{2}}{\sin B}$$

$$= \frac{2 \sin \frac{\pi}{3} \cos \frac{A - C}{2}}{\sin \frac{\pi}{3}}$$

$$= 2 \cos \frac{A - C}{2}$$

213. *(c)* Assume that angles are $4x$, x and x.

As $\quad 4x + x + x = 180° \quad (\because \angle A + \angle B + \angle C = 180°)$

$\Rightarrow \quad x = 30°$

$\therefore$ Angles are 120°, 30° and 30°.

Ratio of sides $= \sin A : \sin B : \sin C$

$$= \sin 120° : \sin 30° : \sin 30° = \sqrt{3} : 1 : 1$$

$$\therefore \text{Required ratio} = \frac{\sqrt{3}}{1 + 1 + \sqrt{3}} = \frac{\sqrt{3}}{2 + \sqrt{3}}$$

214. *(d)* Using sine rule in ΔACD,

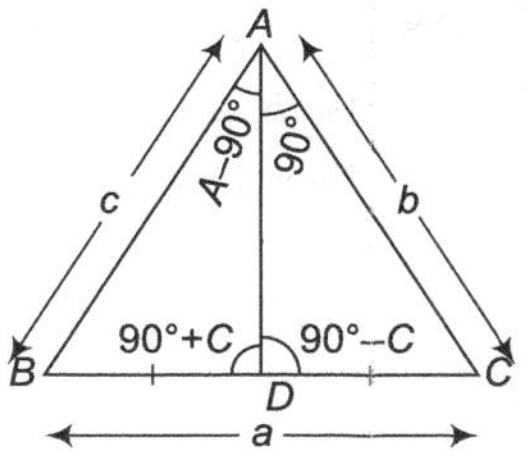

$$\frac{\sin (90° - C)}{b} = \frac{\sin 90°}{\frac{a}{2}}$$

$$\Rightarrow \quad \cos C = \frac{2b}{a}$$

and in ΔABD,

$$\frac{\sin (A - 90°)}{\frac{a}{2}} = \frac{\sin (90° + C)}{c}$$

$$\Rightarrow -\frac{\cos A}{\frac{a}{2}} = \frac{\cos C}{c} \quad \Rightarrow \quad \cos A = -\frac{b}{c}$$

$$\Rightarrow \quad \frac{b^2 + c^2 - a^2}{2bc} = -\frac{b}{c} \quad \Rightarrow \quad c^2 - a^2 = -3b^2$$

$$\therefore \quad \cos A \cos C = -\frac{2b^2}{ac} = \frac{2}{3ac}(c^2 - a^2)$$

215. *(d)* Given, $A - B = 60°$

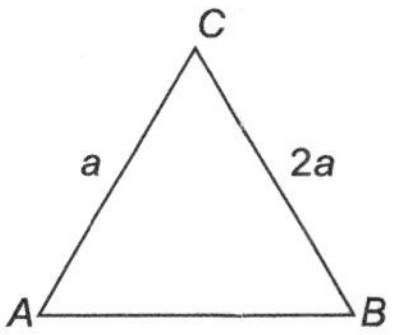

By sine rule, $\frac{2a}{\sin A} = \frac{a}{\sin B}$

$$\Rightarrow \quad \sin A - 2 \sin B = 0$$

$$\Rightarrow \quad \sin (60° + B) - 2 \sin B = 0$$

$$\Rightarrow \quad \frac{\sqrt{3}}{2} \cos B + \frac{1}{2} \sin B - 2 \sin B = 0$$

$$\Rightarrow \quad \frac{\sqrt{3}}{2} \cos B - \frac{3}{2} \sin B = 0$$

$$\Rightarrow \quad \sqrt{3}\left(\frac{1}{2} \cos B - \frac{\sqrt{3}}{2} \sin B\right) = 0$$

$$\Rightarrow \quad \sqrt{3}\,[\cos (60° + B)] = 0$$

$$\Rightarrow \quad 60° + B = 90°$$

$$\Rightarrow \quad B = 30°$$

$$\Rightarrow \quad A = 90°$$

Hence, it is a right angled triangle.

216. *(c)* Using sine rule,

$$\frac{\sin A}{a} = \frac{\sin B}{b} \Rightarrow \frac{\frac{2}{3}}{2} = \frac{\sin B}{3}$$

$$\Rightarrow \quad \sin B = 1 \Rightarrow B = 90°$$

217. *(c)* Given, $\sin A \sin B = \frac{ab}{c^2}$

$$\Rightarrow \quad c^2 = \frac{ab}{\sin A \sin B} = \left(\frac{a}{\sin A}\right)\left(\frac{b}{\sin A}\right)$$

$$\Rightarrow \quad c^2 = \left(\frac{c}{\sin C}\right)^2 \quad \left(\because \frac{a}{\sin A} = \frac{b}{\sin B} = \frac{c}{\sin C}\right)$$

$$\Rightarrow \quad \sin^2 C = 1$$

$$\Rightarrow \quad C = 90°$$

Hence, ΔABC is a right angled triangle.

218. *(a)* Let $a = 7$ cm, $b = 4\sqrt{3}$ cm and $c = \sqrt{13}$ cm

Here, we see that the smallest side is c.

Therefore, the smallest angle will be C.

$$\therefore \quad \cos C = \frac{(7)^2 + (4\sqrt{3})^2 - (\sqrt{13})^2}{2 \times 7 \times 4\sqrt{3}} = \frac{\sqrt{3}}{2}$$

$$\Rightarrow \quad \angle C = \frac{\pi}{6}$$

219. *(b)* Using $\cos C = \dfrac{a^2 + b^2 - c^2}{2ab}$

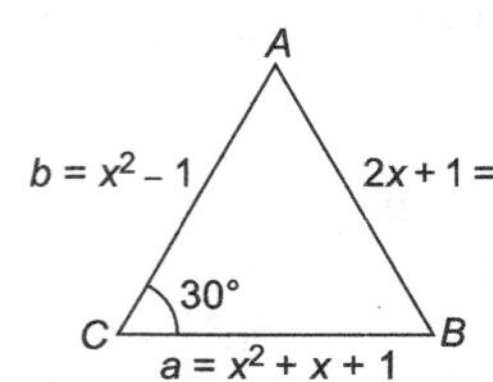

$$\Rightarrow \quad \frac{\sqrt{3}}{2} = \frac{(x^2 + x + 1)^2 + (x^2 - 1)^2 - (2x + 1)^2}{2(x^2 + x + 1)(x^2 - 1)}$$

$$\Rightarrow \quad (x + 2)(x + 1)(x - 1)x + (x^2 - 1)^2$$
$$= \sqrt{3}(x^2 + x + 1)(x^2 - 1)$$

$$\Rightarrow \quad x^2 + 2x + (x^2 - 1) = \sqrt{3}(x^2 + x + 1)$$

$$\Rightarrow \quad (2 - \sqrt{3})x^2 + (2 - \sqrt{3})x - (\sqrt{3} + 1) = 0$$

$$\Rightarrow \quad x = -(2 + \sqrt{3})$$

and $\quad x = 1 + \sqrt{3}$

But $\quad x = -(2 + \sqrt{3}) \Rightarrow c$ is negative.

$\therefore \quad x = 1 + \sqrt{3}$ is the only solution.

220. *(a)* Using sine rule,

$$\frac{a}{\sin A} = \frac{b}{\sin B}$$

$$\Rightarrow \quad \frac{2\sqrt{2}}{\sin 45^\circ} = \frac{6}{\sin B}$$

$$\Rightarrow \quad \sin B = \frac{6}{2\sqrt{2}} \times \frac{1}{\sqrt{2}}$$
$$= \frac{6}{4} = \frac{3}{2} > 1$$

which is not possible.

Hence, no triangle is possible.

221. *(b)* $\because 180^\circ = \pi^c \Rightarrow 1^\circ = \left(\dfrac{\pi}{180}\right)^c$

$$\Rightarrow \quad 25^\circ = \left(\frac{\pi}{180} \times 25\right)^c = \left(\frac{5\pi}{36}\right)^c$$

222. *(a)* Since, 1 radian $= \left(\dfrac{180}{\pi}\right)^\circ$

$$\Rightarrow \quad 11 \text{ radian} = \left(\frac{180}{\pi} \times 11\right)^\circ = \left(\frac{180}{22} \times 7 \times 11\right)^\circ$$
$$= 630^\circ$$

223. *(c)* A. 1 radian $= \dfrac{180^\circ}{\pi} = 57^\circ\, 16'$ (approx.)

B. $\quad 1^\circ = \left(\dfrac{\pi}{180}\right)^c = 0.01746$ radian (approx.)

C. $\quad 3^\circ\, 45' = \left(3\dfrac{45}{60}\right)^\circ = \left(3\dfrac{3}{4}\right)^\circ = \left(\dfrac{15}{4}\right)^\circ$

Also, $180^\circ = \pi$ radian $\Rightarrow 1^\circ = \dfrac{\pi}{180}$ radian

$$\Rightarrow \quad \left(\frac{15}{4}\right)^\circ = \frac{\pi}{180} \times \frac{15}{4} = \frac{\pi}{48} \text{ radian}$$

D. $\quad 50^\circ\, 37'\, 30'' = 50^\circ + \left(37\dfrac{30}{60}\right)'$

$$= 50^\circ + \left(\frac{75}{2}\right)' = 50^\circ + \left(\frac{75}{2 \times 60}\right)^\circ$$

$$= \left(\frac{405}{8}\right)^\circ = \left(\frac{\pi}{180} \times \frac{405}{8}\right)^c = \frac{9\pi}{32} \text{ radian}$$

$$\therefore \quad A \leftrightarrow 2, B \leftrightarrow 1, C \leftrightarrow 4, D \leftrightarrow 3$$

224. *(b)* We know that,

$$\sin(x + y) = \sin x \cos y + \cos x \sin y \qquad \ldots\text{(i)}$$

Now, $\quad \cos^2 x = 1 - \sin^2 x = 1 - \dfrac{9}{25} = \dfrac{16}{25}$

Therefore, $\quad \cos x = \pm \dfrac{4}{5}$

Since, x lies in second quadrant, $\cos x$ is negative.

Hence, $\quad \cos x = -\dfrac{4}{5}$

Now, $\quad \sin^2 y = 1 - \cos^2 y$
$$= 1 - \frac{144}{169} = \frac{25}{169}$$

i.e., $\quad \sin y = \pm \dfrac{5}{13}$

Since, y lies in second quadrant, hence $\sin y$ is positive. Therefore, $\sin y = \dfrac{5}{13}$. Substituting the values of $\sin x, \sin y, \cos x$ and $\cos y$ in Eq. (i), we get

$$\sin(x + y) = \frac{3}{5} \times \left(-\frac{12}{13}\right) + \left(-\frac{4}{5}\right) \times \frac{5}{13}$$
$$= -\frac{36}{65} - \frac{20}{65} = -\frac{56}{65}$$

225. *(b)* We have, $\dfrac{1}{2}\left[2\cos 2x \cos \dfrac{x}{2} - 2\cos \dfrac{9x}{2} \cos 3x\right]$

$$= \frac{1}{2}\left[\cos\left(2x + \frac{x}{2}\right) + \cos\left(2x - \frac{x}{2}\right) - \cos\left(\frac{9x}{2} + 3x\right) - \cos\left(\frac{9x}{2} - 3x\right)\right]$$

$$= \frac{1}{2}\left[\cos \frac{5x}{2} + \cos \frac{3x}{2} - \cos \frac{15x}{2} - \cos \frac{3x}{2}\right]$$

$$= \frac{1}{2}\left[\cos\frac{5x}{2} - \cos\frac{15x}{2}\right]$$

$$= \frac{1}{2}\left[-2\sin\left\{\frac{\frac{5x}{2}+\frac{15x}{2}}{2}\right\}\sin\left\{\frac{\frac{5x}{2}-\frac{15x}{2}}{2}\right\}\right]$$

$$= -\sin 5x \sin\left(-\frac{5x}{2}\right)$$

$$= \sin 5x \sin\frac{5x}{2}$$

226. *(b)* Let $x = \frac{\pi}{8}$. Then, $2x = \frac{\pi}{4}$

Now, $\tan 2x = \frac{2\tan x}{1-\tan^2 x}$

or $\tan\frac{\pi}{4} = \frac{2\tan\frac{\pi}{8}}{1-\tan^2\frac{\pi}{8}}$

Let $y = \tan\frac{\pi}{8}$. Then, $1 = \frac{2y}{1-y^2}$

or $y^2 + 2y - 1 = 0$

Therefore, $y = \frac{-2 \pm 2\sqrt{2}}{2} = -1 \pm \sqrt{2}$

Since $\frac{\pi}{8}$ lies in the first quadrant, $y = \tan\frac{\pi}{8}$ is positive.

Hence, $\tan\frac{\pi}{8} = \sqrt{2} - 1$

227. *(a)* Since $\pi < x < \frac{3\pi}{2}$, $\cos x$ is negative.

Also, $\frac{\pi}{2} < \frac{x}{2} < \frac{3\pi}{4}$

Therefore, $\sin\frac{x}{2}$ is positive and $\cos\frac{x}{2}$ is negative.

Now, $\sec^2 x = 1 + \tan^2 x = 1 + \frac{9}{16} = \frac{25}{16}$

Therefore, $\cos^2 x = \frac{16}{25}$ or $\cos x = -\frac{4}{5}$

Now, $2\sin^2\frac{x}{2} = 1 - \cos x = 1 + \frac{4}{5} = \frac{9}{5}$

Therefore, $\sin^2\frac{x}{2} = \frac{9}{10}$ or $\sin\frac{x}{2} = \frac{3}{\sqrt{10}}$

Again, $2\cos^2\frac{x}{2} = 1 + \cos x = 1 - \frac{4}{5} = \frac{1}{5}$

Therefore, $\cos^2\frac{x}{2} = \frac{1}{10}$

or $\cos\frac{x}{2} = -\frac{1}{\sqrt{10}}$

Hence, $\tan\frac{x}{2} = \frac{\sin\frac{x}{2}}{\cos\frac{x}{2}} = \frac{3}{\sqrt{10}} \times \left(\frac{-\sqrt{10}}{1}\right) = -3$

228. *(d)* We have, $\cos^2 x + \cos^2\left(x + \frac{\pi}{3}\right) + \cos^2\left(x - \frac{\pi}{3}\right)$

$$= \frac{1+\cos 2x}{2} + \frac{1+\cos\left(2x + \frac{2\pi}{3}\right)}{2} + \frac{1+\cos\left(2x - \frac{2\pi}{3}\right)}{2}$$

$$= \frac{1}{2}\left[3 + \cos 2x + \cos\left(2x + \frac{2\pi}{3}\right) + \cos\left(2x - \frac{2\pi}{3}\right)\right]$$

$$= \frac{1}{2}\left[3 + \cos 2x + 2\cos 2x \cos\frac{2\pi}{3}\right]$$

$$= \frac{1}{2}\left[3 + \cos 2x + 2\cos 2x \cos\left(\pi - \frac{\pi}{3}\right)\right]$$

$$= \frac{1}{2}\left[3 + \cos 2x - 2\cos 2x \cos\frac{\pi}{3}\right]$$

$$= \frac{1}{2}[3 + \cos 2x - \cos 2x] = \frac{3}{2}$$

229. *(a)* $2\cos\frac{\pi}{3}\cos\frac{9\pi}{13} + \cos\frac{3\pi}{13} + \cos\frac{5\pi}{13}$

$$= \cos\left(\frac{9\pi}{13} + \frac{\pi}{13}\right) + \cos\left(\frac{9\pi}{13} - \frac{\pi}{13}\right) + \cos\frac{5\pi}{13} + \cos\frac{3\pi}{13}$$

(by formulae)

$$= \cos\frac{10\pi}{13} + \cos\frac{8\pi}{13} + \cos\frac{5\pi}{13} + \cos\frac{3\pi}{13}$$

$$= \left(\cos\frac{10\pi}{13} + \cos\frac{3\pi}{13}\right) + \left(\cos\frac{8\pi}{13} + \cos\frac{5\pi}{13}\right)$$

$$= 2\cos\frac{\frac{10\pi}{13} + \frac{3\pi}{13}}{2}\cos\frac{\frac{10\pi}{13} - \frac{3\pi}{13}}{2} + 2\cos\frac{\frac{8\pi}{13} + \frac{5\pi}{13}}{2}\cos\frac{\frac{8\pi}{13} - \frac{5\pi}{13}}{2}$$

(by formulae)

$$= 2\cos\frac{\pi}{2}\cos\frac{7\pi}{26} + 2\cos\frac{\pi}{2}\cos\frac{3\pi}{26} = 0 + 0 = 0$$

$\left(\because \cos\frac{\pi}{2} = 0\right)$

Note *Remember that* $\cos\frac{10\pi}{13} + \cos\frac{3\pi}{13} \neq \cos\frac{13\pi}{13}$

230. *(c)* I. $(\cos x + \cos y)^2 + (\sin x - \sin y)^2$

$$= \left(2\cos\frac{x+y}{2}\cos\frac{x-y}{2}\right)^2 + \left(2\cos\frac{x+y}{2}\sin\frac{x-y}{2}\right)^2$$

(by formulae)

$$= 4\cos^2\frac{x+y}{2}\cos^2\frac{x-y}{2} + 4\cos^2\frac{x+y}{2}\sin^2\frac{x-y}{2}$$

$$= 4\cos^2\frac{x+y}{2}\left(\cos^2\frac{x-y}{2} + \sin^2\frac{x-y}{2}\right)$$

$$= 4\cos^2\frac{x+y}{2} \qquad (\because \cos^2 x + \sin^2 x = 1)$$

II. $(\sin 3x + \sin x)\sin x + (\cos 3x - \cos x)\cos x$

$$= \left(2\sin\frac{3x+x}{2}\cos\frac{3x-x}{2}\right)\sin x$$
$$-\left(2\sin\frac{3x+x}{2}\sin\frac{3x-x}{2}\right)\cos x \text{ (by formulae)}$$
$$= 2\sin 2x\cos x\sin x - 2\sin 2x\sin x\cos x = 0$$

231. *(a)* $(\cos x - \cos y)^2 + (\sin x + \sin y)^2$

$$= \left(-2\sin\frac{x+y}{2}\sin\frac{x-y}{2}\right)^2 + \left(2\cos\frac{x+y}{2}\sin\frac{x-y}{2}\right)^2$$
(by formulae)
$$= 4\sin^2\frac{x+y}{2}\sin^2\frac{x-y}{2} + 4\cos^2\frac{x+y}{2}\sin^2\frac{x-y}{2}$$
$$= 4\sin^2\frac{x-y}{2}\left(\sin^2\frac{x+y}{2} + \cos^2\frac{x+y}{2}\right)$$
$$= 4\sin^2\frac{x-y}{2} \qquad (\because \cos^2 x + \sin^2 x = 1)$$

232. *(c)* $\sin x + \sin 3x + \sin 5x + \sin 7x$

$$= (\sin 7x + \sin x) + (\sin 5x + \sin 3x)$$
$$= 2\sin\frac{7x+x}{2}\cos\frac{7x-x}{2} + 2\sin\frac{5x+3x}{2}\cos\frac{5x-3x}{2}$$
(by formulae)
$$= 2\sin 4x\cos 3x + 2\sin 4x\cos x$$
$$= 2\sin 4x(\cos 3x + \cos x)$$
$$= 2\sin 4x\, 2\cos\frac{3x+x}{2}\cos\frac{3x-x}{2} \qquad \text{(by formulae)}$$
$$= 4\sin 4x\cos 2x\cos x$$

233. *(c)* $\dfrac{(\sin 7x + \sin 5x) + (\sin 9x + \sin 3x)}{(\cos 7x + \cos 5x) + (\cos 9x + \cos 3x)}$

$$= \frac{2\sin\frac{7x+5x}{2}\cos\frac{7x-5x}{2} + 2\sin\frac{9x+3x}{2}\cos\frac{9x-3x}{2}}{2\cos\frac{7x+5x}{2}\cos\frac{7x-5x}{2} + 2\cos\frac{9x+3x}{2}\cos\frac{9x-3x}{2}}$$
(by formulae)
$$= \frac{2\sin 6x\cos x + 2\sin 6x\cos 3x}{2\cos 6x\cos x + 2\cos 6x\cos 3x}$$
$$= \frac{2\sin 6x(\cos x + \cos 3x)}{2\cos 6x(\cos x + \cos 3x)} = \tan 6x$$

234. *(a)* $\sin 3x + \sin 2x - \sin x = (\sin 3x - \sin x) + \sin 2x$

$$= 2\cos\frac{3x+x}{2}\sin\frac{3x-x}{2} + \sin 2x$$
$$\left(\because \sin C - \sin D = 2\cos\frac{C+D}{2}\sin\frac{C-D}{2}\right)$$
$$= 2\cos 2x\sin x + 2\sin x\cos x \qquad (\because \sin 2x = 2\sin x\cos x)$$
$$= 2\sin x(\cos 2x + \cos x)$$
$$= 2\sin x\, 2\cos\frac{2x+x}{2}\cos\frac{2x-x}{2}$$
$$\left(\because \cos C + \cos D = 2\cos\frac{C+D}{2}\cos\frac{C-D}{2}\right)$$
$$= 4\sin x\cos\frac{3x}{2}\cos\frac{x}{2}$$

235. *(c)* $\tan x = -\dfrac{4}{3}$

Given that, x lies in second quadrant.

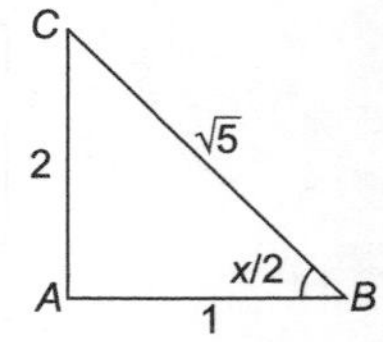

i.e., $\dfrac{\pi}{2} < x < \pi$

$\because$ $$\tan x = \frac{2\tan\frac{x}{2}}{1-\tan^2\frac{x}{2}} = -\frac{4}{3}$$

$\therefore$ $$3\tan\frac{x}{2} = -2\left(1-\tan^2\frac{x}{2}\right)$$

$\Rightarrow$ $$3\tan\frac{x}{2} = -2 + 2\tan^2\frac{x}{2}$$

$\Rightarrow$ $$2\tan^2\frac{x}{2} - 3\tan\frac{x}{2} - 2 = 0$$

$\Rightarrow$ $$2\tan^2\frac{x}{2} - (4-1)\tan\frac{x}{2} - 2 = 0$$

$\Rightarrow$ $$2\tan\frac{x}{2}\left(\tan\frac{x}{2} - 2\right) + 1\left(\tan\frac{x}{2} - 2\right) = 0$$

$\Rightarrow$ $$\left(2\tan\frac{x}{2} + 1\right)\left(\tan\frac{x}{2} - 2\right) = 0$$

$\Rightarrow$ $$\tan\frac{x}{2} = 2 \text{ or } \tan\frac{x}{2} = -\frac{1}{2}$$

$\therefore$ $$\frac{\pi}{2} < x < \pi \Rightarrow \frac{\pi}{4} < \frac{x}{2} < \frac{\pi}{2}$$

i.e., $\dfrac{x}{2}$ lies in first quadrant.

Therefore, $\tan\dfrac{x}{2} = 2 = \dfrac{2}{1} = \dfrac{\text{perpendicular}}{\text{base}} = \dfrac{AC}{AB}$

Using Pythagoras theorem, $(BC)^2 = (AC)^2 + (AB)^2$

$\Rightarrow$ $$(BC)^2 = 4 + 1 = 5$$

$\Rightarrow$ $$BC = \sqrt{5}$$

Now, $$\sin\frac{x}{2} = \frac{\text{perpendicular}}{\text{hypotenuse}} = \frac{2}{\sqrt{5}} = \frac{2\sqrt{5}}{5}$$

$\therefore$ $$\cos\frac{x}{2} = \frac{\text{base}}{\text{hypotenuse}} = \frac{1}{\sqrt{5}} = \frac{\sqrt{5}}{5}$$

Note *Students are advised to use the positive and negative signs carefully according to quadrant system.*

236. *(d)* $\cos x = -\dfrac{1}{3}$

Given that, x is in third quadrant.

i.e., $\pi < x < \dfrac{3\pi}{2}$

($\because$ in third quadrant θ lies between π and $3\pi/2$)

From the formula, $\cos x = 2\cos^2\dfrac{x}{2} - 1$

$\Rightarrow$ $$2\cos^2\frac{x}{2} = 1 + \cos x = 1 - \frac{1}{3}$$

$\Rightarrow \quad 2\cos^2\frac{x}{2}=\frac{2}{3}$

$\Rightarrow \quad \cos^2\frac{x}{2}=\frac{1}{3}$

$\Rightarrow \quad \cos\frac{x}{2}=\pm\frac{1}{\sqrt{3}}$

Now, $\pi < x < \frac{3\pi}{2} \Rightarrow \frac{\pi}{2}<\frac{x}{2}<\frac{3\pi}{4}$

i.e., $\frac{x}{2}$ lies in second quadrant, therefore $\cos\frac{x}{2}$ will be negative.

$\Rightarrow \quad \cos\frac{x}{2}=-\frac{1}{\sqrt{3}}=-\frac{\sqrt{3}}{3}$

Now, $\sin^2\frac{x}{2}=1-\cos^2\frac{x}{2}$

$$\sin^2\frac{x}{2}=1-\left(-\frac{1}{\sqrt{3}}\right)^2$$

$$=1-\frac{1}{3}=\frac{2}{3}$$

$$\sin\frac{x}{2}=\pm\sqrt{\frac{2}{3}}$$

$\Rightarrow \quad \sin\frac{x}{2}=\sqrt{\frac{2}{3}}=\frac{\sqrt{6}}{3}$

(as $\frac{x}{2}$ lies in second quadrant)

$\therefore \quad \tan\frac{x}{2}=\frac{\sin\frac{x}{2}}{-\cos\frac{x}{2}}=\frac{\sqrt{\frac{2}{3}}}{-\frac{1}{\sqrt{3}}}$

$$\tan\frac{x}{2}=-\sqrt{2}$$

237. *(b)* $\sin x=\frac{1}{4}$

Given that, x is in second quadrant.

i.e., $\frac{\pi}{2}<x<\pi$

$\because \quad \sin x=\frac{2\tan\frac{x}{2}}{1+\tan^2\frac{x}{2}}$

$\therefore \quad \frac{1}{4}=\frac{2\tan\frac{x}{2}}{1+\tan^2\frac{x}{2}}$

$\Rightarrow \quad 1+\tan^2\frac{x}{2}=8\tan\frac{x}{2}$

$\Rightarrow \quad \tan^2\frac{x}{2}-8\tan\frac{x}{2}+1=0$

$\Rightarrow \quad \tan\frac{x}{2}=\frac{8\pm\sqrt{64-4}}{2}$

$\Rightarrow \quad \tan\frac{x}{2}=\frac{8\pm 2\sqrt{15}}{2} \Rightarrow \tan\frac{x}{2}=4\pm\sqrt{15}$

$\because \quad \frac{\pi}{2}<x<\pi \Rightarrow \frac{\pi}{4}<\frac{x}{2}<\frac{\pi}{2}$

i.e., $\frac{x}{2}$ lies in Ist quadrant and as

$$\frac{x}{2}\in\left(\frac{\pi}{4},\frac{\pi}{2}\right), \tan\frac{x}{2}>1$$

$\Rightarrow \quad \tan\frac{x}{2}=4+\sqrt{15}$

Now, $\sec^2\frac{x}{2}=1+\tan^2\frac{x}{2}=1+(4+\sqrt{15})^2$

$$=1+16+15+8\sqrt{15}$$

$\Rightarrow \quad \sec^2\frac{x}{2}=32+8\sqrt{15}$

$\Rightarrow \quad \sec^2\frac{x}{2}=8(4+\sqrt{15})$

$\Rightarrow \quad \cos^2\frac{x}{2}=\frac{1}{\sec^2\frac{x}{2}}$

$$=\frac{1}{8(4+\sqrt{15})}\times\frac{(4-\sqrt{15})}{(4-\sqrt{15})}$$

$\Rightarrow \quad \cos^2\frac{x}{2}=\frac{(4-\sqrt{15})}{8}$

$$\sin x=\frac{1}{4}, \cos x=\frac{-\sqrt{15}}{4}$$

$$2\sin^2\frac{x}{2}=1+\frac{\sqrt{15}}{4}=\frac{4+\sqrt{15}}{4}$$

$$\sin^2\frac{x}{2}=\frac{4+\sqrt{15}}{8}$$

$\Rightarrow \quad \sin\frac{x}{2}=\sqrt{\frac{4+\sqrt{15}}{8}}$

238. *(a)* Given that, $a\tan\theta+b\sec\theta=c$ or $a\sin\theta+b=c\cos\theta$

Using the identities,

$$\sin\theta=\frac{2\tan\frac{\theta}{2}}{1+\tan^2\frac{\theta}{2}}$$

and $\cos\theta=\frac{1-\tan^2\frac{\theta}{2}}{1+\tan^2\frac{\theta}{2}}$

We have, $\frac{a\left(2\tan\frac{\theta}{2}\right)}{1+\tan^2\frac{\theta}{2}}+b=\frac{c\left(1-\tan^2\frac{\theta}{2}\right)}{1+\tan^2\frac{\theta}{2}}$

or $\quad (b+c)\tan^2\frac{\theta}{2}+2a\tan\frac{\theta}{2}+b-c=0$

Above equation is quadratic in $\tan\frac{\theta}{2}$ and hence $\tan\frac{\alpha}{2}$ and $\tan\frac{\beta}{2}$ are the roots of this equation.

Therefore, $\tan\frac{\alpha}{2}+\tan\frac{\beta}{2}=\frac{-2a}{b+c}$ and $\tan\frac{\alpha}{2}\tan\frac{\beta}{2}=\frac{b-c}{b+c}$

Using the identity,

$$\tan\left(\frac{\alpha}{2}+\frac{\beta}{2}\right)=\frac{\tan\frac{\alpha}{2}+\tan\frac{\beta}{2}}{1-\tan\frac{\alpha}{2}\tan\frac{\beta}{2}}$$

We have, $\tan\left(\frac{\alpha}{2}+\frac{\beta}{2}\right)=\dfrac{\frac{-2a}{b+c}}{1-\frac{b-c}{b+c}}=\dfrac{-2a}{2c}=\dfrac{-a}{c}$ ...(i)

Again, using another identity,

$$\tan 2\left(\frac{\alpha+\beta}{2}\right)=\frac{2\tan\frac{\alpha+\beta}{2}}{1-\tan^2\frac{\alpha+\beta}{2}}$$

We have, $\tan(\alpha+\beta)=\dfrac{2\left(-\frac{a}{c}\right)}{1-\frac{a^2}{c^2}}=\dfrac{2ac}{a^2-c^2}$ [from Eq. (i)]

Alternate Method

Given that, $a\tan\theta+b\sec\theta=c$

$\Rightarrow$ $(a\tan\theta-c)^2=b^2(1+\tan^2\theta)$

$\Rightarrow a^2\tan^2\theta-2ac\tan\theta+c^2=b^2+b^2\tan^2\theta$

$\Rightarrow$ $(a^2-b^2)\tan^2\theta-2ac\tan\theta+c^2-b^2=0$ ...(i)

Since, α and β are the roots of the Eq. (i), so

$$\tan\alpha+\tan\beta=\frac{2ac}{a^2-b^2}$$

and $\tan\alpha\tan\beta=\dfrac{c^2-b^2}{a^2-b^2}$

Therefore, $\tan(\alpha+\beta)=\dfrac{\tan\alpha+\tan\beta}{1-\tan\alpha+\tan\beta}$

$$=\frac{\frac{2ac}{a^2-b^2}}{\frac{c^2-b^2}{a^2-b^2}}=\frac{2ac}{a^2-c^2}$$

239. *(b)* $2\sin^2\beta+4\cos(\alpha+\beta)\sin\alpha\sin\beta+\cos 2(\alpha+\beta)$

$=2\sin^2\beta+4(\cos\alpha\cos\beta-\sin\alpha\sin\beta)\sin\alpha\sin\beta$
$+(\cos 2\alpha\cos 2\beta-\sin 2\alpha\sin 2\beta)$

$=2\sin^2\beta+4\sin\alpha\cos\alpha\sin\beta\cos\beta-4\sin^2\alpha\sin^2\beta$
$+\cos 2\alpha\cos 2\beta-\sin 2\alpha\sin 2\beta$

$=2\sin^2\beta+\sin 2\alpha\sin 2\beta-4\sin^2\alpha\sin^2\beta$
$+\cos 2\alpha\cos 2\beta-\sin 2\alpha\sin 2\beta$

$=(1-\cos 2\beta)-(2\sin^2\alpha)(2\sin^2\beta)+\cos 2\alpha\cos 2\beta$

$=(1-\cos 2\beta)-(1-\cos 2\alpha)(1-\cos 2\beta)+\cos 2\alpha\cos 2\beta$

$=\cos 2\alpha$

240. *(c)* Let $\theta=\alpha+\beta$. Then, $\tan\alpha=K\tan\beta$

or $\dfrac{\tan\alpha}{\tan\beta}=\dfrac{K}{1}$

Applying componendo and dividendo, we have

$$\frac{\tan\alpha+\tan\beta}{\tan\alpha-\tan\beta}=\frac{K+1}{K-1}$$

or $\dfrac{\sin\alpha\cos\beta+\cos\alpha\sin\beta}{\sin\alpha\cos\beta-\cos\alpha\sin\beta}=\dfrac{K+1}{K-1}$

i.e., $\dfrac{\sin(\alpha+\beta)}{\sin(\alpha-\beta)}=\dfrac{K+1}{K-1}$

Given that, $\alpha-\beta=\phi$ and $\alpha+\beta=\theta$. Therefore,

$$\frac{\sin\theta}{\sin\phi}=\frac{K+1}{K-1}$$

or $\sin\theta=\dfrac{K+1}{K-1}\sin\phi$

241. *(c)* Divide the given equation by 2 to get

$$\frac{\sqrt{3}}{2}\cos\theta+\frac{1}{2}\sin\theta=\frac{1}{\sqrt{2}}$$

or $\cos\frac{\pi}{6}\cos\theta+\sin\frac{\pi}{6}\sin\theta=\cos\frac{\pi}{4}$

or $\cos\left(\frac{\pi}{6}-\theta\right)=\cos\frac{\pi}{4}$

Thus, the solution are given by, $\theta=2m\pi\pm\frac{\pi}{4}+\frac{\pi}{6}$

Hence, the solutions are

$\theta=2m\pi+\frac{\pi}{4}+\frac{\pi}{6}$ and $\theta=2m\pi-\frac{\pi}{4}+\frac{\pi}{6}$

i.e., $\theta=2m\pi+\frac{5\pi}{12}$ and $\theta=2m\pi-\frac{\pi}{12}$

242. *(a)* Since, $2^{\sin\theta}$ and $2^{\cos\theta}$ are positive real numbers, so AM (Arithmetic Mean) of these two numbers is greater or equal to their GM (Geometric Mean) and hence

$$\frac{2^{\sin\theta}+2^{\cos\theta}}{2}\ge\sqrt{2^{\sin\theta}\times 2^{\cos\theta}}=\sqrt{2^{\sin\theta+\cos\theta}}$$

$$\ge 2^{\frac{\sin\theta+\cos\theta}{2}}=2^{\frac{1}{\sqrt{2}}\left(\frac{1}{\sqrt{2}}\sin\theta+\frac{1}{\sqrt{2}}\cos\theta\right)}$$

$$\ge 2^{\frac{1}{\sqrt{2}}\sin\left(\frac{\pi}{4}+\theta\right)}$$

Since, $-1\le\sin\left(\frac{\pi}{4}+\theta\right)\le 1$, we have

$$\frac{2^{\sin\theta}+2^{\cos\theta}}{2}\ge 2^{\frac{-1}{\sqrt{2}}}$$

$\Rightarrow$ $2^{\sin\theta}+2^{\cos\theta}\ge 2^{1-\frac{1}{\sqrt{2}}}$

243. *(d)* We have, $\cos\frac{\pi}{5}\cos\frac{2\pi}{5}\cos\frac{4\pi}{5}\cos\frac{8\pi}{5}$

$$=\frac{1}{2\sin\frac{\pi}{5}}2\sin\frac{\pi}{5}\cos\frac{\pi}{5}\cos\frac{2\pi}{5}\cos\frac{4\pi}{5}\cos\frac{8\pi}{5}$$

$$=\frac{1}{2\sin\frac{\pi}{5}}\sin\frac{2\pi}{5}\cos\frac{2\pi}{5}\cos\frac{4\pi}{5}\cos\frac{8\pi}{5}$$

$$=\frac{1}{4\sin\frac{\pi}{5}}\sin\frac{4\pi}{5}\cos\frac{4\pi}{5}\cos\frac{8\pi}{5}$$

$$= \frac{1}{8\sin\frac{\pi}{5}} \sin\frac{8\pi}{5}\cos\frac{8\pi}{5}$$

$$= \frac{\sin\frac{16\pi}{5}}{16\sin\frac{\pi}{5}}$$

$$= \frac{\sin\left(3\pi + \frac{\pi}{5}\right)}{16\sin\frac{\pi}{5}}$$

$$= \frac{-\sin\frac{\pi}{5}}{16\sin\frac{\pi}{5}}$$

$$= -\frac{1}{16}$$

244. *(a)* We have, $m\sin\theta = n\sin(\theta + 2\alpha)$

$$\Rightarrow \quad \frac{\sin(\theta + 2\alpha)}{\sin\theta} = \frac{m}{n}$$

Using componendo and dividendo, we get

$$\Rightarrow \quad \frac{\sin(\theta + 2\alpha) + \sin\theta}{\sin(\theta + 2\alpha) - \sin\theta} = \frac{m+n}{m-n}$$

$$\Rightarrow \quad \frac{2\sin\left(\frac{\theta + 2\alpha + \theta}{2}\right)\cdot\cos\left(\frac{\theta + 2\alpha - \theta}{2}\right)}{2\cos\left(\frac{\theta + 2\alpha + \theta}{2}\right)\cdot\sin\left(\frac{\theta + 2\alpha - \theta}{2}\right)} = \frac{m+n}{m-n}$$

$$\Rightarrow \quad \frac{2\sin(\theta + \alpha)\cdot\cos\alpha}{2\cos(\theta + \alpha)\cdot\sin\alpha} = \frac{m+n}{m-n}$$

$$\Rightarrow \quad \tan(\theta + \alpha)\cdot\cot\alpha = \frac{m+n}{m-n}$$

245. *(c)* We have, $\tan(A + B) = p$ and $\tan(A - B) = q$

$$\therefore \quad \tan 2A = \tan[(A + B) + (A - B)]$$

$$= \frac{\tan(A + B) + \tan(A - B)}{1 - \tan(A + B)\cdot\tan(A - B)} = \frac{p+q}{1-pq}$$

246. *(c)* We have, $\frac{\sin(x + y)}{\sin(x - y)} = \frac{a+b}{a-b}$

Using componendo and dividendo, we get

$$\frac{\sin(x + y) + \sin(x - y)}{\sin(x + y) - \sin(x - y)} = \frac{a+b+a-b}{a+b-a+b}$$

$$\Rightarrow \quad \frac{2\sin\frac{x+y+x-y}{2}\cdot\cos\frac{x+y-x+y}{2}}{2\cos\frac{x+y+x-y}{2}\cdot\sin\frac{x+y-x+y}{2}} = \frac{2a}{2b}$$

$$\Rightarrow \quad \frac{\sin x\cdot\cos y}{\cos x\cdot\sin y} = \frac{a}{b}$$

$$\Rightarrow \quad \tan x\cdot\cot y = \frac{a}{b}$$

$$\Rightarrow \quad \frac{\tan x}{\tan y} = \frac{a}{b}$$

247. *(c)* We have, $\tan\theta = \frac{\sin\alpha - \cos\alpha}{\sin\alpha + \cos\alpha}$

$$\tan\theta = \frac{\tan\alpha - 1}{\tan\alpha + 1} \left(\begin{array}{l}\text{dividing by } \cos\alpha \text{ in} \\ \text{numerator and denominator}\end{array}\right)$$

$$= \frac{\tan\alpha - \tan\pi/4}{1 + \tan\alpha\cdot\tan\frac{\pi}{4}}$$

$$= \tan\left(\alpha - \frac{\pi}{4}\right)$$

$$\Rightarrow \quad \theta = \alpha - \frac{\pi}{4}$$

$$\Rightarrow \quad \alpha = \left(\theta + \frac{\pi}{4}\right)$$

$$\therefore \sin\alpha + \cos\alpha = \sin\left(\theta + \frac{\pi}{4}\right) + \cos\left(\theta + \frac{\pi}{4}\right)$$

$$= \sin\theta\cdot\frac{1}{\sqrt{2}} + \cos\theta\cdot\frac{1}{\sqrt{2}} + \cos\theta\cdot\frac{1}{\sqrt{2}} - \sin\theta\cdot\frac{1}{\sqrt{2}}$$

$$= \frac{2\cos\theta}{\sqrt{2}} = \sqrt{2}\cos\theta$$

4

Principle *of* Mathematical Induction

The Principle of Mathematical Induction

Level I

1. Application of a general case to a particular case is called

(a) formation (b) deduction
(c) induction (d) None of these

2. In contrast to deduction, which of the following is not true for inductive reasoning?

(a) Inductive reasoning depends on working with each case and developing a conjecture by observing all the incidences
(b) Inductive reasoning is used where collecting and analysing data is not in the norm
(c) Induction is a key aspect of scientific reasoning
(d) None of the above

3. The word induction means

(a) normalisation from particular cases or facts
(b) generalisation from particular cases or facts
(c) Both (a) and (b)
(d) None of the above

4. Principle of mathematical induction is used

(a) to prove any statement
(b) to prove results which are true for all real numbers
(c) to prove that statements which are formulated in terms of n, where n is positive integer
(d) in deductive reasoning

5. Statement I Any subset of R that is an inductive set must contain N.

Statement II A set S is said to be an inductive set, if $1 \in S$ and $x+1 \in S$, whenever $x \in S$.

(a) Statement I is true; Statement II is true; Statement II is a correct explanation for Statement I
(b) Statement I is true; Statement II is true; Statement II is not a correct explanation for Statement I
(c) Statement I is true; Statement II is false
(d) Statement I is false; Statement II is true

6. To prove the formula $1+2+3+\cdots+n=\dfrac{n(n+1)}{2}$, which among the following methods can be used?

(a) The method of mathematical reasoning
(b) The method of binomial theorem
(c) The method of mathematical induction
(d) None of the above

7. If $P(n)$ is a statement involving the natural number n, such that $P(1)$ is true and truth of $P(k)$ implies truth of $P(k+1)$, where k is some positive integer, then according to principle of mathematical induction, which is true?

(a) $P(n)$ is true for all real numbers n
(b) $P(n)$ is true for all natural numbers n
(c) Both (a) and (b) are true
(d) Both (a) and (b) are false

8. A pattern is given below

$1=1^2=1$
$4=2^2=1+3$
$9=3^2=1+3+5$
$16=4^2=1+3+5+7$ etc.

Based on above pattern, we have

$$P(n): 1+3+5+7+\cdots+(2n-1)=n^2.$$

Based on above information, which among the following is true?

(a) $P(1)$ is true
(b) If $P(k)$ is true, then $P(k+1)$ is also true, where k is some positive integer
(c) $P(k+1) = k^2 + 2k + 2$, where k is some positive integer
(d) Both (a) and (b) are true

9. For all positive integers n, which among the following is true?

(a) $2^n < n$ (b) $2^n > n$
(c) $2^n = n$ (d) None of these

10. For every positive integer n, $7^n - 3^n$ is divisible by

I. 8 II. 2 III. 3 IV. 4

(a) Only I is true (b) Only II is true
(c) II and IV are true (d) None of these

11. Which among the following is true for all natural numbers n ?

(a) $(1+x)^n \geq (1+nx)^3$, $x > -1$ (b) $(1+x)^n \geq (1-nx)$, $x > -1$
(c) $(1+x)^n \geq (1+nx)$, $x > -1$ (d) None of these

12. For all $n \in N$,

$1^2 + 2^2 + \cdots + n^2 > \ldots A \ldots$. Here, A equals to

(a) $\frac{n^4}{3}$ (b) $\frac{n^5}{3}$
(c) $\frac{n^3}{3}$ (d) None of these

13. Which among the following is true for all $n \in N$?

(a) $(ab)^n = a^n b^n$ (b) $(ab)^n = a^n b$
(c) $(ab)^n = ab^n$ (d) None of these

14. For all $n \in N$, which among the following is true?

(a) $1 + 3 + 3^2 + \cdots + 3^{n-1} = 3^n - 1$
(b) $1 + 3 + 3^2 + \cdots + 3^{n-1} = 3^n + 1$
(c) $1 + 3 + 3^2 + \cdots + 3^{n-1} = \frac{3^n - 1}{2}$
(d) $1 + 3 + 3^2 + \cdots + 3^{n-1} = \frac{3^n - 1}{3}$

15. For all $n \in N$, $1^3 + 2^3 + 3^3 + \cdots + n^3$ is equal to

(a) $\left[\frac{n(n+1)}{2}\right]^3$ (b) $\left[\frac{n(n+1)}{2}\right]^2$
(c) $\left[\frac{n(n-1)}{2}\right]^2$ (d) None of these

16. For all $n \in N$,

$$1 + \frac{1}{1+2} + \frac{1}{1+2+3} + \cdots + \frac{1}{1+2+3+\cdots+n}$$ is equal to

(a) $\frac{3n}{n+1}$ (b) $\frac{n}{n+1}$ (c) $\frac{2n}{n-1}$ (d) $\frac{2n}{n+1}$

17. For all $n \in N$, which among the following is true?

(a) $1 \cdot 2 \cdot 3 + 2 \cdot 3 \cdot 4 + 3 \cdot 4 \cdot 5 + \cdots + n(n+1)(n+2) = \frac{n(n+1)(n+2)(n+3)}{4}$
(b) $1 \cdot 2 \cdot 3 + 2 \cdot 3 \cdot 4 + 3 \cdot 4 \cdot 5 + \cdots + n(n+1)(n+2) = n(n+1)(n+2)(n+3)$
(c) $1 \cdot 2 \cdot 3 + 2 \cdot 3 \cdot 4 + 3 \cdot 4 \cdot 5 + \cdots + n(n+1)(n+2) = \frac{n(n-1)(n+2)(n+3)}{4}$
(d) None of the above

18. For all $n \in N$, $1 \cdot 3 + 2 \cdot 3^2 + 3 \cdot 3^3 + \cdots + n \cdot 3^n$ is equal to

(a) $\frac{(2n+1)3^{n+1}+3}{4}$ (b) $\frac{(2n-1)3^{n+1}+3}{4}$
(c) $\frac{(2n+1)3^n+3}{4}$ (d) $\frac{(2n-1) \cdot 3^{n+1}+1}{4}$

19. **Statement I** For all $n \in N$,

$$1 \cdot 2 + 2 \cdot 3 + 3 \cdot 4 + \cdots + n \cdot (n+1) = \frac{n(n+1)(n+2)}{3}$$

Statement II For all $n \in N$,

$$1 \cdot 3 + 3 \cdot 5 + 5 \cdot 7 + \cdots + (2n-1)(2n+1) = \frac{n(4n^2 + 6n - 1)}{3}$$

(a) Only Statement I is true (b) Only Statement II is true
(c) Both statements are true (d) Both statements are false

20. Which among the following is true for all $n \in N$?

(a) $1.2 + 2.2^2 + 3.2^3 + \cdots + n \cdot 2^n = (n-1) \cdot 2^{n+1} + 2$
(b) $1.2 + 2.2^2 + 3.2^3 + \cdots + n \cdot 2^n = (n-1) \cdot 2^{n+1} + 1$
(c) $1.2 + 2.2^2 + 3.2^3 + \cdots + n \cdot 2^n = (n+1) \cdot 2^{n+1} + 2$
(d) None of the above

21. For all $n \in N$, consider the following statements

I. $\frac{1}{2} + \frac{1}{4} + \frac{1}{8} + \cdots + \frac{1}{2^n} = 1 - \frac{1}{2^n}$

II. $\frac{1}{2} + \frac{1}{4} + \frac{1}{8} + \cdots + \frac{1}{2^n} = 1 + \frac{1}{2^n}$

III. $\frac{1}{2 \cdot 5} + \frac{1}{5 \cdot 8} + \frac{1}{8 \cdot 11} + \cdots + \frac{1}{(3n-1)(3n+2)} = \frac{n+1}{6n+4}$

IV. $\frac{1}{2 \cdot 5} + \frac{1}{5 \cdot 8} + \frac{1}{8 \cdot 11} + \cdots + \frac{1}{(3n-1)(3n+2)} = \frac{n}{6n+4}$

Choose the correct option.

(a) II and III are not true (b) Only II is not true
(c) Only IV is not true (d) I and IV are not true

22. For all $n \in N$, $a + ar + ar^2 + \ldots + ar^{n-1}$ is equal to

(a) $\frac{a(r^n - 1)}{r+1}$ (b) $\frac{a(r^n + 1)}{r+1}$ (c) $\frac{a(r^n - 1)}{r-1}$ (d) $\frac{(r^n - 1)}{a(r+1)}$

23. For all $n \in N$,

$\left(1+\frac{3}{1}\right)\left(1+\frac{5}{4}\right)\left(1+\frac{7}{9}\right)\cdots\left(1+\frac{(2n+1)}{n^2}\right)$ is equal to

(a) $\frac{(n+1)^2}{2}$ (b) $\frac{(n+1)^3}{3}$
(c) $(n+1)^2$ (d) None of these

24. For all $n \in N$, $\left(1+\frac{1}{1}\right)\left(1+\frac{1}{2}\right)\left(1+\frac{1}{3}\right)\cdots\left(1+\frac{1}{n}\right)$ is equal to

(a) $\frac{(n+1)}{2}$ (b) $(n+1)$ (c) n (d) $\left(n-\frac{1}{n}\right)^2$

25. Which among the following is true for all $n \in N$?

(a) $1^2+3^2+5^2+\cdots+(2n-1)^2=\frac{n(2n-1)(2n+1)}{3}$
(b) $1^2+3^2+5^2+\cdots+(2n-1)^2=\frac{n(2n+1)(2n)}{3}$
(c) $1^2+3^2+5^2+\cdots+(2n-1)^2=\frac{n(2n-1)(2n+3)}{3}$
(d) None of the above

26. For all $n \in N$,

$\frac{1}{1\cdot4}+\frac{1}{4\cdot7}+\frac{1}{7\cdot10}+\cdots+\frac{1}{(3n-2)(3n+1)}$ is equal to

(a) $\frac{n}{2n+1}$ (b) $\frac{n}{3n+1}$ (c) $\frac{n+1}{3n}$ (d) $\frac{n+1}{2n}$

27. **Statement I** $1+2+3+\cdots+n<\frac{1}{8}(2n+1)^2$, $n \in N$.

Statement II $n(n+1)(n+5)$ is a multiple of 3, $n \in N$.

(a) Only Statement I is true (b) Only Statement II is true
(c) Both statements are true (d) Both statements are false

28. For all $n \in N$, $10^{2n-1}+1$ is divisible by

(a) 9 (b) 121 (c) 11 (d) 13

29. For all $n \in N$, $3^{2n+2}-8n-9$ is divisible by

I. 4 II. 6
III. 8 IV. 10

(a) Only I is true (b) Only IV is true
(c) I and III are true (d) All are true

Level II

30. For all $n \in N$,

$\frac{1}{1\cdot2\cdot3}+\frac{1}{2\cdot3\cdot4}+\frac{1}{3\cdot4\cdot5}+\cdots+\frac{1}{n(n+1)(n+2)}$ is equal to [HOTS]

(a) $\frac{(n+3)}{4(n+1)(n+2)}$ (b) $\frac{n(n+3)}{4(n+1)(n+2)}$
(c) $\frac{4(n+1)(n+2)}{n(n+3)}$ (d) None of these

31. For all $n \in N$, which of the following is true? [HOTS]

(a) $\frac{1}{3\cdot5}+\frac{1}{5\cdot7}+\frac{1}{7\cdot9}+\cdots+\frac{1}{(2n+1)(2n+3)}=\frac{n}{2n+3}$
(b) $\frac{1}{3\cdot5}+\frac{1}{5\cdot7}+\frac{1}{7\cdot9}+\cdots+\frac{1}{(2n+1)(2n+3)}=\frac{n}{n+3}$
(c) $\frac{1}{3\cdot5}+\frac{1}{5\cdot7}+\frac{1}{7\cdot9}+\cdots+\frac{1}{(2n+1)(2n+3)}=\frac{n}{3(2n+3)}$
(d) None of the above

32. For all $n \in N$, $x^{2n}-y^{2n}$ is divisible by

(a) $(x+y)$ (b) $(x+y)^3$
(c) $(x+2y)$ (d) $(2x+y)$

33. **Assertion** (A) $P(n): 41^n-14^n$ is a multiple of 27, where $n \in N$.

Reason (R) By principle of mathematical induction, for a given statement $P(n)$, $n \in N$, if $P(1)$ is true and $P(k+1)$ is true whenever $P(k)$ is true, $k \in N$, then for all $n \in N, P(n)$ is true.

(a) Both A and R are correct; R is the correct explanation of A
(b) Both A and R are correct; R is not the correct explanation of A
(c) A is correct; R is incorrect
(d) R is correct; A is incorrect

34. For all $n \in N$, $(2n+7)<$ [HOTS]

(a) $(n+3)$ (b) $(n+3)^2$
(c) $\frac{3(n+3)}{2}$ (d) None of these

35. If $P(n)$ is a statement $(n \in N)$ such that $P(k)$ is true $\Rightarrow P(k+1)$ is true for $k \in N$, then $P(n)$ is true,

(a) for all n (b) for all $n>1$
(c) for all $n>2$ (d) Nothing can be said

36. Let $P(n): n^2+n+1\ (n \in N)$ is an even integer. Therefore, $P(n)$ is true, [HOTS]

(a) for $n>1$ (b) for all n
(c) for $n>2$ (d) None of these

37. If $P(n)$ is a statement $(n \in N)$ such that $P(3)$ is true. If $P(k)$ is true $\Rightarrow P(k+1)$ is true for all $k \geq 3$, then $P(n)$ is true,

(a) for all n (b) for $n \geq 3$
(c) for $n>4$ (d) None of these

38. Let $P(n)$: '$2^n<(1\times2\times3\times\ldots\times n)$'. Then, the smallest positive integer for which $P(n)$ is true, is

(a) 1 (b) 2
(c) 3 (d) 4

39. A student was asked to prove a statement $P(n)$ by induction. He proved that $P(k+1)$ is true, whenever $P(k)$ is true for all $k > 5 \in N$ and also proved that $P(5)$ is true. On the basis of this, he could conclude that $P(n)$ is true

(a) for all $n \in N$
(b) for all $n > 5$
(c) for all $n \geq 5$
(d) for all $n < 5$

40. If $P(n)$: '$2 \cdot 4^{2n+1} + 3^{3n+1}$ is divisible by λ for all $n \in N$' is true, then the value of λ is [NCERT Exemplar]

(a) 8 (b) 9 (c) 10 (d) 11

41. If $P(n)$: '$49^n + 16^n + k$ is divisible by 64 for $n \in N$' is true, then the negative integral value of k is

(a) –2 (b) –3 (c) –4 (d) –1

Miscellaneous Exercise

42. Let $P(k): 1 + 3 + 5 + \cdots + (2k-1) = 3 + k^2$, $k \in N$, then which of the following is true?

(a) $P(1)$ is correct
(b) $P(k) \Rightarrow P(k+1)$
(c) $P(k) \not\Rightarrow P(k+1)$
(d) None of these

43. Statement I For every natural number $n \geq 2$, $\frac{1}{\sqrt{1}} + \frac{1}{\sqrt{2}} + \cdots + \frac{1}{\sqrt{n}} > \sqrt{n}$.

Statement II For every natural number $n \geq 2$, $\sqrt{n(n+1)} < n + 1$. [HOTS]

(a) Statement I is true; Statement II is true; Statement II is a correct explanation for Statement I
(b) Statement I is true; Statement II is true; Statement II is not a correct explanation for Statement I
(c) Statement I is true; Statement II is false
(d) Statement I is false; Statement II is true

44. For all $n \in N$, $\cos\alpha + \cos(\alpha + \beta) + \cos(\alpha + 2\beta) + \cdots + \cos[\alpha + (n-1)\beta]$ is equal to

(a) $\dfrac{\cos\left\{\alpha + \left[\frac{(n-1)}{2}\right]\beta\right\}\sin\left(\frac{n\beta}{2}\right)}{2\sin\frac{\beta}{2}}$

(b) $\dfrac{\cos\left\{\alpha + \left[\frac{(n-1)}{2}\right]\beta\right\}\sin\left(\frac{n\beta}{2}\right)}{\sin\frac{\beta}{2}}$

(c) $\dfrac{\cos\left\{\alpha + \left[\frac{(n+1)}{2}\right]\beta\right\}}{\sin\frac{\beta}{2}}$

(d) None of the above

45. For all $n \in N$, $\cos\theta \cdot \cos 2\theta \cdot \cos 2^2\theta \ldots \cos 2^{n-1}\theta$ is equal to

(a) $\dfrac{\sin 2^n\theta}{2^n \sin\theta}$
(b) $\dfrac{\sin 2^{n+1}\theta}{2^n \sin\theta}$
(c) $\dfrac{\sin 2^{n-1}\theta}{2^n \sin\theta}$
(d) None of these

46. For all $n \in N$, $\sin\theta + \sin 2\theta + \sin 3\theta + \ldots + \sin n\theta$ is equal to [HOTS]

(a) $\dfrac{\sin\frac{n\theta}{2} \cdot \sin\left(\frac{n+1}{2}\right)\theta}{\sin\frac{\theta}{2}}$
(b) $\dfrac{\cos\frac{n\theta}{2} \cdot \cos\left(\frac{n+1}{2}\right)\theta}{\cos\frac{\theta}{2}}$
(c) $\dfrac{\sin n\theta \cdot \sin(n+1)\theta}{\sin\theta}$
(d) None of these

47. For all $n \in N$, the sum of $\frac{n^5}{5} + \frac{n^3}{3} + \frac{7n}{15}$ is [HOTS]

(a) a negative integer
(b) a whole number
(c) a real number
(d) a natural number

48. For any natural number n, $x^n - y^n$ is divisible by ...A..., where x and y are any integers with $x \neq y$. Here, A refers to

(a) $(x-y)^3$
(b) $(x-y)$
(c) $(x-y)^4$
(d) None of these

49. For all $n \in N$, $\sin\alpha + \sin(\alpha + \beta) + \sin(\alpha + 2\beta) + \cdots + \sin(\alpha + (n-1)\beta)$ is equal to [NCERT Exemplar]

(a) $\sin\left(\alpha + \frac{n-1}{2}\beta\right)\sin\frac{n\beta}{2}$
(b) $\dfrac{\sin\left(\alpha + \frac{n-1}{2}\beta\right)\sin\frac{n\beta}{2}}{\sin\frac{\beta}{2}}$
(c) $\sin\left(\alpha + \frac{n-1}{2}\beta\right)\cos\frac{n\beta}{2}$
(d) None of these

50. For all $n \in N$, $1 \times 1! + 2 \times 2! + 3 \times 3! + \cdots + n \times n!$ is equal to [NCERT Exemplar]

(a) $(n+1)! - 2$
(b) $(n+1)!$
(c) $(n+1)! - 1$
(d) $(n+1)! - 3$

51. For given series $1^2 + 2 \times 2^2 + 3^2 + 2 \times 4^2 + 5^2 + 2 \times 6^2 + \ldots$, if S_n is the sum of n terms, then [NCERT Exemplar; HOTS]

(a) $S_n = \frac{n(n+1)^2}{2}$, if n is even
(b) $S_n = \frac{n^2(n+1)}{2}$, if n is odd
(c) Both (a) and (b) are true
(d) Both (a) and (b) are false

Answer *with* Explanations

1. *(b)* Deduction in a nutshell is given a statement to be proven, often called a conjecture or a theorem in Mathematics, valid deductive steps are derived and a proof may or may not be established *i.e.,* deduction is the application of a general case to a particular case.

2. *(b)* In contrast to deduction, inductive reasoning depends on working with each case and developing a conjecture by observing incidences till we have observed each and every case. It is frequently used in Mathematics and is a key aspect of scientific reasoning, where collecting and analysing data is the norm.

Note *Conjecture means opinion or conclusion formed on the basis of slight evidence.*

3. *(b)* The word induction means the generalisation from particular cases or facts.

4. *(c)* In algebra or in other discipline of Mathematics, there are certain results or statements that are formulated in terms of n, where n is a positive integer. To prove such statement the well-suited principle *i.e.*, used-based on the specific technique, is known as the principle of mathematical induction.

5. *(a)* We know, the set of natural numbers N is a special ordered subset of the real numbers. Infact, N is the smallest subset of R with the following property.

A set S is said to be an inductive set, if $1 \in S$ and $x + 1 \in S$ whenever $x \in S$. Since, N is the smallest subset of R which is an inductive set, it follows that any subset of R that is an inductive set must contain N.

So, both the statements are true and Statement II is the correct explanation of Statement I.

6. *(c)* To find the formula for the sum of positive integers 1, 2, 3, ..., n *i.e.*, a formula which will give the value of $1 + 2 + 3$, when $n = 3$, the value $1 + 2 + 3 + 4$, when $n = 4$ and so on and suppose that in some manner we are led to believe that the formula $1 + 2 + 3 + \dots + n = \frac{n(n+1)}{2}$ is the correct one.

How can this formula actually be proved? We can, of course, verify the statement for as many positive integral values of n as we like but this process will not prove the formula for all values of n. What is needed in some kind of chain reaction which will have the effect that once the formula is proved for a particular positive integer the formula will automatically follow for the next positive integer and the next indefinitely. Such a reaction may be considered as produced by the method of mathematical induction.

7. *(b)* Given statement $P(n)$ involving the natural number n such that

(i) The statement is true for $n = 1$, *i.e.,* $P(1)$ is true and

(ii) If the statement is true for $n = k$ (where k is some positive integer), then the statement is also true for $n = k + 1$, *i.e.,* truth of $P(k)$ implies the truth of $P(k + 1)$.

Then, according to principle of mathematical induction, $P(n)$ is true for all natural numbers n.

Note 1. *Property (i) is simply a statement of fact. There may be situations when a statement is true for all $n \geq 4$. In this case, step 1 will start from $n = 4$ and we shall verify the result for $n = 4$, i.e., $P(4)$.*

2. *Property (ii) is a conditional property. It does not assert that the given statement is true for $n = k$, but only that if it is true for $n = k$, then it is also true for $n = k + 1$. So, to prove that the property holds, only prove that conditional proposition.*

If the statement is true for $n = k$, then it is also true for $n = k + 1$.

This is sometimes referred to as the inductive step. The assumption that the given statement is true for $n = k$ in this inductive step is called the inductive hypothesis.

8. *(d)* From the given pattern,

$$1 = 1^2 = 1$$
$$4 = 2^2 = 1 + 3$$
$$9 = 3^2 = 1 + 3 + 5$$
$$16 = 4^2 = 1 + 3 + 5 + 7 \text{ etc.}$$

It is worth to be noted that the sum of the first two odd natural numbers is the square of second natural number, sum of the first three odd natural numbers is the square of third natural number and so on. Thus, from this pattern, it appears that

$$1 + 3 + 5 + 7 + \dots + (2n - 1) = n^2$$

i.e., the sum of the first n odd natural numbers is the square of n.

Let us write

$$P(n): 1 + 3 + 5 + 7 + \dots + (2n - 1) = n^2.$$

To prove that $P(n)$ is true for all n.

The first step in a proof that uses mathematical induction is to prove that $P(1)$ is true. This step is called the basic step. Obviously

$1 = 1^2$, *i.e.,* $P(1)$ is true.

The next step is called the inductive step. Here, we suppose that $P(k)$ is true for some positive integer k, *i.e.,*

$$1 + 3 + 5 + 7 + \dots + (2k - 1) = k^2 \quad \dots(i)$$

Now, we shall prove $P(k + 1)$ is true.

Consider,

$$1 + 3 + 5 + 7 + \dots + (2k - 1) + \{2(k + 1) - 1\} \quad \dots(ii)$$
$$= k^2 + (2k + 1) = (k + 1)^2 \quad \text{[using Eq. (i)]}$$

Therefore, $P(k + 1)$ is true and the inductive proof is now completed. Hence, $P(n)$ is true for all natural numbers n.

9. *(b)* Let $P(n): 2^n > n$

When $n = 1, 2^1 > 1$. Hence, $P(1)$ is true.

Assume that $P(k)$ is true for some positive integer k, *i.e.*,

$$2^k > k \quad \ldots \text{(i)}$$

We shall now prove that $P(k + 1)$ is true whenever $P(k)$ is true. Multiplying both sides of Eq. (i) by 2, we get

$$2 \cdot 2^k > 2k$$

i.e., $\quad 2^{k+1} > 2k = k + k > k + 1$

Therefore, $P(k + 1)$ is true when $P(k)$ is true. Hence, by principle of mathematical induction, $P(n)$ is true for every positive integer n.

10. *(c)* Let $P(n): 7^n - 3^n$ is divisible by 4.

We note that $P(1): 7^1 - 3^1 = 4$ which is divisible by 4. Thus, $P(n)$ is true for $n = 1$.

Let $P(k)$ be true for some natural number k *i.e.*, $P(k): 7^k - 3^k$ is divisible by 4.

We can write $7^k - 3^k = 4d$, where $d \in N$.

Now, we wish to prove that $P(k + 1)$ is true, whenever $P(k)$ is true.

Now, $\quad 7^{(k+1)} - 3^{(k+1)} = 7^{(k+1)} - 7 \cdot 3^k + 7 \cdot 3^k - 3^{(k+1)}$

$$= 7(7^k - 3^k) + (7 - 3)3^k$$

$$= 7(4d) + (7 - 3)3^k$$

$$= 7(4d) + 4 \cdot 3^k = 4(7d + 3^k)$$

From the last line, we see that $7^{(k+1)} - 3^{(k+1)}$ is divisible by 4.Thus, $P(k + 1)$ is true when $P(k)$ is true. Therefore, by principle of mathematical induction, the statement is true, for every positive integer n.

Also, if a number is divisible by 4, it is also divisible by 2.

11. *(c)* Let $P(n): (1 + x)^n \geq (1 + nx)$, for $x > -1$.

We note that $P(n)$ is true when $n = 1$, since $(1 + x) \geq (1 + x)$ for $x > -1$.

Assume that $P(k): (1 + x)^k \geq (1 + kx), x > -1$ is true. $\quad \ldots$(i)

We want to prove that $P(k + 1)$ is true for $x > -1$, whenever $P(k)$ is true. $\quad \ldots$ (ii)

Consider the identity

$$(1 + x)^{k+1} = (1 + x)^k (1 + x)$$

Given that, $x > -1$, so $(1 + x) > 0$.

Therefore, using $(1 + x)^k \geq (1 + kx)$, we have

$$(1 + x)^{k+1} \geq (1 + kx)(1 + x)$$

i.e., $\quad (1 + x)^{k+1} \geq (1 + x + kx + kx^2)$. $\quad \ldots$(iii)

Here, k is a natural number and $x^2 \geq 0$, so that $kx^2 \geq 0$.

Therefore, $(1 + x + kx + kx^2) \geq (1 + x + kx)$ and so we obtain

$$(1 + x)^{k+1} \geq (1 + x + kx)$$

i.e., $\quad (1 + x)^{k+1} \geq [1 + (1 + k)x]$

Thus, the statement in Eq. (ii) is established. Hence, by the principle of mathematical induction, $P(n)$ is true for all natural numbers.

12. *(c)* Let $P(n)$ be the following statement

$$P(n): 1^2 + 2^2 + \cdots + n^2 > \frac{n^3}{3}, n \in N$$

We note that $P(n)$ is true for $n = 1$, since $1^2 > \frac{1^3}{3}$.

Assume that $P(k)$ is true

i.e., $\quad P(k): 1^2 + 2^2 + \ldots + k^2 > \frac{k^3}{3} \quad \ldots$(i)

We shall now prove that $P(k + 1)$ is true, whenever $P(k)$ is true.

We have, $1^2 + 2^2 + 3^2 + \ldots + k^2 + (k + 1)^2$

$$= (1^2 + 2^2 + \ldots + k^2) + (k + 1)^2 > \frac{k^3}{3} + (k + 1)^2$$

[using Eq. (i)]

$$= \frac{1}{3}[k^3 + 3k^2 + 6k + 3]$$

$$= \frac{1}{3}[(k + 1)^3 + 3k + 2] > \frac{1}{3}(k + 1)^3$$

Therefore, $P(k + 1)$ is also true whenever $P(k)$ is true. Hence, by mathematical induction, $P(n)$ is true for all $n \in N$.

13. *(a)* Let $P(n)$ be the following statement

$$P(n): (ab)^n = a^n b^n$$

We note that $P(n)$ is true for $n = 1$, since $(ab)^1 = a^1 b^1$.

Let $P(k)$ be true, *i.e.*,

$$(ab)^k = a^k b^k \quad \ldots \text{(i)}$$

We shall now prove that $P(k + 1)$ is true whenever $P(k)$ is true.

Now, we have

$$(ab)^{k+1} = (ab)^k (ab)$$

$$= (a^k b^k)(ab) \quad \text{[using Eq. (i)]}$$

$$= (a^k \cdot a^1)(b^k \cdot b^1) = a^{k+1} \cdot b^{k+1}$$

Therefore, $P(k + 1)$ is also true whenever $P(k)$ is true. Hence, by principle of mathematical induction, $P(n)$ is true for all $n \in N$.

So, option (a) is true.

14. *(c)* Let the statement $P(n)$ be defined as

$$P(n): 1 + 3 + 3^2 + \ldots + 3^{n-1} = \frac{3^n - 1}{2}$$

Step I For $n = 1$,

$$P(1): 1 = \frac{3^1 - 1}{2} = \frac{3 - 1}{2} = \frac{2}{2} = 1$$

which is true.

Step II Let it is true for $n = k$,

i.e., $$1 + 3 + 3^2 + \ldots + 3^{k-1} = \frac{3^k - 1}{2} \quad \ldots \text{(i)}$$

Step III For $n = k + 1$,

$$(1 + 3 + 3^2 + \ldots + 3^{k-1}) + 3^k = \frac{3^k - 1}{2} + \frac{3^k}{1} \quad \text{[using Eq. (i)]}$$
$$= \frac{3^k - 1 + 2 \times 3^k}{2}$$
$$= \frac{3 \times 3^k - 1}{2}$$
$$= \frac{3^{k+1} - 1}{2}$$

Therefore, $P(k + 1)$ is true when $P(k)$ is true.

Hence, from the principle of mathematical induction, the statement $P(n)$ is true for all natural numbers n.

15. *(b)* Let the statement $P(n)$ be defined as

$$P(n): 1^3 + 2^3 + 3^3 + \ldots + n^3 = \left[\frac{n(n+1)}{2}\right]^2$$

Step I For $n = 1$,

$$P(1): 1^3 = \left[\frac{1(1+1)}{2}\right]^2 = \left[\frac{1 \times 2}{2}\right]^2 = 1^2 = 1 = 1^3$$

which is true.

Step II Let it is true for $n = k$,

i.e., $$1^3 + 2^3 + 3^3 + \ldots + k^3 = \left[\frac{k(k+1)}{2}\right]^2 \quad \ldots \text{(i)}$$

Step III For $n = k + 1$,

$$(1^3 + 2^3 + 3^3 + 4^3 + \ldots + k^3) + (k + 1)^3$$
$$= \left[\frac{k(k+1)}{2}\right]^2 + (k+1)^3 \quad \text{[using Eq. (i)]}$$
$$= \frac{k^2(k+1)^2}{4} + \frac{(k+1)^3}{1}$$
$$= \frac{k^2(k+1)^2 + 4(k+1)^3}{4}$$
$$= \frac{(k+1)^2[k^2 + 4(k+1)]}{4} \quad \left[\begin{array}{l}\text{taking } (k+1)^2 \text{ common} \\ \text{in numerator part}\end{array}\right]$$
$$= \frac{(k+1)^2(k^2 + 4k + 4)}{4}$$
$$= \frac{(k+1)^2(k+2)^2}{4}$$
$$= \frac{(k+1)^2[(k+1)+1]^2}{4}$$
$$= \left[\frac{(k+1)[(k+1)+1]}{2}\right]^2$$

Therefore, $P(k + 1)$ is true when $P(k)$ is true. Hence, from the principle of mathematical induction, the statement is true for all natural numbers n.

16. *(d)* Let the statement $P(n)$ be defined as

$$P(n): 1 + \frac{1}{1+2} + \frac{1}{1+2+3} + \ldots + \frac{1}{1+2+3+\ldots+n} = \frac{2n}{n+1}$$

i.e., $$P(n): 1 + \frac{1}{1+2} + \frac{1}{1+2+3} + \ldots + \frac{2}{n(n+1)} = \frac{2n}{n+1}$$

$$\left[\because \text{sum of } n \text{ natural numbers} = \frac{n(n+1)}{2}\right]$$

Step I For $n = 1$,

$$P(1): 1 = \frac{2 \times 1}{1+1} = \frac{2}{2} = 1$$

which is true.

Step II Let it is true for $n = k$,

i.e., $$1 + \frac{1}{1+2} + \frac{1}{1+2+3} + \ldots + \frac{2}{k(k+1)} = \frac{2k}{k+1} \quad \ldots \text{(i)}$$

Step III For $n = k + 1$,

$$\left(1 + \frac{1}{1+2} + \frac{1}{1+2+3} + \ldots + \frac{2}{k(k+1)}\right) + \frac{2}{(k+1)(k+2)}$$
$$= \frac{2k}{k+1} + \frac{2}{(k+1)(k+2)} \quad \text{[using Eq. (i)]}$$
$$= \frac{2k(k+2) + 2}{(k+1)(k+2)} = \frac{2[k^2 + 2k + 1]}{(k+1)(k+2)}$$

(taking 2 common in numerator part)

$$= \frac{2(k+1)^2}{(k+1)(k+2)} \quad [\because (a+b)^2 = a^2 + 2ab + b^2]$$
$$= \frac{2(k+1)}{k+2} = \frac{2(k+1)}{(k+1)+1}$$

Therefore, $P(k + 1)$ is true when $P(k)$ is true. Hence, from the principle of mathematical induction, the statement is true for all natural numbers n.

Note *If last term of LHS is given in the form of a series, then first, try to write it in simple form (i.e., try to solve the series) using suitable formula.*

17. *(a)* Let the statement $P(n)$ be defined as

$$P(n): 1 \cdot 2 \cdot 3 + 2 \cdot 3 \cdot 4 + \ldots + n(n+1)(n+2) = \frac{n(n+1)(n+2)(n+3)}{4}$$

Step I For $n = 1$,

$$P(1): 1 \cdot 2 \cdot 3 = \frac{1(1+1)(1+2)(1+3)}{4}$$
$$= \frac{1 \times 2 \times 3 \times 4}{4}$$
$$= 1 \cdot 2 \cdot 3$$

which is true.

Step II Let it is true for $n = k$,

i.e., $$1 \cdot 2 \cdot 3 + 2 \cdot 3 \cdot 4 + 3 \cdot 4 \cdot 5 + \ldots + k(k+1)(k+2) = \frac{k(k+1)(k+2)(k+3)}{4} \quad \ldots \text{(i)}$$

Step III For $n = k + 1$.

$[1\cdot2\cdot3 + 2\cdot3\cdot4 + \ldots + k(k+1)(k+2)] + (k+1)(k+2)(k+3)$

$$= \left[\frac{k(k+1)(k+2)(k+3)}{4}\right] + (k+1)(k+2)(k+3) \quad \text{[using Eq. (i)]}$$

$$= \frac{k(k+1)(k+2)(k+3) + 4(k+1)(k+2)(k+3)}{4}$$

$$= \frac{(k+1)(k+2)(k+3)(k+4)}{4} \quad \left[\begin{array}{l}\text{taking } (k+1)(k+2)(k+3) \\ \text{common in numerator part}\end{array}\right]$$

$$= \frac{(k+1)[(k+1)+1][(k+1)+2][(k+1)+3]}{4}$$

Therefore, $P(k+1)$ is true when $P(k)$ is true. Hence, from the principle of mathematical induction, the statement is true for all natural numbers n.

18. *(b)* Let the statement $p(n)$ be defined as

$$P(n): 1\cdot3 + 2\cdot3^2 + 3\cdot3^3 + \ldots + n\cdot3^n = \frac{(2n-1)3^{n+1}+3}{4}$$

Step I For $n = 1$,

$$P(1): 1\cdot3 = \frac{(2\cdot1-1)3^{1+1}+3}{4} = \frac{3^2+3}{4} = \frac{9+3}{4} = \frac{12}{4} = 3 = 1\cdot3$$

which is true.

Step II Let it is true for $n = k$,

i.e., $1\cdot3 + 2\cdot3^2 + 3\cdot3^3 + \ldots + k\cdot3^k = \dfrac{(2k-1)3^{k+1}+3}{4}$ ... (i)

Step III For $n = k + 1$,

$(1\cdot3 + 2\cdot3^2 + 3\cdot3^3 + \ldots + k\cdot3^k) + (k+1)3^{k+1}$

$$= \frac{(2k-1)3^{k+1}+3}{4} + (k+1)3^{k+1} \quad \text{[using Eq. (i)]}$$

$$= \frac{(2k-1)3^{k+1}+3+4(k+1)3^{k+1}}{4}$$

$$= \frac{3^{k+1}(2k-1+4k+4)+3}{4}$$

$\left(\begin{array}{l}\text{taking } 3^{k+1} \text{ common in first and} \\ \text{last term of numerator part}\end{array}\right)$

$$= \frac{3^{k+1}(6k+3)+3}{4}$$

$$= \frac{3^{k+1}\cdot3(2k+1)+3}{4} \quad \left(\begin{array}{l}\text{taking 3 common in first} \\ \text{term of numerator part}\end{array}\right)$$

$$= \frac{3^{(k+1)+1}[2k+2-1]+3}{4}$$

$$= \frac{[2(k+1)-1]3^{(k+1)+1}+3}{4}$$

Therefore, $P(k+1)$ is true when $P(k)$ is true. Hence, from the principle of mathematical induction, the statement is true for all natural numbers n.

19. *(c)* I. Let the statement $P(n)$ be defined as

$$P(n): 1\cdot2 + 2\cdot3 + \ldots + n(n+1) = \frac{n(n+1)(n+2)}{3}$$

Step I For $n = 1$,

$$P(1): 1\cdot2 = \frac{1(1+1)(1+2)}{3} = \frac{1\times2\times3}{3} = 1\cdot2$$

which is true.

Step II Let it is true for $n = k$,

i.e., $1\cdot2 + 2\cdot3 + 3\cdot4 + \ldots + k(k+1)$

$$= \frac{k(k+1)(k+2)}{3} \quad \ldots\text{(i)}$$

Step III For $n = k + 1$,

$[1\cdot2 + 2\cdot3 + \ldots + k(k+1)] + (k+1)(k+2)$

$$= \left[\frac{k(k+1)(k+2)}{3}\right] + (k+1)(k+2) \quad \text{[using Eq. (i)]}$$

$$= \frac{k(k+1)(k+2) + 3(k+1)(k+2)}{3}$$

$$= \frac{(k+1)(k+2)(k+3)}{3} \quad \left[\begin{array}{l}\text{taking } (k+1)(k+2) \\ \text{common in numerator part}\end{array}\right]$$

$$= \frac{(k+1)[(k+1)+1][(k+1)+2]}{3}$$

Therefore, $P(k+1)$ is true when $P(k)$ is true. Hence, from the principle of mathematical induction, the statement is true for all natural numbers n.

II. Let the statement $P(n)$ be defined as

$$P(n): 1\cdot3 + 3\cdot5 + 5\cdot7 + \ldots + (2n-1)(2n+1) = \frac{n(4n^2+6n-1)}{3}$$

Step I For $n = 1$, we have

$$P(1): 1\cdot3 = \frac{1(4\times1^2+6\times1-1)}{3} = \frac{4+6-1}{3} = \frac{9}{3} = 3 = 1\cdot3$$

which is true.

Step II Let it is true for $n = k$,

i.e., $1\cdot3 + 3\cdot5 + 5\cdot7 + \ldots + (2k-1)(2k+1)$

$$= \frac{k(4k^2+6k-1)}{3} \quad \ldots\text{(i)}$$

Step III For $n = k + 1$,

$[1\cdot3 + 3\cdot5 + 5\cdot7 + \ldots + (2k-1)(2k+1)] + (2k-1+2)(2k+1+2)$

$$= \frac{k(4k^2+6k-1)}{3} + (2k+1)(2k+3) \quad \text{[using Eq. (i)]}$$

$$= \frac{k(4k^2+6k-1) + 3(4k^2+6k+2k+3)}{3}$$

$$= \frac{4k^3+6k^2-k+12k^2+24k+9}{3}$$

$$= \frac{4k^3+18k^2+23k+9}{3}$$

$$= \frac{(k+1)(4k^2+14k+9)}{3} \quad \left(\begin{array}{l}\text{factorising it by using}\\ \text{factor theorem}\end{array}\right)$$

$$= \frac{(k+1)(4k^2+8k+4+6k+5)}{3}$$

$$= \frac{(k+1)[4(k^2+2k+1)+6k+6-1]}{3}$$

$$= \frac{(k+1)[4(k+1)^2+6(k+1)-1]}{3}$$

Therefore, $P(k+1)$ is true when $P(k)$ is true. Hence, from the principle of mathematical induction, the statement is true for all natural numbers n.

20. *(a)* Let the statement $P(n)$ be defined as

i.e., $P(n): 1\cdot 2 + 2\cdot 2^2 + 3\cdot 2^3 + \dots + n\cdot 2^n$
$= (n-1)2^{n+1} + 2$

***Step* I** For $n = 1$,

$P(1): 1\cdot 2 = (1-1)2^{1+1} + 2 = 0 + 2 = 2 = 1\cdot 2$

which is true.

***Step* II** Let it is true for $n = k$,

i.e., $1\cdot 2 + 2\cdot 2^2 + 3\cdot 2^3 + \dots + k\cdot 2^k$
$= (k-1)2^{k+1} + 2$ ...(i)

Step III For $n = k+1$,

$(1\cdot 2 + 2\cdot 2^2 + 3\cdot 2^3 + \dots + k\cdot 2^k) + (k+1)2^{k+1}$

$= [(k-1)2^{k+1} + 2] + (k+1)2^{k+1}$ [using Eq. (i)]

$= (k-1)2^{k+1} + 2 + (k+1)2^{k+1}$

$= 2^{k+1}(k-1+k+1) + 2$ $\left(\begin{array}{l}\text{taking } 2^{k+1} \text{common}\\ \text{in first and last term}\end{array}\right)$

$= 2^{k+1}(2k) + 2$

$= 2^{(k+1)+1}k + 2$

$= [(k+1)-1]2^{(k+1)+1} + 2$

Therefore, $P(k+1)$ is true when $P(k)$ is true. Hence, from the principle of mathematical induction, the statement is true for all natural numbers n.

21. *(a)* I. Let the statement $P(n)$ be defined as

i.e., $P(n): \frac{1}{2} + \frac{1}{4} + \frac{1}{8} + \dots + \frac{1}{2^n} = 1 - \frac{1}{2^n}$

***Step* I** For $n = 1$,

$P(1): \frac{1}{2} = 1 - \frac{1}{2^1} = 1 - \frac{1}{2} = \frac{1}{2}$

which is true.

***Step* II** Let it is true for $n = k$,

i.e., $\frac{1}{2} + \frac{1}{2^2} + \frac{1}{2^3} + \dots + \frac{1}{2^k} = 1 - \frac{1}{2^k}$ ...(i)

***Step* III** For $n = k+1$,

$\left(\frac{1}{2} + \frac{1}{2^2} + \frac{1}{2^3} + \dots + \frac{1}{2^k}\right) + \frac{1}{2^{k+1}} = 1 - \frac{1}{2^k} + \frac{1}{2^{k+1}}$

[using Eq. (i)]

$= 1 - \frac{1}{2^k} + \frac{1}{2^k\cdot 2}$

$= 1 - \frac{1}{2^k}\left(1 - \frac{1}{2}\right)$ $\left(\begin{array}{l}\text{taking } \frac{1}{2^k} \text{ common}\\ \text{in last two terms}\end{array}\right)$

$= 1 - \frac{1}{2^k} \times \frac{1}{2} = 1 - \frac{1}{2^{k+1}}$

Therefore, $P(k+1)$ is true when $P(k)$ is true. Hence. from the principle of mathematical induction, the statement is true for all natural numbers n.

So, I is true.

IV. Let the statement $P(n)$ be defined as

i.e., $P(n): \frac{1}{2\cdot 5} + \frac{1}{5\cdot 8} + \frac{1}{8\cdot 11} + \dots + \frac{1}{(3n-1)(3n+2)}$
$= \frac{n}{6n+4}$

***Step* I** For $n = 1$,

$P(1): \frac{1}{2\cdot 5} = \frac{1}{6\times 1 + 4} = \frac{1}{10} = \frac{1}{2\cdot 5}$

which is true.

***Step* II** Let it is true for $n = k$,

i.e., $\frac{1}{2\cdot 5} + \frac{1}{5\cdot 8} + \frac{1}{8\cdot 11} + \dots + \frac{1}{(3k-1)(3k+2)}$
$= \frac{k}{6k+4}$ ...(i)

***Step* III** For $n = k+1$,

$\left[\frac{1}{2\cdot 5} + \frac{1}{5\cdot 8} + \frac{1}{8\cdot 11} + \dots + \frac{1}{(3k-1)(3k+2)}\right]$
$+ \left[\frac{1}{(3k-1+3)(3k+2+3)}\right]$

$= \frac{k}{6k+4} + \frac{1}{(3k+2)(3k+5)}$ [using Eq. (i)]

$= \frac{k}{2(3k+2)} + \frac{1}{(3k+2)(3k+5)}$

$= \frac{k(3k+5)+2}{2(3k+2)(3k+5)} = \frac{3k^2+5k+2}{2(3k+2)(3k+5)}$

$= \frac{3k^2+3k+2k+2}{2(3k+2)(3k+5)} = \frac{3k(k+1)+2(k+1)}{2(3k+2)(3k+5)}$

$= \frac{(k+1)(3k+2)}{2(3k+2)(3k+5)}$

$= \frac{k+1}{2(3k+3+2)} = \frac{k+1}{2[3(k+1)+2]}$

$= \frac{k+1}{6(k+1)+4}$

Therefore, $P(k+1)$ is true when $P(k)$ is true. Hence, from the principle of mathematical induction, the statement is true for all natural numbers n.

So, IV is true.

Hence, II and III are not true.

22. *(c)* Let the statement $P(n)$ be defined as

i.e., $P(n): a + ar + ar^2 + \ldots + ar^{n-1} = \dfrac{a(r^n - 1)}{r-1}$

Step I For $n = 1$,

$P(1): a = \dfrac{a(r^1 - 1)}{r-1} = a$, which is true.

Step II Let it is true for $n = k$,

i.e., $a + ar + ar^2 + \ldots + ar^{k-1} = \dfrac{a(r^k - 1)}{r-1}$ ...(i)

Step III For $n = k + 1$,

$(a + ar + ar^2 + \ldots + ar^{k-1}) + ar^{k+1-1}$

$= \dfrac{a(r^k - 1)}{r-1} + ar^k$ [using Eq. (i)]

$= \dfrac{a(r^k - 1) + ar^k(r-1)}{r-1}$

$= \dfrac{a[r^k - 1 + r^k \cdot r^1 - r^k]}{r-1}$ (taking a common in numerator part)

$= \dfrac{a[r^k - 1 + r^{k+1} - r^k]}{r-1} = \dfrac{a[r^{k+1} - 1]}{r-1}$

Therefore, $P(k+1)$ is true when $P(k)$ is true. Hence, from the principle of mathematical induction, the statement is true for all natural numbers n.

23. *(c)* Let the statement $P(n)$ be defined as

$P(n): \left(1 + \dfrac{3}{1}\right)\left(1 + \dfrac{5}{4}\right)\left(1 + \dfrac{7}{9}\right)\ldots\left(1 + \dfrac{2n+1}{n^2}\right) = (n+1)^2$

Step I For $n = 1$,

i.e., $P(1): \left(1 + \dfrac{3}{1}\right) = (1+1)^2 = 2^2 = 4 = \left(1 + \dfrac{3}{1}\right)$

which is true.

Step II Let it is true for $n = k$,

i.e., $\left(1 + \dfrac{3}{1}\right)\left(1 + \dfrac{5}{4}\right)\left(1 + \dfrac{7}{9}\right)\ldots\left(1 + \dfrac{2k+1}{k^2}\right)$

$= (k+1)^2$ ...(i)

Step III For $n = k + 1$,

$\left\{\left(1 + \dfrac{3}{1}\right)\left(1 + \dfrac{5}{4}\right)\left(1 + \dfrac{7}{9}\right)\ldots\left(1 + \dfrac{2k+1}{k^2}\right)\right\}\left(1 + \dfrac{2k+1+2}{(k+1)^2}\right)$

$= (k+1)^2\left(1 + \dfrac{2k+3}{(k+1)^2}\right)$ [using Eq. (i)]

$= (k+1)^2\left[\dfrac{(k+1)^2 + 2k + 3}{(k+1)^2}\right]$

$= k^2 + 2k + 1 + 2k + 3$

$= (k+2)^2 = [(k+1)+1]^2$

$[\because (a+b)^2 = a^2 + 2ab + b^2]$

Therefore, $P(k+1)$ is true when $P(k)$ is true.Hence, from the principle of mathematical induction, the statement is true for all natural numbers n.

24. *(b)* Let the statement $P(n)$ be defined as

$P(n): \left(1 + \dfrac{1}{1}\right)\left(1 + \dfrac{1}{2}\right)\left(1 + \dfrac{1}{3}\right)\ldots\left(1 + \dfrac{1}{n}\right) = (n+1)$

Step I For $n = 1$,

$P(1): \left(1 + \dfrac{1}{1}\right) = 1 + 1 = 1 + \dfrac{1}{1}$, which is true.

Step II Let it is true for $n = k$,

i.e., $\left(1 + \dfrac{1}{1}\right)\left(1 + \dfrac{1}{2}\right)\left(1 + \dfrac{1}{3}\right)\ldots\left(1 + \dfrac{1}{k}\right) = k + 1$ ...(i)

Step III For $n = k + 1$,

$\left[\left(1 + \dfrac{1}{1}\right)\left(1 + \dfrac{1}{2}\right)\left(1 + \dfrac{1}{3}\right)\ldots\left(1 + \dfrac{1}{k}\right)\right]\left(1 + \dfrac{1}{k+1}\right)$

$= (k+1)\left(1 + \dfrac{1}{k+1}\right)$ [using Eq. (i)]

$= (k+1)\left(\dfrac{k+1+1}{k+1}\right) = (k+1) + 1$

Therefore, $P(k+1)$ is true when $P(k)$ is true. Hence, from the principle of mathematical induction, the statement is true for all natural numbers n.

25. *(a)* Let the statement $P(n)$ be defined as

$P(n): 1^2 + 3^2 + 5^2 + \ldots + (2n-1)^2 = \dfrac{n(2n-1)(2n+1)}{3}$

Step I For $n = 1$,

i.e., $P(1): 1^2 = \dfrac{1(2-1)(2+1)}{3} = \dfrac{1 \times 1 \times 3}{3} = 1 = 1^2$

which is true.

Step II Let it is true for $n = k$,

i.e., $1^2 + 3^2 + 5^2 + \ldots + (2k-1)^2$

$= \dfrac{k(2k-1)(2k+1)}{3}$ ...(i)

Step III For $n = k + 1$,

$[1^2 + 3^2 + 5^2 + \ldots + (2k-1)^2] + (2k - 1 + 2)^2$

$= \dfrac{k(2k-1)(2k+1)}{3} + (2k+1)^2$ [using Eq. (i)]

$= \dfrac{k(2k-1)(2k+1) + 3(2k+1)^2}{3}$

$= \dfrac{(2k+1)[k(2k-1) + 3(2k+1)]}{3}$

[taking $(2k+1)$ common in numerator part]

$= \dfrac{(2k+1)(2k^2 - k + 6k + 3)}{3}$

$= \dfrac{(2k+1)(2k^2 + 5k + 3)}{3}$

$= \dfrac{(2k+1)(2k^2 + 2k + 3k + 3)}{3}$

(factorising the numerator by splitting the middle term)

$$= \frac{(2k+1)[2k(k+1)+3(k+1)]}{3}$$
$$= \frac{(2k+1)(k+1)(2k+3)}{3}$$
$$= \frac{(k+1)(2k+2-1)(2k+2+1)}{3}$$
$$= \frac{(k+1)[2(k+1)-1][2(k+1)+1]}{3}$$

Therefore, $P(k+1)$ is true when $P(k)$ is true. Hence, from the principle of mathematical induction, the statement is true for all natural numbers n.

26. *(b)* Let the statement $P(n)$ be defined as

$$P(n): \frac{1}{1\cdot 4} + \frac{1}{4\cdot 7} + \frac{1}{7\cdot 10} + \dots + \frac{1}{(3n-2)(3n+1)} = \frac{n}{3n+1}$$

***Step* I** For $n = 1$,

$$P(1): \frac{1}{1\cdot 4} = \frac{1}{3\cdot 1+1} = \frac{1}{4} = \frac{1}{1\cdot 4}$$

which is true.

***Step* II** Let it is true for $n = k$,

i.e., $$\frac{1}{1\cdot 4} + \frac{1}{4\cdot 7} + \frac{1}{7\cdot 10} + \dots + \frac{1}{(3k-2)(3k+1)} = \frac{k}{3k+1} \quad \dots \text{(i)}$$

***Step* III** For $n = k+1$,

$$\left[\frac{1}{1\cdot 4} + \frac{1}{4\cdot 7} + \frac{1}{7\cdot 10} + \dots + \frac{1}{(3k-2)(3k+1)}\right] + \frac{1}{(3k-2+3)(3k+1+3)}$$

$$= \frac{k}{3k+1} + \frac{1}{(3k+1)(3k+4)} \quad \text{[using Eq. (i)]}$$
$$= \frac{k(3k+4)+1}{(3k+1)(3k+4)}$$
$$= \frac{3k^2+4k+1}{(3k+1)(3k+4)}$$
$$= \frac{3k^2+3k+k+1}{(3k+1)(3k+4)} \quad \left(\begin{array}{l}\text{factorising the numerator part}\\ \text{by splitting the middle term}\end{array}\right)$$
$$= \frac{3k(k+1)+1(k+1)}{(3k+1)(3k+4)}$$
$$= \frac{(k+1)(3k+1)}{(3k+1)(3k+4)}$$
$$= \frac{k+1}{3k+3+1} = \frac{k+1}{3(k+1)+1}$$

Therefore, $P(k+1)$ is true when $P(k)$ is true. Hence, from the principle of mathematical induction, the statement is true for all natural numbers n.

27. *(c)* I. Let the statement $P(n)$ be defined as

$$P(n): 1+2+3+\dots+n < \frac{1}{8}(2n+1)^2$$

***Step* I** For $n = 1$,

$$P(1): 1 < \frac{1}{8}(2\cdot 1+1)^2 \Rightarrow 1 < \frac{1}{8}\times 3^2 \Rightarrow 1 < \frac{9}{8}$$

which is true.

***Step* II** Let it is true for $n = k$,

$$1+2+3+\dots+k < \frac{1}{8}(2k+1)^2 \quad \dots \text{(i)}$$

***Step* III** For $n = k+1$,

$$(1+2+3+\dots+k)+(k+1) < \frac{1}{8}(2k+1)^2 + (k+1)$$

[using Eq. (i)]

$$= \frac{(2k+1)^2}{8} + \frac{k+1}{1} = \frac{(2k+1)^2+8k+8}{8}$$
$$= \frac{4k^2+1+4k+8k+8}{8}$$
$$= \frac{4k^2+12k+9}{8} = \frac{(2k+3)^2}{8}$$
$$= \frac{(2k+2+1)^2}{8} = \frac{[2(k+1)+1]^2}{8}$$
$$\Rightarrow 1+2+3+\dots+k+(k+1) < \frac{[2(k+1)+1]^2}{8}$$

Therefore, $P(k+1)$ is true when $P(k)$ is true. Hence, from the principle of mathematical induction, the statement is true for all natural numbers n.

II. Let the statement $P(n)$ be defined as

$P(n): n(n+1)(n+5)$ is a multiple of 3.

***Step* I** For $n = 1$,

$$P(1): 1(1+1)(1+5) = 1\times 2\times 6 = 12 = 3\times 4$$

which is a multiple of 3, that is true.

***Step* II** Let it is true for $n = k$,

i.e., $$k(k+1)(k+5) = 3\lambda$$
$$\Rightarrow k(k^2+5k+k+5) = 3\lambda$$
$$\Rightarrow k^3+6k^2+5k = 3\lambda \quad \dots \text{(i)}$$

***Step* III** For $n = k+1$, $(k+1)(k+1+1)(k+1+5)$

$$= (k+1)(k+2)(k+6) = (k^2+2k+k+2)(k+6)$$
$$= (k^2+3k+2)(k+6)$$
$$= k^3+6k^2+3k^2+18k+2k+12$$
$$= k^3+9k^2+20k+12$$
$$= (3\lambda-6k^2-5k)+9k^2+20k+12 \quad \text{[using Eq. (i)]}$$
$$= 3\lambda+3k^2+15k+12$$
$$= 3(\lambda+k^2+5k+4)$$

which is a multiple of 3.

Therefore, $P(k+1)$ is true when $P(k)$ is true. Hence, from the principle of mathematical induction, the statement is true for all natural numbers n.

Hence, both the statements are true.

28. *(c)* Let the statement $P(n)$ be defined as

$$P(n): 10^{2n-1}+1 \text{ is divisible by } 11.$$

Step I For $n=1$,

$$P(1): 10^{2\times 1-1}+1 = 10^{2-1}+1$$
$$= 10^1+1 = 10+1 = 11\times 1$$

which is true.

Step II Let it is true for $n=k$,

i.e., $$10^{2k-1}+1 = 11\lambda \quad \ldots(i)$$

Step III For $n=k+1$,

$$10^{2(k+1)-1}+1 = 10^{2k+2-1}+1$$
$$= 10^{(2k-1)+2}+1$$
$$= 10^{2k-1}10^2+1$$
$$= (11\lambda-1)100+1 \quad \text{[using Eq. (i)]}$$
$$= 11\lambda\times 100-100+1$$
$$= 11\lambda\times 100-99 = 11[100\lambda-9]$$

which is divisible by 11.

Therefore, $P(k+1)$ is true when $P(k)$ is true. Hence, from the principle of mathematical induction, the statement is true for all natural numbers n.

29. *(c)* Let the statement $P(n)$ be defined as

$$P(n): 3^{2n+2}-8n-9 \text{ is divisible by } 8.$$

Step I For $n=1$,

$$P(1): 3^{2\times 1+2}-8\times 1-9 = 3^4-8-9$$
$$= 81-17 = 64 = 8\times 8$$

which is divisible by 8.

Step II Let it is true for $n=k$,

i.e., $$3^{2k+2}-8k-9 = 8\lambda \quad \ldots(i)$$

Step III For $n=k+1$,

$$3^{2(k+1)+2}-8(k+1)-9 = 3^{2k+2+2}-8k-8-9$$
$$= 3^{2k+2}3^2-8k-17$$
$$= (8\lambda+8k+9)3^2-8k-17 \quad \text{[using Eq. (i)]}$$
$$= (8\lambda+8k+9)9-8k-17$$
$$= 72\lambda+72k+81-8k-17$$
$$= 72\lambda+64k+64$$
$$= 8(9\lambda+8k+8)$$

which is divisible by 8.

Therefore, $P(k+1)$ is true when $P(k)$ is true. Hence, from the principle of mathematical induction, the statement is true for all natural numbers n.

Hence, $P(n)$ is divisible by 4 and 8.

30. *(b)* Let the statement $P(n)$ be defined as

$$P(n): \frac{1}{1\cdot 2\cdot 3}+\frac{1}{2\cdot 3\cdot 4}+\frac{1}{3\cdot 4\cdot 5}+\cdots + \frac{1}{n(n+1)(n+2)} = \frac{n(n+3)}{4(n+1)(n+2)}$$

Step I For $n=1$,

$$P(1): \frac{1}{1\cdot 2\cdot 3} = \frac{1(1+3)}{4(1+1)(1+2)} = \frac{4}{4\times 2\times 3} = \frac{1}{1\cdot 2\cdot 3}$$

which is true.

Step II Let it is true for $n=k$.

$$\frac{1}{1\cdot 2\cdot 3}+\frac{1}{2\cdot 3\cdot 4}+\frac{1}{3\cdot 4\cdot 5}+\ldots+\frac{1}{k(k+1)(k+2)} = \frac{k(k+3)}{4(k+1)(k+2)} \quad \ldots(i)$$

Step III For $n=k+1$,

$$\left(\frac{1}{1\cdot 2\cdot 3}+\frac{1}{2\cdot 3\cdot 4}+\frac{1}{3\cdot 4\cdot 5}+\cdots+\frac{1}{k(k+1)(k+2)}\right) + \frac{1}{(k+1)(k+1+1)(k+1+2)}$$
$$= \frac{k(k+3)}{4(k+1)(k+2)}+\frac{1}{(k+1)(k+2)(k+3)} \quad \text{[using Eq. (i)]}$$
$$= \frac{k(k+3)^2+4}{4(k+1)(k+2)(k+3)}$$
$$= \frac{k(k^2+9+6k)+4}{4(k+1)(k+2)(k+3)}$$
$$= \frac{k^3+6k^2+9k+4}{4(k+1)(k+2)(k+3)}$$
$$= \frac{(k+1)(k^2+5k+4)}{4(k+1)(k+2)(k+3)} = \frac{k^2+5k+4}{4(k+2)(k+3)}$$
$$= \frac{k^2+4k+k+4}{4(k+2)(k+3)} = \frac{k(k+4)+(k+4)}{4(k+2)(k+3)}$$
$$= \frac{(k+1)(k+4)}{4(k+2)(k+3)} = \frac{(k+1)[(k+1)+3]}{4[(k+1)+1][(k+1)+2]}$$

Therefore, $P(k+1)$ is true when $P(k)$ is true. Hence, from the principle of mathematical induction, the statement is true for all natural numbers n.

31. *(c)* Let the statement $P(n)$ be defined as

$$P(n): \frac{1}{3\cdot 5}+\frac{1}{5\cdot 7}+\frac{1}{7\cdot 9}+\cdots+\frac{1}{(2n+1)(2n+3)} = \frac{n}{3(2n+3)}$$

Step I For $n=1$,

$$P(1): \frac{1}{3\cdot 5} = \frac{1}{3(2\times 1+3)} = \frac{1}{3\times 5} = \frac{1}{3\cdot 5}, \text{ which is true.}$$

Step II Let it is true for $n=k$, i.e.,

$$\frac{1}{3\cdot 5}+\frac{1}{5\cdot 7}+\frac{1}{7\cdot 9}+\cdots+\frac{1}{(2k+1)(2k+3)} = \frac{k}{3(2k+3)} \quad \ldots(i)$$

Step III For $n=k+1$,

$$\left[\frac{1}{3\cdot 5}+\frac{1}{5\cdot 7}+\frac{1}{7\cdot 9}+\cdots+\frac{1}{(2k+1)(2k+3)}\right] + \frac{1}{(2k+1+2)(2k+3+2)}$$
$$= \frac{k}{3(2k+3)}+\frac{1}{(2k+3)(2k+5)} \quad \text{[using Eq. (i)]}$$

$$= \frac{k(2k+5)+3}{3(2k+3)(2k+5)}$$
$$= \frac{2k^2+5k+3}{3(2k+3)(2k+5)}$$
$$= \frac{2k^2+2k+3k+3}{3(2k+3)(2k+5)}$$
$$= \frac{2k(k+1)+3(k+1)}{3(2k+3)(2k+5)}$$
$$= \frac{(k+1)(2k+3)}{3(2k+3)(2k+5)}$$
$$= \frac{k+1}{3(2k+2+3)} = \frac{k+1}{3[2(k+1)+3]}$$

Therefore, $P(k+1)$ is true when $P(k)$ is true. Hence, from the principle of mathematical induction, the statement is true for all natural numbers n.

32. *(a)* Let the statement $P(n)$ be defined as

$$P(n): x^{2n} - y^{2n} \text{ is divisible by } x+y.$$

Step I For $n = 1$, $P(1): x^2 - y^2 = (x-y)(x+y)$

which is divisible by $(x+y)$, that is true.

Step II Let it is true for $n = k$,

i.e., $$x^{2k} - y^{2k} = \lambda(x+y) \quad \ldots(i)$$

Step III For $n = k+1$,

$$x^{2(k+1)} - y^{2(k+1)} = x^{2k+2} - y^{2k+2}$$
$$= x^{2k}x^2 - y^{2k}y^2$$
$$= [\lambda(x+y) + y^{2k}]x^2 - y^{2k}y^2 \quad \text{[using Eq. (i)]}$$
$$= \lambda(x+y)x^2 + y^{2k}x^2 - y^{2k}y^2$$
$$= \lambda(x+y)x^2 + y^{2k}(x^2-y^2)$$
$$= \lambda(x+y)x^2 + y^{2k}(x-y)(x+y)$$
$$= (x+y)[\lambda x^2 + y^{2k}(x-y)]$$

which is a multiple of $(x+y)$ *i.e.*, divisible by $(x+y)$.

Therefore, $P(k+1)$ is true when $P(k)$ is true. Hence, from the principle of mathematical induction, the statement is true for all natural numbers n.

33. *(a)* Let the statement $P(n)$ be defined as

$P(n): 41^n - 14^n$ is a multiple of 27.

Step I For $n = 1$,

$$P(1): 41^1 - 14^1 = 27 = 1 \times 27$$

which is a multiple of 27 that is true.

Step II Let it is true for $n = k$,

i.e., $$41^k - 14^k = 27\lambda \quad \ldots(i)$$

Step III For $n = k+1$,

$$41^{k+1} - 14^{k+1} = 41^k\, 41 - 14^k\, 14$$
$$= (27\lambda + 14^k)\,41 - 14^k\, 14 \quad \text{[using Eq. (i)]}$$
$$= 27\lambda \times 41 + 14^k \times 41 - 14^k \times 14$$
$$= 27\lambda \times 41 + 14^k(41-14)$$
$$= 27\lambda \times 41 + 14^k \times 27$$
$$= 27(41\lambda + 14^k)$$

which is a multiple of 27.

Therefore, $P(k+1)$ is true when $P(k)$ is true. Hence, from the principle of mathematical induction, the statement is true for all natural numbers n.

Thus, we see that

Both A and R are correct and R is the correct explanation of A.

34. *(b)* Let the statement $P(n)$ be defined as

$P(n): (2n+7) < (n+3)^2$

Step I For $n = 1$,

$$P(1): (2 \times 1 + 7) < (1+3)^2 \Rightarrow 9 < 16$$

which is true.

Step II Let it is true for $n = k$,

i.e., $$2k + 7 < (k+3)^2 \quad \ldots(i)$$

Step III For $n = k+1$,

$$2(k+1) + 7 = 2k + 2 + 7$$
$$= (2k+7) + 2$$
$$< (k+3)^2 + 2 \quad \text{[using Eq. (i)]}$$
$$= k^2 + 9 + 6k + 2$$
$$= k^2 + 8k + 16 - 2k - 5$$
$$= (k+4)^2 - (2k+5) \quad [\because (a+b)^2 = a^2 + 2ab + b^2]$$
$$< (k+4)^2$$
$$= [(k+1)+3]^2 \Rightarrow [2(k+1)+7] < [(k+1)+3]^2$$

Therefore, $P(k+1)$ is true when $P(k)$ is true. Hence, from the principle of mathematical induction, the statement is true for all natural numbers n.

35. *(d)* Unless, we prove $P(1)$ is true, nothing can be said.

36. *(d)* Given, $P(n): n^2 + n + 1$

At $n = 1$, $P(1): 3$, which is not an even integer.

$\therefore$ $P(1)$ is not true.

Also, $n(n+1) + 1$ is always an odd integer.

37. *(b)* Since, $P(3)$ is true.

Assume $P(k)$ is true $\Rightarrow P(k+1)$ is true means, if $P(3)$ is true $\Rightarrow P(4)$ is true $\Rightarrow P(5)$ is true and so on. So, statement $P(n)$ is true for all $n \geq 3$.

38. *(d)* Clearly, $P(1): 2 < 1$ is false

$P(2): 2^2 < 1 \times 2$ is false

$P(3): 2^3 < 1 \times 2 \times 3$ is false

but $P(4): 2^4 < 1 \times 2 \times 3 \times 4$ is true.

So, $P(n)$ is true for $n = 4$.

39. *(c)* Since, $P(5)$ is true and $P(k+1)$ is true, whenever $P(k)$ is true.

40. *(d)* Now, for $n = 1$,

$$2 \cdot 4^{2+1} + 3^{3+1} = 2 \cdot 4^3 + 3^4 = 2 \cdot 64 + 81$$
$$= 128 + 81 = 209$$

For $n = 2$, $2 \cdot 4^5 + 3^7 = 8 \cdot 256 + 2187$

$= 2048 + 2187 = 4235$

Note that the HCF of 209 and 4235 is 11. So, $2 \cdot 4^{2n+1} + 3^{3n+1}$ is divisible by 11.

Hence, λ is 11.

Note *This solution is appropriate for competitive.*

41. *(d)* For $n = 1$, $P(1): 65 + k$ is divisible by 64.

Thus, k should be -1, since $65 - 1 = 64$ is divisible by 64.

42. *(b)* $P(1): 1 = 3 + 1^2$ which is not true.

Suppose $P(k)$ is true, then

$$1 + 3 + 5 + \cdots + (2k - 1) = 3 + k^2$$

On adding $(2k + 1)$ to both the sides, we get

$$1 + 3 + 5 + \cdots + (2k - 1) + (2k + 1) = 3 + k^2 + 2k + 1 = 3 + (k + 1)^2 = P(k + 1)$$

Thus, $P(k) \Rightarrow P(k + 1)$

43. *(b)* I. Let $P(n): \frac{1}{\sqrt{1}} + \frac{1}{\sqrt{2}} + \cdots + \frac{1}{\sqrt{n}} > \sqrt{n}$, for all natural numbers $n \geq 2$.

$\therefore$ $P(2): \frac{1}{\sqrt{1}} + \frac{1}{\sqrt{2}} = 1.707 > \sqrt{2}$

$\Rightarrow P(2)$ is true.

Let us assume that for $n = k$, $P(n)$ is true *i.e.*,

$$P(k): \frac{1}{\sqrt{1}} + \frac{1}{\sqrt{2}} + \cdots + \frac{1}{\sqrt{k}} > \sqrt{k} \text{ is true}$$

Now for $n = k + 1$, we have

$$P(k+1): \frac{1}{\sqrt{1}} + \frac{1}{\sqrt{2}} + \ldots + \frac{1}{\sqrt{k}} + \frac{1}{\sqrt{k+1}} > \sqrt{k} + \frac{1}{\sqrt{k+1}} = \frac{\sqrt{k(k+1)} + 1}{\sqrt{(k+1)}}$$

$$> \frac{k+1}{\sqrt{k+1}} \qquad [\because \sqrt{k(k+1)} > k, \forall k \geq 1]$$

$$= \sqrt{(k+1)}$$

$$\Rightarrow \quad P(k+1): \frac{1}{\sqrt{1}} + \frac{1}{\sqrt{2}} + \ldots + \frac{1}{\sqrt{k+1}} > \sqrt{k+1}$$

$\therefore$ By principle of mathematical induction, Statement I is true, $\forall n \geq 2$.

II. Now, let $\alpha(n): \sqrt{n(n+1)} < n + 1$, for all natural numbers $n \geq 2$.

$\therefore$ $\alpha(2): \sqrt{2(2+1)} = \sqrt{6} < 3$

Let us assume that

$\alpha(k): \sqrt{k(k+1)} < (k+1)$ is true

For $n = k + 1$,

$\alpha(k+1): \sqrt{(k+1)(k+2)} < (k+2) \quad [\because (k+1) < (k+2)]$

By principle of mathematical induction, Statement II is true, for all $n \geq 2$, but Statement II is not a correct explanation for Statement I.

44. *(b)* Let the statement $P(n)$ be defined as

$$P(n): \cos\alpha + \cos(\alpha + \beta) + \cos(\alpha + 2\beta) + \cdots + \cos[\alpha + (n-1)\beta] = \frac{\cos\left[\alpha + \left(\frac{n-1}{2}\right)\beta\right]\sin\left(\frac{n\beta}{2}\right)}{\sin\frac{\beta}{2}}, \text{ for all } n \in N.$$

Step I For $n = 1$,

$$P(1): \cos\alpha = \frac{\cos\left[\alpha + \left(\frac{1-1}{2}\right)\beta\right]\sin\left(\frac{1 \times \beta}{2}\right)}{\sin\frac{\beta}{2}} = \frac{\cos(\alpha + 0)\sin\frac{\beta}{2}}{\sin\frac{\beta}{2}} = \cos\alpha$$

which is true.

Step II Let it is true for $n = k$ *i.e.*,

$$\cos\alpha + \cos(\alpha + \beta) + \cos(\alpha + 2\beta) + \cdots + \cos[\alpha + (k-1)\beta] = \frac{\cos\left[\alpha + \left(\frac{k-1}{2}\right)\beta\right]\sin\frac{k\beta}{2}}{\sin\frac{\beta}{2}} \quad \ldots \text{(i)}$$

Step III For $n = k + 1$,

$$\cos\alpha + \cos(\alpha + \beta) + \cdots + \cos[\alpha + (k-1)\beta] + \cos(\alpha + k\beta)$$

$$= \frac{\cos\left[\alpha + \left(\frac{k-1}{2}\right)\beta\right]\sin\frac{k\beta}{2}}{\sin\frac{\beta}{2}} + \cos(\alpha + k\beta) \qquad \text{[using Eq. (i)]}$$

$$= \frac{\cos\left[\alpha + \left(\frac{k-1}{2}\right)\beta\right]\sin\frac{k\beta}{2} + \cos(\alpha + k\beta)\sin\frac{\beta}{2}}{\sin\frac{\beta}{2}}$$

$$= \frac{2\cos\left[\alpha + \left(\frac{k-1}{2}\right)\beta\right]\sin\frac{k\beta}{2} + 2\cos(\alpha + k\beta)\sin\frac{\beta}{2}}{2\sin\frac{\beta}{2}}$$

(multiply and divide by 2)

$$= \frac{\left\{\sin\left[\alpha + \left(\frac{k-1}{2}\right)\beta + \frac{k\beta}{2}\right] - \sin\left[\left(\alpha + \left(\frac{k-1}{2}\right)\beta\right) - \frac{k\beta}{2}\right] + \sin\left(\alpha + k\beta + \frac{\beta}{2}\right) - \sin\left(\alpha + k\beta - \frac{\beta}{2}\right)\right\}}{2\sin\frac{\beta}{2}}$$

$[\because 2\cos A \sin B = \sin(A + B) - \sin(A - B)]$

$$= \frac{\left\{\sin\left[\alpha + k\beta - \frac{\beta}{2}\right] - \sin\left[\alpha - \frac{\beta}{2}\right] + \sin\left[\alpha + k\beta + \frac{\beta}{2}\right] - \sin\left(\alpha + k\beta - \frac{\beta}{2}\right)\right\}}{2\sin\frac{\beta}{2}}$$

$$= \frac{\sin\left(\alpha + k\beta + \frac{\beta}{2}\right) - \sin\left(\alpha - \frac{\beta}{2}\right)}{2\sin\frac{\beta}{2}}$$

$$= \frac{2\cos\frac{\alpha + k\beta + \frac{\beta}{2} + \alpha - \frac{\beta}{2}}{2}\sin\frac{\alpha + k\beta + \frac{\beta}{2} - \alpha + \frac{\beta}{2}}{2}}{2\sin\frac{\beta}{2}}$$

$$\left(\because \sin C - \sin D = 2\cos\frac{C+D}{2}\sin\frac{C-D}{2}\right)$$

$$= \frac{2\cos\left(\frac{2\alpha + k\beta}{2}\right)\sin\left(\frac{k\beta + \beta}{2}\right)}{2\sin\frac{\beta}{2}}$$

$$= \frac{2\cos\left[\alpha + \frac{\{(k+1)-1\}\beta}{2}\right]\sin\left[\frac{k\beta + \beta}{2}\right]}{2\sin\frac{\beta}{2}}$$

$$= \frac{\cos\left[\alpha + \frac{\{(k+1)-1\}\beta}{2}\right]\sin\left[\frac{\beta(k+1)}{2}\right]}{\sin\frac{\beta}{2}}$$

Therefore, $P(k+1)$ is true when $P(k)$ is true.

Hence, from the principle of mathematical induction, the statement is true for all natural numbers n.

45. *(a)* Let the statement $P(n)$ be defined as

$$P(n): \cos\theta\cos 2\theta\cos 2^2\theta \ldots \cos 2^{n-1}\theta = \frac{\sin 2^n\theta}{2^n\sin\theta}, \text{ for all } n \in N$$

***Step* I** For $n = 1$,

$$P(1): \cos\theta = \frac{\sin 2^1\theta}{2^1\sin\theta} = \frac{2\sin\theta\cos\theta}{2\sin\theta} = \cos\theta$$

Hence, it is true for $n = 1$.

***Step* II** Let it is true for $n = k$, *i.e.*,

$$\cos\theta\cos 2\theta\cos 2^2\theta \ldots \cos 2^{k-1}\theta = \frac{\sin 2^k\theta}{2^k\sin\theta} \quad \ldots(i)$$

***Step* III** For $n = k + 1$,

$$(\cos\theta\cos 2\theta\cos 2^2\theta \ldots \cos 2^{k-1}\theta)(\cos 2^k\theta)$$

$$= \frac{\sin 2^k\theta}{2^k\sin\theta} \times (\cos 2^k\theta) \quad \text{[using Eq. (i)]}$$

$$= \frac{2\sin(2^k\theta)\cos(2^k\theta)}{2^{k+1}\sin\theta}$$

(multiplying numerator and denominator by 2)

$$= \frac{\sin 2^{k+1}\theta}{2^{k+1}\sin\theta}$$

Therefore, $P(k+1)$ is true, when $P(k)$ is true. Hence, from the principle of mathematical induction, the statement is true for all natural numbers n.

46. *(a)* Let the statement $P(n)$ be defined as

$$P(n): \sin\theta + \sin 2\theta + \sin 3\theta + \cdots + \sin n\theta = \frac{\sin\frac{n\theta}{2}\sin\left(\frac{n+1}{2}\right)\theta}{\sin\frac{\theta}{2}}, \text{ for all } n \in N.$$

***Step* I** For $n = 1$,

$$P(1): \sin\theta = \frac{\sin\frac{\theta}{2}\sin\left(\frac{1+1}{2}\right)\theta}{\sin\frac{\theta}{2}} = \sin\theta$$

Hence, it is true for $n = 1$.

***Step* II** Let it is true for $n = k$, *i.e.*,

$$\sin\theta + \sin 2\theta + \sin 3\theta + \cdots + \sin k\theta = \frac{\sin\left(\frac{k\theta}{2}\right)\sin\left(\frac{k+1}{2}\right)\theta}{\sin\frac{\theta}{2}} \quad \ldots(i)$$

***Step* III** For $n = k + 1$,

$$\sin\theta + \sin 2\theta + \sin 3\theta + \cdots + \sin k\theta + \sin(k+1)\theta$$

$$= \frac{\sin\left(\frac{k\theta}{2}\right)\sin\left(\frac{k+1}{2}\right)\theta}{\sin\frac{\theta}{2}} + \sin(k+1)\theta \quad \text{[using Eq. (i)]}$$

$$= \frac{\sin\left(\frac{k\theta}{2}\right)\sin\left(\frac{k+1}{2}\right)\theta}{\sin\frac{\theta}{2}} + 2\sin\left(\frac{k+1}{2}\right)\theta\cos\left(\frac{k+1}{2}\right)\theta$$

$$\left(\because \sin\theta = 2\sin\frac{\theta}{2}\cos\frac{\theta}{2}\right)$$

$$= \sin\left(\frac{k+1}{2}\right)\theta\left[\frac{\sin\frac{k\theta}{2}}{\sin\frac{\theta}{2}} + \frac{2\cos\left(\frac{k+1}{2}\right)\theta}{1}\right]$$

$$\left[\text{taking } \sin\left(\frac{k+1}{2}\right)\theta \text{ common from both the terms}\right]$$

$$= \sin\left(\frac{k+1}{2}\right)\theta\left[\frac{\sin\frac{k\theta}{2} + 2\cos\left(\frac{k+1}{2}\right)\theta\sin\frac{\theta}{2}}{\sin\frac{\theta}{2}}\right]$$

$$= \sin\left(\frac{k+1}{2}\right)\theta\left[\frac{\sin\frac{k\theta}{2} + \sin\left[\left(\frac{k+1}{2}\right)\theta + \frac{\theta}{2}\right] - \sin\left[\left(\frac{k+1}{2}\right)\theta - \frac{\theta}{2}\right]}{\sin\frac{\theta}{2}}\right]$$

$$[\because 2\cos A\sin B = \sin(A+B) - \sin(A-B)]$$

$$= \frac{\sin\left(\frac{k+1}{2}\right)\theta\left[\sin\frac{k\theta}{2}+\sin\left(\frac{k\theta}{2}+\theta\right)-\sin\frac{k\theta}{2}\right]}{\sin\frac{\theta}{2}}$$

$$= \frac{\sin\frac{(k+1)\theta}{2}\sin\left(\frac{k\theta+2\theta}{2}\right)}{\sin\frac{\theta}{2}}$$

$$= \frac{\sin\frac{(k+1)\theta}{2}\sin\left(\frac{(k+1)+1}{2}\right)\theta}{\sin\frac{\theta}{2}}$$

Therefore, $P(k+1)$ is true, when $P(k)$ is true. Hence, from the principle of mathematical induction, the statement is true for all natural numbers n.

47. *(d)* Let the statement $P(n)$ be defined as

$P(n): \frac{n^5}{5}+\frac{n^3}{3}+\frac{7n}{15}$ is a natural number for all $n \in N$.

***Step* I** For $n = 1$,

$$P(1): \frac{1}{5}+\frac{1}{3}+\frac{7}{15} = 1 \in N$$

Hence, it is true for $n = 1$.

***Step* II** Let it is true for $n = k$,

i.e., $\frac{k^5}{5}+\frac{k^3}{3}+\frac{7k}{15} = \lambda \in N$ ...(i)

***Step* III** For $n = k+1$,

$$\frac{(k+1)^5}{5}+\frac{(k+1)^3}{3}+\frac{7(k+1)}{15}$$

$$= \frac{1}{5}(k^5+5k^4+10k^3+10k^2+5k+1) + \frac{1}{3}(k^3+3k^2+3k+1)+\frac{7}{15}k+\frac{7}{15}$$

$$= \left(\frac{k^5}{5}+\frac{k^3}{3}+\frac{7}{15}k\right)+(k^4+2k^3+3k^2+2k)+\frac{1}{5}+\frac{1}{3}+\frac{7}{15}$$

$= \lambda + k^4 + 2k^3 + 3k^2 + 2k + 1$ [using Eq. (i)]

which is a natural number, since $\lambda\, k \in N$.

Therefore, $P(k+1)$ is true, when $P(k)$ is true. Hence, from the principle of mathematical induction, the statement is true for all natural numbers n.

48. *(b)* Let the statement $P(n)$ be defined as

$P(n): (x^n - y^n)$ is divisible by $(x-y)$.

***Step* I** For $n = 1$,

i.e., $P(1): (x-y)$ is divisible by $(x-y)$

which is true.

Hence, $P(n)$ is true for $n = 1$.

***Step* II** Let it is true for $n = k$,

i.e., $P(k): (x^k - y^k)$ is divisible by $(x-y)$

$\Rightarrow$ $(x^k - y^k) = \lambda(x-y)$ ...(i)

***Step* III** For $n = k+1$,

$\Rightarrow x^{k+1} - y^{k+1} = x^k\, x - y^k\, y$

$= [\lambda(x-y)+y^k]\,x - y^k\, y$ [using Eq. (i)]

$= \lambda x(x-y) + y^k\, x - y^k\, y$

$= \lambda x(x-y) + y^k(x-y)$

$= (x-y)[\lambda x + y^k]$

Therefore, $P(k+1)$ is true, when $P(k)$ is true. Hence, from the principle of mathematical induction, the statement is true for all natural numbers n.

49. *(b)* Consider, $P(n): \sin\alpha + \sin(\alpha+\beta)+\sin(\alpha+2\beta)+\ldots + \sin(\alpha+(n-1)\beta)$

$= \dfrac{\sin\left(\alpha+\frac{n-1}{2}\beta\right)\sin\left(\frac{n\beta}{2}\right)}{\sin\left(\frac{\beta}{2}\right)}$, for all natural numbers n.

We observe that $P(1)$ is true, since

$$P(1): \sin\alpha = \frac{\sin(\alpha+0)\sin\frac{\beta}{2}}{\sin\frac{\beta}{2}} = \sin\alpha$$

Assume that $P(n)$ is true for some natural number k, *i.e.,*

$P(k): \sin\alpha + \sin(\alpha+\beta)+\sin(\alpha+2\beta) + \cdots + \sin(\alpha+(k-1)\beta)$

$$= \frac{\sin\left(\alpha+\frac{k-1}{2}\beta\right)\sin\left(\frac{k\beta}{2}\right)}{\sin\left(\frac{\beta}{2}\right)} \quad \ldots\text{(i)}$$

Now, to prove that $P(k+1)$ is true, we have

$P(k+1): \sin\alpha + \sin(\alpha+\beta)+\sin(\alpha+2\beta)+\cdots + \sin(\alpha+(k-1)\beta)+\sin(\alpha+k\beta)$

$$= \frac{\sin\left(\alpha+\frac{k-1}{2}\beta\right)\sin\left(\frac{k\beta}{2}\right)}{\sin\left(\frac{\beta}{2}\right)} + \sin(\alpha+k\beta)$$

[by Eq. (i)]

$$= \frac{\sin\left(\alpha+\frac{k-1}{2}\beta\right)\sin\left(\frac{k\beta}{2}\right)+\sin(\alpha+k\beta)\sin\frac{\beta}{2}}{\sin\frac{\beta}{2}}$$

$$= \frac{\left[\begin{array}{l}\cos\left(\alpha-\frac{\beta}{2}\right)-\cos\left(\alpha+k\beta-\frac{\beta}{2}\right)\\ +\cos\left(\alpha+k\beta-\frac{\beta}{2}\right)-\cos\left(\alpha+k\beta+\frac{\beta}{2}\right)\end{array}\right]}{2\sin\frac{\beta}{2}}$$

$$= \frac{\cos\left(\alpha-\frac{\beta}{2}\right)-\cos\left(\alpha+k\beta+\frac{\beta}{2}\right)}{2\sin\frac{\beta}{2}}$$

$$= \frac{\sin\left(\alpha + \frac{k\beta}{2}\right)\sin\left(\frac{k\beta+\beta}{2}\right)}{\sin\frac{\beta}{2}}$$

$$= \frac{\sin\left(\alpha + \frac{k\beta}{2}\right)\sin(k+1)\left(\frac{\beta}{2}\right)}{\sin\frac{\beta}{2}}$$

Thus, $P(k+1)$ is true, whenever $P(k)$ is true.

Hence, by the principle of mathematical induction, $P(n)$ is true for all natural numbers n.

50. *(c)* Let the statement $P(n)$ be defined as

$P(n): 1\times 1! + 2\times 2! + 3\times 3! \ldots + n\times n! = (n+1)! - 1$ for all natural numbers n.

Note that $P(1)$ is true, since

$$P(1): 1\times 1! = 1 = 2 - 1 = 2! - 1$$

Assume that $P(n)$ is true for some natural number k, *i.e.*,

$$P(k): 1\times 1! + 2\times 2! + 3\times 3! + \ldots + k\times k! = (k+1)! - 1 \quad \ldots(i)$$

To prove $P(k+1)$ is true, we have

$$P(k+1): 1\times 1! + 2\times 2! + 3\times 3! + \cdots + k\times k! + (k+1)\times(k+1)!$$

$$= (k+1)! - 1 + (k+1)!\times(k+1) \quad \text{[by Eq.(i)]}$$

$$= (k+1+1)(k+1)! - 1$$

$$= (k+2)(k+1)! - 1$$

$$= (k+2)! - 1$$

Thus, $P(k+1)$ is true, whenever $P(k)$ is true. Therefore, by the principle of mathematical induction, $P(n)$ is true for all natural numbers n.

51. *(c)* Let $P(n): S_n = \begin{cases} \dfrac{n(n+1)^2}{2}, & \text{when } n \text{ is even} \\ \dfrac{n^2(n+1)}{2}, & \text{when } n \text{ is odd} \end{cases}$

Also, note that any term T_n of the series is given by

$$T_n = \begin{cases} n^2, & \text{if } n \text{ is odd} \\ 2n^2, & \text{if } n \text{ is even} \end{cases}$$

We observe that $P(1)$ is true, since

$$P(1): S_1 = 1^2 = 1 = \frac{1\cdot 2}{2} = \frac{1^2\cdot(1+1)}{2}$$

Assume that $P(k)$ is true for some natural number k, *i.e.*,

Case I When k is odd, then $k+1$ is even. We have,

$$P(k+1): S_{k+1} = 1^2 + 2\times 2^2 + \ldots + k^2 + 2\times(k+1)^2$$

$$= \frac{k^2(k+1)}{2} + 2\times(k+1)^2$$

$$\left[\text{as } k \text{ is odd, } 1^2 + 2\times 2^2 + \cdots + k^2 = k^2\frac{(k+1)}{2}\right]$$

$$= \frac{(k+1)}{2}[k^2 + 4(k+1)]$$

$$= \frac{k+1}{2}[k^2 + 4k + 4]$$

$$= \frac{k+1}{2}(k+2)^2$$

$$= (k+1)\frac{[(k+1)+1]^2}{2}$$

So, $P(k+1)$ is true, whenever $P(k)$ is true, in the case when k is odd.

Case II When k is even, then $k+1$ is odd.

Now, $P(k+1): S_{k+1} = 1^2 + 2\times 2^2 + \ldots + 2.k^2 + (k+1)^2$

$$= \frac{k(k+1)^2}{2} + (k+1)^2$$

$$\left[\text{as } k \text{ is even, } 1^2 + 2\times 2^2 + \ldots + 2k^2 = k\frac{(k+1)^2}{2}\right]$$

$$= \frac{(k+1)^2(k+2)}{2} = \frac{(k+1)^2((k+1)+1)}{2}$$

Therefore, $P(k+1)$ is true, whenever $P(k)$ is true for the case when k is even.

Thus, $P(k+1)$ is true whenever $P(k)$ is true for any natural number k. Hence, $P(n)$ true for all natural numbers n.

5

Complex Numbers *and* Quadratic Equations

Complex Numbers

Level I

1. Which of the following represent quadratic equation?
(a) $ax^2 + bx + c = 0, a \neq 0$
(b) $ax^3 + bx^2 + cx + d = 0, a \neq 0$
(c) $ax + b = 0, a \neq 0$
(d) None of the above

2. In real number system, square of a number is always non-negative.
(a) Correct (b) Incorrect
(c) Not always (d) None of these

3. If $x^2 + 1 = 0 \ \forall \ x \in R$, then roots of the equation is/are
(a) $x = 1$ (b) $x = -1$
(c) $x = \pm 1$ (d) No real roots

4. If $ax^2 + bx + c = 0$ is a quadratic equation, then equation has no real roots, if
(a) $D > 0$
(b) $D = 0$
(c) $D < 0$
(d) None of these

5. Who introduced the symbol i (*iota*) $= \sqrt{-1}$?
(a) Gauss (b) WR Hamilton (c) Euler (d) Cardan

6. Which of the following options define 'imaginary number'?
(a) Square root of any number
(b) Square root of positive number
(c) Square root of negative number
(d) Cube root of number

7. A number $z = a + ib$, where a and b are real numbers, is called
(a) complex number
(b) real number
(c) natural number
(d) integer

8. Which of the following are correct?
I. A number in the form of $x + iy$, where $x, y \in R$, is a complex number.
II. If $z = x + iy$ and $x = 0$, then complex number is purely imaginary.
III. If $z = x + iy$ and $y = 0$, then complex number is purely real.
(a) I and II are correct
(b) I, II and III are correct
(c) II and III are correct
(d) None of the above

9. If $z = a + ib$, then real and imaginary part of z are
(a) $\text{Re}(z) = a, \text{Im}(z) = b$
(b) $\text{Re}(z) = b, \text{Im}(z) = a$
(c) $\text{Re}(z) = a, \text{Im}(z) = ib$
(d) None of the above

10. Two complex numbers are equal, if and only if
(a) their real and imaginary parts are separately equal
(b) their real parts are only equal
(c) their imaginary parts are only equal
(d) None of the above

11. If $4x + i(3x - y) = 3 + i(-6)$, where x and y are real numbers, then the values of x and y are

(a) $x = 3, y = 4$ (b) $x = \frac{3}{4}, y = \frac{33}{4}$
(c) $x = 4, y = 3$ (d) $x = 33, y = 4$

Level II

12. If $x^{2n} + 1 = 0, x \in R$, then root of the equation is/are

(a) $x = 1$ (b) $x = -1$
(c) $x = \pm 1$ (d) No real roots

13. If $3x^2 + 4x + 2 = 0$, then equation has

(a) real roots
(b) imaginary roots
(c) one real and one imaginary root
(d) distinct real roots

14. If $x = \sqrt{-16}$, then

(a) $x = 4i$
(b) $x = 4$
(c) $x = -4$
(d) All of the above

15. Which of the following represent correct form of set of complex numbers?

(a) $C = \{x + iy: x \in R, y \in R, i = \sqrt{-1}\}$
(b) $C = \{x + iy: x \in R, y \in I\}$
(c) $C = \{x + iy: x \in I, y \in I\}$
(d) All of the above

16. $a + ib > c + id$ can be explained only when [HOTS]

(a) $b = 0, c = 0$
(b) $b = 0, d = 0$
(c) $a = 0, c = 0$
(d) $a = 0, d = 0$

17. If $(1 - i)x + (1 + i)y = 1 - 3i$, then (x, y) is equal to

(a) $(2, -1)$ (b) $(-2, 1)$
(c) $(-2, -1)$ (d) $(2, 1)$

18. Which of the following is true?

(a) $1 - i < 1 + i$
(b) $2i + 1 > -2i + 1$
(c) $2i > 1$
(d) None of the above

Algebra of Complex Numbers

Addition of Two Complex Numbers

19. If $z_1 = x_1 + iy_1$ and $z_2 = x_2 + i\,y_2$, then which of the following represent addition of z_1 and z_2?

(a) $(x_1 + y_1) + i\,(x_2 + y_2)$ (b) $(x_1 + x_2) + i\,(y_1 + y_2)$
(c) $(x_1 + x_2) + (y_1 + y_2)$ (d) $(x_1 + y_2) + i\,(x_2 + y_1)$

20. If $z_1 = 2 + 3i$ and $z_2 = 3 + 2i$, then $z_1 + z_2$ equals to

(a) $5 + 5i$ (b) $5 + 10i$ (c) $4 + 6i$ (d) $6 + 4i$

21. Which of the following are correct?

I. Addition of complex numbers satisfies closure property.
II. Addition of complex numbers is commutative.
III. Addition of complex numbers is associative.

(a) I and II are correct (b) I and III are correct
(c) All are correct (d) None of these

22. If $z + 0 = z$, where $z = x + i\,y$ and $0 = 0 + i\,0$, then 0 is called

(a) additive identity (b) additive inverse
(c) closure (d) None of these

23. If z is a complex number and $z + (-z) = 0$, then

(a) $(-z)$ is called additive inverse of z
(b) $-z$ is additive identity of z
(c) $-z$ is closure of z
(d) $-z$ is commutative of z

Difference of Two Complex Numbers

24. If z_1 and z_2 are two complex numbers, then the difference of z_1 and z_2 is

(a) $z_1 - z_2$ or $z_1 + (-z_2)$
(b) $z_1 - (-z_2)$ or $z_1 - z_2$
(c) $-z_1 - z_2$ or $z_1 + z_2$
(d) None of the above

25. If $z_1 = 2 + 3\,i$ and $z_2 == 3 - 2\,i$, then $z_1 - z_2$ is equal to

(a) $-1 + 5i$
(b) $5 - i$
(c) $i + 5$
(d) None of the above

26. Match the terms of Column I with terms of Column II and choose the correct option from the codes given below.

Column I	Column II
A. $3(7 + i\,7) + i(7 + i\,7)$	1. $2 - 7\,i$
B. $(1 - i) - (-1 + 6\,i)$	2. $14 + 28\,i$
C. $\left(\frac{1}{5} + i\frac{2}{5}\right) - \left(4 + i\frac{5}{2}\right)$	3. $\frac{17}{3} + i\,\frac{5}{3}$
D. $\left[\left(\frac{1}{3} + i\frac{7}{3}\right) + \left(4 + i\,\frac{1}{3}\right)\right] - \left(-\frac{4}{3} + i\right)$	4. $-\frac{19}{5} - \frac{21i}{10}$

Codes

	A	B	C	D		A	B	C	D
(a)	1	2	3	4	(b)	2	1	4	3
(c)	3	1	2	4	(d)	1	4	3	2

Multiplication of Two Complex Numbers

27. Statement I If $z_1 = a + ib$ and $z_2 = c + id$ be any two complex numbers, then the product $z_1 z_2 = (ac - bd) + i(ad + bc)$.

Statement II Product of two complex numbers is also a complex number.

Statement III $(3 + i5)(2 + i6) = -24 + i28$.

(a) Statement I is true
(b) Statement II is true
(c) All are true
(d) Neither I nor III is true

28. If z_1 and z_2 are two complex numbers, then the product of z_1 and z_2 is also a complex number, which property it satisfies

(a) closure law of multiplication
(b) commutative law of multiplication
(c) associative law of multiplication
(d) distributive law of multiplication

29. If z_1 and z_2 are two complex numbers and $z_1 \cdot z_2 = z_2 \cdot z_1$, then it satisfies

(a) closure law
(b) commutative law
(c) associative law
(d) distributive law

30. Statement I If $z_1 \cdot z_2 = z_1$, then z_2 is called multiplicative identity.

Statement II If $z \cdot \frac{1}{z} = 1 + 0i$ (the multiplicative identity), then $\frac{1}{z}$ or z^{-1} is called the multiplicative inverse.

(a) Statement I is correct
(b) Statement II is correct
(c) Both are correct
(d) Neither I nor II is correct

31. Match the terms of Column I with the terms of Column II and choose the correct option from the codes given below. (Here, symbol used have their usual meaning.)

Column I	Column II
A. If $z_1 \cdot z_2 = z_2 \cdot z_1$	1. $1 + 0i$
B. If $(z_1\, z_2) \cdot z_3 = = z_1\,(z_2 \cdot z_3)$	2. Associative law
C. If $z \cdot z' = 1 + 0i$ $z \neq 0$, then z' is	3. Commutative law
D. $z \cdot z_1 = z$, then z_1 is	4. Multiplicative inverse

Codes

	A	B	C	D		A	B	C	D
(a)	3	2	4	1	(b)	2	4	1	3
(c)	3	4	2	1	(d)	1	2	3	4

32. Assertion (A) If $(2 + 3i)[(3 + 2i) + (2 + i)] = 1 + 2i$

Reason (R) $z_1\,(z_2 + z_3) = z_1\,z_2 + z_1\,z_3$.

(a) Both A and R are correct; R is the correct explanation of A
(b) Both A and R are correct; R is not the correct explanation of A
(c) A is correct; R is incorrect
(d) R is correct; A is incorrect

33. If $z = 5i\left(-\frac{3}{5}i\right)$, then z is equal to

(a) $0 + 3\,i$
(b) $3 + 0\,i$
(c) $3 - 0\,i$
(d) $-3 + 0\,i$

34. If z is non-zero complex number and $z = a + ib$, then multiplicative inverse of z is

(a) $\frac{a}{a^2 + b^2} + \frac{-bi}{a^2 + b^2}$
(b) $\frac{a}{a^2 - b^2} + \frac{-bi}{a^2 - b^2}$
(c) $\frac{a}{a^2 - b^2} + \frac{ib}{a^2 - b^2}$
(d) $\frac{-a}{a^2 + b^2} + \frac{-bi}{a^2 + b^2}$

35. Match the complex numbers in Column I with their multiplicative inverses in Column II and choose the correct option from the codes given below.

Column I	Column II
A. $4 - 3i$	1. $\frac{\sqrt{5}}{14} - i\,\frac{3}{14}$
B. $\sqrt{5} + 3i$	2. $\frac{4}{25} + i\,\frac{3}{25}$
C. $-i$	3. $0 + i$

Codes

	A	B	C
(a)	1	2	3
(b)	2	1	3
(c)	1	3	2
(d)	2	3	1

Division of Two Complex Numbers

36. If z_1 and z_2 are two complex numbers, then $\frac{z_1}{z_2}$ where $z_2 \neq 0$, is equal to

(a) $z_1 \cdot z_2^{-1}$ (b) $z_1 \cdot z_2$
(c) $\frac{1}{z_1} \cdot z_2$ (d) $\frac{1}{z_1} \cdot \frac{1}{z_2}$

37. If $z_1 = 6 + 3i$ and $z_2 = 2 - i$, then $\frac{z_1}{z_2}$ is equal to

(a) $\frac{1}{5}(9+12i)$ (b) $9 + 12i$
(c) $3 + 2i$ (d) $\frac{1}{5}(12+9i)$

Power of i

38. Find the incorrect option.

(a) $i^3 = -i$ (b) $i^4 = 1$
(c) $i^5 = i$ (d) $i^6 = 1$

39. Match the terms of Column I with the terms of Column II and choose the correct option from the codes given below.

	Column I		Column II
A.	i^{-1}	1.	-1
B.	i^{-2}	2.	$-i$
C.	i^{-3}	3.	i
D.	i^{-4}	4.	1

Codes

	A	B	C	D
(a)	1	2	3	4
(b)	2	1	3	4
(c)	2	3	4	1
(d)	1	4	3	2

40. Statement I If $i = \sqrt{-1}$, then $i^{4k} = 1$, $i^{4k+1} = i$, $i^{4k+2} = -1$ and $i^{4k+3} = -i$.

Statement II $i^{4k} + i^{4k+1} + i^{4k+2} + i^{4k+3} = 0$.

(a) Statement I is correct
(b) Statement II is correct
(c) Both are correct
(d) Neither I nor II is correct

41. If $z = i^9 + i^{19}$, then z is equal to

(a) $0 + 0i$ (b) $1 + 0i$
(c) $0 + i$ (d) $1 + 2i$

42. If $z = i^{-39}$, then simplest form of z is equal to

(a) $1 + 0i$ (b) $0 + i$
(c) $0 + 0i$ (d) $1 + i$

43. Simplify the complex numbers given in Column I and match with Column II and choose the correct option from the codes given below.

Column I	Column II
A. $(1-i)^4$	1. $-\left(\frac{22}{3} + i\frac{107}{27}\right)$
B. $\left(\frac{1}{3} + 3i\right)^3$	2. $-4 + 0i$
C. $\left(-2 - \frac{1}{3}i\right)^3$	3. $-\frac{242}{27} - 26i$

Codes

	A	B	C		A	B	C
(a)	1	2	3	(b)	2	1	3
(c)	3	1	2	(d)	2	3	1

44. Match the terms of Column I with the terms of Column II and choose the correct option from the codes given below.

Column I	Column II
A. $(-5i)\left(\frac{1}{8}i\right)$	1. $-10 - 198i$
B. $(-i)(2i)\left(-\frac{1}{8}i\right)^3$	2. $(-6 + \sqrt{2}) + \sqrt{3}(1 + 2\sqrt{2})i$
C. $(5 - 3i)^3$	3. $0 + \frac{1}{256}i$
D. $(-\sqrt{3} + \sqrt{-2})(2\sqrt{3} - i)$	4. $\frac{5}{8} + 0i$

Codes

	A	B	C	D
(a)	1	2	3	4
(b)	4	3	1	2
(c)	3	4	2	1
(d)	1	4	3	2

The Square Roots of a Negative Real Number

45. If $i^2 = -1$ and $(-i)^2 = i^2 = -1$, then $\sqrt{-1}$ is equal to

(a) i (b) $-i$
(c) i and $-i$ (d) None of these

46. If $x^2 + 1 = 0$, then solution is

(a) $i, -i$ (b) $1, -i$
(c) $-1, i$ (d) None of these

47. Statement I $\sqrt{a} \times \sqrt{b} = \sqrt{ab}\ \forall\ a, b \in (0, \infty)$.

Statement II $\sqrt{a} \times \sqrt{b} = \sqrt{ab}$ when $a > 0$, $b < 0$ or $a < 0$, $b > 0$.

(a) Statement I is true (b) Statement II is true
(c) Both are true (d) Neither I nor II is true

48. If $z = -7 - 24i$, then square root of z is

(a) $3-4i$ and $-3+4i$ (b) $3+4i$ and $-3-4i$
(c) $-3+4i$ and $-3-4i$ (d) None of the above

49. Match the complex numbers given in Column I with their square roots in Column II and choose the correct option from the codes given below.

Column I	Column II
A. $-15-8i$	1. $-1+3i$, $1-3i$
B. $-8-6i$	2. $-1+4i$, $1-4i$
C. $1-i$	3. $\left(\pm\frac{1}{\sqrt{2}}\mp\frac{1}{\sqrt{2}}i\right)$
D. $-i$	4. $\left(\pm\sqrt{\frac{\sqrt{2}+1}{2}}\mp\sqrt{\frac{\sqrt{2}-1}{2}}i\right)$

Codes

	A	B	C	D		A	B	C	D
(a)	2	3	4	1	(b)	2	1	4	3
(c)	1	2	3	4	(d)	2	1	3	4

50. **Statement I** Square root of i is $\left(\pm\frac{1}{\sqrt{2}}\pm\frac{1}{\sqrt{2}}i\right)$.

Statement II Square root of $(1+i)$ is $\left(\pm\sqrt{\frac{\sqrt{2}+1}{3}}\pm\sqrt{\frac{\sqrt{2}-1}{3}}i\right)$.

(a) Statement I is correct (b) Statement II is correct
(c) Both are correct (d) Neither I nor II is correct

Identities

51. If Z_1 and Z_2 are two complex numbers, then $(Z_1 + Z_2)^2$ is equal to

(a) $Z_1^2 + Z_2^2$
(b) $Z_1^2 + Z_2^2 + 2Z_1Z_2$
(c) $Z_1^2 + Z_2^2 + Z_1Z_2$
(d) $Z_1^2 + Z_2^2 - 2Z_1Z_2$

52. Which of the following are correct?

I. $(Z_1 - Z_2)^2 = Z_1^2 + Z_2^2 - 2Z_1Z_2$
II. $(Z_1 + Z_2)^3 = Z_1^3 + 3Z_1^2Z_2 + 3Z_1Z_2^2 + Z_2^3$
III. $(Z_1 - Z_2)^3 = Z_1^3 - 3Z_1^2Z_2 + 3Z_1Z_2^2 - Z_2^3$
IV. $Z_1^2 - Z_2^2 = (Z_1 + Z_2)(Z_1 - Z_2)$

(a) I and II are correct
(b) III and IV are correct
(c) I, II and III are correct
(d) All are correct

53. Expression $\frac{(3+i\sqrt{5})(3-i\sqrt{5})}{(\sqrt{3}+\sqrt{2}\,i)-(\sqrt{3}-i\sqrt{2})}$ in $a+ib$ form is

(a) $0+i$
(b) $0-\sqrt{2}i$
(c) $0+\sqrt{2}i$
(d) $0-\frac{7\sqrt{2}}{2}i$

The Modulus and Conjugate of a Complex Number

Level I

54. If $Z = a + ib$, then modulus of Z, denoted by ...A... is defined to be non-negative real number ...B... . Here, A and B refer to

(a) $|Z|, \sqrt{a^2+b^2}$ (b) Z, a^2+b^2
(c) $Z, \sqrt{a^2+b^2}$ (d) $Z, \sqrt{a^2-b^2}$

55. The conjugate of a complex number $Z = a + ib$, is

(a) $\bar{Z} = a + ib$ (b) $\bar{Z} = a - ib$ (c) $\bar{Z} = ia + b$ (d) $\bar{Z} = ia - b$

56. Which of the following are correct?

I. $|3+i| = \sqrt{10}$; $|2-5i| = \sqrt{29}$
II. $\overline{(3+i)} = 3 - i; \overline{(2-5i)} = 2+5i$ and $\overline{(-3i-5)} = 3i - 5$
III. $Z^{-1} = \frac{\bar{Z}}{|Z|^2}$ or $Z\bar{Z} = |Z|^2, Z \neq 0$

(a) I and III are correct
(b) I and II are correct
(c) All are correct
(d) None of the above

57. Match the terms of Column I with terms of Column II and choose the correct option from the codes given below.

Column I	Column II
A. $\lvert Z_1 Z_2\rvert$	1. $\lvert Z_1\rvert\lvert Z_2\rvert$
B. $\left\lvert\frac{Z_1}{Z_2}\right\rvert$	2. $\bar{Z_1}\,\bar{Z_2}$
C. $\overline{Z_1 Z_2}$	3. $\frac{\lvert Z_1\rvert}{\lvert Z_2\rvert}$
D. $\overline{Z_1 \pm Z_2}$	4. $\bar{Z_1} \pm \bar{Z_2}$
E. $\overline{\left(\frac{Z_1}{Z_2}\right)}$	5. $\frac{\bar{Z_1}}{\bar{Z_2}}$

Codes

	A	B	C	D	E		A	B	C	D	E
(a)	1	2	3	4	5	(b)	1	3	2	4	5
(c)	2	3	1	4	5	(d)	1	2	3	5	4

58. Statement I Multiplicative inverse of $2-3i$ is $2+3i$.

Statement II Simplest form of $\frac{5+\sqrt{2}i}{1-\sqrt{2}i}$ is $1+2\sqrt{2}i$.

Statement III Simplest form of i^{-35} is $-i$.

(a) Statement I is correct (b) Statement II is correct
(c) Statement III is correct (d) All are correct

Level II

59. The real values of x and y for which the equation $(x+iy)+(2-3i)=4+i$ satisfied, are

(a) $x=-2, y=-4$ (b) $x=2, y=4$
(c) $x=2, y=-4$ (d) $x=-2, y=4$

60. Additive inverse of $(1-i)$ is equal to

(a) $0+0i$ (b) $-1+i$
(c) $-1+i$ (d) None of these

61. If $Z_1=1+2i$ and $Z_2=2+3i$, then sum of Z_1 and additive inverse of Z_2 is equal to

(a) $1+2i$ (b) $3+i$
(c) $3+5i$ (d) $-1-i$

62. If Z_1 and Z_2 are two complex numbers such that $Z_1+Z_2=0$, then which of the following option is not possible?

(a) $Z_1=Z_2=0+i0$
(b) Z_1 is additive inverse of Z_2
(c) Z_2 is additive inverse of Z_1
(d) Z_1 is multiplicative inverse of Z_2

63. If Z_1 and Z_2 are two complex numbers, then Z_1-Z_2 represents

(a) a complex number which is sum of Z_1 and additive inverse of Z_2
(b) a complex number which is difference of Z_1 and additive inverse of Z_2
(c) distance between Z_1 and Z_2
(d) None of the above

64. If $Z \neq 0$ is a complex number, then [HOTS]

(a) $\text{Re}(Z)=0 \Rightarrow \text{Im}(Z^2)=0$
(b) $\text{Re}(Z^2)=0 \Rightarrow \text{Im}(Z^2)=0$
(c) $\text{Re}(Z)=0 \Rightarrow \text{Re}(Z^2)=0$
(d) None of the above

65. If $(1-i)^n=2^n$, then n is equal to [RPET 1990]

(a) 1 (b) 0
(c) -1 (d) None of these

66. The value of $(1+i)^5 \times (1-i)^5$ is [KCET 1992]

(a) -8 (b) $8i$ (c) 8 (d) 32

67. The value of $\frac{i^{592}+i^{590}+i^{588}+i^{586}+i^{584}}{i^{582}+i^{580}+i^{578}+i^{576}+i^{574}}-1$ is equal to

(a) -1 (b) -2 (c) -3 (d) -4

68. $i^2+i^4+i^6+\ldots$ upto $(2k+1)$ terms, $k \in N$ is [Kerala CEE 2005]

(a) 0 (b) 1 (c) -1 (d) k

69. If $i^2=-1$, then the value of $\sum_{n=1}^{200} i^n$ is

(a) 50 (b) -50 (c) 0 (d) 100

70. The value of the sum $\sum_{n=1}^{13}(i^n+i^{n+1})$, where $i=\sqrt{-1}$, equals to

(a) i (b) $i-1$ (c) $-i$ (d) 0

71. The value of $i^{1+3+5\ldots+(2n+1)}$ is

(a) i, if n is even; $-i$, if n is odd
(b) 1, if n is even; -1, if n is odd
(c) 1, if n is odd; -1, if n is even
(d) i, if n is even; -1, if n is odd

72. The value of $(1+i)^8+(1-i)^8$ is

(a) 16 (b) -16
(c) 32 (d) -32

73. $\frac{3+2i\sin\theta}{1-2i\sin\theta}$ will be real, if θ equals to [IIT 1976; EAMCET 2002]

(a) $2n\pi$ (b) $n\pi+\frac{\pi}{2}$
(c) $n\pi$ (d) None of these

(where, n is an integer)

74. The value of $\frac{\sqrt{5+12i}+\sqrt{5-12i}}{\sqrt{5+12i}-\sqrt{5-12i}}$ is

(a) $-\frac{3}{2}i$ (b) $\frac{3}{2}i$ (c) $-\frac{3}{2}$ (d) $\frac{3}{2}$

75. $\frac{3+2i\sin\theta}{1-2i\sin\theta}$ will be purely imaginary, if θ equals to

(a) $2n\pi \pm \frac{\pi}{3}$ (b) $n\pi+\frac{\pi}{3}$
(c) $n\pi \pm \frac{\pi}{3}$ (d) None of these

(where, n is an integer)

Argand Plane, Polar Representation and Quadratic Equations

Level I

76. Statement I Every complex number $z = x + iy$ corresponds to the ordered pair (x, y) and can be represented as unique point (x, y) in the xy-plane.

Statement II The plane having a complex number assigned to each of its point is called the complex plane or Argand plane.

(a) Statement I is true (b) Statement II is true
(c) Both are true (d) Neither I nor II is true

77. Which of the following are correct?

I. The modulus of complex number $x + iy = \sqrt{x^2 + y^2}$.

II. If $z = x + iy$, then modulus Z is distance of the point from the origin.

III. Complex numbers which lie on X-axis are in the form of $a + i0$.

IV. Complex numbers which lie on Y-axis are in the form of $0 + bi$.

(a) I and II are correct (b) II and III are correct
(c) I, II and III are correct (d) All are correct

78. Statement I Conjugate of complex number $Z = x + iy$ is $\bar{Z} = x - iy$.

Statement II Geometrically the point $(x, -y)$ is the mirror image of the point (x, y) on the real axis.

(a) Statement I is correct
(b) Statement II is correct
(c) Both are correct
(d) Neither I nor II is correct

Directions (Q. Nos. 79-81) *Let the point P represent the non-zero complex number $Z = x + iy$ and directed line segment $OP = r$ and it is inclined with positive direction of X- axis at an angle θ, then*

79. Polar coordinates of P in ordered pair can be written as

(a) (x, iy) (b) (θ, r)
(c) (r, θ) (d) None of these

80. Complex number $Z = x + iy$ in terms of r and θ can be written as

(a) $Z = r(\cos\theta + i\sin\theta)$ (b) $Z = r(\sin\theta + i\cos\theta)$
(c) $Z = r(\sin\theta - i\cos\theta)$ (d) None of these

81. Representation of $Z = x + iy$ in terms of r and θ is called

(a) polar form of the complex number
(b) cartesian form of the complex number
(c) complex form of the complex number
(d) real form of the complex number

82. If $x = r\cos\theta$, $y = r\sin\theta$ such that $x, y > 0$ and $Z = r(\cos\theta + i\sin\theta)$, then

I. $|Z| = r = \sqrt{x^2 - y^2}$

II. $|Z| = r = \sqrt{x^2 + y^2}$

III. $\arg(Z) = \theta$

(a) I and III are correct (b) II and III are correct
(c) All are correct (d) None of these

83. Statement I For any complex number $Z \neq 0$, there corresponds only one value of θ in $0 \leq \theta < 2\pi$.

Statement II If value of θ lies between $-\pi < \theta \leq \pi$, then θ is called principle argument of Z.

(a) Statement I is correct
(b) Statement II is correct
(c) Both are correct
(d) Neither I nor II is correct

84. Polar form of the complex number $Z = 1 + i\sqrt{3}$, is

(a) $2\left(\cos\frac{\pi}{3} - i\sin\frac{\pi}{3}\right)$ (b) $2\left(\cos\frac{\pi}{3} + i\sin\frac{\pi}{3}\right)$
(c) $2\left(\cos\frac{2\pi}{3} + i\sin\frac{2\pi}{3}\right)$ (d) $2\left(\cos\frac{4\pi}{3} + i\sin\frac{4\pi}{3}\right)$

85. If complex number $Z = \frac{-16}{1 + i\sqrt{3}}$, then

I. $|Z| = 8$ II. $\arg(Z) = \frac{\pi}{3}$

III. $\arg(Z) = \frac{2\pi}{3}$ IV. $Z = 8\left(\cos\frac{\pi}{3} + i\sin\frac{\pi}{3}\right)$

(a) I and II are correct (b) I and III are correct
(c) I, II and III are correct (d) All are correct

86. Identify the correct statements given below and choose the correct code.

I. If $Z = -1 - i\sqrt{3}$, then $|Z| = 2$.

II. $\arg(Z) = -\frac{2\pi}{3}$, where $Z = -1 - i\sqrt{3}$.

III. If $Z = -\sqrt{3} + i$, then $|Z| = 2$.

IV. $\arg(Z) = \frac{5\pi}{6}$, where $Z = -\sqrt{3} + i$.

(a) I and II are correct
(b) II and III are correct
(c) I, II and III are correct
(d) All are correct

87. Match the complex numbers in cartesian form in Column I with its polar form in Column II and choose the correct option from the codes given below.

Column I	Column II
A. $1-i$	1. $\sqrt{2}\left[\cos\left(\frac{3\pi}{4}\right) + i\sin\left(\frac{3\pi}{4}\right)\right]$
B. $-1+i$	2. $\sqrt{2}\left[\cos\left(\frac{-3\pi}{4}\right) + i\sin\left(\frac{-3\pi}{4}\right)\right]$
C. $-1-i$	3. $3[\cos\pi + i\sin\pi]$
D. -3	4. $\sqrt{2}\left[\cos\left(\frac{-\pi}{4}\right) + i\sin\left(\frac{-\pi}{4}\right)\right]$

Codes

	A	B	C	D
(a)	1	2	3	4
(b)	4	1	3	2
(c)	4	1	2	3
(d)	4	3	2	1

88. Statement I If $Z = \sqrt{3} + i$, then polar form of Z is $2\left[\cos\frac{\pi}{6} + i\sin\frac{\pi}{6}\right]$.

Statement II If $Z = i$, then polar form of i is $\left(\cos\frac{\pi}{2} + i\sin\frac{\pi}{2}\right)$.

(a) Statement I is true
(b) Statement II is true
(c) Both are true
(d) Neither I nor II is true

89. If $ax^2 + bx + c = 0$ with real coefficients a, b, c and $a \neq 0$, then

I. Discriminant, $D = b^2 - 4ac$.

II. For $D \geq 0$, equation has real roots.

III. For $D < 0$, equation has imaginary roots.

IV. For $D < 0$, $x = \frac{-b \pm i\sqrt{4ac - b^2}}{2a}$.

(a) I and II are correct
(b) II, III and IV are correct
(c) I, II and IV are correct
(d) All are correct

90. A polynomial equation has

(a) atleast two roots
(b) atleast one root
(c) atmost three roots
(d) atmost two roots

91. If $P(x)$ is a n degree polynomial, then number of roots are

(a) 1
(b) 2
(c) 3
(d) n

92. Roots of $x^2 + 2 = 0$ are

(a) $\pm\sqrt{2}i$
(b) 2
(c) $2i$
(d) None of the above

93. If $x^2 + x + 1 = 0$, then which of the following are correct?

I. $x = \frac{-1 + \sqrt{3}i}{2}$

II. $x = \frac{-1 - \sqrt{3}i}{2}$

III. $x = \frac{1 + \sqrt{3}i}{2}$

IV. $x = \frac{1 - \sqrt{3}i}{2}$

(a) I and II are correct
(b) III and IV are correct
(c) I and III are correct
(d) All are correct

94. If $\sqrt{5}x^2 + x + \sqrt{5} = 0$, then which of the following is/are correct?

I. Discriminant, $D = -19$.

II. $x = \frac{-1 \pm \sqrt{19}i}{2\sqrt{5}}$.

(a) Only I is correct
(b) Only II is correct
(c) Both are correct
(d) Both are incorrect

95. Match the quadratic equations in Column I with their roots in Column II and choose the correct option from the codes given below.

Column I	Column II
A. $x^2 + 3 = 0$	1. $\frac{-3 \pm i\,3\sqrt{3}}{2}$
B. $2x^2 + x + 1 = 0$	2. $\frac{-1 \pm i\sqrt{7}}{-2}$
C. $x^2 + 3x + 9 = 0$	3. $\frac{-1 \pm i\sqrt{7}}{4}$
D. $-x^2 + x - 2 = 0$	4. $\pm i\sqrt{3}$

Codes

	A	B	C	D
(a)	1	2	3	4
(b)	4	3	1	2
(c)	2	1	4	3
(d)	2	3	1	4

96. Statement I Roots of quadratic equation $x^2 + 3x + 5 = 0$ is $x = \frac{-3 \pm i\sqrt{11}}{2}$.

Statement II If $x^2 - x + 2 = 0$ is a quadratic equation, then its roots are $\frac{1 \pm i\sqrt{7}}{2}$.

(a) Statement I is correct (b) Statement II is correct
(c) Both are correct (d) Both are incorrect

97. Match the quadratic equations given in Column I with their roots in Column II and choose the correct option from the codes given below.

Column I	Column II
A. $\sqrt{2}\,x^2 + x + \sqrt{2} = 0$	1. $x = \frac{\sqrt{2} \pm i\sqrt{34}}{2\sqrt{3}}$
B. $\sqrt{3}x^2 - \sqrt{2}x + 3\sqrt{3} = 0$	2. $x = \frac{-1 \pm i\sqrt{7}}{2\sqrt{2}}$
C. $x^2 + x + \frac{1}{\sqrt{2}} = 0$	3. $\frac{-1 \pm i\sqrt{7}}{2\sqrt{2}}$
D. $x^2 + \frac{x}{\sqrt{2}} + 1 = 0$	4. $\frac{-1 \pm i\sqrt{2\sqrt{2} - 1}}{2}$

Codes

	A	B	C	D
(a)	2	1	4	3
(b)	1	2	3	4
(c)	2	1	3	4
(d)	4	3	2	1

Level II

98. Statement I Let z_1 and z_2 be two complex numbers such that $\bar{z}_1 + i\bar{z}_2 = 0$ and $\arg(z_1 \cdot z_2) = \pi$, then $\arg(z_1)$ is $\frac{3\pi}{4}$.

Statement II $\arg(z_1 \cdot z_2) = \arg z_1 + \arg z_2$.

[NCERT Exemplar]

(a) Statement I is correct
(b) Statement II is correct
(c) Both are correct
(d) Neither I nor II is correct

99. Statement I Let z_1 and z_2 be two complex numbers such that $|z_1 + z_2| = |z_1| + |z_2|$, then $\arg(z_1) - \arg(z_2) = 0$.

Statement II $\arg\left(\frac{z_1}{z_2}\right) = \arg(z_1) - \arg(z_2)$.

[NCERT Exemplar]

(a) Statement I is correct
(b) Statement II is correct
(c) Both are correct
(d) Neither I nor II is correct

100. If $z_1 = \sqrt{3} + i\sqrt{3}$ and $z_2 = \sqrt{3} + i$, then $\left(\frac{z_1}{z_2}\right)$ lies in quadrant ...A... . Here, A refers to

(a) I (b) II
(c) III (d) IV

101. If the complex number $z = x + iy$ satisfies the condition $|z + 1| = 1$, then z lies on

(a) X-axis
(b) circle with centre (1, 0) and radius 1
(c) circle with centre (−1, 0) and radius 1
(d) Y-axis

102. The area of the triangle on the complex plane formed by the complex numbers z, iz and $z + iz$, is

[NCERT Exemplar]

(a) $|z|$ (b) $|\bar{z}|^2$
(c) $\frac{1}{2}|z|^2$ (d) $\frac{1}{2}|z + iz|^2$

103. The equation $|z + 1 - i| = |z - 1 + i|$ represents a

[NCERT Exemplar]

(a) straight line (b) circle
(c) parabola (d) hyperbola

104. Number of solutions of the equation $z^2 + |z|^2 = 0$ and $z \neq 0$ is

(a) 1 (b) 2
(c) 3 (d) infinite many

105. Statement I The amplitude of $\sin\frac{\pi}{5} + i\left(1 - \cos\frac{\pi}{5}\right)$ is $\frac{\pi}{10}$.

Statement II If complex number lies on the first quadrant, then $\arg(z) = \theta$.

(a) Statement I is correct (b) Statement II is correct
(c) Both are correct (d) Neither I nor II is correct

106. Match the terms of Column I with terms of Column II and choose the correct option from the codes given below.

Column I	Column II
A. Amplitude of $1 - i$ is	1. $\cos\frac{\pi}{2} - i\sin\frac{\pi}{2}$
B. The polar form of the complex number $(i^{35})^3$ is	2. $\frac{-\pi}{4}$
C. Locus of z, if amplitude of $z - 2 - 3i$ is $\frac{\pi}{4}$, is	3. $x - y + 1 = 0$ 4. $x^2 + y^2 - 1 = 0$

Codes

	A	B	C		A	B	C
(a)	2	1	3	(b)	1	2	3
(c)	3	2	1	(d)	2	4	1

107. If difference in roots of the equation $x^2 - px + 8 = 0$ is 2, then p is equal to [NCERT Exemplar]

(a) ± 6 (b) ± 2 (c) ± 1 (d) ± 5

108. For what value of a the sum of squares of the roots of the equation $x^2 - (a-2)x - (a+1) = 0$ is least? [NCERT Exemplar]

(a) 1 (b) 2 (c) -1 (d) -2

109. The modulus of the complex number z such that $|z + 3 - i| = 1$ and $\arg(z) = \pi$ is equal to

(a) 3 (b) 2 (c) 9 (d) 4

110. The number of complex numbers z such that $|z-1| = |z+1| = |z-i|$, is

(a) 0 (b) 1
(c) 2 (d) ∞

111. If $z_1 = \sqrt{2}\left(\cos\frac{\pi}{4} + i\sin\frac{\pi}{4}\right)$ and $z_2 = \sqrt{3}\left(\cos\frac{\pi}{3} + i\sin\frac{\pi}{3}\right)$, then $|z_1 z_2|$ is equal to

(a) 6 (b) $\sqrt{2}$
(c) $\sqrt{6}$ (d) $\sqrt{3}$

Miscellaneous Exercise

112. Conjugate of $\frac{(3-2i)(2+3i)}{(1+2i)(2-i)}$ is

(a) $\frac{16}{25} + \frac{63}{25}i$ (b) $\frac{63}{25} + \frac{16}{25}i$
(c) $\frac{25}{16} + \frac{25}{63}i$ (d) $\frac{4}{25} + \frac{9}{25}i$

113. Which of the following are correct?

I. Modulus of $\frac{1+i}{1-i}$ is 1.

II. Argument of $\frac{1+i}{1-i}$ is $\frac{\pi}{2}$.

III. Modulus of $\frac{1}{1+i}$ is $\sqrt{2}$.

IV. Argument of $\frac{1}{1+i}$ is $\frac{\pi}{4}$.

(a) I and II are correct
(b) III and IV are correct
(c) I, II and III are correct
(d) All are correct

114. If $x + iy = \frac{a+ib}{a-ib}$, then

(a) $x^2 + y^2 = 1$ (b) $x^2 + y^2 = 2$
(c) $x^2 + y^2 = 0$ (d) $x^2 + y^2 = 4$

115. $Z = \frac{3 + 2i\sin\theta}{1 - 2i\sin\theta}$ is purely real for

(a) all values of θ
(b) when $\theta = (2n+1)\frac{\pi}{2}, n \in Z$
(c) when $\theta = n\pi, n \in Z$
(d) when $\theta = (2n+1)\pi, n \in Z$

116. If $Z = \frac{i-1}{\cos\frac{\pi}{3} + i\sin\frac{\pi}{3}}$, then polar form of Z is

(a) $\sqrt{2}\left(\cos\frac{5\pi}{12} - i\sin\frac{5\pi}{12}\right)$
(b) $\sqrt{2}\left(\cos\frac{5\pi}{12} + i\sin\frac{5\pi}{12}\right)$
(c) $\sqrt{2}\left(\cos\frac{\pi}{4} + i\sin\frac{\pi}{4}\right)$
(d) $\sqrt{2}\left(\cos\frac{\pi}{4} - i\sin\frac{\pi}{4}\right)$

117. Simplest form of $\left[i^{18} + \left(\frac{1}{i}\right)^{25}\right]^3$ is

(a) $2 + 3i$ (b) $2 - 2i$
(c) $2 + 2i$ (d) $2 - 3i$

118. **Statement I** If z_1 and z_2 are two complex numbers, then

$\text{Re}(z_1 z_2) = \text{Re}(z_1)\,\text{Re}(z_2) - \text{Im}(z_1)\,\text{Im}(z_2)$.

Statement II Angle between z and iz is $\frac{\pi}{2}$.

(a) Statement I is true; Statement II is true; Statement II is a correct explanation for Statement I
(b) Statement I is true; Statement II is true; Statement II is not a correct explanation for Statement I
(c) Statement I is true; Statement II is false
(d) Statement I is false; Statement II is true

119. Standard form of $\left(\frac{1}{1-4i} - \frac{2}{1+i}\right)\left(\frac{3-4i}{5+i}\right)$ is

(a) $\frac{307}{442} + i\frac{599}{442}$ (b) $\frac{307}{442} - i\frac{599}{442}$
(c) $\frac{442}{307} + i\frac{599}{307}$ (d) None of these

120. If $x - iy = \sqrt{\dfrac{a - ib}{c - id}}$, then $(x^2 + y^2)^2$ is equal to

(a) $\dfrac{a^2 + b^2}{c^2 + d^2}$ (b) $\dfrac{c^2 + d^2}{a^2 + b^2}$ (c) $\dfrac{a^2 - b^2}{c^2 - d^2}$ (d) $\dfrac{c^2 - d^2}{a^2 - b^2}$

121. Statement I Polar form of $\dfrac{1 + 7i}{(2 - i)^2}$ is $\sqrt{2}\left[\cos\dfrac{3\pi}{4} + i\sin\dfrac{3\pi}{4}\right]$.

Statement II Polar form of $\dfrac{1 + 3i}{1 - 2i}$ is $\sqrt{2}\left[\cos\dfrac{3\pi}{4} + i\sin\dfrac{3\pi}{4}\right]$.

(a) Statement I is correct (b) Statement II is correct
(c) Both are correct (d) Neither I nor II is correct

122. Match the terms of Column I with terms of Column II and choose the correct option from the codes given below.

Column I (Quadratic Equation)		Column II (Roots)
A. $3x^2 - 4x + \dfrac{20}{3} = 0$	1.	$\dfrac{5}{27} \pm \dfrac{\sqrt{2}i}{27}$
B. $x^2 - 2x + \dfrac{3}{2} = 0$	2.	$\dfrac{2}{3} \pm \dfrac{4}{3}i$
C. $27x^2 - 10x + 1 = 0$	3.	$1 \pm \dfrac{i}{\sqrt{2}}$
D. $21x^2 - 28x + 10 = 0$	4.	$\dfrac{2}{3} \pm \dfrac{\sqrt{14}}{21}i$

Codes

	A	B	C	D		A	B	C	D
(a)	2	3	1	4	(b)	1	2	3	4
(c)	2	3	4	1	(d)	1	4	3	2

123. If $z_1 = 2 - i$ and $z_2 = 1 + i$, then value of $\left|\dfrac{z_1 + z_2 + 1}{z_1 - z_2 + 1}\right|$ is

(a) 2 (b) $2i$ (c) $\sqrt{2}$ (d) $\sqrt{2}i$

124. If $a + ib = \dfrac{(x + i)^2}{2x^2 + 1}$, then $a^2 + b^2$ is equal to

(a) $\dfrac{(x^2 + 1)^2}{(2x^2 + 1)^2}$ (b) $\dfrac{(x^2 - 1)^2}{(2x^2 + 1)^2}$ (c) $\dfrac{(x^2 + 1)^2}{(2x^2 - 1)}$ (d) $\dfrac{(x^2 - 1)^2}{(2x^2 - 1)^2}$

125. If $z_1 = 2 - i$ and $z_2 = -2 + i$, then

Statement I $\text{Re}\left(\dfrac{z_1 z_2}{\overline{z_1}}\right) = \dfrac{-2}{5}$.

Statement II $\text{Im}\left(\dfrac{1}{z_1 \overline{z_1}}\right) = 0$.

(a) Statement I is correct (b) Statement II is correct
(c) Both are correct (d) Both are incorrect

126. If $z = \dfrac{1 + 2i}{1 - 3i}$, then

I. Modulus of z is $\dfrac{1}{\sqrt{2}}$.

II. Argument of z is $\dfrac{3\pi}{4}$.

(a) Both I and II are correct (b) Only I is correct
(c) Only II is correct (d) Neither I nor II is correct

127. $(x - iy)(3 + 5i)$ is the conjugate of $(-6 - 24i)$, then x and y are

(a) $x = 3, y = -3$ (b) $x = -3, y = 3$
(c) $x = -3, y = -3$ (d) $x = 3, y = 3$

128. If $z = \dfrac{1 + i}{1 - i} - \dfrac{1 - i}{1 + i}$, then $|z|$ is equal to

(a) 2 (b) 3 (c) 4 (d) 1

129. If $(x + iy)^3 = u + iv$, then $\dfrac{u}{x} + \dfrac{v}{y}$ is equal to

(a) $4x^2$ (b) $4y^2$ (c) $4(x^2 + y^2)$ (d) $4(x^2 - y^2)$

130. α and β are different complex numbers with $|\beta| = 1$, then $\left|\dfrac{\beta - \alpha}{1 - \overline{\alpha}\beta}\right|$ is equal to

(a) 1 (b) 2 (c) 4 (d) 3

131. Number of non-zero integral solutions of the equation $|1 - i|^x = 2^x$, is

(a) 0 (b) 1 (c) 2 (d) 4

132. Statement I If $(a + ib)(c + id)(e + if)(g + ih) = A + iB$, then $(a^2 + b^2)(c^2 + d^2)(e^2 + f^2)(g^2 + h^2) = A^2 + B^2$.

Statement II If $z = x + iy$, then $|z| = \sqrt{x^2 + y^2}$.

(a) Statement I is correct (b) Statement II is correct
(c) Both are correct (d) Neither I nor II is correct

133. If $\left(\dfrac{1 + i}{1 - i}\right)^m = 1$, then least positive integral value of m is

(a) 4 (b) 0 (c) 2 (d) 1

134. Statement I If $|z_1| = |z_2| = |z_3| = \ldots = |z_n| = 1$, then $|z_1 + z_2 + z_3 + \ldots + z_n| = \left|\dfrac{1}{z_1} + \dfrac{1}{z_2} + \dfrac{1}{z_3} + \ldots + \dfrac{1}{z_n}\right|$.

Statement II If $|z| = 1$, then $z = \dfrac{1}{\overline{z}}$. [NCERT Exemplar]

(a) Statement I is correct (b) Statement II is correct
(c) Both are correct (d) Neither I nor II is correct

Answer *with* Explanations

1. *(a)* $ax^2 + bx + c = 0$, $a \neq 0$ represents quadratic equation, since degree of equation is 2.

2. *(a)* In real number system square of every real number is non-negative.

3. *(d)* Here, $x^2 + 1 = 0, \forall x \in R$, then $x^2 = -1$ but we know that in real number system, square of every real number is non-negative. Hence, it has no real roots.

4. *(c)* For a quadratic equation $ax^2 + bx + c = 0$,

$$x = \frac{-b \pm \sqrt{D}}{2a}$$

$$x = \frac{-b \pm \sqrt{b^2 - 4ac}}{2a}$$

For real roots $D \geq 0$. If roots are not real, then $D < 0$.

5. *(c)* Euler was the Mathematician who introduced (i) (*iota*). *i.e.*, $i = \sqrt{-1}$.

6. *(c)* Square root of negative number is imaginary in general $(a)^{\frac{1}{2n}}$, where $a < 0$ and $n \in N$ gives imaginary number.

7. *(a)* A number $z = a + ib$ where $a, b \in R$ is called complex number.

8. *(b)* If $z = x + iy$ is a complex number and if $x = 0$, then $z = iy$, which is purely imaginary and if $y = 0$, then $z = x$ which is purely real.

9. *(a)* Here, $z = a + ib$, then real part of z *i.e.*, $\text{Re}(z) = a$ and imaginary part of z *i.e.*, $\text{Im}(z) = b$.

10. *(a)* Two complex numbers $z_1 = a + ib$ and $z_2 = c + id$ are equal, if $a = c$ and $b = d$ *i.e.*, if their real and imaginary parts are separately equal.

11. *(b)* We have, $4x + i(3x - y) = 3 + i(-6)$ ...(i)

Equating the real and the imaginary parts of Eq. (i), we get $4x = 3, 3x - y = -6$

which on solving simultaneously, give $x = \frac{3}{4}$ and $y = \frac{33}{4}$.

12. *(d)* $x^{2n} + 1 = 0, x \in R, x^{2n} = -1$ but $x^{2n} \geq 0$.

So, above equation has no real roots.

13. *(b)* $3x^2 + 4x + 2 = 0$, $a = 3, b = 4, c = 2$

$$D = b^2 - 4ac$$
$$= 16 - 4(3)(2)$$
$$= 16 - 24$$
$$= -8$$
$$\Rightarrow \quad D < 0$$

$\because b^2 - 4ac < 0$, so above equation has imaginary roots.

14. *(a)* Here, $x = \sqrt{-16}$

$$x = \sqrt{-1 \times 16}$$
$$= \sqrt{-1} \times \sqrt{4 \times 4} = 4i$$

15. *(a)* Set of complex numbers can be represented as

$$C = \{x + iy : x \in R, y \in R, i = \sqrt{-1}\}$$
$$C = \{x + iy : x, y \in R \text{ and } i = \sqrt{-1}\}$$

16. *(b)* Two complex numbers are either equal or not equal. Comparison of complex numbers is not possible *i.e.*, inequality does not exist for complex numbers.

$\therefore$ If $a + ib > c + id$ this comparision is only possible when it is purely real.

Hence, $b = d = 0$.

17. *(a)* $(1 - i)x + (1 + i)y = 1 - 3i$

$$(x + y) + i(y - x) = 1 - 3i$$

Two complex numbers are equal, if their real and imaginary parts are equal.

$$\therefore \quad x + y = 1 \quad \text{and} \quad y - x = -3$$

By simplification $x = 2, y = -1$

Here, (x, y) is $(2, -1)$.

18. *(d)* Since, comparison of complex numbers is not valid. Hence, option (d) is correct.

19. *(b)* If $z_1 = x_1 + iy_1$ and $z_2 = x_2 + iy_2$, then

$$z_1 + z_2 = (x_1 + iy_1) + (x_2 + iy_2)$$
$$\Rightarrow \quad z_1 + z_2 = (x_1 + x_2) + i(y_1 + y_2)$$

20. *(a)* $z_1 + z_2 = (x_1 + x_2) + i(y_1 + y_2)$,

$$z_1 + z_2 = (2 + 3i) + (3 + 2i)$$
$$= (2 + 3) + i(3 + 2) = 5 + 5i$$

21. *(c)* The addition of complex numbers satisfy the closure, commutative and associative properties.

I. **The closure law** The sum of two complex numbers is a complex number, *i.e.*, $z_1 + z_2$ is a complex number for all complex numbers z_1 and z_2.

II. **The commutative law** For any two complex numbers z_1 and z_2, we have $z_1 + z_2 = z_2 + z_1$ *i.e.*, it satisfies commutative law $z_1 + z_2 = z_2 + z_1$.

III. **The associative law** For any three complex numbers z_1, z_2, z_3, we have $(z_1 + z_2) + z_3 = z_1 + (z_2 + z_3)$ *i.e.*, it satisfies associative law.

22. *(a)* For every complex number there exists the complex number $0 + i0$ (denoted as 0) called additive identity or zero complex number such that, for complex number z, $z + 0 = z$.

23. *(a)* To every complex number $z = a + ib$, we have the complex number $-a + i(-b)$ (denoted as $-z$), called the additive inverse or negative of z. We observe that $z + (-z) = 0$ (the additive identity).

24. *(a)* For two complex numbers z_1 and z_2 the difference $z_1 - z_2$ is defined as follows $z_1 - z_2 = z_1 + (-z_2)$.

25. *(a)* Here, $z_1 = 2 + 3i$, $z_2 = 3 - 2i$, then

$$\begin{aligned} z_1 - z_2 &= 2 + 3i - (3 - 2i) \\ &= 2 + 3i - 3 + 2i \\ &= -1 + 5i \end{aligned}$$

26. *(b)* A. We have, $3(7 + i7) + i(7 + i7)$

$$= 21 + 21i + 7i + i^2 7 = 21 + 28i - 7 \quad (\because i^2 = -1)$$
$$= 14 + 28i$$

B. $1 - i + 1 - 6i = (1 + 1) - (i + 6i) = 2 - 7i$

C. $$\frac{1}{5} + i\frac{2}{5} - 4 - \frac{i5}{2} = \left(\frac{1}{5} - \frac{4}{1}\right) + i\left(\frac{2}{5} - \frac{5}{2}\right)$$
$$= \left(\frac{1-20}{5} + i\frac{(4-25)}{10}\right)$$
$$= -\frac{19}{5} - \frac{21i}{10}$$

D. $$\left(\frac{1}{3} + i\frac{7}{3} + 4 + i\frac{1}{3}\right) + \frac{4}{3} - i$$
$$= \left(\frac{1}{3} + 4 + \frac{4}{3}\right) + i\left(\frac{7}{3} + \frac{1}{3} - 1\right)$$
$$= \left(\frac{1}{3} + \frac{4}{1} + \frac{4}{3}\right) + i\left(\frac{7}{3} + \frac{1}{3} - \frac{1}{1}\right)$$
$$= \frac{(1 + 12 + 4)}{3} + i\left(\frac{7 + 1 - 3}{3}\right)$$
$$= \frac{17}{3} + i\frac{5}{3}$$

27. *(c)* I. If $z_1 = a + ib$ and $z_2 = c + id$, then

$$\begin{aligned} z_1 z_2 &= (a + ib)(c + id) \\ &= ac + iad + ibc + i^2 bd \\ &= (ac - bd) + i(ad + bc) \quad (\because i^2 = -1) \end{aligned}$$

Hence, Statement I is correct.

II. Since, product of two complex numbers is also a complex number. This is called closure law of multiplication. *i.e.*,

If z_1 and z_2 are two complex numbers, then $z_1 \cdot z_2$ is also a complex numbers.

III. $(3 + 5i)(2 + i6) = (3 \times 2 - 5 \times 6) + i(3 \times 6 + 5 \times 2)$

$$= -24 + i28$$

Hence, all the statements are correct.

28. *(a)* This property is known as closure law.

29. *(b)* If z_1 and z_2 are two complex numbers and $z_1 z_2 = z_2 z_1$, then it satisfies commutative law.

30. *(c)* $z_1 \cdot z_2 = z_1$

$$\Rightarrow \quad z_2 = 1 + 0i \quad [\because z_1(1 + 0i) = z_1]$$

Then, z_2 is called multiplicative identity of complex number z_1. Hence, Statement I is true.

For every non-zero complex number (z), we have complex number called the multiplicative inverse of z such that $z \cdot \frac{1}{z} = 1 + 0i$

31. *(a)* Multiplication of complex number satisfies commutative law *i.e.*, $z_1 z_2 = z_2 z_1$ as well as associative law *i.e.*, $z_1(z_2 z_3) = (z_1 z_2) z_3$

If z, $z' = 1 + 0i$, $z \neq 0$, then z' is called multipilcative inverse of z.

and if $z \cdot z_1 = z$, then z_1 is called multiplicative identity of z. Hence, $z_1 = 1 + 0i$

32. *(a)* For any three complex numbers z_1, z_2 and z_3, distributive law is $z_1(z_2 + z_3) = z_1 z_2 + z_1 z_3$ and $(z_1 + z_2) z_3 = z_1 z_3 + z_2 z_3$.

$$\therefore 2 + 3i\,[(3 + 2i) + (2 + i)]$$
$$= (2 + 3i)(3 + 2i) + (2 + 3i)(2 + i)$$
$$[\because z_1(z_2 + z_3) = z_1 z_2 + z_1 z_3]$$
$$= (6 - 6) + 13i + (4 - 3) + 8i$$
$$= 1 + 21i$$

33. *(b)* $$5i\left(-\frac{3}{5}i\right) = 5 \times -\frac{3}{5}i^2$$
$$= -3(-1) = 3 = 3 + 0i$$

34. *(a)* $z = a + ib$, let multiplicative inverse of z is z^{-1}.

$$\therefore \quad z^{-1} = \frac{1}{z} = \frac{1}{a + ib}$$
$$= \frac{a - ib}{(a + ib)(a - ib)}$$

[multiplying numerator and denominator by $(a - ib)$]

$$= \frac{a - ib}{a^2 + b^2}$$
$$z^{-1} = \frac{a}{a^2 + b^2} - i\frac{b}{a^2 + b^2}$$

Hence, multiplicative inverse of $a + ib$ is

$$\left(\frac{a}{a^2 + b^2} - i\frac{b}{a^2 + b^2}\right).$$

35. *(b)* A. Let $z = 4 - 3i$

Then, its multiplicative inverse is

$$\frac{1}{z} = \frac{1}{4 - 3i} = \frac{1}{4 - 3i} \times \frac{4 + 3i}{4 + 3i} = \frac{4 + 3i}{16 - 9i^2}$$

[use $(a - b)(a + b) = a^2 - b^2$]

$$= \frac{4 + 3i}{16 + 9} \quad (\because i^2 = -1)$$
$$= \frac{4 + 3i}{25} = \frac{4}{25} + \frac{3i}{25}$$

B. Let $z = \sqrt{5} + 3i$

Then, its multiplicative inverse is

$$\frac{1}{z} = \frac{1}{\sqrt{5}+3i} = \frac{1}{\sqrt{5}+3i} \times \frac{\sqrt{5}-3i}{\sqrt{5}-3i}$$
$$= \frac{\sqrt{5}-3i}{5-9i^2} \quad [\text{use } (a+b)(a-b) = a^2 - b^2]$$
$$= \frac{\sqrt{5}-3i}{5+9} = \frac{\sqrt{5}-3i}{14} \quad (\because i^2 = -1)$$
$$= \frac{\sqrt{5}}{14} - \frac{3i}{14}$$

C. Let $z = -i$

Then, its multiplicative inverse is

$$\frac{1}{z} = -\frac{1}{i} = -\frac{1}{i} \times \frac{i}{i} = \frac{-i}{i^2} = \frac{-i}{-1} = i \quad (\because i^2 = -1)$$
$$= 0 + i$$

36. *(a)* Given, any two complex numbers z_1 and z_2, where $z_2 \neq 0$. Then, quotient $\frac{z_1}{z_2}$ is defined as

$$\frac{z_1}{z_2} = z_1\left(\frac{1}{z_2}\right) = z_1 (z_2^{-1})$$

37. *(a)* Let $z_1 = 6 + 3i$ and $z_2 = 2 - i$

Then, $\frac{z_1}{z_2} = (6+3i)\frac{1}{2-i} = \frac{(6+3i)(2+i)}{(2-i)(2+i)}$

$$= (6+3i)\left(\frac{2}{2^2+(-1)^2} + i\frac{1}{2^2+(-1)^2}\right)$$
$$= (6+3i)\left(\frac{2}{5} + i\frac{1}{5}\right)$$
$$= (6+3i)\frac{(2+i)}{5}$$
$$= \frac{1}{5}[12 - 3 + i(6+6)] = \frac{1}{5}(9 + 12i)$$

38. *(d)* We know that,

$$i = \sqrt{-1}, i^2 = -1$$
$$\therefore \quad i^3 = i^2 \cdot i = (-1)i = -i$$
$$\Rightarrow \quad i^4 = i^2 \times i^2 = -1 \times -1 = 1$$
$$\Rightarrow \quad i^5 = i^2 \times i^2 \cdot i = -1 \times -1 \times i = i$$
$$\Rightarrow \quad i^6 = i^2 \times i^2 \times i = -1 \times -1 \times -1 = -1$$

Hence, option (d) is not correct.

39. *(b)* We know that, $i = \sqrt{-1}, i^2 = -1$

$$\Rightarrow \quad i^{-1} = \frac{1}{i} \times \frac{i}{i} = \frac{i}{-1} = -i$$
$$\Rightarrow \quad i^{-2} = \frac{1}{i^2} = \frac{1}{-1} = -1$$
$$\Rightarrow \quad i^{-3} = \frac{1}{i^3} = \left\{\frac{i}{i^3 \times i}\right\}$$

[multiplying numerator and denominator by i]

$$\Rightarrow \quad = \frac{i}{i^4} = i \quad \Rightarrow i^{-4} = \frac{1}{i^4} = \frac{1}{1} = 1$$

40. *(c)* We know that, $i = \sqrt{-1}$

$$\because \quad i^{4k} = (i^4)^k = 1^k = 1$$
$$\Rightarrow \quad i^{4k+1} = i^{4k} \cdot i = 1 \times i = i$$
$$\Rightarrow \quad i^{4k+2} = i^{4k} \cdot i^2 = 1 \times -1 = -1$$
$$\Rightarrow \quad i^{4k+3} = i^{4k} \cdot i^3 = 1 \times -i = -i$$

Also, $i^{4k} + i^{4k+1} + i^{4k+2} + i^{4k+3}$

$$= i^{4k}(1 + i + i^2 + i^3)$$
$$= i^{4k}(1 + i - 1 - i) = i^{4k}\, 0 = 0$$

Hence, Statements I and II are true.

41. *(a)* $i^9 + i^{19} = i^9 (1 + i^{10}) = i^9 [1 + (i^2)^5]$ (taking i^9 common)

$$= i^9 [(1 + (-1)^5] = i^9 (1 - 1) = 0$$
$$= 0 + 0i$$

42. *(b)* $i^{-39} = \frac{1}{i^{39}}$

Multiplying and dividing by i, we get

$$= \frac{i}{i^{40}} = \frac{i}{(i^4)^{10}} = \frac{i}{(1)^{10}} = \frac{i}{1} = i \quad (\because i^4 = 1)$$
$$= 0 + i$$

43. *(d)* A. $(1-i)^4 = [(1-i)^2]^2$

$$= (1 + i^2 - 2i)^2 \text{use } (a-b)^2 = a^2 + b^2 - 2ab]$$
$$= (1 - 1 - 2i)^2 \quad (\because i^2 = -1)$$
$$= (-2i)^2 = (-2)^2 i^2$$
$$= 4(-1) = -4 + 0i$$

B. Here, we will use the formula $(a+b)^3 = a^3 + b^3 + 3ab$ $(a+b)$ and simplify .

$$\left(\frac{1}{3} + 3i\right)^3 = \left(\frac{1}{3}\right)^3 + (3i)^3 + 3 \times \frac{1}{3} \times 3i\left(\frac{1}{3} + 3i\right)$$
$$= \frac{1}{27} + 27i^3 + 3i\left(\frac{1}{3} + 3i\right)$$
$$= \frac{1}{27} - 27i + 3i \times \frac{1}{3} + 3i \times 3i \quad (\because i^3 = -i)$$
$$= \frac{1}{27} - 27i + i + 9i^2$$
$$= \frac{1}{27} - 27i + i - 9 \quad (\because i^2 = -1)$$
$$= \left(\frac{1}{27} - \frac{9}{1}\right) - i(27 - 1)$$
$$= \left(\frac{1-243}{27}\right) - 26i = -\frac{242}{27} - 26i$$

C. $(-1)^3\left(2 + \frac{1}{3}i\right)^3 = -\left[(2)^3 + \left(\frac{1}{3}i\right)^3 + 3 \times 2 \times \frac{1}{3}i\left(2 + \frac{1}{3}i\right)\right]$

$$= -\left[8 + \frac{1}{27}i^3 + 2i\left(2 + \frac{1}{3}i\right)\right]$$
$$= -\left[8 - \frac{1}{27}i + 4i + \frac{2}{3}i^2\right] \quad (\because i^3 = -i)$$

$$= -\left[8 - \frac{1}{27} i + 4i - \frac{2}{3}\right] \qquad (\because i^2 = -1)$$
$$= -\left[\left(\frac{8}{1} - \frac{2}{3}\right) + i\left(\frac{4}{1} - \frac{1}{27}\right)\right]$$
$$= -\left[\left(\frac{24-2}{3}\right) + i\left(\frac{108-1}{27}\right)\right]$$
$$= -\left[\frac{22}{3} + i\,\frac{107}{27}\right]$$

44. *(b)* A. $(-5i)\left(\frac{1}{8} i\right) = \frac{(-5)}{(8)} i^2 = \frac{(-5)}{8}(-1) = \frac{5}{8} = \frac{5}{8} + i0$

B. $(-i)(2i) \times \left(\frac{-1}{8} i\right)^3 = 2 \times \frac{-1}{8 \times 8 \times 8} \times i^3 = \frac{-1}{256} i^2 \cdot i$
$$= \frac{1}{256} i$$

C. We have,
$$(5-3i)^3 = 5^3 - 3 \times 5^2 \times (3i) + 3 \times 5\,(3i)^2 - (3i)^3$$
$$= 125 - 225i - 135 + 27i = -10 - 198i$$

D. We have, $(\sqrt{3} + \sqrt{-2})(2\sqrt{3} - i) = (-\sqrt{3} + \sqrt{2}i)(2\sqrt{3} - i)$
$$= -6 + \sqrt{3}i + 2\sqrt{6}i - \sqrt{2}i^2 = (-6 + \sqrt{2}) + \sqrt{3}(1 + 2\sqrt{2})i$$

45. *(a)* $(i)^2 = -1$ and $(-i)^2 = i^2 = -1$

Therefore, square roots of -1 are $i, -i$. However, by the symbol $\sqrt{-1}$ we would mean (i) only.

46. *(a)* $x^2 + 1 = 0$

Here, $x = i$ as well as $-i$ are roots of above equation.

Note $\because (\sqrt{3} \cdot i)^2 = (\sqrt{3})^2 i^2 = 3(-1) = -3$

$\therefore (-\sqrt{3}\,i)^2 = (-\sqrt{3})^2 i^2 = 3(-1) = -3$

Therefore, squre root of -3 are $\sqrt{3}i$ and $-\sqrt{3}i$. But symbol $\sqrt{-3}$ meant to represent $\sqrt{3}i$ only i.e, $\sqrt{-3} = \sqrt{3}i$. In general, $\sqrt{-a} = \sqrt{a}\, i$.

47. *(c)* $\sqrt{a} \times \sqrt{b} = \sqrt{ab}$ for all positive number this is also hold true when either $a > 0$ and $b < 0$ or $a < 0$ and $b > 0$.

Hence, above result is only true when atleast one of them is positive.

48. *(a)* Let $x + iy = \sqrt{-7 - 24i}$

Then, $(x + iy)^2 = -7 - 24i$

or $x^2 - y^2 + 2xyi = -7 - 24i$

Equating real and imaginary parts, we have
$$x^2 - y^2 = -7 \qquad \ldots(i)$$
$$2xy = -24$$

We know the identity
$$(x^2 + y^2)^2 = (x^2 - y^2)^2 + (2xy)^2$$
$$= 49 + 576 = 625$$

Thus, $x^2 + y^2 = 25 \qquad \ldots(ii)$

From Eqs. (i) and (ii), we get
$$x^2 = 9 \quad \text{and} \quad y^2 = 16$$

or $x = \pm 3$ and $y = \pm 4$

Since, the product xy is negative, we have
$$x = 3, y = -4 \quad \text{or} \quad x = -3, y = 4$$

Thus, the square roots of $-7 - 24i$ are $3 - 4i$ and $-3 + 4i$.

49. *(b)* A. Let $\sqrt{-15 - 8i} = x + iy$

Then, $(x + iy)^2 = -15 - 8i$

$x^2 - y^2 + 2i\,xy = -15 - 8i \qquad \ldots(i)$

Equating real and imaginary parts, we get
$$x^2 - y^2 = -15$$
$$2xy = -8$$
$$\because \quad (x^2 + y^2)^2 = (x^2 - y^2)^2 + (2xy)^2$$
$$(x^2 + y^2)^2 = (-15)^2 + (-8)^2$$
$$(x^2 + y^2)^2 = 225 + 64$$
$$(x^2 + y^2)^2 = 289$$
$$\therefore \quad x^2 + y^2 = 17 \qquad \ldots(ii)$$

Solving Eqs. (i) and (ii), we get
$$x^2 = 1 \Rightarrow x = \pm 1$$
$$y^2 = 16 \Rightarrow y = \pm 4$$

Since, product of xy is negative.

$\therefore x = 1, y = -4, y = -1, y = 4$

Thus, square root of $-15 - 8i$ is $-1 + 4i$, $1 - 4i$

B. Let $\sqrt{-8 - 6i} = x + iy$
$$x^2 - y^2 + 2x\,yi = -8 - 6i$$

Equating real and imaginary parts, we get
$$x^2 - y^2 = -8,\ 2xy = -6$$
$$x^2 - y^2 = -8 \qquad \ldots(i)$$
$$2xy = -6$$
$$\because \quad (x^2 + y^2)^2 = (x^2 - y^2)^2 + (2xy)^2$$
$$= (-8)^2 + (-6)^2 = 64 + 36$$
$$\Rightarrow \quad (x^2 + y^2)^2 = 100$$
$$\therefore \quad x^2 + y^2 = 10 \qquad \ldots(ii)$$

Solving Eqs. (i) and (ii), we get
$$x = \pm 1, y = \pm 3$$

$\therefore$ Product of xy is negative.
$$\therefore x = 1,\ y = -3 \text{ or } x = -1, y = 3$$

Thus, the square root of complex number $-8 - 6i$ is $-1 + 3i$, $1 - 3i$.

C. Let $x + iy = \sqrt{1 - i}$
$$x^2 - y^2 + 2i\,xy = 1 - i$$
$$\Rightarrow \quad x^2 - y^2 = 1 \qquad \ldots(i)$$
$$\Rightarrow \quad 2xy = -1$$
$$\Rightarrow \quad (x^2 + y^2)^2 = (x^2 - y^2)^2 + (2xy)^2$$
$$= (1)^2 + (-1)^2$$
$$\Rightarrow \quad (x^2 + y^2)^2 = 1 + 1$$
$$\therefore \quad x^2 + y^2 = \sqrt{2} \qquad \ldots(ii)$$

Solving Eqs. (i) and (ii), we get

$$x = \pm\sqrt{\frac{1+\sqrt{2}}{2}}, y = \pm\sqrt{\frac{1-\sqrt{2}}{2}}$$

$\because$ Product of xy is negative.

$$\therefore \quad x = \sqrt{\frac{1+\sqrt{2}}{2}}, \quad y = -\sqrt{\frac{1-\sqrt{2}}{2}}$$

$$x = -\sqrt{\frac{1+\sqrt{2}}{2}}, y = -\frac{\sqrt{1-\sqrt{2}}}{2}$$

Thus, the square root of complex number $1-i$ is $\left(\pm\sqrt{\frac{\sqrt{1+\sqrt{2}}}{2}} \mp i\sqrt{\frac{1-\sqrt{2}}{2}}\right)$.

D. Let $\sqrt{-i} = x + iy - i = (x+iy)^2$

$\Rightarrow \quad x^2 - y^2 + 2i\,xy = -i$

$\Rightarrow \quad x^2 - y^2 = 0 \quad \ldots(i)$

$\Rightarrow \quad 2xy = -1$

$\Rightarrow \quad (x^2+y^2)^2 = (x^2-y^2)^2 + (2xy)^2$

$\Rightarrow \quad (x^2 + y^{2)} = 0 + 1$

$\Rightarrow \quad x^2 + y^2 = 1 \quad \ldots(ii)$

$\therefore \quad x^2 = \frac{1}{2} \Rightarrow x = \pm\frac{1}{\sqrt{2}}$

$y^2 = \frac{1}{2}, y = \pm\frac{1}{\sqrt{2}}$

$\therefore$ Product of xy is negative.

$\because x = \frac{1}{\sqrt{2}}, y = -\frac{1}{\sqrt{2}}, x = -\frac{1}{\sqrt{2}}, y = \frac{1}{\sqrt{2}}$

The square root of complex number $-i$ is $\left(\pm\frac{1}{\sqrt{2}} \mp \frac{1}{\sqrt{2}}\right)$.

50. *(a)* I. Let $x + iy = \sqrt{i}$

$\Rightarrow \quad (x+iy)^2 = i$

$\Rightarrow \quad x^2 - y^2 + 2i\,xy = i$

$\Rightarrow \quad x^2 - y^2 = 0 \quad \ldots(i)$

and $\quad 2xy = 1$

$(x^2+y^2)^2 = (x^2-y^2)^2 + (2xy)^2$

$(x^2 + y^2) = 1 \quad \ldots(ii)$

Solving Eqs. (i) and (ii), we get

$$x = \pm\frac{1}{\sqrt{2}}, y = \pm\frac{1}{\sqrt{2}}$$

$\because$ Product of xy is positive.

$\therefore \quad x = \frac{1}{\sqrt{2}} = y$

$\Rightarrow \quad x = -\frac{1}{\sqrt{2}} = y$

Thus, square root of complex number i is $\left(\pm\frac{1}{\sqrt{2}} \pm \frac{1}{\sqrt{2}}i\right)$.

$\therefore$ Statement I is correct.

II. Let $\quad \sqrt{1+i} = x + iy$

$\Rightarrow \quad x^2 - y^2 + 2i\,xy = 1 + i$

$\Rightarrow \quad x^2 - y^2 = 1 \quad \ldots(i)$

$\Rightarrow \quad 2xy = 1$

$\Rightarrow \quad (x^2+y^2)^2 = (x^2-y^2) + (2xy)^2$

$\Rightarrow \quad x^2 + y^2 = \sqrt{2} \quad \ldots(ii)$

Solving Eqs. (i) and (ii), we get

$$x = \pm\sqrt{\frac{\sqrt{2}+1}{2}}, \quad y = \pm\sqrt{\frac{\sqrt{2}-1}{2}}$$

Thus, square root of complex number $1+i$ is

$$\left(\pm\sqrt{\frac{\sqrt{2}+1}{2}} \pm \sqrt{\frac{\sqrt{2}-1}{2}}i\right).$$

51. *(b)* We have, $(z_1+z_2)^2 = (z_1+z_2)(z_1+z_2)$

$= (z_1+z_2)z_1 + (z_1+z_2)z_2$ (distributive law)

$= z_1^2 + z_2 z_1 + z_1 z_2 + z_2^2$ (distributive law)

$= z_1^2 + z_1 z_2 + z_1 z_2 + z_2^2$

(commutative law, $z_1 z_2 = z_2 z_1$)

$= z_1^2 + 2z_1 z_2 + z_2^2$

52. *(d)* A. We have, $(z_1 - z_2)^2 = [(z_1-z_2)(z_1-z_2)]$

$= (z_1-z_2)z_1 - (z_1-z_2)z_2$ (distributive law)

$= z_1^2 - z_2 z_1 - z_2 z_1 + z_2^2$

$= z_1^2 - 2z_1 z_2 + z_2^2$ (commutative law, $z_1 z_2 = z_2 z_1$)

B. $(z_1+z_2)^3 = [z_1+z_2)(z_1+z_2)](z_1+z_2)$

$= (z_1^2 + 2z_1 z_2 + z_2^2)(z_1+z_2)$

$= (z_1^3 + 3z_1^2 z_2 + 3z_1 z_2^2 + z_2^3)$

C. $(z_1-z_2)^3 = [(z_1-z_2)(z_1-z_2)](z_1-z_2)$

$(z_1^2 - 2z_1 z_2 + z_2^2)(z_1-z_2)$

$z_1^3 - 3z_1^2 z_2 + 3z_1 z_2^2 - z_2^3$

D. $z_1^2 - z_2^2 = (z_1+z_2)(z_1-z_2)$

Hence, all the options are correct.

53. *(d)* $\dfrac{(3+i\sqrt{5})(3-i\sqrt{5})}{(\sqrt{3}+\sqrt{2}i)-(\sqrt{3}-i\sqrt{2})}$

$= \dfrac{3^2 - (i\sqrt{5})^2}{\sqrt{3}+\sqrt{2}i-\sqrt{3}+i\sqrt{2}}$ $[\because (z_1-z_2)(z_1+z_2) = z_1^2 - z_2^2]$

$= \dfrac{9+5}{2\sqrt{2}i}$

$= \dfrac{14i^4}{2\sqrt{2}i}$ $(\because i^4 = 1)$

$= \dfrac{7}{\sqrt{2}}i^3$

$= \dfrac{-7\sqrt{2}}{2}i = 0 - \dfrac{7\sqrt{2}i}{2}$

54. *(a)* Let $z = a + ib$ be a complex number. Then, modulus of z, denoted by $|z|$, is defined to be non-negative real number $\sqrt{a^2+b^2}$ i.e., $|z| = \sqrt{a^2+b^2}$.

55. *(b)* The conjugate of $z = a + ib$ denoted as $\bar{z}$, is the complex number $a - ib$ i.e., $\bar{z} = a - ib$.

56. *(c)* I. $|3+i| = \sqrt{3^2+1^2} = \sqrt{10}, |2-5i| = \sqrt{2^2+(-5)^2} = \sqrt{29}$

II. $\overline{(3+i)} = 3 - i,\ \overline{(2-5i)} = 2 + 5i,\ \overline{(-3i-5)} = 3i - 5$

III. The multiplicative inverse of the non-zero complex number z is given by

$$z^{-1} = \frac{1}{a+ib} = \frac{a}{a^2+b^2} + i\frac{-b}{a^2+b^2} = \frac{a-ib}{a^2+b^2} = \frac{\bar{z}}{|z|^2}$$

$$\therefore\ z^{-1} = \frac{\bar{z}}{|z|^2} \text{ or } z\bar{z} = |z|^2$$

57. *(b)* Let $z_1 = x_1 + iy_1$, $z_2 = x_2 + iy_2$. Then,

A. $z_1 z_2 = (x_1 + iy_1)(x_2 + iy_2)$

$$z_1 z_2 = (x_1x_2 - y_1y_2) + i(x_1y_2 + x_2y_1)$$

$$|z_1 z_2| = \sqrt{(x_1x_2 - y_1y_2)^2 + (x_1y_2 + x_{2+}y_1y_2)^2}$$

$$= \sqrt{x_1^2x_2^2 + y_1^2y_2^2 - 2x_1x_2y_1y_2 + x_1^2y_2^2 + x_2^2y_1^2 + 2x_1x_2y_1y_2}$$

$$= \sqrt{x_1^2(x_2^2+y_2^2) + y_1^2(x_2^2+y_2^2)}$$

$$= \sqrt{(x_1^2+y_1^2)(x_2^2+y_2^2)}$$

$$= \sqrt{(x_1^2+y_1^2)}\sqrt{(x_2^2+y_2^2)}$$

$$|z_1 z_2| = |z_1||z_2|$$

Similarly, we can show that

B. $\left|\frac{z_1}{z_2}\right| = \frac{|z_1|}{|z_2|}$

C. $\overline{z_1 z_2} = \overline{[(x_1+iy_1)(x_2+iy_2)]}$

$$= \overline{((x_1x_2 - y_1y_2) + i(x_1y_2 + x_2y_1))}$$

$$= (x_1x_2 - y_1y_2) - i(x_1y_2 + x_2y_1)$$

$$= x_1(x_2 - iy_2) - yi(x_2 - iy_2)$$

$$= (x_1 - iy_1)(x_2 - iy_2)$$

$$\overline{(z_1 z_2)} = (\overline{z_1}\ \overline{z_2})$$

D. Similarly, $\overline{(z_1 \pm z_2)} = (\overline{z_1} \pm \overline{z_2})$

E. $\overline{\left(\frac{z_1}{z_2}\right)} = \frac{\overline{z_1}}{\overline{z_2}}$

58. *(b)* I. Let $z = 2 - 3i$

Then, $\bar{z} = 2 + 3i$ and $|z|^2 = 2^2 + (-3)^2 = 13$

Therefore, the multiplicative inverse of $2 - 3i$ is

$$z^{-1} = \frac{\bar{z}}{|z|^2} = \frac{2+3i}{13} = \frac{2}{13} + \frac{3}{13}i$$

The above working can be reproduced in the following manner also,

$$z^{-1} = \frac{1}{2-3i} = \frac{2+3i}{(2-3i)(2+3i)}$$

$$= \frac{2+3i}{2^2-(3i)^2} = \frac{2+3i}{13} = \frac{2}{13} + \frac{3}{13}i$$

II. We have,

$$\frac{5+\sqrt{2}i}{1-\sqrt{2}i} = \frac{5+\sqrt{2}i}{1-\sqrt{2}i} \times \frac{1+\sqrt{2}i}{1+\sqrt{2}i}$$

$$= \frac{5+5\sqrt{2}i+\sqrt{2}i-2}{1-(\sqrt{2}i)^2}$$

$$= \frac{3+6\sqrt{2}i}{1+2} = \frac{3(1+2\sqrt{2}i)}{3} = 1 + 2\sqrt{2}i$$

III. $i^{-35} = \frac{1}{i^{35}} = \frac{1}{(i^2)^{17}i} = \frac{1}{-i} \times \frac{i}{i} = \frac{i}{-i^2} = i$

59. *(b)* $(x + iy) + (2 - 3i) = 4 + i$

$\Rightarrow$ $x + 2 + i(y - 3) = 4 + i$

$\Rightarrow$ $x + 2 = 4$ and $y - 3 = 1$

$\Rightarrow$ $x = 2$ and $y = 4$

60. *(c)* Additive inverse of z is $-z$.

$\therefore$ Additive inverse of $(1 - i)$ is

$$-(1-i) = -1 + i$$

61. *(d)* $z_1 = 1 + 2i$; $z_2 = 2 + 3i$

Additive inverse of $z_2 = -2 - 3i$

$$z_1 + (-z_2) = 1 + 2i - 2 - 3i = -1 - i$$

62. *(d)* $\because$ If z_1 is multiplicative inverse of z_2, then $z_1 \cdot z_2 = 1 + 0i$. Hence, (d) is not possible.

63. *(a)* Since, difference of two complex numbers is also a complex number and $z_1 - z_2$ can be written as $z_1 + (-z_2)$, which is sum of z_1 and additive inverse of z_2.

64. *(a)* Let $z = x + iy$

If $\text{Re}(z) = 0$, then $z = iy$

$$z^2 = (iy)^2 = -y^2$$

Thus, $z^2 = -y^2$ (which is real)

$\Rightarrow$ $\text{Im}(z^2) = 0$

65. *(b)* If $(1-i)^n = 2^n$ ...(i)

We know that, if two complex numbers are equal, their modulus must also be equal.

Therefore, from Eq. (i), we have

$$|(1-i)|^n = |2^n|$$

$\Rightarrow$ $|1-i|^n = |2|^n$ $(2^n > 0)$

$\Rightarrow$ $\left[\sqrt{1^2+(-1)^2}\right]^n = 2^n$

$\Rightarrow$ $(\sqrt{2})^n = 2^n$

$\Rightarrow$ $2^{\frac{n}{2}} = 2^n$

$\Rightarrow$ $\frac{n}{2} = n$

$\Rightarrow$ $n = 0$

66. *(d)* $(1+i)^5(1-i)^5 = (1-i^2)^5$

$$= 2^5 = 32$$

67. *(b)* $\dfrac{i^{584}(i^8+i^6+i^4+i^2+1)}{i^{574}(i^8+i^6+i^4+i^2+1)}-1=\dfrac{i^{584}}{i^{574}}-1$

$=i^{10}-1=-1-1=-2$

68. *(c)* $i^2+i^4+i^6+\ldots$ upto $(2k+1)$ terms

$=\dfrac{i^2[1-(i^2)^{2k+1}]}{1-i^2}=\dfrac{-1[1-(-1)^{2k+1}]}{1+1}$

$=\dfrac{-[1-(-1)]}{2}=-1$

69. *(c)* $\sum_{n=1}^{200} i^n=i+i^2+i^3+\ldots+i^{200}$

$=\dfrac{i(1-i^{200})}{1-i}$ (using sum formula of G.P.)

$=\dfrac{i(1-1)}{1-i}=0$

70. *(b)* $\sum_{n=1}^{13}(i^n+i^{n+1})$

$=(i+i^2+i^3+\ldots+i^{13})+(i^2+i^3+\ldots+i^{14})$

$=\dfrac{i(1-i^{13})}{1-i}+\dfrac{i^2(1-i^{13})}{1-i}$

$=i\left(\dfrac{1-i}{1-i}\right)+i^2\dfrac{(1-i)}{(1-i)}=i+i^2=i-1$

71. *(c)* Let $z=i^{[1+3+5+\ldots+(2n+1)]}$

Clearly, series is AP with common difference $=2$

$T_n=2n-1$ and $T_{n+1}=2n+1$

So, number of terms in AP $=n+1$

Now, $S_{n+1}=\dfrac{n+1}{2}$ $[2.1+(n+1-1)2]$

$\Rightarrow$ $S_{n+1}=\dfrac{n+1}{2}[2+2n]=(n+1)^2$ *i.e.*, $i^{(n+1)^2}$

Now, put $n=1,2,3,4,5,\ldots$

$n=1, z=i^4=1, n=2, z=i^6=-1,$

$n=3, z=i^8=1, n=4, z=i^{10}=-1,$

$n=5, z=i^{12}=1,\ldots$

Hence, for odd value of expression it is 1 and for even it is -1.

72. *(c)* $(1+i)^2=1+i^2+2i=2i$

$(1-i)^2=1+i^2-2i=-2i$

$\Rightarrow$ $(1+i)^8+(1-i)^8=[(1+i)^2]^4+[(1-i)^2]^4$

$=(2i)^4+(-2i)^4=2^4(i^4+i^4)=2^4(1+1)=2^5=32$

73. *(c)* $\dfrac{(3+2i\sin\theta)(1+2i\sin\theta)}{(1-2i\sin\theta)(1+2i\sin\theta)}$

$=\left(\dfrac{3-4\sin^2\theta}{1+4\sin^2\theta}\right)+i\left(\dfrac{8\sin\theta}{1+4\sin^2\theta}\right)$

Now, since it is real, therefore $\text{Im}(z)=0$

$\Rightarrow$ $\dfrac{8\sin\theta}{1+4\sin^2\theta}=0$

$\Rightarrow$ $\sin\theta=0$

$\therefore$ $\theta=n\pi$

where, $n=0,1,2,3,\ldots$

74. *(a)* $\dfrac{(\sqrt{5+12i}+\sqrt{5-12i})(\sqrt{5+12i}+\sqrt{5-12i})}{(\sqrt{5+12i}-\sqrt{5-12i})(\sqrt{5+12i}+\sqrt{5-12i})}$

$=\dfrac{5+12i+5-12i+2\sqrt{5+12i}\sqrt{5-12i}}{5+12i-5+12i}$

$=\dfrac{10+2\times(13)}{24i}=-\dfrac{3}{2}i$ $\left[\because \dfrac{1}{i}=-i\right]$

75. *(c)* $\dfrac{3+2i\sin\theta}{1-2i\sin\theta}=\dfrac{3+2i\sin\theta}{1-2i\sin\theta}\times\dfrac{1+2i\sin\theta}{1+2i\sin\theta}$

$=\dfrac{(3-4\sin^2\theta)}{1+4\sin^2\theta}+\dfrac{8i\sin\theta}{1+4\sin^2\theta}$

will be purely imaginary, if the real part vanishes, *i.e.*, $\dfrac{3-4\sin^2\theta}{1+4\sin^2\theta}=0$

$\Rightarrow$ $3-4\sin^2\theta=0$ (only if θ be real)

$\Rightarrow$ $\sin\theta=\pm\dfrac{\sqrt{3}}{2}=\sin\left(\pm\dfrac{\pi}{3}\right)$

$\Rightarrow$ $\theta=n\pi+(-1)^n\left(\pm\dfrac{\pi}{3}\right)$

$\Rightarrow$ $\theta=n\pi\pm\dfrac{\pi}{3}$

76. *(c)* We already know that corresponding to each ordered pair of real numbers (x, y), we get a unique point in the xy-plane and *vice-versa* with reference to a set of mutually perpendicular lines known as the X-axis and the Y-axis. The complex number $x+iy$ which corresponds to the ordered pair (x, y) can be represented geometrically as the unique point $P(x, y)$ in the xy-plane and *vice-versa*.

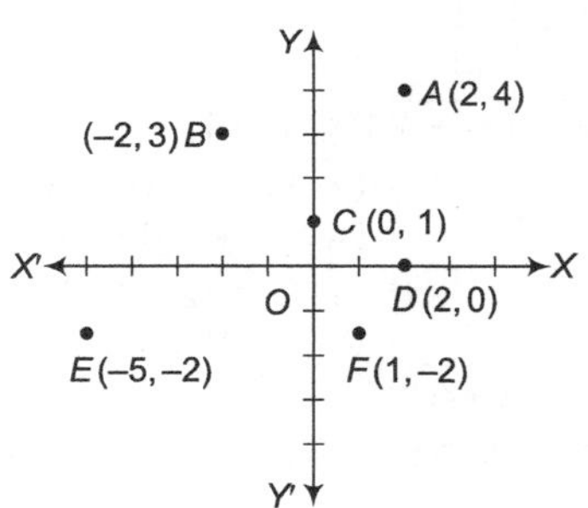

Some complex numbers such as $2+4i$, $-2+3i$, $0+1i$, $2+0i$, $-5-2i$ and $1-2i$ which correspond to the ordered pairs $(2, 4)$, $(-2, 3)$, $(0, 1)$, $(2, 0)$, $(-5, -2)$ and $(1, -2)$ respectively, have been represented geometrically by the points A, B, C, D, E, and F respectively in the plane having a complex number assigned to each of its point is called the complex plane or the Argand plane.

77. *(d)* In the Argand plane, the modulus of the complex number $x+iy=\sqrt{x^2+y^2}$ is the distance between the point $P(x, y)$ and the origin $O(0, 0)$. The points on the X-axis corresponds to the complex numbers of the form $a+i0$ and the points on the Y-axis corresponds to the complex numbers of the form $0+ib$.

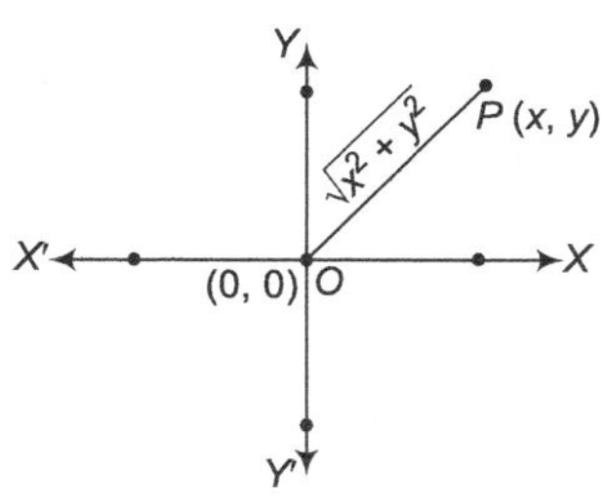

The X-axis and Y-axis in the Argand plane are called, respectively, the real axis and the imaginary axis.

78. *(c)* The representation of a complex number $Z = x + iy$ and its conjugate $\overline{Z} = x - iy$ in the Argand plane are respectively the points $P(x, y)$ and $Q(x, -y)$.

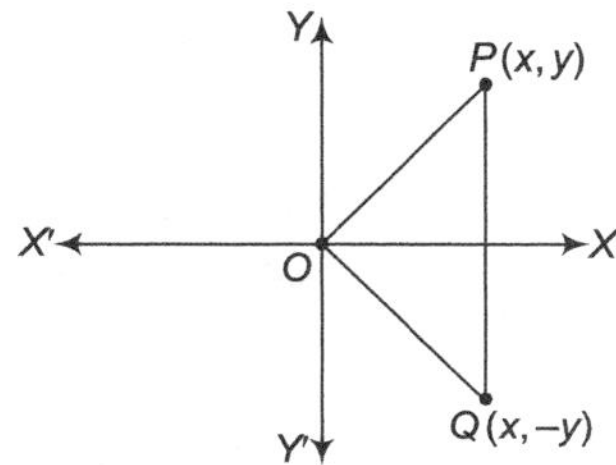

Geometrically, the point $(x, -y)$ is the mirror image of the point (x, y) on the real axis.

79. *(c)* Let the point P represent the non-zero complex number $z = x + iy$. Let the directed line segment OP be of length r and θ be the angle which OP makes with the positive direction of X-axis.

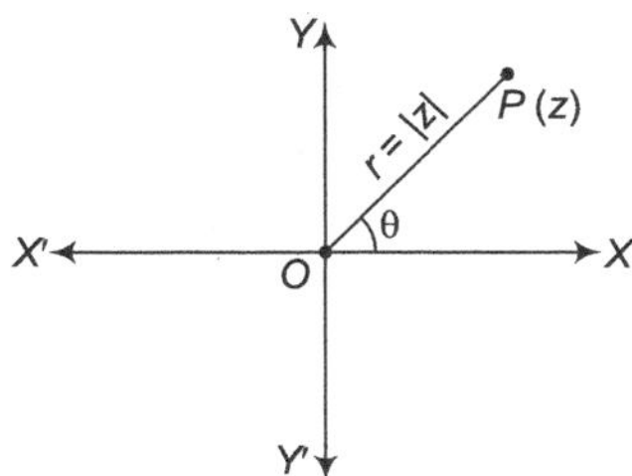

We may note that the point P is uniquely determined by the ordered pair of real numbers (r, θ), called the polar coordinates of the point P.

80. *(a)* Let complex number $z = x + iy$

$\because \quad OP = r, \angle POX = \theta \; x = r\cos\theta;\; y = r\sin\theta.$

Hence, $z = r(\cos\theta + i\sin\theta)$

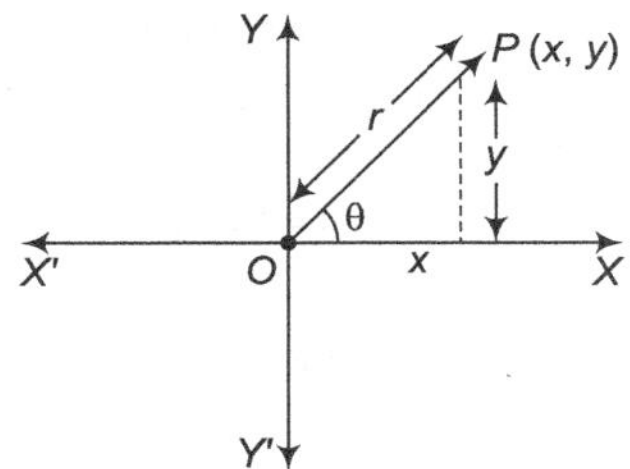

81. *(a)* Representation of z in term of r and θ is called polar form of the complex number *i.e.*, if $z = x + iy$, then $z = r(\cos\theta + i\sin\theta)$ is called polar form.

82. *(b)* If $x = r\cos\theta$; $y = r\sin\theta$, then

$$x^2 + y^2 = r^2(\cos^2\theta + \sin^2\theta)$$

$$\Rightarrow \quad x^2 + y^2 = r^2$$

$$\therefore \quad r = \sqrt{x^2 + y^2} = |z|$$

$\sqrt{x^2 + y^2}$ is the modulus of z and θ is called argument of z which is denoted by arg (z).

83. *(c)* For any complex number $z \neq 0$, there corresponds only one value of θ in $0 \leq \theta \leq 2\pi$. However, any other interval of length 2π, for example $-\pi < \theta \leq \pi$ can be such an interval. We shall take the value of θ such that $-\pi < \theta < \pi$, called principal argument of z and is denoted by arg (z) unless specified otherwise, observe the figure given below.

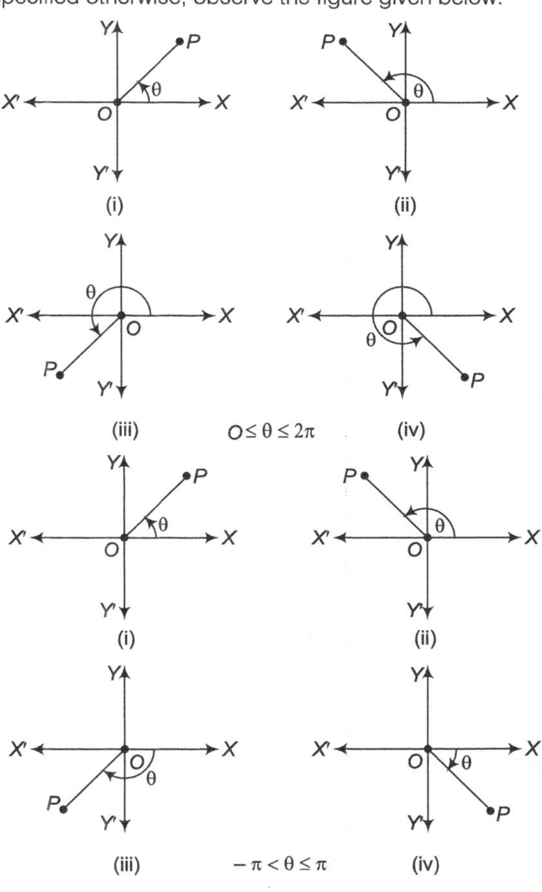

84. *(b)* We have complex number $z = 1 + i\sqrt{3}$

Let $1 = r\cos\theta$, $\sqrt{3} = r\sin\theta$

By squaring and adding, we get

$$r^2(\cos^2\theta + \sin^2\theta) = 4$$

i.e., $r = \sqrt{4} = 2$ (conventionally, $r > 0$)

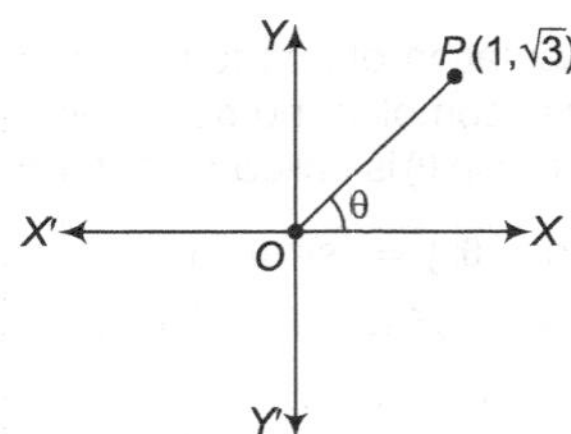

Therefore, $\cos\theta = \frac{1}{2}, \sin\theta = \frac{\sqrt{3}}{2}$, which gives $\theta = \frac{\pi}{3}$.

Therefore, required polar form is $z = 2\left(\cos\frac{\pi}{3} + i\sin\frac{\pi}{3}\right)$.

The complex number $z = 1 + i\sqrt{3}$ is represented as shown in figure.

85. *(b)* The given complex number

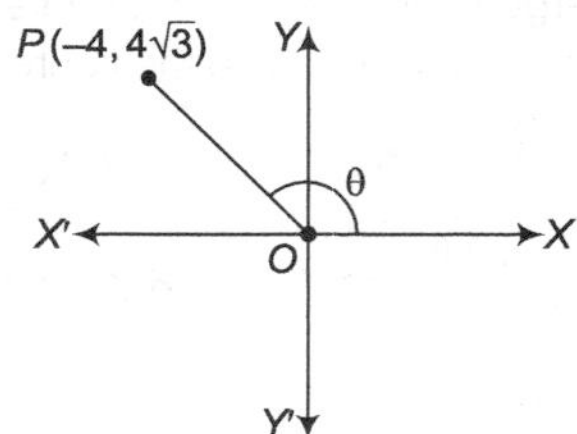

$$\frac{-16}{1+i\sqrt{3}} = \frac{-16}{1+i\sqrt{3}} \times \frac{1-i\sqrt{3}}{1-i\sqrt{3}}$$

$$= \frac{-16(1-i\sqrt{3}}{1-(i\sqrt{3})} = \frac{-16(1-i\sqrt{3})}{1+3}$$

$$= -4(1-i\sqrt{3}) = -4 + i4\sqrt{3}$$

Let $\quad -4 = r\cos\theta,\ 4\sqrt{3} = r\sin\theta$

By squaring and adding, we get

$$16 + 48 = r^2(\cos^2\theta + \sin^2\theta)$$

which gives $\quad r^2 = 64$, *i.e.*, $r = 8$

Hence, $\quad \cos\theta = -\frac{1}{2}, \sin\theta = \frac{\sqrt{3}}{2}$

$$\theta = \pi - \frac{\pi}{3} = \frac{2\pi}{3}$$

Thus, the required polar form is $8\left(\cos\frac{2\pi}{3} + i\sin\frac{2\pi}{3}\right)$.

86. *(d)* If complex number is in the form of $z = a + ib$, then

$$|z| = \sqrt{a^2 + b^2} = \sqrt{(\text{Real part})^2 + (\text{Imaginary part})^2}$$

and argument of z is obtained by $\tan\theta = \left|\frac{b}{a}\right|$ (*i.e.*, dividing imaginary part by real part). We shall take the value of θ such that $-\pi < \theta \le \pi$ (only principal value).

I. $|z| = \sqrt{(-1)^2 + (-\sqrt{3})^2} = \sqrt{1+3} = \sqrt{4} = 2$

II. $\tan\theta = \left|\frac{-\sqrt{3}}{-1}\right| = \sqrt{3} = \tan\frac{\pi}{3} \Rightarrow \theta = \frac{\pi}{3}$

But the point lies in the third quadrant of Argand plane.

$\because$ Real and imaginary parts of z are both negative.

$\therefore \quad \arg(z) = -\pi + \theta = -\pi + \frac{\pi}{3} = \frac{-2\pi}{3}$

III. $|z| = \sqrt{(-\sqrt{3})^2 + (1)^2} = \sqrt{3+1} = \sqrt{4} = 2$

IV. $\tan\theta = \left|\frac{1}{-\sqrt{3}}\right| = \frac{1}{\sqrt{3}} = \tan\frac{\pi}{6}$

$\Rightarrow \quad \theta = \frac{\pi}{6}$

Since, the real part of z is negative and the imaginary part of z is positive, so the points lie in quadrant II.

$\therefore \quad \arg(z) = \pi - \theta = \pi - \frac{\pi}{6} = \frac{5\pi}{6}$

Note *When we determine the argument of z, be careful about the position of z in quadrant.*

87. *(c)* A. Let $z = 1 - i$

and let $1 - i = r\cos\theta + ir\sin\theta$.

On comparing the real and imaginary parts of both sides, we get

$$1 = r\cos\theta \quad \text{...(i)}$$

and $$-1 = r\sin\theta \quad \text{...(ii)}$$

Now, squaring and adding of Eqs. (i) and (ii), we get

$$(1)^2 + (-1)^2 = r^2\cos^2\theta + r^2\sin^2\theta$$

$\Rightarrow r^2(\cos^2\theta + \sin^2\theta) = 1 + 1 \quad (\because \cos^2\theta + \sin^2\theta = 1)$

$\Rightarrow \quad r^2 = 2$

$\Rightarrow \quad r = \sqrt{2}$

Dividing Eq. (ii) by Eq. (i), we get

$$\frac{r\sin\theta}{r\cos\theta} = \left|\frac{-1}{1}\right|$$

$\Rightarrow \quad \tan\theta = |-1| = 1$

$\Rightarrow \quad \tan\theta = \tan\frac{\pi}{4}$

$\Rightarrow \quad \theta = \frac{\pi}{4}$

Since, the real part of z is positive and imaginary part of z is negative, so the point lies in quadrant IV.

$\therefore \quad \arg(z) = -\theta = \left(-\frac{\pi}{4}\right)$

Hence, $\quad z = 1 - i = \sqrt{2}\left[\cos\left(-\frac{\pi}{4}\right) + i\sin\left(-\frac{\pi}{4}\right)\right]$

which is the required polar form of $(1 - i)$.

B. Let $\quad z = -1 + i$

and $\quad -1 + i = r\cos\theta + ir\sin\theta$

On comparing the real and imaginary parts of both sides, we get

$$-1 = r\cos\theta \quad \text{...(i)}$$

and $$1 = r\sin\theta \quad \text{...(ii)}$$

Now, squaring and adding of Eqs. (i) and (ii), we get

$$(-1)^2 + (1)^2 = r^2\cos^2\theta + r^2\sin^2\theta$$

$\Rightarrow \quad r^2(\cos^2\theta + \sin^2\theta) = 1 + 1$

$\Rightarrow \quad r^2 = 2 \quad (\because \cos^2\theta + \sin^2\theta = 1)$

$\therefore \quad r = \sqrt{2}$

Dividing Eq. (ii) by Eq. (i), we get

$$\frac{r\sin\theta}{r\cos\theta} = \left|\frac{1}{-1}\right|$$

$$\Rightarrow \quad \tan\theta = 1 = \tan\frac{\pi}{4}$$

$$\theta = \frac{\pi}{4}$$

Since, the real part of z is negative and imaginary part of z is positive, so the point lies in quadrant II.

$$\therefore \quad \arg(z) = \pi - \theta = \pi - \frac{\pi}{4} = \frac{3\pi}{4}$$

Thus, $z = -1 + i = \sqrt{2}\left[\cos\left(\frac{3\pi}{4}\right) + i\sin\left(\frac{3\pi}{4}\right)\right]$

which is the required polar form of $(-1 + i)$.

C. Let $z = -1 - i$

and let $-1 - i = r\cos\theta + ir\sin\theta$

On comparing the real and imaginary parts of both sides, we get

$$-1 = r\cos\theta \quad \ldots\text{(i)}$$

and $-1 = r\sin\theta \quad \ldots\text{(ii)}$

Now, squaring and adding of Eqs. (i) and (ii), we get

$$r^2\cos^2\theta + r^2\sin^2\theta = (-1)^2 + (-1)^2$$

$\Rightarrow \quad r^2(\cos^2\theta + \sin^2\theta) = 1 + 1$

$\Rightarrow \quad r^2 = 2 \quad (\because \cos^2\theta + \sin^2\theta = 1)$

$\Rightarrow \quad r = \sqrt{2}$

Dividing Eq. (ii) by Eq. (i), we get

$$\frac{r\sin\theta}{r\cos\theta} = \left|\frac{-1}{-1}\right|$$

$\Rightarrow \quad \tan\theta = 1$

$\Rightarrow \quad \tan\theta = \tan\frac{\pi}{4}$

$\Rightarrow \quad \theta = \frac{\pi}{4}$

Since, the real and imaginary parts of z are both negative, so the point lies in quadrant III.

$$\therefore \quad \arg(z) = -\pi + \theta = -\pi + \frac{\pi}{4} = \frac{-3\pi}{4}$$

Thus, $z = -1 - i = \sqrt{2}\left[\cos\left(\frac{-3\pi}{4}\right) + i\sin\left(-\frac{3\pi}{4}\right)\right]$

which is the required polar form of $(-1 - i)$.

D. Let $z = -3 + 0i$

and $-3 + 0i = r\cos\theta + ir\sin\theta$

On comparing the real and imaginary parts of both sides, we get

$$r\cos\theta = -3 \quad \ldots\text{(i)}$$

and $r\sin\theta = 0 \quad \ldots\text{(ii)}$

Now, squaring and adding of Eqs. (i) and (ii), we get

$$r^2\cos^2\theta + r^2\sin^2\theta = (-3)^2 + (0)^2$$

$\Rightarrow \quad r^2(\cos^2\theta + \sin^2\theta) = 9$

$\Rightarrow \quad r^2 = 9 \quad (\because \sin^2\theta + \cos^2\theta = 1)$

$\Rightarrow \quad r = 3$

Dividing Eq. (ii) by Eq. (i), we get

$$\frac{r\sin\theta}{r\cos\theta} = \left|\frac{0}{-3}\right|$$

$\Rightarrow \quad \tan\theta = 0 = \tan 0°$

$\Rightarrow \quad \theta = 0°$

Since, the real part of z is negative and imaginary part of z is positive, so the point lies in quadrant II.

$\therefore \quad \arg z = \pi - \theta = \pi - 0 = \pi$

Thus, $z = -3 + 0i = 3(\cos\pi + i\sin\pi)$

which is the required polar form of (-3).

88. *(c)* I. Let $z = \sqrt{3} + i$ and let $\sqrt{3} + i = r\cos\theta + i\, r\sin\theta$

On comparing real and imaginary parts of both sides, we get

$$r\cos\theta = \sqrt{3} \quad \ldots\text{(i)}$$

and $r\sin\theta = 1 \quad \ldots\text{(ii)}$

Now, squaring and adding of Eqs. (i) and (ii), we get

$$r^2\cos^2\theta + r^2\sin^2\theta = (\sqrt{3})^2 + (1)^2$$

$\Rightarrow \quad r^2(\cos^2\theta + \sin^2\theta) = 3 + 1 = 4 \quad (\because \cos^2\theta + \sin^2\theta = 1)$

$\Rightarrow \quad r^2 = 4 \quad \therefore \quad r = 2$

Dividing Eq. (ii) by Eq. (i), we get

$$\frac{r\sin\theta}{r\cos\theta} = \left|\frac{1}{\sqrt{3}}\right|$$

$$\Rightarrow \quad \tan\theta = \frac{1}{\sqrt{3}} = \tan\frac{\pi}{6}$$

$$\therefore \quad \theta = \frac{\pi}{6}$$

Since, the real and imaginary parts of z are both positive, so the point lies in the first quadrant.

$$\therefore \quad \arg(z) = \theta = \frac{\pi}{6}$$

$$\therefore \quad z = \sqrt{3} + i = 2\left[\cos\frac{\pi}{6} + i\sin\frac{\pi}{6}\right]$$

which is the required polar form of $(\sqrt{3} + i)$.

II. Let $z = 0 + i$ and let $0 + i = r\cos\theta + i\, r\sin\theta$

On comparing real and imaginary parts of both sides, we get

$$r\cos\theta = 0 \quad \ldots\text{(i)}$$

and $r\sin\theta = 1 \quad \ldots\text{(ii)}$

Now, squaring and adding of Eqs. (i) and (ii), we get

$$r^2\cos^2\theta + r^2\sin^2\theta = (0)^2 + (1)^2$$

$\Rightarrow \quad r^2(\cos^2\theta + \sin^2\theta) = 1$

$\Rightarrow \quad r^2 = 1 \quad (\because \cos^2\theta + \sin^2\theta = 1)$

$\Rightarrow \quad r = 1$

Dividing Eq. (ii) by Eq. (i), we get

$$\frac{r\sin\theta}{r\cos\theta} = \left|\frac{1}{0}\right|$$

$$\Rightarrow \quad \tan\theta = \infty = \tan\frac{\pi}{2}$$

$$\Rightarrow \quad \theta = \frac{\pi}{2}$$

Since, real and imaginary parts of z are both positive, so the point lies in the first quadrant.

$$\therefore \quad \arg(z) = \theta = \frac{\pi}{2}$$

$$\therefore \quad z = 0 + i = 1\left[\cos\frac{\pi}{2} + i\sin\frac{\pi}{2}\right]$$

which is the required polar form of i.

89. *(d)* If $ax^2 + bx + c = 0$ is the quadratic equation, a, b and $c \in R, a \neq 0$.
Then, discriminant, $D = b^2 - 4ac$
By quadratic formula,

$$x = \frac{-b \pm \sqrt{D}}{2a}$$

$\therefore$ Root of quadratic equation will be positive, if $D \geq 0$ and roots of quadratic equation will be negative, if $D < 0$

$$x = \frac{-b \pm \sqrt{(b^2 - 4ac)}}{2a}$$

$$x = \frac{-b \pm i\sqrt{4ac - b^2}}{2a}$$

90. *(b)* A polynomial equation has atleast one root.

91. *(d)* A polynomial equation of degree 'n' has 'n' roots.

92. *(a)* $x^2 + 2 = 0$

$$x^2 = -2 \text{ i.e., } x = \pm\sqrt{-2} = \pm\sqrt{2}i$$

$$\therefore \quad x = \pm\sqrt{2}i$$

93. *(a)* $x^2 + x + 1 = 0$

$$\because \quad x = \frac{-b \pm \sqrt{(b^2 - 4ac)}}{2a}$$

$$\Rightarrow \quad x = \frac{-1 \pm \sqrt{1-4}}{2} \Rightarrow x = \frac{-1 \pm i\sqrt{3}}{2}$$

$$\therefore \quad x = \frac{-1 + i\sqrt{3}}{2}, \frac{-1 - i\sqrt{3}}{2}$$

94. *(c)* $\sqrt{5}x^2 + x + \sqrt{5} = 0$

$$D = b^2 - 4ac$$

$$= (1)^2 - 4(\sqrt{5})(\sqrt{5}) = 1 - 20$$

$$D = -19$$

$$x = \frac{-1 \pm \sqrt{-19}}{2\sqrt{5}} = \frac{-1 \pm \sqrt{19}i}{2\sqrt{5}}$$

95. *(b)* A. Given, $x^2 + 3 = 0$

$$\therefore \quad x^2 = -3$$

$$\Rightarrow \quad x = \pm\sqrt{-3} = \pm i\sqrt{3}$$

B. Given, $2x^2 + x + 1 = 0$

On comparing the given equation with $ax^2 + bx + c = 0$, we get

$$a = 2, b = 1, c = 1$$

Now, $D = b^2 - 4ac = (1)^2 - 4 \times 2 \times 1 = 1 - 8 = -7 < 0$

$$\Rightarrow \quad x = \frac{-1 \pm \sqrt{-7}}{2 \times 2} = \frac{-1 \pm i\sqrt{7}}{4} \quad (\because \sqrt{-1} = i)$$

C. Given, $x^2 + 3x + 9 = 0$

On comparing the given equation with $ax^2 + bx + c = 0$, we get
$a = 1, b = 3, c = 9$
Now, $D = b^2 - 4ac = (3)^2 - 4 \times 1 \times 9$

$$= 9 - 36 = -27 < 0$$

$$\Rightarrow \quad x = \frac{-3 \pm \sqrt{-27}}{2 \times 1}, x = \frac{-3 \pm i\sqrt{27}}{2} \quad (\because \sqrt{-1} = i)$$

$$= \frac{-3 \pm i\sqrt{9 \times 3}}{2} = \frac{-3 \pm i\,3\sqrt{3}}{2}$$

D. Given, $-x^2 + x - 2 = 0$

On comparing the given equation with $ax^2 + bx + c = 0$, we get
$a = -1, b = 1, c = -2$
Now, $D = b^2 - 4ac = (1)^2 - 4(-1)(-2)$

$$= 1 - 4 \times 1 \times 2$$

$$= 1 - 8 = -7 < 0$$

$$\Rightarrow \quad x = \frac{-1 \pm \sqrt{-7}}{2 \times (-1)} = \frac{-1 \pm i\sqrt{7}}{-2} \quad (\because \sqrt{-1} = i)$$

96. *(c)* I. Given, $x^2 + 3x + 5 = 0$

On comparing the given equation with $ax^2 + bx + c = 0$, we get
$a = 1, b = 3, c = 5$
Now, $D = b^2 - 4ac = (3)^2 - 4 \times 1 \times 5 = 9 - 20 = -11 < 0$

$$\Rightarrow \quad x = \frac{-3 \pm \sqrt{-11}}{2 \times 1}$$

$$\therefore \quad x = \frac{-3 \pm i\sqrt{11}}{2} \quad (\because \sqrt{-1} = i)$$

II. Given, $x^2 - x + 2 = 0$

On comparing the given equation with $ax^2 + bx + c = 0$, we get
$a = 1, b = -1, c = 2$
Now, $D = b^2 - 4ac = (-1)^2 - 4 \times 1 \times 2 = 1 - 8 = -7 < 0$

$$\Rightarrow \quad x = \frac{-(-1) \pm \sqrt{-7}}{2 \times 1}$$

$$= \frac{1 \pm i\sqrt{7}}{2} \quad (\because \sqrt{-1} = i)$$

97. *(a)* A. Given, $\sqrt{2}\,x^2 + x + \sqrt{2} = 0$

On comparing the given equation with $ax^2 + bx + c = 0$, we get

$a = \sqrt{2}, b = 1, c = \sqrt{2}$

$\therefore\ D = b^2 - 4ac = (1)^2 - 4 \times \sqrt{2} \times \sqrt{2}$

$= 1 - 4 \times 2 = 1 - 8 = -7 < 0$

$\Rightarrow \quad x = \dfrac{-1 \pm \sqrt{-7}}{2 \times \sqrt{2}}$

$x = \dfrac{-1 \pm i\sqrt{7}}{2\sqrt{2}} \qquad (\because \sqrt{-1} = i)$

B. Given, $\sqrt{3}\,x^2 - \sqrt{2}\,x + 3\sqrt{3} = 0$

On comparing the given equation with $ax^2 + bx + c = 0$, we get

$a = \sqrt{3}, b = -\sqrt{2}, c = 3\sqrt{3}$

Now, $D = b^2 - 4ac = (-\sqrt{2})^2 - 4 \times (\sqrt{3}) \times 3\sqrt{3}$

$= 2 - 4 \times 3 \times \sqrt{3} \times \sqrt{3} = 2 - 12 \times 3$

$= 2 - 36 = -34 < 0$

$\Rightarrow \quad x = \dfrac{-(-\sqrt{2}) \pm \sqrt{-34}}{2 \times \sqrt{3}}$

$= \dfrac{\sqrt{2} \pm i\sqrt{34}}{2\sqrt{3}} \qquad (\because \sqrt{-1} = i)$

C. Given, $x^2 + x + \dfrac{1}{\sqrt{2}} = 0$

On comparing the given equation with $ax^2 + bx + c = 0$, we get

$a = 1, b = 1, c = \dfrac{1}{\sqrt{2}}$

Now, $D = b^2 - 4ac = (1)^2 - 4 \times 1 \times \dfrac{1}{\sqrt{2}}$

$= 1 - \dfrac{4}{\sqrt{2}} \times \dfrac{\sqrt{2}}{\sqrt{2}} = 1 - \dfrac{4\sqrt{2}}{2} = 1 - 2\sqrt{2} < 0$

$\Rightarrow x = \dfrac{-1 \pm \sqrt{1 - 2\sqrt{2}}}{2 \times 1} = \dfrac{-1 \pm i\sqrt{2\sqrt{2} - 1}}{2} \qquad (\because \sqrt{-1} = i)$

D. Given, $x^2 + \dfrac{x}{\sqrt{2}} + 1 = 0$

Taking the LCM to make the calculation easier,

$\dfrac{x^2}{1} + \dfrac{x}{\sqrt{2}} + \dfrac{1}{1} = 0, \quad \dfrac{\sqrt{2}x^2 + x + \sqrt{2}}{\sqrt{2}} = 0$

$\Rightarrow \quad \sqrt{2}x^2 + x + \sqrt{2} = 0$

On comparing the above equation with $ax^2 + bx + c = 0$, we get

$a = \sqrt{2}, b = 1, c = \sqrt{2}$

Now, $D = b^2 - 4ac = (1)^2 - 4 \times \sqrt{2} \times \sqrt{2}$

$= 1 - 4 \times 2 = 1 - 8 = -7 < 0$

$\Rightarrow \quad x = \dfrac{-1 \pm \sqrt{-7}}{2 \times \sqrt{2}} = \dfrac{-1 \pm i\sqrt{7}}{2\sqrt{2}} \qquad (\because \sqrt{-1} = i)$

98. *(c)* Given that, $\overline{z_1} + i\,\overline{z_2} = 0$

$\Rightarrow \quad z_1 = iz_2$, i.e., $z_2 = -iz_1$

Thus, $\quad \arg(z_1 z_2) = \arg z_1 + \arg(-iz_1) = \pi$

$\because \quad \arg(z_1 z_2) = \arg z_1 + \arg z_2$

$\Rightarrow \quad \arg(-iz_1^2) = \pi$

$\Rightarrow \quad \arg(-i) + 2\arg(z_1) = \pi$

$\Rightarrow \quad \dfrac{-\pi}{2} + 2\arg(z_1) = \pi \Rightarrow \arg(z_1) = \dfrac{3\pi}{4}$

99. *(c)* Let $z_1 = r_1(\cos\theta_1 + i\sin\theta_1)$ and $z_2 = r_2(\cos\theta_2 + i\sin\theta_2)$

where $\quad r_1 = |z_1|, \arg(z_1) = \theta_1, r_2 = |z_2|, \arg(z_2) = \theta_2$

we have, $|z_1 + z_2| = |z_1| + |z_2|$

$= |r_1(\cos\theta_1 + \cos\theta_2) + r_2(\cos\theta_2 + \sin\theta_2)| = r_1 + r_2$

$r_1^2 + r_2^2 + 2r_1r_2\cos(\theta_1 - \theta_2) = (r_1 + r_2)^2$

$\Rightarrow \quad \cos(\theta_1 - \theta_2) = 1$

$\Rightarrow \quad \theta_1 - \theta_2 = 0 \Rightarrow \theta_1 = \theta_2$ i.e; $\arg z_1 = \arg z_2$

100. *(a)* $\dfrac{z_1}{z_2} = \dfrac{\sqrt{3} + i\sqrt{3}}{\sqrt{3} + i}$

$= \dfrac{\sqrt{3} + i\sqrt{3}}{\sqrt{3} + i} \times \dfrac{\sqrt{3} - i}{\sqrt{3} - i} = \dfrac{3 + 3i - \sqrt{3}i + \sqrt{3}}{4}$

$= \left(\dfrac{3 + \sqrt{3}}{4}\right) + \left(\dfrac{3 - \sqrt{3}}{4}\right)i$

which is represented by a point in I quadrant.

101. *(c)* $|z + 1| = 1$

$\Rightarrow \quad |(x + 1) + iy| = 1$

$\Rightarrow \quad (x + 1)^2 + y^2 = 1$

which is a circle with centre $(-1, 0)$ and radius 1.

102. *(c)* Let $z = x + iy$. Then, $iz = -y + ix$. Therefore,

$z + iz = (x - y) + i(x + y)$

Required area of triangle

= Area of triangle formed by $(x, y)(-y, x)$ and $(x - y, x + y)$

$= \dfrac{1}{2}|x(x - x - y) - y(x + y - y) + (x - y)(y - x)|$

$= \dfrac{1}{2}\left|2xy - 2xy - x^2 - y^2\right|$

$= \dfrac{1}{2}(x^2 + y^2) = \dfrac{|z|^2}{2}$

Geometrically

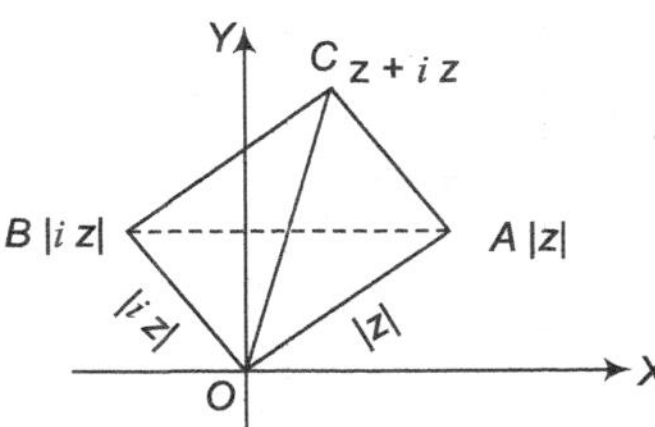

Area of $\Delta ABC = \dfrac{1}{2}$ area of square $OACB$

$= \dfrac{1}{2}|z|^2$

103. *(a)* $|z+1-i| = |z-1+i|$

$\Rightarrow \quad |z-(-1+i)| = |z-(1-i)|$

$\Rightarrow$ $PA = PB$, where A denotes the point $(-1, 1)$, B denotes the point $(1, -1)$ and P denotes the point (x, y).

$\Rightarrow$ z lies on the perpendicular bisector of the line joining A and B and perpendicular bisector is a straight line.

104. *(d)* $z^2 + |z|^2 = 0, z \neq 0$

$\Rightarrow x^2 - y^2 + i2xy + x^2 + y^2 = 0$

$\Rightarrow \quad 2x^2 + i2xy = 0$

$\Rightarrow \quad 2x(x+iy) = 0$

$\Rightarrow \quad x = 0 \quad \text{or} \quad x + iy = 0 \quad$ (not possible)

Therefore, $x = 0, z \neq 0$.

So, y can have any real value. Hence, infinitely many solutions.

105. *(c)* Here, $r\cos\theta = \sin\left(\frac{\pi}{5}\right)$ and $r\sin\theta = 1 - \cos\frac{\pi}{5}$

$\Rightarrow r\cos\theta = 2\sin\frac{\pi}{10}\cos\frac{\pi}{10}$ and $r\sin\theta = 2\sin^2\frac{\pi}{10}$

$\Rightarrow \quad \tan\theta = \tan\left(\frac{\pi}{10}\right) \Rightarrow \theta = \frac{\pi}{10}$

Statements I and II are true.

106. *(a)* A. Let θ be the principle value of amplitude of $1-i$.

Since,

$$\tan\theta = -1$$

$$\Rightarrow \quad \tan\theta = \tan\left(-\frac{\pi}{4}\right)$$

[$\because$ (1, –1) lies in IV quadrant]

$$\Rightarrow \quad \theta = -\frac{\pi}{4}$$

B. $z = (i^{25})^3 = (i)^{75} = i^{4\times18+3} = (i^4)^{18}(i)^3$

$$i^3 = -i = 0 - i$$

Polar form of $z = r(\cos\theta + i\sin\theta)$

$$= 1\left[\cos\left(-\frac{\pi}{2}\right) + i\sin\left(-\frac{\pi}{2}\right)\right]$$

$$= \cos\frac{\pi}{2} - i\sin\frac{\pi}{2}$$

C. Let $z = x + iy$

Then, $z - 2 - 3i = (x-2) + i(y-3)$

Let θ be the amplitude of $(z - 2 - 3i)$

Then, $\quad \tan\theta = \frac{y-3}{x-2}$

$\Rightarrow \quad \tan\frac{\pi}{4} = \frac{y-3}{x-2} \quad \left(\text{since } \theta = \frac{\pi}{4}\right)$

$\Rightarrow \quad 1 = \frac{y-3}{x-2}$

$\Rightarrow \quad x - 2 = y - 3$

$\Rightarrow \quad x - y + 1 = 0$

107. *(a)* Let α, β be the roots of the equation $x^2 - px + 8 = 0$.

Therefore, $\quad \alpha + \beta = p$

and $\quad \alpha \cdot \beta = 8$

Now, $\quad \alpha - \beta = \pm\sqrt{(\alpha+\beta)^2 - 4\alpha\beta}$

Therefore, $\quad 2 = \pm\sqrt{P^2 - 32}$

$\Rightarrow \quad P^2 - 32 = 4$, *i.e.*, $P = \pm 6$

108. *(a)* Let α, β be the roots of the equation.

Therefore, $\alpha + \beta = a - 2$ and $\alpha\beta = -(a+1)$

Now, $\quad \alpha^2 + \beta^2 = (\alpha+\beta)^2 - 2\alpha\beta$

$$= (a-2)^2 + 2(a+1)$$

$$= (a-1)^2 + 5$$

Therefore, $\alpha^2 + \beta^2$ will be minimum, if $(a-1)^2 = 0$, *i.e.*, $a = 1$.

109. *(a)* Let $z = x + iy$

$\therefore |z + 3 - i| = |(x+3) + i(y-1)| = 1$

$\Rightarrow \quad \sqrt{(x+3)^2 + (y-1)^2} = 1 \quad \ldots$(i)

$\because \quad \arg z = \pi$

$\Rightarrow \quad \tan^{-1}\frac{y}{x} = \pi$

$\Rightarrow \quad \frac{y}{x} = \tan\pi = 0$

$\Rightarrow \quad y = 0 \quad \ldots$(ii)

From Eqs. (i) and (ii), we get

$$x = -3, y = 0$$

$\therefore \quad z = -3$

$\Rightarrow \quad |z| = |-3| = 3$

110. *(a)* $|z-1| = |z+1| = |z-i|$

Point which is equidistant from A, B and C is (0, 0).

$\therefore$ $z = 0$ is only complex number which satisfies the given relation.

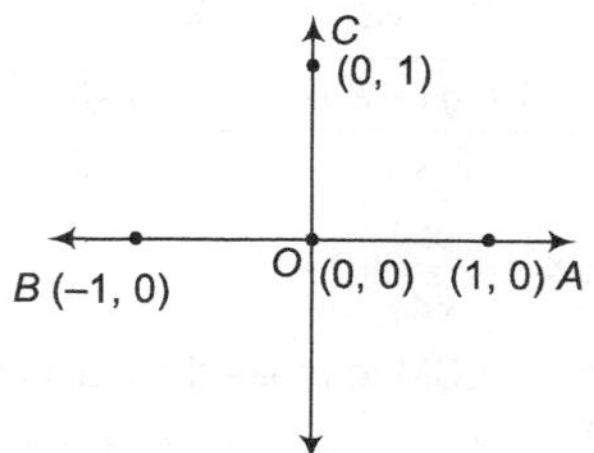

111. *(c)* If $z_1 = \sqrt{2}\left(\cos\frac{\pi}{4} + i\sin\frac{\pi}{4}\right)$

$$z_1 = \sqrt{2}\left(\frac{1}{\sqrt{2}} + i\frac{1}{\sqrt{2}}\right)$$

$\Rightarrow \quad z_1 = (1+i)$

$\Rightarrow \quad |z_1| = \sqrt{2}$

$\Rightarrow \quad z_2 = \sqrt{3}\left(\cos\frac{\pi}{3} + i\sin\frac{\pi}{3}\right)$

$\Rightarrow \quad = \sqrt{3}\left(\frac{1}{2} + i\frac{\sqrt{3}}{2}\right)$

$\because \quad z_2 = \left(\frac{\sqrt{3}}{2} + i\frac{3}{2}\right) \Rightarrow |z_2| = \sqrt{\frac{3}{4} + \frac{9}{4}} = \sqrt{3}$

$\Rightarrow \quad |z_1 z_2| = |z_1||z_2|$

$\Rightarrow \quad |z_1 z_2| = \sqrt{2} \times \sqrt{3} = \sqrt{6}$

Alternate Method

$$z = r(\cos\theta + i\sin\theta) = |z|(\cos\theta + i\sin\theta)$$

Here, $z_1 = \sqrt{2}\left(\cos\frac{\pi}{4} + i\sin\frac{\pi}{4}\right);$

$$z_2 = \sqrt{3}\left(\cos\frac{\pi}{3} + i\sin\frac{\pi}{3}\right)$$

$\therefore \quad |z_1| = \sqrt{2}, |z_2| = \sqrt{3}$

$\because \quad |z_1 z_2| = |z_1||z_2|$

$$|z_1 z_2| = \sqrt{2} \times \sqrt{3} = \sqrt{6}$$

112. *(b)* We have, $\frac{(3-2i)(2+3i)}{(1+2i)(2-i)}$

$$= \frac{6+9i-4i+6}{2-i+4i+2} = \frac{12+5i}{4+3i} \times \frac{4-3i}{4-3i}$$

$$= \frac{48-36i+20i+15}{16+9} = \frac{63-16i}{25} = \frac{63}{25} - \frac{16}{25}i$$

Therefore, conjugate of $\frac{(3-2i)(2+3i)}{(1+2i)(2+i)}$ is $\frac{63}{25} + \frac{16}{25}i$

113. *(a)* I. We have, $\frac{1+i}{1-i} = \frac{1+i}{1-i} \times \frac{1+i}{1+i} = \frac{1-1+2i}{1+1} = i = 0 + i$

Now, let us put $0 = r\cos\theta, 1 = r\sin\theta$

Squaring and adding, $r^2 = 1$ *i.e.*, $r = 1$

So, $\cos\theta = 0$, $\sin\theta = 1$

Therefore, $\theta = \frac{\pi}{2}$

Hence, the modulus of $\frac{1+i}{1-i}$ is 1 and the argument is $\frac{\pi}{2}$.

II. We have, $\frac{1}{1+i} = \frac{1-i}{(1+i)(1-i)} = \frac{1-i}{1+1} = \frac{1}{2} - \frac{i}{2}$

Let $\frac{1}{2} = r\cos\theta, -\frac{1}{2} = r\sin\theta$

Proceeding as

$$r = \frac{1}{\sqrt{2}}; \cos\theta = \frac{1}{\sqrt{2}}, \sin\theta = \frac{-1}{\sqrt{2}}$$

Therefore, $\theta = \frac{-\pi}{4}$

[$\because \cos\theta > 0$ and $\sin\theta < 0$ is in IV quadrant]

Hence, the modulus of $\frac{1}{1+i}$ is $\frac{1}{\sqrt{2}}$ and the argument is $-\frac{\pi}{4}$

114. *(a)* We have, $x + iy = \frac{(a+ib)(a+ib)}{(a-ib)(a+ib)} = \frac{a^2-b^2+2abi}{a^2+b^2}$

$$= \frac{a^2-b^2}{a^2+b^2} + \frac{2ab}{a^2+b^2}i$$

So, $x - iy = \frac{a^2-b^2}{a^2+b^2} - \frac{2ab}{a^2+b^2}i$

$$x^2 + y^2 = (x+iy)(x-iy)$$

$$= \frac{(a^2-b^2)^2}{(a^2+b^2)^2} + \frac{4a^2b^2}{(a^2+b^2)^2}$$

$$= \frac{(a^2+b^2)^2}{(a^2+b^2)^2} = 1$$

115. *(c)* We have, $Z = \frac{3+2i\sin\theta}{1-2i\sin\theta} = \frac{(3+2i\sin\theta)(1+2i\sin\theta)}{(1-2i\sin\theta)(1+2i\sin\theta)}$

$$= \frac{3+6i\sin\theta+2i\sin\theta-4\sin^2\theta}{1+4\sin^2\theta} = \frac{3-4\sin^2\theta}{1+4\sin^2\theta} + \frac{8i\sin\theta}{1+4\sin^2\theta}$$

We are given the complex number to be real. Therefore,

$$\frac{8\sin\theta}{1+4\sin^2\theta} = 0, \text{ i.e., } \sin\theta = 0$$

Thus, $\theta = n\pi, n \in Z$.

116. *(b)* Given that, $Z = \frac{1-i}{\frac{1}{2} + \frac{\sqrt{3}}{2}i}$

$$= \frac{2(i-1)}{1+i\sqrt{(3)}} \times \frac{1-i\sqrt{3}}{1-i\sqrt{3}}$$

$$= \frac{2(i+\sqrt{3}-1+i\sqrt{3})}{1+3}$$

$$= \frac{\sqrt{3}-1}{2} + \frac{\sqrt{3}+1}{2}i$$

Now, put $\frac{\sqrt{3}-1}{2} = r\cos\theta, \frac{\sqrt{3}+1}{2} = r\sin\theta$

Squaring and adding, we obtain

$$r^2 = \left(\frac{\sqrt{3}-1}{2}\right)^2 + \left(\frac{\sqrt{3}+1}{2}\right)^2$$

$$= \frac{2(\sqrt{3})^2+1)}{4} = \frac{2 \times 4}{4} = 2$$

Hence, $r = \sqrt{2}$ which gives $\cos\theta = \frac{\sqrt{3}-1}{2\sqrt{2}}$, $\sin\theta = \frac{\sqrt{3}+1}{2\sqrt{2}}$

Therefore, $\theta = \frac{\pi}{4} + \frac{\pi}{6} = \frac{5\pi}{12}$

Hence, the polar form is $\sqrt{2}\left(\cos\frac{5\pi}{12} + i\sin\frac{5\pi}{12}\right)$.

117. *(b)* $\left[i^{18} + \left(\frac{1}{i}\right)^{25}\right]^3 = \left[(i^2)^9 + \frac{1}{i^{25}}\right]^3$

$$= \left[(i^2)^9 + \frac{1}{i \cdot i^{24}}\right]^3$$

$$= \left[(-1)^9 + \frac{1}{i(i^4)^6}\right]^3 \quad (\because i^2 = -1)$$

$$= \left[-1 + \frac{1}{i}\right]^3 \quad (\because i^4 = 1)$$

$$= \left[-1+\frac{1}{i}\times\frac{i}{i}\right]^3 = \left[-1+\frac{i}{i^2}\right]^3$$

$$= [-1-i]^3 \qquad (\because i^2=-1)$$

$$= (-1)^3\,[1+i]^3$$

$$= -[(1)^3 + i^3 + 3\times 1\times i\,(1+i)]$$

$$[\because (a+b)^3 = a^3+b^3+3ab\,(a+b)]$$

$$= -[1-i+3i+3i^2] \qquad (\because i^3=-i)$$

$$= -[1-i+3i-3] \qquad (\because i^2=-1)$$

$$= -[-2+2i] = 2-2i$$

Note *Students are suggested not to open the cube of expression at initial stage because the expression becomes complex by doing so.*

118. *(b)* I. Let $z_1 = a+ib$

$\Rightarrow$ $\text{Re}(z_1)=a$ and $\text{Im}(z_1)=b$ ...(i)

and $z_2 = c+id$ ...(ii)

$\Rightarrow$ $\text{Re}(z_2)=c$ and $\text{Im}(z_2)=d$

Now, $z_1z_2 = (a+ib)(c+id)$

$$= ac + iad + ibc + i^2bd$$

$$= ac + i\,(ad+bc) - bd \qquad (\because i^2=-1)$$

$\Rightarrow$ $z_1z_2 = (ac-bd) + i\,(ad+bc)$

$\Rightarrow$ $\text{Re}(z_1z_2) = ac - bd$

$\Rightarrow$ $\text{Re}(z_1\,z_2) = \text{Re}(z_1)\,\text{Re}(z_2) - \text{Im}(z_1)\,\text{Im}(z_2)$

[using Eqs. (i) and (ii)]

Here, "Re" stands for real part and "Im" stands for imaginary part.

II. We know that, when a complex number is multiplied by i it rotates 90° anti-clockwise. Hence, angle between z and iz is $\frac{\pi}{2}$.

119. *(a)* $\left(\frac{1}{1-4i}-\frac{2}{1+i}\right)\left(\frac{3-4i}{5+i}\right) = \frac{[1+i-2\,(1-4i)]}{(1-4i)(1+i)}\times\left(\frac{3-4i}{5+i}\right)$

$$= \frac{1+i-2+8i}{1+i-4i-4i^2}\times\frac{3-4i}{5+i}$$

$$= \frac{(1-2)+i\,(1+8)}{1+i\,(1-4)-4(-1)}\times\frac{3-4i}{5+i} \qquad (\because i^2=-1)$$

$$= \frac{-1+9i}{5-3i}\times\frac{3-4i}{5+i}$$

$$= \frac{-3+4i+27i-36i^2}{25+5i-15i-3i^2}$$

$$= \frac{-3+i\,(4+27)-36(-1)}{25+(5-15)\,i-3(-1)} \qquad (\because i^2=-1)$$

$$= \frac{-3+31i+36}{25-10i+3} = \frac{33+31i}{28-10i}\times\frac{28+10i}{28+10i}$$

$$= \frac{924+330i+868i+310i^2}{784-100i^2}$$

$$[\because (a+b)(a-b) = a^2-b^2]$$

$$= \frac{924+(330+868)i-310}{784+100} \qquad (\because i^2=-1)$$

$$= \frac{614+1198i}{884} = \frac{307+599i}{442}$$

120. *(a)* Given, $x-iy = \sqrt{\frac{a-ib}{c-id}}$

$$\Rightarrow \quad x+i\,(-y) = \left(\frac{a-ib}{c-id}\right)^{1/2}$$

Taking modulus both sides, we get

$$|x+i\,(-y)| = \left|\left(\frac{a-ib}{c-id}\right)^{\frac{1}{2}}\right|$$

$$\Rightarrow \quad \sqrt{x^2+(-y)^2} = \left|\frac{a-ib}{c-id}\right|^{\frac{1}{2}} \qquad \left(\because |x+iy| = \sqrt{x^2+y^2} \text{ and } |z^n| = |z|^n\right)$$

$$\Rightarrow \quad \sqrt{x^2+y^2} = \left|\frac{a-ib}{c-id}\right|^{\frac{1}{2}}$$

Squaring both sides, we get

$$x^2+y^2 = \left|\frac{a-ib}{c-id}\right|$$

$$\Rightarrow \quad x^2+y^2 = \frac{|a-ib|}{|c-id|} \qquad \left(\because \left|\frac{z_1}{z_2}\right| = \frac{|z_1|}{|z_2|}\right)$$

$$\Rightarrow \quad (x^2+y^2) = \frac{\sqrt{a^2+b^2}}{\sqrt{c^2+d^2}} \qquad (\because |x-iy| = \sqrt{x^2+y^2})$$

Squaring both sides, we get

$$(x^2+y^2)^2 = \frac{a^2+b^2}{c^2+d^2}$$

121. *(c)* First we will convert the given expressions into $a+ib$ form and then reduce them into polar form

I. Let $z = \frac{1+7i}{(2-i)^2} = \frac{1+7i}{4+i^2-4i}$

$$[\text{use } (a-b)^2 = a^2+b^2-2ab]$$

$$= \frac{1+7i}{4-1-4i} \qquad (\because i^2=-1)$$

$$= \frac{1+7i}{3-4i}\times\frac{3+4i}{3+4i}$$

$$= \frac{3+4i+21i+28i^2}{(3)^2-(4i)^2} \qquad [\because (a-b)(a+b) = a^2-b^2]$$

$$= \frac{3+i\,(4+21)-28}{9-16i^2} = \frac{-25+25i}{9+16} \qquad (\because i^2=-1)$$

$$\Rightarrow z = \frac{-25+25i}{25} = -1+i$$

Now, let $-1+i = r\cos\theta + ir\sin\theta$

On comparing the real and imaginary parts of both sides, we get

$$r\cos\theta = -1 \qquad \text{...(i)}$$

and $$r\sin\theta = 1 \qquad \text{...(ii)}$$

Squaring and adding Eqs. (i) and (ii), we get

$$r^2\cos^2\theta + r^2\sin^2\theta = (-1)^2 + (1)^2$$

$$r^2(\cos^2\theta+\sin^2\theta) = 1+1$$

$$\Rightarrow \quad r^2 = 2 \qquad (\because \cos^2\theta+\sin^2\theta = 1)$$

$\Rightarrow \quad r = \sqrt{2}$

On dividing Eq. (ii) by Eq. (i), we get

$\frac{r\sin\theta}{r\cos\theta} = \left|\frac{1}{-1}\right| \Rightarrow \tan\theta = 1 = \tan\frac{\pi}{4}$

$\Rightarrow \quad \theta = \frac{\pi}{4}$

Since, the real part of z is negative and the imaginary part of z is positive. So, the point lies in II quadrant.

$\therefore \quad \arg(z) = \pi - \theta = \pi - \frac{\pi}{4} = \frac{3\pi}{4}$

$\therefore \quad z = -1 + i = \sqrt{2}\left[\cos\frac{3\pi}{4} + i\sin\frac{3\pi}{4}\right]$

which is the required polar form of $\frac{1+7i}{(2-i)^2}$.

II. Let $z = \frac{1+3i}{1-2i} \times \frac{1+2i}{1+2i} = \frac{1+2i+3i+3i\times 2i}{1^2-(2i)^2}$

$[\because (a+b)(a-b) = a^2 - b^2]$

$= \frac{1+5i+6i^2}{1-4i^2} \quad (\because i^2 = -1)$

$= \frac{1+5i-6}{1+4} = \frac{-5+5i}{5} = -1+i,$

122. *(a)* If $ax^2 + bx + c = 0$ with real coefficients a, b, c and $a \neq 0$, then use

$x = \frac{-b \pm \sqrt{b^2-4ac}}{2a}$, if $b^2 - 4ac > 0$

$x = \frac{-b \pm \sqrt{4ac-b^2}\, i}{2a}$, if $b^2 - 4ac < 0$

A. Given, $3x^2 - 4x + \frac{20}{3} = 0$

On comparing the given equation with $ax^2 + bx + c = 0$, we get

$a = 3, b = -4, c = \frac{20}{3}$

$\therefore \quad D = b^2 - 4ac = (-4)^2 - 4\times 3\times\frac{20}{3}$

$= 16 - 80 = -64 < 0$

$\therefore \quad x = \frac{-(-4) \pm \sqrt{-64}}{2\times 3} = \frac{4 \pm 8i}{2\times 3} = \frac{2(2\pm 4i)}{2\times 3}$

$= \frac{2\pm 4i}{3} = \frac{2}{3} \pm \frac{4}{3}i$

B. Given, $x^2 - 2x + \frac{3}{2} = 0$

On comparing the given equation with $ax^2 + bx + c = 0$, we get

$a = 1, b = -2, c = \frac{3}{2}$

$\therefore \quad D = b^2 - 4ac = (-2)^2 - 4\times 1\times\frac{3}{2}$

$= 4 - 6 = -2 < 0$

$\Rightarrow \quad x = \frac{-(-2) \pm \sqrt{-2}}{2\times 1} = \frac{2\pm\sqrt{2}i}{2}$

$\therefore \quad = \frac{2}{2} \pm i\frac{\sqrt{2}}{2}$

$= 1 \pm i\frac{\sqrt{2}}{2} = 1 \pm \frac{i}{\sqrt{2}}$

C. Given, $27x^2 - 10x + 1 = 0$

On comparing the given equation with $ax^2 + bx + c = 0$, we get

$a = 27, b = -10, c = 1$

$\therefore \quad D = b^2 - 4ac = (-10)^2 - 4\times 27\times 1$

$= 100 - 108 = -8 < 0$

$\Rightarrow \quad x = \frac{-(-10) \pm \sqrt{-8}}{2\times 27} = \frac{10 \pm 2\sqrt{2}\,i}{2\times 27}$

$= \frac{2(5 \pm i\sqrt{2})}{2\times 27} = \frac{5}{27} \pm \frac{i\sqrt{2}}{27}$

D. Given, $21x^2 - 28x + 10 = 0$

On comparing the given equation with $ax^2 + bx + c = 0$, we get

$a = 21, b = -28, c = 10$

$\therefore \quad D = b^2 - 4ac = (-28)^2 - 4\times 21\times 10$

$= 784 - 840 = -56 < 0$

$\Rightarrow \quad = \frac{-(-28) \pm \sqrt{-56}}{2\times 21} = \frac{28 \pm \sqrt{14\times 4}\,i}{2\times 21}$

$= \frac{28}{2\times 21} \pm \frac{2\sqrt{14}\,i}{2\times 21} = \frac{2}{3} \pm \frac{\sqrt{14}}{21}i$

123. *(c)* $\left|\frac{z_1+z_2+1}{z_1-z_2+1}\right| = \left|\frac{2-i+1+i+1}{2-i-(1+i)+1}\right|$

$(\because z_1 = 2-i \text{ and } z_2 = 1+i)$

$= \left|\frac{4}{2-i-1-i+1}\right| = \left|\frac{4}{2-2i}\right| = \left|\frac{2}{1-i}\right| = \frac{2}{|1-i|}$

$\left(\because \left|\frac{z_1}{z_2}\right| = \frac{|z_1|}{|z_2|}\right)$

$= \frac{2}{\sqrt{(1)^2+(-1)^2}} \quad (\because |z| = \sqrt{a^2+b^2})$

$= \frac{2}{\sqrt{2}} = \sqrt{2}$

124. *(a)* Given, $a + ib = \frac{(x+i)^2}{2x^2+1}$

Taking modulus of both sides, we get

$|a+ib| = \left|\frac{(x+i)^2}{2x^2+1}\right|$

$$\Rightarrow \sqrt{a^2+b^2} = \frac{|(x+i)^2|}{|2x^2+1|} = \frac{|x+i|^2}{2x^2+1}$$

$$\left[\because \left|\frac{z_1}{z_2}\right| = \frac{|z_1|}{|z_2|}, |z^n| = |z|^n \text{ and } \mathrm{Re}|(z)| = \mathrm{Re}(z)\right]$$

$$\Rightarrow \sqrt{a^2+b^2} = \frac{\left(\sqrt{x^2+1}\right)^2}{2x^2+1}$$

Squaring on both sides, we get

$$a^2+b^2 = \frac{(x^2+1)^2}{(2x^2+1)^2}$$

125. *(c)* I. $\frac{z_1z_2}{\bar{z}_1} = \frac{(2-i)(-2+i)}{\overline{(2-i)}} = \frac{-(2-i)(2-i)}{(2+i)}$

$$\left(\begin{matrix}\because z_1 = 2-i, z_2 = -2+i \\ \text{and} \\ \because z = a+ib \Rightarrow \bar{z} = a-ib\end{matrix}\right)$$

$$= \frac{-(2-i)^2}{2+i} = \frac{-(4+i^2-4i)}{2+i} = \frac{-(4-1-4i)}{2+i} \quad (\because i^2=-1)$$

$$= \frac{-(3-4i)}{2+i} \times \frac{2-i}{2-i}$$

$$= \frac{-(6-3i-8i+4i^2)}{(2)^2-(i)^2} \quad [\because (a+b)(a-b) = a^2-b^2]$$

$$= \frac{-(6-11i-4)}{4-i^2} \quad (\because i^2=-1)$$

$$= \frac{-(2-11i)}{4+1} \quad (\because i^2=-1)$$

$$= -\left(\frac{2}{5} - \frac{11i}{5}\right) = -\frac{2}{5} + \frac{11}{5}i$$

$$\therefore \mathrm{Re}\left(\frac{z_1z_2}{\bar{z}_1}\right) = -\frac{2}{5}$$

II. $\mathrm{Im}\left(\frac{1}{z_1\bar{z}_1}\right) \quad \because \quad z\bar{z} = |z|^2$

$$\Rightarrow z_1\bar{z}_1 = |z_1|^2 \Rightarrow z_1\bar{z}_1 = |-2+i|^2$$

$$= (\sqrt{4+1})^2 = 5+0i$$

$$\therefore \frac{1}{z_1\bar{z}_1} = \frac{1}{5} + 0i$$

$$\Rightarrow \mathrm{Im}\left(\frac{1}{z_1\bar{z}_1}\right) = 0$$

126. *(a)* Let $z = \frac{1+2i}{1-3i}$

$$\therefore z = \frac{1+2i}{1-3i} \times \frac{1+3i}{1+3i} = \frac{1+3i+2i+6i^2}{1^2-(3i)^2}$$

$$[\because (a+b)(a-b) = a^2-b^2]$$

$$= \frac{1+5i+6(-1)}{1-9i^2} \quad (\because i^2=-1)$$

$$= \frac{1+5i-6}{1+9} = \frac{-5+5i}{10} = \frac{-1+i}{2}$$

$$\Rightarrow z = -\frac{1}{2} + \frac{1}{2}i$$

$$\therefore |z| = \sqrt{\left(-\frac{1}{2}\right)^2 + \left(\frac{1}{2}\right)^2} \quad (\because |a+ib| = \sqrt{a^2+b^2})$$

$$= \sqrt{\frac{1}{4}+\frac{1}{4}} = \sqrt{\frac{2}{4}} = \sqrt{\frac{1}{2}} = \frac{1}{\sqrt{2}}$$

$$\therefore \text{Now,} \quad \tan\theta = \left|\frac{\frac{1}{2}}{-\frac{1}{2}}\right| \quad \left(\because \theta = \tan^{-1}\left|\left(\frac{\mathrm{Im}(z)}{\mathrm{Re}(z)}\right)\right|\right)$$

$$\Rightarrow \tan\theta = 1 = \tan\frac{\pi}{4} \Rightarrow \theta = \frac{\pi}{4}$$

Since, the real part of z is negative and imaginary part of z is positive, so the point lies in IInd quadrant.

$$\therefore \arg(z) = \pi - \theta = \pi - \frac{\pi}{4} = \frac{3\pi}{4}$$

Hence, modulus $= \frac{1}{\sqrt{2}}$ and $\arg(z) = \frac{3\pi}{4}$

127. *(a)* $(x-iy)(3+5i) = 3x+5xi-3yi-5yi^2$

$$= 3x+(5x-3y)i+5y \quad (\because i^2=-1)$$

$$= (3x+5y)+(5x-3y)i \quad \ldots(\text{i})$$

Given, $(x-iy)(3+5i) = -6+24i$

[using Eq. (i) and $z = (a+ib) \Rightarrow \bar{z} = (a-ib)$]

On comparing the real and imaginary parts of both sides, we get

$$3x+5y = -6 \quad \text{and} \quad 5x-3y = 24$$

Solving the above equations by substitution or elimination method, we get

$$x = 3, y = -3$$

128. *(a)* Let $z = \frac{1+i}{1-i} - \frac{1-i}{1+i} = \frac{(1+i)^2-(1-i)^2}{(1-i)(1+i)}$

$$= \frac{(1+i^2+2i)-(1+i^2-2i)}{1-i^2} = \frac{4i}{2} = 2i = 0+2i$$

$$\left[\begin{matrix}\because (a+b)^2 = a^2+b^2+2ab \\ \text{and } (a-b)^2 = a^2+b^2-2ab\end{matrix}\right]$$

$$\therefore |z| = \sqrt{0^2+2^2} = 2$$

129. *(d)* $(x+iy)^3 = u+iv$

$$\Rightarrow x^3+(iy)^3+3xiy(x+iy) = u+iv$$

$$\Rightarrow x^3+i^3y^3+3ix^2y+3xy^2i^2 = u+iv$$

$$\Rightarrow x^3-iy^3+3ix^2y-3xy^2 = u+iv$$

$$(\because i^3 = -i \text{ and } i^2 = -1)$$

$$\Rightarrow (x^3-3xy^2)+i(3x^2y-y^3) = u+iv$$

On comparing real and imaginary parts, we get

$$x^3-3xy^2 = u \quad \text{and} \quad 3x^2y-y^3 = v$$

$$\therefore \frac{u}{x}+\frac{v}{y} = \frac{x^3-3xy^2}{x} + \frac{3x^2y-y^3}{y}$$

$$= x^2-3y^2+3x^2-y^2$$

$$= 4x^2-4y^2 = 4(x^2-y^2)$$

130. *(a)* $\left|\dfrac{\beta-\alpha}{1-\bar{\alpha}\beta}\right| = \left|\dfrac{(\beta-\alpha)\bar{\beta}}{(1-\bar{\alpha}\beta)\bar{\beta}}\right| = \left|\dfrac{(\beta-\alpha)\bar{\beta}}{\bar{\beta}-\bar{\alpha}\,\beta\bar{\beta}}\right|$

(multiply numerator and denominator by $\bar{\beta}$)

$= \left|\dfrac{(\beta-\alpha)\bar{\beta}}{\bar{\beta}-\bar{\alpha}}\right|$ $\quad (\because \beta\bar{\beta} = |\beta|^2 = |1|^2 = 1)$

$= \dfrac{|(\beta-\alpha)\bar{\beta}|}{|\bar{\beta}-\bar{\alpha}|}$ $\quad \left(\because \left|\dfrac{z_1}{z_2}\right| = \dfrac{|z_1|}{|z_2|}\right)$

$= \dfrac{|\beta-\alpha||\bar{\beta}|}{|\beta-\alpha|}$ $\quad (\because |z_1z_2| = |z_1||z_2|$ and $|\bar{z}-\bar{z}_2| = |\overline{z_1-z_2}|)$

$= \dfrac{|\beta-\alpha||\beta|}{|\beta-\alpha|}$ $\quad (\because |\bar{z}| = |z|)$

$= |\beta| = 1$

131. *(a)* $|1-i|^x = 2^x$

$\Rightarrow \left(\sqrt{(1)^2+(-1)^2}\right)^x = 2^x \quad (\because |a+ib| = \sqrt{a^2+b^2})$

$\Rightarrow (\sqrt{1+1})^x = 2^x$

$\Rightarrow (\sqrt{2})^x = 2^x$

$\Rightarrow 2^{\frac{x}{2}} = 2^x$

On comparing the power of 2 on both sides, we get

$\Rightarrow \dfrac{x}{2} = x \Rightarrow x = 2x \Rightarrow 2x - x = 0$

$x = 0$

But we need the non-zero solutions.

Hence, number of solutions is zero.

Note *Here, $x=0$ represents the solution. Students understand the difference between the word solution and number of solutions.*

132. *(c)* $(a+ib)(c+id)(e+if)(g+ih) = A+iB$

Taking modulus on both sides, we get

$|(a+ib)(c+id)(e+if)(g+ih)| = |A+iB|$

$\Rightarrow |a+ib||c+id||e+if||g+ih| = |A+iB|$

$(\because |z_1z_2 \dots z_n| = |z_1||z_2||z_3|\dots|z_n|)$

$\Rightarrow \sqrt{a^2+b^2}\sqrt{c^2+d^2}\sqrt{e^2+f^2}\sqrt{g^2+h^2} = \sqrt{A^2+B^2}$

$(\because$ If $z = a+ib$, then $|z| = \sqrt{a^2+b^2})$

Squaring on both sides, we get

$(a^2+b^2)(c^2+d^2)(e^2+f^2)(g^2+h^2) = A^2+B^2$

133. *(a)* $\left(\dfrac{1+i}{1-i}\right)^m = 1$

$\Rightarrow \left[\dfrac{1+i}{1-i}\times\dfrac{1+i}{1+i}\right]^m = 1$

$\Rightarrow \left[\dfrac{(1+i)^2}{1-i^2}\right]^m = 1$

$\Rightarrow \left[\dfrac{1+i^2+2i}{1+1}\right]^m = 1 \quad \left[\because (a+b)^2 = a^2+b^2+2ab \text{ and } i^2 = -1\right]$

$\Rightarrow \left[\dfrac{1-1+2i}{2}\right]^m = 1$

$\Rightarrow i^m = 1$

$\Rightarrow (\sqrt{-1})^m = 1 \quad (\because i^2 = -1 \Rightarrow i = \sqrt{-1})$

$\Rightarrow (-1)^{\frac{m}{2}} = (-1)^2$

On comparing the power on both sides, we get

$\dfrac{m}{2} = 2$

$\Rightarrow m = 4$

Hence, the least positive integral value of m is 4.

Note *1 can be written as $(-1)^2, (-1)^4, (-1)^6$ .... etc. But for least value we take $(-1)^2$.*

134. *(c)* $\because |z_1| = |z_2| = |z_3| = \dots = |z_n| = 1$

If $|z_1| = 1$

$\Rightarrow |z_1|^2 = 1$

$\Rightarrow z_1\bar{z}_1 = 1 \quad (\because z\bar{z} = |z|^2)$

$\Rightarrow z_1 = \dfrac{1}{\bar{z}_1}$

Similarly, $z_2 = \dfrac{1}{\bar{z}_2}, z_3 = \dfrac{1}{\bar{z}_3} = \dots = z_n = \dfrac{1}{\bar{z}_n}$

Now, $|z_1+z_2+z_3+\dots+z_n|$

$= \left|\dfrac{1}{\bar{z}_1}+\dfrac{1}{\bar{z}_2}+\dfrac{1}{\bar{z}_3}+\dots+\dfrac{1}{\bar{z}_n}\right|$

$= \left|\overline{\dfrac{1}{z_1}+\dfrac{1}{z_2}+\dfrac{1}{z_3}+\dots+\dfrac{1}{z_n}}\right| \quad (\because \bar{z}_1+\bar{z}_2 = \overline{z_1+z_2})$

$= \left|\dfrac{1}{z_1}+\dfrac{1}{z_2}+\dfrac{1}{z_3}+\dots+\dfrac{1}{z_n}\right| \quad (\because |z| = |\bar{z}|)$

6

Linear Inequalities

Inequalities

Level I

1. Consider the following statements

Statement I It is always possible to translate a statement problem in the form of an equation.

Statement II A statement involving a sign '$\neq$' is called inequality.

Choose the correct option.

(a) Statement I is true
(b) Statement II is true
(c) Both are true
(d) Both are false

2. Ravi goes to market with ₹ 200 to buy rice, which is available in packets of 1 kg. The price of one packet of rice is ₹ 30. If x denotes the number of packets of rice which he buys, then the total amount spent by him is ₹ $30x$. The mathematical formulation of the given problem is

(a) $30x > 200$ (b) $30x < 200$
(c) $30x \le 200$ (d) $30x \ge 200$

3. Reshma has ₹ 120 and wants to buy some registers and pens. The cost of one register is ₹ 40 and that of a pen is ₹ 20. The mathematical formulation of the given problem is

(a) $40x + 20y = 120$
(b) $40x + 20y > 120$
(c) $40x + 20y \le 120$
(d) $40x + 20y \ge 120$

4. Two real numbers or two algebraic expressions related by the symbol '<', '>', '≤' or '≥' forms an ...X... . Here, X refers to

(a) equation (b) inequality
(c) set (d) None of these

5. Which of the following is/are the examples of numerical inequalities?

I. $3 < 5$ II. $7 > 5$
III. $x < 5$ IV. $y > 2$
V. $x \ge 3$

(a) I, II (b) III, IV
(c) IV (d) All of these

6. Which of the following is/are double inequalities?

(a) $3 < 5 < 7$ (b) $3 \le x < 5$
(c) $2 < y \le 4$ (d) All of these

7. **Assertion** (A) The inequality $ax + by < 0$ is strict inequality.

Reason (R) The inequality $ax + b \ge 0$ is slack inequality.

(a) Both A and R are correct; R is the correct explanation of A
(b) Both A and R are correct; R is not the correct explanation of A
(c) A is correct; R is incorrect
(d) R is correct; A is incorrect

8. Match the inequalities in Column I with their forms in Column II and choose the correct option from the codes given below.

Column I	Column II
A. $ax + b \geq 0, a \neq 0$	1. Quadratic inequality
B. $ax + b \leq 0, a \neq 0$	2. Linear inequality in two variables
C. $ax^2 + bx + c \leq 0, a \neq 0$	3. Linear inequality in one variable
D. $ax + by \geq c, a \neq 0, b \neq 0$	
E. $ax + by \leq c, a \neq 0, b \neq 0$	

Codes

	A	B	C	D	E
(a)	2	3	1	3	2
(b)	2	2	1	3	3
(c)	3	3	1	2	2
(d)	3	3	1	2	3

Level II

9. Consider the following statements

Statement I The inequality $3x + 2y \geq 5$ is the linear inequality.

Statement II The inequality $2x^2 - 3x \geq 8$ is the linear inequality in one variable.

Choose the correct option.

(a) Statement I is true (b) Statement II is true
(c) Both are true (d) Both are false

10. A shopkeeper sell two types of computer– a desktop model and portable model. He knows that total monthly demand will not exceed 250 units. If x denotes the number of computers of type desktop model and y denotes the number of computers of type portable model, then

(a) $x + y > 250$ (b) $x + y < 250$
(c) $x + y \leq 250$ (d) $x + y \geq 250$

11. Match the inequalities in Column I with their forms in Column II and choose the correct option from the codes given below.

Column I	Column II
A. $3x + 5y > 2$	1. Linear inequality in one variable
B. $x < y$	2. Linear inequality in two variables
C. $3x^2 + 2 \leq 0$	3. Quadratic inequality
D. $5x - 7 > 0$	

Codes

	A	B	C	D		A	B	C	D
(a)	2	2	3	1	(b)	2	1	1	3
(c)	2	1	3	1	(d)	1	2	3	1

Algebraic Solutions of Linear Inequalities in One Variable and Their Graphical Representation

Level I

12. Consider the following statements

Statement I Consider the inequality $30x < 200$ such that x is not a negative integer or fraction. Then, the value of x, which make the inequality a true statement, are 1, 2, 3, 4, 5, 6.

Statement II The solution of an inequality in one variable is the value of that variable which makes it a true statement.

Choose the correct option.

(a) Statement I is true (b) Statement II is true
(c) Both are true (d) Both are false

13. Which of the following is/are true?

I. Solutions of an inequality are found by only trial and error method.

II. Equal numbers may be added to both sides of an equation.

III. Both sides of an equation may be multiplied (or divided) by the same non-zero number.

(a) I and III are true (b) II and III are true
(c) All are true (d) None of these

Directions (Q. Nos. 14-15) *If we multiply (or divide) both sides of an inequality by a negative number, then the sign of inequality changes.*

14. If $-3 < -2$, then 3...A...2. Here, A stands for

(a) > (b) < (c) = (d) ≥

15. If $(-8)(-2) > (-7)(-2)$, then

(a) $-8 > -7$ (b) $-8 < -7$
(c) $-8 \geq -7$ (d) $-8 \leq -7$

16. The rules for solving an inequality is/are

(a) different numbers may be added to both sides of an inequality without affecting the sign of inequality
(b) both sides of an inequality can be multiplied (or divided) by any number without affecting the sign of inequality
(c) Both (a) and (b)
(d) None of the above

17. Match the following inequalities in Column I with their appropriate values in Column II and choose the correct option from the codes given below.

Column I	Column II
A. $30x < 200$, x is a natural number.	1. $\{\ldots -4, -3, -2, -1, 0, 1\}$
B. $30x < 200$, x is an integer.	2. $\{1, 2, 3, 4, 5, 6\}$
C. $5x - 3 < 3x + 1$, x is an integer.	3. $(-\infty, 2)$
D. $5x - 3 < 3x + 1$, x is a real number.	4. $\{\ldots, -3, -2, -1, 0, 1, 2, 3, 4, 5, 6\}$

Codes

	A	B	C	D
(a)	2	1	4	3
(b)	2	3	4	1
(c)	2	4	1	3
(d)	3	1	4	2

18. The solution set of the inequality $4x + 3 < 6x + 7$ is

(a) $[-2, \infty)$ (b) $(-\infty, -2)$
(c) $(-2, \infty)$ (d) None of these

19. The set of real x satisfying the inequality $\dfrac{5-2x}{3} \le \dfrac{x}{6} - 5$ is

(a) $(-\infty, 8)$ (b) $(8, \infty)$
(c) $[8, \infty)$ (d) $(-\infty, 8]$

20. Consider the following statements

Statement I The solution set of $7x + 3 < 5x + 9$ is $(-\infty, 3)$.

Statement II The graph of the solution of above inequality is represented by

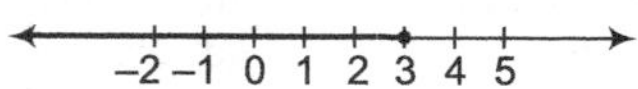

Choose the correct option.

(a) Statement I is true
(b) Statement II is true
(c) Both are true
(d) Both are false

21. The graph of the solutions of inequality $\dfrac{3x-4}{2} \ge \dfrac{x+1}{4} - 1$ on number line is

(a) −1 0 1 2 3
(b) −1 0 1 2 3
(c) −1 0 1 2 3
(d) −1 0 1 2 3

22. The marks obtained by a student of class XI in first and second terminal examinations are 62 and 48, respectively. The minimum marks he should get in the annual examination to have an average of atleast 60 marks, are

(a) 70 (b) 50 (c) 74 (d) 48

23. The number of pairs of consecutive odd natural numbers both of which are larger than 10, such that their sum is less than 40, is

(a) 8 (b) 6 (c) 4 (d) 3

24. Match the following inequalities of Column I with their solutions in Column II and choose the correct option from the codes given below.

Column I	Column II
A. $24x < 100$, x is a natural number.	1. No solution
B. $24x < 100$, x is an integer.	2. $\{1, 2, 3, 4\}$
C. $-12x > 30$, x is a natural number.	3. $\{\ldots, -5, -4, -3\}$
D. $-12x > 30$, x is an integer.	4. $\{\ldots, -3, -2, -1, 0, 1, 2, 3, 4\}$

Codes

	A	B	C	D
(a)	2	1	4	3
(b)	3	1	4	2
(c)	2	4	1	3
(d)	3	4	2	1

25. Consider the following statements

Statement I The solution set of $5x - 3 < 7$, when x is an integer, is $\{\ldots, -3, -2, -1\}$.

Statement II The solution of $5x - 3 < 7$, when x is a real number, is $(-\infty, 2)$.

Choose the correct option.

(a) Statement I is true (b) Statement II is true
(c) Both are true (d) Both are false

26. If $3x + 8 > 2$, then which of the following is true?

(a) $x \in \{-1, 0, 1, 2, \ldots\}$, when x is an integer
(b) $x \in [-2, \infty)$, when x is a real number
(c) Both (a) and (b)
(d) None of the above

27. The solution set of the inequality $4x + 3 < 5x + 7 \ \forall\ x \in R$ is

(a) $(-4, \infty)$ (b) $[-4, \infty)$ (c) $(4, \infty)$ (d) $[4, \infty)$

28. Which of the following is the solution set of $3x - 7 > 5x - 1 \; \forall \; x \in R$?

(a) $(-\infty, -3)$ (b) $(-\infty, -3]$
(c) $(-3, \infty)$ (d) $(-3, 3)$

29. The solution set of the inequality $3(x - 1) \le 2(x - 3)$ is ...A... . Here, A refers to

(a) $(-\infty, 3)$ (b) $(-\infty, -3)$
(c) $(-\infty, 3]$ (d) $(-\infty, -3]$

30. The solution set of the inequality $3(2 - x) \ge 2(1 - x)$ is ...B... . Here, B refers to

(a) $(-\infty, 4)$ (b) $(-\infty, 4]$
(c) $(-\infty, -4)$ (d) $(-\infty, -4]$

31. Consider the following statements

Statement I The solution set of the inequality $x + \frac{x}{2} + \frac{x}{3} < 11$ is $(-\infty, 6)$.

Statement II The solution set of the inequality $\frac{x}{3} > \frac{x}{2} + 1$ is $(-\infty, 6)$.

Choose the correct option.

(a) Statement I is true (b) Statement II is true
(c) Both are true (d) Both are false

32. Consider the following statements

Statement I The solution set of the inequality $\frac{3(x-2)}{5} \le \frac{5(2-x)}{3}$ is $(-\infty, 2)$.

Statement II The solution set of the inequality $\frac{1}{2}\left(\frac{3x}{5} + 4\right) \ge \frac{1}{3}(x - 6)$ is $(-\infty, 120]$.

Choose the correct option.

(a) Statement I is true (b) Statement II is true
(c) Both are true (d) Both are false

33. The solution set of the inequality $2(2x + 3) - 10 < 6(x - 2)$ is

(a) $x \ge 4$ (b) $x \le 4$
(c) $x > 4$ (d) $x < 4$

34. The solution set of the inequality $37 - (3x + 5) \ge 9x - 8(x - 3)$ is

(a) $(-\infty, 2)$ (b) $(-\infty, -2)$
(c) $(-\infty, 2]$ (d) $(-\infty, -2]$

35. Which of the following is the solution set of the inequality $\frac{x}{4} < \frac{(5x-2)}{3} - \frac{(7x-3)}{5}$?

(a) $(4, \infty)$ (b) $(-\infty, 4)$
(c) $[4, \infty)$ (d) $(-\infty, 4]$

36. The solution set of $\frac{2x-1}{3} \ge \left(\frac{3x-2}{4}\right) - \left(\frac{2-x}{5}\right)$ is ...A... . Here, A refers to

(a) $(-\infty, 2)$ (b) $(-\infty, 2]$
(c) $\left(-\infty, -\frac{1}{2}\right)$ (d) $\left(-\infty, -\frac{1}{2}\right]$

37. The graph of the solution on number line of the inequality $3x - 2 < 2x + 1$ is

(a) 1 2 3

(b) 1 2 3 4 5

(c) 1 2 3

(d) 1 2 3 4 5

38. Consider the following statements

Statement I The graph of the solution of inequality $5x - 3 \ge 3x - 5$ does not exist.

Statement II The solution of the inequality $3(1 - x) < 2(x + 4)$ is represented on number line by

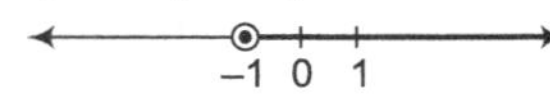

Choose the correct option.

(a) Statement I is true (b) Statement II is true
(c) Both are true (d) Both are false

39. The graph of solution of the inequality $\frac{x}{2} \ge \frac{5x-2}{3} - \frac{7x-3}{5}$ on the number line is

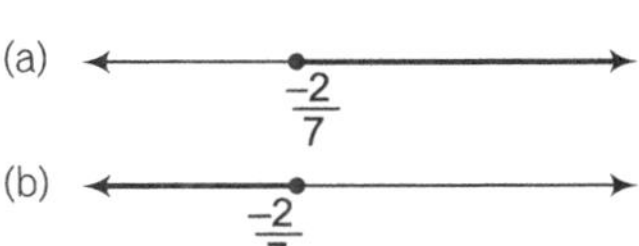

(c) $\frac{-2}{7}$

(d) None of the above

40. Ravi obtained 70 and 75 marks in first two unit tests. Then, the minimum marks he should get in the third test to have an average of atleast 60 marks, are

(a) 45 (b) 35
(c) 25 (d) None of these

41. To receive grade A in a course, one must obtain an average of 90 marks or more in five examinations (each of 100 marks). If Sunita's marks in first four examinations are 87, 92, 94 and 95, the minimum marks that Sunita must obtain in fifth examination to get grade A in the course, are

(a) 72 (b) 82
(c) 90 (d) None of these

42. The number of pairs of consecutive odd positive integers both of which are smaller than 10 such that their sum is more than 11, are

(a) 2 (b) 3
(c) 4 (d) 0

43. The pairs of consecutive even positive integers, both of which are larger than 5 such that their sum is less than 23, are

(a) (4, 6), (6, 8), (8, 10), (10, 12)
(b) (6, 8), (8, 10), (10, 12)
(c) (6, 8), (8, 10), (10, 12), (12, 14)
(d) (8, 10), (10, 12)

44. The longest side of a triangle is 3 times the shortest side and the third side is 2 cm shorter than the longest side. If the perimeter of the triangle is atleast 61 cm, then minimum length of the shortest side is

(a) 7 cm (b) 8 cm
(c) 9 cm (d) 10 cm

45. A man wants to cut three lengths from a single piece of board of length 91 cm. The second length is to be 3 cm longer than the shortest and the third length is to be twice as long as the shortest. The possible length of the shortest board, if the third piece is to be atleast 5 cm longer than the second, is

(a) less than 8 cm
(b) greater than or equal to 8 cm but less than or equal to 22 cm
(c) less than 22 cm
(d) greater than 22 cm

Level II

46. If $3 \leq 3t - 18 \leq 18$, then which one of the following is true? [Kerala CEE 2010]

(a) $15 \leq 2t + 1 \leq 20$ (b) $8 \leq t < 12$
(c) $8 \leq t + 1 \leq 13$ (d) $21 \leq 3t \leq 24$
(e) $t \leq 7$ or $t \geq 12$

47. The solution set of the inequalities $6 \leq -3(2x - 4) < 12$ is

(a) $(-\infty, 1]$ (b) $(0, 1]$
(c) $(0, 1] \cup [1, \infty)$ (d) $[1, \infty)$

48. If $-3 \leq 4 - \frac{7x}{2} \leq 18$, then

(a) $x \in (-\infty, -4]$ (b) $x \in [-4, \infty)$
(c) $x \in [-4, 2]$ (d) $x \in [2, \infty)$

49. Consider the following statements

Statement I The solution set of $2 \leq 3x - 4 \leq 5$ is $(2, 3)$.

Statement II $(-80, -10)$ is the solution set of $-12 < 4 - \frac{3x}{-5} \leq 2$..

Choose the correct option.

(a) Statement I is true (b) Statement II is true
(c) Both are true (d) Both are false

50. If $5x + 1 > -24$ and $5x - 1 < 24$, then $x \in \ldots A \ldots$. Here, A refers to

(a) $(-\infty, -5)$ (b) $(-5, 5)$
(c) $(5, \infty)$ (d) $(-\infty, 5)$

51. Which of the following is the solution set of linear inequalities $2(x - 1) < x + 5$ and $3(x + 2) > 2 - x$?

(a) $(-\infty, -1)$ (b) $(-1, 1)$
(c) $(-1, 7)$ (d) $(1, 7)$

52. Match the statements given in Column I with the sets in Column II and choose the correct option from the codes given below.

Column I	Column II
A. $3x - 5 < x + 7$, x is a natural number.	1. $\{0, 1, 2, 3, 4, 5\}$
B. $3x - 5 < x + 7$, x is an integer.	2. $\{1, 2, 3, 4, 5\}$
C. $3x - 5 < x + 7$, x is a whole number.	3. $\{x : x \in R \text{ and } x < 6\}$
D. $3x - 5 < x + 7$, x is a real number.	4. $\{\ldots, -3, -2, -1, 0, 1, 2, 3, 4, 5\}$

Codes

	A	B	C	D
(a)	2	1	3	4
(b)	4	1	2	3
(c)	2	4	1	3
(d)	2	1	4	3

53. If x satisfies the inequations $2x - 7 < 11$ and $3x + 4 < -5$, then x lies in the interval

(a) $(-\infty, 3)$ (b) $(-\infty, 2)$
(c) $(-\infty, -3)$ (d) $(-\infty, \infty)$

54. If $|x| < 3$ and x is a real number, then [NCERT Exemplar]

(a) $x \geq 3$ (b) $-3 < x < 3$
(c) $x \leq -3$ (d) $-3 \leq x \leq 3$

55. x and b are real numbers. If $b > 0$ and $|x| > b$, then [NCERT Exemplar]

(a) $x \in (-b, \infty)$ (b) $x \in [-\infty, b)$
(c) $x \in (-b, b)$ (d) $x \in (-\infty, -b) \cup (b, \infty)$

56. The solution set of $\frac{x-2}{x+5} > 2$ is

(a) $(-\infty, -12)$ (b) $[-12, -5]$
(c) $(-12, -5)$ (d) $(-5, \infty)$

57. If $|3-4x| \geq 9$, then

(a) $x \in \left(-\infty, -\frac{3}{2}\right)$ (b) $x \in [3, \infty)$
(c) $x \in \left(-\infty, \frac{-3}{2}\right] \cup [3, \infty)$ (d) $x \in (-\infty, \infty)$

58. If $\frac{|x-2|}{x-2} \geq 0$, then [NCERT Exemplar]

(a) $x \in [2, \infty)$ (b) $x \in (2, \infty)$
(c) $x \in (-\infty, 2)$ (d) $x \in (-\infty, 2]$

59. If $a < b$ and $c < 0$, then

(a) $\frac{a}{c} = \frac{b}{c}$ (b) $\frac{a}{c} > \frac{b}{c}$
(c) $\frac{a}{c} < \frac{b}{c}$ (d) None of these

60. Which of the following is/are true?

I. If $x \geq -3$, then $x + 5 \geq 2$.
II. If $-x \leq -4$, then $2x < 8$.
III. If $\frac{1}{x-2} < 0$, then $x < 2$.

(a) Only I is true (b) I and II are ture
(c) I and III are true (d) All are true

61. If $p > 0$ and $q < 0$, then $p + q$ ...A...p. Here, A refers to

(a) $>$ (b) $<$ (c) $\geq$ (d) $\leq$

62. Consider the following statements

Statement I If $|x-1| \leq 2$, then $-1 < x < 3$.

Statement II If $|3x-7| > 2$, then $x < \frac{5}{3}$ or $x > 3$.

Choose the correct option.

(a) Statement I is true (b) Statement II is true
(c) Both are true (d) Both are false

63. If $-3x + 17 < -13$, then [NCERT Exemplar]

(a) $x \in (10, \infty)$ (b) $x \in [-\infty, 10)$
(c) $10x \in (-10, 10)$ (d) $x \in (-\infty, -10) \cup (10, \infty)$

64. The solution set representing by following figure is [NCERT Exemplar]

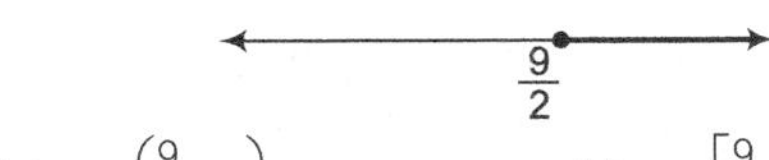

(a) $x \in \left(\frac{9}{2}, \infty\right)$ (b) $x \in \left[\frac{9}{2}, \infty\right)$
(c) $x \in \left[-\infty, \frac{9}{2}\right)$ (d) $x \in \left(-\infty, \frac{9}{2}\right]$

65. The cost and revenue function of a product are given by $C(x) = 20x + 4000$ and $R(x) = 60x + 2000$ respectively, where x is the number of items produced and sold. Then, the number of items must be sold to realise some profit is given by

(a) $x = 50$ (b) $x > 50$
(c) $x < 50$ (d) None of these

66. Which of the following is/are true?

I. If $xy < 0$, then $x > 0, y > 0$.
II. If $xy > 0$, then $x > 0, y < 0$.
III. If $|x| < 5$, then $x \in (-5, 5)$.

(a) Only I is true
(b) Only II is true
(c) Only III is true
(d) I and II are true

67. If $|x+3| \geq 10$, then [NCERT Exemplar]

(a) $x \in (-13, 7]$ (b) $x \in (-13, 7]$
(c) $x \in (-\infty, -13] \cup [7, \infty)$ (d) $x \in [-\infty, -13) \cup (7, \infty)$

68. The length of a rectangle is three times the breadth. If the minimum perimeter of the rectangle is 160 cm, then [NCERT Exemplar]

(a) breadth > 20cm (b) length < 20 cm
(c) breadth ≥ 20 cm (d) length ≤ 20 cm

69. The minimum value of $f(x) = |3-x| + 7$ is [VITEEE 2010]

(a) 0 (b) 6
(c) 7 (d) 8

70. The set of values of x satisfying $2 \leq |x-3| < 4$ is [Kerala CEE 2006]

(a) $(-1, 1] \cup [5, 7)$ (b) $-4 \leq x \leq 2$
(c) $-1 < x < 7$ or $x \geq 5$ (d) $x < 7$ or $x \geq 5$
(e) $-\infty < x \leq 1$ or $5 \leq x < \infty$

71. The number of pairs of consecutive even positive integers, both of which are larger than 8, such that their sum is less than 25, is

(a) 1 (b) 2
(c) 3 (d) 4

72. IQ of a person is given by the formula

$$IQ = \frac{MA}{CA} \times 100$$

where, MA is mental age and CA is chronological age. If $80 \leq IQ \leq 140$ for a group of 12 yr children, then the range of their mental age is

(a) $9.8 \leq MA \leq 16.8$
(b) $10 \leq MA \leq 16$
(c) $9.6 \leq MA \leq 16.8$
(d) $9.6 \leq MA \leq 16.6$

Graphical Solution of Linear Inequalities in Two Variables

Level I

73. Consider the following statements

Statement I Graph of linear inequality in one variable is a visual representation.

Statement II Graph of linear inequality does not represent the solution of the inequality.

Choose the correct option.

(a) Statement I is true (b) Statement II is true
(c) Both are true (d) Both are false

74. A line divides the cartesian plane in ...X... part (s). Here, X refers to

(a) one (b) two
(c) three (d) None of these

75. Each part in which a line divides the cartesian plane, is known as ...Y... plane. Here, Y refers to

(a) half (b) quarter
(c) double (d) None of these

76. Which of the following is/are true?

(a) A vertical line will divide the plane in left and right half planes
(b) A non-vertical line will divide the plane into left and right half planes
(c) Both (a) and (b)
(d) None of the above

77. Consider the following statements

Statement I The graph of a vertical line is represented by

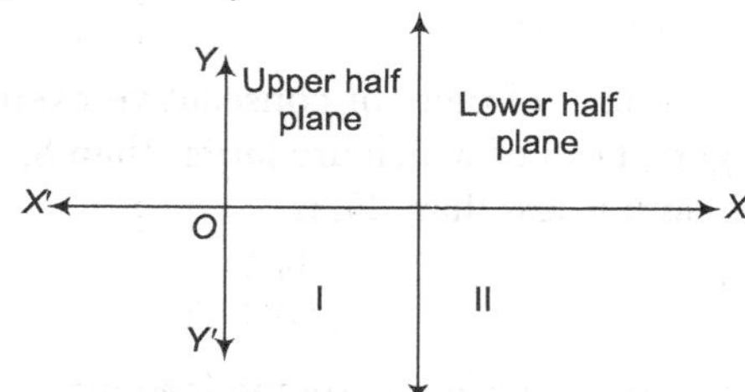

Statement II The visual representation of non-vertical line is represented by

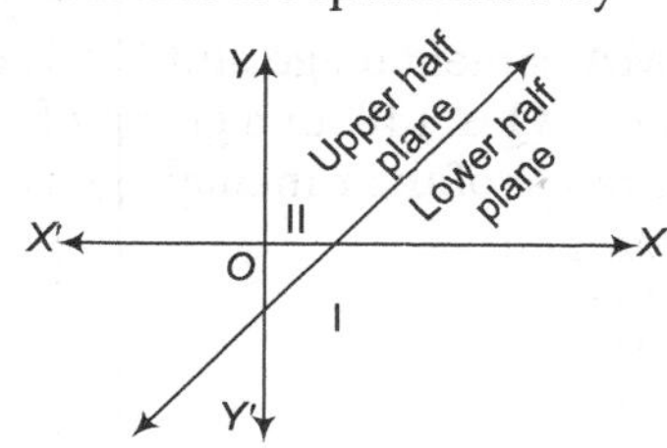

Choose the correct option.

(a) Statement I is true (b) Statement II is true
(c) Both are true (d) Both are false

78. A point in the cartesian plane will either lie on a ...A... or will lie in ...B.... Here, A and B refer to

(a) line and half plane I
(b) line and half plane II
(c) line and either of half plane I or II
(d) None of the above

79. If a point satisfying the line $ax + by = c$, then

(a) it will lie on the line
(b) it will lie in upper half plane
(c) it will lie in lower half plane
(d) None of the above

80. If a point $P(\alpha, \beta)$ on the line $ax + by = c$, then

(a) $a\alpha + b\beta > c$ (b) $a\alpha + b\beta < c$
(c) $a\alpha + b\beta = c$ (d) None of these

81. If $P(\alpha, \beta)$ is a point on the line $ax + by = c, b > 0$ and $Q(\alpha, \gamma)$ is a point in upper half plane, then which of the following is/are true?

I. The visual representation of above situation is

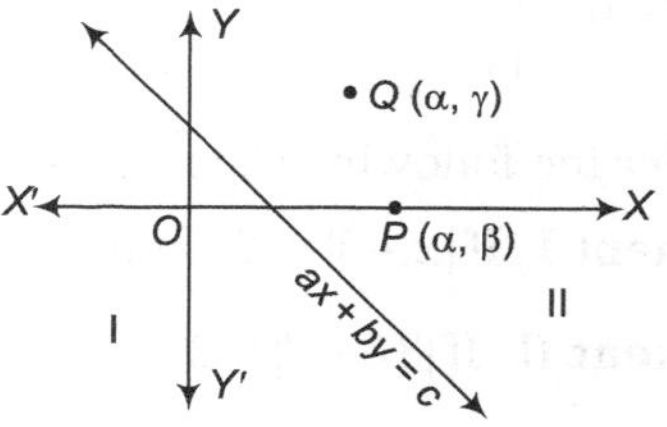

II. $\gamma > \beta$ III. $\gamma < \beta$

(a) I and II are true (b) I and III are true
(c) Only II is true (d) Only III is true

82. By considering the graph of above question 81

Assertion (A) Any point in the upper half plane II satisfies $ax + by > c, b > 0$.

Reason (R) If a point satisfies $ax + by > c, b < 0$, then it lies in the lower half plane I.

(a) Both A and R are correct; R is the correct explanation of A
(b) Both A and R are correct; R is not the correct explanation of A
(c) A is correct; R is incorrect
(d) R is correct; A is incorrect

83. The graph of the inequality $ax + by > c$ is represented by shading in the ...A... . Here, A refers to

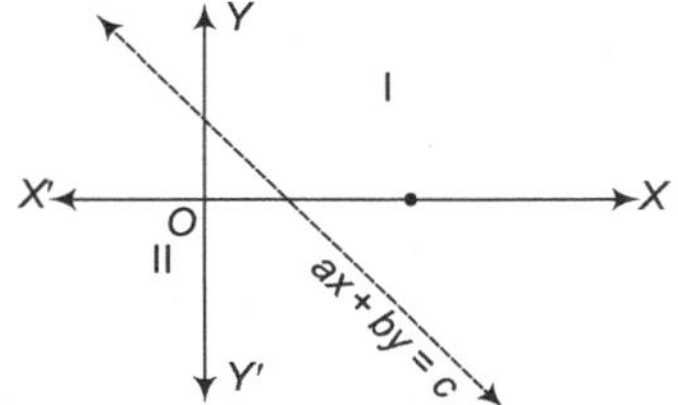

(a) half plane I including the point on $ax + by = c$
(b) half plane II including the point on $ax + by = c$
(c) half plane I not including the point on $ax + by = c$
(d) half plane II not including the point on $ax + by = c$

84. Consider the following statements

Statement I The region containing all the solutions of an inequality is called the solution region.

Statement II The half plane represented by an inequality is checked by taking any point on the line.

Choose the correct option.

(a) Statement I is true (b) Statement II is true
(c) Both are true (d) Both are false

85. Consider the following statements

Statement I If an inequality is of type $ax + by \geq c$ or $ax + by \leq c$, then the points on the line are not to be included in the solution region.

Statement II If an inequality is of the form $ax + by > c$ or $ax + by < c$, then the points on the line $ax + by = c$ are not to be included in the solution region.

Choose the correct option.

(a) Statement I is true (b) Statement II is true
(c) Both are true (d) Both are false

86. Consider the inequality $40x + 20y \leq 120$, where x and y are whole numbers. Then,

(a) (0, 0), (0, 5), (1, 1), (2, 2), (3, 0) are some solutions of given inequality
(b) The graphical representation of given inequality is

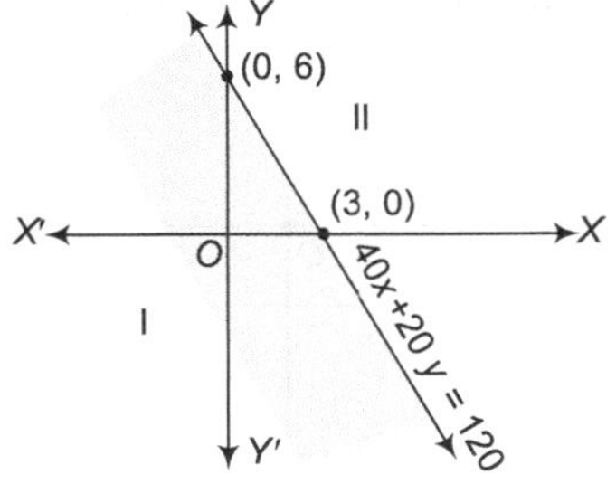

(c) Both (a) and (b) (d) None of these

87. The graph of the inequality $40x + 20y \leq 120$, $x \geq 0$, $y \geq 0$ is

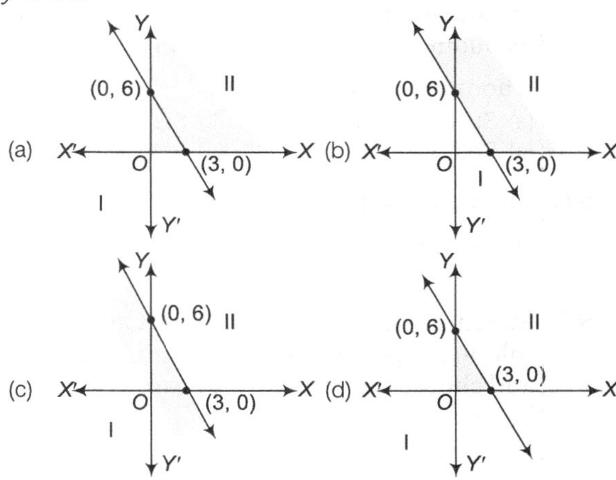

88. The solution of the inequality $3x + 2y > 6$ in the graph is represented by shading the

(a) half plane, which contains origin including the points on the line
(b) half plane, which does not contain origin excluding the points on the line
(c) half plane, which does not contain origin including the points on the line
(d) half plane, which contain origin excluding the points on the line

89. The graphical solution of $3x - 6 \geq 0$ is

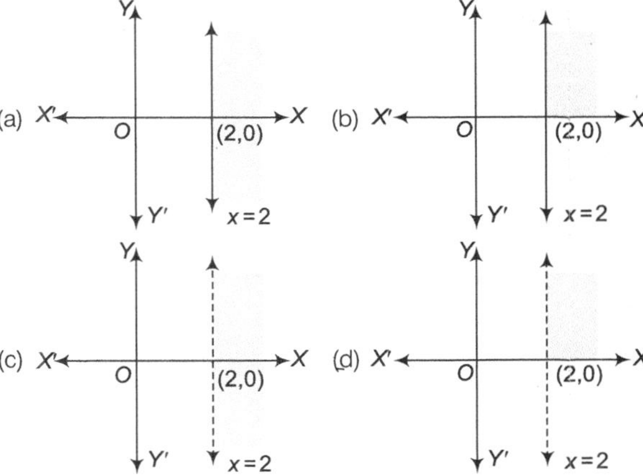

90. Consider the following statements

Statement I The graphical solution of $y < 2$ is

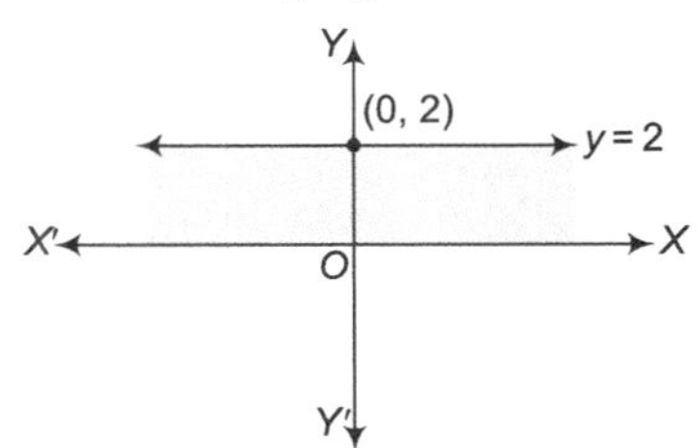

Statement II The graph of $x + y < 5$ is represented by shade the half plane which contains origin excluding the points on the line $x + y = 5$.

Choose the correct option.

(a) Statement I is true (b) Statement II is true
(c) Both are true (d) Both are false

91. The region represented by $2x + y \geq 6$ is

(a) bounded (b) unbounded
(c) Does not exist (d) None of these

92. Match the inequalities given in Column I with the solution representing by the graphs in Column II and choose the correct option from the codes given below.

Column I	Column II
A. $3x + 4y \leq 12$	1. Y, X', X, O, (2, 0), (0, –2), Y'
B. $y + 8 \geq 2x$	2. Y, (0, 3), X', X, O, (4, 0), Y'
C. $x - y \leq 2$	3. Y, (3, 0), X', X, O, (0, –2), Y'
D. $2x - 3y > 6$	4. Y, (4, 0), X', X, O, (0, –8), Y'

Codes

	A	B	C	D
(a)	2	3	4	1
(b)	3	1	4	2
(c)	2	4	1	3
(d)	3	4	2	1

93. The graphical solution of $-3x + 2y \geq -6$ is represented by shade the

(a) half plane II including the points on the line $-3x + 2y = -6$
(b) half plane II excluding the points on the line $-3x + 2y = -6$
(c) half plane I including the points on the line $-3x + 2y = -6$
(d) half plane I excluding the points on the line $-3x + 2y = -6$

94. The solution of $3y - 5x < 30$ is represented by the graph

(a)

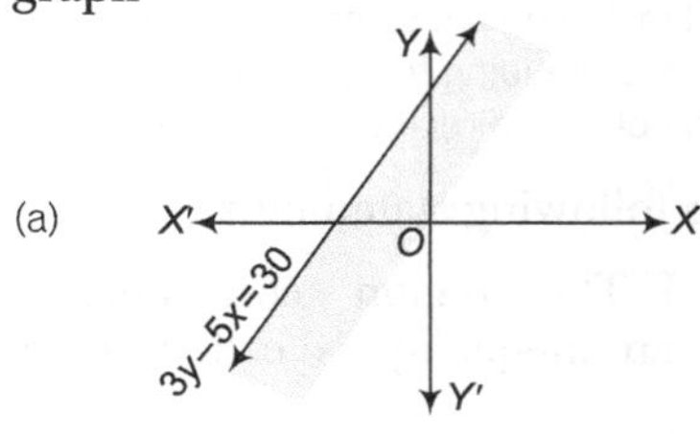

(b)

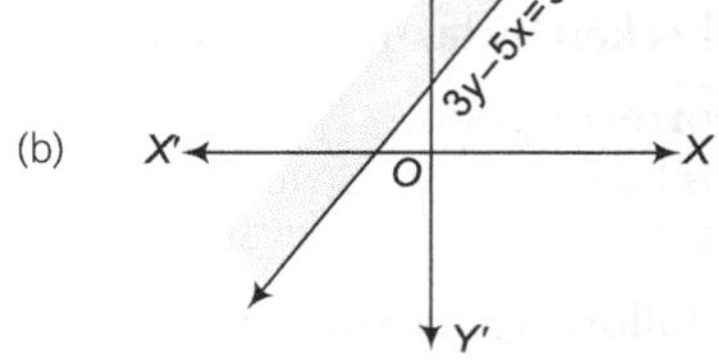

(c)

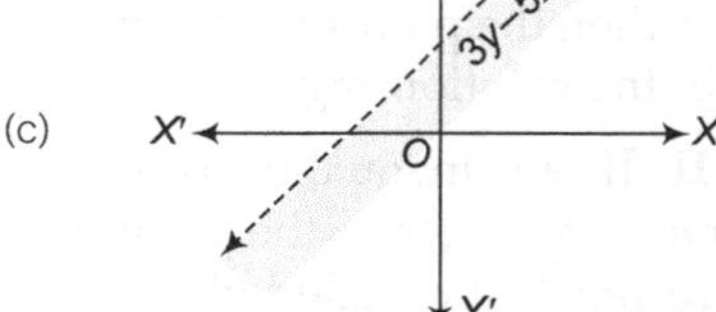

(d)

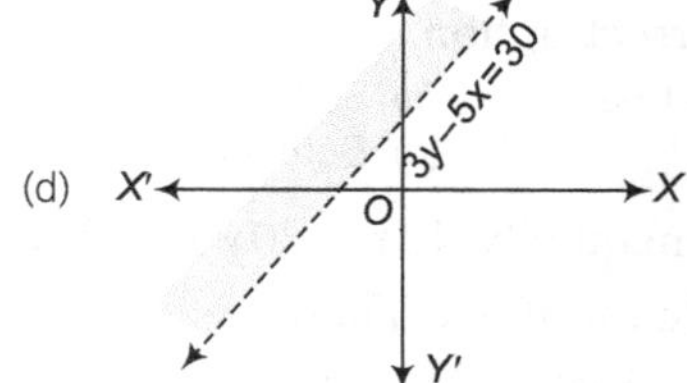

95. Consider the following statements

Statement I The solution of $y < -2$ is represented by

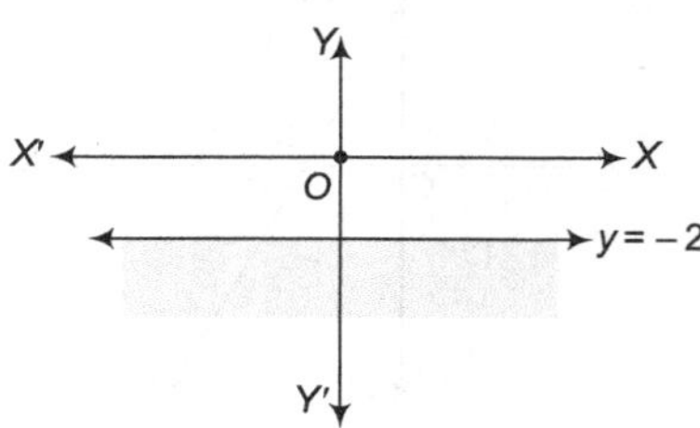

Statement II The graphical solution of $x > -3$ is

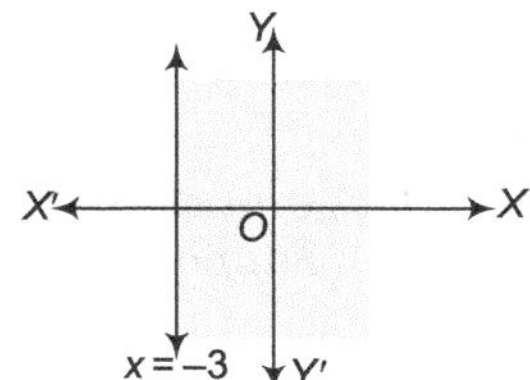

Choose the correct option.

(a) Statement I is true (b) Statement II is true
(c) Both are true (d) Both are false

Level II

96. The graph of $y < 3$ is

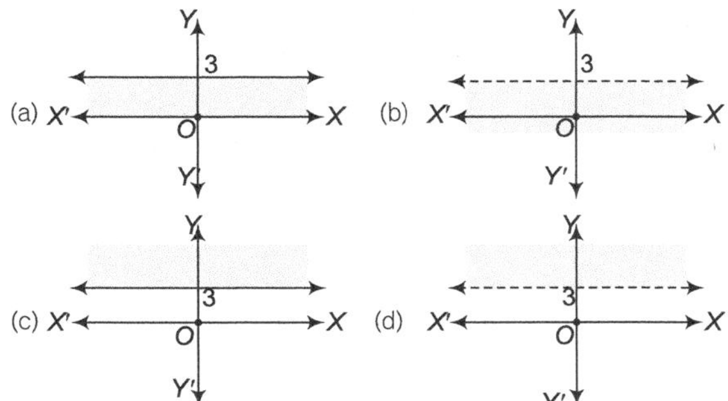

97. The inequality representing the following graph is

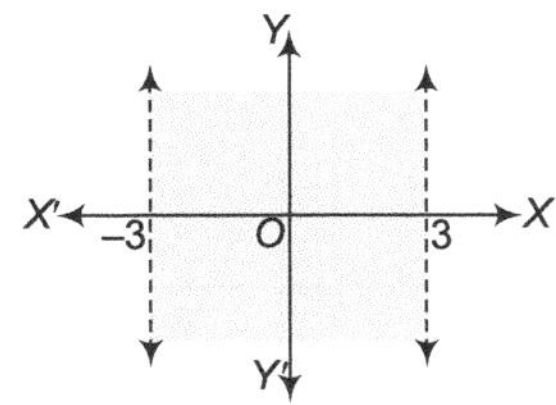

(a) $|x| < 3$ (b) $|x| \le 3$ (c) $|x| > 3$ (d) $|x| \ge 3$

98. Consider the following statements

Statement I The graph of the inequality $2x - y \ge 1$ lies in first quadrant only.

Statement II The graph of $x + y \le 0$ is

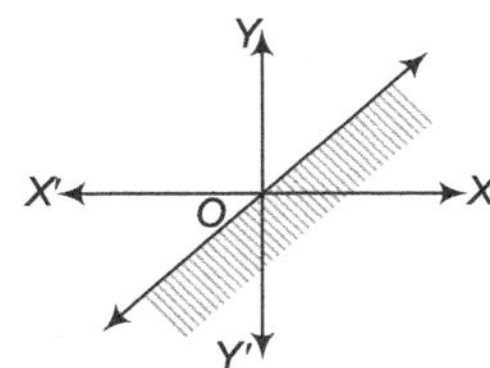

Choose the correct option.

(a) Statement I is true (b) Statement II is true
(c) Both are true (d) Both are false

99. The graph of $2x - 3 \ge 0$ is

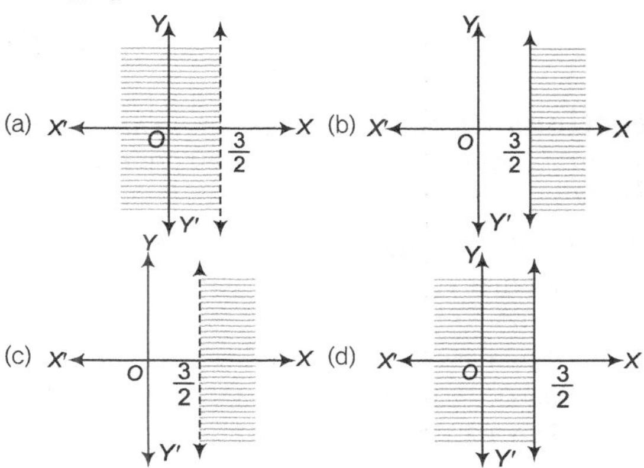

100. Match the inequalities given in Column I with their graph in Column II and choose the correct option from the codes given below.

Column I	Column II
A. $x + y \ge 4$	1.
B. $x + 2y > 1$	2.
C. $x \ge y - 2$	3.
D. $2x - 3y < 4$	4.

Codes

	A	B	C	D		A	B	C	D
(a)	2	1	4	3	(b)	3	2	4	1
(c)	3	4	1	2	(d)	4	1	2	3

Solution of System of Linear Inequalities in Two Variables

Directions (Q. Nos. 101-102) *A furniture dealer deals in only two items — tables and chairs. He has ₹ 15000 to invest and a space to store atmost 60 pieces. A table costs him ₹ 750 and a chair ₹ 150. Suppose he makes x tables and y chairs.*

101. The above data in the form of inequations can be written as

(a) $750x + 150y < 15000$, $x + y < 60$, $x \geq 0$ and $y \geq 0$

(b) $750x + 150y > 15000$, $x + y > 60$, $x \geq 0$ and $y \geq 0$

(c) $750x + 150y \leq 15000$, $x + y \leq 60$, $x \geq 0$ and $y \geq 0$

(d) $750x + 150y \geq 15000$, $x + y \geq 60$, $x \geq 0$ and $y \geq 0$

102. The graphical solution of the inequations representing the given data is

(a)

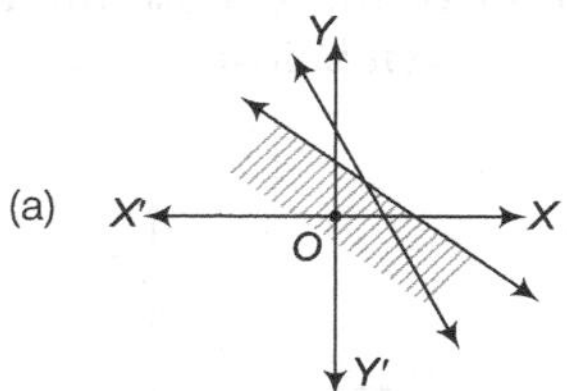

(b)

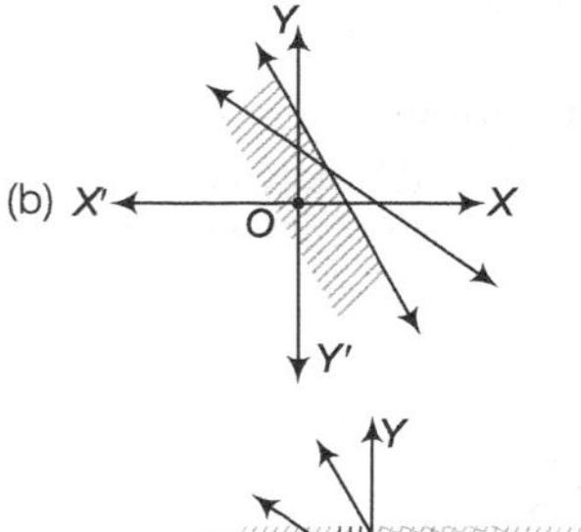

(c)

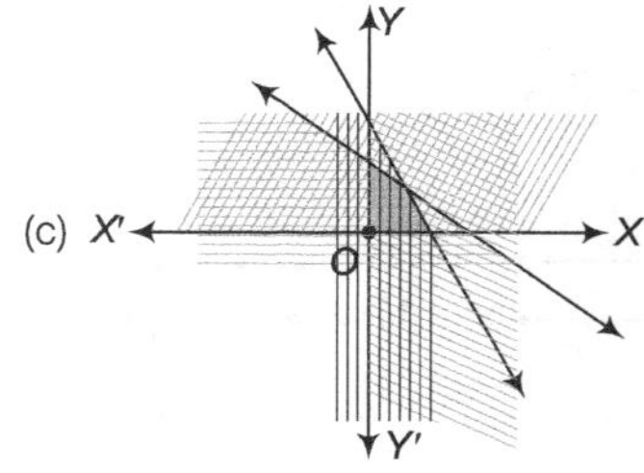

(d) None of these

Level I

103. The graphical solution of linear inequalities $x + y \geq 5$ and $x - y \leq 3$, where $L_1 \equiv x + y = 5$ and $L_2 \equiv x - y = 3$, is

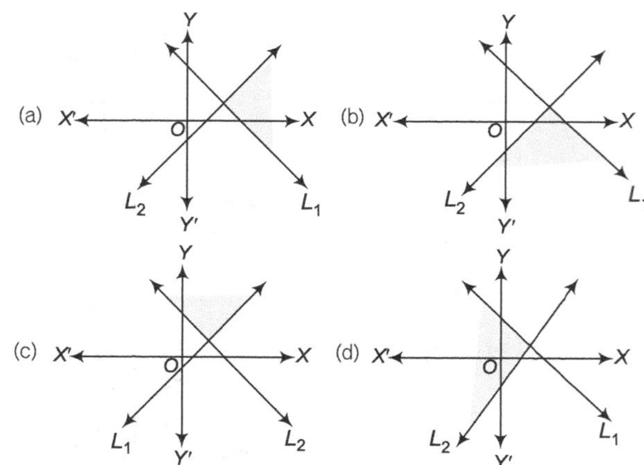

104. The region represented by the solution of system of inequalities $5x + 4y \leq 40$, $x \geq 2$, $y \geq 3$ is

(a) bounded

(b) unbounded

(c) Does not exist

(d) None of the above

105. Which of the following represent the region in the figure?

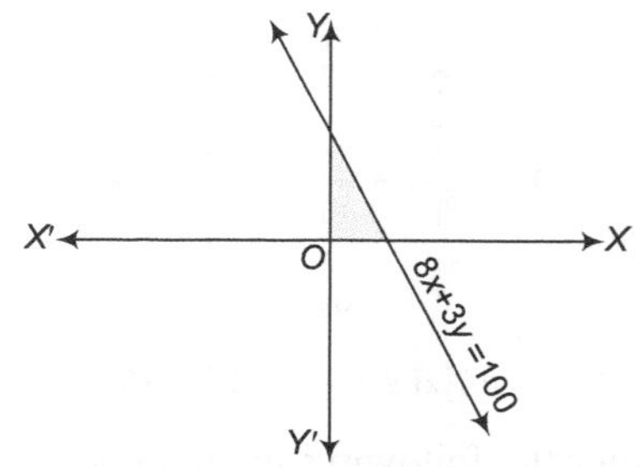

(a) $8x + 3y \geq 100$, $x \geq 0$, $y \geq 0$ (b) $8x + 3y \leq 100$, $x \geq 0$, $y \geq 0$

(c) $8x + 3y \geq 100$ (d) $8x + 3y \leq 100$

106. Consider the following statements

Statement I The solution set of the inequalities $x + 2y \leq 8$, $2x + y \leq 8$, $x \geq 0$, $y \geq 0$ is

Statement II The solution set of the inequalities $x \geq 3$ and $y \geq 2$ is

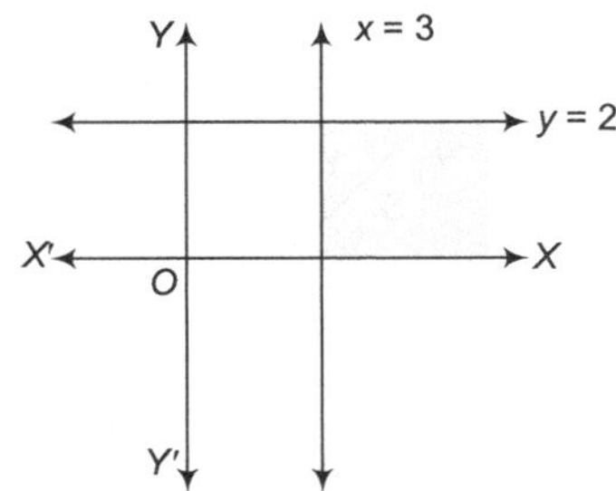

Choose the correct option.

(a) Statement I is true
(b) Statement II is true
(c) Both are true
(d) Both are false

107. Which of the following is/are true?

I. The graphical solution of the system of inequalities $3x + 2y \leq 12$, $x \geq 1$, $y \geq 2$ is

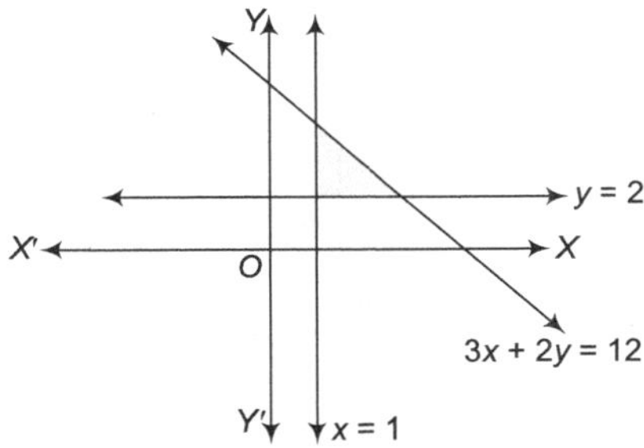

II. The region represented by the solution set of the inequalities $2x + y \geq 6$, $3x + 4y \leq 12$ is bounded.

III. The solution set of the inequalities $x + y \geq 4$, $2x - y > 0$ is

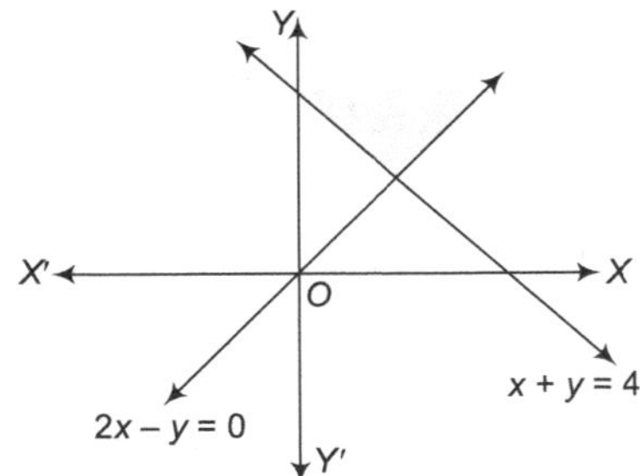

(a) Only I is true
(b) I and II are true
(c) I and III are true
(d) Only III is true

108. Match the linear inequalities given in Column I with solution set representing by graphs in Column II and choose the correct option from the codes given below.

Column I	Column II
A. $2x - y > 1$, $x - 2y < -1$	1. 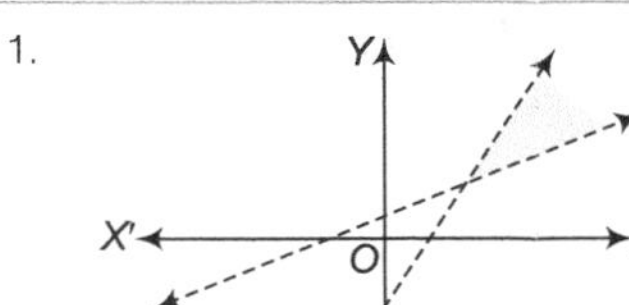
B. $x + y \leq 6$, $x + y \geq 4$	2. 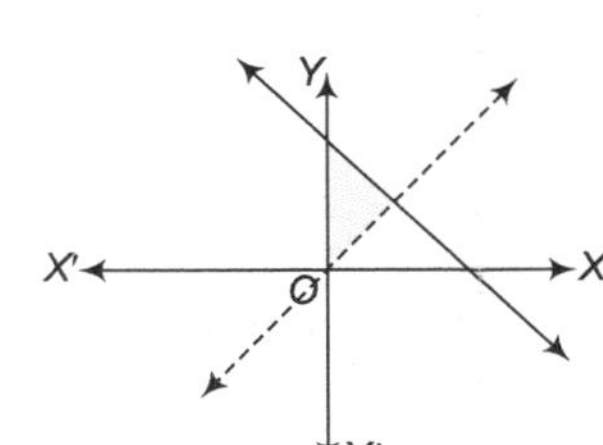
C. $2x + y \geq 8$, $x + 2y \geq 10$	3. 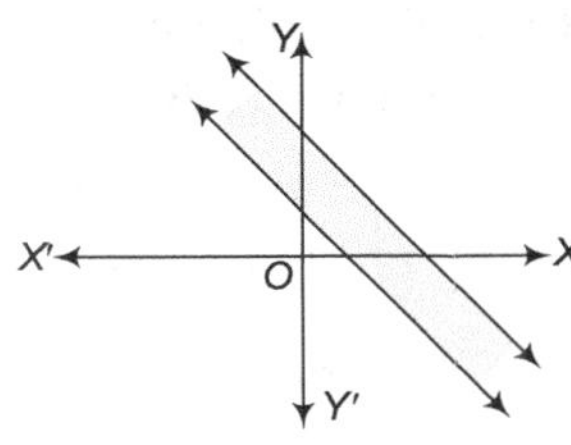
D. $x + y \leq 9$, $y > x$, $x \geq 0$	4. 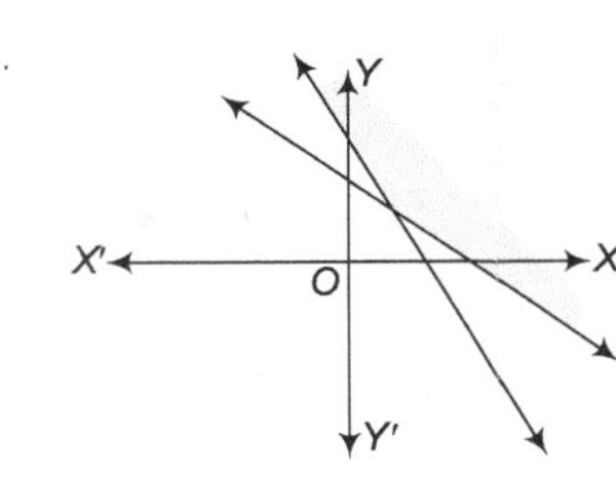

Codes

	A	B	C	D
(a)	4	3	2	1
(b)	2	1	4	3
(c)	1	3	4	2
(d)	3	4	2	1

109. Consider the following statements

Statement I The solution set of the system of inequalities $5x + 4y \leq 20$, $x \geq 1$, $y \geq 2$ is

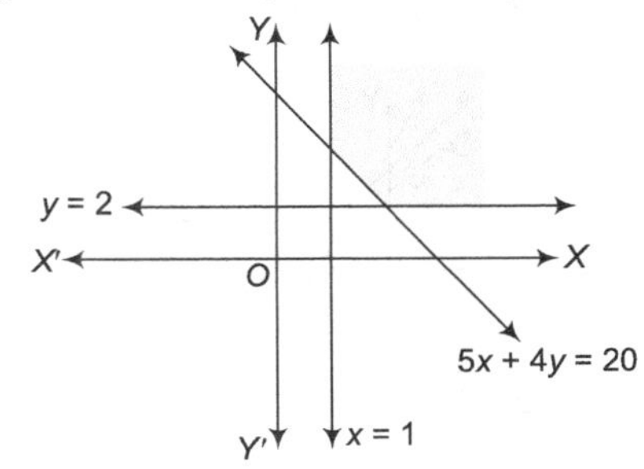

Statement II The solution set of the inequalities $3x + 4y \leq 60$, $x + 3y \leq 30$, $x \geq 0$, $y \geq 0$ is

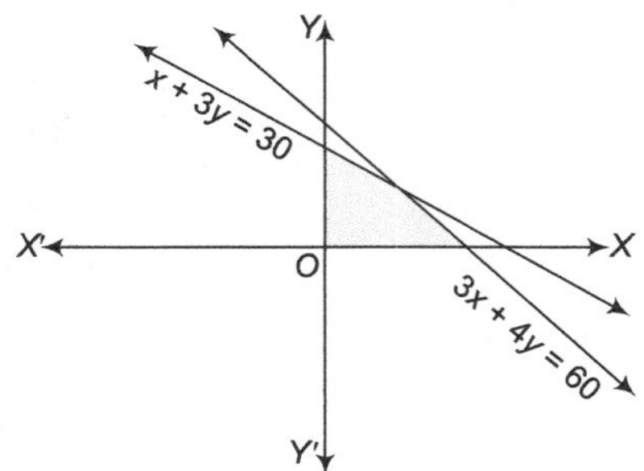

Choose the correct option.

(a) Statement I is true
(b) Statement II is true
(c) Both are true
(d) Both are false

110. The solution set of the inequalities $2x + y \geq 4$, $x + y \leq 3$, $2x - 3y \leq 6$ is

(a)

(b)

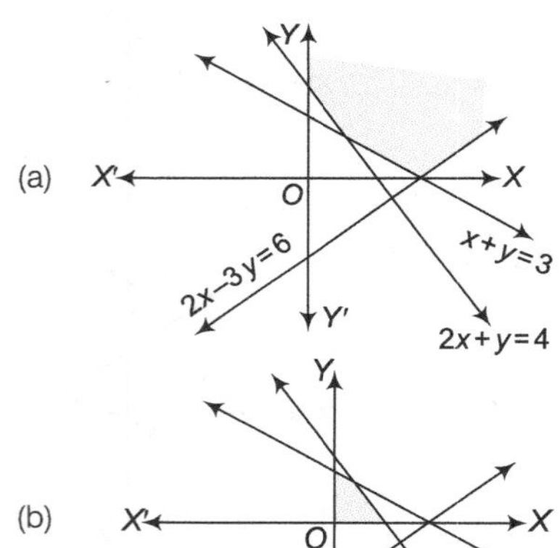

(c)

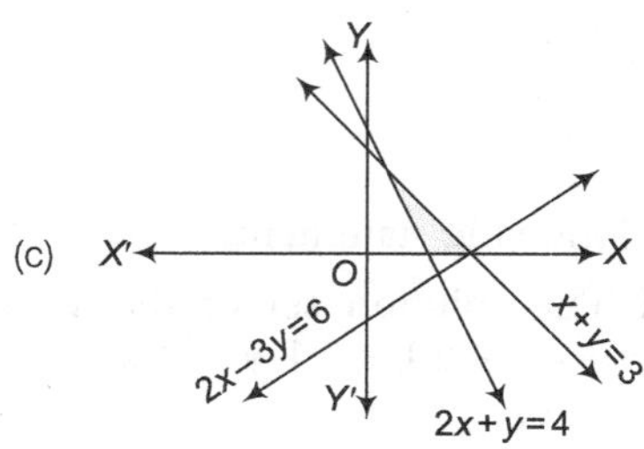

(d)

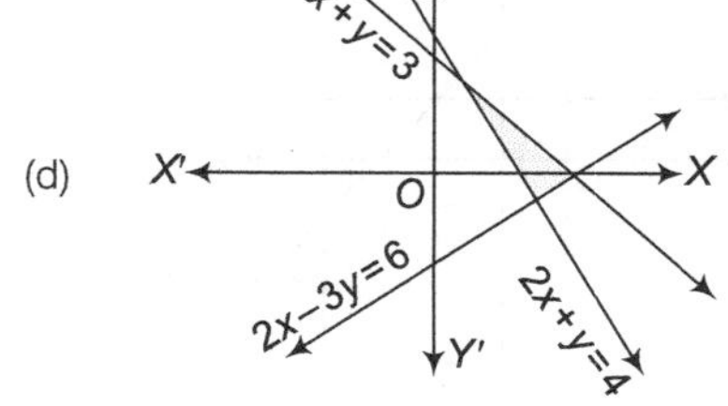

111. The shaded region representing the solution set of the system of inequalities $x - 2y \leq 3$, $3x + 4y \geq 12$, $x \geq 0$, $y \geq 1$ in the following figure is

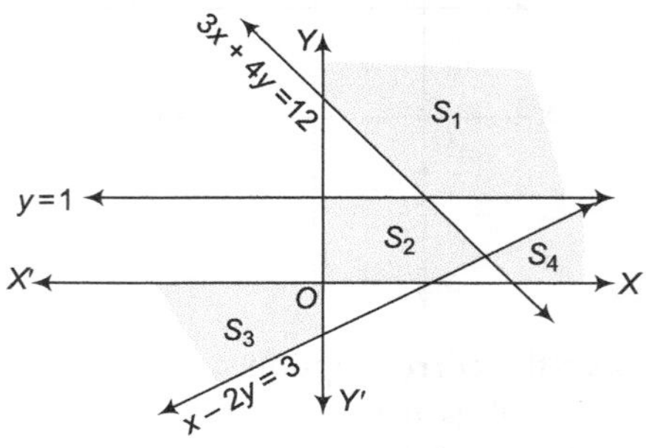

(a) S_1
(b) S_2
(c) S_3
(d) S_4

112. The solution set of the inequalities $4x + 3y \leq 60$, $y \geq 2x$, $x \geq 3$, $x, y \geq 0$ in the figure is ...A... . Here, A refers to

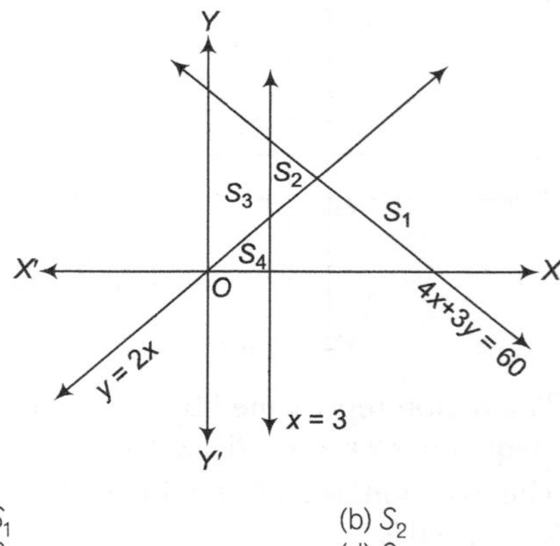

(a) S_1
(b) S_2
(c) S_3
(d) S_4

113. Which of the following represent the solution set in the following figure?

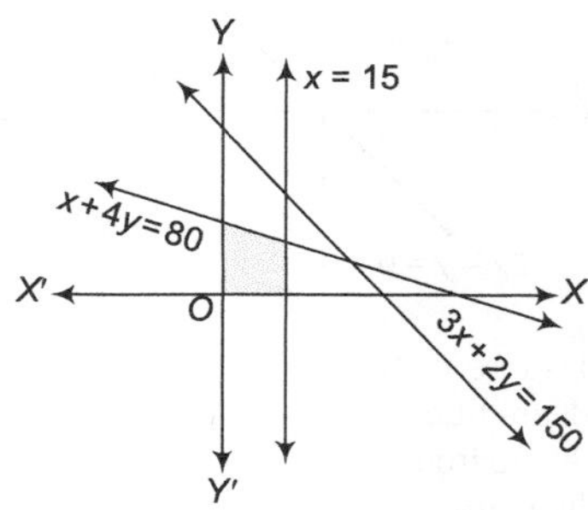

(a) $3x + 2y \geq 150$, $x + 4y \leq 80$, $x \leq 15$, $x, y \geq 0$
(b) $3x + 2y \leq 150$, $x + 4y \geq 80$, $x \geq 15$, $x, y \geq 0$
(c) $3x + 2y \leq 150$, $x + 4y \leq 80$, $x \leq 15$, $x, y \geq 0$
(d) None of the above

114. The graphical solution of the inequalities $x + 2y \le 10, x + y \ge 1, x - y \le 0, x \ge 0, y \ge 0$ is

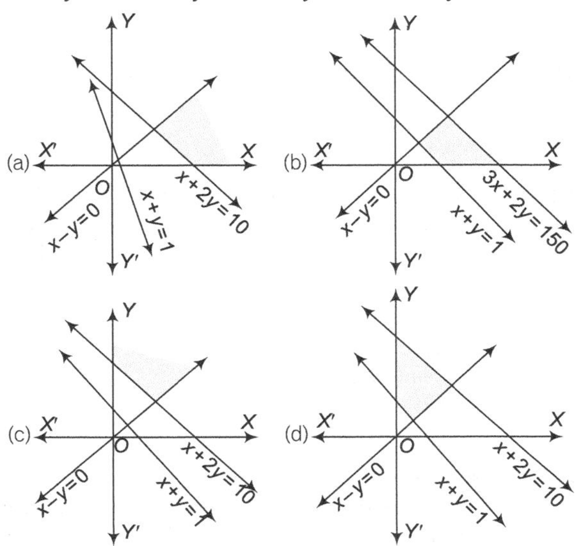

Level II

115. The graphical solution of the system of linear inequalities $3x + 4y \ge 12, y \ge 1, x \ge 0$ is

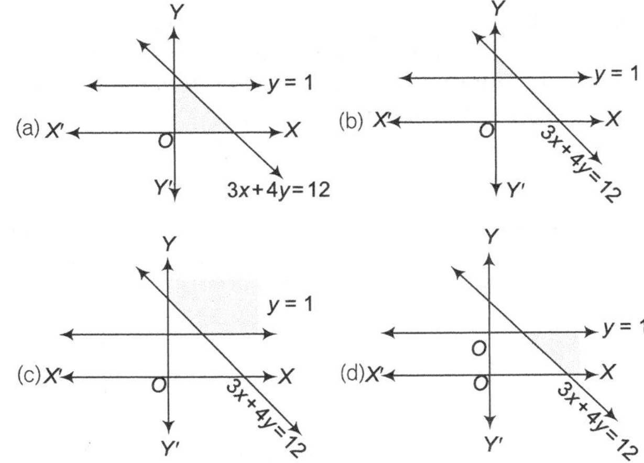

116. The region represented by the solution set of the system of inequalities $x - y \le 1$, $x + 2y \le 8$, $2x + y \ge 2, x \ge 0, y \ge 0$ is

(a) bounded
(b) unbounded
(c) Does not exist
(d) None of these

117. Consider the following statements

Statement I The graphical solution of $|y - x| \le 3$ lies in the first quadrant.

Statement II The solution set of the linear inequations $x - 2y \ge 0, 2x - y \le -2, x \ge 0, y \ge 0$ is empty set.

Choose the correct option.

(a) Statement I is true
(b) Statement II is true
(c) Both are true
(d) Both are false

118. Linear inequalities for which the shaded region for the given figure is the solution set, are [HOTS]

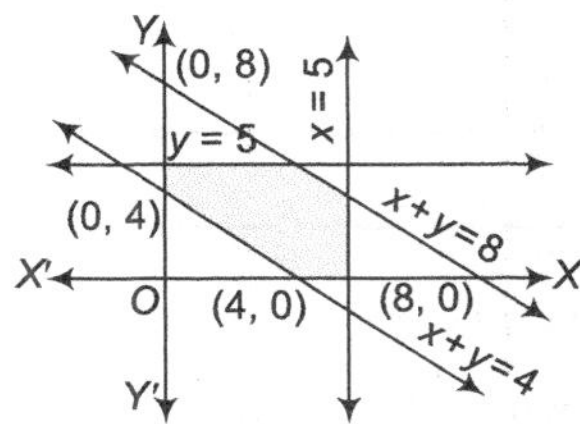

(a) $x + y \le 8, x + y \le 4, x \le 5, y \le 5, x \ge 0, y \ge 0$
(b) $x + y \le 8, x + y \ge 4, x \le 5, y \le 5, x \ge 0, y \ge 0$
(c) $x + y \ge 8, x + y \ge 4, x \ge 5, y \ge 5, x \ge 0, y \ge 0$
(d) None of the above

Directions (Q. Nos. 119-120) *If a young man rides his motorcycle at 25 km/h, he has to spend ₹ 2 per km in petrol, and if he rides it at 40 km/h, the petrol cost rises to ₹ 5 per km. He has ₹ 100 to spend on petrol and wishes to find the maximum distance, he can travel within one hour. If x and y denote the distance travelled by him (in km) at 25 km/h and 40 km/h, respectively.*

119. The inequations represent the data are

(a) $2x + 5y \le 100, \frac{x}{25} + \frac{y}{40} \ge 1, x \ge 0, y \ge 0$
(b) $2x + 5y \ge 100, \frac{x}{25} + \frac{y}{40} \ge 1, x \ge 0, y \ge 0$
(c) $2x + 5y \le 100, \frac{x}{25} + \frac{y}{40} \le 1, x \ge 0, y \ge 0$
(d) $2x + 5y \le 100, 25x + 40y \le 1, x \ge 0, y \ge 0$

120. The solution of the inequalities representing the data is

(a) unique
(b) infinite
(c) No solution
(d) None of the above

Miscellaneous Exercise

121. The solution of the inequality $-8 \leq 5x-3 < 7$ is

(a) $(-1, 2)$ (b) $[-1, \infty)$ (c) $(-\infty, 2]$ (d) $[-1, 2)$

122. If $-5 \leq \frac{5-3x}{2} \leq 8$, then

(a) $-\frac{11}{3} < x < 5$ (b) $-\frac{11}{3} \leq x \leq 5$

(c) $-11 \leq x \leq 5$ (d) $-\frac{11}{3} < x < 5$

123. The solutions of the system of inequalities $3x-7 < 5+x$ and $11-5x \leq 1$ on the number line is

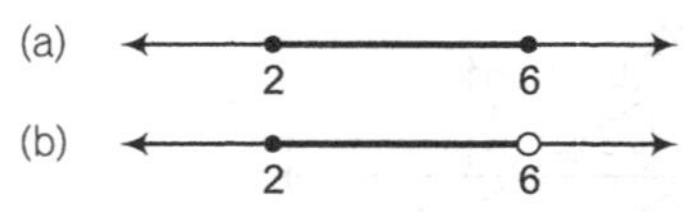

(d) None of the above

124. In an experiment, a solution of hydrochloric acid is to be kept between 30°C and 35° C. The range of temperature in degree Fahrenheit, if conversion formula is given by $C = \frac{5}{9}(F-32)$, where C and F represent temperature in degree Celsius and degree Fahrenheit respectively, is between ...A... and ...B... . Here, A and B refer to

(a) 86°F and 95°F (b) 54°F and 63°F
(c) 54°F and 95°F (d) 63°F and 86°F

125. A manufacturer has 600 L of 12% solution of acid. The volume of 30% acid solution must be added to it, so that acid content in the resulting mixture will be more than 15% but less than 18%, is between ...X... and ...Y... . Here, X and Y refer to

(a) 120 L and 240 L (b) 120 L and 300 L
(c) 60 L and 200 L (d) 240 L and 300 L

126. Match the inequalities given in Column I with their solutions in Column II and choose the correct option from the codes given below.

	Column I		Column II
A.	$-15 < \frac{3(x-2)}{5} \leq 0$	1.	$\left[1, \frac{11}{3}\right]$
B.	$7 \leq \frac{3x+11}{2} \leq 11$	2.	$[-1, 1] \cup [3, 5]$
C.	$1 \leq \lvert x-2 \rvert \leq 3$	3.	$\left[-\frac{34}{3}, \frac{22}{3}\right]$
D.	$-5 \leq \frac{2-3x}{4} \leq 9$	4.	$(-23, 2]$

Codes

	A	B	C	D		A	B	C	D
(a)	3	4	2	1	(b)	4	1	2	3
(c)	4	1	3	2	(d)	2	1	4	3

127. The solution set of the inequalities $3x-7 > 2(x-6)$ and $6-x > 11-2x$, is

(a) $(-5, \infty)$ (b) $[5, \infty)$ (c) $(5, \infty)$ (d) $[-5, \infty)$

128. A solution is to be kept between 68°F and 77°F. The range in temperature in degree Celsius (C), if the Celsius/Fahrenheit (F) conversion formula is given by $F = \frac{9}{5}C + 32$, is

(a) between 20°C and 22°C
(b) between 20°C and 25°C
(c) between 36°C and 45°C
(d) between 4°C and 5°C

129. A solution of 8% boric is to be diluted by adding a 2% boric acid solution to it. The resulting mixture is to be more than 4% but less than 6% boric acid. If we have 640 L of the 8% solution, of the 2% solution will have to be added is

(a) more than 320 and less than 1000
(b) more than 160 and less than 320
(c) more than 320 and less than 1280
(d) more than 320 and less than 640

130. If x L of water will have to be added to 1125 L of the 45% solution of acid, so that the resulting mixture will contain more than 25% but less than 30% acid content, then the value of x is

(a) $562.5 < x < 1125$ (b) $900 < x < 1125$
(c) $525 < x < 900$ (d) $562.5 < x < 900$

131. Solutions of the inequalities comprising a system in variable x are represented on number lines as given below, then

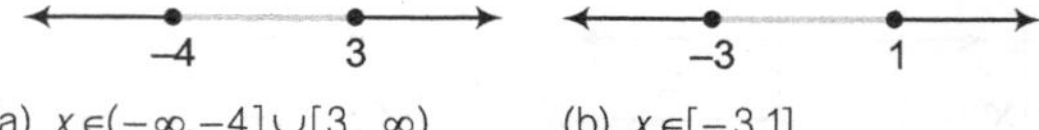

(a) $x \in (-\infty, -4] \cup [3, \infty)$ (b) $x \in [-3, 1]$
(c) $x \in (-\infty, -4) \cup [3, \infty)$ (d) $x \in [-4, 3]$

132. Consider the following statements

Statement I Graph of $x > -2$ is

Statement II Solution set of $x-y\leq 0$ is represented by the following graph

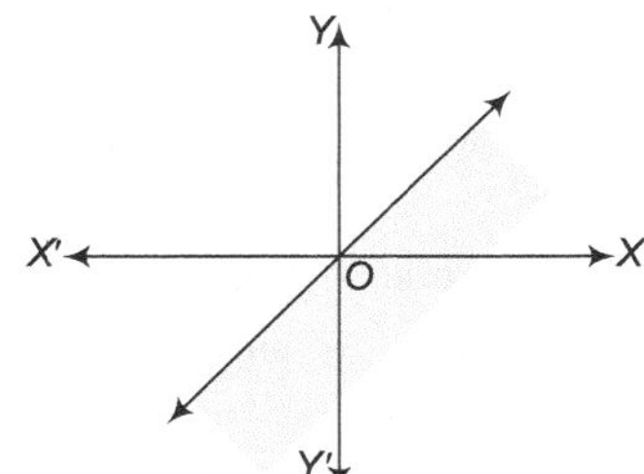

Choose the correct option.

(a) Statement I is true
(b) Statement II is true
(c) Both are true
(d) Both are false

133. The set of admissible values of x such that $\frac{2x+3}{2x-9}<0$, is [Kerala CEE 2008]

(a) $\left(-\infty,\frac{-3}{2}\right)\cup\left(\frac{9}{2},\infty\right)$
(b) $\left(\frac{-3}{2},0\right)$
(c) $\left(0,\frac{9}{2}\right)$
(d) $\left(\frac{-3}{2},\frac{9}{2}\right)$
(e) $(-\infty,0)\cup\left(\frac{9}{2},\infty\right)$

134. The inequality representing the following graph is

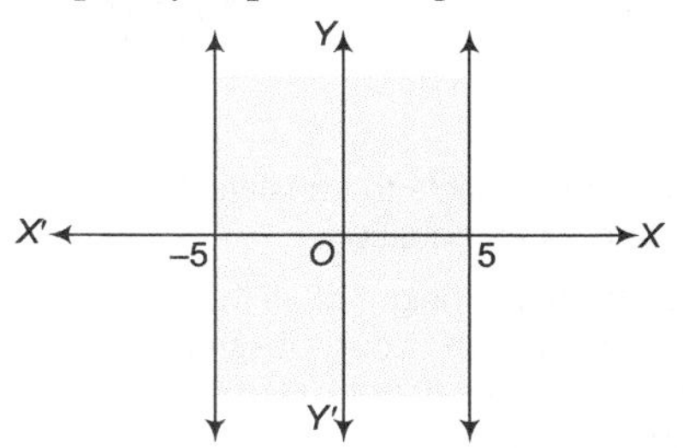

[NCERT Exemplar]

(a) $|x|<5$
(b) $|x|\leq 5$
(c) $|x|>5$
(d) $|x|\geq 5$

135. The solution set of the inequality $|x+1|+|x|>3$ is

(a) $(-\infty,-2)$
(b) $(1,\infty)$
(c) $(-\infty,-2)\cup(1,\infty)$
(d) $(-\infty,-2]\cup[1,\infty)$

136. If $5(2x-7)-3(2x+3)\leq 0$ and $2x+19\leq 6x+47$, then

(a) $x\in(-7,11)$
(b) the solution on the number line is represented by

$-\infty$ — -7 — 11 — ∞

(c) Both (a) and (b)
(d) None of the above

Answer *with* Explanations

1. *(d)* It is not possible to translate every statement problem in the form of an equation.

For example, the height of all the students in your class is less than 160 cm. Your classroom can occupy atmost 60 tables or chairs or both. Here, we get certain statements involving a sign '<' (less than), '>' (greater than), '≤' (less than or equal) and ≥ (greater than or equal) which are known as inequalities.

2. *(b)* Since, Ravi has to buy rice in packets only, he may not be able to spend the entire amount of ₹ 200 because 200 is not a multiple of 30. Hence,

$$30x < 200$$

3. *(c)* Let x denotes the number of registers and y, the number of pens which Reshma buys, then the total amount spent by her is ₹ $(40x + 20y)$. In this case, the total amount spent may be ₹ 120, then

$$40x + 20y \le 120$$

Note *The statement* $40x + 20y \le 120$ *consists of two statements* $40x + 20y < 120$ *and* $40x + 20y = 120$*, where* $40x + 20y < 120$ *is an inequality and* $40x + 20y = 120$ *is an equation.*

4. *(b)* Two real numbers or two algebraic expressions related by the symbol '<', '>', '≤', or '≥' forms an inequality.

For Example, $30x < 200$, $40x + 20y \le 120$, $40x + 20y < 120$ etc.

5. *(a)* $3 < 5$; $7 > 5$ are the examples of numerical inequalities while $x < 5$; $y > 2$; $x \ge 3$ are the example of literal inequalities.

6. *(d)* $3 < 5 < 7$ (read as 5 is greater than 3 and less than 7), $3 \le x < 5$ (read as x is greater than or equal to 3 and less than 5) and $2 < y \le 4$ (read as y is greater than 2 and less than or equal to 4) are the examples of double inequalities.

7. *(b)* Let us consider some inequalities

$$ax + b < 0 \quad \text{...(i)}$$
$$ax + b > 0 \quad \text{...(ii)}$$
$$ax + b \le 0 \quad \text{...(iii)}$$
$$ax + b \ge 0 \quad \text{...(iv)}$$
$$ax + by > c \quad \text{...(v)}$$
$$ax + by \le c \quad \text{...(vi)}$$
$$ax^2 + bx + c > 0 \quad \text{...(vii)}$$
$$ax^2 + bx + c \le 0 \quad \text{...(viii)}$$

Inequalities (i), (ii), (v) and (vii) are strict inequalities, while inequalities (iii), (iv), (vi) and (viii) are slack inequalities.

∴ Both A and R are correct but R cannot explain A.

8. *(c)* Inequalities (A) and (B) are linear inequalities in one variable.

Inequality (C) is quadratic inequality.

Inequalities (D) and (E) are linear inequalities in two variables.

9. *(a)* I. Given inequality is of type $ax + by \ge c$, $a, b \ne 0$, which is a linear inequality in two variables. So, Statement I is true.

II. Given inequality is of type $ax^2 + bx \ge c$, $a \ne 0$, which is a quadratic inequality in one variable. So, Statement II is false.

10. *(c)* Since, maximum number of computers shopkeeper can sell, is 250. Therefore,

$$x + y \le 250$$

11. *(a)* The inequality $3x + 5y > 2$ and $x < y$ contain two variables x and y with degree 1. So, they are linear inequality in two variables whereas $3x^2 + 2 \le 0$ is of the form $ax^2 + bx + c \le 0$, $a \ne 0$, which is a quadratic inequality. Now, $5x - 7 > 0$ is of the form $ax + b > 0$, $a \ne 0$, which is a linear inequality in one variable.

12. *(c)* For $x = 0$, LHS $= 30(0) = 0 < 200$ (RHS), which is true.

For $x = 1$, LHS $= 30(1) = 30 < 200$ (RHS), which is true.

For $x = 2$, LHS $= 30(2) = 60 < 200$, which is true.

For $x = 3$, LHS $= 30(3) = 90 < 200$, which is true.

For $x = 4$, LHS $= 30(4) = 120 < 200$, which is true.

For $x = 5$, LHS $= 30(5) = 150 < 200$, which is true.

For $x = 6$, LHS $= 30(6) = 180 < 200$, which is true.

In the above situation, we find that the values of x, which makes the above inequality a true statement are 0, 1, 2, 3, 4, 5, 6. These values of x, which make above inequality a true statement are called solutions of inequality and the set $\{0, 1, 2, 3, 4, 5, 6\}$ is called its solution set.

Thus, any solution of an inequality in one variable is a value of the variable which makes it a true statement.

13. *(b)* Inequalities can be solved by trial and error as well as other methods. The two important rules of equations are as follows

Rule 1 Equal numbers may be added to (or subtracted from) both sides of an equation.

Rule 2 Both sides of an equation may be multiplied (or divided) by the same non-zero number.

Solutions (Q. Nos. 14-15) In the case of solving inequalities, we follow the same rules except with a difference that in Rule 2, the sign of inequality is reversed (*i.e.*, '<' becomes '>', '≤' becomes '≥' and vice versa) whenever we multiply (or divide) both sides of an inequality by a negative number. It is evident from the facts that

$$3 > 2, \text{ while } -3 < -2,$$

$$-8 < -7, \text{ while } (-8)(-2) > (-7)(-2), \; i.e., 16 > 14$$

14. *(a)*

15. *(b)* In the given inequality $(-8)(-2) > (-7)(-2)$, divide both sides by -2. Since -2 is a negative number, sign of inequality will change. Hence, $-8 < -7$.

16. *(d)* We state the following rules for solving an inequality.

Rule 1 Equal numbers may be added to (or subtracted from) both sides of an inequality without affecting the sign of inequality.

Rule 2 Both sides of an inequality can be multiplied (or divided) by the same positive number. But when both sides are multiplied or divided by a negative number then the sign of inequality is reversed.

17. *(c)* We are given $30x < 200$

or $\dfrac{30x}{30} < \dfrac{200}{30}$, *i.e.*, $x < 20/3$

A. When x is a natural number, in this case the following values of x make the statement true

1, 2, 3, 4, 5, 6

The solution set of the inequality is {1, 2, 3, 4, 5, 6}.

B. When x is an integer, the solutions of the given inequality are

..., –3, –2, – 1, 0, 1, 2, 3, 4, 5, 6

The solution set of the inequality is

{..., –3, –2, – 1, 0, 1,2, 3, 4, 5, 6}.

Again, we have, $5x - 3 < 3x + 1$

or $5x - 3 + 3 < 3x + 1 + 3$

or $5x < 3x + 4$

or $5x - 3x < 3x + 4 - 3x$

or $2x < 4$

or $x < 2$

C. When x is an integer, the solutions of the given inequality are {..., –4, –3, –2, –1, 0, 1}.

D. When x is a real number, the solutions of the inequality are given by $x < 2$, *i.e.,* all real numbers x which are less than 2. Therefore, the solution set of the inequality is $x \in (-\infty, 2)$.

18. *(c)* We have, $4x + 3 < 6x + 7$

or $4x - 6x < 6x + 4 - 6x$

or $-2x < 4$ or $x > -2$

i.e., all the real numbers which are greater than –2, are the solutions of the given inequality. Hence, the solution set is $(-2, \infty)$.

19. *(c)* We have,

$$\frac{5-2x}{3} \le \frac{x}{6} - 5$$

or $2(5 - 2x) \le x - 30$

or $10 - 4x \le x - 30$

or $-5x \le -40$, *i.e.*, $x \ge 8$

Thus, all real numbers x which are greater than or equal to 8 are the solutions of the given inequality, *i.e.*, $x \in [8, \infty)$.

20. *(a)* We have, $7x + 3 < 5x + 9$ or

$2x < 6$ or $x < 3$

$\Rightarrow$ $x \in (-\infty, 3)$

The graphical representation of the solutions are given in figure.

–3 –2 –1 0 1 2 3 4 5 6

21. *(a)* We have,

$$\frac{3x-4}{2} \ge \frac{x+1}{4} - 1$$

or $\dfrac{3x-4}{2} \ge \dfrac{x-3}{4}$

or $2(3x - 4) \ge (x - 3)$

or $6x - 8 \ge x - 3$

or $5x \ge 5$ or $x \ge 1$

The graphical representation of solutions is given in figure.

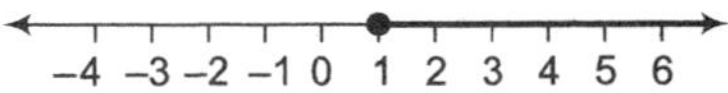

22. *(a)* Let x be the marks obtained by student in the annual examination. Then,

$$\frac{62 + 48 + x}{3} \ge 60$$

or $110 + x \ge 180$

or $x \ge 70$

Thus, the student must obtain a minimum of 70 marks to get an average of atleast 60 marks.

23. *(c)* Let x be the smaller of the two consecutive odd natural number, so that the other one is $x + 2$. Then, we should have

$$x > 10 \qquad \text{...(i)}$$

and $x + (x + 2) < 40$ ...(ii)

Solving (ii), we get

$$2x + 2 < 40$$

i.e., $x < 19$ ...(iii)

From (i) and (iii), we get

$$10 < x < 19$$

Since x is an odd number, x can take the values 11, 13, 15, and 17. So, the required possible pairs will be

(11, 13), (13, 15), (15, 17), (17, 19)

24. *(c)* We have, $24x < 100$

Dividing both sides by 24,

$$\frac{24x}{24} < \frac{100}{24}$$

$$\Rightarrow \quad x < \frac{50}{12} \Rightarrow x < \frac{25}{6}$$

$$\Rightarrow \quad x < 4\frac{1}{6} \quad \left(i.e., x \text{ is less than } 4\frac{1}{6}\right)$$

A. When x is a natural number (only positive integer), in this case, the solution set of the inequality is $\{1, 2, 3, 4\}$.

B. When x is an integer, the solution set of the given inequality is $\{\ldots\ldots -4, -3, -2, -1, 0, 1, 2, 3, 4\}$.

Again, we have $-12x > 30$

Dividing both sides by -12,

$$\frac{-12x}{-12} < \frac{30}{-12}$$

$$\Rightarrow \quad x < -\frac{5}{2}$$

C. When x is a natural number, then there is no solution of given inequality. As natural numbers are positive numbers and there is no positive number which is less than negative number.

D. When x is an integer, the solution of given inequality is $\{\ldots -4, -3\}$ as there are infinite number which are less than $-\frac{5}{2}$.

25. *(b)* We have, $5x - 3 < 7$

Adding 3 on both sides,

$$5x - 3 + 3 < 7 + 3$$

$$\Rightarrow \quad 5x < 10$$

Dividing both sides by 5,

$$\frac{5x}{5} < \frac{10}{5}$$

$$\Rightarrow \quad x < 2$$

I. When x is an integer, the solution of the given inequality is $\{\ldots, -1, 0, 1\}$.

II. When x is a real number, the solution of given inequality is $(-\infty, 2)$ *i.e.*, all the numbers lying between $-\infty$ and 2 but ∞ and 2 are not included as $x < 2$.

26. *(a)* We have, $3x + 8 > 2$

Adding -8 on both sides,

$$3x + 8 - 8 > 2 - 8$$

$$\Rightarrow \quad 3x > -6$$

Dividing by 3 on both sides,

$$\Rightarrow \quad \frac{3x}{3} > \frac{-6}{3}$$

$$\Rightarrow \quad x > -2$$

(i) When x is an integer, the solution of the given inequality is $\{-1, 0, 1, 2 \ldots\}$.

(ii) When x is a real number, the solution of the given inequality is $(-2, \infty)$. *i.e.*, all the numbers lying between -2 and ∞ but -2 and ∞ are not included.

27. *(a)* We have, $4x + 3 < 5x + 7$

Transferring the term $5x$ to LHS and the term 3 to RHS,

$$4x - 5x < 7 - 3 \Rightarrow -x < 4 \Rightarrow x > -4$$

With the help of number line, we can easily look for the numbers greater than -4.

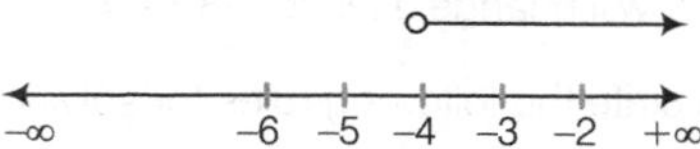

$\therefore$ Solution set is $(-4, \infty)$ *i.e.*, all the numbers lying between -4 and ∞ but -4 and ∞ are not included as $x > -4$.

28. *(a)* We have, $3x - 7 > 5x - 1$

Transferring the term $5x$ to LHS and the term -7 to RHS.

Dividing both sides by 2,

$$3x - 5x > -1 + 7$$

$$\Rightarrow \quad -2x > 6$$

$$\Rightarrow \quad \frac{2x}{2} < -\frac{6}{2}$$

$$\Rightarrow \quad x < -3$$

With the help of number line, we can easily look for the numbers less than -3.

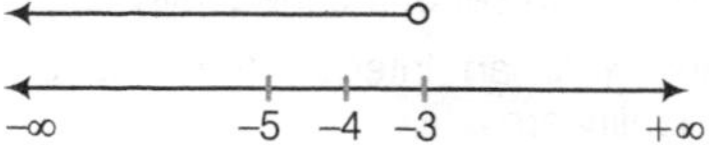

$\therefore$ Solution set is $(-\infty, -3)$ *i.e.*, all the numbers lying between $-\infty$ and -3 but $-\infty$ and -3 are not included as $x < -3$.

29. *(d)* We have, $3x - 3 \le 2x - 6$

Transferring the term $2x$ to LHS and the term -3 to RHS,

$$3x - 2x \le -6 + 3$$

$$\Rightarrow \quad x \le -3$$

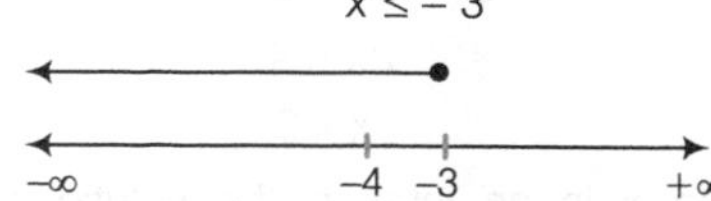

$\therefore$ Solution set is $(-\infty, -3]$.

30. *(b)* We have, $3(2 - x) \ge 2(1 - x) = 6 - 3x \ge 2 - 2x$

Transferring the term 2 to LHS and the term $-3x$ to RHS,

$$6 - 2 \ge -2x + 3x$$

$$\Rightarrow \quad 4 \ge x \Rightarrow x \le 4$$

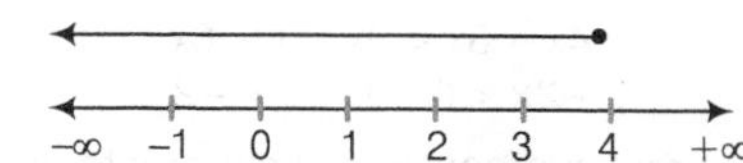

$\therefore$ Solution set is $(-\infty, 4]$.

31. *(a)* I. We have, $x + \frac{x}{2} + \frac{x}{3} < 11$

$$\Rightarrow \quad \frac{x}{1} + \frac{x}{2} + \frac{x}{3} < 11$$

Taking LCM in LHS,

$$\frac{6x + 3x + 2x}{6} < 11 \Rightarrow \frac{11x}{6} < 11$$

Multiplying both sides by $\frac{6}{11}$,

$$\frac{6}{11} \times \frac{11x}{6} < 11 \times \frac{6}{11}$$

$$\Rightarrow \quad x < \frac{6 \times 11}{11} \Rightarrow x < 6$$

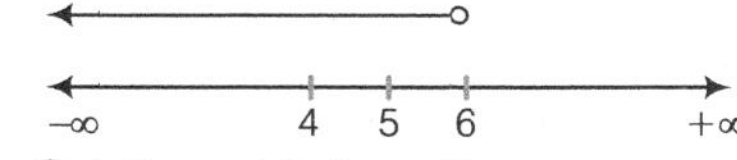

∴ Solution set is $(-\infty, 6)$.

II. Again, we have, $\frac{x}{3} > \frac{x}{2} + 1$

Transferring the term $\frac{x}{2}$ to LHS,

$$\frac{x}{3} - \frac{x}{2} > 1$$

Taking LCM in LHS,

$$\frac{2x - 3x}{6} > 1$$

$$\Rightarrow \quad \frac{-x}{6} > 1$$

Multiplying both sides by 6,

$$6 \times \frac{-x}{6} > 1 \times 6$$

$$\Rightarrow \quad -x > 6 \Rightarrow x < -6$$

∴ Solution set is $(-\infty, -6)$.

32. *(b)* I. We have, $\frac{3(x-2)}{5} \le \frac{5(2-x)}{3}$

$$\Rightarrow \quad \frac{3x - 6}{5} \le \frac{10 - 5x}{3}$$

$$\Rightarrow \quad 9x - 18 \le 50 - 25x$$

Transferring the terms $(-25x)$ to LHS and the term (-18) to RHS,

$$9x + 25x \le 50 + 18$$

$$\Rightarrow \quad 34x \le 68$$

$$\Rightarrow \quad x \le \frac{68}{34} \Rightarrow x \le 2$$

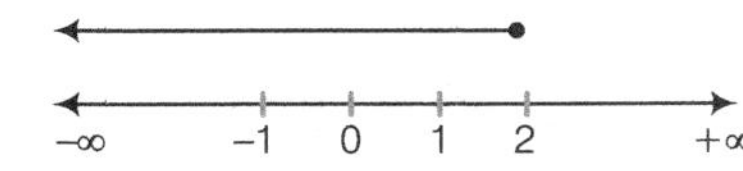

∴ Solution set is $(-\infty, 2]$.

II. We have, $\frac{1}{2}\left(\frac{3x}{5} + 4\right) \ge \frac{1}{3}(x - 6)$

$$\Rightarrow \quad \frac{1}{2}\left(\frac{3x}{5} + \frac{4}{1}\right) \ge \frac{1}{3}(x - 6)$$

Taking LCM in LHS,

$$\frac{1}{2}\left(\frac{3x + 20}{5}\right) \ge \frac{1}{3}(x - 6)$$

$$\Rightarrow \quad \frac{3x + 20}{10} \ge \frac{x - 6}{3}$$

$$\Rightarrow \quad 3(3x + 20) \ge 10(x - 6)$$

$$\Rightarrow \quad 9x + 60 \ge 10x - 60$$

Transferring the term $10x$ to LHS and the term 60 to RHS,

$$9x - 10x \ge -60 - 60 \Rightarrow -x \ge -120$$

Multiplying both sides by -1,

$$x \le 120$$

∴ Solution set is $(-\infty, 120]$.

33. *(c)* We have, $2(2x + 3) - 10 < 6(x - 2)$

$$\Rightarrow \quad 4x + 6 - 10 < 6x - 12$$

$$\Rightarrow \quad 4x - 4 < 6x - 12$$

Transferring the term $6x$ to LHS and (-4) to RHS,

$$4x - 6x < -12 + 4$$

$$\Rightarrow \quad -2x < -8$$

Dividing both sides by -2,

$$\frac{-2x}{-2} > \frac{-8}{-2}$$

$$\Rightarrow \quad x > 4$$

∴ Solution set is $(4, \infty)$.

34. *(c)* We have, $37 - (3x + 5) \ge 9x - 8(x - 3)$

$$(37 - 3x - 5) \ge 9x - 8x + 24$$

$$\Rightarrow \quad 32 - 3x \ge x + 24$$

Transferring the term 24 to LHS and the term $(-3x)$ to RHS,

$$32 - 24 \ge x + 3x$$

$$\Rightarrow \quad 8 \ge 4x$$

$$\Rightarrow \quad 4x \le 8$$

Dividing both sides by 4,

$$\frac{4x}{4} \le \frac{8}{4}$$

$$\Rightarrow \quad x \le 2$$

∴ Solution set is $(-\infty, 2]$.

35. *(a)* We have, $\frac{x}{4} < \frac{(5x-2)}{3} - \frac{(7x-3)}{5}$

$$\frac{x}{4} < \frac{5(5x-2)-3(7x-3)}{15}$$

$$\Rightarrow \quad 15x < 4[(25x-10)-(21x-9)]$$

$$\Rightarrow \quad 15x < 4[(25x-10-21x+9]$$

$$\Rightarrow \quad 15x < 4[4x-1]$$

$$\Rightarrow \quad 15x < 16x-4$$

Transferring the term $16x$ to LHS.

$$15x-16x < -4 \Rightarrow -x < -4$$

Multiplying by -1 both sides, we get

$$x > 4$$

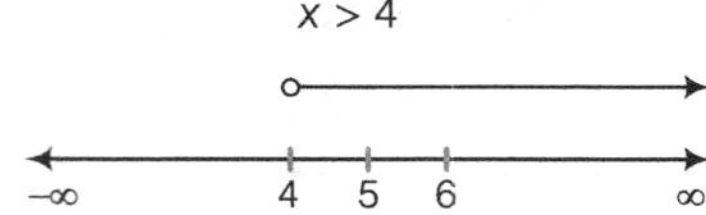

$\therefore$ Solution set is $(4, \infty)$.

36. *(b)* We have, $\frac{2x-1}{3} \geq \left(\frac{3x-2}{4}\right) - \left(\frac{2-x}{5}\right)$

Taking LCM in RHS,

$$\frac{2x-1}{3} \geq \frac{5(3x-2)-4(2-x)}{20}$$

$$\Rightarrow \quad \frac{2x-1}{3} \geq \frac{(15x-10)-(8-4x)}{20}$$

$$\Rightarrow \quad \frac{2x-1}{3} \geq \frac{15x-10-8+4x}{20}$$

$$\Rightarrow \quad \frac{2x-1}{3} \geq \frac{19x-18}{20}$$

$$\Rightarrow \quad 20(2x-1) \geq 3(19x-18)$$

$$\Rightarrow \quad 40x-20 \geq 57x-54$$

Transferring the term $57x$ to LHS and the term -20 to RHS,

$$40x-57x \geq -54+20$$

$$\Rightarrow \quad -17x \geq -34$$

Dividing both sides by -17,

$$\frac{-17x}{-17} \leq \frac{-34}{-17}$$

$$\Rightarrow \quad x \leq \frac{-34}{-17} \Rightarrow x \leq 2$$

$\therefore$ Solution set is $(-\infty, 2]$.

37. *(a)* We have, $3x-2 < 2x+1$

Transferring the term $2x$ to LHS and the term (-2) to RHS,

$$3x-2x < 1+2 \Rightarrow x < 3$$

All the numbers on the left side of 3 will be less than it.

$\therefore$ Solution set is $(-\infty, 3)$.

38. *(b)* I. We have, $5x-3 \geq 3x-5$

Transferring the term $3x$ to LHS and the term (-3) to RHS,

$$5x-3x \geq -5+3$$

$$\Rightarrow \quad 2x \geq -2$$

$$\Rightarrow \quad \frac{2x}{2} \geq -\frac{2}{2}$$

$$\Rightarrow \quad x \geq \frac{-2}{2}$$

$$\Rightarrow \quad x \geq -1$$

All the numbers on the right side of -1 will be greater than it.

$\therefore$ Solution set is $[-1, \infty)$.

II. Again, we have, $3(1-x) < 2(x+4)$

$$\Rightarrow \quad 3-3x < 2x+8$$

Transferring the term $2x$ to the LHS and the term 3 to RHS,

$$-3x-2x < 8-3$$

$$\Rightarrow \quad -5x < 5$$

$$\Rightarrow \quad \frac{-5x}{-5} > \frac{5}{-5}$$

$$\Rightarrow \quad x > \frac{-5}{5} \Rightarrow x > -1$$

$\therefore$ Solution set is $(-1, \infty)$.

39. *(a)* We have, $\frac{x}{2} \geq \frac{(5x-2)}{3} - \frac{(7x-3)}{5}$

Taking LCM in RHS,

$$\frac{x}{2} \geq \frac{5(5x-2)-3(7x-3)}{15}$$

$$\Rightarrow \quad \frac{x}{2} \geq \frac{25x-10-21x+9}{15}$$

$$\Rightarrow \quad \frac{x}{2} \geq \frac{(25x-21x)-(10-9)}{15}$$

$$\Rightarrow \quad \frac{x}{2} \geq \frac{4x-1}{15}$$

$$\Rightarrow \quad 15x \geq 2(4x-1)$$

$$\Rightarrow \quad 15x \geq 8x-2$$

Transferring the term $8x$ to LHS,

$$15x-8x \geq -2 \Rightarrow 7x \geq -2$$

Dividing both sides by 7,

$$\frac{7x}{7} \geq \frac{-2}{7} \Rightarrow x \geq -\frac{2}{7}$$

$\therefore$ Solution set is $\left[\frac{-2}{7}, \infty\right)$.

40. *(b)* Let Ravi got x marks in third unit test.

$\therefore$ Average marks obtained by Ravi

$$= \frac{\text{Sum of marks in all tests}}{\text{Number of tests}}$$

$$= \frac{70 + 75 + x}{3} = \frac{145 + x}{3}$$

Now, it is given that he wants to obtain an average of atleast 60 marks.

Atleast 60 marks means that the marks should be greater than or equal to 60.

i.e., $\frac{145 + x}{3} \geq 60$

$\Rightarrow \quad 145 + x \geq 60 \times 3$

$\Rightarrow \quad 145 + x \geq 180$

Now, transferring the term 145 to RHS,

$x \geq 180 - 145$

$\Rightarrow \quad x \geq 35$

i.e., Ravi should get greater than or equal to 35 marks in third unit test to get an average of atleast 60 marks.

$\therefore$ Minimum marks Ravi should get = 35

41. *(b)* Let Sunita got x marks in the fifth exam.

$\therefore$ Average marks obtained by Sunita

$$= \frac{\text{Sum of marks in all exams}}{\text{Number of exams}}$$

$$= \frac{87 + 92 + 94 + 95 + x}{5} = \frac{368 + x}{5}$$

Now, it is given that Sunita wants to obtain grade A for that her average marks should be greater than or equal to 90.

i.e., $\frac{368 + x}{5} \geq 90$

$\Rightarrow \quad 368 + x \geq 450$

Transferring the term 368 to RHS,

$x \geq 450 - 368$

$\Rightarrow \quad x \geq 82$

i.e., Sunita should get greater than or equal to 82 marks in fifth exam to get grade A.

$\therefore$ Minimum marks = 82

42. *(a)* Let the numbers are $2x + 1$ and $2x + 3$. Then, according to the question,

$2x + 1 < 10$ and $2x + 3 < 10$

$\Rightarrow \quad 2x < 9$ and $2x < 7$

$\Rightarrow \quad x < \frac{9}{2}$ and $x < \frac{7}{2}$

Also, $\quad 2x + 3 + 2x + 1 > 11$

$\Rightarrow \quad 4x + 4 > 11$

$\Rightarrow \quad 4x > 11 - 4$

$\Rightarrow \quad x > \frac{7}{4}$

Now, plotting all the values of x on line.

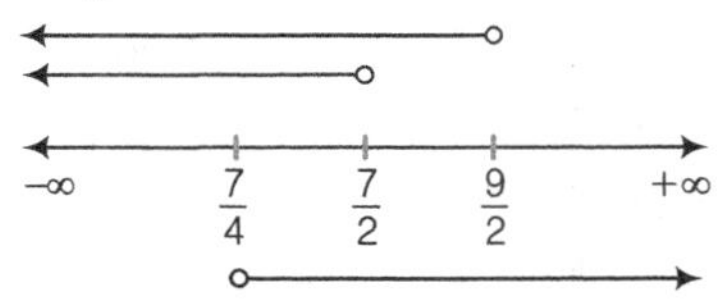

From the graph, it is clear that $x \in \left(\frac{7}{4}, \frac{7}{2}\right)$ in which integer values are $x = 2$ and 3.

When $x = 3$, the numbers are $(2 \times 3 + 1, 2 \times 3 + 3) = (7, 9)$

When $x = 2$, the numbers are $(2 \times 2 + 1, 2 \times 2 + 3) = (5, 7)$

$\Rightarrow$ Required pairs are (5, 7) and (7, 9).

43. *(b)* Let numbers are $2x$ and $2x + 2$

Then, according to the question,

$2x > 5 \Rightarrow x > \frac{5}{2}$

and $\quad 2x + 2 > 5 \Rightarrow 2x > 5 - 2$

$\Rightarrow \quad 2x > 3 \Rightarrow x > 3/2$

and $\quad 2x + 2x + 2 < 23 \Rightarrow 4x < 23 - 2$

$\Rightarrow \quad 4x < 21$

$\Rightarrow \quad x < \frac{21}{4}$

Now, plotting all these values on number line.

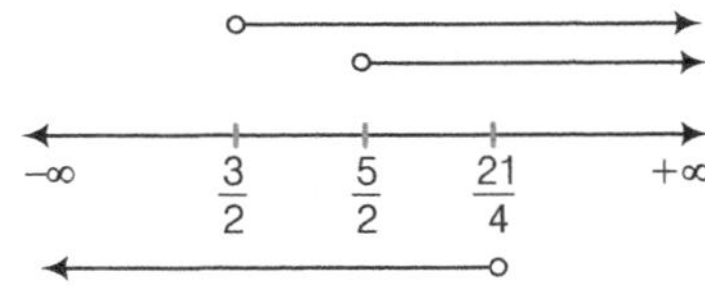

From above graph, it is clear that $x \in \left(\frac{5}{2}, \frac{21}{4}\right)$ in which integer values are $x = 3, 4, 5$

When $x = 3$, pair is $(2 \times 3, 2 \times 3 + 2) = (6, 8)$

When $x = 4$, pair is $(2 \times 4, 2 \times 4 + 2) = (8, 10)$

When $x = 5$, pair is $(2 \times 5, 2 \times 5 + 2) = (10, 12)$

$\therefore$ Required pairs are (6, 8), (8, 10), (10, 12).

44. *(c)* Let the shortest side be x cm.

Then, according to the given condition,

Longest side = $3x$ cm and third side = $(3x - 2)$ cm

Now, perimeter of triangle ≥ 61 *i.e.*, sum of all sides ≥ 61

$\Rightarrow \quad x + 3x + 3x - 2 \geq 61$

$\Rightarrow \quad 2 + 7x - 2 \geq 61 + 2$

$\Rightarrow \quad 7x \geq 63$

$\Rightarrow \quad \frac{7x}{7} \geq \frac{63}{7}$

$\Rightarrow \quad x \geq 9$

$\therefore$ Minimum length of the shortest side is 9 cm.

45. *(b)* Let the shortest side be x cm.

Then, by given condition, second length $= x + 3$ cm

Third length $= 2x$ cm

Also given, total length $= 91$

Hence, sum of all the three lengths should be less than or equal to 91.

$$x + x + 3 + 2x \le 91$$

$$\Rightarrow \quad 4x + 3 \le 91$$

Subtracting (–3) to each term,

$$-3 + 4x + 3 \le 91 - 3$$

$$\Rightarrow \quad 4x \le 88$$

$$\Rightarrow \quad \frac{4x}{4} \le \frac{88}{4}$$

$$\Rightarrow \quad x \le \frac{88}{4}$$

$$\Rightarrow \quad x \le 22 \text{ cm} \quad \ldots\text{(i)}$$

Again, given that

Third length ≥ second length +5

$$\Rightarrow \quad 2x \ge (x + 3) + 5 \Rightarrow 2x \ge x + (3 + 5)$$

Transferring the term x to LHS,

$$2x - x \ge 8 \Rightarrow x \ge 8 \quad \ldots\text{(ii)}$$

From Eqs. (i) and (ii), length of shortest board should be greater than or equal to 8 but less than or equal to 22 *i.e.*, $8 \le x \le 22$.

46. *(c)* Given, $3 < 3t - 18 \le 18$

Adding 18 to each term,

$$3 + 18 < 3t - 18 + 18 \le 18 + 18$$

$$\Rightarrow \quad 21 \le 3t \le 36$$

Dividing by 3 to each term,

$$\frac{21}{3} \le \frac{3t}{3} \le \frac{36}{3}$$

$$\Rightarrow \quad 7 \le t \le 12$$

Adding 1 to each term,

$$7 + 1 \le t + 1 \le 12 + 1$$

$$\Rightarrow \quad 8 \le t + 1 \le 13$$

47. *(b)* The given inequality $6 \le -3(2x - 4) < 12$

$$6 \le -6x + 12 < 12$$

Adding (– 12) to each term,

$$6 - 12 \le -6x + 12 - 12 < 12 - 12$$

$$\Rightarrow \quad -6 \le -6x < 0$$

Dividing by (– 6) to each term,

$$\frac{-6}{-6} \ge \frac{-6x}{-6} > \frac{0}{-6}$$

$$\Rightarrow \quad 1 \ge x > 0$$

$$\Rightarrow \quad 0 < x \le 1$$

∴ Solution set is (0, 1].

48. *(c)* The given inequality $-3 \le 4 - \frac{7x}{2} \le 18$

Adding (– 4) to each term,

$$-3 - 4 \le 4 - \frac{7x}{2} - 4 \le 18 - 4$$

$$\Rightarrow \quad -7 \le \frac{-7x}{2} \le 14$$

Multiplying each term by $\left(\frac{-2}{7}\right)$

$$-7 \times \left(-\frac{2}{7}\right) \ge -\frac{7}{2} x \times \left(-\frac{2}{7}\right) \ge 14 \times \left(-\frac{2}{7}\right)$$

$$\Rightarrow \quad 2 \ge x \ge -4 \Rightarrow -4 \le x \le 2$$

∴ Solution set is [– 4, 2].

49. *(d)* I. The given inequality $2 \le 3x - 4 \le 5$

$$\Rightarrow \quad 2 + 4 \le 3x \le 5 + 4$$

$$\Rightarrow \quad 6 \le 3x \le 9$$

Dividing each term by 3

$$\frac{6}{3} \le \frac{3x}{3} \le \frac{9}{3}$$

$$\Rightarrow \quad 2 \le x \le 3$$

∴ Solution set is [2, 3].

II. The given inequality

$$-12 < 4 - \frac{3x}{-5} \le 2$$

$$\Rightarrow \quad -12 < 4 + \frac{3x}{5} \le 2$$

Adding (– 4) to each term,

$$-12 - 4 < 4 + \frac{3x}{5} - 4 \le 2 - 4$$

$$\Rightarrow \quad -16 < \frac{3x}{5} \le -2$$

Multiplying each term by $\frac{5}{3}$

$$-16 \times \frac{5}{3} < \frac{3x}{5} \times \frac{5}{3} \le -2 \times \frac{5}{3}$$

$$\Rightarrow \quad -\frac{80}{3} < x \le -\frac{10}{3}$$

∴ Solution set is $\left(-\frac{80}{3}, -\frac{10}{3}\right]$ or $\left]-\frac{80}{3}, \frac{-10}{3}\right]$.

50. *(b)* The given inequality is $5x + 1 > -24$.

Adding (–1) on both sides,

$$-1 + 5x + 1 > -24 - 1$$

$$\Rightarrow \quad 5x > -25$$

Dividing both sides by 5,

$$x > \frac{-25}{5}$$

$$\Rightarrow \quad x > -5 \quad \ldots\text{(i)}$$

Also, $5x - 1 < 24$

Adding 1 on both sides,

$$1 + 5x - 1 < 24 + 1$$

$\Rightarrow \quad 5x < 25$

$\Rightarrow \quad x < \frac{25}{5} \Rightarrow x < 5 \quad \ldots(ii)$

Draw the graph of inequations (i) and (ii) on the number line.

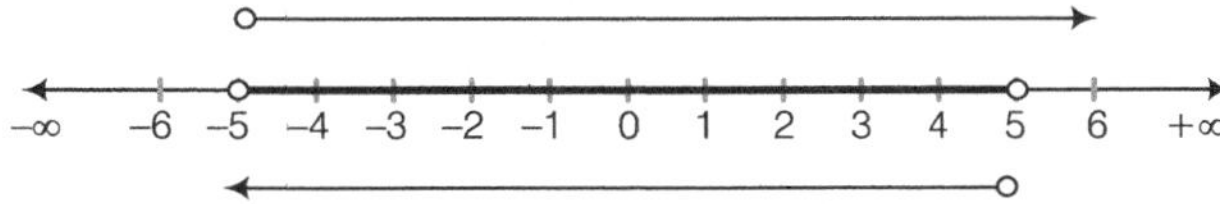

Hence, solutions of the system are real numbers x lying between -5 and 5 excluding -5 and 5 *i.e.*, $-5 < x < 5$.

$\therefore$ Solution set is $(-5, 5)$ or $]-5, 5[$.

51. *(c)* We have the given inequalities as

$$2(x-1) < x + 5 \quad \text{and} \quad 3(x+2) > 2 - x$$

Now, $\quad 2x - 2 < x + 5$

Transferring the term x to LHS and the term -2 to RHS,

$$2x - x < 5 + 2$$

$\Rightarrow \quad x < 7 \quad \ldots(i)$

and $\quad 3(x+2) > 2 - x$

$\Rightarrow \quad 3x + 6 > 2 - x$

Transferring the term $(-x)$ to LHS and the term 6 to RHS,

$\Rightarrow \quad 3x + x > 2 - 6$

$\Rightarrow \quad 4x > -4$

Dividing both sides by 4,

$$x > \frac{-4}{4}$$

$\Rightarrow \quad x > -1 \quad \ldots(ii)$

$\Rightarrow$ Draw the graph of inequalities (i) and (ii) on the number line.

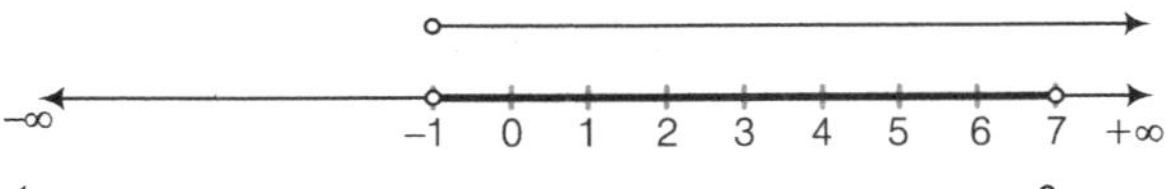

Hence, solution set of the inequalities are real numbers, x lying between -1 and 7 excluding -1 and 7.

i.e., $\quad -1 < x < 7$

$\therefore$ Solution set is $(-1, 7)$ or $]-1, 7[$.

52. *(c)* We have, $3x - 5 < x + 7$

Adding 5 on both sides,

$$3x - 5 + 5 < x + 7 + 5$$

$\Rightarrow \quad 3x < x + 12$

Subtracting x from both sides,

$$3x - x < x + 12 - x$$

$\Rightarrow \quad 2x < 12$

Dividing by 2 both sides,

$$\frac{2x}{2} < \frac{12}{2} \Rightarrow x < 6$$

A. If x is a natural number, solution set = {1, 2, 3, 4, 5}

B. If x is an integer, solution set = {..., –3, –2, –1, 0, 1, 2, 3, 4, 5}

C. If x is a whole number, solution set = {0, 1, 2, 3, 4, 5}

D. If x is a real number, solution set = $\{x : x \in R, x < 6\}$

53. *(c)* We have, $2x - 7 < 11$ and $3x + 4 < -5$

Now, $\quad 2x - 7 < 11$

Adding 7 on both sides,

$$2x < 11 + 7 \Rightarrow 2x < 18$$

Dividing by 2 both sides,

$x < 9 \quad \ldots(i)$

and $\quad 3x + 4 < -5$

Adding (-4) on both sides,

$$3x < -9$$

Dividing by 3 both sides,

$x < -3 \quad \ldots(ii)$

Draw the graph of inequations (i) and (ii) on the number line

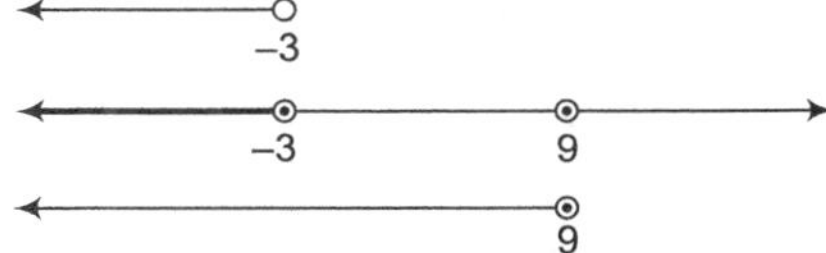

Hence, solution of the system of linear inequations are real numbers less than -3 excluding -3.

$\therefore$ Solution set is $(-\infty, -3)$.

54. *(b)* We have, $|x| < 3$

$\Rightarrow \quad -3 < x < 3 \quad (\because |x| < a \Rightarrow -a < x < a)$

55. *(d)* We have, $|x| > b, b > 0$

$\Rightarrow \quad x < -b$ and $x > b$

$\Rightarrow \quad x \in (-\infty, -b) \cup (b, \infty)$

56. *(c)* We have, $\frac{x-2}{x+5} > 2$

Subtracting 2 from both sides,

$$\frac{x-2}{x+5} - 2 > 0$$

$\Rightarrow \quad \frac{x - 2 - 2(x+5)}{x+5} > 0$

$\Rightarrow \quad \frac{x - 2 - 2x - 10}{x+5} > 0$

$\Rightarrow \quad \frac{-(x+12)}{x+5} > 0$

Multiply by (-1) on both sides,

$$\frac{x+12}{x+5} < 0$$

Given inequality will be less than 0 in two cases.

Case I

$$(x+12)>0 \quad \text{and} \quad x+5<0$$

$$\Rightarrow \quad x>-12 \quad \text{and} \quad x<-5$$

$$\left(\because \frac{a}{b}>0 \Rightarrow a \text{ and } b \text{ are of opposite signs}\right)$$

Case II

$$(x+12)<0 \quad \text{and} \quad x+5>0$$

or $\quad x<-12 \quad \text{and} \quad x>-5 \quad$ (not possible)

$$\therefore \quad -12<x<-5$$

i.e., $\quad x \in(-12,-5)$

57. *(c)* We have, $|3-4x| \geq 9$

$$\Rightarrow \quad 3-4x \leq -9 \quad \text{or} \quad 3-4x \geq 9$$

$$(\because |x| \geq a \Rightarrow x \leq -a \text{ or } x \geq a)$$

Subtracting 3 from both sides of each inequality,

$$3-4x-3 \leq -9-3 \quad \text{or} \quad 3-4x-3 \geq 9-3$$

$$-4x \leq -12 \quad \text{or} \quad -4x \geq 6$$

Dividing by (–4) on both sides of each inequality,

$$\frac{-4x}{-4} \geq \frac{-12}{-4} \quad \text{or} \quad \frac{-4x}{-4} \leq \frac{6}{-4}$$

$$x \geq 3 \quad \text{or} \quad x \leq -\frac{3}{2}$$

$$\therefore \quad x \in\left(-\infty,-\frac{3}{2}\right] \cup[3, \infty)$$

58. *(b)* We have, $\frac{|x-2|}{x-2} \geq 0$

Case I If $x<2$

$$\frac{|x-2|}{x-2} \geq 0$$

$$\Rightarrow \quad \frac{-(x-2)}{x-2} \geq 0 \quad [\because |x-2|=-(x-2), \text{ if } x<2]$$

$$\Rightarrow \quad -1 \geq 0$$

which is not true.

Case II If $x>2$

$$\frac{|x-2|}{x-2} \geq 0$$

$$\Rightarrow \quad \frac{x-2}{x-2} \geq 0 \quad (\because |x-2|=x-2, \text{ if } x>2)$$

$$\Rightarrow \quad 1 \geq 0$$

which is true.

Also, given expression is not defined at $x=2$, since denominator at $x=2$ is 0. There is no solution at $x=2$.

Hence, $x>2$ is the required solution. *i.e.,* $x \in(2, \infty)$

59. *(b)* We have,

$$a<b$$

and $\quad c<0$

Dividing both sides of $a<b$ by c. Since, c is a negative number, sign at inequality will get reversed.

Hence, $\quad \frac{a}{c}>\frac{b}{c}$

60. *(c)* I. Given, $x \geq -3$

Adding 5 on both sides,

$$x+5 \geq -3+5$$

$$\Rightarrow \quad x+5 \geq 2$$

II. We have, $-x \leq -4$

Multiply (–2) on both sides,

$$2x \geq 8$$

III. Given, $\quad \frac{1}{x-2}<0$

$$\Rightarrow \quad x-2<0$$

$$\Rightarrow \quad x<2$$

61. *(b)* We have, $p>0, q<0$

$$\because \quad q<0$$

Adding p on both sides,

$$p+q<p$$

62. *(b)* I. **Case I** If $x<1$, then

$$-(x-1) \leq 2 \quad [\because |x-1|=(x-1), \text{ if } x<1]$$

Subtracting 1 from both sides,

$$-x \leq 1$$

Multiplying (–1) on both sides,

$$x \geq -1$$

$$\Rightarrow \quad x \in[-1,1)$$

Case II If $x \geq 1$, then

$$x-1 \leq 2 \quad (\because |x-1|=x-1, \text{ if } x \geq 1)$$

Adding 1 on both sides,

$$x \leq 3$$

$$\Rightarrow \quad x \in(-\infty, 3]$$

$\therefore$ The solution set of given inequality is

$$[-1,1) \cap(-\infty, 3] \text{ i.e., } [-1,3]$$

Hence, $\quad -1 \leq x \leq 3$

II. **Case I** If $x<\frac{7}{3}$, then

$$-(3x-7)>2$$

$$\left[\because |3x-7|=-(3x-7), \text{ if } x<\frac{7}{3}\right]$$

Multiplying (–1) on both sides,

$$3x-7<-2$$

Adding 7 on both sides,

$$3x<-2+7$$

Dividing both sides by 3,

$$x < \frac{5}{3}$$

***Case* II** $x \geq \frac{7}{3}$

$$3x - 7 > 2 \qquad \left(\because |3x-7| = 3x - 7, \text{if } x \geq \frac{7}{3}\right)$$

Adding 7 on both sides,

$$3x > 9$$

Dividing both sides by 3,

$$x > 3$$

Hence, the solution set of given inequality is

$$x < \frac{5}{3} \quad \text{or} \quad x > 3$$

63. *(a)* We have, $-3x + 17 < -13$

Subtracting 17 from both sides,

$$-3x < -13 - 17$$

$$\Rightarrow \quad -3x < -30$$

Dividing both sides by (-3),

$$x > 10$$

$$\Rightarrow \quad x \in (10, \infty)$$

64. *(b)* The number line represents

$$x \geq \frac{9}{2} \text{ i.e., } x \in \left[\frac{9}{2}, \infty\right)$$

65. *(b)* To earn some profit, we must have

$$R(x) > C(x)$$

$$\Rightarrow \quad 60x + 2000 > 20x + 4000$$

Subtracting $20x$ from both sides,

$$40x + 2000 > 4000$$

Subtracting 2000 from both sides,

$$40x > 2000$$

Dividing both sides by 40,

$$x > \frac{2000}{40} \Rightarrow x > 50$$

66. *(c)* I. If $x > 0$ and $y > 0$, then $xy > 0$ and if $xy < 0$, then one of them must less than zero.

II. If $xy > 0$, either both are positive or both are negative.

i.e., $x > 0$, $y > 0$ or $x < 0$, $y < 0$

III. $|x| < 5$

$$\Rightarrow \quad -5 < x < 5 \qquad (\because |x| < a \Rightarrow -a < x < a)$$

$$\Rightarrow \quad x \in (-5, 5)$$

67. *(c)* We have, $|x + 3| \geq 10$

$$\Rightarrow \quad x + 3 \leq -10 \quad \text{or} \quad x + 3 \geq 10$$

$$(\because |x| \geq a \Rightarrow x < -a \text{ or } x > a)$$

$$\Rightarrow x + 3 - 3 \leq -10 - 3 \quad \text{or} \quad x + 3 - 3 \geq 10 - 3$$

$$\Rightarrow \quad x \leq -13 \quad \text{or} \quad x \geq 7$$

$$\Rightarrow \quad x \in (-\infty, -13] \cup [7, \infty)$$

68. *(c)* Let breadth of rectangle be x cm.

$\therefore$ Length of rectangle $= 3x$

Perimeter of rectangle $= 2$ (Length + Breadth)

$$= 2(x + 3x) = 8x$$

Given, Perimeter ≥ 160 cm

$$8x \geq 160$$

Dividing both sides by 8,

$$x \geq 20 \text{ cm}$$

69. *(c)* We know that,

$$|3 - x| \geq 0 \; \forall \; x \in R$$

Adding 7 on both sides,

$$|3 - x| + 7 \geq 7$$

$$\Rightarrow \quad f(x) \geq 7$$

$\therefore$ Minimum value of $f(x)$ is 7.

70. *(a)* We have, $2 \leq |x - 3| < 4$

***Case* I** If $x < 3$, then

$$2 \leq |x - 3| < 4$$

$$\Rightarrow \quad 2 \leq -(x - 3) < 4$$

$$\Rightarrow \quad 2 \leq -x + 3 < 4$$

Subtracting 3 from both sides,

$$-1 \leq -x < 1$$

Multiplying (-1) on both sides,

$$-1 < x \leq 1$$

$$\Rightarrow \quad x \in (-1, 1]$$

***Case* II** If $x > 3$, then

$$2 \leq |x - 3| < 4$$

$$\Rightarrow \quad 2 \leq x - 3 < 4$$

Adding 3 on both sides,

$$\Rightarrow \quad 5 \leq x < 7$$

Hence, the solution set of given inequality is $x \in (-1, 1] \cup [5, 7)$.

71. *(a)* Let x be the smaller of two consecutive even positive integers. Then, the other even integer is $x + 2$.

Given, $x > 8$ and $x + x + 2 < 25$

$$\Rightarrow \quad x > 8 \text{ and } 2x + 2 < 25$$

$$\Rightarrow \quad x > 8 \text{ and } 2x < 23$$

$$\Rightarrow \quad x > 8 \text{ and } x < \frac{23}{2}$$

$$\Rightarrow \quad x = 10$$

Hence, there exists only one pair of even integer (10, 12).

72. *(c)* We have,

$$IQ = \frac{MA}{CA} \times 100$$

$$\Rightarrow \quad IQ = \frac{MA}{12} \times 100 \qquad (\because CA = 12 \text{ yr})$$

$$= \frac{25}{3} MA$$

Given, $80 \le IQ \le 140$

$\Rightarrow$ $80 \le \frac{25}{3} MA \le 140$

$\Rightarrow$ $240 \le 25MA \le 420$

$\Rightarrow$ $\frac{240}{25} \le MA \le \frac{420}{25}$

$\Rightarrow$ $9.6 \le MA \le 16.8$

73. *(a)* Graph of an inequality in one variable is a visual representation and is a convenient way to represent the solutions of the inequality.

74. *(b)* A line divides the cartesian plane into two parts.

75. *(a)* Each part in which a line divides the cartesian plane is known as half plane.

76. *(a)* A vertical line will divide the plane in left and right half planes and a non-vertical line will divide the plane into lower and upper half planes.

77. *(b)* The graphical representation of a vertical line and a non-vertical line in cartesian plane is represented by the figure given below

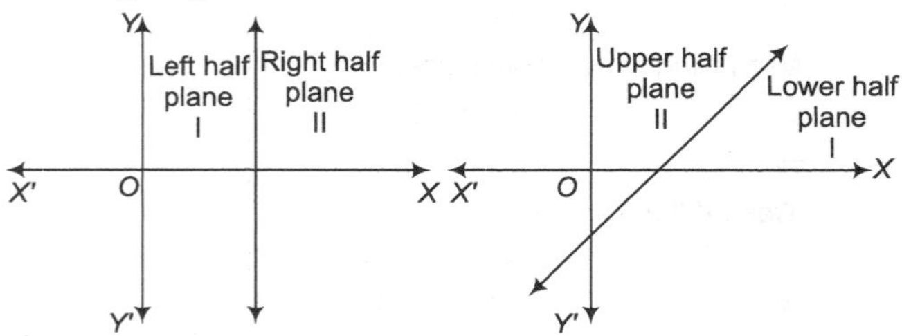

78. *(c)* A point in the cartesian plane will either lie on a line or will lie in either of the half planes I or II.

79. *(a)* All points (x, y) satisfying $ax + by = c$, lies on the line it represent.

80. *(c)* If a point $P(\alpha, \beta)$ lie on the line $ax + by = c$, then it will satisfy $ax + by = c$, *i.e.*, $a\alpha + b\beta = c$. All the points lying on the line $ax + by = c$ are the solution to the given equation.

81. *(c)* Since, $P(\alpha, \beta)$ is a point on the line $ax + by = c$,

then $a\alpha + b\beta = c$ ...(i)

and (α, γ) is a point in upper half plane, so

$a\alpha + b\gamma > c$ ...(ii)

From (i) and (ii), we get

$$a\alpha + b\gamma > a\alpha + b\beta$$

Subtracting $a\alpha$ from both sides,

$\Rightarrow$ $b\gamma > b\beta$

Dividing both sides by b,

$\Rightarrow$ $\gamma > \beta$ $(\because b > 0)$

The graphical representation of both the point is represent by the figure.

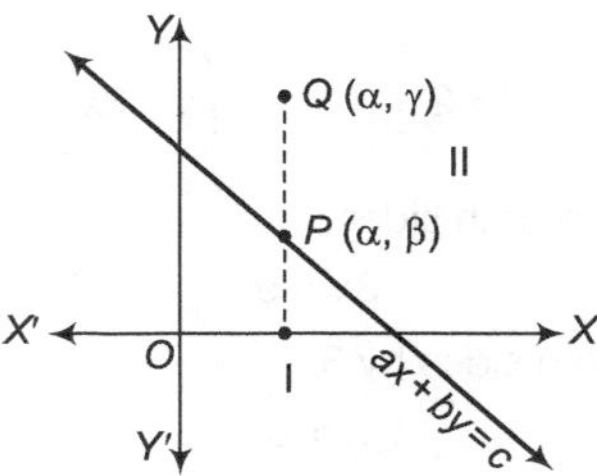

82. *(b)* In case $b > 0$, any point in the half plane II satisfies $ax + by > c$, and conversely any point satisfying the inequality $ax + by > c$ lies in half plane II.

In case $b < 0$, we know that any point satisfying $ax + by > c$ lies in the half plane I, and conversely.

Hence, all points satisfying $ax + by > c$ lies in one of the half planes II or I according as $b > 0$ or $b < 0$, and conversely.

83. *(c)* The graph of the inequality $ax + by > c$ will be half plane *I* (called solution region) and represented by shading half plane *I* not including the point on the line $ax + by = c$.

84. *(a)* I. The region containing all the solutions of an inequality is called the solution region.

II. In order to identify the half plane represented by an inequality, it is just sufficient to take any point (a, b) (not on line) and check whether it satisfies the inequality or not. If it satisfies, then the inequality represents the half plane and shade the region, which contains the point, otherwise,the inequality represents that half plane which does not contains the point within it. For convenience, the point (0, 0) is preferred.

85. *(b)* I. If an inequality is of the type $ax + by \ge c$ or $ax + by \le c$, then the points on the line $ax + by = c$ are also included in the solutions region. So, draw a dark line in the solution region.

II. If an inequality is of the form $ax + by > c$ or $ax + by < c$, then the points on the line $ax + by = c$ are not to be included in the solution region. So, draw a broken or dotted line in the solution region.

86. *(a)* Given, $40x + 20y \le 120$, x and y are whole numbers.

To start with, let $x = 0$. Then, LHS of given inequality is

$$40x + 20y = 40\ (0) + 20y = 20y$$

Thus, we have

$$20y \le 120$$

or $y \le 6$

For $x = 0$, the corresponding values of y can be 0, 1, 2, 3, 4, 5, 6 only. In this case, the solutions of given inequality are (0, 0), (0, 1), (0, 2), (0, 3), (0, 4), (0, 5) and (0, 6).

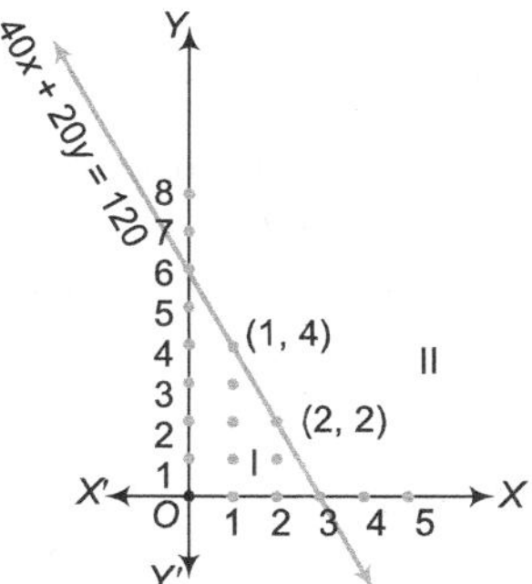

Similarly, other solutions of given inequality when $x = 1, 2$ and 3 are (1, 0), (1, 1), (1, 2), (1, 3), (1, 4), (2, 0), (2, 1), (2, 2), (3, 0).

This is shown in above figure.

Statement (b) is incorrect, since x and y are whole numbers. x and y can only take values $\{0, 1, 2, 3 \ldots\infty\}$. Given shaded region shows negative as well as fractional values of x and y.

87. *(d)* We have, $40x + 20y \le 120, x \ge 0\ y \ge 0$ ...(i)

In order to draw the graph of the inequality (i), we take one point say (0, 0), in half plane I and check whether values of x and y satisfy the inequality or not.

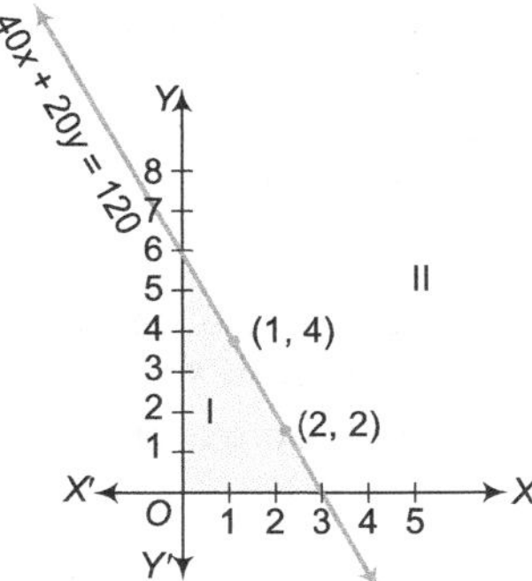

We observe that $x = 0$, $y = 0$ satisfy the inequality. Thus, we say that the half plane I is the graph of the inequality. Since, the points on the line also satisfy the inequality (i) above, the line is also a part of the graph.

Thus, the graph of the given inequality is half plane I including the line itself. Clearly, half plane II is not the part of the graph. Hence, solutions of inequality (i) will consist of all the points of its graph (half plane I including the line).

Also, since it is given $x > 0$, $y > 0$, x and y can only take positive values in half plane I.

88. *(b)* Graph of $3x + 2y = 6$is given as dotted line in the figure.

This line divides the xy-plane in two half planes I and II. We select a point (not on the line), say (0, 0), which lies in one of the half planes (see figure). Now, determine whether this point satisfies the given inequality or not. We note that $3(0) + 2(0) > 6$ or $0 > 6$, which is false.

Hence, half plane I is not the solution region of the given inequality. Clearly, any point on the line does not satisfy the

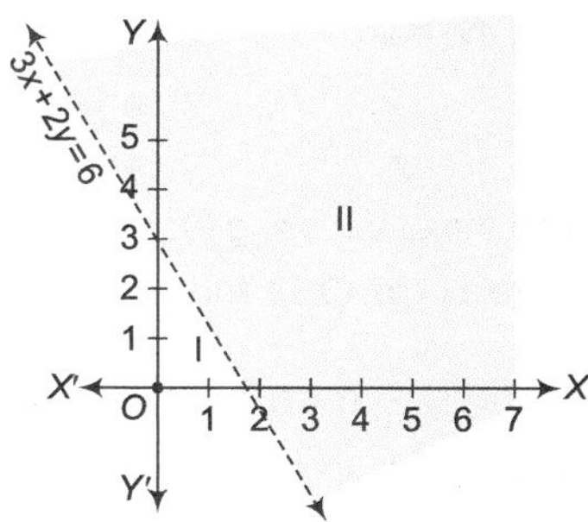

given strict inequality. In other words, the shaded half plane II excluding the points on the line is the solution region of the inequality.

89. *(a)* Graph of $3x - 6 = 0$ is given in the figure.

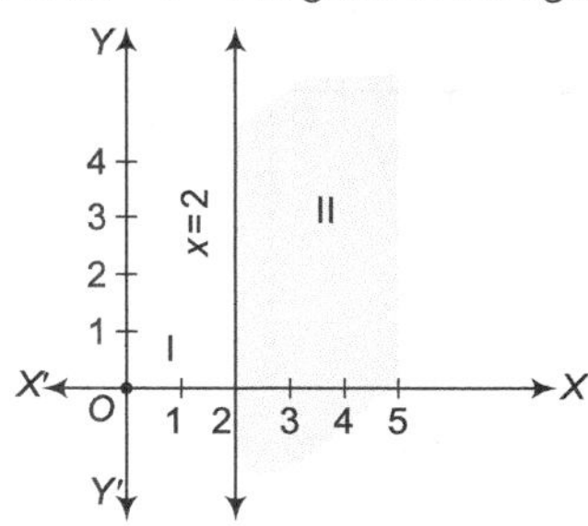

We select a point say (0, 0) and substituting it in given inequality, we see that

$3(0) - 6 \ge 0$ or $-6 \ge 0$ which is false.

Thus, the solution region is the shaded region on the right hand side of the line $x = 2$.

Also, all the points on the line $3x - 6 = 0$ will be included in the solution. Hence, a dark line is drawn in the solution region.

90. *(b)* I. Graph of $y = 2$ is given in the figure.

Let us select a point (0, 0) in lower half plane I and putting $y = 0$ in the given inequality, we see that $1 \times 0 < 2$ or $0 < 2$, which is true.

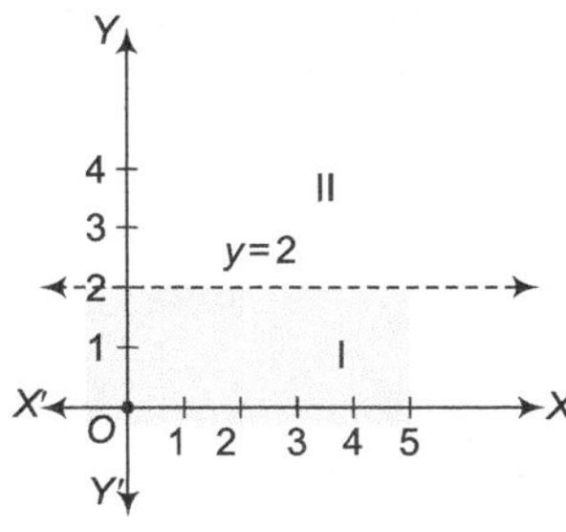

Thus, the solution region is the shaded region below the line $y = 2$. Hence, every point below the line (excluding all the points on the line) determines the solution of the given inequality.

II. We have the given inequality, $x + y < 5$ ...(i)

***Step* I** Consider the inequation as a strict equation *i.e.*, $x + y = 5$

Step II Find the points on X-axis and Y-axis *i.e.*,

x	5	0
y	0	5

Step III Plot the graph using the above table.

Step IV Take a point (0, 0) and put it in given inequation (i), we get

$$0+0<5, 0<5$$

which is true, so shaded region will be towards the origin.

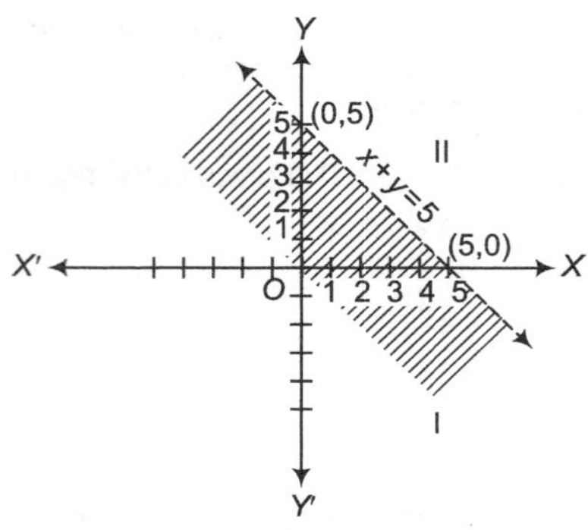

Thus, shaded region shows the inequality $x+y<5$.

Hence, the shaded half plane I excluding the points on the line is the solution region of the inequality.

91. *(b)* We have the given inequality, $2x+y \geq 6$ ... (i)

Step I Consider the inequation as a strict equation *i.e.*,

$$2x+y=6$$

Step II Find the points on X-axis and Y-axis *i.e.*,

x	3	0
y	0	6

Step III Plot the graph using the above table.

Step IV Take a point (0, 0) and put it in the given inequation (i), we get $0+0 \geq 6$, which is false, so shaded region will be away from the origin.

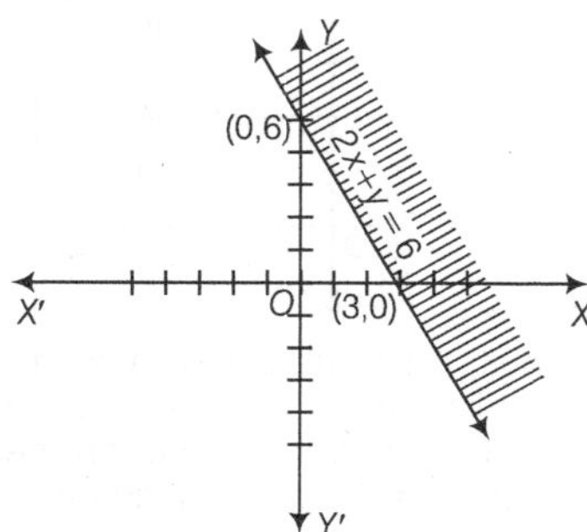

Here, shaded region shows the inequality $2x+y \geq 6$.

Note *The region is said to be bounded, if it is enclosed otherwise it is unbounded.*

92. *(c)* A. We have the given inequality $3x+4y \leq 12$ ...(i)

Step I Consider the inequation as a strict equation *i.e.*, $3x+4y=12$

Step II Find the points on the X-axis and Y-axis *i.e.*,

x	4	0
y	0	3

Step III Plot the graph using the above table.

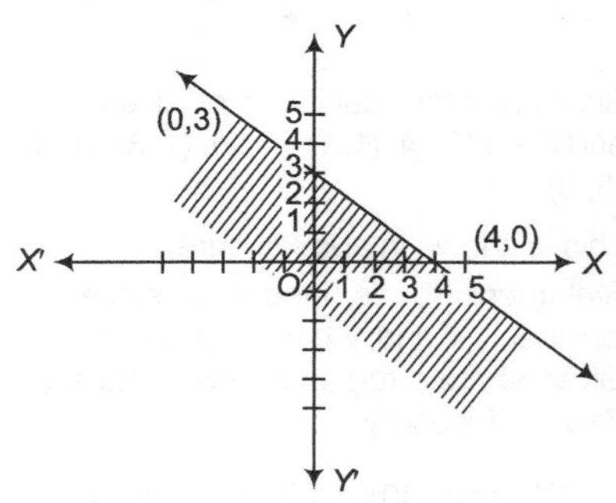

Step IV Take a point (0, 0) and put it in the given inequation (i), we get

$$0+0 \leq 12$$

which is true, so the shaded region will be towards the origin.

Here, shaded region shows the inequality.

B. The given inequation is

$y+8 \geq 2x$ ... (i)

Step I Consider the inequation as strict equation *i.e.*, $y+8=2x$

Step II Find the points on the X-axis and Y-axis *i.e.*,

x	0	4
y	–8	0

Step III Plot the graph using the above table.

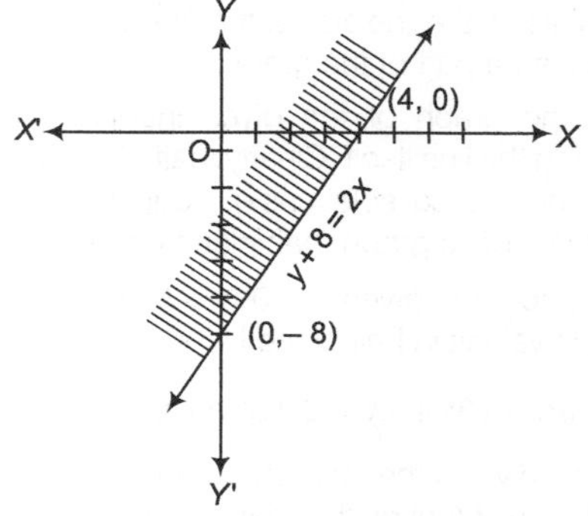

Step IV Take a point (0, 0) and put it in the given inequation (i), we get $0+8 \geq 0$

which is true, so the shaded region will be towards the origin.

Thus, shaded region shows the inequality.

C. The given inequation is

$$x - y \leq 2 \quad \ldots(i)$$

Step I Consider the inequation as a strict equation

i.e., $x - y = 2$

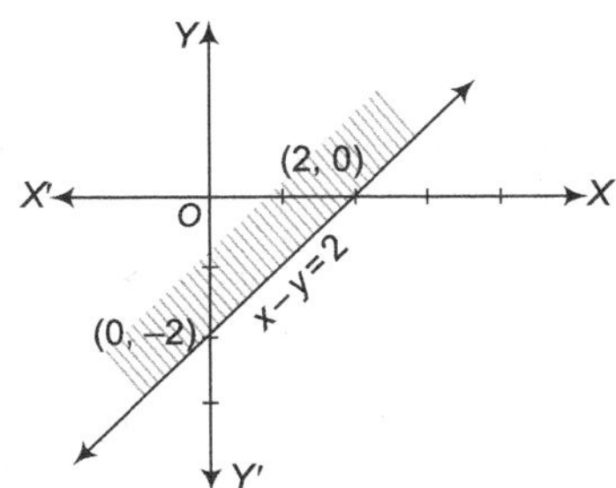

Step II Find the points on the X-axis and Y-axis *i.e.*,

x	2	0
y	0	– 2

Step III Plot the graph using the above table.

Step IV Take a point (0, 0) and put it in the given inequation (i), we get

$0 - 0 \leq 2 \Rightarrow 0 \leq 2$

which is true, so the shaded region will be towards the origin.

Thus, shaded region shows the inequality.

D. The given inequation is

$$2x - 3y > 6 \quad \ldots(i)$$

Step I Consider the inequation as a strict equation *i.e.*, $2x - 3y = 6$

Step II Find the points on the X-axis and Y-axis *i.e.*,

x	0	3
y	– 2	0

Step III Plot the graph using the above table.

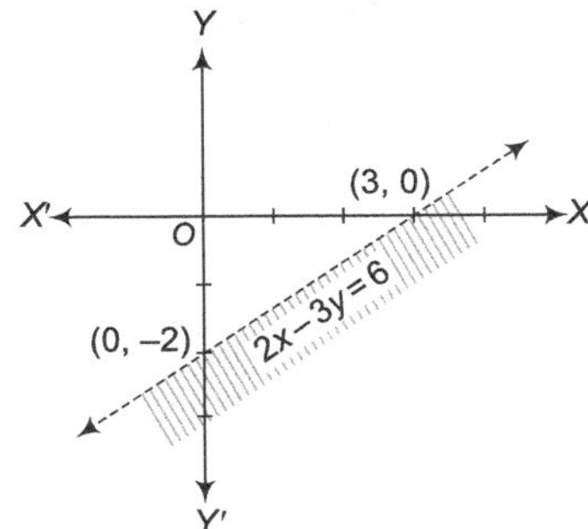

Step IV Take a point (0, 0) and put it in the given inequation (i), we get

$0 - 0 > 6 \Rightarrow 0 > 6$

which is false, so the shaded region will be opposite from the origin.

Thus, shaded region shows the inequality.

93. *(c)* The given inequation is $-3x + 2y \geq -6 \quad \ldots(i)$

Step I Consider the inequation as a strict equation

i.e., $-3x + 2y = -6$

Step II Find the points on the X-axis and Y-axis *i.e.*,

x	0	2
y	– 3	0

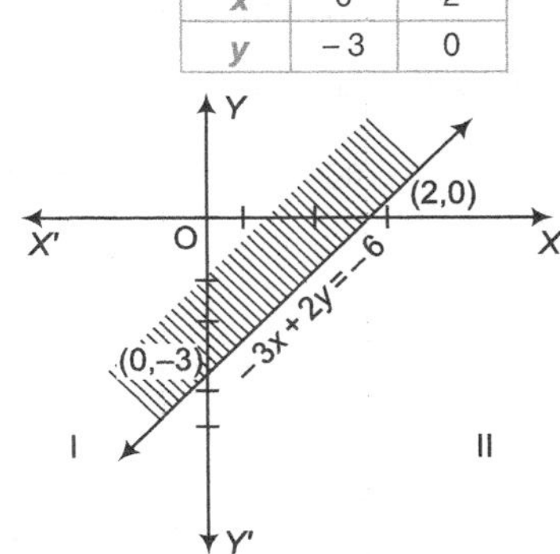

Step III Plot the graph using the above table.

Step IV Take a point (0, 0) and put it in the given inequation (i), we get $0 + 0 \geq -6$

which is true, so the shaded region will be towards the origin.

Hence, the shaded half plane I including the points on the line is the solution of given inequality.

94. *(c)* The given inequation is

$$3y - 5x < 30 \quad \ldots(i)$$

Step I Consider the inequation as a strict equation *i.e.*, $3y - 5x = 30$

Step II Find the points on the X-axis and Y-axis *i.e.*,

x	0	– 6
y	10	0

Step III Plot the graph using the above table.

Step IV Take a point (0, 0) and put it in the given inequation (i), we get

$0 - 0 < 30 \Rightarrow 0 < 30$

which is true, so the shaded region will be towards the origin.

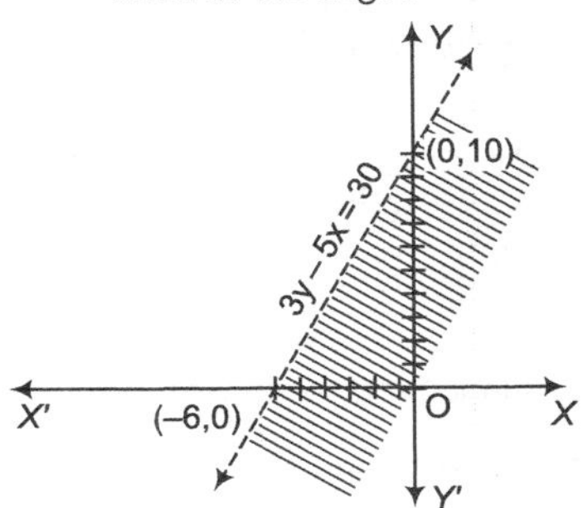

Thus, shaded region shows the inequality.

95. *(d)* I. The given inequation is $y < -2$ ...(i)

***Step* I** Consider the inequation as a strict equation

i.e., $y = -2$

***Step* II** Plot the graph. The graph of $y = -2$ will be a line parallel to X-axis and intersecting the Y-axis at -2.

***Step* III** Take a point (0, 0) and put it in inequation (i), we get $0 < -2$ which is false, so the shaded region will be away from the origin.

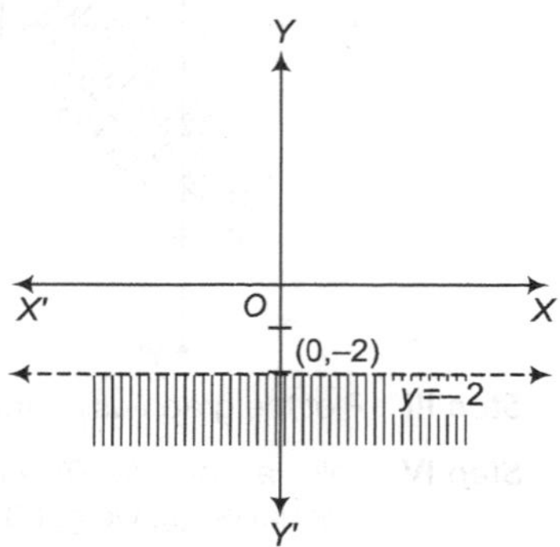

Thus, shaded region shows the inequality. Dotted line will be drawn since it is a strict inequality.

II. The given inequation is

$$x > -3 \quad \text{...(i)}$$

***Step* I** Consider the inequation as a strict equation

i.e., $x = -3$

***Step* II** Plot the graph. The graph of $x = -3$ will be a line parallel to Y-axis and intersecting the X-axis at -3.

***Step* III** Take a point (0, 0) and put it in the given inequation (i), we get $0 > -3$, which is true, so the shaded region will be towards the origin.

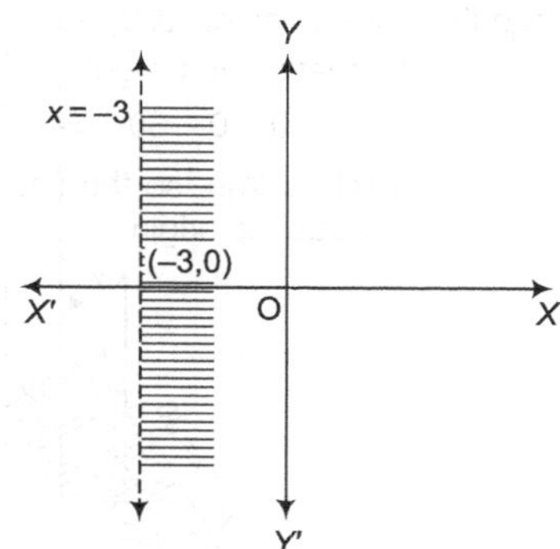

Thus, shaded region shows the inequality.

96. *(b)* First draw the graph of $y = 3$. Since, given inequality is strict inequality, so dotted line is drawn.

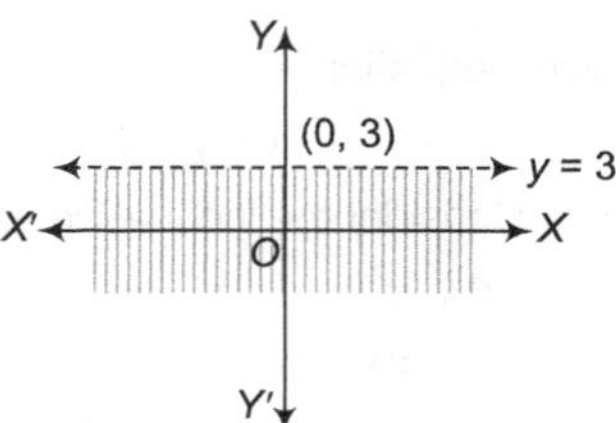

Now, consider the point $O(0, 0)$, putting $y = 0$ in given in equality, we get

$0 < 3$, which is true.

$\therefore$ Solution region is the shaded region below the line $y = 3$.

97. *(a)* The shaded region in the figure lies between $x = -3$ and $x = 3$ not including the line $x = -3$ and $x = 3$ (lines are dotted).

Therefore, $-3 < x < 3$

$\Rightarrow \quad |x| < 3 \quad (\because |x| < a \Leftrightarrow -a < x < a)$

98. *(d)* I. The line corresponding to given inequality is

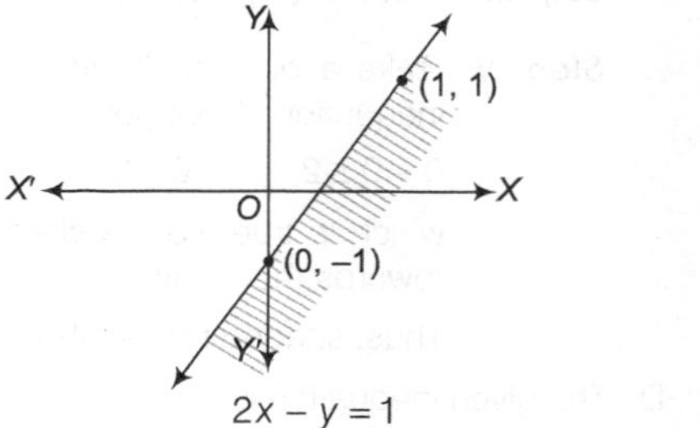

$$2x - y = 1 \quad \text{...(i)}$$

Table for $2x - y = 1$

x	0	1
y	–1	1

Consider the point $O(0, 0)$

Since, $2(0) - (0) \geq 1$

i.e., $0 \geq 1$, where is not true.

Hence, inequality $2x - y \geq 1$ represent the half plane made by the line (*i*) which does not contains origin. Clearly, the graph shows that other quadrants are also included.

II. The line corresponding to given inequality is

$$x + y = 0 \quad \text{...(ii)}$$

Table for $x + y = 0$

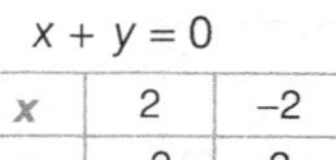

x	2	–2
y	–2	2

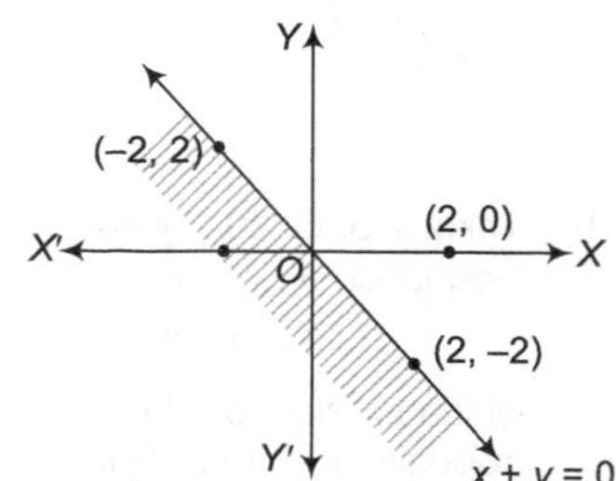

Consider the point (2, 0).

Since, $2 + 0 \leq 0$.

i.e., $2 \leq 0$ which is not true.

Therefore, $x + y \leq 0$ represent the half plane made by the line $x + y = 0$, which does not contains (2, 0).

99. *(b)* Given, $2x - 3 \geq 0$ ...(i)

$\Rightarrow \quad 2x \geq 3$

$\Rightarrow \quad x \geq \frac{3}{2}$

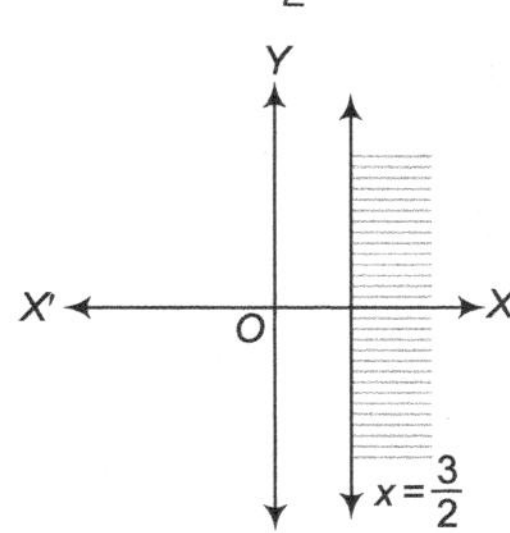

First, draw the graph of the line $x = \frac{3}{2}$. Consider the point O(0, 0). Putting $x = 0$ in Eq. (i),

$$2(0) - 3 \geq 0$$

$\Rightarrow \quad 0 - 3 \geq 0$

$\Rightarrow -3 \geq 0$, which is not true.

Hence, given inequality represent the half plane made by the line $x = \frac{3}{2}$, which does not contains origin.

100. *(c)* A. The line corresponding to $x + y \geq 4$ is $x + y = 4$.

Table for $x + y = 4$

x	0	4
y	4	0

First draw the line

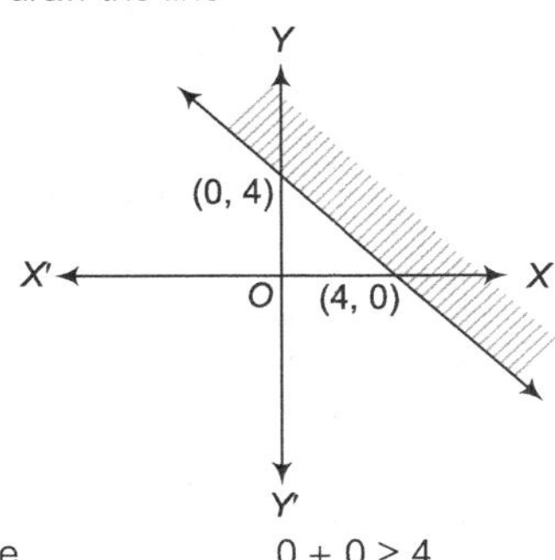

Since, $\quad 0 + 0 \geq 4$

i.e., $\quad 0 \geq 4$

which is not true.

Therefore, given inequality represent the half plane made by the line $x + y = 4$ which does not contain origin.

B. The line corresponding to $x + 2y > 1$ is

$$x + 2y = 1 \quad \text{...(i)}$$

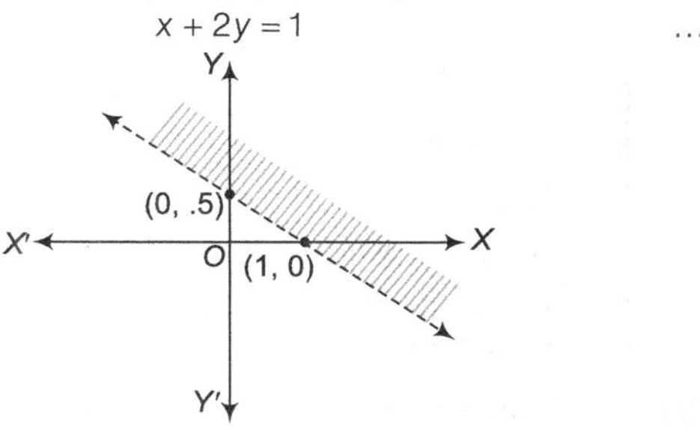

Table for $x + 2y = 1$

x	1	0
y	0	0.5

Since, given inequality is strict. So, the line $x + 2y = 1$ is drawn dotted.

Now, $0 + 2(0) > 1$

i.e., $0 > 1$, which is not true.

Hence, given inequality represent the half made by line (i) which does not contains origin.

C. The line corresponding to $x \geq y - 2$ is

$$x = y - 2$$

Table for $x = y - 2$

x	0	−2
y	2	0

First draw the line

$$x = y - 2$$

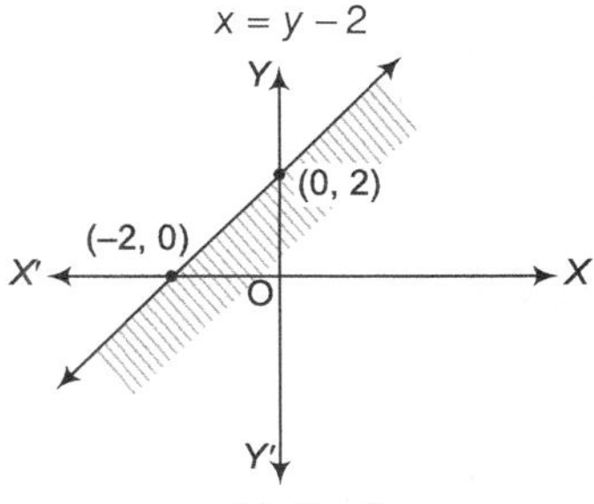

Since, $\quad 0 \geq 0 - 2$

i.e., $0 \geq -2$, which is true.

Therefore, given inequality represent the half plane made by the line $x = y - 2$, which contains origin.

D. The line corresponding to $2x - 3y < 4$ is $2x - 3y = 4$.

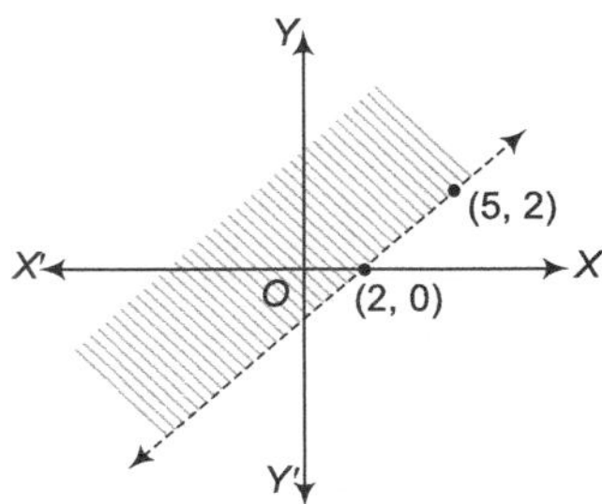

Table for $2x - 3y = 4$

x	2	5
y	0	2

Since, given inequality is strict. So, the line $2x - 3y = 4$ is drawn dotted.

Now, $2(0) - 3(0) < 4$

i.e., $0 < 4$, which is true.

Hence, given inequality represent the half plane which contains origin.

101. *(c)* Total cost of x tables and y chairs $= 750x + 150y$

Since, he has ₹ 15000 to invest, then

$$750x + 150y \leq 15000$$

Again, he has space to store at most 60 pieces

$\therefore \quad x + y \leq 60$

$\because$ x and y are numbers, $x \geq 0$, $y \geq 0$

Hence, linear inequalities representing the above data are

$750x + 150y \leq 15000$, $x + y \leq 60$, $x \geq 0$ and $y \geq 0$.

102. *(c)* The inequalities are

$$750x + 150y \leq 15000$$

i.e., $5x + y \leq 100$ ...(i)

$x + y \leq 60$ ...(ii)

$x \geq 0$ ...(iii)

$y \geq 0$ ...(iv)

The lines corresponding to (i) and (ii) are

$5x + y = 100$ ...(v)

$x + y = 60$ ...(vi)

Table for $5x + y = 100$

x	0	20
y	100	0

Table for $x + y = 60$

x	0	60
y	60	0

First, draw the lines (v) and (vi)

$\because \quad 5(0) + 0 \leq 100$

i.e., $0 \leq 100$, which is true.

Therefore, inequality (i) represent the half plane made by the line (v), which contains the origin.

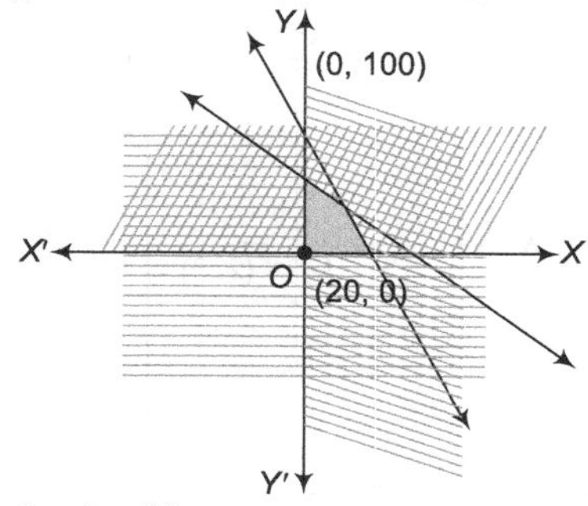

Again, $0 + 0 \leq 60$

i.e., $0 \leq 60$, which is true.

Therefore, inequality (ii) represent the half plane made by the line (vi) which contains origin. Inequality $x \geq 0$ represent the half plane on the right of Y-axis.

Inequality $y \geq 0$ represent the half plane above X-axis.

103. *(c)* The graph of linear equation $x + y = 5$ is drawn in figure. We note that solution of inequality (i) is represented by the shaded region above the line $x + y = 5$, including the points on the line.

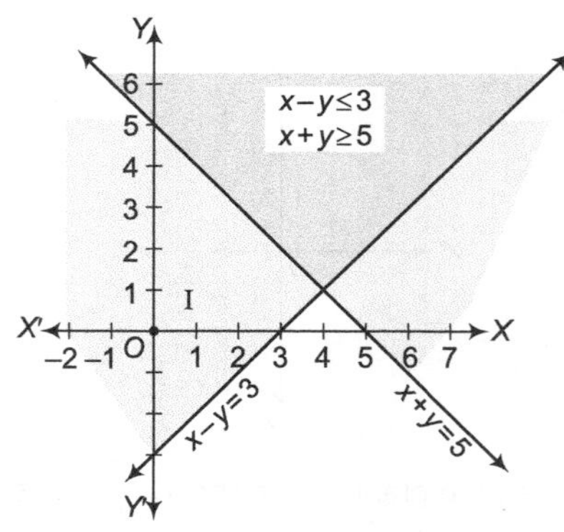

On the same set of axes, we draw the graph of the equation $x - y = 3$ as shown in figure. Then, we note that inequality (ii) represents the shaded region above the line $x - y = 3$, including the points on the line.

Clearly, the double shaded region, common to the above two shaded regions is the required solution region of the given system of inequalities.

104. *(a)* We first draw the graph of the lines

$$5x + 4y = 40,\ x = 2 \text{ and } y = 3$$

Then, we note that the inequality (i) represents shaded region below the line $5x + 4y = 40$ and inequality (ii) represents the shaded region right of line $x = 2$ but inequality (iii) represents the shaded region above the line $y = 3$. Hence, shaded region (in figure) including all the point on the lines are also the solution of the given system of the linear inequalities.

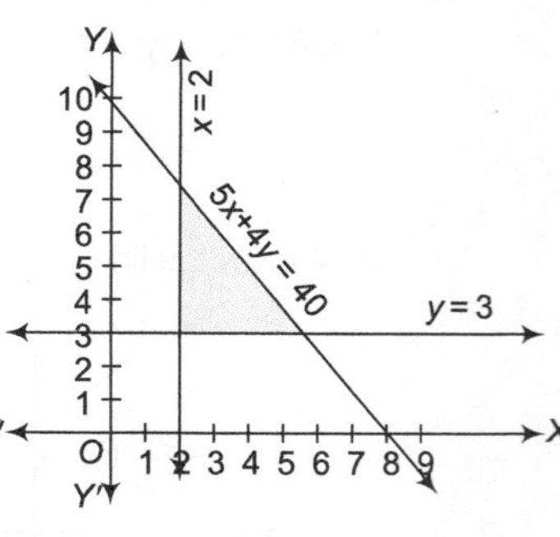

In many practical situations involving system of inequalities the variable x and y often represent quantities that cannot have negative values, for example, number of units produced, number of articles purchased, number of hours worked, etc. Clearly, in such cases, $x \geq 0$, $y \geq 0$ and the solution region lies only in the first quadrant.

Since, the region is enclosed, so it is bounded.

105. *(b)* Consider $8x + 3y = 100$. We observe that the shaded region and the origin lie on the same side of the line and (0, 0) satisfies $8x + 3y \le 100$. Therefore, we must have $8x + 3y \le 100$ as linear inequality corresponding to the line $8x + 3y = 100$. Also, shaded region lies in the first quadrant only. Therefore, $x \ge 0$, $y \ge 0$. Hence, the linear inequalities corresponding to the given solutions are

$8x + 3y \le 100$, $x \ge 0$, $y \ge 0$.

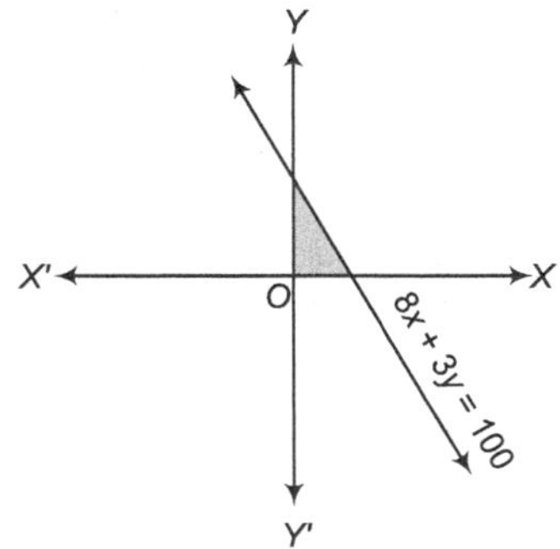

106. *(a)* I. We draw the graphs of the lines $x + 2y = 8$ and $2x + y = 8$. The inequality (1) and (2) represent the region below the two lines, including the point on the respective lines.

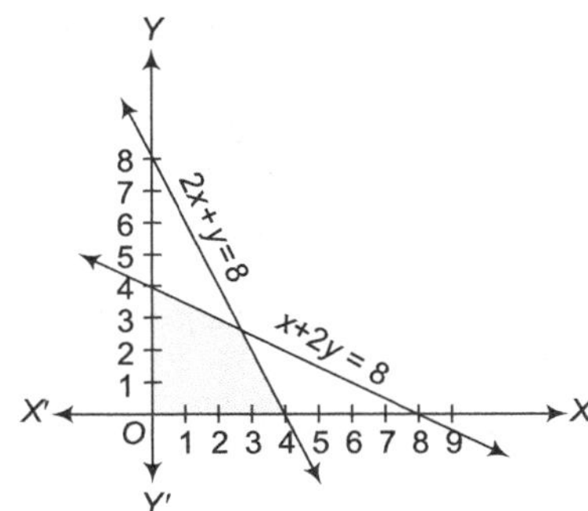

Since, $x \ge 0$, $y \ge 0$, every point in the shaded region in the first quadrant represent a solution of the given system of inequalities.

II. The given system of inequalities

$$x \ge 3 \quad \dots (i)$$

$$y \ge 2 \quad \dots (ii)$$

Step I Consider the inequations as strict equations

i.e., $x = 3$, $y = 2$

Step II Plot the graph. The graph of $x = 3$, (*i.e.*, $y = 0$) will be a line parallel to Y-axis, intersecting the X-axis at 3. The graph of $y = 2$ (*i.e.*, $x = 0$) will be a line parallel to X-axis, intersecting the Y-axis at 2.

Step III Take a point (0, 0) and put it in the given inequations (i) and (ii).

i.e., $0 \ge 3$ (false)

$0 \ge 2$ (false)

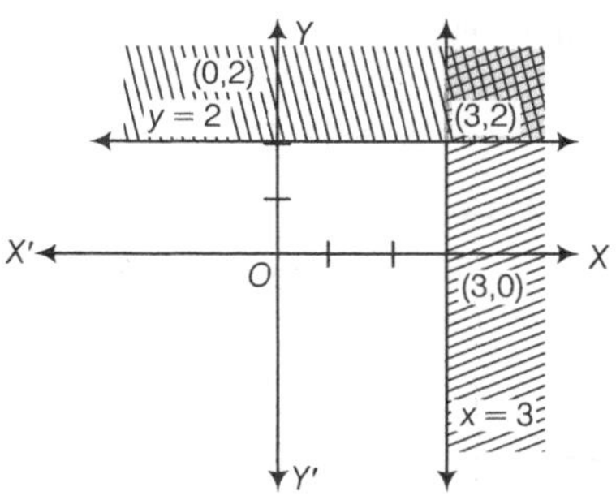

$\therefore$ Both the graphs will be parallel and away from the origin.

Thus, common shaded region shows the solution of the inequalities.

107. *(a)* I. The given system of inequalities

$$3x + 2y \le 12 \quad \dots (i)$$

$$x \ge 1 \quad \dots (ii)$$

$$y \ge 2 \quad \dots (iii)$$

Step I Consider the given inequations as strict equations

i.e., $3x + 2y = 12$, $x = 1$, $y = 2$

Step II Draw the table for $3x + 2y = 12$ (*i.e.*, find the points on X-axis and Y-axis)

Step III Plot the graph.

(i) For $3x + 2y = 12$, use the table.

(ii) Graph of $x = 1$ will be a line parallel to Y-axis cutting X-axis at 1.

(iii) Graph of $y = 2$ will be a line parallel to X-axis cutting Y-axis at 2

Step IV Take a point (0, 0) and put it in the given inequations (i), (ii) and (iii),

i.e., $0 + 0 \le 12$, $0 \le 12$ (true)

So, the shaded region will be towards the origin.

$0 \ge 1$ (false)

So, the shaded region will be away from the origin.

$0 \ge 2$ (false)

So, the shaded region will be away from the origin.

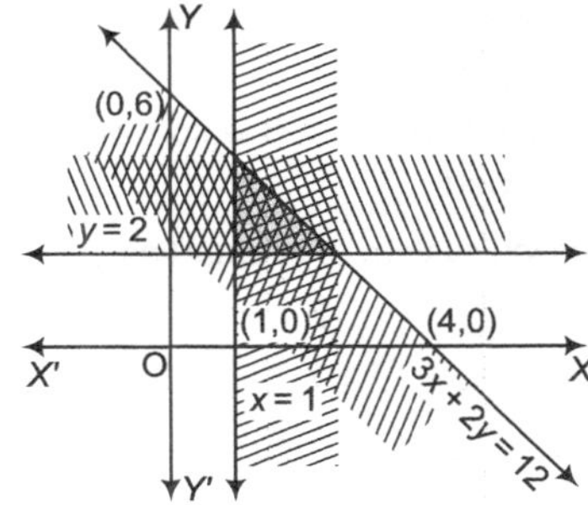

Thus, common shaded region shows the solution of the inequalities.

II. The given system of inequalities

$$2x + y \geq 6 \quad \ldots(i)$$
$$3x + 4y \leq 12 \quad \ldots(ii)$$

Step I Consider the given inequations as strict equations

i.e., $2x + y = 6$

$3x + 4y = 12$

Step II Find the points on the X-axis and Y-axis for

$2x + y = 6$

x	0	3
y	6	0

and $3x + 4y = 12$

x	0	4
y	3	0

Step III Plot the graph using the above tables.

Step IV Take a point (0, 0) and putting the given inequations (i) and (ii),

i.e., $0 + 0 \geq 6$

$0 \geq 6$ (false)

So, the shaded region will be away from the origin.

and $0 + 0 \leq 12$

$0 \leq 12$ (true)

So, the shaded region will be towards the origin.

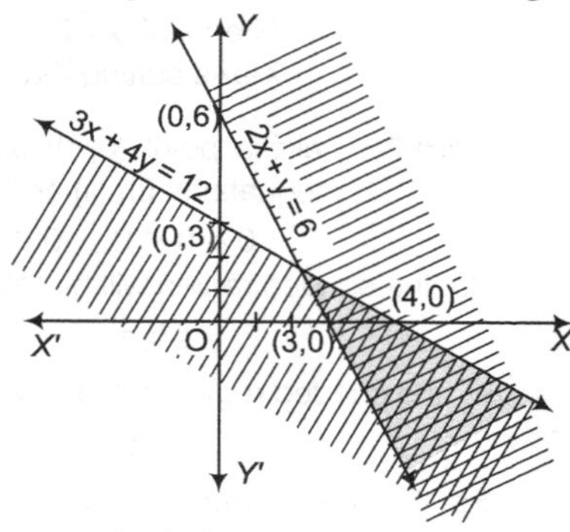

Thus, common shaded region shows the solution of the inequality.

Since, common shaded region is not enclosed. So, it is not bounded.

III. The given system of inequalities

$$x + y \geq 4 \quad \ldots(i)$$
$$2x - y > 0 \quad \ldots(ii)$$

Step I Consider the given inequations as strict equations

i.e., $x + y = 4, 2x - y = 0$

Step II Find the points on the X-axis and Y-axis for $x + y = 4$

x	0	4
y	4	0

and $2x - y = 0$

x	0	1
y	0	2

Step III Plot the graph using the above tables.

Step IV Take a point (0, 0) and put it in the inequation (i)

$0 + 0 \geq 4$ (false)

So, the shaded region will be away from the origin.

Take a point (1, 0) and put it in the inequation (ii)

$2 - 0 > 0$ (true)

So, the shaded region will be towards the point (1, 0)

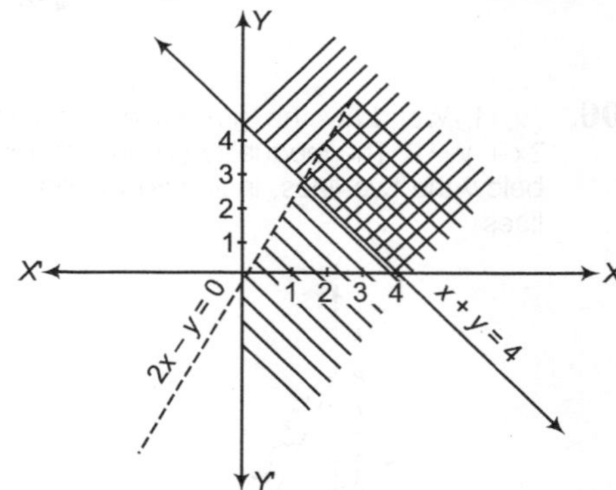

Thus, common shaded region shows the solution of the inequalities.

108. *(c)* A. The given system of inequalities

$$2x - y > 1 \quad \ldots(i)$$
$$x - 2y < -1 \quad \ldots(ii)$$

Step I Consider the inequations as strict equations

i.e., $2x - y = 1$ and $x - 2y = -1$

Step II Find the points on the X-axis and Y-axis for $2x - y = 1$

x	0	1/2
y	−1	0

and $x - 2y = -1$

x	0	−1
y	1/2	0

Step III Plot the graph using the above tables.

Step IV Take a point (0, 0) and put it in the inequations (i) and (ii),

$0 - 0 > 1$ *i.e.,* $0 > 1$ (false)

So, the shaded region will be away from the origin

and $0 - 0 < -1$ *i.e.,* $0 < -1$ (false)

So, the shaded region will be away from the origin.

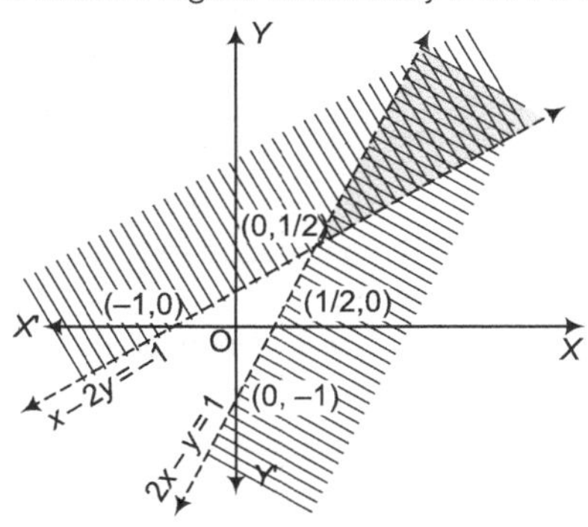

Thus, common shaded region shows the solution of the inequalities.

B. The given system of inequalities

$$x + y \le 6 \quad \ldots\text{(i)}$$
$$x + y \ge 4 \quad \ldots\text{(ii)}$$

***Step* I** Consider the inequations as strict equations *i.e.*, $x + y = 6$ and $x + y = 4$

***Step* II** Find the points on the X-axis and Y-axis for $x + y = 6$

x	0	6
y	6	0

and $x + y = 4$

x	0	4
y	4	0

***Step* III** Plot the graph using the above tables.

***Step* IV** Take a point (0, 0) and put it in the inequations (i) and (ii),

$0 + 0 \le 6 \quad 0 \le 6$ (true)

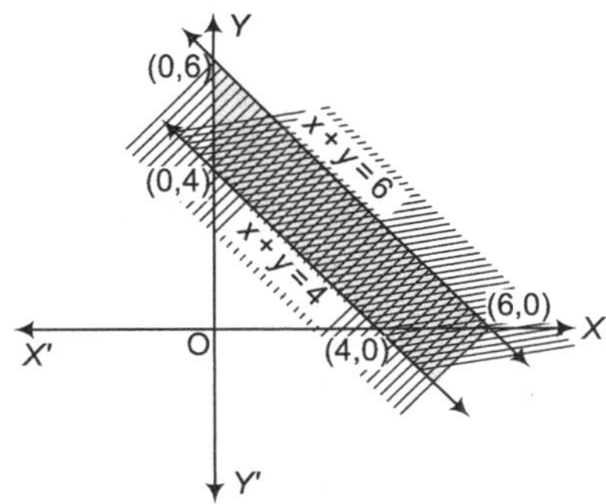

So, the shaded region will be towards the origin.

and $0 + 0 \ge 4 \Rightarrow 0 \ge 4$ (false)

So, the shaded region will be away from the origin.

Thus, common shaded region shows the solution of the inequalities.

C. The given system of inequalities

$$2x + y \ge 8 \quad \ldots\text{(i)}$$
$$x + 2y \ge 10 \quad \ldots\text{(ii)}$$

***Step* I** Consider the inequations as strict equations *i.e.*, $2x + y = 8$ and $x + 2y = 10$

***Step* II** Find the points on the X-axis and Y-axis for $2x + y = 8$

x	0	4
y	8	0

and $x + 2y = 10$

x	0	10
y	5	0

***Step* III** Plot the graph using the above tables.

***Step* IV** Take a point (0, 0) and put it in the given inequations (i) and (ii),

$0 + 0 \ge 8$ *i.e.*, $0 \ge 8$ (false)

So, the shaded region will be away from the origin.

$0 + 0 \ge 10$ *i.e.*, $0 \ge 10$ (false)

So, the shaded region will be away from the origin.

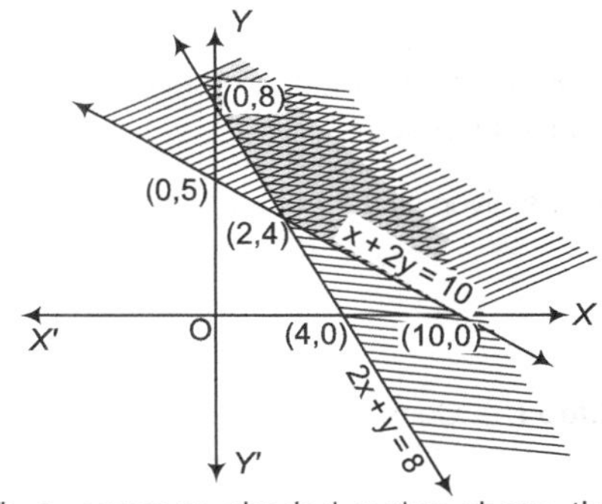

Thus, common shaded region shows the solution of the inequalities.

D. The given system of inequalities

$$x + y \le 9 \quad \ldots\text{(i)}$$
$$y > x \quad \ldots\text{(ii)}$$
$$x \ge 0 \quad \ldots\text{(iii)}$$

***Step* I** Consider the inequations as strict equations *i.e.*, $x + y = 9$, $y = x$, $x = 0$

***Step* II** Find the points on the X-axis and Y-axis for $x + y = 9$

x	0	9
y	9	0

and $y = x$

x	1	2	3
y	1	2	3

***Step* III** Plot the graph using the above tables.

(i) For $x + y = 9$ use the above table.
(ii) For $y = x$ use the table.
(iii) Graph of $x = 0$ will be the Y-axis

Step IV Take a point (0, 0) put it in the inequations (i), (ii) and (iii), we get

$0 + 0 \le 9$ (true)

So, the shaded region will be towards the origin.

Take a point (0, 1) put in $y > x, 1 > 0$ (true)

So, the shaded region will be towards the origin.

Take a point (1, 0) put it in $x \ge 0, 1 \ge 0$ (true)

So, the shaded region will be towards the origin.

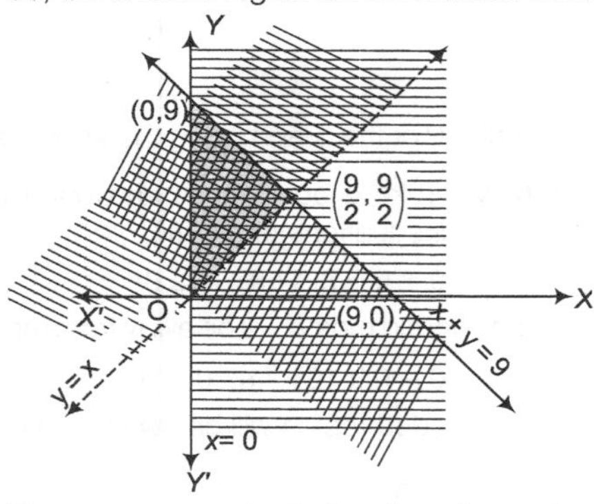

Thus, common shaded region shows the solution of the inequalities.

109. *(b)* I. The given system of inequalites

$5x + 4y \le 20$ ...(i)

$x \ge 1$ ...(ii)

$y \ge 2$ ...(iii)

Step I Consider the inequations as strict equations

i.e., $5x + 4y = 20$

$x = 1, y = 2$

Step II Find the points on the X-axis and Y-axis for $5x + 4y = 20$

x	0	4
y	5	0

Step III Plot the graph using the above tables.

(i) For $5x + 4y = 20$ use the above table.

(ii) Graph of $x = 1$ will be line parallel to Y-axis, intersecting the X-axis at 1.

(iii) Graph of $y = 2$ will be a line parallel to X-axis, intersecting the Y-axis at 2.

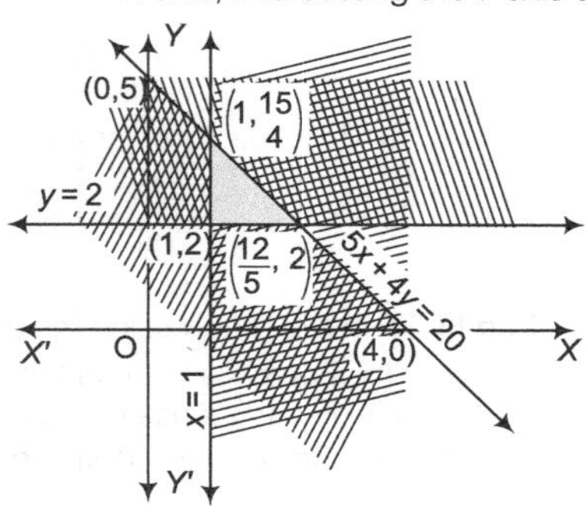

Step IV Take a point (0, 0) and put it in the given inequations (i), (ii) and (iii),

$0 + 0 \le 20$ (true)

$0 \ge 1$ (false)

$0 \ge 2$ (false)

Thus, common shaded region shows the solutions of the inequalities.

II. The given system of inequalities

$3x + 4y \le 60$ ...(i)

$x + 3y \le 30$ ....(ii)

$x \ge 0$ ...(iii)

$y \ge 0$ ...(iv)

Step I Consider the inequations as strict equations

i.e., $3x + 4y = 60, x + 3y = 30$

$x = 0, y = 0$

Step II Find the points on the X-axis and Y-axis for

$3x + 4y = 60$

x	0	20
y	15	0

and $x + 3y = 30$

x	0	30
y	10	0

Step III Plot the graph using the above tables.

(i) For $3x + 4y = 60$ and $x + 3y = 30$ use the above tables.

(ii) Graph of $x = 0$, will be Y-axis.

(iii) Graph of $y = 0$, will be X-axis

Step IV Take a point (0, 0) and put it in inequations (i) and (ii),

$0 + 0 \le 60$ (true)

So, the shaded region will be towards origin.

and $0 + 0 \le 30$ (true)

So, the shaded region will be towards origin.

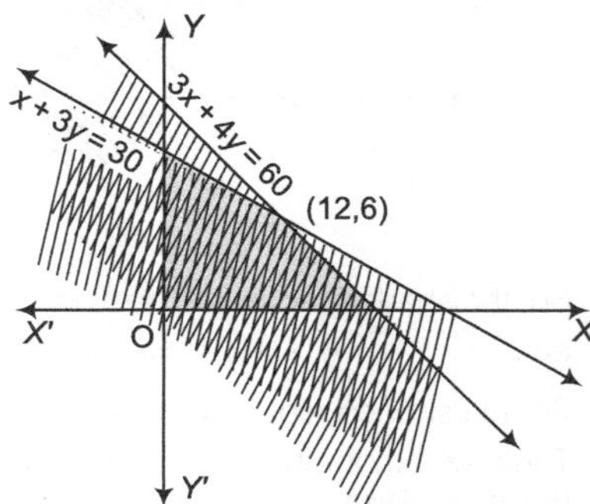

And point (1, 0) in inequation (iii),

$1 \ge 0$ (true)

So, the shaded region will be towards origin.

Point (0, 1) in inequation (iv),

$$1 \geq 0 \quad \text{(true)}$$

So, the shaded region will be towards origin.

Thus, common shaded region shows the solution of the inequalities.

110. *(d)* The given system of inequalities

$$2x + y \geq 4 \quad \ldots(i)$$
$$x + y \leq 3 \quad \ldots(ii)$$
$$2x - 3y \leq 6 \quad \ldots(iii)$$

***Step* I** Consider the inequations as strict equations *i.e.*, $2x + y = 4$

$x + y = 3$

$2x - 3y = 6$

***Step* II** Find the points on the X-axis and Y-axis for, $2x + y = 4$

x	0	2
y	4	0

and $x + y = 3$

x	0	3
y	3	0

and $2x - 3y = 6$

x	0	3
y	−2	0

***Step* III** Plot the graph using the above tables.

***Step* IV** Take a point (0, 0) and put it in the inequations (i), (ii), and (iii), we get

$$0 + 0 \geq 4 \quad \text{(false)}$$

So, the shaded region will be away from origin.

$$0 + 0 \leq 3 \quad \text{(true)}$$

So, the shaded region will be towards origin.

$$0 - 0 \leq 6 \quad \text{(true)}$$

So, the shaded region will be towards origin.

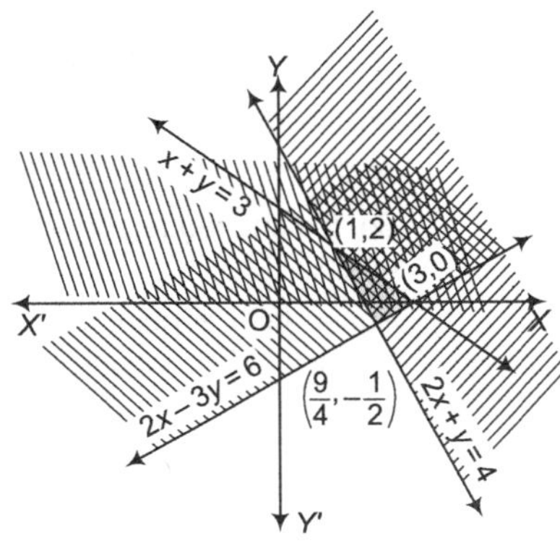

Thus, common shaded region shows the solution of the inequalities.

111. *(a)* The given system of inequalities

$$x - 2y \leq 3 \quad \ldots(i)$$
$$3x + 4y \geq 12 \quad \ldots(ii)$$
$$x \geq 0 \quad \ldots(iii)$$
$$y \geq 1 \quad \ldots(iv)$$

***Step* I** Consider the inequations as strict equations *i.e.*, $x - 2y = 3$, $3x + 4y = 12$ $x = 0$, $y = 1$

***Step* II** Find the point on the X-axis and Y-axis for $x - 2y = 3$

x	0	3
y	− 3/2	0

and $3x + 4y = 12$

x	0	4
y	3	0

***Step* III** Plot the graph using the above tables.

(i) For $x - 2y = 3$ and $3x + 4y = 12$ use the above tables.

(ii) Graph of $x = 0$ will be Y-axis.

(iii) Graph of $y = 1$ will be a line parallel to X-axis, intersecting Y-axis at 1.

***Step* IV** Take a point (0, 0) and put it in the inequations (i) and (ii),

$$0 - 0 \leq 3 \quad \text{(true)}$$

So, the shaded region will be towards origin.

and $\quad 0 + 0 \geq 12 \quad$ (false)

So, the shaded region will be away from the origin.

Again, take a point (1, 0).

Put it in the inequations (iii), we get

$$1 \geq 0 \quad \text{(true)}$$

So, the shaded region will be towards origin.

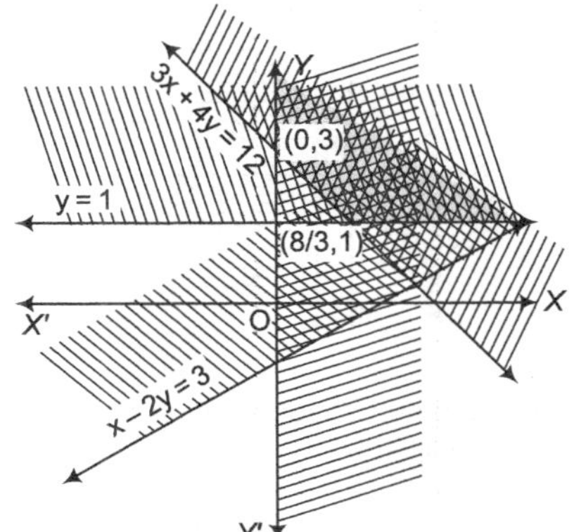

And a point (1, 0) put it in the inequations (iv), we get

$$0 \geq 1 \quad \text{(false)}$$

So, the shaded region will be away from the origin.

Thus, common shaded region shows the solution of the inequalities.

112. *(b)* The given system of inequalities

$$4x + 3y \le 60 \quad \ldots(i)$$
$$y \ge 2x \quad \ldots(ii)$$
$$x \ge 3 \quad \ldots(iii)$$
$$x, y \ge 0 \quad \ldots(iv)$$

***Step* I** Consider the inequations as strict equations

i.e., $4x + 3y = 60$

$y = 2x$

$x = 3$

***Step* II** Find the point on the X-axis and Y-axis for $4x + 3y = 60$

x	0	15
y	20	0

and $y = 2x$

x	0	5
y	0	10

***Step* III** Plot the graph using the above tables.

(i) For $4x + 3y = 60$ use the above tables.

(ii) For $y = 2x$ use the tables.

(iii) Graph of $x = 3$ will be a straight parallel to Y-axis intersecting the X-axis at 3.

***Step* IV** Take a point (0, 0). Put it in the inequations (i) and (iii),

$0 + 0 \le 60$ (true)

$0 \ge 3$ (false)

and a point (1, 1), putting it in the inequation (ii),

$1 \ge 2$ (false)

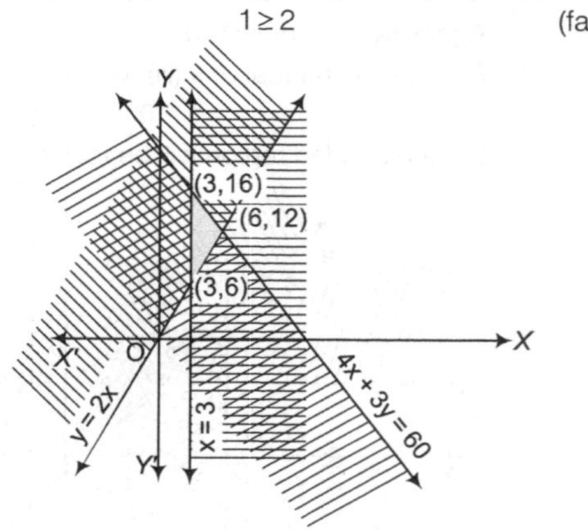

Thus, common shaded region shows the solution of the inequalities.

113. *(c)* (i) Consider, $x + 4y = 80$. We observe that the shaded region and the origin lie on the same side of this line and (0, 0) satisfies $x + 4y \le 80$. Therefore, we must have $x + 4y \le 80$ as linear inequality corresponding to the line $x + 4y = 80$.

(ii) Consider, $3x + 2y = 150$. We observe that the shaded region and the origin lie on the same side of this line and (0, 0) satisfies $3x + 2y \le 150$. Therefore, $3x + 2y \le 150$ is the linear inequality corresponding to the line $3x + 2y = 150$.

(iii) Consider, $x = 15$. We observe that the shaded region and origin lie on the same side and (0, 0) satisfies $x \le 15$. Therefore, $x \le 15$ is the linear inequality corresponding to the line $x = 15$.

(iv) Also, the shaded region lies in the first quadrant only. Therefore, $x \ge 0$, $y \ge 0$.

Hence, in view of (i), (ii), (iii), (iv) above, the linear inequalities corresponding to the given solution set are $3x + 2y \le 150$, $x + 4y \le 80$, $x \le 15$, $x, y \ge 0$.

114. *(d)* The given system of inequalities

$$x + 2y \le 10 \quad \ldots(i)$$
$$x + y \ge 1 \quad \ldots(ii)$$
$$x - y \le 0 \quad \ldots(iii)$$
$$x \ge 0, y \ge 0 \quad \ldots(iv)$$

***Step* I** Consider the given inequations as strict equations *i.e.*,

$x + 2y = 10$, $x + y = 1$, $x - y = 0$

and $x = 0$, $y = 0$

***Step* II** Find the points on the X-axis and Y-axis for

$x + 2y = 10$

x	0	10
y	5	0

and $x + y = 1$

x	0	1
y	1	0

For $x - y = 0$

x	1	2
y	1	2

***Step* III** Plot the graph of $x + 2y = 10$, $x + y = 1$, $x - y = 0$ using the above tables.

***Step* IV** Take a point (0, 0) and put it in the inequations (i) and (ii),

$0 + 0 \le 10$ (true)

So, the shaded region will be towards origin.

And $0 + 0 \ge 1$ (false)

So, the shaded region will be away from the origin.

Again, take a point (2, 2) and put it in the inequation (iv), we get

$2 \ge 0, 2 \ge 0$ (true)

So, the shaded region will be towards point (2, 2).

And take a point (0, 1) and put it in the inequation (iii), we get

$0 - 1 \le 0$ (true)

So, the shaded region will be towards point (0, 1).

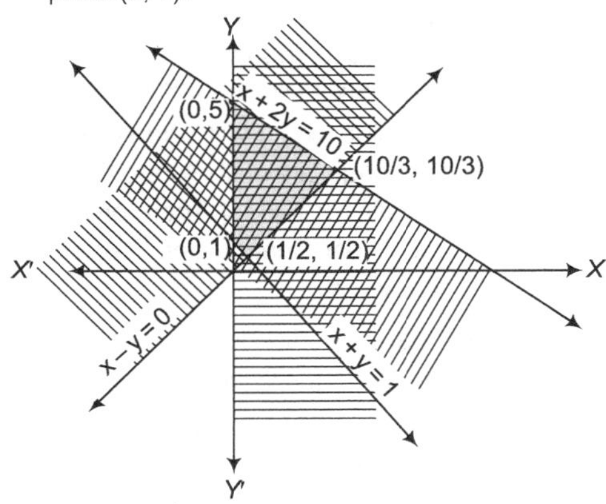

Thus, common shaded region shows the solution of the inequalities.

115. *(c)* Given inequalities are

$$3x + 4y \geq 12 \quad \ldots(i)$$
$$y \geq 1 \quad \ldots(ii)$$
$$x \geq 0 \quad \ldots(iii)$$

The line corresponding to (i) is

$$3x + 4y = 12 \quad \ldots(iv)$$

Table for $3x + 4y = 12$

x	4	0
y	0	3

Draw the graph of the line $3x + 4y = 12$.

$\because 3(0) + 4(0) \geq 12$ *i.e.,* $0 \geq 12$, which is not true.

$\therefore$ Inequality (i) represent the half plane made by the line (iv) which does not contain origin.

Inequality $y \geq 1$ represent the half plane made by the line $y = 1$ which does not contain origin($\because$ $0 \geq 1$ is not true)

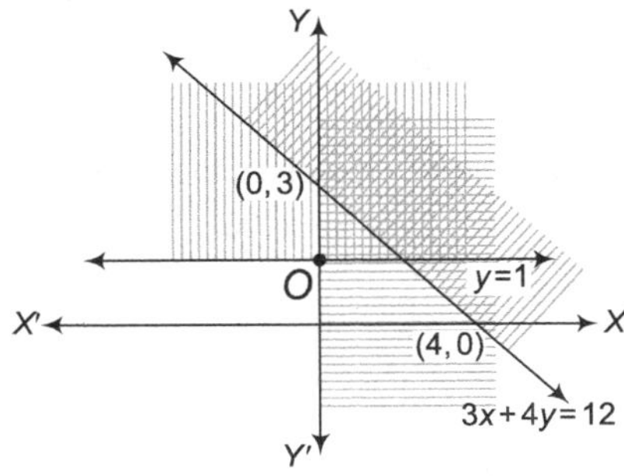

Inequality $x \geq 0$ represent the right side of the Y-axis.

Hence, the shaded part is common to all the given inequalities.

116. *(a)* Given inequalities are

$$x - y \leq 1 \quad \ldots(i)$$
$$x + 2y \leq 8 \quad \ldots(ii)$$
$$2x + y \geq 2 \quad \ldots(iii)$$
$$x \geq 0 \quad \ldots(iv)$$
$$y \geq 0 \quad \ldots(v)$$

The lines corresponding to (i), (ii) and (iii) are

$$x - y = 1 \quad \ldots(vi)$$
$$x + 2y = 8 \quad \ldots(vii)$$
$$2x + y = 2 \quad \ldots(viii)$$

Table for $x - y + 1$

x	1	0
y	0	− 1

Table for $x + 2y = 8$

x	8	0
y	0	4

Table for $2x + y = 2$

x	1	0
y	0	2

Draw the graph of line (vi), (vii) and (viii).

$\because 0 - 0 \leq 1$ *i.e.,* $0 \leq 1$, which is true.

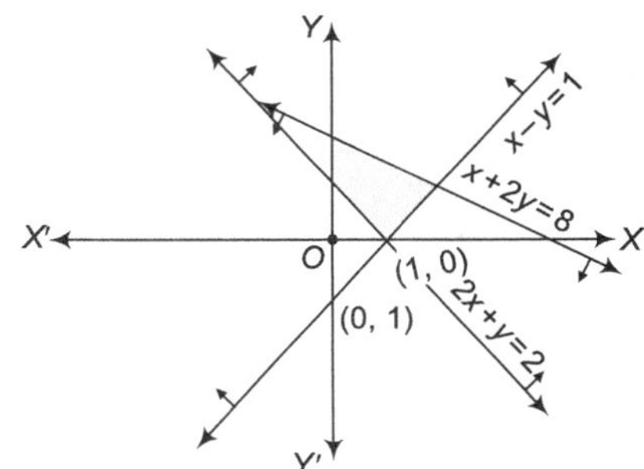

Therefore, inequality (i) represent the half plane made by the line (vi), which contains origin.

Now, $0 + 2(0) \leq 8$ *i.e.,* $0 \leq 8$, which is true. So, inequality (ii) represent the half plane made by line (vii) which contain origin.

Again, $2(0) + 0 \geq 2$ *i.e.,* $0 \geq 2$, which is not true.

$\therefore$ Inequality (iii) represent the half plane made by the line (viii), which does not contain origin.

Also, $x \geq 0$, $y \geq 0$ represent the region in I quadrant.

The region common to (i), (ii), (iii), (iv) and (v) is the shaded region which is bounded.

117. *(b)* I. We have,

$$|y - x| \leq 3$$

$\Rightarrow \quad -3 \leq y - x \leq 3 \quad (\because |x| \leq a, -a \leq x \leq a)$

$\Rightarrow \quad -3 \leq y - x$ and $y - x \leq 3$

$\Rightarrow \quad x - y \leq 3$ and $x - y \geq -3$

The region represented by $|y - x| \leq 3$ is the region represented by $x - y \leq 3$ and $x - y \geq -3$.

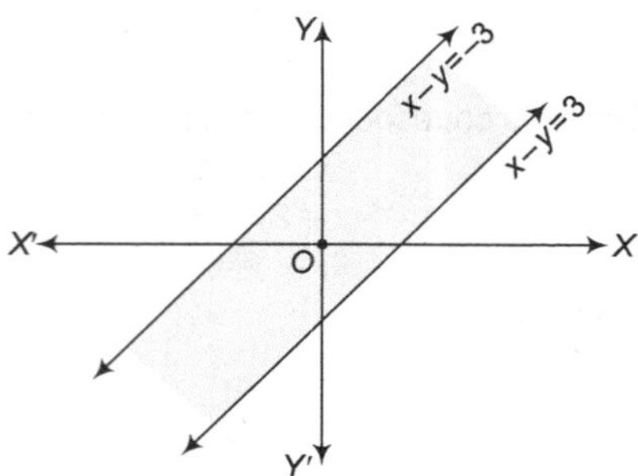

The region represented by $x - y \leq 3$ is the half plane made by the line $x - y = 3$, which contains origin.

The region represented by $x - y \geq -3$ is the half plane made by the line $x - y = -3$, which contains origin.

$\therefore$ Statement is false.

II. Given inequalities are

$$x - 2y \geq 0, 2x - y \leq -2, x \geq 0, y \geq 0$$

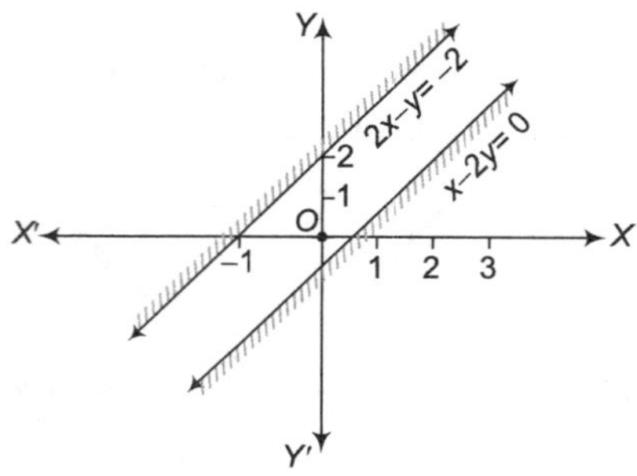

The line corresponding to $x - 2y \geq 0$ is

$$x - 2y = 0$$

i.e., $x = 2y$

Table for $x = 2y$

x	0	4
y	0	2

The line corresponding to $2x - y \leq -2$ is

$$2x - y = -2$$

Table for $2x - y = -2$

x	0	−1
y	2	0

Consider the point (1, 0)

$(1) - 2(0) \geq 1$ *i.e.,* $1 \geq 1$, which is true.

Therefore, inequality $x - 2y \geq 0$ represent the half plane made by the line $x - 2y = 0$, which contain (1, 0).

$\because 2(0) - (0) \leq -2$ *i.e.,* $0 \leq -2$, which is not true.

Therefore, inequality $2x - y \leq -2$ represent the half plane made by the line $2x - y = -2$, which does not contain origin.

It is clear from the graph there is no region common to all inequalities.

Hence, solution set of given inequalities is empty set.

118. *(b)* (i) Consider the line $x + y = 8$. We observe that the shaded region and origin lie on the same side of this line and (0, 0) satisfies $x + y \leq 8$. Therefore, $x + y \leq 8$ is the linear inequality corresponding to the line $x + y = 8$.

(ii) Consider $x + y = 4$. We observe that shaded region and origin are on the opposite side of this line and (0, 0) satisfies $x + y \leq 4$. Therefore, we must have $x + y \geq 4$ as linear inequalities corresponding to the line $x + y = 4$.

(iii) Shaded portion lie below the line $y = 5$. So, $y \leq 5$ is the linear inequality corresponding to $y = 5$.

(iv) Shaded portion lie on the left side of the line $x = 5$. So, $x \leq 5$ is the linear inequality corresponding to $x = 5$.

(v) Also, the shaded region lies in the first quadrant only. Therefore, $x \geq 0, y \geq 0$.

In view of (i), (ii), (iii), (iv) and (v) above the linear inequalities corresponding to the given solutions are

$$x + y \leq 8, x + y \geq 4, y \leq 5, x \leq 5, x \geq 0 \text{ and } y \geq 0$$

119. *(c)* Time taken to cover x km $= \frac{x}{25}$ h

Time taken to cover y km $= \frac{y}{40}$ h

Since, he has only one hour.

$$\therefore \quad \frac{x}{25} + \frac{y}{40} \leq 1$$

Total cost of petrol $= 2x + 5y$

Since he has only ₹100 to spend.

$$\therefore \quad 2x + 5y \leq 100$$

x and y are distances. So, x and y cannot be negative.

$$\therefore \quad x \geq 0, y \geq 0$$

Hence, inequalities represent above data are

$$2x + 5y \leq 100, \frac{x}{25} + \frac{y}{40} \leq 1, x \geq 0, y \geq 0$$

120. *(b)* The inequalities represent the data are

$$2x + 5y \leq 100 \quad \text{...(i)}$$

$$\frac{x}{25} + \frac{y}{40} \leq 1 \quad \text{...(ii)}$$

$$x \geq 0 \quad \text{...(iii)}$$

$$y \geq 0 \quad \text{...(vi)}$$

The line corresponding to (i) and (ii) are

$$2x + 5y = 100 \quad \text{...(v)}$$

$$\frac{x}{25} + \frac{y}{40} = 1 \quad \text{...(vi)}$$

Line (v) cuts X-axis at (50, 0) and Y-axis at (0, 20). Line (vi) cuts X-axis at (25, 0) and Y-axis at (0, 40). Since, (0, 0) satisfies both (i) and (ii). So, both inequality represent the region which contains the origin.

The common region of (i), (ii), (iii), (vi) is the shaded region in the figure.

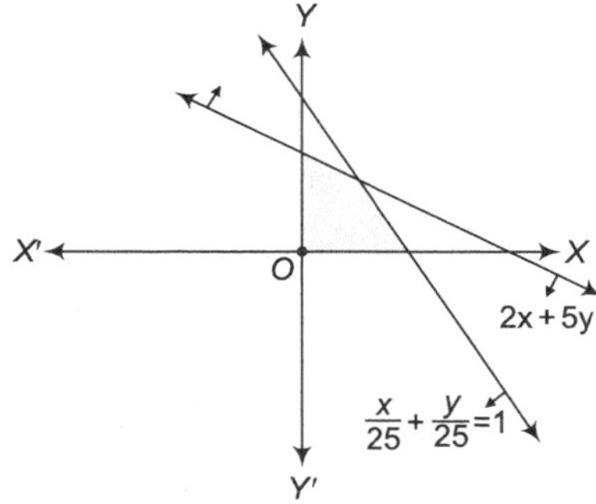

Hence, there are infinite point which satisfy the inequalities (i), (ii), (iii) and (vi).

121. *(d)* In this case, we have two inequalities, $-8 \le 5x - 3$ and $5x - 3 < 7$, which we will solve simultaneously. We have $-8 \le 5x - 3 < 7$ or $-5 \le 5x < 10$ or $-1 \le x < 2$

i.e., $x \in [-1, 2)$.

122. *(b)* We have $-5 \le \dfrac{5-3x}{2} \le 8$

or $\quad -10 \le 5 - 3x \le 16$ or $-15 \le -3x \le 11$

or $$5 \ge x \ge -\frac{11}{3}$$

which can be written as $\dfrac{-11}{3} \le x \le 5$.

123. *(b)* Given inequalities are

$$3x - 7 < 5 + x \quad \text{... (i)}$$

and $$11 - 5x \le 1 \quad \text{...(ii)}$$

From inequality (i), we have

$$3x - 7 < 5 + x$$

or $$x < 6 \quad \text{... (iii)}$$

Also, from inequality (ii), we have

$$11 - 5x \le 1$$

or $$-5x \le -10 \quad \textit{i.e.}, x \ge 2 \quad \text{...(vi)}$$

If we draw the graph of inequalities (iii) and (vi) on the number line, we see that the values of x, which are common to both, are shown by bold line in figure.

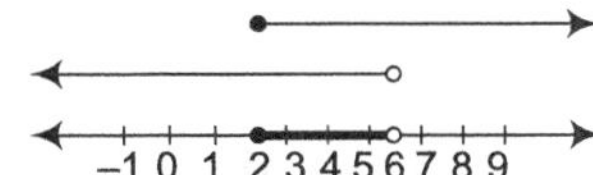

124. *(a)* It is given that $30 < C < 35$.

Putting $C = \dfrac{5}{9}(F - 32)$, we get

$$30 < \frac{5}{9}(F - 32) < 35,$$

Multiplying $\dfrac{9}{5}$ to each term,

or $$\frac{9}{5} \times (30) < (F - 32) < \frac{9}{5} \times (35)$$

or $$54 < (F - 32) < 63$$

or $$86 < F < 95$$

Thus, the required range of temperature is between 86° F and 95° F.

125. *(b)* Let x L of 30% acid solution is required to be added. Then,

Total mixture $= (x + 600)$ L

Therefore 30% x + 12% of 600 > 15% of $(x + 600)$

and 30% x + 12% of 600 < 18% of $(x + 600)$

or $$\frac{30x}{100} + \frac{12}{100}(600) > \frac{15}{100}(x + 600)$$

and $$\frac{30x}{100} + \frac{12}{100}(600) < \frac{18}{100}(x + 600)$$

or $$30x + 7200 > 15x + 9000$$

and $$30x + 7200 < 18x + 10800$$

or $$15x > 1800 \quad \text{and} \quad 12x < 3600$$

or $$x > 120 \quad \text{and} \quad x < 300,$$

i.e., $$120 < x < 300$$

126. *(b)* A. The given inequality $-15 < \dfrac{3(x-2)}{5} \le 0$

Multiplying by 5 to each term,

$$-15 \times 5 < \frac{3(x-2)}{5} \times 5 \le 0 \times 5$$

$\Rightarrow$ $$-75 < 3x - 6 \le 0$$

Adding 6 to each term,

$$-75 + 6 < 3x - 6 + 6 \le 0 + 6$$

$\Rightarrow$ $$-69 < 3x \le 6$$

Dividing by 3 in each term,

$$\frac{-69}{3} < 3\left(\frac{x}{3}\right) \le \frac{6}{3}$$

$\Rightarrow$ $$-23 < x \le 2$$

$\therefore$ Solution set is $(-23, 2]$ or $]-23, 2]$.

B. The given inequality $7 \le \dfrac{3x + 11}{2} \le 11$

Multiplying by 2 to each term,

$$7 \times 2 \le \frac{(3x + 11)}{2} \times 2 \le 11 \times 2$$

$\Rightarrow$ $$14 \le 3x + 11 \le 22$$

Adding (-11) to each term,

$$14 - 11 \le 3x + 11 - 11 \le 22 - 11$$

$\Rightarrow$ $$3 \le 3x \le 11$$

Dividing by 3 in each term,

$$\frac{3}{3} \le \frac{3x}{3} \le \frac{11}{3} \Rightarrow 1 \le x \le \frac{11}{3}$$

$\therefore$ Solution set is $\left[1, \dfrac{11}{3}\right]$.

C. Given inequality can be written as $|x-2| \leq 3$ and $|x-2| \geq 1$

$\because \quad |x-2| \leq 3$

$\Rightarrow \quad -3 \leq x-2 \leq 3 \quad (\because |x| \leq a, -a \leq x \leq a)$

Adding 2 in each term,

$$-1 \leq x \leq 5$$

Again, $\quad |x-2| \geq 1$

$\Rightarrow \quad x-2 \leq -1$ or $x-2 \geq 1$

$(\because |x| \geq a, x \leq -a$ or $x \geq a)$

$\Rightarrow \quad x \leq 1$ or $x \geq 3$

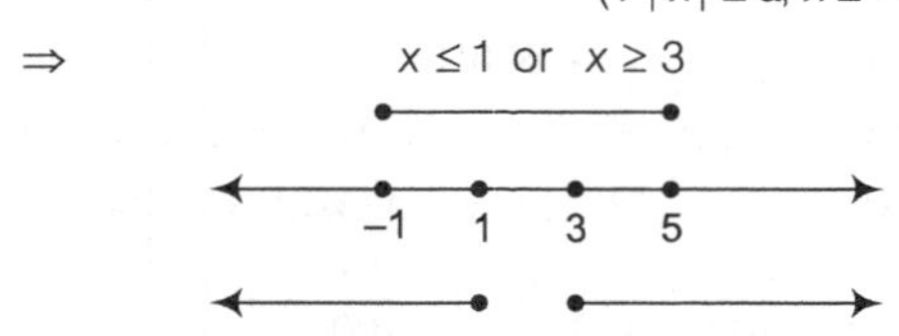

Hence, common values of x are given by

$$x \in [-1, 1] \cup [3, 5]$$

D. Given,

$$-5 \leq \frac{2-3x}{4} \leq 9$$

Multiplying each terms by 4,

$$-5 \times 4 \leq \frac{2-3x}{4} \times 4 \leq 9 \times 4$$

$\Rightarrow \quad -20 \leq 2-3x \leq 36$

Subtracting 2 from each terms,

$\Rightarrow \quad -22 \leq -3x \leq 34$

Multiplying $\left(-\frac{1}{3}\right)$ to each term,

$$(-22)\left(-\frac{1}{3}\right) \geq -3x\left(-\frac{1}{3}\right) \geq 34\left(-\frac{1}{3}\right)$$

$\Rightarrow \quad \frac{22}{3} \geq x \geq \frac{-34}{3}$

$\therefore x \in \left[-\frac{34}{3}, \frac{22}{3}\right]$

127. *(c)* We have, $3x-7 > 2(x-6)$

$\Rightarrow \quad 3x-7 > 2x-12$

Transferring the term $2x$ to LHS and the terrm (–7) to RHS,

$$3x-2x > -12+7$$

$\Rightarrow \quad x > -5 \quad \ldots(i)$

and $\quad 6-x > 11-2x$

Transferring the term $(-2x)$ to LHS and the term 6 to RHS,

$$-x+2x > 11-6$$

$\Rightarrow \quad x > 5 \quad \ldots(ii)$

Draw the graph of inequations (i) and (ii) on the number line,

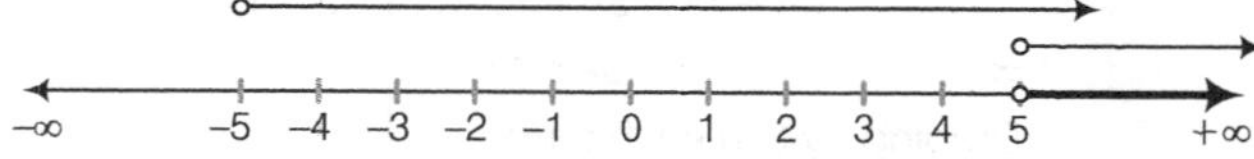

Hence, solution set of the equations are real numbers, x lying on greater than 5 excluding 5.

i.e., $\quad x > 5$

$\therefore$ Solution set is $(5, \infty)$ or $]5, \infty[$.

128. *(b)* Given that $68° < F < 77° \quad \ldots(i)$

Putting $F = \frac{9}{5}C + 32$ in Eq. (i),

$\Rightarrow \quad 68° < \frac{9}{5}C + 32 < 77°$

Adding (– 32) in each term,

$$68° - 32 < \frac{9}{5}C + 32 - 32 < 77° - 32$$

$\Rightarrow \quad 36° < \frac{9}{5}C < 45°$

Multiplying by $\left(\frac{5}{9}\right)$ in each term,

$$36° \times \frac{5}{9} < \frac{9}{5}C \times \frac{5}{9} < 45° \times \frac{5}{9}$$

$\Rightarrow \quad 4° \times 5 < C < 5° \times 5$

$\Rightarrow \quad 20° < C < 25°$

i.e., $\quad C \in (20, 25)$

Hence, temperature in degree Celsius lies between 20 °C and 25°C.

129. *(c)* Let the 2% boric acid solution be x L.

$\therefore \quad$ Mixture $= (640 + x)$ L

Now, according to the question, two conditions arise

I. 2% of x + 8% of 640 > 4% of $(640 + x)$

II. 2% of x + 8% of 640 < 6% of $(640 + x)$

From condition I,

$$\frac{2}{100} \times x + \frac{8}{100} \times 640 > \frac{4}{100} \times (640 + x)$$

Multiplying both sides by 100,

$$100 \times \left[\frac{2x}{100} + \frac{8}{100} \times 640\right] > \frac{4}{100} \times (640 + x) \times 100$$

$\Rightarrow \quad 2x + 8 \times 640 > 4 \times 640 + 4x$

Transferring the term $4x$ to LHS and the term (8×640) to RHS,

$$2x - 4x > 4 \times 640 - 8 \times 640$$

$\Rightarrow \quad -2x > 640\,(4-8)$

$\Rightarrow \quad -2x > -4 \times 640$

Dividing both sides by –2,

$$\frac{-2x}{-2} < \frac{-4 \times 640}{-2}$$

$\Rightarrow \quad x < 2 \times 640$

$\Rightarrow \quad x < 1280 \quad \ldots(i)$

From condition II,

$$\frac{2}{100} \times x + \frac{8}{100} \times 640 < \frac{6}{100} \times (640 + x)$$

$\Rightarrow \quad 100 \times \left[\frac{2x}{100} + \frac{8}{100} \times 640\right] < [6 \times 640 + 6x] \times \frac{100}{100}$

$\Rightarrow \quad 2x + 8 \times 640 < 6 \times 640 + 6x$

Transferring the term $6x$ to LHS and the term (8×640) to RHS,

$$2x - 6x < 6 \times 640 - 8 \times 640$$
$$\Rightarrow \quad -4x < 640(6-8)$$
$$\Rightarrow \quad -4x < -2 \times 640$$

Dividing both sides by -4,

$$\frac{-4x}{-4} > \frac{-2 \times 640}{-4}$$
$$\Rightarrow \quad x > 320 \quad \dots\text{(ii)}$$

Hence, from Eqs. (i) and (ii),

$$320 < x < 1280 \quad i.e., x \in (320, 1280)$$

The number of litres to be added should be greater than 320 L and less than 1280 L.

130. *(d)* Given, x L of water is to be added to 45% solution of acid.

Now, according to the question,

$$25\% \text{ of } (1125 + x) < 45\% \text{ of } 1125 < 30\% \text{ of } (1125 + x)$$
$$\Rightarrow \quad \frac{25}{100} \times (1125 + x) < \frac{45}{100} \times 1125 < \frac{30}{100} \times (1125 + x)$$
$$\Rightarrow \quad 25 \times 1125 + 25x < 45 \times 1125 < 30 \times 1125 + 30x$$

Taking first two parts of above inequality,

$$25 \times 1125 + 25x < 45 \times 1125$$

Transferring the term 25×1125 to the RHS,

$$25x < 45 \times 1125 - 25 \times 1125$$
$$\Rightarrow \quad 25x < 1125(45 - 25)$$
$$\Rightarrow \quad 25x < 1125 \times 20$$

Dividing both sides by 25,

$$\frac{25}{25}x < \frac{1125 \times 20}{25}$$
$$\Rightarrow \quad x < 45 \times 20$$
$$\Rightarrow \quad x < 900 \quad \dots\text{(i)}$$

Now, taking last two parts of the above inequality,

$$\Rightarrow \quad 45 \times 1125 < 30 \times 1125 + 30x$$

Transferring the term (30×1125) to LHS,

$$45 \times 1125 - 30 \times 1125 < 30x$$
$$\Rightarrow \quad 1125(45 - 30) < 30x$$
$$\Rightarrow \quad 1125 \times 15 < 30x$$
$$\Rightarrow \quad 30x > 1125 \times 15$$

Dividing both sides by 30,

$$\frac{30x}{30} > \frac{1125 \times 15}{30}$$
$$\Rightarrow \quad x > \frac{1125}{2}$$
$$\Rightarrow \quad x > 562.5 \quad \dots\text{(ii)}$$

Hence, from Eqs. (i) and (ii),

$$562.5 < x < 900 \; i.e., x \in (562.5, 900).$$

The number of litres of water to be added should be greater than 562.5 L and less than 900 L.

131. *(a)* Common solution of the inequalities is from $-\infty$ to -4 and 3 to ∞.

$$i.e., \quad (-\infty, -4] \cup [3, \infty)$$

132. *(d)* For $x > -2$, the line $x = -2$ has to be dotted *i.e.*, the region does not include the points on the line $x = 2$. Hence, Statement I is false.

Since, (1, 0) is the point in shaded portion, but it does not satisfy $x - y \le 0$. So, Statement II is false.

133. *(d)* Given, $\frac{2x+3}{2x-9} < 0$

$$\Rightarrow \quad 2x + 3 < 0$$
$$\text{and} \quad 2x - 9 > 0 \quad \text{or} \quad 2x + 3 > 0 \quad \text{and} \quad 2x - 9 < 0$$
$$\left(\because \frac{a}{b} < 0 \Rightarrow a < 0 \text{ and } b > 0 \text{ or } a > 0 \text{ and } b < 0\right)$$
$$\Rightarrow x < -\frac{3}{2} \text{ and } x > \frac{9}{2} \text{ or } x > \frac{-3}{2} \text{ and } x < \frac{9}{2} \text{ (not possible)}$$

Hence, $x \in \left(-\frac{3}{2}, \frac{9}{2}\right)$.

134. *(b)* The inequality representing in the graph is $x \ge -5$ and $x \le 5$

$$i.e., \quad |x| \le 5 \quad (\because |x| \le a \Leftrightarrow -a \le x \le a)$$

135. *(c)* On LHS of the given inequality, we have two terms both containing modulus. By equating the expression with in the modulus to zero, we get $x = -1, 0$ as critical points. These critical points divide the real line in three parts as $(-\infty, -1), [-1, 0), [0, \infty)$.

Case I When $-\infty < x < -1$,

$$|x+1| + |x| > 3 \Rightarrow -x - 1 - x > 3 \Rightarrow x < -2$$

Case II When $-1 \le x < 0$,

$$|x+1| + |x| > 3 \Rightarrow x + 1 - x > 3 \Rightarrow 1 > 3 \text{ (not possible)}$$

Case III When $0 \le x < \infty$,

$$|x+1| + |x| > 3 \Rightarrow x + 1 + x > 3 \Rightarrow x > 1.$$

Combining the results of cases (I), (II) and (III), we get

$$x \in (-\infty, -2) \cup (1, \infty)$$

136. *(b)* We have, $5(2x - 7) - 3(2x + 3) \le 0$

$$10x - 35 - 6x - 9 \le 0$$
$$4x - 44 \le 0$$

Adding 44 on both sides,

$$4x - 44 + 44 \le 0 + 44$$
$$\Rightarrow \quad 4x \le 44$$
$$\Rightarrow \quad x \le \frac{44}{4}$$
$$\Rightarrow \quad x \le 11 \quad \dots\text{(i)}$$

Now, $\quad 2x + 19 \le 6x + 47$

Transferring the term $6x$ to LHS and the term 19 to RHS,

$$2x - 6x \le 47 - 19 \Rightarrow -4x \le 28$$

Dividing both sides by 4,

$$\frac{-4x}{4} \le \frac{28}{4}$$

$\Rightarrow \quad -x \le 7$

$\Rightarrow \quad x \ge -7 \quad \ldots\text{(ii)}$

Draw the graph of inequalities (i) and (ii) on the number line,

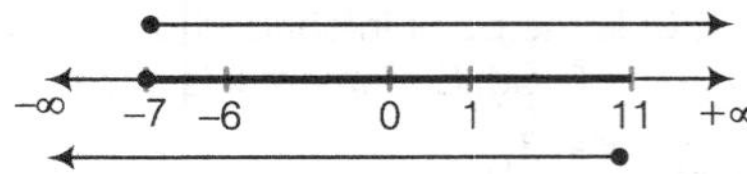

Hence, solution set of the inequalities are real numbers x lying between -7 and 11, including -7 and 11.

i.e., $\quad -7 \le x \le 11$

$\therefore$ Solutions set is $[-7, 11]$.

137. *(b)* Given, $C(x) = 26{,}000 + 30x$

and $\quad R(x) = 43x$

$\therefore \quad \text{Profit} = R(x) - C(x)$

$= 43x - (26{,}000 + 30x)$

$= 13x - 26{,}000$

For some profit, $13x - 26000 > 0$

$\Rightarrow \quad 13x > 26000$

$\Rightarrow \quad x > 2000$

138. *(c)* Let the third pH reading be x.

$\therefore$ Average pH reading of three days $= \dfrac{8.48 + 8.35 + x}{3}$

$$= \frac{16.83 + x}{3}$$

Given, $\quad 8.2 < \dfrac{16.83 + x}{3} < 8.5$

Multiplying each term by 3,

$\Rightarrow \quad 24.6 < 16.83 + x < 25.5$

Subtracting 16.83 from each term,

$\Rightarrow \quad 7.77 < x < 8.67$

139. *(a)* Given, $\dfrac{|x-1|}{x+2} < 1$

i.e., $\quad \dfrac{|x-1|}{x+2} - 1 < 0$

Case I When $x < 1, |x - 1| = -(x - 1) = 1 - x$

$\therefore \quad \dfrac{1-x}{x+2} - 1 < 0 \Rightarrow \dfrac{-2x-1}{x+2} < 0$

$\Rightarrow \quad \dfrac{2x+1}{x+2} > 0$

$\Rightarrow \quad 2x + 1 < 0 \text{ and } x + 2 < 0$

or $\quad 2x + 1 > 0 \text{ and } x + 2 > 0$

$\left(\because \dfrac{a}{b} > 0 \Rightarrow a > 0 \text{ and } b > 0 \text{ or } a < 0 \text{ and } b < 0\right)$

$\Rightarrow \quad x < -\dfrac{1}{2} \text{ and } x < -2$

or $\quad x > -\dfrac{1}{2} \text{ and } x > -2$

$\Rightarrow \quad x < -2 \text{ or } x > -\dfrac{1}{2}$

But $x < 1$

$\therefore \quad x \in (-\infty, -2) \cup \left(-\dfrac{1}{2}, 1\right) \quad \ldots\text{(i)}$

Case II When $x \ge 1, |x - 1| = x - 1$

$\therefore \quad \dfrac{x-1}{x+2} - 1 < 0$

$\Rightarrow \quad -\dfrac{3}{x+2} < 0 \Rightarrow \dfrac{3}{x+2} > 0$

$\Rightarrow \quad x + 2 > 0 \Rightarrow x > -2$

But $\quad x \ge 1$

$\therefore \quad x \in [1, \infty) \quad \ldots\text{(ii)}$

From (i) and (ii), we get

$$x \in (-\infty, -2) \cup \left(-\frac{1}{2}, \infty\right)$$

7

Permutations *and* Combinations

Fundamental Principle of Counting

Level I

1. Suppose you have a suitcase with a number lock. The number lock has 4 wheels each labelled with 10 digits from 0 to 9. The lock can be opened, if 4 specific digits are arranged in a particular sequence with no repetition. Some how, you have forgotten this specific sequence of digits. You remember only the first digit which is 7. In order to open the lock, how many sequences of 3 digits you may have to check with?

(a) 720 (b) 504
(c) 500 (d) None of these

2. Mohan has 3 pants and 2 shirts. The number of different pairs of a pant and shirt he can wear are

(a) 5 (b) 6
(c) 7 (d) None of these

3. Sabnam has 2 school bags, 3 tiffin boxes and 2 water bottles. In how many ways can she carry these items (choosing one each)?

(a) 11 (b) 12 (c) 13 (d) 14

4. Consider the following statements

Statement I "If an event can occur in m different ways, following which another event can occur in n different ways, then the total number of occurrence of the events in the given order is $m \times n$."

Statement II "If an event can occur in m different ways, following which another event can occur in n different ways, following which a third event can occur in p different ways, then the total number of occurrence of the events in the given order is $m \times n \times p$."

Choose the correct option.

(a) Only I is correct (b) Only II is correct
(c) Both I and II are correct (d) Neither I nor II is correct

5. Find the number of 4 letter words, with or without meaning, which can be formed out of the letters of the word ROSE.

Consider the following statements

Statement I When the repetition of the letters is not allowed is 24.

Statement II When the repetition of the letters is allowed is 260.

Choose the correct option.

(a) Statement I is correct (b) Statement II is correct
(c) Both I and II are correct (d) Neither I nor II is correct

6. Given 4 flags of different colours, the number of different signals that can be generated, if a signal requires the use of 2 flags one below the other are

(a) 10 (b) 11 (c) 12 (d) 13

7. The number of 2 digit even numbers that can be formed from the digits 1, 2, 3, 4, 5, if the digits can be repeated are

(a) 10 (b) 11 (c) 12 (d) 14

8. Find the number of different signals that can be generated by arranging atleast 2 flags in order (one below the other) on a vertical staff, if five different flags are available are

(a) 120 (b) 240 (c) 300 (d) 320

9. The number of 3 digit numbers that can be formed from the digits 1, 2, 3, 4 and 5.

Consider the following statements

Statement I When the repetition of numbers is allowed is 125.

Statement II When the repetition of numbers is not allowed is 60.

Which of the above statement(s) is/are true?

(a) Only I (b) Only II
(c) Neither I nor II (d) Both I and II

10. Match the following columns and choose the correct option from the codes given below.

Column I	Column II
A. The number of 3 digit even numbers that can be formed from the digits 1, 2, 3, 4, 5, 6, if the digits can be repeated are	1. 336
B. The number of 4 letter codes that can be formed using the first 10 letters of the English alphabets, if no letter can be repeated, is	2. 108
C. The number of 5 digit telephone numbers that can be constructed using the digits 0 to 9, if each number starts with 67 and no digit appears more than once	3. 5040

Codes

	A	B	C		A	B	C
(a)	1	2	3	(b)	2	3	1
(c)	3	2	1	(d)	2	1	3

11. Consider the following statements

Statement I A coin is tossed 3 times and outcomes are recorded. The number of possible outcome are 6.

Statement II Given 5 flags of different colour, the number of different signals that can be generated, if each signal requires the use of 2 flags, one below the other are 20.

Which of the above statement (s) is /are true?

(a) Only I (b) Only II
(c) Both I and II (d) Neither I nor II

Level II

12. The number of 3 digit numbers having atleast one of their digit as 5 are [HOTS]

(a) 250 (b) 251
(c) 252 (d) 253

13. The following diagram can be coloured in such a way that each of smaller triangle is to be painted with red, blue or green colour and no two adjacent triangles have the same colour.

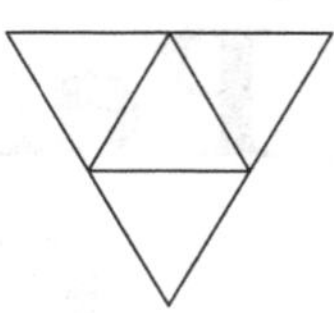

The number of ways in which the diagram can be coloured are. [NCERT Exemplar; HOTS]

(a) 9 (b) 12
(c) 15 (d) 24

14. The number of ways in which 3 prizes can be distributed to 4 children, so that no child gets all the three prizes, are

(a) 64 (b) 62
(c) 60 (d) None of these

15. Match the following columns and choose the correct option from the codes given below.

Column I	Column II
A. The number of ways in which 6 distinct objects can be put into two different boxes, so that no box remains empty are	1. 81
B. The number of ways in which 5 letters can be posted in 4 letter boxes are	2. 62
C. The number of functions from A to B, if $n(A) = 4$ and $n(B) = 3$ are	3. 1024

Codes

	A	B	C
(a)	1	2	3
(b)	2	3	1
(c)	2	1	3
(d)	3	1	2

16. The number of 4-digit numbers that can be formed with the digits 1, 2, 3, 4 and 5, in which atleast 2 digits are identical, is [HOTS]

(a) 505
(b) $4^5 - 5!$
(c) 600
(d) None of the above

Permutations

Level I

17. Find the number of 3 letter words, with or without meaning which can be formed out of the letters of the word 'NUMBER'.

Consider the following statements

Statement I When repetition of letters is not allowed is 120.

Statement II When repetition of letters is allowed is 216.

Choose the correct option.

(a) Only Statement I is correct
(b) Only Statement II is correct
(c) Both I and II are correct
(d) Both I and II are false

18. A permutation is an ...A... in a ...B... order of a number of objects taken some or all at a time. Here, A and B refers to

(a) arrangement, indefinite (b) combination, definite
(c) arrangement, definite (d) combination, indefinite

19. The number of permutations of n ...A... objects taken r at a time, where $0 \le r \le n$ and the objects do not repeat is ...B... which is denoted by ...C... . Here, A, B and C refers to

(a) different , $n(n-1)(n-2)...(n-r+1), {}^nP_r$
(b) similar , $n(n-1)...(n-r-1), {}^nC_r$
(c) different, $n(n-1)...(n-r+1), {}^nC_r$
(d) similar, $n(n-1)...(n-r+1), {}^nP_r$

20. Match the following columns and choose the correct option from the codes given below.

	Column I		Column II
A.	5! equals	1.	5040
B.	7! equals	2.	120
C.	8! equals	3.	4920
D.	$7!-5!$ equals	4.	40320
E.	$4!-3!$ equals	5.	18

Codes

	A	B	C	D	E
(a)	1	2	3	4	5
(b)	2	1	3	4	5
(c)	2	1	4	3	5
(d)	5	4	3	2	1

21. Match the following columns and choose the correct option from the codes given below.

	Column I		Column II
A.	$\frac{7!}{5!}$ equals	1.	28
B.	$\frac{12!}{(10!)(2!)}$ equals	2.	42
C.	$\frac{8!}{6! \times 2!}$ equals	3.	66

Codes

	A	B	C		A	B	C
(a)	1	2	3	(b)	1	3	2
(c)	3	2	1	(d)	2	3	1

22. Consider the following statements

Statement I If $\frac{1}{8!} + \frac{1}{9!} = \frac{x}{10!}$, then x is equal to 100.

Statement II If $\frac{1}{6!} + \frac{1}{7!} = \frac{x}{8!}$, then x is equal to 56.

Which of the above statement (s) is /are true?

(a) Only I (b) Only II
(c) Both I and II (d) Neither I nor II

23. Fill in the blanks.

I. When $n = 6, r = 2$, the value of $\frac{n!}{(n-r)!}$ is ...A... .

II. When $n = 9, r = 5$, the value of $\frac{n!}{(n-r)!}$ is ...B... .

III. When $n = 5$, $r = 2$, the value of $\frac{n!}{r!(n-r)!}$ is ...C... .

IV. $3! + 4! = 7!$ is ...D... .

Here, A, B, C and D refers to

(a) $A \to 10, B \to 15120, C \to 10, D \to$ false
(b) $A \to 30, B \to 15120, C \to 20, D \to$ true
(c) $A \to 30, B \to 15120, C \to 10, D \to$ false
(d) None of the above

24. Consider the following statements

Statement I ${}^nP_r = \frac{n!}{(n-r)!}, 0 \le r \le n.$

Statement II
${}^nP_r = n(n-1)(n-2)........(n-r+1), 0 \le r \le n.$

Which of the above statement (s) is/are true?

(a) Only I (b) Only II
(c) Both I and II (d) Neither I nor II

25. The number of permutations of n different things taken r at a time, when the repetition is allowed, is

(a) r^n (b) n^r
(c) $\frac{n!}{(n-r)!}$ (d) None of these

26. The number of 4 letter words that can be formed from letters of the word 'PART' when

Consider the following statements

Statement I Repetition is not allowed is 24.

Statement II Repetition is allowed is 256.

Which of the above statement (s) is/are ture?

(a) Only I (b) Only II
(c) Both I and II (d) Neither I nor II

27. Match the following columns and choose the correct option form the codes given below.

Column I	Column II
A. The number of 3 letter words that can be formed by the letters of the word 'NUMBER' when the repetition is not allowed are	1. 216
B. The number of 3 letter words that can be formed by the letters of the word 'NUMBER' when the repetition is allowed are	2. 132
C. The number of ways in which a chairman and a vice chairman can be chosen from a group of 12 persons assuming that one person cannot hold more than one position are	3. 120

Codes

	A	B	C		A	B	C
(a)	1	2	3	(b)	3	1	2
(c)	3	2	1	(d)	2	3	1

28. Fill in the blanks.

I. The number of permutations of n objects, where P objects are of the same kind and rest are all different is ...A... .

II. The number or permutations of n objects, where P_1 objects are of one kind, P_2 are of second kind, P_k are of kth kind and the rest, if any, are of different kind is ...B... .

Here, A and B refers to

(a) $\frac{n!}{P!2!}, \frac{n!}{(P_1+P_2+\ldots+P_k)!}$ (b) $\frac{n!}{P!}, \frac{n!}{P_1!P_2!\ldots P_k!}$
(c) $\frac{P!}{n!}, \frac{(P_1+P_2\ldots+P_k)}{n!}$ (d) None of these

29. Identify the correct combination of true (T) and false (F) for the given two statements.

Statement I The number of permutations of the letters of the word 'ROOT' are 12.

Statement II The number of permutation of the letters of the word 'INSTITUTE' are $\frac{9!}{2!\,3!}$.

(a) FT (b) TF (c) FF (d) TT

30. Match the following columns and choose the correct option from the codes given below .

Column I	Column II
A. The number of permutations of the letters of the word ALLAHABAD are	1. 3024
B. The number of 4-digit numbers that can be formed by using the digits from 1 to 9, if repetition of digit is not allowed are	2. 7560
C. The numbers lying between 100 and 1000 that can be formed with the digits 0, 1, 2, 3, 4, 5. If the repetition of the digits is not allowed	3. 100
D. The numbers lying between 100 and 1000 that can be formed with the digits 0, 1, 2, 4, 9, if repetition of digits is allowed	

Codes

	A	B	C	D		A	B	C	D
(a)	3	3	2	1	(b)	1	2	3	3
(c)	2	1	3	3	(d)	3	1	2	3

31. Consider the following statements

Statement I If $5\,{}^4P_r = 6\,{}^5P_{r-1}$, then $r = 8, 3$

Statement II If ${}^5P_r = {}^6P_{r-1}$, then $r = 5$

Which of the above statements (s) is/are true?

(a) Only I (b) Only II
(c) Both I and II (d) Neither I nor II

32. The number of different 8 letter arrangement that can be made from the letters of the word. DAUGHTER so that

(i) all vowels occur together are ...A... .

(ii) all vowels do not occur together are ...B... .

Here, A and B refer to

(a) 4320 , 36000 (b) 3600 , 36000
(c) 4320 , 30000 (d) None of these

33. The number of ways in which 4 red, 3 yellow and 2 green discs can be arranged in a row, if the discs of same colour are identical , are

(a) 1296 (b) 1260
(c) 1200 (d) None of these

34. Match the following columns and choose the correct option from the codes given below.

	Column I		Column II
A.	Number of 3-digit numbers that can be formed by using the digits from 1 to 9, if no digit is repeated are	1.	120
B.	Number of 4-digit numbers that can be formed by using the digits from 0 to 9, if no digit is repeateds is	2.	60
C.	Number of 3-digit even numbers that can be formed by using the digits 1, 2, 3, 4, 6, 7, if no digit is repeated	3.	504
D.	Number of 4-digit numbers that can be formed by using the digits 1, 2, 3, 4, 5, if no digit is repeated	4.	4536
E.	Number of 4-digit even numbers that can be formed by using the digits 1, 2, 3, 4, 5 (repetition not allowed)	5.	48

Codes

	A	B	C	D	E		A	B	C	D	E
(a)	1	2	3	4	5	(b)	3	4	2	1	5
(c)	1	2	4	5	3	(d)	4	3	5	1	2

35. The number of ways of choosing a chairman and a vice chairman from a committee of 8 persons assuming one person cannot hold more than one position are

(a) 56 (b) 60 (c) 64 (d) 28

36. Consider the following statements

Statement I If ${}^{n-1}P_3 : {}^nP_4 = 1:9$, then n equals 9.

Statement II If ${}^5P_r = 2 \cdot {}^6P_{r-1}$, then r equals 3.

Which of the above statement (s) is / are true?

(a) Only I (b) Only II
(c) Both I and II (d) Neither I nor II

37. Consider the following statements

Statement I Number of words with or without meaning, that can be formed using all the letters of the word EQUATION, using each letter exactly once are 40320.

Statement II The number of distinct permutations of the letters in MISSISSIPPI such that four I's do not come together are 33810.

Which of the above statement(s) is /are true?

(a) Only I (b) Only II
(c) Both I and II (d) Neither I nor II

38. Find the number of words with or without meaning that can be made from the letters of the word MONDAY, assuming that no letter is repeated, if

Statement I 4 letters are used at a time equals 360.

Statement II All letters are used at a time equals 720.

Statement III All letters are used but first letters is a vowel equals 120.

Identify the correct combination of true (T) and false (F) for the given statements.

(a) TTT (b) FFF (c) FFT (d) TTF

39. The number of ways, the letters of the word PERMUTATIONS, can be arranged, if

Statement I Words start with P and ends with S equals 1814400.

Statement II Vowels come together equals 2419200.

Statement III There are always 4 letters between P and S equals 25401600.

Identify the correct combination of true (T) and false (F) for the given statements.

(a) TFT (b) FTT (c) TTT (d) FFF

Level II

40. The sum of the digits in unit place of all the numbers formed with the help of 3, 4, 5 and 6 taken all at a time is [NCERT Exemplar]

(a) 432 (b) 108 (c) 36 (d) 18

41. How many words (with or without dictionary meaning) can be made from the letters of the word MONDAY, assuming that no letter is repeated, if [NCERT Exemplar]

	Column I		Column II
A.	4 letters are used at a time	1.	720
B.	All letters are used at a time	2.	240
C.	All letters are used but the first letter is a vowel	3.	360

Match the statements in Column I with Column II and choose the correct option from the codes given below.

Codes

	A	B	C		A	B	C
(a)	1	2	3	(b)	3	1	2
(c)	2	1	3	(d)	3	2	1

42. If the letters of the word KRISNA are arranged in all possible ways and these words are written out as in a dictionary, then the rank of the word KRISNA is

(a) 324 (b) 341
(c) 359 (d) None of these

43. The number of ways in which 10 candidates $A_1, A_2, \ldots, A_{10}$ can be ranked such that A_1 is always above A_{10} is

(a) 5! (b) 2(5!) (c) 10! (d) $\frac{1}{2}(10!)$

44. How many numbers lying between 999 and 10000 can be formed with the help of the digits 0, 2, 3, 6, 7, 8, when the digits are not repeated? [DCE 2008]

(a) 100 (b) 200 (c) 300 (d) 400

45. The ten's digit in $1! + 4! + 7! + 10! + 12! + 13! + 15! + 16! + 17!$ is divisible by [KCET 2005]

(a) 4 (b) 3! (c) 5 (d) 7

46. There are 10 true-false questions in an examination. Then, these questions can be answered in [DCE 2005]

(a) 240 ways (b) 20 ways
(c) 1024 ways (d) 100 ways

47. Using the digits 1, 2, 3, 4, 5, 6, 7, a number of 4 different digits is formed. Find [NCERT Exemplar]

	Column I		Column II
A.	How many numbers are formed?	1.	840
B.	How many numbers are exactly divisible by 2?	2.	200
C.	How many numbers are exactly divisible by 25?	3.	360
D.	How many of these are exactly divisble by 4?	4.	40

Match the questions in Column I with Column II and choose the correct option from the codes given below.

Codes

	A	B	C	D		A	B	C	D
(a)	1	2	3	4	(b)	3	1	4	2
(c)	1	3	4	2	(d)	4	2	3	1

48. The total number of 9-digit numbers which have all different digits is [NCERT Exemplar]

(a) 10! (b) 9!
(c) $9 \times 9!$ (d) $10 \times 10!$

49. If the letters of the word RACHIT are arranged in all possible ways as listed in dictionary. Then, what is the rank of the word RACHIT? [NCERT Exemplar]

(a) 479 (b) 480 (c) 481 (d) 482

50. The number of natural numbers smaller than 10^4 of which all the digits are distinct are

(a) 5000 (b) 5200 (c) 5300 (d) 5274

51. In a class of 10 students there are 3 girls *A, B, C*. The number of different ways they can be arranged in a row such that no two of three girls are consecutive are

(a) $4! \times 300$ (b) $7! \times 336$
(c) $6! \times 300$ (d) $6! \times 336$

52. Consider the following statements

Statement I The number of ways a lawn tennis mixed doubles teams be made up from seven married couples, if no husband and wife play in the same set are 840.

Statement II The number of words beginning with T and ending with E on arranging the letters of the word 'TRIANGLE' are 720.

Choose the correct option.

(a) Only Statement I is correct
(b) Only Statement II is correct
(c) Both I and II are correct
(d) Neither I nor II is correct

53. If the letters of the word SACHIN are arranged in all possible ways and these words are written out as in dictionary, then the word SACHIN appears at serial number [AIEEE 2005]

(a) 602 (b) 603
(c) 600 (d) 601

54. The range of the function $f(x) = {}^{7-x}P_{x-3}$ is [AIEEE 2004]

(a) {1, 2, 3} (b) {1, 2, 3, 4, 5, 6}
(c) {1, 2, 3, 4} (d) {1, 2, 3, 4, 5}

Combinations

Level I

55. The number of ways a team of 2 players can be formed out of a group of 3 lawn tennis players X, Y, Z is

(a) 7 (b) 6
(c) 3 (d) 4

56. Consider the following statements

Statement I Twelve persons meet in a room and each shakes hand with all the others. The number of handshakes can be determined using permutation formula.

Statement II Seven points lie on a circle. The number of chords equals the combinations of 7 different things taken 2 at a time. The number of chords can be determined using permutation formula.

Choose the correct option.

(a) Only I is true (b) Only II is true
(c) Both I and II are true (d) Neither I nor II is true

57. The number of combinations of 4 different objects *A, B, C, D* taken 2 at a time is

(a) 4 (b) 6 (c) 7 (d) 8

58. If there are 5 different objects *A, B, C, D, E*. The number of combinations of 3 different objects is ...A... . Here, A refers to

(a) 9 (b) 12 (c) 11 (d) 10

59. Consider the following statements

Statement I If there are *n* different objects, then $^nP_r = {^nC_r} \cdot r!, 0 \le r \le n.$

Statement II If there are *n* different objects, then $^nC_r = \dfrac{n!}{r!\,(n-r)!}, 0 \le r \le n.$

Statement III The number of combinations of *n* different things taken nothing at all is considered to 1.

Identify the correct combination of true (T) and false (F) of the above statements.

(a) TTT (b) TFT
(c) TFF (d) FFF

60. Consider the following statements

Statement I If there are *n* different objects, then $^nC_r = {^nC_{n-r}}, 0 \le r \le n.$

Statement II If there are *n* different objects, then $^nC_r + {^nC_{r-1}} = {^{n+1}C_r}, 0 \le r \le n.$

Statement III If $^nC_9 = {^nC_8}$, then $^nC_{17} = 2.$

Identify the correct combination of true (T) and false (F) of the above statements.

(a) FFT (b) TTF (c) FTT (d) FFF

61. A group consists of 2 men and 3 women. A committee of 3 persons can be constituted in ...A... different ways and a committee of 3 persons consisting of 1 man and 2 women can be constituted in ...B... different ways. Here, A and B refer to

(a) 10, 5 (b) 5, 10 (c) 6, 10 (d) 10, 6

62. Match the following columns and choose the correct option from the codes given below.

	Column I	Column II
A.	If $^nC_8 = {^nC_2}$, then $^nC_2$ equals	1. 5
B.	If $^{2n}C_3 : {^nC_3} = 12 : 1$, then n equals	2. 45
C.	If $^{2n}C_3 : {^nC_3} = 11 : 1$, then n equals	3. 6

Codes

	A	B	C		A	B	C
(a)	1	2	3	(b)	2	1	3
(c)	3	1	2	(d)	3	2	1

63. The number of chords that can be drawn through 21 points on a circle, is

(a) 200
(b) 190
(c) 210
(d) None of these

64. The number of ways, a team of 3 boys and 3 girls can be selected from 5 boys and 4 girls, is

(a) 20 (b) 40
(c) 30 (d) None of these

65. The number of ways of selecting 9 balls from 6 red balls, 5 white balls and 5 blue balls, if each selection consists of 3 balls of each colour, is

(a) 2000 (b) 1900
(c) 1000 (d) None of these

66. Consider the following statements

Statement I The number of 5 cards combination out of a pack of 52 cards, if there is exactly one ace in each combination is 778320.

Statement II The number of ways of selecting a cricket team of eleven from 17 players in which only 5 players can bowl, if each cricket team of 11 must include exactly 4 bowlers, is 4000.

Which of the above statement(s) is /are true?

(a) Only I (b) Only II
(c) Both I and II (d) Neither I nor II

67. The number of ways in which 2 black and 3 red balls can be selected from a bag containing 5 black and 6 red balls is

(a) 170 (b) 190
(c) 180 (d) 200

68. The number of ways a student can choose a programme out of 5 courses, if 9 courses are available and 2 specific courses are compulsory for every student is

(a) 35 (b) 40
(c) 24 (d) 120

Level II

69. The straight lines l_1, l_2 and l_3 are parallel and lie in the same plane. A total numbers of m points are taken on l_1; n points on l_2, k points on l_3. The maximum number of triangles formed with vertices at these points are [NCERT Exemplar]

(a) ${}^{(m+n+k)}C_3$

(b) ${}^{(m+n+k)}C_3 - {}^{m}C_3 - {}^{n}C_3 - {}^{k}C_3$

(c) ${}^{m}C_3 + {}^{n}C_3 + {}^{k}C_3$

(d) ${}^{m}C_3 \times {}^{n}C_3 \times {}^{k}C_3$

70. Out of 18 points in a plane, no three are in the same line except five points which are collinear. The number of lines that can be formed by joining these points are [NCERT Exemplar]

(a) ${}^{18}C_2 - {}^{5}C_2 + 1$ (b) ${}^{18}C_2 - {}^{5}C_2$

(c) ${}^{18}C_2 - {}^{4}C_2$ (d) None of these

71. The number of triangles that are formed by choosing the vertices from a set of 12 points, seven of which lie on the same line is [NCERT Exemplar]

(a) 105 (b) 15 (c) 175 (d) 185

72. Person wishes to make up as many different parties as he can out of his 20 friends such that each party consists of the same number of persons. The number of friends he should invite are ...A... . Here, A refers to

(a) 8 (b) 9

(c) 10 (d) None of these

73. Consider the following statements

Statement I The number of diagonals of n-sided polygon is ${}^{n}C_2 - n$.

Statement II A polygon has 44 diagonal. The number of its sides are 10.

Choose the correct option from the choices given below.

(a) Only I is true (b) Only II is true

(c) Both I and II are true (d) Both I and II are false

74. A boy has 3 library tickets and 8 books of his interest in library. Out of these 8 books, he does not want to borrow Chemistry part II, unless Chemistry part I is also borrowed. The number of ways he can borrow 3 books equals.

(a) 39 (b) 42 (c) 41 (d) 40

75. The number of ways in which we can choose a committee from four men and six women so that the committee include atleast two men and exactly twice as many women as men is [NCERT Exemplar]

(a) 94 (b) 126

(c) 128 (d) None of these

76. In a shop there are five types of ice-creams available. A child buys six ice-creams.

Consider the following statements

Statement I The number of different ways the child can buy the six ice-creams is ${}^{10}C_5$.

Statement II The number of different ways the child can buy the six ice-creams is equal to the number of different ways of arranging 6 A's and 4 B's in a row.

Choose the correct option.

(a) Statement I is true, Statement II is true; Statement II is correct explanation for Statement I

(b) Statement I is true, Statement II is true; Statement II is not a correct explanation for Statement I

(c) Statement I is true, Statement II is false

(d) Statement I is false, Statement II is true

77. Six X's have to be placed in the square of the figure such that each row contains atleast one 'X'. In how many different ways can this be done?

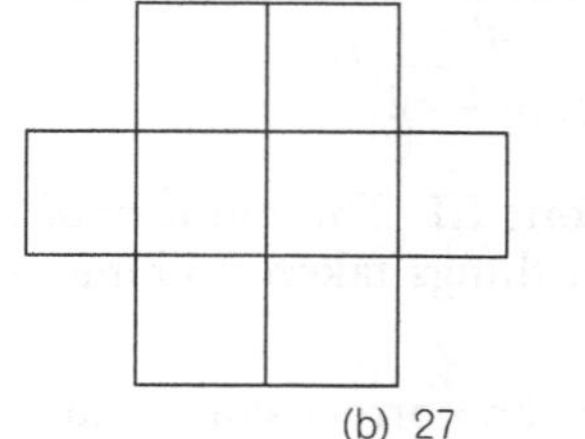

(a) 28 (b) 27

(c) 26 (d) None of these

78. p points are chosen on each of the three coplanar lines. The maximum number of triangles formed with vertices at these points is

(a) $p^3 + 3p^2$ (b) $\frac{1}{2}(p^3 + p)$

(c) $\frac{p^2}{2}(5p - 3)$ (d) $p^2(4p - 3)$

79. A five digit number divisible by 3 is to be formed using the numbers 0, 1, 2, 3, 4 and 5, without repetition. The total number of ways this can be done, is [NCERT Exemplar]

(a) 216 (b) 240 (c) 600 (d) 3125

80. The set $S = \{1, 2, 3, ..., 12\}$ is to be partitioned into three sets, A, B, C of equal size.

Thus, $A \cup B \cup C = S, A \cap B = B \cap C = A \cap C = \phi$

The number of ways to partition S is

(a) $12!/3!(4!)^3$

(b) $12!/3!(3!)^4$

(c) $12!/(4!)^3$

(d) $12!/(3!)^4$

81. There are 10 professors and 20 lecturers out of whom a committee of 2 professors and 3 lecturers is to be formed. Find

Column I	Column II
A. In how many ways committee can be formed	1. ${}^{10}C_2 \times {}^{19}C_3$
B. In how many ways a particular professor is included	2. ${}^{10}C_2 \times {}^{19}C_2$
C. In how many ways a particular lecturer is included	3. ${}^{9}C_1 \times {}^{20}C_3$
D. In how many ways a particular lecturer is excluded	4. ${}^{10}C_2 \times {}^{20}C_3$

Match the columns and choose the correct option from the codes given below.

Codes

	A	B	C	D
(a)	1	2	3	4
(b)	4	3	2	1
(c)	3	2	4	1
(d)	4	2	3	1

82. The number of ways of placing 18 mice in two experimental groups and one control group, with all group equally large, are [NCERT Exemplar]

(a) $\frac{18!}{(6!)^3}$ (b) $\frac{18!}{6!6!}$
(c) $\frac{12!}{(6!)^2}$ (d) None of these

Miscellaneous Exercise

83. The number of words with or without meaning, each of 3 vowels and 2 consonants that can be made from the letters of the word 'INVOLUTE' are

(a) 2800 (b) 2880
(c) 2900 (d) None of these

84. A group consists of 4 girls and 7 boys.

Statement I The number of ways a team of 5 members can be selected, if the team has no girls are 21.

Statement II The number of ways a team of 5 members can be selected, if it has atleast one boy and one girl are 441.

Statement III The number of ways a team of 5 members can be selected, if it has atleast 3 girls are 90.

Identify the correct combination of true (T) and false (F) for the given three statements.

(a) FFT (b) TFF (c) TTF (d) FTT

85. The number of words with or without meaning which can be made using all the letters of the word AGAIN are ...A... and if these words are written as in dictionary, then the 50th word is ...B... . Here, A and B refers to

(a) 60, NAAIG (b) 60, NAAGI
(c) 120, NAAIG (d) 120, NAAGI

86. The numbers greater than 1000000 that can be formed by using the digits 1,2,0,2,4,2,4 are

(a) 420 (b) 400
(c) 360 (d) None of these

87. The number of ways in which 5 girls and 3 boys be seated in a row so that no two boys are together are

(a) 14000 (b) 14300
(c) 14200 (d) 14400

88. Match the following columns and choose the correct option from the codes given below.

Column I	Column II
A. The number of words, with or without meaning, each of 2 vowels and 3 consonants can be formed from the letters of the word DAUGHTER are	1. 50400
B. The number of words, with or without meaning, that can be formed using all the letters of the word EQUATION, so that all vowels and consonants occur together are	2. 3600
C. If the different permutation of all the letters of the word 'EXAMINATION' are listed as in dictionary, the number of words in this list before the first word starting with *E* are	3. 1440
D. The English alphabet has 5 vowels and 21 consonants. The number of words with two different vowels and two different consonants that can be formed from the alphabet are	4. 907200

	A	B	C	D		A	B	C	D
(a)	1	2	3	4	(b)	2	3	4	1
(c)	2	3	1	4	(d)	4	3	2	1

89. A committee of 7 has to be formed from 9 boys and 4 girls.

Statement I The number of ways this can be done, if committee consists of exactly 3 girls equals 504.

Statement II The number of ways this can be done, if committee consists of atleast 3 girls equal 588.

Statement III The number of ways this can be done, if committee consists of atmost 3 girls equal 1632.

Identify the correct combination of true (T) and false (F) for the given three statements

(a) TTT (b) TTF
(c) TFF (d) FTT

90. The number of 6-digit numbers that can be formed from the digits 0, 1, 3, 5, 7, 9 which are divisible by 10 and no digit is repeated equals

(a) 120 (b) 720
(c) 600 (d) None of these

91. In an examination, a question paper consists of 12 questions divided into two parts *i.e.,* part I and part II containing 5 and 7 questions, respectively. A student is required to attempt 8 questions in all, selecting atleast 3 from each part. The number of ways a student can select the questions are

(a) 360 (b) 600
(c) 400 (d) 420

92. The number of 5 cards combinations out of a deck of 52 cards, if each selection of 5 cards has exactly one king are

(a) ${}^4C_3 \times {}^{48}C_1$ (b) ${}^4C_4 \times {}^{48}C_1$
(c) ${}^4C_1 \times {}^{48}C_4$ (d) None of these

93. Consider the following statements

Statement I It is required to seat 5 men and 4 women in a row so that the women occupy the even places. The number of such arrangements equals 2880.

Statement II From a class of 25 students, 10 are to be chosen for an excursion party. There are 3 students who decide that either all of them will join or none of them will join. The number of ways the excursion party can be chosen are ${}^{22}C_7 + {}^{22}C_{10}$.

Which of the above statements is/are true?

(a) Only Statement I is true
(b) Only Statement II is true
(c) Both statements are true
(d) Both statements are false

94. The number of ways the letters of the word ASSASSINATION can be arranged, so that all the S's are together are

(a) 151000 (b) 151200
(c) 150800 (d) None of these

95. Eighteen guests are to be seated, half on each side of a long table. Four particular guests desire to sit on one particular side and three others on the other side of the table. The number of ways in which the seating arrangement can be done equals

(a) ${}^{11}C_4\,(9!)^2$ (b) ${}^{11}C_6(9!)^2$
(c) ${}^6P_0 \times {}^5P_0$ (d) None of these

96. If some or all of n objects are taken at a time, the number of combinations are

(a) 2^n (b) $2^{(n-1)}$
(c) $2^n - 1$ (d) None of these

97. Given five different green dyes, four different blue dyes and three different red dyes, the number of combinations of dyes which can be chosen taking atleast one green and one blue dye are [NCERT Exemplar]

(a) 3600 (b) 3720 (c) 3800 (d) 3600

98. The number of ways of distributing 8 identical balls in 3 distinct boxes, so that none of the box is empty, is [AIEEE 2004]

(a) 5 (b) 21 (c) 3^8 (d) ${}^8C_3$

99. Let l_1 and l_2 be two lines intersecting at P. If A_1, B_1, C_1 are points on l_1 and A_2, B_2, C_2, D_2, E_2 are points on l_2 and if none of these coincides with P, then the number of triangles formed by these nine points, is

(a) 56 (b) 55 (c) 46 (d) 45

100. From 6 different novels and 3 different dictionaries, 4 novels and 1 dictionary are to be selected and arranged in a row on the shelf so that the dictionary is always in the middle. Then, the number of such arrangements are [AIEEE 2009]

(a) atleast 500 but less than 750
(b) atleast 750 but less than 1000
(c) atleast 1000
(d) less than 500

101. At an election, a voter may vote for any number of candidates not greater than the number to be elected. There are 10 candidates and 4 are to be elected. If a voter votes for atleast one candidate, then the number of ways in which he can vote, is [AIEEE 2006]

(a) 6210 (b) 385 (c) 1110 (d) 5040

102. How many ways are there to arrange the letters in the word GARDEN with the vowels in alphabetical order? [AIEEE 2004]

(a) 120 (b) 240
(c) 360 (d) 480

103. The number of parallelograms that can be formed from a set of four parallel lines intersecting another set of three parallel lines is

(a) 18 (b) 12 (c) 9 (d) 6

104. The value of $({}^7C_0 + {}^7C_1) + ({}^7C_1 + {}^7C_2) + \dots\dots +$ $({}^7C_6 + {}^7C_7)$ is equal to

(a) 2^8 (b) $2^7 - 1$
(c) $2^8 - 2$ (d) $2^8 - 1$

105. The greatest possible number of points of intersection of 8 straight lines and 4 circles equals ...A... . Here, A denotes [HOTS]

(a) 32 (b) 64 (c) 76 (d) 104

Directions (Q. Nos. 106-108) *Different words are being formed by arranging the letters of word 'SUCCESS'.*

106. The number of words in which the two C are together but no two S are together, is

(a) 120 (b) 96 (c) 24 (d) 420

107. The number of words in which the consonants appear in alphabetical order is

(a) 42 (b) 40
(c) 420 (d) 280

108. The rank of the word 'SUCCESS' in the dictionary if all the words are written as in a dictionary is

(a) 328 (b) 329
(c) 330 (d) 331

109. Let A and B two sets containing 2 elements and 4 elements respectively. The number of subsets of $A \times B$ having 3 or more elements is [JEE Main 2013]

(a) 256 (b) 220
(c) 219 (d) 211

110. Assuming the balls to be identical except for difference in their colours, the number of ways in which one or more balls can be selected from 10 white, 9 green and 7 black balls is [AIEEE 2012]

(a) 880 (b) 629 (c) 630 (d) 879

111. Let $X = \{1, 2, 3, 4, 5\}$. The number of different ordered pairs (Y, Z) that can be formed such that $Y \subseteq X$, $Z \subseteq X$ and $Y \cap Z$ is empty, is [AIEEE 2012]

(a) 5^2 (b) 3^5 (c) 2^5 (d) 5^3

112. If ${}^nC_r$ denotes the number of combinations of n things taken r at a time, then the expression ${}^nC_{r+1} + {}^nC_{r-1} + 2 \times {}^nC_r$ equals [AIEEE 2003]

(a) ${}^{n+2}C_r$ (b) ${}^{n+2}C_{r+1}$
(c) ${}^{n+1}C_r$ (d) ${}^{n+1}C_{r+1}$

113. A student is to answer 10 out of 13 questions in an examination such that he must choose atleast 4 from the first five questions. The number of choices available to him is [AIEEE 2003]

(a) 140 (b) 196
(c) 280 (d) 346

114. Consider the following statements

Statement I The number of arrangement of letters of the word BANANA are 60.

Statement II The number of permutations of n things, of which p are a like of one kind, q are alike of second kind and remaining all are distinct is $\dfrac{n!}{p!\,q!}$.

Choose the correct option.

(a) Statement I is true; Statement II is true; Statement II is a correct explanation for Statement I
(b) Statement I is true; Statement II is true; Statement II is not a correct explanation for Statement I
(c) Statement I is true; Statement II is false
(d) Statement I is false; Statement II is true

115. ${}^1P_1 + 2 \cdot {}^2P_2 + 3 \cdot {}^3P_3 + \dots\dots + n \cdot {}^nP_n$ is equal to

(a) ${}^{n+1}P_{n+1} - 1$ (b) ${}^{n+1}P_{n+1} - 2$
(c) ${}^{n+1}P_{n+1}$ (d) None of these

116. The number of numbers divisible by 3 that can be formed by four different even digits are [HOTS]

(a) 36 (b) 18
(c) 0 (d) None of these

117. A father with 8 children takes them 3 at a time to the zoological garden, as often as he can without taking the same 3 children together more than once. The number of times he will go the garden is

(a) 56 (b) 100
(c) 112 (d) None of these

Answer *with* Explanations

1. *(b)* We know, if an event can occur in m different ways, following which another event can occur in n different ways, then the total number of occurrence of the events in the given order is $m \times n$. (multiplication principle)

It is given that 1st digit on number lock is 7.

$\therefore$ Number of sequences of 3 digits

$$= 9 \times 8 \times 7 = 504$$

(as for sequence of 3 digits □□□, the first place can be filled in 9 different ways, the second place can be filled in 8 different ways and the third place can be filled in 7 different ways)

2. *(b)* Number of different pairs of a pant and shirt $= 3 \times 2$

$= 6$ (multiplication principle)

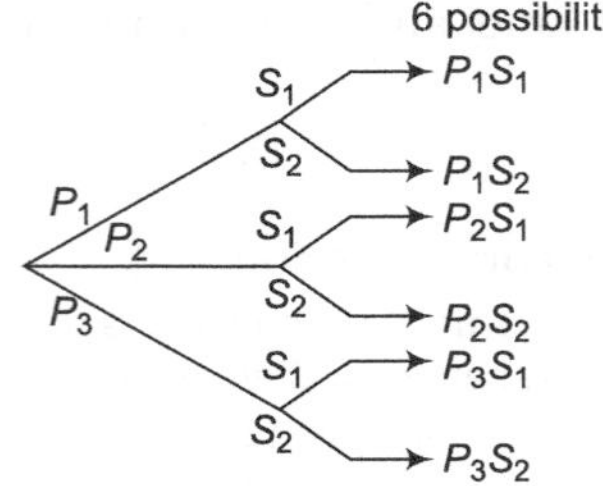

(if the pants are P_1, P_2, P_3 and shirts are S_1 and S_2, then the 6 possibilities are shown in the figure)

3. *(b)* Number of ways she can carry the 3 items $= 2 \times 3 \times 2$

$= 12$ (multiplication principle)

4. *(c)* The fundamental principle of multiplication is as follows "If an event E can occur in m different ways and associated with each way of occurrence of E another event F can occur in n different ways, then the total number of occurrence of the events in the given order in $m \times n$."

Note *The above principle can be generalised for any finite number of events.*

Thus, both the statements are true.

5. *(a)* I. Number of 4 letter words (repetition not allowed)

□□□□

The first place can be filled in 4 different ways by anyone of the 4 letters R, O, S, E. Following which the second, third and the fourth place may be filled in 3, 2, 1 ways respectively.

$= 4 \times 3 \times 2 \times 1$

$= 24$ (multiplication principle)

II. Number of 4 letter words (repetition allowed).

$= 4 \times 4 \times 4 \times 4$

($\because$ each place can be filled in 4 different ways as repetition is allowed)

$= 256$ (multiplication principle)

6. *(c)* Number of signals $= 4 \times 3 = 12$

(because the upper flag can be put up in 4 different ways and lower flag can be chosen in 3 ways)

7. *(a)* Number of 2-digit even numbers = Number of ways of filling 2 vacant places □□ in succession by the five given digits.

Here, we can fill unit's place by 2 or 4 only and the ten's digit can be filled by any of the five numbers as the repetition is allowed. Then, by multiplication principle.

Number of 2 digit even numbers $= 5 \times 2 = 10$

8. *(d)* We know, if an event E can occur in m different ways and another event F can occur in n different ways then both events can occur together in $(m + n)$ways. (addition principle)

A signal can consist of either 2 flags, 3 flags 4 or 5 flags.

Number of signals using 2 flags $= 5 \times 4 = 20$

(because for upper flag in the signal, we have to choose from 5 different flags and for lower, we can choose from 4 different flags)

Similarly,

Number of signals using 3 flags $= 5 \times 4 \times 3 = 60$

Number of signals using 4 flags $= 5 \times 4 \times 3 \times 2 = 120$

Number of signals using 5 flags $= 5 \times 4 \times 3 \times 2 \times 1 = 120$

Hence, the required number of signals $= 20 + 60 + 120 + 120 = 320$ (using addition principle)

Note *Conceptual mix of addition and multiplication principle.*

9. *(d)* I. Number of 3-digit numbers (repetition allowed) $= 5 \times 5 \times 5 = 125$

(as unit's, ten's and hundred's places can be filled in 5 ways)

II. Number of 3-digit numbers (repetition is not allowed)

$= 5 \times 4 \times 3$

$= 60$

(as hundred's, ten's and unit's places can be filled in 5, 4 and 3 ways respectively)

10. *(b)* A. Here, the unit's place can be filled in by 2, 4 or 6 (3 ways), the ten's place can be filled in by 6 ways and the hundred's place can be filled in 6 ways.

$\therefore$ Number of 3-digit even numbers $= 6 \times 6 \times 3$

$= 108$ (multiplication principle)

B. Number of 4 letter codes □ □ □ □

$= 10 \times 9 \times 8 \times 7$

$= 5040$

(as 1st place can be filled in 10 different ways and 2nd, 3rd and 4th places can be filled in 9, 8 and 7 ways respectively)

C. Number of 5-digit telephone numbers = Number of filling five places given that numbers start with 67.

I II III

[6] [7] □ □ □

($\because$ number begins with 67)

$= 8 \times 7 \times 6$

$= 56 \times 6$

$= 336$

(1st, 2nd and 3rd places can be filled in 8, 7 and 6 ways respectively)

Thus, the correct option is

A	B	C
2	3	1

11. *(b)* I. In the first throw of coin we have 2 outcomes (Head or Tail). Similarly, in second and third throw of coin, we have 2 outcomes.

$\therefore$ Number of possible outcomes $= 2 \times 2 \times 2 = 8$

(multiplication principle)

II. Number of 2 flag signals = Number of ways of filling 2 vacant places in succession by the 5 flags available

$= 5 \times 4$

$= 20$ (multiplication principle)

12. *(c)* Total number of 3-digit numbers having atleast one of their digits as 5 = Total number of 3-digit numbers – (Total number of 3-digit numbers in which 5 does not appear at all)

$= 9 \times 10 \times 10 - 8 \times 9 \times 9$

$= 900 - 648 = 252$

13. *(d)* Innermost triangle can be filled using one of the three colours (red, blue or green).

The rest of the triangles can be filled using anyone of the remaining two colours.

By multiplication principle, this can be done in $3 \times 2 \times 2 \times 2 = 24$ ways.

14. *(c)* Each of the three prizes can be given to any of the four children.

$\therefore$ Total number of ways of distributing prizes

$= 4 \times 4 \times 4 = 64$

Number of ways in which one child gets all prizes = 4

$\therefore$ Number of ways in which no child gets all the three prizes $= 64 - 4 = 60$

15. *(b)* A. Number of ways in which 6 distinct objects can be put into two different boxes $= 2 \times 2 \times 2 \times 2 \times 2 \times 2 = 64$

Number of ways in which all objects are put into 1 box = 2

$\therefore$ Number of ways in which no box remains empty

$= 64 - 2 = 62$

B. Each of the letter can be rested in 4 different ways (4 letter boxes).

$\therefore$ Number of ways in which 5 letters can be rested

$= 4 \times 4 \times 4 \times 4 \times 4$

$= 4^5$

$= 1024$

C. If each element of A has a unique image in B, then relation from A to B is a function. Thus, each element in A can have one of the three elements of B as image.

$\therefore$ Number of functions $= 3 \times 3 \times 3 \times 3 = 81$

Correct option is

A	B	C
2	3	1

16. *(a)* Total number of 4-digit numbers $= 5 \times 5 \times 5 \times 5$

$= 625$

(as each place can be filled by anyone of the numbers 1, 2, 3, 4 and 5)

Numbers in which no two digits are identical $= 5 \times 4 \times 3 \times 2$

(*i.e.*, repetition not allowed) $= 120$

(as 1st place can be filled in 5 different ways, 2nd place can be filled in 4 different ways and so on)

Number of 4-digits numbers in which atleast 2 digits are identical $= 625 - 120$

$= 505$

17. *(c)* I. Number of 3 letter words (repetition not allowed)

$= 6 \times 5 \times 4$

$= 120$

(as first place can be filled in 6 different ways, second place can be filled in 5 different ways and third place can be filled in 4 different ways)

II. Number of 3 letter words (repetition is allowed)

$= 6 \times 6 \times 6$

$= 216$

(as each of the place can be filled in 6 different ways)

18. *(c)* A $\rightarrow$ Arrangement

B $\rightarrow$ Definite

19. *(a)* A $\rightarrow$ Different

B $\rightarrow$ $n(n-1)(n-2)\ldots(n-r+1)$

C $\rightarrow$ ${}^nP_r$

There will be as many permutations as there are ways of filling in r vacant places

□ □ □……□

← r vacant places →

by the n objects. The first place can be filled in n ways; following which, the second place can be filled in $(n-1)$ ways, following which the third place can be filled in $(n-2)$ ways,...., the rth place can be filled in $[n-(r-1)]$ ways. Therefore, the number of ways of filling in r vacant places in succession is $n\ (n-1)(n-2)....[n-(r-1)]$ or $n(n-1)(n-2)....(n-r+1)$.

20. *(c)* The notation $n!$ represents the product of first n natural numbers *i.e.*, $n! = 1.2.3.......(n-1).n$

A. $5! = 5 \times 4 \times 3 \times 2 \times 1 = 120$

B. $7! = 7 \times 6 \times 5 \times 4 \times 3 \times 2 \times 1 = 42 \times 120 = 5040$

C. $8! = 8 \times 7! = 8 \times 5040 = 40320$

D. $7! - 5! = 5040 - 120 = 4920$

E. $4! - 3! = 4 \times 3 \times 2 \times 1 - 3 \times 2 \times 1 = 24 - 6 = 18$

21. *(d)* A. $\frac{7!}{5!} = \frac{7 \times 6 \times 5!}{5!} = 42$

B. $\frac{12!}{10!2!} = \frac{12 \times 11 \times 10!}{10! \times 2 \times 1} = 66$

C. $\frac{8!}{6! \times 2!} = \frac{8 \times 7 \times 6!}{6! \times 2 \times 1} = 28$

22. *(a)* I. We have, $\frac{1}{8!} + \frac{1}{9 \times 8!} = \frac{x}{10 \times 9 \times 8!}$

$$\Rightarrow \quad 1 + \frac{1}{9} = \frac{x}{10 \times 9}$$

$$\Rightarrow \quad \frac{10}{9} = \frac{x}{10 \times 9}$$

$$\Rightarrow \quad x = 100$$

Thus, Statement I is true.

II. We have, $\frac{1}{6!} + \frac{1}{7 \times 6!} = \frac{x}{8 \times 7 \times 6!}$

$$\Rightarrow \quad 1 + \frac{1}{7} = \frac{x}{8 \times 7}$$

$$\Rightarrow \quad \frac{8}{7} = \frac{x}{8 \times 7}$$

$$\Rightarrow \quad x = 64$$

Thus, Statement II is false.

23. *(c)* I. $n = 6, r = 2$

$$\frac{n!}{(n-r)!} = \frac{6!}{(6-2)!} = \frac{6!}{4!} = \frac{6 \times 5 \times 4!}{4!} = 30$$

II. $n = 9, r = 5$

$$\frac{n!}{(n-r)!} = \frac{9!}{4!} = \frac{9 \times 8 \times 7 \times 6 \times 5 \times 4!}{4!} = 72 \times 210 = 15120$$

III. $n = 5, r = 2$

$$\frac{n!}{r!(n-r)!} = \frac{5!}{2!(5-2)!} = \frac{5!}{2!3!} = \frac{5 \times 4 \times 3!}{2!3!} = \frac{5 \times 4}{2!} = 10$$

IV. $3! + 4! = 6 + 24 = 30$

$$7! = 7 \times 6 \times 5 \times 4 \times 3 \times 2 \times 1 = 42 \times 120 = 5040$$

$$\Rightarrow \quad 3! + 4! \neq 7!$$

Hence, IV is false.

$A \rightarrow 30, B \rightarrow 15120, C \rightarrow 10, D \rightarrow$ false

24. *(c)* We know, ${}^nP_r = n(n-1)(n-2).....(n-r+1)$

$$= \frac{n(n-1)(n-2).....(n-r+1)(n-r)....3 \times 2 \times 1}{(n-r)(n-r-1)....3 \times 2 \times 1} = \frac{n!}{(n-r)!}$$

Thus, both I and II are true.

25. *(b)* The number of permutations = The number of ways of filling r vacant places by n different things

= □ □------□
r places

(as 1st place can be filled in n different ways,

2nd place can be filled in n different ways...

r^{th} place can be filled in n different ways) (repetition allowed)

∴ Number of permutations

$$= n \times n \times ..... \times n \ (r \text{ times}) = n^r$$

26. *(c)* I. Number of 4 letter words that can be formed from alphabets of the word 'PART'

$$= {}^4P_4 = 4! = 24$$

II. Number of 4 letter words that can be formed when repetition is allowed $= 4^4 = 256$

27. *(b)* A. The number of 3 letter words that can be formed by letters of word 'NUMBER' (repetition not allowed)= ${}^6P_3$

$$= 6 \times 5 \times 4 = 120$$

B. The number of 3 letter words that can be formed by letters of word 'NUMBER' (repetition is allowed)

$$= 6 \times 6 \times 6 = 216$$

C. Number of ways choosing a chairman and vice-chairman from a group of 12 persons assuming that one person cannot hold more than one position

$$= {}^{12}P_2 = \frac{12!}{(12-2)!} = \frac{12!}{10!} = 12 \times 11 = 132$$

28. *(b)* I. The number of permutations of n objects, where P objects are of the same kind and rest are all different $= \frac{n!}{P!}$

II. The number of permutations of n objects, where P_1 objects are of one kind, P_2 are of second kind,...., P_k are of kth kind and the rest if any, are of different kind, is $\frac{n!}{P_1!P_2!\ldots P_k!}$

29. *(d)* I. Here, there are 4 letters of which there are 2 O's and rest are different.

Therefore, the required number of arrangements

$$= \frac{4!}{2!} = 4 \times 3 = 12$$

II. There are 9 letters of which there are 2 I's, 3T's and rest are different.

Therefore, the required number of arrangements $= \frac{9!}{2!3!}$

Alternate Method

I. We have to find the number of ways of rearranging the letters of the word 'ROOT'. In this case, the letters of the word are not all different. There are 2 Os, which are of the same kind. Let us treat, temporarily, the 2 Os as different, say, O_1 and O_2. The number of permutations of 4 different letters, in this case, taken all at a time is 4!. Consider one of these permutations say, RO_1O_2T. Corresponding to this permutation, we have 2! permutations RO_1O_2T and RO_2O_1T which will be exactly the same permutation, if O_1 and O_2 are not treated as different, *i.e.*, if O_1 and O_2 are the same O at both places.

Therefore, the required number of permutations

$$= \frac{4!}{2!} = 3 \times 4 = 12$$

II. Let us now find the number of ways of rerranging the letters of the word 'INSTITUTE'. In this case, there are 9 letters, in which I appears 2 times and T appears 3 times.

Temporarily, let us treat these letters different and name them as I_1, I_2, T_1, T_2 and T_3. The number of permutations of 9 different letters, in this case, taken all at a time is 9!. Consider one such permutation, say, $I_1NT_1SI_2T_2UET_3$. Here, if I_1, I_2 are not same and T_1, T_2, T_3 are not same, then I_1, I_2 can be arranged in 2! ways and T_1, T_2, T_3 can be arranged in 3! ways. Therefore, $2! \times 3!$ permutations will be just the same permutation corresponding to this chosen permutation $I_1NT_1SI_2T_2UET_3$. Hence, total number of different permutations will be $\frac{9!}{2!3!}$.

30. *(c)* A. Here, there are 9 letters of which there are 4A's, 2L's and rest all are different. Therefore, the required number of arrangements $= \frac{9!}{4!2!} = \frac{9 \times 8 \times 7 \times 6 \times 5 \times 4 \times 1}{4 \times 2 \times 1}$

$$= 7560$$

B. The number of 4-digit numbers equals the permutations of 9 different digits taken 4 at a time.

Therefore, the required 4-digit numbers

$$= {}^9P_4 = \frac{9!}{(9-4)!} = \frac{9!}{5!}$$
$$= 9 \times 8 \times 7 \times 6 = 3024$$

C. Every number between 100 and 1000 is a 3-digit number. We, first, have to count the permutations of 6 digits taken 3 at a time. This number would be ${}^6P_3$. But, these permutations will include those also, where 0 is at the 100's place. For example, 092, 042, ..., etc., are such numbers, which are actually 2-digit numbers and hence the number of such numbers has to be subtracted from ${}^6P_3$ to get the required number. To get the number of such numbers, we fix 0 at the 100's place and rearrange the remaining 5 digits taking 2 at a time. This number is ${}^5P_2$.

So, the required number $= {}^6P_3 - {}^5P_2 = \frac{6!}{3!} - \frac{5!}{3!}$

$$= 4 \times 5 \times 6 - 4 \times 5 = 100$$

D. Number lying between 100 and 1000 that can be formed

$$= \square\ \square\ \square$$
$$= 4 \times 5 \times 5$$
$$= 100$$

$\because$ Number of ways to fill hundred's place = 4

and number of ways to fill ten's and unit's place = 5

31. *(d)* I. We have, $5\ {}^4P_r = 6\ {}^5P_{r-1}$

$$\Rightarrow \quad 5 \times \frac{4!}{(4-r)!} = 6 \times \frac{5!}{(5-r+1)!}$$

$$\Rightarrow \quad \frac{5!}{(4-r)!} = \frac{6 \times 5!}{(5-r+1)(5-r)(5-r-1)!}$$

$$\Rightarrow \quad (6-r)(5-r) = 6$$
$$\Rightarrow \quad r^2 - 11r + 24 = 0$$
$$\Rightarrow \quad r^2 - 8r - 3r + 24 = 0$$
$$\Rightarrow \quad (r-8)(r-3) = 0$$
$$\Rightarrow \quad r = 8$$
$$\Rightarrow \quad r = 3$$

But $\quad r \neq 8$ as $r \leq 4$

$\therefore \quad r = 3$

II. We have, ${}^5P_r = {}^6P_{r-1}$

$$\frac{5!}{(5-r)!} = \frac{6!}{[6-(r-1)]!}$$

$$\Rightarrow \quad \frac{5!}{(5-r)!} = \frac{6 \times 5!}{(7-r)!}$$

$$\Rightarrow \quad \frac{1}{(5-r)!} = \frac{6}{(7-r)(6-r)(5-r)!}$$

$$\Rightarrow \quad (7-r)(6-r) = 6$$
$$\Rightarrow \quad 42 - 13r + r^2 = 6$$
$$\Rightarrow \quad r^2 - 13r + 36 = 0$$

$\Rightarrow \quad (r-4)(r-9)=0$

$\Rightarrow \quad r=4, 9$

$r=9$ rejected as ${}^5P_9$ is not defined.

$\therefore \quad r=4$

32. *(a)* I. There are 8 different letters in the word 'DAUGHTER', in which there are 3 vowels, namely, a, u and e. Since, the vowels have to occur together, we can for the time being, assume them as single object (aue). This single object together with 5 remaining letters (objects) will be counted as a 6 objects. Then, we count permutations of these 6 objects taken all at a time. This number would be ${}^6P_6=6!$. Corresponding to each of these permutations, we shall have 3! permutations of the three vowels a, u, e taken all at a time. Hence, by the multiplication principle, the required number of permutaions $=6!\times 3!=4320$.

II. If we have to count those permutations in which all vowels never occur together, we first have to find all possible arrangements of 8 letters taken all at a time, which can be done in 8! ways.Then, we have to subtract from this number, the number of permutations in which the vowels are always together.

Therefore, the required number

$$=8!-6!\times 3!=6!(7\times 8-6)$$
$$=2\times 6!(28-3)$$
$$=50\times 6!=50\times 720$$
$$=36000$$

33. *(b)* Total number of discs are $4+3+2=9$. Out of 9 discs, 4 are of the first kind (red), 3 are of the second kind(yellow) and 2 are of the third kind (green).

Therefore, the number of arrangements $=\dfrac{9!}{4!3!2!}=1260$

34. *(b)* A. Number of 3-digit numbers that can be formed by using digits from 1 to 9

$$={}^9P_3=9\times 8\times 7=72\times 7=504$$

B. Number of 4-digit numbers that can be formed by using digits from 0 to 9

= All possible 4-digit numbers – Numbers starting with zero

$$={}^{10}P_4-{}^9P_3$$
$$=10\times 9\times 8\times 7-9\times 8\times 7$$
$$=9\times 8\times 7(10-1)$$
$$=504\times 9=4536$$

C. Number of 3-digit even numbers = Number of ways of filling 3 places

H T U
= □ □ □

$=4\times 5\times 3=60$

$\because$ Number of ways of filling unit's place = 3

(since, in case of even numbers, units place can only have 2, 4, 6 from the given digits)

Number of ways of filling ten's place = 5

Number of ways of filling hundred's place = 4

D. Number of 4-digit numbers that can be formed using the digits 1, 2, 3, 4 and 5, if no digit is repeated $={}^5P_4$

$$=5\times 4\times 3\times 2=120$$

E. Number of 4-digit even numbers that can be formed using the digits 1, 2, 3, 4, 5 (repetition not allowed)

T H T U
= □ □ □ □

$=2\times 3\times 4\times 2$

$=48$

Unit's place can be filled in 2 ways (2 or 4) ten's, hundred's and thousand's place can be filled in 4, 3 and 2 ways respectively.

35. *(a)* Number of ways of choosing a chairman and a vice-chairman from a committee of 8 persons.

$$={}^8P_2=\frac{8!}{(8-2)!}=\frac{8!}{6!}=\frac{8\times 7\times 6!}{6!}=56$$

36. *(c)* I. Here, $\dfrac{{}^{n-1}P_3}{{}^nP_4}=\dfrac{1}{9}$

$$\Rightarrow \quad \frac{(n-1)(n-2)(n-3)}{n(n-1)(n-2)(n-3)}=\frac{1}{9}$$
$$\Rightarrow \quad \frac{1}{n}=\frac{1}{9}$$
$$\Rightarrow \quad n=9$$

II. Here, $\dfrac{{}^5P_r}{{}^6P_{r-1}}=2$

$$\Rightarrow \quad \frac{\dfrac{5!}{(5-r)!}}{\dfrac{6!}{[6-(r-1)]!}}=2$$
$$\Rightarrow \quad \frac{5!}{(5-r)!}\times\frac{(7-r)!}{6!}=2$$
$$\Rightarrow \quad \frac{5!}{(5-r)!}\times\frac{(7-r)(6-r)(5-r)!}{6\times 5!}=2$$
$$\Rightarrow \quad \frac{(7-r)(6-r)}{6}=2$$
$$\Rightarrow \quad 42-13r+r^2=12$$
$$\Rightarrow \quad r^2-13r+30=0$$
$$\Rightarrow \quad (r-10)(r-3)=0$$
$$\Rightarrow \quad r=10, 3$$

($r=10$ rejected as ${}^5P_{10}$ and ${}^6P_9$ are not defined)

$\therefore \quad r=3$

37. *(c)* I. The word 'EQUATION' has 8 different letters.

Number of ways of arranging 8 different letters at a time $={}^8P_8$

$$=\frac{8!}{(8-8)!}=\frac{8!}{0!}$$
$$=\frac{8\times 7\times 6\times 5\times 4\times 3\times 2\times 1}{1}=40320 \quad (\because 0!=1)$$

II. The word 'MISSISSIPPI' has 11 letters in which

M $\rightarrow$ 1 time
I $\rightarrow$ 4 times
S $\rightarrow$ 4 times
P $\rightarrow$ 2 times

The number of permutations of the word 'MISSISSIPPI' in which 4 I's, 4 S's and 2 P's are alike

$$= \frac{11!}{4!\,4!\,2!} \qquad \ldots(i)$$

If all the I's are together, then it will be considered as one letter together with remaining 7 letters (including 4I's) will be considered as 8 letters. So, the number of permutations is $\frac{8!}{4!\,2!}$.

($\because$ P is repeated twice and S is repeated four times)

Hence, total number of arrangements

$$= \frac{11!}{4!\,4!\,2!} - \frac{8!}{4!\,2!}$$

$$= \frac{11\times10\times9\times8\times7\times6\times5\times4!}{4!\times4\times3\times2\times1\times2\times1} - \frac{8\times7\times6\times5\times4!}{4!\times2\times1}$$

$= 34650 - 840 = 33810$

38. *(d)* In the word MONDAY, all letters are different.

I. Out of 6 different letters, 4 letters can be selected in ${}^6P_4$ ways.

$\therefore$ Required number of words $= {}^6P_4$

$$= \frac{6!}{(6-4)!} = \frac{6!}{2!}$$

$$= \frac{6\times5\times4\times3\times2!}{2!}$$

$= 360$

II. The word 'MONDAY' has 6 different letters.

Number of ways of arranging 6 letters at a time $= {}^6P_6$

$\therefore$ Required number of words $= {}^6P_6$

$$= \frac{6!}{(6-6)!}$$

$$= \frac{6\times5\times4\times3\times2\times1}{0!}$$

$= 720$

III. First, we will fix the vowels.

In the word 'MONDAY', there are two vowels O and A.

$\therefore$ First letter can be chosen by 2 ways.

Number of ways taking 5 different letters from remaining 5 letters $= {}^5P_5$

$\therefore$ Required number of words $= {}^5P_5 = \frac{5!}{(5-5)!} = \frac{5!}{0!}$

$= 5\times4\times3\times2\times1 = 120$

Hence, by multiplication rule, the total number of ways $= 2\times120 = 240$

39. *(c)* The word 'PERMUTATIONS' has

P — 1 time
E — 1 time
R — 1 time
M — 1 time
U — 1 time
T — 2 times
A — 1 time
I — 1 time
O — 1 time
N — 1 time
S — 1 time

I. Word start with P and end with S *i.e.*,

[P] [ ] [ ] [ ] [ ] [ ] [ ] [ ] [ ] [ ] [ ] [S]

First and last position are already filled by P and S, respectively.

Remaining 10 positions can be filled in $\frac{10!}{2!}$ ways

$$= \frac{10\times9\times8\times7\times6\times5\times4\times3\times2!}{2!}$$

$= 720\times7\times6\times5\times4\times3 = 1814400$

II. If all the vowels are together, then it will be considered as one letter, *i.e.*, (A, E, I, O, U) together with remaining 7 letters and 1 vowel (including 5 vowels) will be considered as 8 letters and 5 vowels can be arranged in 5! ways.

$\therefore$ Required number of permutations $= \frac{5!\times8!}{2!}$

($\because$ T is repeated twice)

$$= \frac{5\times4\times3\times2\times1\times8\times7\times6\times5\times4\times3\times2\times1}{2\times1}$$

$= 2419200$

III. Total number of letters to be arranged $= 12$

There are always 4 letters between P and S *i.e.*,

	1	2	3	4	5	6	7	8	9	10	11	12
Ist ways	P					S						
IInd ways	—	P	—	—	—	—	S	—	—	—	—	—
IIIrd ways	—	—	P	—	—	—	—	S	—	—	—	—
IV ways	—	—	—	P	—	—	—	—	S	—	—	—
Vth ways	—	—	—	—	P	—	—	—	—	S	—	—
VIth ways	—	—	—	—	—	P	—	—	—	—	S	—
VIIth ways	—	—	—	—	—	—	P	—	—	—	—	S

Hence, P and S having 4 places between them can be filled in 7 ways.

Hence, P and S or S and P can be filled in $7 + 7 = 14$ ways.

Remaining 10 letters (in which T repeated twice) can be filled in $\frac{10!}{2!}$ ways.

Hence, by FPC, total number of ways

$$= 14 \times \frac{10!}{2!}$$

$$= \frac{14 \times 10 \times 9 \times 8 \times 7 \times 6 \times 5 \times 4 \times 3 \times 2 \times 1}{2}$$

$$= \frac{14 \times 3628800}{2} = 25401600$$

40. *(b)* Number of numbers ending with 6 = □ □ □ □

$$= 1 \times 2 \times 3 \times 1 = 6$$

Similarly, number of numbers ending with 3, 4 or 5 are 6 each.

∴ Sum of digits in unit's place

$$= 6 \times 6 + 6 \times 3 + 6 \times 4 + 6 \times 5$$
$$= 6(6 + 3 + 4 + 5)$$
$$= 6(18) = 108$$

Alternate Method

If $a_1, a_2, \ldots\ldots, a_n$ are numbers, then

Sum of digits at unit's place $= (n-1)!(a_1 + a_2 + \ldots\ldots + a_n)$

$$= (4-1)!(3 + 4 + 5 + 6)$$
$$= 3!(18)$$
$$= 6(8) = 108$$

41. *(b)* A. Number of words using 4 letters out of 6 letters

$$= {}^6P_4 = \frac{6!}{2!} = 6 \times 5 \times 4 \times 3$$
$$= 360$$

B. Number of words using all letters $= {}^6P_6 = 6!$

$$= 720$$

C. Number of words starting with vowel

= Number of ways of choosing first letter (out of O and A) × Number of ways of arranging 5 alphabets

$$= 2 \times 5! = 2 \times 120 = 240$$

42. *(a)* The number of words starting from A are $5! = 120$

The number of words starting from I are $5! = 120$

The number of words starting from KA are $4! = 24$

The number of words starting from KI are $4! = 24$

The number of words starting from KN are $4! = 24$

The number of words starting from KRA are $3! = 6$

The number of words starting from KRIA are $2! = 2$

The number of words starting from KRIN are $2! = 2$

The number of words starting from KRISA are $1! = 1$

The number of words starting from KRISNA are $1! = 1$

Hence, rank of word 'KRISNA'

$$= 2(120) + 3(24) + 6 + 2(2) + 2(1) = 324$$

43. *(d)* The number of ways of ranking

$$A_1, A_2, \ldots, A_{10} = 10!$$

Number of ways in which A_1 is always above $A_{10} = \frac{1}{2}(10!)$

(∵ In exactly half of the total arrangements A_1 will be above A_{10})

44. *(c)* The numbers between 999 and 10000 are all 4-digit numbers. The number of 4-digit numbers formed by digits 0, 2, 3, 6, 7, 8 is ${}^6P_4 = 360$.

But here those numbers are also involved which begin from 0. So, we take those numbers as three-digit numbers. Taking initial digit 0, the number of ways to fill remaining 3 places from five digits 2, 3, 6, 7, 8 are ${}^5P_3 = 60$.

So, the required numbers = 360 − 60 = 300

45. *(b)* As we know, the last two digits of 10! and above factorials will be zero-zero.

$$\therefore\ 1! + 4! + 7! + 10! + 12! + 13! + 15! + 16! + 17!$$
$$= 1 + 24 + 5040 + 10! + 12! + 13! + 15! + 16! + 17!$$
$$= 5065 + 10! + 12! + 13! + 15! + 16! + 17!$$

In this series, the digit in the ten's place is 6 which is divisible by 3!.

46. *(c)* ∵ Each true-false question can be answered in 2 ways.

∴ Number of ways in which 10 questions can be answered

$$= 2^{10} = 1024$$

47. *(c)* A. The number of 4 different digits $= {}^7P_4$

$$= \frac{7!}{(7-4)!}$$
$$= 7 \times 6 \times 5 \times 4 = 840$$

B. The numbers exactly divisible by 2

= Number of ways of filling first 3 places × Number of ways of filling units's place

$$= {}^6P_3 \times 3$$
$$= \frac{6!}{(6-3)!} \times 3 = \frac{6!}{(3!)} \times 3$$
$$= 6 \times 5 \times 4 \times 3 = 360$$

C. Number of 4-digit numbers divisible by 25 = Numbers ending with 25 or 75

5 × 4   25 or 75

= □□ □□

$$= 5 \times 4 \times 2 = 40$$

(∵ when numbers end with 25 or 75, the other two places can be filled in 5 and 4 ways)

D. Number of 4-digit numbers divisible by 4 = Numbers ending with 12, 16, 24, 32, 36, 64, 72, 76, 52, 56

Now, number ending with 12 = □ □ □ □ = 20

4 × 5 × 1 × 1

Similarly, numbers ending with other number (16, 24, …) = 20 each

∴ Required numbers = 10 × 20 = 200

48. *(c)* The total number of 9-digit numbers, having all digit different $= 9\ {}^9P_8$

$$= \frac{9 \times 9!}{1!} = 9 \times 9!$$

(∵ In first place, we select anyone of nine numbers except zero. In rest of the eight places, we select any eight numbers from remaining 9 numbers)

49. *(c)* In the word 'RACHIT', the number of words beginning with A, C,H, I is 5! and the next word we get RACHIT.

$\therefore$ Required number of words

$$= 4 \times 5! + 1 = 4 \times 120 + 1 = 481$$

50. *(d)* The required natural numbers may consist of 4-digit, 3-digit, 2-digit or 1-digit.

Total number of 1-digit numbers = 9

Total number of 2-digit natural numbers = ${}^{10}P_2 - {}^9P_1$

(${}^9P_1$ is the number of numbers which begin with 0)

Similarly,

Total number of 3-digit natural numbers = ${}^{10}P_3 - {}^9P_2$

Total number of 4-digit natural numbers = ${}^{10}P_4 - {}^9P_3$

Hence , the required number of numbers

$$= 9 + ({}^{10}P_2 - {}^9P_1) + ({}^{10}P_3 - {}^9P_2) + ({}^{10}P_4 - {}^9P_3)$$
$$= 9 + (10 \times 9 - 9) + (10 \times 9 \times 8 - 9 \times 8) + (10 \times 9 \times 8 \times 7 - 9 \times 8 \times 7)$$
$$= 5274 \qquad [\because {}^nP_r = n(n-1)\dots(n-(r-1))]$$

51. *(b)* $X B_1 X B_2 X B_3 X B_4 X B_5 X B_6 X B_7 X$

First, we arrange 7 boys in 7! ways.

Now for girls, we have eight places and 3 girls need to be arranged.

This can be done in ${}^8P_3$ ways.

Total ways $= 7! \times {}^8P_3 = 7! \times 8 \times 7 \times 6$

$= 7! \times 336$

52. *(c)* I. We have seven married couples.

$\therefore$ Two husbands can be selected in $7 \times 6 = 42$ ways.

Two wives can be selected in $5 \times 4 = 20$ ways.

(as wives of husband's selected cannot play in same set)

$\therefore$ Required number of ways $= 42 \times 20 = 840$

II. The number of words beginning with T and ending with E

$= 6!$

(rest 6 alphabets can be arranged in 6! ways)

$= 720$ ways

53. *(d)* In the word 'SACHIN' order of alphabets is A, C, H, I, N, S. Number of words start with A = 5!, so are with C, H, I, N. Now, word start with S, and after that ACHIN are in ascending orders of position. So, $5 \cdot 5! = 600$ words are in dictionary before the words that start with S and position of this word is 601.

54. *(a)* Given that, $f(x) = {}^{7-x}P_{x-3}$. The above function is defined, if $7 - x \geq 0$, $x - 3 \geq 0$ and $7 - x \geq x - 3$

$\Rightarrow \quad x \leq 7, x \geq 3$ and $x \leq 5$

$\therefore \quad D_f = \{3, 4, 5\}$

Now, $\quad f(3) = {}^4P_0 = 1$

$f(4) = {}^3P_1 = 3$

and $\quad f(5) = {}^2P_2 = 2$

$\therefore \quad$ Range $= \{1, 2, 3\}$

55. *(c)* There are only 3 possible ways in which the team could be constructed.

These are *XY*, *YZ* and *ZX*.

56. *(d)* I. *X* shaking hand with *Y* and *Y* with *X* will not be two different hand shakes. Here, order is not important. Thus, permutation formula cannot be used.

II. Clearly, the number of chords equals the number of ways of selection of 2 different objects from 7 different objects. Here, also order is not important. Thus, permutation formula cannot be used.

57. *(b)* The combination will be *AB*, *AC*, *AD*, *BC*, *BD* and *CD*. Here, *AB* and *BA* are the same combination as order does not alter the combination.

58. *(d)* We have 5 different objects *A*,*B*,*C*,*D* and *E*. Taking 3 at a time, the combinations will be

ABC, *ABD*, *ABE*, *BCD*
BCE, *CDE*, *ACE*, *ACD*
ADE and *BDE*.

59. *(a)* I. Corresponding to each combination of ${}^nC_r$, we have $r!$ permutations because r objects in every combination can be rearranged in $r!$ ways.

Hence, the total number of permutations of n different things taken r at a time $= {}^nC_r \times r!$

On the other hand, it is ${}^nP_r$. Thus,

$${}^nP_r = {}^nC_r \times r!, \; 0 \leq r \leq n$$

II. From above, we have ${}^nP_r = {}^nC_r \times r!$

$$\Rightarrow \quad {}^nC_r = \frac{{}^nP_r}{r!} = \frac{n!}{r!(n-r)!}$$

III. We know, ${}^nC_r = \dfrac{n!}{r!(n-r)!}$

$$\therefore \quad {}^nC_0 = \frac{n!}{0!(n-0)!} = \frac{n!}{n!} = 1$$

i.e., selecting nothing at all is same as leaving behind all the objects and we know that there is only one way of doing so.

60. *(b)* I. ${}^nC_r = \dfrac{n!}{(n-r)!\,r!} = \dfrac{n!}{(n-r)!\,[n-(n-r)]!} = {}^nC_{n-r}$

II. $${}^nC_r + {}^nC_{r-1} = \frac{n!}{r!(n-r)!} + \frac{n!}{(r-1)!(n-r+1)!}$$
$$= \frac{n!}{r \times (r-1)!(n-r)!} + \frac{n!}{(r-1)!(n-r+1)(n-r)!}$$

$$= \frac{n!}{(r-1)!(n-r)!}\left[\frac{1}{r}+\frac{1}{n-r+1}\right]$$

$$= \frac{n!}{(r-1)!(n-r)!} \times \left[\frac{n-r+1+r}{r(n-r+1)}\right]$$

$$= \frac{(n+1)!}{r!(n+1-r)!} = {}^{n+1}C_r$$

III. ${}^nC_9 = {}^nC_8$

$\Rightarrow \quad {}^nC_{n-9} = {}^nC_8 \quad$ (as ${}^nC_r = {}^nC_{n-r}$)

$\Rightarrow \quad n - 9 = 8$

$\Rightarrow \quad n = 17$

$\therefore \quad {}^nC_{17} = {}^{17}C_{17} = 1$

61. *(d)* Here, order does not matter. Therefore, we need to count combinations. There will be as many committees as there are combinations of 5 different persons taken 3 at a time.

Hence, the required number of ways

$$= {}^5C_3 = \frac{5!}{3!2!} = \frac{4\times5}{2} = 10$$

Now, 1 man can be selected from 2 men in ${}^2C_1$ ways and 2 women can be selected from 3 women in ${}^3C_2$ ways. Therefore, the required number of committees

$$= {}^2C_1 \times {}^3C_2 = \frac{2!}{1!1!} \times \frac{3!}{2!1!} = 6$$

62. *(b)* A. ${}^nC_8 = {}^nC_2$ ($\because {}^nC_x = {}^nC_y \Rightarrow x = y$ or $x + y = n$)

$\Rightarrow \quad n = 10$

$$\therefore {}^nC_2 = {}^{10}C_2 = \frac{10!}{2!(10-2)!} = \frac{10!}{2!8!}$$

$$= \frac{10\times9\times8!}{2!\times8!} = 45$$

B. Here, $\frac{{}^{2n}C_3}{{}^nC_3} = \frac{12}{1}$

$$\therefore \quad \frac{\frac{(2n)(2n-1)(2n-2)}{3\times2\times1}}{\frac{n(n-1)(n-2)}{3\times2\times1}} = \frac{12}{1}$$

$$\Rightarrow \quad \frac{2(2n-1)2(n-1)}{(n-1)(n-2)} = \frac{12}{1}$$

$$\Rightarrow \quad \frac{2n-1}{n-2} = 3$$

$\Rightarrow \quad 2n - 1 = 3n - 6$

$\Rightarrow \quad 5 = n$

C. Given, $\frac{{}^{2n}C_3}{{}^nC_3} = \frac{11}{1}$

$$\frac{\frac{(2n)(2n-1)(2n-2)}{3.2.1}}{\frac{n(n-1)(n-2)}{3\cdot2\cdot1}} = \frac{11}{1}$$

$$\Rightarrow \quad \frac{4n(2n-1)(n-1)}{n(n-1)(n-2)} = \frac{11}{1}$$

$$\Rightarrow \quad \frac{8n-4}{n-2} = \frac{11}{1}$$

$\Rightarrow \quad 8n - 4 = 11n - 22$

$\Rightarrow \quad 18 = 3n$

$\Rightarrow \quad n = 6$

63. *(c)* Number of chords that can be drawn. through 21 points on circle = Number of ways of selecting 2 points from 21 points on circle

$$= {}^{21}C_2 = \frac{21\times20}{2\times1} = 210$$

64. *(b)* The number of ways a team of 3 boys and 3 girls can be selected from 5 boys and 4 girls = ${}^5C_3 \times {}^4C_3$

$$= \frac{5!}{3!2!} \times \frac{4!}{3!1!}$$

$$= \frac{5\times4\times3!}{3!2!} \times \frac{4\times3!}{3!}$$

$$= 10 \times 4 = 40$$

65. *(a)* Since, 3 red balls out of 6 red balls can be selected in ${}^6C_3$ ways, 3 white balls out of 5 white balls can be selected in ${}^5C_3$ ways and 3 blue balls out of 5 blue balls can be selected in ${}^5C_3$ ways.

Hence, by FPC total number of ways selecting 9 balls (3 balls of each colour) is

$$= {}^6C_3 \times {}^5C_3 \times {}^5C_3$$

$$= {}^6C_3 \times {}^5C_2 \times {}^5C_2 \quad (\because {}^nC_r = {}^nC_{n-r})$$

$$= \frac{6\times5\times4}{6} \times \frac{5\times4}{2} \times \frac{5\times4}{2} \quad \left[\because {}^nC_3 = \frac{n(n-1)(n-2)}{6} \text{ and } {}^nC_2 = \frac{n(n-1)}{2}\right]$$

$$= 20 \times 10 \times 10 = 2000$$

66. *(a)* I. Number of ways of selecting one ace from 4 aces = ${}^4C_1$

Number of ways of selecting 4 cards out of 48 cards = ${}^{48}C_4$

Hence, by FPC, 5 cards out of 52 cards with exactly one ace can be selected in ${}^4C_1 \times {}^{48}C_4$ ways

$$= 4 \times \frac{48\times47\times46\times45}{24}$$

$$\left[\because {}^nC_1 = n \text{ and } {}^nC_4 = \frac{n(n-1)(n-2)(n-3)}{24}\right]$$

$$= \frac{48\times47\times46\times45}{6}$$

$$= 8 \times 47 \times 46 \times 45 = 778320 \text{ ways}$$

II.

Cricket team

↓

Total 17 players

↓

5 Bowlers   12 Batsmen

We have to select 11 players including exactly 4 bowlers. Hence, 4 bowlers will be selected from 5 bowlers and remaining 7 players will be selected from 12 batsmen.

Now, 4 bowlers out of 5 bowlers can be selected in ${}^5C_4$ ways.

7 players out of 12 players can be selected in ${}^{12}C_7$ ways.

Hence, by FPC, total number of ways selecting the eleven players

$$= {}^5C_4 \times {}^{12}C_7 = {}^5C_1 \times {}^{12}C_5 \qquad (\because {}^nC_r = {}^nC_{n-r})$$

$$= \frac{5 \times 12 \times 11 \times 10 \times 9 \times 8}{120}$$

$$\left[\because {}^nC_5 = \frac{n(n-1)(n-2)(n-3)(n-4)}{120}\right]$$

$$= 5 \times 11 \times 9 \times 8 = 55 \times 72 = 3960 \text{ ways}$$

67. *(d)* Out of 5 black balls, 2 black balls can be selected in ${}^5C_2$ ways.

Out of 6 red balls, 3 red balls can be selected in ${}^6C_3$ ways.

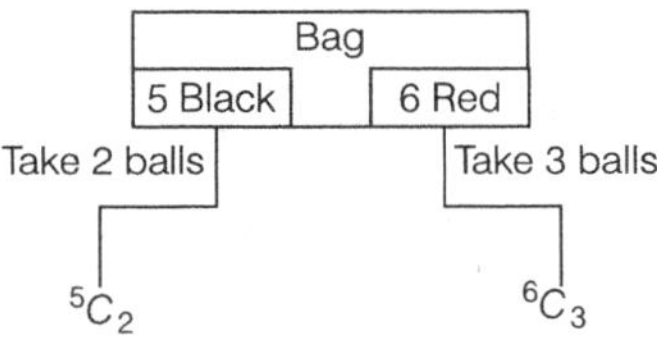

Hence, by FPC, total number of ways

$$= {}^5C_2 \times {}^6C_3 \qquad \left[\because {}^nC_2 = \frac{n(n-1)}{2} \text{ and } {}^nC_3 = \frac{n(n-1)(n-2)}{6}\right]$$

$$= \frac{5 \times 4}{2} \times \frac{6 \times 5 \times 4}{6}$$

$$= 10 \times 20$$

$$= 200 \text{ ways}$$

68. *(a)* Total number of available courses = 9

Out of these 5 courses have to be chosen. But it is given that 2 courses are compulsory for every student *i.e.*, you have to choose only 3 courses, out of 7.

It can be done in ${}^7C_3$ ways $= \frac{7 \times 6 \times 5}{6} = 35$ ways

69. *(b)* Here, the total number of points are $(m + n + k)$ which must give ${}^{m+n+k}C_3$ combinations by taking 3 points at a time, but m points lying on l_1, therefore taking 3 points at a time gives ${}^mC_3$ combinations which produces no triangle. Similarly, ${}^nC_3$ and ${}^kC_3$ number of triangles cannot be formed. Therefore, the required number of triangles is ${}^{(m+n+k)}C_3 - {}^mC_3 - {}^nC_3 - {}^kC_3$.

70. *(a)* Number of lines that can be formed by joining 18 non-collinear points $= {}^{18}C_2$

Number of lines formed by joining 5 collinear points is only 1.

Number of lines if the 5 points have been non-collinear $= {}^5C_2$

Required number of lines $= {}^{18}C_2 - {}^5C_2 + 1$

71. *(d)* Number of triangles that can be formed using 12 non-collinear points $= {}^{12}C_3$

Number of triangles that can be formed using 7 non-collinear points $= {}^7C_3$

But, 7 points are collinear.

$\therefore$ Number of triangles $= {}^{12}C_3 - {}^7C_3$

$$= \frac{12 \times 11 \times 10}{3 \times 2 \times 1} - \frac{7 \times 6 \times 5}{3 \times 2 \times 1}$$

$$= 220 - 35 = 185$$

72. *(c)* Suppose he invites r friends at a time.

Then, number of parties $= {}^{20}C_r$

And we know ${}^nC_r$ is maximum for $r = \frac{n}{2}$, when n is even.

Hence, he should invite 10 friends at a time, in order to have maximum parties.

73. *(a)* I. In n-sided polygon, the number of vertices $= n$

$\therefore$ Number of lines that can be formed using n points $= {}^nC_2$.

Out of these, ${}^nC_2$ lines, n lines form the polygon.

$\therefore$ Number of diagonals $= {}^nC_2 - n$

II. Let the number of sides of a polygon $= n$

Number of diagonal = Number of line segment joining any two vertices of polygon – Number of sides

$$= {}^nC_2 - n$$

$$= \frac{n(n-1)}{2} - n = \frac{n(n-3)}{2}$$

Now, $\frac{n(n-3)}{2} = 44$

$\Rightarrow \quad n^2 - 3n - 88 = 0$

$\Rightarrow \quad (n - 11)(n + 8) = 0$

$\Rightarrow \quad n = 11$

or $\quad n = -8$ rejected.

74. *(c)* There are two possibilities.

(i) Chemistry part I is borrowed. Here, boy may borrow Chemistry part II. So, boy has to select 2 books out of 7 books, which can be done in ${}^7C_2$ ways

$$= \frac{7 \times 6}{2 \times 1} = 21$$

(ii) Chemistry part I is not borrowed. Here, boy will also not borrow Chemistry part II. Thus, he has to select 3 books out of 6 books, which can be done is ${}^6C_3$ ways

$$= \frac{6 \times 5 \times 4}{3 \times 2 \times 1} = 20$$

$\therefore$ Required number of ways $= 21 + 20 = 41$

75. *(a)* There are two possibilities.

Men	Women
(I) 2	4
(II) 3	6

(I) Number of ways of choosing a committee of 2 men and 4 women = ${}^4C_2 \times {}^6C_4$

$$= \frac{4 \times 3}{2 \times 1} \times \frac{6 \times 5}{2 \times 1} = 90$$

(II) Number of ways of choosing a committee of 3 men and 6 women = ${}^4C_3 \times {}^6C_6$

$$= 4 \times 1 = 4$$

$\therefore$ Required number of ways = 94

76. *(d)* I. The number of ways the child can buy ice-cream = The number of integral solutions of $x_1 + x_2 + x_3 + x_4 + x_5 = 6$

$$= {}^{6+5-1}C_{5-1}$$

$$= {}^{10}C_4$$

(number of integral solution of $x_1 + x_2 + \ldots.. + x_r = n$ is ${}^{n+r-1}C_{r-1}$)

II. Number of ways of arranging 6 A's and 4 B's in a row

$$= \frac{10!}{6!\,4!} = {}^{10}C_4$$

77. *(c)* In all , we have 8 squares in which 6 'X's' have to be placed and it can be done in ${}^8C_6 = 28$ ways.

But this includes the possibility that either the top or horizontal row does not have any 'X'. Since, we want that each row must have atleast one 'X', these two possibilities are to be excluded.

Hence, required number of ways $= 28 - 2 = 26$

78. *(d)* Total number of points in a plane is $3p$.

$\therefore$ Maximum number of triangles $= {}^{3p}C_3 - 3 \cdot {}^pC_3$

(here, we subtract those triangles in which 3 points are in a line)

$$= \frac{3p!}{(3p-3)!\,3!} - 3 \cdot \frac{p!}{(p-3)!\,3!}$$

$$= \frac{3p(3p-1)(3p-2)}{3 \times 2} - \frac{3 \times p(p-1)(p-2)}{3 \times 2}$$

$$= \frac{p}{2}[9p^2 - 9p + 2 - (p^2 - 3p + 2)] = p^2(4p - 3)$$

79. *(a)* Since, a five-digit number is formed using digits {0, 1, 2, 3, 4 and 5} divisible by 3 *i.e.*, only possible when sum of digits is multiple of 3 which gives two cases.

Case I { using digits 0, 1, 2, 4, 5}

Number of numbers

$$= {}^4C_1 \times {}^4C_1 \times {}^3C_1 \times {}^2C_1 \times {}^1C_1$$

$$= 4 \times 4 \times 3 \times 2 \times 1 = 96$$

Case II { using digits 1, 2, 3, 4, 5}

Number of numbers

$$= {}^5C_1 \times {}^4C_1 \times {}^3C_1 \times {}^2C_1 \times {}^1C_1$$

$$= 5 \times 4 \times 3 \times 2 \times 1 = 120$$

$\therefore$ Total numbers formed $= 120 + 96 = 216$

80. *(c)* Required number of ways

$$= {}^{12}C_4 \times {}^8C_4 \times {}^4C_4$$

$$= \frac{12!}{8! \times 4!} \times \frac{8!}{4! \times 4!} \times 1 = \frac{12!}{(4!)^3}$$

81. *(b)* A. Number of ways committee can be formed

$$= {}^{10}C_2 \times {}^{20}C_3$$

B. Number of ways a particular professor is included $= {}^9C_1 \times {}^{20}C_3$ ($\because$ only one professor is to be selected)

C. Number of ways a particular lecturer is included $= {}^{19}C_2 \times {}^{10}C_2$

D. Number of ways a particular lecturer is included $= {}^{10}C_2 \times {}^{19}C_3$ ($\because$ of 19 lecturers, 3 need to be selected)

82. *(a)* Number of ways of placing 18 mice in two experimental groups and one control group $= {}^{18}C_6 \times {}^{12}C_6 \times {}^6C_6$

$$= \frac{18!}{12!\,6!} \times \frac{12!}{6! \times 6!} \times 1$$

$$= \frac{18!}{(6!)^3}$$

83. *(b)* In the word 'INVOLUTE', there are 4 vowels, namely, I, O,E,U and 4 consonants namely, N, V, L and T.

The number of ways of selecting 3 vowels out of 4 = ${}^4C_3 = 4$.

The number of ways of selecting 2 consonants out of $4 = {}^4C_2 = 6$

Therefore, the number of combinations of 3 vowels and 2 consonants is $4 \times 6 = 24$.

Now, each of these 24 combinations has 5 letters which can be arranged among themselves in 5! ways. Therefore, the required number of different words is $24 \times 5! = 2880$.

84. *(c)* I. Since, the team will not include any girl, therefore, only boys are to be selected. 5 boys out of 7 boys can be selected in ${}^7C_5$ ways. Therefore, the required number of ways $= {}^7C_5 = \frac{7!}{5!\,2!} = \frac{6 \times 7}{2} = 21$

II. Since, atleast one boy and one girl are to be there in every team. Therefore, the team can consist of

(a) 1 boy and 4 girls
(b) 2 boys and 3 girls
(c) 3 boys and 2 girls
(d) 4 boys and 1 girl.

1 boys and 4 girls can be selected in ${}^7C_1 \times {}^4C_4$ ways.

2 boys and 3 girls can be selected in ${}^7C_2 \times {}^4C_3$ ways.

3 boys and 2 girls can be selected in ${}^7C_3 \times {}^4C_2$ ways.

4 boys and 1 girl can be selected in ${}^7C_4 \times {}^4C_1$ ways.

Therefore, the required number of ways

$= {}^7C_1 \times {}^4C_4 + {}^7C_2 \times {}^4C_3 + {}^7C_3 \times {}^4C_2 + {}^7C_4 \times {}^4C_1$

$= 7 + 84 + 210 + 140 = 441$

III. Since, the team has to consist of atleast 3 girls, the team can consist of

(a) 3 girls and 2 boys, or

(b) 4 girls and 1 boy.

Note that the team cannot have all 5 girls, because the group has only 4 girls.

3 girls and 2 boys can be selected in ${}^4C_3 \times {}^7C_2$ ways.

4 girls and 1 boy can be selected in ${}^4C_4 \times {}^7C_1$ ways.

Therefore, the required number of ways

$$= {}^4C_3 \times {}^7C_2 + {}^4C_4 \times {}^7C_1$$

$$= 84 + 7 = 91$$

85. *(a)* There are 5 letters in the word 'AGAIN', in which A appears 2 times. Therefore, the required number of words $= \frac{5!}{2!} = 60$.

To get the number of words starting with A, we fix the letter A at the extreme left position, we then rearrange the remaining 4 letters taken all at a time. There will be as many arrangements of these 4 letters taken 4 at a time as there are permutations of 4 different things taken 4 at a time. Hence, the number of words starting with A = 4! = 24. Then, starting with G, the number of words $= \frac{4!}{2!} = 12$ as after placing G at the extreme left position, we are left with the letters A, A, I and N. Similarly, there are 12 words starting with the next letter I. Total number of words so far obtained $= 24 + 12 + 12 = 48$.

The 49th word is NAAGI. The 50th word is NAAIG.

86. *(c)* Since, 1000000 is a 7-digit number and the number of digits to be used is also 7. Therefore, the numbers to be counted will be 7 digit only. Also, the numbers have to be greater than 1000000, so they can begin either with 1, 2 or 4.

The number of numbers beginning with $1 = \frac{6!}{3!2!} = \frac{4 \times 5 \times 6}{2} = 60$ as when 1 is fixed at the extreme left position, the remaining digits to be rearranged will be 0, 2, 2, 2, 4, 4 in which there are 3, 2's and 2, 4's.

Total numbers begining with 2

$$= \frac{6!}{2!2!} = \frac{3 \times 4 \times 5 \times 6}{2} = 180$$

and total numbers begining with $4 = \frac{6!}{3!} = 4 \times 5 \times 6 = 120$

Therefore, the required number of numbers $= 60 + 180 + 120 = 360$.

Alternate Method

The number of 7-digit arrangements, clearly, $\frac{7!}{3!2!} = 420$

But this will include those numbers also, which have 0 at the extreme left position. The number of such arrangements $\frac{6!}{3!2!}$ (by fixing 0 at the extreme left position) = 60

Therefore, the required number of numbers $= 420 - 60 = 360$.

87. *(d)* Let us first seat the 5 girls. This can be done in 5! ways. For each such arrangement, the three boys can be seated only at the cross marked places.

$\times G \times G \times G \times G \times G \times$

There are 6 cross marked places and the three boys can be seated in ${}^6P_3$ ways. Hence, by multiplication principle, the total number of ways.

$$= 5! \times {}^6P_3 = 5! \times \frac{6!}{3!} = 5 \times 4 \times 3 \times 2 \times 4 \times 5 \times 6 = 14400$$

88. *(b)* A.

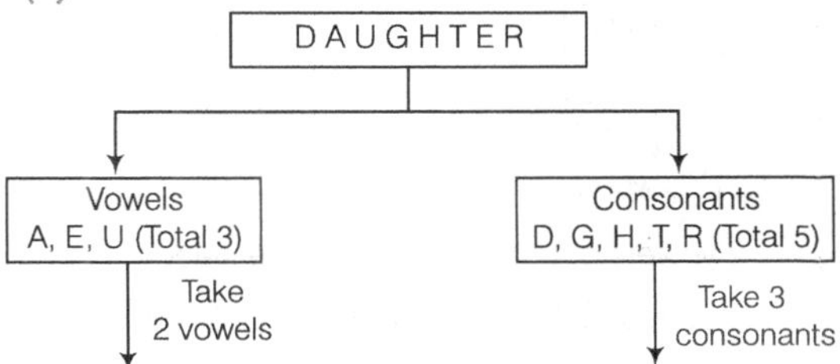

2 vowels out of 3 vowels can be selected = ${}^3C_2$ ways

3 consonants out of 5 consonants can be selected = ${}^5C_3$ ways

and these 5 letters (2 vowels and 3 consonants) can be arranged in 5! ways.

Hence, by FPC, total number of ways

$$= {}^3C_2 \times {}^5C_3 \times 5! = {}^3C_1 \times {}^5C_2 \times 5!$$

$$= \frac{3 \times 5 \times 4 \times 5 \times 4 \times 3 \times 2 \times 1}{2} = 3600 \text{ ways}$$

Note *Students should not forget to take the all possible ways of arrangement.*

B. Given word is → E Q U A T I O N

Vowels → E U A I O

Consonants → T Q N

The vowels can be arranged in 5! ways.

The consonants can be arranged in 3! ways.

These vowels and consonants (both are considered as a single letter) can be arranged in 2! ways *i.e.*, either all vowels and all consonants or all consonants and all vowels.

Hence, by FPC, total number of ways

$= 5! \times 3! \times 2! = 5 \times 4 \times 3 \times 2 \times 1 \times 3 \times 2 \times 1 \times 2 \times 1$

$= 120 \times 6 \times 2 = 120 \times 12 = 1440$

C. The letters of given word are A, A, E, I, I, M, N, N, O, T, X *i.e.*, word starting with A are formed with the letters 2 I's, 2 N's, A, E, X, M, T, O (total 10 letters).

Hence, number of words formed by these letters

$$= \frac{10!}{2!2!}$$

$$= \frac{10 \times 9 \times 8 \times 7 \times 6 \times 5 \times 4 \times 3 \times 2 \times 1}{2 \times 1 \times 2 \times 1}$$

$$= \frac{10 \times 9 \times 8 \times 7 \times 6 \times 5 \times 4 \times 3 \times 2 \times 1}{4}$$

$$= 10 \times 9 \times 8 \times 7 \times 6 \times 5 \times 3 \times 2 \times 1$$

$= 907200$ ways

$\therefore$ The number of words before the first word starts with E are 907200 ways.

D.

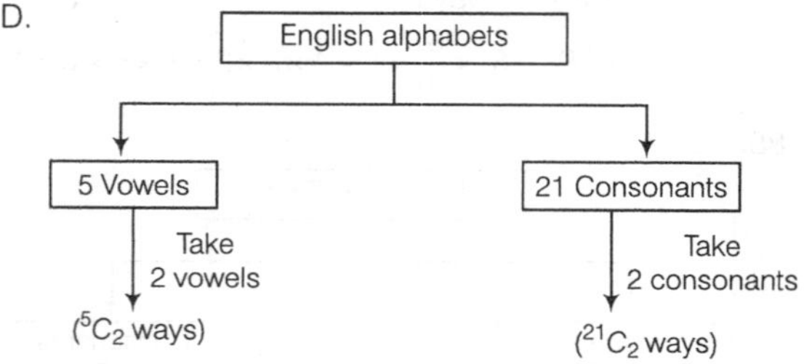

2 vowels out of 5 vowels can be selected in $^5C_2$ ways.

2 consonants out of 21 consonants can be selected in $^{21}C_2$ ways.

These 4 alphabets (2 vowels and 2 consonants) can be arranged in 4! ways.

Hence, by FPC, total number of ways

$$= {}^5C_2 \times {}^{21}C_2 \times 4!$$

$$= \frac{5 \times 4}{2} \times \frac{21 \times 20}{2} \times 4 \times 3 \times 2 \times 1$$

$$= 10 \times 210 \times 24 = 240 \times 210 = 50400$$

89. *(a)* I. Out of 7 persons, we want to select exactly 3 girls *i.e.*, remaining 4 person will be boys.

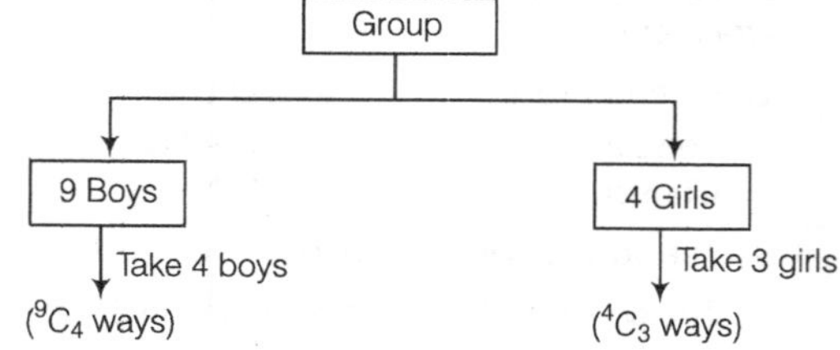

4 boys out of 9 boys can be selected in $^9C_4$ ways.

3 girls out of 4 girls can be selected in $^4C_3$ ways.

Hence, by FPC, total number of ways for making the committee

$$= {}^9C_4 \times {}^4C_3$$

$$= \frac{9 \times 8 \times 7 \times 6}{24} \times {}^4C_1 \qquad (\because {}^nC_r = {}^nC_{n-r})$$

$$= \frac{9 \times 8 \times 7 \times 6 \times 4}{24} \qquad (\because {}^nC_1 = n)$$

$$= 9 \times 8 \times 7 \qquad \left[\because {}^nC_4 = \frac{n(n-1)(n-2)(n-3)}{24}\right]$$

$= 72 \times 7 = 504$ ways

II. Here, we have to select atleast 3 girls.

There are following two possibilities

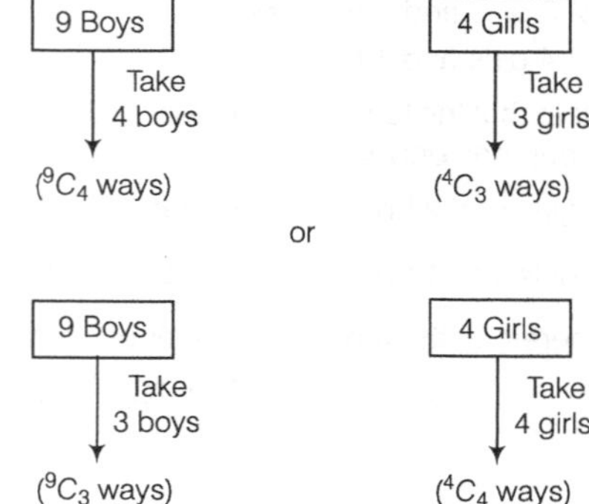

In first case, 4 boys out of 9 boys can be selected in $^9C_4$ ways and 3 girls out of 4 girls can be selected in $^4C_3$ ways.

Hence, by FPC, total number of ways in this case

$$= {}^9C_4 \times {}^4C_3 = \frac{9 \times 8 \times 7 \times 6}{4 \times 3 \times 2 \times 1} \times 4$$

$= 504$ ways

In second case, 3 boys out of 9 boys can be select in $^9C_3$ ways and 4 girls out of 4 girls can be select in $^4C_4$ ways.

Hence, by FPC, total number of ways in this case

$$= {}^9C_3 \times {}^4C_4$$

$$= \frac{9 \times 8 \times 7}{6} \times 1 = 84 \text{ ways}$$

Hence, by fundamental principle of addition, total number of ways $= 504 + 84 = 588$

III. Here, we have to select atmost 3 girls *i.e.*, there are following four possibilities

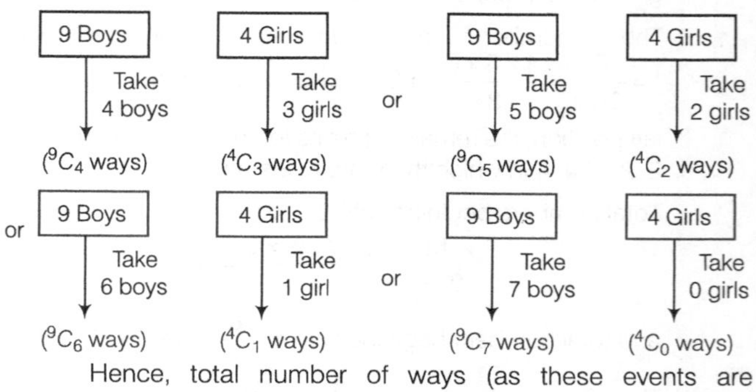

Hence, total number of ways (as these events are dependent to each other)

$$= ({}^9C_4 \times {}^4C_3) + ({}^9C_5 \times {}^4C_2) + ({}^9C_6 \times {}^4C_1) + ({}^9C_7 \times {}^4C_0)$$

$= ({}^9C_4 \times {}^4C_1) + ({}^9C_4 \times {}^4C_2) + ({}^9C_3 \times {}^4C_1) + ({}^9C_2 \times {}^4C_0)$

$(\because {}^nC_r = {}^nC_{n-r})$

$$= \left(\frac{9\times8\times7\times6}{24}\times4\right) + \left(\frac{9\times8\times7\times6}{24}\times\frac{4\times3}{2}\right) + \left(\frac{9\times8\times7}{6}\times4\right) + \left(\frac{9\times8}{2}\times1\right)$$

$= 504 + 756 + 336 + 36$

$= 1632$ ways

90. *(a)* Number of ways to fill the unit place $= 1$ (*i.e.*, 0 in unit place) and remaining 5 numbers (1, 3, 5, 7, 9) can be arranged in 5! ways.

Hence, by FPC, total number of ways

$= 5! \times 1 = 5 \times 4 \times 3 \times 2 \times 1$

$= 120$ ways

91. *(d)*

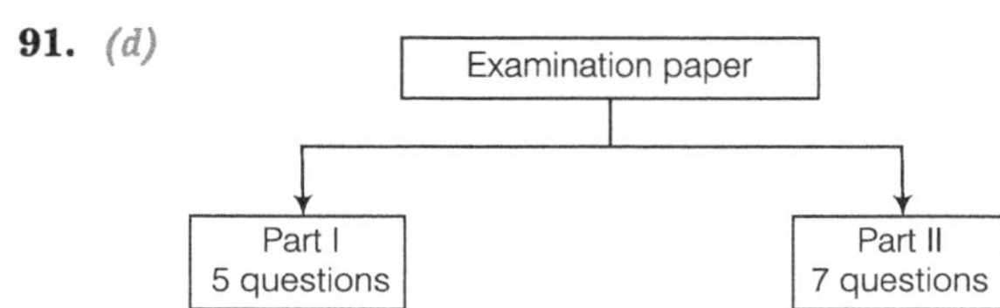

Here, we have to attempt 8 questions keeping in mind that atleast 3 from each part, then following possibility arises :

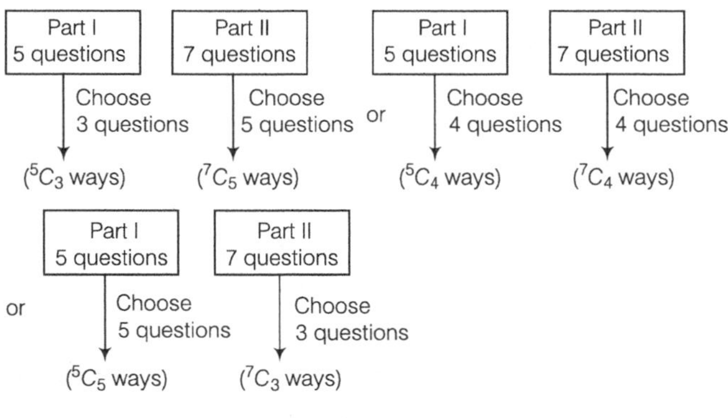

Hence, total number of ways

$= ({}^5C_3 \times {}^7C_5) + ({}^5C_4 \times {}^7C_4) + ({}^5C_5 \times {}^7C_3)$

$= ({}^5C_2 \times {}^7C_2) + ({}^5C_1 \times {}^7C_3) + ({}^5C_5 \times {}^7C_3) \quad (\because {}^nC_r = {}^nC_{n-r})$

$$= \left(\frac{5\times4}{2}\times\frac{7\times6}{2}\right) + \left(5\times\frac{7\times6\times5}{6}\right) + \left(1\times\frac{7\times6\times5}{6}\right)$$

$= (10 \times 21) + 175 + 35$

$= 210 + 210$

$= 420$ ways

92. *(c)* 1 king out of 4 kings can be selected in ${}^4C_1$ ways.

Remaining 4 cards out of remaining 48 cards can be selected in ${}^{48}C_4$ ways.

Hence, by FPC, total number of ways $= {}^4C_1 \times {}^{48}C_4$

93. *(c)* I. Total number of persons = 9

$\boxed{1} \overset{2}{\times} \boxed{3} \overset{4}{\times} \boxed{5} \overset{6}{\times} \boxed{7} \overset{8}{\times} \boxed{9}$

There are 4 even numbers and 4 women want to sit, this can be done in 4! ways.

Remaining 5 seats can be filled by 5 men in 5! ways.

Hence, by FPC, total number of ways

$= 5! \times 4!$

$= 5 \times 4 \times 3 \times 2 \times 1 \times 4 \times 3 \times 2 \times 1$

$= 120 \times 24 = 2880$ ways

II. There are two possibilities.

(a) If all the three students will go in party, then we have to select 7 students out of 22 students.

(b) If all of three students will not go for party then, we have to select 10 students out of 22 students.

Now, 7 students out of 22 students can be selected in ${}^{22}C_7$ ways.

10 students out of 22 students can be selected in ${}^{22}C_{10}$ ways.

$\therefore$ Total number of ways

$= {}^{22}C_7 + {}^{22}C_{10}$

94. *(b)* Given word in A S S A S S I N A T I O N

A	→	3 times
S	→	4 times
I	→	2 times
T	→	1 time
O	→	1 time
N	→	2 times

If all the S's are together, then it will be considered as 1 letter. Now, remaining 9 letters (3A's, 2I's, T, O, 2N's) and 4S's will be considered as 10 letters.

Hence, total number of ways

$$= \frac{10!}{3!2!2!}$$

$$= \frac{10\times9\times8\times7\times6\times5\times4\times3!}{3!\times2\times1\times2\times1}$$

$= 10 \times 9 \times 8 \times 7 \times 6 \times 5$

$= 151200$ ways

95. *(b)* After sending 4 to one side and 3 to other side. We have to select 5 for one side and 6 for other side from remaining.

This can be done in ${}^{11}C_5 \times {}^6C_6$ ways $= {}^{11}C_5$

Now, there are 9 on each side of the long table and each can be arranged in 9! ways.

$\therefore$ Required numbers of ways $= {}^{11}C_5 \times 9! \times 9!$

$= {}^{11}C_6 \times (9!)^2 \quad (\because {}^nC_r = {}^nC_{n-r})$

96. *(c)* 1 objects can be taken out in ${}^nC_1$ ways

2 objects can be taken out in ${}^nC_2$ way

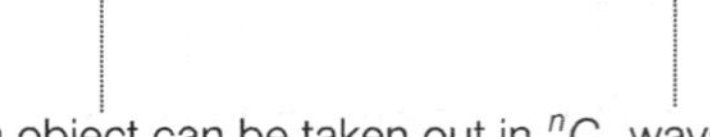

n object can be taken out in ${}^nC_n$ ways

$\therefore$ Required number of ways

$$= {}^nC_1 + {}^nC_2 + \ldots\ldots + {}^nC_n = 2^n - 1$$

(we know ${}^nC_0 + {}^nC_1 + \ldots\ldots\ldots + {}^nC_n = 2^n$)

97. *(b)* Possible number of ways of choosing 5 green dyes

$$= \square\,\square\,\square\,\square\,\square$$
$$= 2 \times 2 \times 2 \times 2 \times 2 = 2^5$$

(dye can be selected or not selected)

Similarly, possible number of ways of choosing 4 blue dyes and 3 red dyes are 2^4 and 2^3 respectively.

Thus, possible number of ways of choosing atleast one green and one blue dye is $2^5 - 1$ and $2^4 - 1$

i.e., 31 and 15.

Required number of ways $= 31 \times 15 \times 2^3$

$$= 31 \times 120, = 3720$$

98. *(b)* Number of ways of distributing 8 identical balls in 3 distinct boxes

= Number of positive integer solutions of $x_1 + x_2 + x_3 = 8$

$= {}^{8-1}C_{3-1}$

$= {}^7C_2 = \dfrac{7 \times 6}{2 \times 1} = 21$

($x + x_2 + \ldots\ldots + x_r = n$, number of positive integer solution equals ${}^{n-1}C_{r-1}$)

99. *(d)* If triangle is formed including point 'P' the other points must be one from l_1 and other point from l_2. Number of triangle formed with $P = {}^3C_1 \times {}^5C_1 = 15$ ways

When P is not included.

Number of triangle formed

$$= {}^3C_2 \times {}^5C_1 + {}^3C_1 \times {}^5C_2$$
$$= 15 + 15 = 30$$

Total number of triangles = 15 + 30 = 45

100. *(c)* The number of ways in which 4 novels can be selected

$$= {}^6C_4 = 15$$

The number of ways in which 1 dictionary can be selected

$$= {}^3C_1 = 3$$

4 novels can be arranged in 4! ways.

$\therefore$ The total number of ways

$$= 15 \times 4! \times 3$$
$$= 15 \times 24 \times 3 = 1080$$

101. *(b)* Total number of ways

$$= {}^{10}C_1 + {}^{10}C_2 + {}^{10}C_3 + {}^{10}C_4$$
$$= 10 + 45 + 120 + 210 = 385$$

102. *(c)* Total number of ways in which all letters can be arranged in alphabetical order = 6!

There are two vowels (A, E) in the word GARDEN. Total number of ways in which these two vowels can be arranged = 2!

$\therefore$ Total number of required ways

$$= \frac{6!}{2!} = 360$$

103. *(a)* Number of ways of choosing 2 parallel lines from a set of four parallel lines $= {}^4C_2$

Number of ways of choosing 2 parallel lines from a set of 3 parallel lines $= {}^3C_2$

$\therefore$ Required number of ways $= {}^4C_2 \times {}^3C_2$

$$= \frac{4 \times 3}{2 \times 1} \times 3$$
$$= 6 \times 3 = 18$$

104. *(c)* $({}^7C_0 + {}^7C_1) + ({}^7C_1 + {}^7C_2) + \ldots + ({}^7C_6 + {}^7C_7)$

$$= 2[{}^7C_0 + {}^7C_1 + {}^7C_2 + \ldots\ldots + {}^7C_7] - {}^7C_0 - {}^7C_7$$
$$= 2 \cdot (2^7) - 1 - 1$$
$$= 2^8 - 2$$

105. *(d)* Maximum points of intersection of two circles = 2

$\therefore$ Maximum point of intersection of 4 circle

$$= {}^4C_2 \times 2 = \frac{12}{2 \times 1} \times 2 = 12$$

Two lines can intersect at only one point

$\therefore$ Number of points of intersection of 8 lines

$$= {}^8C_2 = \frac{8 \times 7}{2 \times 1} = 28$$

A line can intersect circle at two points

$\therefore$ Number of points of intersection of 4 circles and 8 lines

$$= {}^4C_1 \times {}^8C_1 \times 2$$
$$= 4 \times 8 \times 2 = 64$$

$\therefore$ Maximum points of intersection = 12 + 28 + 64 = 104

106. *(c)* Two C's are considered as one alphabet

Now, two C's , U and E can be arranged in 3! = 6 ways.

X C C X U X E X

There are 4 places which S can take

$\therefore$ For placing S, we have ${}^4C_3 = 4$ ways

$\therefore$ Required numbers of words = 6 × 4 = 24

107. *(a)* Since, the consonants have to appear in alphabetical order, we have

X C X C X S X S X S X

Case I U and E are together

Number of words $= {}^6C_1 \times 2! = 12$

***Case* II** U and E are separated.

Number of words $= {}^6C_2 \times 2! = \frac{6 \times 5}{2 \times 1} \times 2 = 30$

$\therefore$ Required number of words = 30 + 12 = 42

108. *(d)* Number of words beginning with C $= \frac{6!}{3!} = 120$

C CESSSU

Number of words beginning with E $= \frac{6!}{3!2!} = 60$

E SSSUCC

Number of words beginning with SC $= \frac{5!}{2!} = 60$

SC UCESS

Number of words beginning with SE $= \frac{5!}{2!2!} = 30$

SE UCC SS

Number of words beginning with SS $= \frac{5!}{2!} = 60$

SS SUCCE

First words beginning with SU is SUCCESS

$\therefore$ The rank of the word SUCCESS is

$$= 120 + 60 + 60 + 30 + 60 + 1 = 331$$

109. *(c)* Number of elements in $A \times B = n(A) \times n(B)$

$= 2 \times 4 = 8$

Number of subsets having 3 or more elements

$$= {}^8C_3 + {}^8C_4 + {}^8C_5 + {}^8C_6 + {}^8C_7 + {}^8C_8$$
$$= 2^8 - ({}^8C_0 + {}^8C_1 + {}^8C_2)$$
$$= 256 - (1 + 8 + 28) = 256 - 37 = 219$$

110. *(d)* If there are m identical objects of one type n identical objects of second type

Then, number of ways of selection of one or more objects
$$= (n + 1)(m + 1) - 1$$

$\therefore$ Number of ways of selection of one or more balls

$$= (10 + 1)(9 + 1)(7 + 1) - 1$$
$$= 11 \times 10 \times 8 - 1 = 879$$

111. *(b)* Here, $X = \{1, 2, 3, 4, 5\}$

Now, the possibilities are

(i) $1 \in Y, 1 \in Z$ (ii) $1 \in Y, 1 \notin Z$

(iii) $1 \notin Y, 1 \in Z$ (iv) $1 \notin Y, 1 \notin Z$

(ii) (iii), (iv) are favourable cases for $Y \cap Z$ to be empty.

Similarly, for other elements we have 3 favourable cases.

$\therefore$ Required number of different ordered pairs $(Y, Z) = 3 \times 3 \times 3 \times 3 \times 3 = 3^5$

112. *(b)* ${}^nC_{r+1} + {}^nC_{r-1} + {}^nC_r + {}^nC_r$

$$= ({}^nC_r + {}^nC_{r-1}) + ({}^nC_r + {}^nC_{r+1})$$
$$= {}^{n+1}C_r + {}^{n+1}C_{r+1}, = {}^{n+2}C_{r+1}$$

113. *(b)* The number of choices available to him

$$= {}^5C_4 \times {}^8C_6 + {}^5C_5 \times {}^8C_5$$
$$= \frac{5!}{4!1!} \times \frac{8!}{6!2!} + \frac{5!}{5!0!} \times \frac{8!}{5!3!}$$
$$= 5 \times \frac{8 \times 7}{2} + 1 \times \frac{8 \times 7 \times 6}{3 \times 2}$$
$$= 5 \times 4 \times 7 + 8 \times 7$$
$$= 140 + 56 = 196$$

114. *(a)* In BANANA, there are 3A's 2N's and 1B

$\therefore$ Number of arrangements $= \frac{6!}{3!2!}$
$$= \frac{6 \times 5 \times 4 \times 3!}{2 \times 3!}$$
$$= 60$$

Also, Statement II is correct explanation of Statement I.

115. *(a)* Here, n^{th} term, $t^n = n \cdot {}^nP_n$

$$= n \cdot n!$$
$$= (n + 1 - 1)n!$$
$$= (n + 1)n! - n!$$
$$t_n = (n + 1)! - n!$$

$\therefore$
$$t_1 = 2! - 1!$$
$$t_2 = 3! - 2!$$
$$t_3 = 4! - 3!$$
$$\vdots \quad \vdots \quad \vdots$$
$$t_n = (n + 1)! - n!$$

$\therefore$ Required sum $= (n + 1)! - 1!$
$$= {}^{n+1}P_{n+1} - 1$$

116. *(a)* Possible even digits are 2, 4, 6, 8, 0.

***Case* I** Number has digits 4, 6, 8, 0

(Here, sum of digits is 18, divisible by 3)

Number of arrangements $= 3 \times 3!$
$= 3 \times 6$
$= 18$ (Ist place can be filled using 4, 6 or 8)

***Case* II** Number has digits 2, 4, 6, 0

(Here, sum of digits is 12, divisible by 3)

Ist place cannot be filled by 0.

Number of arrangements $= 3 \times 3!$
$= 18$

$\therefore$ Number of numbers = 18 + 18 = 36

117. *(a)* Number of times he will go to garden

= Number of ways of selecting 3 children from 8 children

$$= {}^8C_3 = \frac{8 \times 7 \times 6}{3 \times 2} = 56$$

8

Binomial Theorem

Binomial Theorem for Positive Integral Indices

Level I

1. To overcome the difficulty in calculating the higher powers of binomials, we use a theorem known as
(a) principle of mathematical induction
(b) binomial theorem
(c) mean value theorem
(d) None of the above

2. Consider the expansions given below

$(a+b)^0 = 1, a+b \neq 0$

$(a+b)^1 = a+b$

$(a+b)^2 = a^2 + 2ab + b^2$

$(a+b)^3 = a^3 + 3a^2b + 3ab^2 + b^3$

$(a+b)^4 = (a+b)^3 \cdot (a+b) = a^4 + 4a^3b + 6a^2b^2 + 4\,ab^3 + b^4$

Based on the above expansions, which of the following statements is/are true?

I. The total number of terms in the expansion is one more than the index.

II. Powers of the first quantity *a*, go on decreasing by 1, in the successive terms.

III. Powers of the second quantity *b* increase by 1, in the successive terms.

IV. The sum of indices of *a* and *b* is same in each term of the expansion and is not equal to the index of $a+b$.

(a) Only I is true (b) Only III is true
(c) I, II and III are true (d) All are true

3. On arranging the coefficients of $(a+b)^0, (a+b)^1, (a+b)^2, (a+b)^3$ and $(a+b)^4$ respectively, we get the following pattern

Index	Coefficients
0	1
1	1 1
2	1 2 1
3	1 3 3 1
4	1 4 6 4 1

Observing the above pattern, which of the following can be said correct?
(a) Addition of 1's in the row for index 1 gives rise to 2 in the row for index 2
(b) The addition of 1, 2 and 2, 1 in the row for index 2, gives rise to 1 and 1 in the row for index 3
(c) 2 is present at the beginning and at the end of each row
(d) None of the above

Directions (Q. Nos. 4-6) *In the expression of* $(a + b)^n$, *where* $n = 0, 1, 2, 3, 4$, *we get a pattern in the expansions between the index and coefficients that is given below.*

Index	Coefficients
0	1
1	1 1
2	1 2 1
3	1 3 3 1
4	1 4 6 4 1

With the help of above pattern, we get a rule, which helps us to write the expansion of the binomial for any power and get a pattern in the form of combinations as shown below.

Index	Coefficients
0	${}^0C_0$ (= 1)
1	${}^1C_0$ (= 1) ${}^1C_1$ (= 1)
2	${}^2C_0$ (= 1) ${}^2C_1$ (= 2) ${}^2C_2$ (= 1)

Based on the above pattern, answer the questions given below.

4. For the index 3, we will have coefficients as

I. ${}^3C_0, {}^3C_1, {}^3C_2$ II. ${}^3C_0, {}^3C_1, {}^3C_2, {}^3C_3$

III. 1, 2, 3, 1 IV. 1, 3, 3, 1

(a) I and IV are true (b) II and III are true
(c) II and IV are true (d) Only II is true

5. For the index 4, which of the following is not true for all the coefficients?

(a) Coefficients are ${}^4C_0, {}^4C_1, {}^4C_2, {}^4C_3, {}^4C_4$
(b) Coefficients are ${}^4C_1, {}^4C_2, {}^4C_3, {}^4C_4$
(c) 1, 4, 6, 4, 1 are the coefficients
(d) Total number of terms are 5

6. For the index 5, we will have all the coefficients as

(a) ${}^5C_0, {}^5C_1, {}^5C_2, {}^5C_3, {}^5C_4$ (b) 1, 5, 10, 5, 5, 1
(c) 1, 5, 10, 10, 5, 1 (d) 1, 4, 6, 4, 1

7. Based on the pattern discussed in previous problem, which of the following is true?

(a) ${}^7C_0, {}^7C_1, {}^7C_2, {}^7C_3, {}^7C_4, {}^7C_5, {}^7C_6, {}^7C_7$ are the coefficients for index 7
(b) $(a+b)^7 = {}^7C_0\, a^7 + {}^7C_1\, a^6b + {}^7C_2\, a^5 b^2 + {}^7C_3\, a^4 b^3 + {}^7C_4\, a^3 b^4 + {}^7C_5 a^2 b^5 + {}^7C_6\, ab^6 + {}^7C_7\, b^7$
(c) Both (a) and (b) are true
(d) Both (a) and (b) are false

8. For any positive integer n, $(a + b)^n$ is equal to

(a) ${}^nC_1 a^n + {}^nC_2 a^{n-1}b + {}^nC_3 a^{n-2} b^2 + \dots\dots\dots + {}^nC_{n+1}\, b^n$
(b) ${}^nC_0\, b^n + {}^nC_1\, b^{n-3}\, a + {}^nC_2\, b^{n-5}\, a^2 + \dots\dots + {}^nC_n\, a^n$
(c) ${}^nC_0\, a^n + {}^nC_1\, a^{n-1}\, b + {}^nC_2\, a^{n-2}\, b^2 + \dots\dots\dots + {}^nC_{n-1}\, ab^{n-1} + {}^nC_n b^n$
(d) None of the above

9. By using binomial theorem, $(x + 2)^6$ is equal to

(a) $x^6 + 12x^5 + 60\, x^4 + 150\, x^3 + 230\, x^2 + 192\, x + 64$
(b) $x^5 + 12\, x^6 + 60\, x^5 + 160\, x^4 + 240\, x^3 + 192\, x^2 + 62\, x$
(c) $x^6 + 12\, x^5 + 60\, x^4 + 160\, x^3 + 240\, x^2 + 192x + 64$
(d) None of the above

10. Notation form of $(a + b)^n$ is

(a) $\sum_{k=0}^{n} {}^nC_k\, a^{n+k}\, b^k$ (b) $\sum_{k=0}^{n} {}^nC_k\, a^{n-k}\, b^k$
(c) $\sum_{k=0}^{n} {}^nC_k\, b^{n+k}\, a^k$ (d) None of these

11. Which among the following statements is/are true?

(a) In the binomial theorem ${}^nC_r$ is known as binomial coefficients
(b) In the expansion of $(a + b)^n$, number of terms is $(n + 1)$
(c) Both (a) and (b) are true
(d) Both (a) and (b) are false

12. In the expansion of $(a + b)^n$, which of the following is true?

(a) In the successive terms of the expansion, the index of a goes on decreasing by unity
(b) In the successive terms of the expansion, the index of b goes on decreasing by unity
(c) In the first term, index of b is 1
(d) None of the above

13. In every term, the sum of indices of a and b in the expansion of $(a + b)^n$ is

(a) n (b) $n + 1$ (c) $n + 2$ (d) $n - 1$

14. **Statement I** $(x - 2y)^5 = x^5 - 10\, x^4\, y + 40\, x^3\, y^2 - 80\, x^2\, y^3 + 80\, xy^4 - 32\, y^5$

Statement II

$(x - y)^n = {}^nC_0\, x^n - {}^nC_1\, x^{n-1}\, y + {}^nC_2\, x^{n-2}\, y^2 - {}^nC_3\, x^{n-3}\, y^3 + \dots\dots + (-1)^n\, {}^nC_n\, y^n$

(a) Statement I is true; Statement II is true; Statement II is a correct explanation for Statement I
(b) Statement I is true; Statement II is true; Statement II is not a correct explanation for Statement I
(c) Statement I is true; Statement II is false
(d) Statement I is false; Statement II is true

15. **Assertion** (A) $2^n = {}^nC_0 + {}^nC_1 + {}^nC_2 + \ldots\ldots + {}^nC_n$

Reason (R) $(1+x)^n = {}^nC_0 + {}^nC_1\, x + {}^nC_2\, x^2 + \ldots\ldots + {}^nC_n\, x^n$

(a) Both A and R are correct; R is the correct explanation of A
(b) Both A and R are correct; R is not the correct explanation of A
(c) A is correct; R is incorrect
(d) R is correct; A is incorrect

16. Which of the following is/are correct?

(a) $(1-x)^n = {}^nC_0 + {}^nC_1\, x + {}^nC_2\, x^2 + \ldots\ldots + {}^nC_n\, x^n$
(b) ${}^nC_0 + {}^nC_2 + {}^nC_4 + \ldots\ldots = {}^nC_1 + {}^nC_3 + {}^nC_5 + \ldots\ldots$
(c) Both (a) and (b) are true
(d) Both (a) and (b) are false

17. If $x \neq 0$, then $\left(x^2 + \frac{3}{x}\right)^4$ is equal to

(a) $x^8 + 12x^5 + \frac{108}{x} + \frac{81}{x^4}$
(b) $x^8 + 12x^5 + 54x^3 + \frac{180}{x} + \frac{81}{x^3}$
(c) $x^8 + 12x^5 + 54x^2 + \frac{108}{x} + \frac{81}{x^4}$
(d) None of the above

18. The value of $(98)^5$ is

(a) 903920986
(b) 9039207968
(c) 309820968
(d) None of the above

19. Which among the following is true?

(a) $(1.01)^{1000000} > 10000$
(b) $(1.01)^{1000000} < 10000$
(c) $(1.01)^{1000000} < 1000$
(d) None of the above

20. If $6^n - 5n$ is divided by 25, then remainder is

(a) 0 (b) 2
(c) 4 (d) 1

21. Expansion of $\left(\frac{2}{x} - \frac{x}{2}\right)^5$ is equal to

(a) $\frac{32}{x^4} - \frac{40}{x^2} + \frac{20}{x} - 5x + 4x^3 - x^5$
(b) $\frac{32}{x^5} - \frac{40}{x^3} + \frac{20}{x} - 5x + \frac{5x^3}{8} - \frac{x^5}{32}$
(c) $\frac{30}{x^5} - \frac{14}{x^3} + \frac{10}{x} - 5x + \frac{5x^3}{8} - \frac{x^5}{30}$
(d) None of the above

22. Match the terms of Column I with their expansions in Column II and choose the correct option from the codes given below.

	Column I		Column II
A.	$(1-2x)^5$	1.	$\frac{x^5}{243} + \frac{5x^3}{81} + \frac{10x}{27} + \frac{10}{9x} + \frac{5}{3x^3} + \frac{1}{x^5}$
B.	$(2x-3)^6$	2.	$x^6 + 6x^4 + 15x^2 + 20 + \frac{15}{x^2} + \frac{6}{x^4} + \frac{1}{x^6}$
C.	$\left(\frac{x}{3} + \frac{1}{x}\right)^5$	3.	$1 - 10x + 40x^2 - 80x^3 + 80x^4 - 32x^5$
D.	$\left(x + \frac{1}{x}\right)^6$	4.	$64x^6 - 576x^5 + 2160x^4 - 4320x^3 + 4860x^2 - 2916x + 729$

Codes

	A	B	C	D
(a)	1	4	3	2
(b)	2	1	3	4
(c)	3	4	1	2
(d)	4	3	2	1

23. Which among the following is /are true?

I. $(1.1)^{10000} > 100$
II. $(1.1)^{10000} < 100$
III. $(1.1)^{10000} > 1000$
IV. $(1.1)^{10000} < 1000$

(a) I and IV are true
(b) II and IV are true
(c) I and III are true
(d) Only IV is true

24. **Statement I** $(\sqrt{3}+\sqrt{2})^4 - (\sqrt{3}-\sqrt{2})^4 = 40\sqrt{6}$

Statement II $(a+b)^4 - (a-b)^4 = 4ab\,(a^2+b^2)$

(a) Statement I is true; Statement II is true; Statement II is a correct explanation for Statement I
(b) Statement I is true; Statement II is true; Statement II is not a correct explanation for Statement I
(c) Statement I is true; Statement II is false
(d) Statement I is false; Statement II is true

25. Match the terms of Column I with their values in Column II and choose the correct option from the codes given below.

	Column I		Column II
A.	$(96)^3$ equals to	1.	104060401
B.	$(102)^5$ equals to	2.	11040808032
C.	$(101)^4$ equals to	3.	884736
D.	$(99)^5$ equals to	4.	9509900499

Codes

	A	B	C	D		A	B	C	D
(a)	3	4	2	1	(b)	3	2	1	4
(c)	1	3	4	2	(d)	1	2	3	4

26. For a positive integer n, $9^{n+1} - 8n - 9$ is divisible by

(a) 31 (b) 64
(c) 18 (d) None of these

27. If $(1+x)^n = {}^nC_0 + {}^nC_1x + {}^nC_2x^2 + \dots\dots + {}^nC_nx^n$, then $\sum_{r=0}^{n} 3^r \; {}^nC_r$ is equal to

(a) 2^n (b) 3^n
(c) 4^n (d) None of these

28. **Statement I** $(\sqrt{2}+1)^6 + (\sqrt{2}-1)^6 = 190$

Statement II $(x+1)^6 + (x-1)^6 = 2(x^6 + 15x^4 + 15x^2 + 1)$

(a) Statement I is true; Statement II is true; Statement II is a correct explanation for Statement I
(b) Statement I is true; Statement II is true; Statement II is not a correct explanation for Statement I
(c) Statement I is true; Statement II is false
(d) Statement I is false; Statement II is true

Level II

29. **Statement I** ${}^nC_0 + {}^nC_2 + {}^nC_4 + \dots\dots = 2^{n-1}$

Statement II ${}^nC_1 + {}^nC_3 + {}^nC_5 + \dots\dots = 2^{n-1}$ [HOTS]

(a) Statement I is true; Statement II is true; Statement II is a correct explanation for Statement I
(b) Statement I is true; Statement II is true; Statement II is not a correct explanation for Statement I
(c) Statement I is true; Statement II is false
(d) Statement I is false; Statement II is true

30. $(100)^{50} + (99)^{50}$ [NCERT Exemplar]

(a) $<(101)^{50}$ (b) $=101$
(c) $>(101)^{50}$ (d) >101

31. If $(1+x-2x^2)^6 = 1 + a_1x + a_2x^2 + \dots\dots + a_{12}x^{12}$, then the expression $a_2 + a_4 + a_6 + \dots\dots a_{12}$ has the value [HOTS]

(a) 32 (b) 31
(c) 63 (d) 64

32. If for all $n \in N$, $2^{3n+3} - 7n - 8$ is divisible by

(a) 34 (b) 49
(c) 17 (d) None of these

33. The total number of terms in the expansion of $(x+a)^{100} + (x-a)^{100}$ after simplification will be [NCERT Exemplar]

(a) 202 (b) 51
(c) 50 (d) None of these

34. If $(1+ax)^n = 1 + 8x + 24x^2 + \dots$, then the values of a and n are respectively

(a) 2, 4 (b) 2, 3 (c) 3, 6 (d) 1, 2

35. If n is a positive integer, then $(\sqrt{3}+1)^{2n} - (\sqrt{3}-1)^{2n}$ is

(a) an irrational number
(b) an odd positive integer
(c) an even positive integer
(d) a rational number other than positive integers

36. The remainder left out when $8^{2n} - (62)^{2n+1}$ is divided by 9, is [AIEEE 2009]

(a) 0 (b) 2
(c) 7 (d) 8

37. The sum of the series ${}^{20}C_0 - {}^{20}C_1 + {}^{20}C_2 - {}^{20}C_3 + \dots\dots + {}^{20}C_{10}$ is [HOTS]

(a) $-{}^{20}C_{10}$ (b) $\frac{1}{2}\,{}^{20}C_{10}$ (c) 0 (d) ${}^{20}C_{10}$

38. The value of ${}^{50}C_4 + \sum_{r=1}^{6} {}^{56-r}C_3$ is [AIEEE 2005; HOTS]

(a) ${}^{56}C_4$ (b) ${}^{56}C_3$ (c) ${}^{55}C_3$ (d) ${}^{55}C_4$

39. Expansion of $(1-x+x^2)^4$ is

(a) $1 - 4x + 10x^2 - 16x^3 + 19x^4 - 16x^5 + 10x^6 + 4x^7 - x^8$
(b) $1 - 4x + 10x^2 - 16x^3 + 19x^4 - 16x^5 + 10x^6 - 4x^7 + x^8$
(c) $1 - 4x + 16x^2 - 10x^3 + 19x^5 - 16x^4 + 10x^7 - 4x^6 + x^8$
(d) None of the above

40. **Statement I** $2^{4n+4} - 15n - 16$, where $n \in N$ is divisible by 225.

Statement II $2^{4n+4} - 15n - 16 = (1+15)^{n+1} - 15n - 16$ [NCERT Exemplar]

(a) Statement I is true; Statement II is true; Statement II is a correct explanation for Statement I
(b) Statement I is true; Statement II is true; Statement II is not a correct explanation for Statement I
(c) Statement I is true; Statement II is false
(d) Statement I is false; Statement II is true

41. Simplification of $(x^2 - \sqrt{1-x^2})^4 + (x^2 + \sqrt{1-x^2})^4$ is

(a) $2x^8 + 12x^6 - 14x^4 + 4x^2 - 2$
(b) $x^8 + 12x^5 - 14x^3 + 4x^2 + 2$
(c) $2x^8 - 12x^6 + 14x^4 - 4x^2 + 2$
(d) None of the above

42. The total number of terms in the expansion of $(x+a)^{51} - (x-a)^{51}$ after simplification is

(a) 102 (b) 25
(c) 26 (d) None of these

43. The greatest integer which divides the number $(101)^{100} - 1$, is

(a) 100
(b) 1000
(c) 10000
(d) 100000

44. The last digit in 7^{300} is

(a) 7 (b) 3
(c) 9 (d) 1

45. ${}^{10}C_1 + {}^{10}C_3 + {}^{10}C_5 + {}^{10}C_7 + {}^{10}C_9$ is equal to [HOTS]

(a) 2^9 (b) 2^{10}
(c) $2^{10} - 1$ (d) None of these

46. ${}^nC_0 - \frac{1}{2}\,{}^nC_1 + \frac{1}{3}\,{}^nC_2 - \dots\dots + (-1)^n \frac{{}^nC_n}{n+1}$ is equal to [HOTS]

(a) n (b) $\frac{1}{n-1}$ (c) $\frac{1}{n}$ (d) $\frac{1}{n+1}$

General and Middle Terms

Level I

47. In the binomial expansion for $(a+b)^n$, which of the following cannot be said true?

(a) The first term is ${}^nC_0\, a^n$
(b) The second term is ${}^nC_1\, a^{n-1}\, b$
(c) The third term is ${}^nC_2\, a^{n-2}\, b^2$
(d) r^{th} term is ${}^nC_r\, a^{n-r}\, b^r$

48. The general term of the expansion $(a+b)^n$ is

(a) $T_r = {}^nC_r\, a^{n-r}\, b^r$ (b) $T_{r+1} = {}^nC_r\, a^{n-r}\, b^r$
(c) $T_r = {}^nC_r\, a^{n+r}\, b^r$ (d) $T_{r+1} = {}^nC_r\, a^{n+r}\, b^r$

49. In the expansion of $(a+b)^n$, if n is even, then the middle term is said to be

(a) $\left(\frac{n}{2}\right)^{th}$ term
(b) $\left(\frac{n}{2}+1\right)^{th}$ term
(c) $\left(\frac{n+1}{2}\right)^{th}$ term
(d) $\left(\frac{n+1}{2}+1\right)^{th}$ term

50. Statement I In the expansion of $(x+2y)^8$, the middle term is 4th term.

Statement II If n is even in the expansion of $(a+b)^n$, then $\left(\frac{n}{2}+1\right)^{th}$ term is the middle term.

(a) Statement I is true; Statement II is true; Statement II is a correct explanation for Statement I
(b) Statement I is true; Statement II is true; Statement II is not a correct explanation for Statement I
(c) Statement I is true; Statement II is false
(d) Statement I is false; Statement II is true

51. In the expansion of $(a+b)^n$, if n is odd, then middle term(s) is/are

(a) $\left(\frac{n+1}{2}\right)^{th}$ term (b) $\left(\frac{n+1}{2}+1\right)^{th}$ term
(c) Both (a) and (b) (d) Neither (a) nor (b)

52. Statement I In the expansion of $(2x-y)^7$, the middle terms are 4th and 5th terms.

Statement II If n is odd, then there are two middle terms.

(a) Statement I is true; Statement II is true; Statement II is a correct explanation for Statement I
(b) Statement I is true; Statement II is true; Statement II is not a correct explanation for Statement I
(c) Statement I is true; Statement II is false
(d) Statement I is false; Statement II is true

53. In the expansion of $\left(x+\frac{1}{x}\right)^{2n}$, where $x \neq 0$, the middle term is

(a) n^{th} term (b) $(n+1)^{th}$ term
(c) $(n+2)^{th}$ term (d) $\frac{(n+1)^{th}}{2}$ term

54. Which of the following statement is/are true?

(a) In the expansion of $\left(x+\frac{1}{x}\right)^{2n}$, where $x \neq 0$, the middle term is ${}^{2n}C_n$
(b) In the expansion of $\left(x+\frac{1}{x}\right)^{2n}$, where $x \neq 0$, the term independent of x is ${}^{2n}C_n$
(c) Both (a) and (b) are true
(d) Both (a) and (b) are false

55. If 17th and 18th terms of the expansion $(2 + a)^{50}$ are equal, then the value of a is

(a) 0 (b) 1 (c) 18 (d) 91

56. If n is a positive integer, then the middle term in the expansion of $(1 + x)^{2n}$ is

(a) $\frac{2 \cdot 4 \cdot 6 .... 2n}{n!} 2n\, x^n$

(b) $\frac{1 \cdot 3 \cdot 5 .... (2n - 1)}{n!} x^n$

(c) $\frac{3 \cdot 5 \cdot 7 .... (2n + 1)}{n!} 2n\, x^n$

(d) $\frac{1 \cdot 3 \cdot 5 .... (2n - 1)}{n!} 2^n\, x^n$

57. Statement I The coefficient of $x^6 y^3$ in the expansion of $(x + 2y)^9$ is 670.

Statement II General term of the expansion $(x + 2y)^9$ is ${}^9C_r 2^r \cdot x^{9-r} \cdot y^r$.

(a) Statement I is true; Statement II is true; Statement II is a correct explanation for Statement I

(b) Statement I is true; Statement II is true; Statement II is not a correct explanation for Statement I

(c) Statement I is true; Statement II is false

(d) Statement I is false; Statement II is true

58. The second, third and fourth terms in the binomial expansion of $(x + a)^n$ are 240, 720 and 1080 respectively.

Based on the above information, which of the following is true?

(a) The value of x is 3 (b) The value of a is 2

(c) The value of n is 5 (d) None of these

59. If the coefficients of three consecutive terms in the expansion of $(1 + a)^n$ are in the ratio 1 : 7 : 42, then the value of n is

(a) 52 (b) 53

(c) 54 (d) None of these

60. The coefficient of x^5 in $(x + 3)^8$ is

(a) 1412 (b) 1312 (c) 1212 (d) 1512

61. Statement I The coefficient of $a^5\, b^7$ in $(a - 2b)^{12}$ is – 101376.

Statement II In expansion of $(a - 2b)^{12}$, general term is $T_{r+1} = {}^{12}C_r\, a^{12-r}\, b^r\, (-2)^r$.

(a) Statement I is true; Statement II is true; Statement II is a correct explanation for Statement I

(b) Statement I is true; Statement II is true; Statement II is not a correct explanation for Statement I

(c) Statement I is true; Statement II is false

(d) Statement I is false; Statement II is true

62. Which of the following statement(s) is/are true?

(a) The general term in the expansion of $(x^2 - y)^6$ is ${}^6C_r\, x^{10-2r}\, y^r\, (-1)^r$

(b) The general term in the expansion of $(x^2 - yx)^{12}$, $x \neq 0$ is ${}^{12}C_r\, (-1)^r\, x^{24-r}\, y^r$

(c) Both (a) and (b) are true

(d) Both (a) and (b) are false

63. The fourth term in the expansion of $(x - 2y)^{12}$ is

(a) $-176\, x^9 y^3$ (b) $-1860 x^9 y^3$

(c) $-176 x^3 y^9$ (d) None of these

64. The middle terms in the expansion of $\left(3 - \frac{x^3}{6}\right)^7$ are

I. $\frac{-105}{8} x^9$ II. $\frac{48}{35} x^{12}$

III. $\frac{35}{48} x^9$ IV. $\frac{35}{48} x^{12}$

(a) I and II (b) II and III

(c) III and IV (d) IV and I

65. Statement I The 13th term in the expansion of $\left(9x - \frac{1}{3\sqrt{x}}\right)^{18}$, $x \neq 0$ is 18564.

Statement II General term in the expansion of $\left(9x - \frac{1}{3\sqrt{x}}\right)^{18}$, $x \neq 0$ is ${}^{18}C_r\, (9x)^{18+r} \left(\frac{-1}{3\sqrt{x}}\right)^r$.

(a) Statement I is true; Statement II is true; Statement II is a correct explanation for Statement I

(b) Statement I is true; Statement II is true; Statement II is not a correct explanation for Statement I

(c) Statement I is true; Statement II is false

(d) Statement I is false; Statement II is true

66. In the expansion of $(1 + a)^{m+n}$

(a) Coefficient of $a^m = {}^{m+n}C_{m+2}$

(b) Coefficient of a^m >Coefficient of a^n

(c) Coefficient of a^m <Coefficient of a^n

(d) Coefficient of a^m = Coefficient of a^n

67. The middle term in the expansion of $\left(\frac{x}{3} + 9y\right)^{10}$ is

(a) $31236\, x^5\, y^5$ (b) $61236 x^5\, y^5$

(c) $31236\, x^6\, y^6$ (d) $61236\, x^6\, y^6$

68. The positive value of m for which the coefficient of x^2 in the expansion $(1 + x)^m$ is 6, is

(a) 1 (b) 2

(c) 3 (d) 4

69. Statement I The coefficient of x^n in the expansion of $(1+x)^{2n}$ is twice the coefficient of x^n in the expansion of $(1+x)^{2n-1}$.

Statement II Coefficient of x^n in $(1+x)^{2n}$ is $\dfrac{(2n-1)!}{n!(n-1)!}$.

(a) Statement I is true; Statement II is true; Statement II is a correct explanation for Statement I
(b) Statement I is true; Statement II is true; Statement II is not a correct explanation for Statement I
(c) Statement I is true; Statement II is false
(d) Statement I is false; Statement II is true

70. If the coefficients of the $(r-1)^{th}$, r^{th} and $(r+1)^{th}$ terms in the expansion of $(x+1)^n$ are in the ratio 1: 3: 5, then the values of n and r are respectively

(a) 7 and 3
(b) 7 and 4
(c) 4 and 7
(d) 3 and 7

Level II

71. The rth term in the expansion of $\left(x+\dfrac{1}{x}\right)^{2r}$ is [NCERT Exemplar]

(a) $\dfrac{2r!}{(r-1)!\,(r+1)!}x^2$ (b) $\dfrac{2r!}{(2r-1)!\,(2r+1)!}x^2$
(c) $\dfrac{2r!}{(2r+1)!}x^2$ (d) None of these

72. Statement I The fourth term from the end in the expansion of $\left(\dfrac{x^3}{2}-\dfrac{2}{x^2}\right)^9$ is $\dfrac{672}{x^3}$.

Statement II The rth term from the end in the expansion of $(a+b)^n$ is $(n-r+1)^{th}$ term from the beginning. [HOTS ; NCERT Exemplar]

(a) Statement I is true; Statement II is true; Statement II is a correct explanation for Statement I
(b) Statement I is true; Statement II is true; Statement II is not a correct explanation for Statement I
(c) Statement I is true; Statement II is false
(d) Statement I is false; Statement II is true

73. If p is a real number and if the middle term in the expansion of $\left(\dfrac{p}{2}+2\right)^8$ is 1120, then the values of p are

(a) ±3 (b) ±1
(c) ±2 (d) None of these

74. If n is even and the middle term in the expansion of $\left(x^2+\dfrac{1}{x}\right)^n$ is $924\,x^6$, then n is equal to

(a) 10 (b) 12
(c) 14 (d) None of these

75. The coefficient of x^{11} in the expansion of $\left(x^3-\dfrac{2}{x^2}\right)^{12}$ is [NCERT Exemplar]

(a) 21344 (b) –21344
(c) 25344 (d) –25344

76. Statement I The expansion of $\left(x^2-\dfrac{2}{x}\right)^{18}$ does not contain a term containing x^{10}.

Statement II $T_{r+1}={}^{18}C_r\,x^{36-2r}\,2^r(-1)^r$

(a) Statement I is true; Statement II is true; Statement II is a correct explanation for Statement I
(b) Statement I is true; Statement II is true; Statement II is not a correct explanation for Statement I
(c) Statement I is true; Statement II is false
(d) Statement I is false; Statement II is true

77. The term independent of x in the expansion of $\left(3x-\dfrac{2}{x^2}\right)^{15}$ is [NCERT Exemplar; HOTS]

(a) $-3003(3^{10})(2^5)$
(b) $3000(3)^{10}(2)^5$
(c) $-1003(3^{10})(2^5)$
(d) $1003(3^{10})(2^5)$

78. The middle term in the expansion of $\left(x+\dfrac{1}{2x}\right)^{2n}$ is [HOTS]

(a) $\dfrac{1\cdot3\cdot5....(2n-3)}{(n-1)!}$ (b) $\dfrac{1\cdot3\cdot5....(2n-1)}{(n+1)!}$
(c) $\dfrac{1\cdot3\cdot5....(2n-1)}{n!}$ (d) None of these

79. If the middle term of $\left(\dfrac{1}{x}+x\sin x\right)^{10}$ is equal to $7\dfrac{7}{8}$, then value of x is [NCERT Exemplar]

(a) $2n\pi+\dfrac{\pi}{6}$ (b) $n\pi+\dfrac{\pi}{6}$
(c) $n\pi+(-1)^n\dfrac{\pi}{6}$ (d) $n\pi+(-1)^n\dfrac{\pi}{3}$

80. The term independent of x in the expansion of $(1+x+2x^3)\left(\dfrac{3}{2}x^2-\dfrac{1}{3x}\right)^9$ is

(a) $\dfrac{1}{3}$ (b) $\dfrac{19}{54}$
(c) $\dfrac{17}{54}$ (d) $\dfrac{1}{14}$

81. **Statement I** The term independent of x in the expansion of $\left(\frac{\sqrt{x}}{\sqrt{3}}+\frac{\sqrt{3}}{2x^2}\right)^{10}$ is $\frac{5}{12}$.

Statement II 4th term is independent of x in expansion of $\left(\frac{\sqrt{x}}{\sqrt{3}}+\frac{\sqrt{3}}{2x^2}\right)^{10}$. [HOTS]

(a) Statement I is true; Statement II is true; Statement II is a correct explanation for Statement I
(b) Statement I is true; Statement II is true; Statement II is not a correct explanation for Statement I
(c) Statement I is true; Statement II is false
(d) Statement I is false; Statement II is true

82. The middle term in the expansion of $\left(2ax-\frac{b}{x^2}\right)^{12}$ is

(a) $\frac{31596a^6b^6}{x^8}$
(b) $\frac{31596a^6b^6}{x^6}$
(c) $\frac{59136a^6b^6}{x^8}$
(d) $\frac{59136a^6b^6}{x^6}$

83. The coefficient of x in the expansion of $(1-3x+7x^2)(1-x)^{16}$ is

(a) 17
(b) 19
(c) −17
(d) −19

84. If the term free from x in the expansion of $\left(\sqrt{x}-\frac{k}{x^2}\right)^{10}$ is 405, then the values of k are [NCERT Exemplar]

(a) ±1 (b) ±2
(c) ±3 (d) ±4

85. **Statement I** The middle terms in the expansion of $\left(\frac{p}{x}+\frac{x}{p}\right)^9$ are $\frac{126p}{x}$ and $\frac{126x}{p}$.

Statement II If n is odd, then there are two middle terms.

(a) Statement I is true; Statement II is true; Statement II is a correct explanation for Statement I
(b) Statement I is true; Statement II is true; Statement II is not a correct explanation for Statement I
(c) Statement I is true; Statement II is false
(d) Statement I is false; Statement II is true

86. Numerically the greatest term in the expansion of $(2+3x)^9$, where $x=\frac{3}{2}$, is [NCERT Exemplar; HOTS]

(a) $\frac{7}{2}\times 3^{11}$
(b) $\frac{7}{2}\times 3^{13}$
(c) $\frac{5}{2}\times 3^{11}$
(d) $\frac{5}{2}\times 3^{13}$

87. If n is a positive integer, then the coefficient of x^{-1} in the expansion of $(1+x)^n\left(1+\frac{1}{x}\right)^n$ is [NCERT Exemplar; HOTS]

(a) $\frac{n!}{(n-1)!(n+1)!}$ (b) $\frac{(2n)!}{(n-1)!(n+1)!}$
(c) $\frac{(3n)!}{(n-1)!(n+1)!}$ (d) None of these

88. The coefficient of x^{50} after simplifying and collecting the like terms in the expansion of $(1+x)^{1000}+x(1+x)^{999}+x^2(1+x)^{998}+....+x^{1000}$ is [NCERT Exemplar]

(a) $\frac{(1001)!}{(50)!(951)!}$
(b) $\frac{(101)!}{(50)!(49)!}$
(c) $\frac{(1001)!}{(150)!(851)!}$
(d) None of the above

89. It is given that a_1, a_2, a_3 and a_4 are the coefficients of any four consecutive terms in the expansion of $(1+x)^n$.

Based on above information, which among the following is correct? [HOTS; NCERT Exemplar]

(a) $\frac{a_1}{a_1+a_2}=\frac{r+1}{2(n-1)}$
(b) $\frac{a_3}{a_3+a_4}=\frac{2r+1}{n+1}$
(c) $\frac{a_1}{a_1+a_2}+\frac{a_3}{a_3+a_4}=\frac{2a_2}{a_2+a_3}$
(d) All of the above are correct

90. The largest term in the expansion of $(3+2x)^{50}$, where $x=\frac{1}{5}$, is

I. 5th II. 3rd
III. 7th IV. 6th

(a) Only I (b) Only II
(c) Both I and IV (d) Both III and IV

Miscellaneous Exercise

91. The term independent of x in the expansion of $\left(\frac{3}{2}x^2 - \frac{1}{3x}\right)^6$ is

(a) $\frac{12}{5}$ (b) $\frac{3}{5}$
(c) $\frac{5}{12}$ (d) $\frac{7}{12}$

92. If the coefficients of a^{r-1}, a^r and a^{r+1} in the expansion of $(1+a)^n$ are in arithmetic progression, then $n^2 - n(4r+1) + 4r^2 - 2$ is equal to

(a) 0 (b) 1
(c) 2 (d) 3

93. Which of the following is correct?

(a) The coefficient of the middle term in the expansion of $(1+x)^{2n}$ is ${}^{2n}C_{n+1}$
(b) The coefficients of two middle terms in the expansion of $(1+x)^{2n-1}$ are ${}^{2n}C_n$ and ${}^{2n-1}C_n$
(c) Sum of the coefficients of two middle terms in the expansion of $(1+x)^{2n-1}$ is equal to the coefficient of the middle term in the expansion of $(1+x)^{2n}$
(d) None of the above

94. Statement I Using binomial theorem, the coefficient of a^4 in the product $(1+2a)^4(2-a)^5$ is 438.

Statement II $(1+2a)^4(2-a)^5$

$$= (1+8a+24a^2+32a^3+16a^4)(32-80a+80a^2 - 40a^3 + 10a^4 - a^5)$$

(a) Statement I is true; Statement II is true; Statement II is a correct explanation for Statement I
(b) Statement I is true; Statement II is true; Statement II is not a correct explanation for Statement I
(c) Statement I is true; Statement II is false
(d) Statement I is false; Statement II is true

95. The term independent of x in the expansion of $\left(\sqrt[3]{x} + \frac{1}{2\sqrt[3]{x}}\right)^{18}$, $x > 0$, is [HOTS]

(a) ${}^{18}C_r \frac{1}{2^r}$ (b) ${}^{18}C_9 \frac{1}{2^9}$ (c) ${}^{18}C_9$ (d) ${}^{18}C_8 \frac{1}{3^9}$

96. If the coefficients of $(r-5)^{th}$ and $(2r-1)^{th}$ terms in the expansion of $(1+x)^{34}$ are equal, then the value of r is

(a) 12 (b) −4
(c) 13 (d) None of these

97. If the sum of the coefficients of the first three terms in the expansion of $\left(x - \frac{3}{x^2}\right)^m$, $x \neq 0$, m being a natural number, is 559, then the term of the expansion containing x^3 is [HOTS]

(a) $-3940x^3$ (b) $3940x^3$
(c) $-5940x^3$ (d) $5940x^3$

98. Statement I The rth term from the end in the expansion of $(x+a)^n$ is ${}^nC_{n-r+1}\, x^{r-1} \cdot a^{n-r+1}$.

Statement II The rth term from the end in the expansion of $(x+a)^n$ is $(n-r+2)^{th}$ term.

(a) Statement I is true; Statement II is true; Statement II is a correct explanation for Statement I
(b) Statement I is true; Statement II is true; Statement II is not a correct explanation for Statement I
(c) Statement I is true; Statement II is false
(d) Statement I is false; Statement II is true

99. If the coefficients of x^2 and x^3 in the expansion of $(3+ax)^9$ are equal, then the value of a is

(a) $\frac{7}{9}$ (b) $\frac{7}{8}$ (c) $\frac{9}{7}$ (d) $\frac{9}{8}$

100. If the first three terms of the expansion $(a+b)^n$ are 729, 7290 and 30375 respectively, then [HOTS]

(a) $a=3$ (b) $b=6$
(c) $n=5$ (d) None of these

101. By using the binomial theorem, the coefficient of x^5 in the product $(1+2x)^6(1-x)^7$ is

(a) 168 (b) 151
(c) 141 (d) 171

102. Statement I If a and b are distinct integers, then $(a-b)$ is a factor of $a^n - b^n$, whenever n is a positive integer.

Statement II $a^n - b^n = (a-b)[{}^nC_o(a-b)^{n-1} + {}^nC_1 (a-b)^{n-2}b + \ldots]$

(a) Statement I is true; Statement II is true; Statement II is a correct explanation for Statement I
(b) Statement I is true; Statement II is true; Statement II is not a correct explanation for Statement I
(c) Statement I is true; Statement II is false
(d) Statement I is false; Statement II is true

103. The value of $(\sqrt{3}+\sqrt{2})^6 - (\sqrt{3}-\sqrt{2})^6$ is equal to

(a) $386\sqrt{6}$ (b) $396\sqrt{6}$
(c) $196\sqrt{6}$ (d) None of these

104. Using binomial theorem, the expansion of $\left(1+\frac{x}{2}-\frac{2}{x}\right)^4$, $x \neq 0$, is

(a) $\frac{x^4}{16}+\frac{x^3}{2}+\frac{x^2}{2}-4x+\frac{16}{x}+\frac{8}{x^2}+\frac{32}{x^3}-\frac{16}{x^4}$

(b) $\frac{x^4}{16}+\frac{x^3}{2}+\frac{x^2}{2}-4x-5+\frac{16}{x}+8x^2+32x^3-16x^4$

(c) $\frac{x^4}{16}+\frac{x^3}{2}+\frac{x^2}{2}-4x-5+\frac{16}{x}+\frac{8}{x^2}-\frac{32}{x^3}+\frac{16}{x^4}$

(d) None of the above

105. The value of $(a^2+\sqrt{a^2-1})^4+(a^2-\sqrt{a^2-1})^4$ is equal to

(a) $2a^8+12a^6-10a^4-4a^2+2$

(b) $a^8+12a^6+10a^4+4a^2-2$

(c) $2a^{10}+12a^4+10a^6+4a^2+2$

(d) None of the above

106. If the ratio of the fifth term from the beginning to the fifth term from the end in the expansion of $\left(\sqrt[4]{2}+\frac{1}{\sqrt[4]{3}}\right)^n$ is $\sqrt{6}:1$, then

Statement I The value of n is 10.

Statement II $\frac{2^{\frac{n-4}{4}}\cdot 3^{-1}}{2\cdot 3^{\frac{4+n}{4}}}=\sqrt{6}$ [HOTS]

(a) Statement I is true; Statement II is true; Statement II is a correct explanation for Statement I

(b) Statement I is true; Statement II is true; Statement II is not a correct explanation for Statement I

(c) Statement I is true; Statement II is false

(d) Statement I is false; Statement II is true

107. The approximation of $(0.99)^5$ using the first three terms of its expansion is

(a) 0.851

(b) 0.751

(c) 0.951

(d) None of these

108. The expansion of $(3x^2-2ax+3a^2)^3$ by using binomial theorem is

(a) $37x^6-74ax^5+117a^2x^4-116a^3x^3+117a^4x^2-54a^5x+27a^6$

(b) $27x^6-54ax^5+117a^2x^4-116a^3x^3+117a^4x^2-54a^5x+27a^6$

(c) $37x^6+54ax^5-117a^2x^4+116a^3x^3+117a^2x^4-54a^2x^5+27a^6$

(d) None of the above

109. The term independent of x in the expansion of $\left(\frac{3x^2}{2}-\frac{1}{3x}\right)^{15}$, $x \neq 0$, is

[NCERT Exemplar; HOTS]

(a) 6th term (b) 8th term

(c) 10th term (d) None of these

110. In the expansion of $(x+a)^n$, if the sum of odd terms is denoted by O and the sum of even terms by E, then

Statement I $O^2-E^2=(x^2+a^2)^n$

Statement II $4OE=(x+a)^{2n}-(x-a)^{2n}$

[NCERT Exemplar]

(a) Statement I is true; Statement II is true; Statement II is a correct explanation for Statement I

(b) Statement I is true; Statement II is true; Statement II is not a correct explanation for Statement I

(c) Statement I is true; Statement II is false

(d) Statement I is false; Statement II is true

Answer *with* Explanations

1. *(b)* In earlier classes, we have learnt how to find the squares and cubes of binomials like $a + b$ and $a - b$. Using them, we can evaluate the numerical values of numbers like $(98)^2 = (100 - 2)^2$, $(999)^3 = (1000 - 1)^3$, etc. However, for higher powers like $(98)^5$, $(101)^6$, etc., the calculations become difficult by using repeated multiplication. This difficulty was overcome by a theorem known as binomial theorem. It gives an easier way to expand $(a + b)^n$, where n is an integer or a rational number.

2. *(c)* Considering the given expansions as below

$(a + b)^0 = 1, a + b \neq 0$

$(a + b)^1 = a + b$

$(a + b)^2 = a^2 + 2ab + b^2$

$(a + b)^3 = a^3 + 3a^2b + 3ab^2 + b^3$

$(a + b)^4 = (a + b)^3(a + b) = a^4 + 4a^3b + 6a^2b^2 + 4ab^3 + b^4$

In these expansions, we observe that

(i) The total number of terms in the expansion is one more than the index. For example, in the expansion of $(a + b)^2$, number of terms is 3 whereas, the index of $(a + b)^2$ is 2.

(ii) Powers of the first quantity 'a', decreasing by 1 whereas the powers of the second quantity 'b' increases by 1, in the successive terms.

(iii) In each term of the expansion, the sum of the indices of a and b is the same and is equal to the index of $a + b$.

3. *(a)*

Index	Coefficients
0	1
1	1 1
2	1 2 1
3	1 3 3 1
4	1 4 6 4 1

We observe a pattern in this table that helps us to write the next row. It can be seen that the addition of 1's in the row for index 1 gives rise to 2 in the row for index 2. The addition of 1, 2 and 2, 1 in the row for index 2, gives rise to 3 and 3 in the row for index 3 and so on. Also, 1 is present at the beginning and at the end of each row. This can be continued till any index of our interest.

So, option (a) is correct.

Solutions (Q. Nos. 4-6)

In the expansion of $(a + b)^n$, we get the following pattern (using index upto 4) between the index and coefficients.

Index	Coefficients
0	1
1	1 1
2	1 2 1
3	1 3 3 1
4	1 4 6 4 1

The structure given in above figure looks like a triangle with 1 at the top vertex and running down the two slanting sides. This array of numbers is known as Pascal's triangle, after the name of French mathematician Blaise Pascal. It is also known as Meru Prastara by Pingla.

Expansions for the higher powers of a binomial are also possible by using Pascal's triangle. Let us expand $(2x + 3y)^5$ by using Pascal's triangle. The row for index 5 is

1 5 10 10 5 1

Using the result of solution 2, we get

$$(2x + 3y)^5 = (2x)^5 + 5(2x)^4(3y) + 10(2x)^3(3y)^2 + 10(2x)^2(3y)^3 + 5(2x)(3y)^4 + (3y)^5$$

$$= 32x^5 + 240x^4y + 720x^3y^2 + 1080x^2y^3 + 810xy^4 + 243y^5.$$

Now, if we want to find the expansion of $(2x + 3y)^{12}$, we are first required to get the row for index 12. This can be done by writing all the rows of the Pascal's triangle till index 12. This is a slightly lengthy process. The process, as you observe, will become more difficult, if we need the expansions involving still larger powers.

We thus try to find a rule that will help us to find the expansion of the binomial for any power without writing all the rows of the Pascal's triangle, that come before the row of the desired index.

For this, we make use of the concept of combination studied earlier to rewrite the numbers in the Pascal's triangle.

We know that, ${}^nC_r = \dfrac{n!}{r!(n-r)!}$, $0 \leq r \leq n$ and n is a non-negative integer. Also, ${}^nC_0 = 1 = {}^nC_n$.

The Pascal's triangle can now be rewritten as shown in figure.

Index	Coefficients
0	${}^0C_0$ (= 1)
1	${}^1C_0$ (= 1) ${}^1C_1$ (= 1)
2	${}^2C_0$ (= 1) ${}^2C_1$ (= 2) ${}^2C_2$ (= 1)
3	${}^3C_0$ (= 1) ${}^3C_1$ (= 3) ${}^3C_2$ (= 3) ${}^3C_3$ (= 1)
4	${}^4C_0$ (= 1) ${}^4C_1$ (= 4) ${}^4C_2$ (= 6) ${}^4C_3$ (= 4) ${}^4C_4$ (= 1)
5	${}^5C_0$ (= 1) ${}^5C_1$ (= 5) ${}^5C_2$ (= 10) ${}^5C_3$ (= 10) ${}^5C_4$ (= 5) ${}^5C_5$ (= 1)

Pascal's triangle

So, from the obtained values in figure, we get all the coefficients for required index.

4. *(c)*, **5.** *(b)*, **6.** *(c)*

7. *(c)* Observing the pattern discussed in previous problem, we can write the row of the Pascal's triangle for any index without writing the earlier rows. *e.g.*, For the index 7, the row would be

$${}^7C_0 \quad {}^7C_1 \quad {}^7C_2 \quad {}^7C_3 \quad {}^7C_4 \quad {}^7C_5 \quad {}^7C_6 \quad {}^7C_7$$

Thus, using this row and the results of solution 2, we get

$$(a+b)^7 = {}^7C_0a^7 + {}^7C_1a^6b + {}^7C_2a^5b^2 + {}^7C_3a^4b^3 + {}^7C_4a^3b^4 + {}^7C_5a^2b^5 + {}^7C_6ab^6 + {}^7C_7b^7$$

An expansion of a binomial to any positive integral index say n can now be visualised using these observations. We can now write the expansion of a binomial to any positive integral index.

So, option (c) is correct.

8. *(c)* The proof of this result is given below

Let the given statement be

$$P(n) : (a+b)^n = {}^nC_0a^n + {}^nC_1a^{n-1}b + {}^nC_2a^{n-2}b^2 + \dots + {}^nC_{n-1}a\cdot b^{n-1} + {}^nC_nb^n$$

For $n = 1$, we have

$$P(1) : (a+b)^1 = {}^1C_0a^1 + {}^1C_1b^1 = a + b$$

Thus, P (1) is true.

Suppose P (k) is true for some positive integer k, *i.e.*,

$$(a+b)^k = {}^kC_0a^k + {}^kC_1a^{k-1}b + {}^kC_2a^{k-2}b^2 + \dots + {}^kC_kb^k \quad \dots\text{(i)}$$

We shall prove that $P(k+1)$ is also true, *i.e.*,

$$(a+b)^{k+1} = {}^{k+1}C_0a^{k+1} + {}^{k+1}C_1a^kb + {}^{k+1}C_2a^{k-1}b^2 + \dots + {}^{k+1}C_{k+1}b^{k+1}$$

Now, $(a+b)^{k+1} = (a+b)(a+b)^k$

$$= (a+b)({}^kC_0a^k + {}^kC_1a^{k-1}b + {}^kC_2a^{k-2}b^2 + \dots + {}^kC_{k-1}ab^{k-1} + {}^kC_kb^k) \text{ [from Eq. (i)]}$$

$$= {}^kC_0a^{k+1} + {}^kC_1a^kb + {}^kC_2a^{k-1}b^2 + \dots + {}^kC_{k-1}a^2b^{k-1} + {}^kC_kab^k + {}^kC_0a^kb + {}^kC_1a^{k-1}b^2 + {}^kC_2a^{k-2}b^3 + \dots + {}^kC_{k-1}ab^k + {}^kC_kb^{k+1} \text{ (by actual multiplication)}$$

$$= {}^kC_0a^{k+1} + ({}^kC_1 + {}^kC_0)a^kb + ({}^kC_2 + {}^kC_1)a^{k-1}b^2 + \dots + ({}^kC_k + {}^kC_{k-1})ab^k + {}^kC_kb^{k+1} \text{ (grouping like terms)}$$

$$= {}^{k+1}C_0a^{k+1} + {}^{k+1}C_1a^kb + {}^{k+1}C_2a^{k-1}b^2 + \dots + {}^{k+1}C_kab^k + {}^{k+1}C_{k+1}b^{k+1}$$

$$\left(\begin{array}{l}\text{using } {}^{k+1}C_0 = 1, {}^kC_r + {}^kC_{r-1} = {}^{k+1}C_r \\ \text{and } {}^kC_k = 1 = {}^{k+1}C_{k+1}\end{array}\right)$$

Thus, it has been proved that $P(k+1)$ is true whenever $P(k)$ is true. Therefore, by principle of mathematical induction, $P(n)$ is true for every positive integer n.

Hence,

$$(a+b)^n = {}^nC_0a^n + {}^nC_1a^{n-1}b + \dots + {}^nC_{n-1}ab^{n-1} + {}^nC_nb^n$$

9. *(c)* By using binomial theorem, we have

$$(x+2)^6 = {}^6C_0x^6 + {}^6C_1x^5\cdot 2 + {}^6C_2x^42^2 + {}^6C_3x^3\cdot 2^3 + {}^6C_4x^2\cdot 2^4 + {}^6C_5x\cdot 2^5 + {}^6C_6\cdot 2^6$$

$$= x^6 + 12x^5 + 60x^4 + 160x^3 + 240x^2 + 192x + 64$$

Thus,

$$(x+2)^6 = x^6 + 12x^5 + 60x^4 + 160x^3 + 240x^2 + 192x + 64$$

10. *(b)* The notation $\sum_{k=0}^{n} {}^nC_ka^{n-k}b^k$ stands for

$${}^nC_0a^nb^0 + {}^nC_1a^{n-1}b^1 + \dots + {}^nC_ra^{n-r}b^r + \dots + {}^nC_na^{n-n}b^n$$

where, $b^0 = 1 = a^{n-n}$.

Hence, the notation form of $(a+b)^n$ is

$$(a+b)^n = \sum_{k=0}^{n} {}^nC_ka^{n-k}b^k.$$

11. *(c)* (a) The coefficients ${}^nC_r$ occuring in the binomial theorem are known as binomial coefficients.

(b) There are $(n+1)$ terms in the expansion of $(a+b)^n$, *i.e.*, one more than the index. So, both (a) and (b) are true.

12. *(a)* In the successive terms of the expansion the index of a goes on decreasing by unity. It is n in the first term, $(n-1)$ in the second term, and so on ending with zero in the last term. At the same time the index of b increases by unity, starting with zero in the first term, 1 in the second and so on ending with n in the last term.

So, only option (a) is true.

13. *(a)* In the expansion of $(a+b)^n$, the sum of the indices of a and b is $n + 0 = n$ in the first term, $(n-1)+1 = n$ in the second term and so on,

Thus, it can be seen that the sum of the indices of a and b is n in every term of the expansion.

14. *(a)* In the expansion of $(a+b)^n$,

Taking $a = x$ and $b = -y$, we obtain

$$(x-y)^n = [x+(-y)]^n$$
$$= {}^nC_0x^n + {}^nC_1x^{n-1}(-y) + {}^nC_2x^{n-2}(-y)^2 + {}^nC_3x^{n-3}(-y)^3 + \ldots + {}^nC_n(-y)^n$$
$$= {}^nC_0x^n - {}^nC_1x^{n-1}y + {}^nC_2x^{n-2}y^2 - {}^nC_3x^{n-3}y^3 + \ldots + (-1)^n\,{}^nC_ny^n$$

Thus, $(x-y)^n = {}^nC_0x^n - {}^nC_1x^{n-1}y + {}^nC_2x^{n-2}y^2 + \ldots + (-1)^n\,{}^nC_ny^n$

Using this, we have

$$(x-2y)^5 = {}^5C_0x^5 - {}^5C_1x^4(2y) + {}^5C_2x^3(2y)^2 - {}^5C_3x^2(2y)^3 + {}^5C_4x(2y)^4 - {}^5C_5(2y)^5$$
$$= x^5 - 10x^4y + 40x^3y^2 - 80x^2y^3 + 80xy^4 - 32y^5$$

So, both the statements are true and Statement II is the correct explanation of Statement I.

15. *(a)* In the expansion of $(a+b)^n$,

On taking $a = 1$, $b = x$, we obtain

$$(1+x)^n = {}^nC_0(1)^n + {}^nC_1(1)^{n-1}x + {}^nC_2(1)^{n-2}x^2 + \ldots + {}^nC_nx^n$$
$$= {}^nC_0 + {}^nC_1x + {}^nC_2x^2 + {}^nC_3x^3 + \ldots + {}^nC_nx^n$$

Thus, $(1+x)^n = {}^nC_0 + {}^nC_1x + {}^nC_2x^2 + {}^nC_3x^3 + \ldots + {}^nC_nx^n$

In particular, for $x = 1$, we have

$$2^n = {}^nC_0 + {}^nC_1 + {}^nC_2 + \ldots + {}^nC_n$$

So, both the statements are true and Statement II is the correct explanation of Statement I.

16. *(b)* In the expansion of $(a+b)^n$,

On taking $a = 1$, $b = -x$, we obtain

$$(1-x)^n = {}^nC_0 - {}^nC_1x + {}^nC_2x^2 - \ldots + (-1)^n\,{}^nC_nx^n$$

In particular, for $x = 1$, we get

$$0 = {}^nC_0 - {}^nC_1 + {}^nC_2 - \ldots + (-1)^n\,{}^nC_n$$
$$\Rightarrow {}^nC_0 + {}^nC_2 + {}^nC_4 + \ldots = {}^nC_1 + {}^nC_3 + {}^nC_5 + \ldots$$

17. *(c)* By using binomial theorem, we have

$$\left(x^2+\frac{3}{x}\right)^4 = {}^4C_0(x^2)^4 + {}^4C_1(x^2)^3\left(\frac{3}{x}\right) + {}^4C_2(x^2)^2\left(\frac{3}{x}\right)^2 + {}^4C_3(x^2)\left(\frac{3}{x}\right)^3 + {}^4C_4\left(\frac{3}{x}\right)^4$$
$$= x^8 + 4\cdot x^6\cdot\frac{3}{x} + 6\cdot x^4\cdot\frac{9}{x^2} + 4\cdot x^2\cdot\frac{27}{x^3} + \frac{81}{x^4}$$
$$= x^8 + 12x^5 + 54x^2 + \frac{108}{x} + \frac{81}{x^4}.$$

18. *(b)* We express 98 as the sum or difference of two numbers whose powers are easier to calculate and then use binomial theorem.

98 can be written as 100 – 2

Therefore, $(98)^5 = (100-2)^5$

$$= {}^5C_0(100)^5 - {}^5C_1(100)^4\cdot 2 + {}^5C_2(100)^3 2^2 - {}^5C_3(100)^2(2)^3 + {}^5C_4(100)(2)^4 - {}^5C_5(2)^5$$
$$= 10000000000 - 5\times 100000000\times 2 + 10\times 1000000\times 4 - 10\times 10000\times 8 + 5\times 100\times 16 - 32$$
$$= 10040008000 - 1000800032 = 9039207968.$$

19. *(a)* Splitting 1.01 and using binomial theorem to write the first few terms, we have

$$(1.01)^{1000000} = (1+0.01)^{1000000}$$
$$= {}^{1000000}C_0 + {}^{1000000}C_1(0.01) + \text{Other positive terms}$$
$$= 1 + 1000000\times 0.01 + \text{Other positive terms}$$
$$= 1 + 10000 + \text{Other positive terms} > 10000$$

Hence, $(1.01)^{1000000} > 10000$.

20. *(d)* For two numbers a and b, if we can find numbers q and r such that $a = bq + r$, then we say that b divides a with q as quotient and r as remainder. Thus, in order to show that $6^n - 5n$ leaves remainder 1 when divided by 25, we prove that $6^n - 5n = 25k + 1$, where k is some natural number.

We have,

$$(1+a)^n = {}^nC_0 + {}^nC_1a + {}^nC_2a^2 + \ldots + {}^nC_na^n$$

For $a = 5$, we get

$$(1+5)^n = {}^nC_0 + {}^nC_15 + {}^nC_25^2 + \ldots + {}^nC_n5^n$$

i.e., $(6)^n = 1 + 5n + 5^2\cdot{}^nC_2 + 5^3\cdot{}^nC_3 + \ldots + 5^n$

i.e., $6^n - 5n = 1 + 5^2({}^nC_2 + {}^nC_35 + \ldots + 5^{n-2})$

or $6^n - 5n = 1 + 25({}^nC_2 + 5\cdot{}^nC_3 + \ldots + 5^{n-2})$

or $6^n - 5n = 25k + 1$, where $k = {}^nC_2 + 5\cdot{}^nC_3 + \ldots + 5^{n-2}$

This shows that when divided by 25, $6^n - 5n$ leaves remainder 1.

21. *(b)* $$\left(\frac{2}{x}-\frac{x}{2}\right)^5 = {}^5C_0\left(\frac{2}{x}\right)^5 - {}^5C_1\left(\frac{2}{x}\right)^4\frac{x}{2} + {}^5C_2\left(\frac{2}{x}\right)^3\left(\frac{x}{2}\right)^2 - {}^5C_3\left(\frac{2}{x}\right)^2\left(\frac{x}{2}\right)^3 + {}^5C_4\left(\frac{2}{x}\right)\left(\frac{x}{2}\right)^4 - {}^5C_5\left(\frac{x}{2}\right)^5$$

$$= {}^5C_0\frac{32}{x^5} - {}^5C_1\frac{16}{x^4}\times\frac{x}{2} + {}^5C_2\frac{8}{x^3}\times\frac{x^2}{4} - {}^5C_2\frac{4}{x^2}\times\frac{x^3}{8} + {}^5C_1\frac{2}{x}\times\frac{x^4}{16} - {}^5C_0\frac{x^5}{32} \quad (\because {}^nC_r = {}^nC_{n-r})$$

$$= \frac{32}{x^5} - \frac{5\times 16\times x}{2x^4} + \frac{5\times 4}{2}\times\frac{8}{x^3}\times\frac{x^2}{4} - \frac{5\times 4}{2}\times\frac{4}{x^2}\times\frac{x^3}{8} + 5\times\frac{2}{x}\times\frac{x^4}{16} - \frac{x^5}{32}$$

$$= \frac{32}{x^5} - \frac{40}{x^3} + \frac{20}{x} - 5x + \frac{5x^3}{8} - \frac{x^5}{32}$$

22. (c) A. $(1-2x)^5 = {}^5C_0\,(1)^5 - {}^5C_1\,(1)^4\,2x + {}^5C_2\,(1)^3\,(2x)^2 - {}^5C_3\,(1)^2\,(2x)^3 + {}^5C_4\,(1)\,(2x)^4 - {}^5C_5(2x)^5$

$= {}^5C_0 - {}^5C_1\,2x + {}^5C_2\,4x^2 - {}^5C_2\,8x^3 + {}^5C_1\,16x^4 - {}^5C_0\,32x^5 \quad (\because {}^nC_r = {}^nC_{n-r})$

$= 1 - 5\times 2x + \dfrac{5\times 4}{2}\times 4x^2 - \dfrac{5\times 4}{2}\times 8x^3 + 5\times 16\times x^4 - 32x^5$

$= 1 - 10x + 40x^2 - 80x^3 + 80x^4 - 32x^5$

B. $(2x-3)^6 = {}^6C_0\,(2x)^6 - {}^6C_1(2x)^5\times 3 + {}^6C_2\,(2x)^4\,(3)^2 - {}^6C_3\,(2x)^3\,(3)^3 + {}^6C_4\,(2x)^2\,(3)^4 - {}^6C_5\,(2x)\,3^5 + {}^6C_6\,3^6$

$= {}^6C_0\,64x^6 - {}^6C_1\,32x^5\times 3 + {}^6C_2\,16x^4\times 9 - {}^6C_3\,8x^3\times 27 + {}^6C_2\,4x^2\times 81 - {}^6C_1\,2x\times 243 + {}^6C_0\,729 (\because {}^nC_r = {}^nC_{n-r})$

$= 64x^6 - 6\times 32\times 3\times x^5 + \dfrac{6\times 5}{2}\times 16x^4\times 9 - \dfrac{6\times 5\times 4\times 8x^3\times 27}{6} + \dfrac{6\times 5}{2}\times 4x^2\times 81 - 6\times 2x\times 243 + 729$

$= 64x^6 - 576x^5 + 2160x^4 - 4320x^3 + 4860x^2 - 2916x + 729$

C. $\left(\dfrac{x}{3}+\dfrac{1}{x}\right)^5 = {}^5C_0\left(\dfrac{x}{3}\right)^5 + {}^5C_1\left(\dfrac{x}{3}\right)^4\dfrac{1}{x} + {}^5C_2\left(\dfrac{x}{3}\right)^3\left(\dfrac{1}{x}\right)^2 + {}^5C_3\left(\dfrac{x}{3}\right)^2\left(\dfrac{1}{x}\right)^3 + {}^5C_4\left(\dfrac{x}{3}\right)\left(\dfrac{1}{x}\right)^4 + {}^5C_5\left(\dfrac{1}{x}\right)^5$

$= {}^5C_0\,\dfrac{x^5}{243} + {}^5C_1\,\dfrac{x^4}{81}\times\dfrac{1}{x} + {}^5C_2\times\dfrac{x^3}{27}\times\dfrac{1}{x^2} + {}^5C_2\,\dfrac{x^2}{9}\times\dfrac{1}{x^3} + ({}^5C_1)\,\dfrac{x}{3}\times\dfrac{1}{x^4} + ({}^5C_0)\times\dfrac{1}{x^5} \quad (\because {}^nC_r = {}^nC_{n-r})$

$= \dfrac{x^5}{243} + \dfrac{5}{81}\times x^3 + \dfrac{5\times 4}{2}\times\dfrac{1}{27}\times x + \dfrac{5\times 4}{2}\times\dfrac{1}{9}\times\dfrac{1}{x} + 5\times\dfrac{1}{3}\times\dfrac{1}{x^3} + \dfrac{1}{x^5}$

$= \dfrac{x^5}{243} + \dfrac{5x^3}{81} + \dfrac{10x}{27} + \dfrac{10}{9x} + \dfrac{5}{3x^3} + \dfrac{1}{x^5}$

D. $\left(x+\dfrac{1}{x}\right)^6 = {}^6C_0\,x^6 + {}^6C_1\,x^5\left(\dfrac{1}{x}\right) + {}^6C_2x^4\left(\dfrac{1}{x}\right)^2 + {}^6C_3x^3\left(\dfrac{1}{x}\right)^3 + {}^6C_4x^2\left(\dfrac{1}{x}\right)^4 + {}^6C_5\,(x)\left(\dfrac{1}{x}\right)^5 + {}^6C_6\left(\dfrac{1}{x}\right)^6$

$= {}^6C_0\,x^6 + {}^6C_1x^4 + {}^6C_2\,\dfrac{x^4}{x^2} + {}^6C_3\,x^3\times\dfrac{1}{x^3} + {}^6C_2\,x^2\times\dfrac{1}{x^4} + {}^6C_1\,x\times\dfrac{1}{x^5} + {}^6C_0\,\dfrac{1}{x^6} (\because {}^nC_r = {}^nC_{n-r})$

$= x^6 + 6x^4 + \dfrac{6\times 5}{2}\,x^2 + \dfrac{6\times 5\times 4}{6}\times 1 + \dfrac{6\times 5}{2}\,\dfrac{1}{x^2} + 6\times\dfrac{1}{x^4} + \dfrac{1}{x^6}$

$= x^6 + 6x^4 + 15x^2 + 20 + \dfrac{15}{x^2} + \dfrac{6}{x^4} + \dfrac{1}{x^6}$

$\therefore$ A $\leftrightarrow$ 3, B $\leftrightarrow$ 4, C $\leftrightarrow$ 1, D $\leftrightarrow$ 2

23. (c) $(1\cdot 1)^{10000} = (1 + 0\cdot 1)^{10000}$

$= {}^{10000}C_0\,(1)^{10000} + {}^{10000}C_1\,(1)^{9999}(0\cdot 1) + \ldots + \text{Other terms}$

$= 1\times 1 + 10000\times 0\cdot 1 + \ldots \text{Other terms} \quad (\because {}^nC_0 = 1, {}^nC_1 = n)$

$= 1 + 1000 + \text{Other terms}$

$= (1001 + \text{Other terms}) > 1000 \Rightarrow (1\cdot 1)^{10000} > 1000$

24. (c) Now, $(a+b)^4 = {}^4C_0\,a^4 + {}^4C_1\,a^3b + {}^4C_2\,a^2b^2 + {}^4C_3\,ab^3 + {}^4C_4\,b^4$

$= {}^4C_0a^4 + {}^4C_1\,a^3b + {}^4C_2a^2b^2 + {}^4C_1ab^3 + {}^4C_0\,b^4 \quad (\because {}^nC_r = {}^nC_{n-r})$

$= 1\times a^4 + 4\,a^3b + \dfrac{4\times 3}{2}\,a^2b^2 + 4ab^3 + 1\times b^4$

$\Rightarrow (a+b)^4 = a^4 + 4a^3b + 6a^2b^2 + 4ab^3 + b^4 \quad \ldots(\text{i})$

Similarly, $(a-b)^4 = a^4 - 4a^3b + 6a^2b^2 - 4ab^3 + b^4 \quad \ldots(\text{ii})$

Subtracting Eq. (ii) from Eq. (i), we get

$(a+b)^4 - (a-b)^4 = 8a^3b + 8ab^3 = 8\,ab\,(a^2+b^2)$

Now, putting $a = \sqrt{3}$ and $b = \sqrt{2}$

$(\sqrt{3}+\sqrt{2})^4 - (\sqrt{3}-\sqrt{2})^4 = 8\sqrt{3}\sqrt{2}\,[(\sqrt{3})^2 + (\sqrt{2})^2]$

$= 8\sqrt{6}\,(3+2) = 8\sqrt{6}\times 5 = 40\sqrt{6}$

25. (b) A. $(96)^3 = (100-4)^3 = {}^3C_0\,(100)^3 - {}^3C_1(100)^2\,4 + {}^3C_2(100)(4)^2 - {}^3C_3\,(4)^3$

$= {}^3C_0(10^6) - {}^3C_1(10)^4\times 4 + ({}^3C_1)\,10^2\times 16 - ({}^3C_0)\,64 \quad (\because {}^nC_r = {}^nC_{n-r})$

$= 10^6 - 3\times 10^4\times 4 + 3\times 10^2\times 16 - 64$

$= 10^6 - 12\times 10^4 + 48\times 10^2 - 64$

$= (10^6 + 48\times 10^2) - (12\times 10^4 + 64)$

$= (10^6 + 4800) - (120000 + 64) = 1004800 - 120064$

$= 884736$

B. $(102)^5 = (100+2)^5$

$= {}^5C_0\,(100)^5 + {}^5C_1\,(100)^4(2)^1 + {}^5C_2\,(100)^3\,(2)^2 + {}^5C_3\,(100)^2\,(2)^3 + {}^5C_4\,(100)\,(2)^4 + {}^5C_5\,(2)^5$

$= {}^5C_0(10)^{10} + {}^5C_1(10^8)\,2 + {}^5C_2(10^6)\times 4 + ({}^5C_2)\,10^4\times 8 + ({}^5C_1)\,100\times 16 + ({}^5C_0)\times 32 \quad (\because {}^nC_r = {}^nC_{n-r})$

$= 1\times 10^{10} + 5\times 10^8\times 2 + \dfrac{5\times 4}{2}\times 10^6\times 4 + \dfrac{5\times 4}{2}\times 10^4\times 8 + 5\times 100\times 16 + 32$

$= 10000000000 + 1000000000 + 40000000 + 800000 + 8000 + 32$

$= 11040808032$

C. $(100 + 1)^4 = {}^4C_0 (100)^4 + {}^4C_1(100)^3 (1) + {}^4C_2(100)^2 \cdot (1)^2 + {}^4C_3 100 \cdot (1)^3 + {}^4C_4 \cdot (1)^4$

$= {}^4C_0 10^8 + {}^4C_1 10^6 + {}^4C_2 10^4 + {}^4C_1 (100) + {}^4C_0$

$= 10^8 + 4 \times 10^6 + \frac{4 \times 3}{2} \times 10^4 + 4 \times 10^2 + 1$ $(\because {}^nC_r = {}^nC_{n-r})$

$= 100000000 + 4000000 + 60000 + 400 + 1$

$= 104060401$

D. $(99)^5 = (100 - 1)^5$

$= {}^5C_0 (100)^5 - {}^5C_1 (100)^4 \times 1 + {}^5C_2 (100)^3 \times (1)^2 - {}^5C_3 (100)^2 \times 1^3 + {}^5C_4 100 \times 1^4 - {}^5C_5 \times 1^5$

$= {}^5C_0 (10)^{10} - {}^5C_1 (10)^8 + {}^5C_2 (10)^6 - {}^5C_2 (10)^4 + {}^5C_1 \times (10)^2 - {}^5C_0$ $(\because {}^nC_r = {}^nC_{n-r})$

$= 10000000000 - 5 \times 100000000 + \frac{5 \times 4}{2} \times 1000000 - \frac{5 \times 4}{2} \times 10000 + 5 \times 100 - 1$

$= (10000000000 + 10000000 + 500) - (500000000 + 100000 + 1)$

$= 10010000500 - 500100001$

$= 9509900499$

$\therefore$ $A \leftrightarrow 3, \quad B \leftrightarrow 2, \quad C \leftrightarrow 1, \quad D \leftrightarrow 4$

26. *(b)* $9^{n+1} - 8n - 9 = 9^n \times 9 - 8n - 9$

$= (1 + 8)^n \times 9 - 8n - 9$

$= ({}^nC_0 + {}^nC_1 8 + {}^nC_2 8^2 + {}^nC_3 8^3 + \ldots + {}^nC_n 8^n) 9 - 8n - 9$

$= (1 + 8n + {}^nC_2 8^2 + {}^nC_3 8^3 + \ldots + {}^nC_n 8^n) 9 - 8n - 9$

$= 9 + 72n + ({}^nC_2 8^2 + {}^nC_3 8^3 + \ldots + {}^nC_n 8^n) 9 - 8n - 9$

$= (72n - 8n) + 8^2 ({}^nC_2 + {}^nC_3 8 + \ldots + {}^nC_n 8^{n-2}) 9$

$= 64n + 64 ({}^nC_2 + {}^nC_3 8 + \ldots + {}^nC_n 8^{n-2}) 9$

$= 64 [n + ({}^nC_2 + {}^nC_3 8 + \ldots + {}^nC_n 8^{n-2}) 9]$

$= 64 \times$ Some constant numbers

$=$ Divisible by 64

27. *(c)* $\sum_{r=0}^{n} {}^nC_r \times 3^r$

$= {}^nC_0 3^0 + {}^nC_1 3 + {}^nC_2 3^2 + {}^nC_3 3^3 + \ldots + {}^nC_n 3^n$

$= {}^nC_0 + {}^nC_1 3 + {}^nC_2 3^2 + {}^nC_3 3^3 + \ldots + {}^nC_n 3^n$

$= (1 + 3)^n = 4^n$

28. *(d)* Now, $(x + 1)^6 = {}^6C_0 x^6 + {}^6C_1 x^5 \times 1 + {}^6C_2 x^4 \times (1)^2 + {}^6C_3 x^3 \times (1)^3 + {}^6C_4 x^2 \times (1)^4 + {}^6C_5 x \times (1)^5 + {}^6C_6 (1)^6$

$\Rightarrow (x + 1)^6 = {}^6C_0 x^6 + {}^6C_1 x^5 + {}^6C_2 x^4 + {}^6C_3 x^3 + {}^6C_2 x^2 + {}^6C_1 x + {}^6C_0$ $(\because {}^nC_r = {}^nC_{n-r})$

$\Rightarrow (x + 1)^6 = x^6 + 6x^5 + \frac{6 \times 5}{2} x^4 + \frac{6 \times 5 \times 4}{6} x^3 + \frac{6 \times 5}{2} x^2 + 6x + 1$

$\Rightarrow (x + 1)^6 = x^6 + 6x^5 + 15x^4 + 20x^3 + 15x^2 + 6x + 1$ ...(i)

Similarly,

$(x - 1)^6 = x^6 - 6x^5 + 15x^4 - 20x^3 + 15x^2 - 6x + 1$ ...(ii)

Now, adding Eqs. (i) and (ii), we get

$(x + 1)^6 + (x - 1)^6 = 2[x^6 + 15x^4 + 15x^2 + 1]$

Now, put $x = \sqrt{2}$

$\Rightarrow (\sqrt{2} + 1)^6 + (\sqrt{2} - 1)^6 = 2[(\sqrt{2})^6 + 15(\sqrt{2})^4 + 15(\sqrt{2})^2 + 1]$

$= 2 [2^3 + 15 \times 2^2 + 15 \times 2 + 1]$

$= 2 [8 + 15 \times 4 + 30 + 1]$

$= 2 [8 + 60 + 30 + 1] = 2 [99] = 198$

So, only Statement II is true.

29. *(a)* In the binomial expression, we have

$(a + b)^n = {}^nC_0 a^n + {}^nC_1 a^{n-1} b + {}^nC_2 a^{n-2} b^2 + \ldots + {}^nC_n b^n$ ...(i)

The coefficients ${}^nC_0$, ${}^nC_1$, ${}^nC_2$, ..., ${}^nC_n$ are known as binomial or combinatorial coefficients.

Putting $a = b = 1$ in (i), we get

${}^nC_0 + {}^nC_1 + {}^nC_2 + \ldots + {}^nC_n = 2^n$

Thus, the sum of all binomial coefficients is equal to 2^n.

Again, putting $a = 1$ and $b = -1$ in Eq. (i), we get

${}^nC_0 + {}^nC_2 + {}^nC_4 + \ldots = {}^nC_1 + {}^nC_3 + {}^nC_5 + \ldots$

Thus, the sum of all the odd binomial coefficients is equal to the sum of all the even binomial coefficients and each is equal to $\frac{2^n}{2} = 2^{n-1}$.

$\Rightarrow {}^nC_0 + {}^nC_2 + {}^nC_4 + \ldots = {}^nC_1 + {}^nC_3 + {}^nC_5 + \ldots = 2^{n-1}$

30. *(a)* Since, $(101)^{50} = (100 + 1)^{50}$

$= 100^{50} + {}^{50}C_1 100^{49} + {}^{50}C_2 100^{48} + \ldots + 1$ ...(i)

and $(99)^{50} = (100 - 1)^{50}$

$= 100^{50} - {}^{50}C_1 100^{49} + {}^{50}C_2 100^{48} - \ldots + 1$ ...(ii)

On subtracting Eq. (ii) from Eq. (i) we get

$(101)^{50} - (99)^{50} = 2 \{{}^{50}C_1 100^{49} + {}^{50}C_3 100^{47} + \ldots\}$

$= 2 \times {}^{50}C_1 100^{49} + (2 \times {}^{50}C_3 \times 100^{47} + \ldots)$

$= (100 \times 100^{49} + \text{a positive number}) > 100^{50}$

$\Rightarrow (101)^{50} - (99)^{50} > (100)^{50}$

$\Rightarrow (101)^{50} > (100)^{50} + (99)^{50}$

31. *(b)* $(1 + x - 2x^2)^6 = 1 + a_1 x + a_2 x^2 + \ldots + a_{12} x^{12}$

On putting $x = 1$ and $x = -1$ and adding the results

$64 = 2(1 + a_2 + a_4 + \ldots)$

$a_2 + a_4 + a_6 + \ldots + a_{12} = 31$

32. (b) Now, $2^{3n+3}-7n-8$
$=(2^3)^{n+1}-7(n+1)-1$
$=(1+7)^{n+1}-7(n+1)-1$
$={}^{n+1}C_0+{}^{n+1}C_1 7+{}^{n+1}C_2 7^2+{}^{n+1}C_3 7^3+\ldots+{}^{n+1}C_{n+1}7^{n+1}$
$-7(n+1)-1$
$=7^2({}^{n+1}C_2+{}^{n+1}C_3 7^1+\ldots+{}^{n+1}C_{n+1}7^{n-1})$
which is divisible by 49.

33. (b) Since, $(a+b)^n+(a-b)^n$
$=2[{}^nC_0a^nb^0+{}^nC_2a^{n-2}b^2+{}^nC_4a^{n-4}b^4+\ldots]$
$=2$ (Sum of terms at odd places)
So, $(x+a)^{100}+(x-a)^{100}$
$=2({}^{100}C_0x^{100}a^0+{}^{100}C_2x^{98}a^2+\ldots+{}^{100}C_{100}\,a^{100})$
$=51$ terms

34. (a) Given that, $(1+ax)^n=1+8x+24x^2+\ldots$
$\Rightarrow 1+\frac{n}{1!}ax+\frac{n(n-1)}{2!}a^2x^2+\ldots=1+8x+24x^2+\ldots$
On comparing the coefficient of x and x^2, we get
$$na=8,\ \frac{n(n-1)}{2!}a^2=24$$
$\Rightarrow\quad na(n-1)a=48$
$\Rightarrow\quad 8(n-1)a=48$
$\Rightarrow\quad 8na-8a=48$
$\Rightarrow\quad 64-8a=48$
$\Rightarrow\quad a=2\ \Rightarrow\ n=4$

35. (a) $(\sqrt{3}+1)^{2n}={}^{2n}C_0(\sqrt{3})^{2n}+{}^{2n}C_1(\sqrt{3})^{2n-1}+{}^{2n}C_2(\sqrt{3})^{2n-2}+$
$\ldots+{}^{2n}C_{2n}(\sqrt{3})^{2n-2n}$
and $(\sqrt{3}-1)^{2n}={}^{2n}C_0(\sqrt{3})^{2n}(-1)^0+{}^{2n}C_1(\sqrt{3})^{2n-1}(-1)^1$
$+{}^{2n}C_2(\sqrt{3})^{2n-2}(-1)^2+\ldots+{}^{2n}C_{2n}(\sqrt{3})^{2n-2n}(-1)^{2n}$
$\therefore (\sqrt{3}+1)^{2n}-(\sqrt{3}-1)^{2n}=2[{}^{2n}C_1(\sqrt{3})^{2n-1}+{}^{2n}C_3(\sqrt{3})^{2n-3}$
$+{}^{2n}C_5(\sqrt{3})^{2n-5}+\ldots+{}^{2n}C_{2n-1}(\sqrt{3})^{2n-(2n-1)}]$
which is most certainly an irrational number because of odd powers of $\sqrt{3}$ in each of the terms.

36. (b) $8^{2n}-(62)^{2n+1}=(1+63)^n-(63-1)^{2n+1}$
$=(1+63)^n+(1-63)^{2n+1}$
$=[1+{}^nC_1 63+{}^nC_2(63)^2+\ldots+(63)^n]$
$+[1-{}^{2n+1}C_1 63+{}^{(2n+1)}C_2(63)^2-\ldots+(-1)(63)^{(2n+1)}]$
$=2+63[{}^nC_1+{}^nC_2(63)+$
$\ldots+(63)^{n-1}-{}^{(2n+1)}C_1+{}^{(2n+1)}C_2(63)-\ldots+(-1)(63)^{(2n)}]$
$\therefore$ Remainder is 2.

37. (b) We know that,
$(1+x)^{20}={}^{20}C_0+{}^{20}C_1x+\ldots+{}^{20}C_{10}x^{10}+\ldots+{}^{20}C_{20}x^{20}$
On putting $x=-1$, we get
$0={}^{20}C_0-{}^{20}C_1+\ldots-{}^{20}C_9+{}^{20}C_{10}-{}^{20}C_{11}+\ldots+{}^{20}C_{20}$
$\Rightarrow\quad 0={}^{20}C_0-{}^{20}C_1+\ldots-{}^{20}C_9+{}^{20}C_{10}-{}^{20}C_9+{}^{20}C_8$
$-\ldots+{}^{20}C_0\ (\because {}^nC_r={}^nC_{n-r})$
$\Rightarrow\quad 0=2({}^{20}C_0-{}^{20}C_1+\ldots-{}^{20}C_9)+{}^{20}C_{10}$
Adding ${}^{20}C_{10}$ both sides
$\Rightarrow\quad {}^{20}C_{10}=2({}^{20}C_0-{}^{20}C_1+\ldots+{}^{20}C_{10})$
$\Rightarrow {}^{20}C_0-{}^{20}C_1+\ldots+{}^{20}C_{10}=\frac{1}{2}\,{}^{20}C_{10}$

38. (a) ${}^{50}C_4+{}^{55}C_3+{}^{54}C_3+{}^{53}C_3+{}^{52}C_3+{}^{51}C_3+{}^{50}C_3$
$={}^{50}C_3+{}^{50}C_4+{}^{51}C_3+{}^{52}C_3+{}^{53}C_3+{}^{54}C_3+{}^{55}C_3$
$={}^{51}C_4+{}^{51}C_3+{}^{52}C_3+{}^{53}C_3+{}^{54}C_3+{}^{55}C_3$
$(\because {}^nC_r+{}^nC_{r-1}={}^{n+1}C_r)$
$={}^{52}C_4+{}^{52}C_3+{}^{53}C_3+{}^{54}C_3+{}^{55}C_3$
$={}^{53}C_4+{}^{53}C_3+{}^{54}C_3+{}^{55}C_3$
$={}^{54}C_4+{}^{54}C_3+{}^{55}C_3$
$={}^{55}C_4+{}^{55}C_3={}^{56}C_4$

39. (b) Put $1-x=y$.
Then, $(1-x+x^2)^4=(y+x^2)^4$
$={}^4C_0\ y^4(x^2)^0+{}^4C_1\ y^3(x^2)^1+{}^4C_2\ y^2(x^2)^2$
$+{}^4C_3\ y(x^2)^3+{}^4C_4(x^2)^4$
$=y^4+4y^3x^2+6y^2x^4+4yx^6+x^8$
Putting back $y=1-x$ in the above equation
$=(1-x)^4+4x^2(1-x)^3+6x^4(1-x)^2+4x^6(1-x)+x^8$
$=1-4x+10x^2-16x^3+19x^4-16x^5+10x^6-4x^7+x^8$

40. (a) We have,
$2^{4n+4}-15n-16=2^{4(n+1)}-15n-16$
$=16^{n+1}-15n-16$
$=(1+15)^{n+1}-15n-16$
$={}^{n+1}C_0 15^0+{}^{n+1}C_1 15^1+{}^{n+1}C_2 15^2+{}^{n+1}C_3 15^3$
$+\ldots+{}^{n+1}C_{n+1}(15)^{n+1}-15n-16$
$=1+(n+1)15+{}^{n+1}C_2 15^2+{}^{n+1}C_3 15^3$
$+\ldots+{}^{n+1}C_{n+1}(15)^{n+1}-15n-16$
$=1+15n+15+{}^{n+1}C_2 15^2+{}^{n+1}C_3 15^3$
$+\ldots+{}^{n+1}C_{n+1}(15)^{n+1}-15n-16$
$=15^2[{}^{n+1}C_2+{}^{n+1}C_3 15+\ldots+{}^{n+1}C_{n+1}(15)^{n-1}]$
Thus, $2^{4n+4}-15n-16$ is divisible by 225.

41. (c) On putting $\sqrt{1-x^2}=y$, the given expression becomes
$=(x^2-y)^4+(x^2+y)^4$
$=2[x^8+{}^4C_2x^4y^2+{}^4C_4y^4]$

$$=2\left[x^8+\frac{4\times3}{2\times1}x^4\cdot(1-x^2)+(1-x^2)^2\right]$$
$$=2[x^8+6x^4(1-x^2)+(1-2x^2+x^4)]$$
$$=2x^8-12x^6+14x^4-4x^2+2$$

42. *(c)* Since, we know that,

$$(a+b)^n={}^nC_0a^nb^0+{}^nC_1a^{n-1}b^1+\ldots+{}^nC_na^0b^n$$

and $(a-b)^n={}^nC_0a^nb^0-{}^nC_1a^{n-1}b^1+{}^nC_2a^{n-2}b^2-\ldots+(-1)^n\,{}^nC_na^0b^n$

$\Rightarrow (a+b)^n-(a-b)^n=2[{}^nC_1a^{n-1}b^1+{}^nC_3a^{n-3}b^3+\ldots]$

$=2$ [Sum of terms at even places]

Therefore,

$(x+a)^{51}-(x-a)^{51}=2\,[{}^{51}C_1\,x^{50}a+{}^{51}C_3x^{48}a^3+\ldots+{}^{51}C_{51}x^0a^{51}]$

$=26$ terms

43. *(c)* $(1+100)^{100}=1+100\cdot100+\dfrac{100\cdot99}{1\cdot2}\cdot(100)^2+..$

$\Rightarrow\quad (101)^{100}-1=10{,}000\left[1+\dfrac{100\cdot99}{1\cdot2}(1)+\ldots..\right]$

So, it is clear that $(101)^{100}-1$ is divisible by $(100)^2=10000$

44. *(d)* $7^{300}=(7^2)^{150}=(50-1)^{150}$

$={}^{150}C_0(50)^{150}(-1)^0+{}^{150}C_1(50)^{149}(-1)^1+\ldots+{}^{150}C_{150}(50)^{\circ}(-1)^{150}$

Thus, the last digit of 7^{300} is ${}^{150}C_{150}.1\cdot1=1$

45. *(a)* We know that,

$$2^{n-1}={}^nC_0+{}^nC_2+{}^nC_4+\ldots.={}^nC_1+{}^nC_3+{}^nC_5+\ldots$$

$\Rightarrow\quad {}^{10}C_1+{}^{10}C_3+{}^{10}C_5+\ldots+{}^{10}C_9=2^{10-1}=2^9$

46. *(d)* At $n=1$, we have,

$${}^1C_0-\frac{1}{2}{}^1C_1=1-\frac{1}{2}=\frac{1}{2}=\frac{1}{1+1}$$

and at $n=2$, we have,

$${}^2C_0-\frac{1}{2}{}^2C_1+\frac{1}{3}{}^2C_2=1-1+\frac{1}{3}$$
$$=\frac{1}{3}=\frac{1}{2+1}$$

So, option (d) is correct.

47. *(d)* In the binomial expansion for $(a+b)^n$, we observe that the first term is ${}^nC_0\,a^n$, the second term is ${}^nC_1a^{n-1}b$, the third term is ${}^nC_2a^{n-2}b^2$, and so on. Looking at the pattern of the successive terms we can say that the $(r+1)^{th}$ term is ${}^nC_ra^{n-r}b^r$.

48. *(b)* The $(r+1)^{th}$ term is called the general term of the expansion $(a+b)^n$. It is denoted by T_{r+1}.

Thus, $T_{r+1}={}^nC_ra^{n-r}b^r$.

49. *(b)* Regarding the middle term in the expansion $(a+b)^n$, if n is even, then the number of terms in the expansion will be $n+1$. Since, n is even, so $n+1$ is odd. Therefore, the middle term is $\left(\dfrac{n+1+1}{2}\right)^{th}$ *i.e.*, $\left(\dfrac{n}{2}+1\right)^{th}$ term.

50. *(d)* In the expansion of $(x+2y)^8$, the middle term is $\left(\dfrac{8}{2}+1\right)^{th}$ *i.e.*, 5th term.

{consider previous solution for explanation}

So, only Statement II is true.

51. *(c)* Regarding the middle term in the expansion $(a+b)^n$, if n is odd, then $(n+1)$ is even, so there will be two middle terms in the expansion, namely $\left(\dfrac{n+1}{2}\right)^{th}$ term and $\left(\dfrac{n+1}{2}+1\right)^{th}$ term.

52. *(b)* Referring to the previous solution, we see that in the expansion $(2x-y)^7$, the middle terms are $\left(\dfrac{7+1}{2}\right)^{th}$, *i.e.*, 4th and $\left(\dfrac{7+1}{2}+1\right)^{th}$, *i.e.*, 5th term.

So, both the statements are true but Statement II is not a correct explanation of Statement I.

53. *(b)* In the expansion of $\left(x+\dfrac{1}{x}\right)^{2n}$, where $x\neq0$, the middle term is $\left(\dfrac{2n+1+1}{2}\right)^{th}$, *i.e.*, $(n+1)^{th}$ term, as $2n$ is even.

54. *(c)* As $2n$ is even, so in the expansion of $\left(x+\dfrac{1}{x}\right)^{2n}$, where $x\neq0$, the middle term is $\left(\dfrac{2n+1+1}{2}\right)^{th}$ term, *i.e.*, $(n+1)^{th}$ term (as $2n$ is even) and this middle term is given by ${}^{2n}C_n\,x^n\left(\dfrac{1}{x}\right)^n={}^{2n}C_n$ (constant).

This term is called the term independent of x or the constant term.

So, both (a) and (b) are true.

55. *(b)* The $(r+1)^{th}$ term of the expansion $(x+y)^n$ is given by $T_{r+1}={}^nC_rx^{n-r}y^r$.

For the 17th term, we have, $r+1=17$, *i.e.*, $r=16$

Therefore, $T_{17}=T_{16+1}={}^{50}C_{16}(2)^{50-16}a^{16}$

$={}^{50}C_{16}2^{34}\,a^{16}$

Similarly, $T_{18}={}^{50}C_{17}\,2^{33}a^{17}$

Given that, $T_{17}=T_{18}$

So, ${}^{50}C_{16}\,(2)^{34}\,a^{16}={}^{50}C_{17}(2)^{33}\,a^{17}$

Therefore, $\dfrac{{}^{50}C_{16} \cdot 2^{34}}{{}^{50}C_{17} \cdot 2^{33}} = \dfrac{a^{17}}{a^{16}}$

i.e., $a = \dfrac{{}^{50}C_{16} \times 2}{{}^{50}C_{17}}$

$= \dfrac{50!}{16!\,34!} \times \dfrac{17!\,.\,33!}{50!} \times 2 = 1$

56. *(d)* As $2n$ is even, the middle term of the expansion $(1+x)^{2n}$ is $\left(\dfrac{2n}{2}+1\right)^{th}$, *i.e.*, $(n+1)^{th}$ term which is given by

$$T_{n+1} = {}^{2n}C_n (1)^{2n-n} (x)^n = {}^{2n}C_n x^n = \frac{(2n)!}{n!\,n!} x^n$$
$$= \frac{2n(2n-1)(2n-2)\ldots 4\cdot 3\cdot 2\cdot 1}{n!\,n!} x^n$$
$$= \frac{1\cdot 2\cdot 3\cdot 4\ldots(2n-2)(2n-1)(2n)}{n!\,n!} x^n$$
$$= \frac{[1\cdot 3\cdot 5\ldots(2n-1)][2\cdot 4\cdot 6\ldots(2n)]}{n!\,n!} x^n$$
$$= \frac{[1\cdot 3\cdot 5\ldots(2n-1)]2^n[1\cdot 2\cdot 3\ldots n]}{n!\,n!} x^n$$
$$= \frac{[1\cdot 3\cdot 5\ldots(2n-1)]\,n!}{n!\,n!} 2^n x^n$$
$$= \frac{1\cdot 3\cdot 5\ldots(2n-1)}{n!} 2^n x^n$$

57. *(d)* Suppose x^6y^3 occurs in the $(r+1)^{th}$ term of the expansion $(x+2y)^9$.

Now, $T_{r+1} = {}^9C_r x^{9-r}(2y)^r = {}^9C_r 2^r \cdot x^{9-r} \cdot y^r$.

Comparing the indices of x as well as y in x^6y^3 and in T_{r+1}, we get $r = 3$.

Thus, the coefficient of x^6y^3 is

$${}^9C_3\, 2^3 = \frac{9!}{3!\,6!} \cdot 2^3 = \frac{9.8.7}{3.2} \cdot 2^3 = 672$$

So, only Statement II is true.

58. *(c)* Given that, second term $T_2 = 240$

We have, $T_2 = {}^nC_1 x^{n-1} \cdot a$

So, ${}^nC_1 x^{n-1} \cdot a = 240$ ...(i)

Similarly, ${}^nC_2 x^{n-2} \cdot a^2 = 720$ ...(ii)

and ${}^nC_3 x^{n-3} \cdot a^3 = 1080$ ...(iii)

Dividing Eq. (ii) by Eq. (i), we get

$$\frac{{}^nC_2 x^{n-2} a^2}{{}^nC_1 x^{n-1} a} = \frac{720}{240} \text{ i.e., } \frac{(n-1)!}{(n-2)!} \cdot \frac{a}{x} = 6$$

or $\dfrac{a}{x} = \dfrac{6}{(n-1)}$ ...(iv)

Dividing Eq. (iii) by Eq. (ii), we get

$\dfrac{a}{x} = \dfrac{9}{2(n-2)}$ ...(v)

From Eqs. (iv) and (v), we get

$$\frac{6}{n-1} = \frac{9}{2(n-2)}$$

Thus, $n = 5$

Hence, from Eq. (i), $5x^4a = 240$, and from Eq. (iv), $\dfrac{a}{x} = \dfrac{3}{2}$

Solving these equations for a and x, we get $x = 2$ and $a = 3$.

59. *(d)* Suppose the three consecutive terms in the expansion of $(1+a)^n$ are $(r-1)^{th}$, r^{th} and $(r+1)^{th}$ terms.

The $(r-1)^{th}$ term is ${}^nC_{r-2}a^{r-2}$, and its coefficient is ${}^nC_{r-2}$. Similarly, the coefficients of rth and $(r+1)^{th}$ terms are ${}^nC_{r-1}$ and ${}^nC_r$, respectively.

Since, the coefficients are in the ratio 1 : 7 : 42, so we have,

$\dfrac{{}^nC_{r-2}}{{}^nC_{r-1}} = \dfrac{1}{7}$, *i.e.*, $n - 8r + 9 = 0$ ...(i)

and $\dfrac{{}^nC_{r-1}}{{}^nC_r} = \dfrac{7}{42}$, *i.e.*, $n - 7r + 1 = 0$ ...(ii)

On solving Eqs. (i) and (ii), we get $n = 55$.

60. *(d)* The general term of $(x+3)^8$ is

$$T_{r+1} = {}^nC_r\, x^{n-r}\, y^r = {}^8C_r\, x^{8-r}\, 3^r \quad (\because n = 8, r = 3)$$

For coefficient of x^5, put $8 - r = 5$

$\Rightarrow$ $r = 8 - 5 = 3$

$\Rightarrow$ $T_{3+1} = {}^8C_3 x^{8-3}\, 3^3 = {}^8C_3 \times 3^3\, x^5$

$\Rightarrow$ Coefficient of $x^5 = {}^8C_3 \times 3^3$

$$= \frac{8 \times 7 \times 6}{6} \times 27 = 56 \times 27 = 1512$$

61. *(a)* The general term of $(a-2b)^{12}$ is

$T_{r+1} = {}^{12}C_r\,(a)^{12-r}(-2b)^r = {}^{12}C_r\, a^{12-r}(-2)^r b^r$

$T_{r+1} = {}^{12}C_r\, a^{12-r}\, b^r(-2)^r$

For coefficient of a^5b^7, put $12 - r = 5 \Rightarrow r = 7$

$\therefore$ $T_{7+1} = {}^{12}C_7\, a^{12-7}\, b^7(-2)^7$

$\Rightarrow$ $T_8 = {}^{12}C_5\, a^5b^7(-2)^7$ $\quad (\because {}^nC_r = {}^nC_{n-r})$

$\Rightarrow$ Coefficient of $a^5b^7 = {}^{12}C_5(-2)^7$

$$= \frac{12 \times 11 \times 10 \times 9 \times 8}{5 \times 4 \times 3 \times 2 \times 1}(-2)^7$$
$$= -11 \times 9 \times 8 \times 32 \times 4 = -101376$$

So, both the statement are true and Statement II is the correct explanation of Statement I.

62. *(b)* (a) The general term of $(x^2 - y)^6$ is

$$T_{r+1} = {}^6C_r\,(x^2)^{6-r}(-y)^r = {}^6C_r\, x^{12-2r}(-1)^r y^r$$
$$= {}^6C_r\, x^{12-2r}\, y^r(-1)^r$$

(b) The general term of $(x^2 - yx)^{12}$ is

$$T_{r+1} = {}^{12}C_r (x^2)^{12-r} (-yx)^r$$
$$= {}^{12}C_r x^{24-2r} (-1)^r y^r x^r$$
$$= {}^{12}C_r (-1)^r x^{24-2r+r} y^r$$
$$= {}^{12}C_r (-1)^r x^{24-r} y^r$$

So, only (b) is correct.

63. *(d)* General term in the expansion of $(X+Y)^n$ is

$$T_{r+1} = {}^nC_r X^{n-r} Y^r$$

Here, $X = x, \quad Y = -2y \quad$ and $\quad n = 12$

$$T_{r+1} = {}^{12}C_r (x)^{12-r} (-2y)^r$$

On putting $r = 3$, we get

$$T_{3+1} = {}^{12}C_3 x^{12-3} (-2y)^3$$
$$T_4 = \frac{12 \times 11 \times 10}{6} x^9 (-2)^3 y^3$$
$$= 2 \times 11 \times 10 x^9 (-8) y^3$$
$$= 220 \times (-8) x^9 y^3$$
$$= -1760 x^9 y^3$$

64. *(d)* Here, $n = 7$ (odd)

Total terms $= n + 1 = 7 + 1 = 8$ (even)

Middle terms are $\left(\frac{n+1}{2}\right)^{th}$ and $\left(\frac{n+3}{2}\right)^{th}$ term

i.e., $\left(\frac{7+1}{2}\right)^{th}$ term and $\left(\frac{7+3}{2}\right)^{th}$ term

$\Rightarrow$ 4th and 5th terms

$$\therefore \quad T_4 = T_{3+1} = {}^7C_3 (3)^{7-3} \left(-\frac{x^3}{6}\right)^3 \quad (\because T_{r+1} = {}^nC_r x^{n-r} y^r)$$
$$= \frac{7 \times 6 \times 5}{6} \times \frac{3^4 (-1)^3 \times x^9}{6^3}$$
$$= \frac{35 \times 81 \times (-1) \times x^9}{6 \times 6 \times 6}$$
$$= -\frac{35 \times 27 \times x^9}{2 \times 6 \times 6}$$
$$= -\frac{35 \times 3}{2 \times 2 \times 2} \times x^9 = -\frac{105}{8} x^9$$

and $T_5 = T_{4+1} = {}^7C_4 3^{7-4} \left(-\frac{x^3}{6}\right)^4$

$$= {}^7C_3 \frac{3^3 (-1)^4 (x^3)^4}{6^4} \quad (\because {}^nC_r = {}^nC_{r-1})$$
$$= \frac{7 \times 6 \times 5}{6} \times \frac{3 \times 3 \times 3 \times x^{12}}{6 \times 6 \times 6 \times 6} = \frac{35}{48} x^{12}$$

65. *(c)* The general term in the expansion of $(X+Y)^n$ is

$$T_{r+1} = {}^nC_r X^{n-r} Y^r$$

Here, $X = 9x, Y = -\frac{1}{3\sqrt{x}}$ and $n = 18$

The general term in the expansion of $\left(9x - \frac{1}{3\sqrt{x}}\right)^{18}$ is

$$T_{r+1} = {}^{18}C_r (9x)^{18-r} \left(\frac{-1}{3\sqrt{x}}\right)^r$$

On putting $r = 12$, we get

$$T_{12+1} = {}^{18}C_{12} (9x)^{18-12} \left(\frac{-1}{3\sqrt{x}}\right)^{12}$$
$$= {}^{18}C_6 \, 9^6 \times x^6 \frac{1}{3^{12} (\sqrt{x})^{12}} \quad (\because {}^nC_r = {}^nC_{n-r})$$
$$= \frac{18 \times 17 \times 16 \times 15 \times 14 \times 13}{6 \times 5 \times 4 \times 3 \times 2 \times 1} \times \frac{3^{12} \times x^6}{3^{12} \times x^6}$$
$$= 18564$$

So, only Statement I is true.

66. *(d)* In the expansion of $(1+a)^{m+n}$, the general term is

$$T_{r+1} = {}^{m+n}C_r (a)^r$$

For coefficient of a^m, put $r = m$

Then, $\quad T_{m+1} = {}^{m+n}C_m a^m$

$\Rightarrow \quad$ Coefficient of $a^m = {}^{m+n}C_m \quad \ldots$(i)

For coefficient of a^n, put $r = n$, then

$$T_{n+1} = {}^{m+n}C_n a^n$$

$\Rightarrow$ Coefficient of $a^n = {}^{m+n}C_n$

$$= {}^{m+n}C_{m+n-n} \quad (\because {}^nC_r = {}^nC_{n-r})$$
$$= {}^{m+n}C_m \quad \ldots\text{(ii)}$$

From Eqs. (i) and (ii), we get that the coefficient of a^m is equal to coefficient of a^n.

67. *(b)* Here, $n = 10$ (even)

Total terms $= n + 1 = 10 + 1 = 11$ (odd)

There will be one middle term *i.e.*, $\left(\frac{10+2}{2}\right)^{th}$ term or 6th term.

The general term in the expansion of $\left(\frac{x}{3} + 9y\right)^{10}$ is

$$T_{r+1} = {}^{10}C_r \left(\frac{x}{3}\right)^{10-r} (9y)^r \quad (\because T_{r+1} = {}^nC_r x^{n-r} y^r)$$

On putting $r = 5$, we get

$$T_6 = T_{5+1} = {}^{10}C_5 \left(\frac{x}{3}\right)^{10-5} (9y)^5$$

$$= \frac{10 \times 9 \times 8 \times 7 \times 6}{5 \times 4 \times 3 \times 2 \times 1} \left(\frac{x}{3}\right)^{10-5} 9^5 y^5$$

$$T_6 = \frac{9 \times 8 \times 7 \times 2}{4} \times \frac{x^5}{3^5} \times 9^5 \times y^5$$

$$= 61236 x^5 y^5$$

68. *(d)* In the expansion of $(1+x)^m$, the general term is

$$T_{r+1} = {}^mC_r\, x^r$$

For coefficient of x^2, put $r = 2$

$\Rightarrow \quad T_{2+1} = {}^mC_2\, x^2$

$\Rightarrow \quad$ Coefficient of $x^2 = {}^mC_2$

Now given, $\quad {}^mC_2 = 6$

$\Rightarrow \quad \dfrac{m(m-1)}{2} = 6$

$\Rightarrow \quad m^2 - m = 12$

$\Rightarrow \quad m^2 - m - 12 = 0$

On solving it by splitting the middle term,

$\Rightarrow \quad m^2 - (4-3)m - 12 = 0$

$\Rightarrow \quad m^2 - 4m + 3m - 12 = 0$

$\Rightarrow \quad m(m-4) + 3(m-4) = 0$

$\Rightarrow \quad (m-4)(m+3) = 0$

$\Rightarrow \quad m - 4 = 0 \quad$ or $\quad m + 3 = 0$

$\Rightarrow \quad m = 4 \quad$ or $\quad m = -3$

$\Rightarrow$ Only $m = 4 \quad$ ($\because$ we cannot take negative value of m)

69. *(c)* For $(1+x)^{2n}$, the general term is

$$T_{r+1} = {}^{2n}C_r\, x^r$$

For coefficient of x^n, put $r = n$

$$T_{n+1} = {}^{2n}C_n x^n$$

$\Rightarrow \quad$ Coefficient of $x^n = {}^{2n}C_n \quad \ldots$(i)

For $(1+x)^{2n-1}$, the general term is

$$T_{r+1} = {}^{2n-1}C_r\, x^r$$

For coefficient of x^n, put $r = n$

$\Rightarrow \quad T_{n+1} = {}^{2n-1}C_n\, x^n$

$\Rightarrow \quad$ Coefficient of $x^n = {}^{2n-1}C_n = \dfrac{(2n-1)!}{n!(2n-1-n)!}$

$$= \frac{(2n-1)!}{n!(n-1)!} \quad \left(\because {}^nC_r = \frac{n!}{r!(n-r)!}\right)$$

Multiplying and dividing by $2n$, we get

$$= \frac{(2n)(2n-1)!}{(2n)\, n!(n-1)!}$$

$$= \frac{(2n)!}{2n(n-1)!\, n!} \quad (\because n! = n(n-1)!)$$

$$= \frac{(2n)!}{(2)\, n!\, n!} \quad \left(\because {}^{2n}C_n = \frac{2n!}{n!(2n-n)!} = \frac{2n!}{n!\, n!}\right)$$

$$= \frac{1}{2} \times {}^{2n}C_n \quad \ldots\text{(ii)}$$

From Eqs. (i) and (ii),

Coefficient of x^n in $(1+x)^{2n-1} = \dfrac{1}{2} \times$ Coefficient of x^n in $(1+x)^{2n}$

$\Rightarrow$ Coefficient of x^n in $(1+x)^{2n} = 2 \times$ Coefficient of x^n in $(1+x)^{2n-1}$

So, only Statement I is true.

70. *(a)* In the expansion of $(x+1)^n$, the general term is

$$T_{R+1} = {}^nC_R\, x^{n-R} (1)^R \quad \text{or} \quad T_{R+1} = {}^nC_R x^{n-R}$$

For coefficient of $(r-1)^{th}$ term, put $R = r - 2$

$$T_{R+1} = T_{(r-2)+1} = {}^nC_{r-2}\, x^{n-(r-2)}$$

$$T_{(r-1)} = {}^nC_{r-2}\, x^{n-r+2}$$

$\Rightarrow \quad$ Coefficient of $(r-1)^{th}$ term $= {}^nC_{r-2}$

For coefficient of rth term, put $R = r - 1$

$\Rightarrow \quad T_r = T_{r-1+1} = {}^nC_{r-1}\, x^{n-(r-1)}$

$\Rightarrow \quad T_r = {}^nC_{r-1} x^{n-r+1}$

Coefficient of rth term $= {}^nC_{r-1}$

For coefficient of $(r+1)^{th}$ term, put $R = r$

$\Rightarrow \quad T_{r+1} = {}^nC_r\, x^{n-r}$

$\Rightarrow$ Coefficient of $(r+1)^{th}$ term $= {}^nC_r$

Now, according to question it is given that

$${}^nC_{r-2} : {}^nC_{r-1} : {}^nC_r = 1:3:5$$

Taking first two parts, we get

$${}^nC_{r-2} : {}^nC_{r-1} = 1:3 \quad \left(\because {}^nC_r = \frac{n!}{r!(n-r)!}\right)$$

$\Rightarrow \quad \dfrac{n!}{(r-2)!\{n-(r-2)\}!} : \dfrac{n!}{(r-1)!\{n-(r-1)\}!} = 1:3$

$\Rightarrow \quad \dfrac{1}{(r-2)!(n-r+2)!} : \dfrac{1}{(r-1)(r-2)!(n-r+1)!} = 1:3$

$\Rightarrow \quad \dfrac{1}{(n-r+2)(n-r+1)!} : \dfrac{1}{(r-1)(n-r+1)!} = 1:3$

$\Rightarrow \quad \dfrac{1}{n-r+2} : \dfrac{1}{r-1} = 1:3$

$\Rightarrow \quad \dfrac{r-1}{n-r+2} = \dfrac{1}{3}$

$\Rightarrow \quad 3r - 3 = n - r + 2$

$\Rightarrow \quad n - 4r + 5 = 0 \quad \ldots$(i)

Taking last two parts, ${}^nC_{r-1} : {}^nC_r = 3:5$

$$\Rightarrow \frac{n!}{(r-1)!\{n-(r-1)\}!} : \frac{n!}{r!(n-r)!} = 3:5 \left(\because {}^nC_r = \frac{n!}{r!(n-r)!}\right)$$

$$\Rightarrow \frac{1}{(r-1)!\{n-r+1)\}!} : \frac{1}{r(r-1)!(n-r)!} = 3:5$$

$$\Rightarrow \frac{1}{(n-r+1)(n-r)!} : \frac{1}{r(n-r)!} = 3:5$$

$$\Rightarrow \frac{1}{n-r+1} : \frac{1}{r} = 3:5$$

$$\Rightarrow \frac{1}{n-r+1} \times r = \frac{3}{5}$$

$$\Rightarrow 5r = 3n - 3r + 3$$

$$\Rightarrow 3n - 8r + 3 = 0 \quad \ldots\text{(ii)}$$

Multiplying Eq. (i) by 2 and then subtracting from Eq. (ii), we get

$$n - 7 = 0$$

$$\Rightarrow n = 7$$

Putting the value of n in Eq. (i), we get

$$7 - 4r + 5 = 0$$

$$\Rightarrow 4r = 12$$

$$r = 3$$

71. *(a)* We have,

$$T_r = {}^{2r}C_{r-1}(x)^{2r-r+1}\left(\frac{1}{x}\right)^{r-1}$$

$$= \frac{2r!}{(r-1)!(r+1)!} x^{r+1-r+1}$$

$$= \frac{2r!}{(r-1)!(r+1)!} x^2$$

72. *(c)* Since, r^{th} term from the end in the expansion of $(a+b)^n$ is $(n-r+2)^{th}$ term from the beginning. Therefore, 4th term from the end is $9-4+2$, *i.e.*, 7th term from the beginning, which is given by

$$T_7 = {}^9C_6\left(\frac{x^3}{2}\right)^3\left(\frac{-2}{x^2}\right)^6 = {}^9C_3\,\frac{x^9}{8}\cdot\frac{64}{x^{12}} = \frac{9\times8\times7}{3\times2\times1}\times\frac{64}{x^3} = \frac{672}{x^3}$$

So, only Statement I is true.

73. *(c)* Given expression is $\left(\frac{p}{2}+2\right)^8$

Here, $n = 8$ (even)

$$\Rightarrow \text{Middle term} = \left(\frac{8}{2}+1\right)^{th} = 5^{th}\text{ term}$$

$$\therefore T_5 = {}^8C_4\left(\frac{p}{2}\right)^{8-4}\cdot(2)^4$$

$$\Rightarrow \frac{8\times7\times6\times5}{4\times3\times2\times1}\times\frac{p^4}{2^4}\times2^4 = 1120$$

$$\Rightarrow p^4 = 16$$

$$\therefore p = \pm2$$

74. *(b)* Since, n is even , therefore $\left(\frac{n}{2}+1\right)^{th}$ term is the middle term

$$\therefore T_{\frac{n}{2}+1} = {}^nC_{n/2}(x^2)^{n/2}\left(\frac{1}{x}\right)^{n/2} = 924x^6 \quad \text{(given)}$$

$$\Rightarrow x^{n/2} = x^6$$

$$\Rightarrow n = 12$$

75. *(d)* Let the general term, *i.e.*, $(r+1)^{th}$ contain x^{11}.

We have, $T_{r+1} = {}^{12}C_r(x^3)^{12-r}\left(-\frac{2}{x^2}\right)^r$

$$= {}^{12}C_r x^{36-3r-2r}(-1)^r 2^r$$

$$= {}^{12}C_r(-1)^r 2^r x^{36-5r}$$

Now for this to contain x^{11}, we observe that

$$36-5r = 11, \text{ i.e. } r = 5$$

Thus, the coefficient of x^{11} is

$${}^{12}C_5(-1)^5 2^5 = -\frac{12\times11\times10\times9\times8}{5\times4\times3\times2}\times32 = -25344$$

76. *(c)* Let T_{r+1} contain x^{10}. Then,

$$T_{r+1} = {}^{18}C_r(x^2)^{18-r}\left(\frac{-2}{x}\right)^r$$

$$= {}^{18}C_r x^{36-2r}(-1)^r\,.\,2^r\,x^{-r}$$

$$= (-1)^r 2^r\,{}^{18}C_r x^{36-3r}$$

Thus, $36-3r = 10$, *i.e.*, $r = \frac{26}{3}$

Since, r is a fraction, so given expansion cannot have a term containing x^{10}.

So, only Statement I is true.

77. *(a)* The general term of $\left(3x-\frac{2}{x^2}\right)^{15}$ is

$$T_{r+1} = {}^{15}C_r(3x)^{15-r}\left(\frac{-2}{x^2}\right)^r$$

$$= {}^{15}C_r(3)^{15-r}(-2)^r x^{15-3r}$$

For independent term of x,

Put $15-3r = 0$

$$\Rightarrow r = 5$$

$$\therefore \text{Coefficient} = {}^{15}C_5(3)^{15-5}(-2)^5 = -3003(3^{10})(2^5)$$

78. *(c)* Given expression is $\left(x+\frac{1}{2x}\right)^{2n}$.

$$\therefore \text{Middle term} = {}^{2n}C_n(x)^n\left(\frac{1}{2x}\right)^n$$

($\because 2n$ is even $\Rightarrow 2n+1$ is odd)

$$= \frac{2n!}{n!n!2^n} = \frac{1\cdot3\cdot5\ldots(2n-1)}{n!}$$

79. *(c)* $\left(\frac{1}{x}+x\sin x\right)^{10}$

Here, $n = 10$ (even)

$\Rightarrow$ Middle term $=\left(\frac{10}{2}+1\right)^{th} = 6^{th}$

$$T_6 = {}^{10}C_5\left(\frac{1}{x}\right)^{10-5}(x\sin x)^5$$

$\Rightarrow$ $252(\sin x)^5 = \frac{63}{8}$

$\Rightarrow$ $(\sin x)^5 = \frac{1}{32}$

$\Rightarrow$ $\sin x = \frac{1}{2}$

$\Rightarrow$ $\sin x = \sin \pi/6$

$\therefore$ $x = n\pi + (-1)^n\frac{\pi}{6}$

80. *(c)* The general term in the expansion of $\left(\frac{3}{2}x^2-\frac{1}{3x}\right)^9$ is

$$T_{r+1} = {}^9C_r\left(\frac{3}{2}x^2\right)^{9-r}\left(-\frac{1}{3x}\right)^r$$

$$= {}^9C_r\left(\frac{3}{2}\right)^{9-r}\left(-\frac{1}{3}\right)^r x^{18-3r} \quad \ldots\text{(i)}$$

Now, the coefficients of the terms x^0, x^{-1} and x^{-3} in $\left(\frac{3}{2}x^2-\frac{1}{3x}\right)^9$ are

For x^0, $18-3r=0 \Rightarrow r=6$

For x^{-1}, there exists no integer value of r

For x^{-3}, $18-3r=-3 \Rightarrow r=7$

Now, the coefficient of terms independent of x in the expansion of $(1+x+2x^3)\left(\frac{3}{2}x^2-\frac{1}{3x}\right)^9$

$$= 1\cdot {}^9C_6(-1)^6\left(\frac{3}{2}\right)^{9-6}\left(\frac{1}{3}\right)^6 + 0 + 2\cdot{}^9C_7(-1)^7\left(\frac{3}{2}\right)^{9-7}\left(\frac{1}{3}\right)^7$$

$$= \frac{9\cdot 8\cdot 7}{1\cdot 2\cdot 3}\cdot\frac{3^3}{2^3}\cdot\frac{1}{3^6} + 2\cdot\frac{9\cdot 8}{1\cdot 2}(-1)\frac{3^2}{2^2}\cdot\frac{1}{3^7}$$

$$= \frac{7}{18}-\frac{2}{27} = \frac{17}{54}$$

81. *(c)* I. Let $(r+1)^{th}$ term be independent of x which is given by

$$T_{r+1} = {}^{10}C_r\left(\sqrt{\frac{x}{3}}\right)^{10-r}\left(\frac{\sqrt{3}}{2x^2}\right)^r$$

$$= {}^{10}C_r\left(\frac{x}{3}\right)^{\frac{10-r}{2}}3^{r/2}\left(\frac{1}{2^r x^{2r}}\right)$$

$$= {}^{10}C_r\, 3^{\frac{r}{2}-\frac{10-r}{2}}\, 2^{-r}\, x^{\frac{10-r}{2}-2r}$$

Since, the term is independent of x, so

$$\frac{10-r}{2}-2r = 0 \Rightarrow r = 2$$

Hence, 3rd term is independent of x and its value is given by

$$T_3 = {}^{10}C_2\frac{3^{-3}}{4} = \frac{10\times 9}{2\times 1}\times\frac{1}{9\times 12} = \frac{5}{12}$$

Clearly, Statement II is false.

82. *(d)* Since, the power of binomial is even, it has one middle term which is the $\left(\frac{12}{2}+1\right)^{th}$ term and it is given by

$$T_7 = {}^{12}C_6(2ax)^6\left(\frac{-b}{x^2}\right)^6$$

$$= {}^{12}C_6\frac{2^6\, a^6x^6(-b)^6}{x^{12}}$$

$$= {}^{12}C_6\frac{2^6\, a^6\cdot b^6}{x^6} = \frac{59136\, a^6b^6}{x^6}$$

83. *(d)* $(1-3x+7x^2)(1-x)^{16}$

$$= (1-3x+7x^2)({}^{16}C_0 - {}^{16}C_1x + {}^{16}C_2x^2 + \ldots)$$

After multiplying, the term containing x is $-{}^{16}C_1x - 3x\,{}^{16}C_0$.

$\therefore$ Coefficient of $x = -16-3 = -19$

84. *(c)* General term of $\left(\sqrt{x}-\frac{k}{x^2}\right)^{10}$ is

$$T_{r+1} = {}^{10}C_r(\sqrt{x})^{10-r}\left(-\frac{k}{x^2}\right)^r$$

$$= {}^{10}C_r x^{\frac{10-r}{2}-2r}(-k)^r = {}^{10}C_r(-k)^r x^{\frac{10-5r}{2}}$$

For the term free from x,

Put $\frac{10-5r}{2} = 0 \Rightarrow r = 2$

$\therefore$ ${}^{10}C_2(-k)^2 = 405$

$\Rightarrow$ $\frac{10\times 9}{2}\times k^2 = 405$

$\therefore$ $k = \pm 3$

85. *(b)* Since, the power of binomial is odd. Therefore, we have two middle terms which are 5th and 6th terms. These are given by

$$T_5 = {}^9C_4\left(\frac{p}{x}\right)^5\left(\frac{x}{p}\right)^4 = {}^9C_4\frac{p}{x} = \frac{126p}{x}$$

and $$T_6 = {}^9C_5\left(\frac{p}{x}\right)^4\left(\frac{x}{p}\right)^5 = {}^9C_5\frac{x}{p} = \frac{126x}{p}$$

86. *(b)* We have, $(2+3x)^9 = 2^9\left(1+\frac{3x}{2}\right)^9$

Now, $$\frac{T_{r+1}}{T_r} = \frac{2^9\left[{}^9C_r\left(\frac{3x}{2}\right)^r\right]}{2^9\left[{}^9C_{r-1}\left(\frac{3x}{2}\right)^{r-1}\right]}$$

$$= \frac{{}^9C_r}{{}^9C_{r-1}}\left|\frac{3x}{2}\right| = \frac{9!}{r!(9-r)!}\cdot\frac{(r-1)!(10-r)!}{9!}\left|\frac{3x}{2}\right|$$

$$= \frac{10-r}{r}\left|\frac{3x}{2}\right| = \frac{10-r}{r}\left(\frac{9}{4}\right) \qquad \left(\because x = \frac{3}{2}\right)$$

Therefore, $\frac{T_{r+1}}{T_r} \geq 1 \Rightarrow \frac{90-9r}{4r} \geq 1$

$\Rightarrow \quad 90 - 9r \geq 4r$

$\Rightarrow \quad r \leq \frac{90}{13}$

$\Rightarrow \quad r \leq 6\frac{12}{13}$

Thus, the maximum value of r is 6. Therefore, the greatest term is $T_{r+1} = T_7$.

Hence, $T_7 = 2^9\left[{}^9C_6\left(\frac{3x}{2}\right)^6\right]$, where $x = \frac{3}{2}$

$$= 2^9 \cdot {}^9C_6\left(\frac{9}{4}\right)^6$$

$$= 2^9 \cdot \frac{9\times8\times7}{3\times2\times1}\left(\frac{3^{12}}{2^{12}}\right)$$

$$= \frac{7\times3^{13}}{2}$$

87. *(b)* We have,

$$(1+x)^n\left(1+\frac{1}{x}\right)^n = (1+x)^n\left(\frac{x+1}{x}\right)^n = \frac{(1+x)^{2n}}{x^n}$$

Now, to find the coefficient of x^{-1} in $(1+x)^n\left(1+\frac{1}{x}\right)^n$, it is equivalent to finding coefficient of x^{-1} in $\frac{(1+x)^{2n}}{x^n}$, which in turn is equal to the coefficient of x^{n-1} in the expansion of $(1+x)^{2n}$.

Since, $(1+x)^{2n} = {}^{2n}C_0x^0 + {}^{2n}C_1x^1 + {}^{2n}C_2x^2 + \ldots + {}^{2n}C_{n-1}$
$x^{n-1} + {}^{2n}C_{2n}x^{2n}$

Thus, the coefficient of x^{n-1} is

$${}^{2n}C_{n-1} = \frac{(2n)!}{(n-1)!\,(2n-n+1)!} = \frac{(2n)!}{(n-1)!\,(n+1)!}$$

88. *(a)* Since, the given series is a geometric series with the common ratio $\frac{x}{1+x}$, its sum

$$= \frac{(1+x)^{1000}\left[1-\left(\frac{x}{1+x}\right)^{1001}\right]}{\left[1-\left(\frac{x}{1+x}\right)\right]}$$

$$= \frac{(1+x)^{1000} - \frac{x^{1001}}{1+x}}{\frac{1+x-x}{1+x}} = (1+x)^{1001} - x^{1001}$$

Hence, coefficient of x^{50} is given by

$${}^{1001}C_{50} = \frac{1001!}{50!\;951!}$$

89. *(c)* Let a_1, a_2, a_3 and a_4 be the coefficients of four consecutive terms T_{r+1}, T_{r+2}, T_{r+3} and T_{r+4}, respectively.
Then, a_1 = Coefficient of $T_{r+1} = {}^nC_r$

a_2 = Coefficient of $T_{r+2} = {}^nC_{r+1}$

a_3 = Coefficient of $T_{r+3} = {}^nC_{r+2}$

and a_4 = Coefficient of $T_{r+4} = {}^nC_{r+3}$

Thus, $\frac{a_1}{a_1+a_2} = \frac{{}^nC_r}{{}^nC_r + {}^nC_{r+1}}$

$$= \frac{{}^nC_r}{{}^{n+1}C_{r+1}} \qquad (\because {}^nC_r + {}^nC_{r+1} = {}^{n+1}C_{r+1})$$

$$= \frac{n!}{r!\,(n-r)!} \times \frac{(r+1)!\,(n-r)!}{(n+1)!} = \left(\frac{r+1}{n+1}\right)$$

Similarly, $\frac{a_3}{a_3+a_4} = \frac{{}^nC_{r+2}}{{}^nC_{r+2} + {}^nC_{r+3}} = \frac{{}^nC_{r+2}}{{}^{n+1}C_{r+3}} = \frac{r+3}{n+1}$

Hence, $\frac{a_1}{a_1+a_2} + \frac{a_3}{a_3+a_4} = \frac{r+1}{n+1} + \frac{r+3}{n+1}$

$$= \frac{2r+4}{n+1} = \frac{2\,(r+2)}{n+1}$$

Also, $\frac{2a_2}{a_2+a_3} = \frac{2({}^nC_{r+1})}{{}^nC_{r+1} + {}^nC_{r+2}} = \frac{2({}^nC_{r+1})}{{}^{n+1}C_{r+2}}$

$$= 2\frac{n!}{(r+1)!\,(n-r-1)!} \times \frac{(r+2)!\,(n-r-1)!}{(n+1)!}$$

$$= \frac{2(r+2)}{n+1}$$

90. *(d)* $\because (3+2x)^{50} = 3^{50}\left(1+\frac{2x}{3}\right)^{50}$

Here, $T_{r+1} = 3^{50}\;{}^{50}C_r\left(\frac{2x}{3}\right)^r$

and $T_r = 3^{50}\;{}^{50}C_{r-1}\left(\frac{2x}{3}\right)^{r-1}$

But $x = \frac{1}{5}$ (given)

$\therefore \quad \frac{T_{r+1}}{T_r} \geq 1 \Rightarrow \frac{{}^{50}C_r}{{}^{50}C_{r-1}} \cdot \frac{2}{3} \cdot \frac{1}{5} \geq 1$

$\Rightarrow \quad 102 - 2r \geq 15r \Rightarrow r \leq 6$

$\Rightarrow \quad r = 6$

Therefore, there are two greatest terms T_r and T_{r+1} *i.e.*, T_6 and T_7.

91. *(c)* We have, $T_{r+1} = {}^6C_r\left(\frac{3}{2}x^2\right)^{6-r}\left(-\frac{1}{3x}\right)^r$

$$= {}^6C_r\left(\frac{3}{2}\right)^{6-r}(x^2)^{6-r}(-1)^r\left(\frac{1}{x}\right)^r\left(\frac{1}{3^r}\right)$$

$$= (-1)^r\;{}^6C_r\;\frac{(3)^{6-2r}}{(2)^{6-r}}\;x^{12-3r}$$

The term will be independent of x, if the index of x is zero *i.e.*, $12 - 3r = 0$. Thus, $r = 4$, hence 5th term is independent of x and is given by $(-1)^4\;{}^6C_4\;\frac{(3)^{6-8}}{(2)^{6-4}} = \frac{5}{12}$

92. (a) The $(r+1)^{th}$ term in the expansion is ${}^nC_r a^r$. Thus, it can be seen that a^r occurs in the $(r+1)^{th}$ term, and its coefficient is ${}^nC_r$. Hence, the coefficients of a^{r-1}, a^r and a^{r+1} are ${}^nC_{r-1}$, ${}^nC_r$ and ${}^nC_{r+1}$, respectively.

Since, these coefficients are in arithmetic progression. So, we have ${}^nC_{r-1} + {}^nC_{r+1} = 2.{}^nC_r$. This gives

$$\frac{n!}{(r-1)!(n-r+1)!} + \frac{n!}{(r+1)!(n-r-1)!} = 2 \times \frac{n!}{r!(n-r)!}$$

i.e., $$\frac{1}{(r-1)!(n-r+1)(n-r)(n-r-1)!} + \frac{1}{(r+1)(r)(r-1)!(n-r-1)!}$$

$$= 2 \times \frac{1}{r(r-1)!(n-r)(n-r-1)!}$$

or $$\frac{1}{(r-1)!(n-r-1)!}\left[\frac{1}{(n-r)(n-r+1)} + \frac{1}{(r+1)(r)}\right]$$

$$= 2 \times \frac{1}{(r-1)!(n-r-1)![r(n-r)]}$$

i.e., $$\frac{1}{(n-r+1)(n-r)} + \frac{1}{r(r+1)} = \frac{2}{r(n-r)},$$

or $$\frac{r(r+1)+(n-r)(n-r+1)}{(n-r)(n-r+1)r(r+1)} = \frac{2}{r(n-r)}$$

or $$r(r+1)+(n-r)(n-r+1) = 2(r+1)(n-r+1)$$

or $$r^2+r+n^2-nr+n-nr+r^2-r = 2(nr-r^2+r+n-r+1)$$

or $$n^2-4nr-n+4r^2-2=0$$

i.e., $$n^2-n(4r+1)+4r^2-2=0$$

93. (c) As $2n$ is even, so the expansion $(1+x)^{2n}$ has only one middle term which is $\left(\frac{2n}{2}+1\right)^{th}$, i.e., $(n+1)^{th}$ term.

The $(n+1)^{th}$ term is ${}^{2n}C_n x^n$. The coefficient of x^n is ${}^{2n}C_n$.

Similarly, $(2n-1)$ being odd, the other expansion has two middle terms, $\left(\frac{2n-1+1}{2}\right)^{th}$ and $\left(\frac{2n-1+1}{2}+1\right)^{th}$ i.e. n^{th} and $(n+1)^{th}$ terms. The coefficients of these terms are ${}^{2n-1}C_{n-1}$ and ${}^{2n-1}C_n$, respectively.

Now, ${}^{2n-1}C_{n-1} + {}^{2n-1}C_n = {}^{2n}C_n$ (as ${}^nC_{r-1} + {}^nC_r = {}^{n+1}C_r$) as required.

94. (d) We first expand each of the factors of the given product using binomial theorem. We have,

$$(1+2a)^4 = {}^4C_0 + {}^4C_1(2a) + {}^4C_2(2a)^2 + {}^4C_3(2a)^3 + {}^4C_4(2a)^4$$
$$= 1 + 4(2a) + 6(4a^2) + 4(8a^3) + 16a^4$$
$$= 1 + 8a + 24a^2 + 32a^3 + 16a^4$$

and $$(2-a)^5 = {}^5C_0(2)^5 - {}^5C_1(2)^4(a) + {}^5C_2(2)^3(a)^2 - {}^5C_3(2)^2(a)^3 + {}^5C_4(2)(a)^4 - {}^5C_5(a)^5$$
$$= 32 - 80a + 80a^2 - 40a^3 + 10a^4 - a^5$$

Thus, $(1+2a)^4(2-a)^5 = (1+8a+24a^2+32a^3+16a^4)(32-80a+80a^2-40a^3+10a^4-a^5)$

The complete multiplication of the two brackets need not be carried out. We write only those terms which involve a^4. This can be done, if we note that $a^r . a^{4-r} = a^4$. The terms containing a^4 are

$$1(10a^4) + (8a)(-40a^3) + (24a^2)(80a^2) + (32a^3)(-80a) + (16a^4)(32) = -438a^4$$

Thus, the coefficient of a^4 in the given product is -438.

So, only Statement II is correct .

95. (b) We have, $T_{r+1} = {}^{18}C_r (\sqrt[3]{x})^{18-r}\left(\frac{1}{2\sqrt[3]{x}}\right)^r$

$$= {}^{18}C_r x^{\frac{18-r}{3}} . \frac{1}{2^r x^{\frac{r}{3}}} = {}^{18}C_r \frac{1}{2^r} . x^{\frac{18-2r}{3}}$$

Since, we have to find a term independent of x, i.e., term not having x, so take $\frac{18-2r}{3} = 0$, we get $r = 9$.

The required term is ${}^{18}C_9 \frac{1}{2^9}$.

96. (d) The coefficients of $(r-5)^{th}$ and $(2r-1)^{th}$ terms of the expansion $(1+x)^{34}$ are ${}^{34}C_{r-6}$ and ${}^{34}C_{2r-2}$, respectively. Since, they are equal, so ${}^{34}C_{r-6} = {}^{34}C_{2r-2}$.

Therefore, either $r-6 = 2r-2$ or $r-6 = 34-(2r-2)$.

(using the fact that, if ${}^nC_r = {}^nC_p$, then either $r = p$ or $r = n-p$)

So, we get $r = -4$ or $r = 14$.

r being a natural number, $r = -4$ is not possible. So, $r = 14$.

97. (c) The coefficients of the first three terms of $\left(x - \frac{3}{x^2}\right)^m$ are ${}^mC_0$, $(-3)\,{}^mC_1$ and $9.{}^mC_2$. Therefore, by the given condition, we have

$${}^mC_0 - 3\,{}^mC_1 + 9.{}^mC_2 = 559, \text{ i.e., } 1 - 3m + \frac{9m(m-1)}{2} = 559$$

which gives $m = 12$ (m being a natural number).

Now, $$T_{r+1} = {}^{12}C_r x^{12-r}\left(-\frac{3}{x^2}\right)^r = {}^{12}C_r(-3)^r \cdot x^{12-3r}$$

Since, we need the term containing x^3, so put $12-3r=3$, i.e., $r = 3$.

Thus, the required term is ${}^{12}C_3(-3)^3x^3$, i.e., $-5940x^3$.

98. (a) There are $(n+1)$ terms in the expansion of $(x+a)^n$. Observing the terms, we can say that the first term from the end is the last term, i.e., $(n+1)^{th}$ term of the expansion and $n+1 = (n+1)-(1-1)$. The second term from the end is the nth term of the expansion and $n = (n+1)-(2-1)$.

The third term from the end is the $(n-1)^{th}$ term of the expansion and $n-1 = (n+1)-(3-1)$, and so on. Thus, r^{th}

term from the end will be term number $(n+1)-(r-1)=(n-r+2)$ of the expansion and the $(n-r+2)^{th}$ term is ${}^nC_{n-r+1}x^{r-1}a^{n-r+1}$.

So, both the statements are true and Statement II is the correct explanation of Statement I.

99. *(c)* The general term in the expansion of $(3+ax)^9$ is

$$T_{r+1} = {}^9C_r\, 3^{9-r}\, a^r\, x^r$$

For coefficient of x^2, put $r=2$.

$$\therefore \quad T_{2+1} = {}^9C_2\, 3^{9-2}\, a^2\, x^2$$

$$\Rightarrow \text{Coefficient of } x^2 = {}^9C_2\, 3^7\, a^2 \quad \ldots(i)$$

For coefficient of x^3, put $r=3$.

$$\therefore \quad T_{3+1} = {}^9C_3\, 3^{9-3}\, a^3x^3 = {}^9C_3\, 3^6\, a^3x^3$$

$$\Rightarrow \text{Coefficient of } x^3 = {}^9C_3\, 3^6 a^3 \quad \ldots(ii)$$

Given, coefficient of x^2 = Coefficient of x^3

$$\Rightarrow \quad {}^9C_2 3^7a^2 = {}^9C_3 3^6 a^3 \quad \text{[from Eqs. (i) and (ii)]}$$

$$\Rightarrow \quad \frac{9\times 8}{2}\times 3\times 1 = \frac{9\times 8\times 7}{6}\times 1\times a$$

$$\Rightarrow \quad \frac{3}{2} = \frac{7}{6}\times a \Rightarrow a = \frac{9}{7}$$

100. *(a)* Since, $(a+b)^n = {}^nC_0a^n + {}^nC_1\, a^{n-1}b + {}^nC_2\, a^{n-2}b^2 + \ldots + {}^nC_nb^n$

Given that, ${}^nC_0\, a^n = 729 \Rightarrow a^n = 729 \quad \ldots(i)$

Then, ${}^nC_1\, a^{n-1}b = 7290$

$$\Rightarrow \quad na^{n-1}b = 7290 \quad \ldots(ii)$$

$${}^nC_2a^{n-2}b^2 = 30375$$

$$\Rightarrow \quad \frac{n(n-1)}{2}a^{n-2}\, b^2 = 30375 \quad \ldots(iii)$$

Now, multiplying Eqs. (i) and (iii) and divide by square of Eq. (ii), we get

$$\Rightarrow \quad \frac{a^n \times \frac{n(n-1)}{2}\times a^{n-2}\, b^2}{(na^{n-1}b)^2} = \frac{729\times 30375}{7290\times 7290}$$

$$\Rightarrow \quad \frac{a^{2n-2}\frac{n(n-1)}{2}b^2}{n^2\, a^{2n-2}\, b^2} = \frac{30375}{10\times 7290}$$

$$\Rightarrow \quad \frac{n-1}{2n} = \frac{6075}{10\times 1458}$$

$$\Rightarrow \quad \frac{n-1}{2n} = \frac{1215}{2\times 1458}$$

$$\Rightarrow \quad \frac{n-1}{n} = \frac{1215}{1458}$$

$$\Rightarrow \quad 1458n - 1215n = 1458$$

$$\Rightarrow \quad 243n = 1458$$

$$\Rightarrow \quad n = \frac{1458}{243} \Rightarrow n = 6$$

On putting $n=6$ in Eq. (i), we get

$$\Rightarrow \quad a^6 = 729 \Rightarrow a^6 = 3^6 \Rightarrow a = 3$$

From Eq. (ii), ${}^6C_1\times(3)^{6-1}\times b = 7290$

$$\Rightarrow \quad 6\times 3^5\times b = 7290$$

$$\Rightarrow \quad 6\times 243\times b = 7290$$

$$\Rightarrow \quad b = \frac{7290}{6\times 243} = \frac{30}{6} = 5$$

Hence, $a=3, b=5, n=6$

101. *(d)* $(1+2x)^6(1-x)^7$

$$= \{{}^6C_0 + {}^6C_1(2x) + {}^6C_2(2x)^2 + {}^6C_3(2x)^3 + {}^6C_4(2x)^4 + {}^6C_5(2x)^5 + {}^6C_6(2x)^6\}$$
$$\times\{{}^7C_0 - {}^7C_1x + {}^7C_2x^2 - {}^7C_3x^3 + {}^7C_4x^4 - {}^7C_5x^5 + {}^7C_6x^6 - {}^7C_7x^7\}$$

$$= ({}^6C_0 + {}^6C_1\times 2\times x + {}^6C_2\times 4\times x^2 + {}^6C_3\times 8\times x^3 + {}^6C_4\times 2^4\times x^4 + {}^6C_5\times 2^5\times x^5 + {}^6C_6\, 2^6\times x^6)$$
$$\times({}^7C_0 - {}^7C_1x + {}^7C_2x^2 - {}^7C_3x^3 + {}^7C_4x^4 - {}^7C_5x^5 + {}^7C_6x^6 - {}^7C_7x^7)$$

For coefficient of x^5,

$$= {}^6C_0(-{}^7C_5) + {}^6C_1\times 2\times {}^7C_4 + {}^6C_2\times 4\times(-{}^7C_3) + {}^6C_3\times 8\times {}^7C_2 + {}^6C_4\times 2^4\times(-{}^7C_1) + {}^6C_5\times 2^5({}^7C_0)$$

$$= -{}^6C_0\times {}^7C_2 + {}^6C_1\times 2\times {}^7C_3 - {}^6C_2\times 4\times {}^7C_3 + {}^6C_3\times 8\times {}^7C_2 - {}^6C_2\times 16\times {}^7C_1 + {}^6C_1\times 2^5 \quad (\because {}^nC_r = {}^nC_{n-r})$$

$$= -\frac{1\times 7\times 6}{2} + \frac{6\times 2\times 7\times 6\times 5}{6} - \frac{6\times 5}{2}\times 4\times\frac{7\times 6\times 5}{6} + \frac{6\times 5\times 4}{6}\times\frac{8\times 7\times 6}{2} - \frac{6\times 5}{2}\times 16\times 7 + 6\times 32$$

$$= -21 + 12\times 35 - 15\times 28\times 5 + 20\times 24\times 7 - 15\times 16\times 7 + 192$$

$$= -21 + 420 - 420\times 5 + 3360 - 1680 + 192$$

$$= 3972 - 2100 - 21 - 1680$$

$$= 3972 - 3801 = 171$$

102. *(a)* If we take $a^n = (a-b+b)^n$

$$= {}^nC_0(a-b)^n + {}^nC_1(a-b)^{n-1}b + {}^nC_2(a-b)^{n-2}b^2 + \ldots\ldots + {}^nC_nb^n$$

$$\Rightarrow a^n = (a-b)[{}^nC_0(a-b)^{n-1} + {}^nC_1(a-b)^{n-2}b + {}^nC_2(a-b)^{n-3}b^2 + \ldots] + b^n$$

$$\Rightarrow \quad a^n - b^n = (a-b)[{}^nC_0(a-b)^{n-1} + {}^nC_1(a-b)^{n-2}b + \ldots]$$

$\Rightarrow (a-b)$ is a factor of $a^n - b^n$

103. *(b)* Use,

$$(x+y)^6 = {}^6C_0\,x^6 + {}^6C_1x^5y + {}^6C_2\,x^4y^2 + {}^6C_3x^3y^3 + {}^6C_4\,x^2y^4 + {}^6C_5\,xy^5 + {}^6C_6y^6$$

$$= {}^6C_0x^6 + {}^6C_1x^5y + {}^6C_2x^4y^2 + {}^6C_3x^3y^3 + {}^6C_2x^2y^4 + {}^6C_1xy^5 + {}^6C_0y^6 \quad (\because {}^nC_r = {}^nC_{n-r})$$

$$= x^6 + 6x^5y + \frac{6\times5}{2}x^4y^2 + \frac{6\times5\times4}{6}x^3y^3 + \frac{6\times5}{2}x^2y^4 + 6xy^5 + y^6$$

$$\Rightarrow (x+y)^6 = x^6 + 6x^5y + 15x^4y^2 + 20x^3y^3 + 15x^2\,y^4 + 6xy^5 + y^6 \quad \text{...(i)}$$

Similarly, $(x-y)^6 = x^6 - 6x^5y + 15x^4y^2 - 20x^3y^3 + 15x^2y^4 - 6xy^5 + y^6$ ...(ii)

Subtracting Eq. (ii) from Eq. (i), we get

$$(x+y)^6 - (x-y)^6 = 12x^5y + 40x^3y^3 + 12xy^5$$
$$= 4xy\,[3x^4 + 10x^2y^2 + 3y^4]$$

Putting $x=\sqrt{3}$ and $y=\sqrt{2}$, we get

$$(\sqrt{3}+\sqrt{2})^6 - (\sqrt{3}-\sqrt{2})^6$$
$$= 4\sqrt{3}\sqrt{2}\,[3(\sqrt{3})^4 + 10(\sqrt{3})^2(\sqrt{2})^2 + 3(\sqrt{2})^4]$$
$$= 4\sqrt{6}\,[3\times3^2 + 10\times3\times2 + 3\times2^2]$$
$$= 4\sqrt{6}\,[3\times9 + 60 + 3\times4]$$
$$= 4\sqrt{6}\,[27+60+12]$$
$$= 4\sqrt{6}\times(39+60)$$
$$= 4\sqrt{6}\times99 = 396\sqrt{6}$$

104. *(c)* $\left[\left(1+\frac{x}{2}\right)-\frac{2}{x}\right]^4 = {}^4C_0\left(1+\frac{x}{2}\right)^4 - {}^4C_1\left(1+\frac{x}{2}\right)^3\left(\frac{2}{x}\right) + {}^4C_2\left(1+\frac{x}{2}\right)^2\left(\frac{2}{x}\right)^2 - {}^4C_3\left(1+\frac{x}{2}\right)\left(\frac{2}{x}\right)^3 + {}^4C_4\left(\frac{2}{x}\right)^4$

$$= \left(1+\frac{x}{2}\right)^4 - 4\left(1+\frac{x}{2}\right)^3\frac{2}{x} + \frac{4\times3}{2}\left(1+\frac{x}{2}\right)^2\frac{4}{x^2} - 4\left(1+\frac{x}{2}\right)\times\frac{8}{x^3} + \frac{16}{x^4}$$

$$= \left(1+\frac{x}{2}\right)^4 - \frac{8}{x}\left(1+\frac{x}{2}\right)^3 + \frac{24}{x^2}\left(1+\frac{x}{2}\right)^2 - \frac{32}{x^3}\left(1+\frac{x}{2}\right) + \frac{16}{x^4}$$

Now, open the expansion of $\left(1+\frac{x}{2}\right)^4$, $\left(1+\frac{x}{2}\right)^3$ and $\left(1+\frac{x}{2}\right)^2$, we get

$$\left(1+\frac{x}{2}-\frac{2}{x}\right)^4 = \left(1+4\cdot\frac{x}{2}+6\cdot\frac{x^2}{4}+4\cdot\frac{x^3}{8}+\frac{x^4}{16}\right) - 8\cdot\frac{1}{x}\left(1+3\cdot\frac{x}{2}+3\cdot\frac{x^2}{4}+\frac{x^3}{8}\right) + 24\cdot\frac{1}{x^2}\left(1+x+\frac{x^2}{4}\right) - 32\times\frac{1}{x^3}\left(1+\frac{x}{2}\right) + \frac{16}{x^4}$$

$$= \left(1+2x+\frac{3x^2}{2}+\frac{x^3}{2}+\frac{x^4}{16}\right) - \left(\frac{8}{x}+12+6x+x^2\right) + \left(\frac{24}{x^2}+\frac{24}{x}+6\right) - \left(\frac{32}{x^3}+\frac{16}{x^2}\right) + \frac{16}{x^4}$$

$$= \frac{x^4}{16} + \frac{x^3}{2} + x^2\left(\frac{3}{2}-1\right) + x\,(2-6) + (1-12+6) + (24-8)\frac{1}{x} + (24-16)\frac{1}{x^2} - \frac{32}{x^3} + \frac{16}{x^4}$$

$$= \frac{x^4}{16} + \frac{x^3}{2} + \frac{x^2}{2} - 4x - 5 + \frac{16}{x} + \frac{8}{x^2} - \frac{32}{x^3} + \frac{16}{x^4}$$

105. *(a)* Let $a^2 = x$ and $\sqrt{a^2-1} = y$

Now, $(x+y)^4 = {}^4C_0x^4 + {}^4C_1x^3y + {}^4C_2x^2y^2 + {}^4C_3\,xy^3 + {}^4C_4y^4$

$$= {}^4C_0x^4 + {}^4C_1\,x^3y + {}^4C_2\,x^2y^2 + {}^4C_1xy^3 + {}^4C_0y^4 \quad (\because {}^nC_r = {}^nC_{n-r})$$

$$= x^4 + 4x^3y + \frac{4\times3}{2}x^2y^2 + 4xy^3 + y^4$$

$\Rightarrow (x+y)^4 = x^4 + 4x^3y + 6x^2y^2 + 4xy^3 + y^4$ ...(i)

Similarly, $(x-y)^4 = x^4 - 4x^3y + 6x^2y^2 - 4xy^3 + y^4$ ...(ii)

Adding Eqs. (i) and (ii), we get

$$(x+y)^4 + (x-y)^4 = 2x^4 + 12x^2y^2 + 2y^4$$
$$= 2\,[x^4 + 6x^2y^2 + y^4]$$

Putting the values of x and y *i.e.*, $x=a^2$ and $y=\sqrt{a^2-1}$

$\therefore \quad (a^2+\sqrt{a^2-1})^4 + (a^2-\sqrt{a^2-1})^4$

$$= 2\,[(a^2)^4 + 6(a^2)^2(\sqrt{a^2-1})^2 + (\sqrt{a^2-1})^4]$$
$$= 2\,[a^8 + 6a^4(a^2-1) + (a^2-1)^2]$$
$$= 2\,[a^8 + 6a^6 - 6a^4 + a^4 + 1 - 2a^2]$$
$$= 2\,[a^8 + 6a^6 - 5a^4 - 2a^2 + 1]$$
$$= 2a^8 + 12a^6 - 10a^4 - 4a^2 + 2$$

106. *(c)* We know that, in the expansion of $(a+b)^n$, p^{th} term from the end is $(n-p+2)^{th}$ term from the beginning.

So, fifth term from the end is

$= (n-5+2)^{th}$ term from the beginning

$= (n-3)^{th}$ term from the beginning

$= (n-4+1)^{th}$ term from the beginning ...(i)

$\therefore$ We have, $\left(\sqrt[4]{2}+\frac{1}{\sqrt[4]{3}}\right)^n = \left(2^{1/4}+\frac{1}{3^{1/4}}\right)^n$

Now, fifth term from the beginning is

$$T_{4+1} = {}^nC_4(2^{1/4})^{n-4}(3^{-1/4})^4 = {}^nC_4\,2^{\frac{n-4}{4}}\,3^{-1} \quad \text{...(ii)}$$

$(\because {}^nC_r = {}^nC_{n-r})$

And, fifth term from the end is

$$T_{(n-4)+1} = {}^nC_{n-4}\,(2^{1/4})^{n-n+4}\,(3^{-1/4})^{n-4}$$
$$= {}^nC_4\,2\cdot3^{\frac{-n+4}{4}} \quad \text{...(iii)}$$

So, from the given condition, we have

$$\frac{\text{Fifth term from the beginning}}{\text{Fifth term from the end}} = \frac{\sqrt{6}}{1}$$

$$\Rightarrow \quad \frac{{}^nC_4 \cdot 2^{\frac{n-4}{4}} \cdot 3^{-1}}{{}^nC_4 \cdot 2 \cdot 3^{\frac{-n+4}{4}}} = \frac{\sqrt{6}}{1}$$

$$\Rightarrow \quad 2^{\frac{n-4}{4}-1} \cdot 3^{-1-\left(\frac{4-n}{4}\right)} = \sqrt{6}$$

$$\Rightarrow \quad 2^{\frac{n-8}{4}} \cdot 3^{\frac{n-8}{4}} = 6^{1/2}$$

$$\Rightarrow \quad (2 \times 3)^{\frac{n-8}{4}} = (2.3)^{1/2}$$

$$\Rightarrow \quad \frac{n-8}{4} = 1/2$$

$$\Rightarrow \quad n = 2 + 8$$

$$\therefore \quad n = 10$$

107. *(c)* Now, $(0.99)^5 = (1 - 0.01)^5$

$$= {}^5C_0(1)^5 - {}^5C_1(1)^4(0.01) + {}^5C_2(1)^3(0.01)^2$$

(ignore the other terms)

$$= 1 - 5 \times 1 \times 0.01 + \frac{5 \times 4}{2} \times 1 \times 0.01 \times 0.01$$

$$= 1 - 0.05 + 10 \times 0.0001 = 1 - 0.05 + 0.001$$

$$= 1.001 - 0.05 = 0.951$$

108. *(b)* $(3x^2 - 2ax + 3a^2)^3 = [3(x^2 + a^2) - 2ax]^3$

$$= {}^3C_0\{3(x^2 + a^2)\}^3 - {}^3C_1\{3(x^2 + a^2)\}^2\, 2ax + {}^3C_2\, 3(x^2 + a^2)(2ax)^2 - {}^3C_3(2ax)^3$$

$$= 27(x^2 + a^2)^3 - 3 \times 9(x^2 + a^2)^2 \times 2ax + 3 \times 3(x^2 + a^2)\, 4a^2x^2 - 8a^3x^3$$

Now, open the expansion of $(x^2 + a^2)^3$, $(x^2 + a^2)^2$, we get

$$= 27[{}^3C_0(x^2)^3 + {}^3C_1(x^2)^2a^2 + {}^3C_2x^2(a^2)^2 + {}^3C_3(a^2)^3] - 27[{}^2C_0(x^2)^2 + {}^2C_1x^2a^2 + {}^2C_2(a^2)^2] \times 2ax + 9(x^2 + a^2)\, 4a^2x^2 - 8a^3x^3$$

$$= 27[x^6 + 3x^4a^2 + 3x^2a^4 + a^6] - 27[x^4 + 2x^2a^2 + a^4]2ax + 36a^2x^2(x^2 + a^2) - 8a^3x^3$$

$$= 27x^6 + 81x^4a^2 + 81x^2a^4 + 27a^6 - 54ax(x^4 + 2x^2a^2 + a^4) + 36a^2x^4 + 36a^4x^2 - 8a^3x^3$$

$$= 27x^6 + 117x^4a^2 + 117a^4x^2 + 27a^6 - 54ax^5 - 54a^5x - 8a^3x^3 - 108a^3x^3$$

$$= 27x^6 - 54ax^5 + 117a^2x^4 - 116a^3x^3 + 117a^4x^2 - 54a^5x + 27a^6$$

109. *(d)* The general term of $\left(\frac{3x^2}{2} - \frac{1}{3x}\right)^{15}$ is

$$T_{r+1} = {}^{15}C_r\left(\frac{3x^2}{2}\right)^{15-r}\left(\frac{-1}{3x}\right)^r$$

$$= {}^{15}C_r\left(\frac{3}{2}\right)^{15-r}(x^2)^{15-r}\frac{(-1)^r}{3^r x^r}$$

$$= {}^{15}C_r\left(\frac{3}{2}\right)^{15-r}x^{30-2r-r}\frac{(-1)^r}{3^r}$$

$$= {}^{15}C_r\left(\frac{3}{2}\right)^{15-r}x^{30-3r}\frac{(-1)^r}{3^r}$$

For independent of x, *i.e.*, term not having x.

Putting, $30 - 3r = 0 \Rightarrow r = 10$

So, 11th term is independent of x.

Note *Term independent of x means constant term.*

110. *(d)* $(x + a)^n = {}^nC_0x^n + {}^nC_1x^{n-1}a + {}^nC_2x^{n-2}a^2 + \ldots + {}^nC_na^n$

Given, $O = {}^nC_0x^n + {}^nC_2x^{n-2}a^2 + \ldots$

and $E = {}^nC_1x^{n-1}a + {}^nC_3x^{n-3}a^3 + \ldots$

$\therefore \quad O + E = (x + a)^n$ ...(i)

Similarly, $O - E = (x - a)^n$ ...(ii)

Now, multiplying Eqs. (i) and (ii), we get

I. $(O + E)(O - E) = (x + a)^n(x - a)^n$

$$\Rightarrow \quad (O^2 - E^2) = (x^2 - a^2)^n$$

II. Squaring Eqs. (i) and (ii) and then subtracting, we get

$$\Rightarrow \quad (O + E)^2 - (O - E)^2 = [(x + a)^n]^2 - [(x - a)^n]^2$$

$$\Rightarrow \quad O^2 + E^2 + 2OE - O^2 - E^2 + 2OE = (x + a)^{2n} - (x - a)^{2n}$$

$$\Rightarrow \quad 4OE = (x + a)^{2n} - (x - a)^{2n}$$

So, only Statement II is true.

Note *Term independent of x means constant term.*

9

Sequences *and* Series

Introduction of Sequences and Series

Level I

1. The collection of objects listed in a sequence is
(a) random (b) ordered
(c) random or ordered (d) None of these

2. Which of the following is not an example of sequence?
(a) Population of human beings at different times
(b) Amount of money deposited in a bank over a number of years
(c) Living and non-living things in Delhi
(d) Depreciated values of certain commodity

3. Sequences, which follow specific patterns, are called
(a) regressions (b) progressions
(c) reprogressions (d) reversions

4. If there is a generation gap of 30 yr, then

Statement I Total number of generations are 20 over 300 yr.

Statement II The number of person's ancestors for the 10th generation is 1024.
(a) Statement I and Statement II are true
(b) Statement I is true and Statement II is false
(c) Statement I is false and Statement II is true
(d) Statement I and Statement II are false

5. The successive quotients obtained after dividing 10 by 3 form a sequence which is defined as
(a) 3.3, 3.3, 3.3, ...
(b) 3.3, 3.33, 3.333, 3.3333,................
(c) 3. 3.3, 3.3, 3.3, 3.3, ...
(d) 3, 3.3, 3.33, 3.333, 3.3333, ...

6. The various numbers occurring in a sequence are called its
(a) points (b) numbers (c) terms (d) values

7. The terms of a sequence are denoted as $a_1, a_2, a_3, \ldots\ldots a_n \ldots$ etc. Then, subscripts denote the
(a) position of the term
(b) value of the term
(c) number to be added within term
(d) number to be subtracted from term

8. The general term of the sequence is denoted by
(a) a_n (b) a^n (c) n (d) $n \cdot a$

9. What are the terms of the sequence 2, 4, 8, 16, 32, ..., 1024 whose subscripts are multiples of 3?
(a) 8, 16, 32 (b) 8, 64, 512
(c) 64, 2, 1024 (d) 2, 8, 32, 1024

10. If the given sequence is 3,3.3, 3.33, 3.333, 3.3333, 3.33333, then match the statements of Column I with terms of Column II and choose the correct option from the codes given below.

Column I	Column II
A. The terms whose subscripts are odd numbers.	1. 3.3, 3.333, 3.33333
B. The terms whose subscripts are even numbers.	2. 3.3333
C. The terms whose subscripts are multiple of 3.	3. 3, 3.33, 3.3333
D. The terms whose substcripts are multiple of 5.	4. 3.33, 3.33333

Codes

	A	B	C	D
(a)	3	4	1	2
(b)	1	2	3	4
(c)	3	1	4	2
(d)	4	2	1	3

11. A sequence containing finite number of terms is called a/an ...A... sequence. Here, A refers to

(a) indefinite (b) finite
(c) endless (d) indeterminable

12. A sequence is a finite sequence, if it contains n numbers of terms, where n is

(a) odd (b) even
(c) positive (d) fixed

13. A sequence is called a/an ...A... sequence, if it is not a finite sequence. Here, A refers to

(a) bounded (b) limited
(c) infinite (d) None of these

14. The successive quotients 3,3.3, 3.33, 3.333, 3.3333, ... and so on is a/an ...A... sequence, since it ...B.. ends. Here, A and B refers to

(a) finite, never (b) infinite, never
(c) finite, always (d) infinite, always

15. If the sequence of even natural numbers is 2, 4, 6, 8, ..., then what is the nth term of the sequence?

(a) $a_n = n^2$, where $n \in N$ (b) $a_n = 2^n$, where $n \in N$
(c) $a_n = 2n$, where $n \in N$ (d) $a_n = \frac{n}{2}$, where $n \in N$

16. If the sequence of odd natural numbers is 1, 3, 5, 7, ... , then what is the nth term of the sequence?

(a) $a_n = 2n+1$, where $n \in N$ (b) $a_n = 2n-1$, where $n \in N$
(c) $a_n = 2n-2$, where $n \in N$ (d) $a_n = 2n+2$, where $n \in N$

17. If the arrangement of numbers is 1, 1, 2, 3, 5, 8, ... and so on, the sequence is generated by the recurrence relation.

Statement I This sequence is called fibonacci sequence.

Statement II The nth term of the sequence is $a_n = a_{n-2} + a_{n-1}, n > 2$ and $n \in N$

(a) Statement I and Statement II are true
(b) Statement I is true and Statement II is false
(c) Statement I is false and Statement II is true
(d) Statement I and Statement II are false

18. In the sequence of primes 2, 3, 5, 7, ..., the nth term is

(a) $a_n = n+1$, where $n \in N$
(b) $a_n = n+2$, where $n \in N$
(c) Does not exist
(d) None of the above

19. If a sequence is given, then

Statement I We should expect that its terms will necessarily be given by a specific formula.

Statement II We expect a theoretical scheme or a rule for generating the terms $a_1, a_2, \ldots\ldots a_n, \ldots$ in succession.

(a) Statement I and Statement II are true
(b) Statement I is true and Statement II is false
(c) Statement I is false and Statement II is true
(d) Statement I and Statement II are false

20. A sequence can be regarded as a function whose domain is the set of

(a) whole numbers (b) integers
(c) rational numbers (d) natural numbers

21. The domain of a sequence can be subset of natural numbers of the type

(a) $\{3, 4, 5, \ldots\ldots\ldots\ldots, k\}$ (b) $\{1, 2, 3, 4, 5, \ldots\ldots\ldots\ldots, k\}$
(c) $\{0, 1, 2, 3, 4, \ldots\ldots\ldots\ldots, k\}$ (d) None of these

22. The nth term a_n of a sequence can also be denoted as

(a) $n(a)$ (b) $a(n)$
(c) $(n)(a)$ (d) None of these

Series

23. Let $a_1, a_2, a_3, \ldots\ldots, a_n, \ldots\ldots$ be a given sequence. Then, which of the following expressions is called the series associated with the given sequence?

(a) $a_1 - a_2 + a_3 - a_4 \ldots\ldots\ldots + a_n - \ldots\ldots\ldots$
(b) $-a_1 - a_2 - a_3 - a_4 \ldots\ldots\ldots - a_n - \ldots\ldots$
(c) $a_1a_2 + a_2a_3 + \ldots\ldots\ldots + a_{n-1}a_n + \ldots\ldots$
(d) $a_1 + a_2 + a_3 + a_4 \ldots\ldots\ldots + a_n + \ldots\ldots$

24. The series is finite or infinite according as the given sequence is ....

(a) infinite (b) finite
(c) finite or infinite (d) infinite or finite

25. Series are often represented in compact form, called

(a) beta notation (b) nano notation
(c) sigma notation (d) neu notation

26. The ...A... letter ...B... (sigma) is used for indicating the summation. Here, A and B refers to

(a) French, $\sum$ (b) Greek, $\sum$
(c) French, Π (d) Greek, Π

27. The series $a_1 + a_2 + a_3 + \ldots\ldots\ldots\ldots a_n$ is abbreviated as .....

(a) $\sum_{k=1}^{n} a_k$ (b) $\prod_{k=1}^{n} a_k$ (c) $\sum_{k=1}^{n} a_n$ (d) $\prod_{k=1}^{n} a_n$

28. Match the terms of Column I with terms of Column II and choose the correct option from the codes given below.

Column I	Column II
A. Series refers to	1. Finite series with four terms.
B. Sum of a series means	2. 16
C. 1+ 3+ 5+ 7	3. Indicated sum not to the sum itself.
D. Sum of the series 1+ 3+ 5+ 7 is	4. The number that results from adding the terms.

Codes

	A	B	C	D
(a)	3	1	4	2
(b)	4	3	1	2
(c)	2	4	1	3
(d)	3	4	1	2

29. If the nth terms of sequences are given in Column I and corresponding five terms are given in Column II, then match the terms of Column I with the terms of Column II and choose the correct option from the codes given below.

Column I	Column II
A. $a_n = n(n+2)$	1. $\frac{-1}{6}, \frac{1}{6}, \frac{1}{2}, \frac{5}{6}, \frac{7}{6}$
B. $a_n = \frac{n}{n+1}$	2. 2, 4, 8, 16, 32
C. $a_n = 2^n$	3. 3, 8, 15, 24, 35
D. $a_n = \frac{2n-3}{6}$	4. $\frac{1}{2}, \frac{2}{3}, \frac{3}{4}, \frac{4}{5}, \frac{5}{6}$

Codes

	A	B	C	D
(a)	3	4	2	1
(b)	1	4	2	3
(c)	3	2	4	1
(d)	1	3	4	2

30. If the nth term of a sequence is $a_n = 4n-3$, then a_{17} and a_{24} are

(a) 60, 90 (b) 65, 95
(c) 65, 93 (d) 55, 95

31. If the nth term of a sequence is $a_n = (-1)^{n-1} n^3$, then which of the following is 9th term?

(a) 630 (b) 290 (c) 216 (d) 729

32. The first five terms of the sequence, where $a_1 = 3$ and $a_n = 3a_{n-1} + 2$ for all $n > 1$ are

(a) 3, 15, 40, 110, 330 (b) 3, 11, 35, 107, 323
(c) 3, 20, 45, 110, 330 (d) 3, 11, 40, 107, 323

33. If first five terms of a sequence are 2, 2, ...A..., ...B... and ...C... where $a_n = a_{n-1} - 1$, $n > 2$, then A, B and C refers to

(a) 1, −1, −2 (b) 0, −1, −2
(c) 1, 0, −1 (d) 0, −2, −3

Level II

34. If n^{th} term of a sequence is $a_n = 2n^2 - n + 1$, then which of the following is/are true?

I. First and second terms of same sequence are 2 and 7 respectively.
II. Third and fourth terms of same sequence are 16 and 29, respectively.

(a) Both I and II are true (b) Only I is true
(b) Only II is true (d) Both I and II false

35. A sequence may be defined as a [NCERT Exemplar]

(a) relation, whose range $\subseteq N$ (natural number)
(b) function whose range $\subseteq N$
(c) function whose domain $\subseteq N$
(d) progression having real values

36. The first five terms of the sequence whose n^{th} term is $a_n = (-1)^{n-1} 5^{n+1}$ are

(a) 25, −125, 625, −3125, 15625
(b) −25, 125, −625, 3125, −15625
(c) 5, −25, 125, −625, 3125
(d) −5, 25, −125, 625, −1325

37. The first five terms of the sequence are x, y, $\frac{21}{2}$, u, v whose n^{th} term is $a_n = \frac{n(n^2+5)}{4}$. Then,

(a) $x = \frac{3}{2}, y = \frac{5}{2}$ (b) $y = \frac{9}{2}, u = 21$
(c) $u = 21, v = \frac{73}{2}$ (d) $x = \frac{5}{2}, v = \frac{73}{2}$

38. Which of the following is/are true?

I. If n^{th} term of a sequence is $a_n = \frac{n^2}{2^n}$, then its 7th term is $\frac{49}{128}$.
II. If n^{th} term of a sequence is $a_n = \frac{n(n-2)}{n+3}$, then its 20th term is $\frac{323}{22}$.

(a) Both I and II are true (b) I is true and II is false
(c) I is false and II is true (d) Both I and II are false

39. If first term of a sequence is −1 and $a_n = \frac{a_{n-1}}{n}$, $n \geq 2$, then which of the following is/are true?

I. The first five terms of a given sequence are $-1, \frac{-1}{2}, \frac{-1}{6}, \frac{-1}{24}, \frac{-1}{120}$.
II. The sum of first five terms is greater than −2. [HOTS]

(a) OnlyI is true (b) Only II is true
(c) Both I and II are true (d) Both I and II are false

40. The Fibonacci sequence is defined by $a_1 = a_2 = 1$ and $a_n = a_{n-1} + a_{n-2}, n > 2$. Then, match the terms and ratios given in Column I with their values given in Column II and choose the correct option from the codes given below.

	Column I		Column II
A.	a_1, a_2, a_3	1.	2, 3, 5
B.	a_3, a_4, a_5	2.	$\frac{3}{2}, \frac{5}{3}, \frac{8}{5}$
C.	$\frac{a_4}{a_3}, \frac{a_5}{a_4}, \frac{a_6}{a_5}$	3.	$1, 2, \frac{3}{2}$
D.	$\frac{a_2}{a_1}, \frac{a_3}{a_2}, \frac{a_4}{a_3}$	4.	1, 1, 2

Codes

	A	B	C	D
(a)	4	1	3	2
(b)	4	1	2	3
(c)	1	4	3	2
(d)	1	4	2	3

Arithmetic Progression

Level I

41. A sequence $a_1, a_2, a_3, \ldots\ldots, a_n, \ldots\ldots$ is called arithmetic sequence or arithmetic progression, if

(a) $a_{n+1} = a_{n-1} + d,\ n \in N$, where a_1, d are first term and common difference respectively
(b) $a_{n-1} = a_{n+1} + d,\ n \in N$, where a_1, d are first term and common term respectively
(c) $a_{n+1} = a_n + d, n \in N$, where a_1, d are first term and common difference respectively
(d) $a_{n+1} = a_n + d, n \in N$, where a_1, d are first term and common term respectively

42. If the first term is a and the common difference is d, then the arithmetic progression is

(a) $a + d, a + 2d, a + 3d, \ldots\ldots$
(b) $a, a + d, a + 2d, a + 3d, \ldots\ldots$
(c) $a - d, a - 2d, a - 3d, \ldots\ldots$
(d) $a, a - d, a - 2d, a - 3d, \ldots\ldots$

43. If the first term is a and common difference is d, then the nth term (general term) of A.P. is

(a) $a_n = a + (n - 1)d$
(b) $a_n = a + (n + 1)d$
(c) $a_n = a + nd$
(d) $a_n = a - nd$

44. Which of the following is incorrect?

(a) If a constant is added to each term of an A.P., the resulting sequence is also an A.P.
(b) If a constant is subtracted from each term of an A.P., the resulting sequence is also an A.P.
(c) If each term of an A.P. is multiplied by a constant, then the resulting sequence is also an A.P.
(d) If each term of an A.P. is divided by a constant, then the resulting sequence is also an A.P.

45. For an A.P., match the notations in Column I with their meaning in Column II and choose the correct option from the codes given below.

	Column I	Column II
A.	a	1. Number of terms
B.	l	2. Common difference
C.	d	3. Sum to n terms of an A.P.
D.	n	4. First term
E.	S_n	5. Last term

Codes

	A	B	C	D	E
(a)	4	1	2	3	5
(b)	4	5	2	1	3
(c)	5	4	1	2	3
(d)	4	1	2	5	3

46. Let $a, a + d,\ a + 2d, \ldots\ldots, a + (n - 1)d$ be an A.P. Then,

Statement I $l = a + (n - 1)d$

Statement II $S_n = \frac{n}{2}[2a + (n + 1)d]$

(a) Statement I and Statement II are true
(b) Statement I is true and Statement II is false
(c) Statement I is false and Statement II is true
(d) Statement I and Statement II are false

47. Let $a, a + d, a + 2d, \ldots\ldots, a + (n - 1)d$ be an A.P., then $S_n = \frac{n}{2}[2a + (n - 1)d]$ can also be written as

(a) $S_n = \frac{n}{2}[a - l]$
(b) $S_n = n[a + l]$
(c) $S_n = \frac{n}{2}[a + l]$
(d) $S_n = n[a - l]$

48. In an A.P., if m^{th} term is n and the nth term is m, where $m \neq n$, then p^{th} term is

(a) $m+n-p$ (b) $m-n+p$
(c) $n-m+p$ (d) $m+n+p$

49. If the sum of n terms of an A.P. is $nP + \frac{1}{2}n(n-1)Q$, where P and Q are constants, then the common difference is

(a) P (b) Q
(c) $P+Q$ (d) $P-Q$

50. The income of a person is ₹ 300000 in the first year and he receives an increment of ₹ 10000 to his income per year for the next 19 yr. Then, the total amount he received in 20 yr, is

(a) 7900000 (b) 6000000 (c) 8000000 (d) 6900000

51. If A is the arithmetic mean of the numbers a and b, then

(a) $A, a, b,$ are in A.P.
(b) $a, A, b,$ are in A.P.
(c) $A, b, a,$ are in A.P.
(d) None of the above

52. If A is the arithmetic mean of the numbers a and b, then which of the following is not correct?

(a) $A = \frac{a+b}{2}$ (b) $A-a=b-A$
(c) $a-A=A-b$ (d) $A=\frac{a-b}{2}$

53. The arithmetic mean A. M. of two numbers a and b is interpreted as their

(a) addition (b) average
(c) difference (d) multiplication

54. Which of the following is/are true?

I. The arithmetic mean of two numbers 4 and 16 is 10.

II. The sequence 4, 10, 16 is an A.P. .

(a) Only I is true (b) Only II is true
(c) Both I and II are true (d) Neither I nor II is true

55. If x and y are inserted between 4 and 16, so that the resulting sequence becomes an A.P., then

(a) $x=8$, $y=12$ (b) $x=10$, $y=12$
(c) $x=8$, $y=10$ (d) $x=10$, $y=14$

56. If two numbers a and b are given, then we can insert ...X... number(s) between them such that the resulting sequence is an A.P.. Here, X is used for

(a) one (b) two
(c) three (d) many

57. Let $A_1, A_2, \ldots. A_n$ be n numbers between a and b such that $a, A_1, A_2 \ldots. A_n, b$ are in A.P..

Then, match the terms of Column I with terms of Column II and choose the correct option from the codes given below.

Column I	Column II
A. $(n+2)^{th}$ term	1. d
B. $\frac{b-a}{n+1}$	2. A_n
C. $a+\frac{n(b-a)}{n+1}$	3. b

Codes

	A	B	C		A	B	C
(a)	3	1	2	(b)	3	2	1
(c)	1	3	2	(d)	1	2	3

58. The six numbers which will be inserted between 3 and 24 such that resulting sequence become an A.P., are

(a) 6, 9, 12, 15, 19, 22 (b) 4, 6, 8, 10, 12, 14
(c) 6, 8, 10, 12, 14, 16 (d) 6, 9, 12, 15, 18, 21

59. The sum of odd integers from 1 to 2001 is

(a) 1002001 (b) 1030001 (c) 100201 (d) 120001

60. The sum of all natural numbers lying between 100 and 1000, which are multiples of 5, is

(a) 92550 (b) 98450 (c) 93450 (d) 92150

61. If in an A.P., first term is 2 and the sum of first five terms is one-fourth of the next five terms, then the 20th term is

(a) –140 (b) –100 (c) –112 (d) –138

62. If n is the number of terms of an A.P. $-6, \frac{-11}{2}, -5, \ldots..$, which needed to give the sum -25, then the value (s) of n is/are

(a) 6 (b) 5 or 20 (c) 5 (d) 20 or 6

63. If the sum of a certain number of terms of the A.P. 25, 22, 19,..... is 116, then the last term is

(a) 2 (b) 3 (c) 8 (d) 4

64. In an A.P., if K^{th} term is $5K+1$. Then, the sum of first n terms is

(a) $\frac{n}{2}(5n+7)$ (b) $\frac{n}{2}(n+7)$ (c) $\frac{n}{2}(n+5)$ (d) $\frac{n}{2}(7n+5)$

65. If the sum of n terms of an A.P. is $(pn+qn^2)$, where p and q are constants, then the common difference is

(a) $2p$ (b) $2q$
(c) p (d) q

66. The sums of n terms of two arithmetic progressions are in the ratio $5n+4:9n+6$. Then, the ratio of their 18th term is

(a) 123/172 (b) 172/123
(c) 179/321 (d) 321/179

67. If the sum of first p terms of an A.P. is equal to the sum of first q terms, then the sum of first $(p+q)$ terms *i.e.*, S_{p+q} is

(a) 1 (b) 2
(c) 0 (d) 3

68. If the sum of n terms of an A.P. is $3n^2+5n$ and its mth term is 164, then the value of m is

(a) 27 (b) 25
(c) 30 (d) 32

69. If A_1, A_2, A_3, A_4 and A_5 are the numbers to be inserted between 8 and 26 such that the resulting sequence is an A.P., then A_1, A_2, A_3, A_4 and A_5 are respectively

(a) 10, 13, 16, 20, 23 (b) 9, 12, 15, 18, 21
(c) 10, 13, 16, 19, 22 (d) 11, 14, 17, 20, 23

70. Between 1 and 31, m numbers have been inserted in such a way that the resulting sequence is an A.P. and the ratio of 7th and $(m-1)^{th}$ numbers is 5 : 9. Then, the value of m is

(a) 10 (b) 14
(c) 12 (d) 15

71. A man starts repaying a loan as first instalment of ₹ 100. If he increases the instalment by ₹ 5 every month, then the amount he will pay in the 30th instalment is

(a) ₹ 241 (b) ₹ 250
(c) ₹ 245 (d) ₹ 265

Level II

72. If the first term of an A.P. is a, the second term is b and the last term is c, then the sum of the A.P. is

(a) $\frac{2(b+c-a)(c+a)}{(b-a)}$ (b) $\frac{(b+c-2a)(c+a)}{2(b-a)}$
(c) $\frac{(b+c-2a)(c+a)}{b-a}$ (d) $\frac{(b+c-a)(c+a)}{2(b-a)}$

73. The pth term of an A.P. is a and q^{th} term is b. Then, the sum of its $(p+q)$ terms is

(a) $\frac{p+q}{2}\left[a+b+\frac{a-b}{p+q}\right]$ (b) $\frac{p+q}{2}\left[a+b+\frac{a+b}{p+q}\right]$
(c) $\frac{p+q}{2}\left[a+b+\frac{a-b}{p-q}\right]$ (d) $\frac{p+q}{2}\left[a+b+\frac{a+b}{p-q}\right]$

74. In an A.P., if p^{th} term is $\frac{1}{q}$ and q^{th} term is $\frac{1}{p}$, then sum of first pq terms is

(a) $\frac{1}{2}(pq+2)$, where $p \neq q$ (b) $\frac{1}{2}(pq+2)$, where $p = q$
(c) $\frac{1}{2}(pq+1)$, where $p \neq q$ (d) None of these

75. Sum of the first p, q and r terms of an A.P. are a, b and c, respectively. Then,

(a) $\frac{a}{p}(q-r)+\frac{b}{q}(r-p)+\frac{c}{r}(p-q)=0$
(b) $\frac{a}{p}(q+r)+\frac{b}{q}(r+p)+\frac{c}{r}(p+q)=0$
(c) $\frac{a}{p}(q-r)-\frac{b}{q}(r-p)-\frac{c}{r}(p-q)=0$
(d) $\frac{a}{p}(q+r)-\frac{b}{q}(r+p)-\frac{c}{r}(p+q)=0$

76. If the ratio of the sums of m and n terms of an A.P. is $m^2 : n^2$, then the ratio of mth and n^{th} terms is

(a) $(2m+1):(2n+1)$ (b) $(2m+1):(2n-1)$
(c) $(2m-1):(2n+1)$ (d) $(2m-1):(2n-1)$

77. If $\frac{a^n+b^n}{a^{n-1}+b^{n-1}}$ is the A.M. between a and b, then the value of n is [HOTS]

(a) 2 (b) 1 (c) 0 (d) 3

78. Let the difference between any two consecutive interior angles of a polygon is 5°. If the smallest angle is 120°, then the number of sides of polygon is

(a) 9 (b) 8 (c) 7 (d) 16

79. The sum of $(m+n)^{th}$ and $(m-n)^{th}$ terms of an A.P. is equal to ...A... the mth term. Here, A refers to

(a) twice (b) thrice
(c) square (d) cube

80. If the sum of three numbers in A.P. is 24 and their product is 440, then

I. The numbers are 5, 8, 11.

II. The numbers are 11, 8, 5.

(a) Both I and II are true (b) I is true and II is false
(c) I is false and II is true (d) Neither I nor II is true

81. Let the sum of n, $2n$, $3n$ terms of an A.P. be S_1, S_2 and S_3, respectively. Then,

(a) $S_3 = 3(S_2 - S_1)$ (b) $S_3 = 2(S_2 - S_1)$
(c) $S_3 = 3(S_1 - S_2)$ (d) $S_3 = 2(S_1 - S_2)$

82. The sum of all numbers between 200 and 400 which are divisible by 7, is

(a) 8460 (b) 8729 (c) 8520 (d) 8130

83. The sum of integers from 1 to 100 that are divisible by 2 or 5, is

(a) 3000 (b) 3056 (c) 3050 (d) 3060

84. The sum of all two-digit numbers which when divided by 4, yields 1 as remainder, is

(a) 1120 (b) 1310
(c) 1250 (d) 1210

85. The sum of the first four terms of an A.P. is 56. The sum of the last four terms is 112. If its first term is 11, then the number of terms is

(a) 12 (b) 10 (c) 11 (d) 9

86. If the pth, q^{th} and r^{th} terms of an A.P. are *a, b, c* respectively, then

(a) $(q+r)a+(r+p)b+(p+q)c=0$
(b) $(q-r)a+(r-p)b+(p-q)c=0$
(c) $(q+r)a-(r+p)b-(p+q)c=0$
(d) $(q-r)a-(r-p)b-(p-q)c=0$

87. If $a\left(\frac{1}{b}+\frac{1}{c}\right), b\left(\frac{1}{c}+\frac{1}{a}\right), c\left(\frac{1}{a}+\frac{1}{b}\right)$ are in A.P., then *a, b, c*

(a) are in A.P. (b) are not in A.P.
(c) Cannot be determined (d) None of these

88. A farmer buys a used tractor for ₹ 12000. He pays ₹ 6000 cash and agrees to pay the balance in annual instalments of ₹ 500 plus 12% interest on the unpaid amount. Then, the cost of tractor will be

(a) 15850 (b) 16680 (c) 17620 (d) 16280

89. Shamshad Ali buys a scooter for ₹ 22000. He pays ₹ 4000 cash and agrees to pay the balance in annual instalment of ₹ 1000 plus 10% interest on the unpaid amount. The cost of scooter will be

(a) 29200 (b) 34100 (c) 42300 (d) 39100

90. A man deposited ₹ 10000 in a bank at the rate of 5% simple interest annually. Then, the amount after 15th yr and the amount after 20 yr are respectively

(a) 15000, 16000
(b) 17000, 20000
(c) 16000, 17000
(d) 18000, 19000

91. 150 workers were engaged to finish a job in a certain number of days. 4 workers dropped out on second day, 4 more workers dropped out on third day and so on.

They took 8 more days to finish the work. Then, the number of days in which work was completed, is [HOTS]

(a) 22 (b) 23 (c) 24 (d) 25

Geometric Progression, Relationship between A.M. and G.M.

Level I

92. If every term in a progression except the first term bears a constant ratio to the term immediately preceding it, then such a progression is called

(a) arithmetic progression
(b) geometric progression
(c) harmonic progression
(d) None of these

93. Which of the following is not a geometric sequence?

(a) 2, 4, 8, 16, ... (b) $\frac{1}{9}, \frac{-1}{27}, \frac{1}{81}, \frac{-1}{243}, ...$
(c) 0.01, .0001, .000001. ...... (d) None of these

94. Which of the following is/are correct?

I. The common ratio in 2, 4, 8, 16, ... is 2.

II. The common ratio in $\frac{1}{9}, \frac{-1}{27}, \frac{1}{81}, \frac{-1}{243}, ...$ is $\frac{-1}{3}$,

III. The common ratio in 0.01, .0001, .000001, ... is 0.001.

(a) I and II are correct
(b) I and III are correct
(c) II and III are correct
(d) All are correct

95. A sequence $a_1, a_2, a_3, ... a_n, ...$ is called geometric progression, if each term is non-zero and

(a) $\frac{a_{k+1}}{a_k} = r$ (constant), for $k > 1$
(b) $\frac{a_{k-1}}{a_k} = r$ (constant), for $k \geq 1$
(c) $\frac{a_{k+1}}{a_k} = r$ (constant), for $k \geq 1$
(d) $\frac{a_{k+2}}{a_k} = r$ (constant), for $k \geq 1$

96. If $a, ar, ar^2, ar^3, ...$ is a geometric progression, then *a* and *r* are respectively called

(a) common ratio, first term
(b) common difference, first term
(c) first term, common difference
(d) first term, common ratio

97. Match the notations given in Column I which are used in a geometric progression with their meanings given in Column II and choose the correct option from the codes given below.

	Column I		Column II
A.	a	1.	Last term
B.	r	2.	Number of terms
C.	l	3.	First term
D.	n	4.	Sum of first n terms
E.	S_n	5.	Common ratio

Codes

	A	B	C	D	E
(a)	3	5	1	2	4
(b)	3	5	2	1	4
(c)	5	3	1	2	4
(d)	3	5	1	4	2

98. If in a G.P. first non-zero term is a and common ratio is r, then which of the following is incorrect?

(a) $a_3 = ar^2, a_5 = ar^4$ (b) $a_2 = ar^2, a_5 = ar^5$
(c) $a_3 = ar^2, a_6 = ar^5$ (d) $a_5 = ar^4, a_7 = ar^6$

99. The general term of a G.P., where first non-zero term is a and common ratio is r, can be written as

(a) $a_n = ar^n$ (b) $a_{n+1} = ar^{n+1}$
(c) $a_n = ar^{n-1}$ (d) $a_n = ar^{n+1}$

100. Which of the following is correct?

(a) $a, ar, ar^2, ar^3, \ldots, ar^{n-1}$ is a finite G.P.
(b) $a, r, r^2, r^3, \ldots, r^{n-1}, \ldots$ is a finite G.P.
(c) $a + ar + ar^2 + \ldots + ar^{n-1}$ is a finite G.P.
(d) $a + r + r^2 + \ldots + r^{n-1} + \ldots$ is a finite G.P.

101. Which of the following is incorrect?

(a) $a, ar, ar^2, \ldots, ar^{n-1}, \ldots$ is an infinite G.P.
(b) $a + ar + ar^2 + \ldots + ar^{n-1}$ is a finite G.P.
(c) $a + ar + ar^2 + \ldots + ar^{n-1}$ is a finite geometric series
(d) $a + ar + ar^2 + \ldots + ar^{n-1} + \ldots$ is an infinite geometric series

102. Let the first term of a G.P. be a and the common ratio be r. Then, S_n the sum of first n terms of a G.P. is defined as

$$S_n = a + ar + ar^2 + ar^3 + \ldots + ar^{n-1}$$

Statement I If $r = 1$, then $S_n = na$.

Statement II If $r \neq 1$, then $S_n = \frac{a(r^n - 1)}{1 - r}$.

(a) Statement I and Statement II are true
(b) Statement I is true and Statement II is false
(c) Statement I is false and Statement II is true
(d) Statement I and Statement I are false

103. The nth term of a G.P. 5, 25, 125, ... is

(a) 5^n (b) 5^{n-1}
(c) 5^{n+1} (d) 5^{n-2}

104. If the nth term of a G.P. 2, 8, 32, ... is 131072, then the value of n is

(a) 8 (b) 9 (c) 10 (d) 11

105. In a G.P., the 3rd term is 24 and the 6th term is 192. Then, the 10th term is

(a) 1084 (b) 3290
(c) 3072 (d) 2340

106. If geometric series is given as

$$1 + \frac{2}{3} + \left(\frac{2}{3}\right)^2 + \left(\frac{2}{3}\right)^3 + \ldots$$

Then, which of the following is correct?

I. $S_n = 3\left[1 - \left(\frac{2}{3}\right)^n\right]$ II. $S_5 = \frac{211}{81}$

(a) Both I and II are correct
(b) Only I is correct
(c) Only II is correct
(d) I is incorrect and II is correct

107. The sum of first three terms of a G.P. is $\frac{13}{12}$ and their product is -1. Then, which of the following statement is incorrect?

(a) Common ratio is $\frac{-3}{4}$ or $\frac{-4}{3}$
(b) First three terms are $\frac{4}{3}, -1, \frac{3}{4}$ for $r = \frac{-3}{4}$
(c) First three terms are $\frac{3}{4}, -1, \frac{4}{3}$ for $r = \frac{-4}{3}$
(d) Common ratio is $\frac{3}{4}$ or $\frac{4}{3}$

108. A person has 2 parents, 4 grandparent, 8 great grandparent and so on. Then, the number of ancestors during the ten generations preceding his own is

(a) 1084 (b) 2046 (c) 2250 (d) 1024

109. The geometric mean of two positive numbers a and b is

(a) $\frac{a+b}{2}$ (b) $\frac{a-b}{2}$ (c) $\sqrt{ab}$ (d) ab

110. The geometric mean of 2 and 8 is

(a) 4 (b) 6 (c) 7 (d) 5

111. The numbers which can be inserted between two positive numbers a and b to make the resulting sequence a G.P., are

(a) three (b) four
(c) many (d) infinitely many

112. If $G_1, G_2, \ldots, G_n$ be n numbers between positive numbers a and b such that $a, G_1, G_2, \ldots, G_n, b$ is a G.P., then match the terms of Column I with terms of Column II and choose the correct option from the codes given below.

Column I	Column II
A. $(n+2)^{th}$ term	1. $a\left(\frac{b}{a}\right)^{\frac{n}{n+1}}$
B. $\left(\frac{b}{a}\right)^{\frac{1}{n+1}}$	2. b
C. G_n	3. r

Codes

	A	B	C
(a)	2	1	3
(b)	2	3	1
(c)	3	2	1
(d)	1	3	2

113. The three numbers, which can be inserted between 1 and 256, so that the resulting sequence become a G.P., are

(a) 4, 32, 64 (b) 4, 16, 20 (c) 4, 16, 64 (d) 8, 16, 64

Relationship between A.M. and G.M.

114. If A and G be A.M. and G.M. of two given positive real numbers a and b respectively, then A and G are related as

(a) $A \geq G$ (b) $G \geq A$ (c) $A = G$ (d) $A = -G$

115. The 20th term and n^{th} term of the G.P. $\frac{5}{2}, \frac{5}{4}, \frac{5}{8}, \ldots$ are respectively

(a) $\frac{5}{2^{20}}, \frac{5}{2^n}$ (b) $\frac{5}{2^{19}}, \frac{5}{2^{n-1}}$

(c) $\frac{5}{2^{21}}, \frac{5}{2^{n+1}}$ (d) $\frac{5}{2^{18}}, \frac{5}{2^{n-2}}$

116. If the 8th term of a G.P. is 192 with the common ratio 2, then the 12th term is

(a) 1640 (b) 2084 (c) 3072 (d) 3126

117. If the 5th, 8th and 11th terms of a G.P. are p, q and s respectively, then

(a) $p^2 = qs$ (b) $s^2 = pq$ (c) $p = sq$ (d) $q^2 = ps$

118. If the 4th term of a G.P. is square of its second term, and the first term is -3, then the 7th term is

(a) -1096 (b) -2340 (c) -2187 (d) -3120

119. Which of the following is incorrect?

(a) The 13th term of a G.P. $2, 2\sqrt{2}, 4, \ldots$ is 128

(b) The 7th term of a G.P. $5, 5\sqrt{3}, 15, \ldots$ is 135

(c) The number of terms in $\sqrt{3}, 3, 3\sqrt{3}, \ldots, 729$ is 12

(d) The number of terms in $\frac{1}{3}, \frac{1}{9}, \frac{1}{27}, \ldots, \frac{1}{19683}$ is 10

120. If the numbers $\frac{-2}{7}, x, \frac{-7}{2}$ are in G.P., then the values of x are

(a) ± 1 (b) 1, 2 (c) $-1, 2$ (d) ± 2

121. Which of the following is not correct?

(a) The sum of first 20 terms of a G.P. $0.15, 0.015, 0.0015, \ldots$ is $\frac{1}{6}[1-(0.1)^{20}]$

(b) The sum of first n terms of a G.P. $\sqrt{7}, \sqrt{21}, 3\sqrt{7}, \ldots$ is $\frac{\sqrt{7}}{2}(\sqrt{3}+1)(3^{n/2}-1)$

(c) The sum of first n terms of a G.P. $1, -a, a^2, -a^3, \ldots$ is $\frac{[1+(-a)^n]}{1+a}$, where $a \neq -1$

(d) The sum of first n terms of a G.P. $x^3, x^5, x^7, \ldots$ is $\frac{x^3(1-x^{2n})}{1-x^2}$, where $x \neq \pm 1$

122. If the sum of first three terms of a G.P. is $\frac{39}{10}$ and their product is 1, then which of the following is incorrect?

(a) $r = \frac{2}{5}$ or $\frac{5}{2}$ (b) Terms are $\frac{2}{5}, 1, \frac{5}{2}$

(c) Terms are $\frac{5}{2}, 1, \frac{2}{5}$ (d) None of these

123. If n terms of G.P. $3, 3^2, 3^3, \ldots$ are needed to give the sum 120, then the value of n is

(a) 2 (b) 3 (c) 4 (d) 5

124. If the sum of first three terms of a G.P. is 16 and the sum of next three terms is 128, then which of the following are correct?

I. First term $= \frac{16}{7}$

II. Common ratio $= 2$

III. Sum of n terms $= \frac{16}{7}(2^n + 1)$

(a) I and II are correct (b) I and III are correct

(c) II and III are correct (d) All are correct

125. If the first term of a G.P. is 729 and its seventh term is 64, then the sum of first seven terms is

(a) 2080 (b) 2060 (c) 2059 (d) 2120

126. If the sum of first two terms of a G.P. is -4 and the fifth term is 4 times the third term, then the G.P. is

I. $\frac{-4}{3}, \frac{-8}{3}, \frac{-16}{3}, \ldots$

II. $4, -8, 16, -32, 64, \ldots$

(a) Both I and II are correct (b) Only I is correct

(c) Only II is correct (d) Both I and II are incorrect

127. If 4th, 10th and 16th terms of a G.P. are x, y and z respectively, then

(a) y, x, z are in G.P. (b) x, y, z are in G.P.
(c) z, x, y are in G.P. (d) x, z, y are in G.P.

128. The sum of n terms of the sequence 8, 88, 888, 8888, ... is

(a) $\frac{80}{81}(10^n+1)-\frac{8n}{9}$ (b) $\frac{80}{81}(10^n-1)+\frac{8n}{9}$
(c) $\frac{80}{81}(10^n+1)+\frac{8n}{9}$ (d) $\frac{80}{81}(10^n-1)-\frac{8n}{9}$

129. The sum of the products of the corresponding terms of the sequences 2, 4, 8, 16, 32 and 128, 32, 8, 2, $\frac{1}{2}$ is

(a) 496 (b) 390
(c) 450 (d) 466

130. If the products of the corresponding terms of the sequences $a, ar, ar^2, \ldots, ar^{n-1}$ and $A, AR, AR^2, \ldots, AR^{n-1}$ form a G.P., then the common ratio is ...Y... . Here, Y refers to

(a) r/R (b) rR
(c) R (d) r

131. In a G.P., if third term is greater than the first term by 9, and the second term is greater than the fourth term by 18, then the first four terms of this G.P. are

(a) 3, 6, 12, 24 (b) – 3, 6, – 12, 24
(c) 3, – 6, 12, – 24 (d) – 3, – 6, – 12, – 24

132. If the first and the nth term of a G.P. are a and b respectively and P is the product of n terms, then

(a) $P^3=(ab)^n$ (b) $P^2=(ab)^{n-1}$
(c) $P^3=(ab)^{n-1}$ (d) $P^2=(ab)^n$

133. The ratio of the sum of first n terms of a G.P. to the sum of terms from $(n+1)^{th}$ to $(2n)^{th}$ is

(a) $\frac{1}{r^{n-1}}$ (b) $\frac{1}{r^n}$
(c) $\frac{1}{r^{n+1}}$ (d) $\frac{1}{r^{n+2}}$

134. If a, b, c and d are in G.P., then

(a) $(a+b+c)(b+c+d)=(ab+bc+cd)^2$
(b) $(a^2+b^2+c^2)(b^2+c^2+d^2)=(ab+bc+cd)^2$
(c) $(a+b+c)(b+c+d)=(ab+bc+cd)^3$
(d) $(a^2+b^2+c^2)(b^2+c^2+d^2)=(ab+bc+cd)^3$

135. The two numbers are inserted between 3 and 81 such that the resulting sequence is G.P. Then, the numbers are

(a) 15, 27 (b) 9, 16 (c) 9, 27 (d) 9, 15

136. If A and G be A.M. and G.M. respectively between two positive numbers, then the numbers are

(a) $A \pm \sqrt{(A+G)(A-G)}$ (b) $G \pm \sqrt{(A+G)(A-G)}$
(c) $A \pm \sqrt{(G+A)(G-A)}$ (d) $G \pm \sqrt{(G+A)(G-A)}$

137. The principal of ₹ 500 amounts to A in 10 yr after its deposit in bank which pays annual interest rate of 10% compounded annually. Then, the value of A is

(a) $500 \times (1.1)^9$ (b) $500 \times (.1)^{10}$
(c) $500 \times (1.1)^{10}$ (d) $500 \times (.1)^9$

138. If A.M. and G.M. of roots of a quadratic equation are 8 and 5, respectively, then quadratic equation is

(a) $x^2+16x+25=0$ (b) $x^2-16x-25=0$
(c) $x^2+16x-25=0$ (d) $x^2-16x+25=0$

Infinite G.P. and Its Sum

139. G.P. of the form $a, ar, ar^2, ar^3, \ldots$ is called

(a) finite G.P.
(b) infinite G.P.
(c) neither finite nor infinite G.P.
(d) None of the above

140. If a G.P. is $1, \frac{2}{3}, \frac{4}{9}$, ..., then the sum of infinite terms of a G.P. is

(a) 2 (b) 3
(c) 1 (d) 4

141. For a geometric progression $a, ar, ar^2, ar^3, \ldots$, if numerical value of common ratio r is less than 1, then

(a) $S_n \to \frac{a}{1+r}$, as $n \to \infty$ (b) $S_n \to a$, as $n \to \infty$
(c) $S_n \to \frac{1}{1+r}$, as $n \to \infty$ (d) $S_n \to \frac{a}{1-r}$, as $n \to \infty$

142. Sum of infinite terms is denoted by S_∞ or S. Then, for the infinite sequence $a, ar, ar^2, \ldots$, S is equal to

(a) $\frac{a}{1-r}$ (b) $\frac{a}{1+r}$
(c) $\frac{1}{1+r}$ (d) None of these

143. Which of the following is/are true?

I. $1+\frac{1}{2}+\frac{1}{2^2}+\frac{1}{2^3}+\ldots=2$

II. $1-\frac{1}{2}+\frac{1}{2^2}-\frac{1}{2^3}+\ldots=\frac{3}{2}$

(a) Both I and II are true
(b) Only I is true
(c) Only II is true
(d) Both I and II are false

144. Match the geometric progressions given in Column I with the sum of their corresponding infinite terms given in Column II and choose the correct option from the codes given below.

	Column I		Column II
A.	$1, \frac{1}{3}, \frac{1}{9}, \dots$	1.	7.5
B.	6, 1.2, 0.24, ...	2.	1.5
C.	$5, \frac{20}{7}, \frac{80}{49}, \dots$	3.	$\frac{-3}{5}$
D.	$\frac{-3}{4}, \frac{3}{16}, \frac{-3}{64}, \dots$	4.	$\frac{35}{3}$

Codes

	A	B	C	D		A	B	C	D
(a)	2	1	4	3	(b)	2	4	1	3
(c)	1	2	4	3	(d)	2	1	3	4

145. $3^{1/2} \times 3^{1/4} \times 3^{1/8} \times \dots$ upto infinite terms is equal to

(a) 3^2 (b) 3 (c) 3^3 (d) 3^4

146. Let $x = 1 + a + a^2 + \dots$ and $y = 1 + b + b^2 + \dots$, where $|a| < 1$ and $|b| < 1$. Then, which of the following is/are correct?

I. $1 + ab + a^2b^2 + a^3b^3 + \dots = \frac{xy}{x + y + 1}$

II. $1 + ab + a^2b^2 + a^3b^3 + \dots = \frac{xy}{x + y - 1}$

(a) Both I and II are correct
(b) Only I is correct
(c) Only II is correct
(d) Neither I nor II is incorrect

Level II

147. The value of $\sum_{k=1}^{11} (2 + 3^k)$ is

(a) $\frac{3(3^{11} - 1)}{2} + 33$ (b) $\frac{3(3^{10} + 1)}{2} + 22$
(c) $\frac{3(3^{11} + 1)}{2} + 33$ (d) $\frac{3(3^{11} - 1)}{2} + 22$

148. If the pth, q^{th} and r^{th} terms of a G.P. are a, b and c respectively, then $a^{q-r}\, b^{r-p}\, c^{p-q}$ is equal to

(a) 0 (b) 1
(c) – 1 (d) – 2

149. The value of n for which $\frac{a^{n+1} + b^{n+1}}{a^n + b^n}$ may be the geometric mean between a and b, is

(a) 1 (b) $\frac{1}{2}$ (c) $\frac{-1}{2}$ (d) – 1

150. The sum of two numbers is 6 times their geometric mean. Then, the numbers are in the ratio

(a) $(3 - 2\sqrt{2}) : (2 + 3\sqrt{2})$ (b) $(3 + 2\sqrt{2}) : (3 - 2\sqrt{2})$
(c) $(2\sqrt{2} - 3) : (2\sqrt{2} + 3)$ (d) $(2\sqrt{2} + 3) : (2\sqrt{2} - 3)$

151. The number of bacteria in a certain culture doubles every hour. If there were 30 bacteria present in the culture originally, then which of the following is correct?

(a) Bacteria present after 2nd h is 120
(b) Bacteria present after 4th h is 240
(c) Bacteria present after n^{th} h is $(30) \cdot (2)^{n+1}$
(d) None of the above

152. If f is a function satisfying $f(x + y) = f(x) + f(y)$ for all $x, y \in N$ such that $f(1) = 3$ and $\sum_{x=1}^{n} f(x) = 120$, then the value of n is

(a) 2 (b) 3 (c) 4 (d) 5

153. If the sum of terms of a G.P. is 315 whose first term and the common ratio are 5 and 2 respectively, then

I. The last term is 320.

II. The number of terms is 7.

(a) I and II are true (b) I is true and II is false
(c) I is false and II is true (d) I and II are false

154. If the first term of a G.P. is 1 and the sum of the third term and fifth term is 90, then the possible common ratio of the G.P. are

(a) ± 4 (b) ± 2 (c) ± 6 (d) ± 3

155. If the sum of three numbers in G.P. is 56. If we subtract 1, 7, 21 from these numbers in that order, we obtain an Arithmetic Progression (A.P.), then which of the following are true?

I. The numbers are 8, 16, 32.
II. The numbers are 32, 16, 8.
III. The numbers are – 8, 16, – 32.
IV. The numbers are – 32, 16, – 8.

(a) I and II are true (b) III and IV are true
(c) I and III are true (d) II and IV are true

156. A G.P. consists of an even number of terms. If the sum of all the terms is 5 times the sum of terms occupying odd places, then its common ratio is

(a) 2 (b) 3
(c) 4 (d) 5

157. If $\frac{a + bx}{a - bx} = \frac{b + cx}{b - cx} = \frac{c + dx}{c - dx}$ $(x \neq 0)$, then

(a) a, b, c, d are in G.P. (b) a, c, b, d are in G.P.
(c) d, b, c, a are in G.P. (d) d, c, b, a are in G.P.

158. Let S be the sum, P be the product and R be the sum of reciprocals of n terms in a G.P.. Then, [HOTS]
(a) $S^2R^n = P^n$ (b) $R^2P^n = S^n$
(c) $R^2S^n = P^n$ (d) $P^2R^n = S^n$

159. If a, b, c, d are in G.P., then ...P..., ...Q... and ...R... are in G.P.. Here, P, Q and R refers to
(a) $b^n + c^n, a^n + b^n, c^n + d^n$
(b) $c^n + d^n, a^n + b^n, b^n + c^n$
(c) $a^n + b^n, b^n + c^n, c^n + d^n$
(d) $b^n + c^n. c^n + d^n. a^n + b^n$

160. If a and b are the roots of $x^2 - 3x + p = 0$ and c, d are the roots of $x^2 - 12x + q = 0$, where a, b, c, d form a G.P., then [HOTS]
(a) $(q + p) : (q - p) = 17 : 15$
(b) $(q - p) : (q + p) = 17 : 15$
(c) $(p - q) : (p + q) = 17 : 15$
(d) $(p + q) : (p - q) = 17 : 15$

161. If the ratio of the A.M. and G.M. of two positive numbers a and b is $m : n$, then $a : b$ is equal to
(a) $(m + \sqrt{m^2 - n^2}) : (m - \sqrt{m^2 - n^2})$
(b) $(m - \sqrt{m^2 - n^2}) : (m + \sqrt{m^2 - n^2})$
(c) $(m + \sqrt{m^2 + n^2}) : (m - \sqrt{m^2 + n^2})$
(d) $(m - \sqrt{m^2 + n^2}) : (m + \sqrt{m^2 + n^2})$

162. If a, b, c are in A.P., b, c, d are in G.P. and $\frac{1}{c}, \frac{1}{d}, \frac{1}{e}$ are in A.P., then ...X..., ...Y... and ...Z... are in G.P.. Here, X, Y and Z refers to
(a) a, e, c
(b) a, c, e
(c) e, a, c
(d) None of these

163. Which of the following is/are true?
I. The sum of first n terms of the series
$5 + 55 + 555 + \dots\dots\dots$ is $\frac{5}{9}\left[\frac{10(10^n - 1)}{9} + n\right]$.
II. The sum of first n terms of the series
$0.6 + 0.66 + 0.666 + \dots\dots$ is $\frac{2}{3}\left[n + \frac{1}{9}\left\{1 - \frac{1}{10}\right\}^n\right]$.
(a) Both I and II are true
(b) I is true and II is false
(c) I is false and II is true
(d) Neither I nor II is true

164. A manufacturer reckons that the value of a machine, which costs him ₹ 15625, will depreciate each year by 20%. The estimated value at the end of 5 yr is
(a) ₹ 5050 (b) ₹ 5120
(c) ₹ 5920 (d) ₹ 5150

165. A person writes a letter to four of his friends. He asks each one of them, to copy the letter and mail to four different persons with instructions that they move the chain similarly. Assuming that the chain is not broken and that it costs 50 paise to mail one letter. The amount spent on postage when 8th set of letter is mailed, is
(a) 42920 (b) 43650 (c) 43690 (d) 42980

166. If a, b, c, d are in G.P., then $a^2 - b^2, b^2 - c^2, c^2 - d^2$ are in
(a) A.P.
(b) G.P.
(c) neither A.P. nor G.P.
(d) None of the above

167. If a, b, c, d are four distinct positive quantities in G.P., then [NCERT Exemplar]
(a) $a + d > b + c$
(b) $a + d < b + c$
(c) $a + d \geq b + c$
(d) $a + d \leq b + c$

168. If a, b, c are three consecutive terms of an A.P. and x, y, z are three consecutive terms of a G.P., then $x^{b-c}\, y^{c-a}\, z^{a-b}$ is equal to [NCERT Exemplar]
(a) -1
(b) -2
(c) 1
(d) 2

169. If $\sum_{k=1}^{n} f(a + k) = 16(2^n - 1)$, where the function f satisfies $f(x + y) = f(x) \cdot f(y)$ for all natural numbers x, y and $f(1) = 2$, then the natural number a is
(a) 1
(b) 2
(c) 3
(d) 4

170. If x, y, z are positive integers, then value of expression $(x + y)(y + z)(z + x)$ is [NCERT Exemplar]
(a) $= 8xyz$
(b) $\geq 8xyz$
(c) $< 8xyz$
(d) $= 4xyz$

171. In a G.P. of positive terms, if any term is equal to the sum of the next two terms, then the common ratio of the G.P. is [NCERT Exemplar]
(a) $\sin 18°$
(b) $2 \cos 18°$
(c) $\cos 18°$
(d) $2 \sin 18°$

Sum of n Terms of Special Series

Level I

172. Sum of first n natural numbers *i.e.*, $1 + 2 + 3 + \ldots\ldots + n$ is

(a) $S_n = \frac{n(n-1)}{2}$ (b) $S_n = \frac{n(n+1)}{2}$

(c) $S_n = n(n-1)$ (d) $S_n = n(n+1)$

173. Sum of squares of the first n natural numbers *i.e.*, $1^2 + 2^2 + 3^2 + 4^2 + \ldots + n^2$ is

(a) $S_n = \frac{n(n-1)(2n-1)}{6}$ (b) $S_n = \frac{n(n+1)(2n+1)}{4}$

(c) $S_n = \frac{n(n+1)(2n-1)}{4}$ (d) $S_n = \frac{n(n+1)(2n+1)}{6}$

174. Sum of cubes of the first n natural numbers *i.e.*, $1^3 + 2^3 + 3^3 + 4^3 + \ldots + n^3$ is

(a) $\frac{[n(n+2)]^2}{4}$ (b) $\frac{[n(n+1)]^2}{4}$

(c) $\frac{[n(n-1)]^2}{4}$ (d) $\frac{[n(n+1)]^2}{2}$

175. The sum of first n terms of the series $5 + 11 + 19 + 29 + \ldots$ is

(a) $n^2 + 3n + 1$ (b) $n(n+2)(n+4)$

(c) $\frac{n^2 + 3n + 1}{4}$ (d) $\frac{n(n+2)(n+4)}{3}$

176. If n^{th} term is $n(n+3)$, then the sum of first n terms is

(a) $\frac{n(n+1)(n+4)}{3}$ (b) $\frac{n(n+1)(n+2)}{3}$

(c) $\frac{n(n+1)(n+5)}{3}$ (d) $\frac{n(n+1)(n+6)}{3}$

177. The sum of first n terms of the series $1 \times 2 + 2 \times 3 + 3 \times 4 + 4 \times 5 + 5 \times 6 + \ldots$ is

(a) $n(n+1)(n+2)$ (b) $2n(n+1)(n+2)$

(c) $\frac{n(n+1)(n+2)}{6}$ (d) $\frac{n(n+1)(n+2)}{3}$

178. The sum of first n terms of the series $3 \times 1^2 + 5 \times 2^2 + 7 \times 3^2 + \ldots$ is

(a) $\frac{n(n+1)(n^2+5n+1)}{6}$ (b) $\frac{n(n+1)(3n^2+5n+1)}{6}$

(c) $\frac{n(n+1)(n^2+5n+1)}{2}$ (d) $\frac{n(n+1)(n^2+5n+3)}{6}$

179. The sum of the series $5^2 + 6^2 + 7^2 + \ldots + 20^2$ is

(a) 2420 (b) 2620

(c) 2840 (d) 2940

180. The sum of first n terms of the series $3 \times 8 + 6 \times 11 + 9 \times 14 + \ldots$ is

(a) $3n(n+1)(n+3)$

(b) $n(n+1)(n+3)$

(c) $4n(n+1)(n+3)$

(d) $3n(n+1)(n+2)$

181. If n^{th} term is $n(n+1)(n+4)$, then the sum of first n terms of the series is

(a) $\frac{n(n+1)(n^2+23n+34)}{6}$ (b) $\frac{n(n+1)(3n^2+23n+34)}{12}$

(c) $\frac{n(n+1)(n^2+2n+1)}{3}$ (d) $\frac{n(n+1)(3n^2+2n+34)}{6}$

182. If n^{th} term is $n^2 + 2^n$, then the sum of first n terms of the series is

(a) $\frac{n(n+1)(2n+1)+2(2^n-1)}{6}$

(b) $\frac{n(n+1)(2n+1)}{6} + 2(1-2^n)$

(c) $\frac{n(n+1)(2n+1)}{6} + 2(2^n-1)$

(d) $\frac{n(n+1)(2n+1)+2(1-2^n)}{6}$

Level II

183. The sum of first n terms of the series $1 \times 2 \times 3 + 2 \times 3 \times 4 + 3 \times 4 \times 5 + \ldots\ldots$ is

(a) $\frac{n(n+1)(n+2)(n+3)}{2}$ (b) $2n(n+1)(n+2)(n+3)$

(c) $\frac{n(n+1)(n+2)(n+3)}{4}$ (d) $3n(n+1)(n+2)(n+3)$

184. The sum of first n terms of the series $\frac{1}{1 \times 2} + \frac{1}{2 \times 3} + \frac{1}{3 \times 4} + \ldots$ is [HOTS]

(a) $\frac{n}{n+1}$ (b) $\frac{n+1}{n}$

(c) $\frac{n}{n-1}$ (d) $\frac{n-1}{n}$

185. The sum of first n terms of the series $1^2 + (1^2 + 2^2) + (1^2 + 2^2 + 3^2) + \ldots\ldots$ is

(a) $\frac{n(n+1)(n+2)}{12}$ (b) $\frac{n(n+1)^2(n+2)}{12}$

(c) $\frac{n(n+1)(n+2)^2}{12}$ (d) $\frac{n(n+1)^2(n+2)^2}{12}$

186. If n^{th} term is $(2n-1)^2$, then the sum of first n terms of the series is

(a) $\frac{n}{3}(2n+1)(2n-1)$ (b) $\frac{n}{2}(2n+1)(2n-1)$
(c) $\frac{n}{3}(n+1)(n-1)$ (d) $\frac{n}{2}(n+1)(n-1)$

187. The sum upto n terms of the series $3+7+13+21+31+\ldots$ is

(a) $\frac{n(n+1)(n+2)}{3}$ (b) $\frac{n(n^2+3n+5)}{3}$
(c) $\frac{n(n^2+5n+3)}{3}$ (d) $\frac{n(n-1)(n-2)}{3}$

188. If S_1, S_2, S_3 are the sum of first n natural numbers, their squares and their cubes respectively, then

$$aS_2^2 = S_3(1+bS_1)$$

where, a and b respectively are

(a) 8, 9 (b) 8, 8 (c) 9, 8 (d) 9, 9

189. The sum of n terms of the series $\frac{1^3}{1}+\frac{1^3+2^3}{1+3}+\frac{1^3+2^3+3^3}{1+3+5}+\ldots$ is

(a) $\frac{n}{12}[n^2+5n+13]$ (b) $\frac{n}{24}[2n^2+8n+6]$
(c) $\frac{n}{12}[n^2+7n+15]$ (d) $\frac{n}{24}[2n^2+9n+13]$

190. If $\frac{1\times 2^2+2\times 3^2+\ldots+n\times(n+1)^2}{1^2\times 2+2^2\times 3+\ldots+n^2\times(n+1)}=\frac{3n+a}{3n+b}$, then the values of a and b respectively are

(a) 1, 5 (b) 5, 1 (c) 2, 5 (d) 5, 2

191. Which of the following is/are true?

I. Sum of the series $(3^3-2^3)+(5^3-4^3)+(7^3-6^3)+\ldots$ upto n terms is $4n^2+9n+6$.

II. Sum of the series $(3^3-2^3)+(5^3-4^3)+(7^3-6^3)+\ldots$ upto 10 terms is 4960.

(a) Both I and II are true (b) I is true but II is false
(c) I is false but II is true (d) Neither I nor II is true

192. Let S_n denotes the sum of the cubes of the first n natural numbers and s_n denotes the sum of first n natural numbers. Then, $\sum_{r=1}^{n}\frac{S_r}{s_r}$ equals to [NCERT Exemplar]

(a) $\frac{n(n+1)(n+2)}{6}$ (b) $\frac{n(n+1)}{2}$
(c) $\frac{n^2+3n+2}{2}$ (d) None of these

193. If the sum of first n natural numbers is $\frac{1}{5}$ times the sum of their squares, then the value of n is

(a) 5 (b) 6 (c) 7 (d) 8

194. The sum to n terms of the infinite series $1\cdot 3^2+2\cdot 5^2+3\cdot 7^2+\ldots\infty$ is [Jamia Millia Islamia 2009]

(a) $\frac{n}{6}(n+1)(6n^2+14n+7)$ (b) $\frac{n}{6}(n+1)(2n+1)(3n+1)$
(c) $4n^3+4n^2+n$ (d) None of these

195. Sum of n terms of the series $1^3+3^3+5^3+7^3+\ldots$ is [WB JEE 2010]

(a) $n^2(2n^2-1)$ (b) $n^3(n-1)$
(c) n^3+8n+4 (d) $2n^4+3n^2$

196. If the sum of first n terms of an A.P. is cn^2, then the sum of squares of these n terms is [IIT JEE 2009; HOTS]

(a) $\frac{n(4n^2-1)c^2}{6}$ (b) $\frac{n(4n^2+1)c^2}{3}$
(c) $\frac{n(4n^2-1)c^2}{3}$ (d) $\frac{n(4n^2+1)c^2}{6}$

Miscellaneous Exercise

197. The lengths of three unequal edges of a rectangular solid block are in G.P.. The volume of the block is 216 cm^3 and the total surface area is 252 cm^2. The length of the longest edge is [NCERT Exemplar]

(a) 12 cm (b) 6 cm (c) 18 cm (d) 3 cm

198. A person is to count 4500 currency notes. Let a_n denotes the number of notes he counts in the nth min. If $a_1=a_2=\ldots=a_{10}=150$ and $a_{10}, a_{11}, \ldots$ are in A.P. with common difference -2, then the time taken by him to count all notes, is [AIEEE 2012]

(a) 24 min (b) 34 min
(c) 125 min (d) 135 min

199. If a, b, c are in arithmetic progression, then the value of $(a+2b-c)(2b+c-a)(a+2b+c)$ is [WB JEE 2008]

(a) $16abc$ (b) $4abc$ (c) $8abc$ (d) $3abc$

200. If t_n denotes the nth term of the series $2+3+6+11+18+....$, then t_{50} is [NCERT Exemplar]

(a) 49^2-1 (b) 49^2 (c) 50^2+1 (d) 49^2+2

201. If a^2, b^2, c^2 are in A.P., then which of the following is also an A.P.? [DCE 2007]

(a) $\sin A, \sin B, \sin C$ (b) $\tan A, \tan B, \tan C$
(c) $\cot A, \cot B, \cot C$ (d) None of these

202. Let T_r be the rth term of an A.P. whose first term is a and common difference is d. If for some positive integers $m, n, m \neq n$, $T_m = \frac{1}{n}$ and $T_n = \frac{1}{m}$, then $a-d$ equals to [Jamia Millia Islamia 2007]

(a) 0 (b) 1 (c) $\frac{1}{mn}$ (d) $\frac{1}{m}+\frac{1}{n}$

203. Let S_n denotes the sum of the first n terms of an A.P. . If $S_{2n} = 3S_n$, then $S_{3n} : S_n$ is equal to

(a) 4 (b) 6 (c) 8 (d) 10

204. Let $a_1, a_2, a_3, ...$ be the terms of an A.P. . If $\frac{a_1+a_2+....+a_p}{a_1+a_2+....+a_q} = \frac{p^2}{q^2}$, $p \neq q$, then $\frac{a_6}{a_{21}}$ equals to

(a) $\frac{7}{2}$ (b) $\frac{2}{7}$ (c) $\frac{11}{41}$ (d) $\frac{41}{11}$

205. Let $f(x)$ be a polynomial function of second degree. If $f(1) = f(-1)$ and a, b, c are in A.P., then $f'(a), f'(b)$ and $f'(c)$ are in

(a) A.P.
(b) G.P.
(c) H.P.
(d) arithmetico-geometric progression

206. In a cricket tournament, 16 school teams participated. A sum of ₹ 8000 is to be awarded among themselves as prize money. If the last placed team is awarded ₹ 275 in prize money and the award increases by the same amount for successive finishing places, then the amount received by first place team is

(a) 720 (b) 725
(c) 730 (d) 735

207. If a_1, a_2, a_3, a_4, a_5 and a_6 are six arithmetic means between 3 and 31, then $a_6 - a_5$ and $a_1 + a_6$ are respectively [Kerala CEE 2004]

(a) 5 and 34 (b) 4 and 35
(c) 4 and 34 (d) 4 and 36
(e) 6 and 36

208. If $a_1, a_2, a_3, ..., a_n$ are the n arithmetic means between a and b, then $2\sum_{i=1}^{n} a_i$ equals to

(a) ab (b) $n(a+b)$
(c) nab (d) $\frac{(a+b)}{n}$

209. If x, $2y$, $3z$ are in A.P., where the distinct numbers x, y, z are in G.P., then the common ratio of the G.P. is

(a) 3 (b) $\frac{1}{3}$
(c) 2 (d) 1

210. If $1+\sin x+\sin^2 x+.....$ upto $\infty = 4+2\sqrt{3}$, $0 < x < \pi$ and $x \neq \frac{\pi}{2}$, then x is equal to [KCET 2009]

(a) $\frac{\pi}{3}, \frac{5\pi}{6}$ (b) $\frac{2\pi}{3}, \frac{\pi}{6}$
(c) $\frac{\pi}{3}, \frac{2\pi}{3}$ (d) $\frac{\pi}{6}, \frac{\pi}{3}$

211. If G is the G.M. of the product of r sets of observations with geometric means $G_1, G_2, ... G_r$ respectively, then G is equal to [Jamia Millia Islamia 2009]

(a) $\log G_1 + \log G_2 + ... \log G_n$ (b) $G_1 G_2, ... G_r$
(c) $\log G_1, \log G_2, ..., \log G_n$ (d) None of these

212. If A is the arithmetic mean and G_1, G_2 be two geometric means between any two numbers, then which of the following is/are true?

I. $\frac{G_1^2}{G_2}+\frac{G_2^2}{G_1} = A$

II. $\frac{G_1^2}{G_2}+\frac{G_2^2}{G_1} = 2A$

(a) Both I and II are true (b) Only I is true
(c) Only II is true (d) Both I and II are false

213. The first two terms of a geometric progression add upto 12. The sum of the third and the fourth terms is 48. If terms of the geometric progression are alternately positive and negative, then the first term is [AIEEE 2012]

(a) 4 (b) – 4
(c) – 12 (d) 12

214. A man arranges to pay-off a debt of ₹ 3600 by 40 annual instalments, which are in A.P. . When 30 of the instalments are paid, he dies leaving one-third of the debt unpaid. The 8th instalment is

(a) ₹ 60 (b) ₹ 70
(c) ₹ 65 (d) ₹ 75

215. If $0 < \phi < \frac{\pi}{2}$, $x = \sum_{n=0}^{\infty} \cos^{2n} \phi$, $y = \sum_{n=0}^{\infty} \sin^{2n} \phi$ and $z = \sum_{n=0}^{\infty} \cos^{2n} \phi \sin^{2n} \phi$, then [Kerala CEE 2005]

(a) $xyz = xz + y$
(b) $xyz = xy + z$
(c) $xyz = x + y + z$
(d) $xyz = yz + x$
(e) $xyz = x + yz$

216. Which of the following is/are true?

I. The sum of first 23 terms of the A.P. 16, 11, 6, ... is -897.

II. The sum of first 22 terms of the A.P. $x + y, x - y, x - 3y, \ldots$ is $22 [x - 20y]$.

(a) Both I and II are true
(b) Only I is true
(c) Only II is true
(d) Both I and II are false

217. If a, b, c are in A.P., then $2^{ax+1}, 2^{bx+1}, 2^{cx+1}, x \neq 0$ are in [BCECE 2004]

(a) A.P.
(b) G.P. only when $x > 0$
(c) G.P., if $x < 0$
(d) G.P.

218. If the sequence is 54, 51, 48, ..., then the sum of first

I. 18 terms is 513.

II. 19 terms is 513.

(a) Both I and II are true
(b) Only I is true
(c) Only II is true
(d) Both I and II are false

219. If A.M. of two numbers is twice of their G.M., then the ratio of greatest number to smallest number is

(a) $7 - 4\sqrt{3}$
(b) $7 + 4\sqrt{3}$
(c) 21
(d) 5

220. If p, q, r are in G.P. and $\tan^{-1} p, \tan^{-1} q, \tan^{-1} r$ are in A.P., then p, q, r satisfies the relation

(a) $p = q = r$
(b) $p > q > r$
(c) $p + q + r = 0$
(d) $r > q > p$

221. The 20th term of the series $2 \times 4 + 4 \times 6 + 6 \times 8 + \ldots\ldots$ upto n terms is

(a) 1560
(b) 1680
(c) 1640
(d) 1580

Answer *with* Explanations

1. *(b)* By collection of objects listed in a sequence we usually mean that collection is ordered in such a way that it has an identified first member, second member, third member and so on.

2. *(c)* There is no order define for arranging living and non-living things in Delhi.

For example Trees, chairs, animals, ball pens, fungi, fabric, ... etc., are written in random manner. Hence, living and non-living things in Delhi do not form a sequence.

3. *(b)* Sequences, following specific patterns are called progressions.

Regressions, retrogressions and reversions are the antonyms of progressions.

4. *(c)* Total number of generations

$$= \frac{\text{Total number of years}}{\text{Generation gap}}$$

$$= \frac{300}{30} = 10$$

Hence, Statement I is false.

The number of person's ancestors for the first, second, third, ... tenth generations are 2, 4, 8, 16, 32, ... 1024.

Hence, Statement II is true.

5. *(d)* The successive quotients obtained after dividing 10 by 3 are defined as

3, 3.3, 3.33, 3.333, ...

6. *(c)* The various numbers occurring in a sequence are called its terms.

7. *(a)* The subscripts denote the position of the term which starts from 1 and goes on upto infinity.

8. *(a)* The nth term of a sequence is known as general term and can be denoted by a_n.

9. *(b)* The terms whose subscripts are multiples of 3 are a_3, a_6 and a_9 *i.e.*, 8, 64 and 512.

10. *(c)* Total number of given terms = 6

The terms whose subscripts are odd numbers are a_1, a_3 and a_5 *i.e.*, 3, 3.33 and 3.3333.

The terms whose subscripts are even numbers are a_2, a_4 and a_6 *i.e.*, 3.3, 3.333 and 3.33333.

The terms whose subscripts are multiples of 3 are a_3 and a_6 *i.e.*, 3.33 and 3.33333.

The terms whose subscripts are multiples of 5 is a_5 *i.e.*, 3.3333.

11. *(b)* A sequence containing finite number of terms is called a finite sequence. Indefinite, endless and indeterminable are the antonyms of finite.

12. *(d)* A sequence is a finite sequence, if it contains fixed number of terms.

13. *(c)* A sequence is called an infinite sequence, if it is not a finite sequence.

14. *(b)* The successive quotients 3, 3.3, 3.33, ... and so on, is an infinite sequence, since it never ends.

15. *(c)* It is given that

$$a_1 = 2 = 2 \times 1$$
$$a_2 = 4 = 2 \times 2$$
$$a_3 = 6 = 2 \times 3$$
$$a_4 = 8 = 2 \times 4$$
$$\vdots$$

Hence, the nth term of this sequence is $a_n = 2n$, where $n \in N$.

16. *(b)* It is given that

$$a_1 = 1 = 2 - 1$$
$$a_2 = 3 = 2 \times 2 - 1$$
$$a_3 = 5 = 2 \times 3 - 1$$
$$a_4 = 7 = 2 \times 4 - 1$$
$$\vdots$$

Hence, the nth term of this sequence is $a_n = 2n - 1$, where $n \in N$.

17. *(a)* $\because a_1 = 1, a_2 = 1$

Then,
$$a_3 = 2 = a_1 + a_2$$
$$a_4 = 3 = a_2 + a_3$$
$$a_5 = 5 = a_3 + a_4$$
$$a_6 = 8 = a_4 + a_5$$
$$\vdots$$

Hence, the nth term of this sequence is $a_n = a_{n-2} + a_{n-1}$ where $n > 2$ and $n \in N$.

Given, sequence is called Fibonacci sequence.

Hence, Statement I and Statement II are true.

18. *(c)* In the sequence of primes 2,3,5,7, ..., we find that there is no formula for the nth prime. Hence, the nth term does not exist.

19. *(c)* In every sequence, we should not expect that its terms will be given by a specific formula. However, we expect a theoretical scheme or a rule for generating the terms $a_1, a_2, \ldots\ldots\, a_n, \ldots\ldots$ in succession.

Hence, Statement I is false and Statement II is true.

20. *(d)* A sequence can be regarded as a function whose domain is the set of natural numbers.

21. *(b)* The domain of a sequence can be subset of natural numbers of the type $\{1, 2, 3, \ldots, k\}$

22. *(b)* The nth term a_n of a sequence can also be denoted as $a(n)$.

23. *(d)* $a_1, a_2, \ldots, a_n, \ldots\ldots$ be a given sequence.

Then, the expression $a_1 + a_2 + \ldots\ldots + a_n + \ldots\ldots$ is called the series associated with the given sequence.

24. *(c)* The series is finite or infinite according as the given sequence is finite or infinite.

25. *(c)* Series are often represented in compact form called sigma notation.

26. *(b)* The Greek letter Σ (sigma) is used for indicating the summation.

27. *(a)* The series $a_1 + a_2 + a_3 + \ldots\ldots + a_n$ is abbreviated as $\sum_{k=1}^{n} a_k$.

28. *(d)* A. Series refers to the indicated sum not to the sum itself.

B. Sum of a series means the number that results from adding the terms.

C. $1 + 3 + 5 + 7$ is a finite series with four terms.

D. Sum of the series $1 + 3 + 5 + 7$ is 16.

29. *(a)* A. $a_n = n(n+2)$

Then, $a_1 = 3, a_2 = 8, a_3 = 15, a_4 = 24, a_5 = 35$

B. $a_n = \dfrac{n}{n+1}$

Then, $a_1 = \dfrac{1}{2}, a_2 = \dfrac{2}{3}, a_3 = \dfrac{3}{4}, a_4 = \dfrac{4}{5}, a_5 = \dfrac{5}{6}$

C. $a_n = 2^n$

Then, $a_1 = 2, a_2 = 4, a_3 = 8, a_4 = 16, a_5 = 32.$

D. $a_n = \dfrac{2n-3}{6}$

Then, $a_1 = \dfrac{-1}{6}, a_2 = \dfrac{1}{6}, a_3 = \dfrac{1}{2}, a_4 = \dfrac{5}{6}, a_5 = \dfrac{7}{6}$

30. *(c)* $\because a_n = 4n - 3$

Then, $a_{17} = 4(17) - 3 = 65$

and $a_{24} = 4(24) - 3 = 93$

31. *(d)* $\because a_n = (-1)^{n-1} \cdot n^3$

Then,
$$a_9 = (-1)^{9-1} \cdot (9)^3 = (-1)^8 \cdot 729 = 729$$

32. *(b)* $\because a_1 = 3$ and $a_n = 3a_{n-1} + 2$

Then,
$$a_2 = 3a_1 + 2 = 3(3) + 2 = 11$$
$$a_3 = 3a_2 + 2 = 3(11) + 2 = 35$$
$$a_4 = 3a_3 + 2 = 3(35) + 2 = 107$$
$$a_5 = 3a_4 + 2 = 3(107) + 2 = 323$$

33. *(c)* $\because$ Given, $a_1 = 2, a_2 = 2$

and $a_n = a_{n-1} - 1, \quad n > 2$

Then,
$$a_3 = a_2 - 1 = 2 - 1 = 1$$
$$a_4 = a_3 - 1 = 1 - 1 = 0$$
$$a_5 = a_4 - 1 = 0 - 1 = -1$$

Hence, $A = 1, B = 0$ and $C = -1$

34. *(a)* We have, $a_n = 2n^2 - n + 1$

Putting $n = 1$, we get
$$a_1 = 2(1)^2 - 1 + 1 = 2 - 1 + 1 = 2$$

Putting $n = 2$, we get
$$a_2 = 2(2)^2 - 2 + 1 = 8 - 2 + 1 = 7$$

Putting $n = 3$, we get
$$a_3 = 2(3)^2 - 3 + 1 = 18 - 3 + 1 = 16$$

Putting $n = 4$, we get
$$a_4 = 2(4)^2 - 4 + 1 = 32 - 4 + 1 = 29$$

Hence, first four terms are 2, 7, 16 and 29.

35. *(c)* A sequence can be defined as a function whose domain $\subseteq N$.

36. *(a)* We have, $a_n = (-1)^{n-1}\, 5^{n+1}$

Put $n = 1, 2, 3, 4, 5.$

At $n = 1$, $a_1 = (-1)^{1-1}\, 5^{1+1} = (-1)^0\, 5^2 = 25$

At $n=2$, $a_2 = (-1)^{2-1} 5^{2+1} = (-1)^1 5^3 = -125$

At $n=3$, $a_3 = (-1)^{3-1} 5^{3+1} = (-1)^2 5^4 = 625$

At $n=4$, $a_4 = (-1)^{4-1} 5^{4+1} = (-1)^3 5^5 = -3125$

At $n=5$, $a_5 = (-1)^{5-1} 5^{5+1} = (-1)^4 5^6 = 15625$

37. *(b)* We have, $a_n = \dfrac{n(n^2+5)}{4}$

Put $n = 1, 2, 3, 4, 5.$

At $n=1$, $a_1 = \dfrac{1(1^2+5)}{4} = \dfrac{6}{4} = \dfrac{3}{2}$

At $n=2$, $a_2 = \dfrac{2(2^2+5)}{4} = \dfrac{2(4+5)}{4} = \dfrac{9}{2}$

At $n=3$, $a_3 = \dfrac{3(3^2+5)}{4} = \dfrac{3(9+5)}{4} = \dfrac{21}{2}$

At $n=4$, $a_4 = \dfrac{4(4^2+5)}{4} = \dfrac{4(16+5)}{4} = 21$

At $n=5$, $a_5 = \dfrac{5(5^2+5)}{4} = \dfrac{5(25+5)}{4} = \dfrac{75}{2}$

38. *(b)* I. We have, $a_n = \dfrac{n^2}{2^n}$

Putting $n=7$, $a_7 = \dfrac{7^2}{2^7} = \dfrac{49}{128}$

II. We have, $a_n = \dfrac{n(n-2)}{n+3}$

Putting $n=20$, $a_{20} = \dfrac{20(20-2)}{20+3}$

$\Rightarrow \quad a_{20} = \dfrac{20 \times 18}{23}$

$= \dfrac{360}{23}$

39. *(c)* I. We have, $a_1 = -1, a_n = \dfrac{a_{n-1}}{n}, n \geq 2$

Putting $n=2$, $a_2 = \dfrac{a_{2-1}}{2} = \dfrac{a_1}{2} = -\dfrac{1}{2}$

Putting $n=3$, $a_3 = \dfrac{a_{3-1}}{3} = \dfrac{a_2}{3} = \dfrac{-\frac{1}{2}}{3} = -\dfrac{1}{6}$

Putting $n=4$, $a_4 = \dfrac{a_{4-1}}{4} = \dfrac{a_3}{4} = \dfrac{-\frac{1}{6}}{4} = -\dfrac{1}{24}$

Putting $n=5$, $a_5 = \dfrac{a_{5-1}}{5} = \dfrac{a_4}{5} = \dfrac{-\frac{1}{24}}{5} = -\dfrac{1}{120}$

II. Sum of first five terms

$$= -1 - \frac{1}{2} - \frac{1}{6} - \frac{1}{24} - \frac{1}{120}$$
$$= \frac{-120 - 60 - 20 - 5 - 1}{120}$$
$$= \frac{-206}{120} > -2$$

40. *(b)* Here, $1 = a_1 = a_2, a_n = a_{n-1} + a_{n-2}, n > 2$

Putting $n = 3, 4, 5, 6,$

At $n=3$, $a_3 = a_{3-1} + a_{3-2}$
$= a_2 + a_1$
$= 1 + 1 = 2$

At $n=4$, $a_4 = a_{4-1} + a_{4-2}$
$= a_3 + a_2$
$= 2 + 1 = 3$

At $n=5$, $a_5 = a_{5-1} + a_{5-2}$
$= a_4 + a_3 = 3 + 2 = 5$

At $n=6$, $a_6 = a_{6-1} + a_{6-2}$
$= a_5 + a_4 = 5 + 3 = 8$

Now, $\dfrac{a_{n+1}}{a_n}$, for $n = 1, 2, 3\ 4, 5.$

At $n=1$, $\dfrac{a_2}{a_1} = \dfrac{1}{1} = 1$

At $n=2$, $\dfrac{a_3}{a_2} = \dfrac{2}{1} = 2$

At $n=3$, $\dfrac{a_4}{a_3} = \dfrac{3}{2}$

At $n=4$, $\dfrac{a_5}{a_4} = \dfrac{5}{3}$

At $n=5$, $\dfrac{a_6}{a_5} = \dfrac{8}{5}$

Hence, the terms are $1, 2, \dfrac{3}{2}, \dfrac{5}{3}$ and $\dfrac{8}{5}$.

41. *(c)* A sequence $a_1, a_2, a_3, \ldots, a_n, \ldots$ is called arithmetic sequence or arithmetic progression if $a_{n+1} = a_n + d, n \in N$ where a_1 is called the first term and the constant term d is called the common difference of the A.P. .

42. *(b)* If the first term is a and common difference is d, then the arithmetic progression is

$$a, a+d, a+2d, \ldots.$$

43. *(a)* If the first term is a and common difference is d, then the nth term is $a_n = a + (n-1)d$.

44. *(d)* If each term of an A.P. is divided by a non-zero constant, then the resulting sequence is also an A.P. . But in option (d), non-zero constant is not given.

45. *(b)* For an arithmetic progression,

a = First term

l = Last term

d = Common difference

n = Number of terms

S_n = Sum to n terms of an A.P.

46. *(b)* If $a, a+d, a+2d, \ldots, a+(n-1)d$ be an A.P., then

$$l = a + (n-1)d$$

and $$S_n = \frac{n}{2}[2a + (n-1)d]$$

Hence, Statement I is true and Statement II is false.

47. *(c)* $\because S_n = \frac{n}{2}[2a + (n-1)d]$

$$= \frac{n}{2}[a + a + (n-1)d]$$

$$= \frac{n}{2}[a + l] \qquad [\because l = a + (n-1)d]$$

48. *(a)* We have,

$$a_m = a + (m-1)d = n \qquad \ldots(i)$$

and $$a_n = a + (n-1)d = m \qquad \ldots(ii)$$

On solving Eqs. (i) and (ii), we get

$$(m-n)d = n - m$$

$\Rightarrow$ $$d = -1 \qquad \ldots(iii)$$

Now, $$a = n + m - 1 \qquad \ldots(iv)$$

Therefore, $$a_p = a + (p-1)d$$

$$= n + m - 1 + (p-1)(-1)$$

[using Eqs. (iii) and (iv)]

$$= n + m - 1 - p + 1$$

$$= n + m - p$$

49. *(b)* Let $a_1, a_2, a_3, \ldots, a_n$ be the given A.P. .

Then, $$S_n = a_1 + a_2 + a_3 + \ldots + a_n$$

$$= nP + \frac{n}{2}(n-1)Q$$

Therefore, $$S_1 = a_1 = P$$

$$S_2 = a_1 + a_2 = 2P + Q$$

Now, $$a_2 = S_2 - S_1$$

$$= (2P + Q) - (P)$$

$$= P + Q$$

Hence, the common difference is given by

$$d = a_2 - a_1$$

$$= (P + Q) - P$$

$$= Q$$

50. *(a)* Here, we have an A.P. with $a = 300000$

$d = 10000$ and $n = 20$.

Using the sum formula, we get

$$S_{20} = \frac{20}{2}[600000 + 19 \times 10000]$$

$$= 10[600000 + 190000]$$

$$= 10(790000)$$

$$= 7900000$$

51. *(b)* If A is the arithmetic mean of the numbers a and b, then a, A, b are in A.P. .

52. *(d)* If A is the arithmetic mean of the numbers a and b, then

$$A - a = b - A$$

$\Rightarrow$ $$a - A = A - b$$

Also, $$A = \frac{a+b}{2}$$

53. *(b)* The Arithmetic Mean (A.M.) of two numbers a and b is interpreted as their average.

54. *(c)* Given numbers are 4 and 16.

A.M. of 4 and 16 $= \frac{4+16}{2} = \frac{20}{2} = 10$

Hence, Statement I is correct.

Also, 4, 10, 16 is an A.P.

$\because$ $$10 - 4 = 16 - 10$$

Hence, Statement II is correct.

55. *(a)* $\because$ 4, 8, 12, 16 is an A.P. with common difference 4.

Hence, $x = 8$ and $y = 12$

56. *(d)* If two numbers a and b are given, then we can insert as many numbers as we like between them such that resulting sequence is an A.P. .

Hence, X is used for many.

57. *(a)* Let $A_1, A_2, A_3, \ldots A_n$ be n numbers between a and b such that $a, A_1, A_2, A_3, \ldots A_n, b$ is an A.P.

Here, b is the $(n+2)^{th}$ term,

i.e., $$b = a + [(n+2) - 1]d = a + (n+1)d$$

This gives $d = \frac{b-a}{n+1}$

Thus, n numbers between a and b are as follows

$$A_1 = a + d = a + \frac{b-a}{n+1}$$

$$A_2 = a + 2d = a + \frac{2(b-a)}{n+1}$$

$$A_3 = a + 3d = a + \frac{3(b-a)}{n+1}$$

$$\vdots$$

$$A_n = a + nd = a + \frac{n(b-a)}{n+1}$$

58. *(d)* Let A_1, A_2, A_3, A_4, A_5 and A_6 be six numbers between 3 and 24 such that $3, A_1, A_2, A_3, A_4, A_5, A_6, 24$ are in A.P. . Here, $a = 3, b = 24, n = 8$.

Therefore, $24 = 3 + (8-1)d$, so that $d = 3$.

Thus, $$A_1 = a + d = 3 + 3 = 6;$$

$$A_2 = a + 2d = 3 + 2 \times 3 = 9;$$

$A_3 = a + 3d = 3 + 3 \times 3 = 12;$

$A_4 = a + 4d = 3 + 4 \times 3 = 15;$

$A_5 = a + 5d = 3 + 5 \times 3 = 18;$

$A_5 = a + 6d = 3 + 6 \times 3 = 21;$

Hence, six numbers between 3 and 24 are 6, 9, 12, 15, 18 and 21.

59. *(a)* Series of odd integers from 1 to 2001 is $1 + 3 + 5 + 7 + \ldots + 2001$.

Here, first term $(a) = 1$

Common difference $(d) = 2$

We know that,

nth term of a series, $T_n = a + (n-1)d$

$\therefore \quad 2001 = 1 + (n-1)2 \quad (\because a = 1, T_n = 2001, \text{ given})$

$\Rightarrow \quad 2000 = (n-1)2$

$\Rightarrow \quad (n-1) = \frac{2000}{2}$

$\Rightarrow \quad n - 1 = 1000$

$\Rightarrow \quad n = 1001$

Now, required sum $(S_n) = \frac{n}{2}[2a + (n-1)d]$

$\therefore \quad S_{1001} = \frac{1001}{2}[2 \times 1 + (1001 - 1) \times 2]$

$= \frac{1001}{2} \times 2[1 + 1001 - 1]$

$= 1001 \times 1001 = 1002001$

60. *(b)* The numbers are 105, 110, 115,..., 995.

Here, first term $(a) = 105$

and common difference $(d) = 110 - 105 = 5$

Now, $\quad n^{th}$ term $(T_n) = a + (n-1)d$

$\Rightarrow \quad 995 = 105 + (n-1)5 \quad (\because a = 105, T_n = 995)$

$\Rightarrow \quad 995 - 105 = (n-1)5$

$\Rightarrow \quad 890 = (n-1)5$

$\Rightarrow \quad n - 1 = \frac{890}{5}$

$\Rightarrow \quad n - 1 = 178$

$\Rightarrow \quad n = 178 + 1 = 179$

Now, required sum $(S_n) = \frac{n}{2}[2a + (n-1)d]$

$= \frac{179}{2}[2 \times 105 + (179 - 1)5]$

$= \frac{179}{2}[210 + 178 \times 5]$

$= \frac{179}{2}[210 + 890] = \frac{179}{2} \times 1100$

$= 179 \times 550 = 98450$

61. *(c)* Let the A.P. is $a, a + d, a + 2d, a + 3d, \ldots$

Now, given $a = 2$

According to the given condition,

Sum of first five terms $= \frac{1}{4}$ (Sum of next five terms)

$a + (a + d) + (a + 2d) + (a + 3d) + a + 4d$

$= \frac{1}{4}[a + 5d + a + 6d + a + 7d + a + 8d + a + 9d]$

$\Rightarrow \quad 5a + 10d = \frac{1}{4}[5a + 35d]$

$\Rightarrow \quad 4[5a + 10d] = 5a + 35d$

$\Rightarrow \quad 20a + 40d = 5a + 35d$

$\Rightarrow \quad 20a - 5a = 35d - 40d$

$\Rightarrow \quad 15a = -5d$

$\Rightarrow \quad 15 \times 2 = -5d \quad (\because a = 2)$

$\Rightarrow \quad 30 = -5d$

$\Rightarrow \quad d = \frac{-30}{5} = -6$

Now, $\quad T_n = a + (n-1)d$

$\Rightarrow \quad T_{20} = 2 + (20 - 1)(-6) = 2 + 19(-6)$

$= 2 - 19 \times 6 = 2 - 114 = -112$

62. *(b)* Given that the sequence $-6, -\frac{11}{2}, -5 \ldots$ is in A.P.

Here, $a = -6$

$d = -\frac{11}{2} - (-6) = -\frac{11}{2} + 6 = -\frac{11}{2} + \frac{6}{1}$

$= \frac{-11 + 12}{2} = \frac{1}{2}$

Now, $\quad S_n = \frac{n}{2}[2a + (n-1)d]$

$\Rightarrow \quad -25 = \frac{n}{2}\left[2 \times (-6) + (n-1)\frac{1}{2}\right]$

$\Rightarrow \quad -25 \times 2 = n\left[\frac{-12}{1} + \frac{(n-1)}{2}\right]$

$\Rightarrow \quad -50 = n\left[\frac{-24 + n - 1}{2}\right]$

$\Rightarrow \quad -50 \times 2 = n(n - 25)$

$\Rightarrow \quad -100 = n^2 - 25n$

$\Rightarrow \quad n^2 - 25n + 100 = 0$

Now, on factorising it by splitting the middle term, we get

$n^2 - (20 + 5)n + 100 = 0$

$\Rightarrow \quad n^2 - 20n - 5n + 100 = 0$

$\Rightarrow \quad n(n - 20) - 5(n - 20) = 0$

$\Rightarrow \quad n = 5, 20$

63. *(d)* Given, A.P. is 25, 22, 19,... .

Here, $\quad a = 25, d = -3$ and $S_n = 116$

$\because \quad S_n = \frac{n}{2}[2a + (n-1)d]$

$\therefore \quad 116 = \frac{n}{2}[2 \times 25 + (n-1)(-3)]$

$\Rightarrow \quad 116 \times 2 = n[50 - 3n + 3]$

$\Rightarrow \quad 232 = n[53 - 3n]$

$\Rightarrow \quad 232 = 53n - 3n^2$

$\Rightarrow \quad 3n^2 - 53n + 232 = 0$

Now, on factorising it by splitting the middle term, we get

$\Rightarrow \quad 3n^2 - (24 + 29)n + 232 = 0$

$\Rightarrow \quad 3n^2 - 24n - 29n + 232 = 0$

$\Rightarrow \quad 3n(n - 8) - 29(n - 8) = 0$

$\Rightarrow \quad (3n - 29)(n - 8) = 0$

$\Rightarrow \quad n = \frac{29}{3}, n = 8$

$n = \frac{29}{3}$ is not possible as n cannot be fraction, only $n = 8$ is valid.

Now, $T_n = a + (n - 1)d$

$T_8 = 25 + (8 - 1)(-3)$

$= 25 + 7 \times (-3) = 25 - 21 = 4$

64. *(a)* Given, K^{th} term $(T_K) = 5K + 1$

Putting $K = 1, 2$, we get

$T_1 = 5 \times 1 + 1 = 6$

and $T_2 = 5 \times 2 + 1 = 11$

$\Rightarrow \quad a = 6, d = 11 - 6 = 5$

Now, $S_n = \frac{n}{2}[2a + (n - 1)d] = \frac{n}{2}[2 \times 6 + (n - 1)5]$

$= \frac{n}{2}[12 + 5n - 5] = \frac{n}{2}[5n + 7]$

65. *(b)* Given, $S_n = pn + qn^2$

Now, $T_n = S_n - S_{n-1}$

$T_n = pn + qn^2 - [p(n - 1) + q(n - 1)^2]$

$= pn + qn^2 - [pn - p + q(n^2 + 1 - 2n)]$

$= pn + qn^2 - [pn - p + qn^2 + q - 2qn]$

$= pn + qn^2 - pn + p - qn^2 - q + 2qn$

$= p - q + 2qn$

Now, putting $n = 1, 2$, we get

$\Rightarrow \quad T_1 = p - q + 2q \times 1 = p - q + 2q = p + q$

$\Rightarrow \quad T_2 = p - q + 2q \times 2 = p - q + 4q = p + 3q$

Hence, common difference $= (p + 3q) - (p + q) = 2q$

66. *(c)* Let a_1, a_2 be the first terms and d_1, d_2 be the common differences of two A.P.'s.

Now, according to the question,

$$\frac{\text{Sum of } n \text{ terms of first A.P.}}{\text{Sum of } n \text{ terms of second A.P.}} = \frac{5n + 4}{9n + 6}$$

$$\Rightarrow \frac{\frac{n}{2}[2a_1 + (n - 1)d_1]}{\frac{n}{2}[2a_2 + (n - 1)d_2]} = \frac{5n + 4}{9n + 6} \quad \left\{\because S_n = \frac{n}{2}[2a + (n - 1)d]\right\}$$

$$\Rightarrow \frac{2a_1 + (n - 1)d_1}{2a_2 + (n - 1)d_2} = \frac{5n + 4}{9n + 6} \quad \ldots(i)$$

Since, we have to find the ratio of 18th terms, so convert the equation in the form of nth term by taking 2 common.

$$\Rightarrow \frac{2\left[a_1 + \frac{n - 1}{2}d_1\right]}{2\left[a_2 + \frac{n - 1}{2}d_2\right]} = \frac{5n + 4}{9n + 6}$$

$$\Rightarrow \frac{a_1 + \left(\frac{n - 1}{2}\right)d_1}{a_2 + \left(\frac{n - 1}{2}\right)d_2} = \frac{5n + 4}{9n + 6} \quad \ldots(i)$$

We have to find $\frac{a_1 + 17d_1}{a_2 + 17d_2}$.

$\Rightarrow \quad \frac{n - 1}{2} = 17$ [comparing with LHS of Eq. (i)]

$\Rightarrow \quad n - 1 = 2 \times 17$

$\Rightarrow \quad n = 34 + 1$

$\Rightarrow \quad n = 35$

Hence, from Eq. (i), we get

$$\frac{a_1 + 17d_1}{a_2 + 17d_2} = \frac{5 \times 35 + 4}{9 \times 35 + 6} = \frac{175 + 4}{315 + 6} = \frac{179}{321}$$

Alternate Method

Let a_1, a_2 be the first terms and d_1, d_2 be the common differences of two A.P.'s.

Now, according to the question,

$$\frac{\text{Sum of } n \text{ terms of first A.P.}}{\text{Sum of } n \text{ terms of second A.P.}} = \frac{5n + 4}{9n + 6}$$

$$\Rightarrow \frac{\frac{n}{2}[2a_1 + (n - 1)d_1]}{\frac{n}{2}[2a_2 + (n - 1)d_2]} = \frac{5n + 4}{9n + 6} \quad \left\{\because S_n = \frac{n}{2}[2a + (n - 1)d]\right\}$$

$$\Rightarrow \frac{2a_1 + (n - 1)d_1}{2a_2 + (n - 1)d_2} = \frac{5n + 4}{9n + 6} \quad \ldots(i)$$

Now consider,

$$\frac{18^{th} \text{ term of first A.P.}}{18^{th} \text{ term of second A.P.}} = \frac{a_1 + 17d_1}{a_2 + 17d_2}$$

$$= \frac{2a_1 + 34d_1}{2a_2 + 34d_2}$$

$$= \frac{2a_1 + (35 - 1)d_1}{2a_2 + (35 - 1)d_2}$$

$$= \frac{5 \times 35 + 4}{9 \times 35 + 6} \quad \text{[substituting } n = 35 \text{ in Eq. (i)]}$$

$$= \frac{179}{321}$$

67. *(c)* Let a and d be the first term and common difference of an A.P.

$\because$ Sum of the first p terms = Sum of first q terms

$\therefore \quad S_p = S_q$

$$\Rightarrow \frac{p}{2}[2a + (p - 1)d] = \frac{q}{2}[2a + (q - 1)d]$$

$$\left\{\because S_n = \frac{n}{2}[2a + (n - 1)d]\right\}$$

$\Rightarrow \quad p[2a+(p-1)d]=q[2a+(q-1)d]$

$\Rightarrow \quad 2ap+p(p-1)d=2aq+q(q-1)d$

$\Rightarrow \quad 2ap-2aq=q(q-1)d-p(p-1)d$

$\Rightarrow \quad 2a(p-q)=d[q^2-q-p^2+p]$

$\Rightarrow \quad 2a(p-q)=d[(p-q)+(q^2-p^2)]$

$\Rightarrow \quad 2a(p-q)=d[(p-q)+(q-p)(q+p)]$

$\Rightarrow \quad 2a(p-q)=d(p-q)[1-(p+q)]$

$\Rightarrow \quad 2a=d[1-(p+q)]$

$\Rightarrow \quad 2a=-d[(p+q)-1]$

$\Rightarrow \quad 2a+d(p+q-1)=0 \quad \ldots(i)$

Now, $S_{p+q}=\dfrac{p+q}{2}[2a+[p+q-1]d]$

$=\dfrac{p+q}{2}\times 0 \quad$ [from Eq. (i)]

$=0$

68. *(a)* Given, $S_n=3n^2+5n$

$\therefore \quad S_m=3m^2+5m$ and $S_{m-1}=3(m-1)^2+5(m-1)$

Now, using the formula $T_m=S_m-S_{m-1}$, we have

$T_m=(3m^2+5m)-[3(m-1)^2+5(m-1)]$

$=(3m^2+5m)-[3(m^2+1-2m)+5m-5]$

$=(3m^2+5m)-[3m^2+3-6m+5m-5]$

$=3m^2+5m-3m^2-3+6m-5m+5=6m+2$

But given, $T_m=164$

$\therefore \quad 6m+2=164$

$\Rightarrow \quad 6m=164-2$

$\Rightarrow \quad 6m=162$

$\Rightarrow \quad m=\dfrac{162}{6}=27$

69. *(d)* Given, 8, A_1, A_2, A_3, A_4, A_5, 26 are in A.P.

$\because \quad T_n=a+(n-1)d$

$\therefore \quad 26=8+(7-1)d$

($\because n=7$, as there are 5 numbers between 8 and 26)

$\Rightarrow \quad 26-8=6d$

$\Rightarrow \quad 18=6d$

$\Rightarrow \quad d=\dfrac{18}{6} \Rightarrow d=3$

Now, $A_1=a+d=8+3=11$

$\Rightarrow \quad A_2=a+2d=8+2\times 3=14$

$A_3=a+3d=8+3\times 3=17$

$A_4=a+4d=8+4\times 3=20$

and $A_5=a+5d=8+5\times 3=23$

70. *(b)* Let $A_1, A_2, A_3, \ldots, A_m$ are m numbers inserted between 1 and 31.

$\therefore$ We have 1, $A_1, A_2, A_3, \ldots\ldots\ldots, A_m$, 31 are in A.P.

Now, $T_n=a+(n-1)d$

$31=1+(m+2-1)d \quad \left(\begin{array}{l}\because n=m+2, \text{ as there are } m\\ \text{numbers between 1 and 31}\end{array}\right)$

$\Rightarrow \quad 31-1=(m+1)d$

$\Rightarrow \quad 30=(m+1)d$

$\Rightarrow \quad d=\dfrac{30}{m+1} \quad \ldots(i)$

Given, $\dfrac{A_7}{A_{m-1}}=\dfrac{T_8}{T_m}=\dfrac{5}{9}$

$\Rightarrow \quad \dfrac{a+7d}{a+(m-1)d}=\dfrac{5}{9}$

$\Rightarrow \quad \dfrac{1+7\times\dfrac{30}{m+1}}{1+(m-1)\times\dfrac{30}{m+1}}=\dfrac{5}{9} \quad$ [from Eq. (i)]

$\Rightarrow \quad \dfrac{\dfrac{m+1+210}{m+1}}{\dfrac{(m+1)+30(m-1)}{m+1}}=\dfrac{5}{9}$

$\Rightarrow \quad \dfrac{m+211}{m+1+30m-30}=\dfrac{5}{9}$

$\Rightarrow \quad 9(m+211)=5(31m-29)$

$\Rightarrow \quad 9m+1899=155m-145$

$\Rightarrow \quad 1899+145=155m-9m$

$\Rightarrow \quad 146m=2044 \Rightarrow m=\dfrac{2044}{146}=14$

71. *(c)* Given, $a=100, d=5$

$\because T_n=a+(n-1)d$

$\therefore T_{30}=100+(30-1)5=100+29\times 5=100+145=245$

72. *(b)* Let d be the common difference and n be the number of terms of the A.P. .

Since the first term is a and the second term is b.

Therefore, $d=b-a$

Also, the last term is c, so

$c=a+(n-1)(b-a) \quad$ (since, $d=b-a$)

$\Rightarrow \quad n-1=\dfrac{c-a}{b-a}$

$\Rightarrow \quad n=1+\dfrac{c-a}{b-a}=\dfrac{b-a+c-a}{b-a}=\dfrac{b+c-2a}{b-a}$

Therefore, $S_n=\dfrac{n}{2}(a+l)=\dfrac{(b+c-2a)}{2(b-a)}(a+c)$

73. *(c)* Let A be the first term and D be the common difference of the A.P. . It is given that

$t_p=a \Rightarrow A+(p-1)D=a \quad \ldots(i)$

$t_q=b \Rightarrow A+(q-1)D=b \quad \ldots(ii)$

Subtracting Eq. (ii) from Eq. (i), we get

$(p-1-q+1)D=a-b$

$\Rightarrow \quad D=\dfrac{a-b}{p-q} \quad \ldots(iii)$

Adding Eqs. (i) and (ii), we get

$2A+(p+q-2)D=a+b$

$\Rightarrow \quad 2A + (p + q - 1)D = a + b + D$

$\Rightarrow \quad 2A + (p + q - 1)D = a + b + \dfrac{a-b}{p-q}$ ...(iv)

Now, $S_{p+q} = \dfrac{p+q}{2}[2A + (p+q-1)D]$

$= \dfrac{p+q}{2}\left[a + b + \dfrac{a-b}{p-q}\right]$ [using Eq. (iv)]

74. *(c)* Since, $T_p = a + (p-1)d = \dfrac{1}{q}$ ...(i)

and $T_q = a + (q-1)d = \dfrac{1}{p}$ ...(ii)

$\therefore$ On subtracting Eq. (ii) from Eq. (i), we get

$d(p - 1 - q + 1) = \dfrac{1}{q} - \dfrac{1}{p}$

$\Rightarrow \quad d(p-q) = \dfrac{p-q}{pq}$

$\Rightarrow d = \dfrac{1}{pq}$, if $p \neq q$

Putting the value of d in Eq. (i), we get

$a + \dfrac{(p-1)}{pq} = \dfrac{1}{q}$

$\Rightarrow \quad a = \dfrac{1}{q} - \dfrac{p-1}{pq}$

$\Rightarrow \quad a = \dfrac{p-p+1}{pq} = \dfrac{1}{pq}$

Now, $S_{pq} = \dfrac{pq}{2}[2a + (pq-1)d]$

$\left\{\because S_n = \dfrac{n}{2}[2a + (n-1)d]\right\}$

$= \dfrac{pq}{2}\left[2 \times \dfrac{1}{pq} + (pq-1)\dfrac{1}{pq}\right]$

$= \dfrac{pq}{2} \times \dfrac{1}{pq}(2 + pq - 1) = \dfrac{1}{2}(pq+1)$, if $p \neq q$

75. *(a)* Let the first term be A and common difference d.

Given, $S_p = a$

$\Rightarrow \quad \dfrac{p}{2}[2A + (p-1)d] = a$ ...(i)

$S_q = b \Rightarrow \dfrac{q}{2}[2A + (q-1)d] = b$ ...(ii)

and $S_r = c \Rightarrow \dfrac{r}{2}[2A + (r-1)d] = c$ ...(iii)

$\left\{\because S_n = \dfrac{n}{2}[2a + (n-1)d]\right\}$

Consider, $\dfrac{a}{p}(q-r) + \dfrac{b}{q}(r-p) + \dfrac{c}{r}(p-q)$

$= \dfrac{1}{p} \times \dfrac{p}{2}[2A + (p-1)d](q-r) + \dfrac{1}{q} \times \dfrac{q}{2}$

$[2A + (q-1)d](r-p) + \dfrac{1}{r} \times \dfrac{r}{2}[2A + (r-1)d\,](p-q)$

[using Eqs. (i), (ii) and (iii)]

$= \dfrac{1}{2}[\{2A + (p-1)d\}(q-r) + \{2A + (q-1)d\}(r-p)$

$+ \{2A + (r-1)d\}(p-q)]$

$= \dfrac{1}{2}[2A(q-r) + (p-1)d(q-r) + 2A(r-p)$

$+ (q-1)d(r-p) + 2A(p-q) + (r-1)d(p-q)]$

$= \dfrac{1}{2}[2A(q - r + r - p + p - q)$

$+ d[(p-1)(q-r) + (q-1)(r-p) + (r-1)(p-q)]$

$= \dfrac{1}{2}[2A \times (0) + d(pq - pr - q + r$

$+ qr - pq - r + p + rp - rq - p + q]$

$= \dfrac{1}{2}(0 + d \times 0) = 0$

76. *(d)* Let the A.P. be $a, a+d, a+2d, a+3d$ ...

Given, $\dfrac{S_m}{S_n} = \dfrac{m^2}{n^2}$

$\Rightarrow \quad \dfrac{\frac{m}{2}[2a + (m-1)d]}{\frac{n}{2}[2a + (n-1)d]} = \dfrac{m^2}{n^2}\left\{\because S_n = \dfrac{n}{2}[2a + (n-1)d]\right\}$

$\Rightarrow \quad \dfrac{m[2a + (m-1)d]}{n[2a + (n-1)d]} = \dfrac{m^2}{n^2}$

$\Rightarrow \quad \dfrac{2a + (m-1)d}{2a + (n-1)d} = \dfrac{m}{n}$

$\Rightarrow \quad [2a + (m-1)d]n = [2a + (n-1)d]m$

$\Rightarrow \quad 2an + (m-1)dn = 2am + (n-1)dm$

$\Rightarrow \quad 2an - 2am = (n-1)dm - (m-1)dn$

$\Rightarrow \quad 2a(n-m) = d[(n-1)m - (m-1)n]$

$\Rightarrow \quad 2a(n-m) = d[mn - m - mn + n]$

$\Rightarrow \quad 2a(n-m) = d(n-m)$

$\Rightarrow \quad 2a = d$

Now, $\dfrac{T_m}{T_n} = \dfrac{a + (m-1)d}{a + (n-1)d} = \dfrac{a + (m-1)2a}{a + (n-1)2a}$

$[\because T_n = a + (n-1)d]$

Taking a as common in numerator and denominator,

$\Rightarrow \quad \dfrac{T_m}{T_n} = \dfrac{1 + 2(m-1)}{1 + 2(n-1)} = \dfrac{1 + 2m - 2}{1 + 2n - 2} = \dfrac{2m-1}{2n-1}$

77. *(b)* Given, arithmetic mean between a and b is $\dfrac{a^n + b^n}{a^{n-1} + b^{n-1}}$

but we know that, the arithmetic mean between a and b is $\dfrac{a+b}{2}$.

Therefore, $\dfrac{a^n + b^n}{a^{n-1} + b^{n-1}} = \dfrac{a+b}{2}$

$\Rightarrow \quad 2(a^n + b^n) = (a^{n-1} + b^{n-1})(a+b)$

$\Rightarrow \quad 2a^n + 2b^n = a^{n-1}a^1 + a^{n-1}b + b^{n-1}a^1 + b^{n-1}b^1$

$\Rightarrow \quad 2a^n + 2b^n = a^n + a^{n-1}b + b^{n-1}a + b^n$

$\Rightarrow 2a^n + 2b^n - a^n - a^{n-1}b - b^{n-1}a - b^n = 0$

$\Rightarrow \quad a^n + b^n - a^{n-1}b - b^{n-1}a = 0$

$\Rightarrow \quad (a^n - a^{n-1}b) + (b^n - b^{n-1}a) = 0$

$\Rightarrow \quad a^{n-1}(a-b) - b^{n-1}(a-b) = 0$

$\Rightarrow \quad (a-b)(a^{n-1} - b^{n-1}) = 0$

$\Rightarrow \quad a^{n-1} - b^{n-1} = 0 \quad (\because a - b \neq 0)$

$\Rightarrow \quad a^{n-1} = b^{n-1}$

$\Rightarrow \quad \left(\frac{a}{b}\right)^{n-1} = 1$

$\Rightarrow \quad \left(\frac{a}{b}\right)^{n-1} = \left(\frac{a}{b}\right)^0$

On comparing the power of $\left(\frac{a}{b}\right)$ both sides, we get

$\Rightarrow \quad n - 1 = 0$

$\Rightarrow \quad n = 1$

78. *(a)* We know that, $S_n = \frac{n}{2}[2a + (n-1)d]$

Also, we know the sum of interior angles of a polygon having n sides, $S_n = (n-2)\,180°$

Therefore, $(n-2)\,180 = \frac{n}{2}[2 \times 120 + (n-1)(5)]$

$(\because a = 120, d = 5)$

$\Rightarrow \quad (n-2)\,180 \times 2 = n(240 + 5n - 5)$

$\Rightarrow \quad (n-2)\,360 = n\,(5n + 235)$

$\Rightarrow \quad (n-2)\,72 = n\,(n + 47)$

$\Rightarrow \quad 72n - 144 = n^2 + 47n$

$\Rightarrow \quad n^2 + 47n - 72n + 144 = 0$

$\Rightarrow \quad n^2 - 25n + 144 = 0$

Now, on factorising it by splitting the middle term, we get

$\Rightarrow \quad n^2 - (16 + 9)\,n + 144 = 0$

$\Rightarrow \quad n^2 - 16n - 9n + 144 = 0$

$\Rightarrow \quad n(n-16) - 9\,(n-16) = 0$

$\Rightarrow \quad (n-16)\,(n-9) = 0$

$\Rightarrow \quad n = 9, 16$

Only $n = 9$ is the required number of sides.

Note *If* $n = 16$,

$\Rightarrow \quad T_{16} = 120 + (16-1)\,5$
$= 120 + 15 \times 5 = 120 + 75 = 195 > 180°$

which is not possible.
Thus, number of sides in the polygon is 9.

79. *(a)* Consider, $T_{m+n} + T_{m-n}$

$= a + (m + n - 1)d + a + (m - n - 1)d$

$[\because T_n = a + (n-1)d]$

$= 2a + d\,(m + n - 1 + m - n - 1)$

$= 2a + d\,(2m - 2)$

$= 2\,[a + d\,(m-1)] = 2\,T_m$

80. *(a)* Let the three numbers be $a - d, a, a + d$.

Given, sum of the numbers $= 24$

$\therefore \quad a - d + a + a + d = 24$

$\Rightarrow \quad 3a = 24$

$\Rightarrow \quad a = 8$

and product of the numbers $= 440$

$\therefore \quad (a-d)\,a\,(a+d) = 440$

$\Rightarrow \quad a(a^2 - d^2) = 440$

$\Rightarrow \quad 8(64 - d^2) = 440$

$\Rightarrow \quad 64 - d^2 = 55$

$\Rightarrow \quad d^2 = 64 - 55$

$\Rightarrow \quad d^2 = 9$

$\Rightarrow \quad d = \pm 3$

When, $a = 8$ and $d = 3$, then numbers are

$a - d = 8 - 3 = 5$

$a = 8$ and $a + d = 8 + 3 = 11$

$\Rightarrow$ 5, 8, 11

When, $a = 8$ and $d = -3$, then numbers are

$a - d = 8 + 3 = 11$

$a = 8$

and $\quad a + d = 8 - 3$
$= 5$

$\Rightarrow$ 11, 8, 5

Hence, the numbers are 5, 8, 11 or 11, 8, 5.

81. *(a)* Let a and d be the first term and common difference respectively of an A.P.

Given, $S_1 =$ Sum of n terms

$= \frac{n}{2}[2a + (n-1)d] \quad \ldots\text{(i)}$

$S_2 =$ Sum of $2n$ terms

$= \frac{2n}{2}[2a + (2n-1)d] \quad \ldots\text{(ii)}$

and $\quad S_3 =$ Sum of $3n$ terms

$= \frac{3n}{2}[2a + (3n-1)d] \quad \ldots\text{(iii)}$

Consider, $3(S_2 - S_1)$

$= 3\left[\frac{2n}{2}[2a + (2n-1)d] - \frac{n}{2}[2a + (n-1)d]\right]$

[from Eqs. (i) and (ii)]

$= \frac{3n}{2}[2\{2a + (2n-1)d\} - \{2a + (n-1)d\}]$

$= \frac{3n}{2}[4a + 2(2n-1)d - 2a - (n-1)d]$

$= \frac{3n}{2}[(4a - 2a) + d\,(4n - 2 - n + 1)]$

$= \frac{3n}{2}[2a + (3n-1)d]$

$= S_3$ [from Eq. (iii)]

82. *(b)* The numbers which are divisible by 7 between 200 and 400 are 203, 210, 217, ..., 399.

Clearly, they form an A.P.

where, $a = 203, d = 7$ and $T_n = 399$

Now, $T_n = a + (n-1)d$

$\Rightarrow 399 = 203 + (n-1)7$

$\Rightarrow (n-1)7 = 399 - 203$

$\Rightarrow (n-1)7 = 196$

$\Rightarrow n - 1 = \frac{196}{7}$

$\Rightarrow n - 1 = 28$

$\Rightarrow n = 28 + 1$

$\Rightarrow n = 29$

Now, $S_n = \frac{n}{2}(a + l)$

$= \frac{29}{2}(203 + 399)$

$= \frac{29}{2} \times 602$

$= 29 \times 301 = 8729$

83. *(c)* The numbers from 1 to 100 which are divisible by 2 are 2, 4, 6, 8, ...,100.

Clearly, they are in A.P., where $a = 2, d = 4 - 2 = 2$

$\therefore T_n = a + (n-1)d$

$\Rightarrow 100 = 2 + (n-1)2$

$\Rightarrow 100 - 2 = (n-1)2$

$\Rightarrow 98 = (n-1)2$

$\Rightarrow 49 = n - 1 \Rightarrow n = 50$

Therefore, sum of 50 numbers,

$S_{50} = \frac{50}{2}[2 + 100]$ $\left[\because S_n = \frac{n}{2}(a + l)\right]$

$= 25 \times 102$

$= 2550$ ...(i)

Now, the numbers from 1 to 100 which are divisible by 5 are 5, 10, 15, 20, ... 100.

Clearly, they are in A.P., where, $a = 5, d = 10 - 5 = 5$

$\therefore T_n = a + (n-1)d$

$\Rightarrow 100 = 5 + (n-1)5$

$\Rightarrow 100 - 5 = (n-1)5$

$\Rightarrow (n-1) = \frac{95}{5}$

$\Rightarrow n - 1 = 19$

$\Rightarrow n = 19 + 1 = 20$

Now, $S_{20} = \frac{20}{2}[5 + 100]$ $\left[\because S_n = \frac{n}{2}(a + l)\right]$

$= 10 \times 105 = 1050$ ...(ii)

Now, the numbers from 1 to 100 which are divisible by 10 are 10, 20, 30, ..., 100.

Clearly, they are in A.P., where $a = 10, d = 20 - 10 = 10$ and $n = 10$.

Now, $S_{10} = \frac{10}{2}[10 + 100]$ $\left[\because S_n = \frac{n}{2}(a + l)\right]$

$= 5 \times 110 = 550$ ...(iii)

Hence, required sum of integers from 1 to 100 which are divisible by 2 or 5 = 2550 + 1050 − 550

[using Eqs. (i), (ii) and (iii)]

$= 3600 - 550$

$= 3050$

84. *(d)* The two-digit numbers which when divided by 4, leave 1 as remainder, will be the numbers which are one more than the multiple of 4 *i.e.*,

13, 17, 21, ..., 97

Clearly, they form an A.P., where $a = 13, d = 4, T_n = 97$.

$\because T_n = a + (n-1)d$

$\therefore 97 = 13 + (n-1)4$

$\Rightarrow 97 - 13 = (n-1)4$

$\Rightarrow 84 = (n-1)4$

$\Rightarrow 21 = n - 1$

$\Rightarrow n = 21 + 1 = 22$

Now, $S_{22} = \frac{22}{2}(13 + 97)$

$= 11 \times 110 = 1210$

85. *(c)* Let the A.P. be $a, a + d, a + 2d, \ldots\ldots$

Given, sum of first four terms = 56

i.e., $T_1 + T_2 + T_3 + T_4 = 56$

$\Rightarrow a + a + d + a + 2d + a + 3d = 56$

$\Rightarrow 4a + 6d = 56$

$\Rightarrow 4 \times 11 + 6d = 56$ $(\because a = 11)$

$\Rightarrow 6d = 56 - 44 = 12$

$\Rightarrow d = \frac{12}{6}$

$\Rightarrow d = 2$

If last term is T_n, then sum of last four terms

$= T_n + T_{n-1} + T_{n-2} + T_{n-3} = 112$

$\therefore a + (n-1)d + a + (n-1-1)d + a + (n-2-1)d$
$+ a + (n-3-1)d = 112$ $[\because T_n = a + (n-1)d]$

$\Rightarrow 4a + d(n - 1 + n - 2 + n - 3 + n - 4) = 112$

$\Rightarrow 4 \times 11 + 2[4n - 10] = 112$ $(\because a = 11, d = 2)$

$\Rightarrow 44 + 8n - 20 = 112$

$\Rightarrow 8n = 112 - 44 + 20$

$\Rightarrow 8n = 132 - 44$

$\Rightarrow 8n = 88$

$\Rightarrow n = \frac{88}{8}$

$= 11$

Hence, total number of terms in the sequence = 11

86. *(b)* Let the A.P. be $A, A+D, A+2D, A+3D, \ldots\ldots$

Given, p^{th} term $= A + (p-1)D = a$ ...(i)

q^{th} term $= A + (q-1)D = b$ ...(ii)

r^{th} term $= A + (r-1)D = c$ ...(iii)

Consider, $(q-r)a + (r-p)b + (p-q)c$

$= (q-r)[A+(p-1)D] + (r-p)[A+(q-1)D] + (p-q)[A+(r-1)D]$ ...(iv)

[putting the values of a, b and c from Eqs. (i), (ii) and (iii)]

$= (q-r)A + (q-r)(p-1)D + (r-p)A + (r-p)(q-1)D + (p-q)A + (p-q)(r-1)D$

$= A(q-r+r-p+p-q) + D[(q-r)(p-1) + (r-p)(q-1) + (p-q)(r-1)]$

$= A(0) + D[qp - q - rp + r + rq - r - pq + p + pr - p - qr + q]$

$= 0 + 0 = 0$

87. *(a)* Let, $a\left(\frac{1}{b}+\frac{1}{c}\right), b\left(\frac{1}{c}+\frac{1}{a}\right), c\left(\frac{1}{a}+\frac{1}{b}\right)$ are in A.P.

$\Rightarrow \quad a\left(\frac{b+c}{bc}\right), b\left(\frac{a+c}{ac}\right), c\left(\frac{b+a}{ab}\right)$ are in A.P.

$\Rightarrow \quad \frac{ab+ac}{bc}, \frac{ba+bc}{ac}, \frac{cb+ca}{ab}$ are in A.P.

Adding 1 to each term,

$\Rightarrow \frac{ab+ac}{bc}+1, \frac{ba+bc}{ac}+1, \frac{cb+ca}{ab}+1$ are in A.P.

$\Rightarrow \frac{ab+ac+bc}{bc}, \frac{ba+bc+ac}{ac}, \frac{bc+ac+ab}{ab}$ are in A.P.

Dividing each term by $ab+bc+ac$,

$\Rightarrow \frac{1}{bc}, \frac{1}{ac}, \frac{1}{ab}$ are in A.P.

Multiplying each term by abc,

$\Rightarrow \frac{abc}{bc}, \frac{abc}{ac}, \frac{abc}{ab}$ are in A.P.

$\Rightarrow a, b, c$ are in A.P.

88. *(b)* Tractor cost = ₹ 12000

Down payment = ₹ 6000

Balance amount = ₹ 6000

Interest on Ist instalment $= \frac{6000 \times 12 \times 1}{100}$ $\left(\because I = \frac{P \times R \times T}{100}\right)$

= ₹ 720

Now, unpaid amount for second instalment $= 6000 - 500 =$ ₹ 5500

Interest on IInd instalment $= \frac{5500 \times 12 \times 1}{100} =$ ₹ 660

Again, unpaid amount for third instalment $= 5500 - 500$ = ₹ 5000

Interest on IIIrd instalment $= \frac{5000 \times 12 \times 1}{100} =$ ₹ 600

Total interest paid by him

$= 720 + 660 + 600 + \cdots$ upto 12 terms

which is an A.P. with $a = 720, d = 660 - 720 = -60$

Therefore, total interest $= \frac{12}{2}[2 \times 720 + (12-1)(-60)]$

$= 6[1440 - 11 \times 60] = 6[1440 - 660] = 6 \times 780 =$ ₹ 4680

Hence, total amount or actual cost $= 12000 + 4680$ = ₹ 16680

89. *(d)* Scooter cost = ₹ 22000

Down payment = ₹ 4000.

Balance amount = ₹ 18000

Now, interest on Ist instalment $= \frac{18000 \times 10 \times 1}{100} =$ ₹ 1800 $\left(\because I = \frac{P \times R \times T}{100}\right)$

Unpaid amount for second instalment

$= 18000 - 1000 = 17000$

Interest on IInd instalment $= \frac{17000 \times 10 \times 1}{100} =$ ₹ 1700

Unpaid amount for third instalment $= 17000 - 1000 = 16000$

Interest on IIIrd instalment $= \frac{16000 \times 10 \times 1}{100} =$ ₹ 1600

....

....

$\therefore$ Total interest paid by him

$= 1800 + 1700 + 1600 + \ldots$ upto 18 terms

which is an A.P. with $a = 1800, d = 1700 - 1800 = -100$

Therefore, total interest $= \frac{18}{2}[2 \times 1800 + (18-1)(-100)]$

$\left\{\because \text{sum of A.P.} = S_n = \frac{n}{2}[2a + (n-1)d]\right\}$

$= 9(3600 - 1700)$

$= 9 \times 1900 = 17100$

Hence, total amount or actual cost $= 22000 + 17100$ = ₹ 39100

90. *(b)* Interest on ₹ 10000 after 1 yr

$= \frac{10000 \times 5 \times 1}{100}$ $\left(\because I = \frac{P \times R \times T}{100}\right)$

= ₹ 500

Therefore, amount after 1 yr $= 10000 + 500 = 10500$

Now, interest on ₹ 10000 after 2 yr

$= \frac{10000 \times 5 \times 2}{100} =$ ₹ 1000

$\therefore$ Amount after 2 yr $= 10000 + 1000$

= ₹ 11000

Similarly, interest on ₹ 10000 after 3 yr

$= \frac{10000 \times 5 \times 3}{100}$

$= 1500$

$\therefore$ Amount after 3 yr $= 10000 + 1500$
$= ₹11500$

Hence, the amount in the account of man in first, second and third year are ₹10000, 10500, 11000, ...

It is an A.P., where $a = 10000$ and $d = 500$

Amount after 15th yr $= T_{15}$

$$= a + 14d = 10000 + 14 \times 500$$
$$= 10000 + 7000 = ₹17000$$

Amount after 20 yr $= 10000 + 20 \times 500$
$$= 10000 + 10000 = ₹20000$$

91. *(d)* Let the number of days in which the work is completed, be n.

Now, according to question, 4 workers dropped on everyday *i.e.*, 150, 146, 142, 138 ... .

Clearly, work done in both conditions is same.

[Had the workers not dropped then the work would have finished in $(n-8)$ days with 150 workers working on each day. Hence, the total number of workers who would have worked for all the $(n-8)$ days is $150(n-8)$.]

$$\Rightarrow \quad 150(n-8) = \frac{n}{2}[2 \times 150 + (n-1)(-4)]$$
$$\left\{\text{using } S_n = \frac{n}{2}[2a + (n-1)d]\right\}$$
$$\Rightarrow \quad 150n - 1200 = \frac{n}{2} \times 2(150 - 2n + 2)$$
$$\Rightarrow \quad 150n - 1200 = \frac{n}{2} \times 2(152 - 2n)$$

On dividing each term by 2, we get

$$75n - 600 = n(76 - n)$$
$$\Rightarrow \quad 75n - 600 = 76n - n^2$$
$$n^2 - 76n + 75n - 600 = 0$$
$$\Rightarrow \quad n^2 - n - 600 = 0$$

Now, on factorizing it by splitting the middle term, we get

$$\Rightarrow \quad n^2 - (25n - 24n) - 600 = 0$$
$$\Rightarrow \quad n^2 - 25n + 24n - 600 = 0$$
$$\Rightarrow \quad n(n-25) + 24(n-25) = 0$$
$$\Rightarrow \quad (n-25)(n+24) = 0$$

$\Rightarrow n = 25$ and $n \neq -24$, because it is not possible.

Hence, work will be completed in 25 days.

92. *(b)* If every term in a progression except the first term bears a constant ratio to the term immediately preceding it. Then, such a progression is called geometric progression.

93. *(d)* 2, 4, 8, 16, ... is a geometric sequence with common ratio 2.

$\frac{1}{9}, \frac{-1}{27}, \frac{1}{81}, \frac{-1}{243}, \ldots$ is a geometric sequence with common ratio $\frac{-1}{3}$.

.01, .0001, .000001, ... is a geometric sequence with common ratio .01.

Hence, (a), (b) and (c) are the examples of geometric sequence.

94. *(a)* The common ratio in .01, .0001, .000001, ... is 0.01. Hence, I and II are correct.

$$\because \quad \frac{4}{2} = \frac{8}{4} = \frac{16}{8} = \ldots = 2,$$
$$\frac{-1}{27} \times \frac{9}{1} = \frac{1}{81} \times \frac{27}{-1} = \ldots = \frac{-1}{3},$$
$$\text{and} \quad \frac{.0001}{.01} = \frac{.000001}{.0001} = \ldots = .01$$

95. *(c)* A sequence $a_1, a_2, a_3, \ldots, a_n, \ldots$ is called geometric progression, if each term is non-zero and $\frac{a_{k+1}}{a_k} = r$ (constant), for $k \geq 1$.

96. *(d)* If $a, ar, ar^2, ar^3, \ldots,$ is geometric progression, then a and r are called first term and common ratio.

97. *(a)* In geometric progression, following notations are used.

$a =$ First term
$r =$ Common ratio
$l =$ Last term
$n =$ Number of terms
$S_n =$ Sum of first n terms

98. *(b)* If in a G.P. first non-zero term is 'a' and common ratio is 'r', then

Second term $= a_2 = ar$
Third term $= a_3 = ar^2$
Fourth term $= a_4 = ar^3$
Fifth term $= a_5 = ar^4$

99. *(c)* The general term of a G.P. whose first non-zero term is 'a' and common ratio is 'r' can be written as $a_n = ar^{n-1}$.

100. *(a)* $a, ar, ar^2, ar^3, \ldots, ar^{n-1}$ is a finite G.P. .

101. *(b)* $a, ar, ar^2, ar^3, \ldots, ar^{n-1}, \ldots$ is an infinite G.P. .
$a + ar + ar^2 + ar^3 + \ldots ar^{n-1}$ is a finite geometric series.
$a + ar + ar^2 + \ldots + ar^{n-1} + \ldots$ is an infinite geometric series.

102. *(b)* $\because S_n = a + ar + ar^2 + ar^3 + \ldots + ar^{n-1}$ ...(i)

If $r = 1$,

we have $S_n = a + a + \ldots + a$ (n times)
$= na$

Hence, Statement I is true.

If $r \neq 1$, multiplying Eq. (i) by r, we have

$$rS_n = ar + ar^2 + ar^3 + \ldots + ar^n \quad \ldots\text{(ii)}$$

Subtracting Eq. (ii) from Eq. (i), we get

$$\Rightarrow \quad (1-r)S_n = a - ar^n$$

$$\Rightarrow \quad S_n = \frac{a(1-r^n)}{1-r} \quad \text{or} \quad S_n = \frac{a(r^n-1)}{r-1}$$

Hence, Statement II is false.

103. *(a)* Here, $a = 5$ and $r = 5$

Thus, $a_n = ar^{n-1}$
$= 5 \times (5)^{n-1} = 5^n$

104. *(a)* $\because$ 131072 be the nth term of given G.P.

Therefore, $131072 = a_n = 2(4)^{n-1}$ $\quad (\because a = 2 \text{ and } r = 4)$

$\Rightarrow \quad 65536 = (4)^{n-1}$

$\Rightarrow \quad 4^8 = 4^{n-1}$

$\Rightarrow \quad 8 = n - 1$

$\Rightarrow \quad n = 9$

Hence, 131072 is the 9th term of the G.P. .

105. *(c)* Here, $a_3 = ar^2 = 24$ $\quad \ldots(i)$

and $a_6 = ar^5 = 192$ $\quad \ldots(ii)$

On dividing Eq. (ii) by Eq. (i), we get

$$r^3 = \frac{192}{24}$$

$\Rightarrow \quad r^3 = 8 \Rightarrow r = 2$

On substituting $r = 2$ in Eq. (i), we get $a = 6$

Hence, $a_{10} = 6(2)^9 = 3072$

106. *(a)* The geometric series is given as

$$1 + \frac{2}{3} + \left(\frac{2}{3}\right)^2 + \left(\frac{2}{3}\right)^3 + \ldots$$

Here, $a = 1$ and $r = \frac{2}{3}$

Therefore,
$$S_n = \frac{a(1-r^n)}{1-r} = \frac{\left[1-\left(\frac{2}{3}\right)^n\right]}{1-\frac{2}{3}} = 3\left(1-\left(\frac{2}{3}\right)^n\right)$$

In particular, $S_5 = 3\left(1-\left(\frac{2}{3}\right)^5\right)$

$$= 3 \times \frac{211}{243} = \frac{211}{81}$$

Hence, I and II are correct.

107. *(d)* Let $\frac{a}{r}$, a, ar be the first three terms of the G.P. .

Then, $\frac{a}{r} + a + ar = \frac{13}{12}$ $\quad \ldots(i)$

and $\frac{a}{r} \cdot a \cdot ar = -1$ $\quad \ldots(ii)$

From Eq. (ii), we get $a^3 = -1$

$\Rightarrow \quad a = -1$ (considering only real roots)

Substituting $a = -1$ in Eq. (i), we have

$$\frac{-1}{r} - 1 - r = \frac{13}{12}$$

$\Rightarrow \quad r^2 + r + 1 + \frac{13r}{12} = 0$

$\Rightarrow \quad 12r^2 + 25r + 12 = 0$

On solving, we get

$$r = \frac{-3}{4} \quad \text{or} \quad \frac{-4}{3}$$

Thus, the first three terms of G.P. are $\frac{4}{3}, -1, \frac{3}{4}$ for $r = \frac{-3}{4}$ and $\frac{3}{4}, -1, \frac{4}{3}$ for $r = \frac{-4}{3}$.

108. *(b)* Here, $a = 2, r = 2$ and $n = 10$

Since, $S_n = \frac{a(r^n-1)}{r-1}$

We have, $S_{10} = \frac{2(2^{10}-1)}{2-1}$
$= 2(1024 - 1)$
$= 2 \times 1023 = 2046$

Hence, the number of ancestors preceding the person is 2046.

109. *(c)* The geometric mean of two positive numbers a and b is $\sqrt{ab}$.

110. *(a)* The geometric mean of 2 and 8 is $\sqrt{16}$ *i.e.*, 4.

111. *(d)* Infinitely many numbers can be inserted between two positive numbers 'a' and 'b' to make the resulting sequence in a G.P. .

112. *(b)* b is the $(n+2)^{th}$ term of given G.P. .

Hence, $b = ar^{n+1}$

$\Rightarrow \quad r^{n+1} = \frac{b}{a}$

$\Rightarrow \quad r = \left(\frac{b}{a}\right)^{1/n+1}$

Hence, $G_1 = ar = a\left(\frac{b}{a}\right)^{1/n+1}$

$G_2 = ar^2 = a\left(\frac{b}{a}\right)^{2/n+1}$

....

....

....

....

$G_n = ar^n = a\left(\frac{b}{a}\right)^{n/n+1}$

113. *(c)* Let G_1, G_2, G_3 be three numbers between 1 and 256 such that $1, G_1, G_2, G_3, 256$ is a G.P. .

Therefore, $256 = r^4$ giving $r = \pm 4$ (taking real roots only)

For $r = 4$, we have

$$G_1 = ar = 4,$$
$$G_2 = ar^2 = 16,$$
$$G_3 = ar^3 = 64$$

Similarly for $r = -4,$
numbers are −4, 16 and −64.

114. *(a)* Let A and G be A.M and G.M. of two given positive real numbers a and b respectively.

Then, $A = \dfrac{a+b}{2}$ and $G = \sqrt{ab}$

Thus, we have

$$A - G = \frac{a+b}{2} - \sqrt{ab} = \frac{a+b-2\sqrt{ab}}{2}$$
$$= \frac{(\sqrt{a}-\sqrt{b})^2}{2} \geq 0$$

Hence, $A - G \geq 0 \Rightarrow A \geq G$

115. *(a)* The given G.P. is $\frac{5}{2}, \frac{5}{4}, \frac{5}{8}, \ldots$

Here, $a =$ First term $= \frac{5}{2}$ and common ratio $(r) = \dfrac{\frac{5}{4}}{\frac{5}{2}} = \frac{2}{4} = \frac{1}{2}$

Now, $T_{20} = ar^{20-1} = \left(\frac{5}{2}\right) \times \left(\frac{1}{2}\right)^{20-1} = \frac{5}{2} \times \left(\frac{1}{2}\right)^{19}$
$$= \frac{5}{2} \times \frac{1}{2^{19}} = \frac{5}{2^{20}}$$

Again, $T_n = ar^{n-1} = \left(\frac{5}{2}\right) \times \left(\frac{1}{2}\right)^{n-1} = \frac{5}{2} \times \frac{1}{2^{n-1}}$
$$= \frac{5}{2^{1+n-1}} = \frac{5}{2^n}$$

116. *(c)* Given, 8th term, $T_8 = 192$ and common ratio $(r) = 2$

$\Rightarrow \quad ar^{8-1} = 192 \quad (\because T_n = ar^{n-1})$

$\Rightarrow \quad a \times (2)^7 = 192 \Rightarrow a \times 128 = 192$

$\Rightarrow \quad a = \frac{192}{128} \Rightarrow a = \frac{48}{32} = \frac{3}{2}$

Now, $T_{12} = ar^{12-1} = \frac{3}{2} \times (2)^{11}$
$$= \frac{3}{2} \times 2^{11} = 3 \times 2^{10}$$
$$= 3 \times 1024 = 3072$$

117. *(d)* Let a and r be the first term and common ratio of the G.P. .

Given, $T_5 = p \Rightarrow ar^4 = p \quad \ldots$(i)
$T_8 = q \Rightarrow ar^7 = q \quad \ldots$(ii)
and $T_{11} = s \Rightarrow ar^{10} = s \quad \ldots$(iii)

Now, on multiplying Eqs. (i) and (iii), we get
$$ar^4 \times ar^{10} = ps$$
$$a^2 r^{14} = ps$$
$\Rightarrow \quad [(ar)^7]^2 = ps$
$\Rightarrow \quad q^2 = ps \quad$ [from Eq. (ii)]

118. *(c)* Let a and r be the first term and common ratio of the G.P. .

Given, $T_4 = (T_2)^2 \quad (\because T_n = ar^{n-1})$
$\Rightarrow \quad ar^3 = (ar)^2$
$\Rightarrow \quad ar^3 = a^2r^2$
$\Rightarrow \quad r = a$
But it is given that, $a = -3$
$\Rightarrow \quad r = -3$
Now, $T_7 = ar^6 = (-3)(-3)^6 = (-3)^7 = -2187$

119. *(d)* The sequence is $2, 2\sqrt{2}, 4, \ldots$

Here, $a = 2, r = \frac{2\sqrt{2}}{2} = \sqrt{2}$

Now, $T_{13} = ar^{12} = 2(\sqrt{2})^{12} = 2 \cdot 2^6 = 2^7 = 128$

Now, consider the sequence $5, 5\sqrt{3}, 15 \ldots$

Here, $a = 5, r = \sqrt{3}$

$\therefore \quad T_7 = ar^6 = 5(\sqrt{3})^6 = 5 \cdot (3)^3 = 135$

The given sequence is $\sqrt{3}, 3, 3\sqrt{3}, \ldots, 729$

Here, $a = \sqrt{3}, r = \frac{3}{\sqrt{3}} = \sqrt{3}, T_n = 729$

Now, $T_n = ar^{n-1}$
$\Rightarrow \quad 729 = \sqrt{3}(\sqrt{3})^{n-1}$
$\Rightarrow \quad 729 = (\sqrt{3})^n \Rightarrow 3^{n/2} = 3^6$

On comparing the power of 3 on both sides, we get
$$\frac{n}{2} = 6 \Rightarrow n = 12$$

The given sequence is $\frac{1}{3}, \frac{1}{9}, \frac{1}{27}, \ldots, \frac{1}{19683}$

Here, $a = \frac{1}{3}, r = \dfrac{\frac{1}{9}}{\frac{1}{3}} = \frac{3}{9} = \frac{1}{3}$
$$T_n = \frac{1}{19683}$$

Now, $T_n = ar^{n-1}$
$$\frac{1}{19683} = \frac{1}{3} \times \left(\frac{1}{3}\right)^{n-1}$$
$\Rightarrow \quad \frac{1}{6561} = \left(\frac{1}{3}\right)^{n-1} \Rightarrow \left(\frac{1}{3}\right)^{n-1} = \left(\frac{1}{3}\right)^8$

On comparing the power of (1/3) on both sides, we get
$$n - 1 = 8 \Rightarrow n = 8 + 1 \Rightarrow n = 9$$

Hence, option (d) is incorrect.

120. *(a)* Since, $\frac{-2}{7}, x, -\frac{7}{2}$ are in G.P., then

$\Rightarrow \quad x^2 = \left(-\frac{2}{7}\right) \times \left(-\frac{7}{2}\right) \Rightarrow x^2 = \frac{2}{7} \times \frac{7}{2}$
$\Rightarrow \quad x^2 = 1$
$\Rightarrow \quad x = \pm 1$

121. *(c)* Here, $a = 0.15$, $r = \frac{0.015}{0.15} = \frac{15}{1000} \times \frac{100}{15}$

$$r = \frac{1}{10} < 1 \text{ and } n = 20$$

Now, $S_n = \frac{a(1-r^n)}{1-r}$ $(\because r < 1)$

$$\Rightarrow S_{20} = \frac{0.15\left[1-\left(\frac{1}{10}\right)^{20}\right]}{1-\frac{1}{10}} = \frac{0.15\left[1-\frac{1}{10^{20}}\right]}{\frac{10-1}{10}}$$

$$= \frac{0.15 \times 10}{9}\left[1-\frac{1}{10^{20}}\right]$$

$$= \frac{15 \times 10}{900}\left[1-\left(\frac{1}{10}\right)^{20}\right] = \frac{1}{6}[1-(0.1)^{20}]$$

Now, consider the sequence $\sqrt{7}, \sqrt{21}, 3\sqrt{7}\ldots$

Here, $a = \sqrt{7}, r = \frac{\sqrt{21}}{\sqrt{7}} = \frac{\sqrt{7 \times 3}}{\sqrt{7}} = \sqrt{3} > 1$

$\because$ $S_n = \frac{a(r^n - 1)}{r-1}$ $(\because r > 1)$

$$\Rightarrow S_n = \frac{\sqrt{7}\,[(\sqrt{3})^n - 1]}{\sqrt{3}-1}$$

$$= \frac{\sqrt{7}\,[(3^{1/2})^n - 1]}{\sqrt{3}-1} \times \frac{\sqrt{3}+1}{\sqrt{3}+1}$$

$$= \frac{\sqrt{7}\,(\sqrt{3}+1)(3^{n/2}-1)}{3-1}$$

$[\because a^2 - b^2 = (a+b)(a-b)]$

$$= \frac{\sqrt{7}}{2}(\sqrt{3}+1)(3^{n/2}-1)$$

Consider the sequence $1, -a, a^2 \ldots$

Here, $a = 1, r = -\frac{a}{1} = -a$

$\because$ $S_n = \frac{a(1-r^n)}{1-r}$

$$\Rightarrow S_n = \frac{1[1-(-a)^n]}{1-(-a)} = \frac{1-(-a)^n}{1+a}$$

Consider the sequence $x^3, x^5, x^7\ldots$

Here, $a = x^3, r = \frac{x^5}{x^3} = x^2$

$\because$ $S_n = \frac{a(1-r^n)}{1-r}$

$$\therefore S_n = \frac{x^3\,[1-(x^2)^n]}{1-x^2} = \frac{x^3\,[1-x^{2n}]}{1-x^2}$$

Hence, (c) is incorrect.

122. *(d)* Let the first three numbers of a G.P. be $\frac{a}{r}$, a and ar.

$\therefore$ According to the question,

$$\frac{a}{r} + a + ar = \frac{39}{10} \quad \ldots(i)$$

and $\left(\frac{a}{r}\right) \times (a) \times (ar) = 1$

$\Rightarrow$ $a^3 = 1 \Rightarrow a = 1$

Putting the value of $a = 1$ in Eq. (i), we get

$$\frac{1}{r} + 1 + r = \frac{39}{10}$$

$$\Rightarrow \frac{1}{r} + \frac{1}{1} + \frac{r}{1} = \frac{39}{10}$$

$$\Rightarrow \frac{1+r+r^2}{r} = \frac{39}{10}$$

$$\Rightarrow 10 + 10r + 10r^2 = 39r$$

$$\Rightarrow 10r^2 + 10r - 39r + 10 = 0$$

$$\Rightarrow 10r^2 - 29r + 10 = 0$$

Now, on factorizing it by splitting the middle term, we get

$$\Rightarrow 10r^2 - 25r - 4r + 10 = 0$$

$$\Rightarrow 5r(2r-5) - 2(2r-5) = 0$$

$$\Rightarrow (5r-2)(2r-5) = 0$$

$$\Rightarrow 5r - 2 = 0 \text{ and } 2r - 5 = 0$$

$$\Rightarrow r = \frac{2}{5} \text{ and } r = \frac{5}{2}$$

When $a = 1$ and $r = \frac{2}{5}$, then numbers are

$$\frac{a}{r} = \frac{1}{\frac{2}{5}} = \frac{5}{2};\ a = 1 \text{ and } ar = 1 \times \frac{2}{5} = \frac{2}{5}$$

i.e., $\frac{5}{2}, 1, \frac{2}{5}$

When $a = 1$ and $r = \frac{5}{2}$, then numbers are

$$\frac{a}{r} = \frac{1}{\frac{5}{2}} = \frac{2}{5};\ a = 1$$

and $ar = 1 \times \frac{5}{2} = \frac{5}{2}$

i.e., $\frac{2}{5}, 1, \frac{5}{2}$

123. *(c)* Here, $a = 3, r = 3, S_n = 120$

$\because$ $S_n = \frac{a(r^n - 1)}{r-1}$ $(\because r > 1)$

$$\Rightarrow 120 = \frac{3(3^n - 1)}{3-1}$$

$$\Rightarrow 120 = \frac{3(3^n - 1)}{2}$$

$$\Rightarrow 120 \times 2 = 3(3^n - 1)$$

$$\Rightarrow \frac{240}{3} = 3^n - 1$$

$\Rightarrow \quad 3^n - 1 = 80$

$\Rightarrow \quad 3^n = 80 + 1$

$\Rightarrow \quad 3^n = 81$

$\Rightarrow \quad 3^n = 3^4$

On comparing the power of 3 on both sides, we get

$$n = 4$$

124. *(a)* Let the G.P. be $a, ar, ar^2, ar^3, \ldots\ldots$

Given, $\quad a + ar + ar^2 = 16 \quad \ldots(i)$

$ar^3 + ar^4 + ar^5 = 128 \quad \ldots(ii)$

Dividing Eq. (i) by Eq. (ii), we get

$$\frac{a + ar + ar^2}{ar^3 + ar^4 + ar^5} = \frac{16}{128}$$

$$\Rightarrow \quad \frac{a(1 + r + r^2)}{ar^3(1 + r + r^2)} = \frac{1}{8}$$

$$\Rightarrow \quad \frac{1}{r^3} = \left(\frac{1}{2}\right)^3$$

On comparing the base of the power 3 on both sides, we get

$$r = 2 \quad (\because \text{ considering only real roots})$$

Putting $r = 2$ in Eq. (i), we get

$\Rightarrow \quad a + 2a + 4a = 16$

$\Rightarrow \quad 7a = 16$

$\Rightarrow \quad a = \frac{16}{7}$

Now, $\quad S_n = \frac{a(r^n - 1)}{r - 1} \quad (\because r = 2 > 1)$

$$= \frac{\frac{16}{7}(2^n - 1)}{2 - 1}$$

$$= \frac{16}{7}(2^n - 1)$$

Hence, $\quad a = \frac{16}{7}, r = 2$

and $\quad S_n = \frac{16}{7}(2^n - 1)$

125. *(c)* Given, $a = 729, T_7 = 64$

$\Rightarrow \quad ar^{7-1} = 64 \quad (\because T_n = ar^{n-1})$

$\Rightarrow \quad 729r^6 = 64$

$\Rightarrow \quad r^6 = \frac{64}{729} = \left(\frac{2}{3}\right)^6 \quad \ldots(i)$

On comparing base of the power 6 on both sides, we get

$\Rightarrow \quad r = \frac{2}{3} < 1$

Now, $\quad S_n = \frac{a(1 - r^n)}{1 - r} \quad (\because r < 1)$

$$\therefore \quad S_7 = \frac{729\left[1 - \left(\frac{2}{3}\right)^7\right]}{1 - \frac{2}{3}} = \frac{729\left[1 - \frac{2^7}{3^7}\right]}{\frac{1}{1} - \frac{2}{3}}$$

$$= \frac{729\left[\frac{1}{1} - \frac{128}{2187}\right]}{\frac{3 - 2}{3}}$$

$$= \frac{729 \times 3}{1} \times \frac{2187 - 128}{2187}$$

$$= \frac{1}{1} \times 2059$$

$$= 2059$$

126. *(a)* Let the G.P. be $a, ar, ar^2, ar^3, \ldots$

Given, $\quad a + ar = -4 \quad \ldots(i)$

and $\quad T_5 = 4T_3$

$\Rightarrow \quad ar^{5-1} = 4ar^{3-1}$

$\Rightarrow \quad r^4 = 4r^2$

$\Rightarrow \quad r^2 = 4$

$\Rightarrow \quad r = \pm 2$

If $r = 2$, then from Eq. (i), we get

$$a + a(2) = -4$$

$\Rightarrow \quad 3a = -4$

$\Rightarrow \quad a = -\frac{4}{3}$

Hence, G.P. is $-\frac{4}{3}, \left(-\frac{4}{3}\right)(2), \left(-\frac{4}{3}\right)(2)^2 \ldots$ *i.e.*, $-\frac{4}{3}, -\frac{8}{3}, -\frac{16}{3}, \ldots$

If $r = -2$, then from Eq. (i), we get

$$a + a(-2) = -4$$

$\Rightarrow \quad -a = -4$

$\Rightarrow \quad a = 4$

Hence, G.P. is $4, 4(-2), 4(-2)^2, \ldots$ *i.e.*, $4, -8, 16, \ldots$

127. *(b)* Given, $T_4 = x \Rightarrow ar^{4-1} = x \Rightarrow ar^3 = x \quad \ldots(i)$

$T_{10} = y \Rightarrow ar^{10-1} = y \Rightarrow ar^9 = y \quad \ldots(ii)$

$T_{16} = z \Rightarrow ar^{16-1} = z \Rightarrow ar^{15} = z \quad \ldots(iii)$

Now, multiplying Eq. (i) by Eq. (iii), we get

$\Rightarrow \quad ar^3 \times ar^{15} = x \times z$

$\Rightarrow \quad a^2r^{3+15} = xz$

$\Rightarrow \quad a^2r^{18} = xz$

$\Rightarrow \quad (ar^9)^2 = xz$

$\Rightarrow \quad y^2 = xz \quad$ [from Eq. (ii)]

Therefore, x, y, z are in G.P. .

128. *(d)* Let $S = 8 + 88 + 888 + 8888 + \ldots$ upto n terms

$\Rightarrow \quad S = 8(1 + 11 + 111 + 1111 + \ldots$ upto n terms)

$= \frac{8}{9}[9 + 99 + 999 + 9999 + \ldots$ upto n terms]

$= \frac{8}{9}[(10 - 1) + (100 - 1) + (1000 - 1) + \ldots$ upto n terms]

$= \frac{8}{9}[(10 + 100 + 1000 + \ldots\text{upto } n \text{ terms})$

$- (1 + 1 + 1 + \ldots \text{upto } n \text{ terms})]$

$= \frac{8}{9}\left[10\,\frac{(10^n - 1)}{10 - 1} - n\right] \quad \left[\because S_n = \frac{a(r^n - 1)}{r - 1} \text{ as } r > 1\right]$

$= \frac{8}{9}\left[\frac{10\,(10^n - 1)}{9} - n\right]$

$= \frac{8}{9} \times \frac{10}{9} \times (10^n - 1) - \frac{8}{9} \times n$

$= \frac{80}{81}(10^n - 1) - \frac{8n}{9}$

129. *(a)* Given, sequences are 2, 4, 8, 16, 32 ...(i)

and $128, 32, 8, 2, \frac{1}{2}$ ...(ii)

On multiplying the corresponding terms of (i) and (ii) we obtain a new sequence 256, 128, 64, 32, 16.

Let $S = 256 + 128 + 64 + 32 + 16$

Here, $a = 256, r = \frac{1}{2}$

$\therefore$ Required sum $S = \frac{256\left[1 - \left(\frac{1}{2}\right)^5\right]}{1 - \frac{1}{2}}$

$\left[\because S_n = \frac{a(1 - r^n)}{1 - r} \text{ as } r < 1\right]$

$= 256 \times 2\left(1 - \frac{1}{2^5}\right)$

$= 512 \times \left(1 - \frac{1}{32}\right)$

$= 512\left(\frac{32 - 1}{32}\right)$

$= 16 \times 31$

$= 496$

130. *(b)* Given sequences are $a, ar, ar^2, \ldots\ldots, ar^{n-1}$ ...(i)

and $A, AR, AR^2, \ldots, AR^{n-1}$ ...(ii)

Now, on multiplying the corresponding terms of (i) and (ii), we obtain a new sequence.

$aA, ar\, AR, ar^2\, AR^2, \ldots, ar^{n-1}\, AR^{n-1}$

Clearly, common ratio $= \frac{arAR}{aA} = rR$

$\therefore$ Sequence is a G.P. .

131. *(c)* Let the G.P. be $a, ar, ar^2, ar^3, \ldots$

Given, third term = First term + 9

i.e., $T_3 = a + 9$

$\Rightarrow ar^2 = a + 9$

$\Rightarrow ar^2 - a = 9$ ...(i)

Also, second term = Fourth term + 18

i.e., $T_2 = T_4 + 18$

$\Rightarrow ar = ar^3 + 18$

$\Rightarrow ar - ar^3 = 18$ ...(ii)

Dividing Eq. (i) by Eq. (ii), we get

$\frac{ar^2 - a}{ar - ar^3} = \frac{9}{18}$

$\Rightarrow \frac{a(r^2 - 1)}{ar\,(1 - r^2)} = \frac{1}{2}$

$\Rightarrow \frac{-1}{r}\frac{(1 - r^2)}{(1 - r^2)} = \frac{1}{2}$

$\Rightarrow -\frac{1}{r} = \frac{1}{2} \Rightarrow r = -2$

Putting $r = -2$ in Eq. (ii), we get

$a(-2) - a(-2)^3 = 18$

$\Rightarrow -2a + 8a = 18$

$\Rightarrow 6a = 18$

$\Rightarrow a = 3$

G.P. is $3, 3(-2), 3(-2)^2, 3(-2)^3, \ldots\ldots$

i.e., $3, -6, 12, -24, \ldots\ldots$

132. *(d)* Let the G.P. be $A, AR, AR^2, AR^3 \ldots$

Given, first term $A = a$ ...(i)

n^{th} term, $AR^{n-1} = b$ ...(ii)

Now, P = Product of n terms

$\Rightarrow P = A \times AR^1 \times AR^2 \times AR^3 \times \ldots$ upto n terms

$P = A^{1 + 1 + 1 + \ldots \text{upto } n \text{ terms}}\, R^{1 + 2 + 3 + \ldots + (n - 1)}$

$P = A^n\, R^{\frac{n(n-1)}{2}} \quad \left[\because \text{sum of first } n \text{ natural numbers } = \frac{n(n + 1)}{2}\right]$

On squaring both sides, we get

$P^2 = A^{2n}\, R^{n(n-1)}$

$\Rightarrow P^2 = A^n\, A^n\, R^{n(n-1)} = A^n\, (AR^{n-1})^n$

$\Rightarrow P^2 = a^n b^n$ [using Eqs. (i) and (ii)]

$\Rightarrow P^2 = (ab)^n$

133. *(b)* Let the G.P. be

$\underbrace{a, ar, ar^2, ar^3, ar^4, ar^5, \ldots, ar^{n-1}}_{n \text{ terms}},$

$\underbrace{ar^n, ar^{n+1}, \ldots\ldots, ar^{2n-1}}_{n \text{ terms}}, \ldots$

Now, required ratio

$= \frac{\text{Sum of first } n \text{ terms}}{\text{Sum from } (n + 1)^{th} \text{ term to } (2n)^{th} \text{ terms}}$

$= \frac{\frac{a(r^n - 1)}{r - 1}}{\frac{ar^n\,(r^n - 1)}{r - 1}} = \frac{1}{r^n}$

134. *(b)* $\because a, b, c, d$ are in G.P.

$\therefore \quad \frac{b}{a}=\frac{c}{b}=\frac{d}{c}=r$ (say)

$\Rightarrow \quad b=ar, c=br, d=cr$

$\Rightarrow \quad b=ar, c=(ar)r, d=(br)r$

$\Rightarrow \quad b=ar, c=ar^2, d=br^2$

$\Rightarrow \quad b=ar, c=ar^2, d=(ar)r^2=ar^3$ ...(i)

Consider, $(a^2+b^2+c^2)(b^2+c^2+d^2)$

$=(a^2+a^2r^2+a^2r^4)(a^2r^2+a^2r^4+a^2r^6)$

$=a^4r^2(1+r^2+r^4)^2$

$=[a^2r(1+r^2+r^4)]^2$

$=[a^2r+a^2r^3+a^2r^5]^2$

$=[a\cdot ar+ar\cdot ar^2+ar^2\cdot ar^3]^2$

$=[ab+bc+cd]^2$ [from Eq. (i)]

135. *(c)* Let the two numbers be a and b, then 3, a, b, 81 are in G.P. .

$\because n^{th}$ term, $T_n=AR^{n-1}$

$\therefore \quad 81=3R^{4-1}$

$\Rightarrow \quad R^3=\frac{81}{3}$

$\Rightarrow \quad R^3=27 \Rightarrow R^3=3^3 \Rightarrow R=3$

(on comparing the base of power 3 on both sides)

$\Rightarrow \quad a=AR=3\times3=9,\ b=AR^2=3\times3^2=27$

136. *(a)* Let the numbers be α and β.

Given, $\frac{\alpha+\beta}{2}=A$ (arithmetic mean)

$\Rightarrow \quad \alpha+\beta=2A$ ...(i)

and $\sqrt{\alpha\beta}=G$ (geometric mean)

$\Rightarrow \quad \alpha\beta=G^2$ ...(ii)

Now, quadratic equation having roots α and β is

$x^2-(\alpha+\beta)x+\alpha\beta=0$

$\Rightarrow \quad x^2-2Ax+G^2=0$ [using Eqs. (i) and (ii)]

$\Rightarrow x=\frac{2A\pm\sqrt{4A^2-4\times1\times G^2}}{2\times1}$ $\left(\begin{array}{l}\because \text{for } ax^2+bx+c=0\\ x=\frac{-b\pm\sqrt{b^2-4ac}}{2a}\end{array}\right)$

$=\frac{2A\pm2\sqrt{A^2-G^2}}{2}$

$=A\pm\sqrt{(A+G)(A-G)}$ $[\because a^2-b^2=(a+b)(a-b)]$

137. *(c)* Let P be the principal, r be the rate of interest and t be the number of years.

A_1 (Amount in 1 yr) $=P+\frac{P\times r\times t}{100}$

$=500+\frac{500\times10}{100}=500\left(1+\frac{10}{100}\right)$

A_2 (amount in 2 yr) $=500\left(1+\frac{10}{100}\right)^2$

⋮

Clearly, the amount in different years forms a G.P.

$500(1.1), 500(1.1)^2...$

Here, $a=500(1.1)$, $r=1.1$

Now, the amount in 10th yr is given by $T_{10}=ar^9=500(1.1)^{10}$

138. *(d)* Let the roots of the quadratic equation be α and β, then (arithmetic mean) $\frac{\alpha+\beta}{2}=8$ and (geometric mean) $\sqrt{\alpha\beta}=5$

$\Rightarrow \quad \alpha+\beta=16$ and $\alpha\beta=25$

Now, if roots are α and β, then quadratic equation is

x^2- (sum of roots)x + product of roots $=0$

$\Rightarrow \quad x^2-(\alpha+\beta)+\alpha\beta=0$

$\Rightarrow \quad x^2-16x+25=0$

139. *(b)* G.P. of the form $a, ar, ar^2, ar^3, ...$ is called an infinite G.P. .

140. *(b)* Given G.P. is $1, \frac{2}{3}, \frac{4}{9}, ...$

Here, $a=1$ and $r=\frac{2}{3}$

We have, $S_n=\frac{1-\left(\frac{2}{3}\right)^n}{1-\frac{2}{3}}$

$=3\left[1-\left(\frac{2}{3}\right)^n\right]$

Let us study the behaviour of $\left(\frac{2}{3}\right)^n$ as n becomes larger and larger

n	1	5	10	20
$\left(\frac{2}{3}\right)^n$	0.6667	0.1316872428	0.01734152992	0.00030072866

We observe that as n becomes larger and larger, $\left(\frac{2}{3}\right)^n$ becomes closer and closer to zero. Mathematically, we say that as n becomes sufficiently large $\left(\frac{2}{3}\right)^n$ becomes sufficiently small. In other words as $n\to\infty$, $\left(\frac{2}{3}\right)^n\to0$.

Consequently, we find that the sum of infinitely many terms is given by $S_\infty=3$.

141. *(d)* Now, for a gemoetric progression $a, ar, ar^2, \ldots,$ if numerical value of common ratio r is less than 1, then

$$S_n = \frac{a(1-r^n)}{(1-r)} = \frac{a}{1-r} - \frac{ar^n}{1-r}$$

In this case as $n \to \infty$, $r^n \to 0$ since $|r| < 1$.

Therefore, $$S_n \to \frac{a}{1-r}$$

142. *(a)* Symbolically sum to infinity is denoted by S_∞ or S.

Thus, we have, $S = \dfrac{a}{1-r}$.

143. *(b)* I. $1 + \dfrac{1}{2} + \dfrac{1}{2^2} + \dfrac{1}{2^3} + \ldots$

$$= \frac{1}{1-\frac{1}{2}} \qquad \left(\because a = 1 \text{ and } r = \frac{1}{2}\right)$$

$$= \frac{1}{\frac{1}{2}} = 2$$

II. $1 - \dfrac{1}{2} + \dfrac{1}{2^2} - \dfrac{1}{2^3} + \ldots$

$$= 1 + \left(\frac{-1}{2}\right) + \left(\frac{-1}{2}\right)^2 + \left(\frac{-1}{2}\right)^3 + \ldots$$

$$= \frac{1}{1-\left(\frac{-1}{2}\right)} \qquad \left(\because a = 1 \text{ and } r = \frac{-1}{2}\right)$$

$$= \frac{1}{1+\frac{1}{2}} = \frac{1}{\frac{3}{2}} = \frac{2}{3}$$

144. *(a)* A. Given G.P. is $1, \dfrac{1}{3}, \dfrac{1}{9}, \ldots$

Now, $1 + \dfrac{1}{3} + \dfrac{1}{9} + \ldots$

$$= \frac{1}{1-\frac{1}{3}} \qquad \left(\because a = 1 \text{ and } r = \frac{1}{3}\right)$$

$$= \frac{1}{\frac{2}{3}} = \frac{3}{2} = 1.5$$

B. Given, G.P. is 6, 1.2, .24, ...

Now, $= 6 + 1.2 + .24 + \ldots$

$$= 6[1 + 0.2 + (0.2)^2 + \ldots]$$

$$= 6\left[\frac{1}{1-0.2}\right] \qquad (\because a = 1 \text{ and } r = 0.2)$$

$$= \frac{6}{0.8}$$

$$= \frac{60}{8}$$

$$= 7.5$$

C. Given G.P. is

$$5, \frac{20}{7}, \frac{80}{49}, \ldots$$

Now, $$5 + \frac{20}{7} + \frac{80}{49} + \ldots$$

$$= 5\left[1 + \frac{4}{7} + \frac{16}{49} + \ldots\right]$$

$$= 5\left[1 + \frac{4}{7} + \left(\frac{4}{7}\right)^2 + \ldots\right]$$

$$= 5\left[\frac{1}{1-\frac{4}{7}}\right] \qquad \left(\because a = 1 \text{ and } r = \frac{4}{7}\right)$$

$$= \frac{5 \times 7}{3} = \frac{35}{3}$$

D. Given G.P. is

$$\frac{-3}{4}, \frac{3}{16}, \frac{-3}{64}, \ldots$$

Now, $$\frac{-3}{4} + \frac{3}{16} - \frac{3}{64} + \ldots$$

$$= \frac{-3}{4}\left[1 - \frac{1}{4} + \frac{1}{16} \ldots\right]$$

$$= \frac{-3}{4}\left[1 + \left(\frac{-1}{4}\right) + \left(\frac{-1}{4}\right)^2 + \ldots\right]$$

$$= \frac{-3}{4}\left[\frac{1}{1-\left(\frac{-1}{4}\right)}\right] \qquad \left(\because a = 1 \text{ and } r = \frac{-1}{4}\right)$$

$$= \frac{-3}{4} \times \frac{4}{5} = \frac{-3}{5}$$

145. *(b)* $3^{1/2} \times 3^{1/4} \times 3^{1/8} \ldots$

$$= 3^{\frac{1}{2} + \frac{1}{4} + \frac{1}{8} + \ldots}$$

$$= 3^{\frac{1}{2}\left(1 + \frac{1}{2} + \left(\frac{1}{2}\right)^2 + \left(\frac{1}{2}\right)^3 + \ldots\right)}$$

$$= 3^{\frac{1}{2} \times \left(\frac{1}{1-\frac{1}{2}}\right)} \qquad \left(\because a = 1 \text{ and } r = \frac{1}{2}\right)$$

$$= 3^{\frac{1}{2} \times \frac{2}{1}}$$

$$= 3$$

146. *(c)* $\because \quad x = 1 + a + a^2 + a^3 + \ldots$

$\therefore \quad x = \dfrac{1}{1-a}$ (as $|a| < 1$)

$\because \quad y = 1 + b + b^2 + b^3 + \ldots$

$\therefore \quad y = \dfrac{1}{1-b}$ (as $|b| < 1$)

Now, $1 + ab + a^2b^2 + a^3b^3 + \dots$

$$= \frac{1}{1-ab} \quad \dots\text{(i) (as } |ab| = |a||b| < 1)$$

and
$$\frac{xy}{x+y-1} = \frac{\frac{1}{(1-a)(1-b)}}{\frac{1}{1-a} + \frac{1}{1-b} - 1}$$

$$= \frac{1}{(1-b)+(1-a)-(1-a)(1-b)}$$

$$= \frac{1}{1-b+1-a-(1-a-b+ab)}$$

$$= \frac{1}{1-b+1-a-1+a+b-ab}$$

$$= \frac{1}{1-ab} \quad \dots\text{(ii)}$$

[∵ RHS of Eqs. (i) and (ii) are equal]

Hence, LHS of Eqs. (i) and (ii) are also equal.

i.e., $1 + ab + a^2b^2 + \dots = \dfrac{xy}{x+y-1}$

147. *(d)* $\sum_{k=1}^{11} 2 + \sum_{k=1}^{11} 3^k = 2 \times 11 + (3^1 + 3^2 + 3^3 + \dots + 3^{11})$

$$= 22 + \frac{3(3^{11}-1)}{3-1} = 22 + \frac{3(3^{11}-1)}{2}$$

$$\left[\because S_n = \frac{a(r^n - 1)}{r-1} \text{ as } r > 1\right]$$

148. *(b)* Let the first term of G.P. be A and common ratio be R.

Given, p^{th} term $= T_p = a$

$$\Rightarrow \quad AR^{p-1} = a \quad \dots\text{(i)}$$

$$q^{th} \text{ term} = T_q = b \Rightarrow AR^{q-1} = b \quad \dots\text{(ii)}$$

$$r^{th} \text{ term} = T_r = c \Rightarrow AR^{r-1} = c \quad \dots\text{(iii)}$$

Consider, $a^{q-r}\, b^{r-p}\, c^{p-q}$

$= (AR^{p-1})^{q-r}\, (AR^{q-1})^{r-p}\, (AR^{r-1})^{p-q}$

[using Eqs. (i), (ii) and (iii)]

$= A^{q-r}\, R^{(p-1)(q-r)}\, A^{r-p}\, R^{(q-1)(r-p)}\, A^{p-q}\, R^{(r-1)(p-q)}$

$= A^{q-r+r-p+p-q}\, R^{(p-1)(q-r)+(q-1)(r-p)+(r-1)(p-q)}$

$= A^0\, R^{pq-pr-q+r+qr-pq-r+p+rp-rq-p+q}$

$= A^0\, R^0 = 1 \times 1 = 1$

149. *(c)* Given, $\dfrac{a^{n+1}+b^{n+1}}{a^n+b^n} = \sqrt{ab}$

$$\Rightarrow \quad \frac{a^{n+1}+b^{n+1}}{a^n+b^n} = \frac{a^{\frac{1}{2}} b^{\frac{1}{2}}}{1}$$

$$\Rightarrow \quad a^{n+1}+b^{n+1} = (a^n+b^n)(a^{\frac{1}{2}} b^{\frac{1}{2}})$$

$$\Rightarrow \quad a^{n+1}+b^{n+1} = a^{n+\frac{1}{2}} b^{\frac{1}{2}} + a^{\frac{1}{2}} b^{n+\frac{1}{2}}$$

$$\Rightarrow \quad a^{n+1}+b^{n+1} - a^{n+\frac{1}{2}} b^{\frac{1}{2}} - a^{\frac{1}{2}} b^{n+\frac{1}{2}} = 0$$

$$\Rightarrow \quad (a^{n+1} - a^{n+\frac{1}{2}} b^{\frac{1}{2}}) + (b^{n+1} - a^{\frac{1}{2}} b^{n+\frac{1}{2}}) = 0$$

$$\Rightarrow \quad a^{n+\frac{1}{2}}[a^{\frac{1}{2}} - b^{\frac{1}{2}}] - b^{n+\frac{1}{2}}[a^{\frac{1}{2}} - b^{\frac{1}{2}}] = 0$$

$$\Rightarrow \quad (a^{n+\frac{1}{2}} - b^{n+\frac{1}{2}})(a^{\frac{1}{2}} - b^{\frac{1}{2}}) = 0$$

$$\Rightarrow \quad a^{n+\frac{1}{2}} - b^{n+\frac{1}{2}} = 0 \quad (\text{as } a^{\frac{1}{2}} - b^{\frac{1}{2}} \neq 0)$$

$$\Rightarrow \quad a^{n+\frac{1}{2}} = b^{n+\frac{1}{2}}$$

$$\Rightarrow \quad \left(\frac{a}{b}\right)^{n+\frac{1}{2}} = 1$$

$$\Rightarrow \quad \left(\frac{a}{b}\right)^{n+\frac{1}{2}} = \left(\frac{a}{b}\right)^0$$

On comparing the power of base (a/b) on both sides, we get

$$\Rightarrow \quad n + \frac{1}{2} = 0$$

$$\Rightarrow \quad n = -\frac{1}{2}$$

150. *(b)* Here we will use componendo and dividendo.

i.e., if $\dfrac{a}{b} = \dfrac{c}{d} \Rightarrow \dfrac{a+b}{a-b} = \dfrac{c+d}{c-d}$

Let the numbers be a and b.

Given, $a + b = 6\sqrt{ab}$

$$\Rightarrow \quad \frac{a+b}{2\sqrt{ab}} = \frac{3}{1}$$

Now, applying componendo and dividendo, we get

$$\Rightarrow \quad \frac{a+b+2\sqrt{ab}}{a+b-2\sqrt{ab}} = \frac{3+1}{3-1}$$

$$\Rightarrow \quad \frac{(\sqrt{a})^2 + (\sqrt{b})^2 + 2\sqrt{ab}}{(\sqrt{a})^2 + (\sqrt{b})^2 - 2\sqrt{ab}} = \frac{4}{2}$$

$$\Rightarrow \quad \frac{(\sqrt{a}+\sqrt{b})^2}{(\sqrt{a}-\sqrt{b})^2} = \frac{2}{1}$$

$$\Rightarrow \quad \frac{\sqrt{a}+\sqrt{b}}{\sqrt{a}-\sqrt{b}} = \frac{\sqrt{2}}{1}$$

Again, applying componendo and dividendo, we get

$$\Rightarrow \quad \frac{\sqrt{a}+\sqrt{b}+\sqrt{a}-\sqrt{b}}{\sqrt{a}+\sqrt{b}-(\sqrt{a}-\sqrt{b})} = \frac{\sqrt{2}+1}{\sqrt{2}-1}$$

$$\Rightarrow \quad \frac{2\sqrt{a}}{2\sqrt{b}} = \frac{\sqrt{2}+1}{\sqrt{2}-1}$$

$$\Rightarrow \quad \frac{\sqrt{a}}{\sqrt{b}} = \frac{\sqrt{2}+1}{\sqrt{2}-1}$$

Now, squaring both the sides, we get

$\Rightarrow \quad \frac{a}{b} = \frac{(\sqrt{2}+1)^2}{(\sqrt{2}-1)^2}$

$\Rightarrow \quad \frac{a}{b} = \frac{2+1+2\sqrt{2}}{2+1-2\sqrt{2}} \quad \left[\because (a+b)^2 = a^2+b^2+2ab \text{ and } (a-b)^2 = a^2+b^2-2ab\right]$

$\Rightarrow \quad \frac{a}{b} = \frac{3+2\sqrt{2}}{3-2\sqrt{2}}$

$\Rightarrow \quad a:b = (3+2\sqrt{2}):(3-2\sqrt{2})$

151. *(a)* Number of bacteria in the culture, form a G.P. whose first term $a = 30$ and common ratio $r = 2$.

Bacteria present after 2nd hr $= ar^2 = 30 \times (2)^2 = 120$

$(\because T_n = ar^{n-1})$

Bacteria present after 4th hr $= ar^4 = 30 \times (2)^4$

$= 30 \times 16 = 480$

Bacteria present after t^{th} hr $= ar^n = 30 \times 2^n$

$= (30) \cdot (2^n)$

152. *(c)* From given functional relation, first we will find the values of $f(2)$, $f(3)$, $f(4)$ ...... and so on, then we put these values in the summation.

Given, $\quad f(x+y) = f(x) f(y) \quad$ ...(i)

Putting $\quad x = y = 1$ in Eq. (i), we get

$f(1+1) = f(1) f(1)$

$\Rightarrow \quad f(2) = 3 \times 3 \quad [\because f(1) = 3]$

$\Rightarrow \quad f(2) = 9 = 3^2$

Putting $x = 2$, $y = 1$ in Eq. (i), we get

$f(2-1) = f(2) f(1)$

$\Rightarrow \quad f(3) = 9 \times 3 \quad [\because f(2) = 9, f(1) = 3]$

$\Rightarrow \quad f(3) = 27 = 3^3$

Putting $x = 3$, $y = 1$ in Eq. (i), we get

$f(3+1) = f(3) f(1)$

$\Rightarrow \quad f(4) = 27 \times 3 \quad [\because f(3) = 27]$

$\Rightarrow \quad f(4) = 81 = 3^4$

and so on continuing the process, we obtain $f(k) = 3^k$.

Now, $\quad \sum_{x=1}^{n} f(x) = 120$

$\Rightarrow \quad f(1) + f(2) + f(3) + \ldots + f(n) = 120$

$\Rightarrow \quad 3 + 9 + 27 + \ldots$ to n terms $= 120$

Here, $\quad a = 3, r = \frac{9}{3} = 3$

$\because \quad S_n = \frac{a(r^n - 1)}{r-1} = \frac{3(3^n - 1)}{3-1} = 120 \quad (\because r > 1)$

$\Rightarrow \quad 3(3^n - 1) = 120 \times 2 \Rightarrow 3(3^n - 1) = 240$

$\Rightarrow \quad 3^n - 1 = \frac{240}{3} \Rightarrow 3^n - 1 = 80$

$\Rightarrow \quad 3^n = 80 + 1$

$\Rightarrow \quad 3^n = 81$

$\Rightarrow \quad 3^n = 3^4$

On comparing the power of 3 on both sides, we get

$n = 4$

153. *(d)* Let the G.P. of n terms be $a, ar, ar^2, ar^3, \ldots\ldots, ar^{n-1}$

Given, $a = 5, r = 2$ and $S_n = 315$

Therefore, $\quad 315 = \frac{5(2^n - 1)}{2-1}$

$\Rightarrow \quad \frac{315}{5} = 2^n - 1 \quad \left[\because S_n = \frac{a(r^n - 1)}{r-1} \text{ as } r > 1\right]$

$\Rightarrow \quad 2^n - 1 = 63$

$\Rightarrow \quad 2^n = 63 + 1$

$\Rightarrow \quad 2^n = 64 = 2^6$

On comparing the power of 2 on both sides, we get

$n = 6$

Again, for last term we use

$T_n = ar^{n-1} = 5 \times (2)^{6-1} = 5 \times 2^5 = 5 \times 32 = 160 \quad (\because n = 6)$

154. *(d)* Let the G.P. be $a, ar, ar^2, ar^3, \ldots\ldots\ldots$

Given, $\quad a = 1$ and $T_3 + T_5 = 90$

$\therefore \quad ar^2 + ar^4 = 90 \quad (\because T_n = ar^{n-1})$

$\Rightarrow \quad r^2 + r^4 = 90 \quad$ (as $a = 1$)

$\Rightarrow \quad r^4 + r^2 - 90 = 0$

Now, on factorizing it by splitting the middle term, we get

$r^4 + 10r^2 - 9r^2 - 90 = 0$

$\Rightarrow \quad r^2(r^2 + 10) - 9(r^2 + 10) = 0$

$\Rightarrow \quad (r^2 + 10)(r^2 - 9) = 0$

$\because \quad r^2 + 10 \neq 0 \Rightarrow r^2 - 9 = 0$

$\Rightarrow \quad r^2 = 9 \Rightarrow r = \pm 3$

155. *(a)* Let three numbers in G.P. be a, ar, ar^2.

Given, $\quad a + ar + ar^2 = 56 \quad$ ...(i)

Also, $a - 1, ar - 7, ar^2 - 21$ are in A.P.

$\Rightarrow \quad 2(ar - 7) = (a - 1) + (ar^2 - 21)$

$(\because$ if a, b, c are in A.P., then $2b = a + c)$

$\Rightarrow \quad 2ar - 14 = a + ar^2 - 22$

$\Rightarrow \quad a + ar^2 - 2ar = -14 + 22$

$\Rightarrow \quad a + ar^2 - 2ar = 8 \quad$ ...(ii)

Dividing Eq. (i) by Eq. (ii), we get

$\frac{a + ar + ar^2}{a + ar^2 - 2ar} = \frac{56}{8} \Rightarrow \frac{1 + r + r^2}{1 + r^2 - 2r} = \frac{7}{1}$

$\Rightarrow \quad 1 + r + r^2 = 7 + 7r^2 - 14r$

$\Rightarrow \quad 6r^2 - 15r + 6 = 0$

On dividing by 3, we get

$\Rightarrow \quad 2r^2 - 5r + 2 = 0$

Now, on factorizing it by splitting the middle term, we get

$$2r^2 - (4+1)r + 2 = 0$$
$$\Rightarrow \quad 2r^2 - 4r - r + 2 = 0$$
$$\Rightarrow \quad 2r(r-2) - (r-2) = 0$$
$$\Rightarrow \quad (r-2)(2r-1) = 0$$
$$\Rightarrow \quad r = 2, \frac{1}{2}$$

If $r = 2$, then from Eq. (i), we get

$$a + 2a + 4a = 56$$
$$\Rightarrow \quad 7a = 56$$
$$\therefore \quad a = 8$$

Then, numbers are

$$a = 8$$
$$\Rightarrow \quad ar = 8 \times 2 = 16$$
$$\Rightarrow \quad ar^2 = 8 \times 4 = 32$$

If $r = \frac{1}{2}$, then from Eq. (i), we get

$$\frac{a}{1} + \frac{a}{2} + \frac{a}{4} = 56$$
$$\Rightarrow \quad \frac{4a + 2a + a}{4} = 56$$
$$\Rightarrow \quad \frac{7a}{4} = 56$$
$$\therefore \quad a = 32$$

Then, numbers are

$$a = 32$$
$$\Rightarrow \quad ar = 32 \times \frac{1}{2} = 16$$
$$\Rightarrow \quad ar^2 = 32 \times \frac{1}{4} = 8$$

$\Rightarrow$ 32, 16, 8

Hence, required numbers are 8, 16, 32 or 32, 16, 8.

156. *(c)* Let the G.P. be $a, ar, ar^2, ar^3, ar^4, \ldots, ar^{2n-2}, ar^{2n-1}$

where $a, ar^2, ar^4, ar^6, \ldots$ occupy odd places and $ar, ar^3, ar^5, ar^7, \ldots$ occupy even places.

Given, sum of all terms $= 5 \times$ Sum of terms occupying odd places

i.e., $\quad a + ar + ar^2 + \ldots + ar^{2n-1}$

$$= 5 \times (a + ar^2 + ar^4 + \ldots + ar^{2n-2})$$

$$\Rightarrow \quad \frac{a(r^{2n}-1)}{r-1} = \frac{5a[(r^2)^n - 1]}{r^2 - 1} \quad \left[\because S_n = \frac{a(r^n - 1)}{r-1}\right]$$

$$\Rightarrow \quad \frac{r^{2n}-1}{r-1} = \frac{5(r^{2n}-1)}{(r-1)(r+1)}$$

$$\Rightarrow \quad 1 = \frac{5}{r+1} \Rightarrow r + 1 = 5 \Rightarrow r = 4$$

Note *Students should remember that, if a G.P. has 2n terms then there are n even and n odd terms.*

157. *(a)* Here, we will use componendo and dividendo.

i.e., if $\quad \frac{a}{b} = \frac{c}{d} \Rightarrow \frac{a+b}{a-b} = \frac{c+d}{c-d}$

Given, $\quad \frac{a+bx}{a-bx} = \frac{b+cx}{b-cx} = \frac{c+dx}{c-dx}$

Applying componendo and dividendo, we get

$$\frac{a+bx+a-bx}{a+bx-(a-bx)} = \frac{b+cx+b-cx}{b+cx-(b-cx)} = \frac{c+dx+c-dx}{c+dx-(c-dx)}$$

$$\Rightarrow \quad \frac{2a}{2bx} = \frac{2b}{2cx} = \frac{2c}{2dx}$$

$$\Rightarrow \quad \frac{a}{bx} = \frac{b}{cx} = \frac{c}{dx}$$

Multiplying each term by x, we get

$$\frac{a}{b} = \frac{b}{c} = \frac{c}{d} \text{ or } \frac{b}{a} = \frac{c}{b} = \frac{d}{c}$$

Therefore, a, b, c, d are in G.P.

158. *(d)* Let the n terms of G.P. be $a, ar, ar^2, ar^3 \ldots, ar^{n-1}$.

Given, $\quad S =$ Sum of n terms

$$= a + ar + ar^2 + ar^3 + \ldots + ar^{n-1}$$

$$= \frac{a(r^n - 1)}{r-1} \quad \ldots \text{(i)}$$

$R =$ Sum of the reciprocals of n terms

$$= \frac{1}{a} + \frac{1}{ar} + \frac{1}{ar^2} + \ldots + \frac{1}{ar^{n-1}}$$

$$= \frac{\frac{1}{a}\left[1 - \left(\frac{1}{r}\right)^n\right]}{1 - \frac{1}{r}} = \frac{1}{a}\left[\frac{1}{1} - \frac{1}{r^n}\right] \times \frac{1}{\frac{r-1}{r}}$$

$$= \frac{1}{a}\left[\frac{r^n - 1}{r^n}\right] \times \frac{r}{r-1}$$

$$\Rightarrow \quad R = \frac{(r^n - 1)r}{ar^n(r-1)} \quad \ldots \text{(ii)}$$

and $P =$ Product of n terms

$$= a \times ar \times ar^2 \times ar^3 \times \ldots \times ar^{n-1}$$

$$= a^{1+1+1+\ldots \text{upto } n \text{ terms}} r^{1+2+3+\ldots+(n-1) \text{terms}}$$

$$= a^n r^{\frac{n(n-1)}{2}} \quad \left[\because \Sigma n = \frac{n(n+1)}{2}\right]$$

$$\Rightarrow \quad P^2 = a^{2n} r^{n(n-1)} \quad \ldots \text{(iii)}$$

Now, consider $P^2 R^n = a^{2n} r^{n(n-1)} \left[\frac{r(r^n - 1)}{ar^n(r-1)}\right]^n$

[using Eqs. (ii) and (iii)]

$$= a^{2n} r^{n(n-1)} \frac{r^n (r^n - 1)^n}{a^n (r^n)^n (r-1)^n}$$

$$= \frac{a^n r^{n^2} (r^n - 1)^n}{r^{n^2} (r - 1)^n}$$

$$= \frac{a^n (r^n - 1)^n}{(r - 1)^n}$$

$$= \left[\frac{a(r^n - 1)}{r - 1}\right]^n = S^n \quad \text{[using Eq. (i)]}$$

159. *(c)* $\because a, b, c, d$ are in G.P.

$\Rightarrow b = ar, c = ar^2, d = ar^3$ ...(i)

Let $p = a^n + b^n$, $q = b^n + c^n$

and $r = c^n + d^n$

Now, consider $pr = (a^n + b^n)(c^n + d^n)$

$= [a^n + a^n r^n] \times [a^n r^{2n} + a^n r^{3n}]$ [from Eq. (i)]

$= a^n (1 + r^n) a^n r^{2n} (1 + r^n)$

$= a^{2n} r^{2n} (1 + r^n)^2 = [a^n r^n (1 + r^n)]^2 = [a^n r^n + a^n r^{2n}]^2$

$= [(ar)^n + (ar^2)^n]^2 = [b^n + c^n]^2 = q^2$ [from Eq. (i)]

Thus, p, q, r are in G.P. .

160. *(a)* If a quadratic equation, $ax^2 + bx + c = 0$ has roots α and β, then

Sum of roots $= \alpha + \beta = -\frac{b}{a}$ and

Product of roots $= \alpha\beta = \frac{c}{a}$

Given, a and b are roots of $x^2 - 3x + p = 0$

$\therefore$ Sum of roots, $a + b = -\frac{(-3)}{1} = 3$

$\Rightarrow a + b = 3$ ...(i)

and product of roots, $ab = p$ ...(ii)

Also, given that c and d are roots of $x^2 - 12x + q = 0$

$\Rightarrow$ Sum of roots, $c + d = -\frac{(-12)}{1} = 12$

$\Rightarrow c + d = 12$ ...(iii)

and product of roots, $cd = q$ ...(iv)

Also, it is given that a, b, c, d are in G.P.

$\Rightarrow b = ar, c = ar^2, d = ar^3$

Putting these values in Eqs. (i) and (iii), then dividing Eq. (i) by Eq. (iii), we get

$$\frac{a + ar}{ar^2 + ar^3} = \frac{3}{12} \Rightarrow \frac{a(1 + r)}{ar^2(1 + r)} = \frac{1}{4}$$

$$\Rightarrow \frac{1}{r^2} = \frac{1}{4}$$

$$\Rightarrow r^2 = 4 \Rightarrow r = \pm 2$$

Now for $r = 2$, from Eq. (i), we have

$$a + ar = 3 \Rightarrow a + 2a = 3$$

$$\Rightarrow 3a = 3 \Rightarrow a = 1$$

So, the G.P. is

$a = 1$, $b = ar = 1 \times 2 = 2, c = ar^2 = 1 \times 2^2 = 4,$

$d = ar^3 = 1 \times 2^3 = 8$

From Eq. (ii), $p = ab = 1 \times 2 = 2$

From Eq. (iv), $q = cd = 4 \times 8 = 32$

$$\therefore \frac{q + p}{q - p} = \frac{32 + 2}{32 - 2} = \frac{34}{30} = \frac{17}{15}$$

Hence, $(q + p) : (q - p) = 17 : 15$

and for $r = -2$, from Eq. (i), we have $a + ar = 3$

$\Rightarrow a - 2a = 3$

$\Rightarrow a = -3$

So, the G.P. is $a = -3$,

$b = ar = (-3)(-2) = 6$

$c = ar^2 = (-3)(-2)^2 = -12$

$d = ar^3 = (-3)(-2)^3 = 24$

From Eq. (ii), $p = ab = (-3) \cdot (6) = -18$

From Eq. (iv), $q = cd = (-12)(24) = -288$

$$\therefore \frac{q + p}{q - p} = \frac{-288 - 18}{-288 + 18} = \frac{-306}{-270} = \frac{17}{15}$$

Hence, $(q + p) : (q - p) = 17 : 15$

161. *(a)* Let the A.M. of the numbers a and b be A and G.M. of a and b be G, then $A = \frac{a + b}{2}$ and $G = \sqrt{ab}$

Given, $A : G = m : n$

i.e., $\frac{A}{G} = \frac{m}{n} \Rightarrow \frac{a + b}{2\sqrt{ab}} = \frac{m}{n}$

Applying componendo and dividendo rule, we get

$$\frac{a + b + 2\sqrt{ab}}{a + b - 2\sqrt{ab}} = \frac{m + n}{m - n}$$

$$\Rightarrow \frac{(\sqrt{a})^2 + (\sqrt{b})^2 + 2\sqrt{ab}}{(\sqrt{a})^2 + (\sqrt{b})^2 - 2\sqrt{ab}} = \frac{m + n}{m - n}$$

$$\Rightarrow \frac{(\sqrt{a} + \sqrt{b})^2}{(\sqrt{a} - \sqrt{b})^2} = \frac{m + n}{m - n}$$

$$\Rightarrow \frac{\sqrt{a} + \sqrt{b}}{\sqrt{a} - \sqrt{b}} = \frac{\sqrt{m + n}}{\sqrt{m - n}}$$

Again, applying componendo and dividendo rule, we get

$$\frac{(\sqrt{a} + \sqrt{b}) + (\sqrt{a} - \sqrt{b})}{(\sqrt{a} + \sqrt{b}) - (\sqrt{a} - \sqrt{b})} = \frac{\sqrt{m + n} + \sqrt{m - n}}{\sqrt{m + n} - \sqrt{m - n}}$$

$$\Rightarrow \frac{2\sqrt{a}}{2\sqrt{b}} = \frac{\sqrt{m + n} + \sqrt{m - n}}{\sqrt{m + n} - \sqrt{m - n}}$$

$$\Rightarrow \frac{\sqrt{a}}{\sqrt{b}} = \frac{\sqrt{m + n} + \sqrt{m - n}}{\sqrt{m + n} - \sqrt{m - n}}$$

Now, squaring on both the sides, we get

$$\frac{a}{b} = \frac{(\sqrt{m + n} + \sqrt{m - n})^2}{(\sqrt{m + n} - \sqrt{m - n})^2}$$

$$\Rightarrow \frac{a}{b}=\frac{m+n+m-n+2\sqrt{m+n}\sqrt{m-n}}{m+n+m-n-2\sqrt{m+n}\sqrt{m-n}}$$

$$\left[\begin{array}{l}\because (a+b)^2=a^2+b^2+2ab\\ (a-b)^2=a^2+b^2-2ab\end{array}\right]$$

$$\Rightarrow \frac{a}{b}=\frac{2m+2\sqrt{m^2-n^2}}{2m-2\sqrt{m^2-n^2}} \quad [\because (a+b)(a-b)=a^2-b^2]$$

$$\Rightarrow \frac{a}{b}=\frac{m+\sqrt{m^2-n^2}}{m-\sqrt{m^2-n^2}}$$

$$\Rightarrow a:b=(m+\sqrt{m^2-n^2}):(m-\sqrt{m^2-n^2})$$

162. *(b)* Given a, b, c are in A.P.

$$\Rightarrow \quad 2b=a+c \quad \ldots(i)$$

Again, b, c, d are in G.P.

$$\Rightarrow \quad c^2=bd \quad \ldots(ii)$$

Similarly, $\frac{1}{c}, \frac{1}{d}, \frac{1}{e}$ are in A.P.

$$\Rightarrow \quad \frac{2}{d}=\frac{1}{c}+\frac{1}{e} \Rightarrow \frac{2}{d}=\frac{e+c}{ce}$$

$$\Rightarrow \quad d=\frac{2ce}{c+e} \quad \ldots(iii)$$

Putting the values of b and d from Eqs. (i) and (iii) in Eq. (ii), we get

$$c^2=\left(\frac{a+c}{2}\right)\times\left(\frac{2ce}{c+e}\right) \Rightarrow c^2=\frac{a+c}{2}\times\frac{2ce}{c+e}$$

$$\Rightarrow \quad c^2(c+e)=(a+c)ce$$

$$\Rightarrow \quad c(c+e)=(a+c)e$$

$$\Rightarrow \quad c^2+ce=ae+ce \Rightarrow c^2=ae$$

Therefore, a, c, e are in G.P. .

163. *(d)* I. Let $S=5+55+555+\ldots$ upto n terms

$$=5(1+11+111+\ldots \text{ upto } n \text{ terms})$$

$$=\frac{5}{9}(9+99+999+\ldots \text{upto } n \text{ terms})$$

$$=\frac{5}{9}[(10-1)+(100-1)+(1000-1)+\ldots \text{upto } n \text{ terms}]$$

$$=\frac{5}{9}[(10+100+1000+\ldots \text{upto } n \text{ terms}$$

$$-(1+1+1+\ldots \text{ upto } n \text{ terms})]$$

$$=\frac{5}{9}\left[\frac{10(10^n-1)}{10-1}-n\right] \quad \left[\begin{array}{l}\because \text{sum of GP } =\frac{a(r^n-1)}{r-1}, r>1\\ \text{and } \Sigma 1=n\end{array}\right]$$

$$=\frac{5}{9}\left[\frac{10(10^n-1)}{9}-n\right]$$

II. Let $S=0.6+0.66+0.666+\ldots$ upto n terms

$$S=6(0.1+0.11+0.111+\ldots \text{upto } n \text{ terms})$$

$$=\frac{6}{9}(0.9+0.99+0.999+\ldots \text{upto } n \text{ terms})$$

$$=\frac{2}{3}\left[\frac{9}{10}+\frac{99}{100}+\frac{999}{1000}+\ldots \text{upto } n \text{ terms}\right]$$

$$=\frac{2}{3}\left[\left(1-\frac{1}{10}\right)+\left(1-\frac{1}{100}\right)+\left(1-\frac{1}{1000}\right)+\ldots \text{ upto } n \text{ terms}\right]$$

$$=\frac{2}{3}\left[(1+1+1+\ldots \text{upto } n \text{ terms})-\left(\frac{1}{10}+\frac{1}{100}+\frac{1}{1000}+\ldots \text{ upto } n \text{ terms}\right)\right]$$

$$=\frac{2}{3}\left[n-\frac{\frac{1}{10}\left\{1-\left(\frac{1}{10}\right)^n\right\}}{1-\frac{1}{10}}\right]$$

$$\left[\because \text{sum of G.P.}=\frac{a(1-r^n)}{1-r}, r<1\right]$$

$$=\frac{2}{3}\left[n-\frac{\frac{1}{10}\left\{1-\left(\frac{1}{10}\right)^n\right\}}{\frac{9}{10}}\right]=\frac{2}{3}\left[n-\frac{1}{9}\left\{1-\left(\frac{1}{10}\right)^n\right\}\right]$$

164. *(b)* Present cost of machine = ₹ 15625

Now, after 1 yr cost of machine $=15625-\frac{20}{100}\times 15625$

$$=15625-3125$$
$$= ₹\,12500$$

After 2 yr cost of machine $=12500-\frac{20}{100}\times 12500$

$$=12500-2500= ₹\,10000$$

and so on.

Clearly, the cost of machine form a G.P. 15625, 12500, 10000, ... with first term $a=15625$, common ratio $(r)=\frac{4}{5}$

Now, the estimated value at the end of 5 yr is given by

$$T_6=ar^5=15625\left(\frac{4}{5}\right)^5$$

$$=15625\times\frac{4^5}{5^5}=5^6\times\frac{2^{10}}{5^5}$$

$$=5\times 1024= ₹\,5120$$

Note *In depreciation, the cost value decreases every year.*

165. *(c)* Firstly, one person post four letters to four of his friends. Then, those four persons post 4-4 letters *i.e.,* a total $4\times4=16$ letters. Similarly, in next step $4\times4\times4=64$ letters are posted.

Clearly, the number of letters posted form a G.P. 4, 16, 64, 256, ...

where, $\quad a=4$ and $r=4$

Now, total number of letters till 8th set is given by

$\Rightarrow \quad S_8 = \frac{4(4^8 - 1)}{4 - 1}$

$\Rightarrow \quad S_8 = \frac{4}{3}(4^8 - 1) \quad \left[\because S_n = \frac{a(r^n - 1)}{(r - 1)} \text{ as } r > 1\right]$

Cost for one letter = ₹ 0.50

Hence, total cost $= \frac{4}{3}(4^8 - 1) \times 0.50$

$= \frac{4}{3} \times (65536 - 1) \times 0.50$

$= \frac{4}{3} \times 65535 \times 0.50$

= ₹ 43690

166. *(b)* Let r be the common ratio of given G.P. .

Then, $\frac{b}{a} = \frac{c}{b} = \frac{d}{c} = r$

$\Rightarrow \quad b = ar, c = br, d = cr$

$\Rightarrow \quad b = ar, c = ar^2, d = ar^3$

Now, $a^2 - b^2 = a^2 - a^2r^2$

$= a^2(1 - r^2)$

$\Rightarrow \quad b^2 - c^2 = a^2r^2 - a^2r^4$

$= a^2r^2(1 - r^2)$

$\Rightarrow \quad c^2 - d^2 = a^2r^4 - a^2r^6$

$= a^2r^4(1 - r^2)$

Therefore, $\frac{b^2 - c^2}{a^2 - b^2} = \frac{c^2 - d^2}{b^2 - c^2} = r^2$

Hence, $a^2 - b^2, b^2 - c^2, c^2 - d^2$ are in G.P. .

167. *(a)* Given a, b, c, d are in G.P..

$\because$ AM > GM for the first three terms.

Then, $\frac{a + c}{2} > b \quad (\because \sqrt{ac} = b)$

$\Rightarrow \quad a + c > 2b \quad \ldots$(i)

Similarly, for the last three terms

$\Rightarrow \quad \frac{b + d}{2} > c \quad (\because \sqrt{bd} = c)$

$\Rightarrow \quad b + d > 2c \quad \ldots$(ii)

On adding Eqs. (i) and (ii), we get

$(a + c) + (b + d) > 2b + 2c$

$\Rightarrow \quad a + d > b + c$

168. *(c)* We have a, b, c as three consecutive terms of an A.P..

Then, $b - a = c - b = d$ (say)

$c - a = 2d$

$\Rightarrow \quad a - b = -d$

Now consider, $x^{b-c} y^{c-a} z^{a-b}$

$= x^{-d} y^{2d} z^{-d}$

$= x^{-d} (\sqrt{xz})^{2d} z^{-d} \quad \left[\text{since } y = (\sqrt{xz}) \text{ as } x, y, z \text{ are in G.P.}\right]$

$= x^{-d} \cdot x^d \cdot z^d \cdot z^{-d}$

$= x^{d-d} \cdot z^{d-d}$

$= x^0 \cdot z^0 = 1$

169. *(c)* Given that, $f(x + y) = f(x). f(y)$ and $f(1) = 2$

Therefore, $f(2) = f(1 + 1) = f(1) \cdot f(1) = 2^2$

$f(3) = f(1 + 2) = f(1) \cdot f(2) = 2^3$

$f(4) = f(1 + 3) = f(1) \cdot f(3) = 2^4$

and, so on.

Continuing the process, we obtain

$f(k) = 2^k$ and $f(a) = 2^a$

Now consider, $\sum_{k=1}^{n} f(a + k) = \sum_{k=1}^{n} f(a) \cdot (k)$

$= f(a) \sum_{k=1}^{n} f(k)$

$= 2^a(2^1 + 2^2 + 2^3 + \cdots + 2^n)$

$= 2^a \left\{\frac{2 \cdot (2^n - 1)}{2 - 1}\right\} = 2^{a+1}(2^n - 1) \quad \ldots$(i)

But, we are given $\sum_{k=1}^{n} f(a + k) = 16(2^n - 1)$

$2^{a+1}(2^n - 1) = 16(2^n - 1)$

$\Rightarrow \quad 2^{a+1} = 2^4 \Rightarrow a + 1 = 4$

$\therefore \quad a = 3$

170. *(b)* Since, A.M. $\geq$ G.M., therefore we have

$\frac{x + y}{2} \geq \sqrt{xy}, \frac{y + z}{2} \geq \sqrt{yz}$ and $\frac{z + x}{2} \geq \sqrt{zx}$

Multiplying the three inequalities, we get

$\frac{x + y}{2} \cdot \frac{y + z}{2} \cdot \frac{y + z}{2} \geq \sqrt{(xy)(yz)(zx)}$

$\Rightarrow \quad (x + y)(y + z)(z + x) \geq 8xyz$

171. *(d)* Since, $t_n = t_{n+1} + t_{n+2}$

$\Rightarrow \quad ar^{n-1} = ar^n + ar^{n+1}$

$\Rightarrow \quad 1 = r + r^2$

$\Rightarrow \quad r = \frac{-1 \pm \sqrt{5}}{2}$

Since, $r > 0$

Therefore, $r = 2\frac{\sqrt{5} - 1}{4} = 2 \sin 18°$

172. *(b)* $\because 1 + 2 + 3 + \ldots + n$ is a finite arithmetic series.

$\therefore \quad S_n = \frac{n}{2}[2 \times 1 + (n - 1) \times 1]$

($\because$ first term $= a = 1$ and common difference $= d = 1$)

$= \frac{n}{2}[2 + n - 1] = \frac{n(n + 1)}{2}$

173. *(d)* Here, $S_n = 1^2 + 2^2 + 3^2 + \ldots + n^2$

Consider the identity, $K^3 - (K-1)^3 = 3K^2 - 3K + 1$

Putting $K = 1, 2, \cdots, n$ successively, we obtain

$$1^3 - 0^3 = 3(1)^2 - 3(1) + 1$$
$$2^3 - 1^3 = 3(2)^2 - 3(2) + 1$$
$$3^3 - 2^3 = 3(3)^2 - 3(3) + 1$$

....

....

....

....

$$n^3 - (n-1)^3 = 3(n)^2 - 3(n) + 1$$

Adding columnwise, we get

$$n^3 - 0^3 = 3(1^2 + 2^2 + 3^2 + 4^2 + \ldots + n^2) - 3(1 + 2 + \ldots + n) + n$$

$$n^3 = 3\sum_{K=1}^{n} K^2 - 3\sum_{K=1}^{n} K + n$$

Now, $$S_n = \sum_{K=1}^{n} K^2 = \frac{1}{3}\left[n^3 + 3\sum_{K=1}^{n} K - n\right]$$

$$= \frac{1}{3}\left[n^3 + \frac{3n(n+1)}{2} - n\right]$$

$$\left[\because \sum_{K=1}^{n} K = 1 + 2 + 3 + \ldots + n = \frac{n(n+1)}{2}\right]$$

$$= \frac{1}{6}(2n^3 + 3n^2 + 3n - 2n)$$

$$= \frac{1}{6}(2n^3 + 3n^2 + n) = \frac{n(n+1)(2n+1)}{6}$$

174. *(b)* Here, $S_n = 1^3 + 2^3 + 3^3 + \cdots + n^3$

$\because \quad (k+1)^4 - k^4 = 4k^3 + 6k^2 + 4k + 1$

Putting $k = 1, 2, 3, \ldots, n$, we get

$$2^4 - 1^4 = 4(1)^3 + 6(1)^2 + 4(1) + 1$$
$$3^4 - 2^4 = 4(2)^3 + 6(2)^2 + 4(2) + 1$$
$$4^4 - 3^4 = 4(3)^3 + 6(3)^2 + 4(3) + 1$$

....

....

....

....

$$(n-1)^4 - (n-2)^4 = 4(n-2)^3 + 6(n-2)^2 + 4(n-2) + 1$$
$$n^4 - (n-1)^4 = 4(n-1)^3 + 6(n-1)^2 + 4(n-1) + 1$$
$$(n+1)^4 - n^4 = 4n^3 + 6n^2 + 4n + 1$$

Adding columnwise, we get

$$(n+1)^4 - 1^4 = 4(1^3 + 2^3 + \ldots + n^3) + 6(1^2 + 2^2 + 3^3 + \ldots + n^2) + 4(1 + 2 + 3 + \ldots + n) + n$$

$$= 4\sum_{k=1}^{n} k^3 + 6\sum_{k=1}^{n} k^2 + 4\sum_{k=1}^{n} k + n$$

$$\because \sum_{k=1}^{n} k = \frac{n(n+1)}{2} \text{ and } \sum_{k=1}^{n} k^2 = \frac{n(n+1)(2n+1)}{6}$$

$$\therefore 4\sum_{k=1}^{n} k^3 = n^4 + 4n^3 + 6n^2 + 4n - 6\left(\frac{n(n+1)(2n+1)}{6}\right) - 4\left(\frac{n(n+1)}{2}\right) - n$$

$$\Rightarrow \quad 4S_n = n^4 + 4n^3 + 6n^2 + 4n - n(2n^2 + 3n + 1) - 2n(n+1) - n$$

$$= n^4 + 2n^3 + n^2$$
$$= n^2(n^2 + 2n + 1)$$
$$= n^2(n+1)^2$$

Hence, $$S_n = \frac{n^2(n+1)^2}{4}$$

or $$S_n = \frac{[n(n+1)]^2}{4}$$

175. *(d)* Let us write $S_n = 5 + 11 + 19 + 29 + \ldots + a_{n-1} + a_n$

or $S_n = 5 + 11 + 19 + \cdots + a_{n-2} + a_{n-1} + a_n$

On subtracting, we get

$$0 = 5 + [6 + 8 + 10 + 12 + \ldots (n-1) \text{ terms}] - a_n$$

or $$a_n = 5 + \frac{(n-1)[12 + (n-2) \times 2]}{2}$$

$$= 5 + (n-1)(n+4)$$
$$= n^2 + 3n + 1$$

Hence, $$S_n = \sum_{k=1}^{n} a_k = \sum_{k=1}^{n} (k^2 + 3k + 1)$$

$$= \sum_{k=1}^{n} k^2 + 3\sum_{k=1}^{n} k + n$$

$$= \frac{n(n+1)(2n+1)}{6} + \frac{3n(n+1)}{2} + n$$

$$= \frac{n(n+2)(n+4)}{3}$$

176. *(c)* Given that, $a_n = n(n+3)$

$$= n^2 + 3n$$

Thus, sum to n terms is given by

$$S_n = \sum_{k=1}^{n} a_k = \sum_{k=1}^{n} k^2 + 3\sum_{k=1}^{n} k$$

$$= \frac{n(n+1)(2n+1)}{6} + \frac{3n(n+1)}{2}$$

$$= \frac{n(n+1)(2n+1) + 9n(n+1)}{6}$$

$$= \frac{n(n+1)(2n+1+9)}{6}$$

$$= \frac{2n(n+1)(n+5)}{6}$$

$$= \frac{n(n+1)(n+5)}{3}$$

177. *(d)* Let the sum of first n terms of the given series be

$S = 1\times2 + 2\times3 + 3\times4 + 4\times5 + \ldots$ upto n terms

First, we will split the given series into two parts which are 1, 2, 3, 4, ... and 2, 3, 4, 5, ... and find the nth term of each part separately to find the nth term of the given series.

$T_n = (n^{th}$ term of 1, 2, 3, ...) $\times$ (n^{th} term of 2, 3, 4, ...)

$$= [1 + (n-1)1] \times [2 + (n-1)1] \quad [\because T_n = a + (n-1)d]$$

$$= [1 + n - 1][2 + n - 1]$$

$$\Rightarrow \quad T_n = n(n+1)$$

Now, $S = \Sigma T_n = \Sigma (n^2 + n) = \Sigma n^2 + \Sigma n$

$$= \frac{n(n+1)(2n+1)}{6} + \frac{n(n+1)}{2} \quad \left[\because \Sigma n^2 = \frac{n(n+1)(2n+1)}{6},\ \Sigma n = \frac{n(n+1)}{2}\right]$$

$$= \frac{n(n+1)}{2}\left(\frac{2n+1}{3} + \frac{1}{1}\right) = \frac{n(n+1)}{2}\left(\frac{2n+1+3}{3}\right)$$

$$= \frac{n(n+1)}{2}\left(\frac{2n+4}{3}\right) = \frac{n(n+1)(n+2)}{3}$$

178. *(b)* Let the sum of first n terms of the given series be

$S = 3\times1^2 + 5\times2^2 + 7\times3^2 + \ldots$ upto n terms

First, we will split the given series into two parts which are 3, 5, 7, ... and $1^2, 2^2, 3^2, \ldots$ and then find the nth term of each part separately to find the nth term of the given series.

$T_n = (n$th term of 3, 5, 7, ...) $\times$ (n^{th} term of $1^2, 2^2, 3^2, \ldots$)

$$= [3 + (n-1)2](n)^2$$

$$= (3 + 2n - 2)(n)^2 = (2n+1)n^2 = 2n^3 + n^2$$

Now, $S = \Sigma T_n = 2\Sigma n^3 + \Sigma n^2$

$$= 2\left[\frac{n(n+1)}{2}\right]^2 + \frac{n(n+1)(2n+1)}{6}$$

$$= \frac{n(n+1)}{2}\left[2\times\frac{n(n+1)}{2} + \frac{2n+1}{3}\right]$$

$$\left\{\because \Sigma n^3 = \left[\frac{n(n+1)}{2}\right]^2,\ \Sigma n^2 = \frac{n(n+1)(2n+1)}{6}\right\}$$

$$= \frac{n(n+1)}{2}\left[\frac{3n(n+1) + 2n + 1}{3}\right]$$

$$= \frac{n(n+1)}{6} \times (3n^2 + 3n + 2n + 1) = \frac{n(n+1)(3n^2 + 5n + 1)}{6}$$

179. *(c)* Let the given series be

$S = 5^2 + 6^2 + 7^2 + \ldots + 20^2$

$$= (1^2 + 2^2 + 3^2 + 4^2 + 5^2 + 6^2 + \ldots + 20^2) - (1^2 + 2^2 + 3^2 + 4^2)$$

$$= \frac{20\times(20+1)\times(2\times20+1)}{6} - \frac{4(4+1)(2\times4+1)}{6} \quad \left[\because \Sigma n^2 = \frac{n(n+1)(2n+1)}{6}\right]$$

$$= \frac{20\times21\times41}{6} - \frac{4\times5\times9}{6} = \frac{17220}{6} - \frac{180}{6}$$

$$= \frac{17220 - 180}{6} = \frac{17040}{6} = 2840$$

180. *(a)* Let the sum of first n terms of the given series be

$S = 3\times8 + 6\times11 + 9\times14 + \ldots$upto n terms

$T_n = (n^{th}$ term of 3, 6, 9,...) $\times$ (n^{th} term of 8, 11, 14, ...)

$$T_n = [3 + (n-1)3][8 + (n-1)3] \quad [\because T_m = a + (m-1)d]$$

$$= (3 + 3n - 3)(8 + 3n - 3)$$

$$= 3n(3n+5) = 9n^2 + 15n$$

Now, $S = \Sigma T_n$

$$= \Sigma(9n^2 + 15n) = 9\Sigma n^2 + 15\Sigma n$$

$$= \frac{9n(n+1)(2n+1)}{6} + \frac{15n(n+1)}{2}$$

$$\left[\because \Sigma n^2 = \frac{n(n+1)(2n+1)}{6},\ \Sigma n = \frac{n(n+1)}{2}\right]$$

$$= \frac{n(n+1)}{2}\left[\frac{9(2n+1)}{3} + 15\right]$$

$$= \frac{n(n+1)}{2}[3(2n+1) + 15] = \frac{n(n+1)3[2n+1+5]}{2}$$

$$= \frac{3n(n+1)(2n+6)}{2} = 3n(n+1)(n+3)$$

181. *(b)* Let $T_n = n(n+1)(n+4)$

$$= n(n^2 + 4n + n + 4)$$

$$= n(n^2 + 5n + 4)$$

$$= n^3 + 5n^2 + 4n$$

Now, $S = \Sigma T_n = \Sigma(n^3 + 5n^2 + 4n)$

$$= \Sigma n^3 + 5\Sigma n^2 + 4\Sigma n$$

$$= \left[\frac{n(n+1)}{2}\right]^2 + \frac{5n(n+1)(2n+1)}{6} + \frac{4n(n+1)}{2}$$

$$= \frac{n(n+1)}{2}\left[\frac{n(n+1)}{2} + \frac{5(2n+1)}{3} + \frac{4}{1}\right]$$

$$= \frac{n(n+1)}{2}\left[\frac{3n(n+1) + 10(2n+1) + 24}{6}\right]$$

$$= \frac{n(n+1)(3n^2 + 3n + 20n + 10 + 24)}{12}$$

$$= \frac{n(n+1)(3n^2 + 23n + 34)}{12}$$

182. *(c)* Let $T_n = n^2 + 2^n$

$$\Rightarrow S = \Sigma T_n = \Sigma(n^2 + 2^n) = \Sigma n^2 + \Sigma 2^n$$

$$= \frac{n(n+1)(2n+1)}{6} + (2^1 + 2^2 + 2^3 + 2^4 + \ldots + 2^n)$$

$$= \frac{n(n+1)(2n+1)}{6} + \frac{2(2^n - 1)}{2-1} \left[\because S_n = \frac{a(r^n - 1)}{r-1} \text{ as } r > 1\right]$$

$$S = \frac{n(n+1)(2n+1)}{6} + 2(2^n - 1)$$

183. *(c)* Let the sum of first n terms of the given series be

$S = 1\times2\times3 + 2\times3\times4 + 3\times4\times5 + \ldots$ upto n terms

First, we will split the given series into three parts which are 1, 2, 3... and 2, 3, 4, ... and 3, 4, 5... and then find the nth term of each part separately to find the nth term of the given series.

i.e., $T_n = (n^{th}$ term of 1, 2, 3, ...) $\times$ (n^{th} term of 2, 3, 4 ...) $\times$ (n^{th} term of 3, 4, 5, ...)

$= [1 + (n-1)1][2 + (n-1)1][3 + (n-1)1]$ $[\because T_n = a + (n-1)d]$

$= (1 + n - 1)(2 + n - 1)(3 + n - 1)$

$\Rightarrow \quad T_n = n(n+1)(n+2)$

$= n(n^2 + 2n + n + 2) = n(n^2 + 3n + 2)$

$\Rightarrow \quad T_n = n^3 + 3n^2 + 2n$

Now, $\quad S = \Sigma T_n = \Sigma(n^3 + 3n^2 + 2n)$

$= \Sigma n^3 + 3\Sigma n^2 + 2\Sigma n$

$$= \left[\frac{n(n+1)}{2}\right]^2 + \frac{3n(n+1)(2n+1)}{6} + \frac{2n(n+1)}{2}$$

$$= \frac{n(n+1)}{2}\left[\frac{n(n+1)}{2} + (2n+1) + 2\right]$$

$$\left\{\because \Sigma n = \frac{n(n+1)}{2}, \Sigma n^2 = \frac{n(n+1)(2n+1)}{6}, \Sigma n^3 = \left[\frac{n(n+1)}{2}\right]^2\right\}$$

$$= \frac{n(n+1)}{2}\left[\frac{n(n+1)}{2} + \frac{(2n+1)}{1} + \frac{2}{1}\right]$$

$$= \frac{n(n+1)}{2}\left[\frac{n^2 + n + 4n + 2 + 4}{2}\right]$$

$$= \frac{n(n+1)(n^2 + 5n + 6)}{4}$$

$$= \frac{n(n+1)(n^2 + 2n + 3n + 6)}{4}$$

$$= \frac{n(n+1)[n(n+2) + 3(n+2)]}{4}$$

$$= \frac{n(n+1)(n+2)(n+3)}{4}$$

184. *(a)* Let the sum of first n terms of the given series be

$$S = \frac{1}{1\times2} + \frac{1}{2\times3} + \frac{1}{3\times4} + \ldots \text{ upto } n \text{ terms}$$

Here, we will split the denominator of the series into two parts to find the nth term of given series, *i.e.*,

$$T_n = \frac{1}{(n^{th} \text{ term of } 1, 2, 3\ldots\ldots) \times (n^{th} \text{ term of } 2, 3, 4, \ldots.)}$$

$$= \frac{1}{[1 + (n-1)1][2 + (n-1)1]} \quad [\because T_m = a + (m-1)d]$$

$$= \frac{1}{n(n+1)} = \frac{(n+1) - n}{n(n+1)}$$

$$= \frac{n+1}{n(n+1)} - \frac{n}{n(n+1)} = \frac{1}{n} - \frac{1}{n+1}$$

Now, putting $n = 1, 2, 3, 4, \ldots, n$, we get

$$T_1 = \frac{1}{1} - \frac{1}{2}$$

$$T_2 = \frac{1}{2} - \frac{1}{3}$$

$$T_3 = \frac{1}{3} - \frac{1}{4}$$

....

....

....

....

$$T_n = \frac{1}{n} - \frac{1}{n+1}$$

Now, adding these terms, we get

$S = T_1 + T_2 + T_3 + \ldots + T_n$

$$= \left(\frac{1}{1} - \frac{1}{2}\right) + \left(\frac{1}{2} - \frac{1}{3}\right) + \ldots\ldots + \left(\frac{1}{n} - \frac{1}{n+1}\right)$$

$$= 1 - \frac{1}{n+1} = \frac{1}{1} - \frac{1}{n+1} = \frac{n+1-1}{n+1}$$

$$\Rightarrow \quad S = \frac{n}{n+1}$$

185. *(b)* Let the sum of first n terms of the given series be

$S = 1^2 + (1^2 + 2^2) + (1^2 + 2^2 + 3^2) + \ldots$ upto n terms

Here, first term has 1 term, second term has 2 terms, third terms has 3 and so on. Therefore, n^{th} term will have n terms

i.e., $T_n = 1^2 + 2^2 + \ldots + n^2$

$\Rightarrow \quad T_n = \Sigma n^2$

$$\Rightarrow \quad T_n = \frac{n(n+1)(2n+1)}{6}$$

$$\Rightarrow \quad T_n = \frac{n(2n^2 + n + 2n + 1)}{6}$$

$$\Rightarrow \quad T_n = \frac{n(2n^2 + 3n + 1)}{6}$$

$$\Rightarrow \quad T_n = \frac{2n^3 + 3n^2 + n}{6}$$

Now, $\quad S = \Sigma T_n = \frac{1}{6}\Sigma(2n^3 + 3n^2 + n)$

$$= \frac{1}{6}[2\Sigma n^3 + 3\Sigma n^2 + \Sigma n]$$

$$= \frac{1}{6}\left[2\left\{\frac{n(n+1)}{2}\right\}^2 + \frac{3n(n+1)(2n+1)}{6} + \frac{n(n+1)}{2}\right]$$

$$\left\{\because \Sigma n = \frac{n(n+1)}{2}, \Sigma n^2 = \frac{n(n+1)(2n+1)}{6}, \Sigma n^3 = \left[\frac{n(n+1)}{2}\right]^2\right\}$$

$$= \frac{1}{6} \times \frac{n(n+1)}{2}\left[\frac{2n(n+1)}{2} + \frac{2n+1}{1} + \frac{1}{1}\right]$$

$$= \frac{n(n+1)}{12} \times [n^2 + n + 2n + 1 + 1]$$

$$= \frac{n(n+1)}{12}[n^2 + 3n + 2]$$

$$= \frac{n(n+1)(n^2 + 2n + n + 2)}{12}$$

$$= \frac{n(n+1)(n+2)(n+1)}{12}$$

$$= \frac{n(n+1)^2(n+2)}{12}$$

186. *(a)* Let $T_n = (2n-1)^2$

$\Rightarrow \quad T_n = 4n^2 + 1 - 4n$

Now, $S = \Sigma T_n = \Sigma(4n^2 + 1 - 4n)$

$$= 4\Sigma n^2 + \Sigma 1 - 4\Sigma n$$

$$= \frac{4n(n+1)(2n+1)}{6} + n - \frac{4n(n+1)}{2} \quad (\because \Sigma 1 = n)$$

$$= n\left[\frac{2(n+1)(2n+1)}{3} + \frac{1}{1} - \frac{2(n+1)}{1}\right]$$

$$= n\left[\frac{2(2n^2 + n + 2n + 1) + 3 - 6(n+1)}{3}\right]$$

$$= \frac{n[4n^2 + 6n + 2 + 3 - 6n - 6]}{3}$$

$$= \frac{n(4n^2 - 1)}{3}$$

$$= \frac{n}{3}(2n+1)(2n-1)$$

187. *(b)* Let $S = 3 + 7 + 13 + 21 + \ldots + T_n$

or $\quad S = 3 + 7 + 13 + \ldots + T_{n-1} + T_n$

On subtracting, we get

$$0 = (3 + 4 + 6 + 8 + \ldots \text{ to } n \text{ terms}) - T_n$$

$\Rightarrow \quad T_n = 3 + [4 + 6 + 8 + \ldots (n-1) \text{ terms}]$

$$= 3 + \frac{n-1}{2}[2 \times 4 + (n - 1 - 1)2]$$

$$\left\{\because S_n = \frac{n}{2}[2a + (n-1)d]\right\}$$

$$= 3 + \frac{n-1}{2}[8 + (n-2)2]$$

$$= 3 + \frac{n-1}{2}[2n + 4]$$

$$= 3 + \frac{(n-1)}{2} \times 2(n+2)$$

$$= 3 + (n-1)(n+2)$$

$$= 3 + n^2 + 2n - n - 2$$

$\Rightarrow \quad T_n = n^2 + n + 1$

Now, $S = \Sigma T_n = \Sigma(n^2 + n + 1) = \Sigma n^2 + \Sigma n + \Sigma 1$

$$= \frac{n(n+1)(2n+1)}{6} + \frac{n(n+1)}{2} + n$$

$$= n\left[\frac{2n^2 + n + 2n + 1}{6} + \frac{n+1}{2} + \frac{1}{1}\right]$$

$$\left[\because \Sigma n^2 = \frac{n(n+1)(2n+1)}{6}, \; \Sigma n = \frac{n(n+1)}{2}, \Sigma 1 = n\right]$$

$$= n\left[\frac{2n^2 + 3n + 1 + 3n + 3 + 6}{6}\right]$$

$$= n\left[\frac{2n^2 + 6n + 10}{6}\right]$$

$$= \frac{2n(n^2 + 3n + 5)}{6}$$

$$= \frac{n(n^2 + 3n + 5)}{3}$$

188. *(c)* Given, S_1 = Sum of first n natural numbers $= \Sigma n$

$$\Rightarrow \quad S_1 = \frac{n(n+1)}{2} \quad \ldots(i)$$

and S_2 = Sum of squares of first n natural numbers $= \Sigma n^2$

$$\Rightarrow \quad S_2 = \frac{n(n+1)(2n+1)}{6} \quad \ldots(ii)$$

and S_3 = Sum of cubes of first n natural numbers $= \Sigma n^3$

$$\Rightarrow \quad S_3 = \left[\frac{n(n+1)}{2}\right]^2 \quad \ldots(iii)$$

Consider, $S_3(1 + 8S_1)$

$$= \left[\frac{n(n+1)}{2}\right]^2\left[1 + 8 \times \frac{n(n+1)}{2}\right] \quad \text{[using Eqs. (i) and (iii)]}$$

$$= \left[\frac{n(n+1)}{2}\right]^2[1 + 4n(n+1)]$$

$$= \left[\frac{n(n+1)}{2}\right]^2[4n^2 + 4n + 1]$$

$$= \left[\frac{n(n+1)}{2}\right]^2(2n+1)^2$$

On multiplying and dividing by 9, we get

$$= 9 \times \frac{n^2(n+1)^2}{4} \times \frac{(2n+1)^2}{9} = 9\left[\frac{n(n+1)(2n+1)}{6}\right]^2$$

$$= 9 \times S_2^2 = 9S_2^2 \quad \text{[using Eq. (iii)]}$$

189. *(d)* $T_n = \dfrac{n^{th} \text{ term of numerator}}{n^{th} \text{ term of denominator}}$

$$= \frac{1^3 + 2^3 + 3^3 + \ldots + n^3}{1 + 3 + 5 + \ldots \text{ upto } n \text{ terms}}$$

$$= \frac{\Sigma n^3}{\frac{n}{2}[2 \times 1 + (n-1)2]} \quad \left\{\because S_n = \frac{n}{2}[2a + (n-1)d]\right\}$$

$$= \frac{\Sigma n^3}{\frac{n}{2}[2 + 2n - 2]} = \frac{\Sigma n^3}{\frac{n}{2} \times 2n}$$

$$\Rightarrow \quad T_n = \frac{\Sigma n^3}{n^2} = \frac{\left[\frac{n(n+1)}{2}\right]^2}{n^2} = \frac{n^2(n+1)^2}{4n^2}$$

$$\left\{\because \Sigma n^3 = \left[\frac{n(n+1)}{2}\right]^2\right\}$$

$$\Rightarrow \quad T_n = \frac{1}{4}(n+1)^2$$

$$\Rightarrow \quad T_n = \frac{1}{4}(n^2 + 1 + 2n)$$

$$\Rightarrow \quad S = \Sigma T_n = \frac{1}{4}\Sigma(n^2 + 1 + 2n) = \frac{1}{4}(\Sigma n^2 + \Sigma 1 + 2\Sigma n)$$

$$= \frac{1}{4}\left[\frac{n(n+1)(2n+1)}{6} + n + \frac{2 \times n(n+1)}{2}\right]$$

$$\left[\because \Sigma n^2 = \frac{n(n+1)(2n+1)}{6}, \ \Sigma n = \frac{n(n+1)}{2}, \Sigma 1 = n\right]$$

$$= \frac{n}{4}\left[\frac{2n^2 + n + 2n + 1}{6} + 1 + n + 1\right]$$

$$= \frac{n}{4}\left[\frac{2n^2 + 3n + 1}{6} + \frac{n+2}{1}\right]$$

$$= \frac{n}{4}\left[\frac{2n^2 + 3n + 1 + 6n + 12}{6}\right]$$

$$= \frac{n}{24}[2n^2 + 9n + 13]$$

190. *(b)* For numerator, $T_n = n(n+1)^2$

$$= n(n^2 + 1 + 2n) = n^3 + 2n^2 + n$$

$$\Rightarrow \quad S_n = \Sigma T_n = \Sigma(n^3 + 2n^2 + n)$$

$$= \Sigma n^3 + 2\Sigma n^2 + \Sigma n$$

$$= \left[\frac{n(n+1)}{2}\right]^2 + \frac{2n(n+1)(2n+1)}{6} + \frac{n(n+1)}{2}$$

$$\left\{\because \Sigma n^3 = \left[\frac{n(n+1)}{2}\right]^2, \ \Sigma n^2 = \frac{n(n+1)(2n+1)}{6}, \Sigma n = \frac{n(n+1)}{2}\right\}$$

$$= \frac{n(n+1)}{2}\left[\frac{n(n+1)}{2} + \frac{2(2n+1)}{3} + \frac{1}{1}\right]$$

$$= \frac{n(n+1)}{2}\left[\frac{3n(n+1) + 4(2n+1) + 6}{6}\right]$$

$$= \frac{n(n+1)(3n^2 + 11n + 10)}{12}$$

$$= \frac{n(n+1)[3n(n+2) + 5(n+2)]}{12}$$

$$= \frac{n(n+1)(n+2)(3n+5)}{12} \quad \text{...(i)}$$

Again, for denominator

Let $\quad T_n' = n^2(n+1) = n^3 + n^2$

Now, $S_n' = \Sigma T_n' = \Sigma(n^3 + n^2) = \Sigma n^3 + \Sigma n^2$

$$\left\{\because \Sigma n^3 = \left[\frac{n(n+1)}{2}\right]^2, \Sigma n^2 = \frac{n(n+1)(2n+1)}{6}\right\}$$

$$= \left[\frac{n(n+1)}{2}\right]^2 + \frac{n(n+1)(2n+1)}{6}$$

$$= \frac{n(n+1)}{2}\left[\frac{n(n+1)}{2} + \frac{2n+1}{3}\right]$$

$$= \frac{n(n+1)(3n^2 + 3n + 4n + 2)}{12}$$

$$= \frac{n(n+1)(3n^2 + 7n + 2)}{12}$$

$$= \frac{n(n+1)(3n^2 + 6n + n + 2)}{12}$$

$$= \frac{n(n+1)[3n(n+2) + 1(n+2)]}{12}$$

$$S_n' = \frac{n(n+1)(3n+1)(n+2)}{12}$$

Hence, required sum of the series become

$$\frac{S_n}{S_n'} = \frac{\frac{(n(n+1)(n+2)(3n+5))}{12}}{\frac{n(n+1)(3n+1)(n+2)}{12}} \quad \text{[from Eqs. (i) and (ii)]}$$

$$= \frac{3n+5}{3n+1}$$

191. *(c)* $\because a_n = (2n+1)^3 - (2n)^3, \forall n \in N$

$$S_n = \sum_{K=1}^{n} a_K = \sum_{K=1}^{n} [(2K+1)^3 - (2K)^3]$$

$$= \sum_{K=1}^{n} [(2K)^3 + 1 + 6K(2K+1) - (2K)^3]$$

$$= \sum_{K=1}^{n} [12K^2 + 6K + 1]$$

$$= 12\sum_{K=1}^{n} K^2 + 6\sum_{K=1}^{n} K + n$$

$$= \frac{12 \times n(n+1)(2n+1)}{6} + \frac{6 \times n(n+1)}{2} + n$$

$$= n\,[2(n+1)(2n+1) + 3(n+1) + 1]$$

$$= n[2(2n^2 + 3n + 1) + 3n + 3 + 1]$$

$$= n\,[4n^2 + 6n + 2 + 3n + 3 + 1]$$

$$= n\,(4n^2 + 9n + 6)$$

$$= 4n^3 + 9n^2 + 6n$$

When, $n = 10$. Then,

$$S_{10} = 4(10)^3 + 9(10)^2 + 6(10)$$

$$= 4000 + 900 + 60 = 4960$$

192. (a) $\sum_{r=1}^{n}\frac{S_r}{s_r}=\sum_{r=1}^{n}\frac{\frac{r^2(r+1)^2}{4}}{\frac{r(r+1)}{2}}$

$\left\{\begin{array}{l}\because \text{ sum of cubes of the first } n \text{ natural numbers,}\\ \Sigma n^3=\left[\frac{n(n+1)}{2}\right]^2 \text{ and sum of the}\\ \text{first } n \text{ natural numbers, } \Sigma n=\frac{n(n+1)}{2}\end{array}\right\}$

$$=\sum_{r=1}^{n}\frac{r(r+1)}{2}=\frac{1}{2}\sum_{r=1}^{n}(r^2+r)$$

$$=\frac{1}{2}\left[\sum_{r=1}^{n}r^2+\sum_{r=1}^{n}r\right]$$

$$=\frac{1}{2}\left[\frac{n(n+1)(2n+1)}{6}+\frac{n(n+1)}{2}\right]$$

$$=\frac{1}{12}(n)(n+1)(2n+1+3)$$

$$=\frac{n(n+1)2(n+2)}{12}=\frac{n(n+1)(n+2)}{6}$$

193. (c) According to the given condtion

$\Sigma n=\frac{1}{5}(\Sigma n^2)$

$\Rightarrow \quad \frac{n(n+1)}{2}=\frac{1}{5}\frac{n(n+1)(2n+1)}{6}$

$\Rightarrow \quad 2n+1=15$

$\Rightarrow \quad 2n=14 \Rightarrow n=7$

194. (a) Given series is $1\cdot 3^2+2\cdot 5^2+3\cdot 7^2+\ldots\infty$

This is an arithmetico-geometric series whose n^{th} term is equal to

$$T_n=n(2n+1)^2=4n^3+4n^2+n$$

$\therefore \quad S_n=\sum_{1}^{n}T_n=\sum_{1}^{n}(4n^3+4n^2+n)$

$$=4\sum_{1}^{n}n^3+4\sum_{1}^{n}n^2+\sum_{1}^{n}n$$

$$=4\left[\frac{n}{2}(n+1)\right]^2+\frac{4}{6}n(n+1)(2n+1)+\frac{n}{2}(n+1)$$

$$=n(n+1)\left[n^2+n+\frac{4}{6}(2n+1)+\frac{1}{2}\right]$$

$$=\frac{n}{6}(n+1)(6n^2+14n+7)$$

195. (a) $T_n=(2n-1)^3$

$$=8n^3-1^3-3\cdot 2n\cdot 1(2n-1)$$

$$=8n^3-1-12n^2+6n$$

$$=8n^3-12n^2+6n-1$$

$\therefore \quad S_n=\Sigma T_n$

$$=8\Sigma n^3-12\Sigma n^2+6\Sigma n-\Sigma 1$$

$$=8\cdot\left[\frac{n(n+1)}{2}\right]^2-12\cdot\frac{n(n+1)(2n+1)}{6}+6\frac{n(n+1)}{2}-n$$

$$=2n^2(n+1)^2-2n(n+1)(2n+1)+3n(n+1)-n$$

$$=n(n+1)[2n(n+1)-2(2n+1)+3]-n$$

$$=n(n+1)[2n^2+2n-4n-2+3]-n$$

$$=n(n+1)[2n^2-2n+1]-n$$

$$=n(n+1)\cdot 2n(n-1)+n(n+1)-n$$

$$=2n^2(n^2-1)+n^2$$

$$=n^2(2n^2-1)$$

196. (c) Let $S_n=cn^2$

$$S_{n-1}=c(n-1)^2=cn^2+c-2cn$$

$\therefore \quad T_n=2cn-c \quad (\because T_n=S_n-S_{n-1})$

$$T_n^2=(2cn-c)^2=4c^2n^2+c^2-4c^2n$$

$\therefore$ Required sum,

$$\Sigma T_n^2=\frac{4c^2\cdot n(n+1)(2n+1)}{6}+nc^2-2c^2n(n+1)$$

$$=\frac{2c^2n(n+1)(2n+1)+3nc^2-6c^2n(n+1)}{3}$$

$$=\frac{nc^2(4n^2+6n+2+3-6n-6)}{3}$$

$$=\frac{nc^2(4n^2-1)}{3}$$

197. (a) Let the edges of rectangular block are a, ar and ar^2, respectively.

Now, volume $=216\text{ cm}^3$

$\Rightarrow \quad a(ar)(ar^2)=216 \quad \left(\begin{array}{l}\because \text{volume of cuboid}\\ =l\times b\times h\end{array}\right)\ldots$(i)

$\Rightarrow \quad (ar)^3=(6)^3$

$\Rightarrow \quad ar=6\text{ cm}$ (taking cube root) ...(ii)

and total surface area $=252\text{ cm}^2$

$\Rightarrow \quad 2[a(ar)+ar(ar^2)+a(ar^2)]=252$

$\left[\begin{array}{l}\because \text{surface area of cuboid}\\ =2(lb+bh+hl)\end{array}\right]$

From Eq. (ii), we get

$$2(6a+36r+36)=252$$

$\Rightarrow \quad 12(a+6r+6)=252$

$\Rightarrow \quad a+6r=15$ (divide both sides by 12)...(iii)

$\Rightarrow \quad a+6\times\left(\frac{6}{a}\right)=15$ [from Eq. (ii)]

$\Rightarrow \quad a^2-15a+36=0$

$\Rightarrow \quad (a-12)(a-3)=0$

$\Rightarrow \quad a=3, 12$

From Eq. (iii), we get

when $a = 3$, then $3 + 6r = 15 \Rightarrow r = 2$

and when $a = 12$, then $12 + 6r = 15 \Rightarrow r = \frac{1}{2}$

Now, edges are $3, 3 \times 2, 3 \times (2)^2$ or $12, 12 \times \left(\frac{1}{2}\right), 12 \times \left(\frac{1}{2}\right)^2$

i.e., 3, 6, 12 or 12, 6, 3.

Hence, the length of the longest edge is 12 cm.

198. *(b)* Number of notes that the person counts in 10 min

$$= 10 \times 150 = 1500$$

Since $a_{10}, a_{11}, a_{12}, \ldots$ are in A.P. with common difference -2. Let n be the time taken to count remaining 3000 notes, then

$$\frac{n}{2}[2 \times 148 + (n-1) \times -2] = 3000$$

$$\Rightarrow n^2 - 149n + 3000 = 0$$

$$\Rightarrow (n-24)(n-125) = 0$$

$$\Rightarrow n = 24, 125$$

Then, the total time taken by the person to count all notes

$$= 10 + 24 = 34 \text{ min}$$

$\left(\because n \neq 125 \text{ as } a_{125} = -100, \text{ which is not possible}\right)$

199. *(a)* Since, $2b = a + c$

Now, $(a + 2b - c)(2b + c - a)(a + 2b + c)$

$$= (a + a + c - c)(a + c + c - a)(2b + 2b)$$

$$= 2a \cdot 2c \cdot 4b = 16abc$$

200. *(d)* Let $S = 2 + 3 + 6 + 11 + 18 + \ldots + T_n$ ...(i)

or $S = 2 + 3 + 6 + 11 + \ldots + T_{n-1} + T_n$ ...(ii)

On subtracting Eq. (ii) from Eq. (i), we get

$$0 = (2 + 1 + 3 + 5 + 7 + \ldots \text{upto } n \text{ terms}) - T_n$$

$$\Rightarrow T_n = 2 + [1 + 3 + 5 + \ldots (n-1) \text{ terms}]$$

$$\Rightarrow T_n = 2 + \frac{n-1}{2}[2 + (n-2)2]$$

$$\left\{\because S_{n-1} = \frac{n-1}{2}[2a + (n-2)d]\right\}$$

$$\Rightarrow T_n = 2 + (n-1)^2$$

$$\therefore T_{50} = 2 + (50-1)^2 = 49^2 + 2$$

201. *(c)* $\sin^2 B - \sin^2 A = \sin^2 C - \sin^2 B$ ($\because a^2, b^2, c^2$ are in A.P.)

$\Rightarrow \sin(B + A)\sin(B - A) = \sin(C + B)\sin(C - B)$

$\Rightarrow \sin C(\sin B \cos A - \cos B \sin A) = \sin A(\sin C \cos B - \cos C \sin B)$ ($\because A + B + C = \pi$)

$\Rightarrow 2\cot B = \cot A + \cot C$ (divide by $\sin A \sin B \sin C$)

$\Rightarrow \cot A, \cot B, \cot C$ are in A.P.

202. *(a)* Since, $T_m = \frac{1}{n} = a + (m-1)d$ ...(i)

and $T_n = \frac{1}{m} = a + (n-1)d$ ...(ii)

On solving Eqs. (i) and (ii), we get

$$a = \frac{1}{mn} \text{ and } d = \frac{1}{mn}$$

$$\therefore a - d = \frac{1}{mn} - \frac{1}{mn} = 0$$

203. *(b)* Let the first term be a and common difference be d.

Given, $S_{2n} = 3S_n$

$$\therefore \frac{2n}{2}[2a + (2n-1)d] = 3\left\{\frac{n}{2}[2a + (n-1)d]\right\}$$

$$\Rightarrow 4a + (4n-2)d = 6a + (3n-3)d$$

$$\Rightarrow 2a = (n+1)d \quad \ldots(i)$$

Now, $$\frac{S_{3n}}{S_n} = \frac{\frac{3n}{2}[2a + (3n-1)d]}{\frac{n}{2}[2a + (n-1)d]}$$

$$= \frac{3\{(n+1)d + (3n-1)d\}}{(n+1)d + (n-1)d} = \frac{3(4nd)}{2nd} = 6$$

204. *(c)* Given, $$\frac{\frac{p}{2}[2a_1 + (p-1)d]}{\frac{q}{2}[2a_1 + (q-1)d]} = \frac{p^2}{q^2}$$

$$\Rightarrow \frac{(2a_1 - d) + pd}{(2a_1 - d) + qd} = \frac{p}{q} \Rightarrow (2a_1 - d)(p - q) = 0$$

$$\Rightarrow a_1 = \frac{d}{2} \quad (\because p \neq q)$$

Now $$\frac{a_6}{a_{21}} = \frac{a_1 + 5d}{a_1 + 20d} = \frac{\frac{d}{2} + 5d}{\frac{d}{2} + 20d} = \frac{11}{41}$$

205. *(a)* Let $f(x) = Ax^2 + Bx + C$

$$\therefore f(1) = A + B + C$$

and $$f(-1) = A - B + C$$

$$\because f(1) = f(-1)$$

$$\Rightarrow A + B + C = A - B + C$$

$$\Rightarrow B = 0$$

$$\therefore f(x) = Ax^2 + C$$

$$\Rightarrow f'(x) = 2Ax$$

$$\Rightarrow f'(a) = 2Aa, f'(b) = 2Ab$$

and $$f'(c) = 2Ac$$

Also, a, b, c are in A.P. .

$\therefore 2Aa, 2Ab, 2Ac$ are in A.P.

$\Rightarrow f'(a), f'(b), f'(c)$ are in A.P.

206. *(b)* The total number of teams participated in a cricket tournament is $n = 16$

The sum of money awarded as a prize money, $S_n =$ ₹ 8000

Let us consider amount received by last place team as a first term.

$$\therefore a = 275$$

Let common difference be d.

$$\because \quad S_n = \frac{n}{2}[2a + (n-1)d]$$

$$\therefore \quad 8000 = \frac{16}{2}[2 \times 275 + (16-1)d]$$

$$\Rightarrow \quad 1000 = (550 + 15d)$$

$$\Rightarrow \quad 15d = 450$$

$$\Rightarrow \quad d = 30$$

Now, $T_{16} = 275 + (16-1)30 = 275 + 450 = 725$

Hence, the first place team will get the prize ₹ 725.

207. *(c)* Given, A.P. is 3, $a_1, a_2, a_3, a_4, a_5, a_6$, 31

$$\therefore \quad 31 = 3 + 7d$$

$$\Rightarrow \quad d = 4$$

$$\therefore \quad a_1 = 3 + 4 = 7$$

$$a_5 = a + 5d = 3 + 20 = 23$$

$$a_6 = a + 6d = 3 + 24 = 27$$

Now, $\quad a_6 - a_5 = 27 - 23 = 4$

and $\quad a_1 + a_6 = 7 + 27 = 34$

208. *(b)* Since, $a_1, a_2, \ldots, a_n$ are n A.M.'s between a and b, then

$$a_i = a + i\,\frac{(b-a)}{n+1} \qquad \left[\begin{array}{l}\because b = a + (n+1)d \\ \Rightarrow d = \dfrac{b-a}{n+1}\end{array}\right]$$

$$\therefore \quad 2\sum_{i=1}^{n} a_i = 2\left[\sum_{i=1}^{n} a + \frac{(b-a)}{n+1}\sum_{i=1}^{n} i\right]$$

$$= 2\left[na + \frac{(b-a)}{n+1}.\,n\,\frac{(n+1)}{2}\right] = n(a+b)$$

209. *(b)* Since, x, $2y$ and $3z$ are in A.P. .

$$\therefore \quad 4y = x + 3z \qquad \left(\begin{array}{l}\because \text{if } a, b, c \text{ are in A.P.,} \\ \text{then } 2b = a + c\end{array}\right) \quad \ldots\text{(i)}$$

And x, y, z are in G.P.

$\therefore \quad y = rx$ and $z = xr^2$, where r is common ratio.

On putting the value of y and z in Eq. (i), we get

$$4xr = x + 3xr^2$$

$$\Rightarrow \quad 3r^2 - 4r + 1 = 0 \qquad \text{(divide both sides by } x\text{)}$$

$$\Rightarrow \quad 3r^2 - 3r - r + 1 = 0$$

$$\Rightarrow \quad 3r(r-1) - 1(r-1) = 0$$

$$\Rightarrow \quad (3r-1)(r-1) = 0$$

$$\Rightarrow \quad 3r - 1 = 0 \quad \text{or} \quad r - 1 = 0$$

$$\Rightarrow \quad r = \frac{1}{3} \quad \text{or} \quad r = 1$$

$$\therefore \quad r = \frac{1}{3} \qquad (\because r = 1 \text{ is not possible})$$

210. *(c)* Given, $1 + \sin x + \sin^2 x + \ldots \infty = 4 + 2\sqrt{3}$

$$\Rightarrow \quad \frac{1}{1 - \sin x} = 4 + 2\sqrt{3}$$

$$\Rightarrow \quad 1 - \sin x = \frac{1}{4 + 2\sqrt{3}} \times \frac{4 - 2\sqrt{3}}{4 - 2\sqrt{3}}$$

$$\Rightarrow \quad 1 - \sin x = \frac{4 - 2\sqrt{3}}{4}$$

$$\Rightarrow \quad \sin x = \frac{2\sqrt{3}}{4} = \frac{\sqrt{3}}{2}$$

$$\Rightarrow \quad x = \frac{\pi}{3}, \frac{2\pi}{3}$$

211. *(b)* Taking x as the product of variates $x_1, x_2, \ldots, x_r$ corresponding to r set of observation *i.e.*, $x = x_1 x_2 \ldots x_r$, we have

$$\log x = \log x_1 + \log x_2 + \ldots + \log x_r$$

$$\Rightarrow \quad \Sigma \log x = \Sigma \log x_1 + \Sigma \log x_2 + \ldots + \Sigma \log x_r$$

$$\Rightarrow \frac{1}{n}\Sigma \log x = \frac{1}{n}\Sigma \log x_1 + \frac{1}{n}\Sigma \log x_2 + \ldots + \frac{1}{n}\Sigma \log x_r$$

$$\Rightarrow \quad \log G = \log G_1 + \log G_2 + \ldots + \log G_r$$

$$\Rightarrow \quad G = G_1 G_2 \ldots G_r$$

212. *(c)* Let A be the arithmetic mean between two numbers a and b. Again, let G_1 and G_2 be the two geometric mean between two numbers a and b.

Then, $A = \dfrac{a+b}{2}$ and $G_1 = ar$, $G_2 = ar^2$

$$\text{where} \quad r = \left(\frac{b}{a}\right)^{\frac{1}{2+1}} = \left(\frac{b}{a}\right)^{1/3} \qquad \ldots\text{(i)}$$

$$\therefore \quad \frac{G_1^2}{G_2} + \frac{G_2^2}{G_1} = \frac{a^2r^2}{ar^2} + \frac{a^2r^4}{ar}$$

$$= a + ar^3 = a + a\left(\frac{b}{a}\right) \quad \text{[from Eq. (i)]}$$

$$= a + b = 2A \qquad \left(\because A = \frac{a+b}{2}\right)$$

213. *(c)* Since, $a + ar = a(1 + r) = 12 \qquad \ldots$(i)

and $\quad ar^2 + ar^3 = ar^2(1 + r) = 48 \qquad \ldots$(ii)

From Eqs. (i) and (ii), we get

$$r^2 = 4 \quad \Rightarrow \quad r = -2$$

(since the G.P. is in alternate sign, so we take negative value)

On putting the value of r in Eq. (i), we get $a = -12$

214. *(c)* Here, total debt, S = ₹ 3600

and total instalments, $n = 40$

Let a and d be the first instalment and increment in instalment.

Thus, we get an A.P. here.

By using sum of n terms, we get

$$3600 = \frac{40}{2}[2a + (40-1)d]$$

$$\Rightarrow \quad 180 = 2a + 39d \qquad \ldots\text{(i)}$$

Now, after 30 instalments, one-third of the debt is unpaid.

i.e., $\dfrac{3600}{3}$ = ₹ 1200 is unpaid and then paid money

= 3600 − 1200 = ₹ 2400

So, again by using sum of n terms, we get

$$S_{30} = 2400 = \frac{30}{2}[2a + (30-1)d]$$

$$\Rightarrow \quad 160 = 2a + 29d \quad \text{...(ii)}$$

On solving Eqs. (i) and (ii), we get

$$a = 51, d = 2$$

We know that,

$$n^{th} \text{ term } T_n = a + (n-1)d$$

$$\therefore \quad 8^{th} \text{ instalment, } T_8 = a + (8-1)d$$

$$= 51 + 7 \times 2 = 51 + 14 = ₹\,65$$

Hence, the 8th instalment is ₹ 65.

215. *(b)* Since, $x = \sum_{n=0}^{\infty} \cos^{2n}\phi$

$$= 1 + \cos^2\phi + \cos^4\phi + \dots$$

$$= \frac{1}{1-\cos^2\phi} = \frac{1}{\sin^2\phi} \quad (\because |\cos x| < 1)$$

Similarly, $y = \dfrac{1}{1-\sin^2\phi} = \dfrac{1}{\cos^2\phi}$

and $z = \dfrac{1}{1-\sin^2\phi\cos^2\phi} = \dfrac{1}{1-\frac{1}{x}\cdot\frac{1}{y}} = \dfrac{xy}{xy-1}$

$$\Rightarrow \quad xyz = xy + z$$

216. *(a)* I. Given A.P. is 16, 11, 6, ...

Here, $a = 16, d = 11 - 16 = -5$

$$\because \quad S_n = \frac{n}{2}[2a + (n-1)d]$$

$$\therefore \quad S_{23} = \frac{23}{2}[2 \times 16 + (23-1)(-5)]$$

$$= \frac{23}{2}[32 + (22)(-5)] = \frac{23}{2}[32 - 110]$$

$$= \frac{23}{2}[-78] = -897$$

II. Given A.P. is $x + y, x - y, x - 3y, \dots$

Here, $a = x + y$

$$d = (x - y) - (x + y) = -2y$$

$$\because \quad S_n = \frac{n}{2}[2a + (n-1)d]$$

$$\therefore \quad S_{22} = \frac{22}{2}[2 \times (x+y) + (22-1)(-2y)]$$

$$= 11[2x + 2y + (21)(-2y)] = 11[2x + 2y - 42y]$$

$$= 11[2x - 40y]$$

$$= 22[x - 20y]$$

217. *(d)* Let $b = a + d$ and $c = a + 2d$, where d is common difference.

Now, $2^{bx+1} = 2^{(a+d)x+1} = 2^{ax+1} \cdot 2^{dx}$

and $2^{cx+1} = 2^{(a+2d)x+1} = 2^{ax+1} \cdot (2^{dx})^2$

$\therefore$ These numbers are in G.P. for all values of x.

218. *(a)* Given sequence is 54, 51, 48, ...

Here, $a = 54, d = 51 - 54 = -3$

Let n terms are required to give the sum 513.

$$S_n = 513$$

$$\Rightarrow \quad \frac{n}{2}[2a + (n-1)d] = 513$$

$$\Rightarrow \quad \frac{n}{2}[2 \times 54 + (n-1)(-3)] = 513$$

$$\Rightarrow \quad n(108 - 3n + 3) = 513 \times 2$$

$$\Rightarrow \quad -3n^2 + 111n = 1026$$

$$\Rightarrow \quad -(3n^2 - 111n + 1026) = 0$$

$$\Rightarrow \quad n^2 - 37n + 342 = 0$$

$$\Rightarrow \quad (n-18)(n-19) = 0$$

$$\Rightarrow \quad n = 18 \text{ or } = 19$$

Here, the common difference is negative and 19th term

$$= 54 + (19-1) \times (-3) = 0$$

So, the sum of 18 terms as well as that of 19 terms is 513.

219. *(b)* Let the numbers be x and y.

Since, $\dfrac{x+y}{2} = 2\sqrt{xy}$

$$\Rightarrow \quad x + y = 4\sqrt{xy} \quad \text{...(i)}$$

$$\Rightarrow \quad (x+y)^2 = 16xy$$

Also, $(x-y)^2 = (x+y)^2 - 4xy$

$$\therefore \quad (x-y)^2 = 16xy - 4xy = 12xy$$

$$\Rightarrow \quad x - y = 2\sqrt{3xy} \quad \text{...(ii)}$$

On solving Eqs. (i) and (ii), we get

$$x = (2 + \sqrt{3})\sqrt{xy}$$

and $y = (2 - \sqrt{3})\sqrt{xy}$

$$\therefore \quad \text{Required ratio} = \frac{x}{y} = \frac{(2+\sqrt{3})\sqrt{xy}}{(2-\sqrt{3})\sqrt{xy}}$$

$$= (2 + \sqrt{3})^2 = 7 + 4\sqrt{3}$$

220. *(a)* Since, $2\tan^{-1}q = \tan^{-1}p + \tan^{-1}r$

$$\Rightarrow \quad \tan^{-1}\frac{2q}{1-q^2} = \tan^{-1}\frac{p+r}{1-pr}$$

$$\Rightarrow \quad 2q = p + r \quad (\because q^2 = pr)$$

$\Rightarrow$ p, q, r are in A.P.

But p, q, r are in G.P.

$$\Rightarrow \quad p = q = r$$

221. *(b)* Given series is $2 \times 4 + 4 \times 6 + 6 \times 8 + \dots + n$ terms.

Here, we will split the series into two parts which are 2, 4, 6, ... and 4, 6, 8, ... and then find n^{th} term of each part separately to find the nth term of the given series. *i.e.,*

$T_n = (n^{th}$ term of 2, 4, 6 ...) $\times$ (n^{th}th term of 4, 6, 8, ...)

$$= [2 + (n-1)2][4 + (n-1)2] \quad [\because T_n = a + (n-1)d]$$

$$= (2 + 2n - 2)(4 + 2n - 2) = 2n(2n + 2)$$

Putting $n = 20$,

$$T_{20} = 2 \times 20(2 \times 20 + 2) = 40(40 + 2) = 40 \times 42 = 1680$$

10

Straight Lines

Slope of a Line

Level I

1. The two-dimensional coordinate geometry is the combination of

(a) Algebra and Analysis (b) Algebra and Statistics
(c) Algebra and Geometry (d) Geometry and Analysis

2. A systematic study of Geometry by the use of Algebra was first carried out by celebrated French Philosopher and Mathematician ...A..., in his book ...B... published in 1637. Here, A and B refer to

(a) Rene Daniell, La Geometry
(b) Rene Daniell, Le Geometry
(c) Rene Descartes, Le Geometry
(d) Rene Descartes, La Geometry

3. The book 'La Geometry' introduced the notion of the equation of a curve and related analytic methods into the study of

(a) Algebra (b) Statistics (c) Geometry (d) Analysis

4. The resulting combination of analysis and geometry is referred now as

(a) analytical geometry (b) real geometry
(c) complex geometry (d) None of these

5. The study of coordinate geometry include

(a) coordinate axes and coordinate planes
(b) plotting of points in a plane
(c) distance between two points and section formulae
(d) All of the above

6. The point $(6, -4)$ is at 6 units distance from the Y-axis measured along the ...A... and at 4 units distance from the X-axis measured along the ...B... . Here, A and B refer to

(a) positive X-axis, negative Y-axis
(b) positive X-axis, positive Y-axis
(c) negative X-axis, positive Y-axis
(d) negative X-axis, negative Y-axis

7. **Statement I** The point $(3, 0)$ is at 3 units distance from the Y-axis measured along the positive X-axis and has zero distance from the X-axis.

Statement II The point $(3, 0)$ is at 3 units distance from the X-axis measured along the positive Y-axis and has zero distance from the Y-axis.

(a) Both statements are true
(b) Statement I is true and Statement II is false
(c) Statement I is false and Statement II is true
(d) Both statements are false

8. Distance between the points $P(x_1, y_1)$ and $Q(x_2, y_2)$ is

(a) $(x_2 - x_1) + (y_2 - y_1)$
(b) $(x_2 - x_1)^2 + (y_2 - y_1)^2$
(c) $\sqrt{(x_2 - x_1)^2 + (y_2 - y_1)^2}$
(d) $\sqrt[3]{(x_2 - x_1)^2 + (y_2 - y_1)^2}$

9. The distance between the points $(6, -4)$ and $(3, 0)$ is

(a) 7 units (b) 25 units
(c) $\sqrt[3]{25}$ units (d) 5 units

10. The coordinates of a point dividing the line segment joining the points (x_1, y_1) and (x_2, y_2) internally in the ratio $m : n$ are

(a) $\frac{mx_2 + nx_1}{m+n}, \frac{my_2 + ny_1}{m+n}$

(b) $\frac{mx_1 + nx_2}{m+n}, \frac{my_1 + ny_2}{m+n}$

(c) $\frac{mx_2 - nx_1}{m+n}, \frac{my_2 - ny_1}{m+n}$

(d) $\frac{mx_1 - nx_2}{m+n}, \frac{my_1 - ny_2}{m+n}$

11. The coordinates of a point which divides the line segment joining $A(1, -3)$ and $B(-3, 9)$ internally in the ratio $1 : 3$, are given by

(a) $(-2, 6)$ (b) $(0, 0)$
(c) $\left(\frac{-6}{4}, \frac{18}{4}\right)$ (d) $\left(\frac{10}{4}, \frac{-30}{4}\right)$

12. Coordinates of the mid-point of the line segment joining the points (x_1, y_1) and (x_2, y_2), are

(a) $\frac{x_1 + x_2}{2}, \frac{y_1 + y_2}{2}$ (b) $\frac{x_1 + x_2}{3}, \frac{y_1 + y_2}{3}$
(c) $\frac{x_1 - x_2}{2}, \frac{y_1 - y_2}{2}$ (d) $\frac{x_1 - x_2}{3}, \frac{y_1 - y_2}{3}$

13. Area of triangle whose vertices are (x_1, y_1), (x_2, y_2) and (x_3, y_3), is

(a) $| x_1 (y_2 - y_3) + x_2 (y_3 - y_1) + x_3 (y_1 - y_2) |$
(b) $| x_1 (y_2 - y_3) + x_2 (y_1 - y_3) + x_3 (y_1 - y_2) |$
(c) $\frac{1}{2} | x_1 (y_2 - y_3) + x_2 (y_3 - y_1) + x_3 (y_1 - y_2) |$
(d) $\frac{1}{2} | x_1 (y_2 - y_3) + x_2 (y_1 - y_3) + x_3 (y_1 - y_2) |$

14. Area of the triangle whose vertices are $(4, 4)$, $(3, -2)$ and $(-3, 16)$, is

(a) 54 (b) 27 (c) 53 (d) 106

15. If three points A, B and C are collinear, then area of the ΔABC is

(a) one
(b) three
(c) zero
(d) two

16. The simplest geometric figure in coordinate geometry is

(a) straight line
(b) parabola
(c) ellipse
(d) hyperbola

17. For representing the line algebraically, ...M... is most essential. Here, M refers to

(a) location (b) slope
(c) length (d) None of these

18. A line in a coordinate plane forms two angles with the X-axis, which are

(a) supplementary (b) complementary
(c) equal (d) different

19. The angle (say) θ made by the line l with ...A... direction of X-axis and measured ...B... is called the ...C... of the line. Here, A, B and C refer to

(a) negative, clockwise, inclination
(b) positive, clockwise, inclination
(c) negative, anti-clockwise, inclination
(d) positive, anti-clockwise, inclination

20. The inclination of the line lies within

(a) $0°$ to $90°$ (b) $180°$ to $360°$
(c) $90°$ to $180°$ (d) $0°$ to $180°$

21. **Statement I** Lines parallel to X-axis or coinciding with X-axis, have inclination of $0°$.

Statement II The inclination of a vertical line parallel to or coinciding with Y-axis is $90°$.

(a) Both statements are true
(b) Statement I is true and Statement II is false
(c) Statement I is false and Statement II is true
(d) Both statements are false

22. If θ is the inclination of a line l, then the slope or gradient of the line l is

(a) $\sin \theta$ (b) $\cos \theta$
(c) $\tan \theta$ (d) $\cot \theta$

23. The slope of a line whose inclination is $90°$, is

(a) 1 (b) 0
(c) -1 (d) not defined

24. The slope of a line is denoted by m. Thus,

(a) $m = \tan \theta$ (b) $m = \tan \theta, \theta \neq 90°$
(c) $m = \tan \theta, \theta \neq 0°$ (d) None of these

25. **Statement I** Slope of X-axis is zero and slope of Y-axis is not defined.

Statement II Slope of X-axis is not defined and slope of Y-axis is zero.

(a) Both statements are true
(b) Statement I is true and Statement II is false
(c) Statement I is false and Statement II is true
(d) Both statements are false

26. A line is completely determined when we are given ...A... points on it. Here, A refers to

(a) one (b) three
(c) four (d) two

27. I. Let $P(x_1, y_1)$ and $Q(x_2, y_2)$ be two points on a line l. If $x_1 = x_2$, then the line l will become perpendicular to X-axis and its slope will not be defined.

II. The inclination of the line l may be acute or obtuse.

(a) I and II are true
(b) I is true and II is false
(c) I is false and II is true
(d) I and II are false

28. Let $P(x_1, y_1)$ and $Q(x_2, y_2)$ be two points on a non-vertical line l whose inclination is θ. Then,

I. Slope of the line l is $m = \frac{y_2 - y_1}{x_2 - x_1}$, whether θ is acute or obtuse.

II. Slope of the line l is $m = \frac{y_2 - y_1}{x_2 - x_1}$, when θ is acute only.

(a) I and II are true
(b) I is true and II is false
(c) I is false and II is true
(d) I and II are false

29. I. Two non-vertical lines l_1 and l_2 are parallel, if and only if their slopes are equal.

II. Two non-vertical lines l_1 and l_2 are perpendicular to each other, if and only if their slopes are negative reciprocals of each other.

(a) I and II are true
(b) I and II are false
(c) I is true and II is false
(d) II is true and I is false

30. Which of the following is incorrect?

(a) Slope of the line passing through the points (3, – 2) and (– 1, 4) is $\frac{-3}{2}$
(b) Slope of the line passing through the points (3, – 2) and (7, –2) is 0
(c) Slope of the line passing through the points (3, – 2) and (3, 4) is 0
(d) Slope of the line making inclination of 60° with the positive direction of X-axis is $\sqrt{3}$

31. When there are more than one line in a plane, then these lines are

(a) only parallel
(b) only intersecting
(c) intersecting and parallel
(d) either intersecting or parallel

32. When two lines intersect each other, they make two pairs of vertically opposite angles such that sum of any two adjacent angles is

(a) 180°
(b) 90°
(c) 270°
(d) 360°

33. If L_1 and L_2 be two non-vertical lines with slopes m_1 and m_2 respectively. If α_1 and α_2 are the inclinations of lines L_1 and L_2, respectively. Let θ and ϕ be the adjacent angles between the lines L_1 and L_2. Then, match the terms of Column I with terms of Column II and choose the correct option from the codes given below.

Column I	Column II
A. $\tan\theta$	1. $\frac{m_1 - m_2}{1 + m_1 m_2}$ (as $1 + m_1 m_2 \neq 0$)
B. $\tan\phi$	2. $\frac{m_2 - m_1}{1 + m_1 m_2}$ is positive, (as $1 + m_1 m_2 \neq 0$)
C. θ will be acute and ϕ will be obtuse	3. $\frac{m_2 - m_1}{1 + m_1 m_2}$ is negative, (as $1 + m_1 m_2 \neq 0$)
D. θ will be obtuse and ϕ will be acute	4. $\frac{m_2 - m_1}{1 + m_1 m_2}$ (as $1 + m_1 m_2 \neq 0$)

Codes

	A	B	C	D		A	B	C	D
(a)	1	4	2	3	(b)	4	1	2	3
(c)	4	1	3	2	(d)	4	2	1	3

34. The acute angle (θ) between lines L_1 and L_2 with slopes m_1 and m_2 respectively is given by

(a) $\tan\theta = \frac{m_1 - m_2}{1 + m_1 m_2}$, as $1 + m_1 m_2 \neq 0$
(b) $\tan\theta = \frac{m_2 - m_1}{1 + m_1 m_2}$, as $1 + m_1 m_2 \neq 0$
(c) $\tan\theta = \left|\frac{m_2 - m_1}{1 + m_1 m_2}\right|$, as $1 + m_1 m_2 \neq 0$
(d) None of the above

35. The obtuse angle (ϕ) between lines L_1 and L_2 with slopes m_1 and m_2 respectively is given by

(a) $\phi = 180° + \theta$, where $\theta = \tan^{-1}\left|\frac{m_2 - m_1}{1 + m_1 m_2}\right|$, as $1 + m_1 m_2 \neq 0$
(b) $\phi = 180° - \theta$, where $\theta = \tan^{-1}\left|\frac{m_2 - m_1}{1 + m_1 m_2}\right|$, as $1 + m_1 m_2 \neq 0$
(c) $\phi = \theta - 180°$, where $\theta = \tan^{-1}\left(\frac{m_1 - m_2}{1 + m_1 m_2}\right)$, as $1 + m_1 m_2 \neq 0$
(d) $\phi = 180° + \theta$, where $\theta = \tan^{-1}\left(\frac{m_1 - m_2}{1 + m_1 m_2}\right)$, as $1 + m_1 m_2 \neq 0$

36. If the angle between two lines is $\frac{\pi}{4}$ and slope of the lines is $\frac{1}{2}$, then which of the following is not the slope of the other line?

(a) 3
(b) $\frac{-1}{3}$
(c) $\frac{1}{3}$
(d) None of these

37. Lines through the points $(-2, 6)$ and $(4, 8)$ is perpendicular to the line through the points $(8, 12)$ and $(x, 24)$. Then, the value of x is

(a) 2 (b) 6 (c) 8 (d) 4

38. Slopes of two parallel lines are

(a) equal (b) unequal (c) positive (d) negative

39. If two lines having the same slope pass through a common point, then these lines will

(a) non-parallel (b) intersect
(c) coincide (d) None of these

40. If A, B and C are three points in the xy plane, then they will lie on a line or they are collinear, if and only if

(a) Slope of AB = Slope of BC
(b) Slope of AB = – (Slope of BC)
(c) Slope of AB = 2 (Slope of BC)
(d) Slope of $AB = \frac{1}{2}$ (Slope of BC)

41. If the vertices of a quadrilateral are $(-4, 5)$, $(0, 7)$, $(5, -5)$ and $(-4, -2)$, then area of the quadrilateral is

(a) 70 sq units (b) 60 sq units
(c) 75.5 sq units (d) 60.5 sq units

42. The base of an equilateral triangle with side $2a$ lies along the Y-axis such that the mid-point of the base is at the origin. Then, which of the following are not the vertices of triangle?

(a) $(\sqrt{3}a, 0), (0, a), (0, -a)$ (b) $(\sqrt{3}a, 0), (-\sqrt{3}a, 0), (0, a)$
(c) $(-\sqrt{3}a, 0), (0, a), (0, -a)$ (d) None of these

43. I. The distance between $P(x_1, y_1)$ and $Q(x_2, y_2)$ when PQ is parallel to Y-axis, is $|x_2 - x_1|$.

II. The distance between $P(x_1, y_1)$ and $Q(x_2, y_2)$ when PQ is parallel to X-axis, is $|y_2 - y_1|$.

(a) I and II are true
(b) Only I is true
(c) Only II is true
(d) I and II are false

44. The point on X-axis which is equidistant from the points $(7, 6)$ and $(3, 4)$, is

(a) $\left(\frac{13}{2}, 0\right)$ (b) $(13, 0)$ (c) $\left(\frac{15}{2}, 0\right)$ (d) $\left(\frac{17}{2}, 0\right)$

45. The slope of a line, which makes an angle of 30° with the positive direction of Y-axis measured anti-clockwise, is

(a) $\sqrt{3}$ (b) $\frac{-1}{\sqrt{3}}$ (c) $-\sqrt{3}$ (d) $\frac{-1}{\sqrt{3}}$

46. The value of x for which the points $(x, -1)$, $(2, 1)$ and $(4, 5)$ are collinear, will be

(a) 0 (b) 1 (c) 2 (d) 3

47. If $A(-2, -1)$, $B(4, 0)$, $C(3, 3)$ and $D(-3, 2)$ are the vertices of a parallelogram, then

Statement I Slope of AB = Slope of BC and Slope of CD = Slope of AD.

Statement II Mid-point of AC = Mid- point of BD.

(a) Both statements are true
(b) Statement I is true and Statement II is false
(c) Statement I is false and Statement II is true
(d) Both statements are false

48. The angle between the X-axis and the line joining the points $(3, -1)$ and $(4, -2)$ is

(a) 45° (b) 135° (c) 90° (d) 180°

49. The slope of a line is double of the slope of another line. If tangent of the angle between them is $\frac{1}{3}$, then which of the following is not the slope of any line?

(a) 1 (b) $\frac{1}{2}$
(c) 2 (d) None of these

50. A line passes through (x_1, y_1) and (h, k). If slope of the line is m, then

(a) $y_1 - k = m(h - x_1)$ (b) $k - y_1 = m(h - x_1)$
(c) $h - x_1 = m(y_1 - k)$ (d) $h - x_1 = m(k - y_1)$

Level II

51. The line $y - x + 2 = 0$ divides the line joining $(3, -1)$ and $(8, 9)$ in the ratio $\lambda : 1$, where λ is

(a) $\frac{2}{3}$ (b) $\frac{5}{3}$ (c) $\frac{5}{2}$ (d) $\frac{3}{2}$

52. The vertices of a triangle are $(1, 2)$, $(h, -3)$ and $(-4, k)$. If the centroid of the triangle be at the point $(5, -1)$, then the value of $\sqrt{(h+k)^2 + (h+3k)^2}$ is

(a) 10 (b) 20 (c) 30 (d) 40

53. The points $(a, 0)$, $(0, b)$ and $(1, 1)$ are collinear, if $m + n = 1$, where m and n respectively are

(a) $b, -a$ (b) $a, -b$ (c) $\frac{1}{a}, \frac{1}{b}$ (d) $\frac{-1}{a}, \frac{-1}{b}$

54. If the point (x, y) be equidistant from the points $(6, -1)$ and $(2, 3)$, then

(a) $x + y = 3$ (b) $x - y = 3$ (c) $y - x = 3$ (d) $x + y = 6$

55. An equilateral triangle has one vertex at the point $(0, 0)$ and another at $(3, \sqrt{3})$. Then, which of the following are not the coordinates of third vertex?

(a) $(0, 2\sqrt{2})$ (b) $(3, -\sqrt{3})$
(c) $(0, -\sqrt{3})$ (d) None of these

56. Let the opposite angular points of a square be (3, 4) and (1, – 1). Then, which of the following are not the coordinates of remaining angular points?

(a) $\left(\frac{9}{2}, \frac{1}{2}\right)$ (b) $\left(\frac{9}{2}, \frac{5}{2}\right)$

(c) $\left(\frac{-1}{2}, \frac{5}{2}\right)$ (d) None of these

57. In any ΔABC, if D is the middle point of BC, then $AB^2 + AC^2$ is equal to

(a) $AD^2 + BD^2$ (b) $AD^2 - BD^2$

(c) $2(BD^2 - AD^2)$ (d) $2(AD^2 + BD^2)$

58. Let $ABCD$ be a rectangle and P be any point in its plane. Then, $PA^2 + PC^2$ is equal to

(a) $PB^2 + PD^2$ (b) $PB^2 - PD^2$

(c) $PD^2 - PB^2$ (d) $2(PB^2 + PD^2)$

59. If $A(-1, 3)$, $B(1, -1)$ and $C(5, 1)$ are the vertices of triangle, then the length of median through A is

(a) 6 units (b) 3 units (c) 4 units (d) 5 units

60. The line joining the mid-points of any two sides of a triangle is equal to ...A... of the third side. Here, A stands for

(a) twice (b) half (c) equal (d) thrice

61. The coordinates of a point which divides externally the line joining (1, –3) and (– 3, 9) in the ratio 1 : 3 are

(a) (3, – 6) (b) (– 6, 3) (c) (3, – 9) (d) (– 9, 3)

62. The line segment joining $A(6, 3)$ to $B(-1, -4)$ is doubled in length by halving its length added to each end. Then, which of the following are not the coordinates of new ends?

(a) $\left(\frac{19}{2}, \frac{13}{2}\right)$ (b) $\left(\frac{19}{2}, \frac{-15}{2}\right)$

(c) $\left(\frac{-9}{2}, \frac{-15}{2}\right)$ (d) None of these

63. If the point $(2, y)$ divides the line segment joining (4, 3) and (6, 3) in the ratio $\lambda : 1$ and the value of y is ...A... . Here, λ and A respectively are

(a) $\frac{1}{2}, 3$ (b) $3, \frac{1}{2}$

(c) $3, \frac{-1}{2}$ (d) $\frac{-1}{2}, 3$

64. If two vertices of a triangle are (–1, 4) and (5, 2) and its centroid is (0, –3), then the third vertex is

(a) (4, 15)0

(b) (– 4, – 15)

(c) (2, 16)

(d) (– 2, – 16)

65. If G is the centroid of ΔABC, then $AB^2 + BC^2 + CA^2$ is equal to

(a) $GA^2 + GB^2 + GC^2$ (b) $2(GA^2 + GB^2 + GC^2)$

(c) $3(GA^2 + GB^2 + GC^2)$ (d) None of these

66. If the vertices of a triangle are $(1, a)$, $(2, b)$ and $(c^2, -3)$. Then,

I. The centroid cannot lie on the Y-axis.

II. The condition that the centroid may lie on the X-axis is $a + b = 3$.

(a) Both I and II are true (b) Only I is true

(c) Only II is true (d) Both I and II are false

67. The coordinates of A, B, C are (6, 3), (– 3, 5) and (4, – 2) respectively and P is any point (x, y). The ratio of the areas of the ΔPBC and ΔABC is $\frac{m}{n}$, where m and n respectively are

(a) $|x + y - 2|, 7$ (b) $|x + y + 2|, 7$

(c) $|x - y + 2|, 5$ (d) $|x - y - 2|, 5$

68. The coordinates of the vertices of an equilateral triangle cannot be all [HOTS]

(a) irrational (b) rational (c) real (d) complex

69. If A and B are two points having coordinates (3, 4) and (5, – 2) respectively and P is a point such that $PA = PB$ and area of $\Delta PAB = 10$ units, then the coordinates of P are

(a) (7, 4) or (13, 2) (b) (7, 2) or (1, 0)

(c) (2, 7) or (4, 13) (d) None of these

70. Two opposite vertices of a rectangle are (1, 3) and (5, 1). If the rest two vertices lie on the line $y - x + \lambda = 0$, then λ is equal to

(a) 1 (b) – 1

(c) 2 (d) 3

71. The area of the pentagon whose vertices are $A(1, 1)$, $B(7, 21)$, $C(7, -3)$, $D(12, 2)$ and $E(0, -3)$, is

(a) $\frac{137}{2}$ sq units (b) $\frac{137}{3}$ sq units

(c) $\frac{137}{4}$ sq units (d) None of these

72. If points $(K, 2-2K)$, $(-K+1, 2K)$ and $(-4-K, 6-2K)$ are collinear, then K is equal to

(a) 2, 3 (b) 1, 0

(c) $\frac{1}{2}, -1$ (d) 1, 2

73. The distance of the point $(a\cos\alpha, a\sin\alpha)$ from the origin is independent of

(a) α (b) a

(c) a, α (d) None of these

74. If $A(4, 4)$, $B(3, 5)$ and $C(-1, -1)$ are the vertices of right angled triangle at A, then

I. (Slope of AB) $\cdot$ (Slope of BC) $= -1$
II. (Slope of AB) $\cdot$ (Slope of CA) $= -1$
III. (Slope of AC) $\cdot$ (Slope of BC) $= -1$

(a) Only I is true
(b) Only II is true
(c) Only III is true
(d) All are true

75. If three points $(h, 0)$, (a, b) and $(0, k)$ lie on a line, then

(a) $\frac{a}{h}+\frac{b}{k}=1$
(b) $ha+kb=1$
(c) $\frac{a}{h}-\frac{b}{k}=1$
(d) $ha-kb=1$

76. Consider the following population and year graph. [HOTS]

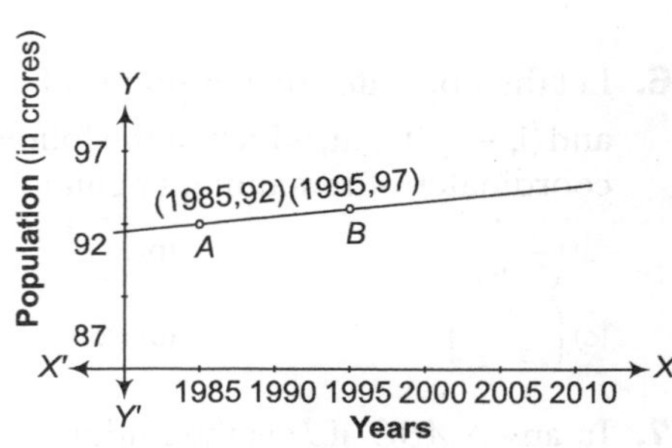

I. Slope of $AB = \frac{1}{2}$.

II. Population in the year 2010 is 104.5 crore.

(a) Both I and II are true
(b) Only I is true
(c) Only II is true
(d) Both I and II are false

77. A ray of light coming from the point (1, 2) is reflected at a point A on the X-axis and then passes through the point (5, 3). Then, the coordinates of the point A are

(a) (13, 0)
(b) $\left(\frac{13}{5}, 0\right)$
(c) $\left(\frac{17}{2}, 0\right)$
(d) (17, 0)

Various Forms of the Equation of a Line

Level I

78. Read the given statements and choose the correct option.

I. For a given line, we should have a definite condition on the points lying on the line.
II. For a given line, we can have any condition on the points lying on the line.

(a) Only I is correct
(b) Only II is correct
(c) Both I and II are incorrect
(d) None of the above

79. Suppose $P(x, y)$ is an arbitrary point on the xy-plane and L is the given line.

Statement I For the equation of L, we wish to construct a statement or condition for the point P that is true, when P is on L, otherwise false.

Statement II Statement for the line is merely an algebraic equation involving the variables x and y.

(a) Statement I is true; Statement II is true; Statement II is a correct explanation for Statement I
(b) Statement I is true; Statement II is true; Statement II is not a correct explanation for Statement I
(c) Statement I is true; Statement II is false
(d) Statement I is false; Statement II is true

80. If a horizontal line L is at a distance a from the X-axis, then ...M... of every point lying on the line is ...N... . Here, M and N respectively are

(a) ordinate, a
(b) ordinate, $-a$
(c) ordinate, either a or $-a$
(d) ordinate, neither a nor $-a$

81. Let a horizontal line L is at a distance 'a' from the X-axis.

I. If line L is above the X-axis, then its equation is $y=-a$.
II. If line L is below the X-axis, then its equation is $y=a$.

(a) Both I and II are true
(b) Only I is true
(c) Only II is true
(d) Both I and II are false

82. The equation of a ...P... line at a distance b from the Y-axis is either $x=b$ or $x=-b$. Here, P stands for

(a) horizontal
(b) vertical
(c) oblique
(d) None of these

83. Which of the following are the equations of the lines parallel to axes and passing through $(-2, 3)$?

(a) $x=2$ and $y=3$
(b) $x=-2$ and $y=-3$
(c) $x=-2$ and $y=3$
(d) $x=2$ and $y=-3$

84. Suppose that $P(x_0, y_0)$ is a fixed point on a non-vertical line L, whose slope is m. Let $P(x, y)$ be any arbitrary point on L. Then, the equation of L is given by

(a) $y - y_0 = m(x - x_0)$ (b) $m = (y - y_0)(x - x_0)$
(c) $x - x_0 = m(y - y_0)$ (d) $(x_0 - x)m = y - y_0$

85. The equation of the line through $(-2, 3)$ with slope -4 is

(a) $x + 4y - 10 = 0$ (b) $4x + y + 5 = 0$
(c) $x + y - 1 = 0$ (d) $3x + 4y - 6 = 0$

86. Let the line L passes through two given points $P_1(x_1, y_1)$ and $P_2(x_2, y_2)$. Let $P(x, y)$ be general point on L. Then, the equation of the line passing through the points (x_1, y_1) and (x_2, y_2) is given by

(a) $y - y_1 = \frac{y_2 - y_1}{x_2 - x_1}(x_1 - x)$ (b) $y + y_1 = \frac{y_1 - y_2}{x_1 - x_2}(x + x_1)$
(c) $y - y_1 = \frac{y_2 - y_1}{x_2 - x_1}(x - x_1)$ (d) $y - y_1 = \frac{y_2 + y_1}{x_2 + x_1}(x - x_1)$

87. **Statement I** The point (x, y) lies on the line with slope m and y-intercept c, if and only if $y = mx + c$.

Statement II Suppose line L with slope m makes x-intercept d. The equation of L is $y = (x + d)$.

(a) Both statements are true (b) Only Statement I is true
(c) Only Statement II is true (d) Both statements are false

88. Suppose a line L make x-intercept a and y-intercept b on the axes. Then, the equation of the line L is

(a) $\frac{x}{a} + \frac{y}{b} = 1$ (b) $\frac{x}{a} - \frac{y}{b} = 1$
(c) $\frac{y}{b} - \frac{x}{a} = 1$ (d) $ax + by = 1$

89. The equation of the line, which makes intercepts a and b on the X and Y-axes respectively, is $2x - 3y + 6 = 0$, where a and b respectively are

(a) 3, 2 (b) $-3, -2$
(c) $3, -2$ (d) $-3, 2$

90. Let L be the line, whose perpendicular distance from origin O be $OA = p$ and the angle between the positive X-axis and OA be $\angle XOA = \omega$. Then, match the terms of Column I with terms of Column II and choose the correct option from the codes given below.

Column I	Column II
A. Number of possible positions of line L according the angle makes the normal with positive direction of X-axis	1. $-\cot\omega$
B. Slope of line L is	2. $-\tan\omega$
C. Point A is	3. four
D. Equation of line L is	4. $(p\cos\omega, p\sin\omega)$
	5. $x\cos\omega + y\sin\omega = p$

Codes

	A	B	C	D
(a)	3	2	4	5
(b)	3	1	4	5
(c)	3	2	5	4
(d)	3	1	2	5

91. The equation of the line whose perpendicular distance from the origin is 4 units and the angle which the normal makes with positive direction of X-axis is 15°, can be written as $ax + by = 8\sqrt{2}$, where a and b respectively are

(a) $\sqrt{3} + 1, \sqrt{3} - 1$
(b) $\sqrt{3} - 1, \sqrt{3} + 1$
(c) $\sqrt{3} + 1, \sqrt{3} + 1$
(d) $\sqrt{3} - 1, \sqrt{3} - 1$

92. The equation contains two constants, namely, m and c. For finding these two constants, we need ...A... conditions satisfied by the equation of line. Here, A stands for

(a) one (b) three
(c) two (d) four

93. Which of the following is incorrect?

(a) The equation for the X-axis and Y-axis is $y = 0$ and $x = 0$ respectively
(b) Equation of line passing through the point $(-4, 3)$ with slope $\frac{1}{2}$ is $x - 2y + 10 = 0$
(c) Equation of line passing through (0, 0) with slope m is $y = mx$
(d) Equation of line passing through $(2, 2\sqrt{3})$ and inclined with X-axis at an angle of 75° is $(2 + \sqrt{3})x + y + 4 = 0$

94. Which of the following is incorrect?

(a) Equation of line intersecting the X-axis at a distance of 3 units to the left of origin with slope -2 is $2x + y + 6 = 0$
(b) Equation of line intersecting the Y-axis at a distance of 2 units above the origin and making an angle of 30° with positive direction of X-axis is $x - \sqrt{3}y + 2\sqrt{3} = 0$
(c) Equation of line passing through the points $(-1, 1)$ and $(2, -4)$ is $5x + 3y + 2 = 0$
(d) None of the above

95. The perpendicular distance between the line and origin is 5 units and the angle made by the perpendicular with the positive X-axis is 30°. Then, the equation of line is

(a) $x + \sqrt{3}y = 10$ (b) $\sqrt{3}x + y = 10$
(c) $x - \sqrt{3}y = 10$ (d) $\sqrt{3}x - y = 10$

96. The vertices of ΔPQR are $P(2, 1)$, $Q(-2, 3)$ and $R(4, 5)$. Then, equation of the median through the vertex R is

(a) $3x + 4y - 32 = 0$
(b) $4x - 3y - 1 = 0$
(c) $3x - 4y + 8 = 0$
(d) $-3x + 4y + 8 = 0$

97. The equation of the line passing through (– 3, 5) and perpendicular to the line through the points (2, 5) and (– 3, 6) is

(a) $5x + y + 30 = 0$ (b) $x + 5y - 22 = 0$
(c) $5x + y + 10 = 0$ (d) $5x - y + 20 = 0$

98. The equation of a line that cuts off equal intercepts on the coordinate axes and passes through the point (2, 3) is

(a) $2x + 3y = 13$(b) $x + y = 5$
(c) $3x + 2y = 12$ (d) $y - x = 1$

99. The equation of the line passing through the point (2, 2) and cutting off intercepts on the axes whose sum is 9. Then, incorrect option is

(a) $2x + y = 6$ (b) $x + 2y = 6$
(c) $2x - y = 2$ (d) None of these

100. The perpendicular from the origin to a line meets it at the point (– 2, 9). Then, the equation of the line is

(a) $2x - 9y + 85 = 0$ (b) $2x + 9y - 77 = 0$
(c) $7x + 3y - 13 = 0$ (d) $5x + 2y - 8 = 0$

101. The length L (in cm) of a copper rod is a linear function of its Celsius temperature C. In an experiment, if $L = 124.942$ when $C = 20$ and $L = 125.134$ when $C = 110$. Then, L in terms of C is $L = \frac{0.192}{90} m + 124.942$. Here, m stands for

(a) $20 - C$ (b) $C - 20$ (c) $C + 20$ (d) $C + 40$

102. If $P(a, b)$ is the mid-point of a line segment between axes, then the equation of line is

(a) $\frac{x}{a} + \frac{y}{b} = 1$ (b) $\frac{x}{a} - \frac{y}{b} = 1$ (c) $\frac{x}{a} - \frac{y}{b} = 2$ (d) $\frac{x}{a} + \frac{y}{b} = 2$

103. Point $R(h, k)$ divides a line segment between the axes in the ratio 1 : 2. Then, the equation of the line is

(a) $2kx + hy = 3hk$ (b) $kx + 2hy = 3hk$
(c) $2hx + ky = 3hk$ (d) $hx + 2ky = 3hk$

104. The three points (3, 0), (– 2, – 2) and (8, 2) are

(a) collinear
(b) vertices of isosceles triangle
(c) vertices of right angled triangle
(d) vertices of scalene triangle

Level II

105. A line perpendicular to the line segment joining the points (1, 0) and (2, 3) divides it in the ratio 1 : n. Then, the equation of line is

(a) $x(n + 1) + 3(n + 1)y = n + 11$
(b) $x(n + 1) + (3n + 1) = n + 11$
(c) $x(n + 1) - 3(n + 1)y = n + 11$
(d) $x(n + 1) - (3n + 1)y = n + 11$

106. L_1 is the equation of line through the point (0, 2) making an angle $\frac{2\pi}{3}$ with the positive X-axis.

L_2 is the equation of line parallel to L_1 and crossing the Y-axis at a distance of 2 units below the origin.

I. $L_1 \equiv \sqrt{3}x + y - 2 = 0$ and $L_2 = \sqrt{3}x + y + 2 = 0$
II. $L_1 \equiv \sqrt{3}x + y + 2 = 0$ and $L_2 = \sqrt{3}x + y - 2 = 0$

(a) Both I and II are true (b) Only I is true
(c) Only II is true (d) Both I and II are false

107. The owner of a milk store finds that he can sell 980 L of milk each week at ₹ 14/L and 1220 L of milk each week at ₹ 16/L. Assuming a linear relationship between selling price and demand. Then, he can sell ...X... weekly at ₹ 17/L. Here , X stands for

(a) 1200 L (b) 1240 L (c) 1340 L (d) 1300 L

108. If the lines $2x + y - 3 = 0$, $5x + Ky - 3 = 0$ and $3x - y - 2 = 0$ are concurrent, then the value of K is

(a) – 1 (b) – 2
(c) – 3 (d) – 4

109. The distance of the line $4x - y = 0$ from the point $P(4, 1)$ measured along the line making an angle of 135° with the positive X-axis is

(a) $\sqrt{2}$ units (b) $2\sqrt{2}$ units
(c) $3\sqrt{2}$ units (d) $4\sqrt{2}$ units

110. If the line $(k - 3)x - (4 - k^2)y + k^2 - 7k + 6 = 0$ is

I. Parallel to the X-axis, then k will be 3.
II. Parallel to the Y-axis, then k will be ± 2.
III. Passing through the origin, then k will be 1, 6.

(a) All are true (b) I and II are true
(c) II and III are true (d) I and III are true

111. The equation of a line, which cut off intercepts on the axes whose sum and product are 1 and – 6, respectively, is

(a) $2x - 3y - 6 = 0$ (b) $3x - 2y + 6 = 0$
(c) Both (a) and (b) (d) Neither (a) nor (b)

112. The equation of line parallel to Y-axis and drawn through the point of intersection of lines $x - 7y + 5 = 0$ and $3x + y = 0$ is

(a) $22x + 5 = 0$ (b) $11x + 5 = 0$
(c) $13x + 5 = 0$ (d) $55x + 11 = 0$

113. The equation of a line which passes through the point (2, 3) and makes an angle of 30° with the positive direction of X-axis, is [NCERT Exemplar]

(a) $x - \sqrt{3}y + (3\sqrt{3} - 2) = 0$
(b) $x + \sqrt{3}y + (3\sqrt{3} + 2) = 0$
(c) $x + \sqrt{3}y + (3\sqrt{3} - 2) = 0$
(d) $x - \sqrt{3}y + (3\sqrt{3} + 2) = 0$

114. Consider the following statements and choose the correct option given below.

I. The variable line $x \cos\alpha + y \sin\alpha = p$ (where p is a constant) intersects X-axis and Y-axis respectively at $\left(\frac{p}{\cos\alpha}, 0\right)$ and $\left(0, \frac{p}{\sin\alpha}\right)$.

II. The locus of the mid-point of distance between the axes of variable line $x \cos\alpha + y \sin\alpha = p$ is $\frac{1}{x^2} + \frac{1}{y^2} = \frac{4}{p^2}$, where p is a constant. [NCERT Exemplar]

(a) Both I and II are true (b) Only I is true
(c) Only II is true (d) Both I and II are false

115. The area of the triangle formed by the lines $y - x = 0$, $x + y = 0$ and $x - k = 0$ is

(a) k sq units (b) k^2 sq units
(c) k^3 sq units (d) k^4 sq units

116. The value of p, so that the three lines $3x + y - 2 = 0$, $px + 2y - 3 = 0$ and $2x - y - 3 = 0$ may intersect at one point, is

(a) 2 (b) 3
(c) 4 (d) 5

117. If the line joining two points A (2, 0) and B (3, 1) is rotated about A in anti-clockwise direction through an angle of 15°, then match the terms of Column I with terms of Column II and choose the correct option from the codes given below. [NCERT Exemplar]

Column I	Column II
A. Slope of line *AB* is	1. $y + \sqrt{3}x - 2\sqrt{3} = 0$
B. Slope of new line after rotation is	2. $y - \sqrt{3}x + 2\sqrt{3} = 0$
C. Equation of new line is	3. $\sqrt{3}$
	4. 1

Codes

	A	B	C
(a)	4	2	3
(b)	3	2	1
(c)	4	3	2
(d)	2	3	1

118. If the slope of a line passing through the point A (3, 2) is $\frac{3}{4}$, then which of the following are the coordinates of the point lie on the line which are 5 units away from the point A? [NCERT Exemplar]

(a) (– 1, – 1) (b) (7, 5)
(c) Both (a) and (b) (d) None of these

119. If three lines whose equations are $y = m_1x + c_1$, $y = m_2x + c_2$ and $y = m_3x + c_3$, are concurrent, then [HOTS]

I. $m_1(c_2 - c_3) + m_2(c_3 - c_1) + m_3(c_1 - c_2) = 0$

II. $m_1(c_1 - c_2) + m_2(c_2 - c_3) + m_3(c_3 - c_1) = 0$

(a) Both I and II are true (b) Only I is true
(c) Only II is true (d) Both I and II are false

General Equation of a Line and Distance of a Point from a Line

Level I

120. The general equation of first degree in two variables is $Ax + By + C = 0$, where

(a) A, B and C are real constants such that $A = B = C = 0$
(b) A, B and C are complex constants such that $A = B = C = 0$
(c) A, B and C are real constants such that A and B are not zero simultaneously
(d) None of the above

121. The graph of the equation $Ax + By + C = 0$ is always a

(a) parabola
(b) straight line
(c) circle
(d) ellipse

122. An equation of the form $Ax + By + C = 0$, where A and B are not zero simultaneously, is called

I. General equation of a line.

II. Normal equation of a line.

(a) Both I and II are true (b) Only I is true
(c) Only II is true (d) Both I and II are false

123. The general equation of line is $Ax + By + C = 0$, where A, B and C are real constants such that A and B are not zero simultaneously. Then, match the terms of Column I with terms of Column II and choose the correct option from the codes given below.

Column I	Column II
A. If $B \neq 0$, then slope of equation is	1. $-\frac{C}{A}$
B. If $B \neq 0$, then y-intercept is	2. $-\frac{A}{B}$
C. If $B = 0$, then slope of equation is	3. $-\frac{C}{B}$
D. If $B = 0$, then x-intercept is	4. Undefined

Codes

	A	B	C	D
(a)	2	4	3	1
(b)	3	2	4	1
(c)	2	3	1	4
(d)	2	3	4	1

124. The general equation of line is $Ax + By + C = 0$, where A, B and C are real constants such that A and B are not zero simultaneously. Then, match the terms of Column I with terms of Column II and choose the correct option from the codes given below.

Column I	Column II
A. If $C \neq 0$, then x-intercept is	1. Zero
B. If $C \neq 0$, then y-intercept is	2. Origin
C. If $C = 0$, then line passes through	3. $-\frac{C}{B}$
D. If $C = 0$, then intercepts on the axes are	4. $-\frac{C}{A}$

Codes

	A	B	C	D
(a)	4	3	2	1
(b)	3	4	1	2
(c)	4	3	1	2
(d)	3	4	2	1

125. Let $x \cos \omega + y \sin \omega = p$ be the normal form of the line represented by the equation $Ax + By + C = 0$. Then, [HOTS]

I. $\cos \omega = \dfrac{\pm B}{\sqrt{A^2 + B^2}}$

II. $\sin \omega = \dfrac{\pm A}{\sqrt{A^2 + B^2}}$

III. $p = \dfrac{\pm C}{\sqrt{A^2 + B^2}}$

(a) I and II are true (b) II and III are true
(c) I and II are false (d) I and III are false

126. Consider the following statements and choose the correct option given below.

I. Slope of line $3x - 4y + 10 = 0$ is $\frac{3}{4}$.

II. x-intercept and y-intercept of $3x - 4y + 10 = 0$ respectively are $\frac{-10}{3}$ and $\frac{5}{2}$.

(a) Both I and II are true (b) Only I is true
(c) Only II is true (d) Both I and II are false

127. If the normal form of the equation $\sqrt{3}x + y - 8 = 0$ is $x \cos \omega + y \sin \omega = p$, then p and ω respectively are

(a) 4, 45° (b) 4, 30°
(c) 3, 45° (d) 3, 30°

128. The angle between the lines $y - \sqrt{3}x - 5 = 0$ and $\sqrt{3}y - x + 6 = 0$ will not be

(a) 30° (b) 150°
(c) 45° (d) None of these

129. The two lines $a_1x + b_1y + c_1 = 0$ and $a_2x + b_2y + c_2 = 0$, where $b_1, b_2 \neq 0$ are

I. Parallel, if $\frac{a_1}{a_2} = \frac{b_1}{b_2}$.

II. Perpendicular, if $a_1a_2 - b_1b_2 = 0$.

(a) Both I and II are true (b) Only I is true
(c) Only II is true (d) Both I and II are false

130. The equation of a line perpendicular to the line $x - 2y + 3 = 0$ and passing through the point $(1, -2)$ is

(a) $y = 2x$ (b) $x = 2y$
(c) $x = -2y$ (d) $y = -2x$

Distance of a Point from a Line

131. Consider the following statements and choose the correct option given below.

I. The distance of a point from a line is the length of any line drawn from the point to the line.

II. The distance of a point from a line is the length of the perpendicular drawn from the point to the line.

(a) Both I and II are true (b) Only I is true
(c) Only II is true (d) Both I and II are false

132. The perpendicular distance (d) of the line $Ax + By + C = 0$ from a point (x_1, y_1) is given by $d = \dfrac{|Ax_1 + By_1 + C|}{m}$, where m stands for [HOTS]

(a) $A^2 + B^2$

(b) $\sqrt{A^2 + B^2}$

(c) $A + B$

(d) $\sqrt{A + B}$

133. The distance d between two parallel lines $y = mx + c_1$ and $y = mx + c_2$ is given by $d = \frac{z}{\sqrt{1+m^2}}$, where z stands for

(a) $c_1 - c_2$ (b) $c_2 - c_1$ (c) $|c_1 + c_2|$ (d) $|c_1 - c_2|$

134. If lines are given in general form, *i.e.*, $Ax + By + C_1 = 0$ and $Ax + By + C_2 = 0$, then the distance d between them is given by $d = \frac{m}{n}$, where m and n respectively are

(a) $|C_1 - C_2|, \sqrt{A^2 + B^2}$ (b) $\sqrt{A^2 + B^2}, C_1 - C_2$
(c) $A^2 + B^2, C_2 - C_1$ (d) $C_2 - C_1, A^2 + B^2$

135. The distance of the point $(3, -5)$ from the line $3x - 4y - 26 = 0$ is

(a) $\frac{3}{7}$ (b) $\frac{2}{5}$ (c) $\frac{7}{5}$ (d) $\frac{3}{5}$

136. The distance between the parallel lines $3x - 4y + 7 = 0$ and $3x - 4y + 5 = 0$, is

(a) $\frac{3}{7}$ (b) $\frac{7}{5}$
(c) $\frac{2}{5}$ (d) $\frac{3}{5}$

137. Consider the following statements and choose the correct option given below.

I. The slope of the line $x + 7y = 0$ is $\frac{1}{7}$ and y-intercept is 0.
II. The slope of the line $6x + 3y - 5 = 0$ is 2 and y-intercept is $\frac{5}{3}$.
III. The slope of the line $y = 0$ is 0.

(a) All are true (b) Only I is true
(c) Only II is true (d) Only III is true

138. The x-intercepts of equations $3x + 2y - 12 = 0$, $4x - 3y = 6$ and $3y + 2 = 0$ respectively are d_1, d_2 and d_3, where

(a) $d_1 = 4, d_2 = \frac{3}{2}$ and d_3 does not exist
(b) $d_1 = 4, d_2 = 3$, and $d_3 = \frac{2}{3}$
(c) $d_1 = 4, d_2 = \frac{3}{2}$ and $d_3 = \frac{-2}{3}$
(d) $d_1 = 4, d_2 = -\frac{3}{2}$ and d_3 does not exist

139. If the equation of line is $x - \sqrt{3}y + 8 = 0$, then

I. The normal form of the equation is $x \cos \alpha + y \sin \alpha = p$, where $\alpha = 120°$ and $p = 4$.
II. The perpendicular distance of the line from the origin is 8.
III. The angle between perpendicular drawn from origin to the line and the positive X-axis is 120°.

(a) All are true (b) I and II are true
(c) I and III are true (d) II and III are true

140. If the equation of line is $y - 2 = 0$, then

I. The normal form of the equation is $x \cos \alpha + y \sin \alpha = p$, where $\alpha = 0°$ and $p = 4$.
II. The perpendicular distance of the line from the origin is 2.
III. The angle between perpendicular and the positive X-axis is 90°.

(a) All are true (b) I and II are true
(c) I and III are true (d) II and III are true

141. If the equation of line is $x - y = 4$, then

I. The normal form of some equation is $x \cos \alpha + y \sin \alpha = p$, where $\alpha = 315°$ and $p = 2\sqrt{2}$.
II. The perpendicular distance of line from the origin is $3\sqrt{2}$.
III. The angle between perpendicular and the positive X-axis is 45°.

(a) All are true (b) Only I is true
(c) I and III are true (d) II and III are true

142. The distance of the point $(-1, 1)$ from the line $12(x + 6) = 5(y - 2)$, is

(a) 3 (b) 4
(c) 5 (d) 6

143. Which of the following is not the point on X-axis, whose distance from the line $\frac{x}{3} + \frac{y}{4} = 1$ is 4 units?

(a) (8, 0) (b) (2, 0)
(c) $(-2, 0)$ (d) None of these

144. The distance between two parallel lines $15x + 8y - 34 = 0$ and $15x + 8y + 31 = 0$ is

(a) $\frac{65}{17}$ (b) $\frac{3}{17}$
(c) $\frac{2}{17}$ (d) $\frac{60}{17}$

145. If $lx + ly + p = 0$ and $lx + ly - r = 0$ are two parallel lines, then distance between them is equal to $\left|\frac{m}{n}\right|$, where m and n respectively are

(a) $p - r, \sqrt{2}l$ (b) $r - p, \sqrt{2}l$
(c) $p + r, \sqrt{2}l$ (d) $p + r, \sqrt{2l}$

146. The equation of the line parallel to the line $3x - 4y + 2 = 0$ and passing through the point $(-2, 3)$ is

(a) $3x - 4y + 18 = 0$ (b) $3x + 4y - 6 = 0$
(c) $4x + 3y - 6 = 0$ (d) $3x + 4y - 18 = 0$

147. The equation of the line perpendicular to the line $x - 7y + 5 = 0$ and having x-intercept 3, is

(a) $x - 7y - 3 = 0$
(b) $7x + y - 21 = 0$
(c) $x + 7y + 3 = 0$
(d) $7x - y - 21 = 0$

148. Which of the following is not the angle between the lines $\sqrt{3}x + y = 1$ and $x + \sqrt{3}y = 1$?

(a) 30° (b) 60°
(c) 150° (d) None of these

149. The line through the points $(h, 3)$ and $(4, 1)$ intersects the line $7x - 9y - 19 = 0$ at right angle. Then, the value of h is

(a) $\frac{11}{7}$ (b) $\frac{22}{7}$
(c) $\frac{22}{9}$ (d) $\frac{11}{9}$

150. The line through the point (x_1, y_1) and parallel to the line $Ax + By + C = 0$ is

(a) $A(x - x_1) - B(y - y_1) = 0$
(b) $A(y - y_1) - B(x - x_1) = 0$
(c) $A(y - y_1) + B(x - x_1) = 0$
(d) $A(x - x_1) + B(y - y_1) = 0$

151. Two lines passing through the point (2, 3) intersects each other at an angle of 60°. If slope of one line is 2, then the equation of other line is [HOTS]

I. $(2 + \sqrt{3})x + (2\sqrt{3} - 1)y - 8\sqrt{3} - 1 = 0$
II. $(2 - \sqrt{3})x - (1 + 2\sqrt{3})y + 8\sqrt{3} - 1 = 0$

(a) Both I and II are possible
(b) Only I is possible
(c) Only II is possible
(d) Neither I nor II is possible

152. The equation of the right bisector of the line segment joining the points (3, 4) and (– 1, 2) is

(a) $2x - y + 1 = 0$
(b) $2x + y - 5 = 0$
(c) $2x - y - 1 = 0$
(d) $2x - y - 5 = 0$

153. The perpendicular from the origin to the line $y = mx + c$ meets at the point $(-1, 2)$. Then, the value of m and c respectively are

(a) $\frac{5}{2}, \frac{1}{2}$ (b) $-\frac{5}{2}, \frac{1}{2}$
(c) $-\frac{1}{2}, \frac{5}{2}$ (d) $\frac{1}{2}, \frac{5}{2}$

154. If p is the length of perpendicular from origin to the line whose intercepts on the axes are a and b, then $p^2(a^2 + b^2) = m^2$, where m stands for

(a) ab (b) a^2b^2 (c) a^3b^3 (d) a^4b^4

Level II

155. The coordinates of the foot of perpendicular from the point $(-1, 3)$ to the line $3x - 4y - 16 = 0$ is

(a) $\left(\frac{68}{25}, \frac{-49}{25}\right)$ (b) $\left(\frac{1}{10}, \frac{-157}{40}\right)$
(c) $\left(\frac{1}{15}, \frac{-79}{60}\right)$ (d) None of these

156. If p and q are lengths of perpendiculars from origin to the lines $x\cos\theta - y\sin\theta = k\cos 2\theta$ and $x\sec\theta + y\operatorname{cosec}\theta = k$ respectively, then

(a) $4q^2 + p^2 = k^2$
(b) $p^2 + 4q^2 = k^2$
(c) $p^2 + q^2 = k^2$
(d) $4p^2 + 4q^2 = k^2$

157. In the ΔABC with vertices $A(2, 3)$, $B(4, -1)$ and $C(1, 2)$,

I. The equation of altitude from the vertex A is $x - y + 1 = 0$.
II. The length of altitude from the vertex A is $\sqrt{2}$. [HOTS]

(a) Both I and II are true
(b) Only I is true
(c) Only II is true
(d) Both I and II are false

158. Assuming that straight lines work as the plane mirror for a point, then the image of the point (1, 2) in the line $x - 3y + 4 = 0$ is

(a) $\left(\frac{6}{3}, \frac{2}{5}\right)$ (b) $\left(\frac{2}{5}, \frac{6}{5}\right)$
(c) $\left(\frac{6}{5}, \frac{7}{5}\right)$ (d) $\left(\frac{7}{5}, \frac{6}{5}\right)$

159. The path of a moving point such that its distances from two lines $3x - 2y = 5$ and $3x + 2y = 5$ are equal, is a/an

(a) circle (b) straight line
(c) ellipse (d) parabola

160. If the equation $x\cos\theta + y\sin\theta = p$ is the normal form of the line $\sqrt{3}x + y + 2 = 0$, then the values of θ and p respectively are

(a) 30°, 1 (b) 210°, 1
(c) 30°, 4 (d) 210°, 4

161. Which of the following is not a point on the Y-axis, whose distance from the line $\frac{x}{3} + \frac{y}{4} = 1$ is 4 units?

(a) $\left(0, -\frac{8}{3}\right)$ (b) $\left(0, \frac{32}{3}\right)$
(c) $\left(0, \frac{8}{3}\right)$ (d) None of these

162. The distance between the lines $3x + 4y = 9$ and $6x + 8y = 15$ is [NCERT Exemplar]

(a) $\frac{3}{10}$ (b) $\frac{2}{25}$ (c) $\frac{7}{10}$ (d) $\frac{3}{25}$

163. The equation of straight line passing through $(2, 1)$ and perpendicular to the line $x + y + 7 = 0$ is [NCERT Exemplar]

(a) $x - y - 1 = 0$ (b) $x + y + 3 = 0$
(c) $x + y - 3 = 0$ (d) $x - y + 1 = 0$

164. The perpendicular distance from the origin to the line joining the points $(\cos\theta, \sin\theta)$ and $(\cos\phi, \sin\phi)$ is [HOTS]

(a) $\cos(\phi + \theta)$ (b) $\cos(\theta - \phi)$ (c) $\cos\left(\frac{\theta + \phi}{2}\right)$ (d) $\cos\left(\frac{\theta - \phi}{2}\right)$

165. The equation of a line drawn perpendicular to the line $\frac{x}{4} + \frac{y}{6} = 1$ through the point, where it meets the Y-axis, is

(a) $2x - 3y + 18 = 0$ (b) $2x + 3y - 18 = 0$
(c) $2x - 3y - 18 = 0$ (d) None of these

166. The equation of a line is $ax + by = 8$, if length of the perpendicular from the origin to the line is 4 units and the inclination of the perpendicular with the positive direction of X-axis is 30°, then

(a) $a = 1, \quad b = \sqrt{3}$ (b) $a = 1, \quad b = 1$
(c) $a = 1, \quad b = \sqrt{3}$ (d) $a = \sqrt{3}, \ b = 1$

167. The equation of the straight line passing through the point of intersection of the lines $5x - 6y - 1 = 0$ and $3x + 2y + 5 = 0$ and perpendicular to the line $3x - 5y + 11 = 0$ is [NCERT Exemplar]

(a) $3x - 5y - 2 = 0$ (b) $3x + 5y + 8 = 0$
(c) $5x + 3y + 8 = 0$ (d) $5x - 3y + 2 = 0$

168. If one diagonal of a square is along the line $8x - 15y = 0$ and one of its vertex is at $(1, 2)$, then the equation of sides of the square passing through the vertex $(1, 2)$ are [NCERT Exemplar]

(a) $23x - 7y - 9 = 0$ and $7x + 23y - 53 = 0$
(b) $7x - 23y + 39 = 0$ and $5x + 23y - 51 = 0$
(c) $5x - 23y + 41 = 0$ and $6x + 7y + 10 = 0$
(d) None of the above

169. The two lines $ax + by = c$ and $a'x + b'y = c'$ are perpendicular, if [NCERT Exemplar]

(a) $aa' + bb' = 0$
(b) $ab' = ba'$
(c) $ab + a'b' = 0$
(d) $ab' + ba' = 0$

170. The coordinates of the foot of the perpendicular from the point $(2, 3)$ on the line $x + y - 11 = 0$ are [NCERT Exemplar]

(a) $(-6, 5)$
(b) $(5, 6)$
(c) $(-5, 6)$
(d) $(6, 5)$

171. The reflection of the point $(4, -13)$ about the line $5x + y + 6 = 0$ is [NCERT Exemplar]

(a) $(-1, -14)$
(b) $(3, 4)$
(c) $(0, 0)$
(d) $(1, 2)$

172. A point moves such that its distance from the point $(4, 0)$ is half that of its distance from the line $x = 16$. The locus of the point is [NCERT Exemplar]

(a) $3x^2 + 4y^2 = 192$
(b) $4x^2 + 3y^2 = 192$
(c) $x^2 + y^2 = 192$
(d) None of the above

Equation of Family of Lines Passing Through the Point of Intersection of Two Lines and Shifting of Origin

Level I

173. Let the two intersecting lines l_1 and l_2 are given by $A_1x + B_1y + C_1 = 0$ and $A_2x + B_2y + C_2 = 0$. Then,

I. $A_1x + B_1y + C_1 + K(A_2x + B_2y + C_2) = 0$
(K is an arbitrary constant) represents a family of lines.

II. $A_1x + B_1y + C_1 + K(A_2x + B_2y + C_2) = 0$
(K is a fixed constant) represents a family of lines.

(a) Both I and II are true (b) Only I is true
(c) Only II is true (d) Neither I nor II is true

174. The equation of line parallel to the Y-axis and drawn through the point of intersection of $x - 7y + 5 = 0$ and $3x + y - 7 = 0$ is

(a) $x = 2$ (b) $x = 3$
(c) $x = 1$ (d) $x = 4$

175. Consider the following statements and choose the correct option given below.

I. The equation of the line through the intersection of lines $3x + 4y = 7$ and $x - y + 2 = 0$ and whose slope is 5 will be $35x - 7y + 18 = 0$.

II. The equation of the line through the intersection of lines $x + 2y - 3 = 0$ and $4x - y + 7 = 0$ and which is parallel to $5x + 4y - 20 = 0$, will be $15x + 12y - 7 = 0$.

(a) Both I and II are true (b) Only I is true
(c) Only II is true (d) Both I and II are false

176. Consider the following statements and choose the correct option given below.

I. The equation of the line through the intersection of the lines $2x + 3y - 4 = 0$ and $x - 5y = 7$ that has its x-intercept equal to -4, is $10x + 93y - 40 = 0$.

II. The equation of the line through the intersection of $5x - 3y = 1$ and $2x + 3y - 23 = 0$ and perpendicular to the line is $5x - 3y - 1 = 0$ is $63x + 105y + 781 = 0$.

(a) Both I and II are true (b) Only I is true
(c) Only II is true (d) Both I and II are false

Shifting of Origin

177. If the origin is shifted to $(1, 2)$ by a translation, then the new coordinates of point $(3, -4)$ are

(a) $(-2, 6)$ (b) $(2, -6)$
(c) $(4, -2)$ (d) None of these

178. The transformed equation of the straight line $2x - 3y + 5 = 0$, when the origin is shifted to the point $(3, -1)$ after translation of axes, is

(a) $2x - 3y - 14 = 0$ (b) $2x + 3y - 14 = 0$
(c) $2x - 3y + 14 = 0$ (d) $2x + 3y + 14 = 0$

179. If the origin is shifted to the point $(-3, -2)$ by translation of axes, then match the old coordinates given in Column I with the new coordinates given in Column II and choose the correct option from the codes given below.

	Column I		Column II
A.	(1, 1)	1.	(8, 2)
B.	(0, 1)	2.	(2, 0)
C.	(5, 0)	3.	(4, 3)
D.	(– 1, – 2)	4.	(3, 3)
E.	(3, – 5)	5.	(6, – 3)

Codes

	A	B	C	D	E		A	B	C	D	E
(a)	3	4	1	2	5	(b)	2	4	1	3	5
(c)	4	3	1	2	5	(d)	3	4	1	5	2

180. If the origin is shifted to the point $(1, 1)$, then

I. The equation $x^2 + xy - 3y^2 - y + 2 = 0$ transformed to $x^2 - 3y^2 + xy + 3x - 6y + 1 = 0$.

II. The equation $xy - y^2 - x + y = 0$ transformed to $xy + y^2 = 0$.

III. The equation $xy - x - y + 1 = 0$ transformed to $xy = 0$.

(a) All are true (b) I and II are true
(c) Only III is true (d) II and III are true

Level II

181. If the origin be shifted to the point (h, k), then the point $(-1, 8)$ become $(-7, 3)$, where h and k respectively are

(a) 6, 5 (b) – 5, – 6 (c) – 6, – 5 (d) – 5, 6

182. The equation of line passing through the point of intersection of the lines $x + 5y + 7 = 0$ and $3x + 2y - 5 = 0$

I. Parallel to the line $7x + 2y - 5 = 0$ is $7x + 2y - 17 = 0$.

II. Perpendicular to the line $7x + 2y - 5 = 0$ is $2x - 7y - 20 = 0$.

(a) Both I and II are true (b) Only I is true
(c) Only II is true (d) Both I and II are false

183. The equation of a curve referred to the new axes, (axes retaining their direction) and origin is $(4, 5)$ is $x^2 + y^2 = 36$.

I. The equation referred to the original axes is $x^2 + y^2 + 8x + 10y + 5 = 0$.

II. The equation referred to the original axes is $x^2 + y^2 - 8x - 10y + 5 = 0$.

(a) Both I and II are true (b) Only I is true
(c) Only II is true (d) Both I and II are false

184. If the origin be shifted to the point (h, k), so that the equation $y^2 + 4y + 8x - 2 = 0$ will not contain term in y and the constant. Then, h and k respectively are

(a) $\frac{3}{4}, 2$ (b) $-\frac{3}{4}, 2$ (c) $-\frac{3}{4}, -2$ (d) $\frac{3}{4}, -2$

185. A variable straight line through the point of intersection of the lines $\frac{x}{a} + \frac{y}{b} = 1$ and $\frac{x}{b} + \frac{y}{a} = 1$ meets the coordinate axes in A and B.

I. The locus of mid-point of AB is the curve $2xy(a + b) = ab(x + y)$.

II. The locus of mid-point of AB is the curve $2xy(x + y) = ab(a + b)$.

(a) Both I and II are true (b) Only I is true
(c) Only II is true (d) Both I and II are false

186. The equation of straight line passing through the point (2, 1) and through the point of intersection of lines $x + 2y - 3 = 0$ and $2x - 3y - 4 = 0$ is $5x + 3y + m = 0$, where m is

(a) 12 (b) 13 (c) – 12 (d) – 13

187. The equation of straight line passing through the intersection of the straight lines $3x - 4y + 1 = 0$ and $5x + y - 1 = 0$ and cuts off equal intercepts on the axes is

(a) $11x + 11y = 23$ (b) $23x + 23y = 11$
(c) $22x + 22y = 23$ (d) $23x + 23y = 15$

188. If t_1 and t_2 are roots of the equation $t^2 + \lambda t + 1 = 0$, where λ is an arbitrary constant, then the line joining the points $(at_1^2, 2at_1)$ and $(at_2^2, 2at_2)$ always passes through a fixed point (m, n), where m and n respectively are

(a) a, 0 (b) 0, a
(c) $-a$, 0 (d) 0, $-a$

Miscellaneous Exercise

189. The equation of the lines through the point (3, 2) which makes an angle of 45° with the line $x - 2y = 3$, is

I. $x + 3y - 9 = 0$ II. $3x - y - 7 = 0$

(a) Both I and II are possible
(b) Only I is possible
(c) Only II is possible
(d) Neither I nor II is possible

190. A straight line moves so that the sum of the reciprocals of its intercepts made on axes is constant (K). Then, the line passes through ...A... point. Here, A stands for [NCERT Exemplar]

(a) variable
(b) origin
(c) fixed
(d) None of the above

191. The equation of the line passing through the point of intersection of the lines $4x + 7y - 3 = 0$ and $2x - 3y + 1 = 0$ that has equal intercepts on the axes, is

(a) $6x + 6y = 13$ (b) $13x + 13y = 6$
(c) Both (a) and (b) (d) None of these

192. If p is the length of perpendicular from the origin on the line $\frac{x}{a} + \frac{y}{b} = 1$ and a^2, p^2, b^2 are in AP, then [NCERT Exemplar; HOTS]

(a) $a^2 + b^2 = 0$
(b) $a^3 + b^3 = 0$
(c) $a^4 + b^4 = 0$
(d) $a^5 + b^5 = 0$

193. The equation of the line passing through the origin and making an angle θ with the line $y = mx + c$ is

(a) $\frac{y}{x} = \frac{m + \tan\theta}{1 + m\tan\theta}$ (b) $\frac{y}{x} = \frac{m - \tan\theta}{1 - m\tan\theta}$
(c) $\frac{y}{x} = \frac{m \pm \tan\theta}{1 \pm m\tan\theta}$ (d) $\frac{y}{x} = \frac{m \pm \tan\theta}{1 \mp m\tan\theta}$

194. A line is drawn through angle θ anti-clockwise from positive side of X-axis. If line passes through point (1, 2) such that its point of intersection with the line $x + y = 4$ is at a distance of $\frac{\sqrt{6}}{3}$ from the given point. Then, the value of θ is equal to [NCERT Exemplar; HOTS]

(a) 30° (b) 65°
(c) 45° (d) 75°

195. The line joining (– 1, 1) and (5, 7) is divided by the line $x + y = 4$ in the ratio $K : 1$, where K stands for

(a) 2 (b) 1 (c) $\frac{1}{3}$ (d) $\frac{1}{2}$

196. A variable line passes through a fixed point P. The algebraic sum of the perpendiculars drawn from the points (2, 0), (0, 2) and (1, 1) on the line is zero. Then, the coordinates of P are [NCERT Exemplar]

(a) (2, 3) (b) (3, 5)
(c) (1, 1) (d) (0, 0)

197. The distance of the line $4x + 7y + 5 = 0$ from the point (1, 2) along the line $2x - y = 0$ is $\frac{23a}{18}$ sq units, where

(a) $a = \sqrt{3}$ (b) $a = \sqrt{5}$
(c) $a = \sqrt{7}$ (d) $a = \sqrt{2}$

198. In an isosceles right angled triangle whose hypotenuse is given by $3x + 4y = 4$ and the opposite vertex of hypotenuse is (2, 2), then the equation of one of the sides other than hypotenuse is [NCERT Exemplar]

I. $x - 7y + 12 = 0$ II. $7x + y - 16 = 0$

(a) Both I and II are true
(b) Only I is true
(c) Only II is true
(d) Neither I nor II is true

199. The straight line must be drawn ...A... to ...B... through the point $(-1, 2)$, so that its point of intersection with the line $x + y = 4$ may be at a distance of 3 units from the point. Here, A and B respectively are

(a) parallel, X-axis (b) perpendicular, X-axis
(c) parallel, Y-axis (d) None of these

200. If the equation of the base of an equilateral triangle is $x + y = 2$ and the vertex is $(2, -1)$, then the length of the side of the triangle is [NCERT Exemplar]

(a) $\sqrt{\frac{3}{2}}$ unit (b) $\sqrt{\frac{2}{3}}$ unit
(c) $\sqrt{2}$ unit (d) $\sqrt{3}$ unit

201. The hypotenuse of a right angled triangle has its ends at the points $(1, 3)$ and $(-4, 1)$. The legs (perpendicular sides) of the triangle are

(a) $x = 1, y = -1$ (b) $x = -1, y = 1$
(c) $x = 1, y = 1$ (d) $x = -1, y = -1$

202. The tangent of an angle between the lines $\frac{x}{a} + \frac{y}{b} = 1$ and $\frac{x}{a} - \frac{y}{b} = 1$ is $\frac{2m}{n}$, where m and n respectively are

[NCERT Exemplar; HOTS]

(a) $a^2 - b^2, ab$ (b) $ab, a^2 - b^2$ (c) $a^2 + b^2, ab$ (d) $ab, a^2 + b^2$

203. Consider the lines $L_1 \equiv x + 3y - 7 = 0$ and $L_2 \equiv 3x - y - 1 = 0$. [HOTS]

(a) L_1 and L_2 are perpendicular to each other
(b) The point of intersection of L_1 and L_2 is $(1, 2)$
(c) The image of the point $(3, 8)$ w.r.t. the line L_1 assuming L_1 to be a plane mirror is $(-1, -4)$
(d) All of the above

204. If $(\sin\theta, \cos\theta)$ and $(3, 2)$ lies on the same side of the line $x + y = 1$, then θ lies between

(a) $\left(0, \frac{\pi}{2}\right)$ (b) $(0, \pi)$
(c) $\left(\frac{\pi}{4}, \frac{\pi}{2}\right)$ (d) $\left(0, \frac{\pi}{4}\right)$

205. If the lines $y = 3x + 1$ and $2y = x + 3$ are equally inclined to the line $y = mx + 4$, then the value(s) of m is (are)

(a) $\frac{1 + 5\sqrt{2}}{-7}$ (b) $\frac{-1 - 5\sqrt{2}}{7}$
(c) $\frac{1 \pm 5\sqrt{2}}{7}$ (d) $\frac{5 \pm \sqrt{2}}{7}$

206. If sum of perpendicular distance of a variable point $P(x, y)$ from the line $x + y - 5 = 0$ and $3x - 2y + 7 = 0$ is always 10, then,

(a) P must move on a circle (b) P must move on a line
(c) Both (a) and (b) are true (d) Neither (a) nor (b) is true

207. The line L given by $\frac{x}{5} + \frac{y}{b} = 1$ passes through the point $(13, 32)$. The line K is parallel to L and has the equation $\frac{x}{c} + \frac{y}{3} = 1$. Then, the distance between L and K is [AIEEE 2011]

(a) $\frac{23}{\sqrt{15}}$ (b) $\sqrt{17}$ (c) $\frac{17}{\sqrt{15}}$ (d) $\frac{23}{\sqrt{17}}$

208. Statement I The equation of the line which is equidistant from parallel line $9x + 6y - 7 = 0$ and $3x + 2y + 6 = 0$ is $18x + 12y + 11 = 0$.

Statement II Any line parallel to $ax + by + c = 0$ is $ax + by + k = 0$ and distance between two lines is $\left|\frac{c - k}{\sqrt{a^2 + b^2}}\right|$.

(a) Statement I is true; Statement II is true; Statement II is a correct explanation for Statement I
(b) Statement I is true; Statement II is true; Statement II is not a correct explanation for Statement I
(c) Statement I is true; Statement II is false
(d) Statement I is false; Statement II is true

209. The point $P(a, b)$ lies on the straight line $3x + 2y = 13$ and the point $Q(b, a)$ lies on the straight line $4x - y = 5$, then equation of the line PQ is [DCE 2008]

(a) $x - 5 = 5$ (b) $x + y = 5$ (c) $x + y = -5$ (d) $x - y = -5$

210. If P_1 and P_2 are the lengths of the perpendiculars drawn from the points $(\sqrt{a^2 - b^2}, 0)$ and $(-\sqrt{a^2 - b^2}, 0)$ respectively, to the line $\frac{x}{a}\cos\theta + \frac{y}{b}\sin\theta = 1$ then,

(a) $P_1P_2 = b^3$ (b) $P_1P_2 = b^2$
(c) Both (a) and (b) are true (c) Neither (a) nor (b) is true

211. The equation of the line bisecting the join of $(3, -4)$ and $(5, 2)$ and having its intercepts on the X-axis and the Y-axis in the ratio $2 : 1$ is

(a) $x + y - 3 = 0$ (b) $2x - y = 9$
(c) $x + 2y = 2$ (d) $2x + y = 7$

212. A person standing at the junction (crossing) of two straight paths represented by the equations $2x - 3y + 4 = 0$ and $3x + 4y - 5 = 0$ wants to reach the path whose equation is $6x - 7y + 8 = 0$ in the least time. The equation of the path that he should follow, is [HOTS]

(a) $119x - 102y + 125 = 0$ (b) $119x + 102y - 125 = 0$
(c) $119x + 102y + 125 = 0$ (d) $119x - 102y - 125 = 0$

Answer *with* Explanations

1. *(c)* The two-dimensional coordinate geometry is the combination of Algebra and Geometry.

2. *(d)* A systematic study of Geometry by the use of Algebra was first carried out by celebrated French Philosopher and Mathematician Rene Descartes, in his book La Geometry published in 1637.

3. *(c)* The book La Geometry introduced the notion of the equation of a curve and related analytical methods into the study of Geometry.

4. *(a)* The resulting combination of analysis and geometry is referred now as analytical geometry.

5. *(d)* The study of coordinate geometry include coordinate axes, coordinate planes, plotting of points in a plane, distance between two points and section formulae.

6. *(a)* The point (6, –4) is at 6 units distance from the *Y*-axis measured along the positive *X*-axis and at 4 units distance from the *X*-axis measured along the negative *Y*-axis.

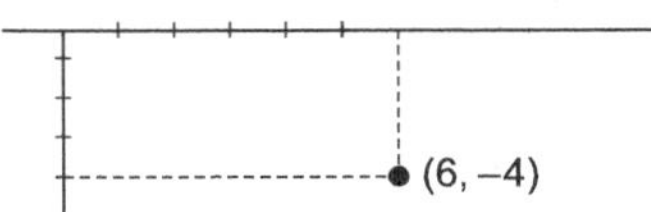

7. *(b)* The point (3, 0) is at 3 units distance from the *Y*-axis measured along the positive *X*-axis and has zero distance from the *X*-axis.

8. *(c)* Distance between the points $P(x_1, y_1)$ and $Q(x_2, y_2)$ is $\sqrt{(x_2 - x_1)^2 + (y_2 - y_1)^2}$.

9. *(d)* The distance between the points (6, – 4) and (3, 0)

$$= \sqrt{(6-3)^2 + (-4-0)^2}$$
$$= \sqrt{(3)^2 + (4)^2}$$
$$= \sqrt{25} = 5 \text{ units}$$

10. *(a)* The coordinates of a point dividing the line segment joining the points (x_1, y_1) and (x_2, y_2) internally in the ratio $m : n$ are $\left(\frac{mx_2 + nx_1}{m+n}, \frac{my_2 + ny_1}{m+n}\right)$.

11. *(b)* The coordinates of the point which divides the line segment joining *A* (1, – 3) and *B* (– 3, 9) internally in the ratio 1 : 3, are given by

$$x = \frac{1\cdot(-3) + 3\cdot 1}{1+3} = 0$$

and $$y = \frac{1.9 + 3\cdot(-3)}{1+3} = 0$$

12. *(a)* In particular, if $m = n$, the coordinates of the mid-point of the line segment joining the points (x_1, y_1) and (x_2, y_2) are $\left(\frac{x_1 + x_2}{2}, \frac{y_1 + y_2}{2}\right)$.

13. *(c)* Area of the triangle, whose vertices are (x_1, y_1), (x_2, y_2) and (x_3, y_3), is $\frac{1}{2}|x_1(y_2 - y_3) + x_2(y_3 - y_1) + x_3(y_1 - y_2)|$

14. *(b)* Area of the triangle, whose vertices are (4, 4), (3, –2) and (–3, 16)

$$= \frac{1}{2}|4(-2-16) + 3(16-4) + (-3)(4+2)|$$
$$= \frac{|-72 + 36 - 18|}{2}$$
$$= \frac{|-54|}{2} = \frac{54}{2} = 27$$

15. *(c)* If three points *A*, *B* and *C* are collinear, then area of the ΔABC is zero.

16. *(a)* The simplest geometric figure in coordinate geometry is straight line.

17. *(b)* For representing the line algebraically, slope is most essential.

18. *(a)* A line in a coordinate plane forms two angles with the *X*-axis which are supplementary.

19. *(d)* The angle (say) θ made by the line *l* with positive direction of *X*-axis and measured anti-clockwise is called the inclination of the line.

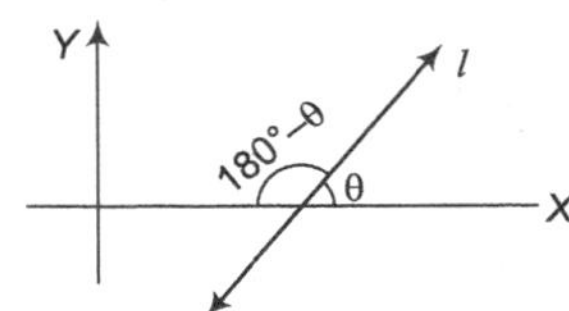

20. *(d)* The inclination of the line (θ) lies within 0° to 180° *i.e.*, $0° \le \theta \le 180°$.

21. *(a)* Lines parallel to *X*-axis or coinciding with *X*-axis, have inclination of 0°. The inclination of a vertical line parallel to or coinciding with *Y*-axis is 90°.

22. *(c)* If θ is the inclination of a line *l*, then tan θ is called the slope or gradient of the line *l*.

23. *(d)* The slope of a line whose inclination is 90°, is not defined.

24. *(b)* The slope of a line is denoted by *m*. Thus, $m = \tan\theta$, $\theta \ne 90°$.

25. *(b)* Slope of X-axis is zero and slope of Y-axis is not defined.

26. *(d)* A line is completely determined when we are given two points on it.

27. *(a)* If two points $P(x_1, y_1)$ and $Q(x_2, y_2)$ lies on non-vertical line 'l' whose inclination is θ, then $x_1 \neq x_2$ otherwise the line will become perpendicular to X- axis and its slope will not be defined, the inclination of a line may be acute or obtuse.

28. *(b)* ***Case* I** When angle θ is acute.

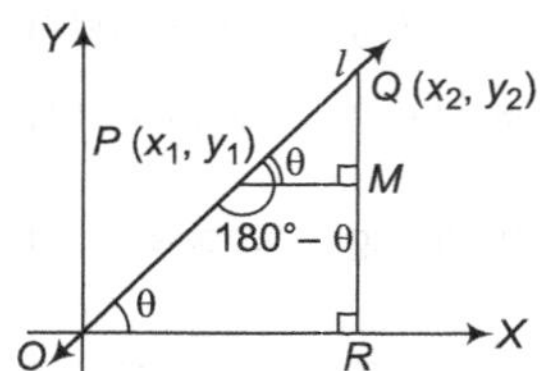

From figure, $\angle MPQ = \theta$

Therefore, slope of line $l = m = \tan\theta$ ...(i)

But in ΔMPQ, we have

$$\tan\theta = \frac{MQ}{MP} = \frac{y_2 - y_1}{x_2 - x_1} \quad \text{...(ii)}$$

From Eqs. (i) and (ii), we have

$$m = \frac{y_2 - y_1}{x_2 - x_1}$$

***Case* II** When angle θ is obtuse.

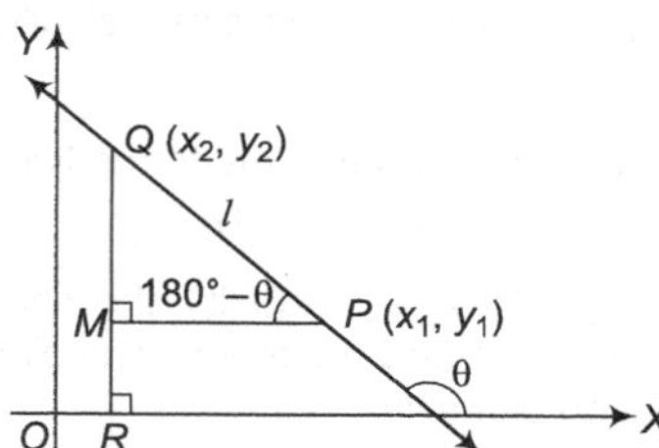

From figure, we have

$$\angle MPQ = 180° - \theta$$

Therefore, $\theta = 180° - \angle MPQ$

Now, slope of the line l is

$$m = \tan\theta$$
$$= \tan(180° - \angle MPQ)$$
$$= -\tan\angle MPQ$$
$$= -\frac{MQ}{MP}$$
$$= -\frac{y_2 - y_1}{x_1 - x_2}$$
$$= \frac{y_2 - y_1}{x_2 - x_1}$$

Consequently, we see that in both the cases, the slope m of the line l through the points (x_1, y_1) and (x_2, y_2) is given by

$$m = \frac{y_2 - y_1}{x_2 - x_1}.$$

29. *(a)* In a coordinate plane, suppose that non-vertical lines l_1 and l_2 have slopes m_1 and m_2, respectively. Let their inclinations be α and β, respectively.

If the line l_1 is parallel to l_2.

Then, their inclinations are equal *i.e.,*

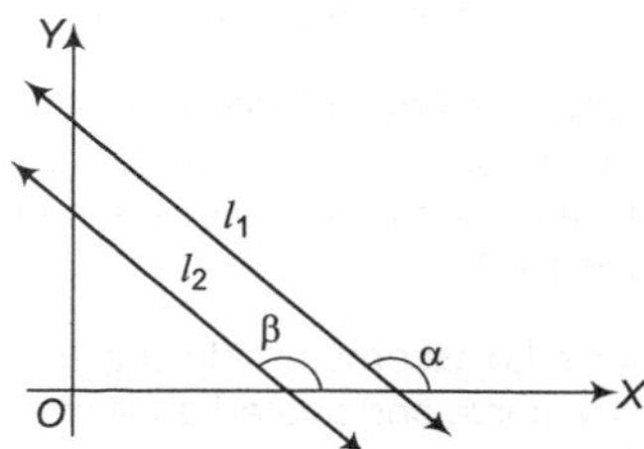

$$\alpha = \beta, \text{ and hence } \tan\alpha = \tan\beta$$

Therefore, $m_1 = m_2$, *i.e.,* their slopes are equal.

Conversely, if the slope of two lines l_1 and l_2 is same *i.e.,*

$$m_1 = m_2$$

Then, $\tan\alpha = \tan\beta$

By the property of tangent function (between 0° and 180°),

$$\alpha = \beta$$

Therefore, the lines are parallel.

Hence, two non-vertical lines l_1 and l_2 are parallel, if and only if their slopes are equal.

If the lines l_1 and l_2 are perpendicular .

Here, $m_1 = \tan\alpha$ ($\because m_1$ is the slope of l_1)

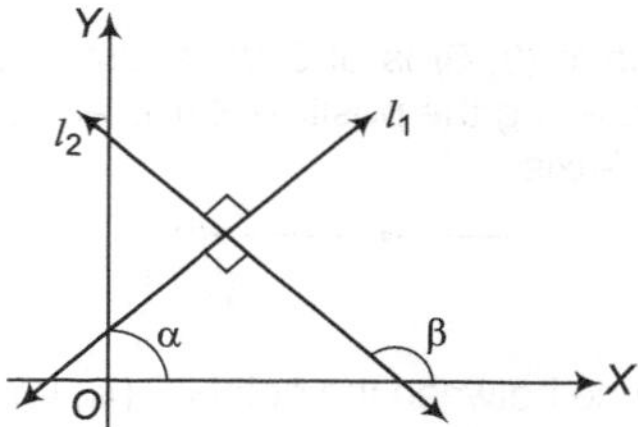

β is the exterior angle of triangle which is equal to the sum of interior opposite angles of triangle *i.e.,* α and 90°.

Therefore, $\tan\beta = \tan(\alpha + 90°)$

$$= -\cot\alpha$$
$$= -\frac{1}{\tan\alpha}$$

i.e., $m_2 = -\frac{1}{m_1}$ or $m_1m_2 = -1$

Conversely, if $m_1m_2 = -1$ *i.e.,* $\tan\alpha\tan\beta = -1$.

Then, $\tan\alpha = -\cot\beta = \tan(\beta + 90°)$ or $\tan(\beta - 90°)$.

Therefore, α and β differ by 90°.

Thus, lines l_1 and l_2 are perpendicular to each other.

Hence, two non-vertical lines are perpendicular to each other, if and only if their slopes are negative reciprocals of each other.

i.e., $m_2 = -\frac{1}{m_1}$ or $m_1m_2 = -1$

30. *(c)* (a) The slope of the line through (3, – 2) and (– 1, 4), is

$$m = \frac{4-(-2)}{-1-3} \qquad \left(\because m = \frac{y_2 - y_1}{x_2 - x_2}\right)$$

$$= \frac{6}{-4} = -\frac{3}{2}$$

(b) The slope of the line through the points (3, – 2) and (7, – 2), is

$$m = \frac{-2-(-2)}{7-3} = \frac{0}{4} = 0$$

(c) The slope of the line through the points (3, –2) and (3, 4), is

$$m = \frac{4-(-2)}{3-3} = \frac{6}{0}, \text{ which is not defined.}$$

(d) Here, inclination of the line $\alpha = 60°$. Therefore, slope of the line is

$$m = \tan 60° = \sqrt{3} \qquad (\because m = \tan\theta)$$

31. *(d)* If there are more than one line in a plane, then we find that these lines are either intersecting or parallel.

32. *(a)* When two lines intersect each other, they make two pairs of vertically opposite angles such that sum of any two adjacent angles is 180°.

33. *(b)* Let L_1 and L_2 be two non-vertical lines with slopes m_1 and m_2, respectively. If α_1 and α_2 are the inclinations of lines L_1 and L_2 respectively, then

$$m_1 = \tan\alpha_1 \text{ and } m_2 = \tan\alpha_2$$

Let θ and ϕ be the adjacent angles between the lines L_1 and L_2. Then,

$\theta = \alpha_2 - \alpha_1$ and $\alpha_1, \alpha_2 \neq 90°$ (exterior angle theorem)

A. Therefore, $\tan\theta = \tan(\alpha_2 - \alpha_1)$

$$= \frac{\tan\alpha_2 - \tan\alpha_1}{1+\tan\alpha_1\tan\alpha_2} = \frac{m_2 - m_1}{1+m_1m_2}, \text{ as } 1+m_1m_2 \neq 0$$

B. Since, $\phi = 180° - \theta$,

$\therefore \quad \tan\phi = \tan(180° - \theta) = -\tan\theta$

$$= -\frac{m_2 - m_1}{1+m_1m_2}, \text{ as } 1+m_1m_2 \neq 0$$

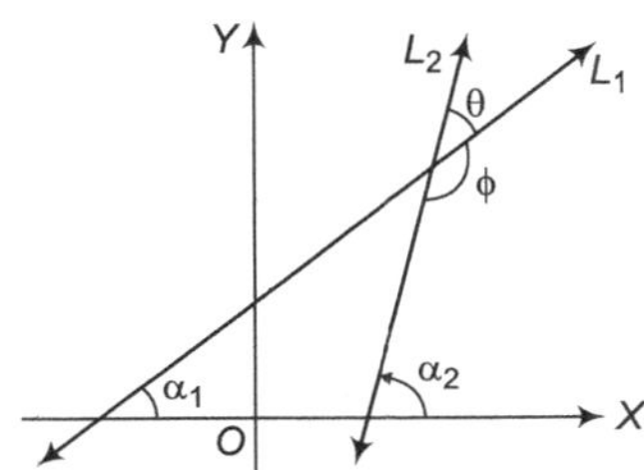

Now, here arise two cases.

C. **Case I** If $\frac{m_2 - m_1}{1+m_1m_2}$ is positive, then $\tan\theta$ will be positive and $\tan\phi$ will be negative, which means θ will be acute and ϕ will be obtuse.

D. **Case II** If $\frac{m_2 - m_1}{1+m_1m_2}$ is negative, then $\tan\theta$ will be negative and $\tan\phi$ will be positive, which means that θ will be obtuse and ϕ will be acute.

34. *(c)* The acute angle (say θ) between lines L_1 and L_2 with slopes m_1 and m_2 respectively, is given by

$$\tan\theta = \left|\frac{m_2 - m_1}{1+m_1m_2}\right|, \text{ as } 1+m_1m_2 \neq 0$$

35. *(b)* The obtuse angle ϕ between lines L_1 and L_2 with slopes m_1 and m_2 respectively, is given by

$$\phi = 180° - \theta$$

where, $$\theta = \tan^{-1}\left|\frac{m_2 - m_1}{1+m_1m_2}\right| \text{ as } 1+m_1m_2 \neq 0$$

36. *(c)* We know that, the acute angle θ between two lines with slopes m_1 and m_2 is given by

$$\tan\theta = \left|\frac{m_2 - m_1}{1+m_1m_2}\right| \qquad \ldots(i)$$

Let $$m_1 = \frac{1}{2}, m_2 = m \text{ and } \theta = \frac{\pi}{4}$$

Now, putting these values in Eq. (i), we get

$$\tan\frac{\pi}{4} = \left|\frac{m - \frac{1}{2}}{1+\frac{1}{2}m}\right|$$

or $$1 = \left|\frac{m - \frac{1}{2}}{1+\frac{1}{2}m}\right|$$

which gives $$\frac{m - \frac{1}{2}}{1+\frac{1}{2}m} = 1$$

or $$\frac{m - \frac{1}{2}}{1+\frac{1}{2}m} = -1$$

Therefore, $m = 3$ or $m = -\frac{1}{3}$

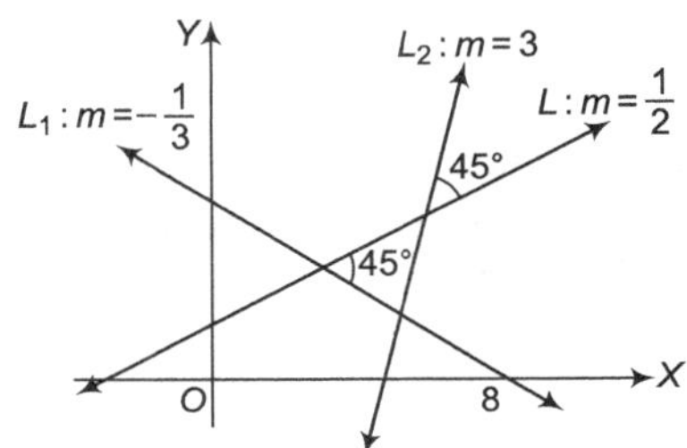

Hence, slope of the other line is 3 or $-\frac{1}{3}$.

37. *(d)* Slope of the line through the points (– 2, 6) and (4, 8) is

$$m_1 = \frac{8-6}{4-(-2)} = \frac{2}{6} = \frac{1}{3}$$

Slope of the line through the points (8, 12) and (*x*, 24) is

$$m_2 = \frac{24-12}{x-8} = \frac{12}{x-8}$$

Since, two lines are perpendicular,

So, $m_1 m_2 = -1$

$\Rightarrow \frac{1}{3} \times \frac{12}{x-8} = -1$

$\Rightarrow 8 - x = 4$ or $x = 4$

38. *(a)* Slopes of two parallel lines are equal.

39. *(c)* If two lines having the same slope pass through a common point, then two lines will coincide.

40. *(a)* If *A*, *B* and *C* are three points in the *xy*-plane, then they will lie on a line or they are collinear, if and only if

Slope of *AB* = Slope of *BC*

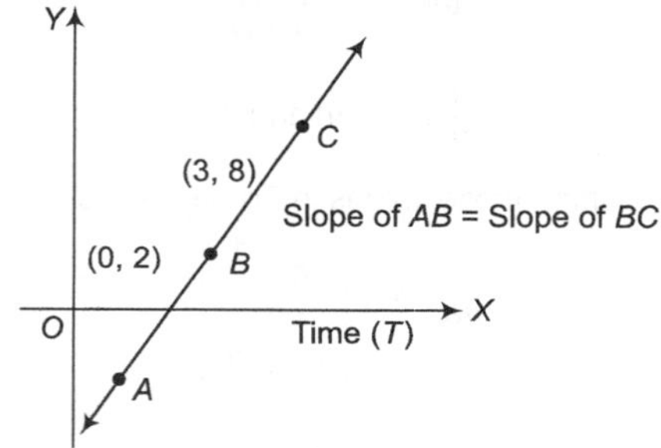

41. *(d)* Given vertices *A*(–4, 5), *B*(0, 7), *C*(5, – 5) and *D*(–4, –2) are shown below in the *xy*-plane. Joining all the vertices a quadrilateral *ABCD* is formed.

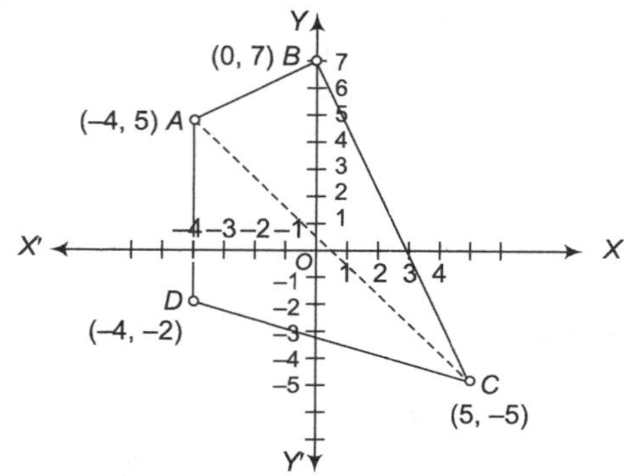

Area of quadrilateral *ABCD*

= Area of ΔADC + Area of ΔABC ...(i)

Now, area of ΔADC

$$= \frac{1}{2} | [-4(-2+5) - 4(-5-5) + 5(5+2)] |$$

$\left[\because \text{Area} = \frac{1}{2} | x_1(y_2 - y_3) + x_2(y_3 - y_1) + x_3(y_1 - y_2) | \text{ and here, } (x_1, y_1) = (-4, 5), (x_2, y_2) = (-4, -2), (x_3, y_3) = (5, -5)\right]$

$$= \frac{1}{2} | [-4 \times 3 - 4(-10) + 5 \times 7] |$$

$$= -\frac{1}{2} | [-12 + 40 + 35] | = \frac{1}{2} | [75 - 12] | = \frac{63}{2}$$

Area of $\Delta ABC = \frac{1}{2} | [-4(7+5) + 0(-5-5) + 5(5-7)] |$

[here, $(x_1, y_1) = (-4, 5), (x_2, y_2) = (0, 7)$ and $(x_3, y_3) = (5, -5)$]

$$= \frac{1}{2} | [-4 \times 12 + 0 \times (-10) + 5(-2)] |$$

$$= \frac{1}{2} | -48 - 10 |$$

$$= \frac{|-58|}{2} = 29$$

$\therefore$ From Eq. (i)

Area of quadrilateral *ABCD*

$$= \frac{63}{2} + 29 = \frac{63 + 58}{2} = \frac{121}{2}$$

= 60.5 sq units

42. *(b)* Let *BC* be the base of a triangle which lies on *Y*-axis and third vertex may be *A* (*h*, 0) or *A'* (– *h*, 0).

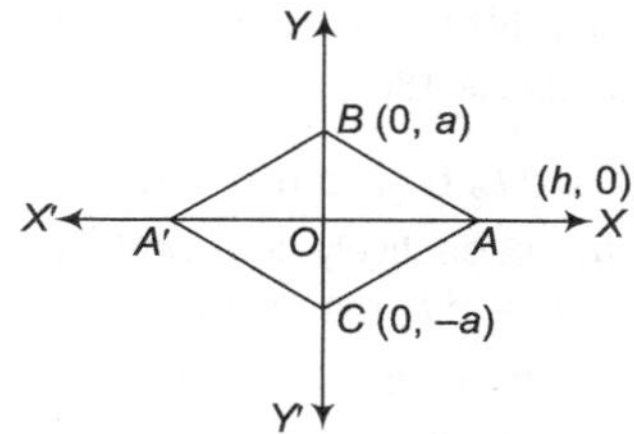

Since, ΔABC is equilateral, then

$$AB = BC$$

$$AB^2 = BC^2$$

$\Rightarrow (h-0)^2 + (0-a)^2 = (2a)^2$

$\left[\because \text{distance between two points } (x_1, y_1) \text{ and } (x_2, y_2) = \sqrt{(x_2 - x_1)^2 + (y_2 - y_1)^2}\right]$

For distance *AB*, $(x_1, y_1) = (a, 0)$ $(x_2, y_2) = (0, a)$

$\Rightarrow h^2 + a^2 = 4a^2$

$\Rightarrow h^2 = 3a^2$

$\Rightarrow h = \pm \sqrt{3} a$ (taking square root)

Hence, the vertices of triangle are $(\sqrt{3} a, 0), (0, a), (0, a)$

or $(-\sqrt{3} a, 0), (0, a), (0, -a)$

43. *(d)* I. When *PQ* is parallel to the *Y*-axis, it means the *x*-coordinates of *P* and *Q* are same *i.e.*, $x_1 = x_2$.

$\therefore$ Distance between two points *PQ*

$$= \sqrt{(x_2 - x_1)^2 + (y_2 - y_1)^2}$$

$$= \sqrt{(x_1 - x_1) + (y_2 - y_1)^2} \quad (\because x_1 = x_2)$$

$$= \sqrt{(y_2 - y_1)^2} = | y_2 - y_1 |$$

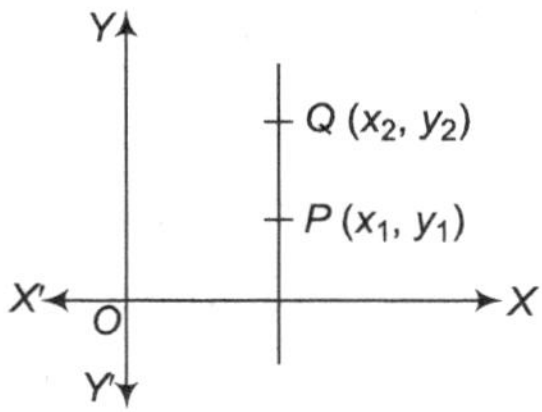

II. When PQ is parallel to X-axis , it means y-coordinates of P and Q are same *i.e.*, $y_2 = y_1$.

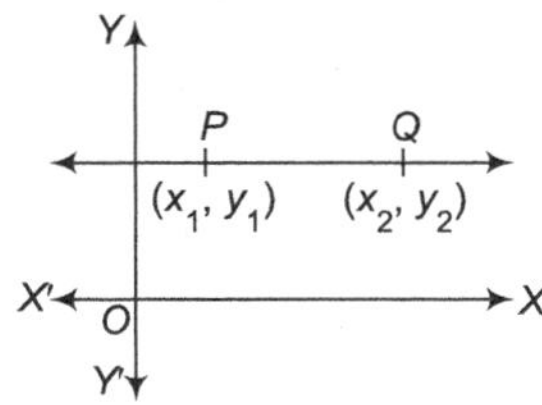

$\therefore$ Distance between two points (PQ)

$$= \sqrt{(x_2 - x_1)^2 + (y_2 - y_1)^2}$$

$$= \sqrt{(x_2 - x_1)^2 + (y_1 - y_1)^2} \qquad (\because y_1 = y_2)$$

$$= \sqrt{(x_2 - x_1)^2} = |x_2 - x_1|$$

44. *(c)* Let any point P on the X-axis is $(x, 0)$ as if a point lie on X-axis, y-coordinate is zero and the given points are $A(7, 6)$ and $B(3, 4)$.

Given, $PA = PB$

$\Rightarrow PA^2 = PB^2$

$\Rightarrow (x-7)^2 + (0-6)^2 = (x-3)^2 + (0-4)^2$

$\left[\because \text{distance between two points} = \sqrt{(x_1 - x_2)^2 + (y_1 - y_2)^2}\right]$

$\Rightarrow x^2 + 49 - 14x + 36 = x^2 + 9 - 6x + 16$

$\Rightarrow -14x + 6x = 25 - 36 - 49$

$\Rightarrow -8x = 25 - 85$

$\Rightarrow -8x = -60$

$\Rightarrow x = \dfrac{60}{8} \Rightarrow x = \dfrac{15}{2}$

$\therefore$ Point P on X-axis $= \left(\dfrac{15}{2}, 0\right)$

45. *(c)* Given, $\angle YPQ = 30°$

To find slope of line AQ.

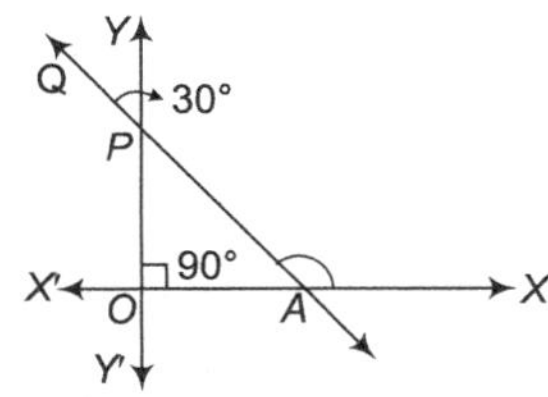

Here, $\angle YPQ = \angle OPA$ (vertically opposite angles)

$\because \angle OPA + \angle POA + \angle PAO = 180°$

($\because$ sum of all angles of a triangle is 180°)

$\Rightarrow 30° + 90° + \angle PAO = 180°$

$\Rightarrow \angle PAO = 180° - 120° = 60°$

$\Rightarrow \angle PAX = 180° - 60° = 120°$

$\therefore$ Slope of line $AQ = m = \tan 120°$ $\left[\begin{array}{l}\because m = \tan\theta \\ \tan(180° - \theta) \\ = -\tan\theta\end{array}\right]$

$= \tan(180° - 60°)$

$= -\tan 60° = -\sqrt{3}$

46. *(b)* Since, points are $A(x, -1)$, $B(2, 1)$ and $C(4, 5)$ are collinear,

Here, $x_1 = x$, $y_1 = -1$, $x_2 = 2$, $y_2 = 1$, $x_3 = 4$ and $y_3 = 5$

Slope of AB = Slope of BC

$\Rightarrow \dfrac{y_2 - y_1}{x_2 - x_1} = \dfrac{y_3 - y_2}{x_3 - x_2}$

$\Rightarrow \dfrac{1+1}{2-x} = \dfrac{5-1}{4-2} \quad \left(\because \text{slope of line} = \dfrac{y_2 - y_1}{x_2 - x_1}\right)$

$\Rightarrow \dfrac{2}{2-x} = \dfrac{4}{2} \Rightarrow 2 - x = 1$

$\Rightarrow x = 2 - 1 = 1$

47. *(c)*

D (–3, 2) C (3, 3)

A (–2, –1) B (4, 0)

$\because$ $ABCD$ is a parallelogram.

$\therefore$ $AB \,||\, CD \Rightarrow$ Slope of AB = Slope of CD

and $BC \,||\, AD \Rightarrow$ Slope of BC = Slope of AD

Hence, Statement I is false.

Now, mid-point of $AC = \left(\dfrac{-2+3}{2}, \dfrac{-1+3}{2}\right)$

$= \left(\dfrac{1}{2}, \dfrac{2}{2}\right) = \left(\dfrac{1}{2}, 1\right)$

and mid-point of $BD = \left(\dfrac{4-3}{2}, \dfrac{0+2}{2}\right) = \left(\dfrac{1}{2}, 1\right)$

$\Rightarrow$ Mid-point of AC = Mid-point of BD

Hence, Statement II is true.

48. *(b)* Given, $(x_1, y_1) = (3, -1)$ and $(x_2, y_2) = (4, -2)$

Slope of line $m = \dfrac{y_2 - y_1}{x_2 - x_1} = \dfrac{-2+1}{4-3}$

$\tan\theta = \dfrac{-1}{1} \Rightarrow \tan\theta = -1$

$\Rightarrow \tan\theta = -\tan 45° \qquad (\because 1 = \tan 45°)$

$\Rightarrow \tan\theta = \tan(180° - 45°)$ $[\because \tan(180° - \theta) = -\tan\theta]$

$\Rightarrow \tan\theta = \tan 135°$

$\Rightarrow \theta = 135°$

49. (c) Let slope of one line is m, then slope of another line is $2m$. Given, the tangent of the angle between them is $\tan\theta = \frac{1}{3}$.

$$\therefore \quad \left|\frac{m_1 - m_2}{1 + m_1 m_2}\right| = \frac{1}{3}$$

$$\Rightarrow \quad \frac{1}{3} = \left|\frac{m - 2m}{1 + m \times 2m}\right|$$

$$\Rightarrow \quad \frac{1}{3} = \left|\frac{-m}{1 + 2m^2}\right|$$

$$\Rightarrow \quad (1 + 2m^2) = 3m$$

$$\Rightarrow \quad 2m^2 - 3m + 1 = 0$$

Factorise it by splitting the middle term

$$\Rightarrow \quad 2m^2 - 2m - m + 1 = 0$$

$$\Rightarrow \quad 2m(m-1) - 1(m-1) = 0$$

$$\Rightarrow \quad (2m-1)(m-1) = 0$$

$$\Rightarrow \quad 2m - 1 = 0$$

or $\quad m - 1 = 0$

$$\Rightarrow \quad m = \frac{1}{2}, m = 1$$

50. (b) Let the given points are $P(x_1, y_1)$ and $Q(h, k)$.

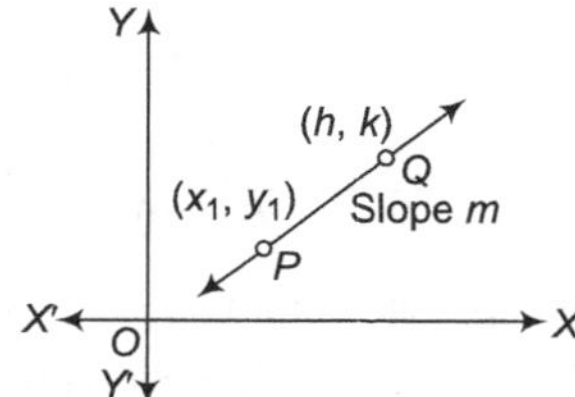

Then, slope $\quad m = \frac{k - y_1}{h - x_1} \quad \left(\because m = \frac{y_2 - y_1}{x_2 - x_1}\right)$

We know, if one point and slope is given, then equation of line is $k - y_1 = m(h - x_1)$.

51. (a) Suppose the line $y - x + 2 = 0$ divides the line segment joining $A(3, -1)$ and $B(8, 9)$ in the ratio $\lambda : 1$ at point P, then the coordinates of the point P are $\left(\frac{8\lambda + 3}{\lambda + 1}, \frac{9\lambda - 1}{\lambda + 1}\right)$.

But P lies on $y - x + 2 = 0$, therefore

$$\left(\frac{9\lambda - 1}{\lambda + 1}\right) - \left(\frac{8\lambda + 3}{\lambda + 1}\right) + 2 = 0$$

$$\Rightarrow \quad 9\lambda - 1 - 8\lambda - 3 + 2\lambda + 2 = 0$$

$$\Rightarrow \quad 3\lambda - 2 = 0$$

or $\quad \lambda = \frac{2}{3}$

So, the required ratio is $\frac{2}{3} : 1$, *i.e.*, $2 : 3$ (internally), since here λ is positive.

52. (b) Here, $\frac{1 + h - 4}{3} = 5$ and $\frac{2 - 3 + k}{3} = -1$

Then, we get $h = 18, k = -2$

$$\therefore \quad \sqrt{(h+k)^2 + (h + 3k)^2} = \sqrt{(18-2)^2 + (18-6)^2}$$

$$= \sqrt{(16^2 + 12^2)}$$

$$= 20$$

53. (c) Let $A \equiv (a, 0)$, $B \equiv (0, b)$ and $C \equiv (1, 1)$

Now points A, B and C will be collinear, if area of $\Delta ABC = 0$

or $\quad \frac{1}{2}\left|\begin{vmatrix} a & 0 \\ 0 & b\end{vmatrix} + \begin{vmatrix} 0 & b \\ 1 & 1\end{vmatrix} + \begin{vmatrix} 1 & 1 \\ a & 0\end{vmatrix}\right| = 0$

$$\Rightarrow \quad |(ab - 0) + (0 - b) + (0 - a)| = 0$$

or $\quad ab - a - b = 0$

or $\quad a + b = ab$

or $\quad \frac{1}{a} + \frac{1}{b} = 1$

54. (b) Let $P \equiv (x, y)$, $A \equiv (6, -1)$ and $B \equiv (2, 3)$

By the given condition, $PA = PB$

$$\Rightarrow \quad \sqrt{(x-6)^2 + (y+1)^2} = \sqrt{(x-2) + (y-3)^2}$$

or $\quad (x - 6) + (y + 1)^2 = (x - 2)^2 + (y - 3)^2$

or $\quad x^2 - 12x + 36 + y^2 + 2y + 1 = x^2 - 4x + 4 + y^2 - 6y + 9$

or $\quad 8x - 8y = 24$

or $\quad x - y = 3$

55. (c) Let $O(0, 0)$ and $A(3, \sqrt{3})$ be the given points and let $B(x, y)$ be the required point. Then,

$$OA = OB = AB \Rightarrow (OA)^2 = (OB)^2 = (AB)^2$$

$$\Rightarrow \quad (3-0)^2 + (\sqrt{3} - 0)^2 = (x-0)^2 + (y-0)^2$$

$$= (x-3)^2 + (y - \sqrt{3})^2$$

$$\Rightarrow \quad 12 = x^2 + y^2 = x^2 + y^2 - 6x - 2\sqrt{3}y + 12$$

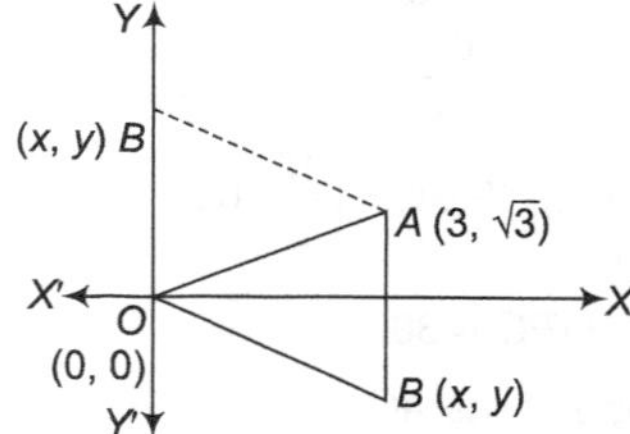

Taking first two equations

$$x^2 + y^2 = 12 \quad \ldots(i)$$

and taking last two equations

$$6x + 2\sqrt{3}y = 12 \quad \text{or} \quad y = \sqrt{3}(2 - x) \quad \ldots(ii)$$

From Eqs. (i) and (ii), we get

$$x^2 + 3(2-x)^2 = 12 \quad \text{or} \quad 4x^2 - 12x = 0$$

$$\Rightarrow \quad x = 0, 3$$

Putting $x = 0$ and 3 respectively in Eq. (ii), we get

$$y = 2\sqrt{3}, -\sqrt{3}$$

Hence, the coordinates of the third vertex B are $(0, 2\sqrt{3})$ or $(3, -\sqrt{3})$.

Alternate Method

If the coordinates of the two vertices of triangle are (x_1, y_1) and (x_2, y_2), then coordinates of third vertices are

$$\left(\frac{x_1 + x_2 \mp \sqrt{3}(y_2 - y_1)}{2}, \frac{y_1 + y_2 \pm \sqrt{3}(x_2 - x_1)}{2}\right)$$

i.e., $$\left(\frac{0 + 3 \mp \sqrt{3}(\sqrt{3} - 0)}{2}, \frac{0 + \sqrt{3} \pm \sqrt{3}(3 - 0)}{2}\right)$$

or $$\left(\frac{3 \mp 3}{2}, \frac{\sqrt{3} \pm 3\sqrt{3}}{2}\right)$$

$\Rightarrow$ $$(0, 2\sqrt{2}) \quad \text{or} \quad (3, -\sqrt{3})$$

56. *(b)* Let $A(3, 4)$ and $C(1, -1)$ be the given opposite angular points of a square $ABCD$ and let $B(x, y)$ be the unknown vertex. Then,

$$AB = BC$$

$\Rightarrow$ $$(AB)^2 = (BC)^2$$

$\Rightarrow$ $$(x - 3)^2 + (y - 4)^2 = (x - 1)^2 + (y + 1)^2$$

$\Rightarrow$ $$4x + 10y - 23 = 0$$

$\Rightarrow$ $$x = \left(\frac{23 - 10y}{4}\right) \quad \ldots(i)$$

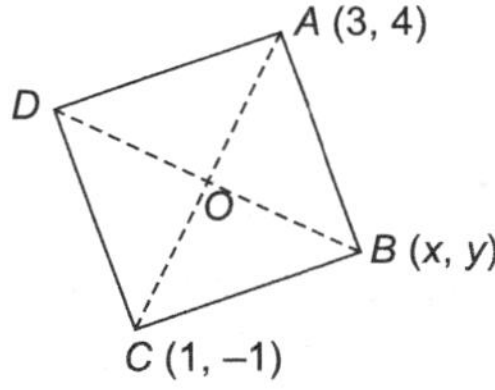

Also in right ΔABC,

$$(AB)^2 + (BC)^2 = (AC)^2$$

$\Rightarrow$ $$(x - 3)^2 + (y - 4)^2 + (x - 1)^2 + (y + 1)^2 = (3 - 1)^2 + (4 + 1)^2$$

$\Rightarrow$ $$x^2 + y^2 - 4x - 3y - 1 = 0 \quad \ldots(ii)$$

Substituting the value of x from Eq. (i) into Eq. (ii), we get

$$\left(\frac{23 - 10y}{4}\right)^2 + y^2 - 4\left(\frac{23 - 10y}{4}\right) - 3y - 1 = 0$$

$\Rightarrow$ $$4y^2 - 12y + 5 = 0$$

or $$(2y - 1)(2y - 5) = 0$$

$\therefore$ $$y = \frac{1}{2} \quad \text{or} \quad = \frac{5}{2}$$

Putting $y = \frac{1}{2}$ in Eq. (i), we get $x = \frac{9}{2}$

and putting $y = \frac{5}{2}$ in Eq. (i), we get $x = -\frac{1}{2}$

Hence, the required vertices of the square are $\left(\frac{9}{2}, \frac{1}{2}\right)$ and $\left(-\frac{1}{2}, \frac{5}{2}\right)$.

57. *(d)* Let D as the origin and DC and DY as the X and Y-axes, respectively. Let $BC = 2a$. Then,

$B(-a, 0)$, $C(a, 0)$ and let coordinate of $A(b, c)$

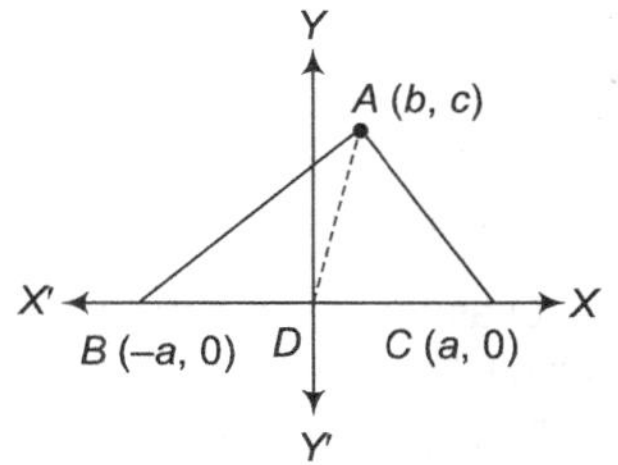

Now, consider

$$= AB^2 + AC^2$$
$$= (b + a)^2 + (c - 0)^2 + (b - a)^2 + (c - 0)^2$$
$$= 2(a^2 + b^2 + c^2) \quad \ldots(i)$$

and $$= 2(AD^2 + BD^2)$$
$$= 2[(b - 0)^2 + (c - 0)^2 + a^2]$$
$$= 2(a^2 + b^2 + c^2) \quad \ldots(ii)$$

From Eqs. (i) and (ii), we get

$$AB^2 + AC^2 = 2(AD^2 + BD^2)$$

58. *(a)* Let A as the origin and AB and AD as the X and Y-axes, respectively. Let $AB = a$ and $AD = b$. Then,

$$B \equiv (a, 0), D \equiv (0, b) \text{ and } C \equiv (a, b)$$

Let $$P \equiv (\alpha, \beta)$$

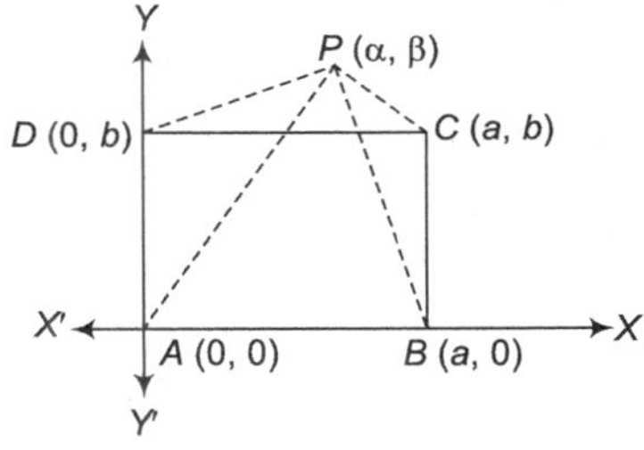

Now, consider $= PA^2 + PC^2$

$$= (\alpha - 0)^2 + (\beta - 0)^2 + (\alpha - a)^2 + (\beta - b)^2$$
$$= 2\alpha^2 + 2\beta^2 - 2a\alpha - 2b\beta + a^2 + b^2 \quad \ldots(i)$$

and $$= PB^2 + PD^2$$
$$= (\alpha - a)^2 + (\beta - 0)^2 + (\alpha - 0)^2 + (\beta - b)^2$$
$$= 2\alpha^2 + 2\beta^2 - 2a\alpha - 2b\beta + a^2 + b^2 \quad \ldots(ii)$$

From Eqs. (i) and (ii), we get

$$PA^2 + PC^2 = PB^2 + PD^2$$

59. *(d)* Let D be the mid-point of BC. Then, coordinates of D are $\left(\frac{1+5}{2}, \frac{-1+1}{2}\right)$ *i.e.*, (3, 0).

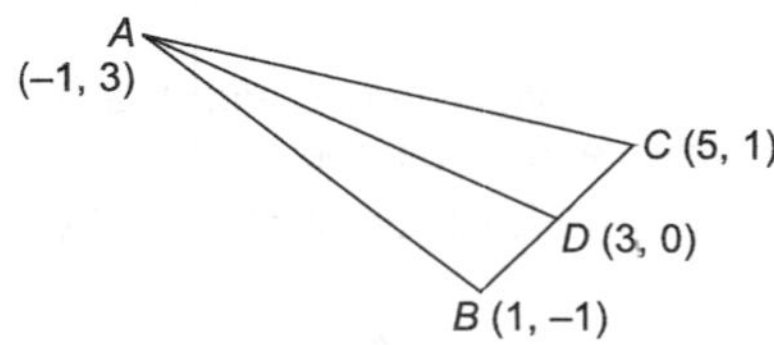

$\therefore$ Median $AD = \sqrt{(3+1)^2 + (0-3)^2}$

$= \sqrt{16+9} = \sqrt{25} = 5$ units

60. *(b)* We take O as the origin and OC and OY as the X and Y-axes, respectively.

Let $BC = 2a$. Then, $B(-a, 0)$, and $C(a, 0)$

Let $A(b, c)$. If E and F are the mid-points of sides AC and AB respectively, then $E\left(\frac{a+b}{2}, \frac{c}{2}\right)$ and $F\left(\frac{b-a}{2}, \frac{c}{2}\right)$

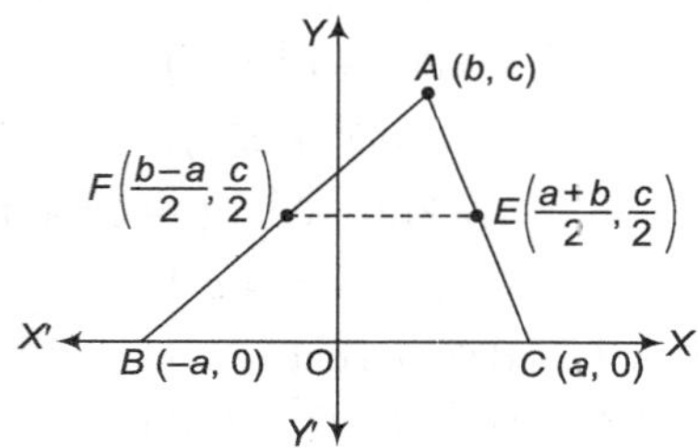

Now, $FE = \sqrt{\left(\frac{a+b}{2} - \frac{b-a}{2}\right)^2 + \left(\frac{c}{2} - \frac{c}{2}\right)^2} = a$

$= \frac{1}{2}(2a) = \frac{1}{2}(BC)$

Hence, the line joining the mid-points of any two sides of a triangle is equal to half of the third side.

61. *(c)* Let the coordinates of the required point be $P(x, y)$.

Then, $x = \left(\frac{1\times(-3) - 3\times1}{1-3}\right)$ and $y = \left(\frac{1\times9 - 3\times(-3)}{1-3}\right)$

i.e., $x = 3$ and $y = -9$

Hence, the required point is $(3, -9)$.

62. *(b)* Let P and Q be the required new ends.

Let the coordinates of P be (x_1, y_1).

Given, $AB = 2AP \Rightarrow \frac{AB}{AP} = \frac{2}{1}$

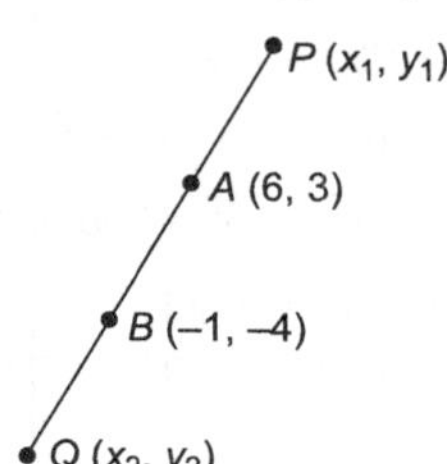

i.e., A divides BP internally in the ratio 2 : 1.

Then, $6 = \frac{2x_1 + 1(-1)}{2+1} \Rightarrow 19 = 2x_1$ or $x_1 = \frac{19}{2}$

and $3 = \frac{2y_1 + 1(-4)}{2+1} \Rightarrow 13 = 2y_1$ or $y_1 = \frac{13}{2}$

$\therefore$ Coordinates of P are $\left(\frac{19}{2}, \frac{13}{2}\right)$.

Also, let coordinates of Q be (x_2, y_2).

Given, $AB = 2BQ$

$\Rightarrow \frac{AB}{BQ} = \frac{2}{1}$

i.e., B divides AQ internally in the ratio 2 : 1.

Then, $-1 = \frac{2x_2 + 1\times6}{2+1} \Rightarrow -9 = 2x_2$ or $x_2 = -\frac{9}{2}$

and $-4 = \frac{2\times y_2 + 1\times3}{2+1} \Rightarrow -15 = 2y_2$ or $y_2 = -\frac{15}{2}$

$\therefore$ Coordinates of Q are $\left(-\frac{9}{2}, -\frac{15}{2}\right)$.

63. *(d)* Let $A \equiv (4, 3)$, $B \equiv (6, 3)$ and $P \equiv (2, y)$.

Let P divides AB internally in the ratio $\lambda : 1$.

Then, $2 = \frac{6\lambda + 4}{\lambda + 1}$

$\Rightarrow 2\lambda + 2 = 6\lambda + 4$

$\Rightarrow -4\lambda = 2$

or $\lambda = -\frac{1}{2}$

A (2, 3) B (4, 3) C (6, 3)

$\therefore$ P divides AB externally in the ratio 1:2. ($\because \lambda$ is negative)

Now, $y = \frac{1\times3 - 2\times3}{1-2} = 3$.

64. *(b)* Let the third vertex be (x, y). Then, the coordinates of the centroid of triangle are

$\left(\frac{-1+5+x}{3}, \frac{4+2+y}{3}\right)$ *i.e.*, $\left(\frac{4+x}{3}, \frac{6+y}{3}\right)$.

Now, $\left(\frac{4+x}{3}, \frac{6+y}{3}\right) = (0, -3)$

$\Rightarrow \frac{4+x}{3} = 0$ and $\frac{6+y}{3} = -3$

$\Rightarrow 4 + x = 0$ and $y + 6 = -9$

or $x = -4$ and $y = -15$

Hence, the third vertex is $(-4, -15)$.

65. *(c)* We take B as the origin and BX and BY as the X and Y-axes, respectively.

Let $BC = a$

then $B \equiv (0, 0)$ and $C \equiv (a, 0)$

and let $A \equiv (h, k)$

Then, coordinates of G will be

$\left(\frac{h+0+a}{3}, \frac{k+0+0}{3}\right)$, *i.e.*, $\left(\frac{h+a}{3}, \frac{k}{3}\right)$.

Take ΔABC as in 1st quadrant (for the sake of exactness).

Now, $\quad \text{LHS} = (AB)^2 + (BC)^2 + (CA)^2$

$$= (h-0)^2 + (k-0)^2 + a^2 + (h-a)^2 + (k-0)^2$$

$$= 2h^2 + 2k^2 - 2ah + 2a^2 \qquad \ldots\text{(i)}$$

$\text{RHS} = 3((GA)^2 + (GB)^2 + (GC)^2)$

$$= 3\left\{\left(\frac{a+h}{3} - h\right)^2 + \left(\frac{k}{3} - k\right)^2 + \left(\frac{a+h}{3} - 0\right)^2 + \left(\frac{k}{3} - 0\right)^2 + \left(\frac{a+h}{3} - a\right)^2 + \left(\frac{k}{3} - 0\right)^2\right\}$$

$$= \frac{3}{9}\{(a-2h)^2 + (-2k)^2 + (a+h)^2 + k^2 + (h-2a)^2 + k^2\}$$

$$= \frac{1}{3}\{6a^2 + 6h^2 + 6k^2 - 6ah\}$$

$$= 2h^2 + 2k^2 - 2ah + 2a^2 \qquad \ldots\text{(ii)}$$

Hence, from Eqs. (i) and (ii), we get

$$AB^2 + BC^2 + CA^2 = 3(GA^2 + GB^2 + GC^2)$$

66. *(a)* Centroid of the triangle is

$$G \equiv \left(\frac{1+2+c^2}{3}, \frac{a+b-3}{3}\right) \text{ i.e.,} \left(\frac{3+c^2}{3}, \frac{a+b-3}{3}\right)$$

I. $\because G$ will lie on Y-axis, then

$$\frac{3+c^2}{3} = 0 \Rightarrow c^2 = -3$$

or $\qquad c \equiv \pm i\sqrt{3}$

$\because$ Both values of c are imaginary.

Hence, G cannot lie on Y-axis.

II. $\because G$ will lies on X-axis, then $\frac{a+b-3}{3} = 0$

$\Rightarrow \qquad a+b-3 = 0$

or $\qquad a+b = 3$

67. *(a)* We have,

$$\frac{\text{Area of } \Delta PBC}{\text{Area of } \Delta ABC} = \frac{\frac{1}{2}|\{x(5+2) - 3(-2-y) + 4(y-5)\}|}{\frac{1}{2}|\{6(5+2) - 3(-2-3) + 4(3-5)\}|}$$

$$= \frac{|7x + 7y - 14|}{|49|}$$

$$= \frac{7|x+y-2|}{49} = \frac{|x+y-2|}{7}$$

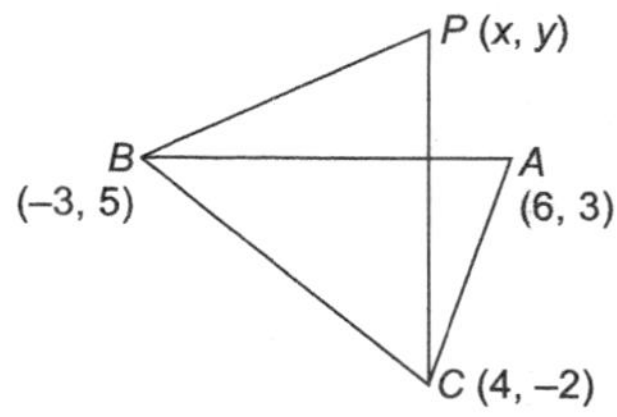

68. *(b)* Let $A(x_1, y_1)$, $B(x_2, y_2)$ and $C(x_3, y_3)$ be the vertices of a ΔABC. If possible let $x_1, y_1, x_2, y_2, x_3, y_3$ be all rational.

Now, area of ΔABC

$$= \frac{1}{2}|x_1(y_2 - y_3) + x_2(y_3 - y_1) + x_3(y_1 - y_2)|$$

$$= \text{Rational} \qquad \ldots\text{(i)}$$

Since, ΔABC is equilateral,

$\therefore \quad$ Area of $\Delta ABC = \frac{\sqrt{3}}{4}(\text{side})^2$

$$= \frac{\sqrt{3}}{4}(AB)^2$$

$$= \frac{\sqrt{3}}{4}\{(x_1 - x_2)^2 + (y_1 - y_2)^2\}$$

$$= \text{Irrational} \qquad \ldots\text{(ii)}$$

From Eqs. (i) and (ii), we get

Rational = Irrational

which is contradiction.

Hence, $x_1, y_1, x_2, y_2, x_3, y_3$ cannot all be rational.

69. *(b)* Let coordinates of P be (h, k).

$\Rightarrow \qquad PA = PB \Rightarrow (PA)^2 = (PB)^2$

$\Rightarrow \qquad (h-3)^2 + (k-4)^2 = (h-5)^2 + (k+2)^2$

$\Rightarrow \quad (h-3)^2 - (h-5)^2 + (k-4)^2 - (k+2)^2 = 0$

$\Rightarrow \quad (2h-8)(2) + (2k-2)(-6) = 0$

$\Rightarrow \qquad (h-4) - 3(k-1) = 0$

$\Rightarrow \qquad h - 3k - 1 = 0 \qquad \ldots\text{(i)}$

Now, area of $\Delta PAB = \frac{1}{2}\begin{vmatrix} h & k & 1 \\ 3 & 4 & 1 \\ 5 & -2 & 1 \end{vmatrix} = 10$

$$6h + 2k - 26 = \pm 20$$

$\Rightarrow \quad 6h + 2k - 46 = 0 \quad \text{or} \quad 6h + 2k - 6 = 0$

$\Rightarrow \quad 3h + k - 23 = 0 \quad \text{or} \quad 3h + k - 3 = 0$

Solving $h - 3k - 1 = 0$ and $3h + k - 23 = 0$, we get

$$h = 7, k = 2$$

Solving $h - 3k - 1 = 0$ and $3h + k - 3 = 0$, we get

$$h = 1, k = 0$$

Hence, the coordinates of P are (7, 2) or (1, 0).

70. *(a)* $\because$ Diagonals of a rectangle bisect each other.

$\therefore$ Mid-point of (1, 3) and (5, 1) satisfies

$$y - x + \lambda = 0 \qquad \ldots\text{(i)}$$

Mid-point of (1, 3) and (5, 1)

$$=\left(\frac{1+5}{2},\frac{3+1}{2}\right)=\left(\frac{6}{2},\frac{4}{2}\right)=(3,2)$$

$\therefore$ Point (3, 2) satisfies Eq. (i)

$\Rightarrow \quad 2-3+\lambda=0 \Rightarrow \lambda=1$

71. *(a)* The required area $=\frac{1}{2}\left\|\begin{matrix} x_1 & y_1\\ x_2 & y_2\\ x_3 & y_3\\ x_4 & y_4\\ x_5 & y_5\\ x_1 & y_1\end{matrix}\right\|$

$$=\frac{1}{2}|[(x_1y_2+x_2y_3+x_3y_4+x_4y_5+x_5y_1]-[y_1x_2+y_2x_3+y_3x_4+y_4x_5+y_5x_1]|$$

$$=\frac{1}{2}\left\|\begin{matrix} 1 & 1\\ 7 & 21\\ 7 & -3\\ 12 & 2\\ 0 & -3\\ 1 & 1\end{matrix}\right\|$$

$$=\frac{1}{2}|[(21-21+14-36+0)-(7+147-36+0-3)]|$$

$$=\frac{137}{2}\text{ sq units}$$

72. *(c)* Since, points are collinear.

$$\therefore \Delta=\frac{1}{2}|[x_1(y_2-y_3)+x_2(y_3-y_1)+x_3(y_1-y_2)]|=0$$

$$\therefore \quad \frac{1}{2}[k\{2k-(6-2k)\}+(1-k)\{6-2k-(2-2k)\}+(-4-k)\{2-2k-2k\}]=0$$

$\Rightarrow k(4k-6)+(1-k)4+(k+4)(4k-2)=0$

$\Rightarrow \quad 8k^2+4k-4=0$

$\Rightarrow \quad 2k^2+k-1=0$

$\Rightarrow \quad (k+1)(2k-1)=0$

$\Rightarrow \quad k=-1,\frac{1}{2}$

73. *(a)* Let $P\equiv(a\cos\alpha, a\sin\alpha)$ and $O\equiv(0,0)$. Then,

$$|OP|=\sqrt{(a\cos\alpha-0)^2+(a\sin\alpha-0)^2}$$

$$=\sqrt{(a^2\cos^2\alpha+a^2\sin^2\alpha)}$$

$$=\sqrt{a^2(\cos^2\alpha+\sin^2\alpha)}=\sqrt{a^2}$$

$=|a|$, which is independent of α.

74. *(b)* Let the given points are $A(4, 4)$, $B(3, 5)$ and $C(-1, -1)$.

Here, $(x_1, y_1)\to A(4, 4), (x_2, y_2)\to B(3, 5)$

and $(x_3, y_3)\to C(-1, 1)$.

$$\text{Now, }\frac{1}{2}\begin{vmatrix} x_1 & y_1 & 1\\ x_2 & y_2 & 1\\ x_3 & y_3 & 1\end{vmatrix}=\frac{1}{2}\begin{vmatrix} 4 & 4 & 1\\ 3 & 5 & 1\\ -1 & -1 & 1\end{vmatrix}$$

$$=\frac{1}{2}[4(5+1)-4(3+1)+1(-3+5)]$$

$$=\frac{1}{2}[24-16+2]=\frac{10}{2}=5\neq 0$$

Hence, A, B and C are the vertices of a triangle.

Slope of AB is $m_1=\frac{y_2-y_1}{x_2-x_1}=\frac{5-4}{3-4}=\frac{1}{-1}$

$$=-1 \qquad \left(\because \text{slope}=\frac{y_2-y_1}{x_2-x_1}\right)$$

Slope of BC is $m_2=\frac{y_3-y_2}{x_3-x_2}=\frac{-1-5}{-1-3}=\frac{-6}{-4}=\frac{3}{2}$

Slope of CA is $m_3=\frac{y_3-y_1}{x_3-x_1}=\frac{-1-4}{-1-4}=\frac{-5}{-5}=1$

$\because$ Slope of $AB\times$ Slope of $CA=(-1)\times 1=(-1)$

$\therefore$ ΔABC is a right angled triangle at A.

75. *(a)* Given points are $A(h, 0)$, $B(a, b)$ and $C(0, k)$. If A, B and C are collinear, then

Slope of AB = Slope of BC = Slope of CA.

Let $(x_1, y_1)\to A(h, 0), (x_2, y_2)\to B(a, b), (x_3, y_3)\to C(0, k)$.

Now, slope of AB = Slope of BC = Slope of CA

$$\therefore \quad \frac{y_2-y_1}{x_2-x_1}=\frac{y_3-y_2}{x_3-x_2}=\frac{y_1-y_3}{x_1-x_3}$$

$$\frac{b-0}{a-h}=\frac{k-b}{0-a}=\frac{0-k}{h-0}$$

Taking first two terms, we get

$$\Rightarrow \quad \frac{b}{a-h}=\frac{k-b}{-a} \Rightarrow -ab=(a-h)(k-b)$$

$$\Rightarrow \quad -ab=ak-ab-hk+bh \Rightarrow ak+bh=hk$$

Dividing each term by (hk), we get

$$\Rightarrow \quad \frac{ak}{hk}+\frac{bh}{hk}=\frac{hk}{hk} \Rightarrow \frac{a}{h}+\frac{b}{k}=1$$

Hence proved.

76. *(a)* Let the population in the year 2010 will be k crore. Since, A, B and C are collinear.

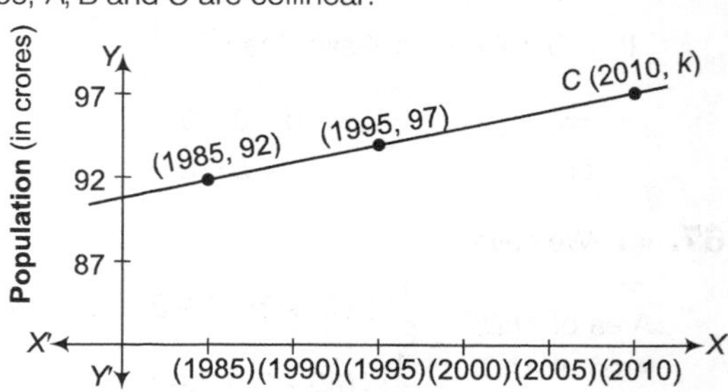

$\therefore$ Slope of AB = Slope of BC

$$\Rightarrow \quad \frac{y_2-y_1}{x_2-x_1}=\frac{y_3-y_2}{x_3-x_2}$$

$$\Rightarrow \quad \frac{97-92}{1995-1985} = \frac{k-97}{2010-1995}$$

$$\left(\because x_1 = 1985, x_2 = 1995, x_3 = 2010, \; y_1 = 92, y_2 = 97, y_3 = k\right)$$

$$\Rightarrow \quad \frac{5}{10} = \frac{k-97}{15}$$

$$\Rightarrow \quad \frac{1}{2} = \frac{k-97}{15}$$

$$\Rightarrow \quad k - 97 = \frac{15}{2}$$

$$\Rightarrow \quad k = 97 + \frac{15}{2} = 97 + 7.5 = 104.5$$

Hence, slope of $AB = \frac{1}{2}$

and population in the year 2010 = 104.5 crore.

77. *(b)* Let the incident ray strike X-axis at the point A whose coordinates be $(x, 0)$. From the figure, the slope of the reflected ray is given by

$$\tan\theta = \frac{3}{5-x} \quad \ldots\text{(i)}$$

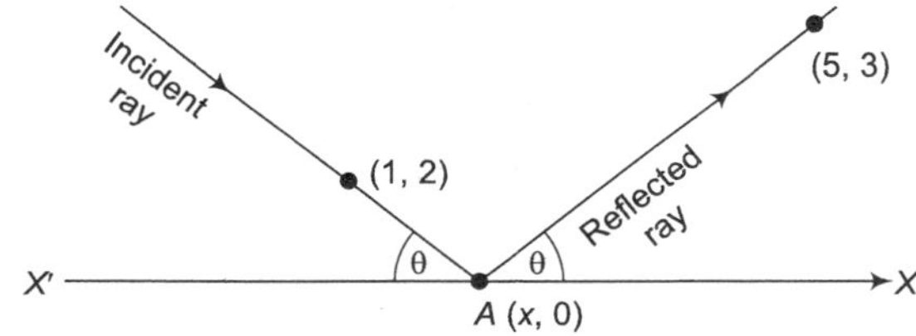

Again, the slope of the incident ray is given by

$$\tan(\pi - \theta) = \frac{-2}{x-1}$$

or

$$-\tan\theta = \frac{-2}{x-1} \quad \ldots\text{(ii)}$$

Solving Eqs. (i) and (ii), we get

$$\frac{3}{5-x} = \frac{2}{x-1}$$

or

$$x = \frac{13}{5}$$

Therefore, the required coordinates of the point A are $\left(\frac{13}{5}, 0\right)$.

78. *(a)* For a given line, we should have a definite condition on the points lying on the line.

79. *(b)* Suppose $P(x, y)$ is an arbitrary point in the xy- plane and L is the given line. For the equation of L, we wish to construct to a statement or condition for the point P that is true, when P is on L, otherwise false. Of course the statement is merely an algebraic equation involving the variables x and y.

80. *(c)* If a horizontal line L is at a distance a from the X-axis, then ordinate of every point lying on the line is either a or $-a$.

81. *(d)* If a horizontal line is at a distance a from the X-axis, then equation of the horizontal line above or below the X-axes is $y = a$ or $y = -a$, respectively.

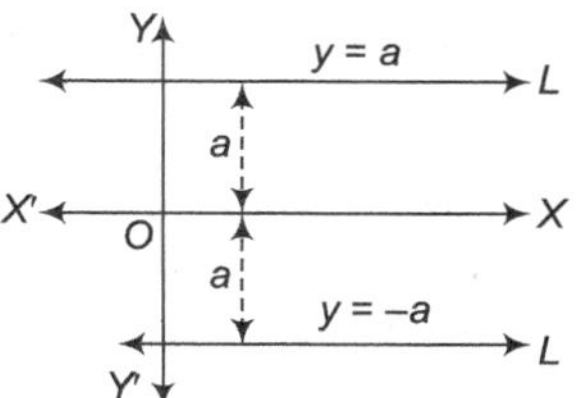

82. *(b)* The equation of a vertical line at a distance b from Y-axis is either $x = b$ or $x = -b$.

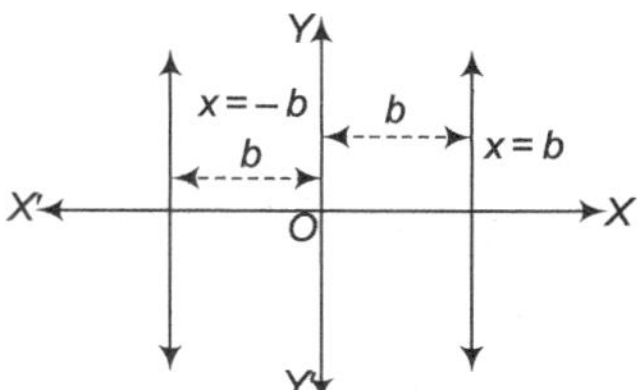

83. *(c)* Since, the y-coordinate of every point on the line parallel to X-axis and passing through $(-2, 3)$ is $y = 3$.

Similarly, equation of the line parallel to Y-axis and passing through $(-2,3)$ is $x = -2$.

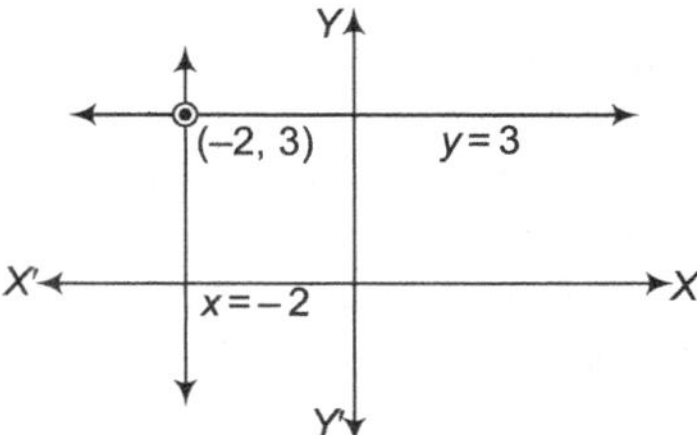

84. *(a)* The slope of L is given by

$$m = \frac{y - y_0}{x - x_0}$$

i.e.,

$$y - y_0 = m(x - x_0) \quad \ldots\text{(i)}$$

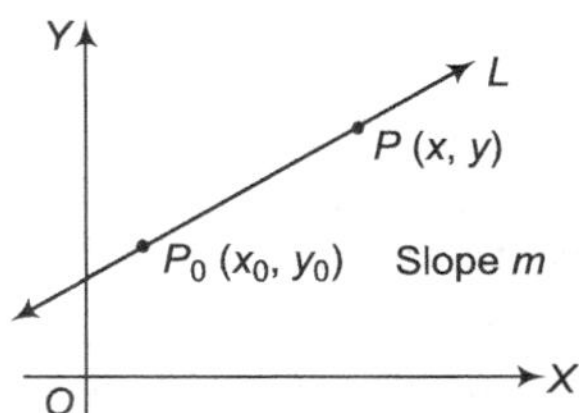

Since, the point $P_0(x_0, y_0)$ along with all points (x, y) on L staisfies Eq. (i) and no other point in the plane satisfies Eq. (i).

$\therefore$ Eq. (i) is indeed the equation for given line L.

Thus, the point (x, y) lies on the line with slope m through the fixed point $(x_0\ y_0)$, if and only if its coordinate satisfy the equation $y - y_0 = m(x - x_0)$.

85. *(b)* Here, $m = -4$ and given point (x_0, y_0) is $(-2, 3)$.

By slope-point form,

Equation of the given line is

$$y - 3 = -4(x + 2)$$

or $4x + y + 5 = 0$, which is the required equation.

86. *(c)* $\because$ The three points P_1, P_2 and P are collinear. Therefore, we have

Slope of P_1P = Slope of P_1P_2

i.e., $$\frac{(y - y_1)}{(x - x_1)} = \frac{(y_2 - y_1)}{(x_2 - x_1)}$$

or $$(y - y_1) = \frac{y_2 - y_1}{x_2 - x_1}(x - x_1)$$

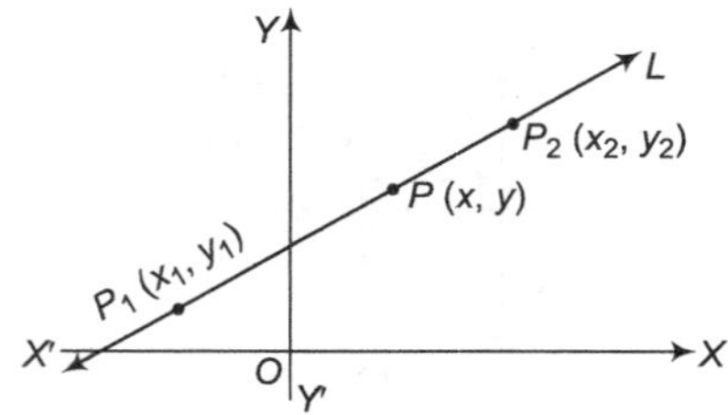

Thus, Equation of the line passing through the points (x_1, y_1) and (x_2, y_2) is given by

$$y - y_1 = \frac{y_2 - y_1}{x_2 - x_1}(x - x_1)$$

87. *(b)* I. Suppose a line L with slope m cuts the Y-axis at a distance c from the origin. The distance c is called the y-intercept of the line L. Obviously, coordinates of the point where the line meet the Y-axis are $(0, c)$. Thus, L has slope m and passes through a fixed point $(0, c)$. Therefore, by point-slope form, the equation of L is

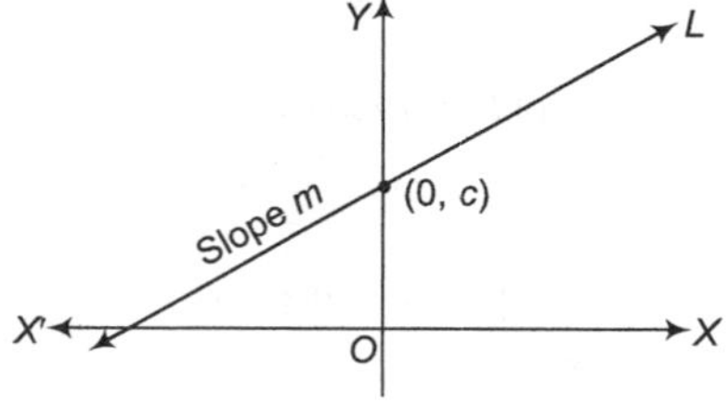

$$y - c = m(x - 0) \quad \text{or} \quad y = mx + c$$

Thus, the point (x, y) on the line with slope m and Y-Intercept c lies on the line, if and only if $y = mx + C$

Note that the value of c will be positive or negative according as the intercept is made on the positive or negative side of the Y-axis, respectively.

II. Suppose line L with slope m makes x-intercept 'd', then coordinate of the point where the line meet the x-axis are (d, o). Thus, L has slope 'm' and passes through fixed point (d, o). Therefore, by point slope form the equation of L is

$$y - 0 = m(x - d) \Rightarrow y = m(x - d)$$

88. *(a)* We know that x-intercept or y-intercept is distance at which line cut the X-axis or Y-axis respectively. Here, x-intercept is 'a' and Y-intercept is 'b'.

Obviously, L meets X-axis at the point $(a, 0)$ and Y-axis at the point $(0, b)$.

By two-point form of the equation of the line, we have

$$y - 0 = \frac{b - 0}{0 - a}(x - a)$$

or $$ay = -bx + ab, \quad i.e., \quad \frac{x}{a} + \frac{y}{b} = 1$$

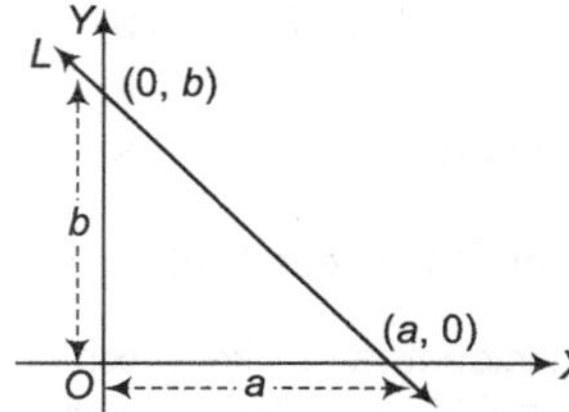

Thus, equation of the line making intercepts 'a' and 'b' on X-axis and Y-axis respectively, is $\frac{x}{a} + \frac{y}{b} = 1$.

89. *(d)* $\because$ Equation of line is $2x - 3y + 6 = 0$

$\Rightarrow$ $$2x - 3y = -6$$

$\Rightarrow$ $$\frac{2x}{-6} - \frac{3y}{-6} = 1$$

$\Rightarrow$ $$\frac{x}{-3} + \frac{y}{2} = 1$$

Hence, $a = -3$

and $b = 2$

90. *(b)* Let L be the line, whose perpendicular distance form origin O be $OA = p$ and the angle between the positive X-axis and OA be $\angle XOA = \omega$. The possible position of line L in the cartesian plane are shown. Now, our purpose is to find slope of L and a point on it. Draw perpendicular AM on the X-axis in each case.

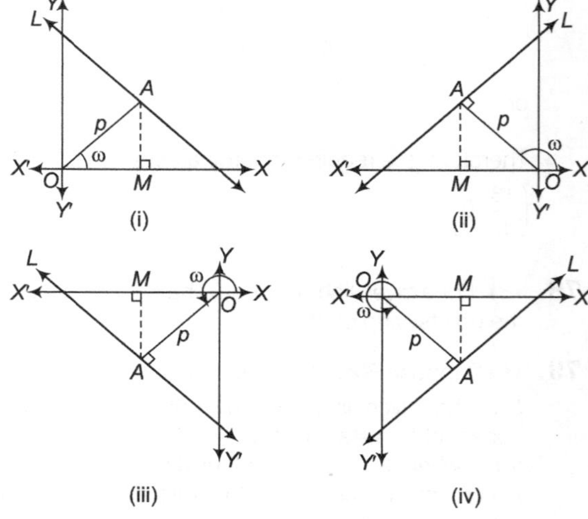

In each case, we have $OM = p\cos\omega$, and $MA = p\sin\omega$, so that the coordinates of the point A are $(p\cos\omega, p\sin\omega)$.

Further, line L is perpendicular to OA. Therefore

The slope of the line $L = \dfrac{-1}{\text{slope of } OA} = \dfrac{-1}{\tan\omega} = \dfrac{-\cos\omega}{\sin\omega}$

Thus, the line L has slope $-\dfrac{\cos\omega}{\sin\omega}$ and point A $(p\cos\omega, p\sin\omega)$ on it. Therefore, by point-slope form, the equation of the line L is

$$y - p\sin\omega = -\frac{\cos\omega}{\sin\omega}(x - p\cos\omega)$$

or $\quad x\cos\omega + y\sin\omega = p(\sin^2\omega + \cos^2\omega)$

or $\quad x\cos\omega + y\sin\omega = p.$

Hence, the equation of the line having normal distance p form the origin and angle ω which the normal makes with positive direction of X-axis is given by

$$x\cos\omega + y\sin\omega = p$$

91. *(a)* Here, we are given $p = 4$ and $\omega = 15°$

Now, $\quad \cos 15° = \dfrac{\sqrt{3}+1}{2\sqrt{2}}$ and $\sin 15° = \dfrac{\sqrt{3}-1}{2\sqrt{2}}$

By the normal form , the equation of the lines is

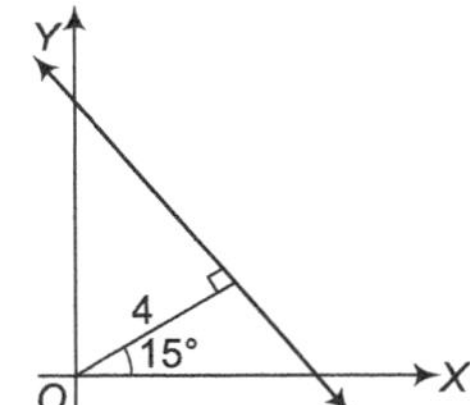

$$x\cos 15° + y\sin 15° = 4$$

or $\quad \dfrac{\sqrt{3}+1}{2\sqrt{2}}x + \dfrac{\sqrt{3}-1}{2\sqrt{2}}y = 4$

or $\quad (\sqrt{3}+1)x + (\sqrt{3}-1)y = 8\sqrt{2}$

This is the required equation.

92. *(c)* We know that the equation $y = mx + c$, contains two constants, namely, m and c. For finding these two constants, we need two conditions satisfied by the equation of line.

93. *(d)* (a) Any point on X-axis is $(x, 0)$, as y-coordinate on X-axis is zero.

Also slope on X-axis is $m = 0$

$\because$ Equation of line is $y - y_1 = m(x - x_1)$

$\therefore \quad y - 0 = 0(x - x) \quad (\because x_1 = x, y_1 = 0)$

$\Rightarrow \quad y = 0$

Hence, equation of X-axis, $y = 0$

Similarly, we can determine the equation of Y-axis, $x = 0$

(b) Equation of line in slope-point form is

$$y - y_1 = m(x - x_1)$$

Here, $\quad (x_1, y_1) = (-4, 3)$ and $m = \dfrac{1}{2}$

$\therefore \quad y - 3 = \dfrac{1}{2}(x + 4)$

$\Rightarrow 2y - 6 = x + 4 \Rightarrow x - 2y + 10 = 0$

(c) Here, $(x_1, y_1) = (0, 0)$, slope $= m$

Equation of line in slope-point form is

$$y - y_1 = m(x - x_1)$$

$\Rightarrow \quad y - 0 = m(x - 0)$

$\Rightarrow \quad y = mx$

(d) Here, $\quad (x_1, y_1) = (2, 2\sqrt{3})$

$\Rightarrow \quad m = \tan 75°$

$= \tan(45° + 30°) = \dfrac{\tan 45° + \tan 30°}{1 - \tan 45° \tan 30°}$

$\left[\because \tan(A+B) = \dfrac{\tan A + \tan B}{1 - \tan A \tan B}\right]$

$\Rightarrow \quad m = \dfrac{1 + \frac{1}{\sqrt{3}}}{1 - \frac{1}{\sqrt{3}}} = \dfrac{\sqrt{3}+1}{\sqrt{3}-1}$

$\Rightarrow \quad m = \dfrac{\sqrt{3}+1}{\sqrt{3}-1} \times \dfrac{\sqrt{3}+1}{\sqrt{3}+1} = \dfrac{(\sqrt{3}+1)^2}{3-1}$

(multiply numerator and denominator by $\sqrt{3}+1$)

$\Rightarrow \quad m = \dfrac{3 + 1 + 2\sqrt{3}}{2} = \dfrac{4 + 2\sqrt{3}}{2}$

$= 2 + \sqrt{3}$

Equation of line in one-point form is

$$y - y_1 = m(x - x_1)$$

$\Rightarrow \quad y - 2\sqrt{3} = (2 + \sqrt{3})(x - 2)$

$\Rightarrow \quad y - 2\sqrt{3} = (2 + \sqrt{3})x - 2(2 + \sqrt{3})$

$\Rightarrow \quad y - 2\sqrt{3} = (2 + \sqrt{3})x - 4 - 2\sqrt{3}$

$\Rightarrow (2 + \sqrt{3})x - y - 4 = 0$

94. *(d)* (a) Since, the line intersect the X-axis to the left of origin. It means it intersect the negative X-axis. Clearly, line AB passes through the point $(-3, 0)$ and $m = -2$.

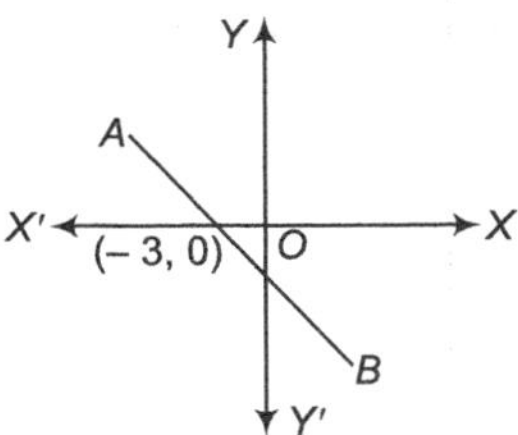

Equation of line in slope-point form is

$$y - y_1 = m(x - x_1)$$

$\Rightarrow \quad y - 0 = -2(x + 3)$

$\Rightarrow \quad y = -2x - 6$

$\Rightarrow \quad 2x + y + 6 = 0$

(b) Since, the line intersect the Y-axis, above the origin. It means it intersects the positive Y-axis. Here, $c = 2$

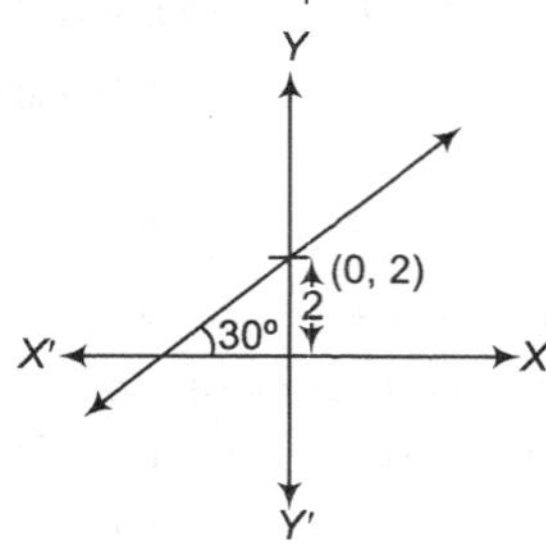

$$m = \tan 30° \Rightarrow m = \frac{1}{\sqrt{3}}$$

$\therefore$ Equation of line is $y = mx + c$

$$\Rightarrow \quad y = \frac{1}{\sqrt{3}} x + 2$$

$$\Rightarrow \quad \sqrt{3}y = x + 2\sqrt{3}$$

$$\Rightarrow \quad x - \sqrt{3}y + 2\sqrt{3} = 0$$

(c) Given points are $A(-1, 1)$ and $B(2, -4)$, then equation of line AB is

$$y - y_1 = \frac{y_2 - y_1}{x_2 - x_1}(x - x_1)$$

$$\Rightarrow \quad y - 1 = \frac{-4-1}{2+1}(x+1) \quad \left(\because x_1 = -1, y_1 = 1, \atop x_2 = 2, y_2 = -4\right)$$

$$\Rightarrow \quad y - 1 = \frac{-5}{3}(x+1)$$

$$\Rightarrow \quad 3y - 3 = -5x - 5$$

$$\Rightarrow \quad 5x + 3y + 2 = 0$$

95. *(b)* Here, $p = 5$ and $\alpha = 30°$

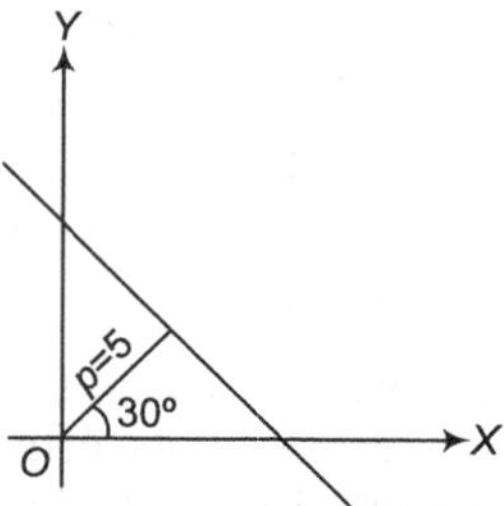

So, equation of line in normal form is

$$x \cos \alpha + y \sin \alpha = p$$

$$\therefore \quad x \cos 30° + y \sin 30° = 5$$

$$\Rightarrow \quad \frac{\sqrt{3}}{2} x + \frac{1}{2} y = 5 \Rightarrow \sqrt{3}x + y = 10$$

96. *(c)* Since, median bisects the opposite side *i.e.,* S is the mid-point of PQ.

$$S = \left(\frac{x_1 + x_2}{2}, \frac{y_1 + y_2}{2}\right)$$

$$= \left(\frac{2-2}{2}, \frac{1+3}{2}\right) = \left(0, \frac{4}{2}\right) = (0, 2)$$

$$(\because x_1 = 2, y_1 = 1, x_2 = -2, y_1 = 3)$$

$\therefore$ Equation of line RS by using

$$y - y_1 = \frac{y_2 - y_1}{x_2 - x_1}(x - x_1)$$

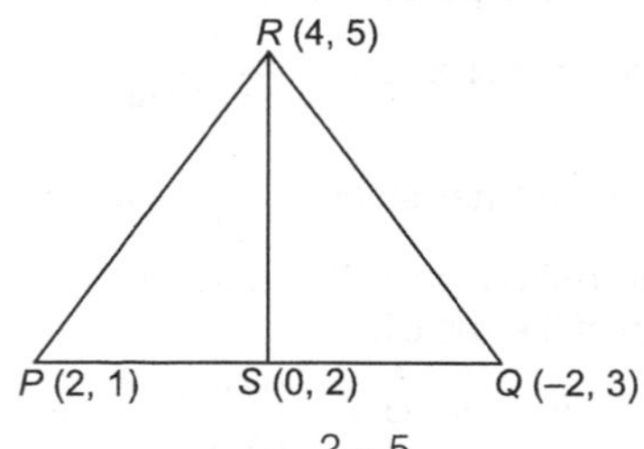

$$\Rightarrow \quad y - 5 = \frac{2-5}{0-4}(x - 4)$$

$$(\because x_1 = 4, y_1 = 5, x_2 = 0, y_2 = 2)$$

$$\Rightarrow \quad y - 5 = \frac{-3}{-4}(x - 4)$$

$$\Rightarrow \quad 4y - 20 = 3x - 12$$

$$\Rightarrow \quad 3x - 4y + 8 = 0$$

97. *(d)* Let the points P and Q are $(2, 5)$ and $(-3, 6)$. We have to find equation of AB.

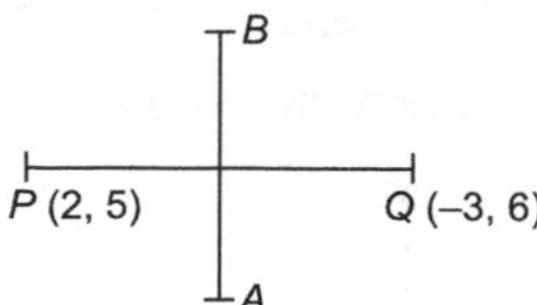

Let slope of AB be m.

Since, $AB \perp PQ$

$\therefore$ Slope of $AB \times$ Slope of $PQ = -1 \qquad (\because m_1 \times m_2 = -1)$

$$\Rightarrow \quad m \times \frac{y_2 - y_1}{x_2 - x_1} = -1$$

$$\Rightarrow \quad m \times \frac{6-5}{-3-2} = -1$$

$$\left(\because \text{slope of } PQ = \frac{6-5}{-3-2}\right)$$

$$\Rightarrow \quad m \times \frac{1}{-5} = -1 \Rightarrow m = 5$$

$\therefore$ Equation of line AB by using

$$y - y_1 = m(x - x_1)$$

$$y - 5 = 5(x + 3) \qquad (\because x_1 = -3, y_1 = 5)$$

$$\Rightarrow \quad y - 5 = 5x + 15$$

$$\Rightarrow \quad 5x - y + 20 = 0$$

98. *(b)* Equation of line in intercept form is

$$\frac{x}{a} + \frac{y}{b} = 1 \qquad \text{...(i)}$$

Since, the line makes equal intercepts on the axes *i.e.,*

$$a = b$$

$$\therefore \text{ From Eq. (i),} \quad \frac{x}{a} + \frac{y}{a} = 1$$

$$\Rightarrow \quad x + y = a \qquad \text{...(ii)}$$

Line (ii) passes through the point (2, 3), it will satisfy the line (ii).

$$\Rightarrow \quad 2 + 3 = a \Rightarrow a = 5$$

Put the value of a in Eq. (ii), we get

$$x + y = 5$$

99. *(c)* Equation of line in intercept form is

$$\frac{x}{a} + \frac{y}{b} = 1 \quad \ldots\text{(i)}$$

Given, $$a + b = 9 \quad \ldots\text{(ii)}$$

and line (i) passes through the point (2, 2), it will satisfy the line (i) *i.e.,* put $x = 2$, $y = 2$ in Eq. (i)

$$\Rightarrow \quad \frac{2}{a} + \frac{2}{b} = 1 \quad \ldots\text{(iii)}$$

Solve Eqs. (ii) and (iii), to find the values of a and b.

From Eq. (ii), $b = 9 - a$ put in Eq. (iii), we get

$$\frac{2}{a} + \frac{2}{9-a} = 1$$

$$\Rightarrow \quad 2(9-a) + 2a = a(9-a)$$

$$\Rightarrow \quad 18 - 2a + 2a = 9a - a^2$$

$$\Rightarrow \quad a^2 - 9a + 18 = 0$$

Factorize it by splitting the middle term,

$$\Rightarrow \quad a^2 - 6a - 3a + 18 = 0$$

$$\Rightarrow \quad a(a-6) - 3(a-6) = 0$$

$$\Rightarrow \quad (a-6)(a-3) = 0$$

$$\Rightarrow \quad a = 6 \text{ or } 3 \Rightarrow b = 3 \text{ or } 6$$

Hence, Eq. (i) becomes

$$\frac{x}{6} + \frac{y}{3} = 1 \quad \text{or} \quad \frac{x}{3} + \frac{y}{6} = 1$$

$$\Rightarrow \quad 3x + 6y = 18 \quad \text{or} \quad 6x + 3y = 18$$

$$\Rightarrow \quad x + 2y = 6 \quad \text{or} \quad 2x + y = 6 \qquad \text{(divide by 3)}$$

100. *(a)* Since, $PQ \perp OR$,

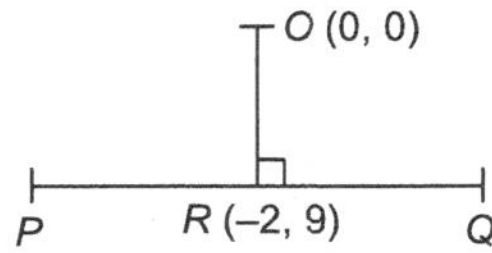

$\therefore$ Slope of $PQ \times$ Slope of $OR = -1$ $\quad (\because m_1 \times m_2 = -1)$

$$\Rightarrow \quad m \times \frac{y_2 - y_1}{x_2 - x_1} = -1 \Rightarrow m \times \frac{9-0}{-2-0} = -1$$

$$(\because x_1 = 0, y_1 = 0, x_2 = -2, y_2 = 9)$$

$$\Rightarrow \quad m \times \frac{9}{-2} = -1$$

$$\Rightarrow \quad m = \frac{2}{9}$$

Now, equation of PQ by using

$$y - y_1 = m(x - x_1)$$

$$y - 9 = \frac{2}{9}(x + 2) \qquad (\because x_1 = -2, y = 9)$$

$$\Rightarrow \quad 9y - 81 = 2x + 4$$

$$\Rightarrow \quad 2x - 9y + 85 = 0$$

101. *(b)* Given, $L = 124.942$ when $C = 20$ and $L = 125.134$ when $C = 110$

$\therefore$ In coordinate form (20, 124.942) and (110, 125.134) are two points.

Equation of linear function *i.e.,* line can be taken as

$$L - L_1 = \frac{L_2 - L_1}{C_2 - C_1}(C - C_1)$$

$$\left[\because y - y_1 = \frac{y_2 - y_1}{x_2 - x_1}(x - x_1)\right]$$

$$\Rightarrow \quad L - 124.942 = \frac{125.134 - 124.942}{110 - 20}(C - 20)$$

$$\Rightarrow \quad L - 124.942 = \frac{0.192}{90}(C - 20)$$

$$\Rightarrow \quad L = \frac{0.192}{90}(C - 20) + 124.942$$

102. *(d)* Let equation of line AB is

$$\frac{x}{h} + \frac{y}{k} = 1 \quad \ldots\text{(i)}$$

where, h and k are the intercepts made by the line on X-axis and Y-axis, respectively.

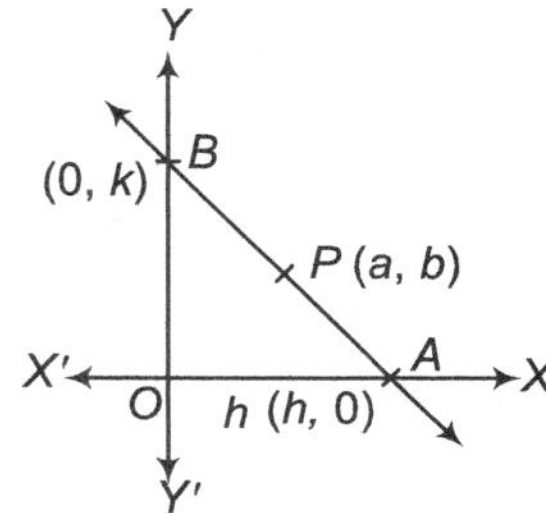

$$\text{Mid-point of } AB = \left(\frac{x_1 + x_2}{2}, \frac{y_1 + y_2}{2}\right) = \left(\frac{h+0}{2}, \frac{0+k}{2}\right)$$

$$(\because x_1 = h, y_1 = 0, x_2 = 0, y_2 = k)$$

$\because$ $P(a, b)$ is the mid-point of AB *i.e.,*

$$\frac{h+0}{2} = a \quad \text{and} \quad \frac{0+k}{2} = b$$

$$\Rightarrow \quad h = 2a \quad \text{and} \quad k = 2b$$

On putting the values of h and k in Eq. (i), we get

$$\frac{x}{2a} + \frac{y}{2b} = 1 \Rightarrow \frac{x}{a} + \frac{y}{b} = 2$$

103. *(a)* Let equation of line AB is

$$\frac{x}{a} + \frac{y}{b} = 1 \quad \ldots\text{(i)}$$

Let a point $R(h, k)$ divide line AB in the ratio 1 : 2.

By using section formula, (internal division)

$$R(h, k) = \left(\frac{1x_2 + 2x_1}{1+2}, \frac{1y_2 + 2y_1}{1+2}\right)$$

$$\Rightarrow \quad h = \frac{1 \times 0 + 2 \times a}{1+2},$$

$$k = \frac{1 \times b + 2 \times 0}{1+2}$$

$\left[\begin{array}{l} \because P(x, y) \text{ divide the line } (x_1, y_1) \text{ and } (x_2, y_2) \\ \text{in the ratio } m : n \text{ internally} \\ \therefore P(x, y) = \left(\dfrac{mx_2 + nx_1}{n+m}, \dfrac{my_2 + ny_1}{n+m}\right) \end{array}\right]$

$$\Rightarrow \quad h = \frac{2a}{3}, k = \frac{b}{3}$$

$$\Rightarrow \quad a = \frac{3h}{2}, b = 3k$$

On putting the values of a and b in Eq. (i), we get

$$\frac{x}{\frac{3h}{2}} + \frac{y}{3k} = 1$$

$$\Rightarrow \quad \frac{2x}{3h} + \frac{y}{3k} = 1$$

$$\Rightarrow \quad \frac{2kx + hy}{3hk} = 1$$

$$\Rightarrow \quad 2kx + hy = 3hk$$

104. *(a)* Let the given points are $A(3, 0)$, $B(-2, -2)$, and $C(8, 2)$.

Equation of line AB is

$$y - y_1 = \frac{y_2 - y_1}{x_2 - x_1}(x - x_1)$$

$$y - 0 = \frac{-2-0}{-2-3}(x - 3) \quad \left(\because x_1 = 3, y_1 = 0, x_2 = -2, y_2 = -2\right)$$

$$\Rightarrow \quad y = \frac{-2}{-5}(x - 3)$$

$$\Rightarrow \quad 5y = 2x - 6$$

$$\Rightarrow 2x - 5y - 6 = 0 \quad \text{...(i)}$$

On putting the point $C(8, 2)$ in Eq. (i), we get

$$2 \times 8 - 5 \times 2 - 6 = 0$$

$$\Rightarrow \quad 16 - 10 - 6 = 0$$

$$\Rightarrow \quad 16 - 16 = 0$$

$$\Rightarrow \quad 0 = 0$$

Hence, points A, B and C are collinear.

105. *(a)* Let the given points are $A(1, 0)$ and $B(2, 3)$. Let the line PQ divide AB in the ratio $1 : n$ at R.

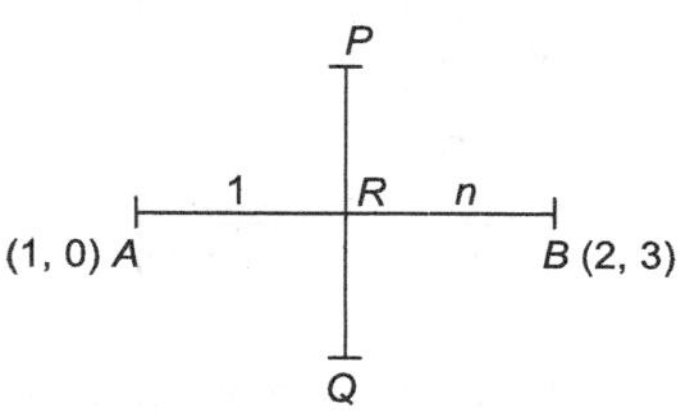

Using section formula for internal division

$$R = \left(\frac{1 \times x_2 + n \times x_1}{1+n}, \frac{1 \times y_2 + n \times y_1}{1+n}\right)$$

$$= \left(\frac{1 \times 2 + n \times 1}{1+n}, \frac{1 \times 3 + n \times 0}{1+n}\right)$$

$$(\because x_1 = 1, y_1 = 0, x_2 = 2, y_2 = 3)$$

$$= \left(\frac{n+2}{n+1}, \frac{3}{1+n}\right)$$

Also, $PQ \perp AB$

Let slope of line PQ is m.

$\therefore$ Slope of line $PQ \times$ Slope of line $AB = -1$ $\quad (\because m_1 m_2 = -1)$

$$\Rightarrow \quad m \times \frac{y_2 - y_1}{x_2 - x_1} = -1$$

$$\Rightarrow \quad m \times \frac{3-0}{2-1} = -1$$

$$\Rightarrow \quad m \times 3 = -1$$

$$\Rightarrow \quad m = -\frac{1}{3}$$

Now, equation of line PQ by using

$$y - y_1 = m(x - x_1)$$

$$\Rightarrow \quad y - \frac{3}{1+n} = \frac{-1}{3}\left(x - \frac{n+2}{n+1}\right)$$

$$\left[\because R\left(\frac{n+2}{n+1}, \frac{3}{1+n}\right) = (x_1, y_1)\right]$$

$$\Rightarrow \quad \frac{3(n+1)y - 9}{1+n} = \frac{-x(n+1) + (n+2)}{n+1}$$

$$\Rightarrow \quad 3(n+1)y - 9 = -x(n+1) + (n+2)$$

$$\Rightarrow \quad x(n+1) + 3(n+1)y = n + 2 + 9$$

$$\Rightarrow \quad x(n+1) + 3(n+1)y = n + 11$$

106. *(b)* $\because m = \tan\theta = \tan\frac{2\pi}{3}$

$$= \tan\left(\pi - \frac{\pi}{3}\right) = -\tan\frac{\pi}{3} \quad [\because \tan(\pi - \theta) = -\tan\theta]$$

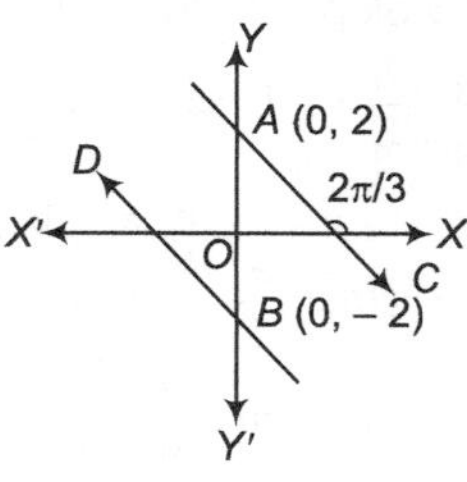

$$m = -\sqrt{3}$$

Hence, equation of line AC through the point $A(0, 2)$ by using $y - y_1 = m(x - x_1)$

$$\Rightarrow \quad y - 2 = -\sqrt{3}(x - 0) \quad (\because x_1 = 0, y_1 = 2)$$

$$\Rightarrow \quad y - 2 = -\sqrt{3}x \Rightarrow \sqrt{3}x + y - 2 = 0$$

Also, the line BD intersect below Y-axis. It means the coordinate of y is negative and x-coordinate is 0. *i.e.,* point is $(0, -2)$.

Hence, equation of line through the point $B(0, -2)$ is

$$y + 2 = -\sqrt{3}(x - 0) \quad (\because y - y_1 = m(x - x_1))$$

$$\Rightarrow \quad \sqrt{3}x + y + 2 = 0$$

107. *(c)* Let price and litre be denoted in ordered pair (x, y) where x denotes the rupees per litre and y denotes the quantity of milk in litre.

Given $(14, 980)$ and $(16, 1220)$ are two points.

Let linear relations *i.e.*, linear equation is

$$y - y_1 = \frac{y_2 - y_1}{x_2 - x_1}(x - x_1)$$

$$\Rightarrow \quad y - 980 = \frac{1220 - 980}{16 - 14}(x - 14)$$

$$\Rightarrow \quad y - 980 = \frac{240}{2}(x - 14) \quad \left(\because \begin{matrix} x_1 = 14, y_1 = 980, \\ x_2 = 16, y_2 = 1220 \end{matrix}\right)$$

$$\Rightarrow \quad y - 980 = 120(x - 14)$$

$$\Rightarrow \quad y - 980 = 120x - 120 \times 14$$

$$\Rightarrow \quad 120x - y = 1680 - 980$$

$$\Rightarrow \quad 120x - y = 700$$

When price $x = 17$,

$$\Rightarrow \quad 120 \times 17 - y = 700$$

$$\Rightarrow \quad y = 2040 - 700$$

$$y = 1340$$

He will sell weekly 1340 L milk at the rate of ₹ 17 per L.

108. *(b)* Three lines are said to be concurrent, if they pass through a common point, *i.e.,* point of intersection of any two lines lies on the third line. Here given lines are

$$2x + y - 3 = 0 \quad \ldots(i)$$

$$5x + ky - 3 = 0 \quad \ldots(ii)$$

$$3x - y - 2 = 0 \quad \ldots(iii)$$

Solving Eqs. (i) and (iii) by cross-multiplication method, we get

$$\frac{x}{-2-3} = \frac{y}{-9+4} = \frac{1}{-2-3} \quad \text{or} \quad x = 1, y = 1$$

Therefore, the point of intersection of two lines is $(1, 1)$. Since, above three lines are concurrent, the point $(1, 1)$ will satisfy Eq. (ii), so that

$$5.1 + k.1 - 3 = 0 \quad \text{or} \quad k = -2$$

109. *(c)* Given line is $4x - y = 0$ $\ldots(i)$

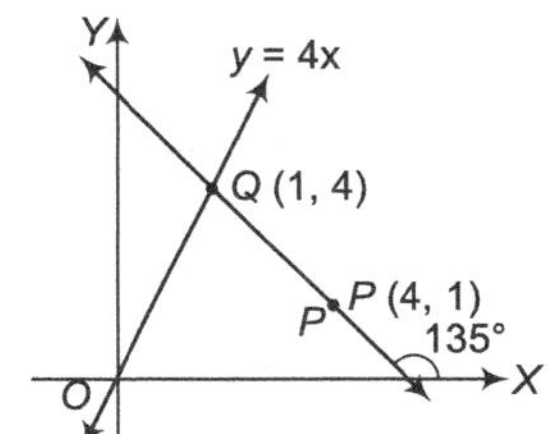

In order to find the distance of the line (i) from the point $P(4, 1)$ along another line, we have to find the point of intersection of both the lines. For this purpose, we will first find the equation of the second line. Slope of second line is $\tan 135° = -1$. Equation of the point with slope -1 through the point $P(4, 1)$ is

$$y - 1 = -1(x - 4) \text{ or } x + y - 5 = 0 \quad \ldots(ii)$$

Solving Eqs. (i) and (ii), we get $x = 1$ and $y = 4$ so that point of intersection of the two lines is $Q(1, 4)$. Now, distance of the line (i) from the point $P(4, 1)$ along the line (ii) is

The distance between the points $P(4, 1)$ and $Q(1, 4)$.

$$= \sqrt{(1-4)^2 + (4-1)^2} = 3\sqrt{2} \text{ units.}$$

110. *(a)* Equation of given line is

$$(k - 3)x - (4 - k^2)y + k^2 - 7k + 6 = 0 \quad \ldots(i)$$

I. When line (i) is parallel to X-axis, then slope = 0

i.e., $$\frac{(k-3)}{(4-k^2)} = 0 \quad \left[\because \text{slope of line (i) is } \frac{k-3}{4-k^2}\right]$$

$$\Rightarrow \quad k - 3 = 0 \Rightarrow k = 3$$

II. When line (i) is parallel to Y-axis, then slope $= \infty = \frac{1}{0}$

i.e., $$\frac{k-3}{4-k^2} = \frac{1}{0} \quad \left[\because \text{slope of line (i) is } \frac{k-3}{4-k^2}\right]$$

$$\Rightarrow \quad 4 - k^2 = 0$$

$$\Rightarrow \quad k = \pm 2$$

III. When line (i) passing through origin then point $(0, 0)$, will satisfy it

$$0 - 0 + k^2 - 7k + 6 = 0$$

[put $x = 0$, $y = 0$ in Eq. (i)]

$$\Rightarrow \quad k^2 - 6k - k + 6 = 0$$

$$\Rightarrow \quad k(k-6) - 1(k-6) = 0$$

$$\Rightarrow \quad (k-6)(k-1) = 0$$

$$\Rightarrow \quad k = 1, 6$$

111. *(c)* Let the intercept form of a line is $\frac{x}{a} + \frac{y}{b} = 1$ $\ldots(i)$

Given, $$a + b = 1 \quad \ldots(ii)$$

and $$ab = -6 \quad \ldots(iii)$$

For solving Eqs. (ii) and (iii), *i.e.*, to find the value of a and b, put $b = \frac{-6}{a}$ from Eq. (iii) in Eq. (ii), we get

$$a - \frac{6}{a} = 1$$

$$\Rightarrow \quad \frac{a}{1} - \frac{6}{a} = 1$$

$$\Rightarrow \quad a^2 - 6 = a$$

$$\Rightarrow \quad a^2 - a - 6 = 0$$

$$\Rightarrow \quad a^2 - (3 - 2)a - 6 = 0$$

$$\Rightarrow \quad a^2 - 3a + 2a - 6 = 0$$

$$\Rightarrow \quad a(a - 3) + 2(a - 3) = 0$$

$$\Rightarrow \quad (a + 2)(a - 3) = 0$$

$$\Rightarrow \qquad a = 3, -2$$
$$b = -2, 3$$

Put $a = 3, -2$ and $b = -2, 3$ respectively in Eq. (i), we get the required equations of lines

$$\frac{x}{3} + \frac{y}{-2} = 1 \quad \text{or} \quad \frac{x}{-2} + \frac{y}{3} = 1$$

$$\Rightarrow \qquad 2x - 3y = 6 \quad \text{or} \quad -3x + 2y = 6$$

$$\Rightarrow \qquad 2x - 3y - 6 = 0 \quad \text{or} \quad 3x - 2y + 6 = 0$$

112. *(a)* Given, equation of lines are

$$x - 7y + 5 = 0 \qquad \ldots\text{(i)}$$

and
$$3x + y = 0 \qquad \ldots\text{(ii)}$$

On solving Eqs. (i) and (ii), to find the point of intersection from Eq. (i)

$$x = 7y - 5$$

On putting the value of x in Eq. (ii), we get

$$3(7y - 5) + y = 0$$

$$\Rightarrow \qquad 21y - 15 + y = 0$$

$$\Rightarrow \qquad 22y = 15$$

$$\Rightarrow \qquad y = \frac{15}{22}$$

$$\therefore \text{From Eq. (i)}, \quad x = \frac{7 \times 15}{22} - 5 = \frac{105 - 110}{22}$$

$$\Rightarrow \qquad x = -\frac{5}{22}$$

Hence, the point is $\left(-\frac{5}{22}, \frac{15}{22}\right)$

Hence, equation of required line by using

$$y - y_1 = m(x - x_1)$$

where $(x_1, y_1) = \left(\frac{-5}{22}, \frac{15}{22}\right)$ and $m = \frac{1}{0}$ as line parallel to Y-axis, is

$$y - \frac{15}{22} = \frac{1}{0}\left(x + \frac{5}{22}\right)$$

$$\Rightarrow \qquad 0 = x + \frac{5}{22} \Rightarrow 22x + 5 = 0$$

113. *(a)* If the line which passes through the point (2, 3) and makes an angle of 30° with the positive direction of X-axis, then

Slope of the line is $m = \tan\theta$

$$= \tan 30° = \frac{1}{\sqrt{3}}$$

Therefore, using point slope formula of the equation of a line, we have

$$y - 3 = \frac{1}{\sqrt{3}}(x - 2)$$

$$\Rightarrow \qquad \sqrt{3}y - 3\sqrt{3} = x - 2$$

$$\Rightarrow \qquad x - \sqrt{3}y + (3\sqrt{3} - 2) = 0$$

114. *(a)* Changing the given equation of the line into intercept form, we have $\dfrac{x}{\frac{p}{\cos\alpha}} + \dfrac{y}{\frac{p}{\sin\alpha}} = 1$ which gives the coordinates $\left(\frac{p}{\cos\alpha}, 0\right)$ and $\left(0, \frac{p}{\sin\alpha}\right)$, where the line intersects X-axis and Y-axis, respectively.

Let (h, k) denote the mid-point of the line segment joining the points $\left(\frac{p}{\cos\alpha}, 0\right)$ and $\left(0, \frac{p}{\sin\alpha}\right)$.

Then, $\quad h = \dfrac{p}{2\cos\alpha}$ and $k = \dfrac{p}{2\sin\alpha}$

This gives $\cos\alpha = \dfrac{p}{2h}$ and $\sin\alpha = \dfrac{p}{2k}$

Squaring and adding, we get

$$\frac{p^2}{4h^2} + \frac{p^2}{4k^2} = 1$$

or
$$\frac{1}{h^2} + \frac{1}{k^2} = \frac{4}{p^2}$$

Therefore, the required locus is $\dfrac{1}{x^2} + \dfrac{1}{y^2} = \dfrac{4}{p^2}$.

115. *(b)* Let the given lines represent the sides of the triangle as

$$AB : y - x = 0 \qquad \ldots\text{(i)}$$
$$BC : x + y = 0 \qquad \ldots\text{(ii)}$$
and
$$CA : x - k = 0 \qquad \ldots\text{(iii)}$$

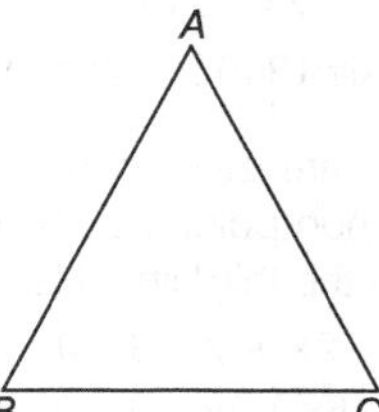

For point A, on solving Eqs. (i) and (iii), we get $A(k, k)$.

For point B, on solving Eqs. (i) and (ii), we get $B(0, 0)$.

For point C, on solving Eqs. (ii) and (iii), we get $C(k, -k)$.

So, the required points are $A(k, k)$, $B(0, 0)$ and $C(k, -k)$.

Hence, area of ΔABC

$$= \frac{1}{2}[k(0 + k) + 0(-k - k) + k(k - 0)]$$

$$\left[\because \text{area of triangle} = \frac{1}{2}\{x_1(y_2 - y_3) + x_2(y_3 - y_1) + x_3(y_1 - y_2)\}\right]$$

$$= \frac{1}{2}[k^2 + k^2] = \frac{2k^2}{2} = k^2 \text{ sq unit}$$

116. *(d)* Given, equation of lines are

$$3x + y - 2 = 0 \qquad \ldots\text{(i)}$$
$$px + 2y - 3 = 0 \qquad \ldots\text{(ii)}$$
and
$$2x - y - 3 = 0 \qquad \ldots\text{(iii)}$$

Adding Eqs. (i) and (iii), we get

$$5x - 5 = 0$$

$$\Rightarrow \quad x = 1$$

$\therefore$ From Eq. (i), $3 \times 1 + y - 2 = 0$ (put $x = 1$)

$$\Rightarrow \quad y = -1$$

So, the point is $(1, -1)$.

$\because$ Lines are concurrent

Point $(1, -1)$ will satisfy Eq. (ii) *i.e.*,

$$p \times 1 + 2(-1) - 3 = 0 \quad \text{(put } x = 1, y = -1\text{)}$$

$$\Rightarrow \quad p = 5$$

117. *(c)* $\because A = (2, 0)$ and $B = (3, 1)$

$\therefore$ Slope of the line AB is $\dfrac{1-0}{3-2} = 1 = \tan 45°$.

After rotation of line AB through an angle of 15° in anti-clockwise direction, the slope of the new line (AC) is $\tan(45° + 15°)$

or $\quad \tan 60° = \sqrt{3}$

Then, the equation of new line AC is

$$y - 0 = \sqrt{3}(x - 2)$$

$$\Rightarrow \quad y - \sqrt{3}x + 2\sqrt{3} = 0$$

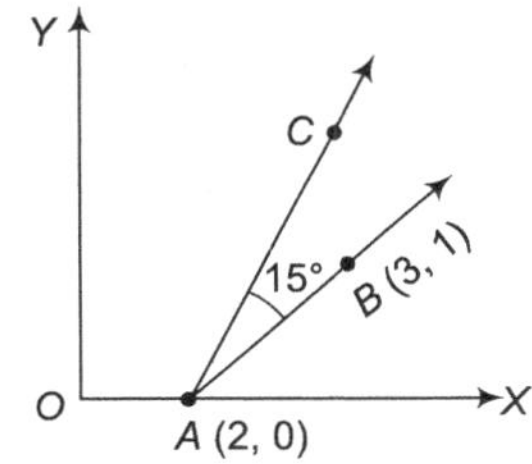

118. *(c)* Equation of a line passing through (3, 2) having slope $\dfrac{3}{4}$ is given by

$$y - 2 = \frac{3}{4}(x - 3)$$

$$\Rightarrow \quad 4y - 8 = 3x - 9$$

$$\Rightarrow \quad 4y - 3x + 1 = 0 \quad \text{...(i)}$$

Let (h, k) be the point on the line which is 5 units away from the point A. Then,

$$\sqrt{(h-3)^2 + (k-2)^2} = 5 \quad \text{...(ii)}$$

Also, we have

$$4k - 3h + 1 = 0 \quad \text{[from Eq. (i)]}$$

$$\Rightarrow \quad k = \frac{3h - 1}{4} \quad \text{...(iii)}$$

Putting the value of k in Eq. (ii) we get

$$\sqrt{(h-3)^2 + \left(\frac{3h-1}{4} - 2\right)^2} = 5$$

$$\Rightarrow \quad 16(h-3)^2 + (3h-9)^2 = 400$$

$$\Rightarrow \quad 16h^2 + 144 - 96h + 9h^2 + 81 - 54h = 400$$

$$\Rightarrow \quad 25h^2 - 150h - 175 = 0$$

$$\Rightarrow \quad h^2 - 6h - 7 = 0$$

$$\Rightarrow \quad (h+1)(h-7) = 0$$

$$\Rightarrow \quad h = -1, \ h = 7$$

Putting these value of h in Eq. (iii), we get

$$k = \frac{3(-1) - 1}{4}$$

$$\Rightarrow \quad k = -1$$

and $\quad k = \dfrac{3(7) - 1}{4}$

$$\Rightarrow \quad k = 5$$

Therefore, the coordinates of the required points are either $(-1, -1)$ or $(7, 5)$.

119. *(b)* Given equation of lines are

$$y = m_1x + c_1 \quad \text{...(i)}$$

$$y = m_2x + c_2 \quad \text{...(ii)}$$

$$y = m_3x + c_3 \quad \text{...(iii)}$$

On solving Eqs. (i) and (ii), we get

$$m_1x + c_1 = m_2x + c_2$$

$$\Rightarrow \quad m_1x - m_2x = c_2 - c_1$$

$$\Rightarrow \quad x = \frac{c_2 - c_1}{m_1 - m_2}$$

From Eq. (i), $\quad y = m_1\left(\dfrac{c_2 - c_1}{m_1 - m_2}\right) + \dfrac{c_1}{1}$

$$\Rightarrow \quad y = \frac{m_1c_2 - m_1c_1 + m_1c_1 - m_2c_1}{m_1 - m_2}$$

$$\Rightarrow \quad y = \frac{m_1c_2 - m_2c_1}{m_1 - m_2}$$

On putting the values of x and y in Eq. (iii), we get

$$\frac{m_1c_2 - m_2c_1}{m_1 - m_2} = m_3\left(\frac{c_2 - c_1}{m_1 - m_2}\right) + \frac{c_3}{1}$$

$$\Rightarrow \quad \frac{m_1c_2 - m_2c_1}{m_1 - m_2} = \frac{m_3c_2 - m_3c_1 + m_1c_3 - m_2c_3}{m_1 - m_2}$$

$$\Rightarrow \quad m_1c_2 - m_2c_1 = m_3c_2 - m_3c_1 + m_1c_3 - m_2c_3$$

$$\Rightarrow \quad m_1c_2 - m_2c_1 - m_3c_2 + m_3c_1 - m_1c_3 + m_2c_3 = 0$$

$$\Rightarrow \quad m_1(c_2 - c_3) + m_2(c_3 - c_1) + m_3(c_1 - c_2) = 0$$

Hence proved.

120. *(c)* The general equation of first degree in two variables, $Ax + By + C = 0$, where A, B and C are real constants such that A and B are not zero simultaneously.

121. *(b)* The graph of the equation $Ax + By + C = 0$ is always a straight line.

122. *(b)* Any equation of the form $Ax + By + C = 0$, where A and B are not zero simultaneously is called general linear equation or general equation of a line.

123. *(d)* If $B \neq 0$, then $Ax + By + C = 0$ can be written as

$$y = -\frac{A}{B}x - \frac{C}{B} \text{ or } y = mx + c \quad \dots(i)$$

where, $m = -\frac{A}{B}$ and $c = -\frac{C}{B}$. We know that Eq. (i) is the slope-intercept form of the equation of a line whose slope is $-\frac{A}{B}$ *i.e.,* *y*-intercept is $-\frac{C}{B}$.

If $B = 0$, then $x = -\frac{C}{A}$, which is a vertical line whose slope is undefined and *x*-intercept is $-\frac{C}{A}$.

124. *(a)* If $C \neq 0$, then $Ax + By + C = 0$ can be written as

$$\frac{x}{\left(-\frac{C}{A}\right)} + \frac{y}{\left(-\frac{C}{B}\right)} = 1 \text{ or } \frac{x}{a} + \frac{y}{b} = 1 \quad \dots(i)$$

where, $a = -\frac{C}{A}$

and $b = -\frac{C}{B}$.

We know that, Eq. (i) is intercept form of the equation of a line whose *x*-intercept is $-\frac{C}{A}$ and *y*-intercept is $-\frac{C}{B}$.

If $C = 0$ then $Ax + By + C = 0$ can be written as $Ax + By + C = 0$, which is a line passing through the origin and, therefore, has zero intercepts on the axes.

125. *(c)* Let $x\cos\omega + y\sin\omega = p$ be the normal form of the line represented by the equation $Ax + By + C = 0$ or $Ax + By = -C$. Thus, both the equation are same and therefore,

$$\frac{A}{\cos\omega} = \frac{B}{\sin\omega} = -\frac{C}{p}$$

which gives $\cos\omega = -\frac{Ap}{C}$

and $\sin\omega = -\frac{Bp}{C}$

Now, $\sin^2\omega + \cos^2\omega = \left(-\frac{Ap}{C}\right)^2 + \left(-\frac{Bp}{C}\right)^2 = 1$

or $p^2 = \frac{C^2}{A^2 + B^2}$

or $p = \pm\frac{C}{\sqrt{A^2 + B^2}}$

Therefore, $\cos\omega = \pm\frac{A}{\sqrt{A^2 + B^2}}$

and $\sin\omega = \pm\frac{B}{\sqrt{A^2 + B^2}}$

Thus, the normal form of the equation $Ax + By + C = 0$ is $x\cos\omega + y\sin\omega = p$, where

$\cos\omega = \pm\frac{A}{\sqrt{A^2 + B^2}}$, $\sin\omega = \pm\frac{B}{\sqrt{A^2 + B^2}}$

and $p = \pm\frac{C}{\sqrt{A^2 + B^2}}$.

126. *(a)* I. Given equation $3x - 4y + 10 = 0$ can be written as

$$y = \frac{3}{4}x + \frac{5}{2} \quad \dots(i)$$

Comparing Eq. (i) with $y = mx + c$, we have slope of the given line as $m = \frac{3}{4}$

II. Equation $3x - 4y + 10 = 0$ can be written as

$$3x - 4y = -10 \text{ or } \frac{x}{-\frac{10}{3}} + \frac{y}{\frac{5}{2}} = 1 \quad \dots(ii)$$

Comparing Eq. (ii) with $\frac{x}{a} + \frac{y}{b} = 1$, we have *x*-intercept as $a = -\frac{10}{3}$ and *y*-intercept as $b = \frac{5}{2}$.

127. *(b)* Given equation is

$$\sqrt{3}x + y - 8 = 0 \quad \dots(i)$$

Dividing Eq. (i) by $\sqrt{(\sqrt{3})^2 + (1)^2} = 2$, we get

$$\frac{\sqrt{3}}{2}x + \frac{1}{2}y = 4 \text{ or } \cos 30°\, x + \sin 30°\, y = 4 \quad \dots(ii)$$

Comparing Eq. (ii) with $x\cos\omega + y\sin\omega = p$, we get $p = 4$ and $\omega = 30°$.

128. *(c)* Given lines are

$$y - \sqrt{3}x - 5 = 0 \text{ or } y = \sqrt{3}x + 5 \quad \dots(i)$$

$$\sqrt{3}y - x + 6 = 0 \text{ or } y = \frac{1}{\sqrt{3}}x - 2\sqrt{3} \quad \dots(ii)$$

Slope of line (i) is $m_1 = \sqrt{3}$ and slope of line (ii) is $m_2 = \frac{1}{\sqrt{3}}$.

The acute angle (say) θ between two lines is given by

$$\tan\theta = \left|\frac{m_2 - m_1}{1 + m_1 m_2}\right| \quad \dots(iii)$$

Putting the values of m_1 and m_2 in Eq. (iii), we get

$$\tan\theta = \left|\frac{\frac{1}{\sqrt{3}} - \sqrt{3}}{1 + \sqrt{3} \times \frac{1}{\sqrt{3}}}\right| = \left|\frac{1-3}{2\sqrt{3}}\right| = \frac{1}{\sqrt{3}}$$

which gives $\theta = 30°$. Hence, angle between two lines is either 30° or 180° – 30° = 150°.

129. *(b)* Given lines can be written as

$$y = -\frac{a_1}{b_1}x - \frac{c_1}{b_1} \quad \dots(i)$$

and $$y = -\frac{a_2}{b_2}x - \frac{c_2}{b_2} \quad \text{...(ii)}$$

Slopes of the lines (i) and (ii) are $m_1 = -\frac{a_1}{b_1}$ and $m_2 = -\frac{a_2}{b_2}$, respectively. Now,

I. Lines are parallel, if $m_1 = m_2$, which gives

$$-\frac{a_1}{b_1} = -\frac{a_2}{b_2}$$

or $$\frac{a_1}{b_1} = \frac{a_2}{b_2}$$

II. Lines are perpendicular, if $m_1 \cdot m_2 = -1$, which gives

$$\frac{a_1}{b_1} \cdot \frac{a_2}{b_2} = -1$$

or $$a_1a_2 + b_1b_2 = 0$$

130. *(d)* Given, line $x - 2y + 3 = 0$ can be written as

$$y = \frac{1}{2}x + \frac{3}{2} \quad \text{...(i)}$$

Slope of the line (i) is $m_1 = \frac{1}{2}$. Therefore, slope of the line perpendicular to line (i) is $m_2 = -\frac{1}{m_1} = -2$.

Equation of the line with slope -2 and passing through the point $(1, -2)$ is

$$y - (-2) = -2(x - 1) \text{ or } y = -2x$$

which is the required equation.

131. *(c)* The distance of a point from a line is the length of the perpendicular drawn from the point to the line.

132. *(b)* Let $L: Ax + By + C = 0$ be a line, whose distance from the point $P(x_1, y_1)$ is d. Draw a perpendicular PM from the point P to the line L.

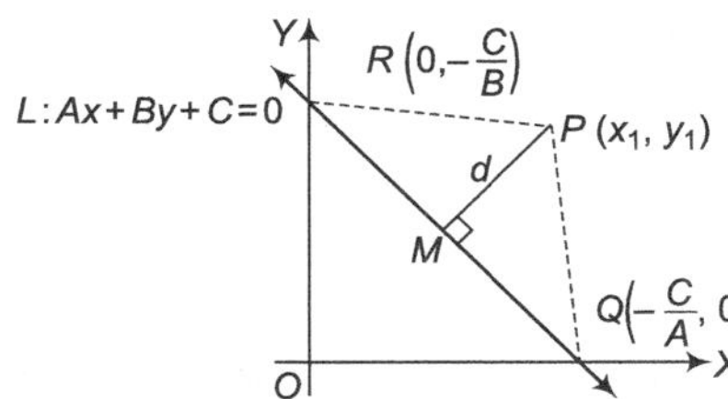

If the line meets the X and Y-axes at the points Q and R respectively, then coordinates of the point are $Q\left(-\frac{C}{A}, 0\right)$ and $R\left(0, -\frac{C}{B}\right)$. Thus, the area of the ΔPQR is given by

area $(\Delta PQR) = \frac{1}{2} PM \cdot QR$, which gives

$$PM = \frac{2 \text{ area}(\Delta PQR)}{QR} \quad \text{...(i)}$$

Also, area (ΔPQR)

$$= \frac{1}{2}\left|x_1\left(0 + \frac{C}{B}\right) + \left(-\frac{C}{A}\right)\left(-\frac{C}{B} - y_1\right) + 0(y_1 - 0)\right|$$

$$= \frac{1}{2}\left|x_1\frac{C}{B} + y_1\frac{C}{A} + \frac{C^2}{AB}\right|$$

or 2 area $(\Delta PQR) = \left|\frac{C}{AB}\right| \cdot |Ax_1 + By_1 + C|$

and $$QR = \sqrt{\left(0 + \frac{C}{A}\right)^2 + \left(\frac{C}{B} - 0\right)^2}$$

$$= \left|\frac{C}{AB}\right|\sqrt{A^2 + B^2}$$

Substituting the values of area (ΔPQR) and QR in Eq. (i), we get

$$PM = \frac{|Ax_1 + By_1 + C|}{\sqrt{A^2 + B^2}}$$

or $$d = \frac{|Ax_1 + By_1 + C|}{\sqrt{A^2 + B^2}}$$

Thus, the perpendicular distance (d) of a line $Ax + By + C = 0$ from a point (x_1, y_1) is given by

$$d = \frac{|Ax_1 + By_1 + C|}{\sqrt{A^2 + B^2}}$$

133. *(d)*

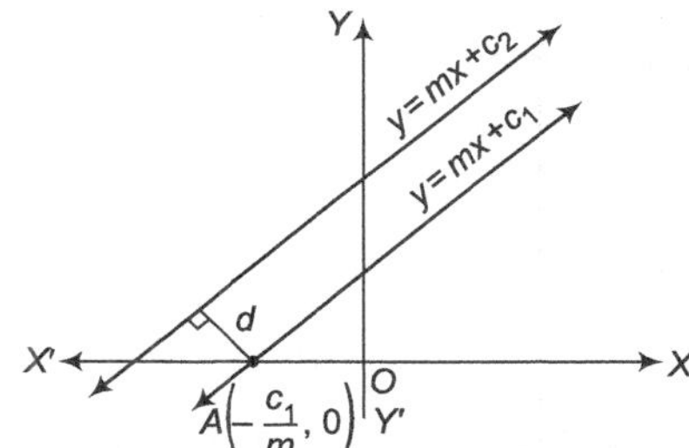

Two parallel lines can be taken in the form

$$y = mx + c_1 \quad \text{...(i)}$$

and $$y = mx + c_2 \quad \text{...(ii)}$$

Line (i) will intersect X-axis at the point $A\left(-\frac{c_1}{m}, 0\right)$.

Distance between two lines is equal to the length of the perpendicular from point A to line (ii). Therefore, distance between the lines (i) and (ii) is

$$d = \frac{\left|(-m)\left(-\frac{c_1}{m}\right) + (-c_2)\right|}{\sqrt{1 + m^2}}$$

or $$d = \frac{|c_1 - c_2|}{\sqrt{1 + m^2}}$$

Thus, the distance d between two parallel lines $y = mx + c_1$ and $y = mx + c_2$ is given by

$$d = \frac{|c_1 - c_2|}{\sqrt{1 + m^2}}.$$

134. *(a)* If lines are given in general form, *i.e.*, $Ax + By + C_1 = 0$ and $Ax + By + C_2 = 0$, then the distance between them is given by

$$d = \frac{\left|\left(\frac{-C_1}{B}\right) - \left(\frac{-C_2}{B}\right)\right|}{\sqrt{1+\left(\frac{-A}{B}\right)^2}} \quad \left(\because \text{ given liens are } y = -\frac{Ax}{B} - \frac{C_1}{B} \text{ and } y = \frac{-Ax}{B} - \frac{C_2}{B}\right)$$

$$= \frac{|C_1 - C_2|}{\sqrt{A^2 + B^2}}$$

135. *(d)* Given line is $3x - 4y - 26 = 0$ ...(i)

On comparing Eq. (i) with general equation of line $Ax + By + C = 0$, we get

$$A = 3, B = -4$$

and $C = -26$

Given point is $(x_1, y_1) = (3, -5)$. The distance of the given point from given line is

$$d = \frac{|Ax_1 + By_1 + C|}{\sqrt{A^2 + B^2}}$$

$$= \frac{|3 \cdot 3 + (-4)(-5) - 26|}{\sqrt{3^2 + (-4)^2}} = \frac{3}{5}$$

136. *(c)* Here, $A = 3$, $B = -4$, $C_1 = 7$, and $C_2 = 5$. Therefore, the required distance is

$$d = \frac{|7-5|}{\sqrt{3^2 + (-4)^2}} = \frac{2}{5}$$

137. *(d)* I. Given equation is $x + 7y = 0$

$$\Rightarrow \quad y = \frac{-x}{7} + 0$$

On comparing with $y = mx + c$, we get

Slope $(m) = \frac{-1}{7}$, y-intercept $= 0$

II. Given equation is $6x + 3y - 5 = 0$

$$\Rightarrow \quad y = -2x + \frac{5}{3}$$

On comparing with $y = mx + c$, we get

Slope $(m) = -2$, y-intercept $= \frac{5}{3}$

III. Given equation is $y = 0$

$$\Rightarrow \quad y = 0 \cdot x + 0$$

On comparing with $y = mx + c$, we get

Slope $(m) = 0$, y-intercept $= 0$

138. *(a)* (i) Given, $3x + 2y = 12$

$$\Rightarrow \quad \frac{3x}{12} + \frac{2y}{12} = 1$$

$$\Rightarrow \quad \frac{x}{4} + \frac{y}{6} = 1$$

On comparing with $\frac{x}{a} + \frac{y}{b} = 1$, we get

$$\Rightarrow \quad a = 4, b = 6$$

$\therefore$ x-intercept $= 4$

(ii) Given, $4x - 3y = 6$

$$\Rightarrow \quad \frac{4x}{6} - \frac{3y}{6} = 1$$

$$\Rightarrow \quad \frac{x}{\frac{6}{4}} - \frac{y}{2} = 1$$

$$\Rightarrow \quad \frac{x}{\frac{3}{2}} + \frac{y}{(-2)} = 1$$

On comparing with $\frac{x}{a} + \frac{y}{b} = 1$, we get

$$x\text{-intercept} = \frac{3}{2}$$

(iii) Given, $3y + 2 = 0$

$$\Rightarrow \quad 3y = -2$$

$$\Rightarrow \quad y = -\frac{2}{3}$$

or $\quad 0 \cdot x + \frac{y}{(-2/3)} = 1$

On comparing with $\frac{x}{a} + \frac{y}{b} = 1$, we get

x-intercept does not exist and y-intercept is $-\frac{2}{3}$.

139. *(c)* I. Given equation of line is

$$x - \sqrt{3}y + 8 = 0$$

$$\Rightarrow \quad x - \sqrt{3}y = -8$$

$$\Rightarrow \quad -x + \sqrt{3}y = 8 \quad \text{...(i)}$$

Now, divide by $\sqrt{(\text{coefficient of } x)^2 + (\text{coefficient of } y)^2}$

i.e., $\sqrt{(-1)^2 + (\sqrt{3})^2} = \sqrt{1+3} = 2$, we get

$$-\frac{1}{2}x + \frac{\sqrt{3}}{2}y = \frac{8}{2}$$

$$\Rightarrow \quad -\cos 60^\circ x + \sin 60^\circ y = 4$$

($\because$ convert in form of $x\cos\alpha + y\sin\alpha = p$)

($\because$ cos x is negative and sin x is positive, it is possible in second quadrant)

$$\Rightarrow x\cos(180^\circ - 60^\circ) + y\sin(180^\circ - 60^\circ) = 4$$

$$\Rightarrow \quad x\cos 120^\circ + y\sin 120^\circ = 4$$

$$\left[\because \cos(180^\circ - \theta) = -\cos\theta \text{ and } \sin(180^\circ - \theta) = \sin\theta\right]$$

On comparing with $x\cos\alpha + y\sin\alpha = p$, we get

$$\alpha = 120^\circ, p = 4$$

Note II. *$p = 4$ is the perpendicular distance of line from the origin.*

III. *$\alpha = 120^\circ$ is the angle between perpendicular and the positive X-axis.*

140. *(d)* Given equation of line is

$$y-2=0 \quad \ldots(i)$$

Now, divide by $\sqrt{(\text{coefficient of } x)^2+(\text{coefficient of } y)^2}$

i.e., $\sqrt{0+1^2}=1$, we get

$$\frac{0\cdot x}{1}+\frac{y}{1}=\frac{2}{1}$$

or $x\cos 90^\circ + y\sin 90^\circ = 2$

On comparing with $x\cos\alpha + y\sin\alpha = p$, we get

$$\alpha = 90^\circ$$

and $p=2$

141. *(b)* Given equation of line is

$$x-y=4$$

On dividing above equation by

$$\sqrt{(\text{coefficient of } x)^2+(\text{coefficient of } y)^2}$$

i.e., $\sqrt{(1)^2+(-1)^2}=\sqrt{2}$, we get

$$\frac{1}{\sqrt{2}}x-\frac{1}{\sqrt{2}}y=\frac{4}{\sqrt{2}}$$

$\Rightarrow$ $\cos 45^\circ\, x - \sin 45^\circ\, y = 2\sqrt{2}$

($\because$ cos x is positive and sin x is negative, it is possible only in fourth quadrant)

$\Rightarrow x\cos(360^\circ-45^\circ)+y\sin(360^\circ-45^\circ)=2\sqrt{2}$

$$\left[\because \cos(360^\circ-\theta)=\cos\theta \text{ and } \sin(360^\circ-\theta)=-\sin\theta\right]$$

$\Rightarrow$ $x\cos 315^\circ + y\sin 315^\circ = 2\sqrt{2}$

On comparing with $x\cos\alpha + y\sin\alpha = p$, we get

$$\alpha = 315^\circ \quad \text{and} \quad p=2\sqrt{2}$$

142. *(c)* Given equation of line is

$$12(x+6)=5(y-2)$$

$\Rightarrow$ $12x+72=5y-10$

$\Rightarrow$ $12x-5y+82=0$

Hence, distance from the point $(-1, 1)$ is

$$d=\left|\frac{12\times(-1)-5\times 1+82}{\sqrt{(12)^2+(-5)^2}}\right|$$

$$\left(\because d=\left|\frac{ax_1+by_1+c}{\sqrt{a^2+b^2}}\right|; x_1=-1, y_1=1; a=12, b=-5, c=82\right)$$

$$=\left|\frac{-12-5+82}{\sqrt{144+25}}\right|$$

$$=\left|\frac{82-17}{\sqrt{169}}\right|=\frac{65}{13}=5$$

143. *(b)* Let any point on the X-axis is $(h, 0)$, whose distance from given line is 4 units.

Given, equation of line is

$$\frac{x}{3}+\frac{y}{4}=1$$

$\Rightarrow$ $4x+3y=12 \quad \ldots(i)$

Distance of $(h, 0)$ from the line (i) is

$$d=\left|\frac{ax_1+by_1+c}{\sqrt{a^2+b^2}}\right|$$

$\Rightarrow$ $4=\left|\dfrac{4\times h+3\times 0-12}{\sqrt{4^2+3^2}}\right|$

$$\left(\because a=4, b=3, c=12, x_1=h, y_1=0\right)$$

$\Rightarrow$ $\left|\dfrac{4h-12}{5}\right|=4$

$\Rightarrow$ $|4h-12|=20$

$\Rightarrow$ $4h-12=\pm 20$

On taking positive sign, we get

$$4h-12=20$$

$\Rightarrow$ $4h=32$

$\Rightarrow$ $h=8$

On taking negative sign, we get

$$4h-12=-20$$

$\Rightarrow$ $4h=-8$

$\Rightarrow$ $h=-2$

Hence, points on the X-axis are $(8, 0)$ and $(-2, 0)$.

144. *(a)* Given, lines $15x+8y-34=0$ and $15x+8y+31=0$ are parallel.

$\therefore$ Distance between two parallel lines

$$=\left|\frac{C_1-C_2}{\sqrt{a^2+b^2}}\right|$$

$$=\left|\frac{-34-31}{\sqrt{15^2+8^2}}\right|$$

$$(\because C_1=-34, C_2=31, a=15, b=8)$$

$$=\frac{|-65|}{\sqrt{225+64}}$$

$$=\frac{65}{\sqrt{289}}$$

$$=\frac{65}{17} \text{ units}$$

145. *(d)* Given, lines $lx + ly + p = 0$ and $lx + ly - r = 0$ are parallel.

$\therefore$ Distance between two parallel lines

$$= \left|\frac{p-(-r)}{\sqrt{(l)^2+(l)^2}}\right| = \left|\frac{p+r}{\sqrt{2l^2}}\right| = \left|\frac{p+r}{\sqrt{2}l}\right|$$

146. *(a)* Given, equation of line is

$$3x - 4y + 2 = 0 \quad \ldots(i)$$

Equation of line parallel to Eq. (i) is

$$3x - 4y + k = 0 \quad \ldots(ii)$$

$\because$ Line (ii) passes through the point $(-2, 3)$.

On putting $x = -2$ and $y = 3$ in Eq. (ii), we get

$$\therefore \quad 3(-2) - 4 \times 3 + k = 0$$

$$\Rightarrow \quad -6 - 12 + k = 0 \Rightarrow k = 18$$

Put $k = 18$ in Eq. (ii), we get

$$3x - 4y + 18 = 0$$

147. *(b)* Given, equation of line is

$$x - 7y + 5 = 0 \quad \ldots(i)$$

Equation of line perpendicular to Eq. (i) is

$$7x + y + k = 0 \quad \ldots(ii)$$

$\because$ Line (ii) passes through the point $(3, 0)$.

On putting $x = 3$ and $y = 0$ in Eq. (ii), we get

$$\therefore \quad 7 \times 3 + 0 + k = 0 \Rightarrow k = -21$$

On putting $k = -21$ in Eq. (ii), we get

$$7x + y - 21 = 0$$

148. *(b)* Given, equation of lines are

$$\sqrt{3}x + y = 1$$

$$\Rightarrow \quad y = -\sqrt{3}x + 1 \text{ and } x + \sqrt{3}y = 1$$

$$\Rightarrow \quad \sqrt{3}y = -x + 1 \Rightarrow y = \frac{-1}{\sqrt{3}}x + \frac{1}{\sqrt{3}}$$

On comparing with $y = m_1x + c_1$ and $y = m_2x + c_2$, we get

$$m_1 = -\sqrt{3} \quad \text{and} \quad m_2 = -\frac{1}{\sqrt{3}}$$

$$\therefore \quad \tan\theta = \left|\frac{-\sqrt{3}-\left(-\frac{1}{\sqrt{3}}\right)}{1+(-\sqrt{3})\left(-\frac{1}{\sqrt{3}}\right)}\right| \quad \left(\because \tan\theta = \left|\frac{m_1-m_2}{1+m_1m_2}\right|\right)$$

$$\Rightarrow \quad \tan\theta = \left|\frac{-\sqrt{3}+\frac{1}{\sqrt{3}}}{1+1}\right| = \left|\frac{\frac{-3+1}{\sqrt{3}}}{2}\right|$$

$$\Rightarrow \quad \tan\theta = \frac{1}{\sqrt{3}} \Rightarrow \tan\theta = \tan 30^\circ$$

$$\Rightarrow \quad \theta = 30^\circ \quad \text{(acute angle)}$$

Now, obtuse angle between the lines $= 180^\circ - 30^\circ = 150^\circ$

149. *(c)* Slope of the line passing through $(h, 3)$ and $(4, 1)$ is

$$m_1 = \frac{1-3}{4-h} = \frac{-2}{4-h} \quad \left(\because m = \frac{y_2-y_1}{x_2-x_1}\right)$$

Given, equation of line is

$$7x - 9y - 19 = 0$$

$$\Rightarrow \quad y = \frac{7}{9}x - \frac{19}{9} \quad \ldots(i)$$

On comparing Eqs. (i) with $y = mx + c$, we get

$$m_2 = \frac{7}{9}$$

$\because$ The lines are perpendicular to each other.

$$\therefore \quad m_1m_2 = -1$$

$$\Rightarrow \quad -\frac{2}{4-h} \times \frac{7}{9} = -1 \Rightarrow 14 = 9(4-h)$$

$$\Rightarrow \quad 14 = 36 - 9h \Rightarrow 9h = 22$$

$$\therefore \quad h = \frac{22}{9}$$

150. *(d)* Given, equation of line is

$$Ax + By + C = 0 \quad \ldots(i)$$

Let equation of line parallel to Eq. (i) is

$$Ax + By + K = 0 \quad \ldots(ii)$$

Line (ii) passes through the point (x_1, y_1) *i.e.*, it will satisfy the line (ii).

$$\therefore \quad Ax_1 + By_1 + K = 0$$

$$\Rightarrow \quad K = -(Ax_1 + By_1)$$

On putting the value of K in Eq. (ii), we get

$$\Rightarrow \quad Ax + By - (Ax_1 + By_1) = 0$$

$$\Rightarrow \quad Ax + By - Ax_1 - By_1 = 0$$

$$\Rightarrow \quad A(x - x_1) + B(y - y_1) = 0$$

151. *(a)* Equation of line l_1 by using $y - y_1 = m(x - x_1)$

where, $(x_1, y_1) = (2, 3)$ and $m = 2$, is

$$y - 3 = 2(x - 2)$$

$$\Rightarrow \quad y - 3 = 2x - 4 \Rightarrow 2x - y - 1 = 0$$

$$\Rightarrow \quad y = 2x - 1 \quad \ldots(i)$$

Here, slope of line l_1 is $m_1 = 2$.

Let slope of line l_2 is m.

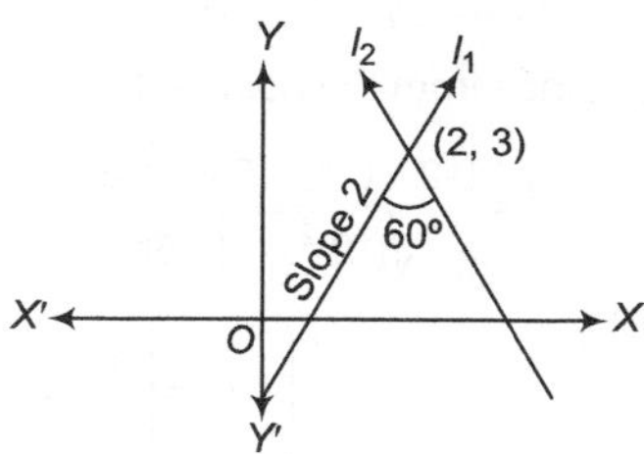

Given, $\theta = 60^\circ$

$$\Rightarrow \quad \tan\theta = \left|\frac{m_1 - m}{1 + m_1m_2}\right| \Rightarrow \tan 60^\circ = \left|\frac{2-m}{1+2m}\right|$$

Taking positive sign, we get

$$\frac{2-m}{1+2m} = \sqrt{3} \Rightarrow 2 - m = \sqrt{3}(1+2m)$$

$$\Rightarrow \quad 2 - m = \sqrt{3} + 2\sqrt{3}m \Rightarrow 2 - \sqrt{3} = 2\sqrt{3}m + m$$

$\Rightarrow \quad m = \dfrac{2-\sqrt{3}}{1+2\sqrt{3}}$

Again taking negative sign, we get

$$\frac{2-m}{1+2m} = -\sqrt{3}$$

$\Rightarrow \quad 2 - m = -\sqrt{3}\,(1+2m)$

$\Rightarrow \quad 2 - m = -\sqrt{3} - 2\sqrt{3}m$

$\Rightarrow \quad m - 2\sqrt{3}m = \sqrt{3} + 2$

$\Rightarrow \quad -m\,(2\sqrt{3}-1) = (\sqrt{3}+2) \Rightarrow m = -\left(\dfrac{\sqrt{3}+2}{2\sqrt{3}-1}\right)$

Now, equation of line l_2 by using

$y - y_1 = m(x - x_1)$ is $y - 3 = \dfrac{2-\sqrt{3}}{1+2\sqrt{3}}(x-2)$

$$\left(\text{when } m = \frac{2-\sqrt{3}}{1+2\sqrt{3}} \text{ and } x_1 = 2, y_1 = 3\right)$$

$\Rightarrow \quad y(1+2\sqrt{3}) - 3 - 6\sqrt{3} = x(2-\sqrt{3}) - 4 + 2\sqrt{3}$

$\Rightarrow x(2-\sqrt{3}) - y(1+2\sqrt{3}) + 8\sqrt{3} - 1 = 0$

Again, $\quad y - 3 = -\left(\dfrac{\sqrt{3}+2}{2\sqrt{3}-1}\right)(x-2)$

$$\left(\text{when } m = -\frac{\sqrt{3}+2}{2\sqrt{3}-1} \text{ and } x_1 = 2, y_1 = 3\right)$$

$\Rightarrow \quad y\,(2\sqrt{3}-1) - 6\sqrt{3} + 3 = -(\sqrt{3}+2)x + 2\sqrt{3} + 4$

$\Rightarrow \quad (2+\sqrt{3})\,x + (2\sqrt{3}-1)y - 8\sqrt{3} - 1 = 0$

152. *(b)* Let the given points are $P\,(3, 4)$ and $Q\,(-1, 2)$. Let, B is the mid-point of PQ then $B = \left(\dfrac{x_1+x_2}{2}, \dfrac{y_1+y_2}{2}\right)$

$$= \left(\frac{3-1}{2}, \frac{4+2}{2}\right) = (1, 3)$$

$$(\because x_1 = 3, y_1 = 4, x_2 = -1, y_2 = 2)$$

Now, $AB \perp PQ$, *i.e.,*

Slope of $AB \times$ Slope of $PQ = -1$

$\Rightarrow \quad m \times \dfrac{y_2 - y_1}{x_2 - x_1} = -1$

$\Rightarrow \quad m \times \dfrac{2-4}{-1-3} = -1$

$\Rightarrow \quad m \times \dfrac{-2}{-4} = -1$

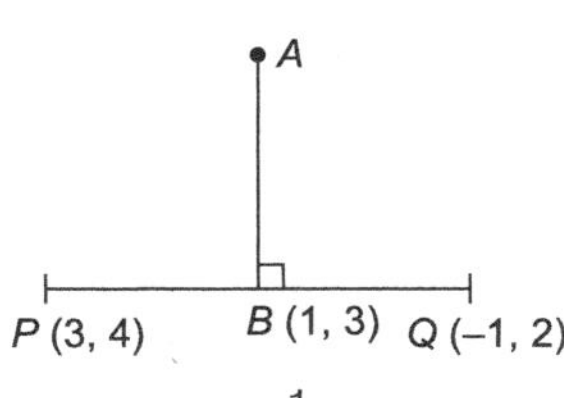

$\Rightarrow \quad m \times \dfrac{1}{2} = -1$

$\Rightarrow \quad m = -2$

Now, equation of line AB by using $y - y_1 = m(x - x_1)$, where $(x_1, y_1) = (1, 3)$ and $m = -2$, is

$y - 3 = -2\,(x-1)$

$\Rightarrow \quad y - 3 = -2x + 2$

$\Rightarrow \quad 2x + y - 5 = 0$

153. *(d)* $\because$ Line $OP \perp$ Line AB

$\therefore$ Slope of $OP \times$ Slope of $AB = -1$

$\Rightarrow \quad \dfrac{2-0}{-1-0} \times m = -1$

$\Rightarrow \quad -2m = -1$

$\Rightarrow \quad m = \dfrac{1}{2}$

Also, point $(-1, 2)$ satisfies the line

$$y = mx + c$$

i.e., $\quad 2 = \dfrac{1}{2}(-1) + c \quad$ (putting $x = -1$, $y = 2$)

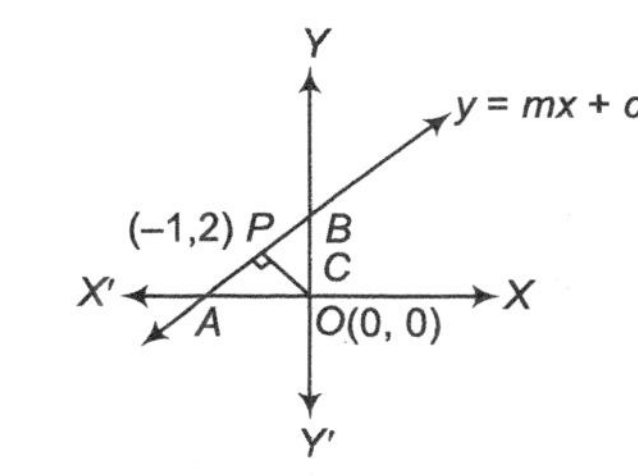

$\Rightarrow \quad 2 = -\dfrac{1}{2} + c$

$\Rightarrow \quad c = 2 + \dfrac{1}{2} = \dfrac{5}{2}$

Hence, $m = \dfrac{1}{2}, \quad c = \dfrac{5}{2}$

154. *(b)* $\because$ Equation of line in intercept form is

$$\frac{x}{a} + \frac{y}{b} = 1$$

$\Rightarrow \quad \dfrac{x}{a} + \dfrac{y}{b} - 1 = 0$

Its distance from origin is

$$\left|\frac{0+0-1}{\sqrt{\frac{1}{a^2}+\frac{1}{b^2}}}\right| = p \qquad \left(\because d = \left|\frac{ax_1 + by_1 + c}{\sqrt{a^2+b^2}}\right|\right)$$

$\Rightarrow \quad p = \dfrac{1}{\sqrt{\dfrac{1}{a^2}+\dfrac{1}{b^2}}}$

$\Rightarrow \quad \dfrac{1}{p} = \sqrt{\dfrac{1}{a^2}+\dfrac{1}{b^2}}$

On squaring both sides, we get

$$\frac{1}{p^2} = \frac{1}{a^2} + \frac{1}{b^2}$$

$\Rightarrow \quad a^2b^2 = p^2(a^2+b^2)$

155. *(a)* The foot is the point of intersection of given line and perpendicular line.

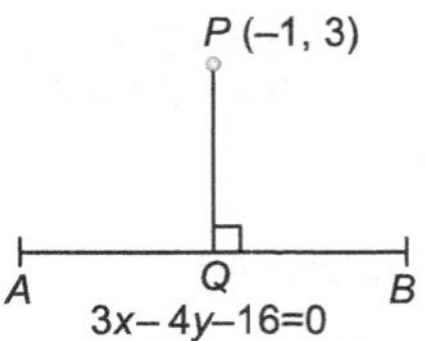

Given, equation of line AB is

$$3x - 4y - 16 = 0$$

$$\Rightarrow \quad y = \frac{3}{4}x - 4 \quad \ldots(\text{i})$$

Slope of line (i) is $m_1 = \frac{3}{4}$

Let slope of line PQ be m.

Since, $PQ \perp AB$

$\therefore$ Slope of PQ × Slope of $AB = -1$

$$\Rightarrow \quad m \times \left(\frac{3}{4}\right) = -1 \quad (\because m_1 m_2 = -1)$$

$$\Rightarrow \quad m = -\frac{4}{3}$$

Hence, equation of line PQ by using $y - y_1 = m(x - x_1)$ where $(x_1, y_1) = (-1, 3)$ and $m = -\frac{4}{3}$, is

$$y - 3 = -\frac{4}{3}(x + 1)$$

$$\Rightarrow \quad 3y - 9 = -4x - 4$$

$$\Rightarrow \quad 4x + 3y - 5 = 0 \quad \ldots(\text{ii})$$

Now, solve Eqs. (i) and (ii), to find the coordinates of Q.

From Eq. (i),

$$y = \frac{3}{4}x - 4$$

On putting the value of y in Eq. (ii), we get

$$4x + 3\left(\frac{3}{4}x - 4\right) - 5 = 0$$

$$\Rightarrow \quad 4x + \frac{9x}{4} - 12 - 5 = 0$$

$$\Rightarrow \quad \frac{25x}{4} = 17$$

$$\Rightarrow \quad x = \frac{68}{25}$$

$\therefore$ From Eq. (i),

$$y = \frac{3}{4} \times \frac{68}{25} - 4 = \frac{3 \times 17}{25} - \frac{4}{1} = \frac{51}{25} - \frac{4}{1} = \frac{51 - 100}{25} = -\frac{49}{25}$$

Hence, required point of intersection is $\left(\frac{68}{25}, -\frac{49}{25}\right)$.

156. *(b)*

O (0,0)
p
(a)

O (0,0)
q
(b)

For figure (a), $x\cos\theta - y\sin\theta = k\cos 2\theta \quad \ldots(\text{i})$

For figure (b), $x\sec\theta + y\,\text{cosec}\,\theta = k \quad \ldots(\text{ii})$

$\therefore$ Perpendicular distance from (0, 0) to the Eq. (i) is

$$p = \left|\frac{0 - 0 - k\cos 2\theta}{\sqrt{\cos^2\theta + \sin^2\theta}}\right| = k\cos 2\theta \quad \ldots(\text{iii})$$

$$\left(\because d = \left|\frac{ax_1 + by_1 + c}{\sqrt{a^2 + b^2}}\right|\right)$$

Again, Eq. (ii) can be written as

$$\frac{x}{\cos\theta} + \frac{y}{\sin\theta} = k$$

$$\Rightarrow \quad x\sin\theta + y\cos\theta = k\cos\theta\sin\theta$$

$$\Rightarrow \quad x\sin\theta + y\cos\theta = \frac{k}{2}(2\cos\theta\sin\theta)$$

$$\Rightarrow \quad x\sin\theta + y\cos\theta = \frac{k}{2}\sin 2\theta$$

$$\Rightarrow \quad x\sin\theta + y\cos\theta - \frac{k}{2}\sin 2\theta = 0 \quad \ldots(\text{iv})$$

Now, perpendicular distance from (0, 0) to Eq. (iv) is

$$q = \left|\frac{0 + 0 - \frac{k}{2}\sin 2\theta}{\sqrt{\sin^2\theta + \cos^2\theta}}\right|$$

$$\left(\because a = \sin\theta, b = \cos\theta, c = -\frac{k}{2}\sin 2\theta, x_1 = 0, y_1 = 0\right)$$

$$\Rightarrow \quad q = \frac{k}{2}\sin 2\theta \quad (\because \sin^2\theta + \cos^2\theta = 1)$$

$$\Rightarrow \quad 2q = k\sin 2\theta \quad \ldots(\text{v})$$

On squaring and adding Eqs. (iii) and (v), we get

$$p^2 + 4q^2 = k^2\cos^2 2\theta + k^2\sin^2 2\theta = k^2(\cos^2 2\theta + \sin^2 2\theta)$$

$$\therefore \quad p^2 + 4q^2 = k^2 \quad (\because \sin^2\theta + \cos^2\theta = 1)$$

157. *(a)* Given, vertices of a ΔABC are $A(2, 3)$, $B(4, -1)$ and $C(1, 2)$.

$\because$ Line $AD \perp$ Line BC

$\therefore$ Slope of AD × Slope of $BC = -1$

$$\Rightarrow \quad m \times \frac{y_2 - y_1}{x_2 - x_1} = -1$$

$$\Rightarrow \quad m \times \frac{2 + 1}{1 - 4} = -1$$

$$(\because x_1 = 4, y_1 = -1, x_2 = 1, y_2 = 2)$$

$$\Rightarrow \quad m \times \frac{3}{-3} = -1 \Rightarrow m = 1$$

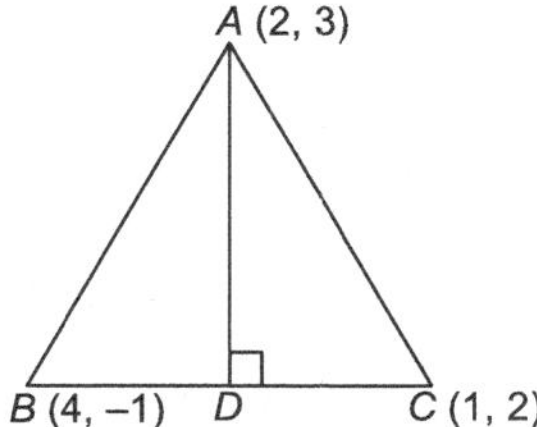

Hence, equation of AD, by using $y - y_1 = m(x - x_1)$ where $(x_1, y_1) = (2, 3)$ and $m = 1$, is

$$y - 3 = 1(x - 2)$$

$$\Rightarrow \quad x - y - 2 + 3 = 0 \Rightarrow x - y + 1 = 0$$

Now, equation of BC by using $y - y_1 = \frac{y_2 - y_1}{x_2 - x_1}(x - x_1)$, is

$$y + 1 = \frac{2 + 1}{1 - 4}(x - 4)$$

$$(\because x_1 = 4, y_1 = -1, x_2 = 1, y_2 = 2)$$

$$\Rightarrow \quad y + 1 = \frac{3}{-3}(x - 4)$$

$$\Rightarrow \quad y + 1 = -x + 4 \Rightarrow x + y - 3 = 0$$

$\therefore$ Length of AD = Perpendicular distance from (2,3) to the line BC

i.e., $x + y - 3 = 0$

$$= \left|\frac{2 + 3 - 3}{\sqrt{1^2 + 1^2}}\right| = \frac{2}{\sqrt{2}} = \sqrt{2} \quad \left(\because p = \left|\frac{ax_1 + by_1 + c}{\sqrt{a^2 + b^2}}\right|\right)$$

158. *(c)* Let $Q(h, k)$ be the image of the point $P(1, 2)$ in the line

$$x - 3y + 4 = 0 \quad \ldots(i)$$

Therefore, the line (i) is the perpendicular bisector of line segment PQ.

Hence, slope of line

$$PQ = \frac{-1}{\text{Slope of line } x - 3y + 4 = 0}$$

So that $\frac{k - 2}{h - 1} = \frac{-1}{\frac{1}{3}}$ or $3h + k = 5$ ...(ii)

and the mid-point of PQ, *i.e.*, point $\left(\frac{h + 1}{2}, \frac{k + 2}{2}\right)$ will satisfy the Eq. (i), so that

$$\frac{h + 1}{2} - 3\left(\frac{k + 2}{2}\right) + 4 = 0 \text{ or } h - 3k = -3 \quad \ldots(iii)$$

Solving Eqs. (ii) and (iii), we get $h = \frac{6}{5}$ and $k = \frac{7}{5}$.

Hence, the image of the point (1, 2) in the line (i) is $\left(\frac{6}{5}, \frac{7}{5}\right)$.

159. *(b)* Given lines are

$$3x - 2y = 5 \quad \ldots(i)$$

and

$$3x + 2y = 5 \quad \ldots(ii)$$

Let (h, k) be any point, whose distances from the lines (i) and (ii) are equal. Therefore,

$$\frac{|3h - 2k - 5|}{\sqrt{9 + 4}} = \frac{|3h + 2k - 5|}{\sqrt{9 + 4}} \text{ or}$$

$$|3h - 2k - 5| = |3h + 2k - 5|,$$

which gives $3h - 2k - 5 = 3h + 2k - 5$

or $-(3h - 2k - 5) = 3h + 2k - 5.$

Solving these two relations, we get $k = 0$ or $h = \frac{5}{3}$. Thus, the point (h, k) satisfies the equations $y = 0$ or $x = \frac{5}{3}$, which represent straight lines. Hence, path of the point equidistant from the lines (i) and (ii) is a straight line.

160. *(b)* Given, equation of line is $\sqrt{3}x + y + 2 = 0$

$$\Rightarrow \quad \sqrt{3}\,x + y = -2$$

$$\Rightarrow \quad -\sqrt{3}x - y = 2$$

On dividing above equation by $\sqrt{(\text{coefficient of } x)^2 + (\text{coefficient of } y)^2}$

i.e., $\sqrt{(-\sqrt{3})^2 + (-1)^2} = \sqrt{3 + 1} = \sqrt{4} = 2$, we get

$$\Rightarrow \quad \frac{-\sqrt{3}}{2}x - \frac{1}{2}y = \frac{2}{2}$$

(for convert in the form of $x \cos\theta + y \sin\theta = p$)

$$\Rightarrow \quad -\cos 30^\circ x - \sin 30^\circ y = 1$$

$$\Rightarrow \quad \cos(180^\circ + 30^\circ)x + \sin(180^\circ + 30^\circ)y = 1$$

($\because$ $\cos\theta$ and $\sin\theta$ both are negative in third quadrant)

$$\Rightarrow \quad (\cos 210^\circ)x + (\sin 210^\circ)y = 1$$

On comparing with $x\cos\theta + y\sin\theta = p$, we get

$$\theta = 210^\circ$$

and

$$p = 1$$

161. *(c)* Let $(0, h)$ be the point on the Y-axis such that its distance from the given line is 4 units.

Equation of given line is

$$\frac{x}{3} + \frac{y}{4} = 1 \text{ or } 4x + 3y = 12 \text{ or } 4x + 3y - 12 = 0$$

Its distance from $(0, h)$ is 4 units. *i.e.*,

$$\left|\frac{0 + 3h - 12}{\sqrt{4^2 + 3^2}}\right| = 4$$

$\left(\because \text{ perpendicular distance, } d = \left|\frac{ax_1 + by_1 + c}{\sqrt{a^2 + b^2}}\right| \right.$ and here $\left. a = 4, b = 3, c = -12, x_1 = 0, y_1 = h\right)$

$$\Rightarrow \quad \left|\frac{3h - 12}{5}\right| = 4$$

$$\Rightarrow \quad |3h - 12| = 20 \Rightarrow 3h - 12 = \pm 20$$

Taking positive sign,

$$3h = 20 + 12$$

$$\Rightarrow \quad 3h = 32 \Rightarrow h = \frac{32}{3}$$

Taking negative sign, $3h = -20 + 12$

$$\Rightarrow \quad 3h = -8 \Rightarrow h = -\frac{8}{3}$$

Hence, points on the Y-axis are $\left(0, -\frac{8}{3}\right)$ or $\left(0, \frac{32}{3}\right)$.

162. *(a)* The equations of lines $3x + 4y = 9$ and $6x + 8y = 15$ may be rewritten as $3x + 4y - 9 = 0$ and $3x + 4y - \frac{15}{2} = 0$

Since, the slope of these lines are same and hence they are parallel to each other. Therefore, the distance between them is given by–

$$\left|\frac{9 - \frac{15}{2}}{\sqrt{3^2 + 4^2}}\right| = \frac{3}{10}$$

163. *(a)* Let m be the slope of the line whose equation is to be found out which is perpendicular to the line $x + y + 7 = 0$. The slope of the given line $y = (-1)x - 7$ is -1. Therefore, using the condition of perpendicularity of lines, we have $m \times (-1) = -1$ or $m = 1$.

Hence, the required equation of the line is

$$y - 1 = (1)(x - 2) \text{ or } y - 1 = x - 2 \Rightarrow x - y - 1 = 0$$

164. *(d)* Equation of line joining the points $(\cos\theta, \sin\theta)$ and $(\cos\phi, \sin\phi)$ is

$$y - \sin\theta = \frac{\sin\phi - \sin\theta}{\cos\phi - \cos\theta}(x - \cos\theta)$$

$$\left(\because y - y_1 = \frac{y_2 - y_1}{x_2 - x_1}(x - x_1) \text{ here, } x_1 = \cos\theta, y_1 = \sin\theta,\ x_2 = \cos\phi, y_2 = \sin\phi\right)$$

$$\Rightarrow \quad y - \sin\theta = \frac{2\cos\frac{\phi+\theta}{2}\sin\frac{\phi-\theta}{2}}{-2\sin\frac{\phi+\theta}{2}\sin\frac{\phi-\theta}{2}}(x - \cos\theta)$$

$$\left(\because \cos C - \cos D = -2\sin\frac{C+D}{2}\sin\frac{C-D}{2} \text{ and } \sin C - \sin D = 2\cos\frac{C+D}{2}\sin\frac{C-D}{2}\right)$$

$$\Rightarrow y - \sin\theta = \frac{\cos\frac{\phi+\theta}{2}}{-\sin\frac{\phi+\theta}{2}}(x - \cos\theta)$$

$$\Rightarrow \quad -y\sin\frac{\phi+\theta}{2} + \sin\theta\sin\frac{\phi+\theta}{2} = x\cos\frac{\phi+\theta}{2} - \cos\theta\cos\frac{\phi+\theta}{2}$$

$$\Rightarrow \quad x\cos\frac{\phi+\theta}{2} + y\sin\frac{\phi+\theta}{2} = \sin\theta\frac{\sin\phi+\theta}{2} + \cos\theta\frac{\cos\phi+\theta}{2}$$

$$\Rightarrow \quad x\cos\frac{\phi+\theta}{2} + y\sin\frac{\phi+\theta}{2} = \cos\left(\theta - \frac{\theta+\phi}{2}\right)$$

$$[\because \cos(A - B) = \cos A\cos B + \sin A\sin B]$$

$$\Rightarrow \quad x\cos\frac{\phi+\theta}{2} + y\sin\frac{\phi+\theta}{2} - \cos\left(\frac{\theta-\phi}{2}\right) = 0$$

Its distance from origin (0, 0) is

$$p = \left|\frac{0 + 0 - \cos\left(\frac{\theta-\phi}{2}\right)}{\sqrt{\cos^2\left(\frac{\phi+\theta}{2}\right) + \sin^2\left(\frac{\phi+\theta}{2}\right)}}\right| \qquad \left(\because p = \left|\frac{ax_1 + by_1 + c}{\sqrt{a^2 + b^2}}\right|\right)$$

$$\Rightarrow \quad p = \cos\left(\frac{\theta-\phi}{2}\right) \qquad (\because \cos^2\theta + \sin^2\theta = 1)$$

165. *(a)* Given, equation of line is

$$\frac{x}{4} + \frac{y}{6} = 1$$

$$\Rightarrow \quad \frac{3x + 2y}{12} = 1$$

$$\Rightarrow \quad 3x + 2y = 12 \qquad \ldots(i)$$

If line (i) meets the Y-axis, putting $x = 0$ in Eq. (i), we get

$$0 + 2y = 12$$

$$\Rightarrow \quad y = 6$$

$\therefore$ Point is (0, 6).

Now, equation of line perpendicular to Eq. (i) is

$$2x - 3y + k = 0 \qquad \ldots(ii)$$

Line (ii) passes through the point (0, 6), so it will satisfy it.

i.e., $\quad 2 \times 0 - 3 \times 6 + k = 0$

$$\Rightarrow \quad k = 18 \qquad (\because \text{put } x = 0, y = 6)$$

Hence, Eq. (ii) becomes

$$2x - 3y + 18 = 0$$

166. *(d)* The normal form of the equation of the line is $x\cos\omega + y\sin\omega = p$. Here, $p = 4$ and $\omega = 30°$. Therefore, the equation of the line is

$$x\cos 30° + y\sin 30° = 4$$

$$x\frac{\sqrt{3}}{2} + y\frac{1}{2} = 4 \quad \text{or} \quad \sqrt{3}x + y = 8$$

167. *(c)* First, we find the point of intersection of lines $5x - 6y - 1 = 0$ and $3x + 2y + 5 = 0$ which is $(-1, -1)$. Also, the slope of the line $3x - 5y + 11 = 0$ is $\frac{3}{5}$. Therefore, the slope of the line perpendicular to this line is $\frac{-5}{3}$. Hence, the equation of the required line is given by

$$y + 1 = \frac{-5}{3}(x + 1) \quad \text{or} \quad 5x + 3y + 8 = 0$$

Alternate Method

The equation of any line through the intersection of lines $5x - 6y - 1 = 0$ and $3x + 2y + 5 = 0$ is

$$5x - 6y - 1 + k(3x + 2y + 5) = 0 \quad \ldots(i)$$

or slope of this line is $\frac{-(5 + 3k)}{-6 + 2k}$.

Also, slope of the line $3x - 5y + 11 = 0$ is $\frac{3}{5}$.

Now, both are perpendicular.

So, $$\frac{-(5 + 3k)}{-6 + 2k} \times \frac{3}{5} = -1$$

or $$k = 45$$

Therefore, equation of required line is given by

$$5x - 6y - 1 + 45(3x + 2y + 5) = 0$$

or $$5x + 3y + 8 = 0$$

168. *(a)* Let $ABCD$ be the given square and the coordinates of the vertex C be $(1, 2)$.

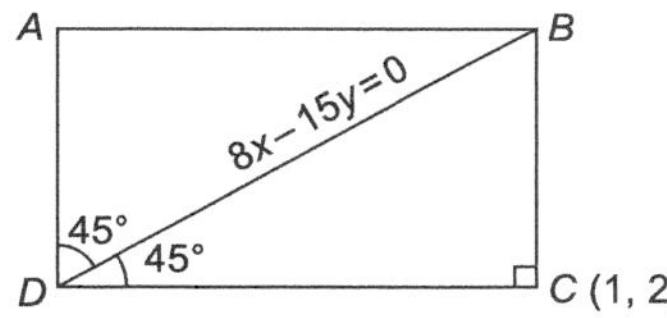

Given that, BD is along the line $8x - 15y = 0$, so its slope is $\frac{8}{15}$.

The angles made by DB with sides DC and CB is 45°. Let the slope of DC be m.

Then, $$\tan 45° = \left|\frac{m - \frac{8}{15}}{1 + \frac{8m}{15}}\right|$$

$$\Rightarrow \quad \pm 1 = \frac{m - \frac{8}{15}}{1 + \frac{8m}{15}}$$

$\Rightarrow 15 + 8m = 15m - 8$ and $-15 - 8m = 15m - 8$

$\Rightarrow \quad 7m = 23$ and $-7 = 23m$

$\Rightarrow \quad m = \frac{23}{7}$ and $m = \frac{-7}{23}$

If slope of DC is $\frac{23}{7}$, then slope of CB is $\frac{-7}{23}$.

Then, equation of DC is given by

$$y - 2 = \frac{23}{7}(x - 1)$$

$$\Rightarrow \quad 23x - 7y - 9 = 0$$

Then, equation of CB is given by

$$y - 2 = \frac{-7}{23}(x - 1)$$

$$\Rightarrow \quad 7x + 23y - 53 = 0$$

169. *(a)* Slope of the line $ax + by = c$ is $\frac{-a}{b}$ and the slope of the line $a'x + b'y = c'$ is $\frac{-a'}{b'}$. The lines are perpendicular, if

$$m_1 \cdot m_2 = -1 \quad \ldots(i)$$

$$\left(\frac{-a}{b}\right)\left(\frac{-a'}{b'}\right) = -1 \quad \text{or} \quad aa' + bb' = 0$$

170. *(b)* Let (h, k) be the coordinates of the foot of the perpendicular from the point $(2, 3)$ on the line $x + y - 11 = 0$. Then, the slope of the perpendicular line is $\frac{k - 3}{h - 2}$. Again, the slope of the given line $x + y - 11 = 0$ is -1.

Using the condition of perpendicularity of lines, we have

$$\left(\frac{k - 3}{h - 2}\right)(-1) = -1 \quad (\because m_1 m_2 = -1)$$

or $$k - h = 1 \quad \ldots(i)$$

Since, (h, k) lies on the given line, so we have

$$h + k - 11 = 0$$

or $$h + k = 11 \quad \ldots(ii)$$

On solving Eqs. (i) and (ii), we get

$$h = 5 \quad \text{and} \quad k = 6$$

Thus, $(5, 6)$ are the required coordinates of the foot of perpendicular.

171. *(a)* Let (h, k) be the point of reflection of the given point $(4, -13)$ about the line $5x + y + 6 = 0$. The mid-point of the line segment joining points (h, k) and $(4, -13)$ is given by

$$\left(\frac{h + 4}{2}, \frac{k - 13}{2}\right)$$

This point lies on the given line, so we have

$$5\left(\frac{h + 4}{2}\right) + \frac{k - 13}{2} + 6 = 0$$

or $$5h + k + 19 = 0 \quad \ldots(i)$$

Again, the slope of the line joining points (h, k) and $(4, -13)$ is given by $\frac{k + 13}{h - 4}$. This line is perpendicular to the given line and hence

$$(-5)\left(\frac{k + 13}{h - 4}\right) = -1$$

This gives $$5k + 65 = h - 4$$

or $$h - 5k - 69 = 0 \quad \ldots(ii)$$

On solving Eqs. (i) and (ii), we get

$$h = -1 \text{ and } k = -14$$

Thus, the point $(-1, -14)$ is the reflection of the given point.

172. *(a)* Let (h, k) be the coordinates of the moving point. Then, we have

$$\sqrt{(h - 4)^2 + k^2} = \frac{1}{2}\left(\frac{h - 16}{\sqrt{1^2 + 0}}\right)$$

$$\Rightarrow \quad (h - 4)^2 + k^2 = \frac{1}{4}(h - 16)^2$$

$$4(h^2 - 8h + 16 + k^2) = h^2 - 32h + 256$$

or $$3h^2 + 4k^2 = 192$$

Hence, the required locus is given by $3x^2 + 4y^2 = 192$

173. *(b)* Let the two intersecting lines l_1 and l_2 be given by

$$A_1x + B_1y + C_1 = 0 \quad \dots(i)$$

and

$$A_2x + B_2y + C_2 = 0 \quad \dots(ii)$$

From the Eqs. (i) and (ii), we can form an equation

$$A_1x + B_1y + C_1 + k(A_2x + B_2y + C_2) = 0 \quad \dots(iii)$$

where, k is an arbitrary constant called parameter. For any value of k, the Eq. (iii) is of first degree in x and y. Hence, it represents a family of lines. A particular member of this family can be obtained for some value of k. This value of k may be obtained from other conditions.

174. *(a)* The equation of any line through the point of intersection of the given lines is of the form

$$x - 7y + 5 + k(3x + y - 7) = 0$$

i.e., $$(1 + 3k)x + (k - 7)y + 5 - 7k = 0 \quad \dots(i)$$

If this line is parallel to Y-axis, then the coefficient of y should be zero, *i.e.,* $k - 7 = 0$ which gives $k = 7$

Substituting this value of k in the Eq. (i), we get

$22x - 44 = 0$, *i.e.,* $x - 2 = 0$, which is the required equation.

175. *(a)* I. The equation of line passing through the intersection of lines $3x + 4y - 7 = 0$ and $x - y + 2 = 0$ is

$$(3x + 4y - 7) + k(x - y + 2) = 0 \quad \dots(i)$$

$$(3 + k)x + (4 - k)y - 7 + 2k = 0 \quad \dots(ii)$$

$\because$ Slope of above line is 5. (given)

$$\therefore \quad -\left(\frac{3+k}{4-k}\right) = 5$$

$$\Rightarrow \quad 3 + k = 5k - 20$$

$$\Rightarrow \quad 4k = 23$$

$$\Rightarrow \quad k = \frac{23}{4}$$

Putting $k = \frac{23}{4}$ in Eq. (i), we get

$$(3x + 4y - 7) + \frac{23}{4}(x - y + 2) = 0$$

$$\Rightarrow 12x + 16y - 28 + 23x - 23y + 46 = 0$$

$$\Rightarrow \quad 35x - 7y + 18 = 0$$

II. The equation of line passing through the intersection of lines $x + 2y - 3 = 0$ and $4x - y + 7 = 0$ is

$$(x + 2y - 3) + k(4x - y + 7) = 0 \quad \dots(i)$$

$$\Rightarrow (1 + 4k)x + (2 - k)y - 3 + 7k = 0$$

$\because$ Above line is parallel to

$$5x + 4y - 20 = 0 \quad \text{(given)} \dots(ii)$$

$$\therefore \quad \frac{-(1+4k)}{2-k} = \frac{-5}{4} \quad \left(\because \text{slope of parallel lines are equal}\right)$$

$$\Rightarrow \quad 4(1 + 4k) = 5(2 - k)$$

$$\Rightarrow \quad 4 + 16k = 10 - 5k$$

$$\Rightarrow \quad 21k = 6$$

$$\Rightarrow \quad k = \frac{6}{21}$$

Putting $k = \frac{6}{21}$ in Eq. (i), we get

$$(x + 2y - 3) + \frac{6}{21}(4x - y + 7) = 0$$

$$\Rightarrow \quad 21x + 42y - 63 + 24x - 6y + 42 = 0$$

$$\Rightarrow \quad 45x + 36y - 21 = 0$$

$$\Rightarrow \quad 15x + 12y - 7 = 0$$

176. *(d)* I. The equation of line passing through the intersection of the lines $2x + 3y - 4 = 0$ and $x - 5y - 7 = 0$ is

$$(2x + 3y - 4) + k(x - 5y - 7) = 0 \quad \dots(i)$$

$$\Rightarrow (2 + k)x + (3 - 5k)y - 4 - 7k = 0$$

$\because$ x-intercept of above line is -4 i.e., it passes through $(-4, 0)$.

Hence, $-4(2 + k) - 4 - 7k = 0$

$$-8 - 4k - 4 - 7k = 0$$

$$\Rightarrow \quad -11k = 12$$

$$\Rightarrow \quad k = \frac{-12}{11}$$

Putting $k = \frac{-12}{11}$ in Eq. (i), we get

$$(2x + 3y - 4) - \frac{12}{11}(x - 5y - 7) = 0$$

$$\Rightarrow 22x + 33y - 44 - 12x + 60y + 84 = 0$$

$$10x + 93y + 40 = 0$$

II. The equation of line passing through the intersection of the lines $5x - 3y - 1 = 0$ and $2x + 3y - 23 = 0$ is

$$(5x - 3y - 1) + k(2x + 3y - 23) = 0 \quad \dots(i)$$

$$\Rightarrow (5 + 2k)x + (-3 + 3k)y + (-1 - 23k) = 0$$

$\because$ Above line is perpendicular to the line

$$5x - 3y - 1 = 0$$

$$\therefore \quad -\left(\frac{5+2k}{-3+3k}\right)\left\{-\left(\frac{5}{-3}\right)\right\} = -1 \quad (\because m_1m_2 = -1)$$

$$\Rightarrow \quad 5(5 + 2k) = 3(-3 + 3k)$$

$$\Rightarrow \quad 25 + 10k = -9 + 9k$$

$$\Rightarrow \quad 10k - 9k = -9 - 25$$

$$\Rightarrow \quad k = -34$$

Putting $k = -34$ in Eq. (i), we get

$$(5x - 3y - 1) + (-34)(2x + 3y - 23) = 0$$

$$\Rightarrow \quad 5x - 3y - 1 - 68x - 102y + 782 = 0$$

$$\Rightarrow \quad -63x - 105y + 781 = 0$$

$$\Rightarrow \quad 63x + 105y - 781 = 0$$

177. *(b)* The coordinates of the new origin are $h = 1$, $k = 2$ and the original coordinates are given to be $x = 3$, $y = -4$.

The transformation relation between the old coordinates (x, y) and the new coordinates (x', y') are given by

$$x = x' + h \text{ i.e., } x' = x - h$$

and $$y = y' + k \text{ i.e., } y' = y - k$$

Substituting the values, we have

$$x' = 3 - 1 = 2 \quad \text{and} \quad y' = -4 - 2 = -6$$

Hence, the coordinates of the point $(3, -4)$ in the new system are $(2, -6)$.

178. *(c)* Let coordinates of a point P changes (x, y) to (x', y') in new coordinate axes whose origin has the coordinates $h = 3$, $k = -1$. Therefore, we can write the transformation formulae as $x = x' + 3$ and $y = y' - 1$. Substituting these values in the given equation of the straight line, we get

$$2(x' + 3) - 3(y' - 1) + 5 = 0$$

or $$2x' - 3y' + 14 = 0$$

Therefore, the equation of the straight line in new system is $2x - 3y + 14 = 0$.

179. *(a)* Suppose old coordinates are (x, y) and the new coordinates are (x', y'). The coordinates of new origin are $h = -3$ and $k = -2$.

A. Given, $(x, y) = (1, 1)$

$$x = x' + h \quad \text{and} \quad y = y' + k$$
$$\Rightarrow \quad 1 = x' - 3 \quad \text{and} \quad 1 = y' - 2$$
$$\Rightarrow \quad x' = 4 \quad \text{and} \quad y' = 3$$

Hence, the new coordinates are (4, 3).

B. Given, $(x, y) = (0, 1)$

$$\because \quad x = x' + h$$
$$\text{and} \quad y = y' + k$$
$$\Rightarrow \quad 0 = x' - 3$$
$$\text{and} \quad 1 = y' - 2$$
$$\Rightarrow \quad x' = 3 \text{ and } y' = 3$$

Hence, the new coordinates are (3, 3).

C. Given, $(x, y) = (5, 0)$

$$\because \quad x = x' + h \text{ and } y = y' + k$$
$$\Rightarrow \quad 5 = x' - 3$$
$$\text{and} \quad 0 = y' - 2$$
$$\Rightarrow \quad x' = 8 \text{ and } y' = 2$$

Hence, the new coordinates are (8, 2).

D. Given, $(x, y) = (-1, -2)$

$$\because \quad x = x' + h \text{ and } y = y' + k$$
$$\Rightarrow \quad -1 = x' - 3$$
$$\text{and} \quad -2 = y' - 2$$
$$\Rightarrow \quad x' = 2 \text{ and } y' = 0$$

Hence, the new coordinates are (2, 0).

E. Given, $(x, y) = (3, -5)$

$$\because \quad x = x' + h$$
$$\text{and} \quad y = y' + k$$
$$\Rightarrow \quad 3 = x' - 3$$
$$\text{and} \quad -5 = y' - 2$$
$$\Rightarrow \quad x' = 6 \text{ and } y' = -3$$

Hence, the new coordinates are (6, –3).

180. *(c)* $\because$ Origin is shifted to the point (1, 1).

$$\therefore \quad h = 1 \text{ and } k = 1$$

I. Given equation is

$$x^2 + xy - 3y^2 - y + 2 = 0$$
$$\because \quad x = x' + h$$
$$\text{and} \quad y = y' + k$$
$$\Rightarrow \quad x = x' + 1$$
$$\text{and} \quad y = y' + 1$$

Now, $(x' + 1)^2 + (x' + 1)(y' + 1) - 3(y' + 1)^2 - (y' + 1) + 2 = 0$

$$\Rightarrow x'^2 + 1 + 2x' + x'y' + x' + y' + 1 - 3y'^2 - 3 - 6y' - y' - 1 + 2 = 0$$
$$\Rightarrow x'^2 - 3y'^2 + x'y' + 3x' - 6y' = 0$$

Therefore, equation in new system is

$$x^2 - 3y^2 + xy + 3x - 6y = 0.$$

II. Given equation is

$$xy - y^2 - x + y = 0$$
$$\because \quad x = x' + 1$$
$$\text{and} \quad y = y' + 1$$

Now, $(x' + 1)(y' + 1) - (y' + 1)^2 - (x' + 1) + (y' + 1) = 0$

$$\Rightarrow x'y' + x' + y' + 1 - y'^2 - 1 - 2y' - x' - 1 + y' + 1 = 0$$
$$\Rightarrow \quad -y'^2 + x'y' = 0 \Rightarrow x'y' - y'^2 = 0$$

Therefore, equation in new system is $xy - y^2 = 0$.

III. Given equation is $xy - x - y + 1 = 0$

$$\because \quad x = x' + 1$$
$$\text{and} \quad y = y' + 1$$

Now, $(x' + 1)(y' + 1) - (x' + 1) - (y' + 1) + 1 = 0$

$$\Rightarrow \quad x'y' + x' + y' + 1 - x' - 1 - y' - 1 + 1 = 0$$
$$\Rightarrow \quad x'y' = 0$$

Therefore, equation in new system is $xy = 0$.

181. *(a)* Let the origin be shifted to the point (h, k) without changing the direction of axes. Then, $x = x' + h$ and $y = y' + k$. Here, given old coordinates and new coordinates are (–1, 8) and (–7, 3), respectively.

$$-1 = -7 + h \text{ and } 8 = 3 + k$$
$$h = 6 \text{ and } k = 5$$

Hence, origin must shifted to (6, 5).

182. *(a)* Any line passing through the point of intersection of the given lines is

$$(x + 5y + 7) + \lambda(3x + 2y - 5) = 0$$
$$\Rightarrow \quad x(1 + 3\lambda) + y(5 + 2\lambda) + (7 - 5\lambda) = 0 \quad \ldots \text{(i)}$$

$$\text{Its slope} = -\frac{(1 + 3\lambda)}{(5 + 2\lambda)}$$

I. If line (i) is to be parallel to $7x + 2y - 5 = 0$.

$$\text{Then,} \quad -\frac{(1 + 3\lambda)}{(5 + 2\lambda)} = -\frac{7}{2}$$
$$\Rightarrow \quad 2 + 6\lambda = 35 + 14\lambda$$
$$\Rightarrow \quad 8\lambda = -33 \Rightarrow \lambda = -\frac{33}{8}$$

Substituting this value of λ in Eq. (i), we get the required equation as $7x + 2y - 17 = 0$.

II. If line (i) is to be perpendicular to $7x + 2y - 5 = 0$.

$$\therefore \quad -\frac{(1 + 3\lambda)}{(5 + 2\lambda)} \times -\left(\frac{7}{2}\right) = -1$$
$$\Rightarrow \quad 7 + 21\lambda = -10 - 4\lambda$$
$$\Rightarrow \quad \lambda = -\frac{17}{25}$$

Substituting this value of λ is Eq. (i), we get the required equation as $2x - 7y - 20 = 0$.

183. *(c)* Here, we want to shift the (4, 5) to the origin without changing the direction of axes. Then, we replace x by $x-4$ and y by $y-5$ in the equation of given curve and the required equation is

$$(x-4)^2+(y-5)^2=36$$

$$\Rightarrow \quad x^2+y^2-8x-10y+5=0$$

184. *(d)* Let the origin be shifted to the point (h, k) without changing the direction of axes. Then, we replace x by $x+h$ and y by $y+k$ in the equation of the given curve and the transformed equation is

$$(y+k)^2+4(y+k)+8(x+h)-2=0$$

$$\Rightarrow y^2+(2k+4)y+8x+(k^2+4k+8h-2)=0$$

Since, this equation is free from the term containing y and the constant, so we have

$$2k+4=0 \quad \text{and} \quad k^2+4k+8h-2=0$$

$$\Rightarrow \quad k=-2 \quad \text{and} \quad h=\frac{3}{4}$$

Hence, the point to which the origin be shifted, is $\left(\frac{3}{4}, -2\right)$.

185. *(b)* Any line through the point of intersection of given lines is

$$\left(\frac{x}{a}+\frac{y}{b}-1\right)+\lambda\left(\frac{x}{b}+\frac{y}{a}-1\right)=0$$

$$x\left(\frac{1}{a}+\frac{\lambda}{b}\right)+y\left(\frac{1}{b}+\frac{\lambda}{a}\right)=(1+\lambda)$$

$$\Rightarrow \quad x\left(\frac{b+a\lambda}{ab}\right)+y\left(\frac{a+b\lambda}{ab}\right)=(1+\lambda)$$

$$\Rightarrow \quad \frac{x}{\left\{\frac{ab(1+\lambda)}{b+a\lambda}\right\}}+\frac{y}{\left\{\frac{ab(1+\lambda)}{a+b\lambda}\right\}}=1$$

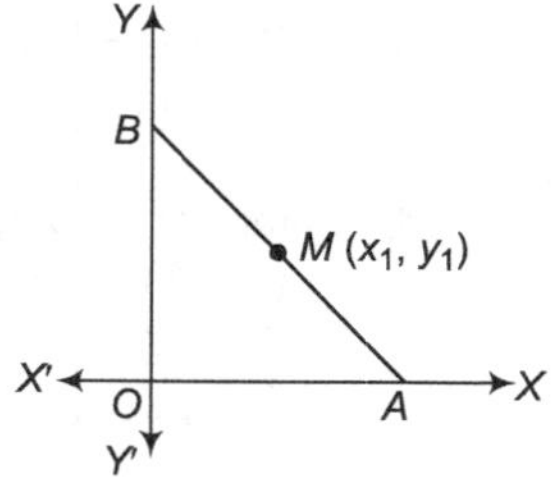

This meets the X-axis at $A \equiv \left(\frac{ab(1+\lambda)}{b+a\lambda}, 0\right)$

and meets the Y-axis at $B \equiv \left(0, \frac{ab(1+\lambda)}{a+b\lambda}\right)$

Let the mid-point of AB is $M(x_1, y_1)$. Then,

$$x_1=\frac{ab(1+\lambda)}{2(b+a\lambda)} \quad \text{and} \quad y_1=\frac{ab(1+\lambda)}{2(a+b\lambda)}$$

$$\therefore \quad \frac{1}{x_1}+\frac{1}{y_1}=\frac{2(b+a\lambda)}{ab(1+\lambda)}+\frac{2(a+b\lambda)}{ab(1+\lambda)}$$

$$=\frac{2}{ab(1+\lambda)}(b+a\lambda+a+b\lambda)$$

$$=\frac{2}{ab(1+\lambda)}(b+a)(1+\lambda)$$

$$\Rightarrow \quad \frac{(x_1+y_1)}{x_1y_1}=\frac{2(a+b)}{ab}$$

$$\Rightarrow \quad 2x_1y_1(a+b)=ab(x_1+y_1)$$

Hence, the locus of mid-point of AB is

$$2xy(a+b)=ab(x+y)$$

186. *(d)* Equation of any straight line passing through the intersection of the lines $x+2y=3$ and $2x-3y=4$ is

$$\lambda(x+2y-3)+(2x-3y-4)=0 \quad \ldots(\text{i})$$

Since, it passes through the point (2, 1),

$$\therefore \quad \lambda(2+2-3)+(4-3-4)=0$$

$$\Rightarrow \quad \lambda-3=0$$

$$\therefore \quad \lambda=3$$

Now substituting this value of λ in Eq. (i), we get

$$3(x+2y-3)+(2x-3y-4)=0$$

i.e.,

$$5x+3y-13=0$$

which is the equation of required line.

187. *(b)* Equation of any line passing through the intersection of the given lines is

$$\Rightarrow \quad (3x-4y+1)+\lambda(5x+y-1)=0$$

$$\Rightarrow \quad x(3+5\lambda)+y(-4+\lambda)+(1-\lambda)=0 \quad \ldots(\text{i})$$

$$\frac{x}{\left(\frac{\lambda-1}{3+5\lambda}\right)}+\frac{y}{\left(\frac{\lambda-1}{\lambda-4}\right)}=1$$

But given, x-intercept = y-intercept

i.e.,

$$\left(\frac{\lambda-1}{3+5\lambda}\right)=\left(\frac{\lambda-1}{\lambda-4}\right)$$

$$\Rightarrow \quad \frac{1}{3+5\lambda}=\frac{1}{\lambda-4}$$

[$\lambda \neq 1$ since, if $\lambda=1$, then line (i) passes through origin]

$$\therefore \quad \lambda-4=3+5\lambda$$

or

$$4\lambda=-7$$

$$\therefore \quad \lambda=-\frac{7}{4}$$

Substituting the value of λ in Eq. (i), we get required equation is $23x+23y=11$.

188. *(c)* $\because t_1$ and t_2 are the roots of the equation $t^2+\lambda t+1=0$

$$\therefore \quad t_1+t_2=-\lambda \quad \text{and} \quad t_1t_2=1 \quad \ldots(\text{i})$$

Equation of the line joining the points $(at_1^2, 2at_1)$ and $(at_2^2, 2at_2)$ is

$$y-2at_1=\frac{2at_2-2at_1}{at_2^2-at_1^2}(x-at_1^2)$$

$$\Rightarrow \quad y-2at_1=\frac{2}{(t_2+t_1)}(x-at_1^2)$$

$$\Rightarrow \quad y(t_1+t_2)-2at_1t_2-2at_1^2=2x-2at_1^2$$

$$\Rightarrow \quad y(t_1+t_2)-2at_1t_2=2x$$

$\Rightarrow \quad y(-\lambda) - 2a = 2x \quad$ [from Eq. (i)]

or $\quad (x + a) + \lambda \left(\frac{y}{2}\right) = 0$

which is equation of a line passing through the point of intersection of the lines $x + a = 0$ and $\frac{y}{2} = 0$.

$\therefore$ Coordinates of fixed point are $(-a, 0)$.

189. *(a)* Equation of line l_1 is

$$x - 2y = 3 \quad \ldots(i)$$

Slope of Eq. (i),

$$m_1 = -\left(\frac{1}{-2}\right) = \frac{1}{2}$$

Angle between two lines is $\tan\theta = \left|\frac{m_1 - m_2}{1 + m_1 m_2}\right|$.

Given, $\theta = 45°$, $m_1 = \frac{1}{2}$, $m_2 = m$ (let)

$$\therefore \quad \tan 45° = \left|\frac{\frac{1}{2} - m}{1 + \frac{1}{2}m}\right| \Rightarrow 1 = \left|\frac{1-2m}{2+m}\right|$$

$$\Rightarrow \quad \frac{1-2m}{2+m} = \pm 1$$

Taking positive sign,

$$\frac{1-2m}{2+m} = 1 \Rightarrow 1 - 2m = 2 + m$$

$$\Rightarrow \quad 1 - 2 = 3m \Rightarrow m = -\frac{1}{3}$$

Taking negative sign, $\frac{1-2m}{2+m} = -1$

$$\Rightarrow \quad 1 - 2m = -2 - m$$

$$\Rightarrow \quad 1 + 2 = 2m - m \Rightarrow m = 3$$

Now, equation of line l_2 by using

$$y - y_1 = m(x - x_1)$$

$$\Rightarrow \quad y - 2 = -\frac{1}{3}(x - 3) \quad \left(\text{when } m = -\frac{1}{3}, x_1 = 3, y_1 = 2\right)$$

$$\Rightarrow \quad 3y - 6 = -x + 3$$

$$\Rightarrow \quad x + 3y - 9 = 0$$

Again, $\quad y - 2 = 3(x - 3) \quad \left(\text{when } m = 3, x_1 = 3, y_1 = 2\right)$

$$\Rightarrow \quad y - 2 = 3x - 9$$

$$\Rightarrow \quad 3x - y - 9 + 2 = 0$$

$$\Rightarrow \quad 3x - y - 7 = 0$$

190. *(c)* Let intercept form of line is

$$\frac{x}{a} + \frac{y}{b} = 1 \quad \ldots(i)$$

where, a and b are intercepts on the axes.

Also given, the sum of the reciprocals of its intercepts made on axes is constant.

$$\Rightarrow \quad \frac{1}{a} + \frac{1}{b} = \frac{1}{k} \quad \text{(where } k \text{ is constant)}$$

$$\Rightarrow \quad \frac{k}{a} + \frac{k}{b} = 1 \quad \ldots(ii)$$

On comparing Eqs. (i) and (ii), we get

$$x = k, y = k$$

Line passes through the fixed point (k, k).

191. *(b)* Given, equation of lines are

$$4x + 7y - 3 = 0 \quad \ldots(i)$$

and $\quad 2x - 3y + 1 = 0 \quad \ldots(ii)$

From Eq. (ii), $\quad 3y = 2x + 1$

$$\Rightarrow \quad y = \frac{2}{3}x + \frac{1}{3}$$

On putting in Eq. (i), we get

$$4x + 7\left(\frac{2}{3}x + \frac{1}{3}\right) - 3 = 0$$

$$\Rightarrow \quad 4x + \frac{14}{3}x + \frac{7}{3} - 3 = 0$$

$$\Rightarrow \quad \frac{12x + 14x}{3} + \frac{7-9}{3} = 0$$

$$\Rightarrow \quad \frac{26x - 2}{3} = 0$$

$$\Rightarrow \quad x = \frac{2}{26}$$

$$\Rightarrow \quad x = \frac{1}{13}$$

On putting the value of x in Eq. (i), we get

$$\frac{4}{13} + 7y - 3 = 0$$

$$\Rightarrow \quad 7y = 3 - \frac{4}{13}$$

$$\Rightarrow \quad 7y = \frac{39 - 4}{13}$$

$$\Rightarrow \quad 7y = \frac{35}{13} \Rightarrow y = \frac{5}{13}$$

Point is $\left(\frac{1}{13}, \frac{5}{13}\right)$.

Now, equation of line in intercept form is

$$\frac{x}{a} + \frac{y}{b} = 1 \quad \ldots(iii)$$

Since, line (iii) has equal intercepts on the axes *i.e.*, $a = b$

$$\Rightarrow \quad \frac{x}{a} + \frac{y}{a} = 1$$

$$\Rightarrow \quad x + y = a \quad \ldots(iv)$$

Above line passes through the point $\left(\frac{1}{13}, \frac{5}{13}\right)$ *i.e.*, point will satisfy it.

$$\therefore \quad \frac{1}{13} + \frac{5}{13} = a \Rightarrow a = \frac{6}{13}$$

Hence, required equation of line (iv) becomes

$$x + y = \frac{6}{13}$$

$$\Rightarrow \quad 13x + 13y = 6$$

192. *(c)* Given equation of line is $\frac{x}{a} + \frac{y}{b} - 1 = 0 \quad \ldots(i)$

At point (0, 0), perpendicular distance from line (i) is

$$p = \left|\frac{0+0-1}{\sqrt{\frac{1}{a^2}+\frac{1}{b^2}}}\right| \quad \left(\because \text{ perpendicular distance, } d = \left|\frac{ax_1+by_1+c}{\sqrt{a^2+b^2}}\right|\right)$$

$$\Rightarrow \quad p = \frac{1}{\sqrt{\frac{1}{a^2}+\frac{1}{b^2}}}$$

$$\Rightarrow \quad \frac{1}{p} = \sqrt{\frac{1}{a^2}+\frac{1}{b^2}}$$

On squaring both sides, we get

$$\Rightarrow \quad \frac{1}{p^2} = \frac{1}{a^2}+\frac{1}{b^2} \quad \ldots(ii)$$

Also, a^2, p^2, b^2 are in AP.

$$\therefore \quad 2p^2 = a^2 + b^2 \quad \ldots(iii)$$

(using formula, if a, b and c are in AP, then $2b = a + c$)

Now, from Eq. (ii), we get

$$\frac{1}{p^2} = \frac{a^2+b^2}{a^2b^2} \Rightarrow p^2 = \frac{a^2b^2}{a^2+b^2}$$

$$\Rightarrow \quad \frac{a^2+b^2}{2} = \frac{a^2b^2}{a^2+b^2} \quad \text{[using Eq. (iii)]}$$

$$\Rightarrow \quad (a^2+b^2)^2 = 2a^2b^2$$

$$\Rightarrow \quad a^4 + b^4 + 2a^2b^2 = 2a^2b^2 \quad [\because (a+b)^2 = a^2+b^2+2ab]$$

$$\Rightarrow \quad a^4 + b^4 = 0$$

193. *(d)* Slope of given line is m and slope of another line is m_2.

$\because$ Angle between two lines is

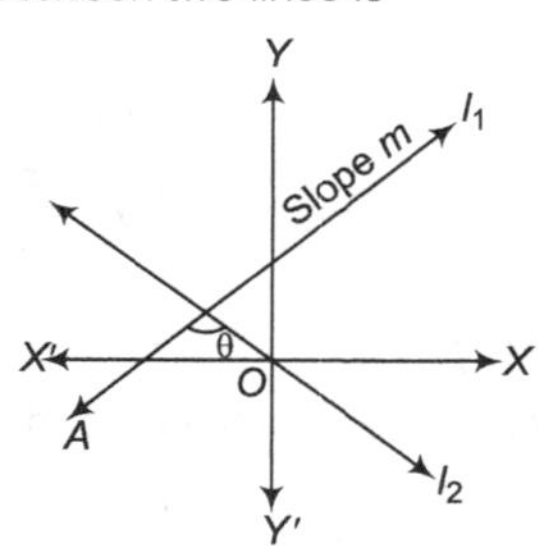

$$\tan\theta = \left|\frac{m_1 - m_2}{1 + m_1m_2}\right|$$

$$\Rightarrow \quad \tan\theta = \left|\frac{m - m_2}{1 + mm_2}\right| \quad (\because m_1 = m)$$

$$\Rightarrow \quad \frac{m - m_2}{1 + mm_2} = \pm\tan\theta$$

Taking positive sign,

$$\frac{m-m_2}{1+mm_2} = \frac{\tan\theta}{1}$$

$$\Rightarrow \quad m - m_2 = \tan\theta + mm_2\tan\theta$$

$$\Rightarrow \quad m - \tan\theta = m_2 + mm_2\tan\theta$$

$$\Rightarrow \quad m_2(1 + m\tan\theta) = m - \tan\theta$$

$$\Rightarrow \quad m_2 = \frac{m - \tan\theta}{1 + m\tan\theta} \quad \ldots(i)$$

Taking negative sign,

$$\frac{m-m_2}{1+mm_2} = \frac{-\tan\theta}{1}$$

$$\Rightarrow \quad m - m_2 = -\tan\theta - mm_2\tan\theta$$

$$\Rightarrow \quad m + \tan\theta = m_2 - mm_2\tan\theta$$

$$\Rightarrow \quad m + \tan\theta = m_2(1 - m\tan\theta)$$

$$\Rightarrow \quad m_2 = \frac{m + \tan\theta}{1 - m\tan\theta} \quad \ldots(ii)$$

$\therefore$ From Eqs. (i) and (ii), we get

$$m_2 = \frac{m \pm \tan\theta}{1 \mp m\tan\theta}$$

Equation of line through (0, 0) by using $y - y_1 = m_2(x - x_1)$ is

$$y - 0 = \frac{m \pm \tan\theta}{1 \mp m\tan\theta}(x - 0)$$

$$\Rightarrow \quad \frac{y}{x} = \frac{m \pm \tan\theta}{1 \mp m\tan\theta}$$

194. *(d)* Let equation of line is

$$\frac{x - x_1}{\cos\theta} = \frac{y - y_1}{\sin\theta} = r$$

$$\therefore \quad \frac{x-1}{\cos\theta} = \frac{y-2}{\sin\theta} = \frac{\sqrt{6}}{3} \quad \left(\because x_1 = 1, y_1 = 2, r = \frac{\sqrt{6}}{3}\right)$$

$$\Rightarrow \quad \frac{x-1}{\cos\theta} = \frac{\sqrt{6}}{3}, \frac{y-2}{\sin\theta} = \frac{\sqrt{6}}{3}$$

$$\Rightarrow \quad x = \frac{\sqrt{6}}{3}\cos\theta + 1, y = \frac{\sqrt{6}}{3}\sin\theta + 2$$

It satisfies $\quad x + y = 4$

$$\therefore \quad \frac{\sqrt{6}}{3}\cos\theta + 1 + \frac{\sqrt{6}}{3}\sin\theta + 2 = 4$$

$$\left(\text{putting } x = \frac{\sqrt{6}}{3}\cos\theta + 1 \text{ and } y = \frac{\sqrt{6}}{3}\sin\theta + 2\right)$$

$$\Rightarrow \quad \frac{\sqrt{6}}{3}\cos\theta + \frac{\sqrt{6}}{3}\sin\theta = 1$$

$$\Rightarrow \quad \sqrt{6}\cos\theta + \sqrt{6}\sin\theta = 3$$

$$\Rightarrow \quad 6\cos^2\theta + 6\sin^2\theta + 2\sqrt{6}\sqrt{6}\cos\theta\sin\theta = 9$$

(squaring on both sides)

$$\Rightarrow \quad 6(\cos^2\theta + \sin^2\theta) + 2 \times 6\cos\theta\sin\theta = 9$$

$$\Rightarrow \quad 6 + 6\sin 2\theta = 9$$

$$(\because \cos^2\theta + \sin^2\theta = 1 \text{ and } 2\sin\theta\cos\theta = \sin 2\theta)$$

$$\Rightarrow \quad 6\sin 2\theta = 9 - 6$$

$$\Rightarrow \quad \sin 2\theta = \frac{3}{6}$$

$$\Rightarrow \quad \sin 2\theta = \frac{1}{2} = \sin 30° \quad \left(\because \sin 30° = \frac{1}{2}\right)$$

$$\Rightarrow \quad 2\theta = 30°$$

$$\Rightarrow \quad \theta = 15°$$

or $\quad \sin 2\theta = \sin(180° - 30°)$

$$[\because \sin\theta = \sin(180° - \theta)]$$

$$\Rightarrow \quad 2\theta = 150°$$

$$\Rightarrow \quad \theta = 75°$$

195. *(d)* Let the point P divides the line joining the points $A(-1, 1)$ and $B(5, 7)$ in the ratio is $K : 1$.

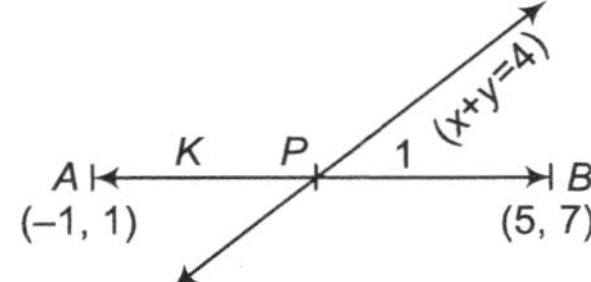

Then,

$$P = \left(\frac{K \times x_2 + 1 \times x_1}{K+1}, \frac{K \times y_2 + 1 \times y_1}{K+1}\right)$$

$$(\because x_1 = -1, y_1 = 1, x_2 = 5, y_2 = 7)$$

$$= \left(\frac{K \times 5 + 1 \times (-1)}{K+1}, \frac{K \times 7 + 1 \times 1}{K+1}\right)$$

$$= \left(\frac{5K - 1}{K+1}, \frac{7K+1}{K+1}\right)$$

Point P will satisfy the line $x + y = 4$

i.e., $$\frac{5K-1}{K+1} + \frac{7K+1}{K+1} = \frac{4}{1}$$

$$\Rightarrow \frac{5K - 1 + 7K + 1}{K+1} = 4$$

$$\Rightarrow 12K = 4K + 4$$

$$\Rightarrow 8K = 4$$

$$\Rightarrow K = \frac{4}{8} = \frac{1}{2}$$

Hence, the required ratio is $K : 1 = 1 : 2$ (internally)

196. *(c)* Let equation of line is $ax + by + c = 0$. Then, according to given condition, the sum of the perpendiculars drawn from the points (2, 0), (0, 2) and (1, 1) on line is zero.

$\therefore$ Perpendicular distance from (2, 0) on the line

+ Perpendicular distance from (0, 2) on the line

+ Perpendicular distance from (1, 1) on the line = 0

$$\Rightarrow \left|\frac{2a + 0b + c}{\sqrt{a^2+b^2}}\right| + \left|\frac{0 + 2b + c}{\sqrt{a^2+b^2}}\right| + \left|\frac{a+b+c}{\sqrt{a^2+b^2}}\right| = 0$$

$$\left(\begin{array}{l}\because \text{perpendicular distance at } (x_1, y_1) \text{ on the line} \\ ax + by + c = 0 \text{ is } d = \left|\dfrac{ax_1 + by_1 + c}{\sqrt{a^2+b^2}}\right|\end{array}\right)$$

$$\Rightarrow \frac{2a+c}{\sqrt{a^2+b^2}} + \frac{2b+c}{\sqrt{a^2+b^2}} + \frac{a+b+c}{\sqrt{a^2+b^2}} = 0$$

$$\Rightarrow 2a + c + 2b + c + a + b + c = 0$$

$$\Rightarrow 3a + 3b + 3c = 0$$

$$\Rightarrow a + b + c = 0$$

Clearly, this line passes through the fixed point (1, 1).

197. *(b)* The equation of line AB is

$$4x + 7y + 5 = 0 \quad \ldots(\text{i})$$

and equation of PQ is

$$2x - y = 0 \quad \ldots(\text{ii})$$

A
(4x+7y+5=0)
(2x–y=0)
Q
P
(1, 2)
B

Now, solve Eqs. (i) and (ii), to obtain the coordinate of Q.

From Eq. (ii), putting $y = 2x$ in Eq. (i), we get

$$4x + 7(2x) + 5 = 0$$

$$\Rightarrow 18x + 5 = 0$$

$$\Rightarrow x = -\frac{5}{18}$$

Putting $x = -\frac{5}{18}$ in Eq. (ii), we get

$$2\left(-\frac{5}{18}\right) - y = 0$$

$$\Rightarrow y = -\frac{5}{9}$$

$$\therefore \text{Coordinate of } Q = \left(-\frac{5}{18}, -\frac{5}{9}\right)$$

$$\text{Length of } QP = \sqrt{(x_2 - x_1)^2 + (y_2 - y_1)^2}$$

$$= \sqrt{\left(1 + \frac{5}{18}\right)^2 + \left(2 + \frac{5}{9}\right)^2}$$

$$\left(\begin{array}{ll}\because x_1 = -\dfrac{5}{18}, & y_1 = -\dfrac{5}{9}, \\ x_2 = 1, & y_2 = 2\end{array}\right)$$

$$= \sqrt{\left(\frac{23}{18}\right)^2 + \left(\frac{23}{9}\right)^2}$$

$$= \frac{23}{9}\sqrt{\frac{1}{4} + 1} = \frac{23}{9} \times \frac{\sqrt{5}}{2}$$

$$= \frac{23}{18}\sqrt{5} \text{ sq units}$$

198. *(a)* Let ABC be an isosceles right angled triangle, with right angle at A and $AB = AC$.

$$\Rightarrow \angle ABC = \angle ACB = 45°$$

The slope of line BC *i.e.*, $3x + 4y = 4$ is

$$m_2 = -\frac{3}{4}$$

Let the slope of the line AC be $m_1 = m$.

$\because$ Angle between two lines is

$$\tan\theta = \left|\frac{m_1 - m_2}{1 + m_1 m_2}\right|$$

$$\Rightarrow \tan 45° = \left|\frac{m - \left(\frac{-3}{4}\right)}{1 + m\left(\frac{-3}{4}\right)}\right|$$

$$\Rightarrow 1 = \left|\frac{m + \frac{3}{4}}{1 - \frac{3m}{4}}\right|$$

$$\Rightarrow \pm 1 = \frac{4m+3}{4-3m}$$

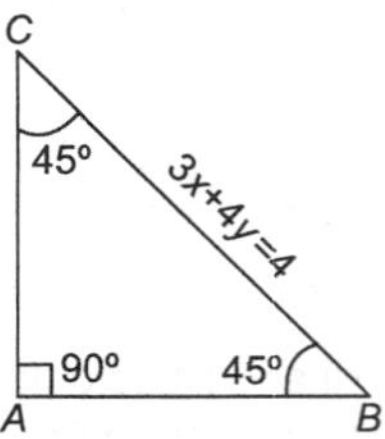

Taking positive sign,

$$1 = \frac{4m+3}{4-3m}$$

$\Rightarrow \quad 4 - 3m = 4m + 3$

$\Rightarrow \quad 4 - 3 = 4m + 3m$

$\Rightarrow \quad 7m = 1 \Rightarrow m = \frac{1}{7}$

Taking negative sign,

$$-1 = \frac{4m+3}{4-3m}$$

$\Rightarrow \quad -4 + 3m = 4m + 3$

$\Rightarrow \quad -4 - 3 = m$

$\Rightarrow \quad m = -7$

Hence, equations of line AB and line AC are

$$y - 2 = \frac{1}{7}(x - 2)$$

and $\quad y - 2 = -7(x - 2) \; [\because y - y_1 = m(x - x_1)]$

$\Rightarrow \quad 7y - 14 = x - 2$

and $\quad y - 2 = -7x + 14$

$\Rightarrow \quad x - 7y + 12 = 0$ and $7x + y - 16 = 0$

199. *(a)* A line passing through point $P(-1, 2)$ and having slope m is

$$y - 2 = m(x+1) \quad \left[\begin{array}{l}\because y - y_1 = m(x - x_1) \\ \text{where, } x_1 = -1, y_1 = 2\end{array}\right]$$

$\Rightarrow \quad y = mx + m + 2 \quad \ldots\text{(i)}$

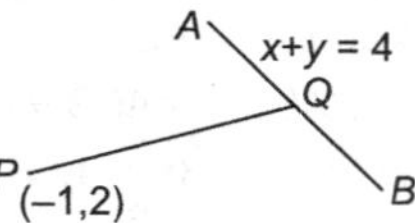

Now, determine the point of intersection of line AB and line PQ.

On putting the value of y from Eq. (i) in $x + y = 4$, we get

$$x + mx + m + 2 = 4$$

$\Rightarrow \quad x(1 + m) = 2 - m \Rightarrow x = \frac{2-m}{1+m}$

$\therefore$ From Eq. (i),

$$y = m\left(\frac{2-m}{1+m}\right) + m + 2$$

$$= \frac{2m - m^2 + m + m^2 + 2 + 2m}{1+m}$$

$$= \frac{2+5m}{1+m}$$

$\therefore$ Coordinates of $Q = \left(\frac{2-m}{1+m}, \frac{2+5m}{1+m}\right)$

$\because$ Distance between $PQ = 3$ (given)

$\Rightarrow \quad (PQ)^2 = 9$

$\Rightarrow \quad (x_2 - x_1)^2 + (y_2 - y_1)^2 = 9$

$\Rightarrow \quad \left(\frac{2-m}{1+m} + 1\right)^2 + \left(\frac{2+5m}{1+m} - 2\right)^2 = 9$

$$\left(\begin{array}{l}\because x_1 = -1, \quad y_1 = 2, \\ x_2 = \frac{2-m}{1+m}, \; y_2 = \frac{2-5m}{1+m}\end{array}\right)$$

$\Rightarrow \quad \left(\frac{3}{1+m}\right)^2 + \left(\frac{3m}{1+m}\right)^2 = 9$

$\Rightarrow \quad 9 + 9m^2 = 9(1+m)^2$

$\Rightarrow \quad 1 + m^2 = 1 + 2m + m^2$

$\Rightarrow \quad 2m = 0 \Rightarrow m = 0$

$\therefore$ Slope of PQ is zero, *i.e.*, it is parallel to X-axis.

200. *(b)* Let ABC be an equilateral triangle. Draw $AD \perp BC$.

Length of AD = Perpendicular distance from $A(2, -1)$ to the line $x + y - 2 = 0$

$$= \left|\frac{2-1-2}{\sqrt{1+1}}\right| = \frac{1}{\sqrt{2}}$$

Since, ΔABC is an equilateral triangle, so $\angle ABC = 60°$.

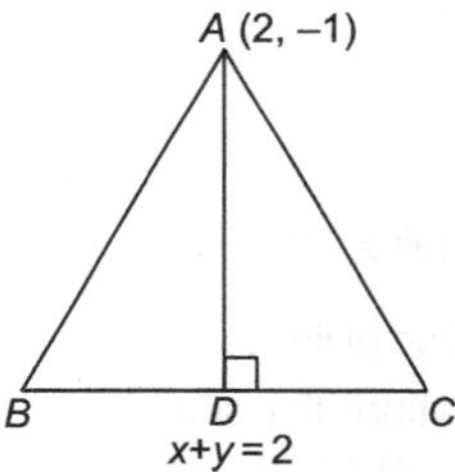

In ΔABD,

$$\sin 60° = \frac{AD}{AB} \Rightarrow \frac{\sqrt{3}}{2} = \frac{\frac{1}{\sqrt{2}}}{AB}$$

$\Rightarrow \quad \frac{\sqrt{3}}{2} = \frac{1}{\sqrt{2}\,AB}$

$\Rightarrow \quad AB = \frac{2}{\sqrt{6}}$

$\Rightarrow \quad AB = \sqrt{\frac{2}{3}}$ unit

Since, ΔABC is an equilateral triangle, so all sides are of equal length.

$\therefore$ Length of AB = Length of BC = Length of $CA = \sqrt{\frac{2}{3}}$

201. *(c)* First we plot the points $A(1, 3)$ and $B(-4, 1)$ in the xy-plane. From the point $A(1, 3)$, we draw a line parallel to Y-axis and from the point $B(-4, 1)$, we draw a line parallel to X-axis. The point of intersection of two lines is on C, which is right angled at C.

$\therefore$ The coordinate of C will be $(1, 1)$.

$\therefore$ Equation of line AC passing through $A(1, 3)$ and $C(1, 1)$ is

$$y - y_1 = \frac{y_2 - y_1}{x_2 - x_1}(x - x_1)$$

$$\therefore \quad y - 3 = \frac{1-3}{1-1}(x - 1)$$

$$\Rightarrow \quad y - 3 = \frac{-2}{0}(x - 1) \Rightarrow x = 1$$

Equation of line BC is

$$y - 1 = \frac{1-1}{1+4}(x - 1) \Rightarrow y - 1 = \frac{0}{1+4}(x - 1)$$

$$\Rightarrow \quad y - 1 = 0 \quad \Rightarrow \quad y = 1$$

Hence, the legs of a triangle are $x = 1$ and $y = 1$.

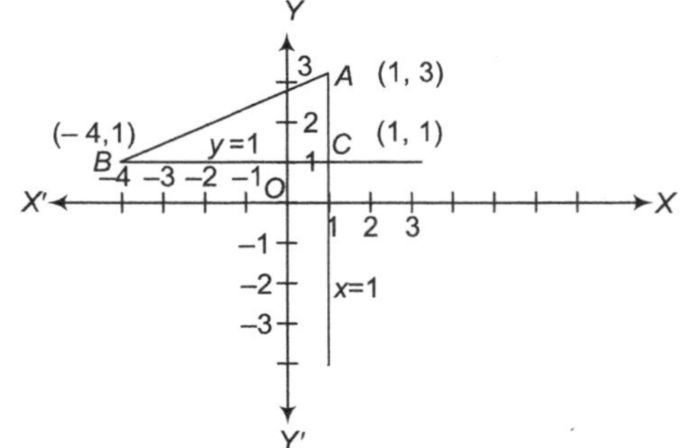

202. *(b)* Given lines are

$$\frac{x}{a} + \frac{y}{b} = 1 \quad \ldots(i)$$

and

$$\frac{x}{a} - \frac{y}{b} = 1 \quad \ldots(ii)$$

Slope of line (i) is $m_1 = -\frac{1/a}{1/b} = -\frac{b}{a}$

Slope of line (ii) is $m_2 = \frac{-\frac{1}{a}}{-\frac{1}{b}} = +\frac{b}{a}$

Let θ be the angle between the lines. Then,

$$\tan\theta = \left|\frac{m_1 - m_2}{1 + m_1 m_2}\right| = \left|\frac{-\frac{b}{a} - \frac{b}{a}}{1 + \left(-\frac{b}{a}\right)\left(\frac{b}{a}\right)}\right|$$

$$\Rightarrow \tan\theta = \left|\frac{-\frac{2b}{a}}{1 - \frac{b^2}{a^2}}\right| = \frac{2b}{a} \times \frac{a^2}{a^2 - b^2} = \frac{2ab}{a^2 - b^2} \quad (\because |-A| = A)$$

$$\Rightarrow \quad \tan\theta = \frac{2ab}{a^2 - b^2}$$

$$\Rightarrow \quad m = ab \text{ and } n = a^2 - b^2.$$

203. *(d)* Let equation of line AB is

$$x + 3y = 7 \quad \ldots(i)$$

Here, slope of line m_1 is $-\frac{1}{3}$.

Let $P(3, 8)$ be the given point, for which image is to be found. Here, $PQ \perp AB$.

Slope of PQ $\times$ Slope of $AB = -1$

Let slope of PQ be m.

$$\therefore \quad m \times \left(\frac{-1}{3}\right) = -1 \quad (\because m_1 m_2 = -1)$$

$$\Rightarrow \quad m = 3$$

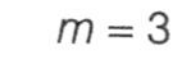

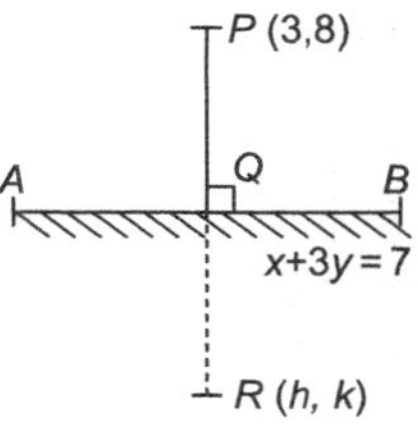

Now, equation of line PQ by using $y - y_1 = m(x - x_1)$ where, $(x_1, y_1) = (3, 8)$ and $m = 3$, is

$$y - 8 = 3(x - 3)$$

$$\Rightarrow \quad y - 8 = 3x - 9$$

$$\Rightarrow \quad 3x - y - 1 = 0 \quad \ldots(ii)$$

To determine point Q, solve Eqs. (i) and (ii).

From Eq. (i), putting $x = 7 - 3y$ in Eq. (ii), we get

$$\Rightarrow \quad 3(7 - 3y) - y - 1 = 0$$

$$\Rightarrow \quad 21 - 9y - y - 1 = 0$$

$$\Rightarrow \quad 20 - 10y = 0$$

$$\Rightarrow \quad y = \frac{20}{10}$$

$$\Rightarrow \quad y = 2$$

$\therefore$ From Eq. (i), $x = 7 - 3 \times 2 = 7 - 6 = 1$

$\therefore$ Coordinates of Q are $(1, 2)$.

Let the image be $R(h, k)$. Since, the line is a plane mirror, so point Q will be the mid-point of P and R.

Now, mid-point of $P(3, 8)$ and $R(h, k)$ is

$$\left(\frac{x_1 + x_2}{2}, \frac{y_1 + y_2}{2}\right) = \left(\frac{3+h}{2}, \frac{8+k}{2}\right)$$

On comparing with coordinate $Q(1, 2)$, we get

$$\frac{3+h}{2} = 1$$

and

$$\frac{8+k}{2} = 2$$

P ———— Q ———— R
(3,8) (1,2) (h, k)

$$\Rightarrow \quad 3 + h = 2 \text{ and } 8 + k = 4$$

$$\Rightarrow \quad h = 2 - 3 \text{ and } k = 4 - 8$$

$$\Rightarrow \quad h = -1 \text{ and } k = -4$$

Hence, image of a point P is $R(-1, -4)$.

204. *(a)* Given points are $(\sin\theta, \cos\theta)$ and $(3, 2)$ and a line

$$x + y - 1 = 0 \quad \ldots(i)$$

Since, $(3, 2)$ lies on Eq. (i),

$$\therefore \quad 3 + 2 - 1 > 0$$

and $(\sin\theta, \cos\theta)$ lies in Eq. (i)

$$\therefore \quad \sin\theta + \cos\theta - 1 > 0$$

$\Rightarrow \quad \sin\theta + \cos\theta > 1$

$\Rightarrow \quad \sqrt{2}\left[\sin\left(\theta + \frac{\pi}{4}\right)\right] > 1$

$\Rightarrow \quad \sin\left(\theta + \frac{\pi}{4}\right) > \frac{1}{\sqrt{2}} = \sin\left(\frac{\pi}{4}\right)$

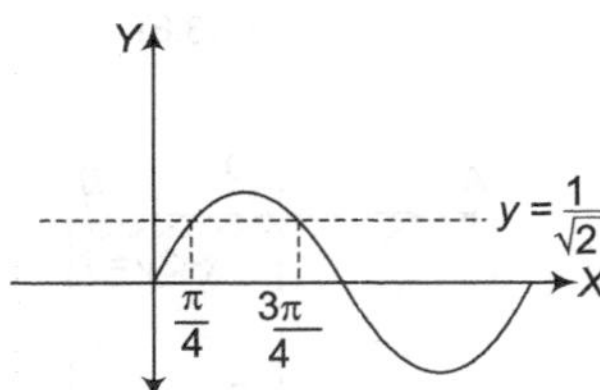

$\Rightarrow \quad \frac{\pi}{4} < \theta + \frac{\pi}{4} < \frac{3\pi}{4} \Rightarrow 0 < \theta < \frac{\pi}{2}$

Hence, $\quad \theta \in \left(0, \frac{\pi}{2}\right)$

205. *(c)* Let the angle between the lines $y = 3x + 1$ and $y = mx + 4$ be θ.

Here, slopes of the lines are $m_1 = 3$ and $m_2 = m$.

$\therefore \quad \tan\theta = \left|\frac{3-m}{1+3m}\right| \quad \ldots(i)$

$\left(\because \tan\theta = \left|\frac{m_1 - m_2}{1 + m_1 m_2}\right|\right)$

Again, it is given that angle between the lines $2y = x + 3$ and $y = mx + 4$ is also θ.

Here, slopes are $m_1 = \frac{1}{2}$ and $m_2 = m$.

$\therefore \quad \tan\theta = \left|\frac{\frac{1}{2} - m}{1 + \frac{1}{2} \times m}\right| \quad \ldots(ii)$

From Eqs. (i) and (ii), we get

$$\left|\frac{3-m}{1+3m}\right| = \left|\frac{\frac{1}{2} - m}{1 + \frac{m}{2}}\right|$$

$\Rightarrow \quad \frac{3-m}{1+3m} = \pm\left(\frac{1-2m}{2+m}\right)$

Taking positive sign,

$(3 - m)(2 + m) = (1 - 2m)(1 + 3m)$

$\Rightarrow \quad 6 + 3m - 2m - m^2 = 1 + 3m - 2m - 6m^2$

$\Rightarrow \quad 6 = 1 - 5m^2$

$\Rightarrow \quad 5m^2 = 1 - 6$

$\Rightarrow \quad m^2 = \frac{-5}{5} \Rightarrow m^2 = -1$

which is not possible as $m^2 \geq 0$.

Again, taking negative sign, we get

$$\frac{3-m}{1+3m} = -\frac{1-2m}{2+m}$$

$\Rightarrow \quad (3 - m)(2 + m) = (2m - 1)(1 + 3m)$

$\Rightarrow \quad 6 + 3m - 2m - m^2 = 2m + 6m^2 - 1 - 3m$

$\Rightarrow \quad 6 + m - m^2 = 6m^2 - m - 1$

$\Rightarrow \quad 6m^2 - m - 1 - 6 - m + m^2 = 0$

$\Rightarrow \quad 7m^2 - 2m - 7 = 0$

$\Rightarrow \quad m = \frac{2 \pm \sqrt{4 + 4 \times 7 \times 7}}{2 \times 7} \quad \left(\because x = \frac{-b \pm \sqrt{b^2 - 4ac}}{2a}\right)$

$\Rightarrow \quad m = \frac{2 \pm 2\sqrt{1 + 49}}{2 \times 7}$

$\Rightarrow \quad m = \frac{1 \pm \sqrt{50}}{7}$

$\Rightarrow \quad m = \frac{1 \pm 5\sqrt{2}}{7}$

206. *(b)* Given, equation of line are

$x + y - 5 = 0 \quad \ldots(i)$

and $\quad 3x - 2y + 7 = 0 \quad \ldots(ii)$

Given, the sum of perpendicular distances of a variable point $P(x, y)$ from lines $x + y - 5 = 0$ and $3x - 2y + 7 = 0$ is 10.

$\therefore \quad \left|\frac{x + y - 5}{\sqrt{(1)^2 + (1)^2}}\right| + \left|\frac{3x - 2y + 7}{\sqrt{3^2 + (-2)^2}}\right| = 10$

$\left(\because \text{perpendicular distance, } d = \left|\frac{ax_1 + by_1 + c}{\sqrt{a^2 + b^2}}\right|\right)$

$\Rightarrow \quad \frac{x + y - 5}{\sqrt{2}} + \frac{3x - 2y + 7}{\sqrt{13}} = 10$

$\Rightarrow \sqrt{13}(x + y - 5) + \sqrt{2}(3x - 2y + 7) = 10\sqrt{2}\sqrt{13}$

$\Rightarrow x(\sqrt{13} + 3\sqrt{2}) + y(\sqrt{13} - 2\sqrt{2}) - 5\sqrt{13} + 7\sqrt{2} = 10\sqrt{26}$

$\Rightarrow x(\sqrt{13} + 3\sqrt{2}) + y(\sqrt{13} - 2\sqrt{2}) + 7\sqrt{2} - 5\sqrt{13} - 10\sqrt{26} = 0$

which is of the form $ax + by + c = 0$.

Hence, it represents a line.

Note *Here, point (x_1, y_1) is given as (x, y), so students should not create any confusion.*

207. *(d)* Since, line L passes through (13, 32).

$\therefore \quad \frac{13}{5} + \frac{32}{b} = 1$

$\Rightarrow \quad \frac{32}{b} = 1 - \frac{13}{5} = -\frac{8}{5}$

$\Rightarrow \quad b = -\frac{32 \times 5}{8} = -20$

$\Rightarrow \quad L: \frac{x}{5} - \frac{y}{20} = 1$

Given, K is parallel to $L = 0$.

$\therefore$ The line K must have equation

$$\frac{x}{5} - \frac{y}{20} = a$$

or $\quad \frac{x}{5a} - \frac{y}{20a} = 1$

Comparing with $\frac{x}{c} + \frac{y}{3} = 1$,

$-20a = 3, c = 5a$

$\Rightarrow \quad a = -\frac{3}{20}, c = -\frac{15}{20}$

$\therefore$ Distance between lines is

$$\left|\frac{a-1}{\sqrt{\frac{1}{25}+\frac{1}{400}}}\right| = \left|\frac{-\frac{3}{20}-1}{\sqrt{\frac{17}{400}}}\right| = \frac{23}{\sqrt{17}}$$

208. *(a)* Equations of given lines are

$$9x + 6y - 7 = 0$$

or $\quad 3x + 2y - \frac{7}{3} = 0 \quad \ldots\text{(i)}$

and $\quad 3x + 2y + 6 = 0 \quad \ldots\text{(ii)}$

Equation of line parallel to either Eq. (i) or Eq. (ii), is

$$3x + 2y + k = 0 \quad \ldots\text{(iii)}$$

Now, given line (iii) is equidistant from lines (i) and (ii).

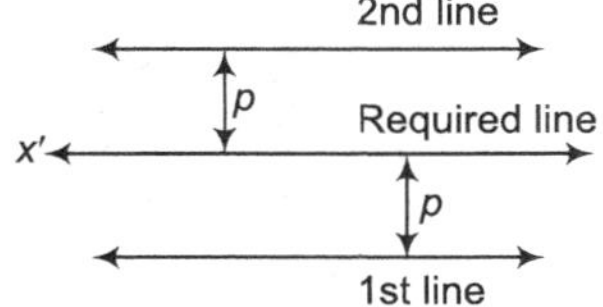

$\therefore$ Distance between lines (i) and (iii)

= Distance between lines (ii) and (iii)

$$\Rightarrow \quad \left|\frac{k+\frac{7}{3}}{\sqrt{3^2+2^2}}\right| = \left|\frac{k-6}{\sqrt{3^2+2^2}}\right|$$

$$\left(\because \text{ distance between parallel lines is } \left|\frac{c-d}{\sqrt{a^2+b^2}}\right|\right)$$

$$\Rightarrow \quad \left|k+\frac{7}{3}\right| = |k-6|$$

$$\Rightarrow \quad k+\frac{7}{3} = \pm(k-6)$$

Taking negative sign,

$$\Rightarrow \quad k+\frac{7}{3} = -k+6 \Rightarrow 2k = 6-\frac{7}{3}$$

$$\Rightarrow \quad 2k = \frac{11}{3} \Rightarrow k = \frac{11}{6}$$

Hence, Eq. (iii) becomes

$$3x + 2y + \frac{11}{6} = 0 \qquad \left(\text{put } k = \frac{11}{6}\right)$$

$\therefore$ $18x + 12y + 11 = 0$ is the required line.

209. *(b)* $\because$ Point $P(a, b)$ lies on $3x + 2y = 13$.

$\therefore \quad 3a + 2b = 13$

and point $Q(b, a)$ lies on $\quad 4x - y = 5$.

$\therefore \quad 4b - a = 5$

On solving Eqs. (i) and (ii), we get

$$a = 3 \text{ and } b = 2$$

Therefore, the coordinates of P and Q are (3, 2) and (2, 3), respectively.

Now, equation of PQ is

$$y - 2 = \frac{3-2}{2-3}(x-3) \Rightarrow x + y = 5$$

210. *(b)* Given, equation of line is

$$\frac{x}{a}\cos\theta + \frac{y}{b}\sin\theta = 1$$

$$\Rightarrow \quad \frac{bx\cos\theta + ay\sin\theta}{ab} = 1$$

$$\Rightarrow \quad bx\cos\theta + ay\sin\theta = ab$$

$$\Rightarrow \quad bx\cos\theta + ay\sin\theta - ab = 0 \quad \ldots\text{(i)}$$

At point $(\sqrt{a^2-b^2}, 0)$, the perpendicular distance from line (i) is

$$p_1 = \left|\frac{Ax_1 + By_1 + C}{\sqrt{A^2+B^2}}\right|$$

$$= \left|\frac{b\sqrt{a^2-b^2}\cos\theta - 0 - ab}{\sqrt{b^2\cos^2\theta + a^2\sin^2\theta}}\right|$$

$$\left(\because A = b\cos\theta, B = a\sin\theta, C = -ab; \; x_1 = \sqrt{a^2-b^2}, y_1 = 0\right)$$

and at point $(-\sqrt{a^2-b^2}, 0)$, the perpendicular distance from line (i) is

$$p_2 = \left|\frac{-b\sqrt{a^2-b^2}\cos\theta - ab}{\sqrt{b^2\cos^2\theta + a^2\sin^2\theta}}\right|$$

$$\left(\because A = b\cos\theta, B = a\sin\theta, \; C = -ab, x_1 = -\sqrt{a^2-b^2}, y_1 = 0\right)$$

$(\sqrt{a^2-b^2}, 0)$ $\qquad$ $(-\sqrt{a^2-b^2}, 0)$

p_1 $\qquad$ p_2

$bx\cos\theta + ay\sin\theta - ab = 0$

$$\therefore \quad p_1p_2 = \left|\frac{b\sqrt{a^2-b^2}\cos\theta - ab}{\sqrt{b^2\cos^2\theta + a^2\sin^2\theta}}\right| \times \left|\frac{-b\sqrt{a^2-b^2}\cos\theta - ab}{\sqrt{b^2\cos^2\theta + a^2\sin^2\theta}}\right|$$

$$= \frac{b\left|\sqrt{a^2-b^2}\cos\theta - a\right| \times b\left|\sqrt{a^2-b^2}\cos\theta + a\right|}{(b^2\cos^2\theta + a^2\sin^2\theta)}$$

$$= \frac{b^2\left|(a^2-b^2)\cos^2\theta - a^2\right|}{(b^2\cos^2\theta + a^2\sin^2\theta)}$$

$$[\because (A+B)(A-B) = A^2 - B^2]$$

$$= \frac{b^2\left|a^2\cos^2\theta - b^2\cos^2\theta - a^2\right|}{(b^2\cos^2\theta + a^2\sin^2\theta)}$$

$$= \frac{b^2\left|a^2(\cos^2\theta - 1) - b^2\cos^2\theta\right|}{(b^2\cos^2\theta + a^2\sin^2\theta)}$$

$$\Rightarrow \quad p_1p_2 = \frac{b^2\left|-a^2\sin^2\theta - b^2\cos^2\theta\right|}{(b^2\cos^2\theta + a^2\sin^2\theta)}$$

$$(\because \sin^2\theta + \cos^2\theta = 1)$$

$$= \frac{b^2 |a^2 \sin^2 \theta + b^2 \cos^2 \theta|}{(b^2 \cos^2 \theta + a^2 \sin^2 \theta)} \quad (\because |-A| = |A|)$$

$\therefore \quad p_1 p_2 = b^2$

211. *(c)* Let the points be $A(3, -4)$ and $B(5, 2)$ and mid-point of $AB = (4, -1)$.

It is given that the bisecting line intersect the coordinate axes in the ratio 2 : 1.

$\therefore$ Points on the coordinate axes are $(2k, 0)$ and $(0, k)$. The equation of line passing through the above point is

$$y - 0 = \frac{k - 0}{0 - 2k}(x - 2k)$$

$$\Rightarrow \quad y = -\frac{1}{2}(x - 2k) \quad \ldots \text{(i)}$$

Since, it passes through the mid-point of AB *i.e.*, $(4, -1)$,

$$\therefore \quad -1 = -\frac{1}{2}(4 - 2k) \Rightarrow k = 1$$

Putting the value of k in Eq. (i), we get

$$y = -\frac{1}{2}(x - 2) \Rightarrow x + 2y = 2$$

212. *(b)* Given, equation of straight path is

$$2x - 3y + 4 = 0 \quad \ldots \text{(i)}$$

and $$3x + 4y - 5 = 0 \quad \ldots \text{(ii)}$$

Now, solve Eqs. (i) and (ii) for crossing point.

From Eq. (i), $\quad 2x = -4 + 3y$

$$\Rightarrow \quad x = \frac{-4 + 3y}{2}$$

Putting $x = \frac{-4 + 3y}{2}$ in Eq. (ii), we get

$$3\left(\frac{-4 + 3y}{2}\right) + 4y - 5 = 0$$

$$\Rightarrow \quad -12 + 9y + 8y - 10 = 0$$

$$\Rightarrow \quad 17y - 22 = 0$$

$$\Rightarrow \quad y = \frac{22}{17}$$

Putting $y = \frac{22}{17}$ in Eq. (i), we get

$$2x - 3\left(\frac{22}{17}\right) + 4 = 0$$

$$\Rightarrow \quad 2x = -4 + \frac{66}{17}$$

$$\Rightarrow \quad 2x = \frac{-68 + 66}{17}$$

$$\Rightarrow \quad x = -\frac{1}{17}$$

$$\therefore \quad x = -\frac{1}{17} \text{ and } y = \frac{22}{17}$$

Suppose the person is standing at point P, then he can reach the path AB in least time, if he goes straight to it *i.e.*, perpendicular to it. It means we have to find the equation of PQ.

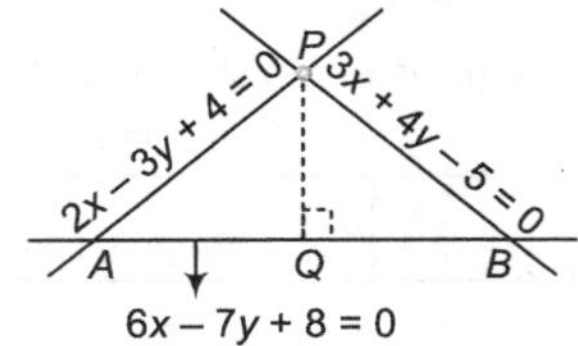

$$\because \quad \text{Line } PQ \perp \text{Line } AB$$

$$\therefore \quad \text{Slope of } PQ \times \text{Slope of } AB = -1$$

$$\Rightarrow \quad m \times \left(\frac{-6}{-7}\right) = -1$$

$$\left[\because \text{slope of line } 6x - 7y + 8 = 0 \text{ is } -\left(\frac{6}{-7}\right)\right]$$

$$\Rightarrow \quad m \times \frac{6}{7} = -1 \Rightarrow m = -\frac{7}{6}$$

Equation of PQ by using $y - y_1 = m(x - x_1)$

where, $(x_1, y_1) = \left(-\frac{1}{17}, \frac{22}{17}\right)$ and $m = -\frac{7}{6}$, is

$$y - \frac{22}{17} = -\frac{7}{6}\left(x + \frac{1}{17}\right)$$

$$\Rightarrow \quad \frac{17y - 22}{17} = -\frac{7}{6}\left(\frac{17x + 1}{17}\right)$$

$$\Rightarrow \quad 6 \times 17y - 22 \times 6 = (-7)17x - 7$$

$$\Rightarrow \quad 102y - 132 = -119x - 7$$

$$\Rightarrow \quad 119x + 102y - 132 + 7 = 0$$

$$\Rightarrow \quad 119x + 102y - 125 = 0$$

11

Conic Sections

Sections of a Cone and Circle

Level I

1. The curves circles, ellipses, parabolas and hyperbolas are known as

(a) conic sections (b) curve sections
(c) line sections (d) plane sections

2. The names parabola and hyperbola are given by

(a) Carl Gauss (b) Apollonius
(c) Andrew Wiles (d) Rene Descartes

3. Conic sections or more commonly conics are obtained by intersections of a ...A... with a double napped ...B... . Here, A and B respectively are

(a) line, right circular cone
(b) cone, plane
(c) line, cone
(d) plane, right circular cone

4. Conic sections have a very wide range of applications in fields such as

(a) planetary motion
(b) design of telescopes and antennas
(c) reflectors in flashlights and automobile headlights
(d) All of the above

5. Let l be a fixed vertical line and m be another line intersecting it at a fixed point V and inclined to it at an angle α. Suppose we rotate the line m around the line l in such a way that the angle α remains constant. Then,

I. The surface generated is right circular hollow cone.

II. The surface generated is extending indefinitely far in both directions.

(a) Both I and II are true (b) Only I is true
(c) Only II is true (d) Both I and II are false

6. The double napped right circular cone is shown in the figure.

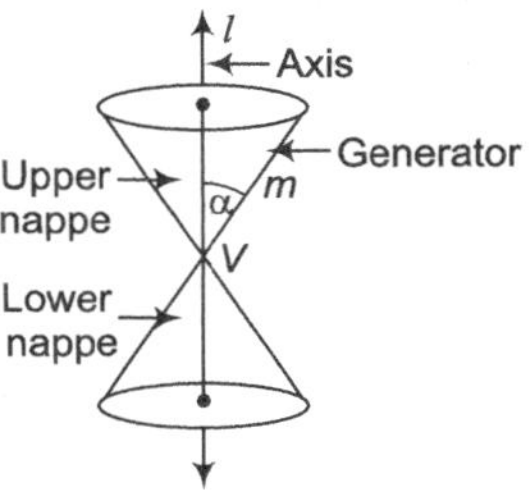

Match the notations used in the figure given in Column I with their meaning given in Column II and choose the correct option from the codes given below.

Column I	Column II
A. V	1. Generator of cone
B. l	2. Slope of cone
C. m	3. Vertex
	4. Axis of cone

Codes

	A	B	C		A	B	C
(a)	3	4	1	(b)	3	4	2
(c)	3	2	1	(d)	2	4	1

7. The vertex separates the cone into ...M... parts called ...N... . Here, M and N respectively stand for

(a) three, nappes (b) two, shapes
(c) two, nappes (d) three, cones

8. I. If we take the intersection of a plane with a cone, the section so obtained is called a conic section.

II. Conic sections are the curves obtained by intersecting a right circular cone by a plane.

(a) Both I and II are true (b) Only I is true
(c) Only II is true (d) Both I and II are false

9. Different kinds of conic sections are obtained depending on

(a) the position of the intersecting plane with respect to the cone
(b) angle made by intersecting plane with the vertical axis of the cone
(c) Both (a) and (b)
(d) Neither (a) nor (b)

10. The intersection of the plane with the cone can take place

(a) at the vertex of the cone
(b) at any part of the nappe below the vertex
(c) at any part of the nappe above the vertex
(d) All of the above

11. When the plane cuts the nappe (other than the vertex) of the cone, we have the following situations

Column I	Column II
A. When $\beta = 90°$	1. Section is a parabola.
B. When $\alpha < \beta < 90°$	2. Section is a circle.
C. When $\beta = \alpha$	3. Section is an ellipse.
D. When $0 \le \beta < \alpha$	4. Plane cuts through both the nappes and the curve of intersection is a hyperbola.

Here, β is the angle made by the intersecting plane with the vertical axis of cone and α is the angle between the generator and vertical axis of cone.

Then, match the terms of Column I with terms of Column II and choose the correct option from the codes given below.

Codes

	A	B	C	D
(a)	2	3	1	4
(b)	3	2	1	4
(c)	2	3	4	1
(d)	2	1	3	4

12. When the plane cuts at the vertex of the cone, then we have the following situations.

Column I	Column II
A. When $\alpha < \beta \le 90°$	1. Section is a pair of intersecting straight lines. (degenerate case of hyperbola)
B. When $\beta = \alpha$	2. Section is a point.
C. When $0 \le \beta < \alpha$	3. Plane contains a generator of the cone and the section is a straight line. (degenerate case of parabola)

Here, β is the angle made by the intersecting plane with the vertical axis of cone and α is the angle between the generator and vertical axis of cone.

Then, match the terms of Column I with terms of Column II and choose the correct option from the codes given below.

Codes

	A	B	C		A	B	C
(a)	2	1	3	(b)	1	3	2
(c)	2	3	1	(d)	3	2	1

Circle

13. A circle is the set of all points in a plane that are equidistant from a ...P... point in the ...Q... . Here, P and Q respectively are

(a) any, space (b) fixed, space
(c) any, plane (d) fixed, plane

14. I. The fixed point in circle is called its centre.

II. The distance from the centre to a point on the circle is called its radius.

(a) Both I and II are true
(b) Only I is true
(c) Only II is true
(d) Both I and II are false

15. The equation of the circle is simplest if the centre of the circle is at the

(a) X-axis (b) origin
(c) Y-axis (d) None of these

16. If $C(h, k)$ be the centre and r be the radius of the circle. Let $P(x, y)$ be any point on the circle, then equation of circle with centre C is

(a) $(x-h)^2 + (y-k)^2 = r$ (b) $(x-h)^2 + (y-k)^2 = r^2$
(c) $\sqrt{(x-h)^2 + (y-k)^2} = r^2$ (d) $\sqrt{(x+h)^2 + (y+k)^2} = r$

17. If the equation of the circle with centre at (h, k) and radius r is $x^2 + y^2 = r^2$. Then, h and k respectively are

(a) $(1, 1)$ (b) $(-1, -1)$
(c) $(0, 0)$ (d) None of these

18. The equation of the circle with centre (–3, 2) and radius 4, is

(a) $(x-3)^2+(y-2)^2=16$ (b) $(x+3)^2+(y+2)^2=16$
(c) $(x-3)^2+(y+2)^2=16$ (d) $(x+3)^2+(y-2)^2=16$

19. If the equation of circle is $x^2+y^2+8x+10y-8=0$, then

I. The centre is (–4, –5).
II. The radius is 7.

(a) Both I and II are true (b) Only I is true
(c) Only II is true (d) Both I and II are false

20. The equation of the circle which passes through the points (2, –2) and (3, 4) and whose centre lies on the line $x+y=2$, is

(a) $(x+0.7)^2+(y-1.3)^2=13.58$
(b) $(x-0.3)^2+(y+0.7)^2=12.58$
(c) $(x-0.7)^2+(y-1.3)^2=12.58$
(d) $(x+0.7)^2+(y+1.3)^2=13.58$

21. If the centre and radius of the circle are denoted by 'C' and 'r' respectively. Then, match the centre and radius of circle given in Column I with their corresponding equation given in Column II and choose the correct option from the codes given below.

Column I	Column II
A. $C(0, 2)$ and $r=2$	1. $x^2+y^2-2x-2y=0$
B. $C(-2, 3)$ and $r=4$	2. $36x^2+36y^2-36x-18y+11=0$
C. $C\left(\frac{1}{2}, \frac{1}{4}\right)$ and $r=\frac{1}{12}$	3. $x^2+y^2+4x-6y-3=0$
D. $C(1, 1)$ and $r=\sqrt{2}$	4. $x^2+y^2-4y=0$

Codes

	A	B	C	D
(a)	3	4	2	1
(b)	4	3	2	1
(c)	4	3	1	2
(d)	4	2	3	1

22. If centre of the circle is $(-a, -b)$ and radius is $\sqrt{a^2-b^2}$, then the equation of circle is

(a) $x^2+y^2+2ax+2by+2b^2=0$
(b) $x^2+y^2-2ax-2by+2b^2=0$
(c) $x^2+y^2-2ax-2by-2b^2=0$
(d) $x^2+y^2+2ax+2by-2b^2=0$

23. If the centre and radius of the circle are denoted by 'C' and 'r' respectively, then match the equation of circle given in Column I with their corresponding centre and radius given in Column II and choose the correct option from the codes given below.

Column I	Column II
A. $(x+5)^2+(y-3)^2=36$	1. $C(2, 4)$ and $r=\sqrt{65}$
B. $x^2+y^2-4x-8y-45=0$	2. $C(4, -5)$ and $r=\sqrt{53}$
C. $x^2+y^2-8x+10y-12=0$	3. $C\left(\frac{1}{4}, 0\right)$ and $r=\frac{1}{4}$
D. $2x^2+2y^2-x=0$	4. $C(-5, 3)$ and $r=6$

Codes

	A	B	C	D		A	B	C	D
(a)	4	2	1	3	(b)	4	1	3	2
(c)	1	4	2	3	(d)	4	1	2	3

24. The equation of the circle passing through the points (4, 1) and (6, 5) and whose centre lies on the line $4x+y=16$ is

(a) $x^2+y^2+6x+8y+15=0$
(b) $x^2+y^2-6x-8y+15=0$
(c) $x^2+y^2+6x+8y-15=0$
(d) $x^2+y^2-6x-8y-15=0$

25. The equation of the circle passing through the points (2, 3) and (–1, 1) and whose centre is on the line $x-3y-11=0$, is $x^2+y^2+ax+by+d=0$, where a, b and d respectively are

(a) –7, –5, 14 (b) 5, –7, –14
(c) –7, 5, –14 (d) –5, 7, 14

26. The number of equation of the circle with radius 5 whose centre lies on X-axis and passes through the point (2, 3), is

(a) one (b) two
(c) three (d) four

27. The equation of the circle passing through (0, 0) and making intercepts a and b on the coordinate axes is $x^2+y^2+mx+ny=0$, where m and n respectively are

(a) a, b (b) $-a, b$
(c) $a, -b$ (d) $-a, -b$

28. The equation of a circle with centre (2, 2) which passes through the point (4, 5), is

(a) $x^2+y^2+4x+4y-77=0$
(b) $x^2+y^2-4x-4y-5=0$
(c) $x^2+y^2+2x+2y-59=0$
(d) $x^2+y^2-2x-2y-23=0$

29. The point (–2.5, 3.5) lies ...A... the circle $x^2+y^2=25$. Here, A stands for

(a) inside (b) outside
(c) on (d) None of these

Level II

30. If the equation of circle is $x^2 + y^2 - 2x + 4y - 8 = 0$, then its centre (C) and radius (r) respectively are

(a) $(-2, 4)$ and $\sqrt{13}$ (b) $(-1, 2)$ and $\sqrt{13}$
(c) $(1, -2)$ and $\sqrt{13}$ (d) $(2, -4)$ and $\sqrt{13}$

31. If the equation of the circle passes through the points (20, 3), (19, 8) and (2, –9), then

(a) centre ≡ (7, 3) and radius = 13
(b) centre ≡ (3, 7) and radius = 12
(c) centre ≡ (3, 5) and radius = 15
(d) centre ≡ (5, 7) and radius = 17

32. The equation of the circle in the first quadrant touching each coordinate axis at a distance of one unit from the origin is [NCERT Exemplar]

(a) $x^2 + y^2 - 2x - 2y + 1 = 0$ (b) $x^2 + y^2 - 2x - 2y - 1 = 0$
(c) $x^2 + y^2 - 2x - 2y = 0$ (d) $x^2 + y^2 - 2x + 2y - 1 = 0$

33. The equation of the circle having centre (1, –2) and passing through the point of intersection of the lines $3x + y = 14$ and $2x + 5y = 18$ is [NCERT Exemplar]

(a) $x^2 + y^2 - 2x + 4y - 20 = 0$
(b) $x^2 + y^2 - 2x - 4y - 20 = 0$
(c) $x^2 + y^2 + 2x - 4y - 20 = 0$
(d) $x^2 + y^2 + 2x + 4y - 20 = 0$

34. If $(2 + 4\cos\theta, -1 + 4\sin\theta)$ are parametric coordinates of the circle, then its cartesian equation is [NCERT Exemplar]

(a) $(x + 2)^2 + (y + 1)^2 = 16$ (b) $(x - 2)^2 + (y - 1)^2 = 16$
(c) $(x + 2)^2 + (y - 1)^2 = 16$ (d) $(x - 2)^2 + (y + 1)^2 = 16$

35. The equation of the circle which passes through the point $(4, -5)$ and has its centre $(-2, -2)$, is

(a) $(x - 2)^2 + (y - 2)^2 = 13$ (b) $(x + 2)^2 + (y + 2)^2 = 85$
(c) Both (a) and (b) (d) Neither (a) nor (b)

36. A circle has radius 3 units and its centre lies on the line $y = x - 1$. If it passes through the point (7, 3), its equation is $x^2 + y^2 + ax + by + c = 0$. Here, a, b and c respectively are [NCERT Exemplar]

(a) –8, –6, 16 (b) –14, –12, 76
(c) Both (a) and (b) (d) Neither (a) nor (b)

37. If a circle passes through the point (0, 0), $(a, 0)$ and $(0, b)$, then its centre is [MNR 1975]

(a) (a, b)
(b) (b, a)
(c) $\left(\frac{a}{2}, \frac{b}{2}\right)$
(d) $\left(\frac{b}{2}, -\frac{a}{2}\right)$

38. The equation of the circle whose centre is (1, –3) and which touches the line $2x - y - 4 = 0$, is

(a) $5x^2 + 5y^2 - 10x + 30y + 49 = 0$
(b) $5x^2 + 5y^2 + 10x - 30y + 49 = 0$
(c) $5x^2 + 5y^2 - 10x + 30y - 49 = 0$
(d) None of the above

39. The lines $2x - 3y - 5 = 0$ and $3x - 4y = 7$ are the diameters of a circle of area $154 (= 49\pi)$ sq uints, then the equation of circle is [HOTS]

(a) $x^2 + y^2 + 2x - 2y - 62 = 0$
(b) $x^2 + y^2 + 2x - 2y - 47 = 0$
(c) $x^2 + y^2 - 2x + 2y - 47 = 0$
(d) $x^2 + y^2 - 2x + 2y - 62 = 0$

40. The equation of the circle which passes through the points (2, 3) and (4, 5) and the centre lies on the straight line $y - 4x + 3 = 0$, is

(a) $x^2 + y^2 + 4x - 10y + 25 = 0$
(b) $x^2 + y^2 - 4x - 10y + 25 = 0$
(c) $x^2 + y^2 - 4x - 10y + 16 = 0$
(d) $x^2 + y^2 - 14x + 8 = 0$

Parabola

Level I

41. I. A parabola is the set of all points in a plane that are equidistant from a fixed line and a fixed point (not on the line) in the plane.

II. A parabola is the set of all points in a plane that are equidistant from a fixed line and a fixed point in the plane.

(a) Both I and II are true
(b) Only I is true
(c) Only II is true
(d) Both I and II are false

42. As all points of parabola are equidistant from a fixed line and fixed point.

I. The fixed line is called the directrix of the parabola.

II. The fixed point is called the focus.

(a) Both I and II are true (b) Only I is true
(c) Only II is true (d) Both I and II are false

43. 'Para' means ...A... and 'bola' means ...B... *i.e.,* the shape described when you throw a ball in the air. Here, A and B respectively are

(a) of, moving (b) for, moving
(c) of, throwing (d) for, throwing

44. If the fixed point lies on the fixed line, then the set of points in the plane, which are equidistant from the fixed point and the fixed line is the straight line through the ...A... point and ...B... to the fixed line. Here, A and B respectively are

(a) any, parallel (b) fixed, perpendicular
(c) any, perpendicular (d) fixed, parallel

45. I. A line through the focus and perpendicular to the directrix is called the axis of the parabola.

II. The point of intersection of parabola with the axis is called the vertex of the parabola.

(a) Both I and II are true (b) Only I is true
(c) Only II is true (d) Both I and II are false

46. The equation of a parabola is simplest, if the vertex is at the ...A... and the axis of symmetry is along the ...B... . Here, A and B respectively are

(a) origin, X-axis (b) origin, Y-axis
(c) Both (a) and (b) (d) None of these

47. The number of possible orientations of parabola is

(a) 1 (b) 2 (c) 3 (d) 4

48. Match the vertex, focus and directrix of parabola given in Column I with their corresponding equation of parabola given in Column II and choose the correct option from the codes given below.

	Column I		Column II
A.	Vertex (0, 0), focus $(a, 0)$, directrix $x = -a$	1.	$x^2 = -4ay$
B.	Vertex (0, 0), focus $(-a, 0)$, directrix $x = a$	2.	$x^2 = 4ay$
C.	Vertex (0, 0), focus $(0, a)$, directrix $y = -a$	3.	$y^2 = -4ax$
D.	Vertex (0, 0), focus $(0, -a)$, directrix $y = a$	4.	$y^2 = 4ax$

Codes

	A	B	C	D
(a)	3	4	2	1
(b)	4	3	2	1
(c)	4	3	1	2
(d)	4	2	3	1

49. The standard equations of parabolas have focus on one of the coordinate axis; vertex at the origin and thereby the ...A... is parallel to the ...B... coordinate axis. Here, A and B respectively are

(a) axis, other (b) directrix, other
(c) Both (a) and (b) (d) Neither (a) nor (b)

50. Parabola is symmetric with respect to the axis of the parabola.

I. If the equation has a term y^2, then the axis of symmetry is along the X-axis.

II. If the equation has a term x^2, then the axis of symmetry is along the Y-axis.

(a) Both I and II are true (b) Only I is true
(c) Only II is true (d) Both I and II are false

51. When the axis of symmetry is along the X-axis the parabola opens to the

(a) right, if the coefficient of x is positive
(b) left, if the coefficient of x is negative
(c) Both (a) and (b)
(d) Neither (a) nor (b)

52. When the axis of symmetry is along the Y-axis the parabola opens

I. upwards, if the coefficient of y is negative.

II. downwards, if the coefficient of y is positive.

(a) Both I and II are true (b) Only I is true
(c) Only II is true (d) Both I and II are false

53. Latusrectum of a parabola is a line segment ...A... to the axis of the parabola, through the ...B... and whose end points lie on the parabola. Here, A and B respectively are

(a) perpendicular, focus (b) parallel, focus
(c) perpendicular, vertex (d) parallel, vertex

54. The length of the latusrectum of the parabola $y^2 = 4ax$, is

(a) a (b) $2a$
(c) $3a$ (d) $4a$

55. If the equation of parabola is $y^2 = 12x$, then

(a) focus = (3, 0)
(b) length of latusrectum = 12
(c) directrix is $x = -3$ and axis of symmetry is X-axis
(d) All of the above

56. I. Foci of parabolas $x^2 = 6y$ and $y^2 = -8x$ are $\left(0, \frac{3}{2}\right)$ and $(-2, 0)$ respectively.

II. Directrices of parabolas $x^2 = 6y$ and $y^2 = -8x$ are $y = \frac{3}{2}$ and $x = -2$ respectively.

III. Lengths of latusrectum of parabolas $x^2 = 6y$ and $y^2 = -8x$ are 6 and 8 respectively.

IV. Axes of $x^2 = 6y$ and $y^2 = -8x$ are Y-axis and X-axis respectively.

(a) I, II and III are correct (b) I, III and IV are correct
(c) II, III and IV are correct (d) I, II and IV are correct

57. Match the equation of parabola given in Column I with their corresponding focus (F), directrix (D), length of latusrectum (L) and axis (A) given in Column II and choose the correct option from the codes given below.

Column I	Column II
A. $x^2 = -16y$	1. $F\left(0, \frac{-9}{4}\right)$, D is $y = \frac{9}{4}$, $L(9)$ and axis is Y-axis.
B. $y^2 = 10x$	2. $F\left(\frac{5}{2}, 0\right)$, D is $x = \frac{-5}{2}$, $L(10)$ and axis is X-axis.
C. $x^2 = -9y$	3. $F(0, -4)$, D is $y = 4$, $L(16)$ and axis is Y-axis.

Codes

	A	B	C		A	B	C
(a)	3	2	1	(b)	2	3	1
(c)	1	2	3	(d)	3	1	2

58. I. If focus (6, 0) and directrix $x = -6$, then the equation of parabola is $y^2 = 24x$.

II. If focus (0, –3) and directrix $y = 3$, then the equation of parabola is $x^2 = 12y$.

(a) Both I and II are true (b) Only I is true
(c) Only II is true (d) Both I and II are false

59. I. If vertex (0, 0) and focus (3, 0), then the equation of parabola is $y^2 = -12x$.

II. If vertex (0, 0) and focus (–2, 0), then the equation of parabola is $y^2 = -8x$.

(a) Both I and II are true (b) Only I is true
(c) Only II is true (d) Both I and II are false

60. If parabola is passing through (2, 3), vertex (0, 0) and axis is along X-axis, then the equation of parabola is

(a) $y^2 = \frac{9}{2}x$ (b) $x^2 = \frac{9}{2}y$ (c) $y^2 = \frac{3}{2}x$ (d) $x^2 = \frac{3}{2}y$

61. If parabola is passing through (5, 2), vertex (0, 0) and symmetric with respect to Y-axis, then the equation of parabola is

(a) $y^2 = \frac{25}{2}x$ (b) $x^2 = \frac{25}{2}y$ (c) $y^2 = \frac{5}{2}x$ (d) $x^2 = \frac{5}{2}y$

Level II

62. The focus of a parabolic mirror as shown in figure is at a distance of 5cm from its vertex. If the mirror is 45 cm deep, then the distance AB is

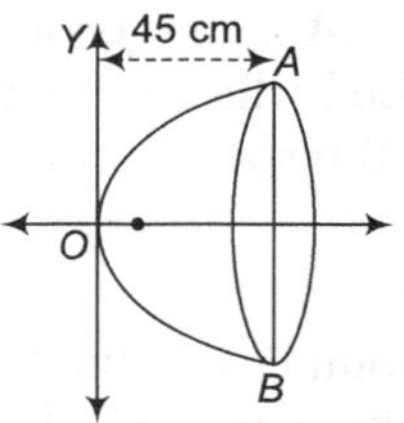

(a) 20 cm (b) 40 cm
(c) 60 cm (d) 80 cm

63. A beam is supported at its ends by supports which are 12 m apart. Since, the load is concentrated at its centre, there is deflection of 3 cm at the centre and the deflected beam is in the shape of a parabola.

The distance between deflection of 1cm and the centre is

(a) $2\sqrt{6}$ m
(b) $3\sqrt{6}$ m
(c) $4\sqrt{6}$ m
(d) $5\sqrt{6}$ m

64. If the equation of the parabola is $x^2 = -8y$, then [NCERT Exemplar]

(a) focus is (0, –2)
(b) directrix is $y = 2$
(c) length of latusrectum is 8
(d) All of the above

65. An equilateral triangle is inscribed in the parabola $y^2 = 4ax$ whose one vertex is at the vertex of the parabola. Then, the length of the side of the triangle is [NCERT Exemplar]

(a) $16\, a\sqrt{2}$
(b) $12\, a\sqrt{3}$
(c) $8\, a\sqrt{3}$
(d) $10\, a\sqrt{2}$

66. The area of the triangle formed by the lines joining the vertex of the parabola $x^2 = 12y$ to the ends of its latusrectum is [NCERT Exemplar]

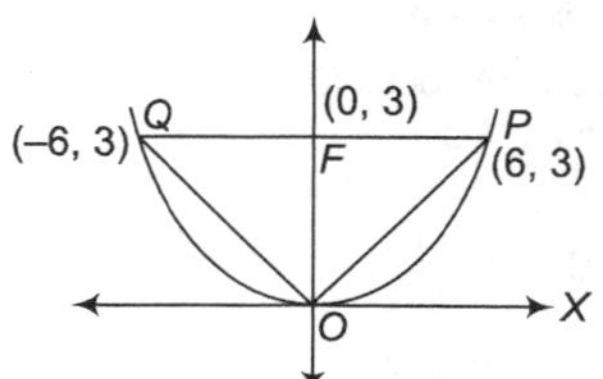

(a) 12 sq units (b) 16 sq units
(c) 18 sq units (d) 24 sq units

67. The equations of the lines joining the vertex of the parabola $y^2 = 6x$ to the points on it which have abscissa 24, are [NCERT Exemplar]

(a) $y \pm 2x = 0$
(b) $2y \pm x = 0$
(c) $x \pm y = 0$
(d) None of these

68. The vertex of parabola $(y-2)^2 = 16(x-1)$, is [Karnataka CET 2001]

(a) (2, 1) (b) (1, –2) (c) (–1, 2) (d) (1, 2)

69. Vertex of the parabola $9x^2 - 6x + 36y + 9 = 0$, is [HOTS]

(a) $\left(\frac{1}{3}, -\frac{2}{9}\right)$ (b) $\left(-\frac{1}{3}, -\frac{1}{2}\right)$
(c) $\left(-\frac{1}{3}, \frac{1}{2}\right)$ (d) $\left(\frac{1}{3}, \frac{1}{2}\right)$

70. The equation of the parabola whose vertex is $(-1, -2)$, axis is vertical and which passes through the point (3, 6), is

(a) $x^2 + 2x - 2y - 3 = 0$ (b) $2x^2 = 3y$
(c) $x^2 - 2x - y + 3 = 0$ (d) None of these

71. The length of the latusrectum of the parabola $9x^2 - 6x + 36y + 19 = 0$, is

(a) 36 (b) 9 (c) 6 (d) 4

72. The equation of the parabola whose focus is the point (2, 3) and directrix is the line $x - 4y + 3 = 0$, is [NCERT Exemplar]

(a) $16x^2 + y^2 + 8xy - 74x - 78y + 212 = 0$
(b) $x^2 + 16y^2 + 8xy - 74x - 78y + 212 = 0$
(c) $x^2 + 16y^2 - 8xy + 74x + 78y + 212 = 0$
(d) $16x^2 + y^2 - 8xy + 74x + 78y + 212 = 0$

Ellipse

Level I

73. An ellipse is the set of all points in a plane, the ...A... of whose distances from two fixed points in the plane is a constant. The two fixed points are called the ...B... of the ellipse. Here, A and B respectively are

(a) sum, vertices (b) difference, vertices
(c) sum, foci (d) difference, foci

74. The constant which is the sum of the distances of a point on the ellipse from the two fixed points is always ...A... the distance between the two fixed points. Here, A stands for

(a) greater than (b) smaller than
(c) equal to (d) None of these

75. Match the centre, major axis, minor axis and vertices of ellipse given in Column I with their respective meaning given in Column II and choose the correct option from the codes given below.

Column I	Column II
A. Centre	1. Line segment through the foci of the ellipse.
B. Major axis	2. Mid-point of line segment joining the foci.
C. Minor axis	3. End points of major axis.
D. Vertices	4. Line segment through the centre and perpendicular to the major axis.

Codes

	A	B	C	D		A	B	C	D
(a)	2	1	4	3	(b)	1	2	4	3
(c)	2	1	3	4	(d)	2	4	1	3

76. A figure of ellipse is shown below. The length of major axis is ...P..., the length of minor axis is ...Q... and distance between the foci is ...R... . Here, P, Q and R respectively are

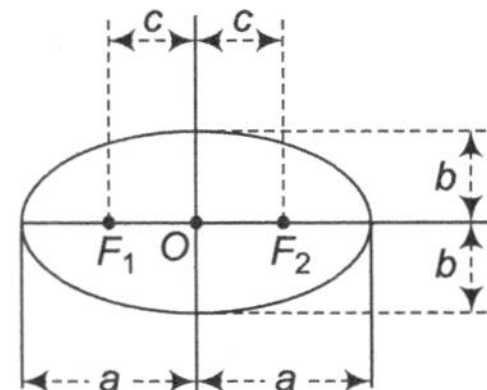

(a) $2b, 2a, 2c$ (b) $2a, 2b, 2c$
(c) $2c, 2a, 2b$ (d) $2c, 2b, 2a$

77. A figure of ellipse is shown below. Then, which of the following is incorrect?

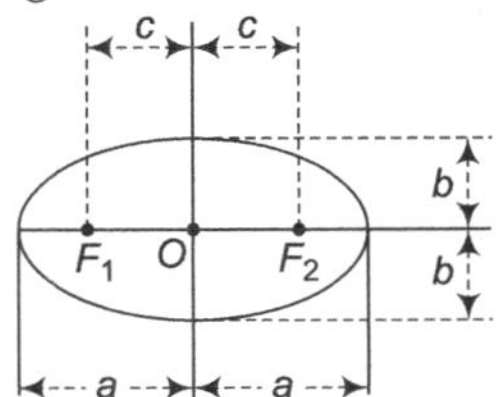

(a) Length of semi-major axis is a
(b) Length of semi-minor axis is b
(c) Both (a) and (b)
(d) Neither (a) nor (b)

78. The relationship between semi-major axis (a), semi-minor axis (b) and the distance of focus from the centre (c) of the ellipse is given as

(a) $c^2 = a^2 + b^2$ (b) $c^2 = b^2 - a^2$
(c) $c^2 = a^2 - b^2$ (d) $c^2 = a^2 \pm b^2$

79. Let lengths of semi-major axis, semi-minor axis and distance of focus from the centre of an ellipse are a, b and c, respectively.

I. When $c = 0$, ellipse becomes circle.
II. When $c = a$, ellipse reduces to the line segment F_1F_2 joining the two foci.

(a) Both I and II are true (b) Only I is true
(c) Only II is true (d) Both I and II are false

80. If length of semi-major axis is 'a' and distance of focus from the centre is 'c', then eccentricity (e) of an ellipse is

(a) $e = ca$ (b) $e = \frac{a}{c}$
(c) $e = \frac{c}{a}$ (d) None of these

81. If 'e' is the eccentricity and 'a' is the length of semi-major axis, then the focus is at a distance ...A... from the centre. Here, A stands for

(a) ae (b) $\frac{e}{a}$
(c) $\frac{a}{e}$ (d) None of these

82. The equation of an ellipse is simplest, if the centre of the ellipse is at the origin and the foci are on ...A... . Here, A stands for

(a) Only X-axis (b) X-axis or Y-axis
(c) Only Y-axis (d) None of these

83. If the centre of ellipse is at origin and major axis is along X-axis, then the equation of ellipse is

(a) $\frac{x^2}{a^2} + \frac{y^2}{b^2} = 1$ (where $a > b$) (b) $\frac{x^2}{a^2} + \frac{y^2}{b^2} = 1$ (where $a < b$)
(c) $\frac{x^2}{a^2} - \frac{y^2}{b^2} = 1$ (where $a > b$) (d) $\frac{x^2}{a^2} - \frac{y^2}{b^2} = 1$ (where $a < b$)

84. If the centre of ellipse is at origin and major axis is along Y-axis, then the equation of ellipse is

(a) $\frac{x^2}{b^2} + \frac{y^2}{a^2} = 1$ (where $b > a$) (b) $\frac{x^2}{b^2} - \frac{y^2}{a^2} = 1$ (where $b > a$)
(c) $\frac{x^2}{b^2} + \frac{y^2}{a^2} = 1$ (where $a > b$) (d) $\frac{x^2}{b^2} - \frac{y^2}{a^2} = 1$ (where $a > b$)

85. The standard equations of ellipse have ...A... at the origin and the major and minor axes are ...B... . Here, A and B respectively are

(a) focus, coordinate axes (b) centre, coordinate axes
(c) Both (a) and (b) (d) Neither (a) nor (b)

86. If the centre of ellipse is at the origin, then

I. Ellipse is not symmetric with respect to X-axis or Y-axis.
II. Ellipse is symmetric with respect to both the coordinate axes.

(a) Both I and II are true (b) Only I is true
(c) Only II is true (d) Both I and II are false

87. I. The foci of an ellipse always lies on major axis.
II. The foci of an ellipse always lies on minor axis.

(a) Both I and II are true (b) Only I is true
(c) Only II is true (d) Both I and II are false

88. Latusrectum of an ellipse is a line segment ...A... to the major axis through any of the ...B... and whose end points lie on the ellipse. Here, A and B respectively are

(a) perpendicular, foci (b) parallel, vertices
(c) parallel, foci (d) perpendicular, vertices

89. If the equation of ellipse is $\frac{x^2}{a^2} + \frac{y^2}{b^2} = 1$ (where $a > b$), then the length of latusrectum is

(a) $\frac{2a^2}{b}$ (b) $\frac{2b^2}{a}$
(c) $\frac{b^2}{a}$ (d) $\frac{a^2}{b}$

90. Match the equation of ellipse given in Column I with their corresponding foci (f), vertices (v), length of major axis (L_1), length of minor axis (L_2), eccentricity (e) and length of latusrectum (l) given in Column II and choose the correct option from the codes given below.

Column I	Column II
A. $\frac{x^2}{36} + \frac{y^2}{16} = 1$	1. $f(0, \pm\sqrt{21})$, $v(0, \pm 5)$, L_1 (10), L_2 (4), $e\left(\frac{\sqrt{21}}{5}\right)$, $l\left(\frac{8}{5}\right)$
B. $\frac{x^2}{4} + \frac{y^2}{25} = 1$	2. $f(0, \pm 5\sqrt{3})$, $v(0, \pm 10)$, L_1 (20), L_2 (10), $e\left(\frac{\sqrt{3}}{2}\right)$, l (5)
C. $\frac{x^2}{16} + \frac{y^2}{9} = 1$	3. $f(\pm 2\sqrt{5}, 0)$, $v(\pm 6, 0)$, L_1 (12), L_2 (8), $e\left(\frac{\sqrt{5}}{3}\right)$, $l\left(\frac{16}{3}\right)$
D. $\frac{x^2}{25} + \frac{y^2}{100} = 1$	4. $f(\pm\sqrt{7}, 0)$, $v(\pm 4, 0)$, L_1 (8), L_2 (6), $e\left(\frac{\sqrt{7}}{4}\right)$, $l\left(\frac{9}{2}\right)$

Codes

	A	B	C	D		A	B	C	D
(a)	3	4	1	2	(b)	3	1	2	4
(c)	3	1	4	2	(d)	1	3	4	2

91. If the equation of ellipse is $\frac{x^2}{49}+\frac{y^2}{36}=1$, then which of the following is not correct?

(a) Foci= $(\pm\sqrt{13}, 0)$, vertices = $(\pm 7, 0)$
(b) Length of major axis is 14 and length of minor axis is 12
(c) Eccentricity is $\sqrt{\frac{13}{7}}$
(d) Length of latusrectum is $\frac{72}{7}$

92. Match the equation of ellipse given in Column I with their corresponding foci (f), vertices (v), length of major axis (L_1), length of minor axis (L_2), eccentricity (e) and length of latusrectum (l) given in Column II and choose the correct option from the codes given below.

	Column I		Column II
A.	$\frac{x^2}{100}+\frac{y^2}{400}=1$	1.	$f(0, \pm 4\sqrt{2})$, $v(0, \pm 6)$, L_1 (12), L_2 (4), $e\left(\frac{2\sqrt{2}}{3}\right)$, $l\left(\frac{4}{3}\right)$
B.	$36x^2+4y^2=144$	2.	$f(0, \pm 10\sqrt{3})$, $v(0, \pm 20)$, L_1 (40), L_2 (20), $e\left(\frac{\sqrt{3}}{2}\right)$, l (10)
C.	$16x^2+y^2=16$	3.	$f(\pm\sqrt{5}, 0)$, $v(\pm 3, 0)$, L_1 (6), L_2 (4), $e\left(\frac{\sqrt{5}}{3}\right)$, $l\left(\frac{8}{3}\right)$
D.	$4x^2+9y^2=36$	4.	$f(0, \pm\sqrt{15})$, $v(0, \pm 4)$, L_1 (8), L_2 (2), $e\left(\frac{\sqrt{15}}{4}\right)$, $l\left(\frac{1}{2}\right)$

Codes

	A	B	C	D		A	B	C	D
(a)	2	1	4	3	(b)	1	2	4	3
(c)	2	1	3	4	(d)	2	4	1	3

93. Match the vertices and foci of ellipse given in Column I with their corresponding equation of ellipse given in Column II and choose the correct option from the codes given below.

	Column I		Column II
A.	Vertices (±5, 0), foci (±4, 0)	1.	$\frac{x^2}{144}+\frac{y^2}{169}=1$
B.	Vertices (0, ± 13), foci (0, ± 5)	2.	$\frac{x^2}{36}+\frac{y^2}{20}=1$
C.	Vertices (±6, 0), foci (±4, 0)	3.	$\frac{x^2}{25}+\frac{y^2}{9}=1$

Codes

	A	B	C		A	B	C
(a)	1	3	2	(b)	3	1	2
(c)	3	2	1	(d)	2	1	3

94. I. In an ellipse, if ends of major axis are (±3, 0) and ends of minor axis are (0, ± 2), then the equation of ellipse is $\frac{x^2}{9}+\frac{y^2}{4}=1$.

II. In an ellipse, if ends of major axis are $(0, \pm\sqrt{5})$ and ends of minor axis are (±1, 0), then the equation of ellipse is $5x^2+y^2=5$.

(a) Both I and II are true
(b) Only I is true
(c) Only II is true
(d) Both I and II are false

95. I. In an ellipse, if length of major axis is 26 and foci are (±5, 0), then the equation of ellipse is $\frac{x^2}{144}+\frac{y^2}{169}=1$.

II. In an ellipse, if length of minor axis is 16 and foci are (0, ±6), then the equation of ellipse is $\frac{x^2}{64}+\frac{y^2}{89}=1$.

(a) Both I and II are true
(b) Only I is true
(c) Only II is true
(d) Both I and II are false

96. In an ellipse, if centre is at origin, foci are (±3, 0) and length of semi-major axis is 4, then the equation of ellipse is

(a) $\frac{x^2}{16}+\frac{y^2}{7}=1$
(b) $\frac{x^2}{7}+\frac{y^2}{16}=1$
(c) $\frac{x^2}{16}+\frac{y^2}{49}=1$
(d) $\frac{x^2}{49}+\frac{y^2}{16}=1$

97. In an ellipse, if centre is at origin, foci are on *X*-axis, length of semi-minor axis is 3 and distance of focus from the centre of ellipse is 4, then the equation of ellipse is

(a) $\frac{x^2}{9}+\frac{y^2}{25}=1$
(b) $\frac{x^2}{25}+\frac{y^2}{9}=1$
(c) $\frac{x^2}{16}+\frac{y^2}{9}=1$
(d) $\frac{x^2}{9}+\frac{y^2}{16}=1$

98. Let the centre of an ellipse is at (0, 0).

I. If major axis is on the *Y*-axis and ellipse passes through the points (3, 2) and (1, 6), then the equation of ellipse is $\frac{x^2}{40}+\frac{y^2}{10}=1$.

II. If major axis is on the *X*-axis and ellipse passes through the points (4, 3) and (6, 2), then the equation of ellipse is $\frac{x^2}{13}+\frac{y^2}{52}=1$.

(a) Both I and II are true
(b) Only I is true
(c) Only II is true
(d) Both I and II are false

Level II

99. If the equation of ellipse is $9x^2 + 25y^2 = 225$, then which of the following is incorrect? [NCERT Exemplar]

(a) Foci $(\pm 4, 0)$, vertices $(\pm 5, 0)$
(b) Major axis and minor axis are X-axis and Y-axis respectively
(c) Eccentricity is $\frac{4}{5}$
(d) None of the above

100. In an ellipse, if foci are $(\pm 5, 0)$ and $x = \frac{36}{5}$ as one of the directrix, then the equation of the ellipse is [NCERT Exemplar]

(a) $\frac{x^2}{36} + \frac{y^2}{11} = 1$ (b) $\frac{x^2}{11} + \frac{y^2}{36} = 1$
(c) $\frac{x^2}{121} + \frac{y^2}{36} = 1$ (d) $\frac{x^2}{36} + \frac{y^2}{121} = 1$

101. The equation of the ellipse whose centre is $(2, -3)$, one of the foci is $(3, -3)$ and the corresponding vertex is $(4, -3)$ is [HOTS]

(a) $\frac{(x-2)^2}{3} + \frac{(y+3)^2}{4} = 1$ (b) $\frac{(x-2)^2}{4} + \frac{(y+3)^2}{3} = 1$
(c) $\frac{x^2}{3} + \frac{y^2}{4} = 1$ (d) None of these

102. The eccentricity of the curve represented by the equation $x^2 + 2y^2 - 2x + 3y + 2 = 0$, is [Roorkee 1998]

(a) 0 (b) 1/2
(c) $1/\sqrt{2}$ (d) $\sqrt{2}$

103. The length of the axes of the conic $9x^2 + 4y^2 - 6x + 4y + 1 = 0$, are [Orissa JEE 2002]

(a) $\frac{1}{2}, 9$ (b) $3, \frac{2}{5}$
(c) $1, \frac{2}{3}$ (d) 3, 2

104. The equation of the ellipse which passes through the point $(-3, 1)$ and has eccentricity $\sqrt{\frac{2}{5}}$, with X-axis as its major axis and centre at the origin, is [NCERT Exemplar]

(a) $3x^2 + 5y^2 = 32$ (b) $5x^2 + 3y^2 = 32$
(c) $5x^2 + 5y^2 = 50$ (d) $3x^2 + 3y^2 = 30$

105. The equation of the ellipse whose centre is at the origin and the X-axis is the major axis, which passes through the points $(-3, 1)$ and $(2, -2)$, is [NCERT Exemplar]

(a) $5x^2 + 3y^2\ 32$ (b) $3x^2 + 5y^2 = 32$
(c) $5x^2 - 3y^2 = 32$ (d) $3x^2 + 5y^2 + 32 = 0$

106. The length of the latusrectum of the ellipse $9x^2 + 4y^2 = 1$, is [MP PET 1999]

(a) $\frac{3}{2}$ (b) $\frac{8}{3}$
(c) $\frac{4}{9}$ (d) $\frac{8}{9}$

107. The locus of a variable point whose distance from $(-2, 0)$ is $\frac{2}{3}$ times its distance from the line $x = -\frac{9}{2}$, is [IIT 1994]

(a) ellipse (b) parabola
(c) hyperbola (d) None of these

108. Equation of the ellipse with eccentricity $\frac{1}{2}$ and foci at $(\pm 1, 0)$, is [MP PET 2002]

(a) $\frac{x^2}{3} + \frac{y^2}{4} = 1$ (b) $\frac{x^2}{4} + \frac{y^2}{3} = 1$
(c) $\frac{x^2}{3} + \frac{y^2}{4} = \frac{4}{3}$ (d) None of these

109. The equation of ellipse whose distance between the foci is equal to 8 and distance between the directrix is 18, is [HOTS]

(a) $5x^2 - 9y^2 = 180$ (b) $9x^2 - 5y^2 = 180$
(c) $x^2 + 9y^2 = 180$ (d) $5x^2 + 9y^2 = 180$

110. If the latusrectum of an ellipse with axis along X-axis and centre at origin is 10 and distance between foci is length of minor axis, then the equation of the ellipse is [NCERT Exemplar]

(a) $x^2 + 2y^2 = 50$ (b) $x^2 + 2y^2 = 100$
(c) $2x^2 + y^2 = 50$ (d) $2x^2 + y^2 = 100$

111. If the foci and vertices of an ellipse be $(\pm 1, 0)$ and $(\pm 2, 0)$ respectively, then the minor axis of the ellipse is

(a) $2\sqrt{5}$ (b) 2 (c) 4 (d) $2\sqrt{3}$

112. The equations of the directrices of the ellipse $16x^2 + 25y^2 = 400$, are

(a) $2x = \pm 25$ (b) $5x = \pm 9$
(c) $3x = \pm 10$ (d) None of these

113. If the eccentricity of an ellipse be $1/\sqrt{2}$, then its latusrectum is equal to its

(a) minor axis (b) semi-minor axis
(c) major axis (d) semi-major axis

114. If the distance between a focus and corresponding directrix of an ellipse be 8 and the eccentricity be 1/2, then length of the minor axis is [HOTS]

(a) 3 (b) $4\sqrt{2}$
(c) 6 (d) None of these

115. The equation of the ellipse whose one of the vertices is $(0, 7)$ and the corresponding directrix is $y = 12$ is

(a) $95x^2 + 144y^2 = 4655$
(b) $144x^2 + 95y^2 = 4655$
(c) $95x^2 + 144y^2 = 13680$
(d) None of the above

116. Eccentricity of the ellipse whose latusrectum is equal to the distance between two focus points, is

(a) $\frac{\sqrt{5}+1}{2}$ (b) $\frac{\sqrt{5}-1}{2}$
(c) $\frac{\sqrt{5}}{2}$ (d) $\frac{\sqrt{3}}{2}$

117. If the length of the major axis of an ellipse is three times the length of its minor axis, then its eccentricity is [EAMCET 1990]

(a) $\frac{1}{3}$ (b) $\frac{1}{\sqrt{3}}$
(c) $\frac{1}{\sqrt{2}}$ (d) $\frac{2\sqrt{2}}{3}$

118. The length of the latusrectum of an ellipse is $\frac{1}{3}$ of the major axis. Its eccentricity is

(a) $\frac{2}{3}$ (b) $\sqrt{\frac{2}{3}}$
(c) $\frac{5 \times 4 \times 3}{7^3}$ (d) $\left(\frac{3}{4}\right)^4$

119. An ellipse is described by using an endless string which is passed over two pins. If the axes are 6 cm and 4 cm, the distance between the pins in centimetre is [HOTS]

(a) 6 (b) $\sqrt{5}$
(c) $2\sqrt{5}$ (d) None of these

120. The equation $\frac{x^2}{2-r} + \frac{y^2}{r-5} + 1 = 0$ represents an ellipse, if [MP PET 1995]

(a) $r > 2$
(b) $2 < r < 5$
(c) $r > 5$
(d) None of the above

Hyperbola

Level I

121. A hyperbola is the set of all points in a plane, the ...A... of whose distances from ...B... fixed points in the plane is a constant. Here, A and B respectively are

(a) sum, two
(b) difference, two
(c) product, two
(d) division, two

122. The term 'difference' that is used in the definition of hyperbola means

(a) the distance to the farther point minus the distance to the closer point
(b) the distance to the farther point plus the distance to the closer point
(c) Both (a) and (b)
(d) Neither (a) nor (b)

123. Match the foci, centre, transverse axis, conjugate axis and vertices of hyperbola given in Column I with their corresponding meaning given in Column II and choose the correct option from the codes given below.

	Column I		Column II
A.	Foci	1.	Mid-point of the line segment joining the foci.
B.	Centre	2.	Points at which the hyperbola intersects the transverse axis.
C.	Transverse axis	3.	Line through the foci.
D.	Conjugate axis	4.	Two fixed points.
E.	Vertices	5.	Line through the centre and perpendicular to the transverse axis.

Codes

	A	B	C	D	E		A	B	C	D	E
(a)	4	3	1	5	2	(b)	1	4	3	5	2
(c)	4	1	5	3	2	(d)	4	1	3	5	2

124. If the distance between the two foci of hyperbola is $2c$ and the distance of its two vertices is $2a$, then

I. The length of transverse axis is $2a$.
II. The length of conjugate axis is $2b$, where $b = \sqrt{c^2 - a^2}$.

(a) Both I and II are true (b) Only I is true
(c) Only II is true (d) Both I and II are false

125. If the distances of foci and vertex of hyperbola from the centre are c and a respectively, then the ...A... is called the eccentricity of the hyperbola. Here, A stands for

(a) product $e = ac$ (b) ratio $e = \frac{a}{c}$
(c) ratio $e = \frac{c}{a}$ (d) None of these

126. If the distances of foci and vertex of hyperbola from the centre are c and a respectively, then

I. Eccentricity is always less than 1.

II. Foci are at a distance of ae from the centre.

(a) Both I and II are true (b) Only I is true
(c) Only II is true (d) Both I and II are false

127. The equation of hyperbola is ...A..., if the centre of hyperbola is at the origin and the foci are on ...B... . Here, A and B respectively are

(a) difficult, X-axis or Y-axis (b) simplest, X-axis and Y-axis
(c) difficult, X-axis and Y-axis (d) simplest, X-axis or Y-axis

128. The number of possible orientations of hyperbola is/are

(a) one (b) two (c) three (d) four

129. If $2a$ and $2b$ are the lengths of transverse axis and conjugate axis respectively, then the equation of hyperbola with origin (0, 0) and transverse axis along X-axis, is

(a) $\frac{x^2}{a^2} - \frac{y^2}{b^2} = 1$ (b) $\frac{x^2}{b^2} - \frac{y^2}{a^2} = 1$
(c) $\frac{y^2}{b^2} - \frac{x^2}{a^2} = 1$ (d) $\frac{y^2}{a^2} - \frac{x^2}{b^2} = 1$

130. A hyperbola in which length of transverse axis is equal to the length of conjugate axis, is called

(a) equilateral hyperbola (b) obtuse hyperbola
(c) acute hyperbola (d) None of these

131. If the equation of hyperbola is $\frac{x^2}{a^2} - \frac{y^2}{b^2} = 1$.

I. Then, no portion of curve lies between the lines $x = +a$ and $x = -a$.

II. Then, no portion of curve lies between the lines $y = +b$ and $y = -b$.

(a) Both I and II are true (b) Only I is true
(c) Only II is true (d) Both I and II are false

132. If $2a$ and $2b$ are the lengths of transverse axis and conjugate axis respectively, then the equation of hyperbola with origin (0, 0) and transverse axis along Y-axis is

(a) $\frac{x^2}{a^2} - \frac{y^2}{b^2} = 1$ (b) $\frac{x^2}{b^2} - \frac{y^2}{a^2} = 1$ (c) $\frac{y^2}{b^2} - \frac{x^2}{a^2} = 1$ (d) $\frac{y^2}{a^2} - \frac{x^2}{b^2} = 1$

133. The standard equations of hyperbolas have transverse and conjugate axes as the coordinate axes and the ...A... at the origin. Here, A stands for

(a) vertex (b) centre
(c) focus (d) None of these

134. Hyperbola is symmetric with respect to

(a) X-axis (b) Y-axis
(c) X-axis or Y-axis (d) X-axis and Y-axis

135. If the equation of hyperbola is $\frac{x^2}{9} - \frac{y^2}{16} = 1$, then

(a) transverse axis is along X-axis of length 6
(b) transverse axis is along Y-axis of length 8
(c) conjugate axis is along Y-axis of length 6
(d) None of the above

136. If the equation of hyperbola is $\frac{y^2}{25} - \frac{x^2}{9} = 1$, then

(a) transverse axis is along X-axis of length 6
(b) transverse axis is along Y-axis of length 10
(c) conjugate axis is along Y-axis of length 10
(d) None of the above

137. Latusrectum of hyperbola is a line segment ...A... to the transverse axis through any of the ...B... and whose end points lie ...C... the hyperbola. Here,

(a) A $\rightarrow$ parallel, B $\rightarrow$ foci, C $\rightarrow$ on
(b) A $\rightarrow$ perpendicular, B $\rightarrow$ vertices, C $\rightarrow$ on
(c) A $\rightarrow$ parallel, B $\rightarrow$ foci, C $\rightarrow$ inside
(d) A $\rightarrow$ perpendicular, B $\rightarrow$ foci, C $\rightarrow$ on

138. In a hyperbola, if length of transverse axis is $2a$ and length of conjugate axis is $2b$, then the length of latusrectum is

(a) $\frac{b^2}{a}$ (b) $\frac{2b^2}{a}$ (c) $\frac{a^2}{b}$ (d) $\frac{2a^2}{b}$

139. Match the equation of hyperbola given in Column I with their corresponding foci (f), vertices (v), eccentricity (e) and length of latusrectum (l) given in Column II and choose the correct option from the codes given below.

Column I	Column II
A. $\frac{x^2}{16} - \frac{y^2}{9} = 1$	1. $f(0, \pm\sqrt{13})$, $v(0, \pm 2)$, $e = \frac{\sqrt{13}}{2}$ and $l = 9$
B. $\frac{y^2}{9} - \frac{x^2}{27} = 1$	2. $f\left(0, \pm\frac{2\sqrt{14}}{\sqrt{5}}\right)$, $v\left(0, \pm\frac{6}{\sqrt{5}}\right)$, $e = \frac{\sqrt{14}}{3}$ and $l = \frac{4\sqrt{5}}{3}$
C. $9y^2 - 4x^2 = 36$	3. $f(\pm 5, 0)$, $v(\pm 4, 0)$, $e = \frac{5}{4}$ and $l = \frac{9}{2}$

Column I	Column II
D. $16x^2 - 9y^2 = 576$	4. $f(0, \pm 6)$, $v(0, \pm 3)$, $e = 2$ and $l = 18$
E. $5y^2 - 9x^2 = 36$	5. $f(0, \pm \sqrt{65})$, $v(0, \pm 4)$, $e = \frac{\sqrt{65}}{4}$ and $l = \frac{49}{2}$
F. $49y^2 - 16x^2 = 784$	6. $f(\pm 10, 0)$, $v(\pm 6, 0)$, $e = \frac{5}{3}$ and $l = \frac{64}{3}$

Codes

	A	B	C	D	E	F
(a)	4	3	1	6	2	5
(b)	3	4	1	6	2	5
(c)	3	4	6	1	2	5
(d)	3	4	1	6	5	2

140. Match the vertices (v) and foci (f) of hyperbola given in Column I with their corresponding equation given in Column II and choose the correct option from the codes given below.

Column I	Column II
A. $v(\pm 2, 0)$, $f(\pm 3, 0)$	1. $\frac{y^2}{25} - \frac{x^2}{39} = 1$
B. $v(0, \pm 5)$, $f(0, \pm 8)$	2. $\frac{y^2}{9} - \frac{x^2}{16} = 1$
C. $v(0, \pm 3)$, $f(0, \pm 5)$	3. $\frac{x^2}{4} - \frac{y^2}{5} = 1$

Codes

	A	B	C		A	B	C
(a)	3	1	2	(b)	1	3	2
(c)	3	2	1	(d)	2	1	3

141. Match the foci (f), length of transverse axis (T), length of conjugate axis (C), length of latusrectum (l) of hyperbola given in Column I with their corresponding equation given in Column II and choose the correct option from the codes given below.

Column I	Column II
A. $f(\pm 5, 0)$, $T(8)$	1. $\frac{x^2}{25} - \frac{y^2}{20} = 1$
B. $f(0, \pm 13)$, $C(24)$	2. $\frac{y^2}{25} - \frac{x^2}{144} = 1$
C. $f(\pm 3\sqrt{5}, 0)$, $l(8)$	3. $\frac{x^2}{4} - \frac{y^2}{12} = 1$
D. $f(\pm 4, 0)$, $l(12)$	4. $\frac{x^2}{16} - \frac{y^2}{9} = 1$

Codes

	A	B	C	D		A	B	C	D
(a)	2	4	1	3	(b)	4	2	3	1
(c)	4	2	1	3	(d)	4	1	2	3

142. If the vertices of hyperbola are $(\pm 7, 0)$ and its eccentricity is $\frac{4}{3}$, then the equation of hyperbola is

(a) $\frac{x^2}{49} - \frac{y^2}{343} = 1$
(b) $\frac{x^2}{49} - \frac{9y^2}{343} = 1$
(c) $\frac{9y^2}{343} - \frac{x^2}{49} = 1$
(d) $\frac{x^2}{343} - \frac{9y^2}{49} = 1$

143. If foci of hyperbola are $(0, \pm \sqrt{10})$ and it is passing through the point $(2, 3)$, then the equation of hyperbola is

(a) $y^2 - x^2 = 5$
(b) $2x^2 - y^2 = -1$
(c) $x^2 - 2y^2 = -14$
(d) $y^2 - 2x^2 = 1$

Level II

144. If the vertices of hyperbola are $(0, \pm 6)$ and its eccentricity is $\frac{5}{3}$, then

I. The equation of hyperbola is $\frac{y^2}{36} - \frac{x^2}{64} = 1$.

II. The foci of hyperbola are $(0, \pm 10)$.

[NCERT Exemplar]

(a) Both I and II are true
(b) Only I is true
(c) Only II is true
(d) Both I and II are false

145. If the vertices of hyperbola are $(\pm 6, 0)$ and one of the directrices is $x = 4$, then the equation of hyperbola is [NCERT Exemplar]

(a) $\frac{y^2}{45} - \frac{x^2}{36} = 1$
(b) $\frac{y^2}{36} - \frac{x^2}{45} = 1$
(c) $\frac{x^2}{36} - \frac{y^2}{45} = 1$
(d) $\frac{x^2}{45} - \frac{y^2}{36} = 1$

146. The length of transverse axis of the hyperbola $3x^2 - 4y^2 = 32$, is [Karnataka CET 2001]

(a) $\frac{8\sqrt{2}}{\sqrt{3}}$
(b) $\frac{16\sqrt{2}}{\sqrt{3}}$
(c) $\frac{3}{32}$
(d) $\frac{64}{3}$

147. One of the directrix of the hyperbola $\frac{x^2}{9} - \frac{y^2}{4} = 1$, is [UPSEAT 2003]

(a) $x = \frac{9}{\sqrt{13}}$
(b) $y = \frac{9}{\sqrt{13}}$
(c) $x = \frac{6}{\sqrt{13}}$
(d) $y = \frac{6}{\sqrt{13}}$

148. The vertices of a hyperbola are at (0, 0) and (10, 0) and one of its foci is at (8, 0). Then, the equation of the hyperbola is

(a) $\frac{x^2}{25} - \frac{y^2}{144} = 1$ (b) $\frac{(x-5)^2}{25} - \frac{y^2}{144} = 1$

(c) $\frac{x^2}{25} - \frac{(y-5)^2}{144} = 1$ (d) $\frac{(x-5)^2}{25} - \frac{(y-5)^2}{144} = 1$

149. Foci of the hyperbola $\frac{x^2}{16} - \frac{(y-2)^2}{9} = 1$, are [HOTS]

(a) (5, 2) (−5, 2) (b) (5, 2) (5, −2)

(c) (5, 2) (−5, −2) (d) None of these

150. The length of the transverse axis along X-axis with centre at origin of a hyperbola is 7 and it passes through the point (5, −2). Then, the equation of the hyperbola is [NCERT Exemplar]

(a) $\frac{4}{49}x^2 - \frac{196}{51}y^2 = 1$ (b) $\frac{49}{4}x^2 - \frac{51}{196}y^2 = 1$

(c) $\frac{4}{49}x^2 - \frac{51}{196}y^2 = 1$ (d) None of these

151. The eccentricity of the hyperbola $\frac{x^2}{a^2} - \frac{y^2}{b^2} = 1$ which passes through the points (3, 0) and $(3\sqrt{2}, 2)$, is [NCERT Exemplar]

(a) $\frac{13}{3}$ (b) $\sqrt{\frac{13}{3}}$ (c) $\frac{\sqrt{13}}{9}$ (d) $\frac{\sqrt{13}}{3}$

152. If the eccentricities of the hyperbolas $\frac{x^2}{a^2} - \frac{y^2}{b^2} = 1$ and $\frac{y^2}{b^2} - \frac{x^2}{a^2} = 1$ be e and e_1, then $\frac{1}{e^2} + \frac{1}{e_1^2}$ is equal to [DCE 2000]

(a) 1 (b) 2

(c) 3 (d) None of these

153. The equation of the hyperbola whose conjugate axis is 5 and the distance between the foci is 13, is

(a) $25x^2 - 144y^2 = 900$ (b) $144x^2 - 25y^2 = 900$

(c) $144x^2 + 25y^2 = 900$ (d) $25x^2 + 144y^2 = 900$

154. The equations of the transverse and conjugate axes of the hyperbola $16x^2 - y^2 + 64x + 4y + 44 = 0$, are respectively [HOTS]

(a) $x = 2,\ y + 2 = 0$ (b) $x = 2,\ y = 2$

(c) $y = 2,\ x + 2 = 0$ (d) None of these

155. The locus of the point of intersection of the lines $\sqrt{3}x - y - 4\sqrt{3}k = 0$ and $\sqrt{3}kx + ky - 4\sqrt{3} = 0$ for different value of k, is

(a) circle (b) parabola (c) hyperbola (d) ellipse

156. The equation of the hyperbola whose foci are (6, 4) and (−4, 4) and eccentricity 2 is given by

(a) $12x^2 - 4y^2 - 24x + 32y - 127 = 0$

(b) $12x^2 + 4y^2 + 24x - 32y - 127 = 0$

(c) $12x^2 - 4y^2 - 24x - 32y + 127 = 0$

(d) $12x^2 - 4y^2 + 24x + 32y + 127 = 0$

157. The distance between the directrices of the hyperbola $x = 8\sec\theta$ and $y = 8\tan\theta$, is [Karnataka CET 2003]

(a) $16\sqrt{2}$ (b) $\sqrt{2}$ (c) $8\sqrt{2}$ (d) $4\sqrt{2}$

158. The eccentricity of the hyperbola $5x^2 - 4y^2 + 20x + 8y = 4$, is [UPSEAT 2004]

(a) $\sqrt{2}$ (b) $\frac{3}{2}$

(c) 2 (d) 3

159. The latusrectum of the hyperbola $9x^2 - 16y^2 + 72x - 32y - 16 = 0$, is [Pb. CET 2004]

(a) $\frac{9}{2}$ (b) $-\frac{9}{2}$ (c) $\frac{32}{3}$ (d) $-\frac{32}{3}$

Miscellaneous Exercise

160. A rod AB of length 15cm rests in between two coordinate axes in such a way that the end point A lies on X-axis and end point B lies on Y-axis. A point $P(x, y)$ is taken on the rod in such a way that $AP = 6$cm. Then, the locus of P is a/an

(a) circle

(b) ellipse

(c) parabola

(d) hyperbola

161. If a parabolic reflector is 20 cm in diameter and 5 cm deep, then the focus is

(a) (2, 0) (b) (3, 0) (c) (4, 0) (d) (5, 0)

162. The points given by $x = \frac{2at}{1+t^2}$ and $y = \frac{a(1-t^2)}{1+t^2}$ lie on a ...A... for all real values of t such that $-1 \le t \le 1$, where a is any given real number. Here, A stands for [NCERT Exemplar]

(a) ellipse (b) circle (c) parabola (d) hyperbola

163. An arch is in the form of a parabola with its axis vertical. The arch is 10 m high and 5m wide at the base. Then, it is 'k' m wide when it is 2m from the vertex of the parabola. Then,

(a) $k = \sqrt{7}$ (b) $k = \sqrt{5}$
(c) $k = \sqrt{3}$ (d) $k = 5$

164. If the eccentricity of an ellipse is $\frac{5}{8}$ and the distance between its foci is 10, then the length of latusrectum of ellipse is [NCERT Exemplar]

(a) $\frac{39}{2}$ (b) $\frac{37}{4}$
(c) $\frac{36}{5}$ (d) $\frac{39}{4}$

165. The cable of a uniformly loaded suspension bridge hangs in the form of a parabola. The roadway which is horizontal and 100m long is supported by vertical wires attached to the cable, the longest wire being 30m and the shortest being 6m. Then, the length of supporting wire attached to the roadway 18m from the middle is

(a) 10.02 m (b) 9.11 m
(c) 10.76 m (d) 12. 06 m

166. The length of the line segment joining the vertex of the parabola $y^2 = 4ax$ and a point on the parabola where the line segment makes an angle θ to the X-axis is $\frac{4am}{n}$. Here, m and n respectively are [NCERT Exemplar]

(a) $\sin\theta, \cos\theta$ (b) $\cos\theta, \sin\theta$
(c) $\cos\theta, \sin^2\theta$ (d) $\sin^2\theta, \cos\theta$

167. An arch is in the form of semi-ellipse. It is 8m wide and 2m high at the centre. Then, the height of the arch at a point 1.5m from one end is

(a) 1.56 m (b) 2.4375 m
(c) 2.056 m (d) 1.086 m

168. If the line $y = mx + 1$ is tangent to the parabola $y^2 = 4x$, then the value of m is [NCERT Exemplar]

(a) 0 (b) 2 (c) 1 (d) 3

169. A rod of length 12cm moves with its ends always touching the coordinate axes. If point P is on the rod, which is 3cm from the end in contact with the X-axis, then the equation of locus of point P is

(a) $\frac{x^2}{9} + \frac{y^2}{81} = 0$ (b) $\frac{x^2}{9} - \frac{y^2}{81} = 0$
(c) $\frac{x^2}{81} + \frac{y^2}{9} = 0$ (d) $\frac{x^2}{81} - \frac{y^2}{9} = 0$

170. The equation of parabola having focus at $(-1, -2)$ and directrix $x - 2y + 3 = 0$, is [NCERT Exemplar]

(a) $4x^2 + y^2 + 4xy + 4x + 32y + 16 = 0$
(b) $x^2 + 4y^2 + 4xy + 32x + 4y + 16 = 0$
(c) $4x^2 + 4y^2 + 4xy + 4x + 32y + 16 = 0$
(d) $4x^2 + 4y^2 + 4xy + 32x + 4y\ 16 = 0$

171. The equation of the set of all points, the sum of whose distances form the points (3, 0) and (9, 0) is 12 can be written as [NCERT Exemplar]

(a) $3x^2 + 4y^2 + 36x = 0$
(b) $3x^2 + 4y^2 - 36x = 0$
(c) $4x^2 + 3y^2 + 36x = 0$
(d) $4x^2 + 3y^2 - 36x = 0$

172. A man running a race course notes that sum of its distances from two flag posts from him is always 10m and the distance between the flag posts is 8 m. Then, the equation of the posts traced by the man is [HOTS]

(a) $\frac{x^2}{25} + \frac{y^2}{9} = 1$ (b) $x^2 + y^2 = 25$
(c) $x^2 + y^2 = 9$ (d) $\frac{x^2}{9} + \frac{y^2}{25} = 1$

173. The equation of the set of all points whose distance from (0, 4) is $\frac{2}{3}$ of their distance from the line $y = 9$ will be written as [NCERT Exemplar]

(a) $8x^2 + 9y^2 = 180$ (b) $9x^2 + 5y^2 = 180$
(c) $8x^2 + 5y^2 = 180$ (d) $9x^2 + 8y^2 = 180$

174. Equation of the circle which touches the lines $x = 0$, $y = 0$ and $3x + 4y = 4$, is

(a) $x^2 - 4x + y^2 + 4y + 4 = 0$
(b) $x^2 - 4x + y^2 - 4y + 4 = 0$
(c) $x^2 + 4x + y^2 + 4y + 4 = 0$
(d) $x^2 + 4x + y^2 - 4y + 4 = 0$

175. Radius of circle $(x - 5)(x - 1) + (y - 7)(y - 4) = 0$, is [HOTS]

(a) 3 (b) 4
(c) $\frac{5}{2}$ (d) $\frac{7}{2}$

176. The equation of the circle passing through the points (0, 0), (0, b) and (a, b) is

(a) $x^2 + y^2 + ax + by = 0$
(b) $x^2 + y^2 - ax + by = 0$
(c) $x^2 + y^2 - ax - by = 0$
(d) $x^2 + y^2 + ax - by = 0$

Answer *with* Explanations

1. *(a)* The curves circles, ellipses, parabolas and hyperbolas are known as conic sections.

2. *(b)* The names parabola and hyperbola are given by Apollonius.

3. *(d)* Conic sections or more commonly conics are obtained by intersections of a plane with a double napped right circular cone.

4. *(d)* Conic sections have a very wide range of applications in fields such as planetary motion, design of telescopes and antennas, reflectors in flashlights and automobile headlights etc.

5. *(c)* Suppose we rotate the line *m* around the line *l* in such a way that the angle α remains constant. Then, the surface generated is a double-napped right circular hollow cone and extending indefinitely far in both directions.

6. *(a)* The point *V* is called the vertex; the line *l* is the axis of the cone; the rotating line is called a generator of the cone.

7. *(c)* The vertex separates the cone into two parts called nappes.

8. *(a)* If we take the intersection of a plane with a cone, the section so obtained is called a conic section.

Thus, conic sections are the curves obtained by intersecting a right circular cone by a plane.

9. *(c)* We obtain different kinds of conic sections depending on the position of the intersecting plane with respect to the cone and the angle made by intersecting plane with the vertical axis of the cone.

10. *(d)* The intersection of the plane with the cone can take place either at the vertex of the cone or at any other part of the nappe either below or above the vertex.

11. *(a)* A. When $\beta = 90°$; the section is a circle.

B. When $\alpha < \beta < 90°$; the section is an ellipse.

C. When $\beta = \alpha$; the section is a parabola.

D. When $0 \le \beta < \alpha$; the plane cuts through both the nappes and the curves of intersection is hyperbola.

(in each of *A*, *B* and *C*, the plane cuts entirely across one nappe of the cone).

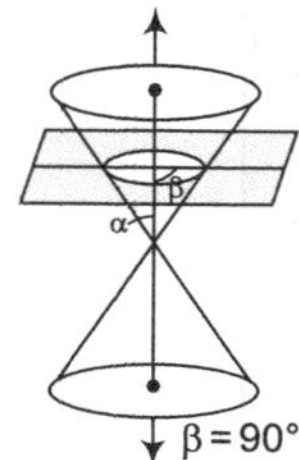

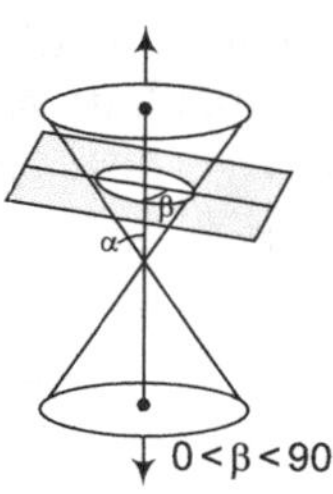

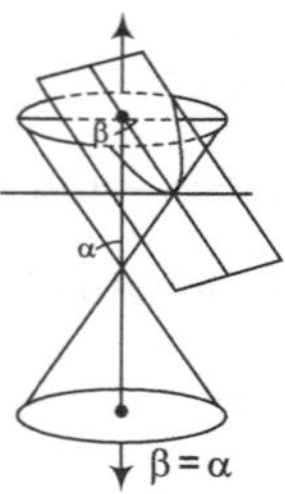

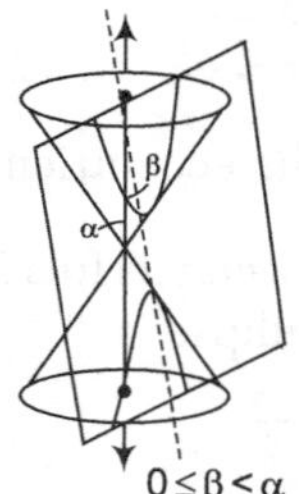

12. *(c)* A. When $\alpha < \beta \le 90°$, then section is a point.

B. When $\beta = \alpha$, the plane contains a generator of the cone and the section is a straight line.

It is the degenerate case of parabola.

C. When $0 \le \beta < \alpha$, the section is a pair of intersecting straight lines. It is the degenerate case of a hyperbola.

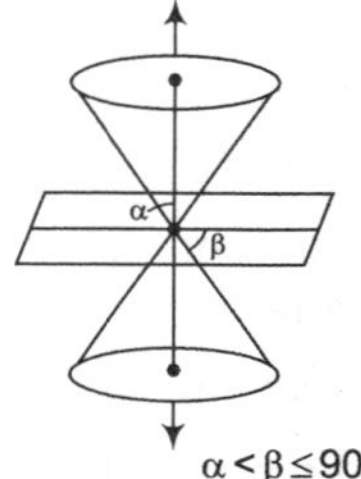

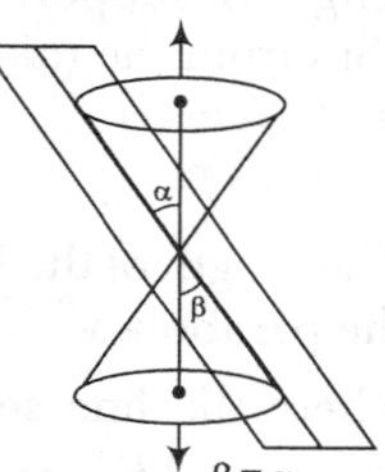

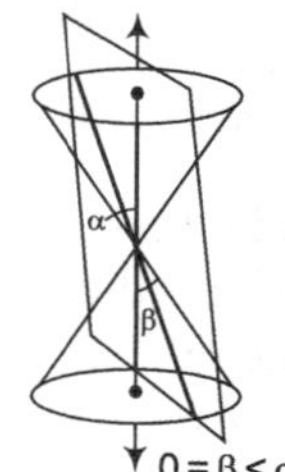

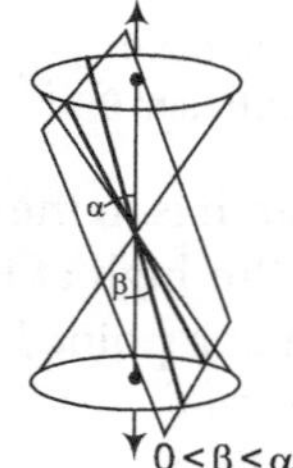

13. *(d)* A circle is the set of all points in a plane that are equidistant from a fixed point in the plane.

14. *(a)* The fixed point in circle is called its centre and the distance from the centre to a point on the circle is called its radius.

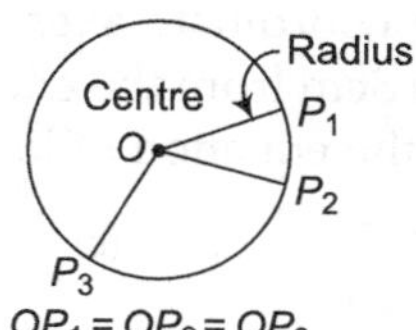

15. *(b)* The equation of the circle is simplest if the centre of the circle is at the origin.

16. *(b)* Given, $C(h, k)$ be the centre and r the radius of circle. Let $P(x, y)$ be any point on the circle. Then, by the definition, $|CP| = r$. By the distance formula, we have

$$\sqrt{(x-h)^2 + (y-k)^2} = r$$

i.e. $$(x-h)^2 + (y-k)^2 = r^2$$

This is the required equation of the circle with centre at (h, k) and radius r.

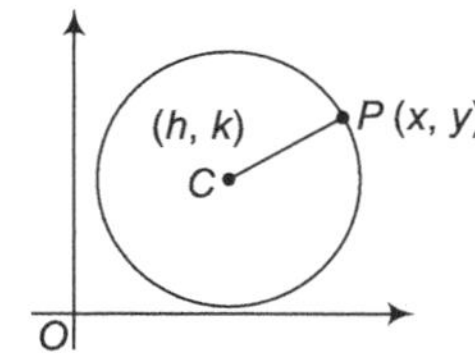

17. *(c)* Suppose $h = k = 0$. Therefore, the equation of the circle is $x^2 + y^2 = r^2$.

18. *(d)* Here, $h = -3$, $k = 2$ and $r = 4$. Therefore, the equation of the required circle is

$$(x+3)^2 + (y-2)^2 = 16$$

19. *(a)* The given equation is $(x^2 + 8x) + (y^2 + 10y) = 8$

Now, completing the squares within the parenthesis, we get

$$(x^2 + 8x + 16) + (y^2 + 10y + 25) = 8 + 16 + 25$$

i.e. $$(x+4)^2 + (y+5)^2 = 49$$

i.e. $$\{x-(-4)\}^2 + \{y-(-5)\}^2 = 7^2$$

Therefore, the given circle has centre at $(-4, -5)$ and radius 7.

20. *(c)* Let the equation of the circle be $(x-h)^2 + (y-k)^2 = r^2$.

Since, the circle passes through $(2, -2)$ and $(3, 4)$, we have

$$(2-h)^2 + (-2-k)^2 = r^2 \quad \ldots\text{(i)}$$

and $$(3-h)^2 + (4-k)^2 = r^2 \quad \ldots\text{(ii)}$$

Also, since the centre lies on the line $x + y = 2$, we have

$$h + k = 2 \quad \ldots\text{(iii)}$$

From Eqs. (i) and (ii), we have

$$(2-h)^2 + (-2-k)^2 = (3-h)^2 + (4-k)^2$$

$$4 + h^2 - 4h + 4 + k^2 + 4k = 9 + h^2 - 6h + 16 + k^2 - 8k$$

$\Rightarrow$ $$4 - 4h + 4 + 4k = 9 - 6h + 16 - 8k$$

$\Rightarrow$ $$-4h + 6h + 4k + 8k = 9 + 16 - 8$$

$\Rightarrow$ $$2h + 12k = 17 \quad \ldots\text{(iv)}$$

Now, multiplying Eq. (iii) by 2 and subtracting it from Eq. (iv), we get

$$(2h + 12k) - (2h + 2k) = 17 - 4$$

$\Rightarrow$ $$10k = 13$$

$\Rightarrow$ $$k = \frac{13}{10} = 1\cdot 3$$

Substituting the value of k in Eq. (iii), we get

$$h + 1\cdot 3 = 2$$

$\Rightarrow$ $$h = 2 - 1\cdot 3 = 0\cdot 7$$

$\Rightarrow$ $$h = 0.7$$

Now, substituting the value of h and k in Eq. (ii), we get

$$(3 - 0\cdot 7)^2 + (4 - 1\cdot 3)^2 = r^2$$

$\Rightarrow$ $$5\cdot 29 + 7\cdot 29 = r^2$$

or $$r^2 = 12\cdot 58$$

$$h = 0.7, k = 1.3$$

and $$r^2 = 12.58$$

Hence, the equation of the required circle is

$$(x - 0.7)^2 + (y - 1.3)^2 = 12.58$$

21. *(b)* A. Given, centre $= (h, k) = (0, 2)$ and radius, $r = 2$

$\therefore$ Equation of circle by using

$(x-h)^2 + (y-k)^2 = r^2$ is

$$(x-0)^2 + (y-2)^2 = 2^2$$

$\Rightarrow$ $$x^2 + y^2 + 4 - 4y = 4 \quad [\because (a-b)^2 = a^2 + b^2 - 2ab]$$

$\Rightarrow$ $$x^2 + y^2 - 4y = 0$$

B. Given, centre $= (h, k) = (-2, 3)$ and radius, $r = 4$

$\therefore$ Equation of circle by using

$$(x-h)^2 + (y-k)^2 = r^2 \text{ is}$$

$$(x+2)^2 + (y-3)^2 = 4^2$$

$\Rightarrow$ $$x^2 + 4 + 4x + y^2 + 9 - 6y = 16$$

$$\left[\begin{array}{l}\because (a+b)^2 = a^2 + b^2 + 2ab \\ \text{and } (a-b)^2 = a^2 + b^2 - 2ab\end{array}\right]$$

$\Rightarrow$ $$x^2 + y^2 + 4x - 6y + 13 - 16 = 0$$

$\Rightarrow$ $$x^2 + y^2 + 4x - 6y - 3 = 0$$

C. Given, centre $= (h, k) = \left(\frac{1}{2}, \frac{1}{4}\right)$ and radius, $r = \frac{1}{12}$

$\therefore$ Equation of circle by using

$$(x-h)^2 + (y-k)^2 = r^2 \text{ is}$$

$$\left(x - \frac{1}{2}\right)^2 + \left(y - \frac{1}{4}\right)^2 = \left(\frac{1}{12}\right)^2$$

$\Rightarrow$ $$x^2 + \frac{1}{4} - 2\cdot\frac{1}{2}x + y^2 + \frac{1}{16} - 2\cdot\frac{y}{4} = \frac{1}{144}$$

$$[\because (a-b)^2 = a^2 + b^2 - 2ab]$$

$\Rightarrow$ $$x^2 + y^2 - x - \frac{y}{2} + \frac{1}{4} + \frac{1}{16} - \frac{1}{144} = 0$$

$\Rightarrow$ $$x^2 + y^2 - x - \frac{y}{2} + \frac{36 + 9 - 1}{144} = 0$$

$\Rightarrow$ $$x^2 + y^2 - x - \frac{y}{2} + \frac{44}{144} = 0$$

$$x^2 + y^2 - x - \frac{y}{2} + \frac{11}{36} = 0$$

$\Rightarrow$ $$36x^2 + 36y^2 - 36x - 18y + 11 = 0$$

D. Given, centre $= (h, k) = (1, 1)$ and radius, $r = \sqrt{2}$

$\therefore$ Equation of circle by using $(x-h)^2 + (y-k)^2 = r^2$ is

$$(x-1)^2 + (y-1)^2 = (\sqrt{2})^2$$

$\Rightarrow$ $$x^2 + 1 - 2x + y^2 + 1 - 2y = 2$$

$\Rightarrow$ $$x^2 + y^2 - 2x - 2y = 0$$

22. *(a)* Given, centre $= (h, k) = (-a, -b)$ and

radius, $r = \sqrt{a^2 - b^2}$

$\therefore$ Equation of circle by using $(x-h)^2 + (y-k)^2 = r^2$ is

$$(x+a)^2 + (y+b)^2 = (\sqrt{a^2 - b^2})^2$$

$$\Rightarrow x^2 + a^2 + 2ax + y^2 + b^2 + 2by = a^2 - b^2$$

$$\Rightarrow x^2 + y^2 + 2ax + 2by + 2b^2 = 0$$

23. *(d)* A. Given, equation of circle is

$$(x+5)^2 + (y-3)^2 = 36 = (6)^2$$

Comparing it with

$(x-h)^2 + (y-k)^2 = r^2$, we get

$$h = -5, k = 3 \quad \text{and} \quad r = 6$$

Hence, centre $= (h, k) = (-5, 3)$ and radius, $r = 6$

B. Given equation of circle is

$$x^2 + y^2 - 4x - 8y - 45 = 0$$

$$\Rightarrow (x^2 - 4x) + (y^2 - 8y) = 45$$

$$\Rightarrow (x^2 - 4x + 4) + (y^2 - 8y + 16) = 45 + 4 + 16$$

$$\Rightarrow (x-2)^2 + (y-4^2) = 65$$

(by completing the perfect square)

$\Rightarrow$ Compring above equation with $(x-h)^2 + (y-k)^2 = r^2$, we have

$$h = 2, k = 4, r^2 = 65 \quad i.e., \quad r = \sqrt{65}$$

Hence, centre (2, 4) and radius $\sqrt{65}$.

Alternate Method

Given equation of circle is

$$x^2 + y^2 - 4x - 8y - 45 = 0$$

Comparing it with

$x^2 + y^2 + 2gx + 2fy + c = 0$, we get

$$2g = -4, 2f = -8, c = -45$$

$$\Rightarrow g = -2, f = -4, c = -45$$

$\because$ Centre and radius of circle

$x^2 + y^2 + 2gx + 2fy + c = 0$ are respectively

$$C(-g, -f) \quad \text{and} \quad r = \sqrt{g^2 + f^2 - c}$$

$$\therefore \quad \text{Centre} = (-g, -f) = (2, 4)$$

and radius $= \sqrt{g^2 + f^2 - c} = \sqrt{(2)^2 + (4)^2 - (-45)}$

$$= \sqrt{4 + 16 + 45} = \sqrt{65}$$

C. Given equation of circle is

$$x^2 + y^2 - 8x + 10y - 12 = 0$$

$$\Rightarrow (x^2 - 8x) + (y^2 + 10y) = 12$$

$$\Rightarrow (x^2 - 8x + 16) + (y^2 + 10y + 25) = 12 + 16 + 25$$

(by completing the square)

$$\Rightarrow (x-4)^2 + (y+5)^2 = 53$$

$$\Rightarrow (x-4)^2 + \{y - (-5)^2\} = 53$$

Comparing above equation with $(x-h)^2 + (y-k)^2 = r^2$, we have

$$h = 4, k = -5, \ r = \sqrt{53}$$

Hence, centre (4,–5) and radius $\sqrt{53}$.

Alternate Method

Given equation of circle is $x^2 + y^2 - 8x + 10y - 12 = 0$

Comparing with $x^2 + y^2 + 2gx + 2fy + c = 0$, we get

$$2g = -8, 2f = 10, c = -12$$

$$\Rightarrow g = -4, f = 5, c = -12$$

$$\therefore \quad \text{Centre} = (-g, -f) = (4, -5)$$

and radius $= \sqrt{g^2 + f^2 - c} = \sqrt{(4)^2 + (-5)^2 - (-12)}$

$$= \sqrt{16 + 25 + 12}$$

$$= \sqrt{53}$$

D. Given equation of circle

$$2x^2 + 2y^2 - x = 0$$

Dividing by 2, we get

$$x^2 + y^2 - \frac{x}{2} = 0$$

$$\Rightarrow \left(x^2 - \frac{1}{2}x\right) + y^2 = 0$$

$$\Rightarrow \left(x^2 - \frac{1}{2}x + \frac{1}{16}\right) + y^2 = \frac{1}{16}$$

(by completing the square within paranthesis)

$$\Rightarrow \left(x - \frac{1}{4}\right)^2 + y^2 = \frac{1}{16}$$

Comparing above equation with $(x-h)^2 + (y-k)^2 = r^2$, we have

$$h = \frac{1}{4}, k = 0, r = \frac{1}{4}$$

Hence, centre $\left(\frac{1}{4}, 0\right)$ and radius $\frac{1}{4}$.

Alternate Method

Given equation of circle is

$$2x^2 + 2y^2 - x = 0$$

On dividing by 2, we get

$$x^2 + y^2 - \frac{x}{2} = 0$$

On Comparing with $x^2 + y^2 + 2gx + 2fy + c = 0$, we get

$$2g = -\frac{1}{2}, 2f = 0, c = 0$$

$$\Rightarrow g = -\frac{1}{4}, f = 0, c = 0$$

$$\therefore \quad \text{Centre} = (-g, -f) = \left(\frac{1}{4}, 0\right)$$

and radius $= \sqrt{g^2 + f^2 - c} = \sqrt{\left(\frac{1}{4}\right)^2 + 0 - 0}$

$$= \sqrt{\frac{1}{16}} = \frac{1}{4}$$

Note *In general equation of circle first make the coefficients of x^2 and y^2 non-negative with unity. Otherwise, it may be error on taking centre.*

24. *(b)* We know that the equation of the circle having centre (h, k) and radius r is

$$(x-h)^2+(y-k)^2=r^2 \quad \ldots(i)$$

$\because$ Circle (i) passes through points (4, 1) and (6, 5).

$\therefore \quad (4-h)^2+(1-k)^2=r^2 \quad \ldots(ii)$

and $\quad (6-h)^2+(5-k)^2=r^2 \quad \ldots(iii)$

From Eqs. (ii) and (iii), we have

$$(4-h)^2+(1-k)^2=(6-h)^2+(5-k)^2$$

$\Rightarrow 16+h^2-8h+1+k^2-2k=36+h^2-12h+25+k^2-10k$

$\Rightarrow \quad -8h+12h-2k+10k=36+25-17$

$\Rightarrow \quad 4h+8k=44 \quad \ldots(iv)$

Also, centre (h, k) lies on the line $4x+y=16$

$\therefore \quad 4h+k=16 \quad \ldots(v)$

Subtracting Eq. (v) from Eq. (iv), we have

$$(4h+8k)-(4h+k)=44-16$$

$\Rightarrow \quad 7k=28$

$\Rightarrow \quad k=4$

Substituting the value of k in Eq. (v), we get

$$4h+4=16$$

$$h=\frac{16-4}{4}=\frac{12}{4}=3$$

Substituting the value of h and k in Eq. (iii), we have

$$(6-3)^2+(5-4)^2=r^2$$

$\Rightarrow \quad 3^2+1^2=r^2 \Rightarrow r^2=10$

Hence, centre (3, 4) and radius $\sqrt{10}$.

$\therefore$ Required equation of circle

$$(x-3)^2+(y-4)^2=10$$

$\Rightarrow \quad x^2+9-6x+y^2+16-8y=10$

$\Rightarrow \quad x^2+y^2-6x-8y+15=0$

Alternate Method

Let equation of circle is

$$x^2+y^2+2gx+2fy+c=0 \quad \ldots(i)$$

Eq. (i), passes through the points (4, 1) and (6, 5) *i.e.*, they will satisfy it.

At point (4, 1),

$$(4)^2+(1)^2+2g(4)+2f(1)+c=0$$

$\Rightarrow \quad 17+8g+2f+c=0$

$\Rightarrow \quad 8g+2f+c+17=0 \quad \ldots(ii)$

At point (6, 5),

$$(6)^2+(5)^2+2g(6)+2f(5)+c=0$$

$\Rightarrow \quad 36+25+12g+10f+c=0$

$\Rightarrow \quad 12g+10f+c+61=0 \quad \ldots(iii)$

As centre $(-g, -f)$ of Eq. (i), lies on the line

$$4x+y=16$$

$\therefore \quad 4(-g)+(-f)=16$

$\Rightarrow \quad 4g+f+16=0 \quad \ldots(iv)$

Now, subtract Eq. (ii) from Eq. (iii), we get

$$4g+8f+44=0 \quad \ldots(v)$$

Now, subtract Eq. (iv) from Eq. (v), we get

$$7f+28=0$$

$\Rightarrow \quad 7f=-28 \Rightarrow f=-4$

From Eq. (iv), we get

$$4g-4+16=0$$

$\Rightarrow \quad 4g+12=0$

$\Rightarrow \quad 4g=-12$

$\Rightarrow \quad g=-3$

From Eq. (ii), we get

$$8(-3)+2(-4)+c+17=0$$

$\Rightarrow \quad -24-8+c+17=0$

$\Rightarrow \quad c-32+17=0$

$\Rightarrow \quad c-15=0$

$\Rightarrow \quad c=15$

On putting the values of g, f and c in Eq. (i), we get required equation of circle

$$x^2+y^2+2(-3)x+2(-4)y+15=0$$

$\Rightarrow \quad x^2+y^2-6x-8y+15=0$

25. *(c)* Let (h, k) be the centre and r be the radius of the circle.

$\therefore$ Equation of circle

$$(x-h)^2+(y-k)^2=r^2 \quad \ldots(i)$$

$\because$ Circle (i) passes through points (2, 3) and (−1, 1).

$\therefore \quad (2-h)^2+(3-k)^2=r^2 \quad \ldots(ii)$

$\quad (-1-h)^2+(1-k)^2=r^2 \quad \ldots(iii)$

From Eqs. (ii) and (iii), we have

$$(2-h)^2+(3-k)^2=(-1-h)^2+(1-k)^2$$

$\Rightarrow 4+h^2-4h+9+k^2-6k=1+h^2+2h+1+k^2-2k$

$\Rightarrow \quad -4h-2h-6k+2k=2-4-9$

$\Rightarrow \quad -6h-4k=-11$

$\Rightarrow \quad 6h+4k=11 \quad \ldots(iv)$

Also, the centre of circle lies on the line $x-3y=11$

$\therefore \quad h-3k=11 \quad \ldots(v)$

Multiplying Eq. (v) by 6 and subtracting from Eq. (iv), we have

$$(6h+4k)-(6h-18k)=11-66$$

$\Rightarrow \quad 22k=-55$

$\Rightarrow \quad k=-\frac{55}{22}=-\frac{5}{2}$

Substituting the value of k in Eq. (v), we get

$$h-3\times\left(-\frac{5}{2}\right)=11$$

$\Rightarrow \quad h+\frac{15}{2}=11$

$\Rightarrow \quad h=11-\frac{15}{2}=\frac{7}{2}$

Now, substituting the value of h and k in Eq. (ii), we get

$$\left(2-\frac{7}{2}\right)^2+\left(3+\frac{5}{2}\right)^2=r^2$$

$$\Rightarrow \left(-\frac{3}{2}\right)^2+\left(\frac{11}{2}\right)^2=r^2$$

$$\Rightarrow \frac{9}{4}+\frac{121}{4}=r^2$$

$$\Rightarrow \frac{130}{4}=r^2$$

Required equation of circle

$$\left(x-\frac{7}{2}\right)^2+\left(y+\frac{5}{2}\right)^2=\frac{130}{4}$$

$$\Rightarrow x^2+\frac{49}{4}-7x+y^2+\frac{25}{4}+5y=\frac{130}{4}$$

$$\Rightarrow x^2+y^2-7x+5y=\frac{130}{4}-\frac{49}{4}-\frac{25}{4}$$

$$\Rightarrow x^2+y^2-7x+5y=\frac{56}{4}$$

$$\Rightarrow x^2+y^2-7x+5y=14$$

or $x^2+y^2-7x+5y-14=0$

Comparing above with $x^2+y^2+ax+by+d$, we get

$$a=-7, b=5$$

and $d=-14$

Alternate Method

Let equation of circle is

$$x^2+y^2+2gx+2fy+c=0 \quad \dots(i)$$

Eq. (i) passes through the points (2, 3) and (– 1, 1) *i.e.*, they will satisfy it.

At point (2, 3),

$$(2)^2+(3)^2+2g(2)+2f(3)+c=0$$

$$\Rightarrow 4g+6f+c+13=0 \quad \dots(ii)$$

and at point (–1, 1),

$$(-1)^2+(1)^2+2g(-1)+2f(1)+c=0$$

$$\Rightarrow -2g+2f+c+2=0 \quad \dots(iii)$$

As centre $(-g, -f)$ lies on the line $x-3y-11=0$

i.e., $(-g)-3(-f)-11=0$

$$\Rightarrow -g+3f-11=0 \quad \dots(iv)$$

Now, subtract Eq. (iii) from Eq. (ii), we get

$$6g+4f+11=0 \quad \dots(v)$$

From Eq. (iv), $g=3f-11$ put in Eq. (v), we get

$$6(3f-11)+4f+11=0$$

$$\Rightarrow 18f-66+4f+11=0$$

$$\Rightarrow 22f-55=0$$

$$\Rightarrow f=\frac{55}{22} \Rightarrow f=\frac{5}{2}$$

From Eq. (iv) we get,

$$g=3\times\frac{5}{2}-11=\frac{15-22}{2}=\frac{-7}{2}$$

From Eq. (ii) we get,

$$4\times\left(-\frac{7}{2}\right)+6\left(\frac{5}{2}\right)+c+13=0$$

$$\Rightarrow -14+15+c+13=0$$

$$\Rightarrow c+14=0$$

$$\Rightarrow c=-14$$

Put the values of g, f and c in Eq. (i) to get the required equation of circle

$$x^2+y^2+2x\left(-\frac{7}{2}\right)+2y\left(\frac{5}{2}\right)-14=0$$

$$\Rightarrow x^2+y^2-7x+5y-14=0$$

26. *(b)* Let centre of circle be $(h, 0)$, as centre of circle is on X-axis, so y-coordinate will be zero.

The distance between any point on the circle from the centre = Radius

$$\therefore \sqrt{(x_2-x_1)^2+(y_2-y_1)^2}=r$$

$$\Rightarrow \sqrt{(h-2)^2+(0-3)^2}=5$$

$$(\because x_1=2, y_1=3, x_2=h, y_2=0)$$

$$\Rightarrow h^2+4-4h+9=25 \quad \text{(squaring both sides)}$$

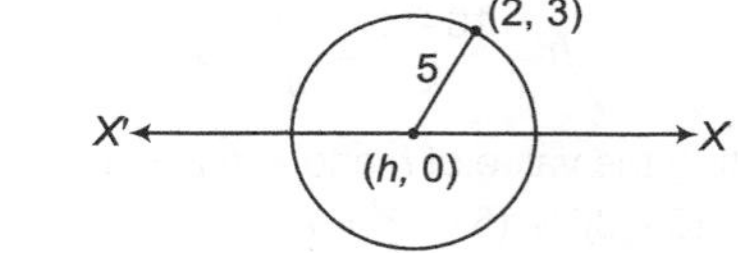

$$\Rightarrow h^2-4h+13-25=0$$

$$\Rightarrow h^2-4h-12=0$$

$$\Rightarrow h^2-6h+2h-12=0$$

$$\Rightarrow (h-6)(h+2)=0$$

$$\Rightarrow h-6=0 \text{ or } h+2=0$$

$$\Rightarrow h=6 \text{ or } -2$$

$\therefore$ Centre is (6, 0) or (– 2, 0).

Hence, equation of circle by using

$$(x-h)^2+(y-k)^2=r^2$$

When $(h, k)=(6, 0)$ and $r=5$, is

$$(x-6)^2+(y-0)^2=5^2$$

$$\Rightarrow x^2+36-12x+y^2=25$$

$$\Rightarrow x^2+y^2-12x+11=0$$

and when $(h, k)=(-2, 0)$ and $r=5$, is

$$\Rightarrow (x+2)^2+(y-0)^2=5^2 \quad [\because (x-h)^2+(y-k)^2=r^2]$$

$$\Rightarrow x^2+4+4x+y^2=25$$

$$\Rightarrow x^2+y^2+4x-21=0$$

27. *(d)* Circle makes intercept a and b at the coordinate axes.

$\therefore$ It intersect X-axis at $(a, 0)$ and Y-axis at $(0, b)$.

Also, circle passes through the points (0, 0), $(a, 0)$ and $(0, b)$.

Let (h, k) be the centre of the circle.

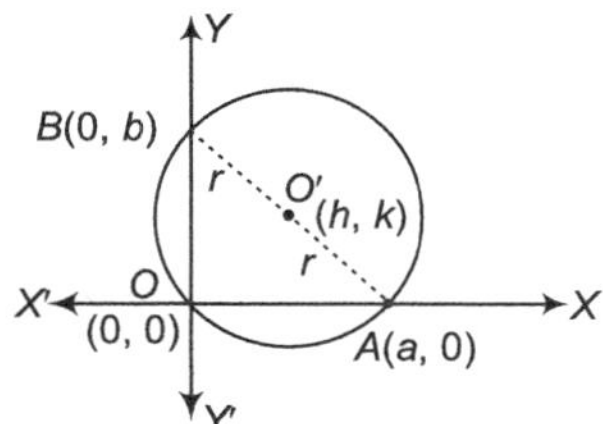

Now, from the figure

$$O'A = O'B = OO' = \text{Radius}$$

$$O'A = \sqrt{(h-a)^2 + k^2} \quad \ldots \text{(i)}$$

$$O'B = \sqrt{h^2 + (k-b)^2} \quad \ldots \text{(ii)}$$

$$OO' = \sqrt{h^2 + k^2} \quad \ldots \text{(iii)}$$

From Eqs. (i) and (ii), we have

$$(O'A)^2 = (O'B)^2$$

$$\Rightarrow \quad (h-a)^2 + k^2 = h^2 + (k-b)^2$$

$$\Rightarrow \quad h^2 + a^2 - 2ha + k^2 = h^2 + k^2 + b^2 - 2bk$$

$$\Rightarrow \quad a^2 - 2ha = b^2 - 2bk$$

$$\Rightarrow \quad -2ha + 2bk = b^2 - a^2$$

$$\text{or} \quad 2ah - 2bk = a^2 - b^2 \quad \ldots \text{(iv)}$$

From Eqs. (ii) and (iii), we have

$$(O'B)^2 = (OO')^2$$

$$\Rightarrow \quad h^2 + (k-b)^2 = h^2 + k^2$$

$$\Rightarrow \quad h^2 + k^2 + b^2 - 2bk = h^2 + k^2$$

$$\Rightarrow \quad b^2 - 2bk = 0$$

$$\Rightarrow \quad -2bk = -b^2$$

$$\Rightarrow \quad k = \frac{-b^2}{-2b} = \frac{b}{2}$$

Substituting the value of k in Eq. (iv), we have

$$2ah - 2b\left(\frac{b}{2}\right) = a^2 - b^2$$

$$\Rightarrow \quad 2ah - b^2 = a^2 - b^2$$

$$\Rightarrow \quad 2ah = a^2$$

$$\Rightarrow \quad h = \frac{a^2}{2a} = \frac{a}{2}$$

$\therefore$ Centre of the circle is $\left(\frac{a}{2}, \frac{b}{2}\right)$.

$\therefore$ Required equation of circle

$$\left(x - \frac{a}{2}\right)^2 + \left(y - \frac{b}{2}\right)^2 = \left[\sqrt{\frac{a^2}{4} + \frac{b^2}{4}}\right]^2$$

$$\Rightarrow x^2 + \frac{a^2}{4} - ax + y^2 + \frac{b^2}{4} - by = \frac{a^2 + b^2}{4}$$

$$\Rightarrow \quad x^2 + y^2 - ax - by = 0$$

Comparing it with $x^2 + y^2 + mx + ny = 0$, we have

$$m = -a, n = -b$$

Alternate Method

Since, the circle makes the intercepts on the coordinate axis at a and b. It means it intersects the X-axis at $(a, 0)$ and the Y-axis at $(0, b)$.

Also, circle passes through the points $O(0, 0)$, $A(a, 0)$ and $B(0, b)$.

Let general equation of circle is

$$x^2 + y^2 + 2gx + 2fy + c = 0 \quad \ldots \text{(i)}$$

Eq. (i) passes through the points $O(0, 0)$, $A(a, 0)$ and $B(0, b)$ *i.e.*, they will satisfy it.

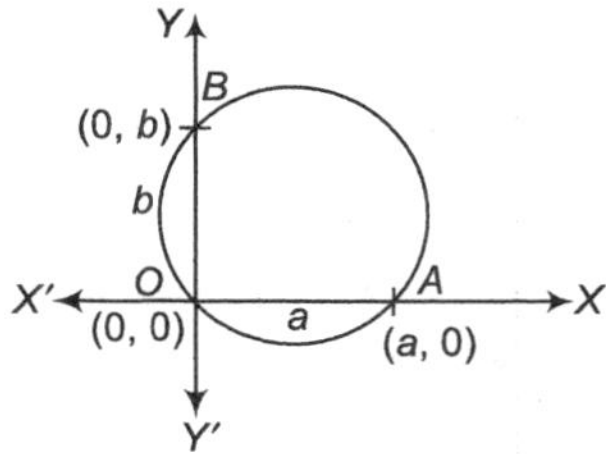

At point (0, 0),

$$0 + 0 + 0 + 0 + c = 0$$

$$\Rightarrow \quad c = 0$$

At point $(a, 0)$,

$$a^2 + 0 + 2ag + 0 + 0 = 0$$

$$\Rightarrow \quad a^2 + 2ag = 0 \Rightarrow g = -\frac{a}{2}$$

and at point $(0, b)$,

$$0 + b^2 + 0 + 2bf + 0 = 0$$

$$\Rightarrow \quad b^2 + 2bf = 0$$

$$\Rightarrow \quad f = -\frac{b}{2}$$

Put the values of g, f and c in Eq. (i) to get the required equation of circle.

$$x^2 + y^2 + 2x\left(-\frac{a}{2}\right) + 2y\left(-\frac{b}{2}\right) + 0 = 0$$

$$\Rightarrow \quad x^2 + y^2 - ax - by = 0$$

Comparing above equation with $x^2 + y^2 + mx + ny = 0$, we get $\quad m = -a \quad$ and $\quad n = -b$

28. *(b)* Here, centre $(h, k) = (2, 2)$ and point on the circle $P(x, y) = (4, 5)$

$$\therefore \quad r = \sqrt{(2-4)^2 + (2-5)^2} = \sqrt{4+9} = \sqrt{13}$$

$$\left[\because r = \sqrt{(x-h)^2 + (y-k)^2}\right]$$

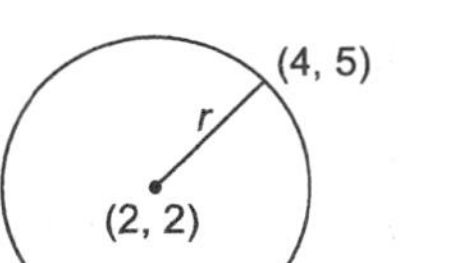

Equation of circle is

$$(x-h)^2+(y-k)^2=r^2$$

$\therefore$ $(x-2)^2+(y-2)^2=(\sqrt{13})^2$

[given, $(h,k)=(2,2)$ and $r=\sqrt{13}$]

$\Rightarrow$ $x^2+4-4x+y^2+4-4y=13$

$\Rightarrow$ $x^2+y^2-4x-4y-5=0$

29. *(a)* Given, equation of circle is

$$x^2+y^2=25$$

Let $S=x^2+y^2-25$ ...(i)

Put $(x,y)=(-2.5,3.5)$ in Eq. (i), we get

$\Rightarrow$ $S=(-2.5)^2+(3.5)^2-25$

$=6.25+12.25-25$

$=18.50-25=-6.50<0$

$\therefore$ Point lies inside the circle.

Note *For the point (x_1,y_1), if $S_1>0$, then point lies outside, if $S_1=0$, then point lies on the circle and if $S_1<0$, then point lies inside.*

30. *(c)* We write the given equation in the standard form as $(x^2-2x)+(y^2+4y)=8$

Now, completing the squares, we get

$$(x^2-2x+1)+(y^2+4y+4)=8+1+4$$

$$(x-1)^2+(y+2)^2=13$$

Comparing it with the standard form of the equation of the circle, we see that the centre of the circle is $(1,-2)$ and radius is $\sqrt{13}$.

31. *(a)* By substitution of coordinates in the general equation of the circle given by

$x^2+y^2+2gx+2fy+c=0$, we have

$$40g+6f+c=-409$$

$$38g+16f+c=-425$$

$$4g-18f+c=-85$$

From these three equations, we get

$$g=-7, f=-3$$

and $c=-111$

Hence, the equation of the circle is

$$x^2+y^2-14x-6y-111=0$$

or $(x-7)^2+(y-3)^2=13^2$

Therefore, the centre of the circle is (7, 3) and radius is 13.

32. *(a)* Since, the circle is in the first quadrant touching each co-ordinate axis at a distance of one unit from origin.

$\therefore$ It passes through the points (1, 0) and (0, 1).

Also, its centre will be (1, 1) and radius is one unit. Hence, the equation of required circle is

$$(x-1)^2+(y-1)^2=1^2$$

$\Rightarrow$ $x^2+y^2-2x-2y+1=0$

Y
(0, 1)
(1, 1)
X′
X
O
(1, 0)
Y′

33. *(a)* The point of intersection of $3x+y-14=0$ and $2x+5y-18=0$ are $x=4$, $y=2$, *i.e.*, the point (4, 2)

Therefore, the radius is $=\sqrt{9+16}=5$ and hence the equation of the circle is given by

$$(x-1)^2+(y+2)^2=25$$

or $x^2+y^2-2x+4y-20=0$

34. *(d)* From given conditions, we have

$x=2+4\cos\theta \Rightarrow (x-2)=4\cos\theta$ ...(i)

and $y=-1+4\sin\theta \Rightarrow (y+1)=4\sin\theta$ ...(ii)

Squaring and adding Eqs. (i) and (ii), we get

$$(x-2)^2+(y+1)^2=16 \quad (\because \sin^2\theta+\cos^2\theta=1)$$

35. *(d)* As the circle is passing through the point $(4,-5)$ and its centre is $(-2,-2)$ so its radius is $\sqrt{(4+2)^2+(-5+2)^2}=\sqrt{45}$

Therefore, the required answer is $(x+2)^2+(y+2)^2=45$

36. *(c)* Let (h,k) be the centre of the circle. Then, $k=h-1$.

Therefore, the equation of the circle is given by

$(x-h)^2+[y-(h-1)]^2=9$ ...(i)

Given that, the circle passes through the point (7, 3) and hence we get

$$(7-h)^2-(3-(h-1))^2=9$$

or $(7-h)^2+(4-h)^2=9$

or $h^2-11h+28=0$

which given $(h-7)(h-4)=0$

$\Rightarrow$ $h=4$ or $h=7$

Therefore, the required equations of the circles are

$$x^2+y^2-8x-6y+16=0$$

or $x^2+y^2-14x-12y+76=0$

Comparing above equations with

$$x^2+y^2+ax+by+c=0$$

where, $a=-8, b=-6, c=16$

and $a=-14, b=-12, c=76$

37. *(c)* Let the equation of circle be

$$x^2+y^2+2gx+2fy+c=0$$

Now, on passing through the given points, we get three equations–

$c=0$ ...(i)

$a^2+2ga+c=0$ ...(ii)

$b^2+2fb+c=0$ ...(iii)

On solving them, we get $g = -\frac{a}{2}, f = -\frac{b}{2}$

Hence, the centre is $\left(\frac{a}{2}, \frac{b}{2}\right)$.

38. *(a)* Since, the circle with centre $(1, -3)$ touches the line $2x - y - 4 = 0$, therefore, the radius of circle is $\left|\frac{2+3-4}{\sqrt{5}}\right| = \frac{1}{\sqrt{5}}$

Therefore, equation is $(x-1)^2 + (y+3)^2 = \frac{1}{5}$

or $\quad x^2 + y^2 - 2x + 6y + 1 + 9 = \frac{1}{5}$

or $5x^2 + 5y^2 - 10x + 30y + 49 = 0$

39. *(c)* The centre of the required circle lies at the intersection of $2x - 3y - 5 = 0$ and $3x - 4y - 7 = 0$. Thus, the coordinates of the centre are $(1, -1)$.

Let r be the radius of the circle. Then, by hypothesis, we have $\pi r^2 = 154 \Rightarrow \frac{22}{7} r^2 = 154 \Rightarrow r = 7$

Hence, the equation of the required circle is

$(x-1)^2 + (y+1)^2 = 7^2 \Rightarrow x^2 + y^2 - 2x + 2y - 47 = 0$

40. *(b)* First find the centre, Let centre be (h, k), then

$$\sqrt{(h-2)^2 + (k-3)^2} = \sqrt{(h-4)^2 + (k-5)^2} \quad \ldots(\text{i})$$

and $\quad k - 4h + 3 = 0 \quad \ldots(\text{ii})$

From Eq. (i), we get

$-4h - 6k + 8h + 10k = 16 + 25 - 4 - 9$

or $\quad 4h + 4k - 28 = 0$ or $h + k - 7 = 0 \quad \ldots(\text{iii})$

From Eqs. (iii) and (ii), we get (h, k) as $(2, 5)$.

Hence, centre is $(2, 5)$ and radius is 2.

Therefore, equation of circle is $x^2 + y^2 - 4x - 10y + 25 = 0$.

41. *(b)* A parabola is the set of all points in a plane that are equidistant from a fixed line and a fixed point (not on the line) in the plane.

42. *(a)* The fixed line is called the directrix of the parabola. The fixed point is called the focus.

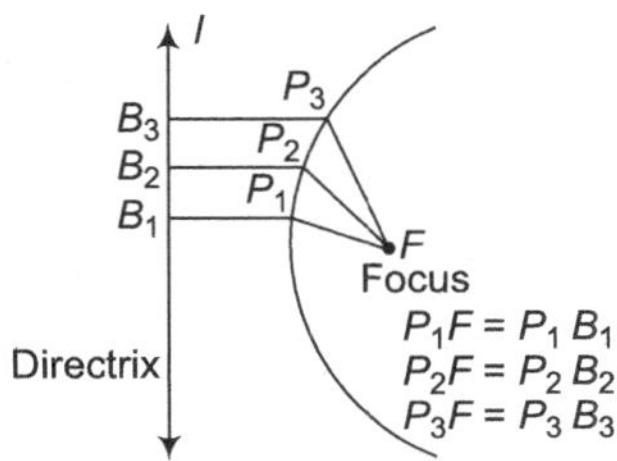

43. *(d)* 'Para' means 'for' and 'bola' means 'throwing' *i.e.,* the shape described when you throw a ball in the air.

44. *(b)* If the fixed point lies on the fixed line, then the set of points in the plane, which are equidistant from the fixed point and the fixed line is the straight line through the fixed point and perpendicular to the fixed line. We call this straight line as degenerate case of parabola.

45. *(a)* A line through the focus and perpendicular to the directrix is called the axis of the parabola. The point of intersection of parabola with the axis is called the vertex of the parabola.

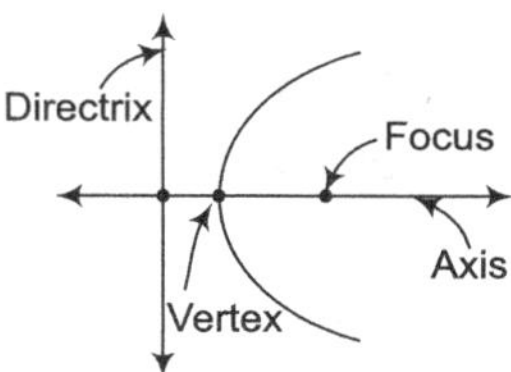

46. *(c)* The equation of a parabola is simplest, if the vertex is at the origin and the axis of symmetry is along the X- axis or Y-axis.

47. *(d)* The number of possible orientations of parabola is four.

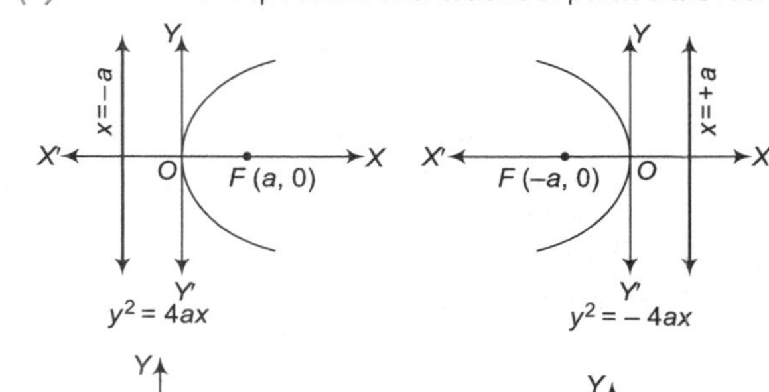

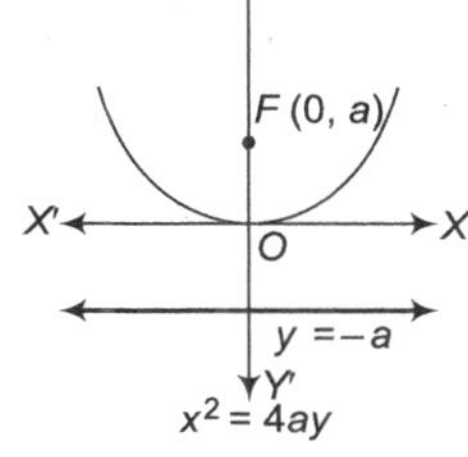

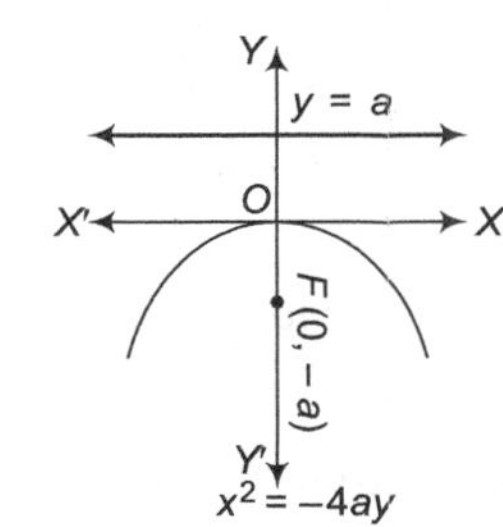

48. *(b)* Let F be the focus and l the directrix. Let FM be perpendicular to the directrix and bisect FM at the point O. Produce MO to X. By the definition of parabola, the mid-point O is on the parabola and is called the vertex of the parabola. Take O as origin, OX the x-axis and OY perpendicular to it as the Y-axis. Let the distance from the directrix to the focus be $2a$. Then, the coordinates of the focus are $(a, 0)$ and the equation of the directrix is $x + a = 0$ as in figure. Let $P(x, y)$ be any point on the parabola such that

$$PF = PB \quad \ldots(\text{i})$$

where, PB is perpendicular to l. The coordinates of B are $(-a, y)$. By the distance formula, we have

$PF = \sqrt{(x-a)^2 + y^2}$ and $PB = \sqrt{(x+a)^2}$

Since, $PF = PB$, we have

$$\sqrt{(x-a)^2 + y^2} = \sqrt{(x+a)^2}$$

i.e. $(x-a)^2 + y^2 = (x+a)^2$

or $x^2 - 2ax + a^2 + y^2 = x^2 + 2ax + a^2$

or $y^2 = 4ax (a > 0)$.

Hence, any point on the parabola satisfies

$$y^2 = 4ax \quad \text{...(ii)}$$

Conversely, let $P(x, y)$ satisfy the Eq. (ii)

$$PF = \sqrt{(x-a)^2 + y^2}$$
$$= \sqrt{(x-a)^2 + 4ax}$$
$$= \sqrt{(x+a)^2} = PB \quad \text{...(iii)}$$

and so $P(x, y)$ lies on the parabola.

Thus, from Eqs. (ii) and (iii) we have proved that the equation to the parabola with vertex at the origin, focus at $(a, 0)$ and directrix $x = -a$ is $y^2 = 4ax$.

Also, in Eq. (ii), since $a > 0$, x can assume any positive value or zero but no negative value and the curve extends indefinitely far into the first and the fourth quadrants. The axis of the parabola is the positive X-axis.

Similarly, we can derive the equations of the parabola as

$$y^2 = -4ax,$$

(vertex at origin, focus $(-a, 0)$ and directrix $x = a$)

$$x^2 = 4ay,$$

(vertex at origin, focus $(0, a)$ and directrix $y = -a$)

$$x^2 = -4ay,$$

(vertex at origin, focus $(0, -a)$ and directrix $y = a$)

These four equations are known as standard equation of parabolas.

49. *(b)* The standard equations of parabolas have focus on one of the coordinate axis; vertex at the origin and thereby the directrix is parallel to the other coordinate axis.

50. *(a)* Parabola is symmetric with respect to the axis of the parabola. If the equation has a y^2 term, then the axis of symmetry is along the X-axis and if the equation has an x^2 term, then the axis of symmetry is along the Y-axis.

51. *(c)* When the axis of symmetry is along the X-axis the parabola opens to the

(i) right, if the coefficient of x is positive.

(ii) left, if the coefficient of x is negative.

52. *(d)* When the axis of symmetry is along the Y-axis the parabola opens

I. upwards, if the coefficient of y is positive.

II. downwards, if the coefficient of y is negative.

53. *(a)* Latusrectum of a parabola is a line segment perpendicular to the axis of the parabola, through the focus and whose end points lie on the parabola.

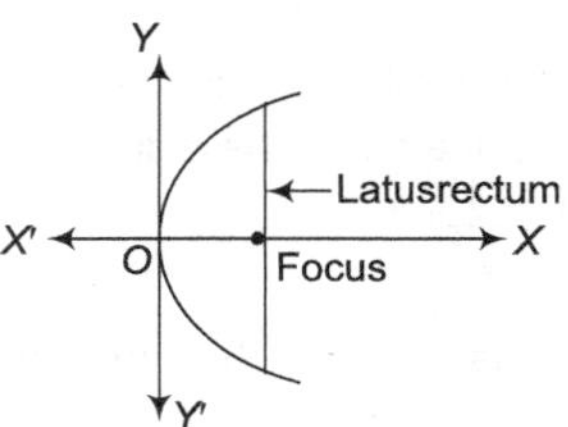

54. *(d)* To find the length of the latusrectum of the parabola $y^2 = 4ax$.

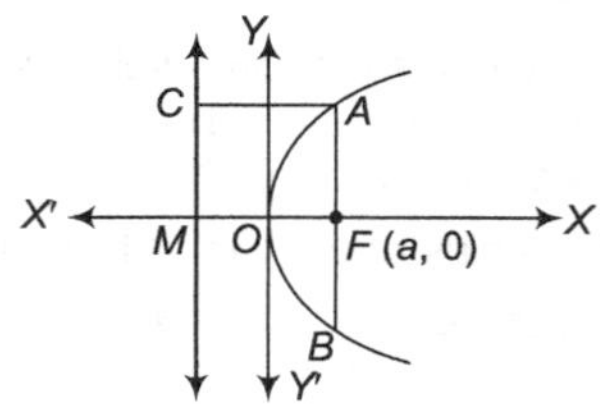

By the definition of the parabola, $AF = AC$.

But $AC = FM = 2a$

Hence, $AF = 2a$

and since the parabola is symmetric with respect to X-axis $AF = FB$ and so

AB = Length of the latusrectum $= 4a$.

55. *(d)* Given, equation of parabola is $y^2 = 12x$

which is of the form $y^2 = 4ax$

i.e., focus lies on the positive direction of X-axis.

Here, $4a = 12$

$\Rightarrow a = 3$

Focus $= (a, 0) = (3, 0)$

Axis $= X$-axis

Directrix, $x = -a$

$\Rightarrow x = -3$

Length of latusrectum $= 4a = 4 \times 3 = 12$

56. *(b)* If equation of parabola is $x^2 = 6y$ which is of the form $x^2 = 4ay$ i.e., focus lies on the positive direction of Y-axis.

Here, $4a = 6 \Rightarrow a = \frac{6}{4} = \frac{3}{2}$

Focus $= (0, a) = \left(0, \frac{3}{2}\right)$

Axis $= Y$-axis

Directrix, $y = -a \Rightarrow y = -\frac{3}{2}$

Length of latusrectum $= 4a = 4 \times \frac{3}{2} = 6$

If equation of parabola is $y^2 = -8x$ which is of the form $y^2 = -4ax$ i.e., focus lies on the negative direction of X-axis.

Here, $4a = 8 \Rightarrow a = 2$

$$\text{Focus} = (-a, 0) = (-2, 0)$$

$$\text{Axis} = X\text{-axis}$$

$$\text{Directrix, } x = a \Rightarrow x = 2$$

$$\text{Length of latusrectum} = 4a = 4 \times 2 = 8$$

57. *(a)* A. If equation of parabola is $x^2 = -16y$, which is of the form $x^2 = -4ay$, *i.e.,* focus lies on the negative direction of Y-axis.

Here, $4a = 16 \Rightarrow a = 4$

$$\text{Focus} = (0, -a) = (0, -4)$$

$$\text{Axis} = Y\text{-axis}$$

$$\text{Directrix, } y = a \Rightarrow y = 4$$

$$\text{Length of latusrectum} = 4a = 4 \times 4 = 16$$

B. If equation of parabola is $y^2 = 10x$, which is of the form $y^2 = 4ax$ *i.e.,* focus lies on the positive direction of X-axis.

Here, $4a = 10 \Rightarrow a = \frac{5}{2}$

$$\text{Focus} = (a, 0) = \left(\frac{5}{2}, 0\right)$$

$$\text{Axis} = X\text{-axis}$$

$$\text{Directrix, } x = -a \Rightarrow x = -\frac{5}{2}$$

$$\text{Length of latusrectum} = 4a = 4 \times \frac{5}{2} = 10$$

C. If equation of parabola is $x^2 = -9y$, which is of the form $x^2 = -4ay$ *i.e.,* focus lies on the negative direction of Y-axis.

Here, $4a = 9 \Rightarrow a = \frac{9}{4}$

$$\text{Focus} = (0, -a) = \left(0, -\frac{9}{4}\right)$$

$$\text{Axis} = Y\text{-axis}$$

$$\text{Directrix, } y = a \Rightarrow y = \frac{9}{4}$$

$$\text{Length of latusrectrum} = 4a = 9$$

58. *(b)* I. Given, focus (6, 0) and directrix $x = -6$

Here, we see that in focus x-coordinate is positive and y-coordinate is zero.

So, focus lies on the positive direction of X-axis *i.e.,* equation of parabola will be of the form $y^2 = 4ax$ with $a = 6$.

Hence, required equation is

$$y^2 = 4 \times 6x \Rightarrow y^2 = 24x$$

II. Given, focus $= (0, -3)$ and directrix $y = 3$

Since, focus lies on the negative direction of Y-axis *i.e.,* equation of parabola will be of the form $x^2 = -4ay$ with $a = 3$.

Hence, required equation is

$$x^2 = -4(3)y \Rightarrow x^2 = -12y$$

59. *(c)* I. Given, vertex $= (0, 0)$

Focus $= (3, 0)$

Since, vertex is (0, 0) and focus lies on positive direction of X-axis. Hence, equation of parabola will be of the form $y^2 = 4ax$ with $a = 3$.

Hence, required equation is

$$y^2 = 4 \times 3x$$

$$\Rightarrow \quad y^2 = 12x$$

II. Given, vertex $= (0, 0)$

Focus $= (-2, 0)$

Since, vertex is (0, 0) and focus lies on the negative direction of X-axis. Hence, equation of parabola will be of the form

$$y^2 = -4ax \text{ with } a = 2$$

Hence, required equation is

$$y^2 = -4(2)x$$

$$= -8x$$

60. *(a)* Given, vertex $= (0, 0)$

Point $= (2, 3)$ and axis $= X$-axis

Since, point (2, 3) lies in first quadrant and axis is X-axis, Hence, equation of parabola will be of the form $y^2 = 4ax$, which passes through (2, 3) *i.e.,*

Put $x = 2, y = 3$ in $y^2 = 4ax$

$$\therefore \quad (3)^2 = 4a \times (2)$$

$$\Rightarrow \quad a = \frac{9}{8}$$

Hence, required equation of parabola is

$$y^2 = 4\left(\frac{9}{8}\right)x$$

$$\Rightarrow \quad y^2 = \frac{9}{2}x$$

61. *(b)* Given, vertex $= (0, 0)$

Point $= (5, 2)$

Since, point (5, 2) lies in first quadrant and axis is Y-axis, therefore, parabola is symmetric with respect to Y-axis. Hence, equation of parabola will be of the form $x^2 = 4ay$, which passes through (5, 2) *i.e.,*

Put $x = 5, y = 2$ in $x^2 = 4ay$

$$\therefore \quad (5)^2 = 4a \times 2$$

$$\Rightarrow \quad a = \frac{25}{8}$$

Hence, required equation of parabola is

$$x^2 = 4 \times \frac{25}{8}y$$

$$\Rightarrow \quad x^2 = \frac{25}{2}y$$

62. *(c)* Since, the distance from the focus to the vertex is 5cm, We have, $a=5$. If the origin is taken at the vertex and the axis of the mirror lies along the positive X-axis, the equation of the parabolic section is

$$y^2 = 4(5)x = 20x$$

Note that $x = 45$. Thus, $y^2 = 900$

Therefore, $y = \pm 30$

Hence, $AB = 2y = 2 \times 30 = 60$ cm.

63. *(a)* Let the vertex be at the lowest point and the axis be vertical.

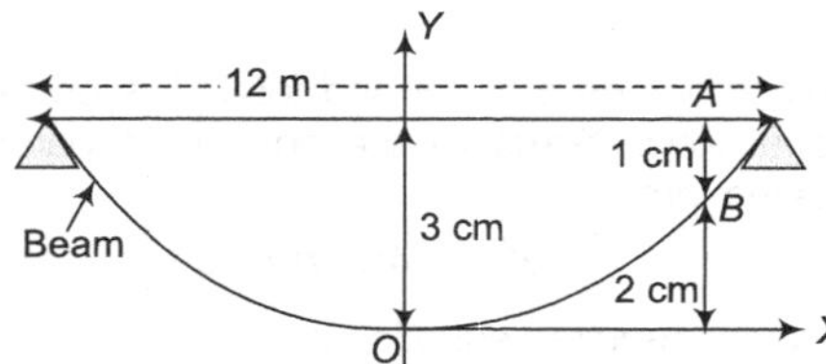

The equation of the parabola takes the form $x^2 = 4ay$. Since, it passes through $\left(6, \frac{3}{100}\right)$, we have $(6)^2 = 4a\left(\frac{3}{100}\right)$

i.e., $a = 36 \times \frac{100}{12} = 300$ m.

Let AB be the deflection of the beam which is $\frac{1}{100}$ m.

Coordinates of B are $\left(x, \frac{2}{100}\right)$.

Therefore, $x^2 = 4 \times 300 \times \frac{2}{100} = 24$

i.e., $x = \sqrt{24} = 2\sqrt{6}$ m

64. *(d)* The given equation is of the form $x^2 = -4ay$, where a is positive.

Therefore, the focus is on Y-axis in the negative direction and parabola opens downwards.

Comparing the given equation with standard form, we get $a = 2$.

Therefore, the coordinates of the focus are $(0, -2)$ and the the equation of directrix is $y = 2$ and the length of the latusrectum is $4a$, *i.e.*, 8.

65. *(c)* As shown in the figure APQ denotes the equilateral triangle with its equal side of length l (say).

Here, $AP = l$ so $AR = l\cos 30° = l\frac{\sqrt{3}}{2}$

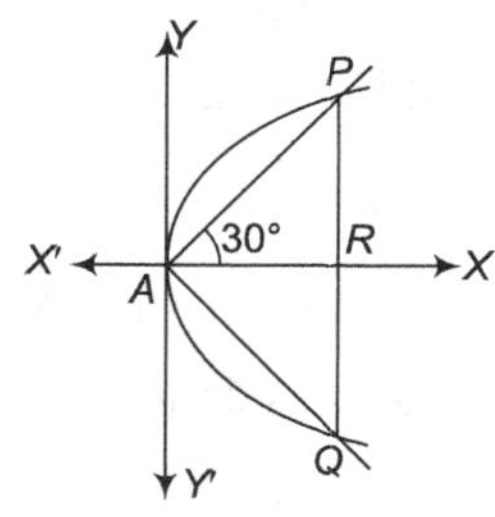

Also, $PR = l\sin 30° = \frac{l}{2}$.

Thus, $\left(\frac{l\sqrt{3}}{2}, \frac{l}{2}\right)$ are the coordinates of the point P lying on the parabola $y^2 = 4ax$.

Therefore, $\frac{l^2}{4} = 4a\left(\frac{l\sqrt{3}}{2}\right) \Rightarrow l = 8a\sqrt{3}$.

Thus, $8a\sqrt{3}$ is the required length of the side of the equilateral triangle inscribed in the parabola $y^2 = 4ax$.

66. *(c)* From the figure, OPQ represents the triangle whose area is to be determined. The area of the triangle

$$= \frac{1}{2}PQ \times OF = \frac{1}{2}(12 \times 3) = 18$$

67. *(b)* Let P and Q be points on the parabola $y^2 = 6x$ and OP, OQ be the lines joining the vertex O to the points P and Q whose abscissa are 24.

Thus, $y^2 = 6 \times 24 = 144$

or $y = \pm 12$

Therefore, the coordinates of the points P and Q are (24, 12) and (24, – 12) respectively. Hence, the lines are

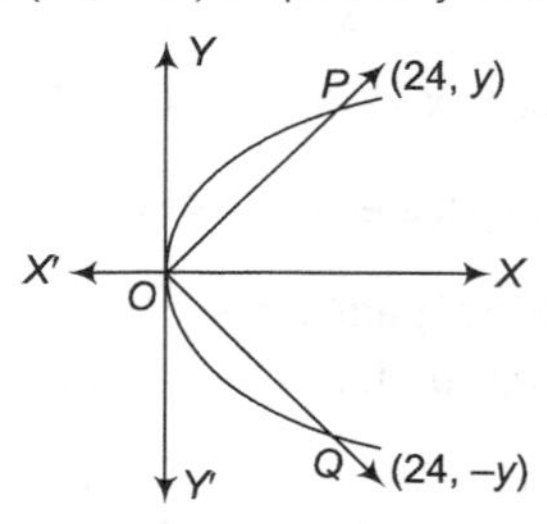

$$y = \pm\frac{12}{24}x \Rightarrow 2y = \pm x.$$

68. *(d)* We know that, the standard equation of parabola is $(y-k)^2 = 4a(x-h)$, where its origin is $(h, k,)$.

Comparing the given equation with the standard equation, we get $h = 1$, $k = 2$ and $4a = 16$

$\Rightarrow a = 4$. Therefore, vertex of the parabola $(h, k) = (1, 2)$.

69. *(a)* The given equation $9x^2 - 6x + 36y + 9 = 0$ can be written as $(3x-1)^2 = -4(9y+2)$.

Hence, the vertex is $\left(\frac{1}{3}, -\frac{2}{9}\right)$.

70. *(a)* The equation of the parabola whose vertex is (–1,–2) is given by

$$(x+1)^2 = 4a(y+2) \qquad \text{...(i)}$$

Since, Eq. (i) passes through the point (3,6),

$\therefore$ $(3+1)^2 = 4a(6+2)$

$\Rightarrow$ $16 = 4a \cdot 8$

$\Rightarrow$ $a = \frac{1}{2}$

Hence, equation of required parabola is

$$(x+1)^2 = 2(y+2)$$

$\Rightarrow \quad x^2 + 2x - 2y - 3 = 0$

71. (d) $9x^2 - 6x + 19 = -36y$

$\Rightarrow \quad (3x-1)^2 = -36y - 18 = -36\left(y + \frac{1}{2}\right)$

$\Rightarrow \quad 9\left(x - \frac{1}{3}\right) = -36\left(y + \frac{1}{2}\right)$

Hence, length of latusrectum is 4.

72. (a) Let $P(x, y)$ be any point on the parabola.

Using the definition of parabola, we have

$$\sqrt{(x-2)^2 + (y-3)^2} = \left|\frac{x - 4y + 3}{\sqrt{17}}\right|$$

Squaring, we get $17(x^2 + y^2 - 4x - 6y + 13)$

$$= x^2 + 16y^2 + 9 - 8xy - 24y + 6x$$

or $\quad 16x^2 + y^2 + 8xy - 74x - 78y + 212 = 0$

73. (c) An ellipse is the set of all points in a plane, the **sum** of whose distances from two fixed points in the plane is a constant. The two fixed points are called the **foci** of the ellipse.

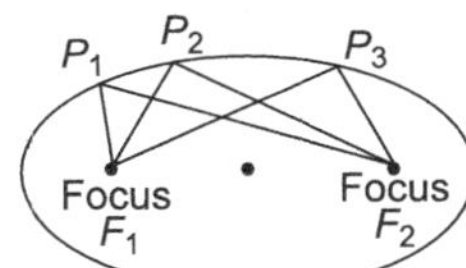

$P_1F_1 + P_1F_2 = P_2F_1 + P_2F_2 = P_3F_1 + P_3F_1$

74. (a) The constant which is the sum of the distances of a point on the ellipse from the two fixed points is always **greater than** the distance between the two fixed points.

75. (a) The mid-point of the line segment joining the foci is called the centre of the ellipse. The line segment through the foci of the ellipse is called the major axis and the line segment through the centre and perpendicular to the major axis is called the minor axis. The end points of the major axis are called the vertices of the ellipse.

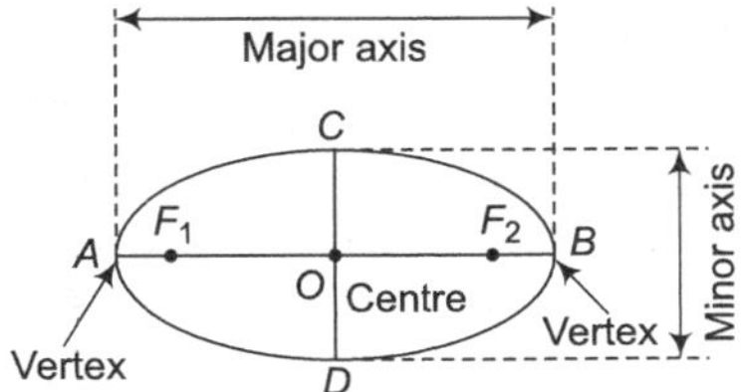

76. (b) We denote the length of the major axis by $2a$, the length of the minor axis by $2b$ and the distance between the foci by $2c$.

77. (d) Length of semi-major and minor axes are half of the lengths of major and minor axes respectively. The length of the semi-major axis is 'a' and semi-minor axis is 'b'.

78. (c) Take a point P at one end of the major axis. Sum of the distances of the point P to the foci is $F_1P + F_2P = F_1O + OP + F_2P$ (since, $F_1P = F_1O + OP$)

$$= c + a + a - c = 2a$$

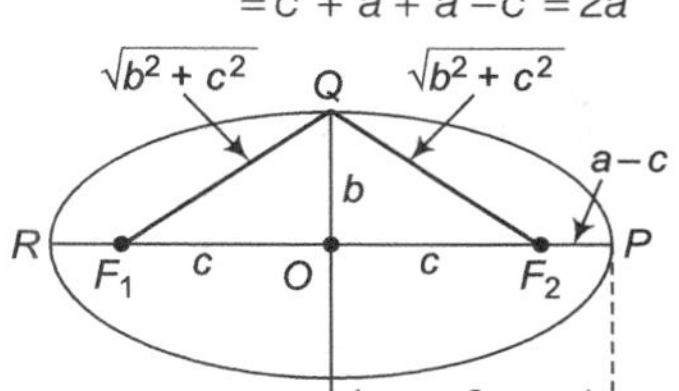

Take a point Q at one end of the minor axis.

Sum of the distances from the point Q to the foci is

$$F_1Q + F_2Q = \sqrt{b^2 + c^2} + \sqrt{b^2 + c^2} = 2\sqrt{b^2 + c^2}$$

Since, both P and Q lie on the ellipse,

By the definition of ellipse, we have

$$2\sqrt{b^2 + c^2} = 2a, \quad i.e., a = \sqrt{b^2 + c^2}$$

or $\quad a^2 = b^2 + c^2, i.e., c = \sqrt{a^2 - b^2}$

79. (a) I. When $c = 0$, both foci merge together with the centre of the ellipse and $a^2 = b^2$, *i.e.*, $a = b$, and so the ellipse becomes circle . Thus, circle is a special case of an ellipse

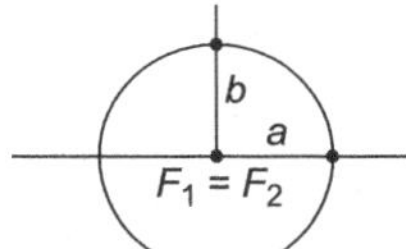

II. When $c = a$, then $b = 0$. The ellipse reduces to the line segment F_1F_2 joining the two foci

a a
F_1 F_2

80. (c) The eccentricity of an ellipse is the ratio of the distances from the centre of the ellipse to one of the foci and to one of the vertices of the ellipse (eccentricity is denoted by e) *i.e.*, $e = \frac{c}{a}$.

81. (a) Since, the focus is at a distance of c from the centre, in terms of the eccentricity the focus is at a distance of ae from the centre.

82. (b) The equation of an ellipse is simplest, if the centre of the ellipse is at the origin and the foci are on the X-axis or Y-axis. The two such possible orientations are shown in figure.

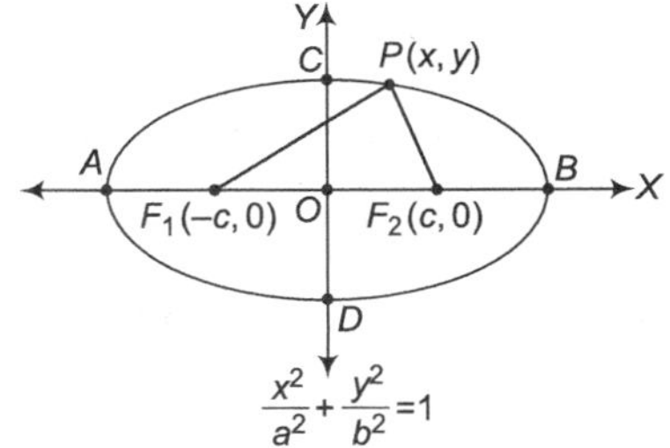

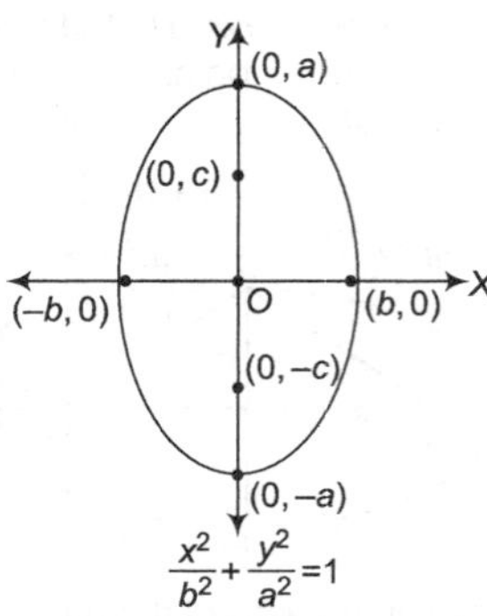

$$\frac{x^2}{b^2}+\frac{y^2}{a^2}=1$$

83. *(a)* We will derive the equation for the ellipse shown in figure with foci on the X-axis.

Let F_1 and F_2 be the foci and O be the mid-point of the line segment F_1F_2. Let O be the origin and the line from O through F_2 be the positive X-axis and the through F_1 as the negative X-axis.

Let, the line through O perpendicular to the X-axis be the Y-axis. Let the coordinates of F_1 be $(-c, 0)$ and F_2 be $(c, 0)$.

Let $P(x, y)$ be any point on the ellipse such that the sum of the distances from P to the two foci be $2a$ so given

$$PF_1 + PF_2 = 2a \quad \ldots(i)$$

Using the distance formula, we have

$$\sqrt{(x+c)^2+y^2}+\sqrt{(x-c)^2+y^2}=2a$$

i.e., $$\sqrt{(x+c)^2+y^2}=2a-\sqrt{(x-c)^2+y^2}$$

Squaring both sides, we get

$$(x+c)^2+y^2=4a^2-4a\sqrt{(x-c)^2+y^2}+(x-c)^2+y^2$$

which on simplification gives $\sqrt{(x-c)^2+y^2}=a-\frac{c}{a}x$

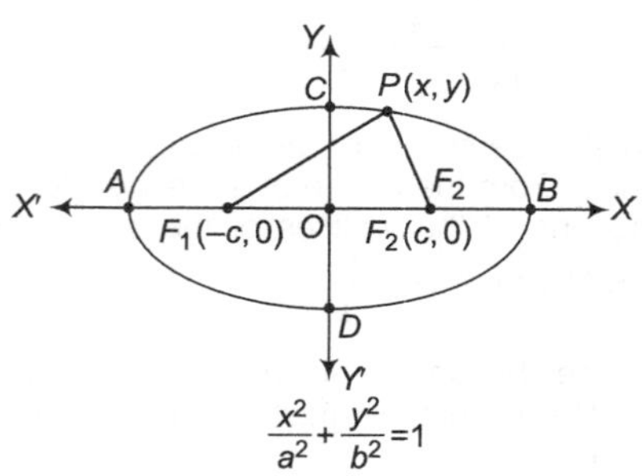

$$\frac{x^2}{a^2}+\frac{y^2}{b^2}=1$$

Squaring again and simplifying, we get

$$\frac{x^2}{a^2}+\frac{y^2}{a^2-c^2}=1$$

i.e., $$\frac{x^2}{a^2}+\frac{y^2}{b^2}=1 \quad (\text{since, } c^2=a^2-b^2)$$

Hence, any point on the ellipse satisfies

$$\frac{x^2}{a^2}+\frac{y^2}{b^2}=1 \quad \ldots(ii)$$

Conversely, let $P(x, y)$ satisfy the Eq. (ii) with $0 < c < a$.

Then, $$y^2=b^2\left(1-\frac{x^2}{a^2}\right)$$

Therefore, $$PF_1=\sqrt{(x+c)^2+y^2}$$

$$=\sqrt{(x+c)^2+b^2\left(\frac{a^2-x^2}{a^2}\right)}$$

$$=\sqrt{(x+c)^2+(a^2-c^2)\left(\frac{a^2-x^2}{a^2}\right)}$$

(since, $b^2=a^2-c^2$)

$$=\sqrt{\left(a+\frac{cx}{a}\right)^2}=a+\frac{c}{a}x$$

Similarly, $PF_2=a-\frac{c}{a}x$

Hence, $$PF_1+PF_2=a+\frac{c}{a}x+a-\frac{c}{a}x=2a \quad \ldots(iii)$$

So, any point that satisfies $\frac{x^2}{a^2}+\frac{y^2}{b^2}=1$, satisfies the geometric condition and so $P(x, y)$ lies on the ellipse.

Hence, from Eqs. (ii) and (iii), we proved that the equation of an ellipse with centre of the origin and major axis along the X-axis is $\frac{x^2}{a^2}+\frac{y^2}{b^2}=1$.

Discussion From the equation of the ellipse obtained above, it follows that for every point $P(x, y)$ on the ellipse, we have

$$\frac{x^2}{a^2}=1-\frac{y^2}{b^2}\le 1, \text{ i.e., } x^2\le a^2, \text{ so } -a\le x\le a$$

Therefore, the ellipse lies between the lines $x=-a$ and $x=a$ and touches these lines.

Similarly, the ellipse lies between the lines $y=-b$ and $y=b$ and touches these lines.

84. *(c)* As, if $a > b$ and major axis is along X-axis.

$\Rightarrow$ The equation of ellipse is $\frac{x^2}{a^2}+\frac{y^2}{b^2}=1$

Similarly, if $a > b$ and major axis is along Y-axis.

$\Rightarrow$ The equation of ellipse is $\frac{x^2}{b^2}+\frac{y^2}{a^2}=1$

85. *(b)* The standard equations of ellipse have **centre** at the origin and the major and minor axis are **coordinate axes**.

86. *(c)* Ellipse is symmetric with respect to both the coordinate axes, since if (x, y) is a point on the ellipse, then $(-x, y)$, $(x, -y)$ and $(-x, -y)$ are also points on the ellipse.

87. *(b)* The foci always lie on the major axis. The major axis can be determined by finding the intercepts on the axes of symmetry. That is, major axis is along the X-axis, if the coefficient of x^2 has the larger denominator and it is along the Y-axis, if the coefficient of y^2 has the larger denominator.

88. *(a)* Latusrectum of an ellipse is a line segment **perpendicular** to the major axis through any of the **foci** and whose end points lie on the ellipse.

89. *(b)* To find the length of the latusrectum of the ellipse $\frac{x^2}{a^2}+\frac{y^2}{b^2}=1$.

Let the length of AF_2 be l.

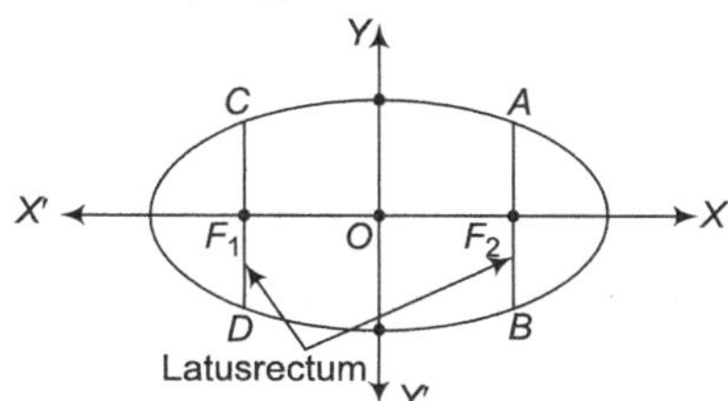

Then, the coordinates of A are (c, l), *i.e.*, (ae, l)

Since, A lies on the ellipse $\frac{x^2}{a^2}+\frac{y^2}{b^2}=1$, we have

$$\frac{(ae)^2}{a^2}+\frac{l^2}{b^2}=1$$

$$\Rightarrow \quad l^2=b^2(1-e^2)$$

But $$e^2=\frac{c^2}{a^2}=\frac{a^2-b^2}{a^2}=1-\frac{b^2}{a^2}$$

$$\Rightarrow \quad \frac{b^2}{a^2}=l-e^2$$

Therefore, $l^2=\frac{b^4}{a^2}$, *i.e.*, $l=\frac{b^2}{a}$

Since, the ellipse is symmetric with respect to Y-axis (of course, it is symmetric w.r.t both the coordinate axes),

$AF_2=F_2B$ and so length of the latusrectum is $\frac{2b^2}{a}$.

90. *(c)* A. Given, $\frac{x^2}{36}+\frac{y^2}{16}=1$

$\because$ Denominator of $\frac{x^2}{36}$ is larger than the denominator of $\frac{y^2}{16}$.

$\therefore$ The major axis is along X-axis. On comparing the given equation with $\frac{x^2}{a^2}+\frac{y^2}{b^2}=1$, we get

$$a^2=36 \quad \text{and} \quad b^2=16$$

$$\Rightarrow \quad a=6 \quad \text{and} \quad b=4$$

Here, a and b are lengths, so we take only positive sign.

Now, $c^2=a^2-b^2=36-16=20$

$$\Rightarrow \quad c=2\sqrt{5}$$

Here, major axis is along X-axis.

$$\text{Foci}=(\pm c, 0)=(\pm 2\sqrt{5}, 0)$$

$$\text{Vertices}=(\pm a, 0)=(\pm 6, 0)$$

Length of major axis $=2a=2\times 6=12$

Length of minor axis $=2b=2\times 4=8$

$$\text{Eccentricity, } e=\frac{c}{a}=\frac{2\sqrt{5}}{6}=\frac{\sqrt{5}}{3}$$

Length of latusrectum $=\frac{2b^2}{a}=\frac{2\times 16}{6}=\frac{16}{3}$

B. Given, equation of ellipse is

$$\frac{x^2}{4}+\frac{y^2}{25}=1$$

$\because$ Denominator of $\frac{x^2}{4}$ is smaller than the denominator of $\frac{y^2}{25}$.

$\therefore$ The major axis is along Y-axis.

On comparing the given equation with $\frac{x^2}{b^2}+\frac{y^2}{a^2}=1$, we get

$$b^2=4 \quad \text{and} \quad a^2=25$$

$$\Rightarrow \quad b=2 \quad \text{and} \quad a=5$$

Here, a and b are lengths, so we take only positive sign.

$\because \quad c^2=a^2-b^2=25-4=21 \Rightarrow c=\sqrt{21}$

Here, major axis is along Y-axis. So,

$$\text{Foci}=(0, \pm c)=(0, \pm\sqrt{21})$$

Vertices $=(0, \pm a)=(0, \pm 5)$

Length of major axis $=2a=2\times 5=10$

Length of minor axis $=2b=2\times 2=4$

Eccentricity, $e=\frac{c}{a}=\frac{\sqrt{21}}{5}$

Length of latusrectum $=\frac{2b^2}{a}=\frac{2\times 4}{5}=\frac{8}{5}$

C. Given, equation of ellipse is $\frac{x^2}{16}+\frac{y^2}{9}=1$

$\because$ Denominator of $\frac{x^2}{16}$ is greater than the denominator of $\frac{y^2}{9}$.

$\therefore$ The major axis is along X-axis.

On comparing given equation with $\frac{x^2}{a^2}+\frac{y^2}{b^2}=1$, we get

$$a^2=16 \quad \text{and} \quad b^2=9$$

$$\Rightarrow \quad a=4 \quad \text{and} \quad b=3$$

Here, a and b are lengths, so we take only positive sign.

Now, $c^2=a^2-b^2$

$$=16-9=7 \Rightarrow c=\sqrt{7}$$

Here, major axis is along X-axis. So,

$$\text{Foci}=(\pm c, 0)=(\pm\sqrt{7}, 0)$$

$$\text{Vertices}=(\pm a, 0)=(\pm 4, 0)$$

Length of major axis $=2a=2\times 4=8$

Length of minor axis $=2b=2\times 3=6$

$$\text{Eccentricity, } e=\frac{c}{a}=\frac{\sqrt{7}}{4}$$

Length of latusrectum $=\frac{2b^2}{a}=\frac{2\times 9}{4}=\frac{9}{2}$

D. Given, equation of ellipse is $\frac{x^2}{25}+\frac{y^2}{100}=1$

$\because$ Denominator of $\frac{x^2}{25}$ is smaller than the denominator of $\frac{y^2}{100}$.

$\therefore$ The major axis is along Y-axis.

On comparing given equation with $\frac{x^2}{b^2}+\frac{y^2}{a^2}=1$, we get

$$b^2=25 \quad \text{and} \quad a^2=100$$

$$\Rightarrow \quad b=5 \quad \text{and} \quad a=10$$

Here, a and b are lengths, so we take only positive sign.

Now, $c^2=a^2-b^2$

$=100-25$

$=75=25\times3$

$\Rightarrow \quad c=5\sqrt{3}$

Here, major axis is along Y-axis. So,

Foci $=(0, \pm c)=(0, \pm 5\sqrt{3})$

Vertices $=(0, \pm a)=(0, \pm 10)$

Length of major axis $=2a=2\times10=20$

Length of minor axis $=2b=2\times5=10$

Eccentricity, $e=\frac{c}{a}=\frac{5\sqrt{3}}{10}=\frac{\sqrt{3}}{2}$

Length of latusrectum $=\frac{2b^2}{a}=\frac{2\times25}{10}=5$

91. *(c)* Given, equation of ellipse is $\frac{x^2}{49}+\frac{y^2}{36}=1$

$\because$ Denominator of $\frac{x^2}{49}$ is larger than the denominator of $\frac{y^2}{26}$.

$\therefore$ The major axis is along X-axis.

On comparing given equation with $\frac{x^2}{a^2}+\frac{y^2}{b^2}=1$, we get

$$a^2=49, b^2=36 \Rightarrow a=7, b=6$$

Here, a and b are lengths, so we take only positive sign.

Now, $c^2=a^2-b^2=49-36=13$

$\Rightarrow \quad c=\sqrt{13}$

Here, major axis is along X-axis. So,

Foci $=(\pm c, 0)=(\pm\sqrt{13}, 0)$

Vertices $=(\pm a, 0)=(\pm 7, 0)$

Length of major axis $=2a=2\times7=14$

Length of minor axis $=2b=2\times6=12$

Eccentricity, $e=\frac{c}{a}=\frac{\sqrt{13}}{7}$

Length of latusrectum $=\frac{2b^2}{a}$

$=\frac{2\times36}{7}=\frac{72}{7}$

92. *(a)* A. Given, equation of ellipse is

$$\frac{x^2}{100}+\frac{y^2}{400}=1$$

$\because$ Denominator of $\frac{x^2}{100}$ is smaller than the denominator of $\frac{y^2}{400}$.

$\therefore$ The major axis is along Y-axis.

On comparing given equation with $\frac{x^2}{b^2}+\frac{y^2}{a^2}=1$,

we get, $b^2=100$ and $a^2=400$

$\Rightarrow \quad b=10$ and $a=20$

Here, a and b are lengths, so we take only positive sign.

Now, $c^2=a^2-b^2=400-100=300$

$\Rightarrow \quad c=10\sqrt{3}$

Here, major axis is along Y-axis. So,

Foci $=(0, \pm c)=(0, \pm 10\sqrt{3})$

Vertices $=(0, \pm a)=(0, \pm 20)$

Length of major axis $=2a=2\times20=40$

Length of minor axis $=2b=2\times10=20$

Eccentricity, $e=\frac{c}{a}=\frac{10\sqrt{3}}{20}=\frac{\sqrt{3}}{2}$

Length of latusrectum $=\frac{2b^2}{a}=\frac{2\times100}{20}=10$

B. Given, equation of ellipse is

$$36x^2+4y^2=144$$

$$\Rightarrow \quad \frac{36x^2}{144}+\frac{4y^2}{144}=1 \quad \text{(divide by 144)}$$

$$\Rightarrow \quad \frac{x^2}{4}+\frac{y^2}{36}=1$$

$\because$ Denominator of $\frac{x^2}{4}$ is smaller than the denominator of $\frac{y^2}{36}$.

$\therefore$ The major axis is along Y-axis.

On comparing given equation with $\frac{x^2}{b^2}+\frac{y^2}{a^2}=1$, we get

$$b^2=4 \quad \text{and} \quad a^2=36$$

$$\Rightarrow \quad b=2 \quad \text{and} \quad a=6$$

Here, a and b are lengths, so we take only positive sign.

and $c^2=a^2-b^2=36-4=32$

$\Rightarrow \quad c=4\sqrt{2}$

Here, major axis is along Y-axis. So,

Foci $=(0, \pm c)=(0, \pm 4\sqrt{2})$

Vertices $=(0, \pm a)=(0, \pm 6)$

Length of major axis $=2a=2\times6=12$

Length of minor axis $=2b=2\times2=4$

$$\text{Eccentricity, } e = \frac{c}{a} = \frac{4\sqrt{2}}{6} = \frac{2\sqrt{2}}{3}$$

$$\text{Length of latusrectum} = \frac{2b^2}{a} = \frac{2 \times 4}{6} = \frac{4}{3}$$

C. Given, equation of ellipse is

$$16x^2 + y^2 = 16$$

$$\Rightarrow \quad \frac{16x^2}{16} + \frac{y^2}{16} = 1 \quad \text{(divide by 16)}$$

$$\Rightarrow \quad \frac{x^2}{1} + \frac{y^2}{16} = 1$$

$\because$ Denominator of $\frac{x^2}{1}$ is smaller than the denominator of of $\frac{y^2}{16}$.

$\therefore$ The major axis is along Y-axis.

On comparing given equation with $\frac{x^2}{b^2} + \frac{y^2}{a^2} = 1$, we get

$$b^2 = 1 \text{ and } a^2 = 16$$

$$\Rightarrow \quad b = 1 \text{ and } a = 4$$

Here, a and b are lengths, so we take only positive sign.

Now, $c^2 = a^2 - b^2$

$$= 16 - 1 = 15$$

$$\Rightarrow \quad c = \sqrt{15}$$

Here, major axis is along Y-axis. So,

$$\text{Foci} = (0, \pm c) = (0, \pm \sqrt{15})$$

$$\text{Vertices} = (0, \pm a) = (0, \pm 4)$$

$$\text{Length of major axis} = 2a = 2 \times 4 = 8$$

$$\text{Length of minor axis} = 2b = 2 \times 1 = 2$$

$$\text{Eccentricity, } e = \frac{c}{a} = \frac{\sqrt{15}}{4}$$

$$\text{Length of latusrectum} = \frac{2b^2}{a} = \frac{2 \times 1}{4} = \frac{1}{2}$$

D. Given, equation of ellipse is

$$4x^2 + 9y^2 = 36$$

$$\Rightarrow \quad \frac{4x^2}{36} + \frac{9y^2}{36} = 1$$

$$\Rightarrow \quad \frac{x^2}{9} + \frac{y^2}{4} = 1$$

$\because$ Denominator of $\frac{x^2}{9}$ is greater than denominator of $\frac{y^2}{4}$.

$\therefore$ The major axis is along X-axis. On comparing given equation with $\frac{x^2}{a^2} + \frac{y^2}{b^2} = 1$, we get

$$a^2 = 9 \text{ and } b^2 = 4$$

$$\Rightarrow \quad a = 3 \text{ and } b = 2$$

Now, $c^2 = a^2 - b^2 = 9 - 4 = 5$

$$\Rightarrow \quad c = \sqrt{5}$$

Here, major axis is along X-axis. So,

$$\text{Foci} = (\pm c, 0) = (\pm \sqrt{5}, 0)$$

$$\text{Vertices} = (\pm a, 0) = (\pm 3, 0)$$

$$\text{Length of major axis} = 2a = 2 \times 3 = 6$$

$$\text{Length of minor axis} = 2b = 2 \times 2 = 4$$

$$\text{Eccentricity, } e = \frac{c}{a} = \frac{\sqrt{5}}{3}$$

$$\text{Length of latusrectum} = \frac{2b^2}{a} = \frac{2 \times 4}{3} = \frac{8}{3}$$

93. *(b)* A. Since, vertices $(\pm 5, 0)$ and foci $(\pm 4, 0)$ are on X-axis, as y-coordinate of both points is zero. Hence, equation of ellipse will be of the form

$$\frac{x^2}{a^2} + \frac{y^2}{b^2} = 1 \quad \dots(i)$$

where, $a = 5$ and $c = 4$

Now, $c^2 = a^2 - b^2$

$$\therefore \quad (4)^2 = (5)^2 - b^2$$

$$\Rightarrow \quad 16 = 25 - b^2$$

$$\Rightarrow \quad b^2 = 25 - 16$$

$$\Rightarrow \quad b^2 = 9$$

Put the values of $a^2 = 25$ and $b^2 = 9$ in Eq. (i), we get

$$\frac{x^2}{25} + \frac{y^2}{9} = 1$$

B. Since, vertices $(0, \pm 13)$ and foci $(0, \pm 5)$ are on Y-axis, x-coordinate of both points is zero.

Hence, equation of ellipse will be of the form

$$\frac{x^2}{b^2} + \frac{y^2}{a^2} = 1 \quad \dots(i)$$

where, $a = 13$ and $c = 5$

$$\because \quad c^2 = a^2 - b^2$$

$$\Rightarrow \quad (5)^2 = (13)^2 - b^2$$

$$\Rightarrow \quad 25 = 169 - b^2$$

$$\Rightarrow \quad b^2 = 169 - 25$$

$$\Rightarrow \quad b^2 = 144$$

Put the values of $a^2 = 169$ and $b^2 = 144$ in Eq. (i), we get

$$\frac{x^2}{144} + \frac{y^2}{169} = 1$$

C. Since, vertices $(\pm 6, 0)$ and foci $(\pm 4, 0)$ are on X-axis, as y-coordinate of both points is zero. Hence, equation of ellipse will be of the form

$$\frac{x^2}{a^2} + \frac{y^2}{b^2} = 1 \quad \dots(i)$$

where, $a = 6$

and $c = 4$

$$\because \quad c^2 = a^2 - b^2$$

$\Rightarrow \quad (4)^2 = (6)^2 - b^2$

$\Rightarrow \quad 16 = 36 - b^2$

$\Rightarrow \quad b^2 = 36 - 16 = 20$

Put the values of $a^2 = 36$ and $b^2 = 20$ in Eq. (i), we get

$$\frac{x^2}{36} + \frac{y^2}{20} = 1$$

94. *(a)* I. Since, ends of major axis ($\pm$ 3, 0) are along X-axis and minor axis (0, $\pm$ 2) are along Y-axis. Hence, equation of ellipse will be of the form

$$\frac{x^2}{a^2} + \frac{y^2}{b^2} = 1 \quad \ldots(i)$$

where, $a = 3$

and $b = 2$

Hence, Eq. (i) becomes $\frac{x^2}{3^2} + \frac{y^2}{2^2} = 1$

or $\frac{x^2}{9} + \frac{y^2}{4} = 1$

II. Since, ends of major axis (0, $\pm \sqrt{5}$) are along Y-axis and minor axis ($\pm$ 1, 0) are along X-axis. Hence, equation of ellipse will be of the form

$$\frac{x^2}{b^2} + \frac{y^2}{a^2} = 1 \quad \ldots(i)$$

where, $a = \sqrt{5}$

and $b = 1$

Hence, Eq. (i) becomes $\frac{x^2}{1^2} + \frac{y^2}{(\sqrt{5})^2} = 1$

or $\frac{x^2}{1} + \frac{y^2}{5} = 1$

$\Rightarrow \quad 5x^2 + y^2 = 5$

95. *(d)* I. Since, foci ($\pm$ 5, 0) lie on X-axis, as y-coordinate is zero. Hence, equation of ellipse will be of the form

$$\frac{x^2}{a^2} + \frac{y^2}{b^2} = 1 \quad \ldots(i)$$

Given that, length of major axis $2a = 26$

$\Rightarrow \quad a = 13$ and $c = 5$

$\because \quad c^2 = a^2 - b^2$

$\Rightarrow \quad (5)^2 = (13)^2 - b^2 \Rightarrow 25 = 169 - b^2$

$\Rightarrow \quad b^2 = 169 - 25 = 144$

Put the values of $a^2 = 169$ and $b^2 = 144$ in Eq. (i), we get

$$\frac{x^2}{169} + \frac{y^2}{144} = 1$$

II. Since, foci (0, $\pm$ 6) lie on Y-axis, as x-coordinate is zero. Hence, equation of ellipse will be of the form

$$\frac{x^2}{b^2} + \frac{y^2}{a^2} = 1 \quad \ldots(i)$$

Given that, length of minor axis $2b = 16$ and $c = 6$

$\Rightarrow \quad b = 8$ and $c = 6$

$\because \quad c^2 = a^2 - b^2 \Rightarrow (6)^2 = a^2 - (8)^2$

$\Rightarrow \quad a^2 = 36 + 64 = 100$

Put the values of $a^2 = 100$ and $b^2 = 64$ in Eq. (i), we get

$$\frac{x^2}{64} + \frac{y^2}{100} = 1$$

96. *(a)* Since, foci ($\pm$ 3, 0) lie on X-axis, as y-coordinate is zero. Hence, equation of ellipse will be of the form

$$\frac{x^2}{a^2} + \frac{y^2}{b^2} = 1 \quad \ldots(i)$$

Given that, foci ($\pm c$, 0) = ($\pm$ 3, 0)

$c = 3$ and $a = 4$

$\because \quad c^2 = a^2 - b^2$

$\Rightarrow \quad (3)^2 = (4)^2 - b^2$

$\Rightarrow \quad 9 = 16 - b^2$

$\Rightarrow \quad b^2 = 16 - 9$

$\Rightarrow \quad b^2 = 7$

Put the values of $a^2 = 16$ and $b^2 = 7$ in Eq. (i), we get

$$\frac{x^2}{16} + \frac{y^2}{7} = 1$$

97. *(b)* Since, foci lie on X-axis, Hence, equation of ellipse will be of the form

$$\frac{x^2}{a^2} + \frac{y^2}{b^2} = 1 \quad \ldots(i)$$

Given that, $b = 3, c = 4$

$\because \quad c^2 = a^2 - b^2$

$\Rightarrow \quad (4)^2 = a^2 - (3)^2$

$\Rightarrow \quad 16 = a^2 - 9 \Rightarrow a^2 = 16 + 9$

$\Rightarrow \quad a^2 = 25$

Put the values of $a^2 = 25$ and $b^2 = 9$ in Eq. (i), we get

$$\frac{x^2}{25} + \frac{y^2}{9} = 1$$

98. *(d)* I. Since, major axis is along Y-axis,

Hence, equation of ellipse will be of the form

$$\frac{x^2}{b^2} + \frac{y^2}{a^2} = 1 \quad \ldots(i)$$

Given that, Eq. (i) passes through the points (3, 2) and (1, 6) *i.e.,* they will satisfy it

$\therefore \quad \frac{3^2}{b^2} + \frac{2^2}{a^2} = 1$

$\Rightarrow \quad \frac{9}{b^2} + \frac{4}{a^2} = 1 \quad \ldots(ii)$

and $\frac{1^2}{b^2} + \frac{6^2}{a^2} = 1$

$\Rightarrow \quad \frac{1}{b^2} + \frac{36}{a^2} = 1 \quad \ldots(iii)$

Multiplying Eq. (ii) by 9 and then subtract the Eq. (iii) from it, we get

$$\frac{80}{b^2} = 8 \Rightarrow b^2 = \frac{80}{8}$$

$$\Rightarrow \quad b^2 = 10$$

From Eq. (ii), we get

$$\frac{9}{10} + \frac{4}{a^2} = 1$$

$$\Rightarrow \quad \frac{4}{a^2} = 1 - \frac{9}{10}$$

$$\Rightarrow \quad \frac{4}{a^2} = \frac{1}{10}$$

$$\Rightarrow \quad a^2 = 40$$

Put the values of $a^2 = 40$ and $b^2 = 10$ in Eq. (i), we get

$$\frac{x^2}{10} + \frac{y^2}{40} = 1$$

II. Since, major axis is along X-axis. Hence, equation of ellipse will be of the form

$$\frac{x^2}{a^2} + \frac{y^2}{b^2} = 1 \quad \ldots(i)$$

Given that, Eq. (i) passes through the points (4, 3) and (6, 2) *i.e.,* they will satisfy it

$$\therefore \quad \frac{(4)^2}{a^2} + \frac{(3)^2}{b^2} = 1$$

$$\Rightarrow \quad \frac{16}{a^2} + \frac{9}{b^2} = 1 \quad \ldots(ii)$$

and

$$\frac{(6)^2}{a^2} + \frac{(2)^2}{b^2} = 1$$

$$\Rightarrow \quad \frac{36}{a^2} + \frac{4}{b^2} = 1 \quad \ldots(iii)$$

Multiply Eq. (ii) by 4 and Eq. (iii) by 9, then subtracting, we get

$$\frac{64}{a^2} - \frac{324}{a^2} = 4 - 9$$

$$\Rightarrow \quad -\frac{260}{a^2} = -5$$

$$\Rightarrow \quad a^2 = \frac{260}{5}$$

$$\Rightarrow \quad a^2 = 52$$

From Eq. (ii), we get

$$\frac{9}{b^2} = 1 - \frac{16}{52}$$

$$\Rightarrow \quad \frac{9}{b^2} = \frac{52-16}{52}$$

$$\Rightarrow \quad \frac{9}{b^2} = \frac{36}{52}$$

$$\Rightarrow \quad b^2 = \frac{9 \times 52}{36} = 13$$

Put the values of $a^2 = 52$ and $b^2 = 13$ in Eq. (i), we get

$$\frac{x^2}{52} + \frac{y^2}{13} = 1$$

99. *(d)* We put the equation in standard form by dividing by 225 and get

$$\frac{x^2}{25} + \frac{y^2}{9} = 1$$

This shows that $a = 5$ and $b = 3$. Hence, $9 = 25(1 - e^2)$, so $e = \frac{4}{5}$. Since, the denominator of x^2 is larger, the major axis is along X-axis and minor axis along Y-axis, foci are (4, 0) and (–4, 0) and vertices are (5, 0) and (–5, 0).

100. *(a)* We have, $ae = 5$, $\frac{a}{e} = \frac{36}{5}$ which give $a^2 = 36$ or $a = 6$.

$\because$ directrix $x = \pm\frac{a}{e}$. Therefore, $e = \frac{5}{6}$.

Now, $b = a\sqrt{1-e^2} = 6\sqrt{1-\frac{25}{36}} = \sqrt{11}$. Thus, the equation of the ellipse is $\frac{x^2}{36} + \frac{y^2}{11} = 1$

101. *(b)* $\because$ The centre of the ellipse is (2, – 3)

and foci = (3,–3) $\Rightarrow$ $ae = 3 - 2 = 1$

Vertex = (4, – 3) $\Rightarrow$ $a = 4 - 2 = 2$ $\Rightarrow$ $e = \frac{1}{2}$

$$\Rightarrow \quad b = a\sqrt{\left(1 - \frac{1}{4}\right)}$$

$$= \frac{2}{2}\sqrt{3} = \sqrt{3}$$

Therefore, equation of ellipse with centre (2, –3) is

$$\frac{(x-2)^2}{4} + \frac{(y+3)^2}{3} = 1$$

102. *(c)* Equation $x^2 + 2y^2 - 2x + 3y + 2 = 0$ can be written as

$$\frac{(x-1)^2}{2} + \left(y + \frac{3}{4}\right)^2 = \frac{1}{16}$$

$$\Rightarrow \quad \frac{(x-1)^2}{(1/8)} + \frac{\left(y + \frac{3}{4}\right)^2}{(1/16)} = 1$$

which is an ellipse with $a^2 = \frac{1}{8}$ and $b^2 = \frac{1}{16}$.

$$\therefore \quad \frac{1}{16} = \frac{1}{8}(1 - e^2)$$

$$\Rightarrow \quad e^2 = 1 - \frac{1}{2}$$

$$\Rightarrow \quad e = \frac{1}{\sqrt{2}}$$

103. *(c)* Given that, the equation of conic

$$9x^2 + 4y^2 - 6x + 4y + 1 = 0$$

$$\Rightarrow \quad (3x-1)^2 + (2y+1)^2 = 1$$

$$\Rightarrow \quad \frac{\left(x - \frac{1}{3}\right)^2}{\frac{1}{9}} + \frac{\left(y + \frac{1}{2}\right)^2}{\frac{1}{4}} = 1$$

Here, $a = \frac{1}{3}, b = \frac{1}{2}$

$\therefore$ $2a = \frac{2}{3}, 2b = 1.$

Therefore, lengths of axes are 1 and $\frac{2}{3}$.

104. *(a)* Let $\frac{x^2}{a^2} + \frac{y^2}{b^2} = 1$ be the equation of the ellipse passing through the point (–3, 1).

Therefore, we have $\frac{9}{a^2} + \frac{1}{b^2} = 1$

or $9b^2 + a^2 = a^2 b^2$

or $9a^2(1-e^2) + a^2 = a^2 a^2 (1-e^2)$ [using $b^2 = a^2(1-e^2)$]

$\Rightarrow 9a^2\left(1-\frac{2}{5}\right) + a^2 = a^4\left(1-\frac{2}{5}\right)$

$\Rightarrow a^2 = \frac{32}{3}$ $\left(\because e = \sqrt{\frac{2}{5}}\right)$

Again, $b^2 = a^2(1-e^2) = \frac{32}{3}\left(1-\frac{2}{5}\right) = \frac{32}{5}$

Hence, the required equation of the ellipse is

$$\frac{x^2}{\frac{32}{3}} + \frac{y^2}{\frac{32}{5}} = 1 \quad \text{or} \quad 3x^2 + 5y^2 = 32$$

105. *(b)* Let $\frac{x^2}{a^2} + \frac{y^2}{b^2} = 1$ be the equation of the ellipse.

Then, according to the given conditions, we have

$$\frac{9}{a^2} + \frac{1}{b^2} = 1$$

and $\frac{1}{a^2} + \frac{1}{b^2} = \frac{1}{4}$

which gives $a^2 = \frac{32}{3}$ and $b^2 = \frac{32}{5}$

Hence, required equation of ellipse is $3x^2 + 5y^2 = 32$.

106. *(c)* Given ellipse is $\frac{x^2}{\left(\frac{1}{3}\right)^2} + \frac{y^2}{\left(\frac{1}{2}\right)^2} = 1$

Here, $b > a$

$\therefore$ Latusrectum $= \frac{2a^2}{b} = \frac{2 \times \frac{1}{9}}{\frac{1}{2}} = \frac{4}{9}$

107. *(a)* Let $P(x_1, y_1)$ be any point

So, $\sqrt{(x_1+2)^2 + y_1^2} = \frac{2}{3}\left(x_1 + \frac{9}{2}\right)$

$\Rightarrow (x_1+2)^2 + y_1^2 = \frac{4}{9}\left(x_1 + \frac{9}{2}\right)^2$

$\Rightarrow 9[x_1^2 + y_1^2 + 4x_1 + 4] = 4\left(x_1^2 + \frac{81}{4} + 9x_1\right)$

$\Rightarrow 5x_1^2 + 9y_1^2 = 45$

$\Rightarrow \frac{x_1^2}{9} + \frac{y_1^2}{5} = 1$

Locus of (x_1, y_1) is $\frac{x^2}{9} + \frac{y^2}{5} = 1$, which is equation of an ellipse.

108. *(b)* Given that, $e = \frac{1}{2}$ and $(\pm ae, 0) = (\pm 1, 0)$

$\Rightarrow ae = 1 \Rightarrow a = 2$

Now, $b^2 = a^2(1-e^2)$

$\Rightarrow b^2 = 4\left(1-\frac{1}{4}\right) \Rightarrow b^2 = 3$

Hence, equation of ellipse is $\frac{x^2}{4} + \frac{y^2}{3} = 1.$

109. *(d)* According to given condition, $2ae = 8, \frac{2a}{e} = 18$

$\Rightarrow a = \sqrt{4 \times 9} = 6$

$\Rightarrow e = \frac{2}{3}, b = 6\sqrt{1-\frac{4}{9}} = \frac{6}{3}\sqrt{5} = 2\sqrt{5}$

Hence, the required equation is $\frac{x^2}{36} + \frac{y^2}{20} = 1$

i.e., $5x^2 + 9y^2 = 180$

110. *(b)* Given that, $\frac{2b^2}{a} = 10$ and $2ae = 2b \Rightarrow b = ae$

Again, we know that $b^2 = a^2(1-e^2)$

or $2a^2e^2 = a^2 \Rightarrow e = \frac{1}{\sqrt{2}}$ (using $b = ae$)

Thus, $a = b\sqrt{2}$

Again, $\frac{2b^2}{a} = 10$ or $b = 5\sqrt{2}$

Thus, we get $a = 10$

Therefore, the required equation of the ellipse is

$$\frac{x^2}{100} + \frac{y^2}{50} = 1$$

$\Rightarrow x^2 + 2y^2 = 100$

111. *(d)* Given that, $ae = 1, a = 2, e = \frac{1}{2}$

$\Rightarrow b = \sqrt{4\left(1-\frac{1}{4}\right)} = \sqrt{3}$

Hence, minor axis $= 2\sqrt{3}$

112. *(d)* $\frac{x^2}{25} + \frac{y^2}{16} = 1$

$\Rightarrow e = \sqrt{1-\frac{16}{25}} = \frac{3}{5}$

Therefore, directrices are $x \pm \frac{5}{3/5} = 0$

or $3x \pm 25 = 0$

113. *(d)* Given that, $e = \frac{1}{\sqrt{2}}$,

latusrectum $= \frac{2b^2}{a} = \frac{2a^2}{a}\left(1 - \frac{1}{2}\right) = a$

i.e., semi-major axis.

114. *(d)* Given that, $\frac{a}{e} - ae = 8$

Also, $e = \frac{1}{2}$

$\Rightarrow \quad a = \frac{8e}{1-e^2} = \frac{8.4}{2(3)} = \frac{16}{3}$

$\therefore \quad b = \frac{16}{3}\sqrt{\left(1 - \frac{1}{4}\right)} = \frac{16}{3}\frac{\sqrt{3}}{2} = \frac{8\sqrt{3}}{3}$

Hence, the length of minor axis is $\frac{16\sqrt{3}}{3}$.

115. *(b)* Given that, vertex (0, 7), directrix $y = 12$ of an ellipse.

$\therefore \quad b = 7$

Also, $\frac{b}{e} = 12$

$\Rightarrow \quad e = \frac{7}{12}, a = 7\sqrt{\frac{95}{144}}$

Hence, equation of ellipse is $144x^2 + 95y^2 = 4655$.

116. *(b)* According to the given condition, $\frac{2b^2}{a} = 2ae$

$\Rightarrow \quad b^2 = a^2e \quad \text{or} \quad e = \frac{b^2}{a^2}$

Also, $e = \sqrt{1 - \frac{b^2}{a^2}} \quad \text{or} \quad e^2 = 1 - e$

or $e^2 + e - 1 = 0$

Therefore, $e = \frac{-1 \pm \sqrt{5}}{2}$

As, $0 < e < 1$

$\therefore \quad e = \frac{\sqrt{5} - 1}{2}$

117. *(d)* Given, major axis = 3 (minor axis)

$\Rightarrow \quad 2a = 3(2b)$

$\Rightarrow \quad a^2 = 9b^2 = 9a^2(1 - e^2) \Rightarrow e = \frac{2\sqrt{2}}{3}$

118. *(b)* Latusrectum = 1/3 (major axis)

$\Rightarrow \quad \frac{2b^2}{a} = \frac{2a}{3} \Rightarrow a^2 = 3b^2 = 3a^2(1 - e^2)$

$\Rightarrow \quad e = \sqrt{\frac{2}{3}}$

119. *(c)* Given, $2a = 6, 2b = 4$ *i.e.*, $a = 3, b = 2$

$e^2 = 1 - \frac{b^2}{a^2} = \frac{5}{9} \Rightarrow e = \frac{\sqrt{5}}{3}$

Distance between the pins $= 2ae = 2\sqrt{5}$ cm

120. *(b)* $\frac{x^2}{2-r} + \frac{y^2}{r-5} + 1 = 0 \Rightarrow \frac{x^2}{r-2} + \frac{y^2}{5-r} = 1$

Hence, $r > 2$ and $r < 5 \Rightarrow 2 < r < 5$.

121. *(b)* A hyperbola is the set of all points in a plane, the difference of whose distances from two fixed points in the plane is a constant.

122. *(a)* The term 'difference' that is used in the definition of hyperbola means the distance to the farther point minus the distance to the closer point.

123. *(d)* The two fixed points are called the foci of the hyperbola. The mid-point of the line segment joining the foci is called the centre of the hyperbola. The line through the foci is called the transverse axis and the line through the centre and perpendicular to the transverse axis is called the conjugate axis. The points at which the hyperbola intersects the transverse axis are called the vertices of the hyperbola.

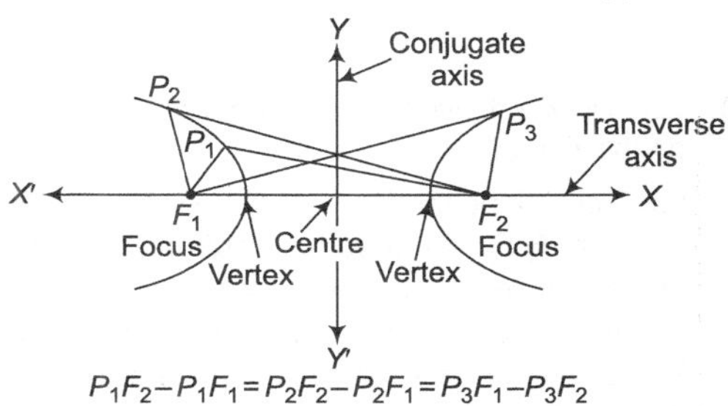

$P_1F_2 - P_1F_1 = P_2F_2 - P_2F_1 = P_3F_1 - P_3F_2$

124. *(a)* In a hyperbola, by taking the point P at A and B in the figure, we have

$BF_1 - BF_2 = AF_2 - AF_1$ (by the definition of the hyperbola)

$$BA + AF_1 - BF_2 = AB + BF_2 - AF_1$$

i.e., $AF_1 = BF_2$

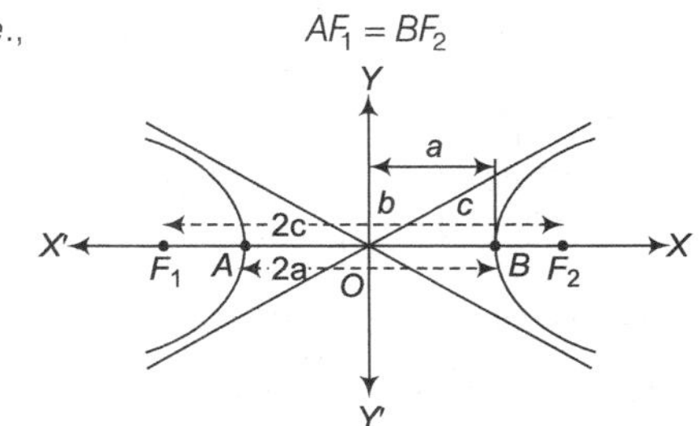

So that, $BF_1 - BF_2 = BA + AF_1 - BF_2 = BA = 2a$

We denote the distance between the two foci by 2c, the distance between two vertices (the length of the transverse axis) by 2a and we define the quantity b as

$$b = \sqrt{c^2 - a^2}$$

Also, 2b is the length of the conjugate axis.

125. *(c)* The ratio $e = \frac{c}{a}$ is called the eccentricity of the hyperbola.

126. *(c)* Since, $c \geq a$, the eccentricity is never less than one. In terms of the eccentricity the foci are at a distance of ae from the centre.

127. *(d)* The equation of a hyperbola is simplest, if the centre of the hyperbola is at the origin and the foci are on the X- axis or Y-axis.

128. *(b)* The number of possible orientations of hyperbola are two which are shown below.

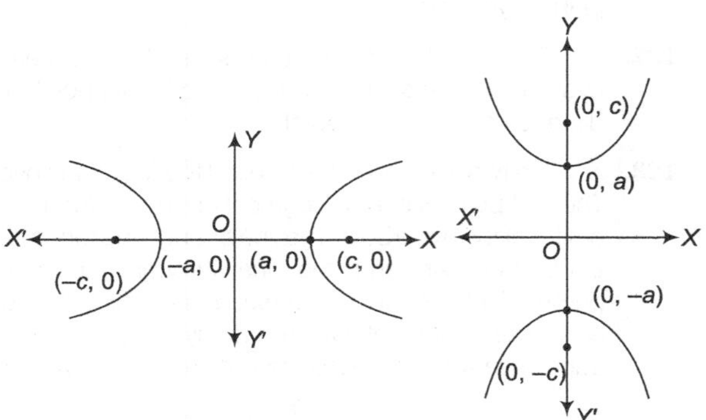

129. *(a)* We will derive the equation for the hyperbola with foci on the X-axis.

Let F_1 and F_2 be the foci and O be the mid-point of the line segment $F_1 F_2$. Let O be the origin and the line through O through F_2 be the positive X- axis and that through F_1 as the negative X-axis. The line through O perpendicular to the X-axis be the Y- axis. Let the coordinates of F_1 be $(-c, 0)$ and F_2 be $(c, 0)$. Let $P(x, y)$ be any point on the hyperbola such that the difference of the distances from P to the farther point minus the closer point be $2a$.

So given, $PF_1 - PF_2 = 2a$

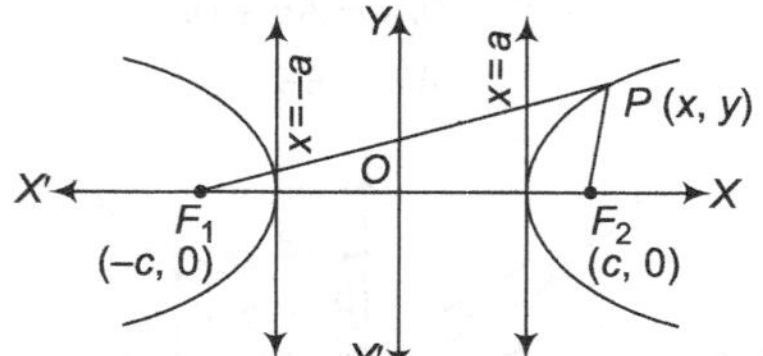

Using the distance formula, we have

$$\sqrt{(x+c)^2+y^2} - \sqrt{(x-c)^2+y^2} = 2a$$

i.e.,
$$\sqrt{(x+c)^2+y^2} = 2a + \sqrt{(x-c)^2+y^2}$$

Squaring both sides, we get

$$(x+c)^2 + y^2 = 4a^2 + 4a\sqrt{(x-c)^2+y^2} + (x-c)^2 + y^2$$

and on simplifying, we get

$$\frac{cx}{a} - a = \sqrt{(x-c)^2+y^2}$$

On squaring again and further simplifying, we get

$$\frac{x^2}{a^2} - \frac{y^2}{c^2-a^2} = 1$$

i.e.,
$$\frac{x^2}{a^2} - \frac{y^2}{b^2} = 1 \qquad (\text{since, } c^2 - a^2 = b^2)$$

Hence, any point on the hyperbola satisfies $\frac{x^2}{a^2} - \frac{y^2}{b^2} = 1$

Conversely, let $P(x, y)$ satisfy the above equation with $0 < a < c$.

Then,
$$y^2 = b^2\left(\frac{x^2-a^2}{a^2}\right)$$

Therefore,
$$PF_1 = +\sqrt{(x+c)^2+y^2}$$
$$= +\sqrt{(x+c)^2 + b^2\left(\frac{x^2-a^2}{a^2}\right)} = a + \frac{c}{a}x$$

Similarly,
$$PF_2 = a - \frac{c}{a}x$$

In hyperbola $c > a$; and since P is to the right of the line $x = a$, $x > a$, $\frac{c}{a}x > a$. Therefore, $a - \frac{c}{a}x$ becomes negative.

Thus,
$$PF_2 = \frac{c}{a}x - a.$$

Therefore,
$$PF_1 - PF_2 = a + \frac{c}{a}x - \frac{cx}{a} + a = 2a$$

Also, note that if P is to the left of the line $x = -a$, then

$$PF_1 = -\left(a + \frac{c}{a}x\right), \quad PF_2 = a - \frac{c}{a}x.$$

In that case $PF_2 - PF_1 = 2a$. So, any point that satisfies $\frac{x^2}{a^2} - \frac{y^2}{b^2} = 1$, lies on the hyperbola.

Thus, we proved that the equation of hyperbola with origin (0, 0) and transverse axis along X-axis is $\frac{x^2}{a^2} - \frac{y^2}{b^2} = 1$.

130. *(a)* If length of transverse axis is equal to the length of conjugate axis, then hyperbola is called equilateral hyperbola.

131. *(b)* From the equation of the hyperbola we have obtained, it follows that, we have for every point (x, y) on the hyperbola, $\frac{x^2}{a^2} = 1 + \frac{y^2}{b^2} \ge 1$.

i.e., $\left|\frac{x}{a}\right| \ge 1$, i.e., $x \le -a$ or $x \le a$. Therefore, no portion of the curve lies between the lines $x = +a$ and $x = -a$ (i.e. no real intercept on the conjugate axis).

132. *(d)* The equation of hyperbola when transverse axis is along X-axis is $\frac{x^2}{a^2} - \frac{y^2}{b^2} = 1$.

Similarly, equation of hyperbola when transverse axis is along Y-axis is $\frac{y^2}{a^2} - \frac{x^2}{b^2} = 1$.

133. *(b)* The standard equation of hyperbolas have transverse and conjugate axes as the coordinate axes and the centre at the origin.

134. *(d)* Hyperbola is symmetric with respect to both the axes, since if (x, y) is a point on the hyperbola, then $(-x, y)$, $(x, -y)$ and $(-x, -y)$ are also points on the hyperbola.

135. *(a)* The foci are always on the transverse axis. It is the positive term whose denominator gives the transverse axis. Thus, $\frac{x^2}{9} - \frac{y^2}{16} = 1$ has transverse axis along X-axis of length 6.

136. *(b)* If equation is $\frac{y^2}{25} - \frac{x^2}{9} = 1$.

Then, transverse axis is along Y-axis and whose length is 10.

137. *(d)* Latusrectum of hyperbola is a line segment perpendicular to the transverse axis through any of the foci and whose end points lie on the hyperbola.

138. *(b)* The length of latusrectum is $\frac{2b^2}{a}$.

139. *(b)* A. Comparing the given equation $\frac{x^2}{16} - \frac{y^2}{9} = 1$ with $\frac{x^2}{a^2} - \frac{y^2}{b^2} = 1$, we get $a^2 = 16$ and $b^2 = 9$

$\Rightarrow \quad a = 4$ and $b = 3$

Now, $\quad c^2 = a^2 + b^2 = 16 + 9 = 25$

$\Rightarrow \quad c = 5 \quad$ ($\because c$ must be positive)

Here, in hyperbolic equation coefficient of x^2 is positive, so transverse axis is along X-axis.

$$\text{Foci} = (\pm c, 0) = (\pm 5, 0)$$

$$\text{Vertices} = (\pm a, 0) = (\pm 4, 0)$$

$$\text{Eccentricity, } e = \frac{c}{a} = \frac{5}{4}$$

$$\text{Length of latusrectum} = \frac{2b^2}{a} = \frac{2 \times 9}{4} = \frac{9}{2}$$

B. Comparing the given equation

$$\frac{y^2}{9} - \frac{x^2}{27} = 1 \text{ with } \frac{y^2}{a^2} - \frac{x^2}{b^2} = 1$$

we get, $\quad a^2 = 9$ and $b^2 = 27$

$\Rightarrow \quad a = 3$ and $b = 3\sqrt{3} \quad (\because a, b > 0)$

$\because \quad c^2 = a^2 + b^2 = 9 + 27 = 36$

$\Rightarrow \quad c = 6 \quad$ ($\because c$ must be positive)

Here, in hyperbolic equation coefficient of y^2 is positive, so major axis is along Y-axis.

$$\text{Foci} = (0, \pm c) = (0, \pm 6)$$

$$\text{Vertices} = (0, \pm a) = (0, \pm 3)$$

$$\text{Eccentricity, } e = \frac{c}{a} = \frac{6}{3} = 2$$

$$\text{Length of latusrectum} = \frac{2b^2}{a} = \frac{2 \times 27}{3} = 18$$

Note *In hyperbolic equation, if the coefficient of x^2 is positive, then its major axis is X-axis and if the coefficient of y^2 is positive, then its major axis is Y-axis.*

C. Given, equation is $9y^2 - 4x^2 = 36$, divide it by 36, we get

$$\frac{9y^2}{36} - \frac{4x^2}{36} = \frac{36}{36}$$

$$\Rightarrow \quad \frac{y^2}{4} - \frac{x^2}{9} = 1$$

Now, comparing $\frac{y^2}{4} - \frac{x^2}{9} = 1$ with $\frac{y^2}{a^2} - \frac{x^2}{b^2} = 1$, we get

$$a^2 = 4 \text{ and } b^2 = 9$$

$\Rightarrow \quad a = 2$ and $b = 3$

$\because \quad c^2 = a^2 + b^2$

$\Rightarrow \quad c^2 = 4 + 9 = 13$

$\Rightarrow \quad c = \sqrt{13} \quad (\because c > 0)$

Here, in hyperbolic equation coefficient of y^2 is positive, so transverse axis is along Y-axis.

$$\text{Foci} = (0, \pm c) = (0, \pm \sqrt{13})$$

$$\text{Vertices} = (0, \pm a) = (0, \pm 2)$$

$$\text{Eccentricity, } e = \frac{c}{a} = \frac{\sqrt{13}}{2}$$

$$\text{Length of latusrectum} = \frac{2b^2}{a} = \frac{2 \times 9}{2} = 9$$

D. Given, equation is $16x^2 - 9y^2 = 576$, divide it by 576, we get

$$\frac{16x^2}{576} - \frac{9y^2}{576} = \frac{576}{576} \Rightarrow \frac{x^2}{36} - \frac{y^2}{64} = 1$$

Now, comparing $\frac{x^2}{36} - \frac{y^2}{64} = 1$ with $\frac{x^2}{a^2} - \frac{y^2}{b^2} = 1$, we get

$$a^2 = 36 \text{ and } b^2 = 64$$

$\Rightarrow \quad a = 6$ and $b = 8$

$\because \quad c^2 = a^2 + b^2 = 36 + 64 = 100$

$\Rightarrow \quad c = 10$

Here, in hyperbolic equation coefficient of x^2 is positive, so transverse axis is along X-axis.

$$\text{Foci} = (\pm c, 0) = (\pm 10, 0)$$

$$\text{Vertices} = (\pm a, 0) = (\pm 6, 0)$$

$$\text{Eccentricity, } e = \frac{c}{a} = \frac{10}{6} = \frac{5}{3}$$

$$\text{Length of latusrectum} = \frac{2b^2}{a} = \frac{2 \times 64}{6} = \frac{64}{3}$$

E. Given, equation is $5y^2 - 9x^2 = 36$, divide it by 36,

we get, $\frac{5y^2}{36} - \frac{9x^2}{36} = \frac{36}{36} \Rightarrow \frac{y^2}{\frac{36}{5}} - \frac{x^2}{4} = 1$

On comparing $\frac{y^2}{\frac{36}{5}} - \frac{x^2}{4} = 1$ with $\frac{y^2}{a^2} - \frac{x^2}{b^2} = 1$, we get

$$a^2 = \frac{36}{5} \text{ and } b^2 = 4$$

$$\Rightarrow \quad a=\frac{6}{\sqrt{5}} \quad \text{and} \quad b=2$$

$$\because \quad c^2=a^2+b^2=\frac{36}{5}+4$$

$$\Rightarrow \quad c^2=\frac{36+20}{5}=\frac{56}{5}$$

$$\Rightarrow \quad c=\frac{2\sqrt{14}}{\sqrt{5}}$$

Here, in hyperbolic equation coefficient of y^2 is positive, so transverse axis is along Y-axis.

$$\text{Foci}=(0, \pm c)=\left(0, \pm \frac{2\sqrt{14}}{\sqrt{5}}\right)$$

$$\text{Vertices}=(0, \pm a)=\left(0, \pm \frac{6}{\sqrt{5}}\right)$$

$$\text{Eccentricity, } e=\frac{c}{a}=\frac{\frac{2\sqrt{14}}{\sqrt{5}}}{\frac{6}{\sqrt{5}}}=\frac{\sqrt{14}}{3}$$

$$\text{Length of latusrectum}=\frac{2b^2}{a}=\frac{2\times 4}{\frac{6}{\sqrt{5}}}=\frac{8\sqrt{5}}{6}=\frac{4\sqrt{5}}{3}$$

F. Given, equation is $49y^2-16x^2=784$, divide it by 784,

we get, $$\frac{49y^2}{784}-\frac{16x^2}{784}=\frac{784}{784}$$

$$\Rightarrow \quad \frac{y^2}{16}-\frac{x^2}{49}=1$$

On comparing $\frac{y^2}{16}-\frac{x^2}{49}=1$ with $\frac{y^2}{a^2}-\frac{x^2}{b^2}=1$, we get

$$a^2=16 \quad \text{and} \quad b^2=49$$

$$\Rightarrow \quad a=4 \quad \text{and} \quad b=7$$

$$\because \quad c^2=a^2+b^2=16+49=65$$

$$\Rightarrow \quad c=\sqrt{65} \qquad (\because c>0)$$

Here, in hyperbolic equation coefficient of y^2 is positive, so transverse axis is along Y-axis.

$$\text{Foci}=(0, \pm c)=(0, \pm \sqrt{65})$$

$$\text{Vertices}=(0, \pm a)=(0, \pm 4)$$

$$\text{Eccentricity, } e=\frac{c}{a}=\frac{\sqrt{65}}{4}$$

$$\text{Length of latusrectum}=\frac{2b^2}{a}=\frac{2\times 49}{4}=\frac{49}{2}$$

140. *(a)* A. Since vertices $(\pm 2, 0)$ and foci $(\pm 3, 0)$ lie on X-axis, as coefficient of Y-axis is zero.

Hence, equation of hyperbola will be of the form

$$\frac{x^2}{a^2}-\frac{y^2}{b^2}=1 \qquad \ldots(\text{i})$$

Here, it is given that $a=2$ and $c=3$

$$\because \quad c^2=a^2+b^2$$

$$\Rightarrow \quad 9=4+b^2 \Rightarrow b^2=5$$

Put the values of $a^2=4$ and $b^2=5$ in Eq. (i), we get

$$\frac{x^2}{4}-\frac{y^2}{5}=1$$

B. Since, vertices $(0, \pm 5)$ and foci $(0, \pm 8)$ lie on Y-axis, as coefficient of X-axis is zero .

Hence, equation of hyperbola will be of the form

$$\frac{y^2}{a^2}-\frac{x^2}{b^2}=1 \qquad \ldots(\text{i})$$

Here, it is given that $(0, \pm a)=(0, \pm 5)$ and foci $(0, \pm c)=(0, \pm 8)$

$$\Rightarrow \quad a=5 \quad \text{and} \quad c=8$$

$$\because \quad c^2=a^2+b^2$$

$$\Rightarrow \quad 64=25+b^2$$

$$\Rightarrow \quad b^2=64-25$$

$$\Rightarrow \quad b^2=39$$

Put $a^2=25$ and $b^2=39$ in Eq. (i), we get

$$\frac{y^2}{25}-\frac{x^2}{39}=1$$

C. Since, vertices $(0, \pm 3)$ and foci $(0, \pm 5)$ lie on Y-axis, as coordinate of X is zero.

Hence, equation of hyperbola will be of the form

$$\frac{y^2}{a^2}-\frac{x^2}{b^2}=1 \qquad \ldots(\text{i})$$

Here, it is given that vertices $(0, \pm 3)=(0, \pm a)$ and foci $(0, \pm 5)=(0, \pm c)$

$$\Rightarrow \quad a=3 \quad \text{and} \quad c=5$$

$$\because \quad c^2=a^2+b^2$$

$$\Rightarrow \quad 25=9+b^2$$

$$\Rightarrow \quad b^2=16$$

Put $a^2=9$ and $b^2=16$ in Eq. (i), we get

$$\frac{y^2}{9}-\frac{x^2}{16}=1$$

141. *(c)* A. Since, foci $(\pm 5, 0)$ lies on X-axis, as coordinate of Y is zero. Hence, equation of hyperbola will be of the form

$$\frac{x^2}{a^2}-\frac{y^2}{b^2}=1 \qquad \ldots(\text{i})$$

where, it is given that foci $(\pm 5, 0)=(\pm c, 0)$ and length of transverse axis is $2a=8$

$$c=5 \quad \text{and} \quad 2a=8$$

$$\Rightarrow \quad a=4$$

$$\because \quad c^2=a^2+b^2$$

$$\Rightarrow \quad 25=16+b^2$$

$$\Rightarrow \quad b^2=9$$

Put $a^2=16$ and $b^2=9$ in Eq. (i), we get

$$\frac{x^2}{16}-\frac{y^2}{9}=1$$

B. Since, foci $(0, \pm 13)$ lies on Y-axis, as x-coordinate is zero. Hence, equation of hyperbola will be of the form

$$\frac{y^2}{a^2} - \frac{x^2}{b^2} = 1 \quad \ldots(i)$$

where, it is given that foci $(0, \pm 13) = (0, \pm c)$ and conjugate length, $2b = 24$

$$c = 13 \quad \text{and} \quad b = 12$$

$\because$ $c^2 = a^2 + b^2$

$\therefore$ $169 = a^2 + 144$

$\Rightarrow$ $a^2 = 169 - 144 \Rightarrow a^2 = 25$

Put $a^2 = 25$ and $b^2 = 144$ in Eq. (i), we get

$$\frac{y^2}{25} - \frac{x^2}{144} = 1$$

C. Since, foci $(\pm 3\sqrt{5}, 0)$ lies on X-axis, as y-coordinate is zero. Hence, equation of hyperbola will be of the form

$$\frac{x^2}{a^2} - \frac{y^2}{b^2} = 1 \quad \ldots(i)$$

where, it is given that foci $(\pm 3\sqrt{5}, 0) = (\pm c, 0)$

$$c = 3\sqrt{5}$$

and length of latusrectum

$$\frac{2b^2}{a} = 8 \Rightarrow b^2 = 4a$$

$\because$ $c^2 = a^2 + b^2$

$\therefore$ $(3\sqrt{5})^2 = a^2 + (4a)$

$\Rightarrow$ $9 \times 5 = a^2 + 4a$

$\Rightarrow$ $a^2 + 4a - 45 = 0$

On splitting the middle term

$\Rightarrow$ $a^2 + 9a - 5a - 45 = 0$

$\Rightarrow$ $a(a + 9) - 5(a + 9) = 0$

$\Rightarrow$ $(a - 5)(a + 9) = 0$

$\Rightarrow$ $a - 5 = 0,\ a + 9 = 0$

$\Rightarrow$ $a = 5, -9$

$\because$ a can't be negative.

So, $a = 5 \Rightarrow b^2 = 4 \times 5 = 20$

Put $a^2 = 25$ and $b^2 = 20$ in Eq. (i), we get

$$\frac{x^2}{25} - \frac{y^2}{20} = 1$$

D. Since, foci $(\pm 4, 0)$ lies on X-axis. Hence, equation of hyperbola will be of the form

$$\frac{x^2}{a^2} - \frac{y^2}{b^2} = 1 \quad \ldots(i)$$

where, it is given that foci $(\pm 4, 0) = (\pm c, 0)$ and length of latusrectum, $\frac{2b^2}{a} = 12$

$\therefore$ $c = 4$ and $b^2 = 6a$ $\quad \ldots(ii)$

$\because$ $c^2 = a^2 + b^2$

$\Rightarrow$ $16 = a^2 + 6a \Rightarrow a^2 + 6a - 16 = 0$

Splitting the middle term

$$a^2 + 8a - 2a - 16 = 0$$

$\Rightarrow$ $a(a + 8) - 2(a + 8) = 0$

$\Rightarrow$ $(a - 2)(a + 8) = 0$

$\Rightarrow$ $a - 2 = 0, a + 8 = 0 \Rightarrow a = 2, -8$

$\because$ a can't be negative.

So, $a = 2$

$\Rightarrow$ $b^2 = 6a = 6 \times 2 = 12$

Put $a^2 = 4$ and $b^2 = 12$ in Eq. (i), we get

$$\frac{x^2}{4} - \frac{y^2}{12} = 1$$

142. *(b)* Since, vertices $(\pm 7, 0)$ lie on X-axis, as y-coordinate is zero. Hence, equation of hyperbola will be of the form

$$\frac{x^2}{a^2} - \frac{y^2}{b^2} = 1 \quad \ldots(i)$$

where, it is given that vertices $(\pm 7, 0) = (\pm a, 0)$

$$a = 7 \quad \text{and} \quad e = \frac{4}{3}$$

$\because$ $c = ae$

$\Rightarrow$ $c = 7 \times \frac{4}{3} = \frac{28}{3}$

Also, $c^2 = a^2 + b^2$

$$\frac{784}{9} = 49 + b^2$$

$\Rightarrow$ $b^2 = \frac{784}{9} - \frac{49}{1}$

$\Rightarrow$ $b^2 = \frac{784 - 441}{9}$

$\Rightarrow$ $b^2 = \frac{343}{9}$

Put $a^2 = 49$ and $b^2 = \frac{343}{9}$ in Eq. (i), we get

$$\frac{x^2}{49} - \frac{y^2}{\frac{343}{9}} = 1 \quad \text{or} \quad \frac{x^2}{49} - \frac{9y^2}{343} = 1$$

143. *(a)* Since, foci $(0, \pm \sqrt{10})$ lies on Y-axis, as x-coordinate is zero. Hence, equation of hyperbola will be of the form

$$\frac{y^2}{a^2} - \frac{x^2}{b^2} = 1 \quad \ldots(i)$$

where, it is given that foci $(0, \pm \sqrt{10}) = (0, \pm c) \Rightarrow c = \sqrt{10}$.

Also, hyperbola passes through the point (2,3) *i.e.,* point will satisfy Eq. (i).

$\therefore$ $\frac{(3)^2}{a^2} - \frac{(2)^2}{b^2} = 1$

$\Rightarrow$ $\frac{9}{a^2} - \frac{4}{b^2} = 1 \quad \ldots(ii)$

$\because$ $c = \sqrt{10} \Rightarrow c^2 = 10$

$\Rightarrow$ $a^2 + b^2 = 10 \quad (\because c^2 = a^2 + b^2)$

$\Rightarrow$ $b^2 = 10 - a^2 \quad \ldots(iii)$

From Eq. (ii), $\frac{9}{a^2} - \frac{4}{10 - a^2} = 1$

$$\Rightarrow \quad \frac{90 - 9a^2 - 4a^2}{a^2(10 - a^2)} = 1$$
$$\Rightarrow \quad 90 - 13a^2 = 10a^2 - a^4$$
$$\Rightarrow \quad a^4 - 23a^2 + 90 = 0$$
$$\Rightarrow \quad a^4 - 18a^2 - 5a^2 + 90 = 0$$
$$\Rightarrow \quad a^2(a^2 - 18) - 5(a^2 - 18) = 0$$
$$\Rightarrow \quad (a^2 - 5)(a^2 - 18) = 0$$
$$\Rightarrow \quad a^2 = 5 \text{ or } a^2 = 18$$

From Eq. (iii), if $a^2 = 5$

$$\Rightarrow \quad b^2 = 10 - 5 = 5$$

If $\quad a^2 = 18$

$$\Rightarrow \quad b^2 = 10 - 18 = -8$$

which is not possible.

So, $\quad a^2 = b^2 = 5$

Put $a^2 = b^2 = 5$ in Eq. (i), we get

$$\Rightarrow \quad \frac{y^2}{5} - \frac{x^2}{5} = 1$$
$$\Rightarrow \quad y^2 - x^2 = 5$$

144. *(a)* Since, the vertices are on the Y- axis (with origin as the mid- point) , the equation is of the form $\frac{y^2}{a^2} - \frac{x^2}{b^2} = 1$.

As vertices are $(0, \pm 6)$

$\therefore a = 6, b^2 = a^2(e^2 - 1) = 36\left(\frac{25}{9} - 1\right) = 64$

So, the required equation of the hyperbola is $\frac{y^2}{36} - \frac{x^2}{64} = 1$ and the foci are $(0, \pm ae) = (0 \pm 10)$.

145. *(c)* As the vertices are on the X- axis and their mid-point is the origin, the equation is of the type $\frac{x^2}{a^2} - \frac{y^2}{b^2} = 1$.

Here, $b^2 = a^2(e^2 - 1)$, vertices are $(\pm a, 0)$ and directrices are given by $x = \pm \frac{a}{e}$.

Thus, $a = 6, \frac{a}{e} = 4$ and so $e = \frac{3}{2}$ which gives

$$b^2 = 36\left(\frac{9}{4} - 1\right) = 45$$

Consequently, the required equation of the hyperbola is

$$\frac{x^2}{36} - \frac{y^2}{45} = 1$$

146. *(a)* The given equation may be written as

$$\frac{x^2}{32/3} - \frac{y^2}{8} = 1 \quad \text{or} \quad \frac{x^2}{(4\sqrt{2}/\sqrt{3})^2} - \frac{y^2}{(2\sqrt{2})^2} = 1$$

Comparing the given equation with $\frac{x^2}{a^2} - \frac{y^2}{b^2} = 1$, we get $a^2 = \left(\frac{4\sqrt{2}}{\sqrt{3}}\right)^2$ or $a = \frac{4\sqrt{2}}{\sqrt{3}}$. Therefore, length of transverse axis of a hyperbola $= 2a = 2 \times \frac{4\sqrt{2}}{\sqrt{3}} = \frac{8\sqrt{2}}{\sqrt{3}}$.

147. *(a)* Directrix of hyperbola $x = \frac{a}{e}$,

where, $\quad e = \sqrt{\frac{b^2 + a^2}{a^2}}$

$\therefore$ Directrix is , $\quad x = \frac{a^2}{\sqrt{a^2 + b^2}} = \frac{9}{\sqrt{9+4}} \Rightarrow x = \frac{9}{\sqrt{13}}$

148. *(b)* $\because 2a = 10 \Rightarrow a = 5$

$$ae - a = 8 \text{ or } e = 1 + \frac{8}{5} = \frac{13}{5}$$

$$\therefore \quad b = 5\sqrt{\frac{13^2}{5^2} - 1} = 5 \times \frac{12}{5} = 12$$

and centre of hyperbola $\equiv (5, 0)$

$\therefore$ Equation is $\frac{(x-5)^2}{5^2} - \frac{(y-0)^2}{12^2} = 1$ or $\frac{(x-5)^2}{25} - \frac{y^2}{144} = 1$

149. *(a)* Here, $a = 4, b = 3 \Rightarrow \frac{9}{16} = (e^2 - 1) \Rightarrow e = \frac{5}{4}$

$[\because b^2 = a^2(e^2 - 1)]$

$\therefore$ Vertex is $(0, 5)$. Hence, focus is $(\pm ae, 2) = (\pm 5, 2)$

150. *(c)* Let $\frac{x^2}{a^2} - \frac{y^2}{b^2} = 1$ represents the hyperbola. Then, according to the given condition, the length of transverse axis, *i.e.*, $2a = 7 \Rightarrow a = \frac{7}{2}$.

Also, the point $(5, -2)$ lies on the hyperbola. So, we have $\frac{4}{49}(25) - \frac{4}{b^2} = 1$, which gives $b^2 = \frac{196}{51}$.

Hence, the equation of the hyperbola is

$$\frac{4}{49}x^2 - \frac{51}{196}y^2 = 1$$

151. *(d)* Given that the hyperbola $\frac{x^2}{a^2} - \frac{y^2}{b^2} = 1$ is passing through the points $(3, 0)$ and $(3\sqrt{2}, 2)$, so we get $a^2 = 9$ and $b^2 = 4$.

Again, we know that $b^2 = a^2(e^2 - 1)$. This gives

$$4 = 9(e^2 - 1)$$

or $\quad e^2 = \frac{13}{9}$

or $\quad e = \frac{\sqrt{13}}{3}$

152. *(a)* $e = \sqrt{1 + \frac{b^2}{a^2}} \Rightarrow e^2 = \frac{a^2 + b^2}{a^2}$

$$e_1 = \sqrt{1 + \frac{a^2}{b^2}}$$

$$\Rightarrow \quad e_1^2 = \frac{b^2 + a^2}{b^2} \Rightarrow \frac{1}{e_1^2} + \frac{1}{e^2} = 1$$

153. *(a)* Conjugate axis is 5 and distance between foci $= 13$
$\Rightarrow 2b = 5$ and $2ae = 13$.

Now, also we know for hyperbola

$$b^2 = a^2(e^2 - 1) \Rightarrow \frac{25}{4} = \frac{(13)^2}{4e^2}(e^2 - 1)$$

$$\Rightarrow \frac{25}{4} = \frac{169}{4} - \frac{169}{4e^2} \text{ or } e^2 = \frac{169}{144}$$

$$\Rightarrow e = \frac{13}{12}$$

or $a = 6, b = \frac{5}{2}$ or hyperbola is $\frac{x^2}{36} - \frac{y^2}{\frac{25}{4}} = 1$

$$\Rightarrow 25x^2 - 144y^2 = 900$$

154. *(c)* The given equation can be written as

$$(4x + 8)^2 - (y - 2)^2 = -44 + 64 - 4$$

$$\Rightarrow \frac{16(x + 2)^2}{16} - \frac{(y - 2)^2}{16} = 1$$

Transverse and conjugate axes are $y = 2$, $x = -2$

$$\Rightarrow x + 2 = 0$$

155. *(c)* Multiplying both, we get $3x^2 - y^2 = 48$

or $\frac{x^2}{\left(\frac{48}{3}\right)} - \frac{y^2}{48} = 1$, which is a hyperbola.

156. *(a)* Foci are (6,4) and (–4,4), $e = 2$ and centre is $\left(\frac{6-4}{2}, \frac{4+4}{2}\right) = (1, 4)$

Since, centre of hyperbola is (1,4),

$\Rightarrow 6 = 1 + ae \Rightarrow ae = 5 \Rightarrow a = \frac{5}{2}$ and $b = \frac{5}{2}(\sqrt{3})$

Hence, the required equation is $\frac{(x-1)^2}{\left(\frac{25}{4}\right)} - \frac{(y-4)^2}{\left(\frac{75}{4}\right)} = 1$

or $12x^2 - 4y^2 - 24x + 32y - 127 = 0$

157. *(c)* Equation of hyperbola is $x = 8\sec\theta$, $y = 8\tan\theta$

$$\Rightarrow \frac{x}{8} = \sec\theta, \frac{y}{8} = \tan\theta$$

$\because \sec^2\theta - \tan^2\theta = 1 \Rightarrow \frac{x^2}{8^2} - \frac{y^2}{8^2} = 1$

Here, $a = 8, b = 8$

Now, $e = \sqrt{1 + \frac{b^2}{a^2}}$

$$= \sqrt{1 + \frac{8^2}{8^2}} = \sqrt{1 + 1}$$

$$\Rightarrow e = \sqrt{2}$$

$\therefore$ Distance between directrices $= \frac{2a}{e} = \frac{2 \times 8}{\sqrt{2}} = 8\sqrt{2}$.

158. *(b)* Given equation of hyperbola is

$$5x^2 - 4y^2 + 20x + 8y = 4$$

$$5(x + 2)^2 - 4(y - 1)^2 = 20$$

$$\Rightarrow \frac{(x + 2)^2}{4} - \frac{(y - 1)^2}{5} = 1$$

From $b^2 = a^2(e^2 - 1)$, we have

$$5 = 4(e^2 - 1)$$

$$\Rightarrow e^2 = \frac{9}{4} \Rightarrow e = \frac{3}{2}$$

159. *(a)* Given equation of hyperbola is

$$9x^2 - 16y^2 + 72x - 32y - 16 = 0$$

$$\Rightarrow 9(x^2 + 8x) - 16(y^2 + 2y) - 16 = 0$$

$$\Rightarrow 9(x + 4)^2 - 16(y + 1)^2 = 144$$

$$\Rightarrow \frac{(x + 4)^2}{16} - \frac{(y + 1)^2}{9} = 1$$

Therefore, latusrectum $= \frac{2b^2}{a} = 2 \times \frac{9}{4} = \frac{9}{2}$

160. *(b)* Let AB be the rod making an angle θ with OX as shown in figure and $P(x, y)$ the point on it such that $AP = 6$ cm.

Since, $AB = 15$ cm, we have

$$PB = 9 \text{ cm}$$

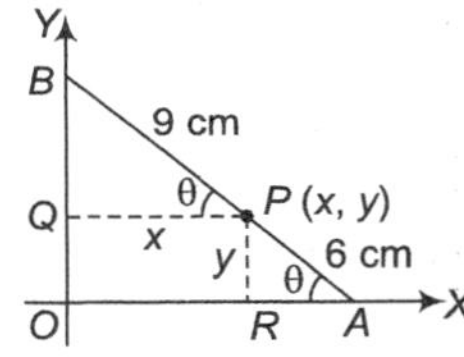

From P, draw PQ and PR perpendiculars on Y-*axis* and X- axis, respectively.

From ΔPBQ, $\cos\theta = \frac{x}{9}$

From ΔPRA, $\sin\theta = \frac{y}{6}$

Since, $\cos^2\theta + \sin^2\theta = 1$,

$$\left(\frac{x}{9}\right)^2 + \left(\frac{y}{6}\right)^2 = 1 \text{ or } \frac{x^2}{81} + \frac{y^2}{36} = 1$$

Thus, the locus of P is an ellipse.

161. *(d)* Taking vertex of the parabolic reflector at origin, X-axis along the axis of parabola. The equation of the parabola is $y^2 = 4ax$. Given depth is 5 cm, diameter is 20 cm.

$\because$ Point $P(5,10)$ lies on parabola.

$\therefore$ $(10)^2 = 4a(5) \Rightarrow a = 5$

Clearly, focus is at the mid-point of given diameter.

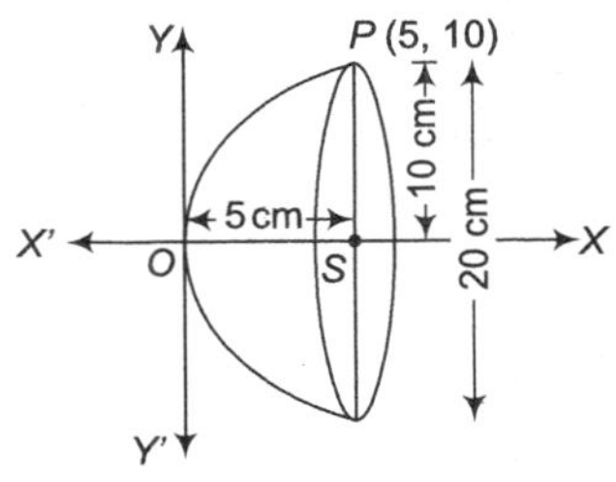

i.e., $S = (5, 0)$

162. *(b)* Given, $x = \frac{2at}{1+t^2}$ and $y = \frac{a(1-t^2)}{1+t^2}$

On squaring and adding, we get

$$x^2 + y^2 = \left(\frac{2at}{1+t^2}\right)^2 + \left[\frac{a(1-t^2)}{1+t^2}\right]^2$$

$$= \frac{4a^2t^2}{(1+t^2)^2} + \frac{a^2(1-t^2)^2}{(1+t^2)^2}$$

$$= \frac{a^2(4t^2+1+t^4-2t^2)}{(1+t^2)^2}$$

$$= \frac{a^2[t^4+1+2t^2]}{(1+t^2)^2} = \frac{a^2(t^2+1)^2}{(t^2+1)^2} = a^2$$

$\Rightarrow \quad x^2 + y^2 = a^2$

which represents a circle.

163. *(b)* Since, axis is vertical, so let arch of parabola is in the form

$$x^2 = 4ay \quad \ldots(i)$$

Given, $OB = 10$ m

and $AC = 5$ m

$\Rightarrow \quad AB = \frac{5}{2}$ m $\quad (\because AC = 5)$

Hence, coordinates of $A = \left(\frac{5}{2}, 10\right)$ will satisfy Eq. (i).

i.e., $\left(\frac{5}{2}\right)^2 = 4a \times 10$

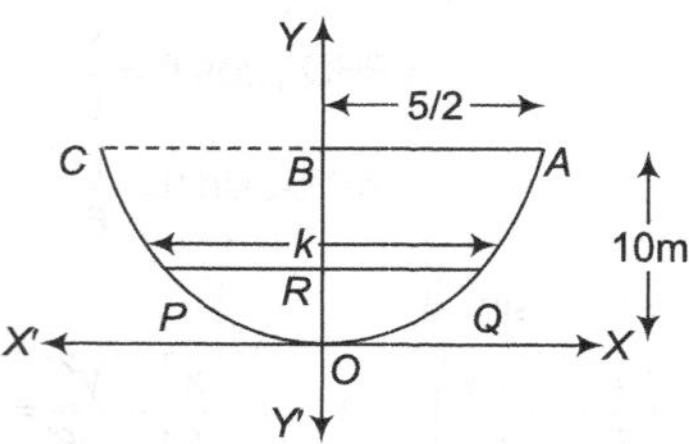

$\Rightarrow \quad \frac{25}{4} = 40a$

$\Rightarrow \quad a = \frac{5}{32}$

From Eq. (i), $\quad x^2 = 4 \times \frac{5}{32} y$

$\Rightarrow \quad x^2 = \frac{5}{8} y$

Now, let $OR = 2$ and $PQ = k \Rightarrow RP = \frac{k}{2}$

Therefore, $P = \left(\frac{k}{2}, 2\right)$ will lie on parabola.

$\therefore \quad \left(\frac{k}{2}\right)^2 = \frac{5}{8} \times 2 \quad \left(\because x^2 = \frac{5}{8}y\right)$

$\Rightarrow \quad \frac{k^2}{4} = \frac{5}{4}$

$\Rightarrow \quad k = \sqrt{5} = 2.23$ m (approx).

164. *(d)* Given, $e = \frac{5}{8}$, distance between foci $2c = 10$

$\Rightarrow \quad c = 5$

$\because \quad c = ae$

$\Rightarrow \quad 5 = a \times \frac{5}{8}$

$\Rightarrow \quad a = 8$

$\because \quad c^2 = a^2 - b^2$

$\Rightarrow \quad 25 = 64 - b^2$

$\Rightarrow \quad b^2 = 64 - 25$

$\Rightarrow \quad b^2 = 39$

Length of latusrectum $= \frac{2b^2}{a} = \frac{2 \times 39}{8} = \frac{39}{4}$

165. *(b)* Since, wires are vertical. Let equation of the parabola is in the form

$$x^2 = 4ay \quad \ldots(i)$$

Focus is at the middle of the cable and shortest and longest vertical supports are 6 m and 30 m and roadway is 100 m long.

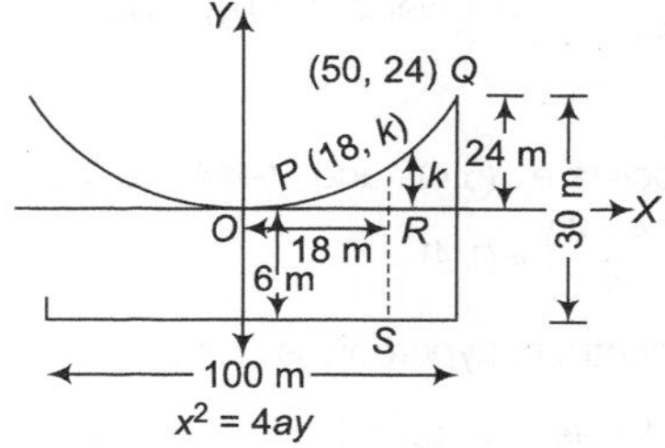

Clearly, coordinate of $Q(50, 24)$ will satisfy Eq. (i)

$$(50)^2 = 4a \times 24$$

$\Rightarrow \quad 2500 = 96a \Rightarrow a = \frac{2500}{96}$

Hence, from Eq. (i), $\quad x^2 = 4 \times \frac{2500}{96} y$

$\Rightarrow \quad x^2 = \frac{2500}{24} y$

Let $\quad PR = k$ m

$\therefore$ Point $P(18, k)$ will satisfy the equation of parabola *i.e.*,

From Eq. (i), $\quad (18)^2 = \frac{2500}{24} \times k$

$\Rightarrow \quad 324 = \frac{2500}{24} k$

$\Rightarrow \quad k = \frac{324 \times 24}{2500} = \frac{324 \times 6}{625} = \frac{1944}{625}$

$\Rightarrow \quad k = 3.11$

$\therefore$ Required length $= 6 + k = 6 + 3.11 = 9.11$ m (approx.)

166. *(c)* Let any point $P(h, k)$ will satisfy

$$y^2 = 4ax \text{ i.e., } k^2 = 4ah \quad \ldots(i)$$

Let a line OP makes an angle θ from the X-axis

$\therefore$ In ΔOAP, $\quad \sin\theta = \frac{PA}{OP}$

$$\sin\theta = \frac{k}{l}$$

$\Rightarrow \quad k = l\sin\theta$

and $\quad \cos\theta = \frac{OA}{OP}$

$\Rightarrow \quad \cos\theta = \frac{h}{l} \Rightarrow h = l\cos\theta$

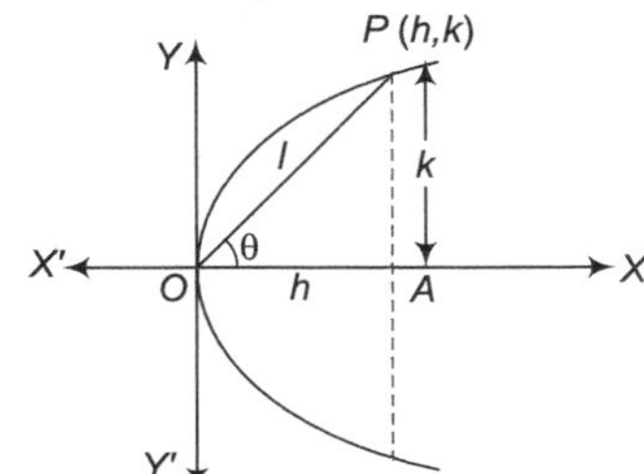

Hence, from Eq. (i),we get

$$l^2\sin^2\theta = 4a \times l\cos\theta \quad (\text{put } k = l\sin\theta, h = l\cos\theta)$$

$\Rightarrow \quad l = \frac{4a\cos\theta}{\sin^2\theta}$

167. *(a)* Clearly, equation of ellipse takes the form

$$\frac{x^2}{a^2} + \frac{y^2}{b^2} = 1 \quad \ldots(i)$$

Here, it is given $2a = 8$ and $b = 2 \Rightarrow a = 4, \quad b = 2$

Put the values of a and b in Eq. (i), we get

$$\frac{x^2}{16} + \frac{y^2}{4} = 1$$

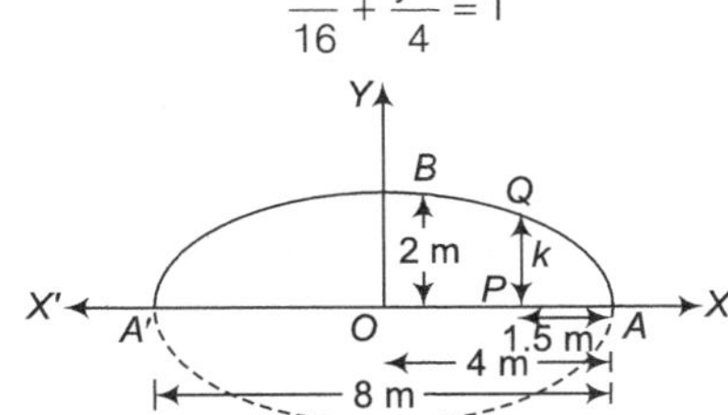

Given, $AP = 1.5$ m

$\Rightarrow \quad OP = OA - AP = 4 - 1.5$

$\Rightarrow \quad OP = 2.5$ m

Let $\quad PQ = k$

$\therefore$ Coordinate $Q = (2.5, k)$ will satisfy the equation of ellipse.

i.e., $\quad \frac{(2.5)^2}{16} + \frac{k^2}{4} = 1 \Rightarrow \frac{6.25}{16} + \frac{k^2}{4} = 1$

$\Rightarrow \quad \frac{k^2}{4} = \frac{1}{1} - \frac{6.25}{16} = \frac{16 - 6.25}{16}$

$\Rightarrow \quad \frac{k^2}{4} = \frac{9.75}{16} \Rightarrow k^2 = \frac{9.75}{4}$

$\Rightarrow \quad k^2 = 2.4375$

$\Rightarrow \quad k = 1.56$ m (approx.)

168. *(c)* Given equation of line is

$$y = mx + 1 \quad \ldots(i)$$

and equation of parabola is $y^2 = 4x \quad \ldots(ii)$

From Eq. (i), $y = mx + 1$ put in Eq. (ii), we get

$$(mx + 1)^2 = 4x$$

$\Rightarrow \quad m^2x^2 + 1 + 2mx - 4x = 0$

$\Rightarrow \quad x^2(m^2) + x(2m - 4) + 1 = 0$

$\because$ For tangent, discriminant should be zero *i.e.*, $D = 0$

$\therefore \quad (2m - 4)^2 - 4m^2 \times 1 = 0 \quad (\because D = b^2 - 4ac = 0)$

$\Rightarrow \quad 4m^2 + 16 - 16m - 4m^2 = 0 \Rightarrow m = 1$

169. *(c)* Let AB be the rod of length 12 cm.

Let $OA = a$ and $OB = b$

Using by Pythagoras theorem in ΔOAB.

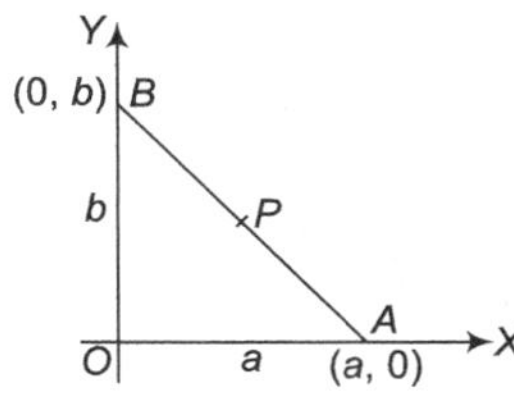

$(\text{Base})^2 + (\text{Perpendicular})^2 = (\text{Hypotenuse})^2$

$\Rightarrow \quad a^2 + b^2 = (12)^2$

$\Rightarrow \quad a^2 + b^2 = 144 \quad \ldots(i)$

Let $\quad AP = 3$ cm

$\Rightarrow \quad PB = AB - AP$

$\quad = 12 - 3 = 9$ cm

$\Rightarrow \quad AP : PB = 3 : 9 = 1 : 3$

Let $P = (h, k)$ for which locus is to be found.

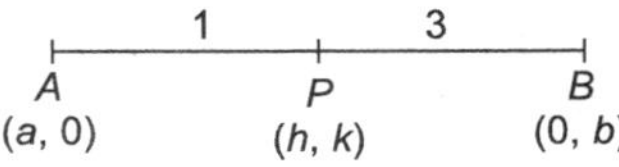

$\therefore$ If $P(h, k)$ divides the points A and B internally in the ratio $m : n$, then

$$h = \frac{mx_2 + nx_1}{m + n}, k = \frac{my_2 + ny_1}{m + n}$$

$\therefore \quad h = \frac{1 \times 0 + 3 \times a}{1 + 3} \Rightarrow h = \frac{3a}{4}$

$\Rightarrow \quad a = \frac{4h}{3}$ and $k = \frac{1 \times b + 3 \times 0}{1 + 3}$

$\Rightarrow \quad k = \frac{b}{4} \Rightarrow b = 4k$

Put the values of $a = \frac{4h}{3}$ and $b = 4k$ in Eq. (i), we get

$$\left(\frac{4h}{3}\right)^2 + (4k)^2 = 144 \Rightarrow \frac{16h^2}{9} + 16k^2 = 144$$

$\Rightarrow \quad \frac{h^2}{9} + k^2 = 9 \Rightarrow \frac{h^2}{81} + \frac{k^2}{9} = 1$

Hence, locus of a point $P(h, k)$ is

$$\frac{x^2}{81} + \frac{y^2}{9} = 1$$

Note *In locus of a point (h, k), we make a relation in h and k as per the given condition and finally, we replace h by x and k by y, we get the required result.*

170. *(a)* By definition of parabola, if P is any point on the parabola, S be the focus and M is a point on directrix.

Then, $SP = PM$.

Given, $S = (-1, -2)$ and $P = (x, y)$

Then, $\sqrt{(x+1)^2 + (y+2)^2} = \dfrac{x - 2y + 3}{\sqrt{1+4}}$

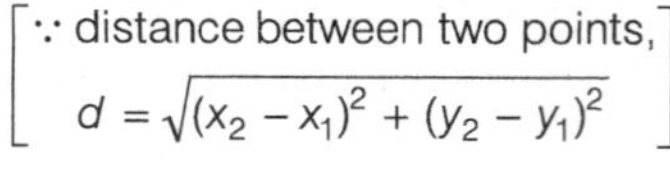

$\left[\because \text{distance between two points, } d = \sqrt{(x_2 - x_1)^2 + (y_2 - y_1)^2}\right]$

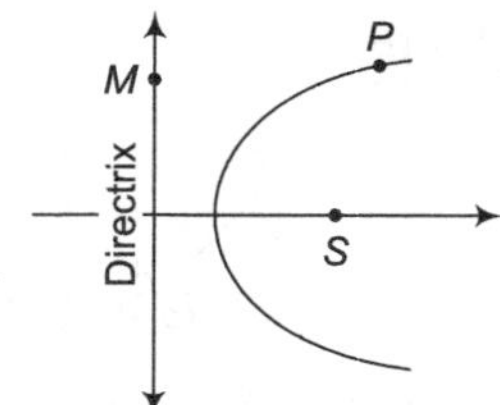

$\Rightarrow (x+1)^2 + (y+2)^2 = \dfrac{(x - 2y + 3)^2}{5}$ (squaring both sides)

$\Rightarrow 5x^2 + 5 + 10x + 5y^2 + 20 + 20y$

$= x^2 + 4y^2 + 9 - 4xy - 12y + 6x$

$\Rightarrow 4x^2 + y^2 + 4xy + 4x + 32y + 16 = 0$

171. *(b)* Let the given points are $A(3, 0)$ and $B(9, 0)$.

Let $P(x, y)$ be any point on the curve.

Given, the sum of distance from point $P(x, y)$ to the points (3, 0) and (9, 0) is equal to 12.

$$PA + PB = 12$$

$\Rightarrow \sqrt{(x-3)^2 + y^2} + \sqrt{(x-9)^2 + y^2} = 12$

$\left[\because \text{distance, } d = \sqrt{(x_2 - x_1)^2 + (y_2 - y_1)^2}\right]$

$\Rightarrow \sqrt{(x-3)^2 + y^2} = 12 - \sqrt{(x-9)^2 + y^2}$

On squaring both sides, we get

$(x-3)^2 + y^2 = 144 + (x-9)^2 + y^2 - 24\sqrt{(x-9)^2 + y^2}$

$\Rightarrow x^2 + 9 - 6x = 144 + x^2 + 81 - 18x - 24\sqrt{(x-9)^2 + y^2}$

$\Rightarrow 12x - 216 = -24\sqrt{(x-9)^2 + y^2}$

$\Rightarrow x - 18 = -2\sqrt{(x-9)^2 + y^2}$ (divide by 12)

Again squaring, we get

$x^2 + 324 - 36x = 4(x^2 + 81 - 18x + y^2)$

$\Rightarrow 3x^2 + 4y^2 - 36x = 0$

172. *(a)* Clearly, path traced by the man will be ellipse.

Given, $SP + S'P = 10$

i.e., $2a = 10$

$\Rightarrow a = 5$

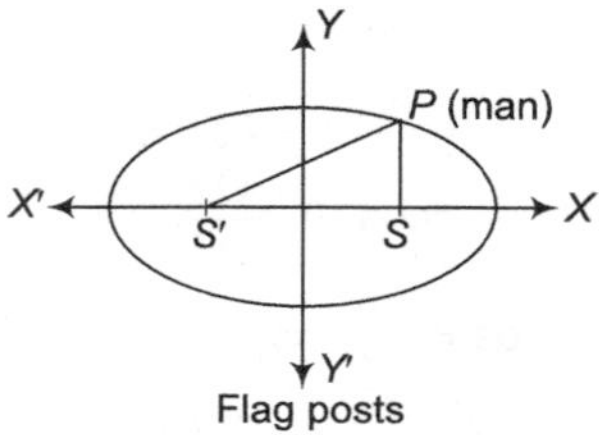

Flag posts

Since, the coordinates of S and S' are $(c, 0)$ and $(-c, 0)$, respectively. Therefore, distance between S and S' is

$2c = 8 \Rightarrow c = 4$

$\because c^2 = a^2 - b^2$

$\Rightarrow 16 = 25 - b^2 \Rightarrow b^2 = 25 - 16$

$\Rightarrow b^2 = 9 \Rightarrow b = 3$

Hence, equation of path (ellipse) is

$\dfrac{x^2}{a^2} + \dfrac{y^2}{b^2} = 1 \Rightarrow \dfrac{x^2}{25} + \dfrac{y^2}{9} = 1$ $(\because a = 5, b = 3)$

173. *(b)* Let point is $P(x, y)$ and another point is $A(0, 4)$.

According to the given condition,

$\sqrt{(x-0)^2 + (y-4)^2} = \dfrac{2}{3} \times \dfrac{y-9}{\sqrt{0+1}}$

$\Rightarrow x^2 + (y-4)^2 = \dfrac{4}{9} \times (y-9)^2$ (on squaring both sides)

$\Rightarrow 9x^2 + 9y^2 + 9 \times 16 - 72y = 4(y^2 + 81 - 18y)$

$\Rightarrow 9x^2 + 9y^2 + 144 - 72y = 4y^2 + 324 - 72y$

$\Rightarrow 9x^2 + 5y^2 = 180$

174. *(b)* Let centre of circle be (h, k). Since, it touches both axes, therefore $h = k = a$.

Hence, equation can be $(x-a)^2 + (y-a)^2 = a^2$

But it also touches the line $3x + 4y = 4$.

Therefore, $\dfrac{3a + 4a - 4}{5} = a \Rightarrow a = 2$

Hence, the required equation of circle is

$$x^2 + y^2 - 4x - 4y + 4 = 0$$

175. *(c)* Extremities of diameter are (5, 7) and (1, 4) and radius is half of the distance between them

$\therefore$ Radius $= \dfrac{1}{2}\sqrt{(4)^2 + (3)^2} = \dfrac{5}{2}$

176. *(c)* Equation of circle passing through (0, 0) is

$x^2 + y^2 + 2gx + 2fy = 0$ ...(i)

Also, circle (i) is passing through (0, b) and (a, b)

$f = -\dfrac{b}{2}$

and $a^2 + b^2 + 2ag + 2\left(-\dfrac{b}{2}\right)b = 0$

$\Rightarrow g = -\dfrac{a}{2}$

Hence, the equation of circle is $x^2 + y^2 - ax - by = 0$.

12

Introduction *to* Three Dimensional Geometry

Coordinate Axes and Coordinate Planes in Three Dimensional Space

Level I

1. To locate the position of a point in a plane, we need two intersecting mutually perpendicular lines in the plane. These lines are called the ...A... and the two numbers are called the ...B... . Here, A and B refers to

(a) 'coordinate geometry' and 'point with respect to origin'
(b) 'point with respect to origin' and 'coordinate geometry'
(c) 'coordinate axes' and 'coordinates of the point with respect to the axes'
(d) None of the above

2. 'In actual life, we do not have to deal with points lying in a plane only'. Which among the following examples satisfy the given statement?

(i) The position of a ball thrown in space at different points of time.

(ii) The position of an aeroplane as it flies from one place to another at different times during its flight.

(iii) The position of the lowest tip of an electric bulb hanging from the ceiling of a room.

(iv) The position of the central tip of the ceiling fan in a room.

(a) Only (i)
(b) (ii) and (iii)
(c) (ii), (iii) and (iv)
(d) All of the above

3. The three numbers representing the three perpendicular distances of the point from three mutually perpendicular planes are called the

(a) coordinates of the point with respect to the two coordinate planes
(b) coordinates of the origin with respect to the three coordinate planes
(c) coordinates of the point with reference to the three coordinate planes
(d) None of the above

Directions (Q. Nos. 4-9) *Consider the figure given below*

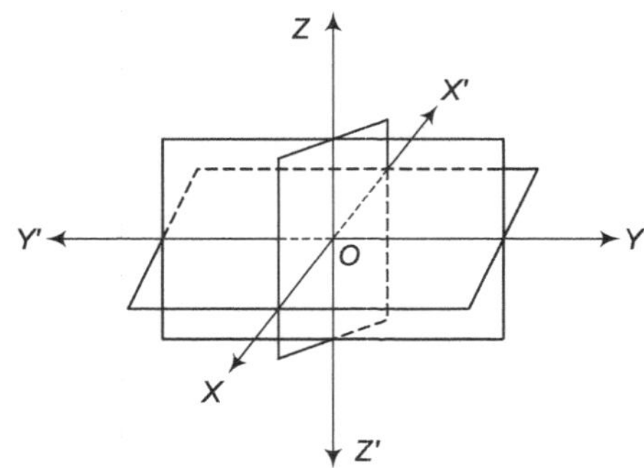

In the above figure, it is shown that three planes are intersecting at a point O, such that these three planes are mutually perpendicular to each other.

Based on the above information, answer the questions given below.

4. The lines $X'OX, Y'OY$ and $Z'OZ$ are constituting the
(a) rectangular non-perpendicular system
(b) rectangular coordinate system
(c) rectangular parallel system
(d) None of the above

5. Which of the following statement is correct?
(a) The plane XOY is called the ZX-plane
(b) The plane YOZ is called the XY-plane
(c) The plane ZOX is called the YZ-plane
(d) The XY-plane, YZ-plane and the ZX-plane are called the three coordinate planes

6. Which of the following is/are incorrect?
I. If we take the XOY plane as the plane of the paper, then the line $Z'OZ$ is perpendicular to the plane XOY.
II. If the plane of the paper is considered as horizontal, then the line $Z'OZ$ will be horizontal.
(a) Only I is incorrect (b) Only II is incorrect
(c) Both I and II are incorrect (d) Both I and II are correct

7. Read the passage given below.
The distances measured from XY-plane upwards in the direction of OZ are taken as ...A... and those measured downwards in the direction of OZ' are taken as ...B... . Similarly, the distance measured to the right of ZX-plane along OY are taken as ...C... to the left of ZX-plane and along OY' as ...D..., in front of the YZ-plane along OX as ...E... and to the back of it along OX' *as* ...F... . In the above passage, A,B,C, D,E and F refers to
(a) A, B, C $\rightarrow$ negative and D, E, F $\rightarrow$ positive
(b) A, B, C $\rightarrow$ positive and D, E, F $\rightarrow$ negative
(c) A, E, C $\rightarrow$ positive and B, D, F $\rightarrow$ negative
(d) A, C, E $\rightarrow$ negative and B, D, F $\rightarrow$ positive

8. Which of the following statement is true?
(a) The point O is called the origin of the coordinate system
(b) The three coordinate planes divide the space into six parts known as hexants
(c) Both (a) and (b) are true
(d) Both (a) and (b) are false

9. Match the terms of Column I with terms of Column II and choose the correct option from the codes given below.

Column I (Octants)	Column II (Name of octants respectively)
A. I, II	1. $XOYZ', X'OYZ'$
B. III, IV	2. $XOYZ, X'OYZ$
C. V, VI	3. $X'OY'Z', XOY'Z'$
D. VII, VIII	4. $X'OY'Z, XOY'Z$

Codes

	A	B	C	D		A	B	C	D
(a)	1	2	3	4	(b)	3	1	4	2
(c)	2	4	1	3	(d)	4	1	2	3

Level II

10. Which of the following statement(s) is correct?
(a) Lines $X'OX, Y'OY$ and $Z'OZ$ are called rectangular axes
(b) Each plane divide the space in two parts
(c) Three coordinate planes together divide the space into eight regions (parts) called octant
(d) All the above are correct

Directions (Q. Nos. 11-12) *Read the following informa -tioncarefully and answer the questions given below it.*

Consider the figure given below

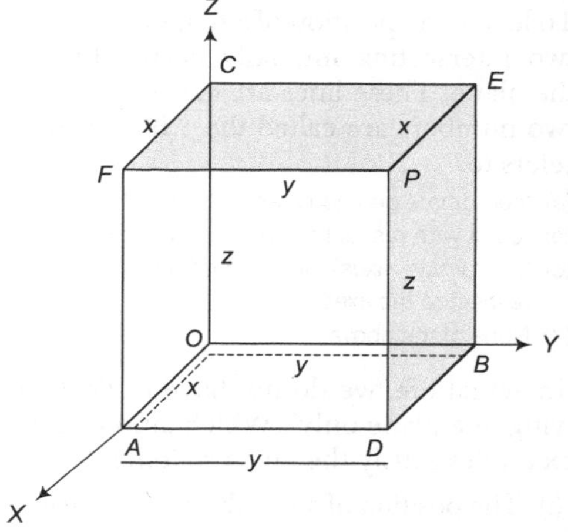

Here, a point P is shown in space, not in a coordinate plane and planes YZ, ZX and XY are meeting at the coordinate axes at the points A, B and C, respectively.

11. Based on above information which among the following is said to be correct?
(a) $ADPF || YZ$-plane
(b) $BDPE || XZ$-plane
(c) $CFPE || XY$-plane
(d) All of the above

12. Which of the following is/are not a pair of rectangular faces of the rectangular parallelopiped (as shown in the figure above)?
(i) (*ADPF, OBEC*) (ii) (*BDPE, CFAO*)
(iii) (*AOEC, FPBD*)
(a) Only (i) (b) Only (ii) (c) Only (iii) (d) Both (i) and (ii)

Coordinates of a Point in Space

Level I

13. Statement I To each point P in the space, there corresponds an ordered triplet (x, y, z) of real numbers.

Statement II There is a one-to-one correspondence between the points in space and ordered triplet (x, y, z) of real numbers.
(a) Statement I is true; Statement II is true; Statement II is a correct explanation for Statement I
(b) Statement I is true; Statement II is true; Statement II is not a correct explanation for Statement I
(c) Statement I is true; Statement II is false
(d) Statement I is false; Statement II is true

14. Statement I If $P(x, y, z)$ is any point in the space, then x, y and z are perpendicular distances from YZ, ZX and XY- planes, respectively.

Statement II If three planes are drawn parallel to YZ, ZX and XY-planes such that they intersect X, Y and Z-axes at $(x, 0, 0)$ $(0, y, 0)$ and $(0, 0, z)$, then the planes meet in space at a point $P(x, y, z)$.
(a) Statement I is true; Statement II is true; Statement II is a correct explanation for Statement I
(b) Statement I is true; Statement II is true; Statement II is not a correct explanation for Statement I
(c) Statement I is true; Statement II is false
(d) Statement I is false; Statement II is true

15. Which of the following statement(s) is correct?
(a) The coordinates of the origin O are $(0, 0, 0)$
(b) The coordinates of any point on the X-axis will be as $(0, y, z)$
(c) The coordinates of any point in the YZ-plane will be $(x, 0, 0)$
(d) All of the above statements are correct

16. Consider the table given below

Octants → Coordinates ↓	I	II	III	IV	V	VI	VII	VIII
x	+	−	−	...A...	+	...B...	...C...	+
y	+	+	−	−	...D...	+	−	−
z	+	+	+	+	...E...	−	−	...F...

Here, the signs of the coordinates in eight octants is shown with the missing part indicated as A,B,C,D, E and F. Here, A, B, C, D,E and F refers to
(a) A, D → +; B, C, E, F → −
(b) A, B, C, D → +; E, F → −
(c) A, B, C → +; D, E, F → −
(d) A, B, E, F → +; C, D → −

17. In the given figure, if P is (2, 4, 5), then the coordinates of F are

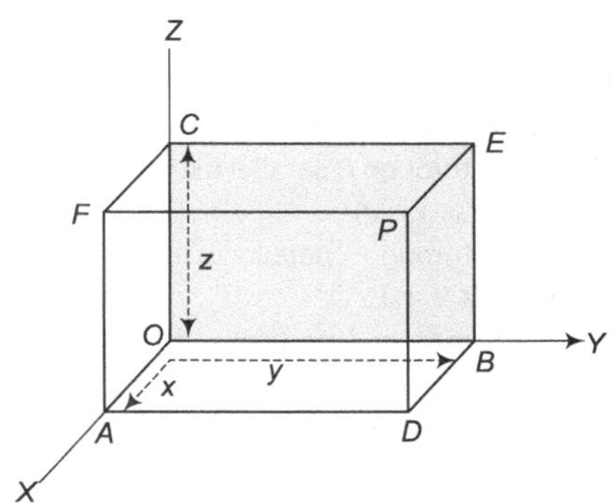

(a) (0, 2, 5) (b) (2, 0, 5)
(c) (0, 5, 2) (d) None of these

18. The octants in which the points (−3, 1, 2) and (−3, 1, − 2) lie, are respectively
(a) II and VI (b) III and V
(c) I and IV (d) II and V

19. A point is on the X-axis. Its y-coordinate and z-coordinate are respectively
(a) y, z (b) y, 0 (c) z, 0 (d) 0, 0

20. If a point is in XZ-plane, then its y-coordinate is
(a) x (b) y (c) o (d) z

21. Match the terms of Column I with the terms of Column II and choose the correct option from the codes given below.

	Column I (Points)		Column II (Name of the octants)
A.	(1, 2, 3)	1.	I- $XOYZ$
B.	(4, –2, 3)	2.	II-$X'OYZ$
C.	(4, –2, –5)	3.	III- $X'OY'Z$
D.	(4, 2, –5)	4.	IV- $XOY'Z$
E.	(–4, 2, –5)	5.	V- $XOYZ'$
F.	(–4, 2, 5)	6.	VI- $X'OYZ'$
G.	(–3, –1, 6)	7.	VII-$X'OY'Z'$
H.	(2, –4, –7)	8.	VIII- $XOY'Z'$

Codes

	A	B	C	D	E	F	G	H
(a)	1	8	4	5	2	3	6	8
(b)	8	1	2	5	4	2	1	5
(c)	1	4	8	5	6	2	3	8
(d)	8	6	2	3	1	4	2	5

22. Which of the following statement is correct?
(a) The X-axis and Y-axis taken together determine a plane known as XY-plane
(b) The coordinates of a point in the XY-plane are of the form $(0, 0, z)$
(c) Coordinate planes divide the space into six octants
(d) All the above are correct

Level II

23. If an arbitrary point P in three-dimensional space is assigned coordinates (x_0, y_0, z_0), then which of the following can be true?
(a) The plane through P parallel to the YZ-plane intersects the X-axis at $(x_0, y_0, 0)$
(b) The plane through P parallel to the XZ-plane intersects the Y-axis at $(0, 0, y_0)$
(c) The plane through P parallel to the XY-plane intersects the Z-axis at $(z_0, 0, 0)$
(d) The space coordinates are called the rectangular coordinates of P.

Directions (Q. Nos. 24-26) *Consider the figure given below and answer the questions that follows.*

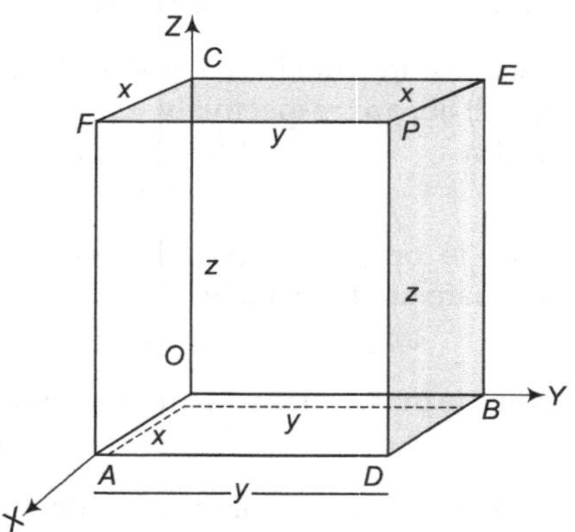

Here, rectangular coordinates of P are given as (x, y, z).

24. Which of the following statement(s) is/are true?
(a) The plane $ADPF$ is parallel to the X-axis
(b) The X-axis is perpendicular to plane $ADPF$
(c) Every line in plane $ADPF$ is perpendicular to line in the plane X-axis
(d) Only (b) and (c) are true

25. Which of the following statement(s) is/are true?
(a) PA is perpendicular to OX
(b) OX is not perpendicular to PA
(c) A is the foot of perpendicular drawn from P on Y-axis
(d) None of the above

26. Which of the following statement(s) is/are correct?
(a) A, B and C are feet of perpendiculars drawn from point P on X, Y and Z-axes, respectively
(b) The coordinates x, y and z of the point P are the perpendicular distance of P from the three coordinate planes YZ, ZX and XY, respectively
(c) Both (a) and (b) are correct
(d) Both (a) and (b) are incorrect

27. Identify the correct combination of the given statements as true (T) and false (F).
(i) The distance measured along or parallel to OX, OY and OZ are negative.
(ii) The distance measured along or parallel to OX', OY', OZ' are positive.
(iii) In first octant all the coordinates are positive and in seventh octant, all coordinates are negative.
(iv) In third octant only x and y-coordinates are negative and in fifth octant, only x and y-coordinates are positive.
(v) In fourth octant, only x and y-coordinates are positive and in sixth octant, only x and y-coordinates are positive.
(vi) In the second octant, only y and z-coordinates are positive.
(a) T, T, F, F, T, T
(b) F, F, T, T, T, F
(c) T, F, T, F, T, F
(d) F, F, T, T, F, T

28. Consider the figure given below

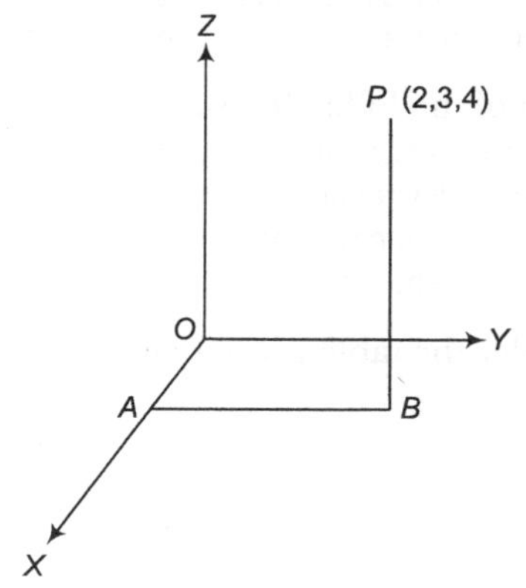

Here, a point in space is shown as P with the coordinates (2, 3, 4).

Based on the above information, which of the following is/are said to be true?

I. A is a point along the positive direction of X-axis.
II. B is a point parallel to positive direction of Y-axis.
III. The coordinates of A are (0, 0, 2).
IV. The coordinates of B are (0, 0, 3). [NCERT Exemplar]

(a) Only I is true
(b) Only II is true
(c) I and II are true
(d) III and IV are true

29. Consider the figure given below

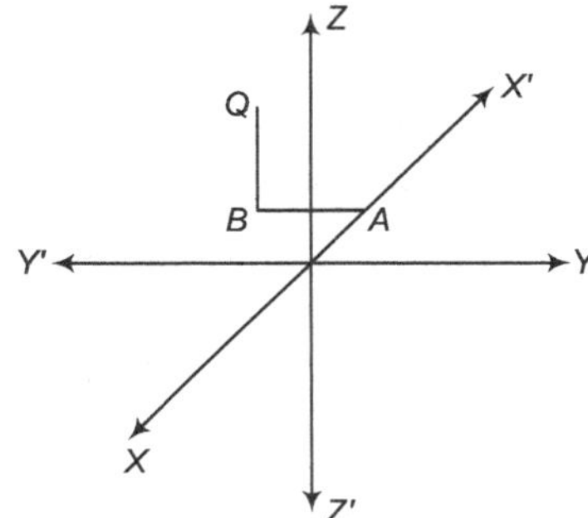

Here, A is at a distance of 2 units from the origin and along the negative direction of X-axis. From A, on moving from 2 units parallel to negative direction of Y-axis is a point B and from B, 3 units parallel to positive direction of Z-axis is a point Q.

Based on the above information, the coordinates of A, B and Q are respectively [HOTS]

(a) (–2, 2, 0), (2, –2, 0) and (2, 2, 3)
(b) (–2, 0, 0), (–2, –2, 0) and (–2, –2, 3)
(c) (2, 0, 0), (2, 2, 0) and (2, 2, 3)
(d) None of the above

30. Which of the following statement(s) is/are correct?

(a) The plane $x = 1$ represents the plane parallel to XZ-plane at a distance 1 unit above YZ-plane
(b) The equation of plane $y = 3$ represents the plane parallel to XZ-plane
(c) The equation of the plane $z = 3$ represents the plane parallel to XY-plane at a distance of 3 units above the XY-plane
(d) All of the above

31. If L, M, N be the feet of the perpendiculars drawn from a point P (3, 4, 5) on the X, Y and Z-axes respectively, then the coordinates of L, M and N are respectively

(a) (0, 0, 5), (0, 4, 0) and (3, 0, 0)
(b) (0, 3, 0), (4, 0, 0) and (0, 0, 5)
(c) (3, 0, 0), (0, 4, 0) and (0, 0, 5)
(d) None of the above

32. If L, M and N be the feet of the perpendicular drawn from a point P (3, 4, 5) on the XY, YZ and ZX-planes respectively, then the coordinates of L, M and N are respectively [NCERT Exemplar]

(a) (3, 0, 0), (0, 4, 5) and (3, 0, 5)
(b) (3, 4, 0), (0, 4, 5) and (3, 0, 5)
(c) (3, 4, 5), (0, 0, 3) and (3, 4, 0)
(d) None of the above

33. L is the foot of the perpendicular drawn from a point P (6, 7, 8) on the XY-plane, then the coordinates of point L are [NCERT Exemplar]

(a) (6, 0, 0)
(b) (6, 7, 0)
(c) (6, 0, 8)
(d) None of the above

34. L is the foot of the perpendicular drawn from a point (6, 7, 8) on X-axis. The coordinates of L are

(a) (6, 0, 0)
(b) (0, 7, 0)
(c) (0, 0, 8)
(d) None of the above

35. What is the locus of a point for which $y = 0$ and $z = 0$?

(a) Equation of X-axis
(b) Equation of Y-axis
(c) Equation of Z-axis
(d) None of the above

36. L is the foot of the perpendicular drawn from a point P (3, 4, 5) on the XZ-plane. What are the coordinates of point L? [NCERT Exemplar]

(a) (3, 0, 0)
(b) (0, 4, 5)
(c) (3, 0, 5)
(d) (3, 4, 0)

Distance between Two Points

Level I

Directions (Q. Nos. 37-40) *Consider the figure given below.*

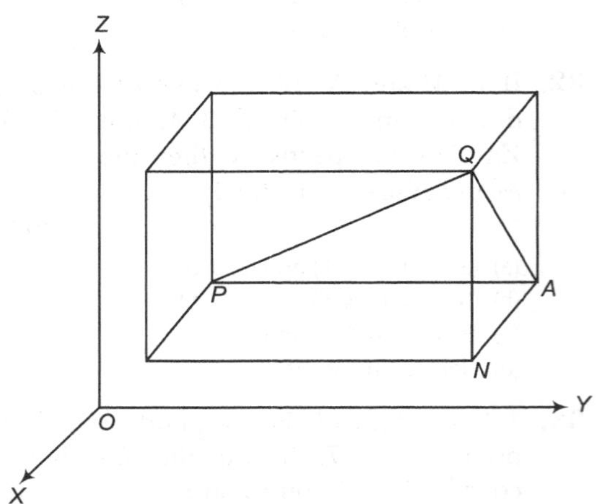

In the figure shown above, $P(x_1, y_1, z_1)$ and $Q(x_2, y_2, z_2)$ are two points referring to a system of rectangular axes OX, OY and OZ. Through the points P and Q, planes are drawn parallel to the coordinate planes forming a rectangular parallelopiped with one diagonal PQ.

Based on the above information, answer the questions given below.

37. PQ^2 is equal to

(a) $PA^2 + AN^2$ (b) $AN^2 + NQ^2$
(c) $PA^2 + NQ^2$ (d) None of these

38. PA, AN and NQ are respectively

(a) $y_2 - y_1, z_2 - z_1, x_2 - x_1$ (b) $y_2 - y_1, x_2 - x_1, z_2 - z_1$
(c) $z_2 - z_1, y_2 - y_1, x_2 - x_1$ (d) None of these

39. The distance PQ between two points $P(x_1, y_1, z_1)$ and $Q(x_2, y_2, z_2)$ is

(a) $\sqrt{(x_2 - x_1)^2 + (y_2 - y_1)^2 + (z_2 - z_1)^2}$
(b) $\sqrt{(x_2 - x_1)^2 - (y_2 - y_1)^2 - (z_2 - z_1)^2}$
(c) $\sqrt{(x_1 + x_2)^2 + (y_1 + y_2)^2 + (z_1 + z_2)^2}$
(d) None of the above

40. The distance OQ between the origin O and any point $Q(x_2, y_2, z_2)$ is

(a) $\sqrt{x_2^2 + y_2^2 + z_2^2}$ (b) $x_2^2 + y_2^2 + z_2^2$
(c) $\sqrt{(x_2 + y_2 + z_2)^2}$ (d) None of these

41. Statement I The distance between the points $P(1, -3, 4)$ and $Q(-4, 1, 2)$ is $\sqrt{5}$ units.

Statement II
$PQ = \sqrt{(x_2 - x_1)^2 + (y_2 - y_1)^2 + (z_2 - z_1)^2}$
where, P and Q are (x_1, y_1, z_1) and (x_2, y_2, z_2).

[NCERT Exemplar]

(a) Statement I is true; Statement II is true; Statement II is a correct explanation for Statement I
(b) Statement I is true; Statement II is true; Statement II is not a correct explanation for Statement I
(c) Statement I is true; Statement II is false
(d) Statement I is false; Statement II is true

42. The equation of set of points P such that $PA^2 + PB^2 = 2k^2$, where A and B are the points (3, 4, 5) and (–1, 3, –7) respectively, is

(a) $2x^2 + 2y^2 + 2z^2 - 4x + 14y + 4z = 2k^2 + 109$
(b) $2x^2 + 2y^2 + 2z^2 - 4x - 14y + 4z = 2k^2 - 109$
(c) $3x^2 + 3y^2 + 3z^2 - 4x - 14y + 4z = 2k^2 + 109$
(d) None of the above

43. Statement I The points $P(-2, 3, 5)$, $Q(1, 2, 3)$ and $R(7, 0, -1)$ are collinear.

Statement II If $PQ + QR = PR$, then P, Q and R will be collinear.

(a) Statement I is true; Statement II is true; Statement II is a correct explanation for Statement I
(b) Statement I is true; Statement II is true; Statement II is not a correct explanation for Statement I
(c) Statement I is true; Statement II is false
(d) Statement I is false; Statement II is true

44. Statement I The points $A(3, 6, 9)$, $B(10, 20, 30)$ and $C(25, -41, 5)$ are the vertices of a right angled triangle.

Statement II ΔABC is not a right angled triangle, if sum of the squares of any two sides is not equal to the square of third side.

(a) Statement I is true; Statement II is true; Statement II is a correct explanation for Statement I
(b) Statement I is true; Statement II is true; Statement II is not a correct explanation for Statement I
(c) Statement I is true; Statement II is false
(d) Statement I is false; Statement II is true

45. Match the terms of Column I with the terms of Column II and choose the correct option from the codes given below.

	Column I (Points)		Column II (Distance between them)
A.	(2, 3, 5) and (4, 3, 1)	1.	$\sqrt{43}$
B.	(–3, 7, 2) and (2, 4, –1)	2.	$2\sqrt{5}$
C.	(–1, 3, –4) and (1, –3, 4)	3.	$2\sqrt{26}$
D.	(2, –1, 3) and (–2, 1, 3)		

Codes

	A	B	C	D
(a)	2	1	3	2
(b)	1	2	3	1
(c)	3	1	2	2
(d)	2	1	3	1

46. **Statement I** Points (–2, 3, 5), (1, 2, 3) and (7, 0, –1) are collinear.

Statement II Three points *A*, *B* and *C* are said to be collinear, if $AB + BC = AC$ (as shown below).

A B C

(a) Statement I is true; Statement II is true; Statement II is a correct explanation for Statement I
(b) Statement I is true; Statement II is true; Statement II is not a correct explanation for Statement I
(c) Statement I is true; Statement II is false
(d) Statement I is false; Statement II is true

47. Which of the following statement(s) is/are correct?
(a) (0, 7, –10), (1, 6, – 6) and (4, 9, – 6) are the vertices of an isosceles triangle
(b) (0, 7, 10), (–1, 6, 6) and (–4, 9, 6) are the vertices of a right angled triangle
(c) (–1, 2, 1), (1, –2, 5) (4, –7, 8) and (2, –3, 4) are the vertices of a parallelogram
(d) All the above are correct

48. The equation of the set of points which are equidistant from the points (1, 2, 3) and (3, 2, –1), is
(a) $x - 2z = 3$
(b) $x - 3z = 2$
(c) $x - 2z = 0$
(d) $x - 2y = 0$

49. The equation of the set of point *P*, the sum of whose distances from *A*(4, 0, 0) and *B*(–4, 0, 0) is equal to 10, is
(a) $3x^2 + 25y^2 + 16z^2 - 225 = 0$
(b) $9x^2 + 25y^2 + 25z^2 - 225 = 0$
(c) $5x^2 + 16y^2 + 18z^2 - 225 = 0$
(d) None of the above

Level II

50. If *L*, *M* and *N* are the feet of perpendiculars drawn from the point *P* (3, 4, 5) on the *XY*, *YZ* and *ZX*-planes respectively, then
(a) distance of the point *L* from the point *P* is 5 units
(b) distance of the point *M* from the point *P* is 3 units
(c) distance of the point *N* from the point *P* is 4 units
(d) All of the above

51. Points *P*(2, 4, 6), *Q*(–2, –2, –2) and *R*(6, 10, 14) are
(a) vertices of a triangle (b) collinear
(c) non-collinear (d) None of these

52. **Statement I** The coordinates of a point P equidistant from the four points *O*(0, 0, 0), *A*(*l*, 0, 0), *B*(0, *m*, 0) and *C*(0, 0, *n*) are $\left(\frac{l}{2}, \frac{m}{2}, \frac{n}{2}\right)$.

Statement II Distance between two points $A(x_1, y_1, z_1)$ and $B(x_2, y_2, z_2)$ is $\sqrt{(x_2 - x_1)^2 + (y_2 - y_1)^2 + (z_2 - z_1)^2}$.

(a) Statement I is true; Statement II is true; Statement II is a correct explanation for Statement I
(b) Statement I is true; Statement II is true; Statement II is not a correct explanation for Statement I
(c) Statement I is true; Statement II is false
(d) Statement I is false; Statement II is true

53. If the point *A* (3, 2, 2) and *B* (5, 5, 4) are equidistant from *P*, which is on *X*-axis, then the coordinates of *P* are [HOTS]
(a) $\left(\frac{39}{4}, 2, 0\right)$ (b) $\left(\frac{49}{4}, 2, 0\right)$ (c) $\left(\frac{39}{4}, 0, 0\right)$ (d) $\left(\frac{49}{4}, 0, 0\right)$

54. The point on *Y*-axis which is at a distance of $\sqrt{10}$ from the point (1, 2, 3), is
(a) (0, 2, 0) (b) (0, 0, 2)
(c) (0, 0, 3) (d) None of these

55. The points (0, 7, 10), (–1, 6, 6) and (–4, 9, 6) form
(a) a right angled isosceles triangle
(b) a scalene triangle
(c) a right angled triangle
(d) an equilateral triangle

56. **Statement I** If a parallelopiped is formed by planes drawn through the points (2, 3, 5) and (5, 9, 7) parallel to the coordinate planes, then the length of the edges of a parallelopiped are 3, 6 and 2.

Statement II Length of a diagonal of above parallelopiped is 7 units.
(a) Only Statement I is true
(b) Only Statement II is true
(c) Both statements are true
(d) Both statements are false

57. The points (5, −1, 1), (7, −4, 7), (1, −6, 10) and (−1, − 3, 4) are [HOTS]

(a) the vertices of a rectangle
(b) the vertices of a square
(c) the vertices of a rhombus
(d) None of these

58. The point in YZ-plane which is equidistant from three points $A(2, 0, 3)$, $B(0, 3, 2)$ and $C(0, 0, 1)$, is

(a) (0, 3, 1)
(b) (0, 1, 3)
(c) (1, 3, 0)
(d) (3, 1, 0)

59. The length of the foot of the perpendicular drawn from the point $P(3, 4, 5)$ on Y-axis is

(a) 10 (b) $\sqrt{34}$ (c) $\sqrt{113}$ (d) $5\sqrt{2}$

60. The perpendicular distance of the point $P(6, 7, 8)$ from XY-plane is

(a) 8 (b) 7
(c) 6 (d) None of these

61. Perpendicular distance of the point $P(3, 5, 6)$ from Y-axis is

(a) $\sqrt{41}$ (b) 6
(c) 7 (d) None of these

Section Formula

Level I

62. Statement I The coordinates of the point R which divides the line segment joining two points $P(x_1, y_1, z_1)$ and $Q(x_2, y_2, z_2)$ internally in the ratio $m : n$ are

$$\left(\frac{mx_2 + nx_1}{m + n}, \frac{my_2 + ny_1}{m + n}, \frac{mz_2 + nz_1}{m + n}\right).$$

Statement II If the point R divides PQ externally in the ratio $m : n$, then its coordinates are $\left(\frac{mx_2 - nx_1}{m - n}, \frac{my_2 - ny_1}{m - n}, \frac{mz_2 - nz_1}{m - n}\right)$

(a) Statement I is true; Statement II is true; Statement II is a correct explanation for Statement I
(b) Statement I is true; Statement II is true; Statement II is not a correct explanation for Statement I
(c) Statement I is true; Statement II is false
(d) Statement I is false; Statement II is true

63. If R is the mid-point of the line segment joining $P(x_1, y_1, z_1)$ and $Q(x_2, y_2, z_2)$, then coordinates of R are

(a) $\left(\frac{x_1 - x_2}{2}, \frac{y_1 - y_2}{2}, \frac{z_1 - z_2}{2}\right)$
(b) $\left(\frac{x_1 + x_2}{2}, \frac{y_1 + y_2}{2}, \frac{z_1 + z_2}{2}\right)$
(c) $\left(\frac{x_2 - x_1}{2}, \frac{y_2 - y_1}{2}, \frac{z_2 - z_1}{2}\right)$
(d) None of the above

64. The coordinates of the point R, which divides the line segment joining $P(x_1, y_1, z_1)$ and $Q(x_2, y_2, z_2)$ in the ratio $k : 1$, are

(a) $\left(\frac{kx_2 - x_1}{1 - k}, \frac{ky_2 - y_1}{1 - k}, \frac{kz_2 - z_1}{1 - k}\right)$
(b) $\left(\frac{kx_2 + x_1}{1 + k}, \frac{ky_2 + y_1}{1 + k}, \frac{kz_2 + z_1}{1 + k}\right)$
(c) $\left(\frac{kx_2 + x_1}{1 - k}, \frac{ky_2 + y_1}{1 - k}, \frac{kz_2 + z_1}{1 - k}\right)$
(d) None of the above

65. Statement I The coordinates of the point which divides the line segment joining the points (1, − 2, 3) and (3, 4, −5) in the ratio 2 : 3 internally, are $\left(\frac{9}{5}, \frac{2}{5}, -\frac{1}{5}\right)$

Statement II The coordinates of the point which divides the line segment joining the points (1,−2, 3) and (3, 4, − 5) in the ratio 2 : 3 externally, are $\left(\frac{19}{2}, \frac{5}{2}, \frac{-1}{2}\right)$.

(a) Statement I is true; Statement II is true; Statement II is a correct explanation for Statement I
(b) Statement I is true; Statement II is true; Statement II is not a correct explanation for Statement I
(c) Statement I is true; Statement II is false
(d) Statement I is false; Statement II is true

66. **Statement I** Points (–4, 6, 10), (2, 4, 6) and (14, 0, – 2) are collinear.

Statement II Point (14, 0, – 2) divides the line segment joining by other two given points in the ratio 3 : 2 internally.

(a) Statement I is true; Statement II is true; Statement II is a correct explanation for Statement I
(b) Statement I is true; Statement II is true; Statement II is not a correct explanation for Statement I
(c) Statement I is true; Statement II is false
(d) Statement I is false; Statement II is true

67. The coordinates of a centroid of a triangle, whose vertices are $(x_1, y_1, z_1), (x_2, y_2, z_2)$ and (x_3, y_3, z_3), are

(a) $\left(\frac{x_1+x_2+x_3}{2}, \frac{y_1+y_2+y_3}{2}, \frac{z_1+z_2+z_3}{2}\right)$
(b) $\left(\frac{x_1+x_2+x_3}{2}, 0, 0\right)$
(c) $\frac{3}{2}(x_1+x_2+x_3), \frac{3}{2}(y_1+y_2+y_3), \frac{3}{2}(z_1+z_2+z_3)$
(d) None of the above

68. The ratio, in which *YZ*-plane divides the line segment joining the points (4, 8, 10) and (6, 10, –8), is

(a) 2 : 3 (externally) (b) 2 : 3 (internally)
(c) 1 : 2 (externally) (d) 1 : 2 (internally)

69. If *P*(3, 2, –4), *Q*(5, 4, –6) and *R*(9, 8, –10) are collinear points, then the ratio in which *Q* divides *PR*, is

(a) 3 : 2 (internally) (b) 3 : 2 (externally)
(c) 1 : 2 (internally) (d) 1 : 2 (externally)

70. **Statement I** The coordinates of the point which divides the line segment joining the points (–2, 3, 5) and (1, –4, 6) in the ratio 2 : 3 internally, are $\left(\frac{-4}{5}, \frac{4}{5}, \frac{27}{5}\right)$.

Statement II The coordinates of the point which divides the line segment joining the points (–2, 3, 5) and (1, –4, 6) in the ratio 2 : 3 externally, are (8, –17, 3).

(a) Only Statement I is true
(b) Only Statement II is true
(c) Both statements are true
(d) Both statements are false

71. The ratio in which *YZ*-plane divides the line segment formed by joining the points (–2, 4, 7) and (3, – 5, 8), is

(a) 2 : 3 (externally)
(b) 2 : 3 (internally)
(c) 1 : 3 (externally)
(d) 1 : 3 (internally)

72. Points *A*(2, –3, 4), *B*(–1, 2, 1) and $C\left(0, \frac{1}{3}, 2\right)$ are

(a) the vertices of a triangle
(b) the vertices of a right angled triangle
(c) the vertices of an equilateral triangle
(d) None of the above

73. The coordinates of the points which trisect the line segment joining the points *P*(4, 2,–6) and *Q*(10, –16, 6), are

(a) (6, –4, –2) and (8, 10, –2) (b) (6, –4, –2) and (8, –10, 2)
(c) (–6, 4, 2) and (–8, 10, 2) (d) None of these

Level II

74. The ratio in which the line segment joining the points (2, 4, 5) and (3, 5, –4) is divided by the *XZ*-plane, is

(a) 4 : 3 (externally) (b) 4 : 3 (internally)
(c) 4 : 5 (externally) (d) 4 : 5 (internally)

75. The coordinates of the point *P* which is five-sixth of the way from *A* (–2 ,0 , 6) to *B* (10 , – 6, – 12), is

(a) (8, 5, 9) (b) (–8, 5, 9)
(c) (8, –5, –9) (d) None of these

76. If *A* (3, 2, 0), *B*(5, 3, 2) and *C*(–9, 6, –3) are three points forming a triangle and *AD*, the bisector of ∠*BAC*, meets *BC* in *D*, then the coordinates of the point *D* are [HOTS]

(a) $\left(\frac{17}{8}, \frac{57}{8}, \frac{17}{8}\right)$ (b) $\left(\frac{19}{8}, \frac{57}{16}, \frac{17}{16}\right)$
(c) $\left(\frac{8}{17}, \frac{8}{19}, \frac{17}{8}\right)$ (d) None of these

77. **Statement I** If three vertices of a parallelogram are *A* (1, 2, 3), *B* (–1, –2, –1) and *C* (2, 3, 2), then the coordinates of fourth vertex are (4, 6, 7).

Statement II Mid-point of *BD* is $\left(\frac{x-1}{2}, \frac{y-2}{2}, \frac{z-1}{2}\right)$, where coordinates of *D* are (x, y, z). [NCERT Exemplar]

(a) Statement I is true; Statement II is true; Statement II is a correct explanation for Statement I
(b) Statement I is true; Statement II is true; Statement II is not a correct explanation for Statement I
(c) Statement I is true; Statement II is false
(d) Statement I is false; Statement II is true

78. The mid-points of the sides of a triangle are (5, 7, 11), (0, 8, 5) and (2, 3, –1), then the vertices are [NCERT Exemplar]

(a) (7, 2, 5), (3, 12, 17), (–3, 4, –7)
(b) (7, 2, 5), (3, 12, 17), (3, 4, 7)
(c) (7, 2, 5), (–3, 11, 15), (3, 4, 8)
(d) None of the above

79. The coordinate of the points which trisect the line segment joining the points $A(2, 1, -3)$ and $B(5, -8, 3)$, are [HOTS]

(a) (3, –2, –1) and (4, –5, 1)
(b) (3, 2, 1) and (4, 5, 2)
(c) (–3, 5, 1) and (3, 4, –5)
(d) None of the above

80. If the origin is the centroid of a ΔABC having vertices $A(a, 1, 3)$, $B(-2, b, -5)$ and $C(4, 7, c)$, then

(a) $a = -2$ (b) $b = 8$
(c) $c = -2$ (d) None of these

81. **Statement I** The XY-plane divides the line joining the points (–1, 3, 4) and (2, –5, 6) externally in the ratio 2 : 3.

Statement II For a point in XY-plane, its z-coordinate should be zero. [HOTS]

(a) Statement I is true; Statement II is true; Statement II is a correct explanation for Statement I
(b) Statement I is true; Statement II is true; Statement II is not a correct explanation for Statement I
(c) Statement I is true; Statement II is false
(d) Statement I is false; Statement II is true

82. A point on XOZ-plane divides the line joining the points (5, –3, –2) and (1, 2, –2) at [MHT CET 2009]

(a) $\left(\frac{13}{5}, 0 -2\right)$
(b) $\left(\frac{13}{5}, 0, 2\right)$
(c) (5, 0, 2)
(d) (5, 0, –2)

83. The ratio in which the line joining the points (2, 4, 5) and (3, 5, –4) is divided by the YZ-plane, is [UPSEE 2007]

(a) 2 : 3 (b) 3 : 2 (c) –2 : 3 (d) 4 : –3

Miscellaneous Exercise

84. It is given that the coordinates of the points A, B, C and D are (1, 2, 3), (–1, –2, –1), (2, 3, 2) and (4, 7, 6), respectively.

Based on the above information, which of the following is correct?

(a) $ABCD$ is a parallelogram (b) $ABCD$ is a rectangle
(c) Both (a) and (b) (d) Neither (a) nor (b)

85. The equation of the set of the point P, such that its distance from the points $A(3, 4 -5)$ and $B(-2, 1, 4)$ is equal, is

(a) $6x + 10y - 18z - 29 = 0$ (b) $10x + 6y - 18z - 29 = 0$
(c) $10y + 6x + 18z + 23 = 0$ (d) None of these

86. If the centroid of a ΔABC is at the point (1, 1, 1) and the coordinates of A and B are (3, –5, 7) and (–1, 7, –6) respectively, then the coordinates of point C are

(a) (1, 2, 1) (b) (1, 1, 2)
(c) (1, 1, 0) (d) (2, 0, 1)

87. If three vertices of a parallelogram $ABCD$ are A (3,–1, 2), B(1, 2, –4) and C(–1, 1, 2), then the coordinates of the fourth vertex are

(a) (–2, 1, 6) (b) (2, 1, 8)
(c) (–1, 2, 8) (d) None of these

88. A triangle is given with the vertices $A(0, 0, 6)$ $B(0, 4, 0)$ and $C(6, 0, 0)$.

Statement I Coordinates of mid-points of side of ΔABC are (3, 0, 2), (0, 3, 3) and (0,2, 3).

Statement II Length of medians of ΔABC are 7, 7 and $\sqrt{34}$. [HOTS]

(a) Only Statement I is true
(b) Only Statement II is true
(c) Both statements are true
(d) Both statements are false

89. If the origin is the centroid of the ΔPQR with vertices $P(2a, 2, 6)$, $Q(-4, 3b, -10)$ and $R(8, 14, 2c)$, then

(a) $a = 2$ (b) $b = \frac{16}{3}$ (c) $c = 2$ (d) $a + b = -4$

90. The coordinates of a point on Y-axis which are at a distance of $5\sqrt{2}$ from the point $P(3, -2, 5)$, are [HOTS]

(a) (0, – 6, 0) (b) (0, 2, 0)
(c) Both (a) and (b) (d) Neither (a) nor (b)

91. If a point R with x-coordinate 4 lies on the line segment joining the points $P(2, -3, 4)$ and $Q(8, 0, 10)$, then the coordinates of the point R are

(a) (4, 2, 6) (b) (4, –2, 6)
(c) (4, 4, –6) (d) None of these

92. If A and B are the points (3, 4, 5) and (–1, 3, –7) respectively, then the equation of set of point P such that $PA^2 + PB^2 = k^2$, where k is constant, is

(a) $2(x^2 + y^2 + z^2) - 4x - 14y + 4z + 109 - k^2 = 0$
(b) $3(x^2 + y^2 + z^2) - 2x - 14y + 4z - 109 + k^2 = 0$
(c) $2(x^2 - y^2 + z^2) + 2x - 14y + 8z + 109 - k^2 = 0$
(d) None of the above

93. If $x^2 + y^2 = 1$, then the point $P(x, y, \sqrt{1 - x^2 - y^2})$ is at a distance of ...A... units from the origin. Here, A refers to [NCERT Exemplar]

(a) 1 (b) 2 (c) 3 (d) 4

94. If $A(2, 2, -3)$, $B(5, 6, 9)$ and $C(2, 7, 9)$ are the vertices of a triangle and the internal bisector of the $\angle A$ meets BC at the point D, then the coordinates of D are [NCERT Exemplar; HOTS]

(a) $\left(\frac{3}{2}, \frac{13}{2}, 9\right)$ (b) $\left(\frac{7}{2}, \frac{13}{2}, 9\right)$

(c) $\left(\frac{5}{2}, \frac{7}{2}, 9\right)$ (d) None of these

95. If the coordinates of the vertices of a cube, whose edge is 2 units, one of whose vertices coincides with the origin and there edges passing through the origin coincides with the positive direction of the axes through the origin, are O, A, G, D, B, F, E and C, then which of the following is correct? [NCERT Exemplar]

(a) O, A, G are (0, 0, 4), (2, 0, 0) and (0, 2, 3) respectively
(b) D, B, F are (0, 4, 2), (2, 2, 0) and (2, 2, 2) respectively
(c) E and C are (2, 0, 2) and (0, 2, 2) respectively
(d) All of the above

96. The centroid of the triangle, if the mid-point of the sides of triangles are $D(1, 2, -3)$, $E(3, 0, 1)$ and $F(-1, 1, -4)$, is [NCERT Exemplar]

(a) (1, 2, −1) (b) (1, 1, −2)
(c) (0, 1, −2) (d) None of these

97. Match questions given under Column I to its correct answer given under Column II and choose the correct option from the codes given below.

Column I	Column II
A. If the centroid of the triangle is origin and two of its vertices are (3, −5, 7) and (−1, 7, −6), then the third vertex is	1. Parallelogram
B. The points $A(3, -1, -1)$, $B(5, -4, 0)$, $C(2, 3, -2)$ and $D(0, 6, -3)$ are the vertices of a	2. An isosceles right angled triangle
C. Points $A(1, -1, 3)$, $B(2, -4, 5)$ and $C(5, -13, 11)$ are	3. (−2, −2, −1)
D. Points $A(2, 4, 3)$, $B(4, 1, 9)$ and $C(10, -1, 6)$ are the vertices of	4. Collinear

Codes

	A	B	C	D		A	B	C	D
(a)	3	1	4	2	(b)	1	3	4	2
(c)	2	1	3	4	(d)	1	2	4	3

98. Identify the statements as true (T) or false (F) and then choose the correct option [HOTS]

(i) A line is parallel to XY-plane, if all the points on the line have equal y-coordinates.
(ii) The equation $x = b$ represents a plane perpendicular to YZ-plane at a unit distance b from the origin.
(iii) The point (4, 5, −6) lies in the VIth octant.
(iv) The X-axis is the intersection of two planes XY-plane and XZ-plane.

(a) T, T, T, F (b) F, F, F, T
(c) T, F, T, F (d) F, T, F, T

99. Which of the following statement(s) is incorrect?

(a) The equation of the plane $z = 6$ represents a plane perpendicular to the XY-plane, having a z-intercept of 6 units
(b) The equation of the plane $x = 0$ represents the YZ-plane
(c) The point on the X-axis with x-coordinate equal to x_0 is written as $(x_0, 0, 0)$
(d) $x = x_0$ represents a plane parallel to the YZ-plane

100. The perimeter of triangle with vertices at (1, 0, 0), (0, 1, 0) and (0, 0, 1) is ...P.... Here, P refers to

(a) 3 (b) 2
(c) $2\sqrt{2}$ (d) $3\sqrt{2}$

101. The distance between the points (1, 4, 5) and (2, 2, 3) is [MP PET 2009]

(a) 5 (b) 4
(c) 3 (d) 2

102. The point in the xy-plane which is equidistant from the points (2, 0, 3), (0, 3, 2) and (0, 0, 1), is [Kerala CEE 2008]

(a) (1, 2, 3) (b) (−3, 2, 0) (c) (3, −2, 0) (d) (3, 2, 0)
(e) (3, 2, 1)

103. The perpendicular distance of the point (6, 5, 8) from Y-axis is ... K .... Here, K refers to [Kerala CEE 2007]

(a) 5 units (b) 6 units (c) 8 units (d) 9 units
(e) 10 units

Answer *with* Explanations

1. *(c)* To locate the position of a point in a plane, we need two intersecting mutually perpendicular lines in the plane. These lines are called the coordinate axes and the two numbers are called the coordinates of the point with respect to the axes.

2. *(d)* In actual life, we do not have to deal with points lying in a plane only. For example, consider the position of a ball thrown in space at different points of time or the position of an aeroplane as it flies from one place to another at different times during its flight.

   Similarly, if we were to locate the position of the lowest tip of an electric bulb hanging from the ceiling of a room or the position of the central tip of the ceiling fan in a room, we will not only require the perpendicular distances of the point to be located from two perpendicular walls of the room but also the height of the point from the floor of the room. Therefore, we need not only two but three numbers representing the perpendicular distances of the point from three mutually perpendicular planes, namely the floor of the room and two adjacent walls of the room.

3. *(c)* The three numbers representing the three perpendicular distances are called the coordinates of the point with reference to the three coordinate planes. So, a point in space has three coordinates.

   Solutions (Q. Nos. 4-9) These problems are based on coordinate axes and coordinate planes in three dimensional space.

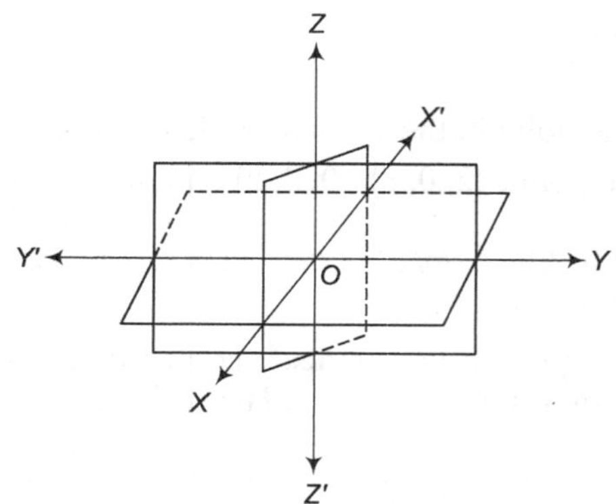

4. *(b)* Considering the three planes intersecting at a point O such that these three planes are mutually perpendicular to each other (see figure). These three planes intersect along the lines $X'OX$, $Y'OY$ and $Z'OZ$, called the X, Y and Z-axes, respectively. We see that these lines are mutually perpendicular to each other. These lines constitute the rectangular coordinate system.

5. *(d)* The planes XOY, YOZ and ZOX, called, respectively the XY-plane, YZ-plane and the ZX-plane, are known as the three coordinate planes.

6. *(b)* If we take the XOY plane as the plane of the paper then line $Z'OZ$ is perpendicular to the plane XOY. If the plane of the paper is considered as horizontal, then the line $Z'OZ$ will be vertical.

7. *(c)* The distances measured from XY-plane upwards in the direction of OZ are taken as positive and those measured downwards in the direction of OZ' are taken as negative. Similarly, the distance measured to the right of ZX-plane and along OY are taken as positive, to the left of ZX-plane and along OY' as negative, in front of the YZ-plane along OX as positive and to the back of it along OX' as negative.

8. *(a)* The point O is called the origin of the coordinate system. The three coordinate planes divide the space into eight parts known as octants.

9. *(c)* These octants are named as

$$XOYZ,\ X'OYZ,\ X'OY'Z,\ XOY'Z$$

   $XOYZ'$, $X'OYZ'$, $X'OY'Z'$ and $XOY'Z'$ and denoted by I, II, III, IV, V, VI, VII and VIII, respectively.

   $\therefore$ A $\leftrightarrow$ 2, B $\leftrightarrow$ 4, C $\leftrightarrow$ 1, D $\leftrightarrow$ 3

10. *(d)* Let $X'OX$, $Y'OY$, and $Z'OZ$ be three mutually perpendicular lines that pass through a point O such that $X'OX$ and $Y'OY$ lies in the plane of the paper and line $Z'OZ$ is perpendicular to the plane of paper. These three lines are called rectangular axes (lines $X'OX$, $Y'OY$ and $Z'OZ$ are called X-axis, Y-axis and Z-axis). We call this coordinate system a three dimensional space, or simply space.

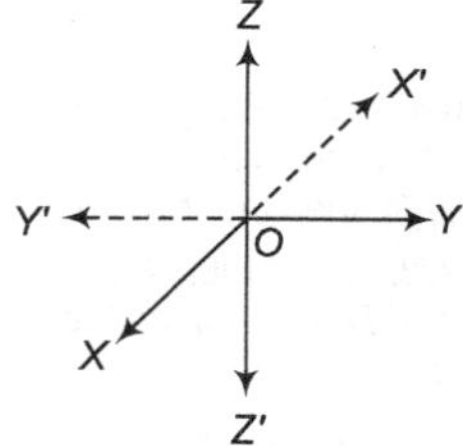

The three axes when taken together in pairs determine XY, YZ and ZX-planes, *i.e.*, three coordinate planes.Each plane divide the space in two parts and the three coordinate planes together divide the space into eight regions (parts) called octants, namely (i) $OXYZ$ (ii) $OX'YZ$ (iii) $OXY'Z$ (iv) $OXYZ'$ (v) $OXY'Z'$ (vi) $OX'YZ'$ (vii) $OX'Y'Z$ (viii) $OX'Y'Z'$.

Solutions (Q. Nos. 11-12) Here, P is any point in the space, not in a coordinate plane, and through P pass planes parallel to the coordinate planes YZ, ZX and XY meeting the coordinate axes in the points A, B, and C, respectively.

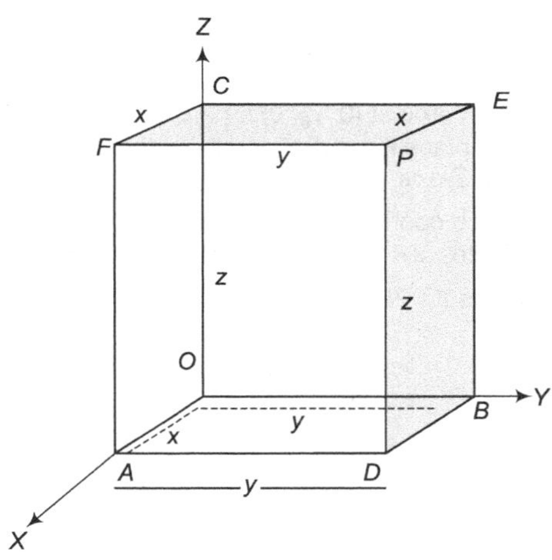

11. *(d)* So, three planes are

(i) *ADPF* || *YZ*-plane (ii) *BDPE* || *XZ*-plane

(iii) *CFPE* || *XY*-plane

12. *(c)* These planes determine a rectangular parallelopiped which has three pairs of rectangular faces.

(*ADPF*, *OBEC*), (*BDPE*, *CFAO*) and (*AOBD*, *FPEC*)

13. *(a)* (This problem is based on coordinates of a point in space)

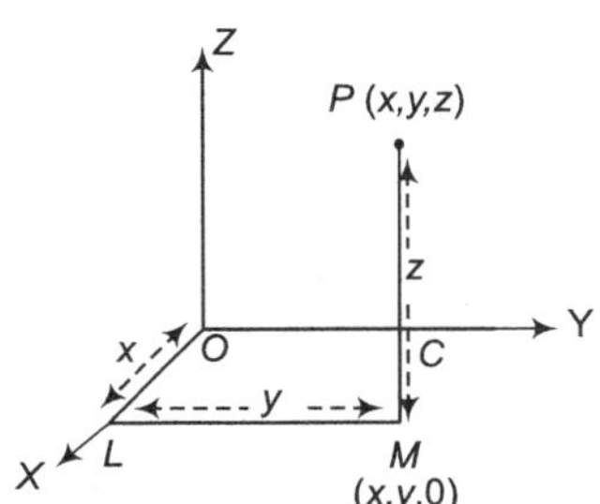

II. Given a point *P* in space, we drop a perpendicular *PM* on the *XY*-plane with *M* as the foot of this perpendicular (see figure). Then, from the point *M*, we draw a perpendicular *ML* to the *X*-axis, meeting it at *L*. Let *OL* be *x*, *LM* be *y* and *MP* be *z*. Then *x* , *y* and *z* are called the *x* , *y* and *z*-coordinates respectively, of the point *P* in the space.

In figure we may note that the point *P* (*x* , *y* , *z*) lies in the octant *XOYZ* and so all *x* , *y* ,and *z* are positive. If *P* was in any other octant, the signs of *x* , *y* and *z* would change accordingly. Thus, to each point *P* in the space there corresponds an ordered triplet (*x* , *y* , *z*) of real numbers.

Conversely, given any triplet (*x* , *y* , *z*), we would first fix the point *L* on the *X*-axis corresponding to *x*, then locate the point *M* in the *XY*-plane such that (*x*, *y*) are the coordinates of the point *M* in the *XY*-plane. Note that *LM* is perpendicular to the *X*-axis or is parallel to the *Y*-axis. Having reached the point *M*, we draw a perpendicular *MP* to the *XY*-plane and locate on it the point *P* corresponding to *z*. The point P so obtained has then the coordinates (*x* , *y* , *z*). Thus, there is a one-to-one correspondence between the points in space and ordered triplet (*x* , *y* , *z*) of real numbers.

So, both the statements are true and Statement II is the correct explanation of Statement I.

14. *(b)* I. Through the point *P* in the space, we draw three planes parallel to the coordinate planes, meeting the *X*-axis, *Y*-axis and *Z*-axis in the points *A*, *B* and *C*, respectively. We observe that, $OA = x, OB = y$ and $OC = z$. Thus, if *P*(*x*, *y*, *z*) is any point in the space, then *x*, *y* and *z* are perpendicular distances from *YZ*, *ZX* and *XY*-planes, respectively.

II. Given *x*, *y* and *z*, we locate the three points *A*, *B* and *C* on the three coordinate axes. Through the points *A*, *B* and *C*, we draw planes parallel to the *YZ*-plane, *ZX*-plane and *XY*-plane, respectively. The point of intersection of these three planes, namely, *ADPF*, *BDPE* and *CEPF* is obviously the point *P*, corresponding to the ordered triplet (*x* , *y z*).

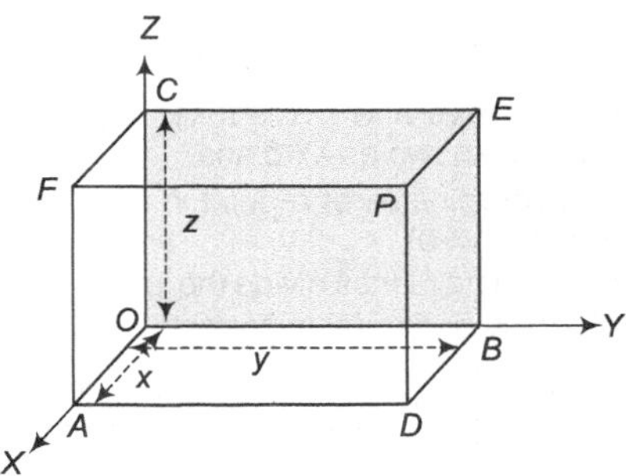

15. *(a)* The coordinates of the origin *O* are (0, 0, 0). The coordinates of any point on the *X*-axis will be as (*x*, 0, 0) and the coordinates of any point in the *YZ*-plane will be as (0, *y* , *z*).

16. *(a)* The sign of the coordinates of a point determine the octant in which the point lies. The following table shows the signs of the coordinates in eight octants.

Octants → Coordinates ↓	I *XOYZ*	II *X'OYZ*	III *X'OY'Z*	IV *XOY'Z*	V *XOYZ'*	VI *X'OYZ'*	VII *X'OY'Z'*	VIII *XOY'Z'*
x	+	−	−	+	+	−	−	+
y	+	+	−	−	+	+	−	−
z	+	+	+	+	−	−	−	−

17. *(b)* For the point F, the distance measured along OY is zero. Therefore, the coordinates of F are (2, 0, 5) .

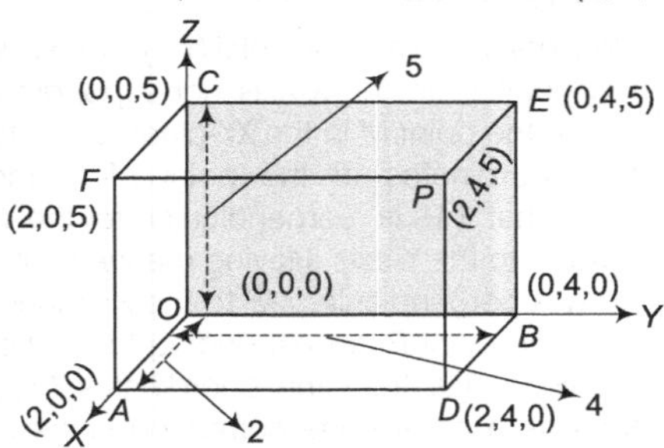

18. *(a)* The point (–3, 1, 2) lies in second octant (*i.e.*, $X'OYZ$) and the point (–3, 1, –2) lies in octant VI (*i.e.*, $X'OYZ'$) (see the table given in solution 16)

19. *(d)* Coordinates of any point on the X-axis is $(x, 0, 0)$. Therefore, both y and z-coordinates are zero.

20. *(c)* Any point on the XZ-plane will have the coordinate $(x, 0, z)$, then its y-coordinate is 0.

21. *(c)*

Point	Octant	Name
A. (1, 2, 3)	I(all the coordinates are positive)	$XOYZ$
B. (4, –2, 3)	IV (y-coordinate is negative)	$XOY'Z$
C. (4, –2, –5)	VIII(y and z-coordinates are negative)	$XOY'Z'$
D. (4, 2, –5)	V (z-coordinate is negative)	$XOYZ'$
E. (–4, 2,–5)	VI(x and z-coordinates are negative)	$X'OYZ'$
F. (–4, 2, 5)	II (x-coordinate is negative)	$X'OYZ$
G. (–3, –1, 6)	III	$X'OY'Z$
H. (2, –4, –7)	VIII (y and z-coordiantes are negative)	$XOY'Z'$

$\therefore$ $A \leftrightarrow 1, B \leftrightarrow 4, C \leftrightarrow 8, D \leftrightarrow 5,$
$E \leftrightarrow 6, F \leftrightarrow 2, G \leftrightarrow 3, H \leftrightarrow 8$

22. *(a)* (i) The X-axis and Y-axis taken together determine a plane known as XY-plane.
(ii) The coordinates of a point in the XY-plane are of the form $(x , y , 0)$.
(iii) Coordinate planes divide the space into eight octants. So, only option (a) is correct.

23. *(d)* An arbitrary point P in three-dimensional space is assigned coordinates (x_0, y_0, z_0) provided that

(i) the plane through P parallel to the YZ-plane intersects the X-axis at $(x_0, 0, 0)$.
(ii) the plane through P parallel to the XZ-plane intersects the Y-axis at $(0, y_0, 0)$.
(iii) the plane through P parallel to the XY-plane intersects the Z-axis at $(0, 0, z_0)$.

The space coordinates (x_0, y_0, z_0) are called the cartesian coordinates of P or simply the rectangular coordinates of P.

Solutions (Q. Nos. 24-26)

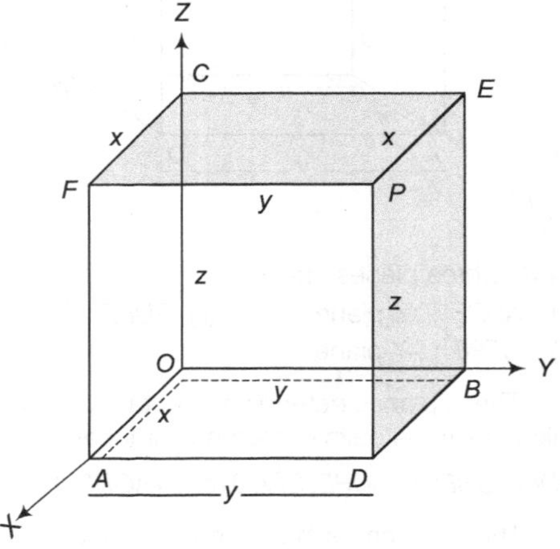

24. *(d)* From the figure shown above, we can say, that plane $ADPF$ is perpendicular to the X-axis or X-axis is perpendicular to the plane $ADPF$ and hence perpendicular to every line in the plane.

25. *(a)* From the above solution, we see that PA is perpendicular to OX and OX is perpendicular to PA. Thus, A is the foot of perpendicular drawn from P on X-axis.

26. *(c)* Distance of foot A from O is x-coordinate of point P. Similarly, we call B and C are the feet of perpendiculars drawn from point P on the Y and Z-axes and distances of these feet B and C from O are the y and z-coordinates of the point P.

Hence, the coordinates x, y and z of a point P are the perpendicular distance of P from the three coordinate planes YZ, ZX and XY, respectively.

27. *(d)* The distance measured along or parallel to OX, OY and OZ will be positive and distance moved along or parallel to OX', OY', OZ' will be negative. The three mutually perpendicular coordinate planes which in turn divide the space into eight parts and each part is known as octant. The sign of the coordinates of a point depend upon the octant, in which it lies. In first octant all the coordinates are positive and in seventh octant, all coordinates are negative. In third octant, x, y-coordinates are negative and z-coordinate is positive. In fifth octant x, y are positive and z-coordinate is negative. In fourth octant, x, z coordinates are positive and y is negative. In sixth octant, x and z are negative y is positive. In the second octant, x is negative and y and z-coordinates are positive.

Octants → Coordinates ↓	I $XOYZ$	II $X'OYZ$	III $X'OY'Z$	IV $XOY'Z$	V $XOYZ'$	VI $X'OYZ'$	VII $X'OY'Z'$	VIII $XOY'Z'$
x	+	–	–	+	+	–	–	+
y	+	+	–	–	+	+	–	–
z	+	+	+	+	–	–	–	–

So, only option (d) is correct.

28. *(c)* A point P is shown in space with the coordinates (2, 3, 4).

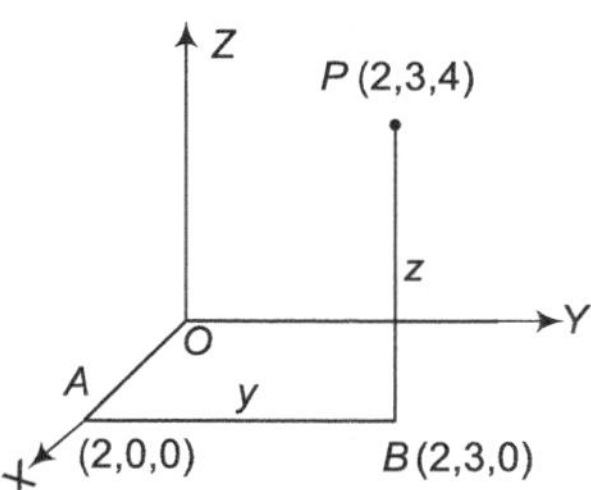

From the figure, it is clear that A is a point along the positive direction of X-axis, so its coordinates are (2, 0, 0) and point B is at a distance of 3 units from x-axis and parallel to positive direction of Y-axis. So, the coordinates of B are (2, 3, 0). Here, P is along the positive direction of Z-axis and at a distance of 4 units from the point B.

29. *(b)*

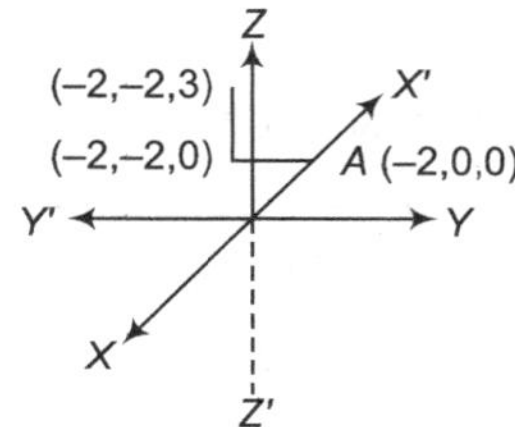

We see that from the origin, on moving 2 units along the negative direction of X-axis, the point A is (–2, 0, 0) and from the point A on moving 2 units parallel to negative direction of Y-axis, we get the coordinates of point B as (–2, –2, 0).

From B, moving 3 units parallel to positive direction of Z-axis, we get the point Q as (–2, –2, 3).

30. *(c)* (a) The equation of the plane $x=0$ represents the YZ-plane and equation of the plane $x=1$ represents the plane parallel to YZ-plane at a distance 1 unit above YZ-plane. A plane parallel to YZ-plane at a distance 1 unit above YZ-plane is shown below.

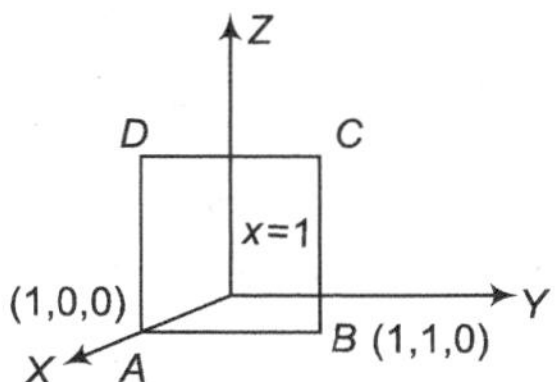

(b) The equation of the plane $y=0$ represents the XZ-plane and the equation of the plane $y=3$ represents the plane parallel to XZ-plane at a distance 3 units above XZ-plane .

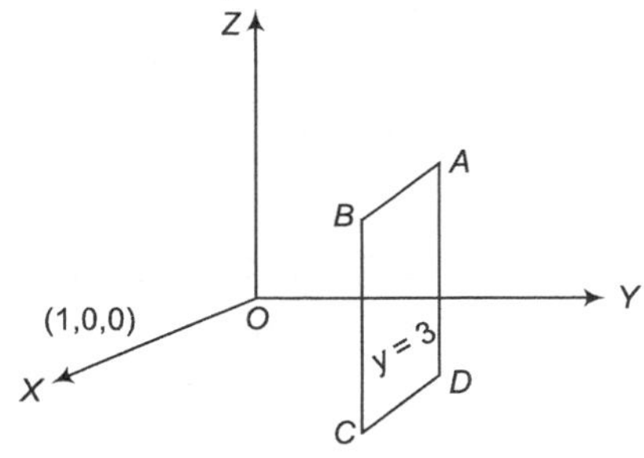

(c) The equation of the plane $z=0$ represents the XY-plane and $z=3$ represents the plane parallel to XY-plane at a distance of 3 units above XY-plane .

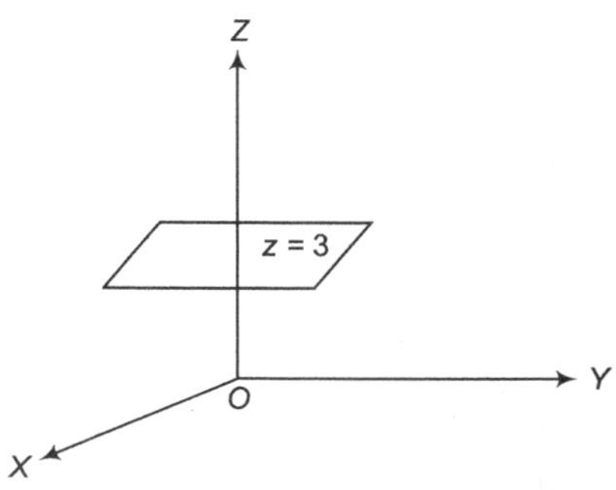

So, only option (c) is correct.

31. *(c)* Since, L is the foot of perpendicular from P on the X-axis, its y and z-coordinates are zero. The coordinates of L is (3, 0, 0). Similarly, the coordinates of M and N are (0, 4, 0) and (0, 0, 5), respectively.

32. *(b)* Since, L is the foot of perpendicular segment from P on the XY-plane, z-coordinate is zero in the XY-plane.

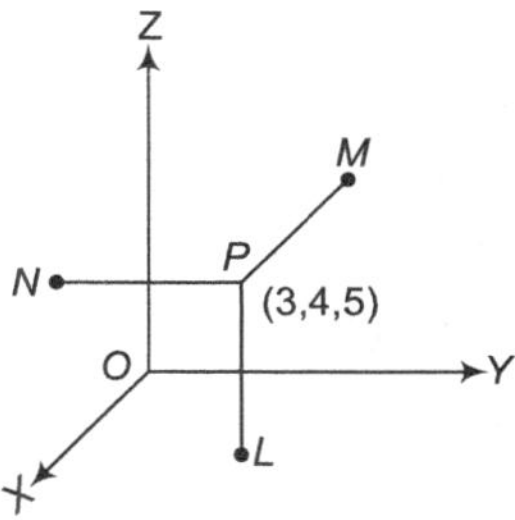

Hence, coordinates of L are (3, 4, 0). Similarly, we can find the coordinates of M (0, 4, 5) and N (3, 0, 5).

33. *(b)* Since, L is the foot of perpendicular from P on the XY-plane, z-coordinate is zero in the XY-plane. Hence, coordinates of L are (6, 7, 0).

34. *(a)* Since, L is the foot of perpendicular from P on the X-axis, y and z-coordinates are zero. Hence, the coordinates of L are (6, 0, 0).

35. *(a)* Locus of the point $y=0$, $z=0$ is X-axis, since on X-axis both $y=0$ and $z=0$.

36. *(c)* Since, L is the foot of perpendicular segment drawn from the point P (3, 4, 5) on the XZ-plane. Since, the y-coordinates of all points in the XZ-plane are zero, coordinate of the foot of perpendicular are (3, 0, 5).

Solutions (Q. Nos. 37-40) Here, $P(x_1, y_1, z_1)$ and $Q(x_2, y_2, z_2)$ are two points referred to a system of rectangular axes OX, OY and OZ. Through the points P and Q, planes are drawn parallel to the coordinate planes so as to form a rectangular parallelopiped with one diagonal PQ .

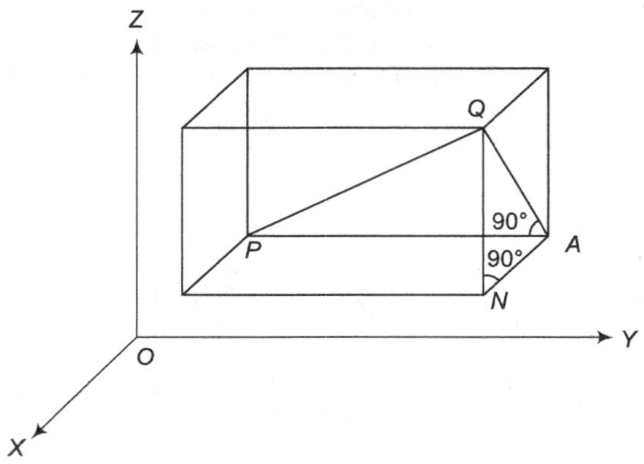

37. *(d)* Now, since $\angle PAQ$ is a right angle, it follows that in triangle PAQ,

$$PQ^2 = PA^2 + AQ^2 \quad \ldots(i)$$

Also, ΔANQ is right angled triangle with $\angle ANQ$ as right angle.

Therefore, $AQ^2 = AN^2 + NQ^2 \quad \ldots(ii)$

From Eqs. (i) and (ii), we have

$$PQ^2 = PA^2 + AN^2 + NQ^2$$

38. *(b)* $PA = y_2 - y_1$, $AN = x_2 - x_1$ and $NQ = z_2 - z_1$

39. *(a)* $\because PQ^2 = PA^2 + AN^2 + NQ^2$

$$PQ^2 = (x_2 - x_1)^2 + (y_2 - y_1)^2 + (z_2 - z_1)^2$$

Therefore, $PQ = \sqrt{(x_2 - x_1)^2 + (y_2 - y_1)^2 + (z_2 - z_1)^2}$

This gives us the distance between two points (x_1, y_1, z_1) and (x_2, y_2, z_2).

40. *(a)* In particular, if $x_1 = y_1 = z_1 = 0$, *i.e.*, point P is origin O, then $OQ = \sqrt{x_2^2 + y_2^2 + z_2^2}$, which gives the distance between the origin O and any point $Q(x_2, y_2, z_2)$.

41. *(d)* The distance PQ between the points $P(1, -3, 4)$ and $Q(-4, 1, 2)$ is

$$PQ = \sqrt{(-4-1)^2 + (1+3)^2 + (2-4)^2}$$
$$= \sqrt{25 + 16 + 4} = \sqrt{45} = 3\sqrt{5} \text{ units}$$

42. *(b)* Let the coordinates of point P be (x, y, z).

Here, $PA^2 = (x-3)^2 + (y-4)^2 + (z-5)^2$

and $PB^2 = (x+1)^2 + (y-3)^2 + (z+7)^2$

By the given condition $PA^2 + PB^2 = 2k^2$, we have

$$(x-3)^2 + (y-4)^2 + (z-5)^2 + (x+1)^2 + (y-3)^2 + (z+7)^2 = 2k^2$$

i.e., $2x^2 + 2y^2 + 2z^2 - 4x - 14y + 4z = 2k^2 - 109$.

43. *(a)* We know that points are said to be collinear, if they lie on a line.

Now, $PQ = \sqrt{(1+2)^2 + (2-3)^2 + (3-5)^2}$

$$= \sqrt{9 + 1 + 4} = \sqrt{14}$$

$$QR = \sqrt{(7-1)^2 + (0-2)^2 + (-1-3)^2} = \sqrt{36 + 4 + 16}$$
$$= \sqrt{56} = 2\sqrt{14}$$

and $PR = \sqrt{(7+2)^2 + (0-3)^2 + (-1-5)^2} = \sqrt{81 + 9 + 36}$

$$= \sqrt{126} = 3\sqrt{14}$$

Thus, $PQ + QR = PR$. Hence, P, Q and R are collinear.

44. *(d)* By the distance formula, we have

$$AB^2 = (10-3)^2 + (20-6)^2 + (30-9)^2$$
$$= 49 + 196 + 441 = 686$$
$$BC^2 = (25-10)^2 + (-41-20)^2 + (5-30)^2$$
$$= 225 + 3721 + 625 = 4571$$
$$CA^2 = (3-25)^2 + (6+41)^2 + (9-5)^2$$
$$= 484 + 2209 + 16 = 2709$$

We find that $CA^2 + AB^2 \neq BC^2$

Hence, ΔABC is not a right angled triangle.

45. *(a)* A. The given points are $A(2, 3, 5)$ and $B(4, 3, 1)$.

$\therefore \quad x_1 = 2, \; y_1 = 3, \; z_1 = 5$

$x_2 = 4, \; y_2 = 3, \; z_2 = 1$

$\therefore$ Required distance

$$AB = \sqrt{(x_2 - x_1)^2 + (y_2 - y_1)^2 + (z_2 - z_1)^2}$$
$$\Rightarrow AB = \sqrt{(4-2)^2 + (3-3)^2 + (1-5)^2}$$
$$= \sqrt{4 + 0 + 16} = \sqrt{20} = 2\sqrt{5}$$

B. The given points are $A(-3, 7, 2)$ and $B(2, 4, -1)$.

Here, $x_1 = -3, \; y_1 = 7, \; z_1 = 2$

$x_2 = 2, \; y_2 = 4, \; z_2 = -1$

$\therefore$ Required distance

$$AB = \sqrt{(x_2 - x_1)^2 + (y_2 - y_1)^2 + (z_2 - z_1)^2}$$
$$= \sqrt{[2-(-3)]^2 + (4-7)^2 + (-1-2)^2}$$
$$= \sqrt{(2+3)^2 + (4-7)^2 + (-1-2)^2}$$
$$= \sqrt{25 + 9 + 9} = \sqrt{43}$$

C. The given points are $A(-1, 3, -4)$ and $B(1, -3, 4)$.

$\therefore$ Required distance

$$AB = \sqrt{(1+1)^2 + (-3-3)^2 + (4+4)^2}$$
$$= \sqrt{4 + 36 + 64} = \sqrt{104} = 2\sqrt{26}$$

D. The given points are $A(2, -1, 3)$ and $B(-2, 1, 3)$.

$\therefore$ Required distance

$$AB = \sqrt{(-2-2)^2 + (1+1)^2 + (3-3)^2}$$
$$= \sqrt{4^2 + 2^2 + 0} = \sqrt{16 + 4} = \sqrt{20} = 2\sqrt{5}$$

46. *(a)* The given points are

$$A(-2, 3, 5), B(1, 2, 3), C(7, 0, -1)$$

Distance between A and B

$$AB = \sqrt{(-2-1)^2 + (3-2)^2 + (5-3)^2}$$
$$= \sqrt{(-3)^2 + (1)^2 + (2)^2} = \sqrt{9 + 1 + 4} = \sqrt{14}$$

Distance between B and C

$$BC = \sqrt{(1-7)^2 + (2-0)^2 + (3+1)^2}$$
$$= \sqrt{(-6)^2 + (2)^2 + (4)^2}$$
$$= \sqrt{36+4+16} = \sqrt{56} = 2\sqrt{14}$$

Distance between A and C

$$AC = \sqrt{(-2-7)^2 + (3-0)^2 + (5+1)^2}$$
$$= \sqrt{(-9)^2 + (3)^2 + (6)^2}$$
$$= \sqrt{81+9+36} = \sqrt{126} = 3\sqrt{14}$$

Clearly, $AB + BC = AC$

Hence, the given points are collinear.

47. *(d)* (a) Let $A(0, 7, -10)$, $B(1, 6, -6)$ and $C(4, 9, -6)$ be the vertices of a triangle. Then,

Side AB = Distance between points A and B

$$= \sqrt{(0-1)^2 + (7-6)^2 + (-10+6)^2}$$
$$= \sqrt{1+1+16} = \sqrt{18} = 3\sqrt{2}$$

and side BC = Distance between points B and C

$$= \sqrt{(1-4)^2 + (6-9)^2 + (-6+6)^2}$$
$$= \sqrt{9+9+0} = \sqrt{18} = 3\sqrt{2}$$

Clearly, $AB = BC$. Hence, triangle is an isosceles triangle.

(b) Let $A(0, 7, 10)$, $B(-1, 6, 6)$ and $C(-4, 9, 6)$ be the vertices of a triangle. Then,

Side $AB = \sqrt{(0+1)^2 + (7-6)^2 + (10-6)^2}$
$$= \sqrt{1+1+16}$$
$$= \sqrt{18} = 3\sqrt{2}$$

Side $BC = \sqrt{(-1+4)^2 + (6-9)^2 + (6-6)^2}$
$$= \sqrt{9+9+0}$$
$$= \sqrt{18} = 3\sqrt{2}$$

and side $CA = \sqrt{(-4-0)^2 + (9-7)^2 + (6-10)^2}$
$$= \sqrt{16+4+16}$$
$$= \sqrt{36} = 6$$

Now, $AB^2 + BC^2 = (3\sqrt{2})^2 + (3\sqrt{2})^2 = 6^2 = CA^2$

$\therefore$ ΔABC is right angled triangle at B.

(c) Let $A(-1, 2, 1)$, $B(1, -2, 5)$, $C(4, -7, 8)$ and $D(2, -3, 4)$ be the vertices of a quadrilateral $ABCD$.

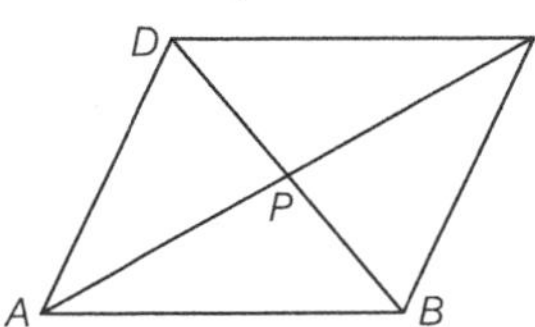

Then, mid-point of $AC = \left(\frac{-1+4}{2}, \frac{2-7}{2}, \frac{1+8}{2}\right)$
$$= \left(\frac{3}{2}, \frac{-5}{2}, \frac{9}{2}\right)$$

Mid-point of $BD = \left(\frac{1+2}{2}, \frac{-2-3}{2}, \frac{5+4}{2}\right)$
$$= \left(\frac{3}{2}, \frac{-5}{2}, \frac{9}{2}\right)$$

Mid-points of both the diagonals are same (*i.e.*, they bisect each other.)

Hence, $ABCD$ is a parallelogram.

48. *(c)* Consider the given points as A and B. Let $P(x, y, z)$ be any point equidistant from A and B.

$\therefore$ $PA = PB$

i.e., Distance between P and A = Distance between P and B

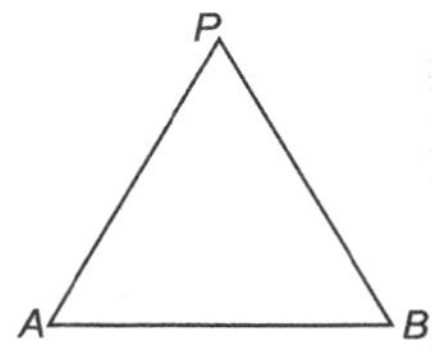

$\Rightarrow$ $\sqrt{(x-1)^2 + (y-2)^2 + (z-3)^2}$
$$= \sqrt{(x-3)^2 + (y-2)^2 + (z+1)^2}$$

$\Rightarrow$ $(x-1)^2 + (y-2)^2 + (z-3)^2$
$$= (x-3)^2 + (y-2)^2 + (z+1)^2$$

$\Rightarrow$ $x^2 + 1 - 2x + y^2 + 4 - 4y + z^2 + 9 - 6z$
$$= x^2 + 9 - 6x + y^2 + 4 - 4y + z^2 + 1 + 2z$$

$\Rightarrow$ $4x - 8z = 0 \Rightarrow x - 2z = 0$

which is the required equation.

49. *(b)* Let the point be $P(x, y, z)$. Then, it is given that $PA + PB = 10$

$\Rightarrow$ $\sqrt{(x-4)^2 + (y-0)^2 + (z-0)^2}$
$$+ \sqrt{(x+4)^2 + (y-0)^2 + (z-0)^2} = 10$$

$\Rightarrow$ $\sqrt{(x-4)^2 + y^2 + z^2} = 10 - \sqrt{(x+4)^2 + y^2 + z^2}$

On squaring both sides, we get

$$(x-4)^2 + y^2 + z^2 = 100 + (x+4)^2 + y^2 + z^2 - 20\sqrt{(x+4)^2 + y^2 + z^2}$$

$\Rightarrow$ $x^2 + 16 - 8x = 100 + x^2 + 16 + 8x - 20\sqrt{(x+4)^2 + y^2 + z^2}$

$\Rightarrow$ $-8x - 8x - 100 = -20\sqrt{(x+4)^2 + y^2 + z^2}$

$\Rightarrow$ $-16x - 100 = -20\sqrt{(x+4)^2 + y^2 + z^2}$

$\Rightarrow$ $4x + 25 = 5\sqrt{(x+4)^2 + y^2 + z^2}$

(dividing both sides by -4)

Again, squaring on both sides, we get

$$(4x+25)^2 = 25[(x+4)^2 + y^2 + z^2]$$

$\Rightarrow$ $16x^2 + 625 + 200x = 25[(x+4)^2 + y^2 + z^2]$

$\Rightarrow 16x^2 + 625 + 200x = 25[x^2 + 16 + 8x + y^2 + z^2]$

$\Rightarrow 16x^2 + 625 + 200x = 25x^2 + 400 + 200x + 25y^2 + 25z^2$

$\Rightarrow 9x^2 + 25y^2 + 25z^2 - 225 = 0$

which is the required equation.

50. *(d)* L is the foot of perpendicular drawn from the point $P(3, 4, 5)$ to the XY-plane. Therefore, the coordinate of the point L is (3, 4, 0). The distance between the point (3, 4, 5) and (3, 4, 0) is 5. Similarly, we can find the lengths of the foot of perpendiculars on YZ and ZX-plane which are 3 and 4 units, respectively.

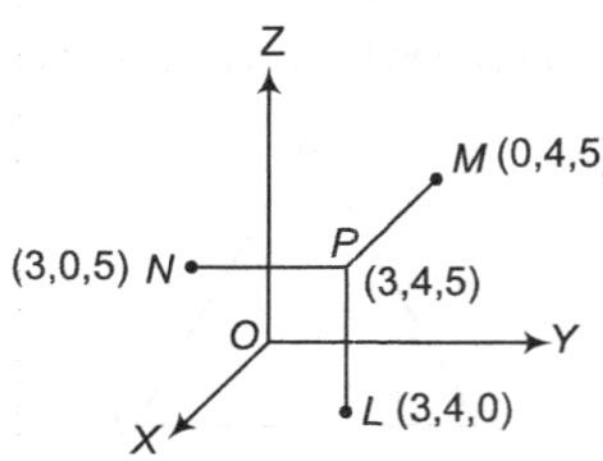

51. *(b)* Three points are collinear if the sum of any two distances is equal to the third distance.

$PQ = \sqrt{(-2-2)^2 + (-2-4)^2 + (-2-6)^2}$

$= \sqrt{16 + 36 + 64} = \sqrt{116} = 2\sqrt{29}$

$QR = \sqrt{(6+2)^2 + (10+2)^2 + (14+2)^2}$

$= \sqrt{64 + 144 + 256} = \sqrt{464} = 4\sqrt{29}$

$PR = \sqrt{(6-2)^2 + (10-4)^2 + (14-6)^2}$

$= \sqrt{16 + 36 + 64} = \sqrt{116} = 2\sqrt{29}$

Since, $QR = PQ + PR$. Therefore, given points are collinear.

52. *(a)* Let $P(x, y, z)$ be the required point.

Then, $OP = PA = PB = PC$.

Now, $OP = PA$

$\Rightarrow OP^2 = PA^2$

$\Rightarrow x^2 + y^2 + z^2 = (x - l)^2 + (y - 0)^2 + (z - 0)^2$

$\Rightarrow x = \frac{l}{2}$

Similarly, $OP = PB \Rightarrow y = \frac{m}{2}$ and $OP = PC \Rightarrow z = \frac{n}{2}$

Hence, the coordinates of the required points are $\left(\frac{l}{2}, \frac{m}{2}, \frac{n}{2}\right)$.

53. *(d)* The point on the X-axis is of the form $P(x, 0, 0)$. Since, the points A and B are equidistant from P. Therefore, $PA^2 = PB^2$, *i.e.*, $(x-3)^2 + (0-2)^2 + (0-2)^2$

$= (x-5)^2 + (0-5)^2 + (0-4)^2$

$\Rightarrow 4x = 25 + 25 + 16 - 17$ *i.e.*, $x = \frac{49}{4}$

Thus, the point P on the X-axis is $\left(\frac{49}{4}, 0, 0\right)$ which is equidistant from A and B.

54. *(a)* Let P be the point on Y-axis. Therefore, it is of the form $P(0, y, 0)$. The point (1, 2, 3) is at a distance of $\sqrt{10}$ from (0, y, 0). Therefore, $\sqrt{(1-0)^2 + (2-y)^2 + (3-0)^2} = \sqrt{10}$

$\Rightarrow y^2 - 4y + 4 = 0 \Rightarrow (y-2)^2 = 0 \Rightarrow y = 2$

Hence, the required point is (0, 2, 0).

55. *(a)* Let $P(0, 7, 10)$, $Q(-1, 6, 6)$ and $R(-4, 9, 6)$ be the vertices of a triangle.

Here, $PQ = \sqrt{1 + 1 + 16} = 3\sqrt{2}$

$QR = \sqrt{9 + 9 + 0} = 3\sqrt{2}$

$PR = \sqrt{16 + 4 + 16} = 6$

Now, $PQ^2 + QR^2 = (3\sqrt{2})^2 + (3\sqrt{2})^2$

$= 18 + 18 = 36 = (PR)^2$

Therefore, ΔPQR is a right angled triangle at Q. Also, $OQ = QR$. Hence, ΔPQR is a right angled isosceles triangle.

56. *(c)* Length of the edges of the parallelopiped are

$(5-2), (9-3), (7-5)$ *i.e.*, 3, 6, 2.

Length of a diagonal is $\sqrt{3^2 + 6^2 + 2^2} = 7$ units.

57. *(c)* Let $A(5, -1, 1)$, $B(7, -4, 7)$, $C(1, -6, 10)$ and $D(-1, -3, 4)$ be the four points of a quadrialateral. Here,

$AB = \sqrt{4 + 9 + 36} = 7,$

$BC = \sqrt{36 + 4 + 9} = 7,$

$CD = \sqrt{4 + 9 + 36} = 7$

$DA = \sqrt{36 + 4 + 9} = 7$

$AC = \sqrt{16 + 25 + 81} = \sqrt{122}$

$BD = \sqrt{64 + 1 + 9} = \sqrt{74}$

Note that $AB = BC = CD = DA$ and $AC \neq BD$. Therefore, $ABCD$ is a rhombus.

58. *(b)* Since x-coordinate of every point in YZ-plane is zero. Let $P(0, y, z)$ be a point on the YZ-plane such that $PA = PB = PC$. Now,

$PA = PB$

$\Rightarrow (0-2)^2 + (y-0)^2 + (z-3)^2$

$= (0-0)^2 + (y-3)^2 + (z-2)^2,$

i.e. $z - 3y = 0$

and $PB = PC$

$\Rightarrow y^2 + 9 - 6y + z^2 + 4 - 4z = y^2 + z^2 + 1 - 2z,$

i.e., $3y + z = 6$

On simplifying the two equations, we get $y = 1$ and $z = 3$.

Here, the coordinates of the point P are (0, 1, 3).

59. *(b)* Let L be the foot of the perpendicular from point P on the Y-axis. Therefore, its x and z-coordinates are zero, *i.e.*, (0, 4, 0). Therefore, distance between the points (0, 4, 0) and (3, 4, 5) is $\sqrt{9 + 25}$ *i.e.*, $\sqrt{34}$.

60. *(a)* Let L be the foot of perpendicular drawn from the point $P(6, 7, 8)$ to the XY-plane and the distance of this foot L from P is z-coordinate of P *i.e.*, 8 units.

61. *(d)* Let M be the foot of perpendicular from P on the Y-axis, therefore, its x and z-coordinates are zero. *The* coordinates of M is (0, 5, 0). Therefore, the perpendicular distance of the point P from Y-axis $\sqrt{3^2 + 6^2} = \sqrt{45}$

62. *(b)* The two given points are $P(x_1, y_1, z_1)$ and $Q(x_2, y_2, z_2)$.

Let the point $R(x, y, z)$ divide PQ in the given ratio $m:n$ internally. Draw PL, QM and RN perpendicular to the XY-plane. Obviously $PL\ |RN|\ QM$ and feet of these perpendiculars lie in XY-plane. The points L, M and N will lie on a line which is the intersection of the plane containing PL, RN and QM with the XY-plane. Through the point R draw a line ST parallel to the line LM. Line ST will intersect the line LP externally at the point S and the line MQ at T, as shown in figure.

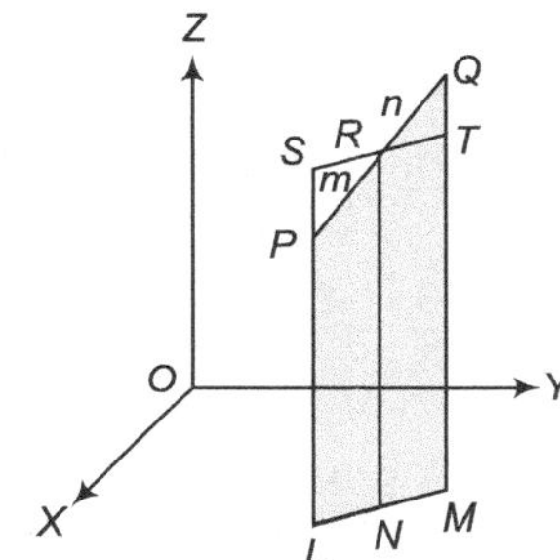

Also, note that quadrilaterals $LNRS$ and $NMTR$ are parallelograms.

The triangles PSR and QTR are similar. Therefore,

$$\frac{m}{n} = \frac{PR}{QR} = \frac{SP}{QT} = \frac{SL - PL}{QM - TM} = \frac{NR - PL}{QM - NR} = \frac{z - z_1}{z_2 - z}$$

This implies $z = \dfrac{mz_2 + nz_1}{m + n}$

Similarly, by drawing perpendiculars to the XZ and YZ-planes, we get

$$y = \frac{my_2 + ny_1}{m + n}$$

and
$$x = \frac{mx_2 + nx_1}{m + n}$$

Hence, the coordinates of the point R which divides the line segment joining two points $P(x_1, y_1, z_1)$ and $Q(x_2, y_2, z_2)$ internally in the ratio $m:n$ are

$$\left(\frac{mx_2 + nx_1}{m + n}, \frac{my_2 + ny_1}{m + n}, \frac{mz_2 + nz_1}{m - n}\right)$$

If the point R divides PQ externally in the ratio $m:n$, then its coordinates are obtained by replacing $-n$ by n, so that coordinates of point R will be

$$\left(\frac{mx_2 - nx_1}{m - n}, \frac{my_2 - ny_1}{m - n}, \frac{mz_2 - nz_1}{m - n}\right)$$

So, both the statements are true but Statement II is not a correct explanation of Statement I.

63. *(b)* Referring to the previous solution,

In case R is the mid-point of PQ, then $m:n = 1:1$, so that

$$x = \frac{x_1 + x_2}{2},\ y = \frac{y_1 + y_2}{2} \text{ and } z = \frac{z_1 + z_2}{2}$$

These are the coordinates of the mid-point of the line segment joining $P(x_1, y_1, z_1)$ and $Q(x_2, y_2, z_2)$.

are
$$\left(\frac{x_1 + x_2}{2}, \frac{y_1 + y_2}{2}, \frac{z_1 + z_2}{2}\right)$$

64. *(b)* The coordinates of the point R which divides PQ in the ratio $k:1$ where coordinates of P and Q are (x_1, y_1, z_1) and (x_2, y_2, z_2) are obtained by taking $k = \dfrac{m}{n}$ in the coordinates of the point R which divides PQ internally in the ratio $m:n$, which are as given below.

$$\left(\frac{kx_2 + x_1}{1 + k}, \frac{ky_2 + y_1}{1 + k}, \frac{kz_2 + z_1}{1 + k}\right)$$

Note *Generally, this result is used in solving problems involving a general point on the line passing through two given points.*

65. *(c)* I. Let $P(x, y, z)$ be the point which divides line segment joining $A(1, -2, 3)$ and $B(3, 4, -5)$ internally in the ratio 2 : 3. Therefore,

$$x = \frac{2(3) + 3(1)}{2 + 3} = \frac{9}{5}$$
$$y = \frac{2(4) + 3(-2)}{2 + 3} = \frac{2}{5}$$
$$z = \frac{2(-5) + 3(3)}{2 + 3} = \frac{-1}{5}$$

Thus, the required point is $\left(\dfrac{9}{5}, \dfrac{2}{5}, \dfrac{-1}{5}\right)$

II. Let $P(x, y, z)$ be the point which divides segment joining $A(1, -2, 3)$ and $B(3, 4, -5)$ externally in the ratio 2 : 3.

Then,
$$x = \frac{2(3) - 3(1)}{2 - 3} = -3$$
$$y = \frac{2(4) - 3(-2)}{2 - 3} = -14$$
$$z = \frac{2(-5) - 3(3)}{2 - 3} = 19$$

Therefore, the required point is $(-3, -14, 19)$.

66. *(c)* Let $A(-4, 6, 10)$, $B(2, 4, 6)$ and $C(14, 0, -2)$ be the given points. Let the point P divides AB in the ratio $k:1$. Then, coordinates of the point P are

$$\frac{2k - 4}{k + 1}, \frac{4k + 6}{k + 1}, \frac{6k + 10}{k + 1}$$

Let us examine whether for some value of k, the point P coincides with point C.

On putting $\dfrac{2k - 4}{k + 1} = 14$, we get $k = -\dfrac{3}{2}$

When $k = -\frac{3}{2}$, then $\frac{4k+6}{k+1} = \frac{4\left(-\frac{3}{2}\right)+6}{-\frac{3}{2}+1} = 0$

and $\frac{6k+10}{k+1} = \frac{6\left(-\frac{3}{2}\right)+10}{-\frac{3}{2}+1} = -2$

Therefore, C (14, 0, −2) is a point which divides AB externally in the ratio 3 : 2 and is same as P. Hence, A, B, C are collinear.

67. *(d)* Let ABC be the triangle. Let the coordinates of the vertices A, B, C be $(x_1, y_1, z_1), (x_2, y_2, z_2)$ and (x_3, y_3, z_3), respectively. Let D be the mid-point of BC. Hence, coordinates of D are

$$\left(\frac{x_2+x_3}{2}, \frac{y_2+y_3}{2}, \frac{z_2+z_3}{2}\right)$$

Let G be the centroid of the triangle. Therefore, it divides the median AD in the ratio 2 : 1. Hence, the coordinates of G are

$$\left(\frac{2\left(\frac{x_2+x_3}{2}\right)+x_1}{2+1}, \frac{2\left(\frac{y_2+y_3}{2}\right)+y_1}{2+1}, \frac{2\left(\frac{z_2+z_3}{2}\right)+z_1}{2+1}\right)$$

or $\left(\frac{x_1+x_2+x_3}{3}, \frac{y_1+y_2+y_3}{3}, \frac{z_1+z_2+z_3}{3}\right)$

68. *(a)* Let YZ-plane divides the line segment joining $A(4, 8, 10)$ and $B(6, 10, -8)$ at $P(x, y, z)$ in the ratio $k : 1$. Then, the coordinates of P are

$$\left(\frac{4+6k}{k+1}, \frac{8+10k}{k+1}, \frac{10-8k}{k+1}\right)$$

Since, P lies on the YZ-plane, its x-coordinates is zero,

i.e., $\frac{4+6k}{k+1} = 0$ or $k = -\frac{2}{3}$

Therefore, YZ- plane divides AB externally in the ratio 2 : 3.

69. *(c)* Let Q divides PR in the ratio $k : 1$.

k 1
P Q R
(3, 2, − 4) (5, 4, − 6) (9, 8, − 10)

Here, the point Q divides the line PR internally, so its coordinates are $\left[\left(\frac{mx_2+nx_1}{m+n}\right), \left(\frac{my_2+ny_1}{m+n}\right), \left(\frac{mz_2+nz_1}{m+n}\right)\right]$

$$Q = \left(\frac{k\times 9+1\times 3}{k+1}, \frac{k\times 8+1\times 2}{k+1}, \frac{k(-10)+1\times(-4)}{k+1}\right)$$

$$= \left(\frac{9k+3}{k+1}, \frac{8k+2}{k+1}, \frac{-10k-4}{k+1}\right)$$

But given, $Q = (5, 4, -6)$

On comparing the corresponding coordinates

$\therefore \quad \frac{9k+3}{k+1} = 5, \frac{8k+2}{k+1} = 4, \frac{-10k-4}{k+1} = -6$

$\Rightarrow \quad 9k+3 = 5k+5, 8k+2 = 4k+4, -10k-4 = -6k-6$

$\Rightarrow \quad 4k = 2$

$\Rightarrow \quad k = \frac{1}{2}$

Hence, point Q divides PR internally in the ratio 1 : 2.

70. *(d)* The given points are $A(-2, 3, 5)$ and $B(1, -4, 6)$.

I. Let the point C divides the line AB internally in the ratio 2 : 3.

2 3
A C B
(−2, 3, 5) (1, −4, 6)

Here, the ratio is 2 : 3.

$\therefore \quad m = 2, n = 3$

The coordinates of point C

$$= \left[\left(\frac{m_1 x_2+nx_1}{m+n}\right), \left(\frac{m_1 y_2+ny_1}{m+n}\right), \left(\frac{m_1 z_2+nz_1}{m+n}\right)\right]$$

$$= \left(\frac{2\times(+1)+3\times(-2)}{(2+3)}, \frac{2\times(-4)+3\times 3}{(2+3)}, \frac{2\times 6+3\times 5}{(2+3)}\right)$$

$$= \left(\frac{2-6}{5}, \frac{-8+9}{5}, \frac{12+15}{5}\right)$$

$$= \left(\frac{-4}{5}, \frac{1}{5}, \frac{27}{5}\right)$$

II. Let the point C divides the line AB externally in the ratio 2 : 3.

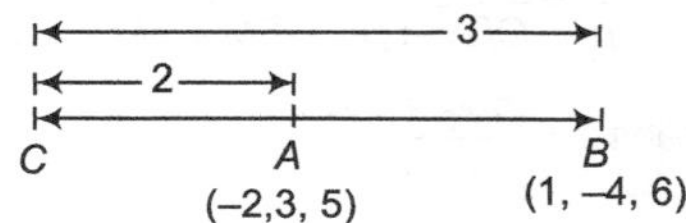

Here, the ratio is 2 : 3.

$\therefore \quad m = 2, n = 3$

The coodinates of point C

$$= \left[\left(\frac{mx_2-nx_1}{m-n}\right), \left(\frac{my_2-ny_1}{m-n}\right), \left(\frac{mz_2-nz_1}{m-n}\right)\right]$$

$$= \left(\frac{2\times(1)-3\times(-2)}{(2-3)}, \frac{2\times(-4)-3\times 3}{(2-3)}, \frac{2\times 6-3\times 5}{(2-3)}\right)$$

$$= \left(\frac{2+6}{(-1)}, \frac{-8-9}{(-1)}, \frac{12-15}{(-1)}\right) = (-8, 17, 3)$$

71. *(b)* The given points are $A(-2, 4, 7)$ and $B(3, -5, 8)$.

Let the point $P(0, y, z)$ in YZ-plane divides AB in the ratio $k : 1$. Then,

P
A k 1 B
(−2, 4, 7) (3, −5, 8)

x-coordinate of point $P = \frac{mx_2+nx_1}{m+n}$

$$\frac{k \times 3 + 1 \times (-2)}{k+1} = 0 \quad (\because \text{x-coordinate of } P \text{ is zero})$$

$$\Rightarrow \quad 3k - 2 = 0$$

$$\Rightarrow \quad k = \frac{2}{3}$$

$$\Rightarrow \quad k : 1 = 2 : 3$$

$\therefore$ *YZ*-plane divides the segment internally in the ratio 2 : 3.

72. *(d)* Let *C* divides *AB* in the ratio *k* : 1.

$A(2, -3, 4) \xrightarrow{k} C(0, 1/3, 2) \xrightarrow{1} B(-1, 2, 1)$

Now, *x*-coordinate of $C = 0$

i.e. $$\frac{k \times (-1) + 1 \times (2)}{k+1} = 0$$

$$\Rightarrow \quad -k + 2 = 0$$

$$\Rightarrow \quad k = 2$$

$$\Rightarrow \quad k : 1 = 2 : 1$$

and *y*-coordinate of $C = \frac{1}{3}$

$$\Rightarrow \quad \frac{k \times 2 + 1 \times (-3)}{k+1} = \frac{1}{3}$$

$$\Rightarrow \quad \frac{2k - 3}{k+1} = \frac{1}{3}$$

$$\Rightarrow \quad 6k - 9 = k + 1$$

$$\Rightarrow \quad 5k = 10$$

$$\Rightarrow \quad k = 2$$

$$\Rightarrow \quad k : 1 = 2 : 1$$

and *z*-coordinate of $C = 2$

$$\Rightarrow \quad \frac{k \times 1 + 1 \times 4}{k+1} = 2$$

$$\Rightarrow \quad k + 4 = 2k + 2$$

$$\Rightarrow \quad k = 2$$

$$\Rightarrow \quad k : 1 = 2 : 1$$

Thus, *C* divides *AB* internally in the ratio 2 : 1. Hence, points *A*, *B* and *C* are collinear.

Note *Sometimes, when we separately find out the values of k in different coordinates, axes are not same, then we can say that it is not internally divided.*

73. *(b)* Let the points R_1 and R_2 trisects the line *PQ* *i.e.*, R_1 divides the line in the ratio 1 : 2.

$P(4, 2, -6) \xrightarrow{1} R_1 \xrightarrow{2} Q(10, -16, 6)$

$$\Rightarrow R_1 = \left(\frac{1 \times 10 + 2 \times 4}{1+2}, \frac{1 \times (-16) + 2 \times 2}{1+2}, \frac{1 \times 6 + 2 \times (-6)}{1+2}\right)$$

$$= \left(\frac{10+8}{3}, \frac{-16+4}{3}, \frac{6-12}{3}\right) = \left(\frac{18}{3}, \frac{-12}{3}, \frac{-6}{3}\right)$$

$$= (6, -4, -2)$$

Again, let the point R_2 divides *PQ* internally in the ratio 2 : 1. Then,

$P(4, 2, -6) \xrightarrow{2} R_2 \xrightarrow{1} Q(10, -16, 6)$

$$\Rightarrow R_2 = \left(\frac{2 \times 10 + 1 \times 4}{2+1}, \frac{2 \times (-16) + 1 \times 2}{2+1}, \frac{2 \times 6 + 1 \times (-6)}{1+2}\right)$$

$$= \left(\frac{20+4}{3}, \frac{-32+2}{3}, \frac{12-6}{3}\right) = \left(\frac{24}{3}, \frac{-30}{3}, \frac{6}{3}\right)$$

$$= (8, -10, 2)$$

Hence, required points are $(6, -4, -2)$ and $(8, -10, 2)$.

74. *(c)* Let the joint of $P(2, 4, 5)$ and $Q(3, 5, -4)$ be divided by *XZ*-plane in the ratio $k : 1$ at the point $R(x, y, z)$. Therefore,

$$x = \frac{3k+2}{k+1}, \; y = \frac{5k+4}{k+1}, \; z = \frac{-4k+5}{k+1}$$

Since, the point $R(x, y, z)$ lies on the *XZ*-plane, the *y*-coordinate should be zero, *i.e.*,

$$\frac{5k+4}{k+1} = 0$$

$$\Rightarrow \quad k = -\frac{4}{5}$$

Hence, the required ratio is $-4 : 5$, *i.e.*, externally in the ratio $4 : 5$.

75. *(c)* Let $P(x, y, z)$ be the required point, *i.e.*, *P* divides *AB* in the ratio 5 : 1. Then,

$$P(x, y, z) = \left(\frac{5 \times 10 + 1 \times -2}{5+1}, \frac{5 \times -6 + 1 \times 0}{5+1}, \frac{5 \times -12 + 1 \times 6}{5+1}\right)$$

$$= (8, -5, -9)$$

76. *(b)* $AB = \sqrt{(5-3)^2 + (3-2)^2 + (2-0)^2}$

$$= \sqrt{4 + 1 + 4} = 3$$

$$AC = \sqrt{(-9-3)^2 + (6-2)^2 + (-3-0)^2} = \sqrt{144 + 16 + 9} = 13$$

Since, *AD* is the bisector of $\angle BAC$, we have $\frac{BD}{DC} = \frac{AB}{AC} = \frac{3}{13}$

i.e., *D* divides *BC* in the ratio 3 : 13.

Hence, the coordinates of *D* are

$$\left(\frac{3(-9) + 13(5)}{3+13}, \frac{3(6) + 13(3)}{3+13}, \frac{3(-3) + 13(2)}{3+13}\right) = \left(\frac{19}{8}, \frac{57}{16}, \frac{17}{16}\right)$$

77. *(d)* We know that, in a parallelogram, diagonals bisect each other.

Now, let the coordinates of *D* be (x, y, z).

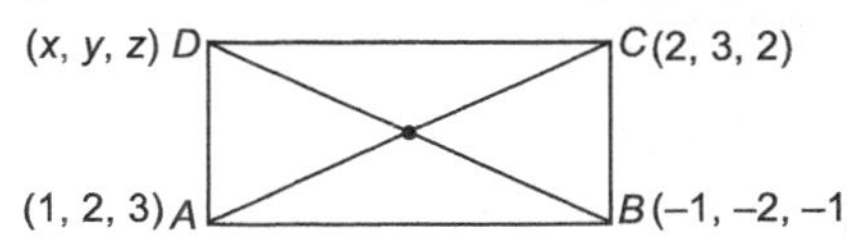

$\therefore$ Mid-point of *BD* is $\left(\frac{x-1}{2}, \frac{y-2}{2}, \frac{z-1}{2}\right)$ *i.e.*, $\left(\frac{3}{2}, \frac{5}{2}, \frac{5}{2}\right)$.

Therefore, $\frac{x-1}{2}=\frac{3}{2} \Rightarrow x=4$

$\frac{y-2}{2}=\frac{5}{2} \Rightarrow y=7$

and $\frac{z-1}{2}=\frac{5}{2} \Rightarrow z=6$

The coordinates of D be (4, 7, 6).

78. *(a)* Let the vertices of a triangle be $A(x_1, y_1, z_1)$, $B(x_2, y_2, z_2)$ and $C(x_3, y_3, z_3)$.

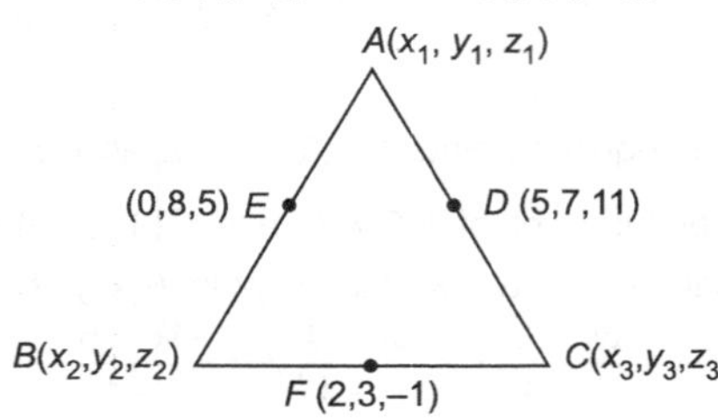

Since, D, E and F are the mid-points of AC, BC and AB.

$\therefore \left(\frac{x_1+x_2}{2}, \frac{y_1+y_2}{2}, \frac{z_1+z_2}{2}\right)=(0, 8, 5)$

$\Rightarrow x_1+x_2=0, y_1+y_2=16, z_1+z_2=10$ ...(i)

$\left(\frac{x_2+x_3}{2}, \frac{y_2+y_3}{2}, \frac{z_2+z_3}{2}\right)=(2, 3, -1)$

$\Rightarrow x_2+x_3=4, y_2+y_3=6, z_2+z_3=-2$ ...(ii)

and $\left(\frac{x_1+x_3}{2}, \frac{y_1+y_3}{2}, \frac{z_1+z_3}{2}\right)=(5, 7, 11)$

$\Rightarrow x_1+x_3=10, y_1+y_3=14, z_1+z_3=22$ ...(iii)

On adding Eqs. (i), (ii) and (iii), we get

$2(x_1+x_2+x_3)=14, 2(y_1+y_2+y_3)=36,$
$2(z_1+z_2+z_3)=30$

$\Rightarrow x_1+x_2+x_3=7, y_1+y_2+y_3=18,$
$z_1+z_2+z_3=15$ ...(iv)

On solving Eqs. (i), (ii), (iii) and (iv), we get

$x_3=7, x_1=3, x_2=-3$

$y_3=2, y_1=12, y_2=4$

and $z_3=5, z_1=17, z_2=-7$

Hence, vertices of a triangle are (7, 2, 5), (3, 12, 17) and (–3, 4, –7).

79. *(a)* Let R_1 and R_2 be the points with coordinates (x_1, y_1, z_1) and (x_2, y_2, z_3) which trisect the line segment AB.

A (2,1,–3) — R_1 (x_1,y_1,z_1) — R_2 (x_2,y_2,z_2) — B (5,–8,3)

Now, $AR_1 : R_1B = 1:2$

$\Rightarrow x_1=\frac{1\times5+2\times2}{1+2}=\frac{9}{3}=3$

$y_1=\frac{1\times(-8)+2\times1}{1+2}=\frac{-6}{3}=-2$

and $z_1=\frac{1\times3+2\times-3}{1+2}=\frac{-3}{3}=-1$

$\therefore R_1=(3, -2, -1)$

Similarly, $AR_2 : R_2B=2:1$

$\Rightarrow x_2=\frac{2\times5+1\times2}{1+2}=\frac{12}{3}=4$

$y_2=\frac{2\times(-8)+1\times1}{1+2}=\frac{-15}{3}=-5$

and $z_2=\frac{2\times3+1\times(-3)}{1+2}=\frac{3}{3}=1$

$\therefore$ Coordinates of R_2 are (4, –5, 1).

80. *(a)* For centroid of ΔABC,

$x=\frac{a-2+4}{3}=\frac{a+2}{3}$

$y=\frac{1+b+7}{3}=\frac{b+8}{3}$

and $z=\frac{3-5+c}{3}=\frac{c-2}{3}$

But, given centroid is (0, 0, 0).

$\therefore \frac{a+2}{3}=0 \Rightarrow a=-2$

$\frac{b+8}{3}=0 \Rightarrow b=-8$

$\frac{c-2}{3}=0 \Rightarrow c=2$

81. *(a)* Suppose xy-plane divides the line joining the given points in the ratio $\lambda:1$. The coordinates of the points of division are $\left[\frac{2\lambda-1}{\lambda+1}, \frac{-5\lambda+3}{\lambda+1}, \frac{6\lambda+4}{\lambda+1}\right]$. Since, the points lies on the XY-plane.

$\therefore \frac{6\lambda+4}{\lambda+1}=0 \Rightarrow \lambda=\frac{-2}{3}$

Alternate Method

The ratio in which XY-plane divides the line segment

$=-z_1 : z_2=-\frac{4}{6}=-\frac{2}{3}$

Note (i) *The ratio in which XY-plane divides the line segment* $=-z_1 : z_2$

(ii) *The ratio in which YZ-plane divides the line segment* $=-x_1 : x_2$

(iii) *The ratio in which ZX-plane divides the line segment* $=-y_1 : y_2$

82. *(a)* Let point $P(x, y, z)$ divides the line joining the points A and B in the ratio $m:1$.

A (5,–3,–2) — m — P — 1 — B (1,2,–2)

Since, point P is on XOZ-plane.

$\therefore$ y-coordinate $=0$

$\Rightarrow \frac{2m-3}{m+1}=0 \Rightarrow m=\frac{3}{2}$

Now, $x=\frac{3+2\times5}{3+2}=\frac{13}{5}$

and $$z = \frac{3\times(-2)+2\times(-2)}{5} = -2$$

$\therefore$ Required point is $\left(\frac{13}{5}, 0, -2\right)$.

83. *(c)* Let the point R divides the line joining the points $P(2, 4, 5)$ and $Q(3, 5, -4)$ in the ratio $m:n$. Then, the coordinates of R are $\left(\frac{3m+2n}{m+n}, \frac{5m+4n}{m+n}, \frac{-4m+5n}{m+n}\right)$.

For yz-plane, x-coordinates will be zero.

$$\therefore \quad \frac{3m+2n}{m+n} = 0 \Rightarrow \frac{m}{n} = \frac{-2}{3}$$

Alternate Method

The ratio in which YZ-plane divides the line segment

$$= -x_1 : x_2 = -2 : 3$$

84. *(a)* To show $ABCD$ is a parallelogram, we need to show that opposite sides are equal.

Now,

$$AB = \sqrt{(-1-1)^2 + (-2-2)^2 + (-1-3)^2} = \sqrt{4+16+16} = 6$$

$$BC = \sqrt{(2+1)^2 + (3+2)^2 + (2+1)^2} = \sqrt{9+25+9} = \sqrt{43}$$

$$CD = \sqrt{(4-2)^2 + (7-3)^2 + (6-2)^2} = \sqrt{4+16+16} = 6$$

$$DA = \sqrt{(1-4)^2 + (2-7)^2 + (3-6)^2} = \sqrt{9+25+9} = \sqrt{43}$$

Since, $AB = CD$ and $BC = AD$, $ABCD$ is a parallelogram.

Now, it is required to prove that $ABCD$ is not a rectangle. For this, we show that diagonals AC and BD are unequal. We have,

$$AC = \sqrt{(2-1)^2 + (3-2)^2 + (2-3)^2} = \sqrt{1+1+1} = \sqrt{3}$$

$$BD = \sqrt{(4+1)^2 + (7+2)^2 + (6+1)^2} = \sqrt{25+81+49} = \sqrt{155}$$

Since, $AC \neq BD$, so $ABCD$ is not a rectangle.

Note *We can also show that ABCD is a parallelogram, using the property that diagonals AC and BD bisect each other.*

85. *(b)* Let $P(x, y, z)$ be any point such that $PA = PB$.

Then, $$\sqrt{(x-3)^2 + (y-4)^2 + (z+5)^2} = \sqrt{(x+2)^2 + (y-1)^2 + (z-4)^2}$$

or $(x-3)^2 + (y-4)^2 + (z+5)^2 = (x+2)^2 + (y-1)^2 + (z-4)^2$

or $\quad 10x + 6y - 18z - 29 = 0$

86. *(b)* Let the coordinates of C be (x, y, z) and given that the coordinates of the centroid G is $(1, 1, 1)$. Then,

$$\frac{x+3-1}{3} = 1, \text{ i.e., } x = 1;$$

$$\frac{y-5+7}{3} = 1$$

i.e., $$y = 1$$

$$\frac{z+7-6}{3} = 1, \text{ i.e., } z = 2$$

Hence, coordinates of C are $(1, 1, 2)$.

87. *(d)* Let $ABCD$ be a parallelogram and coordinates of point D are (α, β, γ) and the diagonals AC and BD intersect at P.

In a parallelogram diagonals bisect each other.

$\therefore$ Coordinates of mid-point of BD

= Coordinates of mid-point of AC

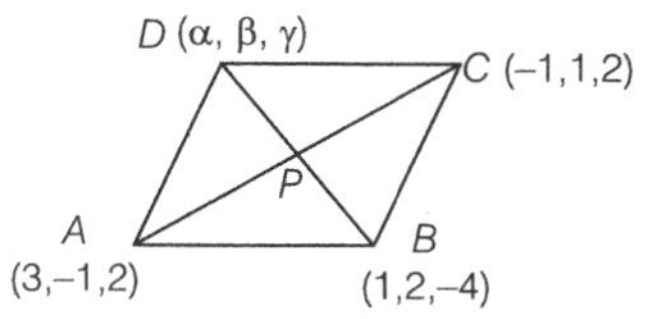

$$\Rightarrow \left(\frac{\alpha+1}{2}, \frac{\beta+2}{2}, \frac{\gamma-4}{2}\right) = \left(\frac{3-1}{2}, \frac{-1+1}{2}, \frac{2+2}{2}\right)$$

$$\Rightarrow \left(\frac{\alpha+1}{2}, \frac{\beta+2}{2}, \frac{\gamma-4}{2}\right) = \left(\frac{2}{2}, 0, \frac{4}{2}\right)$$

$$\Rightarrow \left(\frac{\alpha+1}{2}, \frac{\beta+2}{2}, \frac{\gamma-4}{2}\right) = (1, 0, 2)$$

On comparing the corresponding coordinates, we get

$$\frac{\alpha+1}{2} = 1, \quad \frac{\beta+2}{2} = 0, \quad \frac{\gamma-4}{2} = 2$$

$$\Rightarrow \quad \alpha+1 = 2, \quad \beta+2 = 0, \quad \gamma-4 = 4$$

$$\Rightarrow \quad \alpha = 1, \quad \beta = -2, \quad \gamma = 8$$

$\therefore$ Coordinates of point D are $(1, -2, 8)$.

88. *(b)* ABC is a triangle with vertices $A(0, 0, 6)$, $B(0, 4, 0)$ and $C(6, 0, 0)$.

Let points D, E and F be the mid-points of BC, AC and AB, respectively. So, AD, BE and CF will be the medians of the triangle.

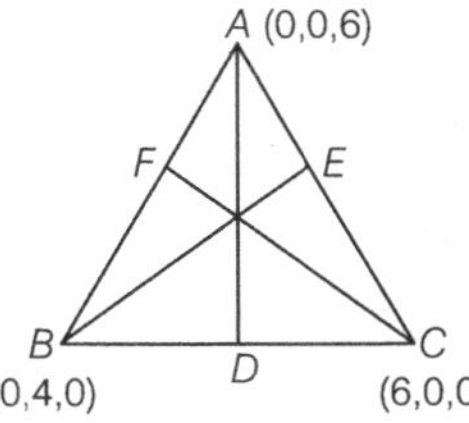

$\Rightarrow$ Coordinates of point D

$$= \left(\frac{0+6}{2}, \frac{4+0}{2}, \frac{0+0}{2}\right) = (3, 2, 0)$$

$$\text{Coordinates of point } E = \left(\frac{0+6}{2}, \frac{0+0}{2}, \frac{6+0}{2}\right) = (3, 0, 3)$$

and coordinates of point F

$$= \left(\frac{0+0}{2}, \frac{0+4}{2}, \frac{6+0}{2}\right) = (0, 2, 3)$$

Now, length of median AD = Distance between A and D

$$AD = \sqrt{(0-3)^2 + (0-2)^2 + (6-0)^2}$$

$$= \sqrt{9+4+36} = \sqrt{49} = 7$$

Similarly, $$BE = \sqrt{(0-3)^2 + (4-0)^2 + (0-3)^2}$$

$$= \sqrt{9+16+9} = \sqrt{34}$$

and $$CF = \sqrt{(6-0)^2 + (0-2)^2 + (0-3)^2}$$

$$= \sqrt{36+4+9} = \sqrt{49} = 7$$

89. (c) Centroid of the ΔPQR is

$$\left(\frac{x_1+x_2+x_3}{3}, \frac{y_1+y_2+y_3}{3}, \frac{z_1+z_2+z_3}{3}\right)$$

i.e., $$\left(\frac{2a-4+8}{3}, \frac{2+3b+14}{3}, \frac{6-10+2c}{3}\right)$$

Origin is the centroid i.e., the coordinates of the centroid are (0, 0, 0). Then,

$$\frac{2a-4+8}{3}=0$$

$$\Rightarrow \quad 2a+4=0 \Rightarrow a=-2$$

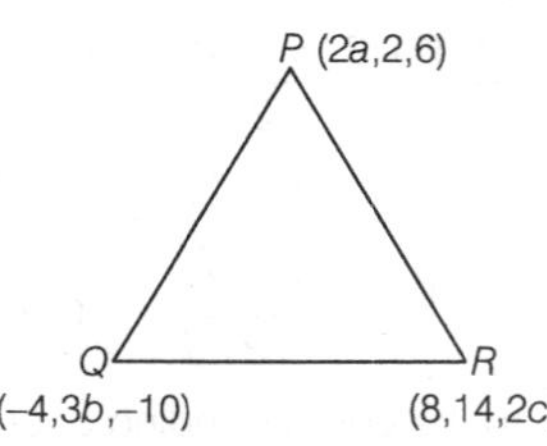

Also, $$\frac{2+3b+14}{3}=0 \Rightarrow 3b+16=0$$

$$\Rightarrow \quad b=-\frac{16}{3}$$

and $$\frac{6-10+2c}{3}=0$$

$$\Rightarrow \quad 2c-4=0 \Rightarrow c=2$$

$$\therefore \quad a=-2, b=\frac{-16}{3}, c=2$$

90. (c) Let any point on the Y-axis be $A(0, y, 0)$.

Given, distance between P and A is $PA=5\sqrt{2}$

$$\Rightarrow \quad \sqrt{(3-0)^2+(-2-y)^2+(5-0)^2}=5\sqrt{2}$$

On squaring both sides, we get

$$(3-0)^2+(-2-y)^2+(5-0)^2=50$$

$$\Rightarrow \quad 9+4+y^2+4y+25=50$$

$$\Rightarrow \quad y^2+4y+38-50=0$$

$$\Rightarrow \quad y^2+4y-12=0$$

Now, factorising it by splitting the middle term, we get

$$y^2+6y-2y-12=0$$

$$\Rightarrow \quad y(y+6)-2(y+6)=0$$

$$\Rightarrow \quad (y+6)(y-2)=0 \Rightarrow y=-6, 2$$

Hence, the points on Y-axis are (0, – 6, 0) or (0, 2, 0).

91. (b) Let the point $R(x, y, z)$ divides PQ in the ratio $k:1$.

k 1
P R Q
(2, –3, 4) (4, y, z) (8, 0, 10)

$$\Rightarrow \quad \frac{k\times 8+1\times 2}{k+1}=4 \qquad \text{(given)}$$

$$\Rightarrow \quad \frac{8k+2}{k+1}=4$$

$$\Rightarrow \quad 8k+2=4k+4$$

$$\Rightarrow \quad 8k-4k=4-2$$

$$\Rightarrow \quad 4k=2$$

$$\Rightarrow \quad k:1=1:2$$

Hence, the point R divides PQ internally in the ratio 1 : 2.

Therefore, y-coordinate of $R=\left(\frac{1\times 0+2\times(-3)}{1+2}\right)$

$$=\left(\frac{-6}{3}\right)=-2$$

and z-coordinate of $R=\left(\frac{1\times 10+2\times 4}{1+2}\right)$

$$=\left(\frac{10+8}{3}\right)=\frac{18}{3}=6$$

Hence, coordinates of point R are (4, – 2, 6).

92. (a) Let the point P be (x, y, z).

Given, $(PA)^2+(PB)^2=k^2$

$$\Rightarrow \quad (x-3)^2+(y-4)^2+(z-5)^2 + (x+1)^2+(y-3)^2+(z+7)^2=k^2$$

$$\Rightarrow \quad x^2+9-6x+y^2+16-8y+z^2+25-10z + x^2+1+2x+y^2+9-6y+z^2+49+14z=k^2$$

$$\Rightarrow \quad 2x^2+2y^2+2z^2-4x-14y+4z+109-k^2=0$$

$$\Rightarrow \quad 2(x^2+y^2+z^2)-4x-14y+4z+109-k^2=0$$

which is the required equation.

93. (a) The coordinates of point P are $(x, y, \sqrt{1-x^2-y^2})$.

Then, $$OP=\sqrt{(x-0)^2+(y-0)^2+(\sqrt{1-x^2-y^2}-0)^2}$$

$$=\sqrt{x^2+y^2+1-x^2-y^2}=\sqrt{1}=1$$

94. (b) If AD is the internal bisector of $\angle A$. Then, we have (by geometry)

$$\frac{BD}{CD}=\frac{AB}{AC}$$

A (2,2,–3)
B (5,6,9) D C (2,7,9)

$$\frac{AB}{AC}=\frac{\sqrt{(2-5)^2+(2-6)^2+(-3-9)^2}}{\sqrt{(2-2)^2+(2-7)^2+(-3-9)^2}}$$

$$=\frac{\sqrt{9+16+144}}{\sqrt{0+25+144}}=\frac{1}{1}$$

Therefore, D is the mid-point of BC.

$$\therefore \quad D=\left(\frac{5+2}{2}, \frac{6+7}{2}, \frac{9+9}{2}\right)=\left(\frac{7}{2}, \frac{13}{2}, 9\right)$$

95. *(c)* Clearly, Points of $O = (0, 0, 0)$
Points of $A = (2, 0, 0)$
Points of $G = (0, 2, 0)$

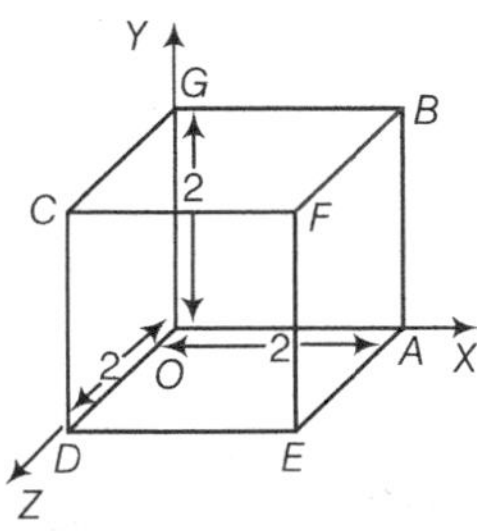

Point of $D = (0, 0, 2)$
Point of $B = (2, 2, 0)$
Point of $F = (2, 2, 2)$
Point of $E = (2, 0, 2)$
Point of $C = (0, 2, 2)$

96. *(b)* Centroid of ΔABC = Centroid of ΔDEF

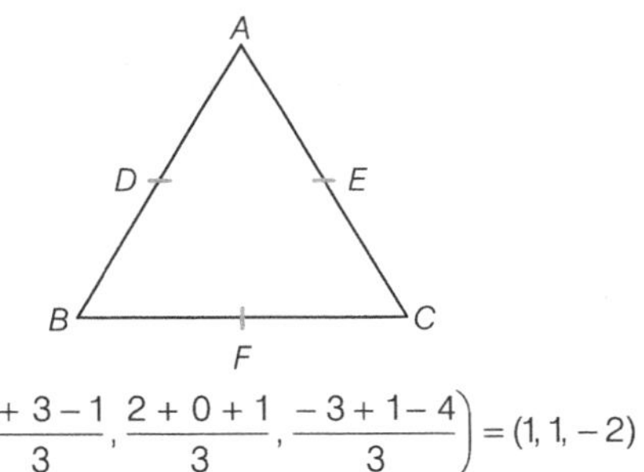

$$= \left(\frac{1+3-1}{3}, \frac{2+0+1}{3}, \frac{-3+1-4}{3}\right) = (1, 1, -2)$$

97. *(a)* A. Let $A(3, -5, 7)$, $B(-1, 7, -6)$, $C(x, y, z)$ be the vertices of a ΔABC with centroid $(0, 0, 0)$

Therefore, $(0, 0, 0) = \left(\frac{3-1+x}{3}, \frac{-5+7+y}{3}, \frac{7-6+z}{3}\right)$.

This implies $\frac{x+2}{3} = 0, \frac{y+2}{3} = 0, \frac{z+1}{3} = 0.$

Hence, $x = -2$, $y = -2$ and $z = -1$. Therefore, (A) $\leftrightarrow$ (3)

B. Mid-point of diagonal AC is

$$\left(\frac{3+2}{2}, \frac{-1+3}{2}, \frac{-1-2}{2}\right) = \left(\frac{5}{2}, 1, \frac{-3}{2}\right)$$

Mid-point of diagonal BD is

$$\left(\frac{5+0}{2}, \frac{-4+6}{2}, \frac{0-3}{2}\right) = \left(\frac{5}{2}, 1, \frac{-3}{2}\right)$$

Diagonals of parallelogram bisect each other. Therefore, (B) $\leftrightarrow$ (1)

C. $|AB| = \sqrt{(2-1)^2 + (-4+1)^2 + (5-3)^2} = \sqrt{14}$

$|BC| = \sqrt{(5-2)^2 + (-13+4)^2 + (11-5)^2} = 3\sqrt{14}$

$|AC| = \sqrt{(5-1)^2 + (-13+1)^2 + (11-3)^2} = 4\sqrt{14}$

Now, $|AB| + |BC| = |AC|$. Hence, points A, B and C are collinear. Hence, (C) $\leftrightarrow$ (4)

D. $AB = \sqrt{4+9+36} = 7$

$BC = \sqrt{36+4+9} = 7$

$CA = \sqrt{64+25+9} = 7\sqrt{2}$

Now, $AB^2 + BC^2 = AC^2$. Hence, ABC is an isosceles right angled triangle and hence (D) $\leftrightarrow$ (2).

98. *(b)* (i) (F), A line parallel to XY-plane, if all the points on line have equal z-coordinates.

(ii) (F), Since $x = 0$ represents YZ-plane. Therefore, $x = b$ represents a plane parallel to YZ-plane at a unit distance b from the origin.

(iii) (F), The point $(4, 5, -6)$ lies in V^{th} octant.

(iv) (T), The X-axis is the intersection of two planes XY-plane and XZ-plane.

99. *(a)* The equation of the plane $z = 6$ represents a plane parallel to the XY-plane, having a z-intercept of 6 units.

100. *(d)* Given $A = (1, 0, 0)$, $B = (0, 1, 0)$ and $C = (0, 0, 1)$

Now, $AB = \sqrt{(1-0)^2 + (0-1)^2 + 0^2} = \sqrt{2}$

$BC = \sqrt{0^2 + (1-0)^2 + (0-1)^2} = \sqrt{2}$

and $CA = \sqrt{(0-1)^2 + 0^2 + (1-0)^2} = \sqrt{2}$

$\therefore$ Perimeter of triangle $= AB + BC + CA$
$= \sqrt{2} + \sqrt{2} + \sqrt{2}$
$= 3\sqrt{2}$

101. *(c)* The distance between given points

$$= \sqrt{(2-1)^2 + (2-4)^2 + (3-5)^2}$$
$$= \sqrt{1+4+4} = 3$$

102. *(d)* Let the point in xy-plane be $P(x_1, y_1, 0)$.

The given points are $A(2, 0, 3)$, $B(0, 3, 2)$ and $C(0, 0, 1)$.

According to the given condition,

$$AP^2 = BP^2 = CP^2$$

$\therefore$ $(x_1 - 2)^2 + y_1^2 + 9 = x_1^2 + (y_1 - 3)^2 + 4$
$= x_1^2 + y_1^2 + 1$

From 1st and 2nd terms,

$$x_1^2 + 4 - 4x_1 + y_1^2 + 9 = x_1^2 + y_1^2 - 6y_1 + 9 + 4$$

$\Rightarrow$ $4x_1 - 6y_1 = 0$ ...(i)

From 2nd and 3rd terms,

$$x_1^2 + y_1^2 + 9 - 6y_1 + 4 = x_1^2 + y_1^2 + 1$$

$\Rightarrow$ $6y_1 = 12$

$\Rightarrow$ $y_1 = 2$

On putting the value of y_1 in Eq. (i), we get

$$x_1 = 3$$

Hence, required point is $(3, 2, 0)$.

103. *(e)* Perpendicular distance of the points $(6, 5, 8)$ from Y-axis

$$= \sqrt{6^2 + 8^2} = 10 \text{ units}$$

13

Limits *and* Derivatives

Intuitive Idea of Derivatives, Limits and Limits of Trigonometric Functions

Level I

1. The branch of Mathematics which mainly deals with the study of change in the value of a function as the points in the domain change, is called
(a) Calculus (b) Analysis (c) Algebra (d) Statistics

2. The distance s in metres covered by a body, dropped from a tall cliff, as a function of time t in seconds is given by
(a) $s = 9.8\,t^2$ (b) $s = 9.8\,t$ (c) $s = 4.9\,t^2$ (d) $s = 4.9\,t$

3. The following table I gives the distance travelled in metres at various intervals of time in seconds of a body dropped from a tall cliff.

Table I

t	s
0	0
1	4.9
1.5	11.025
1.8	15.876
1.9	17.689
1.95	18.63225
2	19.6
2.05	20.59225
2.1	21.609
2.2	23.716
2.5	30.625
3	44.1
4	78.4

Now, the velocity of the body at time $t = 2$ seconds is between
(a) (19, 19.4) (b) (19.5, 19.6)
(c) (19.5, 19.7) (d) None of these

4. Which of the following is/are true?
I. The instantaneous velocity $v(t)$ of a body dropped from a tall cliff at time t_1 is equal to the slope of the tangent of the curve $s = 4.9\,t^2$ at time t_1.
II. The instantaneous velocity $v(t)$ of a body dropped from a tall cliff at time t_1 is equal to the slope of the normal of the curve $s = 4.9\,t^2$ at time t_1.
(a) Both I and II are true
(b) Only I is true
(c) Only II is true
(d) Both I and II are false

Limits

5. The limit of $f(x) = x^2$ as x tends to zero equals
(a) zero (b) one (c) two (d) three

6. As $x \to a$, $f(x) \to l$, then l is called ...A... of the function $f(x)$, where A stands for
(a) value (b) absolute value
(c) limit (d) None of these

7. Consider the following function $g(x) = |x|, x \neq 0$. Then,

I. $g(0)$ is not defined.

II. $\lim_{x \to 0} g(x)$ is not defined.

Which of the following is/are true?

(a) Both I and II are true (b) Only I is true

(c) Only II is true (d) Both I and II are false

8. Consider the function $h(x) = \dfrac{x^2 - 4}{x - 2}, x \neq 2.$

Then,

I. $h(2)$ is not defined.

II. $\lim_{x \to 2} h(x) = 4.$

Which of the following is/are true?

(a) Both I and II are true (b) Only I is true

(c) Only II is true (d) Both I and II are false

9. The right hand limit and left hand limit of a function $f(x)$ at a given point $x = a$ is the value of $f(x)$ which is dictated by the values of $f(x)$ when x tends to a from ...A... and ...B..., respectively. Here, A and B refer to

(a) left, right (b) left, left (c) right, left (d) right, right

10. Consider the function $f(x) = \begin{cases} 1, & x \leq 0 \\ 2, & x > 0 \end{cases}$

Then, left hand limit and right hand limit of $f(x)$ *at* $x = 0$,are respectively

(a) 1, 2 (b) 2, 1 (c) 1, 1 (d) 2, 2

11. Match the terms of Column I with the terms of Column II and choose the correct option from the codes given below.

	Column I		Column II
A.	$\lim_{x \to a^-} f(x)$	1.	left hand limit of f at a
B.	$\lim_{x \to a^+} f(x)$	2.	limit of f at a
C.	$\lim_{x \to a} f(x)$	3.	right hand limit of f at a

Codes

	A	B	C		A	B	C
(a)	3	1	2	(b)	1	3	2
(c)	1	2	3	(d)	2	3	1

12. Consider the function $f(x) = x + 10$. Then,

(a) $\lim_{x \to 5^-} f(x) \neq \lim_{x \to 5^+} f(x)$

(b) $\lim_{x \to 5^-} f(x) = \lim_{x \to 5^+} f(x) = \lim_{x \to 5} f(x)$

(c) $\lim_{x \to 5^-} f(x) = \lim_{x \to 5^+} f(x) \neq \lim_{x \to 5} f(x)$

(d) None of the above

13. Consider the function $f(x) = x^3$. Then, which of the following is not correct?

(a) $\lim_{x \to 1^-} f(x) = 0$ and $\lim_{x \to 1^+} f(x) = 1$

(b) $\lim_{x \to 1^-} f(x) = \lim_{x \to 1^+} f(x) = 1$

(c) $\lim_{x \to 1} f(x) = 1$

(d) None of the above

14. Consider the function $f(x) = 3x$. Then,

(a) $\lim_{x \to 2^-} f(x) = \lim_{x \to 2^+} f(x) = 6$

(b) $\lim_{x \to 2^-} f(x) \neq \lim_{x \to 2^+} f(x)$

(c) $f(2) \neq \lim_{x \to 2^+} f(x)$

(d) $f(2) \neq \lim_{x \to 2^-} f(x)$

15. Consider the function $f(x) = 3$. Then,

(a) $\lim_{x \to 2^-} f(x) \neq \lim_{x \to 2^+} f(x)$

(b) $\lim_{x \to 2^-} f(x) = \lim_{x \to 2^+} f(x) = \lim_{x \to 2} f(x)$

(c) $\lim_{x \to 2^-} f(x) = \lim_{x \to 2^+} f(x) \neq \lim_{x \to 2} f(x)$

(d) None of the above

16. Consider the function $f(x) = x^2 + x$. Then, which of the following is not correct?

(a) $\lim_{x \to 1^-} f(x) = 0$ and $\lim_{x \to 1^+} f(x) = 2$

(b) $\lim_{x \to 1^-} f(x) = \lim_{x \to 1^+} f(x) = 2$

(c) $\lim_{x \to 1} f(x) = 2$

(d) None of the above

17. Consider the function $f(x) = \sin x$. Then,

(a) $\lim_{x \to \frac{\pi^-}{2}} f(x) \neq \lim_{x \to \frac{\pi^+}{2}} f(x)$

(b) $\lim_{x \to \frac{\pi}{2}} f(x) \neq \lim_{x \to \frac{\pi^+}{2}} f(x)$

(c) $\lim_{x \to \frac{\pi^-}{2}} f(x) = \lim_{x \to \frac{\pi^+}{2}} f(x) = 1$

(d) $\lim_{x \to \frac{\pi}{2}} f(x) \neq \lim_{x \to \frac{\pi^-}{2}} f(x)$

18. Consider the function $f(x) = x + \cos x$. Then, which of the following are correct?

(a) $\lim_{x \to 0^-} f(x) = \lim_{x \to 0} f(x) = 1$

(b) $\lim_{x \to 0^+} f(x) = \lim_{x \to 0} f(x) = 1$

(c) $\lim_{x \to 0^+} f(x) = \lim_{x \to 0^-} f(x)$

(d) All of the above

19. Consider the function $f(x) = \dfrac{1}{x^2}$ for $x > 0$. Then,

(a) $\lim_{x \to 0} f(x) = -\infty$ (b) $\lim_{x \to 0} f(x) = +\infty$

(c) $\lim_{x \to 0} f(x)$ is finite (d) $\lim_{x \to 0} f(x)$ is defined

20. Consider the function $f(x)$, where

$$f(x)=\begin{cases}x-2, & x<0\\ 0, & x=0\\ x+2, & x>0\end{cases}.\text{ Then,}$$

(a) $\lim_{x\to0^-} f(x)=\lim_{x\to0^+} f(x)=2$
(b) $\lim_{x\to0^-} f(x)=\lim_{x\to0^+} f(x)=-2$
(c) $\lim_{x\to0^-} f(x)=-2$ and $\lim_{x\to0^+} f(x)=2$
(d) $\lim_{x\to0} f(x)$ exists

21. Consider the function $f(x)$ where

$$f(x)=\begin{cases}x+2, & x\neq1\\ 0, & x=1\end{cases}$$

Then, which of the following are correct?
(a) $\lim_{x\to1^-} f(x)=\lim_{x\to1} f(x)=3$
(b) $\lim_{x\to1^+} f(x)=\lim_{x\to1} f(x)=3$
(c) $\lim_{x\to1^-} f(x)=\lim_{x\to1^+} f(x)$
(d) All of the above

22. The process of finding the limit follows addition, subtraction, multiplication and division as long as the limits and functions under consideration are
(a) undefined
(b) well defined
(c) Both (a) and (b) are correct
(d) Neither (a) nor (b) is correct

23. Let f and g be two functions such that $\lim_{x\to a} f(x)$ and $\lim_{x\to a} g(x)$ exist. Then, which of the following is incomplete?
(a) $\lim_{x\to a}[f(x)+g(x)]=\lim_{x\to a} f(x)+\lim_{x\to a} g(x)$
(b) $\lim_{x\to a}[f(x)-g(x)]=\lim_{x\to a} f(x)-\lim_{x\to a} g(x)$
(c) $\lim_{x\to a}[f(x)\cdot g(x)]=\lim_{x\to a} f(x)\cdot\lim_{x\to a} g(x)$
(d) $\lim_{x\to a}\left[\dfrac{f(x)}{g(x)}\right]=\dfrac{\lim_{x\to a} f(x)}{\lim_{x\to a} g(x)}$

24. If f and g are two functions such that $\lim_{x\to a} f(x)$ and $\lim_{x\to a} g(x)$ exist. When g is the constant function such that $g(x)=\lambda$, for some real number λ, then
(a) $\lim_{x\to a}[(\lambda\cdot f)(x)]=\lambda\cdot\lim_{x\to a} f(x)$
(b) $\lim_{x\to a}\left[\left(\dfrac{f}{\lambda}\right)(x)\right]=\dfrac{1}{\lambda}\cdot\lim_{x\to a} f(x)$
(c) Both (a) and (b) are correct
(d) Neither (a) nor (b) is correct

25. A function f is said to be a polynomial function, if $f(x)$ is zero function or if $f(x)=a_0+a_1x+\ldots..+a_nx^n$, where a_i's are real numbers such that $a_n\neq0$ for some natural number n. Then,
(a) $\lim_{x\to a} x=a\Rightarrow\lim_{x\to a} f(x)\neq f(a)$
(b) $\lim_{x\to a} x=a\Rightarrow\lim_{x\to a} f(x)=f(0)$
(c) $\lim_{x\to a} x=a\Rightarrow\lim_{x\to a} f(x)=f(a)$
(d) $\lim_{x\to a} x=a\Rightarrow\lim_{x\to a} f(x)$ does not exist

26. A function f is said to be a rational function, if $f(x)=\dfrac{g(x)}{h(x)}$, where $g(x)$ and $h(x)$ are polynomials such that $h(x)\neq0$, then
(a) $h(a)\neq0\Rightarrow\lim_{x\to a} f(x)=\dfrac{g(a)}{h(a)}$
(b) $h(a)=0$ and $g(a)\neq0\Rightarrow\lim_{x\to a} f(x)$ does not exist
(c) Both (a) and (b) are true
(d) Both (a) and (b) are false

27. For any positive integer n, $\lim_{x\to a}\left[\dfrac{x^n-a^n}{x-a}\right]$ is equal to
(a) na^n
(b) $(n-1)a^{n-1}$
(c) na^{n-1}
(d) $(n-1)a^n$

28. Which of the following is/are true?

I. $\lim_{x\to1}\left[\dfrac{x^{15}-1}{x^{10}-1}\right]=\dfrac{3}{2}$ II. $\lim_{x\to0}\left[\dfrac{\sqrt{1+x}-1}{x}\right]=\dfrac{1}{2}$

(a) Both I and II are true
(b) Only I is true
(c) Only II is true
(d) Both I and II are false

Limits of Trigonometric Functions

29. Let f and g be two real valued functions with the same domain such that $f(x)\leq g(x)$ for all x in the domain. For some a, if both $\lim_{x\to a} f(x)$ and $\lim_{x\to a} g(x)$ exist, then
(a) $\lim_{x\to a} f(x)>\lim_{x\to a} g(x)$
(b) $\lim_{x\to a} f(x)<\lim_{x\to a} g(x)$
(c) $\lim_{x\to a} f(x)\geq\lim_{x\to a} g(x)$
(d) $\lim_{x\to a} f(x)\leq\lim_{x\to a} g(x)$

30. Let f, g and h be real functions such that $f(x)\leq g(x)\leq h(x)$ for all x in the common domain. For some real number a, if
(a) $\lim_{x\to a} f(x)=l=\lim_{x\to a} h(x)$, then $\lim_{x\to a} g(x)=l$
(b) $\lim_{x\to a} f(x)=l=\lim_{x\to a} g(x)$, then $\lim_{x\to a} h(x)=l$
(c) $\lim_{x\to a} g(x)=l=\lim_{x\to a} h(x)$, then $\lim_{x\to a} f(x)=l$
(d) None of the above

31. Which of the following is/are true?

I. $\lim_{x \to 0} \frac{\sin x}{x} = 1$ II. $\lim_{x \to 0} \frac{1 - \cos x}{x} = 0$

(a) Both I and II are true (b) Only I is true
(c) Only II is true (d) Both I and II are false

32. Which of the following is/are true?

I. $\lim_{x \to 0} \frac{\sin 4x}{\sin 2x} = 2$ II. $\lim_{x \to 0} \frac{\tan x}{x} = 0$

(a) Both I and II are true (b) Only I is true
(c) Only II is true (d) Both I and II are false

33. Let $f(x)$ and $g(x)$ be two functions such that $f(x) = f_1(x) f_2(x)$ with $f_1(a) = 0$, $f_2(a) \neq 0$ and $g(x) = g_1(x) g_2(x)$ with $g_1(a) = 0$, $g_2(a) \neq 0$. If on cancelling out the common factors from $f(x)$ and $g(x)$, $\frac{f(x)}{g(x)} = \frac{p(x)}{q(x)}$, where $q(x) \neq 0$, then

(a) $\lim_{x \to a} \frac{f(x)}{g(x)}$ does not exist
(b) $\lim_{x \to a} \frac{f(x)}{g(x)} = \frac{p(a)}{q(a)}$
(c) Both (a) and (b) are true
(d) Both (a) and (b) are false

34. Match the limits of the functions given in Column I with their corresponding values given in Column II and choose the correct option from the codes given below.

	Column I		Column II
A.	$\lim_{x \to 3} x + 3$	1.	π
B.	$\lim_{x \to \pi} \left(x - \frac{22}{7}\right)$	2.	6
C.	$\lim_{r \to 1} \pi r^2$	3.	$\frac{19}{2}$
D.	$\lim_{x \to 4} \left(\frac{4x + 3}{x - 2}\right)$	4.	$\frac{-1}{2}$
E.	$\lim_{x \to -1} \left(\frac{x^{10} + x^5 + 1}{x - 1}\right)$	5.	$\pi - \frac{22}{7}$

Codes

	A	B	C	D	E
(a)	5	2	1	4	3
(b)	2	5	1	3	4
(c)	5	2	1	3	4
(d)	2	5	3	1	4

35. Match the limits of the functions given in Column I with their corresponding values given in Column II and choose the correct option from the codes given below.

	Column I		Column II
A.	$\lim_{x \to 0} \frac{(x + 1)^5 - 1}{x}$	1.	$\frac{11}{4}$
B.	$\lim_{x \to 2} \frac{(3x^2 - x - 10)}{x^2 - 4}$	2.	b
C.	$\lim_{x \to 3} \frac{x^4 - 81}{2x^2 - 5x - 3}$	3.	5
D.	$\lim_{x \to 0} \left(\frac{ax + b}{cx + 1}\right)$	4.	2
E.	$\lim_{z \to 1} \left(\frac{z^{\frac{1}{3}} - 1}{z^{\frac{1}{6}} - 1}\right)$	5.	$\frac{108}{7}$

Codes

	A	B	C	D	E		A	B	C	D	E
(a)	3	1	5	2	4	(b)	1	3	5	2	4
(c)	3	1	5	4	2	(d)	3	5	1	2	4

36. Which of the following is/are true?

I. $\lim_{x \to 1} \frac{ax^2 + bx + c}{cx^2 + bx + a}$ (where, $a + b + c \neq 0$) is 1.

II. $\lim_{x \to -2} \frac{\frac{1}{x} + \frac{1}{2}}{x + 2}$ is $\frac{1}{4}$.

(a) Both I and II are true (b) Only I is true
(c) Only II is true (d) Both I and II are false

37. Which of the following is/are true?

I. $\lim_{x \to 0} \frac{\sin ax}{bx} = \frac{a}{b}$

II. $\lim_{x \to 0} \frac{\sin ax}{\sin bx} = \frac{a}{b}$ (where a, $b \neq 0$)

(a) Both I and II are true (b) Only I is true
(c) Only II is true (d) Both I and II are false

38. Match the limits of the functions given in Column I with their corresponding values given in Column II and choose the correct option from the codes given below.

Column I		Column II
A. $\lim_{x \to \pi} \frac{\sin(\pi - x)}{\pi(\pi - x)}$	1.	4
B. $\lim_{x \to 0} \frac{\cos x}{\pi - x}$	2.	$\frac{1}{\pi}$
C. $\lim_{x \to 0} \frac{\cos 2x - 1}{\cos x - 1}$	3.	$\frac{a + 1}{b}$
D. $\lim_{x \to 0} \frac{ax + x\cos x}{b\sin x}$	4.	0

Column I	Column II
E. $\lim_{x \to 0} x \sec x$	5. 1
F. $\lim_{x \to 0} \frac{\sin ax + bx}{ax + \sin bx}$ $(a, b, a + b \neq 0)$	

Codes

	A	B	C	D	E	F
(a)	2	2	1	3	5	4
(b)	2	2	3	1	4	5
(c)	2	2	1	4	3	5
(d)	2	2	1	3	4	5

39. Which of the following is/are true?

I. $\lim_{x \to 0} (\text{cosec } x - \cot x) = 0$

II. $\lim_{x \to \frac{\pi}{2}} \frac{\tan 2x}{x - \frac{\pi}{2}} = 1$

(a) Both I and II are true (b) Only I is true
(c) Only II is true (d) Both I and II are false

40. If $f(x) = \begin{cases} 2x + 3, & x \leq 0 \\ 3(x+1), & x > 0 \end{cases}$, then

(a) $\lim_{x \to 0} f(x) = 3$ (b) $\lim_{x \to 1} f(x) = 6$
(c) Both (a) and (b) are true (d) Neither (a) nor (b) is true

41. If $f(x) = \begin{cases} \frac{|x|}{x}, & x \neq 0 \\ 0, & x = 0 \end{cases}$, then

(a) LHL = RHL = 0, at $x = 0$
(b) LHL ≠ RHL, at $x = 0$
(c) LHL = RHL = 1, at $x = 0$
(d) None of the above

42. If $f(x) = |x| - 5$, then

(a) $\lim_{x \to 5} f(x) = 0$ (b) $\lim_{x \to 5} f(x) = 10$
(c) $\lim_{x \to 5} f(x) = -10$ (d) None of these

43. If the function $f(x)$ satisfies $\lim_{x \to 1} \frac{f(x) - 2}{x^2 - 1} = \pi$

(a) $\lim_{x \to 1} f(x) = 1$ (b) $\lim_{x \to 1} f(x) = 2$
(c) $\lim_{x \to 1} f(x) = 0$ (d) $\lim_{x \to 1} f(x) = \pi$

Level II

44. If $f(x) = \begin{cases} x^2 - 1, & x \leq 1 \\ -x^2 - 1, & x > 1 \end{cases}$, then

(a) LHL ≠ RHL, at $x = 1$
(b) LHL = RHL = −2, at $x = 1$
(c) LHL = RHL = −1, at $x = 1$
(d) None of the above

45. If $f(x) = \begin{cases} \frac{x}{|x|}, & x \neq 0 \\ 0, & x = 0 \end{cases}$, then

(a) LHL = RHL = 1, at $x = 0$ (b) LHL = RHL = −1, at $x = 0$
(c) LHL ≠ RHL, at $x = 0$ (d) None of the above

46. Suppose $f(x) = \begin{cases} a + bx, & x < 1 \\ 4, & x = 1 \\ b - ax, & x > 1 \end{cases}$

and if $\lim_{x \to 1} f(x) = f(1)$, then

(a) $a = 4, b = 0$ (b) $a = 0, b = 4$
(c) $a = 0, b = 0$ (d) $a = 4, b = 4$

47. Let $a_1, a_2, a_3, \ldots, a_n$ be fixed real numbers and define a function

$f(x) = (x - a_1)(x - a_2) \ldots (x - a_n)$, then

(a) $\lim_{x \to a_1} f(x) = 0$
(b) $\lim_{x \to a} f(x) = (a - a_1)(a - a_2) \ldots (a - a_n)$, for some $a \neq a_1, a_2, a_3, \ldots, a_n$
(c) Both (a) and (b) are true
(d) Either (a) or (b) is true

48. If $f(x) = \begin{cases} |x| + 1, & x < 0 \\ 0, & x = 0 \\ |x| - 1, & x > 0 \end{cases}$, then $\lim_{x \to a} f(x)$ exists for all

(a) $a \neq 1$ (b) $a \neq 0$ (c) $a \neq -1$ (d) $a \neq 2$

49. If $f(x) = \begin{cases} mx^2 + n, & x < 0 \\ nx + m, & 0 \leq x \leq 1 \\ nx^3 + m, & x > 1 \end{cases}$, then

(a) $\lim_{x \to 0} f(x)$ exists when $m = n$
(b) $\lim_{x \to 1} f(x)$ exists for any integral value of m and n
(c) Both (a) and (b) are true
(d) Either (a) or (b) is true

50. $\lim_{x \to 2} \left[\frac{1}{x - 2} - \frac{2(2x - 3)}{x^3 - 3x^2 + 2x} \right]$ is equal to [NCERT Exemplar]

(a) $\frac{1}{2}$ (b) $\frac{-1}{2}$ (c) 1 (d) −2

51. $\lim_{x \to 0} \frac{\sqrt{2 + x} - \sqrt{2}}{x}$ is equal to [NCERT Exemplar]

(a) $\frac{1}{2}$ (b) $\frac{1}{\sqrt{2}}$ (c) $\frac{1}{2\sqrt{2}}$ (d) $\sqrt{2}$

52. If $\lim_{x \to 3} \frac{x^n - 3^n}{x - 3} = 108$, the positive integer n is equal to [NCERT Exemplar]

(a) 3 (b) 5 (c) 2 (d) 4

53. $\lim_{x \to \frac{\pi}{2}} (\sec x - \tan x)$ is equal to [NCERT Exemplar]

(a) 1 (b) 2
(c) 0 (d) Not defined

54. $\lim_{x \to 0} \frac{\sin(2+x) - \sin(2-x)}{x}$ is equal to $p \cos q$, where p and q are respectively [NCERT Exemplar]

(a) 1 , 2 (b) 2, 1 (c) 1, 1 (d) 2, 2

55. $\lim_{x \to \frac{\pi}{6}} \frac{2\sin^2 x + \sin x - 1}{2\sin^2 x - 3\sin x + 1}$ is equal to [NCERT Exemplar]

(a) –1 (b) 2 (c) –3 (d) 3

56. $\lim_{x \to 0} \frac{\tan x - \sin x}{\sin^3 x}$ is equal to [NCERT Exemplar]

(a) $\frac{1}{2}$ (b) 0
(c) 1 (d) Not defined

57. $\lim_{x \to a} \frac{\sqrt{a+2x} - \sqrt{3x}}{\sqrt{3a+x} - 2\sqrt{x}}$ is equal to $\frac{2m}{9}$, where m is equal to [NCERT Exemplar]

(a) $\sqrt{3}$ (b) $\sqrt{2}$ (c) $\sqrt{5}$ (d) $\sqrt{7}$

58. $\lim_{x \to 0} \frac{\cos ax - \cos bx}{\cos cx - 1}$ is equal to $\frac{m}{n}$, where m and n are respectively [NCERT Exemplar]

(a) $a^2 + b^2, c^2$ (b) $c^2, a^2 + b^2$
(c) $a^2 - b^2, c^2$ (d) $c^2, a^2 - b^2$

59. $\lim_{h \to 0} \frac{(a+h)^2 \sin(a+h) - a^2 \sin a}{h}$ is equal to

I. $a^2 \sin a + 2a\cos a$ II. $a^2 \cos a + 2a \sin a$ [NCERT Exemplar]

(a) Both I and II are true (b) Only I is true
(c) Only II is true (d) Both I and II are false

60. $\lim_{x \to 0} \frac{\sin x}{x(1+\cos x)}$ is equal to [NCERT Exemplar]

(a) 0 (b) $\frac{1}{2}$ (c) 1 (d) –1

61. $\lim_{x \to \frac{\pi}{2}} \frac{1 - \sin x}{\cos x}$ is equal to [NCERT Exemplar]

(a) 0 (b) –1
(c) 1 (d) Does not exist

62. $\lim_{x \to 1} [x-1]$, where $[\cdot]$ is greatest integer function, is equal to [NCERT Exemplar]

(a) 1 (b) 2
(c) 0 (d) Does not exist

63. $\lim_{x \to 0} x \sin \frac{1}{x}$ is equal to [NCERT Exemplar]

(a) 0 (b) 1
(c) $\frac{1}{2}$ (d) Does not exist

64. $\lim_{n \to \infty} \frac{1 + 2 + 3 + \ldots\ldots + n}{n^2}$, $n \in N$ is equal to [NCERT Exemplar]

(a) 0 (b) 1 (c) $\frac{1}{2}$ (d) $\frac{1}{4}$

Derivatives

Level I

65. "Differentiation is the rate of change in the dependent variable with respect to independent variable". Then, which of the following statement is related with above concept?

(a) People maintaining a reservoir need to know when will a reservoir overflow knowing the depth of the water at several instances of time

(b) Rocket scientists need to compute the precise velocity with which the satellite needs to be shot out from the rocket knowing the height of the rocket at various times

(c) Financial institutions need to predict the changes in the value of a particular stock knowing its present value.

(d) All of the above

66. Let f is a real valued function and a is a point in its domain of definition, then the derivative of f at a is defined by

(a) $\lim_{h \to 0} \frac{f(a+h) - f(a)}{h}$ (b) $\lim_{h \to 0} \frac{f(a) - f(a+h)}{h}$

(c) $\lim_{h \to 0} h f(a+h) - f(a)$ (d) $\lim_{h \to 0} \frac{h}{f(a+h) - f(a)}$

67. Which of the following is/are true?

I. $f'(a)$ quantifies the change in $f(x)$ at x with respect to a.

II. $f'(a)$ quantifies the change in $f(x)$ at a with respect to x.

(a) Both I and II are true (b) Only I is true
(c) Only II is true (d) Both I and II are false

68. The derivative of the function $f(x) = 3x$ at $x = 2$ is

(a) 0 (b) 1 (c) 2 (d) 3

69. If f be a function given by $f(x) = 2x^2 + 3x - 5$. Then, $f'(0) = m f'(-1)$, where m is equal to

(a) –1 (b) –2 (c) –3 (d) –4

70. The derivative of $\sin x$ at $x = 0$ is

(a) 0 (b) 2 (c) 1 (d) 3

71. The derivative of $f(x) = 3$ at $x = 0$ and at $x = 3$ are

(a) negative (b) zero
(c) different (d) not defined

72. Let $y = f(x)$ be a function and let $P = (a, f(a))$ and $Q(a + h, f(a + h))$ be two points close to each other on the given graph of this function.

I. $\lim_{h \to 0} \dfrac{f(a+h) - f(a)}{h} = \lim_{Q \to P} \dfrac{QR}{PR}$

II. $f'(a) = \tan \Psi$

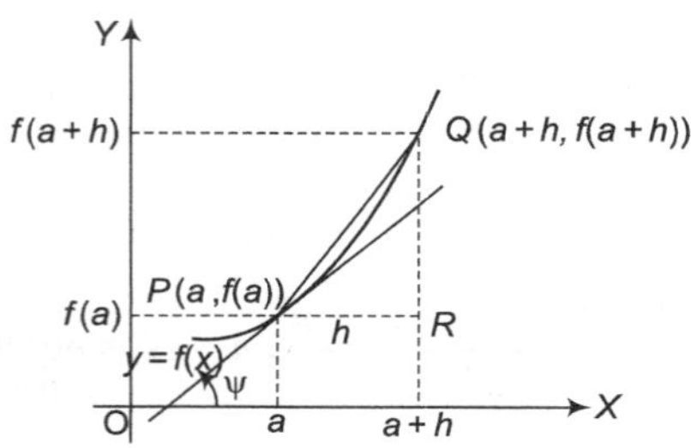

Which of the following is/are true?

(a) Both I and II are true (b) Only I is true
(c) Only II is true (d) Both I and II are false

73. If the derivative of a function f exists at ...A... point, it defines a new function called the ...B... of f. Here, A and B refer to

(a) particular, derivative (b) every, derivative
(c) particular, limit (d) every, limit

74. Suppose f is a real valued function, the derivative of f at x is given by

(a) $f'(x) = \lim_{h \to x} \dfrac{f(x+h) - f(x)}{h}$

(b) $f'(x) = \lim_{h \to \infty} \dfrac{f(x+h) - f(x)}{h}$

(c) $f'(x) = \lim_{h \to 0} \dfrac{f(x+h) - f(x)}{h}$

(d) None of the above

75. If $y = f(x)$ is the function, then derivative of f at any x is denoted by

(a) $f'(x)$ (b) $\dfrac{dy}{dx}$
(c) $D(f(x))$ (d) All of these

76. Derivative of f at $x = a$ is denoted by

(a) $\dfrac{d}{dx} f(x)\Big|_a$ (b) $\dfrac{df}{dx}\Big|_a$ (c) $\left(\dfrac{df}{dx}\right)_{x=a}$ (d) All of these

77. Match the functions given in Column I with their corresponding derivatives given in Column II and choose the correct option from the codes given below.

Column I	Column II
A. $f(x) = 10x$	1. $2x$
B. $f(x) = x^2$	2. $-\dfrac{1}{x^2}$
C. $f(x) = a$ for fixed real number a	3. 0
D. $f(x) = \dfrac{1}{x}$	4. 10

Codes

	A	B	C	D		A	B	C	D
(a)	4	1	3	2	(b)	1	4	3	2
(c)	4	1	2	3	(d)	4	3	1	2

78. Let f and g be two functions such that their derivatives are defined in a common domain. Then, which of the following is incorrect?

(a) $\dfrac{d}{dx}[f(x) + g(x)] = \dfrac{d}{dx} f(x) + \dfrac{d}{dx} g(x)$

(b) $\dfrac{d}{dx}[f(x) - g(x)] = \dfrac{d}{dx} f(x) - \dfrac{d}{dx} g(x)$

(c) $\dfrac{d}{dx}[f(x) \cdot g(x)] = f(x) \cdot \dfrac{d}{dx} g(x) + g(x) \cdot \dfrac{d}{dx} f(x)$

(d) $\dfrac{d}{dx}\left[\dfrac{f(x)}{g(x)}\right] = \dfrac{g(x) \dfrac{d}{dx} f(x) - f(x) \dfrac{d}{dx} g(x)}{(g(x))^2}$

79. Which of the following is/are true?

Let $u = f(x)$ and $v = g(x)$. Then,

I. $(uv)' = u'v + uv'$ is a Leibnitz rule or product rule.

II. $\left(\dfrac{u}{v}\right)' = \dfrac{u'v - uv'}{v^2}$ is a Leibnitz rule or quotient rule.

Which of the following is/are true?

(a) Both I and II are true (b) Only I is true
(c) Only II is true (d) Both I and II are false

80. The derivative of the function $f(x) = x$ is

(a) 0 (b) 1
(c) ∞ (d) None of these

81. Which of the following is/are true?

I. The derivative of $f(x) = 10x$ is 10.

II. The derivative of $f(x) = x^2$ is $2x$.

(a) Both I and II are true (b) Only I is true
(c) Only II is true (d) Both I and II are false

82. For any positive integer n, the derivative of $f(x) = x^n$ is

(a) n^{x^n} (b) $(n-1)x^n$ (c) nx^{n-1} (d) $(n-1)x^{n-1}$

83. Let $f(x) = a_n x^n + a_{n-1} x^{n-1} + \ldots\ldots + a_1 x + a_0$

be a polynomial function, where a_i's are all real numbers and $a_n \neq 0$. Then, the derivative of function is given by

(a) $(n+1) a_n x^{n+1} + n a_{n-1} x^n + \ldots\ldots + 3 a_2 x^2 + 2a_1 x + a_0$

(b) $n a_n x^{n-1} + (n-1) a_{n-1} x^{n-2} + \ldots\ldots + 2a_2 x + a_1$

(c) $(n-1) a_n x^{n-1} + (n-2) a_{n-2} x^{n-1} + \ldots\ldots + 2a_2 x^2 + a_1 x + a_0$

(d) None of the above

84. The derivative of function $6x^{100} - x^{55} + x$ is

(a) $600x^{100} - 55x^{55} + x$ (b) $600x^{99} - 55x^{54} + 1$

(c) $99x^{99} - 54x^{54} + 1$ (d) $99x^{99} - 54x^{54}$

85. Which of the following is/are true?

I. The derivative of $f(x) = 1 + x + x^2 + \ldots\ldots + x^{50}$ at $x = 1$ is 1250.

II. The derivative of $f(x) = \dfrac{x+1}{x}$ is $\dfrac{1}{x^2}$.

(a) Both I and II are true (b) Only I is true

(c) Only II is true (d) Both I and II are false

86. Match the functions given in Column I with their corresponding derivatives given in Column II and choose the correct option from the codes given below.

	Column I		Column II
A.	$f(x) = \sin x$	1.	$\sin 2x$
B.	$f(x) = \tan x$	2.	$\cos x$
C.	$f(x) = \sin^2 x$	3.	$\sec^2 x$

Codes

	A	B	C		A	B	C
(a)	2	3	1	(b)	2	1	3
(c)	1	3	2	(d)	3	2	1

87. Which of the following is/are true?

I. The derivative of $x^2 - 2$ at $x = 10$ is 18.

II. The derivative of $99x$ at $x = 100$ is 99.

III. The derivative of x at $x = 1$ is 1.

(a) I , II and III are true

(b) I and II are true

(c) II and III are true

(d) I and III are true

88. Match the functions given in Column I with their corresponding derivatives given in Column II and choose the correct option from the codes given below.

	Column I		Column II
A.	$x^3 - 27$	1.	$2x - 3$
B.	$(x-1)(x-2)$	2.	$\dfrac{-2}{x^3}$
C.	$\dfrac{1}{x^2}$	3.	$3x^2$
D.	$\dfrac{x+1}{x-1}$	4.	$\dfrac{-2}{(x-1)^2}$

Codes

	A	B	C	D		A	B	C	D
(a)	1	3	2	4	(b)	3	1	2	4
(c)	3	1	4	2	(d)	3	2	1	4

89. For the function

$$f(x) = \frac{x^{100}}{100} + \frac{x^{99}}{99} + \ldots\ldots\ldots \frac{x^2}{2} + x + 1,$$

$f'(1) = mf'(0)$, where m is equal to

(a) 50 (b) 0 (c) 100 (d) 200

90. If a is some fixed real number, then the derivative of $x^n + ax^{n-1} + a^2x^{n-2} + \ldots. + a^{n-1}x + a^n$ is

(a) $nx^{n-1} + a(n-1)x^{n-2} + a^2(n-2)x^{n-3} + \ldots. + a^{n-1}$

(b) $(n-1)x^{n-1} + a(n-2)x^{n-2} + a^2(n-3)x^{n-3} + \ldots. + a^{n-1}$

(c) $nx^n + a(n-1)x^{n-1} + a^2(n-2)x^{n-2} + \ldots. + a^n$

(d) None of the above

91. For some constants a and b, match the functions given in Column I with their corresponding derivatives given in Column II and choose the correct option from the codes given below.

	Column I		Column II
A.	$(x-a)(x-b)$	1.	$\dfrac{a-b}{(x-b)^2}$
B.	$(ax^2+b)^2$	2.	$2x - a - b$
C.	$\dfrac{x-a}{x-b}$	3.	$4a^2x^3 + 4abx$

Codes

	A	B	C		A	B	C
(a)	2	1	3	(b)	1	3	2
(c)	3	2	1	(d)	2	3	1

92. For some constant a, the derivative of $\dfrac{x^n - a^n}{x-a}$ is

(a) $\dfrac{nx^n + anx^{n-1} + x^n + a^n}{(x-a)^2}$ (b) $\dfrac{nx^n - anx^{n-1} - x^n + a^n}{(x-a)^2}$

(c) $\dfrac{anx^{n-1} - nx^n - a^n + x^n}{(x-a)^2}$ (d) $\dfrac{anx^{n-1} - nx^n - x^n + a^n}{(x-a)^2}$

93. Which of the following is/are true?

I. The derivative of $y = 2x - \dfrac{3}{4}$ is 2.

II. The derivative of $y = (5x^3 + 3x - 1)(x-1)$ is $20x^3 + 15x^2 + 6x - 4$

(a) Both I and II are true (b) Only I is true

(c) Only II is true (d) Both I and II are false

94. Match the functions given in Column I with their corresponding derivatives given in Column II and choose the correct option from the codes given below.

	Column I		Column II
A.	$\cos x$	1.	$5\sec x \tan x - 4\sin x$
B.	$\sin x \cos x$	2.	$\cos 2x$
C.	$\sec x$	3.	$-\sin x$
D.	$5\sec x + 4\cos x$	4.	$\tan x \sec x$

Codes

	A	B	C	D
(a)	3	2	4	1
(b)	2	3	4	1
(c)	3	2	1	4
(d)	3	4	2	1

95. Match the functions given in Column I with their corresponding derivatives given in Column II and choose the correct option from the codes given below.

	Column I		Column II
A.	$x^{-3}(5 + 3x)$	1.	$-12x^{-5} + 36x^{-10}$
B.	$x^5(3 - 6x^{-9})$	2.	$\frac{-2}{(x+1)^2} - \frac{x(3x-2)}{(3x-1)^2}$
C.	$x^{-4}(3 - 4x^{-5})$	3.	$-15x^{-4} - 6x^{-3}$
D.	$\frac{2}{x+1} - \frac{x^2}{3x-1}$	4.	$15x^4 + 24x^{-5}$

Codes

	A	B	C	D
(a)	3	4	1	2
(b)	4	3	1	2
(c)	3	1	4	2
(d)	3	4	2	1

96. Match the functions given in Column I with their corresponding derivatives given in Column II and choose the correct option from the codes given below.

	Column I		Column II
A.	$\operatorname{cosec} x$	1.	$5\cos x + 6\sin x$
B.	$3\cot x + 5\operatorname{cosec} x$	2.	$-3\operatorname{cosec}^2 x - 5\operatorname{cosec} x \cot x$
C.	$5\sin x - 6\cos x + 7$	3.	$2\sec^2 x - 7\sec x \tan x$
D.	$2\tan x - 7\sec x$	4.	$-\cot x \operatorname{cosec} x$

Codes

	A	B	C	D
(a)	4	1	2	3
(b)	4	2	3	1
(c)	2	4	1	3
(d)	4	2	1	3

97. Which of the following is/are true?

I. The derivative of $f(x) = \sin 2x$ is $2(\cos^2 x - \sin^2 x)$.

II. The derivative of $g(x) = \cot x$ is $-\operatorname{cosec}^2 x$.

(a) Both I and II are true (b) Only I is true
(c) Only II is true (d) Both I and II are false

98. Which of the following is/are true?

I. The derivative of $f(x) = \frac{x^5 - \cos x}{\sin x}$ is $\frac{x^5 \cos x + 5x^4 \sin x + 1}{(\sin x)^2}$.

II. The derivative of $h(x) = \frac{x + \cos x}{\tan x}$ is $\frac{(1 - \sin x)\sec^2 x - (x + \cos x)\tan x}{(\tan x)^2}$

(a) Both I and II are true (b) Only I is true
(c) Only II is true (d) Both I and II are false

Level II

99. Which of the following is/are true?

I. If a and b are non-zero constants, then the derivative of $f(x) = ax + b$ is a.

II. If a, b and c are non-zero constants, then the derivative of $f(x) = ax^2 + bx + c$ is $ax + b$. [NCERT Exemplar]

(a) Both I and II are true (b) Only I is true
(c) Only II is true (d) Both I and II are false

100. Which of the following is/are true?

I. The derivative of $f(x) = x^3$ is x^2.

II. The derivative of $f(x) = \frac{1}{x^3}$ is $\frac{-1}{x^2}$.

(a) Both I and II are true (b) Only I is true
(c) Only II is true (d) Both I and II are false

101. If $f(x) = |\log_e |x||$, then $f'(x)$ equals to

(a) $\frac{1}{|x|}, x \neq 0$

(b) $\frac{1}{x}$ for $|x| > 1$ and $-\frac{1}{x}$ for $|x| < 1$

(c) $-\frac{1}{x}$ for $|x| > 1$ and $\frac{1}{x}$ for $|x| < 1$

(d) $\frac{1}{x}$ for $x > 0$ and $-\frac{1}{x}$ for $x < 0$

102. The derivative of $f(x) = \tan(ax + b)$ is [NCERT Exemplar]

(a) $\sec^2(ax + b)$ (b) $b\sec^2(ax + b)$
(c) $a\sec^2(ax + b)$ (d) $ab\sec^2(ax + b)$

103. The derivative of $f(x) = \sqrt{\sin x}$ is [NCERT Exemplar]

(a) $\frac{1}{2}\cos x\sqrt{\sin x}$ (b) $2\cos x\sqrt{\sin x}$

(c) $2\cot x\sqrt{\sin x}$ (d) $\frac{1}{2}\cot x\sqrt{\sin x}$

104. The derivative of $f(x) = \sqrt{\cos x}$ is

(a) $-\frac{\tan x}{2}\sqrt{\cos x}$ (b) $-2\tan x\sqrt{\cos x}$

(c) $-2\sin x\sqrt{\cos x}$ (d) $-\frac{1}{2}\sin x\sqrt{\cos x}$

105. Let f be an even function and $f'(0)$ exists, then $f'(0)$ is

(a) 1 (b) 2

(c) 0 (d) −1

106. If $f(x) = x\sin x$, then $f'\left(\frac{\pi}{2}\right)$ is equal to [NCERT Exemplar]

(a) 0 (b) 1 (c) −1 (d) $\frac{1}{2}$

107. The functions $u = e^x \sin x$, $v = e^x \cos x$ satisfy the equation

(a) $v\frac{du}{dx} - u\frac{dv}{dx} = u^2 + v^2$ (b) $\frac{d^2u}{dx^2} = 2v$

(c) $\frac{d^2v}{dx^2} = -2u$ (d) All of these

108. If $f(x) = |\cos x - \sin x|$, then $f'\left(\frac{\pi}{4}\right)$ is equal to

(a) $\sqrt{2}$ (b) $-\sqrt{2}$

(c) 0 (d) None of these

109. The derivative of $\sqrt{ax+b}$ is

(a) $\frac{a}{2\sqrt{ax+b}}$ (b) $\frac{2}{a\sqrt{ax+b}}$ (c) $\frac{2a}{\sqrt{ax+b}}$ (d) $\frac{1}{2a\sqrt{ax+b}}$

110. If $f(x) + f(y) = f\left(\frac{x+y}{1-xy}\right)$ for all $x, y \in R$ $(xy \neq 1)$ and $\lim_{x \to 0} \frac{f(x)}{x} = 2$. Then, $f'\left(\frac{1}{\sqrt{3}}\right)$ is

(a) $\frac{3}{4}$ (b) $\frac{4}{3}$ (c) $\frac{3}{6}$ (d) $\frac{3}{2}$

111. The derivative of $\cot\sqrt{x}$ is

(a) $\frac{-\sin^2\sqrt{x}}{2\sqrt{x}}$ (b) $\frac{\sin^2\sqrt{x}}{2\sqrt{x}}$

(c) $\frac{-\operatorname{cosec}^2\sqrt{x}}{2\sqrt{x}}$ (d) $\frac{-\sec^2\sqrt{x}}{2\sqrt{x}}$

112. If $\frac{d}{dx}\left\{\frac{1}{f(x)}\right\} = \frac{p}{q}$, where p and q respectively are

(a) $f'(x), \{f(x)\}^2$ (b) $-f'(x), \{f(x)\}^2$

(c) $-f'(x), f(x)$ (d) $f(x), f'(x)$

Miscellaneous Exercise

113. Which of the following is/are true?

I. The derivative of $-x$ is -1.

II. The derivative of $(-x)^{-1}$ is $\frac{1}{x^2}$

(a) Both I and II are true

(b) Only I is true

(c) Only II is true

(d) Both I and II are false

114. Which of the following is/are true?

I. The derivative of $\sin(x+1)$ is $\cos(x+1)$.

II. The derivative of $\cos\left(x - \frac{\pi}{8}\right)$ is $\frac{-\pi}{8}\sin\left(x - \frac{\pi}{8}\right)$.

(a) Both I and II are true (b) Only I is true

(c) Only II is true (d) Both I and II are false

115. If a is a non-zero constant, then the derivative of $x + a$ is

(a) 1 (b) 0

(c) a (d) None of these

116. If p, q, r and s are fixed non-zero constants, then the derivative of $(px+q)\left(\frac{r}{x}+s\right)$ is

(a) $ps + \frac{qr}{x^2}$ (b) $ps - \frac{qr}{x}$

(c) $pr - \frac{qs}{x^2}$ (d) $ps - \frac{qr}{x^2}$

117. If a, b, c and d are fixed non-zero constants, then the derivative of $(ax+b)(cx+d)^2$ is

(a) $(cx+d)(3acx + bc + ad)$

(b) $(cx+d)(3acx + 2bc + ad)$

(c) $(cx+d)(acx + 2bc + ad)$

(d) $(cx+d)(acx + bc + ad)$

118. If a, b, c and d are non-zero fixed constants, then the derivative of $\frac{ax+b}{cx+d}$ is $\frac{m}{(cx+d)^n}$, where m and n respectively are

(a) $ad + bc, 2$ (b) $ac - bd, 3$ (c) $ad - bc, 2$ (d) $bc - ad, 2$

119. The derivative of $\dfrac{1+\frac{1}{x}}{1-\frac{1}{x}}$ is

(a) $\dfrac{2}{(1+x)^2}$ (b) $\dfrac{-2}{(1-x)^2}$ (c) $\dfrac{-1}{(1-x)^2}$ (d) $\dfrac{3}{(1-x)^2}$

120. If a, b and c are fixed non-zero constants, then the derivative of $\dfrac{1}{ax^2+bx+c}$ is

(a) $\dfrac{(ax+b)}{(ax^2+bx+c)^2}$ (b) $\dfrac{-(ax+b)}{(ax^2+bx+c)^2}$

(c) $\dfrac{-(2ax+b)}{(ax^2+bx+c)^2}$ (d) $\dfrac{2ax+b}{(ax^2+bx+c)^2}$

121. If a, b, p, q and r are fixed non-zero constants, then the derivative of $\dfrac{ax+b}{px^2+qx+r}$ is

(a) $\dfrac{-apx^2-2bpx+ra-bq}{(px^2+qx+r)^2}$ (b) $\dfrac{apx^2+2bpx-ra+bq}{(px^2+qx+r)^2}$

(c) $\dfrac{apx^2+bpx-2ra+bq}{(px^2+qx+r)^2}$ (d) $\dfrac{-apx^2-bpx+2ra-bq}{(px^2+qx+r)^2}$

122. If a, b, p, q and r are fixed non-zero constants, then the derivative of $\dfrac{px^2+qx+r}{ax+b}$ is

(a) $\dfrac{apx^2+2bpx+bq+ra}{(ax+b)^2}$ (b) $\dfrac{apx^2+bpx+2bq+ra}{(ax+b)^2}$

(c) $\dfrac{apx^2+bpx+2bq-ra}{(ax+b)^2}$ (d) $\dfrac{apx^2+2bpx+bq-ra}{(ax+b)^2}$

123. If a, b are fixed non-zero constants, then the derivative of $\dfrac{a}{x^4}-\dfrac{b}{x^2}+\cos x$ is $ma+nb-p$, where

(a) $m=4x^3, n=\dfrac{-2}{x^3}, p=\sin x$

(b) $m=\dfrac{-4}{x^5}, n=\dfrac{2}{x^3}, p=\sin x$

(c) $m=\dfrac{-4}{x^5}, n=\dfrac{-2}{x^3}, p=-\sin x$

(d) $m=4x^3, n=\dfrac{2}{x^3}, p=-\sin x$

124. The derivative of $4\sqrt{x}-2$ is

(a) $\dfrac{1}{\sqrt{x}}$ (b) $2\sqrt{x}$

(c) $\dfrac{2}{\sqrt{x}}$ (d) $\sqrt{x}$

125. If a and b are fixed non-zero constants, then the derivative of $(ax+b)^n$ is

(a) $n(ax+b)^{n-1}$ (b) $na(ax+b)^{n-1}$

(c) $nb(ax+b)^{n-1}$ (d) $nab(ax+b)^{n-1}$

126. If a, b, c and d are fixed non-zero constants, then the derivative of $(ax+b)^n(cx+d)^m$ is $(ax+b)^{n-1}(cx+d)^{m-1}[p(ax+b)+q(cx+d)]$, where p and q respectively are

(a) na, mc (b) $-na, mc$ (c) $mc, -na$ (d) mc, na

127. Which of the following is/are true?

I. The derivative of $\sin(x+a)$ is $\cos(x+a)$, where a is a fixed non-zero constant.

II. The derivative of $\operatorname{cosec} x \cot x$ is $\operatorname{cosec}^3 x-\cot^2 x \operatorname{cosec} x$

(a) Both I and II are true (b) Only I is true

(c) Only II is true (d) Both I and II are false

128. The derivative of $\dfrac{\sin x+\cos x}{\sin x-\cos x}$ is $\dfrac{m}{(\sin x-\cos x)^2}$, where m is equal to

(a) 1 (b) −1 (c) 2 (d) −2

129. The derivative of $\dfrac{\sec x-1}{\sec x+1}$ is $\dfrac{p}{(\sec x+1)^2}$, where p is equal to

(a) $2\sec x\tan x$ (b) $2\sin x\tan x$

(c) $2\operatorname{cosec} x\cot x$ (d) $2\cos x\cot x$

130. The derivative of $\sin^n x$

(a) $n\sin^{n-1}x$ (b) $n\cos^{n-1}x$

(c) $n\sin^{n-1}x\cos x$ (d) $n\cos^{n-1}x\sin x$

131. If a, b, c and d are fixed non-zero constants, then the derivative of $\dfrac{a+b\sin x}{c+d\cos x}$ is

(a) $\dfrac{bc\cos x-ad\sin x+bd}{(c+d\cos x)^2}$ (b) $\dfrac{bc\cos x-ad\sin x-bd}{(c+d\cos x)^2}$

(c) $\dfrac{ad\cos x+bc\sin x+bd}{(c+d\cos x)^2}$ (d) $\dfrac{bc\cos x+ad\sin x+bd}{(c+d\cos x)^2}$

132. If a is a fixed non-zero constant, then the derivative of $\dfrac{\sin(x+a)}{\cos x}$ is

(a) $\dfrac{\cos a}{\cos^2 x}$ (b) $\dfrac{-\cos a}{\cos^2 x}$

(c) $\dfrac{\sin a}{\cos^2 x}$ (d) $\dfrac{-\sin a}{\cos^2 x}$

133. If the derivative of $x^4(5\sin x-3\cos x)$ is $x^3[mx\cos x+nx\sin x+p\sin x+q\cos x]$, then m, n, p and q respectively are

(a) 5, −3, −20, 12 (b) 5, 3, 20, −12

(c) 5, 3, 12, −20 (d) 5, −3, −12, 20

134. The derivative of $(x^2+1)\cos x$ is

(a) $-x^2\sin x-\sin x-2x\cos x$ (b) $-x^2\sin x-\sin x+2\cos x$

(c) $-x^2\sin x-x\sin x+2\cos x$ (d) $-x^2\sin x-\sin x+2x\cos x$

135. If a, p and q are fixed non-zero constants, then the derivative of $(ax^2 + \sin x)(p + q\cos x)$ is $A(ax^2 + \sin x) + B(p + q\cos x)$, where A and B respectively are

(a) $-q\sin x, 2ax + \cos x$ (b) $2ax + \cos x, -q\sin x$
(c) $2ax - \cos x, q\sin x$ (d) $q\sin x, 2ax - \cos x$

136. If the derivative of $(x + \cos x)(x - \tan x)$ is $A(x + \cos x) + B(x - \tan x)$, then A and B respectively are

(a) $\tan^2 x, 1 - \sin x$ (b) $-\tan^2 x, 1 - \sin x$
(c) $\tan^2 x, \sin x - 1$ (d) $-\tan^2 x, \sin x - 1$

137. If the derivative of $\dfrac{4x + 5\sin x}{3x + 7\cos x}$ is $\dfrac{m + nx\cos x + p(\cos x + x\sin x) + q\sin x}{(3x + 7\cos x)^2}$, where m, n, p and q respectively are

(a) 35, 15, 28, 15 (b) 35, 28, 15, –15
(c) 35, 15, 28, –15 (d) 35, –15, 28, 15

138. The derivative of $\dfrac{x^2 \cos\left(\frac{\pi}{4}\right)}{\sin x}$ is

(a) $\dfrac{x\sin\left(\frac{\pi}{4}\right)[2\sin x + x\cos x]}{\sin^2 x}$ (b) $\dfrac{x\cos\left(\frac{\pi}{4}\right)[2\sin x - x\cos x]}{\sin^2 x}$

(c) $\dfrac{x\sin\left(\frac{\pi}{4}\right)[2\sin x - x\cos x]}{\cos^2 x}$ (d) $\dfrac{x\cos\left(\frac{\pi}{4}\right)[2\sin x + x\cos x]}{\cos^2 x}$

139. The derivative of $\dfrac{x}{1 + \tan x}$ is

(a) $\dfrac{1 + \tan x + \sec^2 x}{(1 + \tan x)^2}$ (b) $\dfrac{1 + \tan x + x\sec^2 x}{(1 + \tan x)^2}$

(c) $\dfrac{1 + \tan x - x\sec^2 x}{(1 + \tan x)^2}$ (d) $\dfrac{1 - x\tan x + \sec^2 x}{(1 + \tan x)^2}$

140. If the derivative of $(x + \sec x)(x - \tan x)$ is $A(x + \sec x) + B(x - \tan x)$, where A and B respectively are

(a) $\sec^2 x - 1, 1 + \sec x \tan x$ (b) $1 - \sec^2 x, 1 - \sec x \tan x$
(c) $1 - \sec^2 x, 1 + \sec x \tan x$ (d) $\sec^2 x - 1, 1 - \sec x \tan x$

141. The derivative of $\dfrac{x}{\sin^n x}$ is

(a) $\dfrac{\sin^n x + nx\sin^{n-1} x\cos x}{\sin^{2n} x}$

(b) $\dfrac{\sin^n x - nx\sin^{n-1} x\cos x}{\sin^n x}$

(c) $\dfrac{\sin^n x + nx\sin^{n-1} x\cos x}{\sin^n x}$

(d) $\dfrac{\sin^n x - nx\sin^{n-1} x\cos x}{\sin^{2n} x}$

142. Let $f(x) = x - [x]; x \in R$, where $[\cdot]$ represents the greatest integer function, then $f'\left(\frac{1}{2}\right)$ is equal to [NCERT Exemplar]

(a) 1 (b) 0 (c) –1 (d) 2

143. Let $f(x) = \begin{cases} x^2 - 1, & 0 < x < 2 \\ 2x + 3, & 2 \le x < 3 \end{cases}$

If $\lim_{x \to 2^-} f(x)$ and $\lim_{x \to 2^+} f(x)$ are the roots of a quadratic equation, then the equation is [NCERT Exemplar]

(a) $x^2 - 10x + 21 = 0$ (b) $x^2 + 10x + 21 = 0$
(c) $x^2 + 4x - 21 = 0$ (d) $x^2 - 4x - 21 = 0$

144. If $f(x) = \begin{cases} \dfrac{\sin[x]}{[x]}, & [x] \ne 0 \\ 0, & [x] = 0 \end{cases}$ where $[\cdot]$ denotes the greatest integer function, then $\lim_{x \to 0} f(x)$ is equal to [NCERT Exemplar]

(a) 1 (b) 0
(c) –1 (d) Does not exist

145. $\lim_{x \to 0} \dfrac{\operatorname{cosec} x - \cot x}{x}$ is equal to [NCERT Exemplar]

(a) $\frac{1}{2}$ (b) 0 (c) –1 (d) $\frac{-1}{2}$

146. Let $f(x) = \begin{cases} x + 2, & x \le -1 \\ cx^2, & x > -1 \end{cases}$

If $\lim_{x \to -1} f(x)$ exists, then c is equal to [NCERT Exemplar]

(a) 0 (b) 1 (c) –1 (d) 2

147. Let $f(x) = \begin{cases} \dfrac{k\cos x}{\pi - 2x}, & \text{when } x \ne \dfrac{\pi}{2} \\ 3, & \text{when } x = \dfrac{\pi}{2} \end{cases}$

If $\lim_{x \to \frac{\pi}{2}} f(x) = f\left(\frac{\pi}{2}\right)$, then k is equal to [NCERT Exemplar]

(a) 2 (b) 4 (c) 6 (d) 8

148. $\lim_{x \to 4} \dfrac{|x - 4|}{x - 4}$ is equal to [NCERT Exemplar]

(a) 1 (b) 0
(c) –1 (d) Does not exist

149. $\lim_{x \to \pi} \dfrac{1 - \sin\frac{x}{2}}{\cos\frac{x}{2}\left(\cos\frac{x}{4} - \sin\frac{x}{4}\right)}$ is equal to [NCERT Exemplar]

(a) $\sqrt{2}$ (b) $\frac{1}{\sqrt{2}}$ (c) $\sqrt{3}$ (d) $\frac{1}{\sqrt{3}}$

Answer *with* Explanations

1. *(a)* Calculus is that branch of Mathematics which mainly deals with the study of change in the value of a function with respect to the change in points of its domain.

2. *(c)* We know that,

$$s = ut + \frac{1}{2}gt^2, \text{ where}$$

s = distance, u = initial velocity, t = time,
g = gravitational acceleration

Here, $u = 0, g = 9.8 \text{ m/s}^2$

$$\Rightarrow \quad s = 0 \times t + \frac{1}{2} \times 9.8t^2$$

$$\Rightarrow \quad s = 4.9t^2 \text{ m in } t \text{ s}$$

3. *(c)* The adjoining table I gives the distance travelled in metres at various intervals of time in seconds of a body dropped from a tall cliff.

The objective is to find the velocity of the body at time $t = 2$ s from this data. One way to approach this problem is to find the average velocity for various intervals of time ending at $t = 2$ s and hope that these throw some light on the velocity at $t = 2$ s.

Average velocity between $t = t_1$ and $t = t_2$ equals to distance travelled between $t = t_1$ and $t = t_2$ divided by $(t_2 - t_1)$. Hence, the average velocity in the first two seconds

$$= \frac{\text{Distance travelled between } t_2 = 2 \text{ and } t_1 = 0}{\text{Time interval } (t_2 - t_1)}$$

$$= \frac{(19.6 - 0)\text{m}}{(2 - 0)\text{s}} = 9.8 \text{ m/s}$$

Similarly, the average velocity between $t = 1$ and $t = 2$ is

$$\frac{(19.6 - 4.9)\text{m}}{(2 - 1)\text{ s}} = 14.7 \text{ m/s}$$

Table I

t	s
0	0
1	4.9
1.5	11.025
1.8	15.876
1.9	17.689
1.95	18.63225
2	19.6
2.05	20.59225
2.1	21.609
2.2	23.716
2.5	30.625
3	44.1
4	78.4

Likewise, we compute the average velocity between $t = t_1$ and $t = 2$ for various t_1. The following table II gives the average velocity (v), $t = t_1$ s and $t = 2$ s

Table II

t_1	0	1	1.5	1.8	1.9	1.95	1.99
v	9.8	14.7	17.15	18.62	19.11	19.355	19.551

From table II, we observe that the average velocity is gradually increasing. As we make the time intervals ending at $t = 2$ smaller, we see that we get a better idea of the velocity at $t = 2$. Hoping that nothing really dramatic happens between 1.99 s and 2 s, we conclude that the average velocity at $t = 2$ s is just above 19.551 m/s.

This conclusion is somewhat strengthened by the following set of computation. Compute the average velocities for various time intervals starting at $t = 2$ s. As before the average velocity v between $t = 2$ s and $t = t_2$ s is

$$= \frac{\text{Distance travelled between 2 s and } t_2 \text{ s}}{t_2 - 2}$$

$$= \frac{\left[\begin{array}{l}\text{Distance travelled in } t_2 \text{ s} \\ \quad - \text{Distance travelled in 2 s}\end{array}\right]}{t_2 - 2}$$

$$= \frac{\text{Distance travelled in } t_2 \text{ s} - 19.6}{t_2 - 2}$$

The following table III gives the average velocity v in metres per second between $t = 2$ s and t_2 s.

Table III

t_2	4	3	2.5	2.2	2.1	2.05	2.01
v	29.4	24.5	22.05	20.58	20.09	19.845	19.649

Here, again we note that if we take smaller time intervals starting at $t = 2$, we get better idea of the velocity at $t = 2$.

In the first set of computations, what we have done is to find average velocities in increasing time intervals ending at $t = 2$ and then hope that nothing dramatic happens just before $t = 2$. In the second set of computations, we have found the average velocities decreasing in time intervals ending at $t = 2$ and then hope that nothing dramatic happens just after $t = 2$. Purely on the physical grounds, both these sequences of average velocities must approach a common limit.

We can safely conclude that the velocity of the body at $t = 2$ is between 19.551 m / s and 19.649 m / s. Technically, we say that the instantaneous velocity at $t = 2$ is between 19.551 m/s and 19.649 m / s. As is well-known, velocity is the rate of change of displacement. Hence, what we have accomplished is the following:

From the given data of distance covered at various time instants we have estimated the rate of change of the distance at a given instant of time. We say that the derivative of the distance function $s = 4.9t^2$ at $t = 2$ is between 19.551 and 19.649.

4. *(b)* The figure given below is a plot of distance s of the body from the top of the cliff versus the time t elapsed. In the limit as the sequence of time intervals $h_1, h_2, \ldots,$ approaches zero, the sequence of average velocities approaches the same limit as does the sequence of ratios

$$\frac{C_1B_1}{AC_1}, \frac{C_2B_2}{AC_2}, \frac{C_3B_3}{AC_3}, \ldots$$

where, $C_1B_1 = s_1 - s_0$ is the distance travelled by the body in the time interval $h_1 = AC_1$, etc. From the figure it is safe to conclude that this latter sequence approaches the slope of the tangent to the curve at point A. In other words, The instantaneous velocity $v(t)$ of a body at time $t = 2$ (say) is equal to the slope of the tangent of the curve $s = 4.9t^2$ at $t = 2$ (say).

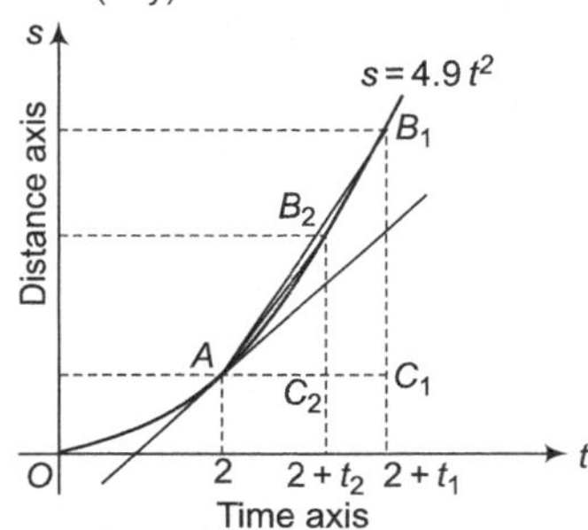

5. *(a)* Given function $f(x) = x^2$. Observe that as x takes values very close to 0, the value of $f(x)$ also approaches towards 0.

We say $\lim_{x \to 0} f(x) = 0$

(*i.e.*, the limit of $f(x)$ as x tends to zero equals zero). The limit of $f(x)$ as x tends to zero is to be thought of as the value $f(x)$ should assume at $x = 0$.

6. *(c)* In general as $x \to a$, $f(x) \to l$, then l is called limit of the function $f(x)$ which is symbolically written as $\lim_{x \to a} f(x) = l$.

7. *(b)* Given function $g(x) = |x|$, $x \neq 0$. Observe that $g(0)$ is not defined.

Now, on computing the value of $g(x)$ for values of x very near to 0, we see that the value of $g(x)$ moves towards 0. So, $\lim_{x \to 0} g(x) = 0$. This is intuitively clear from the graph of $y = |x|$ for $x \neq 0$.

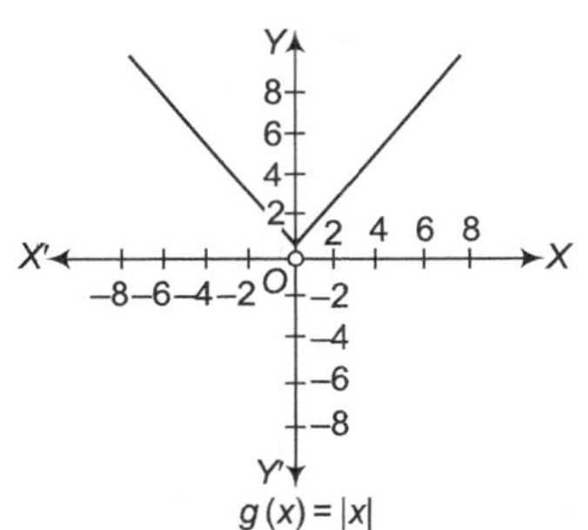

$g(x) = |x|$

8. *(a)* Given, the following function.

$$h(x) = \frac{x^2 - 4}{x - 2}, \; x \neq 2$$

Now, on computing the value of $h(x)$ for values of x very near to 2 (but not at $x = 2$), we get all these values are near to 4. This is somewhat strengthened by considering the graph of the function $y = h(x)$,

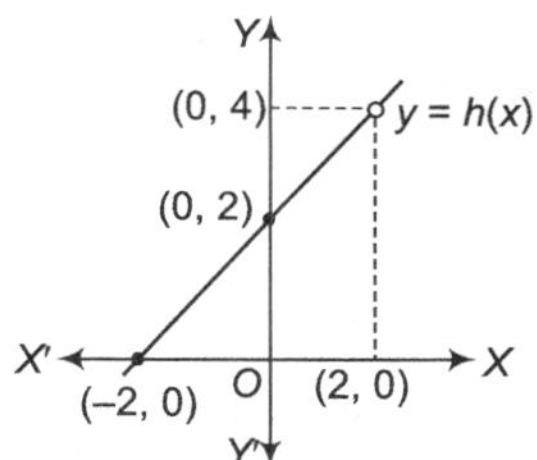

9. *(c)* There are essentially two ways x could approach a number a either from left or from right, *i.e.*, all the values of x near a could be less than a or could be greater than a. This naturally leads to two limits – the right hand limit and the left hand limit. Right hand limit and left hand limit of a function $f(x)$ is that value of $f(x)$ which is dictated by the values of $f(x)$ when x tends to a from the right and left.

10. *(a)* Given function $f(x) = \begin{cases} 1, & x \leq 0 \\ 2, & x > 0 \end{cases}$

Graph of this function is shown below. It is clear that the value of f at 0 dictated by values of $f(x)$ with $x \leq 0$ equals 1, *i.e.*, the left hand limit of $f(x)$ at $x = 0$ is

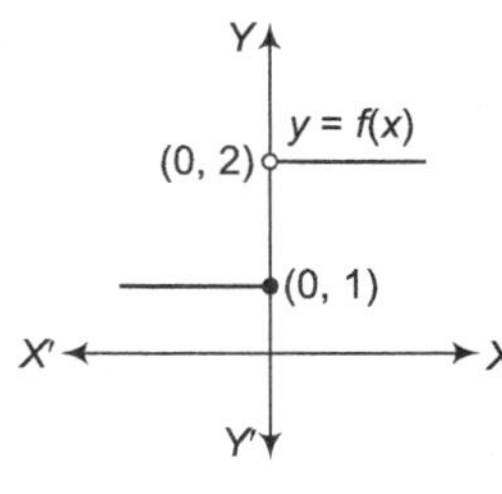

$$\lim_{x \to 0^-} f(x) = 1$$

Similarly, the value of f at $x = 0$ dictated by values of $f(x)$ with $x > 0$ equals 2, *i.e.*, the right hand limit of $f(x)$ at $x = 0$ is

$$\lim_{x \to 0^+} f(x) = 2$$

In this case the right and left hand limits are different, and hence we say that the limit of $f(x)$ as x tends to zero does not exist (even though the function is defined at 0).

11. *(b)* We say $\lim_{x \to a^-} f(x)$ is the expected value of f at $x = a$ given the values of f near x to the left of a. This value is called the left hand limit of f at a.

Now, $\lim_{x \to a^+} f(x)$ is the expected value of f at $x = a$ given the values of f near x to the right of a. This value is called the right hand limit of $f(x)$ at a and

If the right and left hand limits coincide, we call that common value as the limit of $f(x)$ at $x = a$ and denote it by $\lim_{x \to a} f(x)$.

12. *(b)* Given, function $f(x) = x + 10$. The limit of this function at $x = 5$ is given by the value of the function $f(x)$ for x very near to 5. Some of the points near and to the left of 5 are 4.9, 4.95, 4.99, 4.995,...etc, values of the function at these points are tabulated below. Similarly, the real number 5.001, 5.01, 5.1 are also points near and to the right of 5. Values of the function at these points are also given in the table.

x	4.9	4.95	4.99	4.995	5.001.	5.01	5.1
$f(x)$	14.9	14.95	14.99	14.995	15.001	15.01	15.1

From the table, we deduce that the value of $f(x)$ at $x = 5$ should be greater than 14.995 and less than 15.001 assuming nothing dramatic happens between $x = 4.995$ and 5.001. Now, assume that the value of the $f(x)$ at $x = 5$ as dictated by the numbers to the left of 5 is 15, *i.e.,*

$$\lim_{x \to 5^-} f(x) = 15$$

Similarly, when x approaches 5 from the right, $f(x)$ should be taking value 15, *i.e.,*

$$\lim_{x \to 5^+} f(x) = 15$$

Hence, the left hand limit of $f(x)$ and the right hand limit of $f(x)$ are both equal to 15. Thus,

$$\lim_{x \to 5^-} f(x) = \lim_{x \to 5^+} f(x) = \lim_{x \to 5} f(x) = 15$$

As x approaches 5 from either right or left, the graph of the function $f(x) = x + 10$ approaches the point (5, 15).

Then the value of the function at $x = 5$ also happens to be equal to 15.

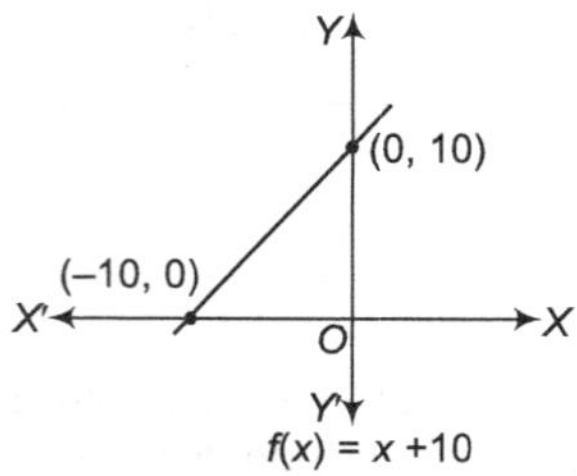

$f(x) = x + 10$

13. *(a)* Given, function $f(x) = x^3$. Let us find the limit of this function at $x = 1$. Now, tabulate the value of $f(x)$ at x near 1. This is given in the table

x	$f(x)$
0.9	0.729
0.99	0.970299
0.999	0.997002999
1.001	1.003003001
1.01	1.030301
1.1	1.331

From this table, we deduce the value of $f(x)$ at $x = 1$ should be greater than 0.997002999 and less than 1.003003001 assuming nothing dramatic happens between $x = 0.999$ and 1.001. It is reasonable to assume that the value of the $f(x)$ at $x = 1$ as dictated by the numbers to the left of 1 is 1, *i.e.,*

$$\lim_{x \to 1^-} f(x) = 1$$

Similarly, when x approaches 1 form the right, $f(x)$ should be taking value 1, *i.e.,*

$$\lim_{x \to 1^+} f(x) = 1$$

Hence, it is likely that the left hand limit of $f(x)$ and the right hand limit of $f(x)$ are both equal to 1. Thus,

$$\lim_{x \to 1^-} f(x) = \lim_{x \to 1^+} f(x) = \lim_{x \to 1} f(x) = 1$$

This conclusion about the limit being equal to 1 is some what strengthened by seeing the graph of this function. In the figure given below, we note that as x approaches 1 from either right or left, the graph of the function $f(x) = x^3$ approaches the point (1, 1).

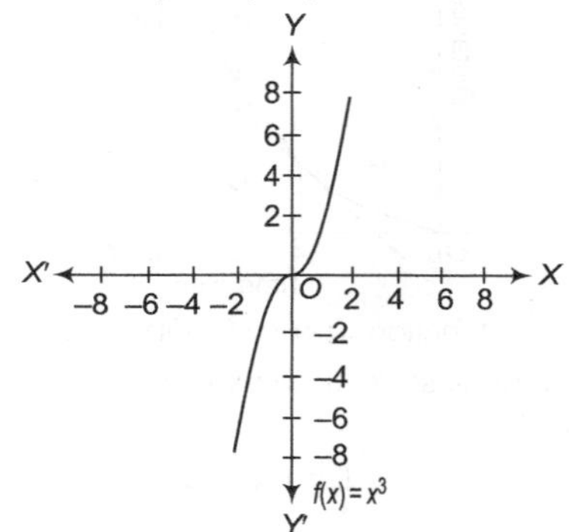

14. *(a)* Given, function $f(x) = 3x$. Let us find the limit of this function at $x = 2$. The following table is now self-explanatory.

x	1.9	1.95	1.99	1.999	2.001	2.01	2.1
$f(x)$	5.7	5.85	5.97	5.997	6.003	6.03	6.3

As x approaches 2 from either left or right, then value of $f(x)$ seems to approach 6. Therefore,

$$\lim_{x \to 2^-} f(x) = \lim_{x \to 2^+} f(x) = \lim_{x \to 2} f(x) = 6$$

and its graph strengthens this fact.

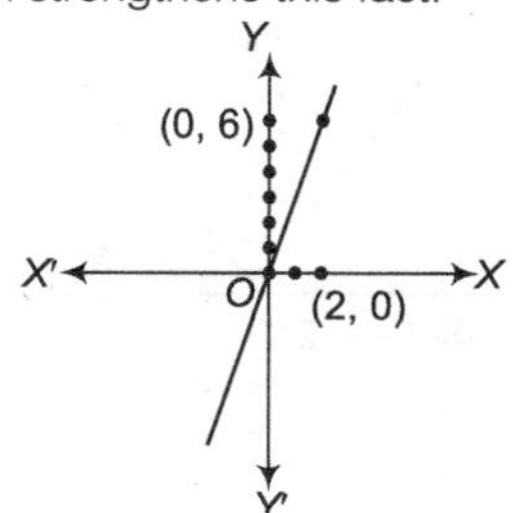

Note that the value of the function at $x = 2$ coincides with the limit at $x = 2$.

15. *(b)* Given, constant function $f(x)=3$. Let us find its limit at $x=2$. This function being the constant function takes the same value (3, in this case) everywhere, *i.e.,* its value at points close to 2 is 3. Hence,

$$\lim_{x\to 2^-} f(x)=\lim_{x\to 2^+} f(x)=\lim_{x\to 2} f(x)=3$$

Graph of $f(x)=3$ is anyway the line parallel to X-axis passing through (0, 3).

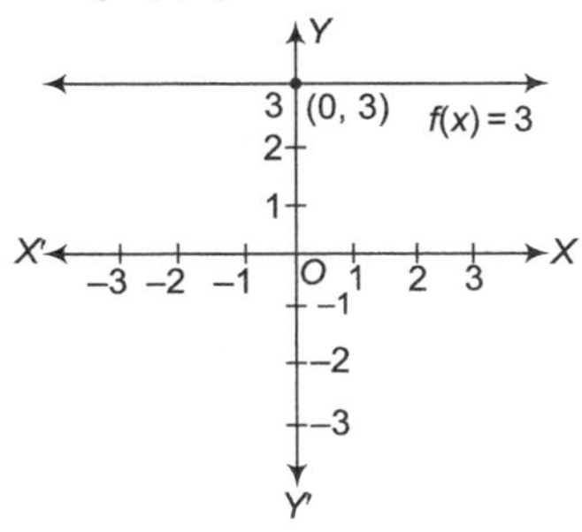

It is clear that the required limit is 3. Hence, $\lim_{x\to a} f(x)=3$ for any real number a.

16. *(a)* Given, function $f(x)=x^2+x$. Let us to find the $\lim_{x\to 1} f(x)$.

Tabulate the values of $f(x)$ near $x=1$ in the table

x	0.9	0.99	0.999	1.01	1.1	1.2
$f(x)$	1.71	1.9701	1.997001	2.0301	2.31	2.64

From the table, it is reasonable to deduce that

$$\lim_{x\to 1^-} f(x)=\lim_{x\to 1^+} f(x)=\lim_{x\to 1} f(x)=2$$

and from the graph of $f(x)=x^2+x$ shown below.

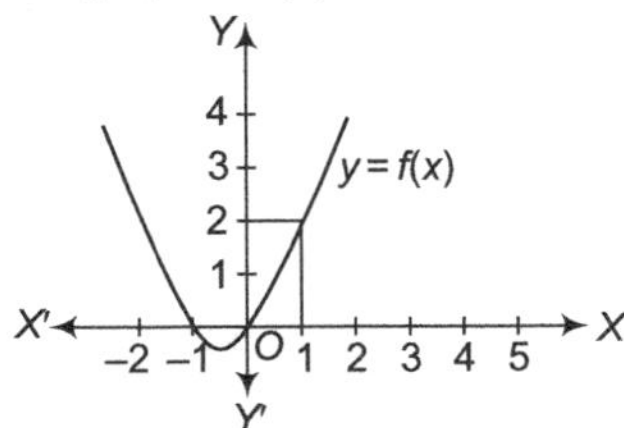

it is clear that as x approaches to 1, the graph approaches to (1, 2).

Hence, $\lim_{x\to 1} f(x)=f(1)$

Now, convince yourself of the following three facts

$$\lim_{x\to 1} x^2=1,\ \lim_{x\to 1} x=1 \text{ and } \lim_{x\to 1}(x+1)=2$$

Then, $\lim_{x\to 1} x^2+\lim_{x\to 1} x=1+1=2=\lim_{x\to 1}[x^2+x]$

Also $\lim_{x\to 1} x\cdot\lim_{x\to 1}(x+1)=1\cdot 2=2=\lim_{x\to 1}[x(x+1)]=\lim_{x\to 1}[x^2+x]$

17. *(c)* Given, function $f(x)=\sin x$. To find the value of $\lim_{x\to \frac{\pi}{2}} \sin x$, where the angle x is measured in radians, let us tabulate the value of $f(x)$ near $\frac{\pi}{2}$ (table).

x	$\frac{\pi}{2}-0.1$	$\frac{\pi}{2}-0.01$	$\frac{\pi}{2}+0.01$	$\frac{\pi}{2}+0.1$
$f(x)$	0.9950	0.9999	0.9999	0.9950

From this, we may deduce that

$$\lim_{x\to \frac{\pi}{2}^-} f(x)=\lim_{x\to \frac{\pi}{2}^+} f(x)=\lim_{x\to \frac{\pi}{2}} f(x)=1$$

In this case, the $\lim_{x\to \pi/2} \sin x=1$

18. *(d)* Given, function $f(x)=x+\cos x$. Then, to find the $\lim_{x\to 0} f(x)$ let us tabulate the value of $f(x)$ near 0 (table).

x	$f(x)$
– 0.1	0.9850
– 0.01	0.98995
– 0.001	0.9989995
0.001	1.0009995
0.01	1.00995
0.1	1.0950

From the table, it is clear that

$$\lim_{x\to 0^-} f(x)=\lim_{x\to 0^+} f(x)=\lim_{x\to 0} f(x)=1$$

In this case, we observe that $\lim_{x\to 0} f(x)=f(0)=1$.

19. *(b)* Given, function $f(x)=\frac{1}{x^2}$ for $x>0$. Let us find the $\lim_{x\to 0} f(x)$.

Here, the domain of the function is given to be all positive real numbers. Therefore, when we tabulate values of $f(x)$, it does not make sense to talk of x approaching 0 from the left. Now, let us tabulate the values of the function for positive x close to 0 (in this table n denotes any positive integer).

From the table given below, as x tends to 0, $f(x)$ becomes larger and larger. What we mean here is that the value of $f(x)$ may be made larger than any given number.

x	1	0.1	0.01	10^{-n}
$f(x)$	1	100	10000	10^{2n}

Hence,

$$\lim_{x\to 0} f(x)=+\infty$$

20. *(c)* Here, to find $\lim_{x\to 0} f(x)$, where

$$f(x)=\begin{cases} x-2, & x<0 \\ 0 & , x=0 \\ x+2, & x>0 \end{cases}$$

Let us tabulate the values of $f(x)$ near 0. Observe that for negative values of x we need to evaluate $x-2$ and for positive values, we need to evaluate $x+2$.

x	−0.1	−0.01	−0.001	0.001	0.01	0.1
$f(x)$	−2.1	−2.01	−2.001	2.001	2.01	2.1

From the first three entries of the table, we deduce that the value of the function is decreasing to −2 and hence,

$$\lim_{x \to 0^-} f(x) = -2$$

From the last three entries of the table, we deduce that the value of the function is increasing from 2 and hence,

$$\lim_{x \to 0^+} f(x) = 2$$

Since, the left and right hand limits at 0 do not coincide, therefore, the limit of the function at $x = 0$ does not exist.

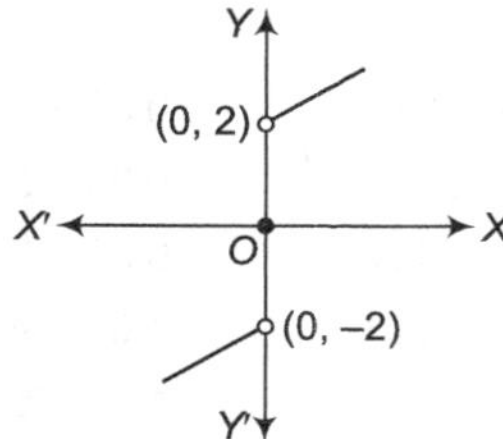

From the graph of this function as shown above, we observe that the value of the function at $x = 0$ is well defined and is, equal to 0, but the limit of the function at $x = 0$ is not even defined.

21. *(d)* We find the $\lim_{x \to 1} f(x)$,

Given, function $f(x) = \begin{cases} x + 2, & x \neq 1 \\ 0, & x = 1 \end{cases}$

x	0.9	0.99	0.999	1.001	1.01	1.1
$f(x)$	2.9	2.99	2.999	3.001	3.01	3.1

From the values of $f(x)$ for x less than 1, it seems that the function should take value 3 at x is less than 1, *i.e.,*

$$\lim_{x \to 1^-} f(x) = 3$$

Similarly, the value of $f(x)$ should be 3 as dictated by values of $f(x)$ at x greater than 1. *i.e.,* $\lim_{x \to 1^+} f(x) = 3$

But the left and right hand limits coincide and hence

$$\lim_{x \to 1^-} f(x) = \lim_{x \to 1^+} f(x) = \lim_{x \to 1} f(x) = 3$$

Graph of this function as shown below, strengthens our deduction about the limit.

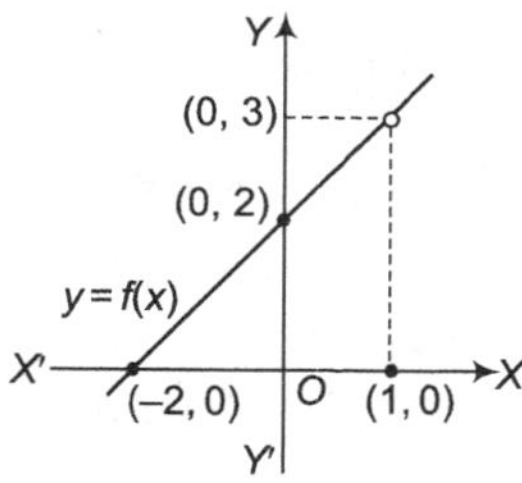

Note that, in general the value of the function at a given point and its limit may be different (even when both are defined).

22. *(b)* The limiting process respects addition, subtraction, multiplication and division as long as the limits and functions under consideration are well defined.

23. *(d)* Let f and g be two functions such that both $\lim_{x \to a} f(x)$ and $\lim_{x \to a} g(x)$ exist. Then,

(i) Limit of sum of two functions is sum of the limits of the functions, *i.e.,*

$$\lim_{x \to a} [f(x) + g(x)] = \lim_{x \to a} f(x) + \lim_{x \to a} g(x).$$

(ii) Limit of difference of two functions is difference of the limits of the functions, *i.e.,*

$$\lim_{x \to a} [f(x) - g(x)] = \lim_{x \to a} f(x) - \lim_{x \to a} g(x).$$

(iii) Limit of product of two functions is product of the limits of the functions, *i.e.,*

$$\lim_{x \to a} [f(x) \cdot g(x)] = \lim_{x \to a} f(x) \cdot \lim_{x \to a} g(x).$$

(iv) Limit of quotient of two functions is quotient of the limits of the functions (whenever the denominator is non-zero), *i.e.,*

$$\lim_{x \to a} \frac{f(x)}{g(x)} = \frac{\lim_{x \to a} f(x)}{\lim_{x \to a} g(x)}, \text{if } \lim_{x \to a} g(x) \neq 0$$

24. *(a)* Given, g is the constant function such that $g(x) = \lambda$, for some real number λ. Then,

$$\lim_{x \to a} [(\lambda \cdot f)(x)] = \lambda \cdot \lim_{x \to a} f(x)$$

(The limit of product of functions is product of the limits of the function.)

25. *(c)* A function f is said to be a polynomial function, if $f(x)$ is zero function or if $f(x) = a_0 + a_1x + a_2x^2 + \ldots + a_nx^n$, where a_i's are real numbers such that $a_n \neq 0$ for some natural number n.

As $\lim_{x \to a} x = a$

$$\therefore \quad \lim_{x \to a} x^2 = \lim_{x \to a} (x \cdot x)$$

$$= \lim_{x \to a} x \cdot \lim_{x \to a} x = a \cdot a = a^2$$

Now, induction on n tells us that

$$\lim_{x \to a} x^n = a^n$$

Since, $f(x) = a_0 + a_1x + a_2x^2 + \ldots + a_nx^n$ be a polynomial function, therefore thinking of each of $a_0, a_1x, a_2x^2, \ldots, a_nx^n$ as a function of x, we have

$$\lim_{x \to a} f(x) = \lim_{x \to a} [a_0 + a_1x + a_2x^2 + \ldots + a_nx^n]$$

$$= \lim_{x \to a} a_0 + \lim_{x \to a} a_1x + \lim_{x \to a} a_2x^2 + \ldots + \lim_{x \to a} a_nx^n$$

$$= a_0 + a_1 \lim_{x \to a} x + a_2 \lim_{x \to a} x^2 + \ldots + a_n \lim_{x \to a} x^n$$

$$= a_0 + a_1a + a_2a^2 + \ldots + a_na^n$$

$$= f(a)$$

Hence, $\lim_{x \to a} f(x) = f(a)$.

26. *(c)* A function f is said to be a rational function, if $f(x) = \frac{g(x)}{h(x)}$, where $g(x)$ and $h(x)$ are polynomials such that $h(x) \neq 0$. Then,

$$\lim_{x\to a} f(x) = \lim_{x\to a} \frac{g(x)}{h(x)} = \frac{\lim_{x\to a} g(x)}{\lim_{x\to a} h(x)} = \frac{g(a)}{h(a)}$$

If $h(a) = 0$, there are two scenarios – (i) when $g(a) \neq 0$ and (ii) when $g(a) = 0$. In case I, the limit does not exist. In case II, we can write $g(x) = (x - a)^k g_1(x)$, where k is the maximum of powers of $(x - a)$ in $g(x)$. Similarly, $h(x) = (x - a)^l h_1(x)$ as $h(a) = 0$. Now, if $k > l$, then

$$\lim_{x\to} f(x) = \frac{\lim_{x\to a} g(x)}{\lim_{x\to} h(x)} = \frac{\lim_{x\to a} (x - a)^k g_1(x)}{\lim_{x\to a} (x - a)^l h_1(x)}$$

$$= \frac{\lim_{x\to a} (x - a)^{k-l} g_1(x)}{\lim_{x\to a} h_1(x)} = \frac{0 \cdot g_1(a)}{h_1(a)} = 0$$

If $k < l$, the limit is not defined.

27. *(c)* For any positive integer n,

$$\lim_{x\to a} \frac{x^n - a^n}{x - a} = na^{n-1}$$

Dividing $(x^n - a^n)$ by $(x - a)$, we get

$$x^n - a^n = (x - a)(x^{n-1} + x^{n-2}a + x^{n-3}a^2 + \dots + xa^{n-2} + a^{n-1})$$

Now, $\lim_{x\to a} \frac{x^n - a^n}{x - a} = \lim_{x\to a} (x^{n-1} + x^{n-2}a + x^{n-3}a^2 + \dots + xa^{n-2} + a^{n-1})$

$$= a^{n-1} + aa^{n-2} + \dots + a^{n-2}(a) + a^{n-1}$$

$$= a^{n-1} + a^{n-1} + \dots + a^{n-1} + a^{n-1} \qquad (n \text{ terms})$$

$$= na^{n-1}$$

Note *The above expression is true even if n is any rational number and a is positive.*

28. *(a)* I. Given,

$$\lim_{x\to 1} \frac{x^{15} - 1}{x^{10} - 1} = \lim_{x\to 1} \left[\frac{x^{15} - 1}{x - 1} \div \frac{x^{10} - 1}{x - 1}\right]$$

$$= \lim_{x\to 1} \left[\frac{x^{15} - 1}{x - 1}\right] \div \lim_{x\to 1} \left[\frac{x^{10} - 1}{x - 1}\right]$$

$$= 15(1)^{14} \div 10(1)^9 = 15 \div 10 = \frac{3}{2}$$

II. Put $y = 1 + x$, so that $y \to 1$ as $x \to 0$.

Then, $\lim_{x\to 0} \frac{\sqrt{1 + x} - 1}{x} = \lim_{y\to 1} \frac{\sqrt{y} - 1}{y - 1}$

$$= \lim_{y\to 1} \frac{y^{\frac{1}{2}} - 1^{\frac{1}{2}}}{y - 1}$$

$$= \frac{1}{2}(1)^{\frac{1}{2} - 1} = \frac{1}{2}$$

29. *(d)* We have f and g be two real valued functions with the same domain such that $f(x) \le g(x)$ for all x in the domain. Now, if for some a, both $\lim_{x\to a} f(x)$ and $\lim_{x\to a} g(x)$ exist, then $\lim_{x\to a} f(x) \le \lim_{x\to a} g(x)$. This is illustrated in figure.

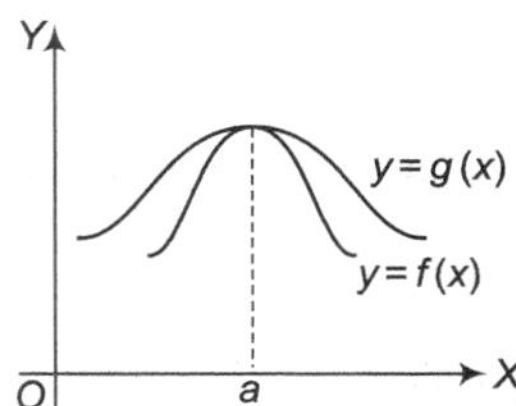

30. *(a)* **Sandwich theorem** Let f, g and h be real functions such that $f(x) \le g(x) \le h(x)$ for all x in the common domain. For some real number a, if $\lim_{x\to a} f(x) = l = \lim_{x\to a} h(x)$, then $\lim_{x\to a} g(x) = l$. This is illustrated in the following figure.

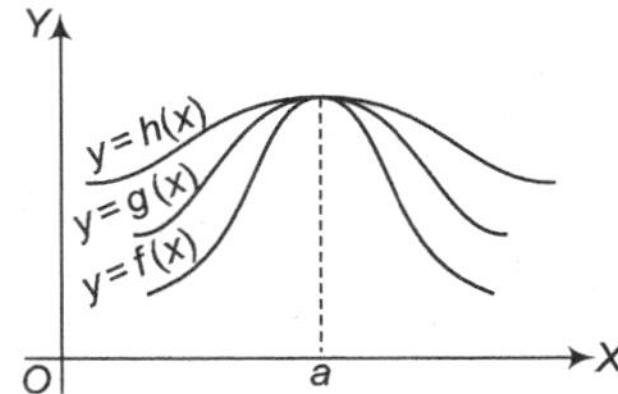

31. *(a)* Let us first consider the geometric proof of the following important inequality relating trigonometric functions.

$\cos x < \frac{\sin x}{x} < 1$ for $0 < |x| < \frac{\pi}{2}$

We know that, $\sin(-x) = -\sin x$ and $\cos(-x) = \cos x$. Hence, it is sufficient to prove the inequality for $0 < x < \frac{\pi}{2}$.

In the figure, O is the centre of the unit circle such that the $\angle AOC$ is x radians and $0 < x < \frac{\pi}{2}$. Line segment BA and

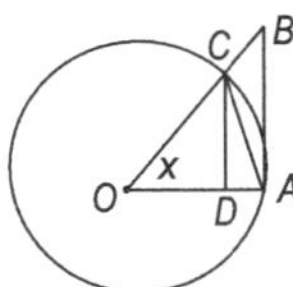

CD are perpendicular to OA. Further, join AC, Then,

Area of ΔOAC < Area of sector OAC < Area of ΔOAB.

i.e., $\frac{1}{2} OA \cdot CD < \frac{x}{2\pi} \cdot \pi \cdot (OA)^2 < \frac{1}{2} OA \cdot AB$.

i.e., $CD < x \cdot OA < AB$.

From ΔOCD,

$\sin x = \frac{CD}{OA}$ ($\because OC = OA$) and hence

$CD = OA \sin x$. Also, $\tan x = \frac{AB}{OA}$ and

hence $AB = OA \cdot \tan x$. Thus

$$OA \sin x < OA \cdot x < OA \cdot \tan x.$$

Since, length OA is positive, we have

$$\sin x < x < \tan x.$$

Since $0 < x < \frac{\pi}{2}$, $\sin x$ is positive and thus by dividing throughout by $\sin x$, we have

$1 < \frac{x}{\sin x} < \frac{1}{\cos x}$.

Taking reciprocals throughout, we have

$$\cos x < \frac{\sin x}{x} < 1$$

I. The inequality $\cos x < \frac{\sin x}{x} < 1$ for $0 < |x| < \frac{\pi}{2}$ says that the function $\frac{\sin x}{x}$ is sandwiched between the function $\cos x$ and the constant function which takes value 1

Since, $\lim_{x\to 0} \cos x = 1$, therefore the proof of I is completed by sandwich theorem.

II. Let us recall the trigonometric identity $1-\cos x = 2\sin^2\left(\frac{x}{2}\right)$.

Then, $\lim_{x\to 0}\frac{1-\cos x}{x} = \lim_{x\to 0}\frac{2\sin^2\left(\frac{x}{2}\right)}{x}$

$$= \lim_{x\to 0}\frac{\sin\left(\frac{x}{2}\right)}{\frac{x}{2}}\cdot\sin\left(\frac{x}{2}\right)$$

$$= \lim_{x\to 0}\frac{\sin\left(\frac{x}{2}\right)}{\frac{x}{2}}\cdot\lim_{x\to 0}\sin\left(\frac{x}{2}\right)$$

$$= 1\cdot 0 = 0$$

Observe that, we have implicitly used the fact that $x \to 0$ is equivalent to $\frac{x}{2} \to 0$. This may be justified by putting $y = \frac{x}{2}$.

32. *(b)* I. $\lim_{x\to 0}\frac{\sin 4x}{\sin 2x}$

$$= \lim_{x\to 0}\left[\frac{\sin 4x}{4x}\cdot\frac{2x}{\sin 2x}\cdot 2\right]$$

$$= 2\cdot\left(\lim_{x\to 0}\left[\frac{\sin 4x}{4x}\right]\div\left[\frac{\sin 2x}{2x}\right]\right)$$

$$= 2\cdot\left(\lim_{4x\to 0}\left[\frac{\sin 4x}{4x}\right]\div\lim_{2x\to 0}\left[\frac{\sin 2x}{2x}\right]\right)$$

(as $x \to 0$, $4x \to 0$ and $2x \to 0$)

$$= 2\,(1 \div 1) = 2$$

II. We have, $\lim_{x\to 0}\frac{\tan x}{x} = \lim_{x\to 0}\frac{\sin x}{x\cos x}$

$$= \lim_{x\to 0}\frac{\sin x}{x}\cdot\lim_{x\to 0}\frac{1}{\cos x} = 1\cdot 1 = 1$$

33. *(b)* Given that, $f(x) = f_1(x)f_2(x)$ so that $f_1(a) = 0$, and $f_2(a) \neq 0$ and $g(x) = g_1(x)g_2(x)$, where $g_1(a) = 0$ and $g_2(a) \neq 0$. Cancel out the common factors from $f(x)$ and $g(x)$, we have

$$\frac{f(x)}{g(x)} = \frac{p(x)}{q(x)}, \text{ where } q(x) \neq 0.$$

$$\therefore \quad \lim_{x\to a}\frac{f(x)}{g(x)} = \frac{p(a)}{q(a)}$$

34. *(b)* A. $\lim_{x\to 3} x + 3 = 3 + 3 = 6$

B. $\lim_{x\to\pi}\left(x - \frac{22}{7}\right) = \pi - \frac{22}{7}$

C. $\lim_{r\to 1}\pi r^2 = \pi\times(1)^2 = \pi$

D. $\lim_{x\to 4}\frac{4x+3}{x-2} = \frac{4\times 4+3}{4-2} = \frac{19}{2}$

E. $\lim_{x\to -1}\frac{x^{10}+x^5+1}{x-1} = \frac{(-1)^{10}+(-1)^5+1}{-1-1} = \frac{1-1+1}{-2} = \frac{-1}{2}$

35. *(a)* A. $\lim_{x\to 0}\frac{(x+1)^5-1}{x}$ $\left(\frac{0}{0}\text{ form}\right)$

Let $x + 1 = h$

when $x \to 0$, then $h \to 1$

$\therefore$ Given limit $= \lim_{h\to 1}\frac{h^5-1}{h-1} = \lim_{h\to 1}\frac{h^5-1^5}{h-1}$

$$= 5(1)^{5-1} = 5\times 1^4 = 5$$

Note *Students should not forget to change the limit.*

B. **Hint :** It is not directly solvable by putting the value of $x = 2$. Firstly factorise the numerator and denominator by splitting the middle term and cancel the common factor and put the value of x.

$\lim_{x\to 2}\frac{3x^2-x-10}{x^2-4}$ $\left(\frac{0}{0}\text{ form}\right)$

$$= \lim_{x\to 2}\frac{3x^2-(6-5)x-10}{(x-2)(x+2)} = \lim_{x\to 2}\frac{3x^2-6x+5x-10}{(x-2)(x+2)}$$

$$= \lim_{x\to 2}\frac{3x(x-2)+5(x-2)}{(x-2)(x+2)}$$

$$= \lim_{x\to 2}\frac{(x-2)(3x+5)}{(x-2)(x+2)} = \lim_{x\to 2}\frac{3x+5}{x+2}$$

$$= \frac{3\times 2+5}{2+2} = \frac{6+5}{4} = \frac{11}{4}$$

C. Given, $\lim_{x\to 3}\frac{x^4-81}{2x^2-5x-3}$ $\left(\frac{0}{0}\text{ form}\right)$

$$= \lim_{x\to 3}\frac{(x^2-9)(x^2+9)}{2x^2-6x+x-3}$$

$$= \lim_{x\to 3}\frac{(x-3)(x+3)(x^2+9)}{2x(x-3)+1(x-3)}$$

$$= \lim_{x\to 3}\frac{(x-3)(x+3)(x^2+9)}{(x-3)(2x+1)}$$

$$= \lim_{x\to 3}\frac{(x+3)(x^2+9)}{2x+1}$$

$$= \frac{(3+3)(3^2+9)}{2\times 3+1}$$

$$= \frac{6\times 18}{7} = \frac{108}{7}$$

D. $\lim_{x\to 0} \frac{ax+b}{cx+1} = \frac{a\times 0+b}{c\times 0+1} = \frac{b}{1} = b$

E. Given, $\lim_{z\to 1} \frac{z^{1/3}-1}{z^{1/6}-1} = \lim_{z\to 1} \frac{z^{1/3}-1}{z-1} \times \frac{z-1}{z^{1/6}-1}$

$$= \frac{\left(\frac{1}{3}\right)(1)^{\frac{1}{3}-1}}{\left(\frac{1}{6}\right)(1)^{\frac{1}{6}-1}}$$

$$= \frac{6}{3}\times 1 = 2$$

36. *(b)* I. Given, $\lim_{x\to 1} \frac{ax^2+bx+c}{cx^2+bx+a}$

$$= \frac{a\times(1)^2+b\times 1+c}{c\times(1)^2+b\times 1+a} = \frac{a+b+c}{c+b+a} = 1$$

Note *If a + b + c = 0, then given limit is of the form 0/0, which is indeterminate.*

II. $\lim_{x\to -2} \frac{\frac{1}{x}+\frac{1}{2}}{x+2}$

$$= \lim_{x\to -2} \frac{(2+x)}{2x(x+2)} = \lim_{x\to -2} \frac{1}{2x} = \frac{1}{2(-2)} = -\frac{1}{4}$$

37. *(a)* I. Given $\lim_{x\to 0} \frac{\sin ax}{bx} = \lim_{x\to 0} \frac{(a)\sin ax}{b(ax)}$

(dividing and multiplying by a)

$$= \frac{a}{b}\times 1 = \frac{a}{b} \qquad \left(\because \lim_{x\to 0} \frac{\sin ax}{ax} = 1\right)$$

II. Given, $\lim_{x\to 0} \frac{\sin ax}{\sin bx}$

Multiplying and dividing by (ax) and (bx), we get

$$= \lim_{x\to 0} \frac{\sin ax}{ax} \times \frac{bx}{\sin bx} \times \frac{ax}{bx}$$

$$= 1\times 1\times \frac{a}{b} = \frac{a}{b} \qquad \left(\because \lim_{x\to 0} \frac{\sin ax}{ax} = \lim_{x\to 0} \frac{bx}{\sin bx} = 1\right)$$

38. *(d)* A. Given, $\lim_{x\to \pi} \frac{\sin(\pi-x)}{\pi(\pi-x)}$

Let $\pi - x = h$, As $x\to\pi$, then $h\to 0$

$$\therefore \lim_{x\to\pi} \frac{\sin(\pi-x)}{\pi(\pi-x)} = \lim_{h\to 0} \frac{\sin h}{\pi h} = \lim_{h\to 0} \frac{1}{\pi}\times\frac{\sin h}{h}$$

$$= \frac{1}{\pi}\times 1 = \frac{1}{\pi} \qquad \left(\because \lim_{h\to 0} \frac{\sin h}{h} = 1\right)$$

B. Given $\lim_{x\to 0} \frac{\cos x}{\pi - x}$

Put the limit directly, we get $\frac{\cos 0}{\pi - 0} = \frac{1}{\pi}$

C. Given, $\lim_{x\to 0} \frac{\cos 2x-1}{\cos x-1} = \lim_{x\to 0} \frac{1-\cos 2x}{1-\cos x}$

$$= \lim_{x\to 0} \frac{2\sin^2 x}{2\sin^2 \frac{x}{2}}$$

$$\left(\because 1-\cos 2x = 2\sin^2 x \text{ and } 1-\cos x = 2\sin^2 \frac{x}{2}\right)$$

Multiplying and dividing by x^2 and then multiplying by $\frac{4}{4}$ in the numerator. we get

$$= \lim_{x\to 0} \frac{\sin^2 x}{x^2} \times \frac{4\times\frac{x^2}{4}}{\sin^2\frac{x}{2}}$$

$$= \lim_{x\to 0} \left(\frac{\sin x}{x}\right)^2 \times \left(\frac{\frac{x}{2}}{\sin\frac{x}{2}}\right)^2 \times 4$$

$$= 1\times 1\times 4 = 4$$

D. Given, $\lim_{x\to 0} \frac{ax + x\cos x}{b\sin x}$

Dividing each term by x, we get

$$= \lim_{x\to 0} \frac{\frac{ax}{x}+\frac{x\cos x}{x}}{\frac{b\sin x}{x}}$$

$$= \lim_{x\to 0} \frac{a+\cos x}{b\left(\frac{\sin x}{x}\right)}$$

$$= \frac{a+\cos 0}{b\times 1} = \frac{a+1}{b} \qquad \left(\because \lim_{x\to 0} \frac{\sin x}{x} = 1\right)$$

Note *In most of the times, when we put the limit, it becomes an indeterminate form. Firstly, simplify the function by expanding or by using the formula such that we get the factor which is causing the function to indeterminate form and then remove the cause.*

E. $\lim_{x\to 0} x\sec x = 0\times\sec 0 = 0\times 1 = 0$

F. $\lim_{x\to 0} \frac{\sin ax + bx}{ax + \sin bx}$

Dividing each term by x, we get

$$= \lim_{x\to 0} \frac{\frac{\sin ax}{x}+\frac{bx}{x}}{\frac{ax}{x}+\frac{\sin bx}{x}} = \lim_{x\to 0} \frac{\frac{a\sin ax}{ax}+b}{a+\frac{b\sin bx}{bx}}$$

$$= \frac{a\times 1+b}{a+b\times 1} = \frac{a+b}{a+b} = 1 \qquad \left(\because \lim_{x\to 0} \frac{\sin x}{x} = 1\right)$$

39. *(b)* I. Given, $\lim_{x\to 0} (\operatorname{cosec} x - \cot x) = \lim_{x\to 0} \left(\frac{1}{\sin x} - \frac{\cos x}{\sin x}\right)$

$$= \lim_{x\to 0} \frac{1-\cos x}{\sin x}$$

$$= \lim_{x \to 0} \frac{2\sin^2 \frac{x}{2}}{2\sin\frac{x}{2}\cos\frac{x}{2}} = \lim_{x \to 0} \frac{\sin\frac{x}{2}}{\cos\frac{x}{2}}$$

$$= \lim_{x \to 0} \tan\frac{x}{2} = \tan 0 = 0$$

II. Given, $\lim_{x \to \frac{\pi}{2}} \frac{\tan 2x}{x - \frac{\pi}{2}}$

Let $x - \frac{\pi}{2} = h$, when $x \to \frac{\pi}{2}$, then $h \to 0$

Therefore, given limit $= \lim_{h \to 0} \frac{\tan 2\left(\frac{\pi}{2} + h\right)}{h}$

$$= \lim_{h \to 0} \frac{\tan(\pi + 2h)}{h}$$

$$= \lim_{h \to 0} \frac{\tan 2h}{h} \qquad [\because \tan(\pi + \theta) = \tan\theta]$$

$$= \lim_{h \to 0} \frac{2\tan 2h}{2h}$$

$$= 2 \times 1 = 2 \qquad \left(\because \lim_{x \to 0} \frac{\tan x}{x} = 1\right)$$

Note *Students should not forget to change the limit, while changing the variable.*

40. *(c)* We have, $f(x) = \begin{cases} 2x + 3, & x \le 0 \\ 3(x + 1), & x > 0 \end{cases}$

At $x = 0$,

RHL $= \lim_{x \to 0^+} f(x) = \lim_{h \to 0} f(0 + h)$

$$= \lim_{h \to 0} 3(0 + h + 1)$$

$$= 3(0 + 0 + 1) = 3 \times 1 = 3$$

LHL $= \lim_{x \to 0^-} f(x) = \lim_{h \to 0} f(0 - h)$

$$= \lim_{h \to 0} 2(0 - h) + 3$$

$$= 2(0 - 0) + 3 = 3$$

$\Rightarrow$ RHL = LHL $= \lim_{x \to 0} f(x)$

Hence, at $x = 0$ limit exists.

At $x = 1$, RHL $= \lim_{x \to 1^+} f(x) = \lim_{h \to 0} f(1 + h)$

$$= \lim_{h \to 0} 3(1 + h + 1) \qquad \text{(putting } x = 1 + h\text{)}$$

$$= 3(1 + 0 + 1) = 6$$

LHL $= \lim_{x \to 1^-} f(x) = \lim_{h \to 0} f(1 - h)$ (putting $x = 1 - h$)

$$= \lim_{h \to 0} 3(1 - h + 1)$$

$$= 3(1 + 1) = 6$$

Therefore, RHL = LHL and hence $\lim_{x \to 1} f(x)$ exists and equal to 6.

41. *(b)* Given, $f(x) = \begin{cases} \frac{|x|}{x}, & x \ne 0 \\ 0, & x = 0 \end{cases}$

At $x = 0$,

RHL $= \lim_{x \to 0^+} f(x) = \lim_{h \to 0} f(0 + h)$

$$= \lim_{h \to 0} \frac{|0 + h|}{(0 + h)} = \lim_{h \to 0} \frac{h}{h} = 1$$

LHL $= \lim_{x \to 0^-} f(x) = \lim_{h \to 0} f(0 - h)$

$$= \lim_{h \to 0} \frac{|0 - h|}{(0 - h)} = \lim_{h \to 0} \frac{h}{-h} = \lim_{h \to 0} (-1) = -1$$

$\Rightarrow$ RHL $\ne$ LHL

Hence, at $x = 0$, limit does not exist.

Note *Students should be careful while putting the limits, sometimes LHL ≠ RHL. So, check both LHL and RHL.*

42. *(a)* Given, $f(x) = |x| - 5$

We know that, $|x| = \begin{cases} x, & x \ge 0 \\ -x, & x < 0 \end{cases}$

At $x = 5$, RHL $= \lim_{x \to 5^+} f(x) = \lim_{h \to 0} f(5 + h)$

$$= \lim_{h \to 0} |5 + h| - 5$$

$$= \lim_{h \to 0} (5 + h) - 5$$

$$= \lim_{h \to 0} h = 0$$

and LHL $= \lim_{x \to 5^-} f(x) = \lim_{h \to 0} f(5 - h)$

$$= \lim_{h \to 0} |5 - h| - 5$$

$$= \lim_{h \to 0} (5 - h) - 5$$

$$= \lim_{h \to 0} (-h) = 0$$

Hence, RHL = LHL $= \lim_{x \to 5} f(x) = 0$

43. *(b)* Given, $\lim_{x \to 1} \frac{f(x) - 2}{x^2 - 1} = \pi$

$$\Rightarrow \frac{\lim_{x \to 1} [f(x) - 2]}{\lim_{x \to 1} (x^2 - 1)} = \pi$$

$$\Rightarrow \lim_{x \to 1} [f(x) - 2] = \pi \lim_{x \to 1} (x^2 - 1)$$

$$\Rightarrow \lim_{x \to 1} f(x) - 2 = \pi(1^2 - 1)$$

$$\Rightarrow \lim_{x \to 1} f(x) - 2 = \pi \times 0$$

$$\Rightarrow \lim_{x \to 1} f(x) - 2 = 0$$

$$\Rightarrow \lim_{x \to 1} f(x) = 2$$

44. *(a)* Given, $f(x) = \begin{cases} x^2 - 1, & x \le 1 \\ -x^2 - 1, & x > 1 \end{cases}$

At $x = 1$,

RHL $= \lim_{x \to 1^+} f(x) = \lim_{h \to 0} f(1 + h) = \lim_{h \to 0} -(1 + h)^2 - 1$

(put $x = 1 + h$)

$$= -(1 + 0)^2 - 1 = -1 - 1 = -2$$

$$\text{LHL} = \lim_{x \to 1^-} f(x) = \lim_{h \to 0} f(1-h) \qquad \text{(put } x = 1-h)$$
$$= \lim_{h \to 0} (1-h)^2 - 1$$
$$= (1-0)^2 - 1$$
$$= 1 - 1 = 0$$
$$\Rightarrow \quad \text{RHL} \neq \text{LHL}$$

Hence, at $x = 1$, limit doesn't exist.

45. *(c)* Given, $f(x) = \begin{cases} \dfrac{x}{|x|}, & x \neq 0 \\ 0, & x = 0 \end{cases}$

At $x = 0$, $\text{RHL} = \lim_{x \to 0^+} f(x) = \lim_{h \to 0} f(0+h) = \lim_{h \to 0} \dfrac{0+h}{|0+h|}$

$$= \lim_{h \to 0} \frac{0+h}{(0+h)} = 1$$
$$\text{LHL} = \lim_{x \to 0^-} f(x) = \lim_{h \to 0} f(0-h)$$
$$= \lim_{h \to 0} \frac{(0-h)}{|0-h|} = \lim_{h \to 0} \frac{(0-h)}{-(0-h)} = -1$$
$$\Rightarrow \quad \text{RHL} \neq \text{LHL}$$

Hence, at $x = 0$ limit does not exist.

46. *(b)* We have, $f(x) = \begin{cases} a + bx, & x < 1 \\ 4, & x = 1 \\ b - ax, & x > 1 \end{cases}$

Given, $\lim_{x \to 1} f(x) = f(1)$

i.e., $\text{RHL} = \text{LHL} = f(1)$

$$\Rightarrow \quad \lim_{x \to 1^+} f(x) = \lim_{x \to 1^-} f(x) = 4$$
$$\Rightarrow \quad \lim_{h \to 0} f(1+h) = \lim_{h \to 0} f(1-h) = 4$$
$$\Rightarrow \quad \lim_{h \to 0} b - a(1+h) = \lim_{h \to 0} a + b(1-h) = 4$$
$$\Rightarrow \quad b - a(1+0) = a + b(1-0) = 4$$
$$\Rightarrow \quad b - a = a + b = 4$$
$$\Rightarrow \quad b - a = 4$$
$$\text{and} \quad b + a = 4$$

On solving, we get $b = 4$, $a = 0$.

47. *(c)* Given, $f(x) = (x - a_1)(x - a_2) \ldots\ldots (x - a_n)$

$$\therefore \quad \lim_{x \to a_1} f(x) = \lim_{x \to a_1} (x - a_1)(x - a_2) \ldots\ldots (x - a_n)$$
$$= (a_1 - a_1)(a_1 - a_2) \ldots\ldots (a_1 - a_n)$$
$$= 0 \times (a_1 - a_2) \ldots\ldots (a_1 - a_n) = 0$$

Again, $\lim_{x \to a} f(x) = \lim_{x \to a} (x - a_1)(x - a_2) \ldots\ldots (x - a_n)$

$$= (a - a_1)(a - a_2) \ldots\ldots (a - a_n)$$

48. *(b)* Given, $f(x) = \begin{cases} |x| + 1, & x < 0 \\ 0, & x = 0 \\ |x| - 1, & x > 0 \end{cases} = \begin{cases} -x + 1, & x < 0 \\ 0, & x = 0 \\ x - 1, & x > 0 \end{cases}$

Let us first check the existence of limit of $f(x)$ at $x = 0$.

At $x = 0$,

$$\text{RHL} = \lim_{x \to 0^+} f(x) = \lim_{h \to 0} f(0+h) = \lim_{h \to 0} (0+h) - 1$$
$$= \lim_{h \to 0} h - 1 = 0 - 1 = -1$$
$$\text{LHL} = \lim_{x \to 0^-} f(x) = \lim_{h \to 0} f(0-h)$$
$$= \lim_{h \to 0} -(0-h) + 1$$
$$= \lim_{h \to 0} h + 1 = 0 + 1 = 1$$
$$\Rightarrow \quad \text{RHL} \neq \text{LHL}$$
$$\Rightarrow \quad \text{At } x = 0, \text{ limit does not exist.}$$

Note that for any $a < 0$ or $a > 0$, $\lim_{x \to a} f(x)$ exists, as for $a < 0$, $\lim_{x \to a} f(x) = \lim_{x \to a} -x + 1 = -a + 1$ exists and for $a > 0$, $\lim_{x \to a} f(x) = \lim_{x \to a} x - 1 = a - 1$ exists. Hence, $\lim_{x \to a} f(x)$ exists for all $a \neq 0$.

49. *(c)* $\because$ At $x = 0$, limit exists

Therefore, RHL = LHL

$$\Rightarrow \quad \lim_{x \to 0^+} f(x) = \lim_{x \to 0^-} f(x)$$
$$\Rightarrow \quad \lim_{h \to 0} f(0+h) = \lim_{h \to 0} f(0-h)$$
$$\Rightarrow \quad \lim_{h \to 0} n(0+h) + m = \lim_{h \to 0} m(0-h)^2 + n$$
$$\Rightarrow \quad n(0+0) + m = m(0-0)^2 + n$$
$$\Rightarrow \quad m = n \qquad \ldots\text{(i)}$$

Again, at $x = 1$ limit exists

$$\Rightarrow \quad \text{RHL} = \text{LHL}$$
$$\Rightarrow \quad \lim_{x \to 1^+} f(x) = \lim_{x \to 1^-} f(x)$$
$$\Rightarrow \quad \lim_{h \to 0} f(1+h) = \lim_{h \to 0} f(1-h)$$
$$\Rightarrow \quad \lim_{h \to 0} n(1+h)^3 + m = \lim_{h \to 0} n(1-h) + m$$
$$\Rightarrow \quad n(1+0)^3 + m = n(1-0) + m$$
$$\Rightarrow \quad n + m = m + n \qquad \ldots\text{(ii)}$$

Hence, from Eqs. (i) and (ii), we get $\lim_{x \to 0} f(x)$ exist, when $m = n$ and $\lim_{x \to 1} f(x)$ exists for any integral value of m and n.

50. *(b)* We have,

$$\lim_{x \to 2} \left[\frac{1}{x-2} - \frac{2(2x-3)}{x^3 - 3x^2 + 2x} \right] = \lim_{x \to 2} \left[\frac{1}{x-2} - \frac{2(2x-3)}{x(x-1)(x-2)} \right]$$
$$= \lim_{x \to 2} \left[\frac{x(x-1) - 2(2x-3)}{x(x-1)(x-2)} \right]$$
$$= \lim_{x \to 2} \left[\frac{x^2 - 5x + 6}{x(x-1)(x-2)} \right]$$
$$= \lim_{x \to 2} \left[\frac{(x-2)(x-3)}{x(x-1)(x-2)} \right] \qquad (x - 2 \neq 0)$$
$$= \lim_{x \to 2} \left[\frac{x-3}{x(x-1)} \right] = \frac{-1}{2}$$

51. (c) Put $y = 2 + x$ so that when $x \to 0$, $y \to 2$. Then,

$$\lim_{x\to 0}\frac{\sqrt{2+x}-\sqrt{2}}{x} = \lim_{y\to 2}\frac{y^{\frac{1}{2}}-2^{\frac{1}{2}}}{y-2}$$

$$= \frac{1}{2}(2)^{\frac{1}{2}-1} = \frac{1}{2}\cdot 2^{-\frac{1}{2}} = \frac{1}{2\sqrt{2}}$$

52. (d) We have, $\lim_{x\to 3}\frac{x^n-3^n}{x-3} = n(3)^{n-1}$

Therefore, $n(3)^{n-1} = 108 = 4(27) = 4(3)^{4-1}$

On comparing, we get $n = 4$

53. (c) Put $y = \frac{\pi}{2} - x$. Then, $y \to 0$ as $x \to \frac{\pi}{2}$. Therefore,

$$\lim_{x\to\frac{\pi}{2}}(\sec x - \tan x) = \lim_{y\to 0}\left[\sec\left(\frac{\pi}{2}-y\right) - \tan\left(\frac{\pi}{2}-y\right)\right]$$

$$= \lim_{y\to 0}(\operatorname{cosec} y - \cot y)$$

$$= \lim_{y\to 0}\left(\frac{1}{\sin y} - \frac{\cos y}{\sin y}\right) = \lim_{y\to 0}\left(\frac{1-\cos y}{\sin y}\right)$$

$$= \lim_{y\to 0}\frac{2\sin^2\frac{y}{2}}{2\sin\frac{y}{2}\cos\frac{y}{2}} \quad \left(\text{form of } \frac{0}{0}\right)$$

$$= \lim_{\frac{y}{2}\to 0}\tan\frac{y}{2} = 0$$

54. (d) We have, $\lim_{x\to 0}\frac{\sin(2+x)-\sin(2-x)}{x}$

$$= \lim_{x\to 0}\frac{2\cos\frac{(2+x+2-x)}{2}\sin\frac{(2+x-2+x)}{2}}{x}$$

$$= \lim_{x\to 0}\frac{2\cos 2\sin x}{x}$$

$$\left(\because \sin C - \sin D = 2\cos\frac{C+D}{2}\sin\frac{C-D}{2}\right)$$

$$= 2\cos 2\lim_{x\to 0}\frac{\sin x}{x} = 2\cos 2 \quad \left(\because \lim_{x\to 0}\frac{\sin x}{x} = 1\right)$$

Hence, $p = 2$ and $q = 2$

55. (c) We know that,

$2\sin^2 x + \sin x - 1 = (2\sin x - 1)(\sin x + 1)$

$2\sin^2 x - 3\sin x + 1 = (2\sin x - 1)(\sin x - 1)$

Given, $\lim_{x\to\frac{\pi}{6}}\frac{2\sin^2 x + \sin x - 1}{2\sin^2 x - 3\sin x + 1} = \lim_{x\to\frac{\pi}{6}}\frac{(2\sin x - 1)(\sin x + 1)}{(2\sin x - 1)(\sin x - 1)}$

$$= \lim_{x\to\frac{\pi}{6}}\frac{\sin x + 1}{\sin x - 1} = \frac{\sin\frac{\pi}{6}+1}{\sin\frac{\pi}{6}-1} = -3$$

56. (a) We have, $\lim_{x\to 0}\frac{\tan x - \sin x}{\sin^3 x} = \lim_{x\to 0}\frac{\sin x\left(\frac{1}{\cos x}-1\right)}{\sin^3 x}$

$$= \lim_{x\to 0}\frac{1-\cos x}{\cos x\sin^2 x} = \lim_{x\to 0}\frac{2\sin^2\frac{x}{2}}{\cos x\left(4\sin^2\frac{x}{2}\cdot\cos^2\frac{x}{2}\right)} = \frac{1}{2}$$

57. (a) We have, $\lim_{x\to a}\frac{\sqrt{a+2x}-\sqrt{3x}}{\sqrt{3a+x}-2\sqrt{x}}$

On rationalising, we get

$$= \lim_{x\to a}\frac{\sqrt{a+2x}-\sqrt{3x}}{\sqrt{3a+x}-2\sqrt{x}}\times\frac{\sqrt{a+2x}+\sqrt{3x}}{\sqrt{a+2x}+\sqrt{3x}}$$

$$= \lim_{x\to a}\frac{a+2x-3x}{(\sqrt{3a+x}-2\sqrt{x})(\sqrt{a+2x}+\sqrt{3x})}$$

$$= \lim_{x\to a}\frac{(a-x)(\sqrt{3a+x}+2\sqrt{x})}{(\sqrt{a+2x}+\sqrt{3x})(\sqrt{3a+x}-2\sqrt{x})(\sqrt{3a+x}+2\sqrt{x})}$$

$$= \lim_{x\to a}\frac{(a-x)[\sqrt{3a+x}+2\sqrt{x}]}{(\sqrt{a+2x}+\sqrt{3x})(3a+x-4x)}$$

$$= \frac{4\sqrt{a}}{3\times 2\sqrt{3a}} = \frac{2}{3\sqrt{3}} = \frac{2\sqrt{3}}{9}$$

Hence, m is $\sqrt{3}$.

58. (c) We have, $\lim_{x\to 0}\frac{\cos ax - \cos bx}{\cos cx - 1}$

$$= \lim_{x\to 0}\frac{2\sin\left(\frac{(a+b)}{2}x\right)\sin\left(\frac{(a-b)x}{2}\right)}{2\sin^2\left(\frac{cx}{2}\right)}$$

$$= \lim_{x\to 0}\frac{\sin\frac{(a+b)x}{2}\cdot\sin\frac{(a-b)x}{2}}{x^2}\cdot\frac{x^2}{\sin^2\frac{cx}{2}}$$

$$= \lim_{x\to 0}\frac{\sin\frac{(a+b)x}{2}}{\frac{(a+b)x}{2}\cdot\left(\frac{2}{a+b}\right)}\cdot\frac{\sin\frac{(a-b)x}{2}}{\frac{(a-b)x}{2}\cdot\frac{2}{a-b}}\cdot\frac{\left(\frac{cx}{2}\right)^2\times\frac{4}{c^2}}{\sin^2\frac{cx}{2}}$$

$= \left(\frac{a+b}{2}\times\frac{a-b}{2}\times\frac{4}{c^2}\right) = \frac{a^2-b^2}{c^2}$. Hence, m and n are $a^2 - b^2$ and c^2 respectively.

59. (c) We have, $\lim_{h\to 0}\frac{(a+h)^2\sin(a+h) - a^2\sin a}{h}$

$$= \lim_{h\to 0}\frac{(a^2+h^2+2ah)[\sin a\cos h + \cos a\sin h] - a^2\sin a}{h}$$

$$= \lim_{h\to 0}\left[\frac{a^2\sin a(\cos h - 1)}{h} + \frac{a^2\cos a\sin h}{h} + (h+2a)(\sin a\cos h + \cos a\sin h)\right]$$

$$= \lim_{h\to 0}\left[\frac{a^2 \sin a\left(-2\sin^2 \frac{h}{2}\right)}{\frac{h^2}{2}}\cdot\frac{h}{2}\right] + \lim_{h\to 0}\frac{a^2 \cos a \sin h}{h}$$

$$+ \lim_{h\to 0}(h + 2a)\sin(a + h)$$

$$= a^2 \sin a \times 0 + a^2 \cos a(1) + 2a\sin a = a^2 \cos a + 2a\sin a.$$

60. *(b)* We have, $\lim_{x\to 0}\frac{\sin x}{x(1+\cos x)} = \lim_{x\to 0}\frac{2\sin\frac{x}{2}\cos\frac{x}{2}}{x\left(2\cos^2\frac{x}{2}\right)}$

$$= \frac{1}{2}\lim_{x\to 0}\frac{\tan\frac{x}{2}}{\frac{x}{2}} = \frac{1}{2}$$

61. *(a)* We have, $\lim_{x\to\frac{\pi}{2}}\frac{1-\sin x}{\cos x} = \lim_{y\to 0}\left[\frac{1-\sin\left(\frac{\pi}{2}-y\right)}{\cos\left(\frac{\pi}{2}-y\right)}\right]$

$\left(\text{taking } \frac{\pi}{2} - x = y\right)$

$$= \lim_{y\to 0}\frac{1-\cos y}{\sin y} = \lim_{y\to 0}\frac{2\sin^2\frac{y}{2}}{2\sin\frac{y}{2}\cos\frac{y}{2}} = \lim_{y\to 0}\tan\frac{y}{2} = 0$$

62. *(d)* Since, RHL $= \lim_{x\to 1^+}[x-1] = 0$

and LHL $= \lim_{x\to 1^-}[x-1] = -1$

Hence, at $x = 1$ limit does not exist.

63. *(a)* Since,

$\lim_{x\to 0} x = 0$ and $-1 \le \sin\frac{1}{x} \le 1$, therefore $\lim_{x\to 0} x\sin\frac{1}{x} = 0$

64. *(c)* We have, $\lim_{n\to\infty}\frac{1+2+3+\ldots+n}{n^2}$

$$= \lim_{n\to\infty}\frac{n(n+1)}{2n^2} = \lim_{n\to\infty}\frac{1}{2}\left(1+\frac{1}{n}\right) = \frac{1}{2}$$

65. *(d)* People maintaining a reservoir need to know when will a reservoir overflow knowing the depth of the water at several instances of time. Rocket Scientists need to compute the precise velocity with which the satellite needs to be shot out from the rocket knowing the height of the rocket at various times.

Financial institutions need to predict the changes in the value of a particular stock knowing its present value. In these and many such cases it is desirable to know how a particular parameter is changing with respect to some other parameter.

66. *(a)* Suppose f is a real valued function and a is a point in its domain of definition. The derivative of f at a is defined by

$\lim_{h\to 0}\frac{f(a+h)-f(a)}{h}$ provided this limit exists. Derivative of $f(x)$ at a is denoted by $f'(a)$.

67. *(c)* $f'(a)$ quantifies the change in $f(x)$ at a with respect to x.

68. *(d)* We have,

$$f'(2) = \lim_{h\to 0}\frac{f(2+h)-f(2)}{h} = \lim_{h\to 0}\frac{3(2+h)-3(2)}{h}$$

$$= \lim_{h\to 0}\frac{6+3h-6}{h} = \lim_{h\to 0}\frac{3h}{h} = \lim_{h\to 0} 3 = 3.$$

The derivative of the function $f(x) = 3x$ at $x = 2$ is 3.

69. *(c)* We first find the derivatives of $f(x)$ at $x = -1$ and at $x = 0$. We have,

$$f'(-1) = \lim_{h\to 0}\frac{f(-1+h)-f(-1)}{h}$$

$$= \lim_{h\to 0}\frac{[2(-1+h)^2 + 3(-1+h) - 5] - [2(-1)^2 + 3(-1) - 5]}{h}$$

$$= \lim_{h\to 0}\frac{2h^2 - h}{h} = \lim_{h\to 0}(2h-1) = 2(0) - 1 = -1$$

and $f'(0) = \lim_{h\to 0}\frac{f(0+h)-f(0)}{h}$

$$= \lim_{h\to 0}\frac{[2(0+h)^2 + 3(0+h) - 5] - [2(0)^2 + 3(0) - 5]}{h}$$

$$= \lim_{h\to 0}\frac{2h^2 + 3h}{h} = \lim_{h\to 0}(2h+3) = 2(0) + 3 = 3$$

Clearly, $f'(0) = -3f'(-1)$

70. *(c)* Let $f(x) = \sin x$. Then,

$$f'(0) = \lim_{h\to 0}\frac{f(0+h)-f(0)}{h}$$

$$= \lim_{h\to 0}\frac{\sin(0+h)-\sin(0)}{h} = \lim_{h\to 0}\frac{\sin h}{h} = 1$$

71. *(b)* Since, the derivative measures the change in the function, intuitively it is clear that the derivative of the constant function must be zero at every point. This is indeed, supported by the following computation.

$$f'(0) = \lim_{h\to 0}\frac{f(0+h)-f(0)}{h} = \lim_{h\to 0}\frac{3-3}{h} = \lim_{h\to 0}\frac{0}{h} = 0$$

Similarly, $f'(3) = \lim_{h\to 0}\frac{f(3+h)-f(3)}{h}$

$$= \lim_{h\to 0}\frac{3-3}{h} = 0$$

72. *(a)* We know that, $f'(a) = \lim_{h\to 0}\frac{f(a+h)-f(a)}{h}$

From the ΔPQR, it is clear that the ratio whose limit we are taking is precisely equal to $\tan(QPR)$ which is the slope of the chord PQ. In the limiting process, as h tends to 0, the point Q tends to P and we have

$$\lim_{h\to 0}\frac{f(a+h)-f(a)}{h} = \lim_{Q\to P}\frac{QR}{PR}$$

This is equivalent to the fact that the chord PQ tends to the tangent at P of the curve $y = f(x)$. Thus, the limit turns out to be equal to the slope of the tangent. Hence,

$$f'(a) = \tan\psi.$$

73. *(b)* For a given function f, we can find the derivative at every point. If the derivative exists at every point, it defines a new function called the derivative of f.

74. *(c)* Suppose f is a real valued function, the function defined by $\lim\limits_{h \to 0} \frac{f(x+h)-f(x)}{h}$

wherever the limit exists is defined to be the derivative of f at x and is denoted by $f'(x)$. This definition of derivative is also called the first principle of derivative.

Thus, $$f'(x) = \lim_{h \to 0} \frac{f(x+h)-f(x)}{h}$$

75. *(d)* There are different notations to denote the derivative of a function. Sometimes $f'(x)$ is denoted by $\frac{d}{dx}(f(x))$ or if $y = f(x)$, it is denoted by $\frac{dy}{dx}$. This is referred to as derivative of $f(x)$ or y with respect to x. It is also denoted by $D(f(x))$.

76. *(d)* The derivative of f at $x = a$ is denoted by

$$\frac{d}{dx} f(x)\bigg|_a \text{ or } \frac{df}{dx}\bigg|_a \text{ or even } \left(\frac{df}{dx}\right)_{x=a}$$

77. *(a)* A. Since, $f'(x) = \lim\limits_{h \to 0} \frac{f(x+h)-f(x)}{h}$

$$= \lim_{h \to 0} \frac{10(x+h)-10(x)}{h}$$

$$= \lim_{h \to 0} \frac{10h}{h} = \lim_{h \to 0} (10) = 10$$

B. We have, $f'(x) = \lim\limits_{h \to 0} \frac{f(x+h)-f(x)}{h}$

$$= \lim_{h \to 0} \frac{(x+h)^2-(x)^2}{h} = \lim_{h \to 0} (h+2x) = 2x$$

C. We have, $f'(x) = \lim\limits_{h \to 0} \frac{f(x+h)-f(x)}{h}$

$$= \lim_{h \to 0} \frac{a-a}{h} = \lim_{h \to 0} \frac{0}{h} = 0 \quad (\text{as } h \neq 0)$$

D. We have, $f'(x) = \lim\limits_{h \to 0} \frac{f(x+h)-f(x)}{h}$

$$= \lim_{h \to 0} \frac{\frac{1}{(x+h)} - \frac{1}{x}}{h}$$

$$= \lim_{h \to 0} \frac{1}{h}\left[\frac{x-(x+h)}{x(x+h)}\right]$$

$$= \lim_{h \to 0} \frac{1}{h}\left[\frac{-h}{x(x+h)}\right]$$

$$= \lim_{h \to 0} \frac{-1}{x(x+h)} = \frac{-1}{x^2}$$

78. *(d)* Let f and g be two functions such that their derivatives are defined in a common domain. Then,

(i) Derivative of sum of two functions is sum of the derivatives of the functions.

$$\frac{d}{dx}[f(x)+g(x)] = \frac{d}{dx} f(x) + \frac{d}{dx} g(x)$$

(ii) Derivative of difference of two functions is difference of the derivatives of the functions.

$$\frac{d}{dx}[f(x)-g(x)] = \frac{d}{dx} f(x) - \frac{d}{dx} g(x)$$

(iii) Derivative of product of two functions is given by the following product rule.

$$\frac{d}{dx}[f(x)\cdot g(x)] = \frac{d}{dx} f(x)\cdot g(x) + f(x)\cdot \frac{d}{dx} g(x)$$

(iv) Derivative of quotient of two functions is given by the following quotient rule (whenever the denominator is non-zero).

$$\frac{d}{dx}\left(\frac{f(x)}{g(x)}\right) = \frac{\frac{d}{dx} f(x)\cdot g(x) - f(x)\frac{d}{dx} g(x)}{(g(x))^2}, \; g(x) \neq 0$$

79. *(b)* Let $u = f(x)$ and $v = g(x)$. Then,

$$(uv)' = u'v + uv'$$

This is referred to a Leibnitz rule or the product rule for differentiating product of functions. Similarly, the quotient rule is

$$\left(\frac{u}{v}\right)' = \frac{u'v - uv'}{v^2}$$

80. *(b)* It is easy to see that the derivative of the function $f(x) = x$ is the constant function 1. This is because $f'(x) = \lim\limits_{h \to 0} \frac{f(x+h)-f(x)}{h} = \lim\limits_{h \to 0} \frac{x+h-x}{h} = \lim\limits_{h \to 0} 1 = 1$

81. *(a)* The derivative of $f(x) = 10x = x + \ldots\ldots + x$ (ten terms) is $\frac{df(x)}{dx} = \frac{d}{dx}(x + \ldots + x)$ (ten terms)

$$= \frac{d}{dx} x + \ldots + \frac{d}{dx} x \quad \text{(ten terms)}$$

$$= 1 + \ldots + 1 \quad \text{(ten terms)}$$

$$= 10$$

We note that this limit may be evaluated using product rule too. Write $f(x) = 10x = uv$, where u is the constant function taking value 10 everywhere and $v(x) = x$. We know that the derivative of u equals 0. Also, derivative of $v(x) = x$ equals to 1. Thus, by the product rule, we have

$$f'(x) = (10x)' = (uv)' = u'v + uv' = 0\cdot x + 10\cdot 1 = 10$$

On similar lines the derivative of $f(x) = x^2$ may be evaluated We have $f(x) = x^2 = x.x$ and hence

$$\frac{df}{dx} = \frac{d}{dx}(x.x) = \frac{d}{dx}(x).x + x.\frac{d}{dx}(x)$$

$$= 1\cdot x + x \cdot 1 = 2x$$

82. *(c)* By definition of the derivative function, we have

$$f'(x) = \lim_{h \to 0} \frac{f(x+h)-f(x)}{h} = \lim_{h \to 0} \frac{(x+h)^n - x^n}{h}$$

By binomial theorem,

$(x+h)^n = ({}^nC_0)x^n + ({}^nC_1)x^{n-1}h + \ldots + ({}^nC_n)h^n$ and hence $(x+h)^n - x^n = h(nx^{n-1} + \ldots + h^{n-1})$. Thus,

$$\frac{df(x)}{dx} = \lim_{h \to 0} \frac{(x+h)^n - x^n}{h}$$

$$= \lim_{h \to 0} \frac{h(nx^{n-1} + \ldots + h^{n-1})}{h}$$

$$= \lim_{h \to 0} (nx^{n-1} + \ldots + h^{n-1}) = nx^{n-1}$$

Alternatively, we may also prove this by induction on n and the product rule as follows: The result is true for $n = 1$, (which has been proved earlier). We have

$$\frac{d}{dx}(x^n) = \frac{d}{dx}(x \cdot x^{n-1})$$

$$= \frac{d}{dx}(x) \cdot (x^{n-1}) + x \cdot \frac{d}{dx}(x^{n-1}) \quad \text{(by product rule)}$$

$$= 1 \cdot x^{n-1} + x \cdot ((n-1)x^{n-2})$$

(by induction hypothesis)

$$= x^{n-1} + (n-1)x^{n-1} = nx^{n-1}$$

83. *(b)* Let $f(x) = a_n x^n + a_{n-1}x^{n-1} + \ldots\ldots + a_1x + a_0$ be a polynomial function, where a_i's are all real numbers and $a_n \neq 0$. Then, the derivative function is given by

$$\frac{df(x)}{dx} = na_n x^{n-1} + (n-1)a_{n-1}x^{n-2} + \ldots\ldots + 2a_2x + a_1$$

84. *(b)* If given function is $6x^{100} - x^{55} + x$.Then, the derivative of function is $6 \cdot 100 \cdot x^{99} - 55 \cdot x^{54} + 1$

or $\quad 600x^{99} - 55x^{54} + 1$

85. *(d)* I. The derivative of the function is $1 + 2x + 3x^2 + \ldots\ldots\ldots\ldots + 50x^{49}$. At $x = 1$ the value of this function equals to

$$1 + 2(1) + 3(1)^2 + \ldots\ldots + 50(1)^{49} = 1 + 2 + 3 + \ldots\ldots + 50$$

$$= \frac{(50)(51)}{2} = 1275$$

II. Clearly, this function is defined everywhere except at $x = 0$. We use the quotient rule with $u = x + 1$ and $v = x$. Hence, $u' = 1$ and $v' = 1$. Therefore,

$$\frac{df(x)}{dx} = \frac{d}{dx}\left(\frac{x+1}{x}\right) = \frac{d}{dx}\left(\frac{u}{v}\right) = \frac{u'v - uv'}{v^2}$$

$$= \frac{1(x) - (x+1)1}{x^2} = -\frac{1}{x^2}$$

86. *(a)* A. We have, $f(x) = \sin x$. Then,

$$\frac{df(x)}{dx} = \lim_{h \to 0} \frac{f(x+h) - f(x)}{h} = \lim_{h \to 0} \frac{\sin(x+h) - \sin(x)}{h}$$

$$= \lim_{h \to 0} \frac{2\cos\left(\frac{2x+h}{2}\right)\sin\left(\frac{h}{2}\right)}{h}$$

$$= \lim_{h \to 0} \cos\left(x + \frac{h}{2}\right) \cdot \lim_{h \to 0} \frac{\sin\frac{h}{2}}{\frac{h}{2}} = \cos x \cdot 1 = \cos x$$

B. We have, $f(x) = \tan x$. Then,

$$\frac{df(x)}{dx} = \lim_{h \to 0} \frac{f(x+h) - f(x)}{h} = \lim_{h \to 0} \frac{\tan(x+h) - \tan(x)}{h}$$

$$= \lim_{h \to 0} \frac{1}{h}\left[\frac{\sin(x+h)}{\cos(x+h)} - \frac{\sin x}{\cos x}\right]$$

$$= \lim_{h \to 0}\left[\frac{\sin(x+h)\cos x - \cos(x+h)\sin x}{h\cos(x+h)\cos x}\right]$$

$$= \lim_{h \to 0} \frac{\sin(x+h-x)}{h\cos(x+h)\cos x}$$

$$\left[\begin{array}{l}\text{using formula for } \sin(A-B) \\ = \sin A \cos B - \cos A \sin B\end{array}\right]$$

$$= \lim_{h \to 0} \frac{\sin h}{h} \cdot \lim_{h \to 0} \frac{1}{\cos(x+h)\cos x}$$

$$= 1 \cdot \frac{1}{\cos^2 x} = \sec^2 x$$

C. We use the Leibnitz product rule to evaluate this.

$$\frac{df(x)}{dx} = \frac{d}{dx}(\sin x \sin x)$$

$$= (\sin x)' \sin x + \sin x (\sin x)'$$

$$= (\cos x)\sin x + \sin x (\cos x)$$

$$= 2\sin x \cos x = \sin 2x$$

87. *(c)* I. Let $f(x) = x^2 - 2$, we have

$$f'(x) = \lim_{h \to 0} \frac{f(x+h) - f(x)}{h}$$

$$\Rightarrow \quad f'(x) = \lim_{h \to 0} \frac{[(x+h)^2 - 2)] - (x^2 - 2)}{h}$$

$$= \lim_{h \to 0} \frac{x^2 + h^2 + 2xh - 2 - x^2 + 2}{h}$$

$$= \lim_{h \to 0} \frac{h(h + 2x)}{h}$$

$$= 0 + 2x = 2x$$

At $\quad x = 10, f'(10) = 2 \times 10 = 20$

II. Let $f(x) = 99x$

We have, $f'(x) = \lim_{h \to 0} \frac{f(x+h) - f(x)}{h}$

$$\Rightarrow \quad f'(x) = \lim_{h \to 0} \frac{99(x+h) - 99x}{h}$$

$$= \lim_{h \to 0} \frac{99x + 99h - 99x}{h}$$

$$= \lim_{h \to 0} \frac{99h}{h} = 99$$

At $\quad x = 100, f'(100) = 99$

III. Let $f(x) = x$

We have, $f'(x) = \lim_{h \to 0} \frac{f(x+h) - f(x)}{h}$

$$\Rightarrow \quad f'(x) = \lim_{h \to 0} \frac{x + h - x}{h} \Rightarrow f'(x) = \lim_{h \to 0} \frac{h}{h} = 1$$

At $\quad x = 1, f'(1) = 1$

88. *(b)* A. Let $f(x) = x^3 - 27$

We have, $\quad f'(x) = \lim_{h \to 0} \frac{f(x+h) - f(x)}{h}$

$$\Rightarrow \quad f'(x) = \lim_{h \to 0} \frac{[(x+h)^3 - 27] - (x^3 - 27)}{h}$$

$[\because f(x) = x^3 - 27]$

$$= \lim_{h \to 0} \frac{(x+h)^3 - x^3}{h}$$

$$= \lim_{h \to 0} \frac{x^3 + h^3 + 3xh(x+h) - x^3}{h}$$

$$= \lim_{h \to 0} \frac{h^3 + 3xh(x+h)}{h}$$

$$= \lim_{h \to 0} \frac{h[h^2 + 3x(x+h)]}{h}$$

$$= 0 + 3x(x+0) = 3x^2$$

$$\Rightarrow \quad f'(x) = 3x^2$$

B. Let $f(x) = (x-1)(x-2) = x^2 - 2x - x + 2$

$$\Rightarrow \quad f(x) = x^2 - 3x + 2$$

We have, $f'(x) = \lim_{h \to 0} \frac{f(x+h) - f(x)}{h}$

$$\Rightarrow f'(x) = \lim_{h \to 0} \frac{[(x+h)^2 - 3(x+h) + 2] - (x^2 - 3x + 2)}{h}$$

$[\because f(x) = x^2 - 3x + 2]$

$$= \lim_{h \to 0} \frac{x^2 + h^2 + 2xh - 3x - 3h + 2 - x^2 + 3x - 2}{h}$$

$$= \lim_{h \to 0} \frac{h^2 + 2xh - 3h}{h}$$

$$= \lim_{h \to 0} \frac{h(h + 2x - 3)}{h}$$

$$= \lim_{h \to 0} h + 2x - 3$$

$$= 0 + 2x - 3 = 2x - 3$$

$$\therefore \quad f'(x) = 2x - 3$$

C. Let $f(x) = \frac{1}{x^2}$

We have, $f'(x) = \lim_{h \to 0} \frac{f(x+h) - f(x)}{h}$

$$= \lim_{h \to 0} \frac{\frac{1}{(x+h)^2} - \frac{1}{x^2}}{h}$$

$$= \lim_{h \to 0} \frac{x^2 - (x+h)^2}{x^2 h (x+h)^2}$$

$$= \lim_{h \to 0} \frac{x^2 - x^2 - h^2 - 2xh}{x^2 h (x+h)^2}$$

$$= - \lim_{h \to 0} \frac{(h + 2x)h}{x^2 h (x+h)^2}$$

$$= - \lim_{h \to 0} \frac{h + 2x}{x^2 (x+h)^2} = \frac{-2}{x^3}$$

D. Let $f(x) = \frac{x+1}{x-1}$

We have, $f'(x) = \lim_{h \to 0} \frac{f(x+h) - f(x)}{h}$

$$\Rightarrow \quad f'(x) = \lim_{h \to 0} \frac{\frac{x+h+1}{x+h-1} - \frac{x+1}{x-1}}{h}$$

$$= \lim_{h \to 0} \frac{(x+h+1)(x-1) - (x+1)(x+h-1)}{(x+h-1)(x-1)h}$$

$$= \lim_{h \to 0} \frac{(x^2 - x + hx - h + x - 1) - (x^2 + hx - x + x + h - 1)}{(x+h-1)(x-1)h}$$

$$= \lim_{h \to 0} \frac{x^2 + hx - h - 1 - x^2 - hx - h + 1}{(x+h-1)(x-1)h}$$

$$= \lim_{h \to 0} \frac{-2h}{(x+h-1)(x-1)h}$$

$$= \frac{-2}{(x+0-1)(x-1)} = \frac{-2}{(x-1)^2}$$

89. *(c)* Given, $f(x) = \frac{x^{100}}{100} + \frac{x^{99}}{99} + \ldots + \frac{x^2}{2} + x + 1$

$$\Rightarrow \quad f'(x) = \frac{100x^{99}}{100} + \frac{99x^{98}}{99} + \ldots + \frac{2x}{2} + 1 + 0$$

$[\because f(x) = x^n \Rightarrow f'(x) = nx^{n-1}]$

$$\Rightarrow \quad f'(x) = x^{99} + x^{98} + \ldots + x + 1 \quad \ldots(i)$$

Putting $x = 1$, we get

$$f'(1) = \underbrace{(1)^{99} + 1^{98} + \ldots + 1 + 1}_{100 \text{ times}}$$

$$= \underbrace{1 + 1 + 1 \ldots + 1 + 1}_{100 \text{ times}}$$

$$\Rightarrow \quad f'(1) = 100 \quad \ldots(ii)$$

Again, putting $x = 0$, we get

$$f'(0) = 0 + 0 + \ldots + 0 + 1$$

$$\Rightarrow \quad f'(0) = 1 \quad \ldots(iii)$$

From Eqs. (ii) and (iii), we get

$$f'(1) = 100 f'(0)$$

Hence, $m = 100$

90. *(a)* Let $f(x) = x^n + ax^{n-1} + a^2x^{n-2} + \ldots + a^{n-1}x + a^n$

Differentiating w.r.t. x, we get

$$f'(x) = nx^{n-1} + a(n-1)x^{n-1-1} + a^2(n-2)x^{n-2-1} + \ldots + a^{n-1}(1) + 0$$

$$\Rightarrow \quad f'(x) = nx^{n-1} + a(n-1)x^{n-2} + a^2(n-2)x^{n-3} + \ldots + a^{n-1}$$

91. *(d)* A. Let $f(x) = (x-a)(x-b) = x^2 - ax - bx + ab$

$$= x^2 - (a+b)x + ab$$

Differentiating w.r.t. x, we get

$$f'(x) = 2x - (a+b)1 + 0$$

$$= 2x - a - b$$

B. Let $f(x) = (ax^2 + b)^2$

$$\Rightarrow f(x) = a^2x^4 + b^2 + 2abx^2$$

Differentiating w.r.t x, we get

$$f'(x) = 4a^2x^3 + 0 + 2ab(2x)$$

$$= 4a^2x^3 + 4abx = 4ax(ax^2 + b)$$

C. Let $f(x) = \frac{x-a}{x-b}$

Differentiating w.r.t. x, we get

$$f'(x) = \frac{(x-b)\frac{d}{dx}(x-a) - (x-a)\frac{d}{dx}(x-b)}{(x-b)^2}$$

$$\left[\because f(x) = \frac{f_1(x)}{f_2(x)} \quad \therefore f'(x) = \frac{f_2(x) f_1'(x) - f_1(x) f_2'(x)}{\{f_2(x)\}^2}\right]$$

$$= \frac{(x-b)(1-0) - (x-a)(1-0)}{(x-b)^2}$$

$$= \frac{(x-b) - (x-a)}{(x-b)^2}$$

$$= \frac{x-b-x+a}{(x-b)^2}$$

$$= \frac{a-b}{(x-b)^2}$$

92. *(b)* Let $y = \frac{x^n - a^n}{x-a}$

Differentiating y w.r.t x, we get

$$\frac{dy}{dx} = \frac{(x-a)\frac{d}{dx}(x^n - a^n) - (x^n - a^n)\frac{d}{dx}(x-a)}{(x-a)^2}$$

$$= \frac{(x-a)[nx^{n-1} - 0] - (x^n - a^n)(1-0)}{(x-a)^2}$$

$$= \frac{(x-a)\, nx^{n-1} - x^n + a^n}{(x-a)^2}$$

$$= \frac{x \times nx^{n-1} - anx^{n-1} - x^n + a^n}{(x-a)^2}$$

$$= \frac{nx^n - anx^{n-1} - x^n + a^n}{(x-a)^2}$$

93. *(b)* I. We have, $y = 2x - \frac{3}{4}$

Differentiating y w.r.t. x, we get

$$\frac{dy}{dx} = 2 \times 1 - 0 = 2$$

II. We have, $y = (5x^3 + 3x - 1)(x-1)$

Differentiating y w.r.t x, we get

$$\frac{dy}{dx} = (5x^3 + 3x - 1)\frac{d}{dx}(x-1) + (x-1)\frac{d}{dx}(5x^3 + 3x - 1)$$

$$= (5x^3 + 3x - 1)(1-0) + (x-1)(5 \times 3x^2 + 3 \times 1 - 0)$$

$$= (5x^3 + 3x - 1) + (x-1)(15x^2 + 3)$$

$$= 5x^3 + 3x - 1 + 15x^3 + 3x - 15x^2 - 3$$

$$= 20x^3 - 15x^2 + 6x - 4$$

94. *(a)* A. Let $f(x) = \cos x$

We have, $f'(x) = \lim_{h \to 0} \frac{f(x+h) - f(x)}{h}$ (by first principle)

$$\Rightarrow \quad f'(x) = \lim_{h \to 0} \frac{\cos(x+h) - \cos x}{h}$$

$$= \lim_{h \to 0} \frac{-2\sin\left(\frac{x+h+x}{2}\right)\sin\left(\frac{x+h-x}{2}\right)}{h}$$

$$= \lim_{h \to 0} \frac{-2\sin\frac{2x+h}{2}\sin\frac{h}{2}}{2 \times \frac{h}{2}}$$

$$= \lim_{h \to 0}\left\{-\sin\left(\frac{2x+h}{2}\right)\right\}\left\{\frac{\sin\frac{h}{2}}{\frac{h}{2}}\right\}$$

$$= -\lim_{h \to 0} \sin\frac{2x+h}{2} \lim_{h \to 0} \frac{\sin h/2}{h/2}$$

$$= -\sin\frac{2x}{2} \times 1 \qquad \left(\because \lim_{h \to 0} \frac{\sin\frac{h}{2}}{\frac{h}{2}} = 1\right)$$

$$\Rightarrow \quad f'(x) = -\sin x$$

B. Let $y = \sin x \cos x$

Differentiating y w.r.t. x, we get

$$\frac{dy}{dx} = \sin x \frac{d}{dx}\cos x + \cos x \frac{d}{dx}\sin x$$

(by product rule)

$$= \sin x(-\sin x) + \cos x \cos x = -\sin^2 x + \cos^2 x$$

$$= \cos 2x$$

C. Let $y = \sec x$

$$\Rightarrow \quad y = \frac{1}{\cos x}$$

Differentiating y w.r.t x, we get

$$\frac{dy}{dx} = \frac{\cos x \frac{d}{dx}(1) - 1 \times \frac{d}{dx}\cos x}{\cos^2 x} \quad \text{(by quotient rule)}$$

$$= \frac{\cos x \times 0 - 1 \times (-\sin x)}{\cos^2 x} = \frac{\sin x}{\cos x} \times \frac{1}{\cos x}$$

$$\Rightarrow \quad \frac{dy}{dx} = \tan x \sec x$$

D. Let $y = 5\sec x + 4\cos x$

Differentiating y w.r.t. x, we get

$$\frac{dy}{dx} = 5\sec x \tan x - 4\sin x$$

95. *(a)* A. Let $y = x^{-3}(5+3x) \Rightarrow y = 5x^{-3} + 3x^{-2}$

Differentiating y w.r.t. x, we get

$$\frac{dy}{dx} = 5(-3)x^{-3-1} + 3(-2)\,x^{-2-1}$$

$$= -15x^{-4} - 6x^{-3}$$

B. Let $y = x^5(3 - 6x^{-9}) = 3x^5 - 6x^{5-9} = 3x^5 - 6x^{-4}$

Differentiating y w.r.t. x, we get

$$\frac{dy}{dx} = 3 \times 5x^{5-1} - 6(-4)\,x^{-4-1}$$

$$= 15x^4 + 24x^{-5}$$

C. Let $y = x^{-4}(3 - 4x^{-5})$

$\Rightarrow \quad y = 3x^{-4} - 4x^{-4-5}$

$\Rightarrow \quad y = 3x^{-4} - 4x^{-9}$

Differentiating y w.r.t. x, we get

$$\frac{dy}{dx} = 3(-4)x^{-4-1} - 4(-9)x^{-9-1} = -12x^{-5} + 36x^{-10}$$

D. Let $y = \frac{2}{x+1} - \frac{x^2}{3x-1}$

Differentiating y w.r.t. x, we get

$$\frac{dy}{dx} = \frac{(x+1)\frac{d}{dx}(2) - 2\frac{d}{dx}(x+1)}{(x+1)^2} - \frac{(3x-1)\frac{d}{dx}(x^2) - x^2\frac{d}{dx}(3x-1)}{(3x-1)^2}$$

$$\Rightarrow \frac{dy}{dx} = \frac{(x+1)\times 0 - 2\times 1}{(x+1)^2} - \frac{(3x-1)(2x) - x^2(3)}{(3x-1)^2}$$

$$= -\frac{2}{(x+1)^2} - \frac{(6x^2 - 2x - 3x^2)}{(3x-1)^2}$$

$$= -\frac{2}{(x+1)^2} - \frac{(3x^2 - 2x)}{(3x-1)^2}$$

$$\Rightarrow \frac{dy}{dx} = \frac{-2}{(x+1)^2} - \frac{x(3x-2)}{(3x-1)^2}$$

96. *(d)* A. Let $y = \text{cosec } x = \frac{1}{\sin x}$

Differentiating y w.r.t. x, we get

$$\frac{dy}{dx} = \frac{\sin x \frac{d}{dx}(1) - (1)\frac{d}{dx}(\sin x)}{\sin^2 x}$$

$$= \frac{\sin x \times 0 - 1 \times \cos x}{\sin^2 x}$$

$$= \frac{0 - \cos x}{\sin^2 x} = \frac{-\cos x}{\sin x} \times \frac{1}{\sin x}$$

$$\Rightarrow \frac{dy}{dx} = -\cot x \text{ cosec } x$$

B. Let $y = 3\cot x + 5\text{cosec } x$

Differentiating y w.r.t. x, we get

$$\frac{dy}{dx} = -3\text{cosec}^2 x - 5\text{cosec } x \cot x$$

C. Let $y = 5\sin x - 6\cos x + 7$

Differentiating y w.r.t. x, we get

$$\frac{dy}{dx} = 5\cos x - 6(-\sin x) + 0$$

$$= 5\cos x + 6\sin x$$

D. Let $y = 2\tan x - 7\sec x$

Differentiating y w.r.t. x, we get

$$\frac{dy}{dx} = 2\sec^2 x - 7\sec x \tan x$$

97. *(a)* I. Recall the trigonometric rule $\sin 2x = 2\sin x \cos x$. Thus,

$$\frac{df(x)}{dx} = \frac{d}{dx}(2\sin x \cos x) = 2\frac{d}{dx}(\sin x \cos x)$$

$$= 2[(\sin x)' \cos x + \sin x (\cos x)']$$

$$= 2[(\cos x)\cos x + \sin x(-\sin x)]$$

$$= 2(\cos^2 x - \sin^2 x)$$

II. $g(x) = \cot x = \frac{\cos x}{\sin x}$

$$\Rightarrow \frac{d}{dx}(g(x)) = \frac{d}{dx}(\cot x) = \frac{d}{dx}\left(\frac{\cos x}{\sin x}\right)$$

$$= \frac{(\cos x)'(\sin x) - (\cos x)(\sin x)'}{(\sin x)^2}$$

$$= \frac{(-\sin x)(\sin x) - (\cos x)(\cos x)}{(\sin x)^2}$$

$$= \frac{-(\sin^2 x + \cos^2 x)}{\sin^2 x}$$

$$= -\text{cosec}^2 x$$

Alternatively, this may be computed by noting that $\cot x = \frac{1}{\tan x}$. Here, we use the fact that the derivative of $\tan x$ is $\sec^2 x$ and also that the derivative of the constant function is 0.

$$\frac{d}{dx}(g(x)) = \frac{d}{dx}(\cot x) = \frac{d}{dx}\left(\frac{1}{\tan x}\right)$$

$$= \frac{(1)'(\tan x) - (1)(\tan x)'}{(\tan x)^2}$$

$$= \frac{(0)(\tan x) - (\sec x)^2}{(\tan x)^2}$$

$$= \frac{-\sec^2 x}{\tan^2 x} = -\text{cosec}^2 x$$

98. *(d)* I. We have, $f(x) = \frac{x^5 - \cos x}{\sin x}$. We use the quotient rule on this function wherever it is defined.

$$f'(x) = \frac{(x^5 - \cos x)' \sin x - (x^5 - \cos x)(\sin x)'}{(\sin x)^2}$$

$$= \frac{(5x^4 + \sin x)\sin x - (x^5 - \cos x)\cos x}{\sin^2 x}$$

$$= \frac{-x^5 \cos x + 5x^4 \sin x + 1}{(\sin x)^2}$$

II. We use quotient rule on the function $h(x) = \frac{x + \cos x}{\tan x}$ wherever it is defined.

$$h'(x) = \frac{(x + \cos x)' \tan x - (x + \cos x)(\tan x)'}{(\tan x)^2}$$

$$= \frac{(1 - \sin x)\tan x - (x + \cos x)\sec^2 x}{(\tan x)^2}$$

99. *(b)* I. By definition,

$$f'(x)=\lim_{h\to0}\frac{f(x+h)-f(x)}{h}$$
$$=\lim_{h\to0}\frac{a(x+h)+b-(ax+b)}{h}=\lim_{h\to0}\frac{ah}{h}=a$$

II. By definition,

$$f'(x)=\lim_{h\to0}\frac{f(x+h)-f(x)}{h}$$
$$=\lim_{h\to0}\frac{a(x+h)^2+b(x+h)+c-ax^2-bx-c}{h}$$
$$=\lim_{h\to0}\frac{bh+ah^2+2axh}{h}=\lim_{h\to0} ah+2ax+b=b+2ax$$

100. *(d)* I. $f'(x)=\lim_{h\to0}\frac{f(x+h)-f(x)}{h}$

$$=\lim_{h\to0}\frac{(x+h)^3-x^3}{h}$$
$$=\lim_{h\to0}\frac{x^3+h^3+3xh(x+h)-x^3}{h}$$
$$=\lim_{h\to0}(h^2+3x(x+h))=3x^2$$

II. $f'(x)=\lim_{h\to0}\frac{f(x+h)-f(x)}{h}$

$$=\lim_{h\to0}\frac{\frac{1}{(x+h)^3}-\frac{1}{x^3}}{h}\qquad\left[\because f(x)=\frac{1}{x^3}\right]$$
$$=\lim_{h\to0}\frac{x^3-(x+h)^3}{(x+h)^3x^3h}$$
$$=\lim_{h\to0}\frac{x^3-[x^3+h^3+3xh(x+h)]}{(x+h)^3x^3h}$$
$$=\lim_{h\to0}\frac{-h^3-3xh(x+h)}{(x+h)^3x^3h}$$
$$=\lim_{h\to0}\frac{-h[h^2+3x(x+h)]}{(x+h)^3x^3h}=\frac{-3}{x^4}$$

101. *(b)* For $x>1$, we have $f(x)=|\log|x||=\log x$

$$\Rightarrow\qquad f'(x)=\frac{1}{x}$$

For $x<-1$, we have

$$f(x)=|\log|x||=\log(-x)$$
$$\Rightarrow\qquad f'(x)=\frac{1}{x}$$

For $0<x<1$, we have

$$f(x)=|\log|x||=-\log x\quad\Rightarrow\quad f'(x)=-\frac{1}{x}$$

For $-1<x<0$, we have

$$f'(x)=-\log(-x)\quad\Rightarrow\quad f'(x)=-\frac{1}{x}$$

Hence, $$f'(x)=\begin{cases}\frac{1}{x}, & |x|>1\\ -\frac{1}{x}, & |x|<1\end{cases}$$

102. *(c)* We have, $f'(x)=\lim_{h\to0}\frac{f(x+h)-f(x)}{h}$

$$=\lim_{h\to0}\frac{\tan(a(x+h)+b)-\tan(ax+b)}{h}$$
$$=\lim_{h\to0}\frac{\frac{\sin(ax+ah+b)}{\cos(ax+ah+b)}-\frac{\sin(ax+b)}{\cos(ax+b)}}{h}$$
$$=\lim_{h\to0}\frac{\frac{\sin(ax+ah+b)\cos(ax+b)-\sin(ax+b)\cos(ax+ah+b)}{}}{h\cos(ax+b)\cos(ax+ah+b)}$$
$$=\lim_{h\to0}\frac{a\sin(ah)}{a.h\cos(ax+b)\cos(ax+ah+b)}$$
$$=\lim_{h\to0}\frac{a}{\cos(ax+b)\cos(ax+ah+b)}$$
$$\lim_{h\to0}\frac{\sin ah}{ah}\ [\text{as } h\to0,\ ah\to0]$$
$$=\frac{a}{\cos^2(ax+b)}=a\sec^2(ax+b)$$

103. *(d)* By definition,

$$f'(x)=\lim_{h\to0}\frac{f(x+h)-f(x)}{h}$$
$$=\lim_{h\to0}\frac{\sqrt{\sin(x+h)}-\sqrt{\sin x}}{h}$$
$$=\lim_{h\to0}\frac{(\sqrt{\sin(x+h)}-\sqrt{\sin x})(\sqrt{\sin(x+h)}+\sqrt{\sin x})}{h(\sqrt{\sin(x+h)}+\sqrt{\sin x})}$$
$$=\lim_{h\to0}\frac{\sin(x+h)-\sin x}{h(\sqrt{\sin(x+h)}+\sqrt{\sin x})}$$
$$=\lim_{h\to0}\frac{2\cos\left(\frac{2x+h}{2}\right)\sin\frac{h}{2}}{2.\frac{h}{2}(\sqrt{\sin(x+h)}+\sqrt{\sin x})}$$
$$=\frac{\cos x}{2\sqrt{\sin x}}=\frac{1}{2}\cot x\sqrt{\sin x}$$

104. *(a)* By definition,

$$f'(x)=\lim_{h\to0}\frac{f(x+h)-f(x)}{h}$$
$$=\lim_{h\to0}\frac{\sqrt{\cos(x+h)}-\sqrt{\cos x}}{h}$$
$$=\lim_{h\to0}\frac{(\sqrt{\cos(x+h)}-\sqrt{\cos x})(\sqrt{\cos(x+h)}+\sqrt{\cos x})}{h(\sqrt{\cos(x+h)}+\sqrt{\cos x})}$$
$$=\lim_{h\to0}\frac{\cos(x+h)-\cos x}{h(\sqrt{\cos(x+h)}+\sqrt{\cos x})}$$
$$=\lim_{h\to0}\frac{-2\sin\left(\frac{2x+h}{2}\right)\sin\left(\frac{h}{2}\right)}{2.\frac{h}{2}(\sqrt{\cos(x+h)}+\sqrt{\cos x})}$$
$$=\frac{-\sin x}{2\sqrt{\cos x}}$$
$$=-\frac{\tan x}{2}\sqrt{\cos x}$$

105. *(c)* Since, f is an even function.

So, $f(-x) = f(x)$ for all x.

Also, $f'(0)$ exists.

So, $Rf'(0) = Lf'(0)$

$$\Rightarrow \lim_{h\to 0} \frac{f(0+h)-f(0)}{h} = \lim_{h\to 0} \frac{f(0-h)-f(0)}{-h}$$

$$\Rightarrow \lim_{h\to 0} \frac{f(h)-f(0)}{h} = \lim_{h\to 0} \frac{f(-h)-f(0)}{-h}$$

$$\Rightarrow \lim_{h\to 0} \frac{f(h)-f(0)}{h} = \lim_{h\to 0} \frac{f(h)-f(0)}{-h}$$

[using Eq. (i), $f(-h) = f(h)$]

$$\Rightarrow \lim_{h\to 0} \frac{f(h)-f(0)}{h} = -\lim_{h\to 0} \frac{f(h)-f(0)}{h}$$

$$\Rightarrow 2\lim_{h\to 0} \frac{f(h)-f(0)}{h} = 0$$

$$\Rightarrow 2f'(0) = 0 \Rightarrow f'(0) = 0$$

106. *(b)* $\because f(x) = x\sin x$

$$\Rightarrow f'(x) = \frac{d}{dx}(x\sin x)$$

$$= \sin x \frac{d}{dx}x + x\frac{d}{dx}\sin x = \sin x + x\cos x$$

$$\Rightarrow f'\left(\frac{\pi}{2}\right) = \sin\frac{\pi}{2} + \frac{\pi}{2}\cos\frac{\pi}{2} = 1$$

107. *(d)* We have, $u = e^x \sin x$

$$\Rightarrow \frac{du}{dx} = e^x \sin x + e^x \cos x = u + v$$

$$v = e^x \cos x$$

$$\Rightarrow \frac{dv}{dx} = e^x \cos x - e^x \sin x = v - u$$

$$\therefore v\frac{du}{dx} - u\frac{dv}{dx} = v(u+v) - u(v-u) = u^2 + v^2$$

$$\frac{d^2u}{dx^2} = \frac{du}{dx} + \frac{dv}{dx} = u + v + v - u = 2v$$

and $$\frac{d^2v}{dx^2} = \frac{dv}{dx} - \frac{du}{dx} = (v-u) - (v+u) = -2u$$

108. *(d)* We have, $f(x) = |\cos x - \sin x|$

$$\Rightarrow f(x) = \begin{cases} \cos x - \sin x, & \text{for } 0 < x \le \frac{\pi}{4} \\ \sin x - \cos x, & \text{for } \frac{\pi}{4} < x < \frac{\pi}{2} \end{cases}$$

Clearly, $$Lf'\left(\frac{\pi}{4}\right) = \left\{\frac{d}{dx}(\cos x - \sin x)\right\}_{at\ x=\frac{\pi}{4}}$$

$$= (-\sin x - \cos x)_{x=\frac{\pi}{4}} = -\sqrt{2}$$

and $$Rf'\left(\frac{\pi}{4}\right) = \left\{\frac{d}{dx}(\sin x - \cos x)\right\}_{at\ x=\frac{\pi}{4}}$$

$$= (\cos x + \sin x)_{x=\frac{\pi}{4}} = \sqrt{2}$$

$$\because Lf'\left(\frac{\pi}{4}\right) \ne Rf'\left(\frac{\pi}{4}\right)$$

$\therefore f'\left(\frac{\pi}{4}\right)$ doesn't exist.

109. *(a)* Let $f(x) = \sqrt{ax+b}$. Then,

$$f(x+h) = \sqrt{a(x+h)+b} = \sqrt{(ax+b)+ah}$$

$$\therefore \frac{d}{dx}(f(x)) = \lim_{h\to 0} \frac{f(x+h)-f(x)}{h}$$

$$\Rightarrow \frac{d}{dx}(f(x)) = \lim_{h\to 0} \frac{\sqrt{(ax+b)+ah} - \sqrt{ax+b}}{h}$$

$$\Rightarrow \frac{d}{dx}(f(x)) = \lim_{h\to 0} \frac{\sqrt{(ax+b)+ah} - \sqrt{ax+b}}{h} \times \frac{\sqrt{(ax+b)+ah} + \sqrt{ax+b}}{\sqrt{(ax+b)+ah} + \sqrt{ax+b}}$$

$$\Rightarrow \frac{d}{dx}(f(x)) = \lim_{h\to 0} \frac{(ax+b)+ah-(ax+b)}{h\{\sqrt{(ax+b)+ah} + \sqrt{ax+b}\}}$$

$$\Rightarrow \frac{d}{dx}(f(x)) = \lim_{h\to 0} \frac{ah}{h\{\sqrt{ax+b+ah} + \sqrt{ax+b}\}}$$

$$\Rightarrow \frac{d}{dx}(f(x)) = \lim_{h\to 0} \frac{a}{\sqrt{(ax+b)+ah} + \sqrt{ax+b}}$$

$$\Rightarrow \frac{d}{dx}(f(x)) = \frac{a}{\sqrt{ax+b} + \sqrt{ax+b}}$$

Hence, $$\frac{d}{dx}(\sqrt{ax+b}) = \frac{a}{2\sqrt{ax+b}}$$

110. *(d)* $$f(x) + f(y) = f\left(\frac{x+y}{1-xy}\right) \quad \ldots\text{(i)}$$

Putting $x = y = 0$, we get $f(0) = 0$

Putting $y = -x$, we get $f(x) + f(-x) = f(0) = 0$

$$\Rightarrow f(-x) = -f(x) \quad \ldots\text{(ii)}$$

Also, $$\lim_{x\to 0} \frac{f(x)}{x} = 2 \quad \ldots\text{(iii)}$$

Now, $$f'(x) = \lim_{h\to 0} \frac{f(x+h)-f(x)}{h}$$

$$= \lim_{h\to 0} \frac{f(x+h)+f(-x)}{h}$$

[using Eq. (ii) $-f(x) = f(-x)$]

$$f'(x) = \lim_{h\to 0} \frac{f\left(\frac{x+h-x}{1-(x+h)(-x)}\right)}{h} \quad \text{[using Eq. (i)]}$$

$$\Rightarrow f'(x) = \lim_{h\to 0} \left[\frac{f\left(\frac{h}{1+x(x+h)}\right)}{h}\right]$$

$$\Rightarrow f'(x) = \lim_{h\to 0} \frac{f\left(\frac{h}{1+xh+x^2}\right)}{\left(\frac{h}{1+xh+x^2}\right)} \times \left(\frac{1}{1+xh+x^2}\right)$$

$$\Rightarrow f'(x) = \lim_{h\to 0} \frac{f\left(\frac{h}{1+xh+x^2}\right)}{\left(\frac{h}{1+xh+x^2}\right)} \times \lim_{h\to 0} \frac{1}{1+xh+x^2}$$

$$\left[\text{using } \lim_{x\to 0} \frac{f(x)}{x} = 2\right]$$

$$\Rightarrow \quad f'(x)=2\times\frac{1}{1+x^2}$$

$$\Rightarrow \quad f'(x)=\frac{2}{1+x^2}$$

$$\Rightarrow \quad f'\left(\frac{1}{\sqrt{3}}\right)=\frac{2}{1+\frac{1}{3}}$$

$$=\frac{2}{4/3}=\frac{6}{4}=\frac{3}{2}$$

111. *(c)* Let $f(x)=\cot\sqrt{x}$.

Then, $f(x+h)=\cot\sqrt{x+h}$

$$\therefore \quad \frac{d}{dx}f(x)=\lim_{h\to 0}\frac{f(x+h)-f(x)}{h}$$

$$\Rightarrow \quad \frac{d}{dx}f(x)=\lim_{h\to 0}\frac{\cot\sqrt{x+h}-\cot\sqrt{x}}{h}$$

$$\Rightarrow \quad \frac{d}{dx}(f(x))=\lim_{h\to 0}\frac{-\sin(\sqrt{x+h}-\sqrt{x})}{h\sin\sqrt{x+h}\sin\sqrt{x}}$$

$$\Rightarrow \quad \frac{d}{dx}f(x)=\lim_{h\to 0}\frac{-\sin(\sqrt{x+h}-\sqrt{x})}{[(x+h)-x]\sin\sqrt{x+h}\sin\sqrt{x}]}$$

$$\Rightarrow \quad \frac{d}{dx}f(x)=\lim_{h\to 0}\frac{-\sin(\sqrt{x+h}-\sqrt{x})}{(\sqrt{x+h}-\sqrt{x})(\sqrt{x+h}+\sqrt{x})\times\sin\sqrt{x+h}\sin\sqrt{x}}$$

$$\Rightarrow \quad \frac{d}{dx}f(x)=\lim_{h\to 0}\frac{-\sin(\sqrt{x+h}-\sqrt{x})}{\sqrt{x+h}-\sqrt{x}}\times\lim_{h\to 0}\frac{1}{(\sqrt{x+h}+\sqrt{x})\sin\sqrt{x+h}\sin\sqrt{x}}$$

$$\Rightarrow \quad \frac{d}{dx}f(x)=\frac{-1}{2\sqrt{x}\sin\sqrt{x}\sin\sqrt{x}}$$

$$\Rightarrow \quad \frac{d}{dx}f(x)=\frac{-\operatorname{cosec}^2\sqrt{x}}{2\sqrt{x}}$$

112. *(b)* Let $\phi(x)=\frac{1}{f(x)}$. Then, $\phi(x+h)=\frac{1}{f(x+h)}$

$$\therefore \quad \frac{d}{dx}\{\phi(x)\}=\lim_{h\to 0}\frac{\phi(x+h)-\phi(x)}{h}$$

$$\Rightarrow \quad \frac{d}{dx}\{\phi(x)\}=\lim_{h\to 0}\frac{\frac{1}{f(x+h)}-\frac{1}{f(x)}}{h}$$

$$\Rightarrow \quad \frac{d}{dx}\{\phi(x)\}=\lim_{h\to 0}\frac{f(x)-f(x+h)}{hf(x)f(x+h)}$$

$$\Rightarrow \quad \frac{d}{dx}\{\phi(x)\}=\lim_{h\to 0}\frac{f(x)-f(x+h)}{h}\times\lim_{h\to 0}\frac{1}{f(x)f(x+h)}$$

$$\Rightarrow \quad \frac{d}{dx}\{\phi(x)\}=-\lim_{h\to 0}\frac{f(x+h)-f(x)}{h}\times\lim_{h\to 0}\frac{1}{f(x)f(x+h)}$$

$$\Rightarrow \quad \frac{d}{dx}\{\phi(x)\}=-f'(x)\times\frac{1}{f(x)f(x)}$$

$\because$ $f(x)$ is differentiable $\Rightarrow$ $f(x)$ is continuous

$\left[\because f(x) \text{ is differentiable} \Rightarrow f(x) \text{ is continuous} \Rightarrow \lim_{h\to 0} f(x+h)=f(x)\right]$

$$\Rightarrow \quad \frac{d}{dx}\{\phi(x)\}=\frac{-f'(x)}{\{f(x)\}^2}$$

113. *(a)* I. Let $f(x)=-x$

We have, $f'(x)=\lim_{h\to 0}\frac{f(x+h)-f(x)}{h}$ (by first principle)

$$=\lim_{h\to 0}\frac{-(x+h)-(-x)}{h}$$

$$=\lim_{h\to 0}\frac{-x-h+x}{h}$$

$$\Rightarrow \quad f'(x)=\lim_{h\to 0}\frac{-h}{h}=-1$$

II. Let $f(x)=(-x)^{-1}$

$$\Rightarrow \quad f(x)=-\frac{1}{x}$$

We have, $f'(x)=\lim_{h\to 0}\frac{f(x+h)-f(x)}{h}$ (by first principle)

$$\Rightarrow \quad f'(x)=\lim_{h\to 0}\frac{-\frac{1}{x+h}+\frac{1}{x}}{h}$$

$$\Rightarrow \quad f'(x)=\lim_{h\to 0}\frac{\frac{1}{x}-\frac{1}{x+h}}{h}=\lim_{h\to 0}\frac{x+h-x}{x(x+h)h}$$

$$=\lim_{h\to 0}\frac{h}{x(x+h)h}=\frac{1}{x(x+0)}=\frac{1}{x^2}$$

114. *(b)* I. $f(x)=\sin(x+1)$

We have, $f'(x)=\lim_{h\to 0}\frac{f(x+h)-f(x)}{h}$ (by first principle)

$$f'(x)=\lim_{h\to 0}\frac{\sin(x+h+1)-\sin(x+1)}{h}$$

$[\because f(x)=\sin(x+1)]$

$\Rightarrow f'(x)$

$$=\lim_{h\to 0}\frac{2\cos\left(\frac{x+h+1+x+1}{2}\right)\sin\left(\frac{x+h+1-x-1}{2}\right)}{h}$$

$$=\lim_{h\to 0}\frac{2\cos\left(\frac{2x+h+2}{2}\right)\sin\frac{h}{2}}{2\times\frac{h}{2}}$$

$$=\cos\left(\frac{2x+0+2}{2}\right)\times 1 \quad \left(\because \lim_{h\to 0}\frac{\sin\frac{h}{2}}{\frac{h}{2}}=1\right)$$

$$\Rightarrow \quad f'(x)=\cos(x+1)$$

II. Let $f(x)=\cos\left(x-\frac{\pi}{8}\right)$

We have, $f'(x)=\lim_{h\to 0}\frac{f(x+h)-f(x)}{h}$ (by first principle)

$$\Rightarrow \quad f'(x)=\lim_{h\to 0}\frac{\cos\left(x+h-\frac{\pi}{8}\right)-\cos\left(x-\frac{\pi}{8}\right)}{h}$$

$$=\lim_{h\to 0}\frac{-2\sin\frac{x+h-\frac{\pi}{8}+x-\frac{\pi}{8}}{2}\sin\frac{x+h-\frac{\pi}{8}-x+\frac{\pi}{8}}{2}}{h}$$

$$= \lim_{h \to 0} \frac{-2\sin\dfrac{2x - 2\left(\dfrac{\pi}{8}\right) + h}{2} \sin\dfrac{h}{2}}{2 \times \dfrac{h}{2}}$$

$$= -\sin\frac{2x - 2\left(\dfrac{\pi}{8}\right) + 0}{2} \times 1 \qquad \left(\because \lim_{x \to 0} \frac{\sin\dfrac{h}{2}}{\dfrac{h}{2}} = 1\right)$$

$$= -\sin\frac{2\left(x - \dfrac{\pi}{8}\right)}{2} \Rightarrow f'(x) = -\sin\left(x - \frac{\pi}{8}\right)$$

115. *(a)* Let $y = x + a$

Differentiating y w.r.t. x, we get

$$\frac{dy}{dx} = 1 + 0 = 1$$

116. *(d)* Let $y = (px + q)\left(\frac{r}{x} + s\right)$

Differentiating y w.r.t. x, we get

$$\frac{dy}{dx} = (px + q)\frac{d}{dx}\left(\frac{r}{x} + s\right) + \left(\frac{r}{x} + s\right)\frac{d}{dx}(px + q)$$

$$= (px + q)\left[-\frac{r}{x^2} + 0\right] + \left(\frac{r}{x} + s\right)(p \times 1 + 0)$$

$$\left[\because \frac{d}{dx}\left(\frac{1}{x}\right) = \frac{d}{dx}(x^{-1}) = (-1)\,x^{-1-1} = -\frac{1}{x^2}\right]$$

$$= (px + q)\left(-\frac{r}{x^2}\right) + \left(\frac{r}{x} + s\right)p$$

$$= -\frac{pxr}{x^2} - \frac{qr}{x^2} + \frac{pr}{x} + ps = -\frac{pr}{x} - \frac{qr}{x^2} + \frac{pr}{x} + ps$$

$$\Rightarrow \frac{dy}{dx} = ps - \frac{qr}{x^2}$$

117. *(b)* Let $y = (ax + b)(cx + d)^2$

Differentiating y w.r.t. x, we get

$$\frac{dy}{dx} = (ax + b)\frac{d}{dx}(cx + d)^2 + (cx + d)^2\frac{d}{dx}(ax + b)$$

$$= (ax + b)\frac{d}{dx}(c^2x^2 + d^2 + 2cxd) + (cx + d)^2(a \times 1 + 0)$$

$$= (ax + b)(c^2(2x) + 0 + 2c \times 1 \times d) + (cx + d)^2 \times a$$

$$= (ax + b)(2c^2x + 2cd) + a(cx + d)^2$$

$$= (ax + b)2c(cx + d) + a(cx + d)^2$$

$$= (cx + d)[2c(ax + b) + a(cx + d)]$$

$$= (cx + d)(2acx + 2bc + acx + ad)$$

$$= (cx + d)(3acx + 2bc + ad)$$

118. *(c)* Let $y = \frac{ax + b}{cx + d}$

Differentiating y w.r.t. x, we get

$$\frac{dy}{dx} = \frac{(cx + d)\dfrac{d}{dx}(ax + b) - (ax + b)\dfrac{d}{dx}(cx + d)}{(cx + d)^2}$$

$$= \frac{(cx + d)(a \times 1 + 0) - (ax + b)(c \times 1 + 0)}{(cx + d)^2}$$

$$= \frac{a(cx + d) - (ax + b)c}{(cx + d)^2}$$

$$= \frac{acx + ad - acx - bc}{(cx + d)^2}$$

$$\Rightarrow \frac{dy}{dx} = \frac{ad - bc}{(cx + d)^2}$$

119. *(b)* Let $y = \frac{1 + \dfrac{1}{x}}{1 - \dfrac{1}{x}} \Rightarrow y = \frac{x + 1}{x - 1}$

Differentiating y w.r.t. x, we get

$$\frac{dy}{dx} = \frac{(x - 1)\dfrac{d}{dx}(x + 1) - (x + 1)\dfrac{d}{dx}(x - 1)}{(x - 1)^2}$$

$$= \frac{(x - 1)(1 + 0) - (x + 1)(1 - 0)}{(x - 1)^2} = \frac{x - 1 - x - 1}{(x - 1)^2}$$

$$\Rightarrow \frac{dy}{dx} = \frac{-2}{(x - 1)^2} = \frac{-2}{(1 - x)^2}$$

120. *(c)* Let $y = \frac{1}{ax^2 + bx + c}$

Differentiating y w.r.t. x, we get

$$\frac{dy}{dx} = \frac{(ax^2 + bx + c)\dfrac{d}{dx}(1) - 1 \times \dfrac{d}{dx}(ax^2 + bx + c)}{(ax^2 + bx + c)^2}$$

$$= \frac{(ax^2 + bx + c) \times 0 - 1 \times (2ax + b)}{(ax^2 + bx + c)^2}$$

$$\Rightarrow \frac{dy}{dx} = -\frac{(2ax + b)}{(ax^2 + bx + c)^2}$$

121. *(a)* Let $y = \frac{ax + b}{px^2 + qx + r}$

Differentiating y w.r.t. x, we get

$$\frac{dy}{dx} = \frac{(px^2 + qx + r)\dfrac{d}{dx}(ax + b) - (ax + b)\dfrac{d}{dx}(px^2 + qx + r)}{(px^2 + qx + r)^2}$$

$$= \frac{(px^2 + qx + r)(a \times 1 + 0) - (ax + b)(2px + q)}{(px^2 + qx + r)^2}$$

$$= \frac{(px^2 + qx + r)a - (ax + b)(2px + q)}{(px^2 + qx + r)^2}$$

$$= \frac{(apx^2 + aqx + ra) - (2apx^2 + aqx + 2bpx + qb)}{(px^2 + qx + r)^2}$$

$$= \frac{apx^2 + aqx + ra - 2apx^2 - aqx - 2bpx - bq}{(px^2 + qx + r)^2}$$

$$= \frac{-apx^2 - 2bpx + ra - bq}{(px^2 + qx + r)^2}$$

122. *(d)* Let $y = \frac{px^2 + qx + r}{(ax + b)}$

Differentiating y w.r.t. x, we get

$$\frac{dy}{dx} = \frac{(ax + b)\dfrac{d}{dx}(px^2 + qx + r) - (px^2 + qx + r)\dfrac{d}{dx}(ax + b)}{(ax + b)^2}$$

$$= \frac{(ax + b)(2px + q + 0) - (px^2 + qx + r)(a \times 1 + 0)}{(ax + b)^2}$$

$$= \frac{(ax + b)(2px + q) - (px^2 + qx + r)a}{(ax + b)^2}$$

$$= \frac{(2apx^2 + aqx + 2bpx + bq) - (apx^2 + aqx + ra)}{(ax + b)^2}$$

$$= \frac{2apx^2 + aqx + 2bpx + bq - apx^2 - aqx - ra}{(ax + b)^2}$$

$$\Rightarrow \quad \frac{dy}{dx} = \frac{apx^2 + 2bpx + bq - ra}{(ax + b)^2}$$

123. *(b)* Let $y = \frac{a}{x^4} - \frac{b}{x^2} + \cos x$

$$\Rightarrow \quad y = a\,x^{-4} - bx^{-2} + \cos x$$

Differentiating y w.r.t. x, we get

$$\frac{dy}{dx} = a\frac{d}{dx}(x^{-4}) - b\frac{d}{dx}(x^{-2}) + \frac{d}{dx}(\cos x)$$

$$= a(-4)x^{-4-1} - b(-2)x^{-2-1} - \sin x$$

$$= -\frac{4a}{x^5} + \frac{2b}{x^3} - \sin x \qquad \left[\because \frac{d}{dx}(x^n) = nx^{n-1}\right]$$

124. *(c)* Let $y = 4\sqrt{x} - 2 \Rightarrow y = 4x^{1/2} - 2$

Differentiating y w.r.t. x, we get

$$\frac{dy}{dx} = 4\cdot\frac{1}{2}x^{\frac{1}{2}-1} - 0$$

$$= 2x^{\frac{-1}{2}} = \frac{2}{\sqrt{x}}$$

125. *(b)* Let $y = (ax + b)^n$

Differentiating y w.r.t. x, we get

$$\Rightarrow \quad \frac{dy}{dx} = n(ax + b)^{n-1}\frac{d}{dx}(ax + b)$$

$$= n(ax + b)^{n-1}a$$

$$\Rightarrow \quad \frac{dy}{dx} = na(ax + b)^{n-1}$$

126. *(d)* Let $y = (ax + b)^n (cx + d)^m$

Differentiating y w.r.t. x, we get

$$\frac{dy}{dx} = (ax + b)^n \frac{d}{dx}(cx + d)^m + (cx + d)^m \frac{d}{dx}(ax + b)^n$$

$$= (ax + b)^n m(cx + d)^{m-1}\frac{d}{dx}(cx + d) + (cx + d)^m n(ax + b)^{n-1}\frac{d}{dx}(ax + b)$$

$$= m(ax + b)^n (cx + d)^{m-1}(c \times 1 + 0) + n(cx + d)^m (ax + b)^{n-1}(a \times 1 + 0)$$

$$= m(ax + b)^n (cx + d)^{m-1}c + n(cx + d)^m (ax + b)^{n-1}a$$

$$\Rightarrow \quad \frac{dy}{dx} = (ax + b)^{n-1}(cx + d)^{m-1}[mc(ax + b) + na(cx + d)]$$

127. *(b)* I. Let $y = \sin(x + a)$

$$y = \sin x\cos a + \cos x\sin a$$

$$[\because \sin(A + B) = \sin A\cos B + \cos A\sin B]$$

Differentiating y w.r.t. x, we get

$$\frac{dy}{dx} = \cos a\frac{d}{dx}(\sin x) + \sin a\frac{d}{dx}(\cos x)$$

$$= \cos a\cos x - \sin a\sin x = \cos(x + a)$$

II. Let $y = \operatorname{cosec} x\cot x$

Differentiating y w.r.t. x, we get

$$\frac{dy}{dx} = \operatorname{cosec} x\frac{d}{dx}(\cot x) + \cot x\frac{d}{dx}(\operatorname{cosec} x)$$

$$= -\operatorname{cosec} x\operatorname{cosec}^2 x + \cot x(-\operatorname{cosec} x\cot x)$$

$$= -\operatorname{cosec}^3 x - \cot^2 x\operatorname{cosec} x$$

128. *(d)* Let $y = \frac{\sin x + \cos x}{\sin x - \cos x}$

Differentiating y w.r.t. x, we get

$$\frac{dy}{dx} = \frac{\left[(\sin x - \cos x)\frac{d}{dx}(\sin x + \cos x) - (\sin x + \cos x)\frac{d}{dx}(\sin x - \cos x)\right]}{(\sin x - \cos x)^2}$$

(by quotient rule)

$$= \frac{\left[(\sin x - \cos x)(\cos x - \sin x) - (\sin x + \cos x)(\cos x + \sin x)\right]}{(\sin x - \cos x)^2}$$

$$= \frac{-(\cos x - \sin x)(\cos x - \sin x) - (\cos x + \sin x)^2}{(\sin x - \cos x)^2}$$

$$= \frac{-(\cos x - \sin x)^2 - (\cos x + \sin x)^2}{(\sin x - \cos x)^2}$$

$$= \frac{-[(\cos^2 x + \sin^2 x - 2\cos x\sin x) + (\cos^2 x + \sin^2 x + 2\cos x\sin x)]}{(\sin x - \cos x)^2}$$

$$= \frac{-[1 + 1]}{(\sin x - \cos x)^2} = \frac{-2}{(\sin x - \cos x)^2}$$

129. *(a)* Let $y = \frac{\sec x - 1}{\sec x + 1}$

Differentiating y w.r.t. x, we get

$$\frac{dy}{dx} = \frac{(\sec x + 1)\frac{d}{dx}(\sec x - 1) - (\sec x - 1)\frac{d}{dx}(\sec x + 1)}{(\sec x + 1)^2}$$

$$= \frac{\left[(\sec x + 1)(\sec x\tan x - 0) - (\sec x - 1)(\sec x\tan x + 0)\right]}{(\sec x + 1)^2}$$

$$= \frac{(\sec x + 1)(\sec x\tan x) - (\sec x - 1)(\sec x\tan x)}{(\sec x + 1)^2}$$

$$= \frac{\sec x\tan x[\sec x + 1 - \sec x + 1]}{(\sec x + 1)^2} = \frac{2\sec x\tan x}{(\sec x + 1)^2}$$

130. *(c)* Let $y = \sin^n x \Rightarrow y = (\sin x)^n$

Differentiating y w.r.t.x, we get

$$\frac{dy}{dx} = n(\sin x)^{n-1}\frac{d}{dx}(\sin x)$$

$$\Rightarrow \quad \frac{dy}{dx} = n(\sin x)^{n-1}\cos x$$

131. *(d)* Let $y = \dfrac{a + b\sin x}{c + d\cos x}$

Differentiating y w.r.t. x, we get

$$\frac{dy}{dx} = \frac{\left[\begin{array}{l}(c + d\cos x)\dfrac{d}{dx}(a + b\sin x)\\ -(a + b\sin x)\dfrac{d}{dx}(c + d\cos x)\end{array}\right]}{(c + d\cos x)^2}$$

$$= \frac{(c + d\cos x)[0 + b\cos x] - (a + b\sin x)(0 - d\sin x)}{(c + d\cos x)^2}$$

$$= \frac{(c + d\cos x)(b\cos x) - (a + b\sin x)(-d\sin x)}{(c + d\cos x)^2}$$

$$= \frac{bc\cos x + bd\cos^2 x + ad\sin x + bd\sin^2 x}{(c + d\cos x)^2}$$

$$= \frac{bc\cos x + ad\sin x + bd(\cos^2 x + \sin^2 x)}{(c + d\cos x)^2}$$

$$\Rightarrow \frac{dy}{dx} = \frac{bc\cos x + ad\sin x + bd}{(c + d\cos x)^2}$$

132. *(a)* Let $y = \dfrac{\sin(x + a)}{\cos x} = \dfrac{\sin x\cos a + \cos x\sin a}{\cos x}$

$(\because \sin(A + B) = \sin A\cos B + \cos A\sin B)$

$$= \frac{\sin x\cos a}{\cos x} + \frac{\cos x\sin a}{\cos x} = \cos a\tan x + \sin a$$

Differentiating y w.r.t. x, we get

$$\frac{dy}{dx} = \cos a\frac{d}{dx}(\tan x) + \frac{d}{dx}(\sin a)$$

$$= \cos a\sec^2 x + 0 = \frac{\cos a}{\cos^2 x}$$

133. *(b)* Let $y = x^4(5\sin x - 3\cos x)$

Differentiating y w.r.t. x, we get

$$\frac{dy}{dx} = x^4\frac{d}{dx}(5\sin x - 3\cos x) + (5\sin x - 3\cos x)\frac{d}{dx}(x^4)$$

(by product rule)

$$= x^4(5\cos x + 3\sin x) + (5\sin x - 3\cos x)4x^3$$

$$= x^3[x(5\cos x + 3\sin x) + 4(5\sin x - 3\cos x)]$$

$$= x^3[5x\cos x + 3x\sin x + 20\sin x - 12\cos x]$$

134. *(d)* Let $y = (x^2 + 1)\cos x$,

Differentiating y w.r.t x, we get

$$\frac{dy}{dx} = (x^2 + 1)\frac{d}{dx}(\cos x) + \cos x\frac{d}{dx}(x^2 + 1) \text{ (by product rule)}$$

$$= (x^2 + 1)(-\sin x) + \cos x(2x)$$

$$= -x^2\sin x - \sin x + 2x\cos x$$

135. *(a)* Let $y = (ax^2 + \sin x)(p + q\cos x)$

Differentiating y w.r.t. x, we get

$$\frac{dy}{dx} = (ax^2 + \sin x)\frac{d}{dx}(p + q\cos x) + (p + q\cos x)\frac{d}{dx}(ax^2 + \sin x)$$

(by product rule)

$$= (ax^2 + \sin x)(0 - q\sin x) + (p + q\cos x)(2ax + \cos x)$$

$$= -q\sin x(ax^2 + \sin x) + (p + q\cos x)(2ax + \cos x)$$

136. *(b)* Let $y = (x + \cos x)(x - \tan x)$

Differentiating y w.r.t. x, we get

$$\frac{dy}{dx} = (x + \cos x)\frac{d}{dx}(x - \tan x) + (x - \tan x)\frac{d}{dx}(x + \cos x)$$

(by product rule)

$$= (x + \cos x)(1 - \sec^2 x) + (x - \tan x)(1 - \sin x)$$

$$= -(x + \cos x)(\sec^2 x - 1) + (x - \tan x)(1 - \sin x)$$

$$= -(x + \cos x)\tan^2 x + (x - \tan x)(1 - \sin x)$$

$(\because \sec^2 x - \tan^2 x = 1)$

137. *(c)* Let $y = \dfrac{4x + 5\sin x}{3x + 7\cos x}$

Differentiating y w.r.t. x, we get

$$\frac{dy}{dx} = \frac{\left[\begin{array}{l}(3x + 7\cos x)\dfrac{d}{dx}(4x + 5\sin x)\\ -(4x + 5\sin x)\dfrac{d}{dx}(3x + 7\cos x)\end{array}\right]}{(3x + 7\cos x)^2}$$

(by quotient rule)

$$= \frac{(3x + 7\cos x)(4 \times 1 + 5\cos x) - (4x + 5\sin x)(3 - 7\sin x)}{(3x + 7\cos x)^2}$$

$$= \frac{\left[\begin{array}{r}12x + 15x\cos x + 28\cos x + 35\cos^2 x - 12x\\ + 28x\sin x - 15\sin x + 35\sin^2 x\end{array}\right]}{(3x + 7\cos x)^2}$$

$$= \frac{\left[\begin{array}{r}35(\cos^2 x + \sin^2 x) + 15x\cos x + 28\cos x\\ + 28x\sin x - 15\sin x\end{array}\right]}{(3x + 7\cos x)^2}$$

$$= \frac{35 + 15x\cos x + 28\cos x + 28x\sin x - 15\sin x}{(3x + 7\cos x)^2}$$

138. *(b)* Let $y = \dfrac{x^2\cos\left(\dfrac{\pi}{4}\right)}{\sin x} \Rightarrow y = \left(\cos\dfrac{\pi}{4}\right)\dfrac{x^2}{\sin x}$

Differentiating y w.r.t x, we get

$$\frac{dy}{dx} = \cos\frac{\pi}{4}\cdot\frac{d}{dx}\left(\frac{x^2}{\sin x}\right)$$

$$\Rightarrow \frac{dy}{dx} = \left(\cos\frac{\pi}{4}\right)\frac{\sin x\dfrac{d}{dx}(x^2) - x^2\left(\dfrac{d}{dx}\sin x\right)}{\sin^2 x}$$

(by quotient rule)

$$= \left(\cos\frac{\pi}{4}\right)\frac{(\sin x)(2x) - x^2(\cos x)}{\sin^2 x}$$

$$= \frac{x\cos\dfrac{\pi}{4}[2\sin x - x\cos x]}{\sin^2 x}$$

139. *(c)* Let $y = \dfrac{x}{1 + \tan x}$

Differentiating y w.r.t. x, we get

$$\frac{dy}{dx} = \frac{(1 + \tan x)\dfrac{d}{dx}(x) - x\dfrac{d}{dx}(1 + \tan x)}{(1 + \tan x)^2}$$

$$= \frac{(1 + \tan x)(1) - x(0 + \sec^2 x)}{(1 + \tan x)^2} = \frac{1 + \tan x - x\sec^2 x}{(1 + \tan x)^2}$$

140. *(c)* Let $y = (x + \sec x)(x - \tan x)$

Differentiating y w.r.t. x, we get

$$\frac{dy}{dx} = (x + \sec x)\frac{d}{dx}(x - \tan x) + (x - \tan x)\frac{d}{dx}(x + \sec x)$$
(by product rule)
$$= (x + \sec x)(1 - \sec^2 x) + (x - \tan x)(1 + \sec x \tan x)$$

141. *(d)* Let $y = \dfrac{x}{\sin^n x}$

Differentiating y w.r.t x, we get

$$\frac{dy}{dx} = \frac{\sin^n x \frac{d}{dx}(x) - x\frac{d}{dx}(\sin^n x)}{(\sin^n x)^2}$$ (by quotient rule)

$$= \frac{\sin^n x \times 1 - x\frac{d}{dx}(\sin x)^n}{\sin^{2n} x}$$

$$= \frac{\sin^n x - nx(\sin x)^{n-1}\frac{d}{dx}(\sin x)}{\sin^{2n} x}$$

[differentiating $(\sin x)^n$ by chain rule]

$$= \frac{\sin^n x - nx\sin^{n-1} x\cos x}{\sin^{2n} x}$$

142. *(a)* We have, $f(x) = x - [x],\ x \in R$

Now $$Lf'\left(\frac{1}{2}\right) = \lim_{h\to 0}\frac{f\left(\frac{1}{2} - h\right) - f\left(\frac{1}{2}\right)}{-h}$$

$$= \lim_{h\to 0}\frac{\left(\frac{1}{2} - h\right) - \left[\frac{1}{2} - h\right] - \left(\frac{1}{2} - \left[\frac{1}{2}\right]\right)}{-h}$$

$$= \lim_{h\to 0}\frac{\frac{1}{2} - h - 0 - \frac{1}{2} + 0}{-h}$$

$$= \lim_{h\to 0}\left(\frac{-h}{-h}\right) = \lim_{h\to 0} 1 = 1$$

and $$Rf'\left(\frac{1}{2}\right) = \lim_{h\to 0}\frac{f\left(\frac{1}{2} + h\right) - f\left(\frac{1}{2}\right)}{h}$$

$$= \lim_{h\to 0}\frac{\left(\frac{1}{2} + h\right) - \left[\frac{1}{2} + h\right] - \left(\frac{1}{2} - \left[\frac{1}{2}\right]\right)}{h}$$

$$= \lim_{h\to 0}\frac{\frac{1}{2} + h - 0 - \frac{1}{2} + 0}{h}$$

$$= \lim_{h\to 0}\left(\frac{h}{h}\right) = \lim_{h\to 0} 1 = 1$$

$$\therefore f'\left(\frac{1}{2}\right) = 1$$

143. *(a)* Given, $f(x) = \begin{cases} x^2 - 1, & 0 < x < 2 \\ 2x + 3, & 2 \le x < 3 \end{cases}$

At $x = 2$, $\text{RHL} = \lim_{x\to 2^+} f(x) = \lim_{h\to 0} f(2 + h)$

$$= \lim_{h\to 0} 2(2 + h) + 3$$
$$= 2(2 + 0) + 3$$
$$= 4 + 3 = 7 = \alpha \text{ (say)}$$

and $$\text{LHL} = \lim_{x\to 2^-} f(x) = \lim_{h\to 0} f(2 - h)$$
$$= \lim_{h\to 0}(2 - h)^2 - 1 = (2 - 0)^2 - 1$$
$$= 4 - 1 = 3 = \beta\text{(say)}$$

If a quadratic equation has roots α and β, then the equation is

$$x^2 - (\text{Sum of roots})\,x + \text{Product of roots} = 0$$

i.e., $x^2 - (\alpha + \beta)x + \alpha\beta = 0$

i.e., $x^2 - (7 + 3)x + 7\cdot 3 = 0$

$\Rightarrow$ $x^2 - 10x + 21 = 0$

144. *(d)* We have,

$$f(x) = \begin{cases} \dfrac{\sin[x]}{[x]}, & [x] \ne 0 \\ 0, & [x] = 0 \end{cases}$$

At $x = 0$, $\text{RHL} = \lim_{x\to 0^+} f(x) = \lim_{h\to 0} f(0 + h) = 0$

$$\left(\because \lim_{h\to 0}[0 + h] = 0\right)$$

and $$\text{LHL} = \lim_{x\to 0^-} f(x) = \lim_{x\to 0} f(0 - h)$$
$$= \lim_{h\to 0}\frac{\sin[0 - h]}{[0 - h]} \quad \left(\because \lim_{h\to 0}[0 - h] \ne 0\right)$$
$$= \frac{\sin(-1)}{(-1)}$$
$$= -\frac{\sin(1)}{-1} = \sin(1)$$

$\because$ RHL $\ne$ LHL

$\therefore$ $\lim_{x\to 0} f(x)$ does not exist.

145. *(a)* We have, $\lim_{x\to 0}\dfrac{\operatorname{cosec} x - \cot x}{x}$

$$= \lim_{x\to 0}\frac{\frac{1}{\sin x} - \frac{\cos x}{\sin x}}{x} = \lim_{x\to 0}\frac{1 - \cos x}{x\sin x}$$

$$= \lim_{x\to 0}\frac{2\sin^2\frac{x}{2}}{x\left(2\sin\frac{x}{2}\cos\frac{x}{2}\right)} \quad \left(\because 1 - \cos x = 2\sin^2\frac{x}{2} \text{ and } \sin x = 2\sin\frac{x}{2}\cos\frac{x}{2}\right)$$

$$= \lim_{x\to 0}\frac{\sin\frac{x}{2}}{x\cos\frac{x}{2}} = \lim_{x\to 0}\frac{\tan\frac{x}{2}}{2\cdot\frac{x}{2}} = \frac{1}{2}\times 1$$

$$= \frac{1}{2} \quad \left(\because \lim_{x\to 0}\frac{\tan x}{x} = 1\right)$$

146. *(b)* Given, $f(x) = \begin{cases} x + 2, & x \le -1 \\ cx^2, & x > -1 \end{cases}$

At $x = -1$, limit exists

$\therefore$ RHL = LHL

$$\Rightarrow \lim_{x\to -1^+} f(x) = \lim_{x\to -1^-} f(x)$$

$\Rightarrow \quad \lim_{h\to 0} f(-1+h) = \lim_{h\to 0} f(-1-h)$

$\Rightarrow \quad \lim_{h\to 0} c\,(-1+h)^2 = \lim_{h\to 0} (-1-h+2)$

$\Rightarrow \quad c\,(-1+0)^2 = 1-0$

$\Rightarrow \quad c = 1$

147. *(c)* Given, $f(x) = \begin{cases} \dfrac{k\cos x}{\pi - 2x}, & x \neq \dfrac{\pi}{2} \\ 3, & x = \dfrac{\pi}{2} \end{cases}$

Since, $\lim_{x\to\frac{\pi}{2}} f(x) = f\left(\frac{\pi}{2}\right)$

$\Rightarrow \quad \lim_{x\to\frac{\pi}{2}} f(x) = 3$

$\Rightarrow \quad \lim_{h\to 0} f\left(\frac{\pi}{2}+h\right) = 3$

$\Rightarrow \quad \lim_{h\to 0} \dfrac{k\cos\left(\frac{\pi}{2}+h\right)}{\pi - 2\left(\frac{\pi}{2}+h\right)} = 3$

$\Rightarrow \quad \lim_{h\to 0} \dfrac{-k\sin h}{\pi - \pi - 2h} = 3$

$\Rightarrow \quad \lim_{h\to 0} \dfrac{-k\sin h}{-2h} = 3$

$\Rightarrow \quad \dfrac{k}{2} \times \lim_{h\to 0} \dfrac{\sin h}{h} = 3$

$\Rightarrow \quad \dfrac{k}{2} \times 1 = 3$

$\Rightarrow \quad k = 6 \qquad \left(\because \lim_{h\to 0} \dfrac{\sin h}{h} = 1\right)$

148. *(d)* Let $f(x) = \dfrac{|x-4|}{x-4}$

At $x = 4$,

$\text{RHL} = \lim_{x\to 4^+} f(x) = \lim_{h\to 0} f(4+h) = \lim_{h\to 0} \dfrac{|4+h-4|}{(4+h-4)}$

$= \lim_{h\to 0} \dfrac{(4+h-4)}{(4+h-4)} = 1$

At $x = 4$, $\text{LHL} = \lim_{x\to 4^-} f(x) = \lim_{h\to 0} f(4-h)$

$= \lim_{h\to 0} \dfrac{|4-h-4|}{(4-h-4)} = \lim_{h\to 0} \dfrac{-(4-h-4)}{(4-h-4)} = -1$

$\because$ RHL $\neq$ LHL

$\therefore$ Hence, at $x = 4$, limit does not exist.

Note *In an absolute function, sometimes it may be possible that LHL ≠ RHL. Please be careful about it.*

149. *(b)* Given $\lim_{x\to\pi} \dfrac{1-\sin\frac{x}{2}}{\cos\frac{x}{2}\left(\cos\frac{x}{4} - \sin\frac{x}{4}\right)}$

Multiplying and dividing by $\left(1+\sin\frac{x}{2}\right)$, we get

$$\lim_{x\to\pi} \frac{\left(1-\sin\frac{x}{2}\right)}{\cos\frac{x}{2}\left(1+\sin\frac{x}{2}\right)} \frac{\left(1+\sin\frac{x}{2}\right)}{\left(\cos\frac{x}{4}-\sin\frac{x}{4}\right)}$$

Multiplying and dividing by $\left(\cos\frac{x}{4}+\sin\frac{x}{4}\right)$, we get

$$\lim_{x\to\pi} \frac{\left(1-\sin^2\frac{x}{2}\right)}{\cos\frac{x}{2}\left(\cos\frac{x}{4}-\sin\frac{x}{4}\right)\left(1+\sin\frac{x}{2}\right)} \cdot \frac{\left(\cos\frac{x}{4}+\sin\frac{x}{4}\right)}{\left(\cos\frac{x}{4}+\sin\frac{x}{4}\right)}$$

$$= \lim_{x\to\pi} \frac{\left(\cos^2\frac{x}{2}\right)\left(\cos\frac{x}{4}+\sin\frac{x}{4}\right)}{\cos\frac{x}{2}\left(\cos^2\frac{x}{4}-\sin^2\frac{x}{4}\right)\left(1+\sin\frac{x}{2}\right)}$$

$(\because 1-\sin^2 x = \cos^2 x)$

$$= \lim_{x\to\pi} \frac{\cos\frac{x}{2}\left(\cos\frac{x}{4}+\sin\frac{x}{4}\right)}{\cos\frac{2x}{4}\left(1+\sin\frac{x}{2}\right)}$$

$$= \frac{\cos\frac{\pi}{4}+\sin\frac{\pi}{4}}{1+\sin\frac{\pi}{2}} = \frac{\frac{1}{\sqrt{2}}+\frac{1}{\sqrt{2}}}{1+1}$$

$$= \frac{1}{\sqrt{2}}$$

14

Mathematical Reasoning

Statements

Level I

1. In mathematical language, the reasoning is of ...A... types. Here, A refers to

(a) two (b) three
(c) one (d) four

2. The basic unit involved in mathematical reasoning is a mathematical ...A... .

Here, A refers to

(a) sentence
(b) statement
(c) identity
(d) equation

3. Consider the following sentences

I. In 2003, the President of India was a woman.
II. An elephant weighs more than a human being.
III. Women are more intelligent than men.

Which of the following holds good?

(a) Sentence I is true
(b) Sentence II is false
(c) Sentence III is ambiguous
(d) None of the above

4. A...B... is mathematically acceptable statement, if it is either true or false but not both.

Here, B refers to

(a) statement
(b) sentence
(c) not a sentence
(d) None of these

5. Match the given sentences of Column I with Column II and choose the correct option from the codes given below.

Column I	Column II
A. Two plus two equals four.	1. True
B. The sum of two positive numbers is positive.	2. False
C. All prime numbers are odd numbers.	3. Ambiguous
D. The sum of x and y is greater than 0.	

Codes

	A	B	C	D
(a)	1	1	2	3
(b)	2	3	1	1
(c)	3	1	2	2

(d) None of the above

6. Consider the following sentences

I. How beautiful!
II. Open the door.
III. Where are you going?
IV. Tomorrow is Friday.

Then, which one is a statement?

(a) I is a statement
(b) II and III are statements
(c) IV is a statement
(d) None of the above

7. Match the given sentences of Column I with Column II and choose the correct option from the codes given below.

Column I	Column II
A. She is a Mathematics graduate.	1. Statement
B. Kashmir is far from here.	2. Not a statement
C. There are 40 days in a month.	

Codes

	A	B	C
(a)	2	1	1
(b)	2	2	1
(c)	1	1	2
(d)	2	2	2

8. The statements are always denoted by the ...A... . Here, A refers to

(a) small letters (b) capital letters
(c) numbers (d) None of these

9. Assertion (A) The sentence "8 is less than 6" is a statement.

Reason (R) A sentence is called a statement, if it is either true or false but not both.

(a) Both A and R are correct; R is the correct explanation of A
(b) Both A and R are correct; R is not the correct explanation of A
(c) A is correct; R is incorrect
(d) R is correct; A is incorrect

10. Match the given sentences of Column I with Column II and choose the correct option from the codes given below.

Column I	Column II
A. Every set is a finite set.	1. Statement
B. The Sun is a star.	2. Not a statement
C. Mathematics is fun.	
D. There is no rain without clouds.	
E. How far is Chennai from here?	

Codes

	A	B	C	D	E
(a)	2	1	2	2	2
(b)	1	1	2	1	2
(c)	1	2	1	2	1
(d)	1	1	2	2	2

11. The sentence "There are 35 days in a month" is

(a) a statement (b) not a statement
(c) may be statement or not (d) None of the above

12. Assertion (A) "Mathematics is difficult", is a statement.

Reason (R) A sentence is a statement, if it is either true or false but not both.

(a) Both A and R are correct; R is the correct explanation of A
(b) Both A and R are correct; R is not the correct explanation of A
(c) A is correct; R is incorrect
(d) R is correct; A is incorrect

13. Match the given sentences of Column I with Column II and choose the correct option from the codes given below.

Column I	Column II
A. The sum of 5 and 7 is greater than 10.	1. Statement
B. The square of a number is an even number.	2. Not a statement
C. The sides of a quadrilateral have equal length.	
D. Answer this question.	

Codes

	A	B	C	D
(a)	1	2	1	1
(b)	1	2	2	1
(c)	2	2	1	2
(d)	1	2	2	2

14. Consider the following sentence and then choose the answer.

"The product of (−1) and 8 is 8".

(a) True (b) False
(c) May be true or false (d) None of these

15. The sentence "The sum of all interior angles of a triangle is 180°", is

(a) true (b) false
(c) may be true or false (d) None of these

16. Assertion (A) The sentence, "Today is a windy day", is not a statement.

Reason (R) A sentence is called a statement, if it is either true or false.

(a) Both A and R are correct; R is the correct explanation of A
(b) Both A and R are correct; R is not the correct explanation of A
(c) A is correct; R is incorrect
(d) R is correct; A is incorrect

17. The sentence "All real numbers are complex numbers" is ...A... . Here, A refers to

(a) true (b) false
(c) may be true or false (d) None of these

18. Which of the following is a statement?

(a) Everyone in this room is bold
(b) She is an engineering student
(c) $\sin^2 \theta$ is greater than 1/2
(d) Three plus three equals six

Level II

19. Which of the following is a statement? [NCERT Exemplar]

(a) x is a real number (b) Switch off the fan
(c) 6 is a natural number (d) Let me go

20. Which of the following is not a statement? [NCERT Exemplar]

(a) Smoking is injurious to health
(b) $2+2=4$
(c) 2 is the only even prime number
(d) Come here

21. Match the sentences of Column I with Column II and choose the correct option from the codes given below.

Column I	Column II
A. The angles opposite to equal sides of a triangle are equal.	1. Statement
B. The Moon is a satellite of Earth .	2. Not a statement
C. May God bless you!	
D. Asia is a continent.	
E. How are you?	

[NCERT Exemplar; HOTS]

Codes

	A	B	C	D	E		A	B	C	D	E
(a)	1	1	2	1	2	(b)	1	2	1	1	2
(c)	2	2	1	2	1	(d)	1	2	1	2	1

22. The sentence "New Delhi is in India", is [NCERT Exemplar]

(a) a statement
(b) not a statement
(c) may be statement or not
(d) None of the above

23. Assertion (A) "Every rectangle is a square" is a statement.

Reason (R) "Close the door" is not a statement. [NCERT Exemplar]

(a) Both A and R are correct; R is the correct explanation of A
(b) Both A and R are correct; R is not the correct explanation of A
(c) A is correct; R is incorrect
(d) R is correct; A is incorrect

24. Assertion (A) "How old are you" is not a statement.

Reason (R) "x is a natural number" is a statement.

(a) Both A and R are correct; R is the correct explanation of A
(b) Both A and R are correct; R is not the correct explanation of A
(c) A is correct; R is incorrect
(d) R is correct; A is incorrect

25. Which of the following is/are not a statement? [HOTS]

I. Earth is a planet.
II. Plants are living objects.
III. $\sqrt{-3}$ is a rational number.
IV. $x^2 - 5x + 6 < 0$, when $x \in -R$.

(a) I and II (b) II and IV
(c) III and I (d) None of these

26. Which of the following statement is false? [HOTS]

(a) A quadratic equation has always a real root
(b) The number of ways of seating 2 persons in two chairs out of n persons is $P(n,2)$
(c) The cube roots of unity are in GP
(d) None of the above

27. Assertion (A) "Washington DC is in America" is a statement.

Reason (R) "Moon revolves around the Earth" is not a statement.

(a) Both A and R are correct; R is the correct explanation of A
(b) Both A and R are correct; R is not the correct explanation of A
(c) A is correct; R is incorrect
(d) R is correct; A is incorrect

28. Assertion (A) "Two plus three is five" is not a statement.

Reason (R) "Every square is a rectangle" is a statement.

(a) Both A and R are correct; R is the correct explanation of A
(b) Both A and R are correct; R is not the correct explanation of A
(c) A is correct; R is incorrect
(d) R is correct; A is incorrect

29. " The sun is a star" is a ...X... . Here, X refers to

(a) sentence (b) statement
(c) Both (a) and (b) (d) None of these

30. Match the sentences of Column I with Column II and choose the correct option from the codes given below.

Column I	Column II
A. New Delhi is in Nepal.	1. Not a statement
B. Every relation is a function.	2. Statement
C. Do your home work.	
D. Please do me a favour.	

Codes

	A	B	C	D		A	B	C	D
(a)	1	1	2	2	(b)	2	2	1	1
(c)	1	2	1	2	(d)	1	2	2	1

31. "Paris is in England" is a ...X... . Here, X refers to

(a) sentence (b) not a sentence
(c) statement (d) Both (a) and (c)

New Statements from Old

Level I

32. The denial of a statement is called the ...A... of the statement. Here, A refers to

(a) sentence (b) ambiguous
(c) negation (d) None of these

33. Consider the statement : New Delhi is a city.
Then, which of the following is not a negation of this statement?

(a) It is not the case that New Delhi is a city
(b) It is false that New Delhi is a city
(c) New Delhi is not a city
(d) It is true that New Delhi is a city

34. The negation of a statement is said to be a ...X... . Here, X refers to

(a) sentence (b) statement
(c) not a sentence (d) ambiguous

35. **Assertion** (A) 'It is false that both the diagonals in a rectangle have the same length' is a negation of the statement "Both the diagonal of a rectangle have the same length".

Reason (R) The negation of a statement is also a statement.

(a) Both A and R are correct; R is the correct explanation of A
(b) Both A and R are correct; R is not the correct explanation of A
(c) A is correct; R is incorrect
(d) R is correct; A is incorrect

36. **Assertion** (A) " $\sqrt{7}$ is not rational" is a negation of the statement "$\sqrt{7}$ is rational".

Reason (R) The negation of a statement is not a statement.

(a) Both A and R are correct; R is the correct explanation of A
(b) Both A and R are correct; R is not the correct explanation of A
(c) A is correct; R is incorrect
(d) R is correct; A is incorrect

37. Many mathematical statements are obtained by combining one or more statements using some ...X... words like "and" "or", etc. Here, X refers to

(a) separating (b) non-connecting
(c) connecting (d) adding

38. A statement which is made up of two or more statements is called a

(a) sentence (b) component statement
(c) compound statement (d) None of these

39. **Assertion** (A) "Chennai is not the capital of Tamil Nadu" is a negation of the statement "Chennai is the capital of Tamil Nadu".

Reason (R) The denial of a statement is called the negation of the statement.

(a) Both A and R are correct; R is the correct explanation of A
(b) Both A and R are correct; R is not the correct explanation of A
(c) A is correct; R is incorrect
(d) R is correct; A is incorrect

40. The negation of the statement "$\sqrt{2}$ is not a complex number" is

(a) $\sqrt{2}$ is a rational number (b) $\sqrt{2}$ is an irrational number
(c) $\sqrt{2}$ is a complex number (d) None of the above

41. The negation of the statement "All triangles are not equilateral triangle" is

(a) all triangles are closed curve
(b) all triangles are isosceles triangle
(c) all triangles are similar
(d) all triangles are equilateral triangle

42. **Assertion** (A) The following pairs of statements are negations of each other

The numbers x is not a rational number.
The number x is not an irrational number.

Reason (R) The denial of a statement is called the negation of the statement.

(a) Both A and R are correct; R is the correct explanation of A
(b) Both A and R are correct; R is not the correct explanation of A
(c) A is correct; R is incorrect
(d) R is correct; A is incorrect

43. The negation of the statement "the number x is rational number" is ...P... . Here, P refers to

(a) the number x is a real
(b) the number x is not real
(c) the number x is odd
(d) the number x is an irrational number

44. Match the compound statements in Column I with their component statements in Column II and choose the correct option from the codes given below.

Column I	Column II
A. Number 3 is prime or it is odd.	1. p : All integers are positive. q : All integers are negative.
B. All integers are positive or negative.	2. p : Number 3 is prime. q : Number 3 is odd.

	Column I		Column II
C.	100 is divisible by 3, 11 and 5.	3.	p : 100 is divisible by 3. q : 100 is divisible by 11. r : 100 is divisible by 5.

Codes

	A	B	C
(a)	1	2	3
(b)	3	1	2
(c)	1	1	2
(d)	2	1	3

Level II

45. The negation of the statement "A circle is an ellipse" is [NCERT Exemplar]

(a) an ellipse is a circle
(b) an ellipse is not a circle
(c) a circle is not an ellipse
(d) a circle is an ellipse

46. The negation of the statement " 7 is greater than 8" is [NCERT Exemplar]

(a) 7 is equal to 8
(b) 7 is not greater than 8
(c) 8 is less than 7
(d) None of these

47. Let S be a non-empty subset of R. Consider the following statement

p : There is a rational number $x \in S$ such that $x > 0$. Which of the following statement(s) is the negation of the statement p ?

(a) There is a rational number $x \in S$ such that $x \leq 0$
(b) There is no rational number $x \in S$ such that $x \leq 0$
(c) Every rational number $x \in S$ satisfies $x \leq 0$
(d) $x \in S$ and $x \leq 0 \Rightarrow x$ is not rational

48. Match the statements of Column I with their negations in Column II and choose the correct option from the codes given below:

	Column I (Statements)		Column II (Negation of statements)
A.	Bengaluru is the capital of Karnataka.	1.	It did not rain on July 4, 2005.
B.	It rained on July 4, 2005.	2.	Ravish is not honest.
C.	Ravish is honest.	3.	Bengaluru is not the capital of Karnataka.
D.	The Earth is round.	4.	The Earth is not round.

Codes

	A	B	C	D
(a)	1	2	3	4
(b)	3	1	2	4
(c)	2	3	1	4
(d)	4	3	2	1

49. The negation of the statement "101 is not a multiple of 3" is [NCERT Exemplar]

(a) 101 is a multiple of 3
(b) 101 is a multiple of 2
(c) 101 is an odd number
(d) 101 is an even number

50. Which of the following is not a negation of the statement "A natural number is greater than zero". [NCERT Exemplar; HOTS]

(a) A natural number is not greater than zero.
(b) It is false that a natural number is greater than zero
(c) It is false that a natural number is not greater than zero
(d) None of the above

51. "Roses are red and violets are blue" is a ...X... . Here, X refers to

(a) compound statement
(b) component statement
(c) Both (a) and (b)
(d) None of these

52. "The school works or a holiday is declared" is a ...P... . Here, P refers to

(a) component statement (b) compound statement
(c) Both (a) and (b) (d) None of these

53. The negation of the statement "The product of 3 and 4 is 9" is

(a) it is false that the product of 3 and 4 is 9
(b) the product of 3 and 4 is 12
(c) the product of 3 and 4 is not 12
(d) it is false that the product of 3 and 4 is not 9

54. Match the compound statements of Column I with their component statements in Column II and choose the correct option from the codes given below.

	Column I		Column II
A.	125 is a multiple of 7 or 8.	1.	p : $\sqrt{13}$ is a rational number. q : $\sqrt{13}$ is an irrational number.
B.	Mumbai is the capital of Gujarat or Maharashtra.	2.	p : 125 is a multiple of 7. q : 125 is a multiple of 8.
C.	$\sqrt{13}$ is a rational or an irrational number.	3.	p : Mumbai is the capital of Gujarat. q : Mumbai is the capital of Maharashtra.
D.	A rectangle is a quadrilateral or a 5- sided polygon.	4.	p : A rectangle is a quadrilateral. q : A rectangle is a 5-sided polygon.

Codes

	A	B	C	D		A	B	C	D
(a)	1	2	3	4	(b)	4	3	2	1
(c)	2	3	1	4	(d)	3	2	4	1

55. Consider the following statements

Statement I The negation of the statement "The number 2 is greater than 7" is "The number 2 is not greater than 7".

Statement II The negation of the statement "Every natural number is an integer" is "Every natural number is not an integer".

Choose the correct option.

(a) Only Statement I is true
(b) Only Statement II is true
(c) Both statements are true
(d) Both statements are false

Special Words/Phrases

Level I

56. Which of the following is/are connectives?

(a) Today (b) Yesterday
(c) Tomorrow (d) "And", "or"

57. **Assertion** (A) The compound statement with 'And' is true if all its component statements are true.

Reason (R) The compound statement with 'And' is false if any of its component statements is false.

(a) Both A and R are correct; R is the correct explanation of A
(b) Both A and R are correct; R is not the correct explanation of A
(c) A is correct; R is incorrect
(d) R is correct; A is incorrect

58. Match the compound statements in Column I with component statements in Column II and choose the correct option from the codes given below.

Column I	Column II
A. A line is straight and extends indefinitely in both directions.	1. Cannot be considered as a compound statement.
B. 0 is less than every positive integer and every negative integer.	2. p : A line is straight. q : A line extends indefinitely in both directions.
C. All living things have two legs and two eyes.	3. p : 0 is less than every positive integer. q : 0 is less than every negative integer.
D. A mixture of alcohol and water can be separated by chemical methods.	4. p : All living things have two legs. q : All living things have two eyes.

Codes

	A	B	C	D		A	B	C	D
(a)	1	2	3	4	(b)	2	3	4	1
(c)	3	4	2	1	(d)	4	1	2	3

59. Consider the statement "Two lines in a plane either intersect at one point or they are parallel". Then, this statement is true, if

(a) two lines do not intersect
(b) lines are not parallel
(c) either lines intersect or parallel
(d) None of the above

60. **Statement I** For the statement "An ice-cream or pepsi is available with thali in a restaurant", "or" is exclusive.

Statement II For the statement "A student who has taken Biology or Chemistry can apply for M.Sc. Microbiology programme","or" is inclusive.

(a) Statement I is true; Statement II is true; Statement II is a correct explanation for Statement I
(b) Statement I is true; Statement II is true; Statement II is not a correct explanation for Statement I
(c) Statement I is true; Statement II is false
(d) Statement I is false; Statement II is true

61. Match the statements of Column I with Column II and choose the correct option from the codes given below.

Column I	Column II
A. To enter a country, you need a passport or a voter registration card.	1. Exclusive "or"
B. The school is closed, if it is a holiday or a Sunday.	2. Inclusive "or"
C. Two lines intersect at a point or are parallel.	
D. Students can take French or Sanskrit as their third language.	

Codes

	A	B	C	D		A	B	C	D
(a)	1	1	2	2	(b)	1	2	1	2
(c)	2	2	1	1	(d)	1	2	2	2

62. **Statement I** A compound statement with an "or" is true when one component statement is true or both the component statements are true.

Statement II A compound statement with an "or" is false when both the component statements are false.

(a) Statement I is true; Statement II is true; Statement II is a correct explanation for Statement I
(b) Statement I is true; Statement II is true; Statement II is not a correct explanation for Statement I
(c) Statement I is true; Statement II is false
(d) Statement I is false; Statement II is true

63. Match the compound statements of Column I with Column II and choose the correct option from the codes given below.

	Column I		Column II
A.	The school is closed, if there is a holiday or Sunday.	1.	False
B.	Mumbai is the capital of Kolkata or Karnataka.	2.	True

Codes

	A	B
(a)	1	1
(b)	2	1
(c)	2	2
(d)	1	2

64. Match the statements of Column I with Column II and Column III and choose the correct option from the codes given below.

	Column I		Column II	Column III
(i)	$\sqrt{2}$ is a rational number or an irrational number.	A.	Inclusive "or"	(p) False
(ii)	To enter into a public library, children need an identity card from the school or a letter from the school authorities.	B.	Exclusive "or"	(q) True
(iii)	A rectangle is a quadrilateral or a 5-sided polygon.			

Codes

	(i)	(ii)	(iii)
(a)	A,p	B,q	A,q
(b)	B,q	A,q	B,q
(c)	A,p	A,q	B,p
(d)	B,p	B,q	A,p

65. Which of the following is an example of quantifiers?

(a) p : There exists a rectangle whose all sides are equal
(b) q: For every prime number P, $\sqrt{P}$ is an irrational number
(c) Both (a) and (b)
(d) None of the above

66. Consider the following statements

Statement I For every positive number x there exists a positive number y such that $y < x$.

Statement II There exists a positive number y such that for every positive number x, we have $y < x$.

Choose the correct option.

(a) Only Statement I is true (b) Only Statement II is true
(c) Both statements are true (d) Both statements are false

67. Consider the following statements

Statement I The words "And" and "or" are connectives.

Statement II "There exists" and "For all" are called quantifiers.

Choose the correct option.

(a) Only Statement I is true
(d) Only Statement II is true
(c) Both statements are true
(d) Both statements are false

68. Match the statements of Column I with Column II and Column III appropriately and choose the correct option from the codes given below.

	Column I (Compound statements)	Column II (Connecting words)	Column III (Component statements)
(i)	All rational numbers are real and all real numbers are not complex.	A. "And"	1. p : The sand heats up quickly in the sun. q : The sand does not cool down fast at night.
(ii)	Square of an integer is positive or negative.	B. "Or"	2. p : $x = 2$ is the root of equation $3x^2 - x - 10 = 0$. q : $x = 3$ is the root of equation $3x^2 - x - 10 = 0$.
(iii)	The sand heats up quickly in the sun and does not cool down fast at night.		3. p : All rational numbers are real. q : All real numbers are not complex.
(iv)	$x = 2$ and $x = 3$ are the roots of the equation $3x^2 - x - 10 = 0$.		4. p: The square of an integer is positive. q : The square of an integer is negative.

Codes

	(i)	(ii)	(iii)	(iv)
(a)	A,1	B, 2	A,3	B, 4
(b)	B,1	B, 3	A,1	A, 2
(c)	A,3	B, 4	A,1	A, 2
(d)	B,4	B, 3	A,2	A, 3

69. Match the statements of Column I with Column II and Column III and choose the correct option from the codes given below.

Column I [Statements (p)]	Column II (Quantifier)	Column III (Negation of statements)
(i) There exists a number which is equal to its square.	A "There exists"	1. There does not exist a capital for every state of India.
(ii) For every real number x, x is less than $x+1$.	B. "For every"	2. There does not exist a number which is equal to its square.
(iii) There exists a capital for every state in India.		3. There exists a real number such that x is not less than $(x+1)$.

Codes

	(i)	(ii)	(iii)		(i)	(ii)	(iii)
(a)	A,2	B, 3	A,1	(b)	B,1	B, 2	A,1
(c)	A,3	A, 1	A,2	(d)	A,1	B, 1	B,2

70. Consider the following statements

I. $x+y=y+x$ is true for every real numbers x and y.

II. There exists real numbers x and y for which $x+y=y+x$.

(a) I is negation of II
(b) II is negation of I
(c) I is not negation of II
(d) Both (a) and (b) are true

71. For the statement "Sun rises or moon sets", "or" is ...X... . Here, X refers to

(a) exclusive
(b) inclusive
(c) Both (a) and (b)
(d) None of these

72. **Assertion** (A) For the statement "To apply for a driving licence, you should have a ration card or a passport", "or" is inclusive.

Reason (R) For the statement "All integers are positive or negative", "or" is exclusive.

(a) Both A and R are correct; R is the correct explanation of A
(b) Both A and R are correct; R is not the correct explanation of A
(c) A is correct; R is incorrect
(d) R is correct; A is incorrect

Level II

73. The component statements of the statement "The sky is blue or the grass is green" are

(a) p : The sky is blue, q : The sky is not blue
(b) p : The sky is blue, q : The grass is green
(c) Both (a) and (b)
(d) None of the above

74. The component statements of the statement " The earth is round or the sun is cold" are

(a) p : The earth is round, q : The sun is cold
(b) p : The sun is cold, q : The sun is hot
(c) Both (a) and (b)
(d) None of the above

75. Match the compound statements of Column I with Column II and Column III and choose the correct option from the codes given below. [NCERT Exemplar]

Column I (Compound statements)	Column II (Component statements)	Column III (Connecting words)
(i) It is raining or the sun is shining.	A. p : 2 is positive number. q : 2 is negative number.	1. "Or"
(ii) 2 is a positive number or a negative number.	B. p: It is raining. q : The sun is shining.	2. "And"

Codes

	(i)	(ii)		(i)	(ii)
(a)	A,1	B, 2	(b)	B,1	A, 1
(c)	B,1	B, 2	(d)	A,2	B, 1

76. **Assertion** (A) Component statements of the compound statement " 2 and 3 are prime numbers" are p : 2 is a prime number, q : 3 is a prime number and this statement is connected by " and ".

Reason (R) "Tigers are found in Gir Forest or Rajaji National Park." Then, component statements of this statement are

p : Tigers are found in Gir Forest.

q : Tigers are found in Rajaji National Park.

And the given statement is connected by "or".

(a) Both A and R are correct; R is the correct explanation of A
(b) Both A and R are correct; R is not the correct explanation of A
(c) A is correct; R is incorrect
(d) R is correct; A is incorrect

77. For the statement " A lady gives birth to a baby boy or a baby girl", the "or" is

(a) exclusive
(b) inclusive
(c) Only (a)
(d) Both (a) and (b)

78. For the statement "17 is a real number or a positive integer", the "or" is

(a) inclusive
(b) exclusive
(c) Only (a)
(d) None of the above

79. Match the statements of Column I with Column II and Column III and choose the correct option from the codes given below.

Column I (Statements)	Column II (Quantifier)	Column III (Negation of statements)
(i) There exists a number which is equal to its square.	A. "For all"	1. There does not exist a number which is a multiple of both 6 and 9.
(ii) For all even integers x, x^2 is also even.	B. "There exists"	2. There exists an even integer x such that x^2 is not even.
(iii) There exists a number which is a multiple of 6 and 9.		3. There does not exist a number which is equal to its square.

[NCERT Exemplar]

Codes

	(i)	(ii)	(iii)
(a)	A,1	B, 3	A,2
(b)	A,1	B, 2	A,3
(c)	B,3	A, 2	B,1
(d)	B,1	B, 2	A,3

80. Match the statements of Column I with their quantifiers in Column II and choose the correct option from the codes given below.

Column I	Column II
A. There exists a triangle which is not equilateral.	1. "For all"
B. For all real numbers x and y, $xy = yx$.	2. "For every"
C. There exists a real number which is not a rational number.	3. "There exists"
D. For every natural number x, $x + 1$ is also a natural number.	

Codes

	A	B	C	D
(a)	1	2	3	1
(b)	3	1	3	2
(c)	2	3	1	2
(d)	1	3	1	2

81. **Statement I** The quantifier for the statement "For all real numbers x with $x > 3$, x^2 is greater than 9" is "there exists".

Statement II The quantifier for the statement "There exists a triangle which is not an isosceles triangle" is "there exists".

(a) Statement I is true; Statement II is true; Statement II is a correct explanation for Statement I
(b) Statement I is true; Statement II is true; Statement II is not a correct explanation for Statement I
(c) Statement I is true; Statement II is false
(d) Statement I is false; Statement II is true

Implications

Level I

82. The component statements of the compound statement "If you are born in some country, then you are a citizen of that country" are

p : You are born in some country.

q : You are a citizen of that country.

Then, the correct option is

(a) If p is true, then q must be true
(b) If p is true, then q is not true
(c) Both p and q are false
(d) None of the above

83. Consider the following statements

r : If a number is a multiple of 9, then it is a multiple of 3.

Let p and q denote the statements

p : A number is a multiple of 9.

q : A number is a multiple of 3.

Then, "if p then q" is the same as

(a) p implies q
(b) p not implies q
(c) p is sufficient condition for q
(d) Both (a) and (c)

84. Consider the following statements

r : If a number is a multiple of 9, then it is a multiple of 3.

Let p and q denote the statements

p : A number is a multiple of 9.

q : A number is a multiple of 3.

Then, "if p then q" is the same as

(a) p only if q
(b) q is a necessary condition for p
(c) $\sim q$ implies $\sim p$
(d) All (a), (b) and (c)

85. Those statements which can be formed from a given statement with "if then" are called ....X... . Here, X refers to

(a) compound statements (b) contrapositive statements
(c) converse statements (d) Both (b) and (c)

86. Match the statements of Column I with statements of Column II and choose the correct option from the codes given below.

Column I (Statements)	Column II (Contrapositive statements)
A. If a number is divisible by 9, then it is divisible by 3.	1. If you are not a citizen of India, then you were not born in India.
B. If you are born in India, then you are a citizen of India.	2. If a triangle is not isosceles, then it is not equilateral.
C. If a triangle is equilateral, then it is isosceles.	3. If a number is not divisible by 3, it is not divisible by 9.

Codes

	A	B	C		A	B	C
(a)	3	1	2	(b)	1	2	3
(c)	1	1	2	(d)	3	2	1

87. Consider the following statements

p : If a number is divisible by 10, then it is divisible by 5.

q : If a number is divisible by 5, then it is divisible by 10.

Then, the correct option is

(a) q is converse of p (b) p is converse of q
(c) p is not converse of q (d) Both (a) and (b)

88. Match the statements of Column I with statements of Column II and choose the correct option from the codes given below.

Column I (Statements)	Column II (Converse statements)
A. If a number n is even, then n^2 is even.	1. If two integers a and b are such that $a-b$ is always a positive integer, then $a > b$.
B. If you do all the exercises in the book, you get an A grade in the class.	2. If a number n^2 is even, then n is even.
C. If two integers a and b are such that $a > b$, then integer $a-b$ is always positive.	3. If you get an A grade in the class, then you have done all the exercises of the book.

Codes

	A	B	C		A	B	C
(a)	1	2	3	(b)	3	2	1
(c)	2	3	1	(d)	1	3	2

89. Statement I If a ΔABC is equilateral, then it is isosceles, is true.

Statement II If a and b are integers, then ab is a rational number, is true.

(a) Statement I is true; Statement II is true; Statement II is a correct explanation for Statement I
(b) Statement I is true; Statement II is true; Statement II is not a correct explanation for Statement I
(c) Statement I is true; Statement II is false
(d) Statement I is false; Statement II is true

90. Which of the following are equivalent forms of the symbol $\Leftrightarrow$ for the statements p and q

(i) p if and only if q
(ii) q if and only if p
(iii) p is necessary and sufficient condition for q and *vice-versa*
(iv) $p \Leftrightarrow q$

(a) (i) and (ii)
(b) (ii) and (iii)
(c) (iii) and (iv)
(d) All of the above

91. Statement I p : If a rectangle is a square, then all its four sides are equal.

q : If all the four sides of a rectangle are equal, then the rectangle is a square.

Then, the combination form of p if and only if q is "A rectangle is a square if and only if all its four sides are equal."

Statement II p : If the sum of digits of a number is divisible by 3, then the number is divisible by 3.

q : If a number is divisible by 3, then the sum of its digits is divisible by 3.

Then, the combination form of p if and only if q is "A number is divisible by 3 if and only if the sum of its digits is divisible by 3".

(a) Statement I is true; Statement II is true; Statement II is a correct explanation for Statement I
(b) Statement I is true; Statement II is true; Statement II is not a correct explanation for Statement I
(c) Statement I is true; Statement II is false
(d) Statement I is false; Statement II is true

92. The statement "If a natural number is odd, then its square is also odd" can be written with 'if- then' in following different ways.

(i) A natural number is odd implies that its square is odd.
(ii) A natural number is odd only, if its square is odd.
(iii) If the square of a natural number is not odd, then the natural number is also not odd.

(iv) For a natural number to be odd, it is necessary that its square is odd.
(v) For a square of a natural number to be odd, it is sufficient that the number is odd.

Then, which of the following is/are true from of the given statements?

(a) (i) and (ii)
(b) (iii) and (iv)
(c) Only (v)
(d) All are true

93. **Statement I** The contrapositive statement of the statement "If x is prime number, then x is odd" is "If a number x is not odd, then x is not a prime number."

Statement II The converse statement of the statement "If x is a prime number, then x is odd" is "If x is odd, then x is a prime number."

(a) Statement I is true; Statement II is true; Statement II is a correct explanation for Statement I
(b) Statement I is true; Statement II is true; Statement II is not a correct explanation for Statement I
(c) Statement I is true; Statement II is false
(d) Statement I is false; Statement II is true

94. Match the statements of Column I with the statements of Column II and Column III and choose the correct option from the codes given below.

Column I (Statements)	Column II (Contrapositive)	Column III (Converse)
(i) If the two lines are parallel, then they do not intersect in the same plane.	A. If the temperature of something is not low, then it is not cold.	1. If x is divisible by 4, then x is an even number.
(ii) Something is cold implies that it has low temperature.	B. If you know how to reason deductively, then you can comprehend geometry.	2. If you do not know how to reason deductively, then you cannot comprehend geometry.
(iii) You cannot comprehend geometry, if you do not know how to reason deductively.	C. If x is not divisible by 4, then x is not an even number.	3. If something has low temperature, then it is cold.
(iv) x is an even number implies that x is divisible by 4.	D. If two straight lines intersect in a plane, then the lines are not parallel.	4. If two lines do not intersect in same plane, then two lines are parallel.

Codes

	(i)	(ii)	(iii)	(iv)
(a)	A,1	C, 4	B,2	D,3
(b)	D,4	A, 3	B,2	C,1
(c)	A,4	B, 1	C,2	D,3
(d)	B,4	D, 3	C,1	A,2

95. Match the statements of Column I with the statements in the form "if-then" in Column II and choose the correct option from the codes given below.

Column I	Column II
A. You get a job implies that your credentials are good.	1. If you get on A^+ in the class, then you do all the exercises of the book.
B. The Banana trees will bloom, if it stays warm for a month.	2. If you get a job, then your credentials are good.
C. A quadrilateral is a parallelogram if its diagonals bisect each other.	3. If it stays warm for a month, then the banana trees will bloom.
D. To get an A^+ in the class, it is necessary that you do all the exercises of the book.	4. If the diagonals of a quadrilateral bisect each other, then it is a parallelogram.

Codes

	A	B	C	D
(a)	1	2	3	4
(b)	4	1	2	3
(c)	3	2	4	1
(d)	2	3	4	1

96. Let p :If you live in Delhi, then you have winter clothes.

Statement I If you do not have winter clothes, then you do not live in Delhi.

Statement II If you have winter clothes, then you live in Delhi.

Choose the correct option.

(a) Statement I is contrapositive of p
(b) Statement II is converse of p
(c) Both (a) and (b) are true
(d) None of the above

97. Let q : If a quadrilateral is a parallelogram, then its diagonals bisect each other.

Statement I If the diagonals of a quadrilateral do not bisect each other, then the quadrilateral is not a parallelogram.

Statement II If the diagonals of a quadrilateral bisect each other, then it is a parallelogram.

Choose the correct option.

(a) Statement I is converse of q
(b) Statement II is contrapositive of q
(c) Statement II is converse of q
(d) Both (b) and (c) are true

Level II

98. Which of the following is the conditional $p \rightarrow q$? [NCERT Exemplar]

(a) q is sufficient for p (b) p is necessary for q
(c) p only if q (d) if q, then p

99. The contrapositive of the statement
"If 7 is greater than 5, then 8 is greater than 6" is [NCERT Exemplar]

(a) If 8 is greater than 6, then 7 is greater than 5
(b) If 8 is not greater than 6, then 7 is greater than 5
(c) If 8 is not greater than 6, then 7 is not greater than 5
(d) If 8 is greater than 6, then 7 is not greater than 5

100. The converse of the statement
" If $x > y$, then $x + a > y + a$" is [NCERT Exemplar]

(a) If $x < y$, then $x + a < y + a$ (b) If $x + a > y + a$, then $x > y$
(c) If $x < y$, then $x + a > y + a$ (d) If $x > y$, then $x + a < y + a$

101. Assertion (A) p : It is raining today, q : $2 + 3 > 4$. Then, the statement in the form of "if p then q" is "If it is raining today, then $2 + 3 > 4$."

Reason (R) The contrapositive form of the statement "If my car is in the repair shop, then I cannot go to market" is "If I can go to the market, then my car is not in the repair shop".

(a) Both A and R are correct; R is the correct explanation of A
(b) Both A and R are correct; R is not the correct explanation of A
(c) A is correct; R is incorrect
(d) R is correct; A is incorrect

102. The converse of the statement "If sun is not shining, then sky is filled with clouds" is [NCERT Exemplar]

(a) If sky is filled with clouds, then the sun is not shining
(b) If sun is shining, then sky is filled with clouds
(c) If sky is clear, then sun is shining
(d) If sun is not shining, then sky is not filled with clouds

103. The contrapositive of "If two triangles are identical, then these are similar" is ...X.... Here, X refers to

(a) If two triangles are not similar, then these are not identical
(b) If two triangles are not identical, then these are not similar
(c) Both (a) and (b)
(d) None of the above

104. The contrapositive of the statement
" If p, then q", is [NCERT Exemplar]

(a) If q, then p (b) If p ,then $\sim q$
(c) If $\sim q$, then $\sim p$ (d) If $\sim p$, then $\sim q$

105. The statement " If x^2 is not even, then x is not even" is converse of the statement [NCERT Exemplar]

(a) If x^2 is odd, then x is even
(b) If x is not even, then x^2 is not even
(c) If x is even, then x^2 is even
(d) If x is odd, then x^2 is even

106. For the statement "If you drive over 80 km/h, then you will get a fine", the sufficient condition is ...X.... Here, X refers to

(a) getting a fine (b) driving over 80 km/h
(c) Both (a) and (b) (d) None of these

107. Match the implication of the statements in Column I with their symbolic forms in Column II and choose the correct option from the codes given below. [HOTS]

	Column I		Column II
A.	If ABC is a triangle, then $AB + BC > AC$	1.	$x = 3 \Rightarrow x^2 = 9$
B.	If $3 > 5$, then $4 > 6$	2.	ABC is triangle, $\Rightarrow AB + BC > AC$
C.	If $x = 3$, then $x^2 = 9$	3.	$3 > 5 \Rightarrow 4 > 6$

Codes

	A	B	C		A	B	C
(a)	3	1	2	(b)	2	3	1
(c)	1	2	3	(d)	1	3	2

Validating Statements

Level I

108. The validity of a statement, depends upon

(a) special words (b) phrases
(c) implication, quantifiers (d) All of these

109. If 'p and q' is a true statement, then

(a) only 'p' is true
(b) only 'q' is true
(c) both 'p' and 'q' are true
(d) both 'p' and 'q' are not true

110. By assuming 'p is false and q is true' or 'q is false and p is true', then

(a) 'p or q' is true (b) 'p or q' is false
(c) 'p and q' is true (d) None of these

111. In order to prove the statement " If p then q" we can use ..A... . Here, A refers to

(a) direct method (b) contrapositive method
(c) Either (a) or (b) (d) None of these

112. In order to prove the statement "p if and only if q" we need to show that

(a) if p is true, then q is true (b) if q is true, then p is true
(c) Both (a) and (b) (d) None of these

113. Consider the following statements
$p : x, y \in Z$ such that x and y are odd.
$q : xy$ is odd. Then,

(a) $p \Rightarrow q$ is true (b) $\sim q \Rightarrow p$ is true
(c) Both (a) and (b) (d) None of these

114. For statement "If $x, y \in Z$ such that xy is odd, then both x and y are odd", which of the following is a correct option?

(a) The validity of above statement can be checked by contrapositive method
(b) The above statement is true
(c) The above statement is false
(d) Both (a) and (b)

115. Consider the following statements

p : A tumbler is half empty.

q : A tumbler is half full.

Then, the combination form of "p if and only if q" is

(a) a tumbler is half empty and half full
(b) a tumbler is half empty if and only if it is half full
(c) Both (a) and (b)
(d) None of the above

116. By the method of contradiction to prove "$\sqrt{7}$ is irrational" we assume that the given statement is ... P... . Here, P refers to

(a) $\sqrt{7}$ is real (b) $\sqrt{7}$ is not real
(c) $\sqrt{7}$ is rational (d) None of these

117. Consider the statement

"If n is an odd integer, then n is prime".

Then, which of the following is a correct option?

(a) The validity of the above statement can be checked by giving a counter example
(b) The above statement is false
(c) Both (a) and (b)
(d) None of the above

118. The statement p : 'If x is a real number such that $x^3 + 4x = 0$, then x is 0' is true by

(a) direct method (b) method of contradiction
(c) method of contrapositive (d) All of these

119. Consider the following statements

Statement I The statement "For any real numbers a and b, $a^2 = b^2$ implies that $a = b$" is not true by giving a counter example.

Statement II The following statement is true by the method of contrapositive, p : If x is an integer and x^2 is even, then x is also even.

Choose the correct option.

(a) Only Statement I is true
(b) Only Statement II is true
(c) Both statements are true
(d) Both statements are false

120. Consider the following statements

Statement I By giving a counter example, the following statement is not true.

p : If all the angles of a triangle are equal, then the triangle is an obtuse angled triangle.

Statement II By giving a counter example, the following statement is true.

q : The equation $x^2 - 1 = 0$ does not have a root lying between 0 and 2.

Choose the correct option.

(a) Only Statement I is true (b) Only Statement II is true
(c) Both statements are true (d) Both statements are false

121. Match the statements of Column I with the terms of Column II and choose the correct option from the codes given below.

	Column I	Column II
A.	p : Each radius of a circle is a chord of the circle.	1. True
B.	q : The centre of a circle bisects each chord of the circle.	2. False
C.	r : Circle is a particular case of an ellipse.	
D.	s : If x and y are integers such that $x > y$, then $-x < -y$.	
E.	t : $\sqrt{11}$ is a rational number.	

Codes

	A	B	C	D	E
(a)	1	2	1	2	1
(b)	1	1	2	1	2
(c)	2	2	1	1	2
(d)	2	1	2	2	1

Level II

122. The statement p : For any real numbers x, y if $x = y$, then $2x + a = 2y + a$ when $a \in Z$ [NCERT Exemplar]

(a) is true
(b) is false
(c) its contrapositive is not true
(d) None of the above

123. Let p : "100 is a multiple of 4 and 5".

Statement I p is true.

Statement II Both the component statements of p are true.

Choose the correct option. [NCERT Exemplar]

(a) Statement I is true; Statement II is true; Statement II is a correct explanation for Statement I
(b) Statement I is true; Statement II is true; Statement II is not a correct explanation for Statement I
(c) Statement I is true; Statement II is false
(d) Statement I is false; Statement II is true

124. Let q: "60 is a multiple of 3 or 5."

Statement I q is true.

Statement II Both the component statements of q are true.

Choose the correct option. [NCERT Exemplar]

(a) Statement I is true; Statement II is true; Statement II is a correct explanation for Statement I
(b) Statement I is true; Statement II is true; Statement II is not a correct explanation for Statement I
(c) Statement I is true; Statement II is false
(d) Statement I is false; Statement II is true

125. Assertion (A) The validity of the statement p : 125 is divisible by 5 and 7 is false.

Reason (R) The validity of the statement q : 131 is a multiple of 3 or 11 is false. [NCERT Exemplar]

(a) Both A and R are correct; R is the correct explanation of A
(b) Both A and R are correct; R is not the correct explanation of A
(c) A is correct; R is incorrect
(d) R is correct; A is incorrect

126. Consider the following statements

p : 80 is a multiple of 5.

q : 80 is a multiple of 4.

Assertion (A) The compound statement "80 is multiple of 5 and 4" is a true statement.

Reason (R) Both "p" and "q" are true statements.

(a) Both A and R are correct; R is the correct explanation of A
(b) Both A and R are correct; R is not the correct explanation of A
(c) A is correct; R is incorrect
(d) R is correct; A is incorrect

127. Let p : 25 is a multiple of 5.

q : 25 is a multiple of 8.

Statement I The compound statement "p and q" is false.

Statement II The compound statement "p or q" is false.

Choose the correct option.

(a) Only Statement I is true
(b) Only Statement II is true
(c) Both statements are true
(d) Both statements are false

128. Statement "For any real number a and b, $a^4 = b^4$ implies that $a = b$" is ...K... by giving a counter example. Here, K refers to

(a) true
(b) false
(c) Cannot say
(d) None of these

129. Validity of statement with "if-then" such as r : "if p then q" is true can be shown by the [NCERT Exemplar]

(a) direct method
(b) contrapositive method
(c) contradiction method and by giving a counter example
(d) All of the above

130. The statement p: "The equation $x^3 - 1 = 0$ does not have a real root lying between 0 and 2"

(a) is true
(b) is false
(c) equation does not have any real roots
(d) None of the above

Miscellaneous Exercise

131. Which of the following sentences are statements?

(i) $\sin^2 x + \cos^2 x = 0$
(ii) A leap year has 366 days.
(iii) A cow has four legs.
(iv) $15 + 7 > 23$

(a) Only (i)
(b) Only (ii)
(c) Only (iii)
(d) All of the above

132. The negation of the statement

p : "for every positive real number x, the number $x - 1$ is also positive" is [NCERT Exemplar]

(a) "there exists atleast one positive real number x for which $(x - 1)$ is not positive"
(b) "for every positive real number x, the number $(x + 1)$ is also positive"
(c) Both (a) and (b)
(d) Neither (a) nor (b)

133. **Statement I** The negation of the statement q : "all cats scratch" is "all cats do not scratch."

Statement II The negation of the statement r : "For every real number x, either $x > 1$ or $x < 1$" is "there exists one number x such that neither $x > 1$ nor $x < 1$."

(a) Statement I is true; Statement II is true; Statement II is a correct explanation for Statement I
(b) Statement I is true; Statement II is true; Statement II is not a correct explanation for Statement I
(c) Statement I is true; Statement II is false
(d) Statement I is false; Statement II is true

134. Negation of the statement S : "there exists a number x such that $0 < x < 1$" is ...P... . Here, P refers to

(a) there does not exist a number such that $0 < x < 2$
(b) there does not exist a number such that $0 < x < 1$
(c) Both (a) and (b)
(d) None of the above

135. Match the statements of Column I with the statements of Column II and choose the correct option from the codes given below.

Column I (Statements)	Column II (Contrapositive statements)
A. p : A positive integer is prime only, if it has no divisor other than 1 and itself.	1. If it is not a sunny day, then I do not go to beach.
B. q : I go to a beach whenever it is a sunny day.	2. If you do not feel thirsty, then it is not hot outside.
C. r : If it is hot outside, then you feel thirsty.	3. If a positive integer has divisor other than 1 and itself, then it is not prime.

Codes

	A	B	C		A	B	C
(a)	1	2	3	(b)	3	1	2
(c)	2	3	1	(d)	1	3	2

136. Match the statements of Column I with the statements of Column II and choose the correct option from the codes given below.

Column I (Statements)	Column II (Converse statements)
A. p : A positive integer is prime only, if it has no divisor other than 1 and itself.	1. If you feel thirsty, then it is hot outside.
B. q : I go to a beach whenever it is a sunny day.	2. If a positive integer has no divisor other than 1 and itself, then it is prime.
C. r : If it is hot outside, then you feel thirsty.	3. If it is a sunny day, then I go to beach.

Codes

	A	B	C
(a)	1	3	2
(b)	2	1	3
(c)	2	3	1
(d)	1	2	3

137. Consider the statement

p : "It is necessary to have a password to log on to the server".

Then, "if-then" form of this statement is

(a) if you log on to sever, then you have a password
(b) if you log on to server, then you do not have a password
(c) Both (a) and (b)
(d) None of the above

138. **Statement I** The "if p then q" form of the statement "There is traffic jam whenever it rains" is "If it rains, then there is a traffic jam".

Statement II The "if p then q" form of the statement "You can access the website only if you pay a subscription fee" is "If you pay a subscription fee, then you can access the website."

(a) Statement I is true; Statement II is true; Statement II is a correct explanation for Statement I
(b) Statement I is true; Statement II is true; Statement II is not a correct explanation for Statement I
(c) Statement I is true; Statement II is false
(d) Statement I is false; Statement II is true

139. Match the statements of Column I with their "p if and only if q" form in Column II and choose the correct option from the codes given below.

Column I	Column II
A. p : If you watch a television, then your mind is free and if your mind is free, then you watch a television.	1. A quadrilateral is equiangular, if and only if it is a rectangle.
B. q : For you to get grade A, it is necessary and sufficient that you do all the homework regularly.	2. You get grade A, if and only if you do all the homework regularly.
C. r : If a quadrilateral is equiangular, then it is a rectangle and if a quadrilateral is a rectangle, then it is equiangular.	3. You watch a television, if and only if your mind is free.

Codes

	A	B	C
(a)	3	2	1
(b)	1	2	3
(c)	2	1	3
(d)	3	1	2

140. **Statement I** p : The sum of an irrational number and a rational number is irrational (by contradiction method) is true.

Statement II q : If n is a real number with $n > 3$, then $n^2 > 9$ (by contradiction method) is a valid statement.

(a) Statement I is true; Statement II is true; Statement II is a correct explanation for Statement I
(b) Statement I is true; Statement II is true; Statement II is not a correct explanation for Statement I
(c) Statement I is true; Statement II is false
(d) Statement I is false; Statement II is true

141. The contrapositive of statement "If Chandigarh is capital of Punjab, then Chandigarh is in India" is [NCERT Exemplar]

(a) "If Chandigarh is not in India, then Chandigarh is not the capital of Punjab"
(b) "If Chandigarh is in India, then Chandigarh is capital of Punjab"
(c) "If Chandigarh is not capital of Punjab, then Chandigarh is not capital of India"
(d) "If Chandigarh is capital of Punjab, then Chandigarh is not in India"

142. The statement p : "If a triangle is equiangular, then it is an obtuse angled triangle" can be written in five different forms such as

(i) A triangle is equiangular implies that it is an obtuse angled triangle.
(ii) A triangle is equiangular only if it is an obtuse angled triangle.
(ii) For a triangle to be equiangular, it is necessary that it is an obtuse angled triangle.
(iv) For a triangle to be obtuse angled triangle, it is sufficient that it is equiangular.
(v) If a triangle is not obtuse angled triangle, then it is not an equiangular triangle.

Then, which of following convey the same meaning of the given statement

(a) Only (i)
(b) (ii) and (iii)
(c) (iv) and (v)
(d) All of these

Answer *with* Explanations

1. *(a)* In mathematical language, there are two kinds of reasoning-inductive and deductive.

2. *(b)* The basic unit involved in mathematical reasoning is a mathematical statement.

3. *(c)* The given two sentences are

I. In 2003, the President of India was a woman.

II. An elephant weighs more than a human being.

When we read these sentences, we immediately decide that the first sentence is false and the second is true. There is no confusion regarding these. In Mathematics, such sentences are called statements. On the other hand, the third sentence is

III. Women are more intelligent than men.

For this sentence, we see that

Some people may think it is true, while others may disagree. Regarding this sentence, we cannot say whether it is always true or false. That means this sentence is ambiguous. Such a sentence is not acceptable as a statement in Mathematics.

4. *(b)* A sentence is called a mathematically acceptable statement, if it is either true or false but not both. Whenever we mention a statement, it is a "mathematically acceptable" statement.

5. *(a)* The given three sentences are

A. Two plus two equals four.

B. The sum of two positive numbers is positive.

C. All prime numbers are odd numbers.

Out of these sentences, the first two are true and the third one is false. There is no ambiguity regarding these sentences. Therefore, they are statements.

Now, the given fourth sentence is

D. The sum of x and y is greater than 0.

Here, we are not in a position to determine whether it is true or false, unless we know what x and y are. For example, it is false where $x = 1$, $y = -3$ and true when $x = 1$ and $y = 0$. Therefore, this sentence is not a statement. In other words, this sentence is vague or ambiguous. But the sentence "For any natural number x and y, the sum of x and y is greater than 0" is a statement.

6. *(d)* For the given three sentences

I. How beautiful!

II. Open the door.

III. Where are you going?

They are not statements because the first one is an exclamation, the second is an order and the third is a question.

None of these is considered as a statement in mathematical language.

Now, for the given fourth sentence *i.e.*, Tomorrow is Friday. Sentences involving variable time such as "today" "tomorrow" or "yesterday' are not statements. This is because it is not known what time is referred here.

So, this sentence is not a statement. The sentence is correct (true) on a Thursday but not on other days. The same argument holds for sentences with pronouns unless a particular person is referred to and for variable places such as "here", "there" etc.

7. *(b)* For the given sentences

A. She is a Mathematics graduate.

B. Kashmir is far from here.

We know that, the sentences with pronouns unless a particular person is referred to and for variable places such as "here", "there" etc. are not statements.

Hence, the sentences (i) and (ii) are not statements.

Now, the sentence

C. There are 40 days in a month.

Note that the period mentioned in the sentence above is a variable time that is any of 12 months. But we know that the sentence is always false (irrespective of the month), since the maximum number of days in a month can never exceed 31. Therefore, this sentence is a statement. So, what makes a sentence a statement is the fact that the sentence is either true or false but not both.

8. *(a)* While dealing with statements ,we usually denote them by small letters $p, q, r, \dots$. For example, we denote the statement " Fire is always hot" by p. This is also written as p: Fire is always hot.

9. *(a)* We know that, 8 is greater than 6. Hence, the given sentence is false.

$\therefore$ This is a statement.

10. *(b)* A. This sentence is false since there are sets which are not finite. Hence, it is a statement.

B. It is a scientifically established fact that sun is a star and therefore, this sentence is always true. Hence, it is a statement.

C. This sentence is subjective in the sense that for those who like Mathematics, it may be fun but for others it may not be. This means that this sentence is not always true. Hence, it is not a statement.

D. It is a scientifically established natural phenomenon that cloud is formed before it rains. Therefore, this sentence is always true. Hence, it is a statement.

E. This is a question which also contains the word "here". Hence, it is not a statement.

11. *(a)* We know that a month has 30 or 31 days. It is false to say that a month has 35 days. Hence, it is a statement.

12. *(d)* Mathematics may be difficult for one but may be easy for other. Hence, it is not a statement.

13. *(d)* A. It is true that sum of 5 and 7 is greater than 10. Hence, it is a statement.

B. The square of a number may be even or it may be odd. Square of an odd number is always odd and square of an even number is always even. Hence it is not a statement.

C. A quadrilateral may have equal lengths, as it may be a rhombus or a square or the quadrilateral may have unequal sides (like parallelogram). Hence, it is not a statement.

D. It is an order. Hence, it is not a statement.

14. *(b)* It is false because product of (-1) and 8 is not 8. It is -8. Hence, it is a statement.

15. *(a)* It is true that sum of interior angles of a triangle is $180°$. Hence, it is a statement.

16. *(c)* Today is a windy day. It is not clear that about which day it is said. Thus, it cannot be concluded whether it is true or false. Hence, it is not a statement.

17. *(a)* It is true that all real numbers are complex numbers. All real numbers can be expressed as $a + ib$. Hence, it is a statement.

18. *(d)* (a) Everyone in this room is bold. This is not a statement because from the context, it is not clear which room is referred here and the term 'bold' is not clearly defined.

(b) She is an engineering student. This is also not a statement because it is not clear, who she is?

(c) $\sin^2 \theta$ is greater than 1/2. This is not a statement because we cannot say whether the sentence is true or not.

(d) We know that, $3 + 3 = 6$. It is true. Hence, this sentence is a statement.

19. *(c)* (a) "x is a real number" is an open statement.

So, this is not a statement.

(b) "Switch off the fan" is not a statement, it is an order.

(c) "6 is a natural number" is a true sentence. So, it is a statement.

(d) "Let me go!" (optative sentence). So, it is not a statement.

20. *(d)* (a) "Smoking is injurious to health". It is true. So, it is a statement.

(b) $2 + 2 = 4$ is true. So it is a statement.

(c) "2 is the only even prime number" is true. So, it is a statement.

(d) "Come here" is an order. So, it is not a statement.

21. *(a)* A. "The angles opposite to equal sides of a triangle are equal" is true. So, it is a statement.

B. "The Moon is a satellite of Earth" is true. Hence, it is a statement.

C. "May God bless you!" is an exclamation. So, it is not a statement.

D. "Asia is a continent" is true. So, it is a statement.

E. "How are you?" is an interrogative sentence. So, it is not a statement.

22. *(a)* "New Delhi is in India" is true. So, it is a statement.

23. *(b)* "Every rectangle is a square" is false. So, it is a statement.

"Close the door" is an order. So, it is not a statement.

24. *(c)* "How old are you?" is an interrogative sentence. So, it is not a statement.

"x is a natural number" is not define. So, it is not a statement.

25. *(d)* I. "Earth is planet" is true. So, it is a statement.

II. "Plants are living objects" is true. So, it is a statement.

III. "$\sqrt{-3}$ is a rational number" is false. So, it is a statement.

IV. "$x^2 - 5x + 6 < 0$" when $x \in -R$ is false. So, it is a statement.

26. *(a)* (a) It is false. (b) It is true. (c) It is true.

27. *(c)* "Washington DC is in America" is true. So, it is a statement and "Moon revolves around the Earth" is also true. So, it is also a statement.

28. *(d)* $2 + 3 = 5$ is true. So, it is a statement and every square is a rectangle is also true. So, it is a statement.

29. *(c)* "The Sun is a Star" is true. So, it is a statement.

30. *(b)* A. It is false. So, it is a statement.

B. It is false. So, it is a statement.

C. It is an order. So, it is not a statement.

D. It is a request. So, it is not a statement.

31. *(c)* It is false. So, it is a statement.

32. *(c)* The denial of a statement is called the negation of the statement.

33. *(d)* Let us consider the statement

p : New Delhi is a city.

The negation of this statement is

"It is not the case that New Delhi is a city."

This can also be written as

"It is false that New Delhi is a city."

This can simply be expressed as

"New Delhi is not a city."

and the statement "It is true that New Delhi is a city" is not a negation of the given statement.

34. *(b)* If p is a statement, then the negation of p is also a statement and is denoted by $\sim p$, and read as "not p"

Here is an example to illustrate how, by looking at the negation of a statement, we may improve our understanding of it.

Let us consider the statement

p : Everyone in Germany speaks German.

The denial of this sentence tells us that not everyone in Germany speaks German. This does not mean that no person in Germany speaks German. It says merely that at least one person in Germany does not speak German.

Note *While forming the negation of a statement, phrases like, "It is not the case" or " It is false that" are also used.*

35. *(b)* This statement says that in a rectangle, both the diagonals have the same length. This means that if you take any rectangle, then both the diagonals have the same length . The negation of this statement is

"It is false that both the diagonals in a rectangle have the same length."

This means the statement

There is atleast one rectangle whose both diagonals do not have the same length.

36. *(c)* The given statement is "$\sqrt{7}$ is rational". Now, the negation of this statement is "$\sqrt{7}$ is not a rational".

37. *(c)* Many mathematical statements are obtained by combining one or more statements using some connecting words like "and" , "or" , etc. Consider the following statements

p : There is something wrong with the bulb or with the wiring.

This statement tells us that there is something wrong with the bulb or there is something wrong with the wiring. That means the given statement is actually made up of two smaller statements.

q : there is something wrong with the bulb.

r : there is something wrong with the wiring.

connected by "or"

Now, suppose two statements are given below.

p : 7 is an odd number.

q : 7 is a prime number.

These two statements can be combined with "and"

r : 7 is both odd and prime number.

This is a compound statement.

38. *(c)* A compound statement is a statement which is made up of two or more statements. In this case, each statement is called a component statement.

39. *(a)* Negation of the statement "Chennai is the capital of Tamil Nadu" is "Chennai is not the capital of Tamil Nadu."

40. *(c)* In negative statement, if word 'not' is 'not' given in the statement, then insert word 'not' in the statement. If word 'not' is given in the statement, then remove word 'not' from the statement.

$\therefore$ The negation of the given statement is "$\sqrt{2}$ is a complex number".

41. *(d)* The negation of the given statement is "all the triangles are equilateral triangles."

42. *(a)* The negation of the statement "The number x is not a rational number" is "the number x is a rational number". The second statement is the same as "x is not irrational". Hence, both statements are negations of each other.

43. *(d)* The negation of the statement "The number x is a rational number" is "The number x is not a rational number" or we can say "x is an irrational number". Therefore, both are negations of given statement.

44. *(d)* Firstly separate the given compound statement into two component statements. After that check whether they are true or false.

A. p : number 3 is prime.
q : number 3 is odd.
p and q are true.

B. p : All integers are positive.
q : All integers are negative.
p and q both are false.

C. p : 100 is divisible by 3.
q : 100 is divisible by 11.
r : 100 is divisible by 5.
p is false, q is false and r is true.
$\therefore$ p, q are false statements and r is a true statement.

45. *(c)* The negation of statement "A circle is an ellipse" is "A circle is not an ellipse".

46. *(b)* The negation is "7 is not greater than 8".

47. *(c)* Given statement is

p : there is a rational number $x \in S$ such that $x > 0$.

$\sim p$: every rational number $x \in S$ satisfies $x \le 0$.

48. *(b)* A. Negation is "Bengaluru is not the capital of Karnataka"
B. Negation is "It did not rain on July 4, 2005"
C. Negation is "Ravish is not honest"
D. Negation is "The Earth is not round"

49. *(a)* The given statement is

p : 101 is not a multiple of 3.

$\sim p$: 101 is a multiple of 3.

50. *(c)* The negation of given statement can be

(i) A natural number is not greater than zero.

(ii) It is false that a natural number is greater than zero.

$\therefore$ "It is false that a natural number is not greater than zero" is not a negation of the given statement.

51. *(a)* Roses are red and violets are blue is a compound statement as it is composed of the statements "Roses are red" and "violets are blue".

52. *(b)* "The school works or a holiday is declared" is a compound statement as it is a combination of the statement "The school works" and "A holiday is declared".

53. *(a)* The negation of the statement "The product of 3 and 4 is 9" is "It is false that the product of 3 and 4 is 9."

54. *(c)* The components of the given statements are

A. p : 125 is a multiple of 7.
   q : 125 is a multiple of 8.

B. p : Mumbai is the capital of Gujarat.
   q : Mumbai is the capital of Maharashtra.

C. p : $\sqrt{13}$ is a rational number.
   q : $\sqrt{13}$ is an irrational number.

D. p : a rectangle is a quadrilateral.
   q : a rectangle is a 5- sided polygon.

55. *(c)* I. The given statement is "The number 2 is greater than 7". Its negation is "The number 2 is not greater than 7".

II. The given statement is "Every natural number is an integer". Its negation is "Every natural number is not an integer".

56. *(d)* Some of the connecting words which are found in compound statement like "And", "or", etc, are often used in mathematical statements. These are called connectives.

57. *(b)* Consider the following compound statements

p : A point occupies a position and its location can be determined.

The statement can be broken into two component statements as

q : a point occupies a position.
r : Its location can be determined.

Here, we observe that both the statements are true.

Let us look at another statement.

p : 42 is divisible by 5, 6 and 7.

This statement has following component statements

q : 42 is divisible by 5.
r : 42 is divisible by 6.
s : 42 is divisible by 7.

Here, we know that the first is false, while the other two are true.

$\therefore$ p is false in this case.

Thus we can conclude that

1. The compound statement with 'and' is true, if all its component statements are true.
2. The compound statement with 'and' is false, if any of its component statements is false (this includes the case that some of its component statements are false or all of its component statements are false.)

58. *(b)* A. The component statements are

p : a line is straight.
q : a line extends indefinitely in both directions.

Both these statements are true. Therefore, the compound statement is true.

B. The component statements are

p : 0 is less than every positive integer.
q : 0 is less than every negative integer.

The second statement is false. Therefore, the compound statement is false.

C. The two component statements are

p : all living things have two legs.
q : all living things have two eyes.

Both these statements are false. Therefore, the compound statement is false.

Now, consider the following statement

p : A mixture of alcohol and water can be separated by chemical methods. This sentence cannot be considered as a compound statement with "and". Here, the word "and" refers to two things-alcohol and water.

59. *(c)* Consider the given statement

p : two lines in a plane either intersect at one point or they are parallel.

We know that, this is a true statement. This means that if two lines in a plane intersect, then they are not parallel. Alternatively, if the two lines are not parallel, then they intersect at a point. That is this statement is true in both the situations.

60. *(b)* I. p : an ice-cream or pepsi is available with a thali in a restaurant.

This means that a person who does not want ice cream can have a pepsi alongwith thali or one does not want pepsi can have an ice-cream alongwith thali. That is, who do not want a pepsi can have an ice cream. A person cannot have both ice-cream and pepsi. This is called an exclusive "Or".

II. A student who has taken Biology or Chemistry can apply for M.Sc. Microbiology Programme.

Here, we mean that the students who have taken both Biology and Chemistry can apply for the Microbiology Programme as well as the students who have taken only one of these subjects. In this case, we are using inclusive "Or".

61. *(c)* A. Here, "or" is inclusive, since a person can have both a passport and a voter registration card to enter a country.

B. Here, also "or" is inclusive since school is closed on holiday as well as on Sunday.

C. Here, "or" is exclusive because it is not possible for two lines to intersect and parallel together.

D. Here, also "or" is exclusive because a student cannot take both French and Sanskrit.

62. *(b)* I. A compound statement with an "or" is true when one component statement is true or both the component statements are true.

II. A compound statement with an "or" is false when both the component statements are false.

For example, consider the following statement.

p : two lines intersect at a point or they are parallel.

The component statements are

q : two lines intersect at a point.

r : two lines are parallel.

Then, when q is true, r is false and when r is true, q is false. Therefore, the compound statement p is true.

Consider another statement.

p : 125 is a multiple of 7 or 8.

Its component statements are

q : 125 is a multiple of 7.

r : 125 is a multiple of 8

Both q and r are false. Therefore, the compound statement p is false.

63. *(b)* A. p : the school is closed, if there is a holiday or Sunday.

The component statements are

q : school is closed, if there is a holiday.

r : school is closed, if there is a Sunday.

Both q and r are true. Therefore, the compound statement is true.

B. p : Mumbai is the capital of Kolkata or Karnataka.

The component statements are

q : Mumbai is the capital of Kolkata.

r : Mumbai is the capital of Karnataka.

Both these statements are false. Therefore, the compound statement is false.

64. *(b)* (i) The component statements are

p : $\sqrt{2}$ is a rational number.

q : $\sqrt{2}$ is an irrational number.

Here, we know that the first statement is false and the second is true and "or" because is exclusive. Therefore, the compound statement is true.

(ii) The component statements are

p : to get into a public library children need an identity card.

q : to get into a public library children need a letter from the school authorities.

Children can enter the library if they have either of the two , an identity card or the letter, as well as when they have both. Therfore, it is inclusive "Or" because the compound statement is also true when children have both the card and the letter.

(iii) Here, "Or" is exclusive. When we look at the component statements, we get that the statement is true.

65. *(c)* Quantifiers are phrases like, "There exists" and "For all" Another phrase which appears in mathematical statements is "there exists". For example, consider the statement, p : There exists a rectangle whose all sides are equal. This means that there is atleast one rectangle whose all sides are equal.

A word closely connected with "There exists" is "For every" (or for all). Consider a statement.

p : for every prime number p, $\sqrt{p}$ is an irrational number.

This means that if S denotes the set of all prime numbers, then for all the members p of the set S, $\sqrt{p}$ is an irrational number.

In general, a mathematical statement that says "For every" can be interpreted as saying that all the members of the given set S where the property applies must satisfy that property.

66. *(a)* Given statements are

I. For every positive number x there exists a positive number y such that $y < x$.

II. There exists a positive number y such that every positive number x, we have $y < x$.

Although these statements may look similar, they do not say the same thing. As a matter of fact, (1) is true and (2) is false. Thus, in order for a piece of mathematical writing to make sense, all of the symbols must be carefully introduced and each symbol must be introduced precisely at the right place - not too early and not too late.

Note *We should also observe that it is important to know precisely where in the sentence a given connecting word is introduced.*

67. *(c)* The words "And" and "or" are called connectives and "There exists" and "For all" are called quantifiers.

68. *(c)* (i) Connecting word is 'and'.

p : all rational numbers are real.

q : all real numbers are not complex.

(ii) The connecting word is 'or'.

p : the square of an integer is positive.

q : the square of an integer is negative.

(iii) Connecting word is 'and'.

p : the sand heats up quickly in the sun.

q : the sand does not cool down fast at night.

(iv) Connecting word is 'and'.

p : $x = 2$ is the root of equation $3x^2 - x - 10 = 0$.

q : $x = 3$ is the root of equation $3x^2 - x - 10 = 0$.

69. *(a)* (i) Quantifier is "there exists."

p : there exists a number which is equal to its square.

not p $(\sim p)$: there does not exist a number which is equal to its square.

(ii) Quantifier is "For every."

p : for every real number x, x is less than $(x + 1)$.

$\sim p$: there exists a real number x such that x is not less than $(x + 1)$.

(iii) Quantifier is "there exists."

p : there exists a capital for every state of India.

$\sim p$: there does not exist a capital for every state of India.

70. *(c)* Consider the given statements

I. $x + y = y + x$ is true for every real numbers x and y.

II. There exists real numbers x and y for which $x + y = y + x$.

$\therefore$ It is clear that

Statement I and II are not negation of each other.

71. *(a)* For the given statement "Sun rises or moon sets"

We know that, when Sun rises, the Moon sets. One of the happenings will take place. Hence, 'or' is exclusive.

72. *(b)* We know that, if one of the statements is true, then 'or' is exclusive and if both the statements are true, then 'or' is inclusive.

(i) To apply for a driving licence either a ration card or a passport or both can be used.

∴ 'or' is inclusive.

(ii) All integers are positive or negative. An integer cannot be both positive or negative at a time.

∴ Here, 'or' is exclusive.

73. *(b)* The component statements are

p : the sky is blue.

q : the grass is green.

74. *(a)* The component statements are

p : the Earth is round .

q : the Sun is cold .

75. *(b)* (i) The component statement are given by

p : it is raining .

q : the sun is shining .

The connective is "or".

(ii) The component statements are given by

p : 2 is a positive number.

q : 2 is a negative number.

The connective is 'or'.

76. *(b)* The component statements of the given statements are

p : 2 is a prime number

q : 3 is a prime number

This is connected by "and".

The component statements of the given statement are

p : tigers are found in Gir Forest.

q : tigers are found in Rajaji National Park.

This is connected by "or".

77. *(a)* Exclusive "or" . A lady cannot give birth to a baby who is both a boy and a girl.

78. *(a)* Inclusive "or". 17 is a real number or a positive integer or both.

79. *(c)* (i) The quantifier is "there exists" and the negation is "There does not exist a number which is equal to its square".

(ii) The quantifier is "For all" and the negation is "there exists an even integer x such that x^2 is not even".

(iii) The quantifier is "there exists" and the negation is "There does not exist a number which is multiple of both 6 and 9".

80. *(b)* The quantifiers in the given statements are

A. There exists

B. For all

C. There exists

D. For every

81. *(d)* The quantifier of Statement I is "for all"

and the quantifier of Statement II is "there exists".

82. *(a)* r : if you are born in some country, then you are citizen of that country.

When we look at this statement, we observe that it corresponds to two statements p and q given by

p : you are born in some country.

q : you are citizen of that country.

Then, the sentence "if p then q" says that in the event, if p is true, then q must be true.

One of the most important facts about the sentence "if p then q" is that it does not say anything (or places no demand) on q when p is false. For example, if you are not born in the country, then you cannot say anything about q. To put it in other words not happening of p has no effect on happening of q.

Another point to be noted for the statement "if p then q" is that the statement does not imply that p happens.

83. *(d)* "If p then q" is the same as the following

1. p implies q is denoted by $p \Rightarrow q$. The symbol $\Rightarrow$ stands for implies.

   This says that a number is a multiple of 9 implies that it is a multiple of 3.

2. p is a sufficient condition for q.

   This says that knowing that a number is a multiple of 9 is sufficient to conclude that it is a multiple of 3.

84. *(d)* (a) p only if q. This says that number is a multiple of 9 only, if it is a multiple of 3.

(b) q is a necessary condition for p.

This says that when a number is a multiple of 9, it is necessarily a multiple of 3.

(c) $\sim q$ implies $\sim p$.

This says that if a number is not a multiple of 3, then it is not a multiple of 9.

85. *(d)* Contrapositive and converse are certain other statements which can be formed from a given statement with "if - then".

For example, consider the following "if-then" statement.

If the physical environment changes, then the biological environment changes.

Then, the contrapositive of this statement is

If the biological environment does not change, then the physical environment does not change.

Note that both these statements convey the same meaning.

86. *(a)* The contrapositive of the given statements are

A. If a number is not divisible by 3, it is not divisible by 9.

B. If you are not a citizen of India, then you were not born in India.

C. If a triangle is not isosceles, then it is not equilateral.

The above examples show the contrapositive of the statement if "p then q" is "if $\sim q$ then $\sim p$".

87. *(d)* The converse of the statement "if p then q" is "if q then p". Given statements are

p : if a number is divisible by 10, it is divisible by 5.

q : if a number is divisible by 5, then it is divisible by 10.

It is clear that q is converse of p and p is converse of q.

88. *(c)* The converse of these statemets are

A. If a number n^2 is even, then n is even.

B. If you get an A grade in the class, then you have done all the exercises of the book.

C. If two integers a and b are such that $a-b$ is always a positive integer, then $a > b$.

89. *(b)* I. The component statements are given by

$p : \Delta ABC$ is equilateral.

$q : \Delta ABC$ is isosceles.

Since, an equilateral triangle is isosceles, we infer that the given compound statement is true.

II. The component statements are given by

p : a and b are integers

q : ab is a rational number.

Since, the product of two integers is an integer and therefore a rational number, the compound statement is true.

90. *(d)* 'If and only if' represented by the symbol '$\Leftrightarrow$' means the following equivalent forms for the given statements p and q

(i) p, if and only if q.

(ii) q, if and only if p.

(iii) p, is necessary and sufficient condition for q and *vice-versa*.

(iv) $p \Leftrightarrow q$.

91. *(b)* I. A rectangle is a square, if and only if all its four sides are equal.

II. A number is divisible by 3, if and only if the sum of its digits is divisible by 3.

92. *(d)* Given statement can be written in the following ways.

(i) A natural number is odd implies that its square is odd.

(ii) A natural number is odd only, if its square is odd.

(iii) If the square of a natural number is not odd, then the natural number is also not odd.

(iv) For a natural number to be odd, it is necessary that its square is odd.

(v) For a square of a natural number to be odd, if it is sufficient that the number is odd.

93. *(b)* I. **Contrapositive statement** If a number x is not odd, then x is not a prime number.

II. **Converse statement** If x is odd, then x is a prime number.

94. *(b)* (i) **Contrapositive statement** If two straight lines intersect in a plane, then the lines are not parallel.

Converse statement If two lines do not intersect in the same plane, then two lines are parallel.

(ii) **Contrapositive statement** If the temperature of something is not low, then it is not cold.

Converse statement If something has low temperature, then it is cold.

(iii) **Contrapositive statement** If you know how to reason deductively, then you can comprehend geometry.

Converse statement If you do not know how to reason deductively, then you cannot comprehend geometry.

(iv) **Contrapositive statement** If x is not divisible by 4, then x is not an even number.

Converse statement If x is divisible by 4, then x is an even number.

95. *(d)* A. If you get a job, then your credentials are good.

B. If it stays warm for a month, then the banana trees will bloom.

C. If the diagonals of a quadrilateral bisect each other, then it is a parallelogram.

D. If you get on A^+ in the class, then you do all the exercises of the book.

96. *(c)* I. Contrapositive statement.

II. Converse statement.

97. *(d)* I. Contrapositive statement.

II. Converse statement.

98. *(c)* $p \rightarrow q$ is the same as "p only if q".

99. *(c)* Contrapositive statement is

"If 8 is not greater than 6, then 7 is not greater than 5".

100. *(b)* Converse statement is

"If $x + a > y + a$, then $x > y$".

101. *(b)* (i) The required conditional statement is

"If it is raining today, then $2 + 3 > 4$".

(ii) The contrapositive of the given statement is

"If I can go to the market, then my car is not in the repair shop".

102. *(a)* The converse statement is

"If sky is filled with clouds, then the sun is not shining."

103. *(a)* The contrapositive statement is

"If two triangles are not similar, then these are not identical."

104. *(c)* Contrapositive statement is

"If $\sim q$, then $\sim p$."

105. *(b)* Converse statement is
"If x is not even then x^2 is not even".

106. *(b)* Let p and q denote the statements
p : you drive over 80 km.
q : you will get a fine.

We know that, the implication of "if p then q" indicates that p is sufficient for q.

$\therefore$ Sufficient condition is "Driving over 80 km per hour."

107. *(b)* The given statements can be written as
A. ABC is a triangle $\Rightarrow AB + BC > AC$
B. $3 > 5 \Rightarrow 4 > 6$
C. $x = 3 \Rightarrow x^2 = 9$

108. *(d)* To answer this question, one must answer all the following questions.

What does the statement mean? What would it mean to say that this statement is true and when this statement is not true?

The answer to these questions depend upon which of the special words and phrases "and", "or", and which of the implications "if and only", "if-then", and which of the quantifiers "for every", "there exists", appear in the given statement.

109. *(c)* If p and q are mathematical statements, then in order to show that the statement "p and q" is true, the following steps are followed.

***Step* I** Show that the statement p is true.

***Step* II** Show that the statement q is true.

110. *(a)* If p and q are mathematical statements, then in order to show that the statement "p or q" is true, one must consider the following

***Case* I** By assuming that p is false, show that q must be true.

***Case* II** By assuming that q is false, show that p must be true.

111. *(c)* In order to prove the statement "if p then q" we need to show that anyone of the following case is true.

***Case* I** By assuming that p is true, prove that q must be true. (direct method)

***Case* II** By assuming that q is false, prove that p must be false. (contrapositive method)

112. *(c)* In order to prove the statement "p if and only if q" we need to show
(i) If p is true, then q is true and
(ii) If q is true, then p is true.

113. *(a)* Let $p : x, y \in Z$ such that x and y are odd.
$q : xy$ is odd.

To check the validity of the given statement, assume that if p is true, then q is true.

p is true means that x and y are odd integers. Then,
$x = 2m + 1$, for some integer m.
$y = 2n + 1$, for some integer n.

Thus, $xy = (2m + 1)(2n + 1)$
$= 2(2mn + m + n) + 1$

This shows that xy is odd. Therefore, then given statement is true.

Also, if we assume that q is not true. This implies that we need to consider the negation of the statement q. This gives the statement

$\sim q$: product xy is even.

This is possible only, if either x or y is even. This shows that p is not true. Thus, we have shown that

$$\sim q \Rightarrow \sim p$$

Note *The above problem illustrates that to prove $p \Rightarrow q$, it is enough to show $\sim q \Rightarrow \sim p$ which is the contrapositive of the statement $p \Rightarrow q$.*

114. *(d)* Let us name the statements as below:
p : xy is odd.
q : both x and y are odd.

We have to check whether the statement $p \Rightarrow q$ is true or not, *i.e.*, by checking its contrapositive statement i,e., $\sim q \Rightarrow \sim p$.

Now, $\sim q$: it is false that both x and y are odd. This implies that x (or y) is even.

Then, $x = 2n$ for some integer n.

Therefore, $xy = 2ny$ for some integer n. This shows that xy is even. That is $\sim p$ is true.

Thus, we have shown that $\sim q \Rightarrow \sim p$ and hence the given statement is true.

115. *(b)* The given statements are
p : a tumbler is half empty.
q : a tumbler is half full.

We know that, if the first statement happens, then the second happens and also if the second happens, then the first happens. We can express this fact as

If a tumbler is half empty, then it is half full.

If a tumbler is half full, then it is half empty.

We combine these two statements and get the following

A tumbler is half empty, if and only if it is half full.

116. *(c)* Here, to check whether a statement p is true, we assume that p is not true *i.e.*, $\sim p$ is true. Then, we arrive at some result which contradicts our assumption. Therefore, we conclude that p is true.

We assume that $\sqrt{7}$ is rational . This means that there exists positive integers a and b such that $\sqrt{7} = \frac{a}{b}$, where a and b

have no common factors. Squaring the equation, we get, $7=\frac{a^2}{b^2}\Rightarrow a^2=7b^2\Rightarrow 7$ divides a. Therefore, there exists an integer c such that $a=7c$. Then, $a^2=49c^2$ and $a^2=7b^2$

Hence, $7b^2=49c^2\Rightarrow b^2=7c^2\Rightarrow 7$ divides b. But, we have already shown that 7 divides a. This implies that 7 is common factor of both of a and b which contradicts our earlier assumption that a and b have no common factors. This shows that the assumption $\sqrt{7}$ is rational is wrong. Hence, the statement $\sqrt{7}$ is irrational, is true.

117. *(c)* To show that the given statement is false, we use the method involves giving an example of a situation, where the statement is not valid *i.e.*, counter example. Now, the given statement is in the form "if p, then q" we have to show that this is false. For this purpose, we need to show that if p then $\sim q$.

To show this, we look for an odd integer n which is not a prime number. 9 is one such number. So, $n=9$ is a counter example. Thus, we conclude that the given statement is false.

Note *In Mathematics, counter examples are used to disprove the statement. However, generating examples in favour of a statement do not provide validity of the statement.*

118. *(d)* (a) **Direct method**

$$x^3+4x=0 \quad \text{or} \quad x(x^2+4)=0$$

$$\Rightarrow \quad x=0,\ x^2+4\neq 0,\ x\in R$$

(b) **Method of contradiction**

Let $x\neq 0$ and let it be $x=p$, $p\in R$ and p is a root of $x^3+4x=0$.

$$\therefore \quad p^3+4p=0$$

$$\Rightarrow \quad p(p^2+4)=0$$

$$\Rightarrow \quad p^2+4=0 \quad (\because p\neq 0)$$

which is a contradiction, as $p^2+4\neq 0$, $p\in R$

$$\Rightarrow \quad x=0$$

(c) **Method of contrapositive**

Let $x=0$ is not true and $x=p\neq 0$.

$\therefore p^3+4p=0$, p being the root of $x^3+4x=0$

$$\therefore \quad p(p^2+4)=0$$

$$\Rightarrow \quad p=0 \quad \text{or} \quad p^2+4=0$$

$$\Rightarrow \quad p(p^2+4)\neq 0 \quad \left(\because p\neq 0 \text{ and } p^2+4\neq 0 \text{ as } p\in R\right)$$

$$\Rightarrow \quad x^3+4x\neq 0 \text{ for } x\in R$$

$\therefore x=0$ is the root of $x^3+4x=0$

119. *(c)* I. For the statement

"for any real number a and b"

$a^2=b^2$ implies that $a=b$

Let $\quad a=1$

and $\quad b=-1$

Then $\quad a^2=b^2=1$

but $\quad a\neq b$.

Hence, the given statement is not true.

II. Now, for the statement

p : if x is an integer and x^2 is even, then x is also even.

Let x is not even *i.e.*, $x=2n+1$

$$\therefore \quad x^2=(2n+1)^2 = 4n^2+4n+1 = 4(n^2+n)+1$$

$\therefore 4(n^2+n)+1$ is odd

$\Rightarrow n^2$ is odd.

i.e., 'if q is not true, then p is not true' is proved.

Hence, the given statement is true.

120. *(a)* I. Let an angle of triangle be $90°+\alpha$.

$\therefore$ Sum of the angles $=3(90°+\alpha)=270°+3\alpha$ which is greater than 180°.

$\therefore$ A triangle having equal angles cannot be obtuse angled triangle.

II. The equation $x^2-1=0$ has the root $x=1$, which lies between 0 and 2.

$\therefore$ The given statement is not true.

121. *(c)* A. **False** The end points of radius do not lie on the circle, therefore it is not a chord.

B. **False** Only diameters are bisected at the centre. Other chords do not pass through the centre. Therefore, centre cannot bisect them.

C. **True** Equation of ellipse is $\frac{x^2}{a^2}+\frac{y^2}{b^2}=1$.

When $b=a$, the equation becomes $\frac{x^2}{a^2}+\frac{y^2}{a^2}=1$

$$\Rightarrow \quad x^2+y^2=a^2$$

which is the equation of circle.

D. **True** By the rule of inequality, if x and y are integers and $x>y$, then $-x<-y$. *e.g.*, $3>2\Rightarrow -3<-2$.

E. **False** Since, 11 is prime number, therefore, $\sqrt{11}$ is irrational. ($\because$ prime numbers are not perfect squares)

122. *(a)* We prove the statement p is true by contrapositive method and by direct method.

Direct method For any real number x and y,

$$x=y$$

$$\Rightarrow \quad 2x=2y$$

$$\Rightarrow \quad 2x+a=2y+a \text{ for some } a\in Z$$

Contrapositive method The contrapositive statement of p is "For any real numbers x, y if $2x+a\neq 2y+a$, where $a\in Z$, then $x\neq y$."

Given, $\quad 2x+a\neq 2y+a$

$$\Rightarrow \quad 2x\neq 2y$$

$$\Rightarrow \quad x\neq y$$

Hence, the given statement is true.

123. *(a)* (i) Let p : 100 is a multiple of 4 and 5.

Let r : "100 is a multiple of 4"
q : "100 is a multiple of 5"
Then, r and q both are true.
Hence, p is true.

124. *(a)* Let q : 60 is a multiple of 3 and 5.

Let r : "60 is a multiple of 3"
s : "60 is a multiple of 5".
Then, r and s both are true.
Hence, q is true.

125. *(b)* For the statement
p : 125 is divisible by 5 and 7.
Let q : 125 is divisible by 5.
r : 125 is divisible by 7.
Now, q is true and r is false.
$\therefore$ $p = q$ and r is false.
Hence, p is not valid.
Now for the statement
q : 131 is a multiple of 3 or 11.
Let r : 131 is a multiple of 3.
s : 131 is a multiple of 11.
Clearly, r is false and s is also false.
$\therefore$ $q = r$ or s is not true.
Hence, $q = r$ or s is not valid
$\therefore$ Both A and R are correct but R is not a correct explanation of A.

126. *(a)* The compound statement is "80 is multiple of 5 and 4". We know that, 80 is multiple of 5 as well as 4. So, p and q are true statements.

Hence, the compound statement is also true, *i.e.* the compound statement "p and q" is a valid statement.

127. *(a)* I. Compound statement with 'and' is
25 is a multiple of 5 and 8.
This is a false statement. Since, p is true but q is false.
($\because$ 25 is divisible by 5 but not divisible by 8)

II. Compound statement with 'or' is
25 is a multiple of 5 or it is a multiple of 8.
This is a true statement. Since p is true and q is false.

128. *(b)* We observe that
$(-2)^4 = (2)^4 = 16$ but $-2 \neq 2$
So, the given statement is not true.

129. *(d)* To show statement r : "If p then q is true", we can adopt the following methods

(a) **Direct method** Assume p is true and show q is true, *i.e.*, $p \Rightarrow q$.
(b) **Contrapositive method** Assume $\sim q$ is true and show $\sim p$ is true, *i.e.*, $\sim q \Rightarrow \sim p$.
(c) **Contradiction method** Assume that p is true and q is false and obtain a contradiction from assumption.
(d) **By giving a counter example** To prove the given statement r is false, we give a counter example.

130. *(b)* We observe that $x = 1$ is a root of $x^3 - 1 = 0$ and $x = 1$ lies between 0 and 2. So, the given statement is not true.

131. *(d)* (i) $\sin^2 x + \cos^2 x = 0$, it is false.
So, it is a statement.
(ii) A leap year has 366 days. It is true. So, it is a statement.
(iii) A cow has four legs. It is true. So it is a statement.
(iv) $15 + 7 > 23 \Rightarrow 22 > 23$. It is false. So, it is a statement.

132. *(a)* $\sim p$: there exists atleast one positive real number x for which $(x - 1)$ is not positive.

133. *(b)* $\sim q$: all cats do not scratch and $\sim r$: there exists atleast one number x such that neither $x > 1$ nor $x < 1$.

134. *(b)* The negation is "There does not exist a number such that $0 < x < 1$".

135. *(b)* A. **Contrapositive** If a positive integer has divisor other than 1 and itself, then it is not prime.
B. **Contrapositive** If it is not a sunny day, then I do not go to beach.
C. **Contrapositive** If you do not feel thirsty, then it is not hot outside.

136. *(c)* A. **Converse** If a positive integer has no divisor other than 1 and itself, then it is a prime.
B. **Converse** If it is a sunny day, then I go to beach.
C. **Converse** If you feel thirsty, then it is hot outside.

137. *(a)* If you log on to server, then you have a password.

138. *(b)* I. If it rains, then there is a traffic jam.
II. If you pay a subscription fee, then you can access the website.

139. *(a)* A. You watch a television, if and only if your mind is free.
B. You get grade A, if and only if you do all the homework regularly.
C. A quadrilateral is equiangular, if and only if it is a rectangle.

140. *(b)* I. Let $\sqrt{p}$ be irrational number and q be rational number.
Their sum $= q + \sqrt{p}$
Let it is not irrational. Then, it is a rational number.

i.e., $$q + \sqrt{p} = \frac{a}{b}$$

where a and b are coprime.

$$\Rightarrow \quad \sqrt{p} = \frac{a}{b} - q$$

$\therefore$ LHS $= \sqrt{p}$ = An irrational number

$$\text{RHS} = \frac{a}{b} - q = \text{A rational number}$$

It is a contradiction.

Hence, the sum of rational and irrational number is irrational.

II. Let $n > 3$ and $n^2 \le 9$

Putting $n = 3 + p,\ p \in R^+$

$\Rightarrow \quad n^2 = (3 + p)^2$

$\Rightarrow \quad n^2 = 9 + 6p + p^2 = 9 + p(6 + p)$

$\Rightarrow \quad n^2 > 9$

which is a contradiction.

$\therefore \quad$ if $n > 3$, then $n^2 > 9$.

141. *(a)* The contrapositive statement is "If Chandigarh is not in India, then Chandigarh is not the capital of Punjab"

142. *(d)* The given statement can be written in the following ways.

(i) A triangle is equiangular implies that it is an obtuse angled triangle.

(ii) A triangle is equiangular only, if it is an obtuse angled triangle.

(iii) For a triangle to be equiangular, it is necessary that it is an obtuse angled triangle.

(iv) For a triangle to be obtuse angled triangle, it is sufficient that it is equiangular.

(v) If a triangle is not obtuse angled triangle, then it is not an equiangular triangle.

15

Statistics

Measures of Dispersion
(Range and Mean Deviation)

Level I

1. The method used in Statistics to find a representative value for the given data is called
(a) measure of skewness
(b) measure of central tendency
(c) measure of dispersion
(d) None of the above

2. The value which represents the measure of central tendency, is/are
(a) mean (b) median (c) mode (d) All of these

3. Consider the following data which represents the runs scored by two batsmen in their last ten matches as

Batsman A 30, 91, 0, 64, 42, 80, 30, 5, 117, 71

Batsman B 53, 46, 48, 50, 53, 53, 58, 60, 57, 52

Which of the following is/are true about the data?

I. Mean of batsman A runs is 53.
II. Median of batsman A runs is 42.
III. Mean of batsman B runs is 53.
IV. Median of batsman B runs is 53.

(a) Only I is true
(b) I and III are true
(c) I, III and IV are true
(d) All are true

4. Which of the following is/are true about the result obtained from the previous question?

I. Mean of the runs scored by both the batsmen are same.
II. Median of the runs scored by both the batsmen are different.
III. Performance of both players is same.
IV. Performance of both players is different.

(a) Only III is true
(b) I and IV are true
(c) I, II and III are true
(d) All are true

5. The number which indicates variability of data or observations, is called
(a) measure of central tendency
(b) mean
(c) median
(d) measure of dispersion

6. Which of the following is/are used for the measures of dispersion?
(a) Range
(b) Quartile deviation
(c) Standard deviation
(d) All of these

Range

7. Which of the following is/are true about the range of the data?

I. It helps to find the variability in the observations on the basis of maximum and minimum value of observations.

II. Range of series = Minimum value − Maximum value.

III. It tells us about the dispersion of the data from a measure of central tendency.

(a) Only I is true (b) II and III are true
(c) I and II are true (d) All are true

Mean Deviation

8. Statement I In order to find the dispersion of values of x from mean $\bar{x}$, we take absolute measure of dispersion.

Statement II Sum of the deviations from mean $(\bar{x})$ is zero.

(a) Statement I is true; Statement II is true; Statement II is a correct explanation for Statement I
(b) Statement I is true; Statement II is true; Statement II is not a correct explanation for Statement I
(c) Statement I is true; Statement II is false
(d) Statement I is false; Statement II is true

9. The formula of mean deviation used to find the dispersion of values from a central value is

(a) Mean of deviations $= \dfrac{\text{Sum of deviations}}{\text{Number of observations}}$

(b) Mean of deviations $= \dfrac{\text{Product of deviations}}{\text{Number of observations}}$

(c) Mean of deviations $= \dfrac{\text{Sum of absolute values of deviations}}{\text{Number of observations}}$

(d) Mean of deviations $= \dfrac{\text{Product of absolute values of deviations}}{\text{Number of observations}}$

10. If $x_1, x_2, \ldots, x_n$ be the n observations of ungrouped data.

Statement I The mean deviation about mean is given by

$$\text{MD}(\bar{x}) = \frac{1}{n}\sum_{i=1}^{n}|x_i - \bar{x}|, \text{ where } \bar{x} = \text{mean}.$$

Statement II The mean deviation about median is given by

$$\text{MD}(M) = \frac{1}{n}\sum_{i=1}^{n}|x_i - M|, \text{ where } M = \text{median}.$$

(a) Only Statement I is true (b) Only Statement II is true
(c) Both statements are true (d) Both statements are false

11. Statement I Mean deviation about the mean for the data 6, 7, 10, 12, 13, 4, 8, 12 is 2.75.

Statement II Mean deviation about the mean for the data 12, 3, 18, 17, 4, 9, 17, 19, 20, 15, 8, 17, 2, 3, 16, 11, 3, 1, 0, 5 is 3.1.

(a) Only Statement I is true (b) Only Statement II is true
(c) Both statements are true (d) Both statements are false

12. Mean deviation about the median for the data 3, 9, 5, 3, 12, 10, 18, 4, 7, 19, 21 is

(a) 4.27 (b) 5.24 (c) 5.27 (d) 4.24

13. We can grouped data into ...A... ways. Here, A refers to

(a) three (b) four
(c) two (d) None of these

14. If the data consists of n distinct values $x_1, x_2, \ldots, x_n$ occurring with frequencies $f_1, f_2, \ldots, f_n$ respectively.

I. The mean $\bar{x}$ of the given data is calculated by using the formula

$$\bar{x} = \frac{1}{N}\sum_{i=1}^{n}\frac{x_i}{f_i}$$

II. The mean deviation about the mean is

$$\text{MD}(\bar{x}) = \frac{1}{N}\sum_{i=1}^{n} f_i|x_i - \bar{x}|$$

III. The mean deviation about median is

$$\text{MD}(M) = \frac{1}{N}\sum_{i=1}^{n} f_i|x_i + M|$$

Which of the following is/are true?

(a) Only II is true (b) I and II are true
(c) II and III are true (d) Only III is true

15. Statement I Consider the following data

x_i	2	5	6	8	10	12
f_i	2	8	10	7	8	5

Then, the mean deviation about the mean is 2.3.

Statement II Consider the following data

x_i	3	6	9	12	13	15	21	22
f_i	3	4	5	2	4	5	4	3

Then, the mean deviation about the median is 4.90.

(a) Only Statement I is true (b) Only Statement II is true
(c) Both statements are true (d) Both statements are false

16. Consider the following data

Marks obtained	10-20	20-30	30-40	40-50	50-60	60-70	70-80
Number of students	2	3	8	14	8	3	2

Then, the mean deviation about the mean is
(a) 20 (b) 10 (c) 30 (d) 15

17. Mean deviation about median for continuous frequency distribution is calculated by using the formula

(a) MD $(M)=\frac{1}{N}\sum_{i=1}^{n}f_i|x_i+M|$ (b) MD $(M)=\frac{1}{N}\sum_{i=1}^{n}f_i|x_i-M|$

(c) MD $(M)=\frac{1}{N}\sum_{i=1}^{n}\frac{f_i}{|x_i-M|}$ (d) MD $(M)=\frac{1}{N}\sum_{i=1}^{n}\frac{|x_i-M|}{f_i}$

18. Consider the following data

Class	0-10	10-20	20-30	30-40	40-50	50-60
Frequency	6	7	15	16	4	2

Then, mean deviation about the median is
(a) 10.14 (b) 10.15 (c) 10.17 (d) 10.16

19. **Statement I** The mean deviation about the mean for the data 4, 7, 8, 9, 10, 12, 13, 17 is 3.5.

Statement II The mean deviation about the mean for the data 38, 70, 48, 40, 42, 55, 63, 46, 54, 44 is 8.5.

(a) Only Statement I is true (b) Only Statement II is true
(c) Both statements are true (d) Both statements are false

20. Mean deviation about the median for the data

13, 17, 16, 14, 11, 13, 10, 16, 11, 18, 12, 17 is
(a) 2.44 (b) 2.33
(c) 1.44 (d) 1.33

21. Consider the following data

36, 72, 46, 42, 60, 45, 53, 46, 51, 49

Then, the mean deviation about the median for the data is
(a) 6 (b) 8
(c) 7 (d) None of these

22. **Statement I** Consider the following data.

x_i	5	10	15	20	25
f_i	7	4	6	3	5

Then, the mean deviation about the mean is 6.32.

Statement II Consider the following data.

x_i	10	30	50	70	90
f_i	4	24	28	16	8

Then, the mean deviation about the mean is 15.
(a) Only Statement I is true (b) Only Statement II is true
(c) Both statements are true (d) Both statements are false

23. Consider the following data

x_i	5	7	9	10	12	15
f_i	8	6	2	2	2	6

Then, the mean deviation about the median for the data is
(a) 3.15 (b) 3.23 (c) 3.21 (d) 3.17

24. Consider the following data

x_i	15	21	27	30	35
f_i	3	5	6	7	8

Then, the mean deviation about the median for the data is
(a) 5 (b) 5.3 (c) 5.1 (d) 5.2

25. Match the following continuous series given in Column I with their corresponding mean deviations about mean or median in Column II and choose the correct option from the codes given below.

Column I									Column II
A. The mean deviation about the mean for the data									1. 10.34
Income per day	0-100	100-200	200-300	300-400	400-500	500-600	600-700	700-800	
Number of persons	4	8	9	10	7	5	4	3	
B. The mean deviation about the mean for the data									2. 157.92
Height (in cm)	95-105	105-115	115-125	125-135	135-145	145-155			
Number of boys	9	13	26	30	12	10			
C. The mean deviation about median for the data									3. 11.288
Marks	0-10	10-20	20-30	30-40	40-50	50-60			
Number of girls	6	8	14	16	4	2			

Codes

	A	B	C		A	B	C
(a)	2	3	1	(b)	3	2	1
(c)	3	1	2	(d)	1	2	3

26. The mean deviation about median age for the age distribution of 100 persons given below is

Age	16-20	21-25	26-30	31-35	36-40	41-45	46-50	51-55
Number of persons	5	6	12	14	26	12	16	9

[**Hint** Convert the given data into continuous frequency distribution by subtracting 0.5 from the lower limit and adding 0.5 to the upper limit of each class interval.]

(a) 7.38 (b) 7.42
(c) 7.72 (d) 7.35

27. **Assertion** (A) The mean deviation about median calculated for series, where variability is very high, cannot be fully relied.

Reason (R) The median is not a representative of central tendency for the series where degree of variability is very high.

(a) Both A and R are correct; R is the correct explanation of A
(b) Both A and R are correct; R is not the correct explanation of A
(c) A is correct; R is incorrect
(d) R is correct; A is incorrect

28. **Assertion** (A) The mean deviation about the mean to find measure of dispersion has certain limitations.

Reason (R) The sum of deviations from the mean is more than the sum of deviations from median. Therefore, the mean deviation about the mean is not very scientific, where degree of variability is very high.

(a) Both A and R are correct; R is the correct explanation of A
(b) Both A and R are correct; R is not the correct explanation of A
(c) A is correct; R is incorrect
(d) R is correct; A is incorrect

Level II

29. The mean deviation of the data 2, 9, 9, 3, 6, 9, 4 from the mean is [NCERT Exemplar]

(a) 2.23 (b) 2.57
(c) 3.23 (d) 3.57

30. The mean deviation about the mean of the following data is

Size (x)	1	3	5	7	9	11	13	15
Frequency (f)	3	3	4	14	7	4	3	4

(a) 1.95 (b) 3.95
(c) 2.95 (d) 4.95

31. The mean deviation of the data 3, 10, 10, 4, 7, 10, 5 from the mean is [NCERT Exemplar]

(a) 2 (b) 2.57
(c) 3 (d) 3.75

32. Consider the following data

Size	20	21	22	23	24
Frequency	6	4	5	1	4

Which of the following are true?

I. Mean of the data is 22.65.
II. Mean deviation of the data is 1.25.
III. Mean of the data is 21.65.
IV. Mean deviation of the data is 2.25.

(a) I and II are true (b) II and III are true
(c) I and IV are true (d) III and IV are true

33. When tested, the lives (in hours) of 5 bulbs were noted as follows

1357, 1090, 1666, 1494, 1623

The mean deviations (in hours) from their mean is [NCERT Exemplar]

(a) 178 (b) 179
(c) 220 (d) 356

34. Following are the marks obtained by 9 students in a Mathematics test.

50, 69, 20, 33, 53, 39, 40, 65, 59

The mean deviation from the median is [NCERT Exemplar]

(a) 9 (b) 10.5
(c) 12.67 (d) 14.76

35. Consider the following data

Marks obtained	10	11	12	14	15
Number of students	2	3	8	3	4

Which of the following are true?

I. Median of the data is 11.
II. Median of the data is 12.
III. Mean deviation about the median is 2.25.
IV. Mean deviation about the median is 1.25.

(a) I and III are true (b) I and IV are true
(c) II and III are true (d) II and IV are true

36. The mean deviation about the mean for the following frequency distribution is

Class interval	0-4	4-8	8-12	12-16	16-20
Frequency	4	6	8	5	2

(a) 2.84 (b) 4.84
(c) 3.84 (d) 1.84

37. The mean deviation from the median of the following data is

Class interval	0-6	6-12	12-18	18-24	24-30
Frequency	4	5	3	6	2

(a) 14 (b) 10 (c) 5 (d) 7

38. Statement I The average marks of boys in a class is 52 and that of girls is 42. The average marks of boys and girls combined is 50. The percentage of boys in the class is 80%.

Statement II Mean marks scored by the students of a class is 53. The mean marks of the girls is 55 and the mean marks of the boys is 50. The percentage of girls in the class is 64%.

(a) Only Statement I is true (b) Only Statement II is true
(c) Both statements are true (d) Both statements are false

39. The mean deviation about median from the following data is

Value	7	8	10	9	11	12	13
Frequency	2	1	4	5	6	1	3

(a) 3.6 (b) 1.36 (c) 2.36 (d) 2.5

40. The median of a set of 9 distinct observations is 20.5. If each of the largest 4 observations of the set is increased by 2, then the median of the new set is [HOTS]

(a) increased by 2
(b) decreased by 2
(c) two times the original median
(d) remains the same as that of original set

41. Statement I The weights (in kg) of 15 students are as follows

31, 35, 27, 29, 32, 43, 37, 41, 34, 28, 36, 44, 45, 42, 30

If the weight 44 kg is replaced by 46 kg and 27 kg is by 25 kg, then new median is 35.

Statement II The mean deviation from the median of the weights (in kg) 54, 50, 40, 42, 51, 45, 47, 57 is 4.78.

(a) Only Statement I is true
(b) Only Statement II is true
(c) Both statements are true
(d) Both statements are false

42. The mean age of a combined group of men and women is 25 yrs. If the mean age of the group of men is 26 and that of the group of women is 21, then the ratio of men and women in the group is [HOTS]

(a) 1 : 4 (b) 4 : 1
(c) 3 : 1 (d) 1 : 3

43. The mean deviation from the mean of the set of observations -1, 0 and 4 is

(a) 3 (b) 1
(c) -2 (d) 2

44. In a class of 100 students, there are 70 boys whose average marks in a subject are 75. If the average marks of the whole class is 72, then the average marks of the girls is

(a) 73 (b) 65
(c) 68 (d) 74

Variance and Standard Deviation

Level I

45. If $x_1, x_2, x_3, \ldots, x_n$ are n observations and $\bar{x}$ is their mean, then

$$(x_1 - \bar{x})^2 + (x_2 - \bar{x})^2 + \cdots + (x_n - \bar{x})^2 = \sum_{i=1}^{n}(x_i - \bar{x})^2$$

Which of the following is/are true based on above information?

I. If $\sum_{i=1}^{n}(x_i - \bar{x})^2 = 0$, then each $(x_i - \bar{x})$ has to be zero.

II. If $\sum_{i=1}^{n}(x_i - \bar{x})^2$ is small, then there is a lower degree of dispersion of the observations from the mean $\bar{x}$.

III. If $\sum_{i=1}^{n}(x_i - \bar{x})^2$ is large, then there is a higher degree of dispersion of the observations from the mean $\bar{x}$.

IV. $\sum_{i=1}^{n}(x_i - \bar{x})^2$ is a reasonable indicator of the degree of dispersion.

(a) Only I is true (b) II and III are true
(c) I, II and III are true (d) All are true

46. The quantity which leads to a proper measure of dispersion, is

(a) $\sum(x_i - \bar{x})^2$ (b) $\frac{1}{n}\sum(x_i - \bar{x})$
(c) $\frac{1}{n}\sum(x_i - \bar{x})^2$ (d) $\sum(x_i - \bar{x})$

47. Number which is mean of the squares of deviations from mean, is called ...A... . Here, A refers to

(a) standard deviation
(b) variance
(c) median
(d) None of the above

48. The variance of n observations $x_1, x_2, \ldots, x_n$ is given by

(a) $\sigma^2 = \frac{1}{n}\sum_{i=1}^{n}(x_i - \bar{x})$
(b) $\sigma^2 = \frac{1}{n}\sum_{i=1}^{n}(x_i - \bar{x})^2$
(c) $\sigma^2 = \frac{1}{n}\sum_{i=1}^{n}(x_i + \bar{x})$
(d) $\sigma^2 = \frac{1}{n}\sum_{i=1}^{n}(x_i + \bar{x})^2$

49. Statement I The proper measure of dispersion about the mean of a set of observations *i.e.*, standard deviation is expressed as positive square root of the variance.

Statement II The units of individual observations x_i and the unit of their mean are different from that of variance, since variance involves sum of squares of $(x - \bar{x})$.

(a) Statement I is true; Statement II is true; Statement II is a correct explanation for Statement I
(b) Statement I is true; Statement II is true; Statement II is not a correct explanation for Statement I
(c) Statement I is true; Statement II is false
(d) Statement I is false; Statement II is true

50. The standard deviation of n observations $x_1, x_2, \ldots, x_n$ is given by

(a) $\sigma = \sqrt{\sum_{i=1}^{n}(x_i - \bar{x})^2}$
(b) $\sigma = \sqrt{\sum_{i=1}^{n}(x_i + \bar{x})^2}$
(c) $\sigma = \sqrt{\frac{1}{n}\sum_{i=1}^{n}(x_i - \bar{x})^2}$
(d) $\sigma = \sqrt{\frac{1}{n}\sum_{i=1}^{n}(x_i + \bar{x})^2}$

51. Consider the following data

6, 8, 10, 12, 14, 16, 18, 20, 22, 24

I. The variance of the data is 33.
II. The standard deviation of the data is 4.74.

(a) Only Statement I is true
(b) Only Statement II is true
(c) Both statements are true
(d) Both statements are false

52. Consider the following data

x_i	4	8	11	17	20	24	32
f_i	3	5	9	5	4	3	1

Statement I The variance of the data is 45.8.

Statement II The standard deviation of the data is 6.77.

(a) Only Statement I is true
(b) Only Statement II is true
(c) Both statements are true
(d) Both statements are false

53. Consider the following distribution

Class	30-40	40-50	50-60	60-70	70-80	80-90	90-100
Frequency	3	7	12	15	8	3	2

Which of the following is/are true?

I. Mean of data is 31.
II. Variance of the data is 201.
III. Standard deviation of the data is 14.10.

(a) Only II is true
(b) II and III are true
(c) I and III are true
(d) All are true

54. The standard deviation for the following data is

x_i	3	8	13	18	23
f_i	7	10	15	10	6

(a) 6.21
(b) 6.31
(c) 6.12
(d) 6.13

55. Consider the following distribution

Class	30-40	40-50	50-60	60-70	70-80	80-90	90-100
Frequency	3	7	12	15	8	3	2

Then, match the following terms for the data in Column I with their corresponding values in Column II and choose the correct option from the codes given below.

Column I		Column II	
A.	Mean of the data	1.	14.18
B.	Variance of the data	2.	62
C.	Standard deviation of the data	3.	201

Codes

	A	B	C
(a)	3	1	2
(b)	2	1	3
(c)	2	3	1
(d)	3	2	1

56. The mean and variance for the data 6, 7, 10, 12, 13, 4, 8, 12 respectively are

(a) 9, 9.50
(b) 8, 8.50
(c) 9, 9.25
(d) 8, 8.25

57. Statement I The mean and variance for first n natural numbers are $\frac{n+1}{2}$ and $\frac{n^2+1}{12}$, respectively.

Statement II The mean and variance for first 10 positive multiples of 3 are 16.5 and 74.25, respectively. [HOTS]

(a) Only Statement I is true
(b) Only Statement II is true
(c) Both statements are true
(d) Both statements are false

58. Consider the following data

x_i	6	10	14	18	24	28	30
f_i	2	4	7	12	8	4	3

Statement I The mean of the data is 19.

Statement II The variance of the data is 43.4.

(a) Only Statement I is true (b) Only Statement II is true
(c) Both statements are true (d) Both statements are false

59. The mean and variance of the following data respectively are

x_i	92	93	97	98	102	104	109
f_i	3	2	3	2	6	3	3

(a) 100, 16.09 (b) 100, 29.09
(c) 50, 16.09 (d) 50, 29.09

60. Consider the following data

x_i	60	61	62	63	64	65	66	67	68
f_i	2	1	12	29	25	12	10	4	5

Statement I The mean of the data using shortcut method is 32.

Statement II The standard deviation of the data using shortcut method is 1.69.

(a) Only Statement I is true
(b) Only Statement II is true
(c) Both statements are true
(d) Both statements are false

61. Consider the following frequency distribution

Class	0-30	30-60	60-90	90-120	120-150	150-180	180-210
Frequency	2	3	5	10	3	5	2

Which of the following are true for the above data?

I. Mean of the data is 107.
II. Variance of the data is 2276.
III. Mean of the data is 105.
IV. Variance of the data is 2280.

(a) I and IV are true (b) III and IV are true
(c) III and II are true (d) I and II are true

62. Consider the following frequency distribution

Class	0-10	10-20	20-30	30-40	40-50
Frequency	5	8	15	16	6

Which of the following are true?

I. Mean of the data is 27.
II. Mean of the data is 32.
III. Variance of the data is 132.
IV. Variance of the data is 164.

(a) II and IV are true (b) I and IV are true
(c) II and III are true (d) I and III are true

63. Consider the following frequency distribution

Height (in cm)	70-75	75-80	80-85	85-90	90-95	95-100	100-105	105-110	110-115
Number of children	3	4	7	7	15	9	6	6	3

Match the following measures in Column I with their corresponding values in Column II and choose the correct option from the codes given below.

Column I	Column II
A. Mean of the data	1. 105.58
B. Variance of the data	2. 10.27
C. Standard deviation of the data	3. 93

Codes

	A	B	C		A	B	C
(a)	3	2	1	(b)	2	3	1
(c)	3	1	2	(d)	2	1	3

64. The diameters of circles (in mm) drawn in a design are given below.

Diameters	33-36	37-40	41-44	45-48	49-52
Number of circles	15	17	21	22	25

The standard deviation and mean diameter of the circles respectively are

(a) 5.55, 44.5
(b) 5.45, 43.5
(c) 5.55, 43.5
(d) 5.45, 44.5

Level II

65. Consider the following data

57, 64, 43, 67, 49, 59, 44, 47, 61, 59

Match the following measures of data in Column I with their corresponding values in Column II and choose the correct option from the codes given below.

Column I	Column II
A. Mean of the data	1. 66.2
B. Variance of the data	2. 8.13
C. Standard deviation of the data	3. 55

Codes

	A	B	C			A	B	C
(a)	3	2	1		(b)	3	1	2
(c)	2	3	1		(d)	2	1	3

66. Find the variance of the following data

Class interval	4-8	8-12	12-16	16-20
Frequency	3	6	4	7

(a) 13 (b) 18 (c) 19 (d) 20

67. The standard deviation of the data 6, 5, 9, 13, 12, 8, 10 is [NCERT Exemplar]

(a) $\sqrt{\frac{52}{7}}$ (b) $\frac{52}{7}$ (c) $\sqrt{6}$ (d) 6

68. Consider the following frequency distribution

Class	1-10	11-20	21-30	31-40	41-50	51-60
Frequency	11	29	18	4	5	3

Which of the following is/are true?

I. Mean of the data is 21.5.
II. Variance of the data is 161.14.
III. Standard deviation of the data is 12.

(a) Only II is true (b) II and III are true
(c) I and II are true (d) All are true

69. Variance of the data 2, 4, 5, 6, 8, 17 is 23.33. Then, variance of 4, 8, 10, 12, 16, 34 will be [NCERT Exemplar]

(a) 23.33 (b) 25.33 (c) 93.32 (d) 98.32

70. Consider the following frequency distribution

x	A	$2A$	$3A$	$4A$	$5A$	$6A$
f	2	1	1	1	1	1

Where, A is a positive integer and has variance 160. Then, the value of A is

(a) 5 (b) 6 (c) 7 (d) 8

71. The mean of 100 observations is 50 and their standard deviation is 5. The sum of squares of all observations is [NCERT Exemplar]

(a) 50000 (b) 250000 (c) 252500 (d) 255000

72. If there are 60 students in a class and the following is the frequency distribution of the marks obtained by the students in a test.

Marks	0	1	2	3	4	5
Frequency	$x-2$	x	x^2	$(x+1)^2$	$2x$	$x+1$

where x is a positive integer.

Which of the following are true?

I. Mean of the marks is 3.
II. Mean of the marks is 2.8.
III. Variance of the marks is 1.12.
IV. Variance of the marks is 1.14.

(a) I and III are true (b) II and III are true
(c) I and IV are true (d) II and IV are true

73. Consider the following data

1, 2, 3, 4, 5, 6, 7, 8, 9, 10

If 1 is added to each number, then variance of the numbers so obtained is

(a) 6.5 (b) 2.87 (c) 3.87 (d) 8.25

74. **Statement I** For the following frequency distribution, the standard deviation is 1.38.

x	2	3	4	5	6	7
f	4	9	16	14	11	6

Statement II The mean life of a sample of 60 bulbs was 650 h and the standard deviation was 8 h. A second sample of 80 bulbs has a mean life of 660 h and standard deviation 7 h. Then, the overall standard deviation is 8.9.

(a) Only Statement I is true (b) Only Statement II is true
(c) Both statements are true (d) Both statements are false

75. Consider the first 10 positive integers. If we multiply each number by (–1) and then add 1 to each number, the variance of the numbers so obtained is [NCERT Exemplar]

(a) 8.25 (b) 6.5 (c) 3.87 (d) 2.87

Analysis of Frequency Distributions

Level I

76. The measure of variability which is independent of units, is called

(a) mean deviation
(b) variance
(c) standard deviation
(d) coefficient of variation

77. The coefficient of variation (CV) is defined as

(a) $CV = \frac{\bar{x}}{\sigma} \times 100$ (b) $CV = \frac{\sigma^2}{\bar{x}} \times 100$

(c) $CV = \frac{(\sigma)^{1/2}}{\bar{x}} \times 100$ (d) $CV = \frac{\sigma}{\bar{x}} \times 100$

where, σ and $\bar{x}$ are the standard deviation and mean of the data.

78. **Statement I** The series having greater CV is said to be less variable than the other.

Statement II The series having lesser CV is said to be more consistent than the other.

(a) Only Statement I is true
(b) Only Statement II is true
(c) Both statements are true
(d) Both statements are false

79. If $\bar{x}_1$ and σ_1 are the mean and standard deviation of the first distribution and $\bar{x}_2$ and σ_2 are the mean and standard deviation of the second distribution.

Which of the following is/are true?

I. CV (1st distribution) $= \frac{\sigma_1}{\bar{x}_1} \times 100$

II. CV (2nd distribution) $= \frac{\bar{x}_2}{\sigma_2} \times 100$

III. For $\bar{x}_1 = \bar{x}_2$, the series with lesser value of standard deviation is said to be more variable than the other.

IV. For $\bar{x}_1 = \bar{x}_2$, the series with greater value of standard deviation is said to be more consistent than other.

(a) Only I is true (b) III and IV are true
(c) I, III and IV are true (d) All are true

80. Two plants A and B of a factory show the following results about the number of workers and the wages paid to them.

	A	B
Number of workers	5000	6000
Average monthly wages	₹ 2500	₹ 2500
Variance of distribution of wages	81	100

Which of the following is/are true?

I. The plant A has greater variability in the individual wages.

II. The plant B has greater variability in the individual wages.

III. Both plants have equal variability in the individual wages.

(a) Only II is true (b) I and III are true
(c) II and III are true (d) All are true

81. If the coefficient of variation of two distributions are 60 and 70 and their standard deviations are 21 and 16 respectively, then their arithmetic means are

(a) 35, 22.80
(b) 32, 22.80
(c) 35, 22.85
(d) 32, 22.85

82. Consider the following values which are calculated in respect of heights and weights of students of a section of class XI.

	Height	Weight
Mean	162.6 cm	52.36 kg
Variance	127.69 cm^2	23.1361 kg^2

Which of the following is/are true?

I. Weights show greater variations than the heights.

II. Heights show greater variations than the weights.

III. Both heights and weights show equal variations.

IV. To compare the variations, we have to calculate their coefficients of variation.

(a) Only I is true (b) I and IV are true
(c) I, II and IV are true (d) All are true

83. Consider the following distribution

Marks	10-20	20-30	30-40	40-50	50-60	60-70	70-80
Group *A*	9	17	32	33	40	10	9
Group *B*	10	20	30	25	43	15	7

More variable group is ...P.... Here, P refers to

(a) A (b) B
(c) Data insufficient (d) None of these

84. Table below shows the prices of shares X and Y.

X	35	54	52	53	56	58	52	50	51	49
Y	108	107	105	105	106	107	104	103	104	101

The prices of shares more stable in value is

(a) X (b) Cannot say
(c) Y (d) Data insufficient

Directions (Q. Nos. 85-86) *Read the following information carefully and answer the questions given below it.*

An analysis of monthly wages paid to workers in two firms A and B, belonging to the same industry, gives the following results

	Firm A	Firm B
Number of wages earners	586	648
Mean of monthly wages	₹ 5253	₹ 5253
Variance of the distribution of wages	100	121

85. The firm which pays larger amount as monthly wages, is

(a) A (b) B
(c) Cannot say (d) Data insufficient

86. The firm which shows greater variability in individual wages, is

(a) A (b) B
(c) Cannot say (d) Data insufficient

87. The following is the record of goals scored by team A in a football session.

Number of goals scored	0	1	2	3	4
Number of matches	1	9	7	5	3

For team B, mean number of goals scored per match was 2 with standard deviation 1.25 goals.

Statement I Team A is more consistent than team B.

Statement II CV of team A is less than CV of team B.

(a) Statement I is true; Statement II is true; Statement II is a correct explanation for Statement I
(b) Statement I is true; Statement II is true; Statement II is not a correct explanation for Statement I
(c) Statement I is true; Statement II is false
(d) Statement I is false; Statement II is true

88. If the sum and sum of squares corresponding to length x (in cm) and weight y (in grams) of 50 plant products are

$\sum_{i=1}^{50} x_i = 212$, $\sum_{i=1}^{50} x_i^2 = 902.8$, $\sum_{i=1}^{50} y_i = 261$ and $\sum_{i=1}^{50} y_i^2 = 1457.6$, then

(a) length is more varying than weight
(b) weight is more varying than length
(c) both are equally varying
(d) Cannot say

Level II

89. Coefficient of variation of two distributions are 50 and 60 and their arithmetic means are 30 and 25, respectively. Then, difference of their standard deviations is [NCERT Exemplar]

(a) 0 (b) 1
(c) 1.5 (d) 2.5

90. If the coefficient of variation is 45% and the mean is 12, then its standard deviation is

(a) 5.2 (b) 5.3
(c) 5.4 (d) None of these

91. The coefficient of variation for the given data is

Class interval	0-10	10-20	20-30	30-40	40-50
Frequency	2	10	8	4	6

(a) 48.1
(b) 52.1
(c) 49.1
(d) 50.1

92. We have the following results of two samples of polythene bags from two manufacturer A and B tested by a prospective buyer for bursting pressure.

Bursting pressure (in kg)	Number of bags manufactured	
	A	B
5-10	2	9
10-15	9	11
15-20	29	18
20-25	54	32
25-30	11	27
30-35	5	13

Which of the following are true?

I. Bags manufactured by B have higher bursting pressure.
II. Bags manufactured by A have higher bursting pressure.
III. Bags manufactured by A have more uniform pressure.
IV. Bags manufactured by B have more uniform pressure.

(a) I and III are true (b) III and IV are true
(c) II and IV are true (d) I and IV are true

93. Coefficient of variation of two distributions are 50% and 60% and their standard deviations are 10 and 15, respectively. Then, difference of their arithmetic means is

(a) 3 (b) 4 (c) 5 (d) 6

94. The sum of the squares of deviations for 10 observations taken from their mean 50 is 250. Then, the coefficient of variation is

(a) 10% (b) 40%
(c) 50% (d) None of these

95. If $n = 10, \bar{x} = 12$ and $\Sigma x_i^2 = 1530$, then the coefficient of variation is

(a) 35% (b) 42% (c) 30% (d) 25%

96. If runs of two players A and B in cricket matches are such that player A has mean 50 and variance 36 and player B has mean 60 and variance 81 of runs, then match the following measures in Column I with their corresponding values in Column II and choose the correct option from the codes given below.

	Column I		Column II
A.	Coefficient of variation of player A	1.	A
B.	Coefficient of variation of player B	2.	B
C.	The more consistent player is	3.	15
D.	Less consistent player is	4.	12

Codes

	A	B	C	D
(a)	2	1	4	3
(b)	4	3	1	2
(c)	3	4	2	1
(d)	2	4	1	3

97. Following is the statement of marks obtained by two students A and B in 10 examination papers.

Paper	1	2	3	4	5	6	7	8	9	10
Marks scored by A	44	80	76	48	52	72	68	56	60	54
Marks scored by B	48	75	54	60	63	69	72	51	57	66

If the consistency of performance is the criterion for awarding the prize, then the prize winner is [HOTS]

(a) student A (b) student B
(c) either student A or B (d) None of these

Miscellaneous Exercise

98. The variance of 20 observations is 5. If each observation is multiplied by 2, then the new variance of the resulting observations is [HOTS]

(a) $2^3 \times 5$ (b) $2^2 \times 5$
(c) 2×5 (d) $2^4 \times 5$

99. The mean of 5 observations is 4.4 and their variance is 8.24. If three of the observations are 1, 2 and 6, then the other two observations are

(a) 5, 5
(b) 4, 9
(c) 6, 5
(d) 7, 4

100. **Statement I** If each of the observations $x_1, x_2, \ldots, x_n$ is increased by a, where a is a negative or positive number, then the variance remains unchanged.

Statement II Adding or subtracting a positive or negative number to (or from) each observation of a group does not affect the variance.

(a) Statement I is true; Statement II is true; Statement II is a correct explanation for Statement I
(b) Statement I is true; Statement II is true; Statement II is not a correct explanation for Statement I
(c) Statement I is true; Statement II is false
(d) Statement I is false; Statement II is true

101. The mean and standard deviation of 100 observations were calculated as 40 and 5.1 respectively by a student who took by mistake 50 instead of 40 for one observation.

Statement I The correct mean is 39.9.

Statement II The correct standard deviation is 5.

(a) Only Statement I is true
(b) Only Statement II is true
(c) Both statements are true
(d) Both statements are false

102. The mean and standard deviation of six observations are 8 and 4, respectively. If each observation is multiplied by 3, then which of the following is/are true?

I. The new mean of the resulting observations is 24.
II. New standard deviation of resulting observations is 16.
III. Difference of new mean and original mean is 12.
IV. Difference of new standard deviation and original standard deviation is 8.

(a) Only I is true (b) I and IV are true
(c) II, III and IV are true (d) All are true

103. **Statement I** The mean and variance of eight observations are 9 and 9.25, respectively. If six of the observations are 6, 7, 10, 12, 12 and 13, then the remaining two observations are 4 and 8.

Statement II The mean and variance of 7 observations are 8 and 16, respectively. If five of the observations are 2, 4, 10, 12 and 14, then the remaining two observations are 6 and 5.

(a) Only Statement I is true (b) Only Statement II is true
(c) Both statements are true (d) Both statements are false

104. If $\bar{x}$ is the mean and σ^2 is the variance of n observations $x_1, x_2, \ldots, x_n$, then which of the following are true for the observations $ax_1, ax_2, ax_3, \ldots, ax_n$?

I. Mean of the observations is $\frac{\bar{x}}{a}$.
II. Variance of the observations is $\frac{\sigma^2}{a^2}$.
III. Mean of the observations is $a\bar{x}$.
IV. Variance of the observations is $a^2\sigma^2$.

(a) I and II are true
(b) I and IV are true
(c) II and III are true
(d) III and IV are true

105. The mean and standard deviation of marks obtained by 50 students of a class in three subjects, Mathematics, Physics and Chemistry are given below.

Subjects	Mathematics	Physics	Chemistry
Mean	42	32	40.9
Standard deviation	12	15	20

Arrangement of these subjects from the highest variability in marks to the lowest variability is

(a) CV of Chemistry > CV of Mathematics > CV of Physics
(b) CV of Physics > CV of Chemistry > CV of Mathematics
(c) CV of Chemistry > CV of Physics > CV of Mathematics
(d) CV of Mathematics > CV of Physics > CV of Chemistry

106. The mean and standard deviation of 20 observations are found to be 10 and 2, respectively. On rechecking, it was found that an observation 8 was incorrect.

Match the following cases in Column I with their corresponding values in Column II and choose the correct option from the codes given below.

	Column I		Column II
A.	Correct mean, if wrong item is omitted	1.	10.2
B.	Correct standard deviation, if wrong item is omitted	2.	10.10
C.	Correct mean, if it is replaced by 12	3.	1.98
D.	Correct standard deviation, if it is replaced by 12	4.	1.99

Codes

	A	B	C	D		A	B	C	D
(a)	4	2	1	3	(b)	2	4	1	3
(c)	3	1	2	4	(d)	2	1	4	3

107. The mean and standard deviation of a group of 100 observations were found to be 20 and 3, respectively. Later on, it was found that three observations were incorrect, which were recorded as 21, 21 and 18, then the mean and standard deviation, if the incorrect observations are omitted, are [HOTS]

(a) 21, 3.02 (b) 20, 3.02 (c) 21, 3.03 (d) 20, 3.03

108. Life of bulbs produced by two factories A and B are given below.

Length of life (in hours)	Number of bulbs	
	Factory A	Factory B
550-650	10	8
650-750	22	60
750-850	52	24
850-950	20	16
950-1050	16	12
	120	**120**

From the point of view of length of life, more consistent bulbs are from the factory

(a) A (b) B
(c) either A or B (d) Data insufficient

109. Consider the frequency distribution table given below

x	$1 \le x < 3$	$3 \le x < 5$	$5 \le x < 7$	$7 \le x < 10$
f	6	4	5	1

Which of the following are true?

I. Mean of the distribution is 4.15.
II. Variance of the distribution is 4.043.
III. Mean of the distribution is 4.16.
IV. Variance of the distribution is 4.042.

(a) I and IV are true (b) I and II are true
(c) II and III are true (d) III and IV are true

110. Let a, b, c, d and e be the observations with mean m and standard deviation s. The standard deviation of the observations $a+k$, $b+k$, $c+k$, $d+k$ and $e+k$ is [NCERT Exemplar]

(a) s (b) ks
(c) $s+k$ (d) s/k

111. Let x_1, x_2, x_3, x_4 and x_5 be the observations with mean m and standard deviations. Then, standard deviation of the observations kx_1, kx_2, kx_3, kx_4 and kx_5 is [NCERT Exemplar]

(a) $k+5$ (b) π/k (c) ks (d) s

112. Following are the marks obtained, out of 100 by two students Raju and Sita in 10 tests.

Raju	25	50	45	30	70	42	36	48	35	60
Sita	10	70	50	20	95	55	42	60	48	80

Which of the following are true?

I. Raju is more intelligent.
II. Sita is more intelligent.
III. Raju is more consistent.
IV. Sita is more consistent.

(a) I and IV are true (b) II and III are true
(c) I and III are true (d) II and IV are true

113. Let $x_1, x_2, \ldots, x_n$ be n observations. Let $w_i = lx_i + k$ for $i = 1, 2, \ldots, n$, where l and k are constants. If the mean of x_i's is 48 and their standard deviation is 12, the mean of w_i's is 55 and standard deviation of w_i's is 15. The values of l and k should be [NCERT Exemplar]

(a) $l = 1.25, k = -5$
(b) $l = -1.25, k = 5$
(c) $l = 2.5, k = -5$
(d) $l = 2.5, k = 5$

114. Two formulae for the standard deviation $(\sigma)\ \sqrt{\dfrac{\Sigma(x_i - \bar{x})^2}{n}}$ and $\sqrt{\dfrac{\Sigma x_i^2}{n} - \bar{x}^2}$ of ungrouped data are ...K... . Here, K refers to

(a) proportional
(b) different
(c) equivalent
(d) dependent

115. The mean and standard deviation of a set of n_1 observations are $\bar{x}_1$ and s_1 respectively, while the mean and standard deviation of another set of n_2 observations are $\bar{x}_2$ and s_2, respectively. Then standard deviation of the combined set of $(n_1 + n_2)$ observations is given by [HOTS]

(a) $\text{SD} = \sqrt{\dfrac{n_1(s_1)^2 + n_2(s_2)^2}{n_1 + n_2} + \dfrac{n_1(\bar{x}_1 - \bar{x}_2)^2}{n_2(n_1 + n_2)^2}}$

(b) $\text{SD} = \sqrt{\dfrac{n_2(s_1)^2 + n_1(s_2)^2}{n_1 + n_2} + \dfrac{n_1 n_2(\bar{x}_1 - \bar{x}_2)^2}{(n_1 + n_2)^2}}$

(c) $\text{SD} = \sqrt{\dfrac{n_1(s_1)^2 + n_2(s_2)^2}{n_1 + n_2} + \dfrac{n_1 n_2(\bar{x}_1 - \bar{x}_2)^2}{(n_1 + n_2)^2}}$

(d) None of the above

116. If for a distribution $\Sigma(x - 5) = 3$, $\Sigma(x - 5)^2 = 43$ and the total number of items is 18.

Statement I Mean of the distribution is 4.1666.

Statement II Standard deviation of the distribution is 1.54.

(a) Only Statement I is true (b) Only Statement II is true
(c) Both statements are true (d) Both statements are false

117. The mean and standard deviation of some data for the time taken to complete a test, calculated with the following results

Number of observations = 25,

Mean = 18.2,

Standard deviation = 3.25

Further, another set of 15 observations $x_1, x_2, \ldots, x_n$ also in seconds, is now available and we have $\sum_{i=1}^{n=15} x_i = 279$ and $\sum_{i=1}^{15} x_i^2 = 5524$. The standard deviation based on all 40 observations is [HOTS]

(a) 2.87 (b) 4.87
(c) 3.87 (d) 5.87

118. Consider the first n terms of an AP whose first term is a and common difference is d.

Which of the following are true?

I. Mean of first n terms of an AP is $\left[a + \dfrac{d(n+1)}{2}\right]$.

II. Standard deviation of first n terms of an AP is $\left[d\sqrt{\dfrac{n^2+1}{2}}\right]$.

III. Mean of first n terms of an AP is $\left[a + \dfrac{d(n-1)}{2}\right]$.

IV. Standard deviation of first n terms of an AP is $\left[d\sqrt{\dfrac{n^2-1}{12}}\right]$.

(a) I and II are true
(b) I and IV are true
(c) II and III are true
(d) III and IV are true

119. Mean and standard deviation of 100 observations were found to be 40 and 10, respectively. If at the time of calculation, two observations were wrongly taken as 30 and 70 in place of 3 and 27 respectively, then the correct standard deviation is

(a) 29.1 (b) 10.24 (c) 30.1 (d) 27.1

120. While calculating the mean and variance of 10 readings, a student wrongly used the reading 52 for the correct reading 25. He obtained the mean and variance as 45 and 16, respectively. The correct mean and the variance is [HOTS]

(a) 42.1, 43.81
(b) 42.3, 43.82
(c) 42.3, 43.81
(d) 42.1, 43.82

121. Statement I The variance of first n even natural numbers is $\dfrac{n^2 - 1}{4}$.

Statement II The sum of first n natural numbers is $\dfrac{n(n+1)}{2}$ and the sum of squares of first n natural numbers is $\dfrac{n(n+1)(2n+1)}{6}$. [AIEEE 2009]

(a) Statement I is true; Statement II is true; Statement II is a correct explanation for Statement I
(b) Statement I is true; Statement II is true; Statement II is not a correct explanation for Statement I
(c) Statement I is true; Statement II is false
(d) Statement I is false; Statement II is true

122. If the mean deviation of numbers, $1, 1+d, 1+2d, \ldots, 1+100d$ from their mean is 255, then d is equal to

(a) 10.0
(b) 20.0
(c) 10.1
(d) 20.2

123. Which of the following is/are true?

I. The standard deviation of the numbers 31, 32, 33, ..., 46, 47 is $4\sqrt{3}$.

II. If the variance of 1, 2, 3, 4, 5, ..., 10 is $\frac{99}{12}$, then the standard deviation of 3, 6, 9, 12, ..., 30 is $\frac{3}{2}\sqrt{33}$.

III. If the standard deviation of the scores 1, 2, 3, 4, 5, 6 and 7 is 2, then the standard deviation of 12, 23, 34, 45, 56, 67 and 78 is 22.

IV. If the standard deviation of observations –5, –4, –3, –2, –1, 0, 1, 2, 3, 4 and 5 is $\sqrt{10}$.

The standard deviation of observations 15, 16, 17, 18, 19, 20, 21, 22, 23, 24 and 25 will be $\sqrt{10}+10$.

(a) Only II is true
(b) II and III are true
(c) I, II and III are true
(d) All are true

1244. The standard deviation of first n natural numbers is

(a) $\frac{n^2+1}{12}$
(b) $\sqrt{\frac{n^2-1}{12}}$
(c) $\frac{(n+1)(2n+1)}{6}$
(d) $\left[\frac{n(n+1)}{2}\right]^2$

125. If $x_1, x_2, \ldots, x_{18}$ are observations such that $\sum_{i=1}^{18}(x_i-8)=9$ and $\sum_{i=1}^{18}(x_i-8)^2=45$, then standard deviation of these observations is

(a) $\sqrt{\frac{81}{34}}$
(b) 5
(c) $\sqrt{5}$
(d) $\frac{3}{2}$

126. The mean and variance of n observations $x_1, x_2, x_3, \ldots, x_n$ are 5 and 0, respectively. If $\sum_{i=1}^{n} x_i^2 = 400$, then the value of n is

(a) 80
(b) 16
(c) 25
(d) 20

127. **Statement I** If the mean of n observations $1^2, 2^2, 3^2, \ldots, n^2$ is $\frac{46n}{11}$, then n is equal to 11.

Statement II For two data sets each of size 5, the variances are given to be 4 and 5 and the corresponding means are given to be 2 and 4, respectively. The variance of combined data set is $\frac{11}{2}$.

(a) Only Statement I is true
(b) Only Statement II is true
(c) Both statements are true
(d) Both statements are false

Answer *with* Explanations

1. *(b)* We know that, Statistics deals with data collected for specific purposes. We can make decisions about the data by analysing and interpreting it. We have studied methods of representing data graphically and in tabular form. This representation reveals certain salient features or characteristics of the data. We have also studied the methods of finding a representative value for the given data. This value is called the measure of central tendency.

2. *(d)* Mean (arithmetic mean), median and mode are three measures of central tendency. A measure of central tendency gives us a rough idea, where data points are centred. But in order to make better interpretations from the data, we should also have an idea how the data are scattered or how much they are bunched around a measure of central tendency.

3. *(c)* The runs scored by two batsmen in their last ten matches are as follows:

Batsman A 30, 91, 0, 64, 42, 80, 30, 5, 117, 71

Batsman B 53, 46, 48, 50, 53, 53, 58, 60, 57, 52

Clearly, the mean and median of the data are

	Batsman A	Batsman B
Mean	53	53
Median	53	53

We calculate the mean of a data (denoted by $\bar{x}$) by dividing the sum of the observations by the number of observations, *i.e.,*

$$\bar{x} = \frac{1}{n}\sum_{i=1}^{n} x_i$$

Also, the median is obtained by first arranging the data in ascending or descending order and applying the following rule.

If the number of observations is odd, then the median is $\left(\frac{n+1}{2}\right)^{th}$ observation.

If the number of observations is even, then median is the mean of $\left(\frac{n}{2}\right)^{th}$ and $\left(\frac{n}{2}+1\right)^{th}$ observations.

Now, we show how we arrive at the result of mean and median.

Mean for bastman A

$$= \frac{30 + 91 + 0 + 64 + 42 + 80 + 30 + 5 + 117 + 71}{10}$$

$$= \frac{530}{10} = 53$$

Mean for batsman B

$$= \frac{53 + 46 + 48 + 50 + 53 + 53 + 58 + 60 + 57 + 52}{10}$$

$$= \frac{530}{10} = 53$$

To apply the formula to obtain median first arrange the data in ascending order

For batsman A	0	5	30	30	42	64	71	80	91	117
For batsman B	46	48	50	52	53	53	53	57	58	60

Here, we have $n = 10$ which is even number. So median is the mean of 5th and 6th observations.

Median for batsman A $= \frac{42 + 64}{2} = \frac{106}{2} = 53$

Median for batsman B $= \frac{53 + 53}{2} = \frac{106}{2} = 53$

4. *(b)* We find the mean and median of the runs scored by both the batsmen A and B are same *i.e.,* 53. Can we say that the performance of two players is same? Clearly no, because the variability in the scores of batsman A is from 0 (minimum) to 117 (maximum). Whereas, the range of the runs scored by batsman B is from 46 to 60.

Let us now plot the above scores as dots on a number line. We find the following diagrams

For batsman A

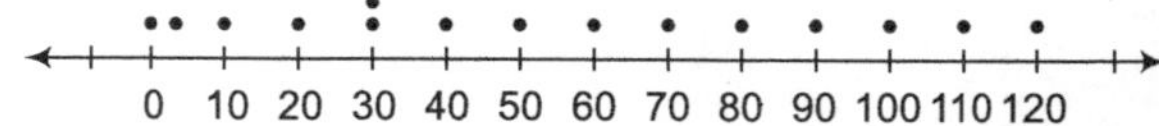

For batsman B

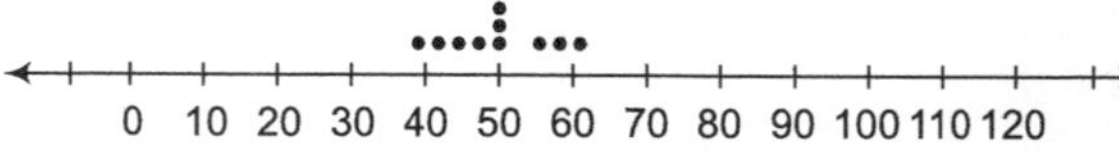

We can see that the dots corresponding to batsman B are close to each other and are clustering around the measure of central tendency (mean and median), while those corresponding to batsman A are scattered or more spread out.

5. *(d)* Variability is another factor which is required to be studied under Statistics. Like *'measures of central tendency'* we want to have a single number to describe variability. This single number is called a *'measure of dispersion'*.

6. *(d)* The dispersion or scatter in a data is measured on the basis of the observations and the types of the measure of central tendency used there. There are following measures of dispersion

(i) Range (ii) Quartile deviation

(iii) Mean deviation (iv) Standard deviation

7. *(a)* Consider the data [given in question 3] of runs scored by two batsmen A and B, we had some idea of variability in the scores on the basis of minimum and maximum runs in each series.

To obtain a single number for this, we find the difference of maximum and minimum values of each series. This difference is called the range of the data.

In case of batsman A,

$$\text{Range} = 117 - 0 = 117$$

and for batsman B,

$$\text{Range} = 60 - 46 = 14$$

Clearly, range of A > range of B. Therefore, the scores are scattered or dispersed in case of A, while for B these are close to each other.

Thus, range of a series = Maximum value – Minimum value

The range of data gives us a rough idea of variability or scatter but does not tell about the dispersion of the data from a measure of central tendency.

For this purpose, we need some other measures of variability. Clearly, such measure must depend upon the difference (or deviation) of the values from the central tendency.

8. *(a)* The deviation of an observation x from a fixed value 'a' is the difference $(x - a)$. In order to find the dispersion of values of x from a central value a, we find the deviations about a. An absolute measure of dispersion is the mean of these deviations.

To find the mean, we must obtain the sum of the deviations. But, we know that a measure of central tendency lies between the maximum and the minimum values of the set of observations.

Therefore, some of the deviations will be negative and some positive. Thus, the sum of deviations may vanish. Moreover, the sum of the deviations from mean $(\bar{x})$ is zero. Also,

$$\text{Mean of deviations} = \frac{\text{Sum of deviations}}{\text{Number of observations}} = \frac{0}{n} = 0$$

Thus, finding the mean of deviations about mean is not of any use for us, as far as the measure of dispersion is concerned.

9. *(c)* In finding a suitable measure of dispersion, we require the distance of each value from a central tendency or a fixed number a. The absolute value of the difference of two numbers gives the distance between the numbers when represented on a number line. Thus, to find the measure of dispersion from a fixed number a, we may take the mean of the absolute values of the deviations from the central value.

This mean is called the mean deviation. Thus, mean deviation about a central value a is the mean of the absolute values of the deviations of the observations from a. The mean deviations from a is denoted as MD (a). Therefore,

$$\text{MD}(a) = \frac{\text{Sum of absolute values of deviations from } a}{\text{Number of observations}}$$

10. *(c)* Let n observations be $x_1, x_2, x_3, \dots, x_n$. The following steps are involved in the calculation of mean deviation about mean or median.

***Step* I** Calculate the measure of central tendency about which we are to find the mean deviation. Let it be a.

***Step* II** Find the deviation of each x_i from a, *i.e.*, $x_1 - a, x_2 - a, x_3 - a, \dots x_n - a$.

***Step* III** Find the absolute values of the deviations, *i.e.*, drop the minus sign (–), if it is there *i.e.*,

$$|x_1 - a|, |x_2 - a|, |x_3 - a|, \dots, |x_n - a|$$

***Step* IV** Find the mean of the absolute values of the deviations. This mean is the mean deviation about a, *i.e.*,

$$\text{MD}(a) = \frac{\sum_{i=1}^{n} |x_i - a|}{n}$$

Thus, $\text{MD}(\bar{x}) = \frac{1}{n}\sum_{i=1}^{m} |x_i - \bar{x}|$, where $\bar{x}$ = mean

and $\text{MD}(M) = \frac{1}{n}\sum_{i=1}^{n} |x_i - M|$, where M = median.

11. *(a)* I. We proceed stepwise and get the following

***Step* I** Mean of the given data is

$$\bar{x} = \frac{6 + 7 + 10 + 12 + 13 + 4 + 8 + 12}{8} = \frac{72}{8} = 9$$

***Step* II** The deviations of the respective observations from the mean $\bar{x}$, *i.e.*, $x_i - \bar{x}$ are

$$6 - 9, 7 - 9, 10 - 9, 12 - 9, 13 - 9, 4 - 9, 8 - 9, 12 - 9,$$

or $\quad -3, -2, 1, 3, 4, -5, -1, 3.$

***Step* III** The absolute values of the deviations, *i.e.*, $|x_i - \bar{x}|$ are 3, 2, 1, 3, 4, 5, 1, 3.

***Step* IV** The required mean deviation about the mean is

$$\text{MD}(\bar{x}) = \frac{\sum_{i=1}^{8} |x_i - \bar{x}|}{8}$$

$$= \frac{3 + 2 + 1 + 3 + 4 + 5 + 1 + 3}{8} = \frac{22}{8} = 2.75$$

Note *Instead of carrying out the steps every time, we can carry on calculation, stepwise without referring to steps.*

II. We have to first find the mean of the given data

$$(\bar{x}) = \frac{1}{20}\sum_{i=1}^{20} x_i = \frac{200}{20} = 10$$

The respective absolute values of the deviations from mean, *i.e.*, $|x_i - \bar{x}|$ are 2, 7, 8, 7, 6, 1, 7, 9, 10, 5, 2, 7, 8, 7, 6, 1, 7, 9, 10, 5.

Therefore, $\quad \sum_{i=1}^{20} |x_i - \bar{x}| = 124$

and $\quad \text{MD}(\bar{x}) = \frac{124}{20} = 6.2$

12. *(c)* Here, the number of observations is 11 which is odd. Arranging the data into ascending order we have 3, 3, 4, 5, 7, 9, 10, 12, 18, 19, 21.

Now, median $= \left(\frac{11+1}{2}\right)^{th}$ or 6th observation = 9

The absolute values of the respective deviations from the median, *i.e.*, $|x_i - M|$ are 6, 6, 5, 4, 2, 0, 1, 3, 9, 10, 12.

Therefore, $\sum_{i=1}^{11} |x_i - M| = 58$

and $\text{MD}(M) = \frac{1}{11}\sum_{i=1}^{11}|x_i - M|$

$= \frac{1}{11} \times 58 = 5.27$

13. *(c)* We know that, data can be grouped into two ways

(i) Discrete frequency distribution.

(ii) Continuous frequency distribution.

14. *(a)* Let the given data consist of n distinct values $x_1, x_2, \ldots, x_n$ occurring with frequencies $f_1, f_2, \ldots, f_n$, respectively. This data can be represented in the tabular form as given below and is called discrete frequency distribution.

x	f
x_1	f_1
x_2	f_2
x_3	f_3
x_4	f_4
⋮	⋮
x_n	f_n

I. We find the mean $\bar{x}$ of the given data by using the formula

$$\bar{x} = \frac{\sum_{i=1}^{n} x_i f_i}{\sum_{i=1}^{n} f_i} = \frac{1}{N}\sum_{i=1}^{n} x_i f_i$$

where, $\sum_{i=1}^{n} x_i f_i$ denotes the sum of the products of observations x_i with their respective frequencies f_i and $N = \sum_{i=1}^{n} f_i$ is the sum of the frequencies.

II. **Mean deviation about mean** First, we find the deviations of observations x_i from the mean $\bar{x}$ and then take their absolute values, *i.e.*, $|x_i - \bar{x}|$ for all $i = 1, 2, \ldots, n$.

After this, find the mean of the absolute values of the deviations, which is the required mean deviation about the mean. Thus,

$$\text{MD}(\bar{x}) = \frac{\sum_{i=1}^{n} f_i|x_i - \bar{x}|}{\sum_{i=1}^{n} f_i} = \frac{1}{N}\sum_{i=1}^{n} f_i|x_i - \bar{x}|$$

III. **Mean deviation about median** To find mean deviation about median, we find the median of the given discrete frequency distribution. For this, the observations are arranged in ascending order. After this, the cumulative frequencies are obtained. Then, we identify the observation whose cumulative frequency is equal to or just greater than $\frac{N}{2}$, where N is the sum of frequencies. This value of the observation lies in the middle of the data. Therefore, it is the required median. After finding median, we obtain the mean of the absolute values of the deviation from median. Thus,

$$\text{MD}(M) = \frac{1}{N}\sum_{i=1}^{n} f_i|x_i - M|$$

15. *(a)* I. We have the following data

x_i	2	5	6	8	10	12
f_i	2	8	10	7	8	5

Let us make a table of the given data and append other columns after calculations.

x_i	f_i	$f_i x_i$	$\|x_i - \bar{x}\|$	$f_i\|x_i - \bar{x}\|$
2	2	4	5.5	11
5	8	40	2.5	20
6	10	60	1.5	15
8	7	56	0.5	3.5
10	8	80	2.5	20
12	5	60	4.5	22.5
	40	**300**		**92**

$$N = \sum_{i=1}^{6} f_i = 40, \sum_{i=1}^{6} f_i x_i = 300,$$

Therefore, $\bar{x} = \frac{1}{N}\sum_{i=1}^{6} f_i x_i = \frac{1}{40} \times 300 = 7.5$

Now, $\sum_{i=1}^{6} f_i|x_i - \bar{x}| = 92$

and $\text{MD}(\bar{x}) = \frac{1}{N}\sum_{i=1}^{6} f_i|x_i - \bar{x}| = \frac{1}{40} \times 92 = 2.3$

II. We have the following data

x_i	3	6	9	12	13	15	21	22
f_i	3	4	5	2	4	5	4	3

Here, the given observations are already in ascending order. Adding a row corresponding to cumulative frequencies to the given data, we get (table).

x_i	3	6	9	12	13	15	21	22
f_i	3	4	5	2	4	5	4	3
cf	3	7	12	14	18	23	27	30

Now, $N = 30$ which is even.

Median is the mean of the 15th and 16th observations. Both of these observations lie in the cumulative frequency 18, for which the corresponding observation is 13.

Therefore, $M = \frac{15^{th} \text{ observation} + 16^{th} \text{ observation}}{2}$

$= \frac{13 + 13}{2} = 13$

Now, absolute values of the deviations from median, *i.e.*, $|x_i - M|$ are shown in table.

$\|x_i - M\|$	10	7	4	1	0	2	8	9
f_i	3	4	5	2	4	5	4	3
$f_i\|x_i - M\|$	30	28	20	2	0	10	32	27

We have, $\sum_{i=1}^{8} f_i = 30$ and $\sum_{i=1}^{8} f_i|x_i - M| = 149$

Therefore, MD $(M) = \frac{1}{N}\sum_{i=1}^{8} f_i|x_i - M| = \frac{1}{30} \times 149 = 4.97$

16. *(b)* We have the following data

Marks obtained	10-20	20-30	30-40	40-50	50-60	60-70	70-80
Number of students	2	3	8	14	8	3	2

We make the following table from the given data

Marks obtained	Numbers of students (f_i)	Mid value (x_i)	$f_i x_i$	$\|x_i - \bar{x}\|$	$f_i\|x_i - \bar{x}\|$
10-20	2	15	30	30	60
20-30	3	25	75	20	60
30-40	8	35	280	10	80
40-50	14	45	630	0	0
50-60	8	55	440	10	80
60-70	3	65	195	20	60
70-80	2	75	150	30	60
	40		**1800**		**400**

Here, $N = \sum_{i=1}^{7} f_i = 40, \sum_{i=1}^{7} f_i x_i = 1800, \sum_{i=1}^{7} f_i|x_i - \bar{x}| = 400$

Therefore, $\bar{x} = \frac{1}{N}\sum_{i=1}^{7} f_i x_i = \frac{1800}{40} = 45$

and MD $(\bar{x}) = \frac{1}{N}\sum_{i=1}^{7} f_i|x_i - \bar{x}| = \frac{1}{40} \times 400 = 10$

Alternate Method

Take the assumed mean $a = 45$ and $h = 10$, and form the following table

Marks obtained	Number of students (f_i)	Mid value (x_i)	$d_i = \frac{x_i - 45}{10}$	$f_i d_i$	$\|x_i - \bar{x}\|$	$f_i\|x_i - \bar{x}\|$
10-20	2	15	−3	−6	30	60
20-30	3	25	−2	−6	20	60
30-40	8	35	−1	−8	10	80
40-50	14	45	0	0	0	0
50-60	8	55	1	8	10	80
60-70	3	65	2	6	20	60
70-80	2	75	3	6	30	60
	40			**0**		**400**

Therefore, $\bar{x} = a + \frac{\sum_{i=1}^{7} f_i d_i}{N} \times h = 45 + \frac{0}{40} \times 10 = 45$

and $MD(\bar{x}) = \frac{1}{N}\sum_{i=1}^{7} f_i|x_i - \bar{x}| = \frac{400}{40} = 10$

Note *The step deviation method is applied to compute $\bar{x}$. Rest of the procedure is same.*

17. *(b)* **Mean deviation about median** The process of finding the mean deviation about median for a continuous frequency distribution is similar as we did for mean deviation about the mean. The only difference lies in the replacement of the mean by median while taking deviations.

To find the median for a continuous frequency distribution. The data is first arranged in ascending order. Then, the median of continuous frequency distribution is obtained by first identifying the class in which median lies (median class) and then applying the formula.

$$\text{Median} = l + \frac{\frac{N}{2} - C}{f} \times h$$

where, median class is the class interval whose cumulative frequency is just greater than or equal to $\frac{N}{2}$, N is the sum of frequencies, l, f, h and C are respectively the lower limit, the frequency, the width of the median class and C the cumulative frequency of the class just preceding the median class. After finding the median, the absolute values of the deviations of mid-point x_i of each class from the median *i.e.*, $|x_i - M|$ are obtained.

Then, $\quad \text{MD}(M) = \frac{1}{N}\sum_{i=1}^{n} f_i|x_i - M|$

18. *(d)* We have the following data

Class	0-10	10-20	20-30	30-40	40-50	50-60
Frequency	6	7	15	16	4	2

Form the following table from the given data

Class	Frequency (f_i)	Cumulative frequency (cf)	Mid value (x_i)	$\|x_i - M\|$	$f_i\|x_i - M\|$
0-10	6	6	5	23	138
10-20	7	13	15	13	91
20-30	15	28	25	3	45
30-40	16	44	35	7	112
40-50	4	48	45	17	68
50-60	2	50	55	27	54
	50				**508**

The class interval containing $\frac{N^{th}}{2}$ or 25th item is 20-30.

Therefore, 20-30 is the median class. We know that,

$$\text{Median} = l + \frac{\frac{N}{2} - C}{f} \times h$$

Here, $l = 20, C = 13, f = 15, h = 10$ and $N = 50$

Therefore, median $= 20 + \frac{25 - 13}{15} \times 10 = 20 + 8 = 28$

Thus, mean deviation about median is given by

$$\text{MD}(M) = \frac{1}{N}\sum_{i=1}^{6} f_i|x_i - M| = \frac{1}{50} \times 508 = 10.16$$

19. *(d)* I. Mean of the given series

$$\bar{x} = \frac{\text{Sum of terms}}{\text{Number of terms}} = \frac{\Sigma x_i}{n}$$

$$= \frac{4+7+8+9+10+12+13+17}{8} = 10$$

x_i	$\lvert x_i - \bar{x} \rvert$
4	$\lvert 4-10 \rvert = 6$
7	$\lvert 7-10 \rvert = 3$
8	$\lvert 8-10 \rvert = 2$
9	$\lvert 9-10 \rvert = 1$
10	$\lvert 10-10 \rvert = 0$
12	$\lvert 12-10 \rvert = 2$
13	$\lvert 13-10 \rvert = 3$
17	$\lvert 17-10 \rvert = 7$
$\Sigma x_i = \mathbf{80}$	$\Sigma \lvert x_i - \bar{x} \rvert = \mathbf{24}$

$\therefore$ Mean deviation about mean $= \frac{\Sigma \lvert x_i - \bar{x} \rvert}{n} = \frac{24}{8} = 3$

II. Mean of the given series

$$\bar{x} = \frac{\text{Sum of terms}}{\text{Number of terms}} = \frac{\Sigma x_i}{n}$$

$$= \frac{38+70+48+40+42+55+63+46+54+44}{10}$$

$= 50$

x_i	$\lvert x_i - \bar{x} \rvert$
38	$\lvert 38-50 \rvert = 12$
70	$\lvert 70-50 \rvert = 20$
48	$\lvert 48-50 \rvert = 02$
40	$\lvert 40-50 \rvert = 10$
42	$\lvert 42-50 \rvert = 08$
55	$\lvert 55-50 \rvert = 05$
63	$\lvert 63-50 \rvert = 13$
46	$\lvert 46-50 \rvert = 04$
54	$\lvert 54-50 \rvert = 04$
44	$\lvert 44-50 \rvert = 06$
$\Sigma x_i = \mathbf{500}$	$\Sigma \lvert x_i - \bar{x} \rvert = \mathbf{84}$

$\therefore$ Mean deviation about mean $= \frac{\Sigma \lvert x_i - \bar{x} \rvert}{n} = \frac{84}{10} = 8.4$

20. *(b)* The given data is 13, 17, 16, 14, 11, 13, 10, 16, 11, 18, 12, 17

Arranging in ascending order,

10, 11, 11, 12, 13, 13, 14, 16, 16, 17, 17, 18

Number of observations = 12 (even)

$$\text{Median } M = \frac{\left(\frac{N}{2}\right)^{th} \text{observation} + \left(\frac{N}{2}+1\right)^{th} \text{observation}}{2}$$

$$= \frac{\left(\frac{12}{2}\right)^{th} \text{observation} + \left(\frac{12}{2}+1\right)^{th} \text{observation}}{2}$$

$$= \frac{6^{th} \text{ observation} + 7^{th} \text{ observation}}{2}$$

$$= \frac{13+14}{2} = \frac{27}{2}$$

$\Rightarrow \quad M = 13.5$

x_i	$\lvert x_i - M \rvert$
10	$\lvert 10-13.5 \rvert = 3.5$
11	$\lvert 11-13.5 \rvert = 2.5$
11	$\lvert 11-13.5 \rvert = 2.5$
12	$\lvert 12-13.5 \rvert = 1.5$
13	$\lvert 13-13.5 \rvert = 0.5$
13	$\lvert 13-13.5 \rvert = 0.5$
14	$\lvert 14-13.5 \rvert = 0.5$
16	$\lvert 16-13.5 \rvert = 2.5$
16	$\lvert 16-13.5 \rvert = 2.5$
17	$\lvert 17-13.5 \rvert = 3.5$
17	$\lvert 17-13.5 \rvert = 3.5$
18	$\lvert 18-13.5 \rvert = 4.5$
	$\Sigma \lvert x_i - M \rvert = \mathbf{28}$

$\therefore$ Mean deviation about median $= \frac{\Sigma \lvert x_i - M \rvert}{n}$

$$= \frac{28}{12} = 2.33$$

21. *(c)* The given data is 36, 72, 46, 42, 60, 45, 53, 46, 51, 49

Arranging the data in ascending order,

36, 42, 45, 46, 46, 49, 51, 53, 60, 72

Number of observations = 10 (even)

$$\text{Median } M = \frac{\left(\frac{N}{2}\right)^{th} \text{observation} + \left(\frac{N}{2}+1\right)^{th} \text{observation}}{2}$$

$$= \frac{\left(\frac{10}{2}\right)^{th} \text{observation} + \left(\frac{10}{2}+1\right)^{th} \text{observation}}{2}$$

$$= \frac{5^{th} \text{ observation} + 6^{th} \text{ observation}}{2}$$

$$= \frac{46+49}{2} = 47.5$$

x_i	$\lvert x_i - M \rvert$
36	$\lvert 36-47.5 \rvert = 11.5$
42	$\lvert 42-47.5 \rvert = 5.5$
45	$\lvert 45-47.5 \rvert = 2.5$
46	$\lvert 46-47.5 \rvert = 1.5$
46	$\lvert 46-47.5 \rvert = 1.5$
49	$\lvert 49-47.5 \rvert = 1.5$
51	$\lvert 51-47.5 \rvert = 3.5$
53	$\lvert 53-47.5 \rvert = 5.5$
60	$\lvert 60-47.5 \rvert = 12.5$
72	$\lvert 72-47.5 \rvert = 24.5$
	$\Sigma \lvert x_i - M \rvert = \mathbf{70}$

$\therefore$ Mean deviation about median $= \frac{\Sigma \lvert x_i - M \rvert}{n} = \frac{70}{10} = 7$

22. *(a)* I.

x_i	f_i	$f_i x_i$	$\|x_i - \bar{x}\|$	$f_i\|x_i - \bar{x}\|$
5	7	35	$\|5-14\| = 9$	63
10	4	40	$\|10-14\| = 4$	16
15	6	90	$\|15-14\| = 1$	06
20	3	60	$\|20-14\| = 6$	18
25	5	125	$\|25-14\| = 11$	55
Total	$\Sigma f_i = 25$	**350**		**158**

$$\text{Mean } (\bar{x}) = \frac{\Sigma f_i x_i}{\Sigma f_i} = \frac{350}{25} = 14$$

$$\therefore \text{ Mean deviation about mean} = \frac{\Sigma f_i |x_i - \bar{x}|}{\Sigma f_i} = \frac{158}{25} = 6.32$$

II.

x_i	f_i	$f_i x_i$	$\|x_i - \bar{x}\|$	$f_i\|x_i - \bar{x}\|$
10	4	40	$\|10-50\| = 40$	160
30	24	720	$\|30-50\| = 20$	480
50	28	1400	$\|50-50\| = 00$	000
70	16	1120	$\|70-50\| = 20$	320
90	8	720	$\|90-50\| = 40$	320
Total	$\Sigma f_i = 80$	$\Sigma f_i x_i = 4000$		**1280**

$$\text{Mean} = \frac{\Sigma f_i x_i}{\Sigma f_i} = \frac{4000}{80} = 50$$

$$\therefore \text{ Mean deviation about mean} = \frac{\Sigma f_i |x_i - \bar{x}|}{\Sigma f_i} = \frac{1280}{80} = 16$$

23. *(b)*

x_i	f_i	cf	$\|x_i - M\|$	$f_i\|x_i - M\|$
5	8	8	$\|5-7\| = 2$	16
7	6	14	$\|7-7\| = 0$	00
9	2	16	$\|9-7\| = 2$	04
10	2	18	$\|10-7\| = 3$	06
12	2	20	$\|12-7\| = 5$	10
15	6	26	$\|15-7\| = 8$	48
Total	$\Sigma f_i = 26$			$\Sigma f_i\|x_i - M\| = 84$

Here, $N = \Sigma f_i = 26$ (even)

$$\text{Median } (M) = \frac{\left(\frac{N}{2}\right)^{th} \text{observation} + \left(\frac{N}{2}+1\right)^{th} \text{observation}}{2}$$

$$= \frac{\left(\frac{26}{2}\right)^{th} \text{observation} + \left(\frac{26}{2}+1\right)^{th} \text{observation}}{2}$$

$$= \frac{13^{th} \text{ observation} + 14^{th} \text{ observation}}{2}$$

$$= \frac{7+7}{2} = \frac{14}{2} = 7$$

$$\therefore \text{Mean deviation about median} = \frac{\Sigma f_i |x_i - M|}{\Sigma f_i}$$

$$= \frac{84}{26} = 3.23$$

24. *(c)*

x_i	f_i	cf	$\|x_i - M\|$	$f_i\|x_i - M\|$
15	3	3	$\|15-30\| = 15$	45
21	5	8	$\|21-30\| = 9$	45
27	6	14	$\|27-30\| = 3$	18
30	7	21	$\|30-30\| = 0$	0
35	8	29	$\|35-30\| = 5$	40
Total	$\Sigma f_i = 29$			$\Sigma f_i\|x_i - M\| = 148$

Here, $N = \Sigma f = 29$ (odd)

$$\therefore \text{ Median } M = \left(\frac{N+1}{2}\right)^{th} \text{observation}$$

$$= \left(\frac{29+1}{2}\right)^{th} \text{observation} = 15^{th} \text{ observation}$$

$\Rightarrow$ $M = 30$

$\therefore$ Mean deviation about median

$$= \frac{\Sigma f_i |x_i - M|}{\Sigma f_i} = \frac{148}{29} = 5.1$$

25. *(a)* A.

Class	f_i	Mid value (x_i)	$d_i = \frac{x_i - A}{h}$, $A = 350, h = 100$	$f_i d_i$	$\|x_i - \bar{x}\|$	$f_i\|x_i - \bar{x}\|$
0-100	4	50	−3	−12	308	1232
100-200	8	150	−2	−16	208	1664
200-300	9	250	−1	−9	108	972
300-400	10	350	0	0	8	80
400-500	7	450	1	7	92	644
500-600	5	550	2	10	192	960
600-700	4	650	3	12	292	1168
700-800	3	750	4	12	392	1176
Total	$\Sigma f_i = 50$			**4**		**7896**

$$\text{Mean } \bar{x} = A + \frac{\Sigma f d_i}{\Sigma f_i} \times h = 350 + \frac{4}{50} \times 100 = 350 + 8$$

$$\bar{x} = 358$$

Mean deviation about the mean

$$= \frac{\Sigma f_i |x_i - \bar{x}|}{\Sigma f_i} = \frac{7896}{50} = 157.92$$

B.

Class	f_i	Mid value (x_i)	$d_i = \frac{x_i - A}{h}$, $A = 130, h = 10$	$f_i d_i$	$\|x_i - \bar{x}\|$	$f_i\|x_i - \bar{x}\|$
95-105	9	100	−3	−27	25.3	227.7
105-115	13	110	−2	−26	15.3	198.9
115-125	26	120	−1	−26	5.3	137.8
125-135	30	130	0	0	4.7	141.0
135-145	12	140	1	12	14.7	176.4
145-155	10	150	2	20	24.7	247.0
Total	**100**			**−47**		**1128.8**

$$\text{Mean } (\bar{x}) = A + \frac{\Sigma f_i d_i}{\Sigma f_i} \times h$$

$$= 130 + \frac{(-47)}{100} \times 10 = 130 - 4.7 = 125.3$$

$$\text{Mean deviation about mean} = \frac{\Sigma f_i |x_i - \bar{x}|}{\Sigma f_i} = \frac{1128.8}{100}$$

$$= 11.288$$

C.

Class	f_i	cf	Mid value (x_i)	$\|x_i - M\|$	$f_i\|x_i - M\|$
0-10	6	6	5	$\|5 - 27.86\| = 22.86$	137.16
10-20	8	(14)C	15	$\|15 - 27.86\| = 12.86$	102.88
(20-30)	(14)	28	25	$\|25 - 27.86\| = 2.86$	40.04
30-40	16	44	35	$\|35 - 27.86\| = 7.14$	114.24
40-50	4	48	45	$\|45 - 27.86\| = 17.14$	68.56
50-60	2	50	55	$\|55 - 27.86\| = 27.14$	54.28
Total	$\Sigma f_i = 50$				**517.16**

Here, $\frac{N}{2} = \frac{50}{2} = 25$

$\Rightarrow$ $C = 14, f = 14, l = 20, h = 10$

$$\text{Median } (M) = l + \frac{\frac{N}{2} - C}{f} \times h$$

$$= 20 + \frac{25 - 14}{14} \times 10$$

$$= 20 + \frac{11 \times 10}{14}$$

$$= 20 + 7.86 = 27.86$$

$$\therefore \text{Mean deviation about median} = \frac{\Sigma f_i |x_i - M|}{\Sigma f_i}$$

$$= \frac{517.16}{50} = 10.34$$

26. *(d)*

Class	f_i	cf	Mid value (x_i)	$\|x_i - M\|$	$f_i\|x_i - M\|$
15.5-20.5	5	5	18	$\|18 - 38\| = 20$	100
20.5-25.5	6	11	23	$\|23 - 38\| = 15$	90
25.5-30.5	12	23	28	$\|28 - 38\| = 10$	120
30.5-35.5	14	37	33	$\|33 - 38\| = 5$	70
35.5-40.5	26	63	38	$\|38 - 38\| = 0$	0
40.5-45.5	12	75	43	$\|43 - 38\| = 5$	60
45.5-50.5	16	91	48	$\|48 - 38\| = 10$	160
50.5-55.5	9	100	53	$\|53 - 38\| = 15$	135
Total	**100**				**735**

Here, $N = 100$

$\therefore$ $\frac{N}{2} = 50$

$\therefore$ Median class is 35.5 - 40.5.

$\Rightarrow$ $l = 35.5, C = 37, f = 26, h = 5$

$$\because \quad M = l + \frac{\frac{N}{2} - C}{f} \times h$$

$$\therefore \quad M = 35.5 + \frac{50 - 37}{26} \times 5$$

$$= 35.5 + \frac{13}{26} \times 5$$

$$= 35.5 + 2.5 = 38$$

$$\therefore \text{Mean deviation about median} = \frac{\Sigma f_i |x_i - M|}{\Sigma f_i}$$

$$= \frac{735}{100} = 7.35$$

27. *(a)* In a series, where the degree of variability is very high, the median is not a representative of central tendency. Thus, the mean deviation about median calculated for such series can not be fully relied.

28. *(a)* The sum of the deviations from the mean (minus signs ignored) is more than the sum of the deviations from median. Therefore, the mean deviation about the mean is not very scientific. Thus, in many cases, mean deviation may give unsatisfactory results. Also mean deviation is calculated on the basis of absolute values of the deviations and therefore, cannot be subjected to further algebraic treatment. This implies that we must have some other measure of dispersion. Standard deviation is such a measure of dispersion.

29. *(b)* Mean $(\bar{x}) = \frac{42}{7} = 6$

$$\text{MD } (\bar{x}) = \frac{\Sigma |x_i - \bar{x}|}{n}$$

$$= \frac{4 + 3 + 3 + 3 + 0 + 3 + 2}{7} = 2.57$$

30. *(c)* Mean $= \bar{x} = \frac{\Sigma f_i x_i}{\Sigma f_i}$

$$= \frac{3 + 9 + 20 + 98 + 63 + 44 + 39 + 60}{42}$$

$$= \frac{336}{42} = 8$$

$$\text{MD } (\bar{x}) = \frac{\Sigma f_i |x_i - \bar{x}|}{\Sigma f_i}$$

$$\frac{3(7) + 3(5) + 4(3) + 14(1) + 7(1) + 4(3) + 3(5) + 4(7)}{42}$$

$$= \frac{21 + 15 + 12 + 14 + 7 + 12 + 15 + 28}{42}$$

$$= \frac{62}{21} = 2.95$$

31. *(b)* We have, 3, 10, 10, 4, 7, 10, 5

$$\text{Mean } (\bar{x}) = \frac{3 + 10 + 10 + 4 + 7 + 10 + 5}{7}$$

$$= \frac{49}{7} = 7$$

Mean deviation $= \frac{1}{n}\sum_{i=1}^{7}|x_i - \bar{x}|$

$$= \frac{\left[\begin{array}{r}|3-7| + |10-7| + |10-7| + |4-7| + |7-7| \\ + |10-7| + |5-7|\end{array}\right]}{7}$$

$$= \frac{4+3+3+3+0+3+2}{7} = \frac{18}{7} = 2.57$$

32. *(b)* Let us write the given data in tabular form and calculate the required values to find mean deviation about the mean as

x_i	f_i	$f_i x_i$	$\lvert d_i\rvert = \lvert x_i - \bar{x}\rvert = \lvert x_i - 21.65\rvert$	$f_i\lvert d_i\rvert$
20	6	120	1.65	9.90
21	4	84	0.65	2.60
22	5	110	0.35	1.75
23	1	23	1.35	1.35
24	4	96	2.35	9.40
Total	**20**	**433**		**25.00**

Mean $(\bar{x}) = \frac{\Sigma f_i x_i}{\Sigma f_i} = \frac{433}{20} = 21.65$

Hence, mean of the data is 21.65

Mean deviation $= \frac{\Sigma f_i|x_i - \bar{x}|}{\Sigma f_i} = \frac{25}{20} = 1.25$

Hence, mean deviation of the data is 1.25.

33. *(a)* Mean $(\bar{x}) = \frac{1357 + 1090 + 1666 + 1494 + 1623}{5}$

$$= \frac{7230}{5} = 1446$$

Mean deviation $= \sum_{i=1}^{5}|x_i - \bar{x}|$

$$= \frac{\left[\begin{array}{r}|1357-1446| + |1090-1446| + |1666-1446| \\ + |1494-1446| + |1623-1446|\end{array}\right]}{5}$$

$$= \frac{89 + 356 + 220 + 48 + 177}{5}$$

$$= \frac{890}{5} = 178$$

34. *(c)* First arrange the data in ascending order as

20, 33, 39, 40, 50, 53, 59, 65, 69

Median $= \left(\frac{n+1}{2}\right)^{th}$ term

$$= \left(\frac{10}{2}\right)^{th} \text{ term} = 5^{th} \text{ term} = 50$$

Deviations are $|20-50|, |33-50|, |39-50|, |40-50|, |50-50|, |53-50|, |59-50|, |65-50|, |69-50|$

Mean deviation $= \frac{30 + 17 + 11 + 10 + 0 + 3 + 9 + 15 + 19}{9}$

$$= \frac{114}{9} = 12.67$$

35. *(d)* Total number of students $(n) = 2 + 3 + 8 + 3 + 4 = 20$

Median of numbers $= \frac{1}{2}\left[\left(\frac{n}{2}\right)^{th} \text{term} + \left(\frac{n}{2}+1\right)^{th} \text{term}\right]$

$$= \frac{1}{2}\left[\left(\frac{20}{2}\right)^{th} \text{term} + \left(\frac{20}{2}+1\right)^{th} \text{term}\right]$$

$$= \frac{1}{2}(10^{th} \text{ term} + 11^{th} \text{ term})$$

Marks obtained	f_i	cf	$\lvert d_i\rvert = \lvert x_i - 12\rvert$	$f_i\lvert d_i\rvert$
10	2	2	2	4
11	3	5	1	3
12	8	13	0	0
14	3	16	2	6
15	4	20	3	12
Total	**20**		$\Sigma f_i\lvert d_i\rvert = 25$	**25**

Median $= \frac{12 + 12}{2} = 12$

Mean Deviation $= \frac{\Sigma f_i|d_i|}{\Sigma f_i} = \frac{25}{20} = 1.25$

36. *(c)*

Class interval	f_i	x_i	$f_i x_i$	$\lvert d_i\rvert = \lvert x_i - \bar{x}\rvert = \lvert x_i - 9.2\rvert$	$f_i\lvert d_i\rvert$
0-4	4	2	8	7.2	28.8
4-8	6	6	36	3.2	19.2
8-12	8	10	80	0.8	6.4
12-16	5	14	70	4.8	24.0
16-20	2	18	36	8.8	17.6
	$\Sigma f_i = 25$		$\Sigma f_i x_i = 230$		**96.0**

Mean $(\bar{x}) = \frac{\Sigma f_i x_i}{\Sigma f_i} = \frac{230}{25} = 9.2$

Mean deviation $= \frac{\Sigma f_i|d_i|}{\Sigma f_i} = \frac{96}{25} = 3.84$

Hence, the mean deviation about the mean is 3.84.

37. *(d)*

Class interval	f_i	x_i	cf	$\lvert d_i\rvert = \lvert x_i - 14\rvert$	$f_i\lvert d_i\rvert$
0-6	4	3	4	11	44
6-12	5	9	9	5	25
12-18	3	15	12	1	3
18-24	6	21	18	7	42
24-30	2	27	20	13	26
	$\Sigma f_i = 20$				$\Sigma f_i\lvert d_i\rvert = 140$

Median number $= \frac{n}{2} = \frac{20}{2} = 10$ ($\because$ n is even number)

$\therefore$ Median class $= 12 - 18$

Median $= l + \frac{h}{f}\left(\frac{n}{2} - C\right) = 12 + \frac{6}{3}(10 - 9) = 12 + 2 = 14$

Mean deviation $= \frac{\Sigma f_i|d_i|}{\Sigma f_i} = \frac{140}{20} = 7$

Hence, mean deviation about the median is 7.

38. *(a)* I. Let the number of boys and girls be x and y.

$\therefore \quad 52x + 42y = 50(x + y)$

$\Rightarrow \quad 2x = 8y$

$\Rightarrow \quad x = 4y$

$\therefore$ Total number of students in the class $= x + y = 5y$

$\therefore$ Required percentage of boys

$$= \frac{4y}{5y} \times 100\%$$

$$= 80\%$$

II. Let the number of boys be x and number of girls be y.

$\therefore \quad 53(x + y) = 55y + 50x$

$\Rightarrow \quad 3x = 2y$

$\Rightarrow \quad x = \frac{2y}{3}$

$\therefore$ Total number of students $= x + y = \frac{2y}{3} + y = \frac{5}{3}y$

Hence, required percentage $= \frac{y}{5y/3} \times 100\%$

$$= \frac{3}{5} \times 100\% = 60\%$$

39. *(b)* Arranging the terms in increasing order

Value (x)	Frequency (f)	cf	$\lvert x_i - M \rvert$	$f_i \lvert x_i - M \rvert$
7	2	2	+3	6
8	1	3	+2	2
9	5	8	+1	5
10	4	12	0	0
11	6	18	1	6
12	1	19	2	2
13	3	22	3	9
	$N = 22$			30

$\therefore$ Median number $= \frac{N+1}{2} = 11.5$

which comes under the cumulative frequency 12. The corresponding value of x will be the median, *i.e.*,

$$\text{Median} = 10$$

$$\text{MD} = \frac{\Sigma f_i |x_i - M|}{\Sigma f_i}$$

$$= \frac{30}{22} = 1.36$$

40. *(d)* After arranging the terms in ascending order median is the $\left(\frac{n+1}{2}\right)^{th}$ term, *i.e.*, 5th term.

Here, we increase largest four observations of the set which will come after 5th term.

Hence, median remains the same as that of original set.

41. *(c)* I. Since, 44 kg is replaced by 46 kg and 27 kg is replaced by 25 kg, then the given series becomes 31, 35, 25, 29, 32, 43, 37, 41, 34, 28, 36, 46, 45, 42, 30.

On arranging this series in ascending order, we get 25, 28, 29, 30, 31, 32, 34, 35, 36, 37, 41, 42, 43, 45, 46.

Total number of students are 15, therefore middle term is 8th whose corresponding value is 35.

II. On arranging the terms in increasing order of magnitude

40, 42, 45, 47, 50, 51, 54, 55, 57

Number of terms, $N = 9$

$\therefore$ Median $= \left(\frac{9+1}{2}\right)^{th}$ term $= 5^{th}$ term $= 50$ kg

Weight (in kg)	Deviation from median (d)	$\lvert d \rvert$
40	−10	10
42	−8	8
45	−5	5
47	−3	3
50	0	0
51	1	1
54	4	4
55	5	5
57	7	7
		$\lvert d \rvert = 43$

MD from median $= \frac{43}{9} = 4.78$ kg

42. *(b)* Let n_1 and n_2 be the numbers of men and women in a group.

According to the given condition,

$$\frac{n_1 \times 26 + n_2 \times 21}{n_1 + n_2} = 25$$

$\Rightarrow \quad 26n_1 + 21n_2 = 25n_1 + 25n_2$

$\Rightarrow \quad n_1 = 4n_2$

$\Rightarrow \quad \frac{n_1}{n_2} = \frac{4}{1}$

43. *(d)* Mean $(\bar{x}) = \frac{-1+0+4}{3} = 1$

$\therefore \quad \text{MD} = \frac{\Sigma |x_i - \bar{x}|}{n}$

$$= \frac{|-1-1| + |0-1| + |4-1|}{3} = 2$$

44. *(b)* Total number of students $= 100$

Average marks of the class $= 72$

Total marks of the class $= 72 \times 100 = 7200$

and total marks of the boys $= 70 \times 75 = 5250$

So, total marks of the girls $= 7200 - 5250 = 1950$

Hence, average marks of girls $= \frac{1950}{30} = 65$

45. *(c)* If $x_1, x_2, x_3, \ldots, x_n$ be n observations and $\bar{x}$ is their mean. Then,

$$(x_1 - \bar{x})^2 + (x_2 - \bar{x})^2 + \ldots + (x_n - \bar{x})^2 = \sum_{i=1}^{n} (x_i - \bar{x})^2$$

I. If this sum is zero, then each $(x_i - \bar{x})$ has to be zero. This implies that there is no dispersion at all as all observations are equal to the mean $\bar{x}$.

II. If $\sum_{i=1}^{n} (x_i - \bar{x})^2$ is small, this indicates that the observations $x_1, x_2, x_3, \ldots, x_n$ are close to the mean $\bar{x}$ and therefore, there is a lower degree of dispersion.

III. If this sum is large, there is a higher degree of dispersion of the observations from the mean $\bar{x}$.

IV. Can we thus say that the sum $\sum_{i=1}^{n} (x_i - \bar{x})^2$ is reasonable indicator of the degree of dispersion or scatter?

Now we can understand above statements with the help of an example.

Let us take the set A of six observations 5, 15, 25, 35, 45, 55. The mean of the observations is $\bar{x} = 30$. The sum of squares of deviations from $\bar{x}$ for this set is

$$\sum_{i=1}^{6} (x - \bar{x})^2 = (5-30)^2 + (15-30)^2 + (25-30)^2 + (35-30)^2 + (45-30)^2 + (55-30)^2$$

$$= 625 + 225 + 25 + 25 + 225 + 625 = 1750$$

Let us now take another set B of 31 observations 15, 16, 17, 18, 19, 20, 21, 22, 23, 24, 25, 26, 27, 28, 29, 30, 31, 32, 33, 34, 35, 36, 37, 38, 39, 40, 41, 42, 43, 44, 45. The mean of these observations is $\bar{y} = 30$.

Note that both the sets A and B of observations have a mean of 30.

Now, the sum of squares of deviations of observations for set B from the mean $\bar{y}$ is given by

$$\sum_{i=1}^{31} (y_i - \bar{y}) = (15 - 30)^2 + (16 - 30)^2 + (17 - 30)^2 + \ldots + (44 - 30)^2 + (45 - 30)^2$$

$$= (-15)^2 + (-14)^2 + \ + (-1)^2 + 0^2 + 1^2 + 2^2 + 3^2 + \ldots + 14^2 + 15^2$$

$$= 2[15^2 + 14^2 + \ldots + 1^2]$$

$$= 2 \times \frac{15 \times (15 + 1)(30 + 1)}{6} = 5 \times 16 \times 31 = 2480$$

[Because sum of squares of first n natural numbers $= \frac{n(n+1)(2n+1)}{6}$, here $n = 15$]

If $\sum_{i=1}^{n} (x_i - \bar{x})^2$ is simply our measure of dispresion or scatter about mean, we will tend to say that the set A of six observations has a lesser dispersion about the mean than the set B of 31 observations,even though the observations in set A are more scattered from the mean (the range of deviations being from −25 to 25) than in the set B (where the range of deviations is from −15 to 15).

This is also clear from the following diagrams.

For the set A, we have

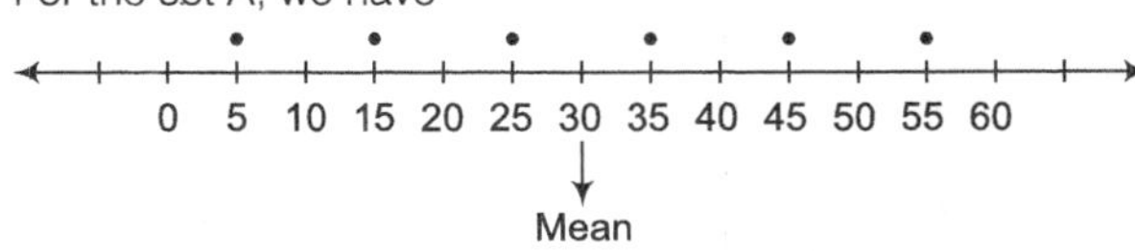

For the set B, we have

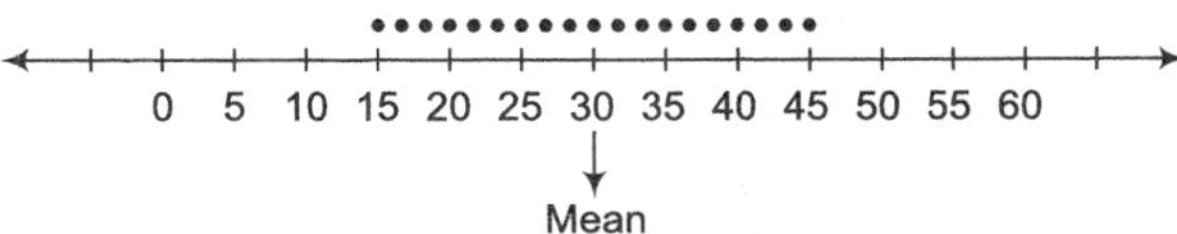

Thus, we can say that the sum of squares of deviations from the mean is not a proper measure of dispersion. To overcome this difficulty we take the mean of the squares of the deviations, i.e., we take $\frac{1}{n} \sum_{i=1}^{n} (x_i - \bar{x})^2$. In case of the set A, we have

$$\text{Mean} = \frac{1}{6} \times 1750 = 291.67$$

and in case of the set B,

$$\text{Mean} = \frac{1}{31} \times 2480 = 80$$

This indicates that the scatter or dispersion is more in set A than the scatter or dispersion in set B, which confirms with the geometrical representation of the two sets.

46. *(c)* We can take $\frac{1}{n} \sum_{i=1}^{n} (x_i - \bar{x})^2$ as a quantity which leads to a proper measure of dispersion.

47. *(b)* This number, *i.e.*, mean of the squares of the deviations from mean is called the variance and is denoted by σ^2 (read as sigma square).

48. *(b)* The variance of n observations $x_1, x_2, \ldots x_n$ is given by

$$\sigma^2 = \frac{1}{n} \sum_{i=1}^{n} (x_i - \bar{x})^2$$

49. *(a)* In the calculation of variance, we find that the units of individual observations x_i and the unit of their mean $\bar{x}$ are different from that of variance, since variance involves the sum of squares of $(x_i - \bar{x})$. For this reason, the proper measure of dispersion about the mean of a set of observations is expressed as positive square-root of the variance and is called standard deviation.

50. *(c)* The standard deviation, usually denoted by σ, is given by

$$\sigma = \sqrt{\frac{1}{n} \sum_{i=1}^{n} (x_i - \bar{x})^2}$$

51. *(a)* From the given data we can form the following table. The mean is calculated by step-deviation method taking 14 as assumed mean. The number of observations is $n = 10$.

x_i	$d_i = \frac{x_i - 14}{2}$	Deviations from mean $(x_i - \bar{x})$	$(x_i - \bar{x})^2$
6	−4	−9	81
8	−3	−7	49
10	−2	−5	25
12	−1	−3	9
14	0	−1	1
16	1	1	1
18	2	3	9
20	3	5	25
22	4	7	49
24	5	9	81
	5		330

$$\therefore \quad \text{Mean } (\bar{x}) = \text{Assumed mean} + \frac{\sum_{i=1}^{n} d_i}{n} \times h$$

$$= 14 + \frac{5}{10} \times 2$$

$$= 15$$

$$\text{Variance } (\sigma^2) = \frac{1}{n}\sum_{i=1}^{10}(x_i - \bar{x})^2 = \frac{1}{10} \times 330 = 33$$

$$\therefore \text{ Standard deviation } (\sigma) = \sqrt{33}$$

$$= 5.74$$

52. *(c)* Presenting the data in tabular form, we get

x_i	f_i	$f_i x_i$	$x_i - \bar{x}$	$(x_i - \bar{x})^2$	$f_i(x_i - \bar{x})^2$
4	3	12	−10	100	300
8	5	40	−6	36	180
11	9	99	−3	9	81
17	5	85	3	9	45
20	4	80	6	36	144
24	3	72	10	100	300
32	1	32	18	324	324
	30	420			1374

$$N = 30, \sum_{i=1}^{7} f_i x_i = 420, \sum_{i=1}^{7} f_i(x_i - \bar{x})^2 = 1374$$

$$\text{Therefore,} \quad \bar{x} = \frac{\sum_{i=1}^{7} f_i x_i}{N} = \frac{1}{30} \times 420 = 14$$

$$\therefore \text{ Variance } (\sigma^2) = \frac{1}{N}\sum_{i=1}^{7} f_i(x_i - \bar{x})^2$$

$$= \frac{1}{30} \times 1374$$

$$= 45.8$$

$$\text{Standard deviation } (\sigma) = \sqrt{45.8} = 6.77$$

53. *(a)* From the given data, we construct the following table

Class	Frequency (f_i)	Mid value (x_i)	$f_i x_i$	$(x_i - \bar{x})^2$	$f_i(x_i - \bar{x})^2$
30-40	3	35	105	729	2189
40-50	7	45	315	289	2023
50-60	12	55	660	49	588
60-70	15	65	975	9	135
70-80	8	75	600	169	1352
80-90	3	85	255	529	1587
90-100	2	95	190	1089	2178
	50		3100		10050

$$\therefore \quad \text{Mean } (\bar{x}) = \frac{1}{N}\sum_{i=1}^{7} f_i x_i = \frac{3100}{50} = 62$$

$$\text{Variance } (\sigma)^2 = \frac{1}{N}\sum_{i=1}^{7} f_i(x_i - \bar{x})^2 = \frac{1}{50} \times 10050 = 201$$

$$\text{Standard deviation } (\sigma) = \sqrt{201} = 14.18$$

54. *(c)* Let us form the following table

x_i	f_i	$f_i x_i$	x_i^2	$f_i x_i^2$
3	7	21	9	63
8	10	80	64	640
13	15	195	169	2535
18	10	180	324	3240
23	6	138	529	3174
	48	614		9652

Now, we have

$$\sigma = \frac{1}{N}\sqrt{N\Sigma f_i x_i^2 - (\Sigma f_i x_i)^2}$$

$$= \frac{1}{48}\sqrt{48 \times 9652 - (614)^2}$$

$$= \frac{1}{48}\sqrt{463296 - 376996}$$

$$= \frac{1}{48} \times 293.77 = 6.12$$

$$\therefore \text{Standard deviation } (\sigma) = 6.12$$

55. *(c)* Let the assumed mean $A = 65$. Here, $h = 10$

We obtain the following table from the given data.

Class	Frequency (f_i)	Mid value (x_i)	$y_i = \frac{x_i - 65}{10}$	y_i^2	$f_i y_i$	$f_i y_i^2$
30-40	3	35	−3	9	−9	27
40-50	7	45	−2	4	−14	28
50-60	12	55	−1	1	−12	12
60-70	15	65	0	0	0	0
70-80	8	75	1	1	8	8
80-90	3	85	2	4	6	12
90-100	2	95	3	9	6	18
	$N = 50$				−15	105

$$\text{Therefore,} \quad \bar{x} = A + \frac{\Sigma f_i y_i}{50} \times h = 65 - \frac{15}{50} \times 10 = 62$$

Variance, $\sigma^2 = \frac{h^2}{N^2}[N\Sigma f_i y_i^2 - (\Sigma f_i y_i)^2]$

$$= \frac{(10)^2}{(50)^2}[50 \times 105 - (-15)^2]$$

$$= \frac{1}{25}[5250 - 225] = 201$$

Standard deviation $(\sigma) = \sqrt{201} = 14.18$

56. *(c)*

x_i	x_i^2
6	36
7	49
10	100
12	144
13	169
4	16
8	64
12	144
Total = 72	**722**

$$\text{Mean} = \frac{\Sigma x_i}{n} = \frac{72}{8} = 9$$

$$\text{Variance} = \frac{\Sigma x_i^2}{n} - \left(\frac{\Sigma x_i}{n}\right)^2 = \frac{722}{8} - \left(\frac{72}{8}\right)^2$$

$$= 90.25 - 81 = 9.25$$

57. *(b)* I. First n natural numbers are 1, 2, 3, 4, ..., n.

x_i	x_i^2
1	1^2
2	2^2
3	3^2
4	4^2
⋮	⋮
n	n^2
Total = $\frac{n(n+1)}{2}$	$\frac{n(n+1)(2n+1)}{6}$

$\therefore$ $$\text{Mean} = \frac{\Sigma x_i}{n}$$

$\therefore$ $$\bar{x} = \frac{n(n+1)}{2n} = \frac{n+1}{2}$$

$$\text{Variance} = \frac{\Sigma x_i^2}{n} - \left(\frac{\Sigma x_i}{n}\right)^2$$

$$= \frac{n(n+1)(2n+1)}{6n} - \left[\frac{n(n+1)}{2n}\right]^2$$

$$= \frac{(n+1)(2n+1)}{6} - \frac{(n+1)^2}{4}$$

$$= \frac{(n+1)}{2}\left[\frac{2n+1}{3} - \frac{n+1}{2}\right]$$

$$= \frac{(n+1)}{2}\left[\frac{4n+2-3n-3}{6}\right]$$

$$= \left(\frac{n+1}{2}\right)\left[\frac{n-1}{6}\right] = \frac{n^2-1}{12}$$

II. First 10 positive multiples of 3 are 3, 6, 9, 12, 15, 18, 21, 24, 27, 30.

x_i	x_i^2
3	9
6	36
9	81
12	144
15	225
18	324
21	441
24	576
27	729
30	900
Total = 165	**3465**

$$\text{Mean } (\bar{x}) = \frac{\Sigma x}{n} = \frac{165}{10} = 16.5$$

$$\text{Variance} = \frac{\Sigma x^2}{n} - \left(\frac{\Sigma x}{n}\right)^2 = \frac{3465}{10} - \left(\frac{165}{10}\right)^2$$

$$= 346.5 - (16.5)^2 = 346.5 - 272.25 = 74.25$$

58. *(c)*

x_i	f_i	x_i^2	$f_i x_i$	$f_i x_i^2$
6	2	36	12	72
10	4	100	40	400
14	7	196	98	1372
18	12	324	216	3888
24	8	576	192	4608
28	4	784	112	3136
30	3	900	90	2700
Total = 130	**40**		**760**	**16176**

$$\text{Mean } (\bar{x}) = \frac{\Sigma f_i x_i}{\Sigma f_i} = \frac{760}{40} = 19$$

$$\text{Variance} = \frac{\Sigma f_i x_i^2}{\Sigma f_i} - \left(\frac{\Sigma f_i x_i}{\Sigma f_i}\right)^2$$

$$= \frac{16176}{40} - \left(\frac{760}{40}\right)^2$$

$$= 404.4 - (19)^2$$

$$= 404.4 - 361 = 43.4$$

59. *(b)*

x_i	f_i	x_i^2	$f_i x_i$	$f_i x_i^2$
92	3	8464	276	25392
93	2	8649	186	17298
97	3	9409	291	28227
98	2	9604	196	19208
102	6	10404	612	62424
104	3	10816	312	32448
109	3	11881	327	35643
Total	**22**		**2200**	**220640**

$$\text{Mean } (\bar{x}) = \frac{\Sigma f_i x_i}{\Sigma f_i} = \frac{2200}{22} = 100$$

$$\text{Variance} = \frac{\Sigma f_i x_i^2}{\Sigma f_i} - \left(\frac{\Sigma f_i x_i}{\Sigma f_i}\right)^2$$

$$= \frac{220640}{22} - (100)^2$$

$$= 10029.09 - 10000 = 29.09$$

60. *(b)*

Mid value (x_i)	Frequency (f_i)	Deviation from mean $d_i = x_i - A$, $A = 64$	d_i^2	$f_i d_i$	$f_i d_i^2$
60	2	−4	16	−8	32
61	1	−3	9	−3	9
62	12	−2	4	−24	48
63	29	−1	1	−29	29
64	25	0	0	0	0
65	12	1	1	12	12
66	10	2	4	20	40
67	4	3	9	12	36
68	5	4	16	20	80
Total	**100**	**0**		**0**	**286**

$$\text{Mean } (\bar{x}) = A + \frac{\Sigma f d_i}{\Sigma f_i}$$

$$(\bar{x}) = 64 + \frac{0}{100} = 64$$

$$\text{Standard deviation } (\sigma) = \sqrt{\frac{\Sigma f d_i^2}{\Sigma f_i} - \left(\frac{\Sigma f d_i}{\Sigma f_i}\right)^2}$$

$$= \sqrt{\frac{286}{100} - \left(\frac{0}{100}\right)^2} = \sqrt{2.86} = 1.69$$

61. *(d)*

Class	Frequency (f_i)	Mid value (x_i)	Deviation from mean $d_i = \frac{x_i - A}{h}$, $A = 105$	d_i^2	$f_i d_i$	$f_i d_i^2$
0-30	2	15	−3	9	−6	18
30-60	3	45	−2	4	−6	12
60-90	5	75	−1	1	−5	5
90-120	10	105	0	0	0	0
120-150	3	135	1	1	3	3
150-180	5	165	2	4	10	20
180-210	2	195	3	9	6	18
Total	**30**				**2**	**76**

$$\text{Mean } (\bar{x}) = A + \frac{\Sigma f d_i}{\Sigma f_i} \times h$$

$$= 105 + \frac{2}{30} \times 30 \qquad (\because h = 30)$$

$$= 105 + 2 = 107$$

$$\text{Variance} = \left[\frac{\Sigma f d_i^2}{\Sigma f_i} - \left(\frac{\Sigma f d_i}{\Sigma f_i}\right)^2\right] \times h^2$$

$$= \left[\frac{76}{30} - \left(\frac{2}{30}\right)^2\right] \times (30)^2 = \left[\frac{76}{30} - \frac{4}{30 \times 30}\right] \times 900$$

$$= \left[\frac{76 \times 30 - 4}{900}\right] \times 900 = \left(\frac{2280 - 4}{900}\right) \times 900$$

$$= \frac{2276}{900} \times 900 = 2276$$

62. *(d)*

Class	Frequency (f_i)	Mid value (x_i)	Deviation from mean $d_i = \frac{x_i - 25}{10}$	d_i^2	$f_i d_i$	$f_i d_i^2$
0-10	5	5	−2	4	−10	20
10-20	8	15	−1	1	−8	8
20-30	15	25	0	0	0	0
30-40	16	35	1	1	16	16
40-50	6	45	2	4	12	24
Total	**50**				**10**	**68**

$$\text{Mean } (\bar{x}) = A + \frac{\Sigma f d_i}{\Sigma f_i} \times h = 25 + \frac{10}{50} \times 10$$

$$= 25 + \frac{100}{50} = 25 + 2 = 27$$

$$\text{Variance} = \left[\frac{\Sigma f d_i^2}{\Sigma f_i} - \left(\frac{\Sigma f d_i}{\Sigma f_i}\right)^2\right] \times h^2$$

$$= \left[\frac{68}{50} - \left(\frac{10}{50}\right)^2\right] \times (10)^2 = \frac{[68 \times 50 - 100]}{50 \times 50} \times 100$$

$$= \frac{(3400 - 100)}{50} \times 2 = \frac{3300 \times 2}{50} = \frac{6600}{50} = 132$$

63. *(c)*

Class	Frequency (f_i)	Mid value (x_i)	Deviation from mean $d_i = \frac{x_i - 92.5}{5}$	d_i^2	$f_i d_i$	$f_i d_i^2$
70-75	3	72.5	−4	16	−12	48
75-80	4	77.5	−3	9	−12	36
80-85	7	82.5	−2	4	−14	28
85-90	7	87.5	−1	1	−7	7
90-95	15	92.5	0	0	0	0
95-100	9	97.5	1	1	9	9
100-105	6	102.5	2	4	12	24
105-110	6	107.5	3	9	18	54
110-115	3	112.5	4	16	12	48
Total	**60**				**6**	**254**

Let assumed mean, $A = 92.5$, $h = 5$

$$\text{Mean } (\bar{x}) = A + \frac{\Sigma f d_i}{\Sigma f_i} \times h = 92.5 + \frac{6}{60} \times 5$$

$$= 92.5 + \frac{30}{60} = 92.5 + 0.5 = 93$$

$$\text{Variance} = \left[\frac{\Sigma f d_i^2}{\Sigma f_i} - \left(\frac{\Sigma f d_i}{\Sigma f_i}\right)^2\right] \times h^2$$

$$= \left[\frac{254}{60} - \left(\frac{6}{60}\right)^2\right] \times (5)^2$$

$$= \left[\frac{254 \times 60 - 36}{60 \times 60}\right] \times 25 = \frac{(15240 - 36)}{60 \times 12} \times 5$$

$$= \frac{15204 \times 5}{720} = 105.58$$

Standard deviation $(\sigma) = \sqrt{\text{Variance}} = \sqrt{105.58} = 10.27$

64. *(c)*

Class	f_i	Mid value (x_i)	Deviation from mean $d_i = \frac{x_i - 42.5}{4}$, $A = 42.5$, $h = 4$	d_i^2	$f_i d_i$	$f_i d_i^2$
32.5-36.5	15	34.5	−2	4	−30	60
36.5-40.5	17	38.5	−1	1	−17	17
40.5-44.5	21	42.5	0	0	0	0
44.5-48.5	22	46.5	1	1	22	22
48.5-52.5	25	50.5	2	4	50	100
Total	**100**				**25**	**199**

$$\text{Mean } (\bar{x}) = A + \frac{\Sigma f d_i}{\Sigma f_i} \times h$$

$$= 42.5 + \frac{25}{100} \times 4$$

$$= 42.5 + 1 = 43.5$$

$$\text{Standard deviation } (\sigma) = \sqrt{\frac{\Sigma f d_i^2}{\Sigma f_i} - \left(\frac{\Sigma f d_i}{\Sigma f_i}\right)^2} \times h$$

$$= \sqrt{\frac{199}{100} - \left(\frac{25}{100}\right)^2} \times 4$$

$$= \sqrt{\frac{199 \times 100 - 625}{100 \times 100}} \times 4$$

$$= \frac{4}{100}\sqrt{19900 - 625}$$

$$= \frac{1}{25} \times \sqrt{19275} = \frac{138.83}{25} = 5.55$$

65. *(b)* Mean $(\bar{x})$

$$= \frac{57 + 64 + 43 + 67 + 49 + 59 + 61 + 59 + 44 + 47}{10}$$

$$= \frac{550}{10} = 55$$

$$\text{Variance } (\sigma^2) = \frac{\Sigma (x_i - \bar{x})^2}{n}$$

$$= \frac{2^2 + 9^2 + 12^2 + 12^2 + 6^2 + 4^2 + 6^2 + 4^2 + 11^2 + 8^2}{10}$$

$$= \frac{662}{10} = 66.2$$

Standard deviation $(\sigma) = \sqrt{\sigma^2} = \sqrt{662} = 8.13$

66. *(c)*

Class interval	Mid value (x_i)	f_i
4-8	6	3
8-12	10	6
12-16	14	4
16-20	18	7

$$\text{Mean } (\bar{x}) = \frac{\Sigma f_i x_i}{\Sigma f_i} = \frac{3 \times 6 + 6 \times 10 + 4 \times 14 + 7 \times 18}{20} = 13$$

$$\text{Variance } (\sigma^2) = \frac{\Sigma f_i (x_i - \bar{x})^2}{\Sigma f_i} = \frac{3(-7)^2 + 6(-3)^2 + 4(1)^2 + 7(5)^2}{20}$$

$$= \frac{147 + 54 + 4 + 175}{20} = 19$$

67. *(a)*

x_i	$d_i = (x_i - 9)$	d_i^2
5	−4	16
6	−3	9
8	−1	1
9	0	0
10	1	1
12	3	9
13	4	16
$\Sigma x_i = 63$	$\Sigma \lvert d_i \rvert = 0$	$\Sigma d_i^2 = 52$

Total number of data $(n) = 7$

$$\text{Mean } (\bar{x}) = \frac{63}{7} = 9$$

$$\text{SD} = \sqrt{\frac{\Sigma d_i^2}{\Sigma f_i}} = \sqrt{\frac{52}{7}}$$

68. *(c)* Let A, the assumed mean, be 25.5. Here, $h = 10$.

Class	x_i	$y_i = \frac{x_i - 25.5}{10}$	f_i	$f_i y_i$	$f_i y_i^2$
1-10	5.5	−2	11	−22	44
11-20	15.5	−1	29	−29	29
21-30	25.5	0	18	0	0
31-40	35.5	1	4	4	4
41-50	45.5	2	5	10	20
51-60	55.5	3	3	9	27
			70	**−28**	**124**

$$\frac{\Sigma f_i y_i}{\Sigma f_i} = \frac{-28}{70} = -0.4$$

$$\bar{x} = A + \frac{\Sigma f_i y_i}{\Sigma f_i} \times h$$

$$\text{Mean} = \bar{x} = 25.5 + (10)(-0.4) = 21.5$$

$$\text{Variance } (\sigma^2) = \left[\frac{h}{N}\sqrt{N\Sigma f_i y_i^2 - (\Sigma f_i y_i)^2}\right]^2$$

$$= \frac{10 \times 10}{70 \times 70}[70(124) - (-28)^2]$$

$$= \frac{70(124)}{7 \times 7} - \frac{28 \times 28}{7 \times 7}$$

$$= \frac{1240}{7} - 16 = 161.14$$

$$\text{SD }(\sigma) = \sqrt{161 \cdot 14} = 12.69 \approx 12.7$$

69. *(c)* When each observation is multiplied by 2, then variance is also multiplied by 2.

We are given, 2, 4, 6, 8, 17.

When each observation multiplied by 2, we get 4, 8, 12, 16, 34.

$\therefore$ Variance of new series $= 2^2 \times$ Variance of given data

$$= 4 \times 23.33 = 93.32$$

70. *(c)*

x_i	f_i	$f_i x_i$	$f_i x_i^2$
A	2	$2A$	$2A^2$
$2A$	1	$2A$	$4A^2$
$3A$	1	$3A$	$9A^2$
$4A$	1	$4A$	$16A^2$
$5A$	1	$5A$	$25A^2$
$6A$	1	$6A$	$36A^2$
Total	**7**	**$22A$**	**$92A^2$**

$\because \quad \sigma^2 = \dfrac{\Sigma f_i x_i^2}{\Sigma f_i} - \left(\dfrac{\Sigma f_i x_i}{\Sigma f_i}\right)^2$

$\Rightarrow \quad 160 = \dfrac{92A^2}{7} - \left(\dfrac{22A}{7}\right)^2$

$\Rightarrow \quad 160 = \dfrac{92A^2}{7} - \dfrac{484A^2}{49}$

$\Rightarrow \quad 160 = \dfrac{92 \times 7A^2 - 484A^2}{49}$

$\Rightarrow \quad 160 \times 49 = 644A^2 - 484A^2$

$\Rightarrow \quad 160A^2 = 7840$

$\Rightarrow \quad A^2 = \dfrac{7840}{160}$

$\Rightarrow \quad A^2 = 49$

$\Rightarrow \quad A = \pm 7$

$A = 7$ as A is a positive integer.

71. *(c)* We have, $n = 100$, $\bar{x} = 50$, $\sigma = 5$, $\sigma^2 = 25$

We know that,

$$\sigma^2 = \frac{\Sigma x_i^2}{n} - \left(\frac{1}{n}\Sigma x_i\right)^2$$

$\Rightarrow \quad 25 = \dfrac{\Sigma x_i^2}{100} - (50)^2$

$\Rightarrow \quad 2500 = \Sigma x_i^2 - 250000$

$\Rightarrow \quad \Sigma x_i^2 = 252500$

72. *(b)* First find x.

As given total number of students is 60 *i.e.*,

$x - 2 + x + x^2 + (x+1)^2 + 2x + x + 1 = 60$

$\Rightarrow \quad 5x - 2 + x^2 + x^2 + 1 + 2x + 1 = 60$

$\Rightarrow \quad 2x^2 + 7x = 60$

$\Rightarrow \quad 2x^2 + 7x - 60 = 0$

$\Rightarrow \quad 2x^2 + 15x - 8x - 60 = 0$

$\Rightarrow \quad x(2x + 15) - 4(2x + 15) = 0$

$\Rightarrow \quad (2x + 15)(x - 4) = 0$

$\Rightarrow \quad x = -\dfrac{15}{2}, x = 4$

$\Rightarrow \quad x = 4 \quad (\because x \text{ is positive})$

Marks (x_i)	Frequency (f_i)	x_i^2	$f_i x_i$	$f_i x_i^2$
0	2	0	0	0
1	4	1	4	4
2	16	4	32	64
3	25	9	75	225
4	8	16	32	128
5	5	25	25	125
Total	**60**		**168**	**546**

$$\text{Mean }(\bar{x}) = \frac{\Sigma f_i x_i}{N} = \frac{168}{60} = 2.8$$

$\therefore$ Variance $(\sigma^2) = \sqrt{\dfrac{\Sigma f_i x_i^2}{N} - (\bar{x})^2} = \sqrt{\dfrac{546}{60} - (2.8)^2}$

$$= \sqrt{9.1 - 7.84} = \sqrt{1.26} = 1.12$$

73. *(d)* We have the following numbers

1, 2, 3, 4, 5, 6, 7, 8, 9, 10

If 1 is added to each number, we get

2, 3, 4, 5, 6, 7, 8, 9, 10, 11

Sum of these numbers, $\Sigma x_i = 2 + 3 + \ldots + 11 = 65$

Sum of squares of these numbers,

$$\Sigma x_i^2 = 2^2 + 3^2 + \ldots + 11^2 = 505$$

$$\text{Variance }(\sigma^2) = \frac{\Sigma x_i^2}{n} - \left(\frac{\Sigma x_i}{n}\right)^2$$

$$= \frac{505}{10} - (6.5)^2 = 50.5 - 42.25 = 8.25$$

74. *(c)* I.

x_i	f_i	$d_i = x_i - 4$	$f_i d_i$	$f_i d_i^2$
2	4	−2	−8	16
3	9	−1	−9	9
4	16	0	0	0
5	14	1	14	14
6	11	2	22	44
7	6	3	18	54
	$\Sigma f_i = 60$		$\Sigma f_i d_i = 37$	$\Sigma f_i d_i^2 = 137$

Standard deviation $(\sigma) = \sqrt{\frac{\Sigma f_i d_i^2}{\Sigma f_i} - \left(\frac{\Sigma f_i d_i}{\Sigma f_i}\right)^2}$

$= \sqrt{\frac{137}{60} - \left(\frac{37}{60}\right)^2}$

$= \sqrt{2.2833 - (0.616)^2}$

$= \sqrt{2.2833 - 0.3794}$

$= \sqrt{1.9039} = 1.38$

II. We have, $n_1 = 60, \bar{x}_1 = 650, \sigma_1 = 8$

$n_2 = 80, \bar{x}_2 = 660, \sigma_2 = 7$

Combined standard deviation (σ)

$= \sqrt{\frac{n_1\sigma_1^2 + n_2\sigma_2^2}{n_1 + n_2} + \frac{n_1 n_2(\bar{x}_1 - \bar{x}_2)^2}{(n_1 + n_2)^2}}$

$= \sqrt{\frac{60 \times 64 + 80 \times 49}{60 + 80} + \frac{60 \times 80 \times (660 - 650)^2}{(60 + 80)^2}}$

$= \sqrt{\frac{6 \times 64 + 8 \times 49}{14} + \frac{60 \times 80 \times 100}{140 \times 140}}$

$= \sqrt{\frac{3 \times 64 + 4 \times 49}{7} + \frac{6 \times 8 \times 100}{14 \times 14}}$

$= \sqrt{\frac{388}{7} + \frac{1200}{49}} = \frac{\sqrt{2716 + 1200}}{49}$

$= \frac{\sqrt{3916}}{7} = \frac{62.58}{7} = 8.9$

75. *(a)* First ten positive integers are 1, 2, 3, 4, 5, 6, 7, 8, 9 and 10.

Sum of these numbers $(\Sigma x_i) = 1 + 2 + \ldots + 10 = 55$

Sum of squares of these numbers $(\Sigma x_i^2) = 1^2 + 2^2 + \ldots + 10^2 = 385$

Standard deviation $(\sigma) = \sqrt{\frac{\Sigma x_i^2}{n} - \left(\frac{1}{n}\Sigma x_i\right)^2}$

$= \sqrt{\frac{385}{10} - (5.5)^2} = \sqrt{38.5 - 30.25} = \sqrt{8.25}$

$\therefore$ Variance $(\sigma^2) = 8.25$

76. *(d)* We have studied about some types of measures of dispersion. The mean deviation and the standard deviation have the same units in which the data are given. Whenever we want to compare the variability of two series with same mean, which are measured in different units, we do not merely calculate the measures of dispersion but we require such measures which are independent of the units. The measure of variability which is independent of units, is called coefficient of variation (denoted as CV).

77. *(d)* The coefficient of variation is defined as

$$CV = \frac{\sigma}{\bar{x}} \times 100, \bar{x} \neq 0$$

where, σ and $\bar{x}$ are the standard deviation and mean of the data, respectively.

78. *(b)* For comparing the variability or dispersion of two series, we calculate the coefficient of variance for each series. The series having greater CV is said to be more variable than the other. The series having lesser CV is said to be more consistent than the other.

79. *(a)* If $\bar{x}_1$ and σ_1 are the mean and standard deviation of the first distribution, and $\bar{x}_2$ and σ_2 are the mean and standard deviation of the second distribution.

Then, CV (1st distribution) $= \frac{\sigma_1}{\bar{x}_1} \times 100$

and CV (2nd distribution) $= \frac{\sigma_2}{\bar{x}_2} \times 100$

Given, $\bar{x}_1 = \bar{x}_2 = \bar{x}$ (say)

Therefore, CV (1st distribution) $= \frac{\sigma_1}{\bar{x}} \times 100$ ...(i)

and CV (2nd distribution) $= \frac{\sigma_2}{\bar{x}} \times 100$ ...(ii)

It is clear from Eqs. (i) and (ii) that the two CVs can be compared on the basis of values of σ_1 and σ_2 only.

Thus, we say that for two series with equal means, the series with greater standard deviation (or variance) is called more variable or dispersed than the other. Also, the series with lesser value of standard deviation (or variance) is said to be more consistent than the other.

80. *(a)* The variance of the distribution of wages in plant A $(\sigma_1^2) = 81$.

Therefore, standard deviation of the distribution of wages in plant A $(\sigma_1) = 9$.

Also, the variance of the distribution of wages in plant B $(\sigma_2^2) = 100$.

Therefore, standard deviation of the distribution of wages in plant B $(\sigma_2) = 10$.

Since, the average monthly wages in both the plants is same, *i.e.*, ₹ 2500. Therefore, the plant with greater standard deviation will have more variability.

Thus, the plant B has greater variability in the individual wages.

81. *(c)* Given, CV (1st distribution) = 60, $\sigma_1 = 21$

and CV (2nd distribution) = 70, $\sigma_2 = 16$

Let $\bar{x}_1$ and $\bar{x}_2$ be the means of 1st and 2nd distributions, respectively. Then,

CV (1st distribution) $= \frac{\sigma_1}{\bar{x}_1} \times 100$

Therefore, $60 = \frac{21}{\bar{x}_1} \times 100 \Rightarrow \bar{x}_1 = \frac{21}{60} \times 100 = 35$

and CV (2nd distribution) $= \frac{\sigma_2}{\bar{x}_2} \times 100$

i.e., $70 = \frac{16}{\bar{x}_2} \times 100 \Rightarrow \bar{x}_2 = \frac{16}{70} \times 100 = 22.85$

82. *(b)* To compare the variability, we have to calculate their coefficients of variation.

Given, variance of height = 127.69 cm^2

Therefore, standard deviation of height = $\sqrt{127.69} = 11.3$ cm

Also, variance of weight = 23.1361 kg^2

Therefore, standard deviation of weight = $\sqrt{23.1361} = 4.81$ kg

Now, the coefficient of variations (CV) are given by

$$\text{CV (in heights)} = \frac{\text{Standard deviation}}{\text{Mean}} \times 100$$

$$= \frac{11.3}{162.6} \times 100 = 6.95$$

$$\text{and CV (in weights)} = \frac{4.81}{52.36} \times 100 = 9.18$$

Clearly, CV in weights is greater than the CV in heights.

Therefore, we can say that weights show more variability than heights.

83. *(b)* **For group A**

Class	f_i	x_i	$d_i = \frac{x_i - 45}{10}$	d_i^2	$f_i d_i$	$f_i d_i^2$
10-20	9	15	−3	9	−27	81
20-30	17	25	−2	4	−34	68
30-40	32	35	−1	1	−32	32
40-50	33	45	0	0	0	0
50-60	40	55	1	1	40	40
60-70	10	65	2	4	20	40
70-80	9	75	3	9	27	81
Total	**150**				**−6**	**342**

$$\text{Mean } (\bar{x}) = A + \frac{\Sigma f d_i}{\Sigma f_i} \times h = 45 + \frac{(-6)}{150} \times 10$$

$$= 45 - \frac{60}{150}$$

$$= 45 - 0.4 = 44.6$$

$$\text{Standard deviation } (\sigma) = \sqrt{\frac{\Sigma f d_i^2}{\Sigma f_i} - \left(\frac{\Sigma f d_i}{\Sigma f_i}\right)^2} \times h$$

$$= \sqrt{\frac{342}{150} - \left(-\frac{6}{150}\right)^2} \times 10$$

$$= \sqrt{\frac{342 \times 150 - 36}{150 \times 150}} \times 10$$

$$= \frac{10}{150}\sqrt{51300 - 36}$$

$$= \frac{1}{15} \times \sqrt{51264} = \frac{226.41}{15} = 15.09$$

$$\therefore \quad \text{CV for group A} = \frac{\sigma}{\bar{x}} \times 100$$

$$= \frac{15.09}{44.6} \times 100$$

$$= \frac{1509}{44.6} = 33.83$$

For group B

Class	f_i	x_i	$d_i = \frac{x_i - 45}{10}$	d_i^2	$f_i d_i$	$f_i d_i^2$
10-20	10	15	−3	9	−30	90
20-30	20	25	−2	4	−40	80
30-40	30	35	−1	1	−30	30
40-50	25	45	0	0	00	00
50-60	43	55	1	1	43	43
60-70	15	65	2	4	30	60
70-80	7	75	3	9	21	63
Total	**150**				**−6**	**366**

$$\text{Mean } \bar{x} = A + \frac{\Sigma f d_i}{\Sigma f_i} \times h = 45 + \frac{(-6)}{150} \times 10$$

$$= 45 - \frac{60}{150} = 45 - 0.4 = 44.6$$

$$\text{Standard deviation } (\sigma) = \sqrt{\frac{\Sigma f d_i^2}{\Sigma f_i} - \left(\frac{\Sigma f d_i}{\Sigma f_i}\right)^2} \times h$$

$$= \sqrt{\frac{366}{150} - \left(\frac{-6}{150}\right)^2} \times 10$$

$$= \sqrt{\frac{366 \times 150 - 36}{150 \times 150}} \times 10$$

$$= \frac{10}{150}\sqrt{54900 - 36}$$

$$= \frac{1}{15} \times \sqrt{54864} = \frac{234.23}{15} = 15.61$$

$$\text{CV for group B} = \frac{\sigma}{\bar{x}} \times 100$$

$$= \frac{15.61}{44.6} \times 100 = \frac{1561}{44.6} = 35$$

$\because$ Coefficient of variation of group B is greater than the coefficient of variation of group A. Hence, group B is more variable than group A.

84. *(c)*

For share X			For share Y		
X	$d_1 = X - A$	d_1^2	Y	$d_2 = Y - A$	d_2^2
35	−17	289	108	3	9
54	2	4	107	2	4
52 (*A*)	0	0	105 (*A*)	0	0
53	1	1	105	0	0
56	4	16	106	1	1
58	6	36	107	2	4
52	0	0	104	−1	1
50	−2	4	103	−2	4
51	−1	1	104	−1	1
49	−3	9	101	−4	16
Total	**−10**	**360**		**0**	**40**

where, *A* = assumed mean

For share X

$$\text{Mean } (\bar{x}) = A + \frac{\Sigma d_1}{n} = 52 - \frac{10}{10}$$

$$= 52 - 1 = 51$$

$$\text{Standard deviation } (\sigma) = \sqrt{\frac{\Sigma d_1^2}{n} - \left(\frac{\Sigma d_1}{n}\right)^2}$$

$$= \sqrt{\frac{360}{10} - \left(\frac{-10}{10}\right)^2} = \sqrt{36 - 1}$$

$$= \sqrt{35} = 5.92$$

$$\text{CV for share X} = \frac{\sigma}{\bar{x}} \times 100 = \frac{5.92}{51} \times 100$$

$$= \frac{592}{51} = 11.60$$

For share Y

$$\text{Mean } (\bar{y}) = A + \frac{\Sigma d_2}{n} = 105 + \frac{0}{10} = 105$$

$$\text{Standard deviation } (\sigma) = \sqrt{\frac{\Sigma d_2^2}{n} - \left(\frac{\Sigma d_2}{n}\right)^2}$$

$$= \sqrt{\frac{40}{10} - 0} = \sqrt{4} = 2$$

$$\Rightarrow \quad \text{CV for share Y} = \frac{\sigma}{\bar{y}} \times 100 = \frac{2}{105} \times 100$$

$$= \frac{200}{105} = 1.90$$

∵ CV of share X > CV of share Y

∴ Share Y is more stable than share X.

Solutions (Q. Nos. 85-86)

For firm A

Number of wages earners = 586

Mean of monthly wages $(\bar{x})$ = ₹ 5253

Amount paid by firm A = 586 × 5253 = ₹ 3078258 ...(i)

Variance of distribution of wages = 100

Standard deviation $(\sigma) = \sqrt{100} = 10$

$$\text{Coefficient of variation} = \frac{\sigma}{\bar{x}} \times 100$$

$$= \frac{10}{5253} \times 100 = 0.19 \quad \text{...(ii)}$$

For firm B

Number of wages earners = 648

Mean of monthly wages $(\bar{x})$ = ₹ 5253

Amount paid by firm B = 648 × 5253 = ₹ 3403944 ...(iii)

Variance of distribution of wages = 121

Standard deviation $(\sigma) = \sqrt{\text{Variance}} = \sqrt{121} = 11$

$$\text{Coefficient of variation} = \frac{\sigma}{\bar{x}} \times 100 = \frac{11}{5253} \times 100$$

$$= \frac{1100}{5253} = 0.21 \quad \text{...(iv)}$$

85. *(b)* From Eqs. (i) and (iii),

Monthly wages paid by firm A = ₹ 3078258

and from firm B = ₹ 3403944

Thus, firm B pays out larger amount as monthly wages.

86. *(b)* From Eqs. (ii) and (iv),

∵ CV of firm B > CV of firm A as 0.21 > 0.19

∴ Firm B shows greater variability in individual wages.

87. *(a)*

Number of goals scored (x_i)	Number of matches (f_i)	x_i^2	$f_i x_i$	$f_i x_i^2$
0	1	0	0	0
1	9	1	9	9
2	7	4	14	28
3	5	9	15	45
4	3	16	12	48
Total	**25**		**50**	**130**

$$\text{Mean } (\bar{x}) = \frac{\Sigma f_i x_i}{\Sigma f_i} = \frac{50}{25} = 2$$

$$\text{Standard deviation } (\sigma) = \sqrt{\frac{\Sigma f_i x_i^2}{\Sigma f_i} - \left(\frac{\Sigma f_i x_i}{\Sigma f_i}\right)^2}$$

$$= \sqrt{\frac{130}{25} - \left(\frac{50}{25}\right)^2} = \sqrt{\frac{130}{25} - 4}$$

$$= \sqrt{\frac{130 - 100}{25}} = \sqrt{\frac{30}{25}}$$

$$= \frac{1}{5}\sqrt{30} = \frac{5.48}{5} = 1.096$$

$$\text{CV for team } A = \frac{\sigma}{\bar{x}} \times 100 = \frac{1.096}{2} \times 100$$

$$= \frac{109.6}{2} = 54.8$$

For team B

Given, $\bar{x} = 2$ and $\sigma = 1.25$

$$\therefore \quad \text{CV for team B} = \frac{\sigma}{\bar{x}} \times 100 = \frac{1.25}{2} \times 100 = \frac{125}{2} = 62.5$$

∵ CV of team A < CV of team B as 54.8 < 62.5

Hence, team A is more consistent.

88. *(b)* **For length**

$$\text{Mean } (\bar{x}) = \frac{\Sigma x_i}{n} = \frac{212}{50} = 4.24$$

$$\text{Standard deviation } (\sigma) = \sqrt{\frac{\Sigma x_i^2}{n} - \left(\frac{\Sigma x_i}{n}\right)^2}$$

$$= \sqrt{\frac{902.8}{50} - \left(\frac{212}{50}\right)^2}$$

$$= \sqrt{\frac{902.8 \times 50 - (212)^2}{50 \times 50}}$$

$$= \frac{1}{50}\sqrt{45140 - 44944}$$

$$= \frac{1}{50}\sqrt{196} = \frac{14}{50} = \frac{7}{25} = 0.28$$

$\therefore$ Coefficient of variation for length

$$= \frac{\sigma}{\bar{x}} \times 100 = \frac{0.28}{4.24} \times 100 = 6.604$$

For weight

$$\text{Mean } (\bar{y}) = \frac{\Sigma y_i}{n} = \frac{261}{50} = 5.22$$

$$\text{Standard deviation } (\sigma) = \sqrt{\frac{\Sigma y_i^2}{n} - \left(\frac{\Sigma y_i}{n}\right)^2}$$

$$= \sqrt{\frac{1457.6}{50} - \left(\frac{261}{50}\right)^2}$$

$$= \sqrt{\frac{1457.6 \times 50 - (261)^2}{50 \times 50}}$$

$$= \frac{1}{40}\sqrt{72880 - 68121}$$

$$= \frac{1}{50}\sqrt{4759} = \frac{1}{50} \times 68.98 = 1.37$$

$$\therefore \text{ CV of weight} = \frac{\sigma}{\bar{y}} \times 100 = \frac{1.37}{5.22} \times 100 = \frac{137}{5.22} = 26.24$$

$\because$ CV of weight (26.24) > CV of length (6.6)

$\therefore$ Weight is more varying than length.

89. *(a)* We know that,

$$\text{Coefficient of variation} = \frac{\sigma}{\bar{x}} \times 100$$

$$\therefore \text{ CV of 1st distribution} = \frac{\sigma_1}{30} \times 100$$

$$\Rightarrow \quad 50 = \frac{\sigma_1}{30} \times 100$$

[CV of 1st distribution = 50(given)]

$$\Rightarrow \quad \sigma_1 = 15$$

$$\text{Also, CV of 2nd distribution} = \frac{\sigma_2}{25} \times 100$$

$$\Rightarrow \quad 60 = \frac{\sigma_2}{25} \times 100$$

$$\Rightarrow \quad \sigma_2 = \frac{60 \times 25}{100}$$

$$\Rightarrow \quad \sigma_2 = 15$$

Thus, $\sigma_1 - \sigma_2 = 15 - 15 = 0$

Hence, option (a) is correct.

90. *(c)* Since, coefficient of variation

$$= \frac{\text{Standard deviation}}{\text{Mean}} \times 100$$

$$\therefore \quad 45 = \frac{\sigma}{12} \times 100$$

$$\Rightarrow \quad \sigma = \frac{45 \times 12}{100} = 5.4$$

91. *(a)*

Class	Mid value (x)	f	fx	$d = x_i - \bar{x}$	fd	fd^2
0-10	5	2	10	−20.7	−41.4	856.98
10-20	15	10	150	−10.7	−107	1144.9
20-30	25	8	200	−0.7	−5.6	3.92
30-40	35	4	140	9.3	37.2	345.96
40-50	45	6	270	19.3	115.8	2234.94
		$\Sigma f = 30$	$\Sigma fx = 770$		$\Sigma fd = -1$	$\Sigma fd^2 = 4586.70$

$$\bar{x} = \frac{770}{30} = 25.7$$

$$\text{SD } (\sigma) = \sqrt{\frac{\Sigma fd^2}{\Sigma f} - \left(\frac{\Sigma fd}{\Sigma f}\right)^2}$$

$$= \sqrt{\frac{4586.7}{30} - \left(\frac{-1}{30}\right)^2}$$

$$= \sqrt{152.89 - 0.005} = 12.365$$

$$\therefore \text{Coefficient of SD} = \frac{\sigma}{\bar{x}} = \frac{12.365}{25.7} = 0.481$$

and coefficient of variation = Coefficient of SD × 100

$$= 0.481 \times 100 = 48.1$$

92. *(a)* To find out the correct statements, we compute coefficient of variation.

I. **For manufacturer A**

Bursting pressure	Mid value	f_i	$u_i = \frac{x_i - 17.5}{5}$	f_iu_i	$f_iu_i^2$
5-10	7.5	2	−2	−4	8
10-15	12.5	9	−1	−9	9
15-20	17.5	29	0	0	0
20-25	22.5	54	1	54	54
25-30	27.5	11	2	22	44
30-35	32.5	5	3	15	45
		$N = \Sigma f_i = 110$	$\Sigma u_i = 3$	$\Sigma f_iu_i = 78$	$\Sigma f_iu_i^2 = 160$

$$\bar{x}_A = a + h\left(\frac{\Sigma fu_i}{N}\right)$$

$$\Rightarrow \quad \bar{x}_A = 17.5 + 5 \times \frac{78}{110}$$

$$\Rightarrow \quad \bar{x}_A = 17.5 + 3.5$$

$$\Rightarrow \quad \bar{x}_A = 21 \quad \ldots \text{(i)}$$

$$\sigma_A^2 = h^2\left[\frac{1}{N}\Sigma fu_i^2 - \left(\frac{1}{N}\Sigma fu_i\right)^2\right]$$

$$\sigma_A^2 = 25\left[\frac{160}{110} - \left(\frac{78}{110}\right)^2\right]$$

$$\Rightarrow \quad \sigma_A^2 = 25\left[\frac{17600 - 6084}{110 \times 110}\right]$$

$$\Rightarrow \quad \sigma_A^2 = 23.79$$

$$\Rightarrow \quad \sigma_A = \sqrt{23.79} = 4.87$$

$\therefore$ Coefficient of variation $= \dfrac{\sigma_A}{\bar{x}_A} \times 100 = \dfrac{4.87}{21} \times 100$

$= 23.19$ ...(ii)

II. **For manufacturer B**

Bursting pressure	Mid value	f_i	$u_i = \dfrac{x_i - 17.5}{5}$	$f_i u_i$	$f_i u_i^2$
5-10	7.5	9	−2	−18	36
10-15	12.5	11	−1	−11	11
15-20	17.5	18	0	0	0
20-25	22.5	32	1	32	32
25-30	27.5	27	2	54	108
30-35	32.5	13	3	39	117
		$N = \Sigma f_i = 110$	$\Sigma u_i = 3$	$\Sigma f_i u_i = 96$	$\Sigma f_i u_i^2 = 304$

$$\bar{x}_B = a + h\left(\frac{\Sigma f_i u_i}{N}\right)$$

$\Rightarrow \quad \bar{x}_B = 17.5 + 5 \times \dfrac{96}{110}$

$\Rightarrow \quad \bar{x}_B = 17.5 + 4.36$

$\Rightarrow \quad \bar{x}_B = 21.86$ ...(iii)

$$\sigma_B^2 = h^2\left[\frac{1}{N}(\Sigma f_i u_i^2) - \left(\frac{1}{N}\Sigma f_i u_i\right)^2\right]$$

$\Rightarrow \quad \sigma_B^2 = 25\left[\dfrac{304}{110} - \left(\dfrac{96}{110}\right)^2\right]$

$\Rightarrow \quad \sigma_B^2 = 25\left[\dfrac{33440 - 9216}{110 \times 110}\right]$

$\Rightarrow \quad \sigma_B^2 = 50.04$

$\Rightarrow \quad \sigma_B = \sqrt{50.04}$

$\Rightarrow \quad \sigma_B = 7.07$

$\therefore$ Coefficient of variation $= \dfrac{\sigma_B}{\bar{x}_B} \times 100$

$= \dfrac{7.07}{21.86} \times 100 = 32.34$ ...(iv)

We find that the average bursting pressure is higher for manufacturer B [from Eqs. (i) and (iii)].

Also, the coefficient of variation is less for manufacturer A [from Eqs. (ii) and (iv)]. So, bags manufactured by A have more uniform pressure.

93. *(c)* We have,

CV of 1st distribution (CV_1) = 50

CV of 2nd distribution (CV_2) = 60

$\sigma_1 = 10$ and $\sigma_2 = 15$

We know that, $\quad CV = \dfrac{\sigma}{\bar{x}} \times 100$

$\therefore \quad CV_1 = \dfrac{\sigma_1}{\bar{x}_1} \times 100 \Rightarrow 50 = \dfrac{10}{\bar{x}_1} \times 100$

$\Rightarrow \quad \bar{x}_1 = \dfrac{10 \times 100}{50} = 20$

Also, $\quad CV_2 = \dfrac{\sigma_2}{\bar{x}_2} \times 100$

$\Rightarrow \quad 60 = \dfrac{15 \times 100}{\bar{x}_2}$

$\Rightarrow \quad \bar{x}_2 = \dfrac{15 \times 100}{60}$

$\Rightarrow \quad \bar{x}_2 = 25$

Thus, $\quad \bar{x}_2 - \bar{x}_1 = 25 - 20 = 5$

94. *(a)* We have,

$$\Sigma(x_i - \bar{x})^2 = 250$$

$$n = 10 \text{ and } \bar{x} = 50$$

$\therefore \quad \sigma^2 = \left[\dfrac{\sum_{i=1}^{10}(x_i - \bar{x})^2}{n}\right]$

$\Rightarrow \quad \sigma^2 = \left[\dfrac{250}{10}\right]$

$\Rightarrow \quad \sigma = \sqrt{25} = 5$

Coefficient of variation $= \dfrac{\sigma}{\bar{x}} \times 100 = \dfrac{5}{50} \times 100 = 10\%$

95. *(d)* We have, $n = 10, \bar{x} = 12$ and $\Sigma x_i^2 = 1530$

$\therefore \quad \sigma^2 = \dfrac{1}{10}\left(\sum_{i=1}^{10} x_i^2\right) - \left(\dfrac{1}{10}\sum_{i=1}^{10} x_i\right)^2$

$\Rightarrow \quad \sigma^2 = \dfrac{1530}{10} - (12)^2$

$\Rightarrow \quad \sigma^2 = 153 - 144$

$\Rightarrow \quad \sigma^2 = 9$

$\Rightarrow \quad \sigma = 3$

Coefficient of variation $= \dfrac{\sigma}{\bar{x}} \times 100$

$= \dfrac{3}{12} \times 100 = 25\%$

96. *(b)* Given that,

Mean and variance of players A and B are

$\bar{x}_1 = 50, \sigma_1^2 = 36$, or $\sigma_1 = 6$

$\bar{x}_2 = 60, \sigma_2^2 = 81$ or $\sigma_2 = 9$

From the following formula, we can check consistency.

Coefficient of variation $= \dfrac{\text{SD}}{\text{Mean}} \times 100$

Therefore, coefficient of variation of player A

$= \dfrac{\sigma_1}{\bar{x}_1} \times 100 = \dfrac{6}{50} \times 100 = 12\%$

and coffecient of variation of player B

$= \dfrac{\sigma_2}{\bar{x}_2} \times 100 = \dfrac{9}{60} \times 100 = 15\%$

Since, coefficient of variation of player A is less, hence it is more consistent in comparison of player B.

97. *(b)* For consistency of performance, we need to know the coefficient of variation of marks obtained by two students A and B. Therefore, from given data, we make the following table.

Coefficient of variation of marks obtained by student A

x	$d = x - A, A = 72$	d^2
44	−28	784
80	+8	64
76	+4	16
48	−24	576
52	−20	400
72 = A	0	0
68	−4	16
56	−16	256
60	−12	144
54	−18	324
	$\Sigma d = -110$	$\Sigma d^2 = 2580$

Hence, $\sigma_A = \sqrt{\frac{\Sigma d^2}{n} - \left(\frac{\Sigma d}{n}\right)^2} = \sqrt{\frac{2580}{10} - \left(\frac{-110}{10}\right)^2}$

$$= \sqrt{258 - 121} = \sqrt{137} = 11.7$$

$\therefore \quad \bar{x}_A = A + \frac{\Sigma d}{n} = 72 - \frac{110}{10} = 61$

Therefore, coefficient of variation

$$= \frac{SD_A}{\bar{x}_A} \times 100 = \frac{11.7}{61} \times 100 = 19.18$$

Again from given data, we make the following table.

Coefficient of variation of marks obtained by student B

y	$d = y - A, A = 69$	d^2
48	−21	441
75	+6	36
54	−15	225
60	−9	81
63	−6	36
69 = A	0	0
72	+3	9
51	−18	324
57	−12	144
66	−3	9
	$\Sigma d = -75$	$\Sigma d^2 = 1305$

$\therefore \quad \sigma_B = \sqrt{\frac{\Sigma d^2}{n} - \left(\frac{\Sigma d}{n}\right)^2} = \sqrt{\frac{1305}{10} - \left(\frac{-75}{10}\right)^2}$

$$= \sqrt{130.5 - 56.25} = \sqrt{74.25} = 8.6168$$

Now, $\bar{x}_B = A + \frac{\Sigma d}{n} = 69 - \frac{75}{10} = \frac{690 - 75}{10} = 61.5$

$\therefore$ Coefficient of variation $= \frac{\sigma_B}{\bar{x}_B} \times 100$

$$= \frac{8.6168}{61.5} \times 100 = 14.01$$

Since, coefficient of variation is less in case of student B. Therefore, B is more consistent.

Thus, the prize winner is student B.

98. *(b)* Let the observations be $x_1, x_2, \ldots x_{20}$ and $\bar{x}$ be their mean. Given that, variance $= 5$ and $n = 20$. We know that,

$$\text{Variance } (\sigma^2) = \frac{1}{n}\sum_{i=1}^{20}(x_i - \bar{x})^2,$$

i.e., $5 = \frac{1}{20}\sum_{i=1}^{20}(x_i - \bar{x})^2$ or $\sum_{i=1}^{20}(x_i - \bar{x})^2 = 100$ ...(i)

If each observation is multiplied by 2 and the new resulting observations are y_i, then

$$y_i = 2x_i \quad i.e., \quad x_i = \frac{1}{2}y_i$$

Therefore, $\bar{y} = \frac{1}{n}\sum_{i=1}^{20} y_i = \frac{1}{20}\sum_{i=1}^{20} 2x_i = 2 \cdot \frac{1}{20}\sum_{i=1}^{20} x_i$

i.e., $\bar{y} = 2\bar{x}$ or $\bar{x} = \frac{1}{2}\bar{y}$

On substituting the values of x_i and $\bar{x}$ in Eq. (i), we get

$$\sum_{i=1}^{20}\left(\frac{1}{2}y_i - \frac{1}{2}\bar{y}\right)^2 = 100$$

i.e., $\sum_{i=1}^{20}(y_i - \bar{y})^2 = 400$

Thus, the variance of new observations

$$= \frac{1}{20} \times 400 = 20 = 2^2 \times 5$$

Note *The reader may note that if each observation is multiplied by a constant k, the variance of the resulting observations becomes k^2 times the original variance.*

99. *(b)* Let the other two observations be x and y.

Therefore, the series is 1,2,6, x,y.

Now, mean $(\bar{x}) = 4.4 = \frac{1 + 2 + 6 + x + y}{5}$ or $22 = 9 + x + y$

Therefore, $x + y = 13$ ...(i)

Also, variance $(\sigma^2) = 8.24 = \frac{1}{n}\sum_{i=1}^{5}(x_i - \bar{x})^2$

i.e., $8.24 = \frac{1}{5}[(3.4)^2 + (2.4)^2 + (1.6)^2 + x^2 + y^2 - 2 \times 4.4(x + y) + 2 \times (4.4)^2]$

or $41.20 = 11.56 + 5.76 + 2.56 + x^2 + y^2 - 8.8 \times 13 + 38.72$

Therefore, $x^2 + y^2 = 97$ ...(ii)

But from Eq. (i), we have

$$x^2 + y^2 + 2xy = 169 \quad \text{...(iii)}$$

From Eqs. (ii) and (iii), we have

$$2xy = 72 \quad \text{...(iv)}$$

On subtracting Eq. (iv) from Eq. (ii), we get

$$x^2 + y^2 - 2xy = 97 - 72$$

i.e., $(x - y)^2 = 25$ or $x - y = \pm 5$ ...(v)

So, from Eqs. (i) and (v), we get

$x = 9, y = 4$ when $x - y = 5$

or $x = 4, y = 9$ when $x - y = -5$

Thus, the remaining observations are 4 and 9.

100. *(a)* Let $\bar{x}$ be the mean of $x_1, x_2, \ldots, x_n$. Then, variance is given by

$$\sigma_1^2 = \frac{1}{n}\sum_{i=1}^{n}(x_i - \bar{x})^2$$

If a is added to each observation, the new observations will be

$$y_i = x_i + a \quad \ldots(i)$$

Let the mean of the new observations be $\bar{y}$. Then,

$$\bar{y} = \frac{1}{n}\sum_{i=1}^{n} y_i = \frac{1}{n}\sum_{i=1}^{n}(x_i + a)$$

$$= \frac{1}{n}\left[\sum_{i=1}^{n} x_i + \sum_{i=1}^{n} a\right] = \frac{1}{n}\sum_{i=1}^{n} x_i + \frac{na}{n} = \bar{x} + a$$

i.e., $$\bar{y} = \bar{x} + a \quad \ldots(ii)$$

Thus, the variance of the new observations is

$$\sigma_2^2 = \frac{1}{n}\sum_{i=1}^{n}(y_i - \bar{y})^2 = \frac{1}{n}\sum_{i=1}^{n}(x_i + a - \bar{x} - a)^2$$

[using Eqs. (i) and (ii)]

$$= \frac{1}{n}\sum_{i=1}^{n}(x_i - \bar{x})^2 = \sigma_1^2$$

Thus, the variance of the new observations is same as that of the original observations.

Note *We may note that adding (or subtracting) a positive number to (or from) each observation of a group does not affect the variance.*

101. *(c)* Given that, number of observations $(n) = 100$

Incorrect mean $(\bar{x}) = 40$
Incorrect standard deviation $(\sigma) = 5.1$

I. We know that, $\bar{x} = \frac{1}{n}\sum_{i=1}^{n} x_i$

i.e., $$40 = \frac{1}{100}\sum_{i=1}^{100} x_i \text{ or } \sum_{i=1}^{100} x_i = 4000$$

i.e., Incorrect sum of observations = 4000

Thus, the correct sum of observations

= Incorrect sum − 50 + 40

= 4000 − 50 + 40 = 3990

$$\therefore \text{ Correct mean } = \frac{\text{Correct sum}}{100} = \frac{3990}{100} = 39.9$$

II. Also, standard deviation $(\sigma) = \sqrt{\frac{1}{n}\sum_{i=1}^{n} x_i^2 - \frac{1}{n^2}\left(\sum_{i=1}^{n} x_i\right)^2}$

$$= \sqrt{\frac{1}{n}\sum_{i=1}^{n} x_i^2 - (\bar{x})^2}$$

i.e., $$5.1 = \sqrt{\frac{1}{100} \times \text{Incorrect} \sum_{i=1}^{n} x_i^2 - (40)^2}$$

or $$26.01 = \frac{1}{100} \times \text{Incorrect} \sum_{i=1}^{n} x_i^2 - 1600$$

Therefore, incorrect $\sum_{i=1}^{n} x_i^2 = 100\,(26.01 + 1600) = 162601$

Now, correct $\sum_{i=1}^{n} x_i^2 = \text{Incorrect} \sum_{i=1}^{n} x_i^2 - (50)^2 + (40)^2$

$$= 162601 - 2500 + 1600 = 161701$$

Therefore, correct standard deviation

$$= \sqrt{\frac{\text{Correct } \Sigma x_i^2}{n} - (\text{Correct mean})^2}$$

$$= \sqrt{\frac{161701}{100} - (39.9)^2}$$

$$= \sqrt{1617.01 - 1592.01} = \sqrt{25} = 5$$

102. *(b)* Given, $\bar{x} = 8$, $\sigma = 4$ and $n = 6$

Let the observations are x_1, x_2, x_3, x_4, x_5 and x_6. Then,

Mean $(\bar{x}) = 8$ (given)

i.e., $$\text{Mean} = \frac{x_1 + x_2 + x_3 + x_4 + x_5 + x_6}{6} = 8$$

$$\Rightarrow \quad x_1 + x_2 + x_3 + x_4 + x_5 + x_6 = 48$$

Now, if and each observation is multiplied by 3, then mean is given by

$$3x_1 + 3x_2 + 3x_3 + 3x_4 + 3x_5 + 3x_6 = 48 \times 3$$

$$\Rightarrow \quad \Sigma x_i = 144$$

Now, new mean $(\bar{x}) = \frac{144}{6} = 24$

Difference of mean = New mean − Original mean

= 24 − 8 = 16

$\because$ Variance $= 4^2$ (given)

i.e., $$\frac{\Sigma x_i^2}{n} - \left(\frac{\Sigma x_i}{n}\right)^2 = 4^2$$

$$\Rightarrow \quad \frac{\Sigma x_i^2}{6} - (8)^2 = 4^2$$

$$\Rightarrow \quad \frac{\Sigma x_i^2}{6} = 16 + 64$$

$$\Rightarrow \quad \Sigma x_i^2 = 80 \times 6$$

$$\Sigma x_i^2 = 480 \quad \ldots(i)$$

Now, new $\Sigma x^2 = (3x_1)^2 + (3x_2)^2 + (3x_3)^2 + (3x_4)^2 + (3x_5)^2 + (3x_6)^2$

$$= 9(x_1^2 + x_2^2 + x_3^2 + x_4^2 + x_5^2 + x_6^2) = 9 \times 480$$

[from Eq. (i)]

$$\Sigma x^2 = 4320$$

$$\therefore \quad \text{New variance } = \frac{\Sigma x^2}{n} - (\bar{x})^2$$

$$= \frac{4320}{6} - (24)^2 = 720 - 576 = 144$$

$\therefore$ New standard deviation $= \sqrt{\text{New variance}} = \sqrt{144} = 12$

Difference of standard deviations = 12 − 4 = 8

Note *Mean depends on the change of origin and scale. Standard deviation does not depend on the change of origin but it depends on change of scale.*

103. *(a)* I. Let the remaining two observations be a and b.

Given, $\bar{x} = 9$ and $\sigma^2 = 9.25$

$\because$ $\bar{x} = 9$

$\therefore$ $\dfrac{\text{Sum of all observations}}{\text{Number of observations}} = 9$

$\Rightarrow$ Sum of all observations $= 9 \times 8$

$\Rightarrow 6 + 7 + 10 + 12 + 12 + 13 + a + b = 72$

$\Rightarrow$ $60 + a + b = 72$

$\Rightarrow$ $a + b = 72 - 60$

$\Rightarrow$ $a + b = 12$ ...(i)

Again, $\sigma^2 = \dfrac{\Sigma x_i^2}{n} - \left(\dfrac{\Sigma x_i}{n}\right)^2$ or $\sigma^2 = \dfrac{\Sigma x_i^2}{n} - (\bar{x})^2$

$\Rightarrow$ $9.25 = \dfrac{36 + 49 + 100 + 144 + 144 + 169 + a^2 + b^2}{8} - (9)^2$

$\Rightarrow$ $9.25 = \dfrac{642 + a^2 + b^2}{8} - 81$

$\Rightarrow$ $9.25 + 81 = \dfrac{642 + a^2 + b^2}{8}$

$\Rightarrow$ $90.25 \times 8 = 642 + a^2 + b^2$

$\Rightarrow$ $a^2 + b^2 = 722 - 642$

$\Rightarrow$ $a^2 + b^2 = 80$ ...(ii)

Now, putting $b = 12 - a$ in Eq. (ii), we get

$a^2 + (12 - a)^2 = 80$ [from Eq. (i)]

$\Rightarrow$ $a^2 + 144 + a^2 - 24a = 80$

$\Rightarrow$ $2a^2 - 24a + 144 - 80 = 0$

$\Rightarrow$ $2a^2 - 24a + 64 = 0$

$\Rightarrow$ $a^2 - 12a + 32 = 0$

$\Rightarrow$ $a^2 - 8a - 4a + 32 = 0$

$\Rightarrow$ $a(a - 8) - 4(a - 8) = 0$

$\Rightarrow$ $(a - 4)(a - 8) = 0$

$\Rightarrow$ $a = 4$ or $a = 8$

From Eq. (i), $b = 8$ or $= 4$

Hence, observations are 4 and 8.

II. Let the remaining two observations be a and b.

Given, $\bar{x} = 8$ and $\sigma^2 = 16$

$\because$ $\bar{x} = 8$

$\therefore$ $\dfrac{\text{Sum of all observations}}{\text{Number of observations}} = 8$

$\Rightarrow$ $\dfrac{2 + 4 + 10 + 12 + 14 + a + b}{7} = 8$

$\Rightarrow$ $42 + a + b = 56$

$\Rightarrow$ $a + b = 14$ ...(i)

Again, $\sigma^2 = \dfrac{\Sigma x_i^2}{n} - \left(\dfrac{\Sigma x_i}{n}\right)^2$

$\Rightarrow$ $\sigma^2 = \dfrac{\Sigma x_i^2}{n} - (\bar{x})^2$

$\Rightarrow$ $16 = \dfrac{4 + 16 + 100 + 144 + 196 + a^2 + b^2}{n} - (8)^2$

$\Rightarrow$ $16 + 64 = \dfrac{460 + a^2 + b^2}{7}$

$\Rightarrow$ $7 \times (80) = 460 + a^2 + b^2$

$\Rightarrow$ $a^2 + b^2 = 560 - 460$

$\Rightarrow$ $a^2 + b^2 = 100$ ...(ii)

Now, from Eq. (i), putting $b = 14 - a$ in Eq. (ii), we get

$a^2 + (14 - a)^2 = 100$

$\Rightarrow$ $a^2 + 196 + a^2 - 28a = 100$

$\Rightarrow$ $2a^2 - 28a + 196 - 100 = 0$

$\Rightarrow$ $2a^2 - 28a + 96 = 0$

$\Rightarrow$ $a^2 - 14a + 48 = 0$

$\Rightarrow$ $a^2 - 8a - 6a + 48 = 0$

$\Rightarrow$ $a(a - 8) - 6(a - 8) = 0$

$\Rightarrow$ $(a - 6)(a - 8) = 0$

$\Rightarrow$ $a = 6$ or 8

From Eq. (i), $b = 8$ or 6

Hence, the observations are 6 and 8.

104. *(d)* Mean of $ax_1, ax_2, \ldots, ax_n$

$= \dfrac{ax_1 + ax_2 + \ldots + ax_n}{n}$

$= a\left(\dfrac{x_1 + x_2 + x_3 + \ldots + x_n}{n}\right)$

$= a\bar{x}$ $\left(\because \bar{x} = \dfrac{x_1 + x_2 + \ldots + x_n}{n}\right)$

Variance of $ax_1, ax_2, \ldots, ax_n$

$= \dfrac{\Sigma(ax_i - a\bar{x})^2}{n}$

$= \dfrac{(ax_1 - a\bar{x})^2 + (ax_2 - a\bar{x})^2 + \ldots + (ax_n - a\bar{x})^2}{n}$

$= \dfrac{a^2[(x_1 - \bar{x})^2 + (x_2 - \bar{x})^2 + \ldots + (x_n - \bar{x})^2]}{n}$

$= a^2 \dfrac{\Sigma(x_i - \bar{x})^2}{n} = a^2\sigma^2$ $\left[\because \sigma^2 = \dfrac{\Sigma(x_i - \bar{x})^2}{n}\right]$

105. *(c)* Here, $n = 50$

For Mathematics, CV $= \dfrac{\sigma}{\bar{x}} \times 100$

$= \dfrac{12}{42} \times 100 = \dfrac{2}{7} \times 100 = \dfrac{200}{7}$

$= 28.57$ ...(i)

For Physics, CV $= \dfrac{\sigma}{\bar{x}} \times 100 = \dfrac{15}{32} \times 100$

$= \dfrac{1500}{32} = 46.87$ ...(ii)

For Chemistry, $CV = \frac{\sigma}{\bar{x}} \times 100$

$$= \frac{20}{40.9} \times 100 = \frac{2000}{40.9}$$

$$= 48.89 \quad \ldots(iii)$$

From Eqs. (i), (ii) and (iii), we have

$\because$ CV of Chemistry > CV of Physics > CV of Mathematics

$\therefore$ Chemistry shows the highest variability and Mathematics shows the lowest variability.

106. *(b)* A. If wrong item is omitted.

Given, $\bar{x} = 10$ and $\sigma = 2, n = 20$

$$\Rightarrow \frac{\Sigma x_i}{20} = 10$$

$$\Rightarrow \Sigma x_i = 20 \times 10$$

$$\Rightarrow \Sigma x_i = 200$$

If observation 8 is omitted, then

$$\Sigma x_i = 200 - 8 = 192$$

Now, remaining observations = 19

$\therefore$ Correct mean $= \frac{\Sigma x_i}{n} = \frac{192}{19} = 10.10$

B. Again, $\sigma = 2 \Rightarrow \sigma^2 = 4$

$$\Rightarrow \frac{\Sigma x_i^2}{n} - (\bar{x})^2 = 4$$

$$\Rightarrow \frac{\Sigma x_i^2}{20} - (10)^2 = 4$$

$$\Rightarrow \Sigma x_i^2 = (4 + 100) \times 20$$

$$= 104 \times 20 = 2080$$

If observation 8 is omitted, then

$$\Sigma x_i^2 = 2080 - 64 = 2016$$

Now, correct standard deviation $\sigma = \sqrt{\frac{2016}{19} - \left(\frac{192}{19}\right)^2}$

$$= \sqrt{\frac{2016 \times 19 - (192)^2}{19 \times 19}}$$

$$= \frac{1}{19}\sqrt{38304 - 36864}$$

$$= \frac{1}{19}\sqrt{1440} = \frac{37.95}{19} = 1.99$$

C. If it is replaced by 12, then

Given, $\bar{x} = 10, \sigma = 2, n = 20$

$$\Rightarrow \frac{\Sigma x_i}{20} = 10$$

$$\Rightarrow \Sigma x_i = 200$$

If observation 8 is replaced by 12, then

$$\Sigma x_i = 200 - 8 + 12$$

$$= 192 + 12 = 204$$

Correct mean $= \frac{204}{20} = \frac{102}{10} = 10.2$

D. Again, $\sigma = 2 \Rightarrow \sigma^2 = 4$

$$\Rightarrow \frac{\Sigma x_i^2}{n} - (\bar{x})^2 = 4$$

$$\Rightarrow \frac{\Sigma x_i^2}{20} - (10)^2 = 4$$

$$\Rightarrow \Sigma x_i^2 = 2080$$

If observation 8 is replaced by 12, then

$$\Sigma x_i^2 = 2080 - (8)^2 + (12)^2$$

$$= 2080 - 64 + 144 = 2224 - 64 = 2160$$

Now, correct standard deviation (σ)

$$= \sqrt{\frac{\Sigma x^2}{n} - (\bar{x})^2} = \sqrt{\frac{2160}{20} - \left(\frac{204}{20}\right)^2}$$

$$= \sqrt{\frac{2160 \times 20 - (204)^2}{20 \times 20}}$$

$$= \frac{1}{20}\sqrt{43200 - 41616}$$

$$= \frac{1}{20}\sqrt{1584} = \frac{39.79}{20} = 1.98$$

107. *(d)* Given, $n = 100, \bar{x} = 20$ and $\sigma = 3$

$$\because \bar{x} = 20$$

$$\therefore \frac{\Sigma x_i}{100} = 20$$

$$\Rightarrow \Sigma x_i = 100 \times 20$$

$$\Rightarrow \Sigma x_i = 2000$$

Now, incorrect observations 21, 21 and 18 are omitted, then correct sum is

$$\Sigma x_i = 2000 - 21 - 21 - 18 = 2000 - 60 = 1940$$

Now, correct mean of remaining 97 observations is

$$\bar{x} = \frac{1940}{97} = 20$$

Again, $\sigma = 3 \Rightarrow \sqrt{\frac{\Sigma x_i^2}{n} - (\bar{x})^2} = 3$

$$\Rightarrow \frac{\Sigma x_i^2}{100} - (20)^2 = 9 \Rightarrow \frac{\Sigma x_i^2}{100} = 9 + 400$$

$$\Rightarrow \Sigma x_i^2 = 409 \times 100 = 40900$$

Now, correct Σx_i^2 is

$$= 40900 - (21)^2 - (21)^2 - (18)^2$$

$$= 40900 - 441 - 441 - 324$$

$$= 40900 - 1206 = 39694$$

Now, correct SD for remaining 97 observations is

$$\sigma = \sqrt{\frac{39694}{97} - \left(\frac{1940}{97}\right)^2}$$

$$= \sqrt{409.2 - (20)^2}$$

$$= \sqrt{409.2 - 400}$$

$$= \sqrt{9.2} = 3.03$$

108. *(a)* Here, $h = 100$, let A (assumed mean) = 800

Length of life (in hours)	Mid value (x_i)	$y_i = \frac{x_i - A}{100}$	Factory A			Factory B		
			f_i	f_iy_i	$f_iy_i^2$	f_i	f_iy_i	$f_iy_i^2$
550-650	600	−2	10	−20	40	8	−16	32
650-750	700	−1	22	−22	22	60	−60	60
750-850	800	0	52	0	0	24	0	0
850-950	900	1	20	20	20	16	16	16
950- 1050	1000	2	16	32	64	12	24	48
			120	**10**	**146**	**120**	**−36**	**156**

For factory A

$$\text{Mean } (\bar{x}) = 800 + \frac{10}{120} \times 100 = 808.33$$

$$\text{SD} = \frac{100}{120}\sqrt{120(146) - 100} = 109.98$$

Therefore, coefficient of variation (CV)

$$= \frac{\text{SD}}{\bar{x}} \times 100 = \frac{109.54}{808.33} \times 100 = 13.60$$

For factory B

$$\text{Mean } (\bar{x}) = 800 + \left(\frac{-36}{120}\right) \times 100 = 770$$

$$\text{SD} = \frac{100}{120}\sqrt{120(156) - (-36)^2} = 110$$

Therefore, coefficient of variation

$$= \frac{\text{SD}}{\text{Mean}} \times 100 = \frac{110}{770} \times 100 = 14.28$$

Since, CV of factory B > CV of factory A,

⇒ Factory B has more variability which means bulbs of factory A are more consistent.

109. *(b)*

x	f_i	x_i	f_ix_i	$f_ix_i^2$
1-3	6	2	12	24
3-5	4	4	16	64
5-7	5	6	30	180
7-10	1	8.5	8.5	72.25
	$\Sigma f_i = 16$		**66.5**	**340.25**

$$\text{Mean } (\bar{x}) = \frac{\Sigma f_ix_i}{\Sigma f_i} = \frac{66.5}{16} = 4.15$$

$$\text{Variance } (\sigma^2) = \frac{\Sigma f_ix_i^2}{\Sigma f_i} - \left(\frac{\Sigma f_ix_i}{\Sigma f_i}\right)^2 = \frac{340.25}{16} - (4.15)^2$$

$$= 21.2656 - 17.2225 = 4.043$$

110. *(a)* We know that, if any constant is added in each observation, then standard deviation remains same.

∴ The standard deviation of the observations $a + k, b + k, c + k, d + k, e + k$ is s.

111. *(c)* Standard deviation is dependent on change of scale. Therefore, the standard deviation of $kx_1, kx_2, kx_3, kx_4, kx_5$ is ks.

112. *(b)* **For Raju**

x_i	$d_i = x_i - 45$	d_i^2
25	−20	400
50	5	25
45	0	0
30	−15	225
70	25	625
42	−3	9
36	−9	81
48	3	9
35	−10	100
60	15	225
	−9	**1699**

$$\text{Standard deviation } (\sigma) = \sqrt{\frac{1699}{10} - \left(\frac{-9}{10}\right)^2}$$

$$= \sqrt{169.9 - 0.81}$$

$$= \sqrt{169.09} = 13$$

$$\text{Mean } (\bar{x}) = 45 - \left(\frac{9}{10}\right) = 45 - 0.9 = 44.1$$

For Sita

x_i	$d_i = x_i - 55$	d_i^2
10	−45	2025
70	15	225
50	−5	25
20	−35	1225
95	40	1600
55	0	0
42	−13	169
60	5	25
48	−7	49
80	25	625
	− 20	**5968**

Mean $(\bar{x}) = 55 - \frac{20}{10} = 53$

Standard deviation $(\sigma) = \sqrt{\frac{5968}{10} - (-10)^2}$

$= \sqrt{496.8} = 22.28$

Coefficient of variation of both Raju and Sita are

For Raju

$$= \frac{\sigma}{\bar{x}} \times 100 = \frac{13}{44.1} \times 100 = \frac{1300}{44.1} = 29.47$$

For Sita

$$= \frac{\sigma}{\bar{x}} \times 100 = \frac{22.28}{53} \times 100 = \frac{2228}{53} = 42.04$$

Since, CV of Sita > CV of Raju

Also, mean of Sita > mean of Raju

Hence, Raju is more consistent but Sita is more intelligent.

113. *(a)* Given, $w_i = lx_i + k$

$$M(x_i) = \bar{x} = 48, \sigma(x_i) = 12$$

$$M(w_i) = \bar{w} = 55 \quad \text{and} \quad \sigma(w) = 15$$

$$M(w_i) = lM(x_i) + M(k)$$

$$55 = l \times 48 + k \quad \ldots \text{(i)}$$

and $\sigma(w_i) = l\sigma(x_i) + \sigma(k)$

$\Rightarrow$ $15 = l(12) + 0$

$\Rightarrow$ $l = \frac{15}{12}$

$= 1.25$

From Eq. (i),

$55 = 1.25 \times 48 + k$

$\Rightarrow$ $k = 55 - 60$

$\therefore$ $k = -5$

114. *(c)* We have,

$$\Sigma(x_i - \bar{x})^2 = \Sigma(x_i^2 - 2\bar{x}\, x_i + \bar{x}^2)$$

$$= \Sigma x_i^2 + \Sigma - 2\bar{x}\, x_i + \Sigma \bar{x}^2$$

$$= \Sigma x_i^2 - 2\bar{x}\Sigma x_i + (\bar{x})^2 \Sigma 1$$

$$= \Sigma x_i^2 - 2\bar{x}\,(n\bar{x}) + n\bar{x}^2$$

$$= \Sigma x_i^2 - n\bar{x}^2$$

On dividing both sides by n and taking their square root, we get $\sigma = \sigma'$.

Hence, both formulae are equivalent.

115. *(c)* Let x_{1i}, $i = 1, 2, \ldots, n_1$ and x_{2j}, $j = 1, 2, \ldots n_2$ be the two series. Then,

For first set

$$\bar{x}_1 = \frac{1}{n} \sum_{i=1}^{n_1} x_{1i} \text{ and } \bar{x}_2 = \frac{1}{n_2} \sum_{j=1}^{n_2} x_{2j}$$

$$\sigma_1^2 = \frac{1}{n_1} \sum_{i=1}^{n_1} (x_{1i} - \bar{x}_1)^2$$

and $$\sigma_2^2 = \frac{1}{n_2} \sum_{j=1}^{n_2} (x_{2j} - \bar{x}_2)^2$$

Mean $\bar{x}$ of the combined series is given by

$$\bar{x}_1 = \frac{1}{n_1 + n_2} \left[\sum_{i=1}^{n_1} x_{1i} + \sum_{i=1}^{n_2} x_{2j} \right] = \frac{n_1\bar{x}_1 + n_2\bar{x}_2}{n_1 + n_2}$$

The variance σ^2 of the combined series is given by

$$\sigma^2 = \frac{1}{n_1 + n_2} \left[\sum_{i=1}^{n_1} (x_{1i} - \bar{x})^2 + \sum_{j=1}^{n_2} (x_{2j} - \bar{x})^2 \right]$$

Now, $\sum_{i=1}^{n_1} (x_{1i} - \bar{x})^2 = \sum_{i=1}^{n_1} (x_{1i} - \bar{x}_1 + \bar{x}_1 - \bar{x})^2$

$$= \sum_{i=1}^{n_1} (x_{1i} - \bar{x}_1)^2 + n_1 (\bar{x}_1 - \bar{x})^2 + 2(\bar{x}_1 - \bar{x}) \sum_{i=1}^{n_1} (x_{1i} - \bar{x}_1)$$

But $\sum_{i=1}^{n_1} (x_{1i} - \bar{x}_1) = 0$ as it is the algebraic sum of the deviation of the values of first series from their mean.

Also, $\sum_{i=1}^{n_1} = (x_{1i} - \bar{x})^2 = n_1 s_1^2 + n_1 (\bar{x}_1 - \bar{x})^2 = n_1 s_1^2 + n_1 d_1^2$

where, $d_1 = \bar{x}_1 - \bar{x}$

Similarly, we get

$$\sum_{j=1}^{n_2} (x_{1j} - \bar{x})^2 = \sum_{j=1}^{n_2} (x_{1j} - \bar{x}_2 + \bar{x}_2 - \bar{x})^2$$

$$= n_2 s_2^2 + n_2 d_2^2$$

where, $d_2 = \bar{x}_2 - \bar{x}$

Now, from above,

$$\sigma^2 = \frac{1}{n_1 + n_2} [n_1(s_1^2 + d_1^2) + n_2 (s_2^2 + d_2^2)]$$

Now $$d_1 = \bar{x}_1 - \bar{x} = \bar{x}_1 - \frac{n_1\bar{x}_1 + n_2\bar{x}_2}{n_1 + n_2} = \frac{n_2(\bar{x}_1 - \bar{x}_1)}{n_1 + n_2}$$

$$d_2 = \bar{x}_2 - \bar{x} = \bar{x}_2 - \frac{n_1\bar{x}_1 + n_2\bar{x}_2}{n_1 + n_2} = \frac{n_1(\bar{x}_2 - \bar{x}_1)}{n_1 + n_2}$$

Thus, $$\sigma^2 = \frac{1}{n_1 + n_2} \left[n_1 s_1^2 + n_2 s_2^2 + \left\{ \frac{n_1 n_2 (\bar{x}_1 - \bar{x}_2)^2}{(n_1 + n_2)^2} + \frac{n_2 n_1 (\bar{x}_2 - \bar{x}_1)^2}{(n_1 + n_2)^2} \right\} \right]$$

$$\sigma = \sqrt{\frac{n_1 s_1^2 + n_2 s_2^2}{n_1 + n_2} + \frac{n_1 n_2 (\bar{x}_1 - \bar{x}_2)^2}{(n_1 + n_2)^2}}$$

116. *(b)* Given, $\Sigma (x - 5) = 3$

$\therefore$ $\Sigma x - \Sigma 5 = 3$

$\Rightarrow$ $\Sigma x - 5 \times 18 = 3$ $\quad (\because n = 18)$

$\Rightarrow$ $\Sigma x = 3 + 90$

$\Rightarrow$ $\Sigma x = 93$

Now, $\Sigma (x - 5)^2 = 43$

$\Rightarrow \quad \Sigma(x^2 + 25 - 10x) = 43$

$\Rightarrow \quad \Sigma x^2 + \Sigma 25 - 10\Sigma x = 43$

$\Rightarrow \quad \Sigma x^2 + 25 \times 18 - 10 \times 93 = 43$

$\Rightarrow \quad \Sigma x^2 = 43 + 930 - 450$

$\Rightarrow \quad \Sigma x^2 = 973 - 450$

$\Rightarrow \quad \Sigma x^2 = 523$

Now, $\quad \text{mean} = \dfrac{\Sigma x}{n} = \dfrac{93}{18} = 5.16$

and $\quad \text{SD}(\sigma) = \sqrt{\dfrac{\Sigma x^2}{n} - \left(\dfrac{\Sigma x}{n}\right)^2} = \sqrt{\dfrac{523}{18} - \left(\dfrac{93}{18}\right)^2}$

$= \sqrt{\dfrac{523 \times 18 - 93 \times 93}{18 \times 18}}$

$= \dfrac{1}{18}\sqrt{9414 - 8649}$

$= \dfrac{1}{18}\sqrt{765} = \dfrac{27.66}{18} = 1.54$

117. *(c)* $n = 25, \bar{x}_1 = 18.2, \sigma_1 = 3.25$ and

$n = 15, \sum_{i=1}^{15} x_i = 279, \sum_{i=1}^{15} x_i^2 = 5524$

For first set

$\Sigma x_i = 25 \times 18.2 = 455$

$\sigma_1^2 = \dfrac{\Sigma x_i^2}{25} - (18.2)^2$

$\Rightarrow \quad (3.25)^2 = \dfrac{\Sigma x_i^2}{25} - 331.24$

$\Rightarrow \quad 10.5625 + 331.24 = \dfrac{\Sigma x_i^2}{25}$

$\Rightarrow \quad \Sigma x_i^2 = 25 \times 341.8025 = 8545.0625$

For required standard deviation, $n = 40$

Combined sum of squares of two data

$\Sigma x_i^2 = 8545.0625 + 5524 = 14069.0625$

Combined sum of two data, $\Sigma x_i = 455 + 279 = 734$

$\therefore$ Standard deviation $(\sigma) = \sqrt{\dfrac{14069.0625}{40} - \left(\dfrac{734}{40}\right)^2}$

$= \sqrt{351.726 - (18.35)^2} = \sqrt{351.726 - 336.7225}$

$= \sqrt{15.0035} = 3.87$

118. *(d)*

x_i	$d_i = x - a$	d_i^2
a	0	0
$a + d$	d	d^2
$a + 2d$	$2d$	2^2d^2
$a + 3d$	$3d$	3^2d^2
:	:	:
:	:	:
$a + (n-1)d$	$(n-1)d$	$(n-1)^2d^2$
$\Sigma x_i = \dfrac{n}{2}[2a + (n-1)d]$		

$\text{Mean}(\bar{x}) = \dfrac{\Sigma x_i}{n} = \dfrac{1}{n}\left[\dfrac{n}{2}\{2a + (n-1)d\}\right]$

$= \dfrac{1}{2}[2a + (n-1)d] = a + \dfrac{d(n-1)}{2}$

$\Sigma d_i = d[1 + 2 + 3 + \ldots + (n-1)] = \dfrac{d(n-1)n}{2}$

$\left[\because 1 + 2 + \ldots + n = \dfrac{n(n+1)}{2}\right]$

$\Sigma d_i^2 = d^2[1^2 + 2^2 + 3^2 + \ldots + (n-1)^2]$

$= \dfrac{d^2(n-1)(n)(2n-1)}{6}$

$\left[\because 1^2 + 2^2 + \ldots + n^2 = \dfrac{n(2n+1)(n+1)}{6}\right]$

Standard deviation

$(\sigma) = \sqrt{\dfrac{\Sigma d_i^2}{n} - \left(\dfrac{\Sigma d_i}{n}\right)^2}$

$= \sqrt{\dfrac{d^2(n-1)(n)(2n-1)}{6n} - \left[\dfrac{d(n)(n-1)}{2n}\right]^2}$

$= \sqrt{\dfrac{d^2(n-1)(2n-1)}{6} - \dfrac{d^2(n-1)^2}{4}}$

$= d\sqrt{\dfrac{(n-1)(2n-1)}{6} - \dfrac{(n-1)^2}{4}}$

$= d\sqrt{\dfrac{(n-1)}{2}\left(\dfrac{2n-1}{3} - \dfrac{n-1}{2}\right)}$

$= d\sqrt{\dfrac{n-1}{2}\left(\dfrac{4n-2-3n+3}{6}\right)}$

$= d\sqrt{\dfrac{(n-1)(n+1)}{12}} = d\sqrt{\dfrac{n^2-1}{12}}$

119. *(b)* We have, $n = 100, \bar{x} = 40, \sigma = 10$

$\text{Mean}(\bar{x}) = \dfrac{\Sigma x_i}{n} = 40$

$\Rightarrow \quad \dfrac{\Sigma x_i}{100} = 40 \Rightarrow \Sigma x_i = 4000$

Corrected $\Sigma x_i = 4000 - 30 - 70 + 3 + 27$

$= 4030 - 100 = 3930$

Corrected mean $= \dfrac{3930}{100} = 39.3$

Variance $(\sigma^2) = \dfrac{\Sigma x_i^2}{n} - (40)^2$

$\Rightarrow \quad 100 = \dfrac{\Sigma x_i^2}{100} - 1600 \Rightarrow \Sigma x_i^2 = 170000$

Corrected $\Sigma x_i^2 = 170000 - (30)^2 - (70)^2 + (3)^2 + (27)^2$

$= 170000 - 900 - 4900 + 9 + 729$

$= 164938$

Corrected standard deviation (σ)

$= \sqrt{\dfrac{164938}{100} - (39.3)^2} = \sqrt{1649.38 - 1544.49}$

$= \sqrt{104.48} = 10.24$

120. *(c)* We have, $n = 10, \bar{x} = 45, \sigma^2 = 16$

Now, mean $(\bar{x}) = \frac{\Sigma x_i}{10} = 45 \Rightarrow \Sigma x_i = 450$

Corrected $\Sigma x_i = 450 - 52 + 25 = 423$

Corrected mean $(\bar{x}) = \frac{423}{10} = 42.3$

$$\text{Variance } (\sigma^2) = \frac{\Sigma x_i^2}{n} - \left(\frac{\Sigma x_i}{n}\right)^2$$

$$\Rightarrow \quad 16 = \frac{\Sigma x_i^2}{10} - (45)^2$$

$$\Rightarrow \quad \frac{\Sigma x_i^2}{10} = 2025 + 16$$

$$\Rightarrow \quad \Sigma x_i^2 = 2041 \times 10 = 20410$$

$$\text{Corrected } \Sigma x_i^2 = 20410 - (52)^2 + (25)^2 = 20410 - 2704 + 625 = 18331$$

$$\text{Corrected variance } (\sigma^2) = \frac{18331}{10} - (42.3)^2 = 1833.10 - 1789.29 = 43.81$$

121. *(d)* Statement II is true.

I. Sum of n even natural numbers $= n(n+1)$

$$\text{Mean } (\bar{x}) = \frac{n(n+1)}{n} = n + 1$$

$$\text{Variance} = \left[\frac{1}{n}\Sigma (x_i)^2\right] - (\bar{x})^2$$

$$= \frac{1}{n}[2^2 + 4^2 + \ldots + (2n)^2] - (n+1)^2$$

$$= \frac{1}{n}2^2[1^2 + 2^2 + \ldots + n^2] - (n+1)^2$$

$$= \frac{4}{n}\frac{n(n+1)(2n+1)}{6} - (n+1)^2$$

$$= \frac{(n+1)[2(2n+1) - 3(n+1)]}{3}$$

$$= \frac{(n+1)(n-1)}{3} = \frac{n^2-1}{3}$$

∴ Statement I is false.

122. *(c)* $\bar{x} = \frac{\text{Sum of quantities}}{n} = \frac{\frac{n}{2}(a+l)}{n}$

$$= \frac{1}{2}[1 + 1 + 100d] = 1 + 50d$$

$$\therefore \quad \text{MD} = \frac{1}{n}\Sigma |x_i - \bar{x}|$$

$$\Rightarrow \quad 255 = \frac{1}{101}[50d + 49d + \ldots + d + 0 + d + \ldots + 50d]$$

$$\Rightarrow \quad 255 = \frac{2d}{101}\left[\frac{50 \times 51}{2}\right]$$

$$\Rightarrow \quad d = \frac{255 \times 101}{50 \times 51} = 10.1$$

123. *(b)* I. $\bar{x} = \frac{31 + 32 + 33 + \ldots + 47}{17} = \frac{663}{17} = 39$

$$\therefore \sum_{i=1}^{17}(x_i - \bar{x})^2 = (31-39)^2 + (32-39)^2 + (33-39)^2 + (34-39)^2 + (35-39)^2 + (36-39)^2 + (37-39)^2 + (38-39)^2 + (39-39)^2 + (40-39)^2 + (41-39)^2 + (42-39)^2 + (43-39)^2 + (44-39)^2 + (45-39)^2 + (46-39)^2 + (47-39)^2$$

$$= 64 + 49 + 36 + 25 + 16 + 9 + 4 + 1 + 0 + 1 + 4 + 9 + 16 + 25 + 36 + 49 + 64$$

$$= 408$$

Hence, standard deviation $= \sqrt{\frac{408}{17}} = \sqrt{24} = 2\sqrt{6}$

II. Given, $\sigma_{10}^2 = \frac{99}{12} = \frac{33}{4}$

$$\Rightarrow \quad \sigma_{10} = \frac{\sqrt{33}}{2}$$

Required SD $= 3\sigma_{10} = \frac{3\sqrt{33}}{2}$

III. Let $x_1 = i$, where $i = 1, 2, 3, 4, 5, 6, 7$

and $y_1 = 12$, $y_2 = 23$, $y_3 = 34$, $y_4 = 45$, $y_5 = 56$
$y_6 = 67$, $y_7 = 78$

Clearly, $y_i = 11x_i + 1$, $\forall \ i = 1, 2, \ldots, 7$

Now, SD of y_i's $= 11 \times$ SD of x_1's
$= 11 \times 2$
$= 22$

IV. The new observations are obtained by adding 20 to each. Hence, σ does not change.

124. *(b)* $\bar{x} = \frac{1 + 2 + 3 \ldots + n}{n} = \frac{(n+1)}{2}$

$$\therefore \quad \sigma^2 = \frac{\Sigma (x_i)^2}{n} - (\bar{x})^2 = \frac{\Sigma n^2}{n} - \left(\frac{n+1}{2}\right)^2$$

$$\sigma^2 = \frac{n(n+1)(2n+1)}{6n} - \left(\frac{n+1}{2}\right)^2 = \frac{n^2-1}{12} \Rightarrow \sigma = \sqrt{\frac{n^2-1}{12}}$$

125. *(d)* Standard deviation

$$= \sqrt{\frac{\sum_1^{18}(x_i - 8)^2}{n} - \left(\frac{\sum_1^{18}(x_i - 8)}{n}\right)^2}$$

$$= \sqrt{\frac{45}{18} - \left(\frac{9}{18}\right)^2}$$

$$= \sqrt{\frac{45}{18} - \frac{1}{4}} = \sqrt{\frac{81}{36}} = \frac{9}{6} = \frac{3}{2}$$

126. *(b)* $\bar{x} = 5$

$$\text{Variance} = \frac{1}{h_1}\Sigma x_i^2 - (\bar{x})^2$$

$$0 = \frac{1}{n}.\ 400 - 25 \Rightarrow n = \frac{400}{25} = 16$$

127. *(c)* I. Mean of $1^2, 2^2, 3^2, \ldots, n^2$ is

$$\frac{1^2 + 2^2 + 3^2 + \ldots + n^2}{n} = \frac{\Sigma n^2}{n}$$

$$\therefore \quad \frac{46n}{11} = \frac{n(n+1)(2n+1)}{6n}$$

$$\Rightarrow 22n^2 + 33n + 11 - 276n = 0$$

$$\Rightarrow (n - 11)(22n - 1) = 0$$

$$\Rightarrow n = 11 \text{ and } n \neq \frac{1}{22}$$

II. $\because \sigma_x^2 = 4$ and $\sigma_y^2 = 5$

Also, $\bar{x} = 2$ and $\bar{y} = 4$

Now, $\frac{\Sigma x_i}{5} = 2 \Rightarrow \Sigma x_i = 10$

$\frac{\Sigma y_i}{5} = 4 \Rightarrow \Sigma y_i = 20$

Since, $\sigma_x^2 = \frac{1}{5}(\Sigma x_i^2) - (\bar{x})^2$

$\Rightarrow \Sigma x_i^2 = 40$

Similarly, $\Sigma y_i^2 = 105$

$$\therefore \quad \sigma_z^2 = \frac{1}{10}(\Sigma x_i^2 + \Sigma y_i^2) - \left(\frac{\bar{x} + \bar{y}}{2}\right)^2$$

$$= \frac{1}{10}(40 + 105) - 9$$

$$= \frac{55}{10} = \frac{11}{2}$$

16

Probability

Random Experiments

Level I

1. An experiment is called random experiment, if it ...X... . Here, X refers to

(a) has more than one possible outcome
(b) is not possible to predict the outcome in advance
(c) Both (a) and (b)
(d) None of the above

2. Statement I A possible result of a random experiment is called outcome.

Statement II The set of outcomes is called the sample space of the experiment.

(a) Statement I is true; Statement II is true; Statement II is a correct explanation for Statement I
(b) Statement I is true; Statement II is true; Statement II is not a correct explanation for Statement I
(c) Statement I is true; Statement II is false
(d) Statement I is false; Statement II is true

3. Two coins (a ₹ 1 coin and a ₹ 2 coin) are tossed once, then the sample space is

(a) $\{HH, HT, TH, TT\}$ (b) $\{TT, HH\}$
(c) $\{HT, TH, HH\}$ (d) None of these

4. When a pair of dice (one is blue and the other is red) is rolled once, then

(a) sample space is $\{(1, 1), (1, 2), (1, 3), (1, 4), (1, 5), (1, 6), (2, 1), (2, 2), (2, 3), (2, 4), (2, 5), (2, 6), (3, 1), (3, 2), (3, 3), (3, 4), (3, 5), (3, 6), (4, 1), (4, 2), (4, 3) (4, 4), (4, 5), (4, 6), (5, 1), (5, 2), (5, 3), (5, 4), (5, 5), (5, 6), (6, 1), (6, 2), (6, 3), (6, 4), (6, 5), (6, 6)\}$
(b) total number of elements in sample space is 36
(c) Both (a) and (b)
(d) None of the above

5. Statement I A boy has a ₹1 coin, a ₹2 coin and a ₹5 coin in his pocket. He takes out two coins out of his pocket, one after the other.

The sample space for the experiment is

$$S = (QH, QR, HQ, HR, RH, RQ)$$

where, Q denotes a ₹1 coin, H denotes ₹2 coins and R denotes ₹5 coin.

Statement II A person is noting down the number of accidents along a busy highway during a year.

The sample space for the experiment is

$$S = \{0, 1, 2 ......\}$$

(a) Statement I is true; Statement II is true; Statement II is a correct explanation for Statement I
(b) Statement I is true; Statement II is true; Statement II is not a correct explanation for Statement I
(c) Statement I is true; Statement II is false
(d) Statement I is false; Statement II is true

6. Statement I A coin is tossed. If it shows head, we draw a ball from a bag consisting of 3 blue and 4 white balls; if it shows tail we throw a die, then the sample space of this experiment is

$$S = \{HB_1, HB_2, HB_3, HW_1, HW_2, HW_3, HW_4, T1, T2, T3, T4, T5, T6\}$$

Statement II Consider the experiment in which a coin is tossed repeatedly until a head comes up, then the sample space is

$$S = \{H, TH, TTH, TTTH, ...\}$$

where B_1, B_2, B_3 denote blue balls and W_1, W_2, W_3, W_4 denote the white balls.

(a) Statement I is true; Statement II is true; Statement II is a correct explanation for Statement I
(b) Statement I is true; Statement II is true; Statement II is not a correct explanation for Statement I
(c) Statement I is true; Statement II is false
(d) Statement I is false; Statement II is true

7. Match the given experiments in Column I with their respective sample spaces in Column II and choose the correct option from the codes given below.

Column I	Column II
A. A coin is tossed three times	1. S = {$HHHH$, $HHHT$, $HHTH$, $HHTT$, $HTHH$, $HTHT$, $HTTH$, $HTTT$, $THHH$, $THHT$, $THTH$, $THTT$, $TTHH$, $TTHT$, $TTTH$ $TTTT$} where, H and T represent head and tail of a coin
B. A die is thrown two times	2. S = {HHH, HHT, THH, THT, HTH, TTH, TTT, HTT}
C. A coin is tossed four times	3. S = {(1, 1), (1, 2) ... (6, 5), (6, 6)}
D. A coin is tossed and a die is thrown	4. S = {H1, H2, H3, H4, H5, H6, T1, T2, T3, T4, T5, T6}

Codes

	A	B	C	D
(a)	2	3	1	4
(b)	1	2	3	4
(c)	4	3	2	1
(d)	2	3	4	1

8. Statement I A coin is tossed and then a die is rolled only in case a head is shown on the coin.

The sample space for the experiment is

$$S = \{H1, H2, H3, H4, H5, H6, T\}$$

Statement II 2 boys and 2 girls are in Room X, and 1 boy and 3 girls are in Room Y. Then, the sample space for the experiment in which a room is selected and then a person, is $S = \{XB_1, XB_2, XG_1, XG_2, YB_3, YG_3, YG_4, YG_5\}$

where B_i, denote the boys and G_j, denote the girls.

(a) Statement I is true; Statement II is true; Statement II is a correct explanation for Statement I
(b) Statement I is true; Statement II is true; Statement II is not a correct explanation for Statement I
(c) Statement I is true; Statement II is false
(d) Statement I is false; Statement II is true

9. One die of red colour(R), one of white colour(W) and one of blue colour(B) are placed in a bag. One die is selected at random and rolled, its colour and the number on its uppermost face is noted. Then, the sample space is

(a) $S = \begin{Bmatrix} R1, & R2, & R3, & R4, & R5, & R6 \\ W1, & W2, & W3, & W4, & W5, & W6 \\ B1, & B2, & B3, & B4, & B5, & B6 \end{Bmatrix}$.

(b) $S = \begin{Bmatrix} R1, & R2 & R3, & R4, & R5 & R6 \\ W1, & W2, & W4, & W5 & & \\ B1, & B3, & B4, & B5 & B6 & \end{Bmatrix}$

(c) Both (a) and (b)
(d) None of the above

10. An experiment consists of recording boy-girl composition of families with 2 children.

Statement I Then, the sample space, if we are interested in knowing whether it is a boy or girl in the order of their births, is

$$S = \{Bb, Bg, Gb, Gg\}$$

where B and b denote the first and second boy and G and g denote the first and second girl respectively.

Statement II The sample space, if we are interested in the number of girls in the family, is

$$S = \{0, 1, 2\}$$

(a) Both I and II are correct (b) Both I and II are incorrect
(c) I is correct; II is incorrect (d) II is correct; I is incorrect

11. Match the given experiments in Column I with their respective sample spaces in Column II and choose the correct option from the codes given below.

Column I	Column II
A. A box contains 1 red and 3 identical white balls. Two balls are drawn at random in succession without replacement. Then, the sample space is	1. S = {HH, HT, T1, T2, T3, T4, T5, T6}
B. An experiment consists of tossing a coin and then throwing it second time, if a head occurs. If a tail occurs on the first toss, then a die is rolled once. Then, the sample space is	2. S = {DDD, DDN, DND, NDD, NDN, NND, NNN, DNN}
C. Suppose 3 bulbs are selected at random from a lot. Each bulb is tested and classified as defective (D) or non-defective (N). Then, the sample space of this experiment is	3. S = {T, H1, H3, H5, H21, H22, H23, H24, H25, H26, H41, H42, H43, H44, H45, H46, H61, H62, H63, H64, H65, H66}
D. A coin is tossed. If the outcome is a head, a die is thrown. If the die shows up an even number, the die is thrown again. Then, the sample space for the experiment is	4. S = {RW, WW, WR}

Codes

	A	B	C	D		A	B	C	D
(a)	1	2	3	4	(b)	4	1	2	3
(c)	3	2	4	1	(d)	2	3	4	1

12. **Statement I** The numbers 1, 2, 3 and 4 are written separately on four slips of paper. The slips are put in a box and mixed thoroughly. A person draws two slips from the box, one after the other, without replacement. Then, the sample space for the experiment is

$S = \{(1, 2), (1, 3), (1, 4), (2, 1), (2, 3), (2, 4), (3, 1), (3, 2), (3, 4), (4, 1), (4, 2), (4, 3)\}$

Statement II An experiment consists of rolling a die and then tossing a coin once, if the number on the die is even. If the number on the die is odd, the coin is tossed twice. Then, the sample space for this experiment is

$S = \{1HH, 1HT, 1TH, 1TT, 2H, 2T, 3HH, 3HT, 3TT, 3TH, 4H, 4T, 5HH, 5HT, 5TH, 5TT, 6H, 6T\}$

(a) Statement I is true; Statement II is true; Statement II is a correct explanation for Statement I
(b) Statement I is true; Statement II is true; Statement II is not a correct explanation for Statement I
(c) Statement I is true; Statement II is false
(d) Statement I is false; Statement II is true

13. A coin is tossed. If it shows a tail, we draw a ball from a box which contains 2 red and 3 black balls. If it shows head, we throw a die. Then, the sample space for this experiment is
(a) $S = \{TR_1, TR_2, TB_1, TB_2, TB_3, H1, H2, H3, H4, H5, H6\}$
(b) $S = \{TR, TR_2, TB_1, H1, H4\}$
(c) Both (a) and (b)
(d) None of the above

14. A die is thrown repeatedly until a six comes up. Then, the sample space for this experiment is
(a) $S = \{6, (1, 6)\ (2, 6), (3, 6)\}$
(b) $S = \{(6, 1)\ (6, 2)\ (6, 4)\ (6, 6)\}$
(c) infinite number of possibilities occur
(d) None of the above

Level II

15. A coin is tossed once, then the sample space is
(a) $\{H\}$ (b) $\{T\}$
(c) $\{H, T\}$ (d) None of these

16. The total number of elementary events associated to the random experiment of throwing three dice together is
(a) 210 (b) 216 (c) 215 (d) 220

17. A coin is tossed twice. If the second throw result in a tail, a die is thrown. Then, the total number of the outcomes of this experiment is
(a) 11 (b) 13 (c) 14 (d) 16

18. A coin is tossed repeatedly until a tail comes up for the first time. Then, the sample space for this experiment is
(a) $\{T, HT, HTT\}$
(b) $\{TT, TTT, HTT\ THH\}$
(c) $\{T, HT, HHT, HHHT, HHHHT, \ldots\}$
(d) None of the above

19. A pair of dice is rolled. If the outcome is a doublet, a coin is tossed. Then, the total number of outcomes for this experiment is ...K... . Here, K refers to [HOTS]
(a) 40 (b) 42 (c) 41 (d) 43

20. A bag contains 4 identical red balls and 3 identical black balls. The experiment consists of drawing one ball, then putting it into the bag and again drawing a ball. Then, the possible outcome of this experiment is [HOTS]
(a) $\{RR, BB\}$ (b) $\{RR, B, B, RR\}$
(c) $\{BB, R\}$ (d) $\{RR, RB, BR, BB\}$

21. A bag contains one white and one red ball. A ball is drawn from the bag. If the ball drawn is white, it is replaced in the bag and again a ball is drawn. Otherwise a die is tossed, then which of the following is true?
I. Total number of outcomes is 8.
II. Sample space $S = \{(W, W), (W, R), (R, 1), (R, 2), (R, 3), (R, 4), (R, 5), (R, 6)\}$.
(a) Only I is true (b) Only II is true
(c) Both I and II are true (d) Both I and II are false

22. Which of following is true?

Statement I A letter is chosen at random from the word 'ASSASSINATION'. Then, the total number of outcomes is 13.

Statement II A letter is chosen at random from the word "NAGATATION". Then, the total number of outcomes is 10.

(a) Statement I is true (b) Statement II is true
(c) Both statements are true (d) Both statements are false

23. A box contains 1 white and 3 identical black balls. Two balls are drawn at random in succession without replacement. Then, the sample space S for this experiment is
(a) $\{WB, BW, BB\}$ (b) $\{W, B, W\}$
(c) $\{B, W, B\}$ (d) None of these

24. In a random sampling, three items are selected from a lot. Each item is tested and classified as defective (D) or non-defective (N). Then, the total possible outcomes for this experiment is
(a) 6 (b) 7 (c) 5 (d) 8

Event

Level I

25. Assertion (A) A subset of a sample space is associated with an event and an event is associated with a subset of sample space.

Reason (R) Any subset E of a sample space S is called an event.

(a) Both A and R are correct; R is the correct explanation of A
(b) Both A and R are correct; R is not the correct explanation of A
(c) A is correct, R is incorrect
(d) R is correct; A is incorrect

26. Statement I The event E of a sample space S occurred, if the outcome ω of the experiment is such that $\omega \in E$.

Statement II The event E has not occurred, if the outcome ω is such that $\omega \notin E$.

(a) Statement I is true; Statement II is true; Statement II is a correct explanation for Statement I
(b) Statement I is true; Statement II is true; Statement II is not a correct explanation for Statement I
(c) Statement I is true; Statement II is false
(d) Statement I is false; Statement II is true

27. Event can be classified into various types on the basis of the ...K... they have. Here, K refers to

(a) experiment
(b) sample space
(c) elements
(d) None of the above

28. Which of the following is true?

I. If the empty set ϕ and the sample space describe events, then ϕ is called an impossible event.

II. In the above statement, the whole sample space S is called the sure event.

(a) Only I is true
(b) Only II is true
(c) Both I and II are true
(d) Both I and II are false

29. Consider the experiment of rolling a die. Then, sample space is

$$S = \{1, 2, 3, 4, 5, 6\}$$

Statement I The event E: "the number appears on the die is a multiple of 7", is an impossible event.

Statement II The event F : "the number turns up is odd or even", is a sure event.

(a) Statement I is true; Statement II is true; Statement II is a correct explanation for Statement I
(b) Statement I is true; Statement II is true; Statement II is not a correct explanation for Statement I
(c) Statement I is true; Statement II is false
(d) Statement I is false; Statement II is true

30. A event which has only ...X... sample point of a sample space, is called simple event. Here, X refers to

(a) two (b) three (c) one (d) zero

31. If an event has more than one sample point, then it is called a/an

(a) simple event (b) elementary event
(c) compound event (d) None of these

32. For every event A, there corresponds another event A' called the ...K... to A. Here, K refers to

(a) complementary event (b) simple event
(c) not complementary event (d) None of these

33. When the sets A and B are two events associated with a sample space. Then, event '$A \cup B$' denotes

(a) A and B (b) only A (c) A or B (d) only B

34. If A and B are two events, then the set $A \cap B$ denotes the event

(a) A or B (b) A and B (c) only A (d) only B

35. The set $A - B$ denotes the event

(a) A and B (b) A or B (c) only A (d) A but not B

36. Consider the experiment of rolling a die. Let A be the event 'getting a prime number' and B be the event 'getting an odd number'.

Then, which of the following is true?

I. A or $B = A \cup B = \{1, 2, 3\}$

II. A and $B = A \cap B = \{3, 5\}$

III. A but not $B = A - B = \{2\}$

IV. Not $A = A' = \{1, 5, 6\}$

(a) Only I is true (b) Only II is true
(c) II and III are true (d) Only IV is true

37. Statement I Two events A and B are called mutually exclusive events, if the occurrence of any one of them excludes the occurrence of the other event.

Statement II In the above statement, if they can not occur simultaneously, then the events A and B are not mutually exclusive.

(a) Statement I is true; Statement II is true; Statement II is a correct explanation for Statement I
(b) Statement I is true; Statement II is true; Statement II is not a correct explanation for Statement I
(c) Statement I is true; Statement II is false
(d) Statement I is false; Statement II is true

38. **Statement I** The events $E_1, E_2, \ldots, E_n$ are said to be exhaustive, if atleast one of them necessarily occurs, whenever the experiment is performed.

Statement II $E_i \cap E_j = \phi$ for $i \neq j$ *i.e.*, the events E_i and E_j are pairwise disjoint and $\bigcup_{i=1}^{n} E_i = S$, then events $E_1, E_2, \ldots, E_n$ are called mutually exclusive and exhaustive events.

(a) Statement I is true; Statement II is true; Statement II is a correct explanation for Statement I
(b) Statement I is true; Statement II is true; Statement II is not a correct explanation for Statement I
(c) Statement I is true; Statement II is false
(d) Statement I is false; Statement II is true

39. Two dice are thrown and the sum of the numbers which come up on the dice is noted. Let us consider the following events associated with this experiment

A: "the sum is even".
B: "the sum is a multiple of 3".
C: "the sum is less than 4".
D: "the sum is greater than 11".

Which pair of these events is mutually exclusive?

(a) A and B (b) B and C (c) C and D (d) A and C

40. A coin is tossed three times, consider the following events

A: no head appears.
B: exactly one head appears.
C: atleast two heads appear.

Then,

(a) A, B and C are mutually exclusive events
(b) A, B and C are exhaustive events
(c) Only (a)
(d) Both (a) and (b)

41. A die is rolled. Let E be the event "die shows 4" and F be the event "die shows even number". Then, E and F are

(a) mutually exclusive
(b) exhaustive
(c) mutually exclusive and exhaustive
(d) None of the above

42. A die is thrown. Then, match the events of Column I with their respective sample points in Column II and choose the correct option from the codes given below.

Column I	Column II
A. A : a number less than 7.	1. {3, 4, 5, 6}
B. B : a number greater than 7.	2. {6}
C. C : a multiple of 3.	3. {1, 2, 3}
D. D : a number less than 4.	4. {3, 6}
E. E : an even number greater than 4.	5. {}
F. F : a number not less than 3.	6 {1, 2, 3, 4, 5, 6}

Codes

	A	B	C	D	E	F
(a)	6	5	4	3	2	1
(b)	1	2	3	4	5	6
(c)	5	6	4	3	2	1
(d)	3	4	5	6	2	1

43. A die is thrown. If A, B, C, D, E and F are events described in above question. Then, match the events of Column I with their respective sample points in Column II and choose the correct option from the codes given below.

Column I	Column II
A. $A \cup B$	1. {1, 2}
B. $A \cap B$	2. ϕ
C. $B \cup C$	3. {1, 2, 3}
D. $E \cap F$	4. {1, 2, 4, 5}
E. $D \cap E$	5. {6}
F. $A - C$	6. {3, 6}
G. $D - E$	7. {1, 2, 3, 4, 5, 6}
H. $E \cap F'$	
I. F'	

Codes

	A	B	C	D	E	F	G	H	I
(a)	1	2	7	3	4	5	6	4	2
(b)	7	2	6	5	2	4	3	2	1
(c)	1	2	4	7	5	3	4	2	1
(d)	6	5	4	1	2	3	6	5	3

44. An experiment involves rolling a pair of dice and recording the numbers that come up. Describe the following events

A: the sum is greater than 8.
B: 2 occurs on either die.
C: the sum is atleast 7 and a multiple of 3.

Which pair of these events is mutually exclusive?

(a) A and B (b) B and C
(c) Both (a) and (b) (d) None of these

45. Three coins are tossed once. Let A denotes the event "three heads show", B denotes the event "two heads and one tail show", C denotes the event "three tails show" and D denotes the event "a head shows on the first coin". Then, match the terms of Column I with the terms of Column II and choose the correct option from the codes given below.

Column I	Column II
A. Mutually exclusive events	1. A and B, A and C, B and C, C and D, A, B and C
B. Simple events	2. B, D
C. Compound events	3. A and C

Codes

	A	B	C
(a)	2	3	1
(b)	3	2	1
(c)	1	2	3
(d)	1	3	2

46. "Three coins are tossed." Then, match the terms of Column I with the terms of Column II and choose the correct option from the codes given below.

Column I	Column II
A. Two events which are mutually exclusive.	1. A, B and C, where A be the event 'three heads show'. B be the event 'two heads shows', and C be the event 'three tails show'.
B. Two events which are mutually exclusive and exhaustive.	2. A and B, where A be the event 'three heads show' and B be event 'atleast two heads show'.
C. Two events which are not mutually exclusive.	3. A and B where A be\` the event atleast one head and B be the event getting three tails.
D. Two events which are mutually exclusive but not exhaustive.	4. A and B where A be the event 'three heads show' and B be the event 'three tails show'.
E. Three events which are mutually exclusive but not exhaustive.	

Codes

	A	B	C	D	E
(a)	4	3	2	2	1
(b)	1	2	3	4	4
(c)	2	3	1	4	2
(d)	3	2	1	4	3

47. Two dice are thrown. The events A, B and C are as follows: [HOTS]

A : getting an even number on the first die.

B: getting an odd number on the first die.

C: getting the sum of the numbers on the dice ≤ 5.

Then, which of following is not correct?

(i) $A' = B$

(ii) Not $B = A'$

(iii) A or $B = A \cup B =$ All elements of A and B

(iv) A and $B = \phi$

(v) A but not $C = \{(2, 4), (2, 5), (2, 6), (4,2), (4, 3), (4,4), (4, 5), (4, 6), (6, 1), (6, 2), (6, 3), (6, 4), (6, 5), (6, 6)\}$

(vi) B or $C = \{(1, 1), (1, 2), (1, 3), (1, 4), (1, 5), (1, 6), (3, 1), (3, 2), (3, 3), (3, 4), (3, 5), (3, 6), (5, 1), (5, 2), (5, 3), (5, 4), (5, 5), (5, 6), (2, 1), (2, 2), (2, 3), (4, 1)\}$

(vii) B and $C = \{(1, 1), (1, 2), (1, 3), (1\ 4), (3, 1), (3, 2)\}$

(viii) $A \cap B' \cap C' = \{(2, 4), (2, 5), (2, 6), (4, 2), (4, 3), (4, 4)\ (4, 5), (4, 6), (6, 1), (6, 2), (6, 3), (6, 4), (6, 5), (6, 6)\}$

(a) (ii) (b) (i), (ii), (iii) (c) (iii), (iv), (v) (d) All of these

48. Two dice are thrown. The events A, B and C are as follows:

A : getting an even number on the first die.

B : getting on odd number on the first die.

C: getting the sum of the numbers on the dice ≤ 5.

Then, which of the following is true?

(i) A and B are mutually exclusive.

(ii) A and B are mutually exclusive and exhaustive.

(iii) $A = B'$

(iv) A and C are mutually exclusive.

(v) A and B' are mutually exclusive.

(vi) A', B' and C are mutually exclusive and exhaustive.

(a) (i) (b) (ii), (iv), (v)
(c) (vi), (iii) (d) (i), (ii) and (iii)

Level II

49. An ordinary deck of cards contains 52 cards divided into four suits. The red suits are diamonds and hearts and black suits are clubs and spades. The cards J, Q and K are called face cards. Suppose we pick one card from the deck at random. Then, the event that the chosen card is a black face card, is (where E be a event that a black card is chosen)

(a) $E = \{J, Q, K$ of spade and club$\}$
(b) $E = \{J, Q$ of spade and club$\}$
(c) Both (a) and (b)
(d) None of the above

50. Suppose that each child born is equally likely to be a boy or a girl. Consider a family with exactly three children.

Then, match the events of Column I with respective expressions in Column II and choose the correct option from the codes given below.

Column I	Column II
A. The event that exactly one child is a girl.	1. {BBG, BGB, GBB}
B. The event that atleast two children are girls.	2. {BBB}
C. The event that no child is a girl.	3. {GGB, GBG, BGG, GGG}

Codes [NCERT Exemplar]

	A	B	C		A	B	C
(a)	1	3	2	(b)	3	2	1
(c)	1	2	3	(d)	2	1	3

51. An experiment consists of rolling a die until a 2 appears.

Then, the number of elements of the sample space correspond to the event that the 2 appears on the kth roll of the die is

(a) 5^{k-1} (b) 5^{k-2}
(c) 5^{k+1} (d) 5^{k+2}

52. An experiment consists of rolling a die until a 2 appears. Then, number of elements of the sample space correspond to the event that the 2 appears not later than the kth roll of the die is

(a) $\frac{5^k}{4}$ (b) 5^{k-1} (c) $\frac{5^k-1}{4}$ (d) $\frac{5^k-1}{5}$

53. Let $S = \{1, 2, 3, 4, 5, 6\}$ and $E = \{1, 3, 5\}$, then E' or $\overline{E}$ is

(a) {2, 4} (b) {3, 6}
(c) {1, 2, 4} (d) {2, 4, 6}

54. Match the terms of Column I with the terms of Column II and choose the correct option from the codes given below.

Column I	Column II
A. If E_1 and E_2 are the two mutually exclusive events, then	1. $E_1 \cap E_2 = E_1$
B. If E_1 and E_2 are the mutually exclusive and exhaustive events, then	2. $(E_1 - E_2) \cup (E_1 \cap E_2) = E_1$
C. If E_1 and E_2 have common outcomes, then	3. $E_1 \cap E_2 = \phi$, $E_1 \cup E_2 = S$
D. If E_1 and E_2 are two events such that $E_1 \subset E_2$, then	4. $E_1 \cap E_2 = \phi$

Codes

	A	B	C	D		A	B	C	D
(a)	1	2	3	4	(b)	4	3	2	1
(c)	2	3	4	1	(d)	1	4	2	3

55. Two dice are thrown simultaneously such as the event E: "an even number as the sum." Then, the set of this event is

(a) {(1, 1), (1, 3) (3, 1), (2, 2) (1, 5) (5, 1) (3, 3), (2, 4), (4, 2), (3, 5), (5, 3), (4, 4), (6, 2), (2, 6), (5, 5), (6, 4), (4, 6), (6, 6)}
(b) {(1, 1) (2, 2), (4, 4), (2, 6)}
(c) {(1, 1), (1, 3), (1, 5), (3, 1), (6, 2)}
(d) None of the above

56. Two dice are thrown simultaneously. Then, the set of the event E "the sum as a prime number", is

(a) E = {(1, 1), (1, 2), (2, 1), (1, 4), (4, 1), (2, 3), (3, 2), (1, 6), (6, 1), (2, 5), (5, 2), (3, 4), (4, 3), (6, 5), (5, 6)}
(b) E = {(1, 1), (1, 2), (1, 5), (1, 6), (2, 4)}
(c) E = {(1, 1), (1, 2), (2, 1), (3, 6), (2, 6)}
(d) None of the above

Directions (Q. Nos. 57-59) *Two dice are thrown. The events A, B and C are described as follows:*
A : getting an even number on the first die.
B : getting an odd number on the first die.
C : getting atmost 5 as sum of the numbers on two dice.

57. Then, the set of the event A and B is

(a) ϕ (b) {1, 2}
(c) {2, 4} (d) {1, 3}

58. The total number of outcomes in the event 'B or C' is

(a) 20 (b) 22
(c) 21 (d) 23

59. The set of the event 'B and C' is

(a) {(1, 1), (1, 2), (1, 3) (1, 4)}
(b) {(1, 2), (1, 3), (1, 4), (1, 5)}
(c) {(1, 1), (1, 2), (1, 3), (1, 4), (3, 1), (3, 2)}
(d) None of the above

60. Two dice are thrown simultaneously. Then, the set of event 'a total of atleast 10' is

(a) {(6, 4), (4, 6), (5, 5), (6, 5), (5, 6), (6, 6)}
(b) {(5, 5), (6, 4), (4, 6)}
(c) {(6, 5), (5, 6), (6, 6)}
(d) {(4, 4), (4, 5), (6, 5), (5, 5)}

61. Ticket numbered from 1 to 20 are mixed up together and then a ticket is drawn at random, then the total number of elementary elements is

(a) 18
(b) 20
(c) 19
(d) 17

62. A set containing the numbers from 1 to 25. Then, the set of event getting a prime number, when each of the given number is equally likely to be selected, is

(a) {2, 3, 7, 11, 13, 17}
(b) {1, 2, 3, 7, 11, 19}
(c) {2, 5, 7, 9, 11, 13, 17, 19, 23}
(d) {2, 3, 5, 7, 11, 13, 17, 19, 23}

63. Four cards are drawn at random from a deck of 52 playing cards. Then, the total numbers of elementary elements is ...K... . Here, K refers to

(a) ${}^{50}C_4$
(b) ${}^{51}C_4$
(c) ${}^{52}C_4$
(d) ${}^{53}C_4$

Axiomatic Approach to Probability

Level I

64. Let S be a sample space containing outcomes $\omega_1, \omega_2, \omega_3, \ldots, \omega_n$ *i.e.*, $S = \{\omega_1, \omega_2, \ldots, \omega_n\}$.

Then, which of the following is true?

I. $0 \le P(\omega_i) \le 1$ for each $\omega_i \in S$

II. $P(\omega_1) + P(\omega_2) + \ldots + P(\omega_n) = 1$

III. For any event A, $P(A) = \Sigma P(\omega_i)$, $\omega_i \in A$

(a) Only I (b) Only II
(c) Only III (d) All of these

65. Statement I In a coin tossing experiment, we can assign the number $\frac{1}{2}$ to each of the outcomes H and T.

i.e., $P(H) = \frac{1}{2}$ and $P(T) = \frac{1}{2}$. Then, this assignment satisfy the condition of axiomatic approach.

Statement II In Statement I, if we take $P(H) = \frac{1}{4}$ and $P(T) = \frac{3}{4}$, then this assignment does not satisfy the condition of axiomatic approach.

(a) Statement I is true; Statement II is true; Statement II is a correct explanation for Statement I
(b) Statement I is true; Statement II is true; Statement II is not a correct explanation for Statement I
(c) Statement I is true; Statement II is false
(d) Statement I is false; Statement II is true

66. Let a sample space be $S = \{\omega_1, \omega_2, \ldots \omega_6\}$. Which of the following assignments of probabilities to each outcome are valid?

Outcomes

	ω_1	ω_2	ω_3	ω_4	ω_5	ω_6
(i)	$\frac{1}{6}$	$\frac{1}{6}$	$\frac{1}{6}$	$\frac{1}{6}$	$\frac{1}{6}$	$\frac{1}{6}$
(ii)	1	0	0	0	0	0
(iii)	$\frac{1}{8}$	$\frac{2}{3}$	$\frac{1}{3}$	$\frac{1}{3}$	$-\frac{1}{4}$	$-\frac{1}{3}$
(iv)	$\frac{1}{12}$	$\frac{1}{12}$	$\frac{1}{6}$	$\frac{1}{6}$	$\frac{1}{6}$	$\frac{3}{2}$
(v)	0.1	0.2	0.3	0.4	0.5	0.6

(a) Only (i) (b) Only (ii) (c) Only (v) (d) (i) and (ii)

67. Let S be a sample space associated with the experiment examining three consecutive pens produced by a machine and classified as good (non-defective) and bad (defective). We may get 0, 1, 2 or 3 defective pens as result of this examination and consider event A : there is exactly one defective pen and event B : there are atleast two defective pens. Then,

Statement I $P(A) = \Sigma P(\omega_i)$, $\forall\, \omega_i \in A = \frac{3}{8}$

Statement II $P(B) = \Sigma P(\omega_i)$, $\forall\, \omega_i \in B = \frac{1}{2}$

(a) Statement I is true; Statement II is true; Statement II is a correct explanation for Statement I
(b) Statement I is true; Statement II is true; Statement II is not a correct explanation for Statement I
(c) Statement I is true; Statement II is false
(d) Statement I is false; Statement II is true

68. Consider the experiment "tossing a coin twice" and probability of outcomes are as $P\ (HH) = \frac{1}{4}$, $P\ (HT) = \frac{1}{7}$, $P(TH) = 2/7$ and $P(TT) = \frac{9}{28}$, then

Statement I Probability of the event E : 'both the tosses yield the same result' is $\frac{4}{7}$.

Statement II Probability of the event F : 'exactly two heads' is $\frac{1}{4}$.

(a) Statement I is true; Statement II is true; Statement II is a correct explanation for Statement I
(b) Statement I is true; Statement II is true; Statement II is not a correct explanation for Statement I
(c) Statement I is true; Statement II is false
(d) Statement I is false; Statement II is true

69. Let a sample space of an experiment be $S = \{\omega_1, \omega_2, \omega_3 \ldots\ldots \omega_n\}$. Then, $\sum_{i=1}^{n} P(\omega_i)$ is equal to

(a) 0 (b) 1
(c) $\frac{1}{2}$ (d) $\frac{1}{3}$

70. If A and B are two events, then which of the following is true?

(a) $P(A \cup B) = P(A) + P(B)$
(b) $P(A \cup B) = P(A) + P(B) - \Sigma P(\omega_i), \forall\, \omega_i \in A \cap B$
(c) $P(A \cup B) = P(A) + P(B) - P(A \cap B)$
(d) Both (b) and (c)

71. If A and B are any two events associated with random experiment, then by the definition of probability, which of following is true?

(i) $P(A \cup B) = \Sigma P(\omega_i)\ \forall\, \omega_i \in A \cup B$
(ii) $P(A) + P(B) = P(A \cup B) + P(A \cap B)$

(a) Only (i) (b) Only (ii)
(c) Both (i) and (ii) (d) None of these

72. If A and B be two events associated with a random experiment, then

Statement I $A \cup B = A \cup (B - A)$, where A and $B - A$ are mutually exclusive and $B = (A \cap B) \cup (B - A)$, where $A \cap B$ and $B - A$ are mutually exclusive.

Statement II $P(A \cup B) = P(A) + P(B)$, if $P(A \cap B) = P(\phi) = 0$.

(a) Both I and II are correct
(b) Both I and II are incorrect
(c) I is correct, II is incorrect
(d) II is correct, I is incorrect

73. Consider the event $A = \{2, 4, 6, 8\}$ associated with the experiment of drawing a card from a deck of ten cards numbered from 1 to 10. Clearly, the sample space is $S = \{1, 2, 3, \ldots\ldots, 10\}$.
Then, which of following is true?

(i) $P(A') = 1 - P(A)$ (ii) $P(A') = \frac{3}{5}$

(iii) $P(A) + P(A') = 1$

(a) Only (i) (b) Only (ii)
(c) All are true (d) (i) and (ii)

74. One card is drawn from a well-shuffled deck of 52 cards. If each outcome is equally likely, then match the probability that the card will be of Column I to the Column II and choose the correct option from the codes given below.

Column I	Column II
A. A diamond	1. $\frac{1}{2}$
B. Not an ace	2. $\frac{3}{4}$
C. A black card (*i.e.*, a club or a spade)	3. $\frac{12}{13}$
D. Not a diamond	4. $\frac{1}{4}$
E. Not a black card	

Codes

	A	B	C	D	E		A	B	C	D	E
(a)	4	3	1	2	1	(b)	3	2	4	2	1
(c)	2	3	1	4	4	(d)	1	2	3	4	3

75. A bag contains 9 discs. Of which 4 are red, 3 are blue and 2 are yellow. The discs are similar in shape and size. A disc is drawn at random from the bag.

Let the events A, B and C be defined as

A : the disc drawn is red.

B : the disc drawn is yellow.

C : the disc drawn is blue.

Then, which of the following is correct?

(i) $P(A) = \frac{2}{9}$

(ii) $P(B) = \frac{4}{9}$

(iii) $P(C) = \frac{1}{9}$

(iv) $P(\text{not } C) = \frac{2}{3}$

(v) $P(A \text{ or } C) = \frac{7}{9}$

(a) (i), (ii) and (iii) (b) Only (ii)
(c) Only (iv) (d) (iv) and (v)

76. Two students Anil and Ashima appeared in an examination. The probability that Anil will qualify the examination is 0.05 and that Ashima will qualify the examination is 0.10. The probability that both will qualify the examination is 0.02. Then, which of the following is correct?

I. Probability that Anil and Ashima will not qualify the examination is 0.87.
II. Probability that atleast one of them will not qualify the examination is 0.98.
III. Probabiity that only one of them will qualify the examination is 0.11.

(a) Only I is correct
(b) Only II is correct
(b) Only III is correct
(d) All are correct

77. A committee of two persons is selected from two men and two women. The probabilities that the committee will have (i) no man (ii) one man (iii) two men are respectively

(a) $\frac{1}{2}, \frac{1}{3}$ and $\frac{1}{3}$
(b) $\frac{1}{6}, \frac{2}{3}$ and $\frac{1}{6}$
(c) $\frac{1}{2}, \frac{1}{6}$ and $\frac{1}{5}$
(d) $\frac{1}{6}, \frac{1}{2}$ and $\frac{1}{3}$

78. Which of the following cannot be valid assignment of probabilities for outcomes of sample space $S = \{\omega_1, \omega_2, \omega_3, \omega_4, \omega_5, \omega_6, \omega_7\}$?

Assignment

	ω_1	ω_2	ω_3	ω_4	ω_5	ω_6	ω_7
(i)	0.1	0.01	0.05	0.03	0.01	0.2	0.6
(ii)	$\frac{1}{7}$	$\frac{1}{7}$	$\frac{1}{7}$	$\frac{1}{7}$	$\frac{1}{7}$	$\frac{1}{7}$	$\frac{1}{7}$
(iii)	0.1	0.2	0.3	0.4	0.5	0.6	0.7
(iv)	– 0.1	0.2	0.3	0.4	– 0.2	0.1	0.3
(v)	$\frac{1}{14}$	$\frac{2}{14}$	$\frac{3}{14}$	$\frac{4}{14}$	$\frac{5}{14}$	$\frac{6}{14}$	$\frac{15}{14}$

(a) (i) and (ii)
(b) (iii) and (iv)
(c) (iv) and (v)
(d) Only (iii)

79. A coin is tossed twice. Then, the probability that atleast one tail occurs is

(a) $\frac{1}{4}$
(b) $\frac{1}{2}$
(c) $\frac{1}{3}$
(d) $\frac{3}{4}$

80. A die is thrown. Then, match the events of Column I with their respective probabilities in Column II and choose the correct option from the codes given below.

Column I	Column II
A. A prime number will appear.	1. $\frac{5}{6}$
B. A number greater than or equal to 3 will appear.	2. 0
C. A number less than or equal to one will appear.	3. $\frac{1}{6}$
D. A number more than 6 will appear.	4. $\frac{2}{3}$
E. A number less than 6 will appear.	5. $\frac{1}{2}$

Codes

	A	B	C	D	E
(a)	1	2	3	4	5
(b)	5	4	3	2	1
(c)	2	3	4	5	1
(d)	1	2	5	4	3

81. A card is selected from a deck of 52 cards. Then, match the events of Column I with their respective values in Column II and choose the correct option from the codes given below.

Column I	Column II
A. Number of points are there in the sample space.	1. $\frac{1}{13}$
B. The probability that the card is an ace of spade.	2. $\frac{1}{52}$
C. The probability that the card is an ace.	3. 52
D. The probability that the card is black card.	4. $\frac{1}{2}$

Codes

	A	B	C	D
(a)	1	2	3	1
(b)	2	4	3	3
(c)	3	2	4	2
(d)	3	2	1	4

82. Statement I A fair coin with 1 marked on one face and 6 on the other and a fair die are both tossed. Then,

(i) Probability that sum of numbers gives 3 is $\frac{1}{12}$.

(ii) Probability that sum of numbers gives 12 is $\frac{1}{13}$.

Statement II There are four men and six women in the city council. If one council member is selected for a committee at random. Then, the probability that it is a woman, is $\frac{3}{5}$.

(a) Statement I is true; Statement II is true; Statement II is a correct explanation for Statement I
(b) Statement I is true; Statement II is true; Statement II is not a correct explanation for Statement I
(c) Statement I is true; Statement II is false
(d) Statement I is false; Statement II is true

83. **Statement I** A fair coin is tossed four times and a person win ₹ 1 for each head and lose ₹ 1.50 for each tail that turns up. From the sample space, how many different amounts of money you can have after four tosses and then the probability of having each of these amounts is $\frac{1}{16}, \frac{1}{4}, \frac{3}{8}, \frac{1}{4}, \frac{1}{16}$ respectively?

Statement II If $\frac{2}{11}$ is the probability of an event, then the probability of the event 'not A' is $\frac{9}{11}$.

(a) Statement I is true; Statement II is true; Statement II is a correct explanation for Statement I
(b) Statement I is true; Statement II is true; Statement II is not a correct explanation for Statement I
(c) Statement I is true; Statement II is false
(d) Statement I is false; Statement II is true

84. Three coins are tossed once. Then, match the probability of getting the events in Column I with their respective values in Column II and choose the correct option from the codes given below.

Column I		Column II	
A.	3 heads	1.	$\frac{1}{8}$
B.	2 heads	2.	$\frac{7}{8}$
C.	atleast 2 heads	3.	$\frac{1}{2}$
D.	atmost 2 heads	4.	$\frac{3}{8}$
E.	no head		

Codes

	A	B	C	D	E
(a)	3	2	4	1	4
(b)	1	2	3	4	2
(c)	1	4	3	2	1
(d)	1	3	2	4	4

85. If three coins are tossed once, then which of following is true?

(i) Probability of getting 3 tails is $\frac{1}{7}$.

(ii) Probability of getting exactly two tails is $\frac{3}{8}$.

(iii) Probability of getting no tail is $\frac{1}{8}$.

(iv) Probability of getting atmost two tails is $\frac{3}{8}$.

(a) (i) and (iv)
(b) Only (ii)
(c) Only (iii)
(d) (ii) and (iii)

86. **Statement I** A letter is chosen at random from the word 'ASSASSINATION'. Then, the probability that letter is (i) a vowel (ii) a consonant are $\frac{6}{13}$ and $\frac{7}{13}$, respectively.

Statement II In a lottery, a person chooses six different natural numbers at random from 1 to 20, and if these six numbers match with the six numbers already fixed by the lottery committee, he wins the prize. Then, the probability of winning the prize in the game is $\frac{1}{38760}$.

(a) Statement I is true; Statement II is true; Statement II is a correct explanation for Statement I
(b) Statement I is true; Statement II is true; Statement II is not a correct explanation for Statement I
(c) Statement I is true; Statement II is false
(d) Statement I is false; Statement II is true

87. Check whether the following probabilities $P(A)$ and $P(B)$ are consistently defined.

(i) $P(A) = 0.5, P(B) = 0.7, P(A \cap B) = 0.6$
(ii) $P(A) = 0.5, P(B) = 0.4, P(A \cup B) = 0.8$

(a) (i) is consistent
(b) (ii) is consistent
(c) Both (i) and (ii) are consistent
(d) None of these

88. Consider the following table

	$P(A)$	$P(B)$	$P(A \cap B)$	$P(A \cup B)$
(i)	$\frac{1}{3}$	$\frac{1}{5}$	$\frac{1}{15}$	p
(ii)	0.35	q	0.25	0.6
(iii)	0.5	0.35	r	0.7

where p, q and r refers to (respectively)

(a) $\frac{7}{15}$, 0.5 and 0.15
(b) $\frac{7}{12}$, 0.4 and 0.14
(c) $\frac{3}{15}$, 0.14 and 0.42
(d) $\frac{7}{15}$, 0.5 and 0.13

89. **Statement I** Given, $P(A) = \frac{3}{5}$ and $P(B) = \frac{1}{5}$. Then, P (A or B), if A and B are mutually exclusive events, is $\frac{4}{5}$.

Statement II If E and F are events such that $P(E) = \frac{1}{4}$, $P(F) = \frac{1}{2}$ and $P(E \text{ and } F) = \frac{1}{8}$, then (i) $P(E \text{ or } F)$ and (ii) P (not E and not F) are respectively $\frac{5}{8}$ and $\frac{3}{8}$.

(a) Statement I is true; Statement II is true; Statement II is a correct explanation for Statement I
(b) Statement I is true; Statement II is true; Statement II is not a correct explanation for Statement I
(c) Statement I is true; Statement II is false
(d) Statement I is false; Statement II is true

90. **Statement I** Events E and F are such that P (not E or not F) $= 0.25$, then E and F are not mutually exclusive.

Statement II A and B are events such that $P(A) = 0.42$, $P(B) = 0.48$ and $P(A \text{ and } B) = 0.16$ then $P(\text{not } A)$, $P(\text{not } B)$ and $P(A \text{ or } B)$ are 0.58, 0.52 and 0.74, respectively.

(a) Statement I is true; Statement II is true; Statement II is a correct explanation for Statement I
(b) Statement I is true; Statement II is true; Statement II is not a correct explanation for Statement I
(c) Statement I is true; Statement II is false
(d) Statement I is false; Statement II is true

91. In Class XI of a school 40% of the students study Mathematics and 30% study Biology. 10% of the class study both Mathematics and Biology. If a student is selected at random from the class, then the probability that he will be studying Mathematics or Biology, is

(a) 0.4 (b) 0.5
(c) 0.7 (d) 0.6

92. **Statement I** In an entrance test that is graded on the basis of two examinations, the probability of a randomly chosen student passing the first examination is 0.8 and the probability of passing the second examination is 0.7. The probability of passing atleast one of them is 0.95. Then, the probability of passing both is 0.55.

Statement II The probability that a student will pass the final examination in both English and Hindi is 0.5 and the probability of passing neither is 0.1. If the probability of passing the English examination is 0.75, then the probability of passing the Hindi examination is 0.65.

(a) Statement I is true; Statement II is true; Statement II is a correct explanation for Statement I
(b) Statement I is true; Statement II is true; Statement II is not a correct explanation for Statement I
(c) Statement I is true; Statement II is false
(d) Statement I is false; Statement II is true

93. In a class of 60 students, 30 opted for NCC, 32 opted for NSS and 24 opted for both NCC and NSS. If one of these student is selected at random, then match the probability of events of Column I with their respective values in Column II and choose the correct option from the codes given below.

Column I	Column II
A. The student opted for NCC or NSS.	1. $\frac{2}{15}$
B. The student has opted neither NCC nor NSS.	2. $\frac{11}{30}$
C. The student has opted NSS but not NCC.	3. $\frac{19}{30}$

Codes

	A	B	C
(a)	1	2	3
(b)	2	3	1
(c)	2	1	3
(d)	3	2	1

Level II

94. The probability that a randomly chosen two-digit positive integer is a multiple of 3, is

(a) $\frac{1}{2}$ (b) $\frac{1}{3}$ (c) $\frac{1}{4}$ (d) $\frac{1}{5}$

95. A typical PIN (personal identification number) is a sequence of any four symbols chosen from the 26 letters in the alphabet and the ten digits. If all PINs are equally likely, then probability that a randomly chosen PIN contains a repeated symbol is

(a) 0.5142 (b) 0.1583 (c) 0.1582 (d) 0.1483

96. An experiment has four possible outcomes A, B, C and D, that are mutually exclusive. Then, which of the following assignments of probabilities are permissible.

(i) $P(A) = 0.12$, $P(B) = 0.63$, $P(C) = 0.45$, $P(D) = -0.20$

(ii) $P(A) = \frac{9}{120}$, $P(B) = \frac{45}{120}$, $P(C) = \frac{27}{120}$, $P(D) = \frac{46}{120}$

(a) (i) is permissible
(b) (ii) is permissible
(c) Both (i) and (ii) are permissible
(d) Both (i) and (ii) are not permissible

97. If M and N are any two events, the probability that atleast one of them occurs is ...K... . Here, K refers to

(a) $P(M) + P(N) - 2P(M \cap N)$
(b) $P(M) + P(N) - P(M \cap N)$
(c) $P(M) + P(N) + P(M \cap N)$
(d) $P(M) + P(N) + 2P(M \cap N)$

98. If A and B are mutually exclusive events, then
(a) $P(A) \le P(\bar{B})$ (b) $P(A) \ge P(\bar{B})$
(c) $P(A) < P(\bar{B})$ (d) None of these

99. If $P(A \cup B) = P(A \cap B)$ for any two events A and B, then
(a) $P(A) = P(B)$ (b) $P(A) > P(B)$
(c) $P(A) < P(B)$ (d) None of these

100. **Statement I** If the letters of the word 'ALGORITHM' are arranged at random in a row, then the probability that the letters GOR must remain together as a unit, is $\frac{1}{72}$.

Statement II Six new employees, two of whom are married to each other, are to be assigned six desks that are lined up in a row. If the assignment of employees to desks is made randomly, then the probability that the married couple will have non-adjacent desks, is $\frac{2}{3}$

(a) Statement I is true; Statement II is true; Statement II is a correct explanation for Statement I
(b) Statement I is true; Statement II is true; Statement II is not a correct explanation for Statement I
(c) Statement I is true; Statement II is false
(d) Statement I is false; Statement II is true

101. A die is loaded in such a way that each odd number is twice as likely to occur as each even number. Then $P(G)$, when G is the event that a number greater than 3 occurs on a single roll of the die, is ...K... . Here, K refers to
(a) $\frac{4}{5}$ (b) $\frac{5}{7}$ (c) $\frac{2}{3}$ (d) $\frac{4}{9}$

102. If A and B are mutually exclusive events, $P(A) = 0.35$, and $P(B) = 0.45$, then match the terms of Column I with their respective values in Column II and choose the correct option from the codes given below.

	Column I		Column II
A.	$P(A') + P(B')$	1.	0.2
B.	$P(A \cup B)$	2.	0.35
C.	$P(A \cap B)$	3.	0
D.	$P(A \cap B')$	4.	0.8
E.	$P(A' \cap B')$	5.	1.20

Codes

	A	B	C	D	E		A	B	C	D	E
(a)	1	2	3	4	5	(b)	5	1	3	4	2
(c)	3	2	1	4	5	(d)	5	4	3	2	1

103. In a non-leap year, the probability of having 53 Tuesdays or 53 Wednesdays is [NCERT Exemplar]
(a) $\frac{1}{7}$ (b) $\frac{2}{7}$
(c) $\frac{3}{7}$ (d) None of these

104. Three numbers are chosen from 1 to 20. Find the probability that they are not consecutive.
(a) $\frac{186}{190}$ (b) $\frac{187}{190}$ (c) $\frac{188}{190}$ (d) $\frac{18}{^{20}C_3}$

105. While shuffling a pack of 52 playing cards, 2 are accidentally dropped. Find the probability that the missing cards to be of different colours. [HOTS]
(a) $\frac{29}{52}$ (b) $\frac{1}{2}$ (c) $\frac{26}{51}$ (d) $\frac{27}{51}$

106. Seven persons are to be seated in a row. The probability that two particular persons sit next to each other, is
(a) $\frac{1}{3}$ (b) $\frac{1}{6}$ (c) $\frac{2}{7}$ (d) $\frac{1}{2}$

107. Match the proposed probability under Column I with the appropriate written description under Column II and choose the correct option from the codes given below.

	Column I (Probability)		Column II (Written description)
A.	0.95	1.	An incorrect assignment
B.	0.02	2.	No chance of happening
C.	− 0.3	3.	As much chance of happening as not
D.	0.5	4.	Very likely to happen
E.	0	5.	Very little chance of happening

Codes

	A	B	C	D	E
(a)	4	5	1	3	2
(b)	1	2	3	4	5
(c)	3	2	4	5	1
(d)	5	2	3	4	1

108. A box contains 6 nails and 10 nuts. Half of the nails and half of the nuts are rusted. If one item is chosen at random, then the probability that it is rusted or is a nail, is ...K... . Here, K refers to
(a) $\frac{3}{16}$ (b) $\frac{5}{16}$ (c) $\frac{11}{16}$ (d) $\frac{14}{16}$

109. From 10000 lottery tickets numbered from 1 to 10000. One ticket is drawn at random. Then, the probability that the number marked on the drawn ticket is divisible by 20, is ...K... . Here, K refers to

(a) $\frac{1}{100}$ (b) $\frac{1}{50}$ (c) $\frac{1}{20}$ (d) $\frac{1}{10}$

110. Without repetition of the numbers, four-digit numbers are formed with the numbers 0, 2, 3, 5. The probability of such a number divisible by 5 is

(a) $\frac{1}{5}$ (b) $\frac{4}{5}$ (c) $\frac{1}{30}$ (d) $\frac{5}{9}$

Miscellaneous Exercise

111. The probabilities that when a hand of 7 cards is drawn from a well-shuffled deck of 52 cards, it contains all kings, 3 kings and atleast 3 kings are respectively

(a) $\frac{1}{7735}, \frac{9}{1547}$ and $\frac{46}{7735}$ (b) $\frac{1}{7734}, \frac{4}{1547}$ and $\frac{16}{1775}$

(c) $\frac{1}{7770}, \frac{4}{7151}$ and $\frac{45}{1794}$ (d) None of these

112. On her vacations, Veena visits four cities (A, B, C and D) in a random order. Then, match the probability that she visit in Column I with their respective values in Column II and choose the correct option from the codes given below.

Column I	Column II
A. A before B	1. $\frac{1}{4}$
B. A before B and B before C	2. $\frac{1}{2}$
C. A first and B last	3. $\frac{1}{12}$
D. A either first or second	4. $\frac{1}{6}$
E. A just before B	

Codes

	A	B	C	D	E
(a)	2	4	3	2	1
(b)	1	2	3	4	2
(c)	4	3	2	2	1
(d)	2	1	4	3	4

113. If A, B and C are three events associated with a random experiment, then $P(A \cup B \cup C)$ is equal to ...K... . Here, K refers to

(a) $P(A) + P(B) + P(A \cap B) + P(A \cap C)$

(b) $P(A) + P(B) + P(C) + P(A \cap B) + P(B \cap C) + P(C \cap A) - P(A \cap B \cap C)$

(c) $P(A) + P(B) + P(C) - P(A \cap B) - P(A \cap C) - P(B \cap C) + P(A \cap B \cap C)$

(d) None of the above

114. Statement I In a relay race, there are five teams A, B, C, D and E. The probability that A, B and C finish first, second and third respectively, is $\frac{1}{60}$.

Statement II The probability that A, B and C are first three to finish (in any order) (Assume that all finishing orders are equally likely), is $\frac{1}{10}$.

(a) Both statements are true
(b) Both statements are false
(c) Statement I is true; Statement II is false
(d) Statement I is false; Statement II is true

115. A box contains 10 red marbles, 20 blue marbles and 30 green marbles. 5 marbles are drawn from the box. What is the probability that

(i) all will be blue?
(ii) atleast one will be green?

(a) $\frac{^{20}C_4}{^{60}C_5}, 1 - \frac{^{20}C_4}{^{30}C_5}$ (b) $\frac{^{20}C_5}{^{60}C_4}, 1 - \frac{^{20}C_4}{^{30}C_5}$

(c) $\frac{^{20}C_3}{^{60}C_5}, 1 - \frac{^{20}C_3}{^{28}C_4}$ (d) $\frac{^{20}C_5}{^{60}C_5}, 1 - \frac{^{30}C_5}{^{60}C_5}$

116. 4 cards are drawn from a well-shuffled deck of 52 cards. Thus, the probability of obtaining 3 diamonds and one spade is

(a) $\frac{^{13}C_1 \times {}^{12}C_1}{^{52}C_1}$ (b) $\frac{^{13}C_3 \times {}^{13}C_1}{^{52}C_4}$

(c) $\frac{^{13}C_1 \times {}^{13}C_2}{^{52}C_2}$ (d) $\frac{^{13}C_3 \times {}^{13}C_2}{^{51}C_4}$

117. The number lock of a suitcase has 4 wheels, each labelled with ten digits, *i.e.*, from 0 to 9. The lock opens with a sequence of four digits with no repeats. Then, the probability of a person getting the right sequence to open the suitcase is

(a) $\frac{1}{5041}$ (b) $\frac{1}{5040}$ (c) $\frac{1}{5042}$ (d) $\frac{1}{4050}$

118. If 4-digit numbers greater than 5000 are randomly formed from the digits 0, 1, 3, 5 and 7, then the probabilities of forming a number divisible by 5, when (i) the digits are repeated (ii) the repetition of digits is not allowed, are respectively

(a) $\frac{1}{5}, \frac{2}{5}$ (b) $\frac{2}{5}, \frac{1}{8}$ (c) $\frac{2}{5}, \frac{3}{8}$ (d) $\frac{1}{5}, \frac{1}{8}$

119. A die has two faces each with number '1', three faces each with number '2' and one face with number '3'. If die is rolled once, then match the probabilities of Column I with their respective values in Column II and choose the correct option from the codes given below.

Column I	Column II
A. $P(2)$	1. $\frac{5}{6}$
B. $P(1 \text{ or } 3)$	2. $\frac{1}{2}$
C. $P(\text{not } 3)$	

Codes [NCERT Exemplar]

	A	B	C		A	B	C
(a)	2	1	1	(b)	2	2	1
(c)	1	1	2	(d)	1	2	2

120. In a certain lottery, 10000 tickets are sold and ten equal prizes are awarded. Then, the probabilities of not getting a prize, if you buy (i) one ticket (ii) two tickets (iii) 10 tickets, are respectively

(a) $\frac{999}{1000}, \frac{^{9990}C_2}{^{10000}C_2}, \frac{^{9990}C_{10}}{^{10000}C_{10}}$ (b) $\frac{999}{1000}, \frac{^{999}C_2}{^{10000}C_1}, \frac{99901}{100000}$

(c) $\frac{99}{1000}, \frac{999}{10000}, \frac{^{99}C_2}{^{10000}C_1}$ (d) None of these

121. From the employees of a company, 5 persons are selected to represent them in the managing committee of the company. Particulars of five persons are as follows

S. No.	Name	Sex	Age (in years)
1.	Harish	M	30
2.	Rohan	M	33
3.	Sheetal	F	46
4.	Alis	F	28
5.	Salim	M	41

A person is selected at random from this group to act as a spokesperson. Then, the probability that the spokesperson will be either male or over 35 yr, is

(a) $\frac{2}{5}$ (b) $\frac{4}{5}$ (c) $\frac{1}{5}$ (d) $\frac{3}{5}$

122. Out of 100 students, two sections of 40 and 60 are formed. If you and your friend are among the 100 students.

Statement I The probability that you both enter the same section, is $\frac{17}{33}$.

Statement II The probability that you both enter the different sections, is $\frac{16}{33}$

(a) Statement I is true; Statement II is true; Statement II is a correct explanation for Statement I
(b) Statement I is true; Statement II is true; Statement II is not a correct explanation for Statement I
(c) Statement I is true; Statement II is false
(d) Statement I is false; Statement II is true

123. Three letters are dictated to three persons and an envelope is addressed to each of them, the letters are inserted into the envelopes at random, so that each envelope contains exactly one letter. Then, the probability that atleast one letter is in its proper envelope, is

(a) $\frac{2}{5}$ (b) $\frac{1}{5}$
(c) $\frac{2}{3}$ (d) $\frac{1}{3}$

124. A and B are two events such that $P(A) = 0.54$, $P(B) = 0.69$ and $P(A \cap B) = 0.35$.

Then, match the terms of Column I with terms of Column II and choose the correct option from the codes given below.

Column I	Column II
A. $P(A \cup B)$	1. 0.34
B. $P(A' \cap B')$	2. 0.19
C. $P(A \cap B')$	3. 0.12
D. $P(B \cap A')$	4. 0.88

Codes

	A	B	C	D		A	B	C	D
(a)	4	3	2	1	(b)	1	2	3	4
(c)	2	3	4	1	(d)	3	2	1	4

125. Probability that a truck stopped at a roadblock will have faulty brakes and badly worn tires, are 0.23 and 0.24, respectively. Also, the probability is 0.38 that a truck stopped at the roadblock will have faulty brakes or badly working tires. Then, the probability that a truck stopped at this roadblock will have faulty breaks as well as badly worn tires, is ...K... . Here, K refers to

(a) 0.04 (b) 0.07
(c) 0.06 (d) 0.09

126. In a leap year, the probability of having 53 Sundays or 53 Mondays is

(a) $\frac{2}{7}$ (b) $\frac{3}{7}$

(c) $\frac{4}{7}$ (d) $\frac{5}{7}$

127. Three-digit numbers are formed using the digits 0, 2, 4, 6, 8. A number is chosen at random out of these numbers. What is the probability that this number has the same digits?

(a) $\frac{1}{16}$ (b) $\frac{16}{25}$

(c) $\frac{1}{645}$ (d) $\frac{1}{25}$

128. One mapping (function) is selected at random from all the mappings of the set $A = \{1, 2, 3, \ldots, n\}$ into itself. The probability that the mapping selected is one-to-one, is

(a) $\frac{1}{n^n}$ (b) $\frac{1}{n!}$

(c) $\frac{(n-1)!}{n^{n-1}}$ (d) None of these

129. If A, B and C are three mutually exclusive and exhaustive events of an experiment such that $3P(A) = 2P(B) = P(C)$, then $P(A)$ is equal to ...K... . Here, K refers to

(a) $\frac{1}{11}$

(b) $\frac{2}{11}$

(c) $\frac{5}{11}$

(d) $\frac{6}{11}$

130. Three of the six vertices of a regular hexagon are chosen at random. Then, the probability that the triangle formed with these vertices is equilateral, is

(a) $\frac{3}{10}$

(b) $\frac{3}{20}$

(c) $\frac{1}{20}$

(d) $\frac{1}{10}$

Answer *with* Explanations

1. *(c)* In our day-to-day life, we perform many activities which have a fixed result no matter any number of times they are repeated. Such as, given any triangle, without knowing the three angles, we can definitely say that the sum of measure of angles is 180°.

We also perform many experimental activities, where the result may not be same, when they are repeated under identical conditions. Such as when a coin is tossed it may turn up a head or a tail, but we are not sure which one of these results will actually be obtained. Such experiments are called random experiments.

An experiment is called random experiment, if it satisfies the following two conditions:

(i) It has more than one possible outcomes.

(ii) It is not possible to predict the outcome in advance.

2. *(b)* A possible result of a random experiment is called its outcome.

Consider the experiment of rolling a die. The outcomes of this experiment are 1, 2, 3, 4, 5, or 6, if we are interested in the number of dots on the upper face of the die.

The set of outcomes {1, 2, 3, 4, 5, 6} is called the sample space of the experiment.

Thus, the set of all possible outcomes of a random experiment is called the sample space associated with the experiment. Sample space is denoted by the symbol 'S'.

Each element of the sample space is called a sample point. In other words, each outcome of the random experiment is also called sample point.

3. *(a)* Clearly, the coins are distinguishable in the sense that we can speak of the first coin and the second coin. Since, either coin can turn up Head (H) or Tail (T), the possible outcomes may be

Heads on both coins $= (H, H) = HH$

Head on first coin and tail on the other $= (H, T) = HT$

Tail on first coin and head on the other $= (T, H) = TH$

Tail on both coins $= (T, T) = TT$

Thus, the sample space is $S = \{HH, HT, TH, TT\}$

Note *The outcomes of this experiment are ordered pairs of H and T. For the sake of simplicity, the commas are omitted from the ordered pairs.*

4. *(c)* Suppose 1 appears on blue die and 2 on the red die. We denote this outcome by an ordered pair (1, 2). Similarly, if 3 appears on blue die and 5 on red, the outcome is denoted by the ordered pair (3, 5).

In general, each outcome can be denoted by the ordered pair (x, y), where x is the number appeared on the blue die and y is the number appeared on the red die. Therefore, this sample space is given by

$S = \{(x, y)$: x is the number on the blue die and y is the number on the red die$\}$.

The number of elements of this sample space is $6 \times 6 = 36$ and the sample space is given below.

{(1, 1), (1, 2), (1, 3), (1, 4), (1, 5), (1, 6), (2, 1), (2, 2), (2, 3), (2, 4), (2, 5), (2, 6), (3, 1), (3, 2), (3, 3), (3, 4), (3, 5), (3, 6), (4, 1), (4, 2), (4, 3), (4, 4), (4, 5), (4, 6), (5, 1), (5, 2), (5, 3), (5, 4), (5, 6), (6, 1), (6, 2), (6, 3), (6, 4), (6, 5), (6, 6)}

5. *(b)* I. Let Q denotes a ₹ 1 coin, H denotes a ₹ 2 coin and R denotes a ₹ 5 coin. The first coin he takes out of his pocket may be anyone of the three coins Q, H or R. Corresponding to Q, the second draw may be H or R. So, the result of two draws may be QH or QR. Similarly, corresponding to H, the second draw may be Q or R.

Therefore, the outcomes may be HQ or HR. Lastly, corresponding to R, the second draw may be H or Q.

So, the outcomes may be RH or RQ.

Thus, the sample space is $S = \{QH, QR, HQ, HR, RH, RQ\}$

II. The number of accidents along a busy highway during the year of observation can be either 0 (for no accident) or 1 or 2, or some other positive integer.

Thus, a sample space associated with this experiment is $S = \{0, 1, 2, \ldots\}$.

6. *(b)* I. Let us denote blue balls by B_1, B_2, B_3 and the white balls by W_1, W_2, W_3, W_4. Then, a sample space of the experiment is

$S = \{HB_1, HB_2, HB_3, HW_1, HW_2, HW_3, HW_4,$
$T1, T2, T3, T4, T5, T6\}$.

Here, HB_i means head on the coin and ball B_i is drawn, HW_i means head on the coin and ball W_i is drawn. Similarly, T_i means tail on the coin and the number i on the die.

II. In the experiment, head may come up on the 1st toss, or the 2nd toss, or the 3rd toss and so on till head is obtained. Hence, the desired sample space is

$S = \{H, TH, TTH, TTTH, TTTTH, \ldots\}$

7. *(a)* A. We know that, sample space means a set of all possible outcomes occurring in an experiment. If a coin is tossed three times, then possible outcomes are

$\{HHH, HHT, HTT, THH, THT, HTH, TTH, TTT\}$

where, H represents head and T represents tail.

Hence, sample space

$$S = \{HHH, HHT, HTT, THH, THT, HTH, TTH, TTT\}$$

B. When die is thrown then, the outcome may be any number from 1 to 6 in any of the two dice. Then, sample space, $S = \{(1, 1), (1, 2), \dots, (6, 5), (6, 6)\}$

↓ \ →	1	2	3	4	5	6
1	1, 1	1, 2	1, 3	1, 4	1, 5	1, 6
2	2, 1	2, 2	2, 3	2, 4	2, 5	2, 6
3	3, 1	3, 2	3, 3	3, 4	3, 5	3, 6
4	4, 1	4, 2	4, 3	4, 4	4, 5	4, 6
5	5, 1	5, 2	5, 3	5, 4	5, 5	5, 6
6	6, 1	6, 2	6, 3	6, 4	6, 5	6, 6

C. The sample space

$S = \{HHHH, HHHT, HHTH, HHTT, HTHH, HTHT, HTTH, HTTT, THHH, THHT, THTH, THTT, TTHH, TTHT, TTTH, TTTT\}$

where, H and T represent head and tail of a coin respectively.

D. The sample space is

$$S = \{H1, H2, H3, H4, H5, H6, T1, T2, T3, T4, T5, T6\}$$

where, H and T represent head and tail of a coin.

8. *(b)* I. The sample space is $S = \{H1, H2, H3, H4, H5, H6, T\}$

where, H represents head of a coin.

II.

X	Y
B_1, B_2 G_1, G_2	B_3 G_3, G_4, G_5

When room X is selected, then there are four possibilities of selection of members which are B_1, B_2, G_1, G_2. Similarly, there will be four possibilities for room Y.

So, the sample space is

$$S = \{XB_1, XB_2, XG_1, XG_2, YB_3, YG_3, YG_4, YG_5\}$$

9. *(a)* Let R, W and B denote the red, white and blue die respectively. When a die will be selected, then there will be possibilities for colour R, B or W. But there will be 6 possibilities for 6 numbers (1 to 6) with each colour. So, the sample space is

$$S = \begin{Bmatrix} R1, & R2, & R3, & R4, & R5, & R6, \\ W1, & W2, & W3, & W4, & W5, & W6, \\ B1, & B2, & B3, & B4, & B5, & B6 \end{Bmatrix}$$

10. *(a)* I. Let B and b denote the first and second boy and G and g denote the first and second girl, respectively. The different possibilities are, both the babies are boys, 1st baby is boy and 2nd baby is girl, 1st baby is girl and 2nd is boy, 1st baby is girl and 2nd is also a girl.

So, the sample space is $S = \{Bb, Bg, Gb, Gg\}$

II. Sample space for number of girls, $S = \{0, 1, 2\}$.

($\because$ the different possibilities are no girl, one girl and two girls.)

11. *(b)* A. Here, all three white balls are identical. So, all are denoted by W and red ball is denoted by R.

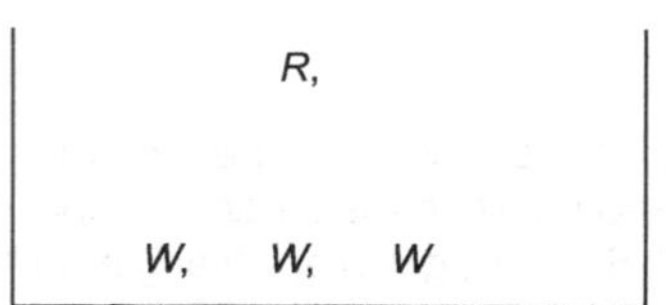

The sample space is $S = \{RW, WW, WR\}$

Note *If white balls are not identical, then total number of sample point is 6.*

B.

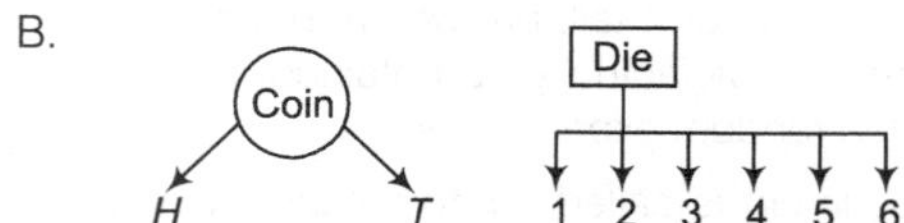

Different possibilities on tossing the coin, when head occurs, then tossed again which may result in head or tail.

If tail occurs, then die is thrown which may have any number from 1 to 6.

$\therefore \quad S = \{HH, HT, T1, T2, T3, T4, T5, T6\}$

where, H and T represent a head and tail of the coin.

C. Let D represents defective and N represents non-defective. Then, the sample space is

$S = \{DDD, DDN, DND, DNN, NDD, NDN, NND, NNN\}$

D. The sample space is shown below the chart.

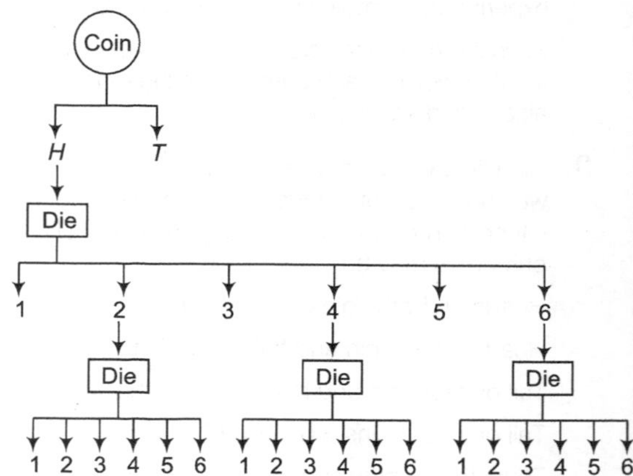

The sample space is

$S = \{T, H1, H3, H5, H21, H22, H23, H24, H25, H26, H41, H42, H43, H44, H45, H46, H61, H62, H63, H64, H65, H66\}$

12. *(b)* I. Since, in a slip box, four slips marked as 1, 2, 3, 4.

Slip box

The sample space is

$S = \{(1, 2), (1, 3), (1, 4), (2, 1), (2, 3), (2, 4), (3, 1), (3, 2), (3, 4),$
$(4, 1), (4, 2), (4, 3)\}$

Note *If above experiment is with replacement, then additional sample points in space are* $\{(1, 1), (2, 2), (3, 3), (4, 4)\}$.

II. Let H and T denote the head and tail. The sample space is shown below in the chart.

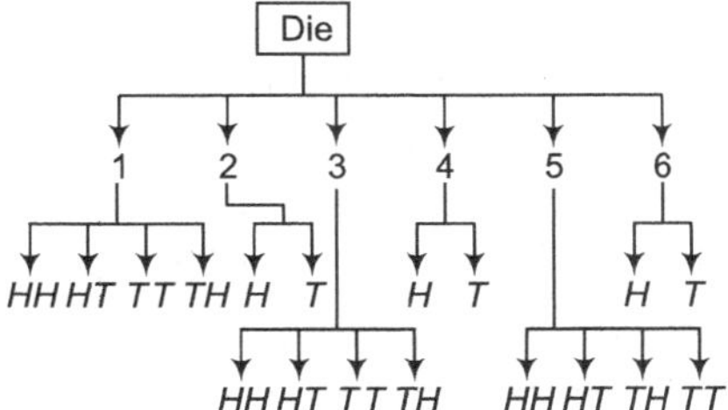

Hence, the sample space is

$S = \{1HH, 1HT, 1TT, 1TH, 2H, 2T, 3HH, 3HT, 3TT, 3TH, 4H, 4T, 5HH, 5HT, 5TH, 5TT, 6H, 6T\}$

13. *(a)* Let H and T represent a head and tail.

Let red ball be represented by R_1, R_2 and black ball be represented by B_1, B_2 and B_3.

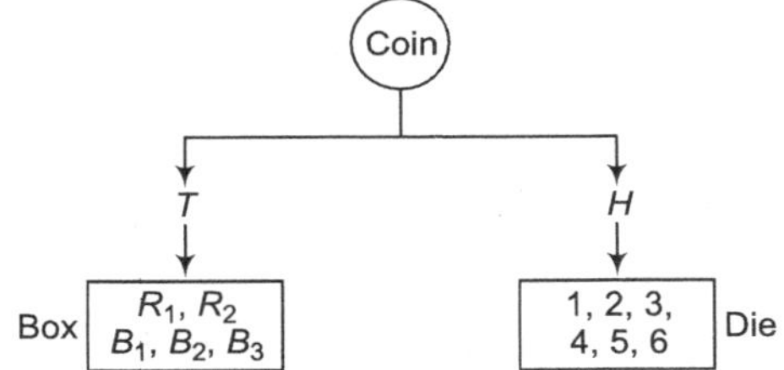

The sample space is

$S = \{TR_1, TR_2, TB_1, TB_2, TB_3, H1, H2, H3, H4, H5, H6\}$

14. *(c)* The sample space is

$S = \{6, (1, 6), (2, 6), (3, 6), (4, 6), (5, 6), (1, 1, 6), (1, 2, 6), (1, 3, 6) \ldots\}$

Hence, infinite number of possibilities occur.

15. *(c)* A coin is tossed once, then the sample space is $S = \{H, T\}$.

16. *(b)* When three dice tossed together, then the total number of outcomes $= 6^3$

$$= 6 \times 6 \times 6 = 216$$

17. *(c)* The sample space S is

$S = \{HH, TH, (HT, 1), (HT, 2), (HT, 3), (HT, 4), (HT, 5), (HT, 6), (TT, 1), (TT, 2), (TT, 3), (TT, 4), (TT, 5), (TT, 6)\}$

Hence, total number of outcomes $= 14$

18. *(c)* The sample space is

$$S = \{T, HT, HHT, HHHT, HHHHT, \ldots\}$$

19. *(b)* The sample space associated with the given random experiment is given by

$S = \{(1, 2), (1, 3), (1, 4), (1, 5), (1, 6), (2, 1), (2, 3), (2, 4), (2, 5), (2, 6), (3, 1), (3, 2), (3, 4), (3, 5), (3, 6), (4, 1), (4, 2), (4, 3), (4, 5), (4, 6), (5, 1), (5, 2), (5, 3), (5, 4), (5, 6), (6, 1), (6, 2), (6, 3), (6, 4), (6, 5)\}$

Now, when outcomes is doublet, then coin is tossed. So, sample space

$= \{(11, H)\ (11, T)$
$(22, H)\ (22, T)$
$(33, H)\ (33, T)$
$(44, H)\ (44, T)$
$(55, H)\ (55, T)$
$(66, H)\ (66, T)\}$

Hence, total number of sample points $= 42$

20. *(d)* The sample space for this experiment

$$S = \{RR, RB, BR, BB\}$$

where, R denotes the red ball and B denotes the black ball.

21. *(c)* Let W denotes the white ball and R denotes the red ball.

We know that, when a die is rolled, then the sample space $= \{1, 2, 3, 4, 5, 6\}$

Now, according to the question, required sample space is

$S = \{(W, W), (W, R), (R, 1), (R, 2), (R, 3), (R, 4), (R, 5), (R, 6)\}$

22. *(c)* I. There are 13 letters in the word 'ASSASSINATION'.

II. There are 10 letters in the word 'NAGATATION'.

23. *(a)* Let white ball is denoted by W and black ball is denoted by B. Now, two balls are drawn at random in succession without replacement.

Then, the sample space is

$$S = \{WB, BW, BB\}$$

24. *(d)* The defective items denoted by D and non-defective items denoted by N, then the sample space is

$$S = \{DDD, DDN, DND, NDD, DNN, NDN, NND, NNN\}$$

25. *(b)* We have studied about random experiment and sample space associated with an experiment. The sample space serves as an universal set for all questions concerned with the experiment.

Consider the experiment of tossing a coin two times. An associated sample space is $S = \{HH, HT, TH, TT\}$.

Now, suppose that we are interested in those outcomes which correspond to the occurrence of exactly one head. We find that HT and TH are the only elements of S corresponding to the occurrence of this happening (event). These two elements form the set $E = \{HT, TH\}$.

We know that, the set E is a subset of the sample space S. Similarly, we find the following correspondence between events and subsets of S.

Description of events	Corresponding subset of S
Number of tails is exactly 2	$A = \{TT\}$
Number of tails is atleast one	$B = \{HT, TH, TT\}$
Number of heads is atmost one	$C = \{HT, TH, TT\}$
Second toss is not head	$D = \{HT, TT\}$
Number of tails is atmost two	$S = \{HH, HT, TH, TT\}$
Number of tails is more than two	ϕ

The above discussion suggests that a subset of sample space is associated with an event and an event is associated with a subset of sample space. In the light of this, we define an event as follows:

Any subset E of a sample space S is called an event.

26. *(b)* Consider the experiment of throwing a die. Let E denotes the event "a number less than 4 appears". If actually 1 had appeared on the die, then we say that event E has occurred. As a matter of fact, if outcomes are 2 or 3, we say that event E has occured.

Thus, the event E of a sample space S is said to have occurred, if the outcome ω of the experiment is such that $\omega \in E$. If the outcome ω is such that $\omega \notin \text{E}$, we say that the event E has not occurred.

27. *(c)* Events can be classified into various types on the basis of the elements they have.

28. *(c)* The empty set ϕ and the sample space S describe events. Infact ϕ is called an impossible event and S, *i.e.*, the whole sample space is called the sure event.

29. *(b)* Let E be the event "the number appears on the die is a multiple of 7". Can you write the subset associated with the event E?

Clearly, no outcome satisfies the condition given in the event, *i.e.*, no element of the sample space ensures the occurrence of the event E. Thus, we say that the empty set only correspond to the event E. In other words, we can say that it is impossible to have a multiple of 7 on the upper face of the die. Thus, the event $E = \phi$ is an impossible event.

Now, let us take up another event F "the number turns up is odd or even". Clearly, $F = \{1, 2, 3, 4, 5, 6\} = S$, *i.e.*, all outcomes of the experiment ensure the occurrence of the event F. Thus, the event $F = S$ is a sure event.

30. *(c)* If an event E has only one sample point of a sample space, then it is called a simple (or elementary) event.

In a sample space containing n distinct elements, there are exactly n simple events.

For example, in the experiment of tossing two coins, a sample space is

$$S = \{HH, HT, TH, TT\}$$

There are four simple events corresponding to this sample space. These are $E_1 = \{HH\}$, $E_2 = \{HT\}$, $E_3 = \{TH\}$ and $E_4 = \{TT\}$.

31. *(c)* If an event has more than one sample point, then it is called a compound event.

For example, in the experiment of "tossing a coin thrice" the events

E: 'exactly one head appeared'

F: 'atleast one head appeared'

G: 'atmost one head appeared' etc.

are all compound events. The subsets of S associated with these events are

$$E = \{HTT, THT, TTH\}$$
$$F = \{HTT, THT, TTH, HHT, HTH, THH, HHH\}$$
$$G = \{TTT, THT, HTT, TTH\}$$

Each of the above subsets contain more than one sample point, hence they are all compound events.

32. *(a)* For every event A, there corresponds another event A' called the complementary event to A. It is also called the event 'not A'.

For example, take the experiment 'of tossing three coins'. An associated sample space is

$$S = \{HHH, HHT, HTH, THH, HTT, THT, TTH, TTT\}$$

Let $A = \{HTH, HHT, THH\}$ be the event 'only one tail appears'

Clearly, for the outcome HTT, the event A has not occurred. But we may say that the event 'not A' has occurred. Thus, with every outcome which is not in A, we say that 'not A' occurs.

Thus, the complementary event 'not A' to the event A is

$$A' = \{HHH, HTT, THT, TTH, TTT\}$$

or $\quad A' = \{\omega : \omega \in S \text{ or } \omega \notin A\} = S - A$

33. *(c)* Recall that union of two sets A and B denoted by $A \cup B$ contains all those elements which are either in A or in B or in both. When the sets A and B are two events associated with a sample space, then '$A \cup B$' is the event 'either A or B or both'. This event '$A \cup B$' is also called 'A or B'.

Therefore, event 'A or B' $= A \cup B$

$$= \{\omega : \omega \in A \text{ or } \omega \in B\}$$

34. *(b)* We know that, intersection of two sets $A \cap B$ is the set of those elements which are common to both A and B, *i.e.*, which belong to both 'A and B'.

If A and B are two events, then the set $A \cap B$ denotes the event 'A and B'.

Thus, $\quad A \cap B = \{\omega : \omega \in A \text{ and } \omega \in B\}$

For example, in the experiment of 'throwing a die twice' Let A be the event 'score on the first throw is 6' and B is the event 'sum of two scores is atleast 11'. Then,

$$A = \{(6, 1), (6, 2), (6, 3), (6, 4), (6, 5), (6, 6)\}$$

and $\quad B = \{(5, 6), (6, 5), (6, 6)\}$

So, $A \cap B = \{(6, 5), (6, 6)\}$

Note that the set $A \cap B = \{(6, 5), (6, 6)\}$ may represent the event 'the score on the first throw is 6 and the sum of the scores is alteast 11'.

35. *(d)* We know that, $A - B$ is the set of all those elements which are in A but not in B. Therefore, the set $A - B$ may denote the event 'A but not B'.

We know that,

$$A - B = A \cap B'$$

36. *(c)* Here, $S = \{1, 2, 3, 4, 5, 6\}$, $= A = \{2, 3, 5\}$ and $B = \{1, 3, 5\}$

Obviously,

I. 'A or B' $= A \cup B = \{1, 2, 3, 5\}$

II. 'A and B' $=$ ' $A \cap B = \{3, 5\}$

III. 'A but not B' $= A - B = \{2\}$

IV. 'not A' $= A' = \{1, 4, 6\}$

37. *(b)* In the experiment of rolling a die, a sample space is $S = \{1, 2, 3, 4, 5, 6\}$.

Consider events A 'an odd number appears' and B 'an even number appears'.

Clearly, the event A excludes the event B and *vice-versa*. In other words, there is no outcome which ensures the occurrence of events A and B simultaneously. Here,

$$A = \{1, 3, 5\} \text{ and } B = \{2, 4, 6\}$$

Clearly, $A \cap B = \phi$, *i.e.*, A and B are disjoint sets.

In general, two events A and B are called mutually exclusive events, if the occurrence of anyone of them excludes the occurrence of the other event, *i.e.*, they cannot occur simultaneously. In this case, the sets A and B are disjoint.

Again, in the experiment of rolling a die, consider the events A 'an odd number appears' and event B 'a number less than 4 appears'.

Obviously, $A = \{1, 3, 5\}$ and $B = \{1, 2, 3\}$

Now, $3 \in A$ as well as $3 \in B$

Therefore, A and B are not mutually exclusive events.

Remark Simple events of a sample space are always mutually exclusive.

38. *(b)* Consider the experiment of throwing a die.

We have, $S = \{1, 2, 3, 4, 5, 6\}$.

Let us define the following events.

A : 'a number less than 4 appears'.

B : 'a number greater than 2 but less than 5 appears' and

C : 'a number greater than 4 appears'

Then, $A = \{1, 2, 3\}$, $B = \{3, 4\}$ and $C = \{5, 6\}$

We observe that,

$$A \cup B \cup C = \{1, 2, 3\} \cup \{3, 4\} \cup \{5, 6\} = S$$

Such events A, B and C are called exhaustive events. In general, if $E_1, E_2, \ldots, E_n$ are n events of a sample space S and if

$$E_1 \cup E_2 \cup E_3 \cup \ldots \cup E_n = \bigcup_{i=1}^{n} E_i = S$$

Then $E_1, E_2, \ldots, E_n$ are called exhaustive events. In other words, events $E_1, E_2, \ldots, E_n$ are said to be exhaustive, if atleast one of them necessarily occurs whenever the experiment is performed.

Further, if $E_i \cap E_j = \phi$ for $i \neq j$, *i.e.*, events, E_i and E_j are pairwise disjoint and $\bigcup_{i=1}^{n} E_i = S$, then $E_1, E_2, \ldots, E_n$ are called mutually exclusive and exhaustive events.

39. *(c)* There are 36 elements in the sample space $S = \{(x, y) : x, y = 1, 2, 3, 4, 5, 6\}$.

Then,

$A = \{(1, 1), (1, 3), (1, 5), (2, 2), (2, 4), (2, 6), (3, 1), (3, 3,), (3, 5), (4, 2), (4, 4), (4, 6), (5, 1), (5, 3), (5, 5), (6, 2), (6,4), (6,6)\}$

$B = \{(1, 2), (2, 1), (1, 5), (5, 1), (3, 3), (2, 4), (4, 2), (3, 6), (6, 3), (4, 5), (5, 4), (6, 6)\}$

$C = \{(1, 1), (2, 1), (1, 2)\}$ and $D = \{(6, 6)\}$

We find that

$A \cap B = \{(1, 5), (2, 4), (3, 3), (4, 2), (5, 1), (6, 6)\} \neq \phi$

Therefore, A and B are not mutually exclusive events.

Similarly, $A \cap C \neq \phi$, $A \cap D \neq \phi$, $B \cap C \neq \phi$ and $B \cap D \neq \phi$

Thus, the pairs, (A, C), (A, D), (B, C), (B, D) are not mutually exclusive events.

Also, $C \cap D = \phi$ and so C and D are mutually exclusive events.

40. *(d)* The sample space of the experiment is

$S = \{HHH, HHT, HTH, THH, HTT, THT, TTH, TTT\}$

and $A = \{TTT\}$,

$B = \{HTT, THT, TTH\}$,

$C = \{HHT, HTH, THH, HHH\}$

Now,

$A \cup B \cup C = \{TTT, HTT, THT, TTH, HHT, HTH, THH, HHH\}$
$= S$

Therefore, A, B and C are exhaustive events.

Also, $A \cap B = \phi$, $A \cap C = \phi$ and $B \cap C = \phi$

Therefore, the events are pair-wise disjoint, *i.e.*, they are mutually exclusive.

Hence, A, B and C form a set of mutually exclusive and exhaustive events.

41. *(d)*

Let E = The die shows $4 = \{4\}$

Let F = The die shows even number
$= \{2, 4, 6\}$

$\therefore \quad E \cap F = \{4\} \neq \phi$

Hence, E and F are not mutually exclusive.

42. *(a)*

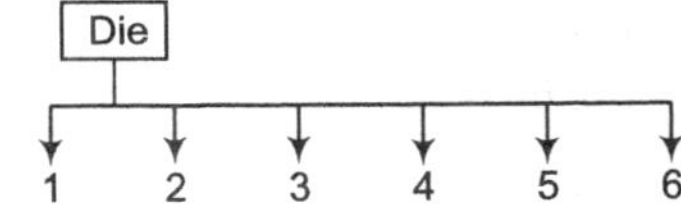

A : a number less than 7 = {1, 2, 3, 4, 5, 6}
B : a number greater than 7 = { } = ϕ ($\because$ the maximum number on a die is 6, so there cannot be a number on die greater than 7)
C : a multiple of 3 = {3, 6}
D : a number less than 4 = {1, 2, 3}
E : an even number greater than 4 = {6}
F : a number not less than 3 = {3, 4, 5, 6}

43. *(b)* A. Now, $A \cup B$ = The elements which are in A or B

$= \{1, 2, 3, 4, 5, 6\} \cup \phi = \{1, 2, 3, 4, 5, 6\}$

B. $A \cap B$ = The elements which are common in both A and B.

$= \{1, 2, 3, 4, 5, 6\} \cap \phi = \phi$

C. $B \cup C$ = The elements which are in both B and C.

$= \{\ \} \cup \{3, 6\} = \{3, 6\}$

D. $E \cap F$ = The elements which are common in both E and F.

$= \{6\} \cap \{3, 4, 5, 6\} = \{6\}$

E. $D \cap E$ = The elements which are common in both D and E.

$= \{1, 2, 3\} \cap \{6\} = \phi$

F. $A - C$ = The elements which are in A but not in C

$= \{1, 2, 3, 4, 5, 6\} - \{3, 6\} = \{1, 2, 4, 5\}$

G. $D - E$ = The elements which are in D but not in E

$= \{1, 2, 3\} - \{6\} = \{1, 2, 3\}$

H. $E \cap F' = E \cap (U - F) = E \cap [\{1, 2, 3, 4, 5, 6\} - \{3, 4, 5, 6\}]$

$[\because U = \{1, 2, 3, 4, 5, 6\}] = \{6\} \cap \{1, 2\} = \phi$

I. and $F' = (U - F) = \{1, 2, 3, 4, 5, 6\} - \{3, 4, 5, 6\} = \{1, 2\}$

44. *(c)* When two dice are thrown, then there are 36 possible outcomes.

↓ \ →	1	2	3	4	5	6
1	1, 1	1, 2	1, 3	1, 4	1, 5	1, 6
2	2, 1	2, 2	2, 3	2, 4	2, 5	2, 6
3	3, 1	3, 2	3, 3	3, 4	3, 5	3, 6
4	4, 1	4, 2	4, 3	4, 4	4, 5	4, 6
5	5, 1	5, 2	5, 3	5, 4	5, 5	5, 6
6	6, 1	6, 2	6, 3	6, 4	6, 5	6, 6

A = The sum is greater than 8.
= {(3, 6), (4, 5), (5, 4), (6, 3), (4, 6), (5, 5), (6, 4), (5, 6), (6, 5), (6, 6)}

B = 2 occurs on the either die
= {(1, 2), (2, 1), (2, 2), (2, 3), (2, 4), (2, 5), (2, 6), (3, 2), (4, 2), (5, 2), (6, 2)}

C = The sum is atleast 7 and a multiple of 3 *i.e.*, the sum is 9 and 12.
= {(3, 6), (4, 5), (5, 4), (6, 3), (6, 6)}

$\therefore$ $A \cap B = \phi \Rightarrow A$ and B are mutually exclusive.
$B \cap C = \phi \Rightarrow B$ and C are mutually exclusive.

45. *(d)* When three coins are tossed, then there are $2^3 = 8$ possible outcomes.

i.e., S = {*HHH, HHT, HTH, HTT, THH, THT, TTH, TTT*}

A = Three heads shows = {*HHH*}
B = Two heads and one tail show = {*HHT, HTH, THH*}
C = Three tails show = {*TTT*}
D = A head shows on the first toss = {*HHH, HHT, HTH, HTT*}

A. Here, $A \cap B = \phi$
$A \cap C = \phi$
$B \cap C = \phi$
$C \cap D = \phi$
$A \cap B \cap C = \phi$

Here, A and B, A and C, B and C, C and D and A, B and C are mutually exclusive.

B. Since, events A, C have only one sample point, Hence, A and C are simple events.

C. Since, events B, D have two or more sample points, Hence, B and D are compound events.

46. *(a)* If three coins are tossed, then possible number of outcomes = $2^3 = 8$

Sample space S = {*HHH, HHT, HTH, HTT,* THH, THT, TTH, TTT}

A. Let A be the event. 'three heads show'.
$\Rightarrow$ A = {*HHH*}
and B be the event 'three tails show'.
$\Rightarrow$ B = {*TTT*}
$\Rightarrow$ $A \cap B = \phi$
Hence, A and B are mutually exclusive events.

B. Let A be the event 'atleast one head'.
$\Rightarrow A$ = {*HHH, HHT, HTH, HTT, THH, THT, TTH*}
and B be the event getting three tails.
$\Rightarrow$ B = {*TTT*}
$\Rightarrow$ $A \cap B = \phi$ and $A \cup B = S$
(Two events are exhaustive, if $A \cup B = S$)
Hence, A and B are mutually exclusive and exhaustive.

C. Let A be the event 'three heads show' and B be the event 'atleast two heads show'
$\Rightarrow$ A = {*HHH*}
and B = {*HHT, HTH, THH, HHH*}
$\Rightarrow$ $A \cap B$ = {*HHH*} $\neq \phi$
Hence, A and B are not mutually exclusive.

D. Same as in C, A and B are not mutually exclusive. Also, $A \cup B \neq S$.

E. Let A be the event 'three heads show'.

B be the event 'two heads shows', and C be the event 'three tails show'.

$\Rightarrow \quad A = \{HHH\}$

$\Rightarrow \quad B = \{HHT, HTH, THH\}$

and $\quad C = \{TTT\}$

$\Rightarrow \quad A \cap B \cap C = \phi$ and $A \cup B \cup C \neq S$

$\Rightarrow$ A, B and C are mutually exclusive but not exhaustive.

47. *(a)* If two dice are thrown then, total number of possible outcomes, $S = 6 \times 6 = 36$ which are as follows

S.No.	1	2	3	4	5	6
1	1, 1	1, 2	1, 3	1, 4	1, 5	1, 6
2	2, 1	2, 2	2, 3	2, 4	2, 5	2, 6
3	3, 1	3, 2	3, 3	3, 4	3, 5	3, 6
4	4, 1	4, 2	4, 3	4, 4	4, 5	4, 6
5	5, 1	5, 2	5, 3	5, 4	5, 5	5, 6
6	6, 1	6, 2	6, 3	6, 4	6, 5	6, 6

A = Getting an even number on the first die

$= \{(2, 1), (2, 2), (2, 3), (2, 4), (2, 5), (2, 6), (4, 1), (4, 2), (4, 3), (4, 4), (4, 5), (4, 6), (6, 1), (6, 2), (6, 3), (6, 4), (6, 5), (6, 6\}$

B = Getting an odd number on the first die

$= \{(1, 1), (1, 2), (1, 3), (1, 4), (1, 5), (1, 6), (3, 1), (3, 2), (3, 3), (3, 4), (3, 5), (3, 6), (5, 1), (5, 2), (5, 3), (5, 4), (5, 5), (5, 6)\}$

$C = \{(1, 1), (1, 2), (1, 3), (1, 4), (2, 1), (2, 2), (2, 3), (3, 1), (3, 2), (4, 1)\}$

(i) A' = which are not in $A = B$ (if A is an event getting an even number on the first die, so A' will be the event getting an odd number on the first die *i.e.*, B)

(ii) not $B = B'$ = which are not in $B = A$ (B is an event getting an odd number on the first, so B' will be the event getting an even number on the first die *i.e.*, A)

(iii) A or $B = A \cup B$

$A \cup B$ includes all elements of A and B *i.e.*, it will include both the events A and B *i.e.*, event having an even number on the first die and event having odd number on the first die *i.e.*, $A \cup B$ will include all possible events *i.e.*, S (sample space)

$\therefore \quad A \cup B = S$

(iv) A and $B = A \cap B$ *i.e.*, elements which are common in both A and B. There is no element common in A and B.

So, $\quad A \cap B = \phi$ (null set)

(v) A but not $C = A - C$ = Elements which are in A but not in C.

$= \{(2, 4), (2, 5), (2, 6), (4, 2), (4, 3), (4, 4), (4, 5), (4, 6), (6, 1), (6, 2), (6, 3), (6, 4), (6, 5), (6, 6)\}$

(vi) B or $C = B \cup C$ *i.e.*, elements which are in both B or C.

$= \{(1, 1), (1, 2), (1, 3), (1, 4), (1, 5), (1, 6), (3, 1), (3, 2), (3, 3), (3, 4), (3, 5), (3, 6), (5, 1), (5, 2), (5, 3), (5, 4), (5, 5), (5, 6), (2, 1), (2, 2), (2, 3), (4, 1)\}$

(vii) B and $C = B \cap C$ *i.e.*, elements which are common in both B and C.

$= \{(1, 1), (1, 2), (1, 3), (1, 4), (3, 1), (3, 2)\}$

(viii) $A \cap B' \cap C' = \{(2, 4), (2, 5), (2, 6), (4, 2), (4, 3), (4, 4), (4, 5), (4, 6). (6, 1), (6, 2), (6, 3), (6, 4), (6, 5), (6, 6)\}$

48. *(d)* (i) True

$\because \quad A$ = Getting an even number on the first die

B = Getting an odd number on the first die

$\Rightarrow \quad A \cap B = \phi$

$\therefore$ A and B are mutually exclusive events.

(ii) True

$\therefore \quad A \cup B = S$ *i.e.*, exhaustive. Also, $A \cap B = \phi$

(iii) True

$\therefore \quad B$ = Getting an odd number on the first die

$\Rightarrow B'$ = Getting an even number on first die = A

$\therefore \quad A = B'$

(iv) False

$\because A \cap C = \{(2, 1), (2, 2), (2, 3), (4, 1)\} \neq \phi$, so A and C are not mutually exclusive.

(v) False

$\because \quad B' = A$

$\therefore \quad A \cap B' = A \cap A = A \neq \phi \qquad (\because B' = A)$

So, A and B' are not mutually exclusive.

(vi) False

$\because \quad A' \cap B' = \phi$

$\therefore \quad A' \cap B' \cap C = \phi$ and $A' \cup B' \cup C = S$

But $\quad A' \cap C = B \cap C = \{(1, 1), (1, 2), (1, 3), (1, 4), (3, 1), (3, 2)\} \neq \phi \qquad (\because A' = B)$

and $\quad B' \cap C = A \cap C = \{(2, 1), (2, 2), (2, 3), (4, 1)\} \neq \phi \qquad (\because B' = A)$

$\therefore$ A', B' and C are not mutually exclusive.

49. *(a)* Let E be the event that a black face card is chosen. The outcomes in E are Jack, Queen, King of spade and club. Symbolically, E = {J, Q, K, of spade and club} or E = {J♣, Q♣, K♣, J♠, Q♠, K♠}.

50. *(a)* All possible genders are expressed as: S = {BBB, BBG, BGB, BGG, GBB, GBG, GGB, GGG}

A. Let A denotes the event: 'exactly one child is a girl'

A = {BBG, BGB, GBB}

B. Let B denotes the event that atleast two children are girls. B = {GGB, GBG, BGG, GGG}

C. Let C denotes the event: 'no child is a girl'.

C = {BBB}

51. *(a)* A die has 6 faces *i.e.,* 1, 2, 3, 4, 5, 6 let first $(k-1)$ rolls have 5 outcomes each and k^{th} roll should result in 1 outcomes, Now, 5^{k-1} elements correspond to the event that 2 appears on the kth roll of the die.

52. *(c)* After $(k-1)^{th}$ roll *i.e.*, 1 outcome 5 outcomes appears in $(k+1)^{th}$ roll, 5^2 outcomes in $(k+1)^{th}$ roll etc.

$\therefore$ Total elements $= 1 + 5 + 5^2 + \ldots\ldots + 5^{k-1}$

$$= \frac{1(5^k - 1)}{5-1} = \frac{5^k - 1}{4}$$

53. *(d)* Given that, $S = \{1, 2, 3, 4, 5, 6\}$ and $E = \{1, 3, 5\}$

Then, $\bar{E} = S - E = \{2, 4, 6\}$.

54. *(b)* A. If E_1 and E_2 are two mutually exclusive events, then

$$E_1 \cap E_2 = \phi$$

B. It E_1 and E_2 are the mutually exclusive and exhaustive events, then $E_1 \cap E_2 = \phi$ and $E_1 \cup E_2 = S$

where, S is the sample space for the events E_1 and E_2

C. If E_1 and E_2 have common outcomes, then $(E_1 - E_2) \cup (E_1 \cap E_2) = E_1$

D. If E_1 and E_2 are two events such that $E_1 \subset E_2$ and $E_1 \cap E_2 = E_1$

55. *(a)* The sample space associated with the given random experiment is given by

$S = \{(1, 1), (1, 2), (1, 3), (1, 4), (1, 5), (1, 6)$
$(2, 1), (2, 2), (2, 3), (2, 4), (2, 5), (2, 6)$
$(3, 1), (3, 2), (3, 3), (3, 4), (3, 5), (3, 6)$
$(4, 1), (4, 2), (4, 3), (4, 4), (4, 5), (4, 6)$
$(5, 1), (5, 2), (5, 3), (5, 4), (5, 5), (5, 6)$
$(6, 1), (6, 2), (6, 3), (6, 4), (6, 5), (6, 6)\}$

Now, let E = An even number as the sum

Then, $E = \{(1, 1), (1, 3), (3, 1), (2, 2), (1, 5), (5, 1)$
$(3, 3), (2, 4), (4, 2), (3, 5), (5, 3), (4, 4)$
$(6, 2), (2, 6), (5, 5), (6, 4), (4, 6), (6, 6)\}$

56. *(a)* The event E = The sum as a prime number

then event $E = \{(1, 1), (1, 2), (2, 1), (1, 4), (4, 1), (2, 3),$
$(3, 2), (1, 6), (6, 1), (2, 5), (5, 2), (3, 4, (4, 3), (6, 5), (5, 6)\}$

Solutions (Q.Nos. 57-59)

The sample space associated with the given random experiment is given by

$S = \{(1, 1), (1, 2), (1, 3), (1, 4), (1, 5), (1, 6)$
$(2, 1), (2, 2), (2, 3), (2, 4), (2, 5), (2, 6)$
$(3, 1), (3, 2), (3, 3), (3, 4), (3, 5), (3, 6)$
$(4, 1), (4, 2), (4, 3), (4, 4), (4, 5), (4, 6)$
$(5, 1), (5, 2), (5, 3), (5, 4), (5, 5), (5, 6)$
$(6, 1), (6, 2), (6, 3), (6, 4), (6, 5), (6, 6)\}$

57. *(a)* Now,

A = Getting an even number or the first die

$A = \{(2, 1), (2, 2), (2, 3), (2, 4), (2, 5), (2, 6)$
$(4, 1), (4, 2), (4, 3), (4, 4), (4, 5), (4, 6),$
$(6, 1), (6, 2), (6, 3), (6, 4), (6, 5), (6, 6)\}$

B = Getting an odd number on first die

$B = \{(1, 1), (1, 2), (1, 3), (1, 4), (1, 5), (1, 6),$
$(3, 1), (3, 2), (3, 3), (3, 4), (3, 5), (3, 6)$
$(5, 1), (5, 2), (5, 3), (5, 4), (5, 5), (5, 6)\}$

A and $B = A \cap B = \phi$

58. *(b)* Event B given above

and C = Getting atmost 5 as sum of the numbers on the two dice

$C = \{(1, 1), (1, 2), (1, 3), (1, 4), (2, 1),$
$(2, 2), (2, 3), (3, 1), (3, 2), (4, 1)\}$

B or $C = B \cup C = \{(1, 1), (1, 2), (1, 3), (1, 4), (1, 5), (1, 6),$
$(2, 1), (2, 2), (2, 3), (3, 1), (3, 2), (3, 3),$
$(3, 4), (3, 5), (3, 6), (4, 1), (5, 1), (5, 2),$
$(5, 3), (5, 4), (5, 5), (5, 6)\}$

$n(B \cup C) = 22$

59. *(c)* The set of event B as above and the set of event C as above

Now, set of event B and C

$= B \cap C = \{(1, 1), (1, 2), (1, 3), (1, 4), (3, 1), (3, 2)\}$

60. *(a)* The sample space associated with the given random experiment is given by

$S = \{(1, 1), (1, 2), (1, 3), (1, 4), (1, 5), (1, 6)$
$(2, 1), (2, 2), (2, 3), (2, 4), (2, 5), (2, 6)$
$(3, 1), (3, 2), (3, 3), (3, 4), (3, 5), (3, 6)$
$(4, 1), (4, 2), (4, 3), (4, 4), (4, 5), (4, 6)$
$(5, 1), (5, 2), (5, 3), (5, 4), (5, 5), (5, 6)$
$(6, 1), (6, 2), (6, 3), (6, 4), (6, 5), (6, 6)\}$

Let event E = a total of atleast 10

$E = \{(6, 4), (4, 6), (5, 5), (6, 6), (5, 6), (6, 5)\}$

61. *(b)* The sample space for the experiment of drawing a ticket from ticket numbered from 1 to 20 is S.

Then, $S = \{1, 2, 3, 4, 5, 6, 7, 8, 9, 10, 11, 12, 13, 14, 15, 16, 17, 18, 19, 20\}$

Then, the total number of elementary events = 20.

62. *(d)* Let the set be S

Then, $S = \{1, 2, 3, 4, 5, 6, 7, 8, 9, 10, 11, 12, 13, 14, 15, 16, 17, 18, 19, 20, 21, 22, 23, 24, 25\}$

Now, let the event E = Getting a prime number when each of the given number is equally likely to be selected.

$E = \{2, 3, 5, 7, 11, 13, 17, 19, 23,\}$.

63. *(c)* We know that, r objects can be selected from n objects in ${}^nC_r$ ways

$\because$ Four cards can be drawn from a pack of 52 cards in ${}^{52}C_4$ ways.

64. *(d)* Let S be the sample space of a random experiment. The probability P is a real valued function whose domain is the power set of S and range is the interval [0, 1] satisfying the following axioms.

I. For any event E, $P(E) \geq 0$

II. $P(S) = 1$

III. If E and F are mutually exclusive events, then $P(E \cup F) = P(E) + P(F)$.

It follows from (iii) that $P(\phi) = 0$. To prove this, we take $F = \phi$ and note that E and ϕ are disjoint events. Therefore, from axiom (iii), we get

$$P(E \cup \phi) = P(E) + P(\phi)$$

or $$P(E) = P(E) + P(\phi) \text{ i.e., } P(\phi) = 0.$$

Let S be a sample space containing outcomes $\omega_1, \omega_2, \ldots \omega_n$, i.e., $S = \{\omega_1, \omega_2, \ldots \omega_n,\}$

It follows from the axiomatic definition of probability that

(i) $0 \leq P(\omega_i) \leq 1$ for each $\omega_i \in S$

(ii) $P(\omega_1) + P(\omega_2) + \ldots + P(\omega_n) = 1$

(iii) For any event A, $P(A) = \sum P(\omega_i)$, $\omega_i \in A$.

Note *It may be noted that the singleton $\{\omega_i\}$ is called elementary event and for notational convenience, we write $P(\omega_i)$ for $P(\{\omega_i\})$.*

65. *(b)* In 'a coin tossing' experiment, we can assign the number $\frac{1}{2}$ to each of the outcomes H and T

i.e., $$P(H) = \frac{1}{2} \text{ and } P(T) = \frac{1}{2} \quad \ldots\text{(i)}$$

Clearly, this assignment satisfies both the conditions, i.e., each number is neither less than zero nor greater than 1 and

$$P(H) + P(T) = \frac{1}{2} + \frac{1}{2} = 1$$

Therefore, in this case we can say that probability of $H = \frac{1}{2}$, and probability of $T = \frac{1}{2}$

If we take $$P(H) = \frac{1}{4} \text{ and } P(T) = \frac{3}{4} \quad \ldots\text{(ii)}$$

Does this assignment satisfy the conditions of axiomatic approach?

Yes, in this case, probability of $H = \frac{1}{4}$ and probability of $T = \frac{3}{4}$.

We find that both the assignments (i) and (ii) are valid for probability of H and T.

In fact, we can assign the numbers p and $(1-p)$ to both the outcomes such that $0 \leq p \leq 1$

and $$P(H) + P(T) = p + (1-p) = 1$$

This assignment, too, satisfies both conditions of the axiomatic approach of probability. Hence, we can say that there are many ways (rather infinite) to assign probabilities to outcomes of an experiment.

66. *(d)* (i) **Condition I** Each of the number $p(\omega_i)$ is positive and less than one.

Condition II Sum of probabilities

$$= \frac{1}{6} + \frac{1}{6} + \frac{1}{6} + \frac{1}{6} + \frac{1}{6} + \frac{1}{6} = 1$$

Therefore, the assignment is valid

(ii) **Condition I** : Each of the number $p(\omega_i)$ is either 0 or 1.

Condition II : Sum of the probabilities

$= 1 + 0 + 0 + 0 + 0 + 0 = 1$

Therefore, the assignment is valid

(iii) **Condition** (I) Two of the probabilities $p(\omega_5)$ and $p(\omega_6)$ are negative, the assignment is not valid.

(iv) Since, $p(\omega_6) = \frac{3}{2} > 1$, the assignment is not valid,

(v) Since, sum of probability

$= 0.1 + 0.2 + 0.3 + 0.4 + 0.5 + 0.6 = 2.1$,

the assignment is not valid.

67. *(b)* A sample space associated with given experiment is

S = {BBB, BBG, BGB, GBB, BGG, GBG, GGB, GGG}

where, B stands for a defective or bad pen and G for a non-defective or good pen.

Let the probability assigned to the outcomes be as follows:

Sample point	BBB	BBG	BGB	GBB	BGG	GBG	GGB	GGG
Probability	$\frac{1}{8}$	$\frac{1}{8}$	$\frac{1}{8}$	$\frac{1}{8}$	$\frac{1}{8}$	$\frac{1}{8}$	$\frac{1}{8}$	$\frac{1}{8}$

Let event A : there is exactly one defective pen

and event B : there are atleast two defective pens.

Hence, A = {BGG, GBG, GGB}

and B = {BBG, BGB, GBB, BBB}

Now, $P(A) = \sum P(\omega_i), \forall\, \omega_i \in A$

$= P(BGG) + P(GBG) + P(GGB)$

$= \frac{1}{8} + \frac{1}{8} + \frac{1}{8} = \frac{3}{8}$

and $P(B) = \Sigma P(\omega_i), \forall\, \omega_i \in B$

$= P(BBG) + P(BGB) + P(GBB) + P(BBB)$

$= \frac{1}{8} + \frac{1}{8} + \frac{1}{8} + \frac{1}{8} = \frac{4}{8} = \frac{1}{2}$

68. *(b)* Let us consider an experiment of 'tossing a coin "twice". The sample space of this experiment is S = {HH, HT, TH, TT}

Let the following probability be assigned to the outcomes

$$P(HH) = \frac{1}{4}, P(HT) = \frac{1}{7}, P(TH) = \frac{2}{7}, P(TT) = \frac{9}{28}$$

Clearly, this assignment satisfies the conditions of axiomatic approach. Now, let us find the probability of the event E : 'Both the tosses yield the same result'.

Here, $E = \{HH, TT\}$

Now, $P(E) = \sum P(\omega_i)$, for all $\omega_i \in E$

$$= P(HH) + P(TT) = \frac{1}{4} + \frac{9}{28} = \frac{4}{7}$$

For the event F : 'exactly two heads', we have $F = \{HH\}$

and $P(F) = P(HH) = \frac{1}{4}$

69. *(b)* Let a sample space of an experiment be $S = \{\omega_1, \omega_2, \ldots \omega_n\}$.

Let all the outcomes are equally likely to occur, *i.e.*, the chance of occurrence of each simple event must be same.

i.e., $P(\omega_i) = p$, for all $\omega_i \in S$, where $0 \le p \le 1$

Since $\sum_{i=1}^{n} P(\omega_i) = 1$, *i.e.*, $p + p + \ldots + p$ (n times) $= 1$

or $np = 1$ *i.e.*, $p = \frac{1}{n}$

If S be a sample space and E be an event, such that $n(S) = n$ and $n(E) = m$. If each outcome is equally likely, then if follows that

$$P(E) = \frac{m}{n} = \frac{\text{Number of outcomes favourable to } E}{\text{Total possible outcomes}}$$

70. *(d)* To find the probability of event 'A or B', *i.e.*, $P(A \cup B)$.

If S is sample space for tossing of three coins, then

$S = \{HHT, HHH, HTH, HTT, THH, THT, TTH, TTT\}$

Let $A = \{HHT, HTH, THH\}$ and $B = \{HTH, THH, HHH\}$ be two events associated with 'tossing of a coin thrice'

Clearly, $A \cup B = \{HHT, HTH, THH, HHH\}$

Now, $P(A \cup B) = P(HHT) + P(HTH) + P(THH) + P(HHH)$

If all the outcomes are equally likely, then

$$P(A \cup B) = \frac{1}{8} + \frac{1}{8} + \frac{1}{8} + \frac{1}{8} = \frac{4}{8} = \frac{1}{2}$$

Also, $P(A) = P(HHT) + P(HTH) + P(THH) = \frac{3}{8}$

and $P(B) = P(HTH) + P(THH) + P(HHH) = \frac{3}{8}$

Therefore, $P(A) + P(B) = \frac{3}{8} + \frac{3}{8} = \frac{6}{8}$

It is clear that $P(A \cup B) \ne P(A) + P(B)$

The points HTH and THH are common to both A and B. In the computation of $P(A) + P(B)$ the probabilities of points HTH and THH, *i.e.*, the elements of $A \cap B$ are included twice. Thus, to get the probability $P(A \cup B)$ we have to subtract the probabilities of the sample points in $A \cap B$ from $P(A) + P(B)$.

i.e., $P(A \cup B) = P(A) + P(B) - \Sigma P(\omega_i), \forall \omega_i \in A \cap B$

$$= P(A) + P(B) - P(A \cap B)$$

Thus, we observe that,

$$P(A \cup B) = P(A) + P(B) - P(A \cap B)$$

71. *(c)* In general, if A and B are any two events associated with a random experiment, then by the definition of probability of an event, we have

$$P(A \cup B) = \Sigma P(\omega_i), \forall \omega_i \in A \cup B$$

Since $A \cup B = (A - B) \cup (A \cap B) \cup (B - A)$,

We have

$$P(A \cup B)$$
$$= [\Sigma P(\omega_i) \forall \omega_i \in (A - B)] + [\Sigma P(\omega_i) \forall \omega_i \in A \cap B] + [\Sigma P(\omega_i) \forall \omega_i \in B - A]$$

(because $A - B$, $A \cap B$ and $B - A$ are mutually exclusive) ....(i)

Also, $P(A) + P(B) = [\Sigma p(\omega_i) \forall \omega_i \in A] + [\Sigma P(\omega_i) \forall \omega_i \in B]$

$$= [\Sigma P(\omega_i) \forall \omega_i \in (A - B) \cup (A \cap B)] + [\Sigma P(\omega_i) \forall \omega_i \in (B - A) \cup (A \cap B)]$$

$$= [\Sigma P(\omega_i) \forall \omega_i \in (A - B)] + [\Sigma P(\omega_i) \forall \omega_i \in (A \cap B)] + [\Sigma P(\omega_i) \forall \omega_i \in (B - A)] + [\Sigma P(\omega_i) \forall \omega_i \in (A \cap B)]$$

$= P(A \cup B) + [\Sigma P(\omega_i) \forall \omega_i \in A \cap B]$ [using Eq. (i)]

$= P(A \cup B) + P(A \cap B)$

Hence, $P(A) + P(B) = P(A \cup B) + P(A \cap B)$

72. *(a)* $A \cup B = A \cup (B - A)$, where A and $B - A$ are mutually exclusive, and $B = (A \cap B) \cup (B - A)$, where $A \cap B$ and $B - A$ are mutually exclusive.

Using axiom (iii) of probability, we get

$$P(A \cup B) = P(A) + P(B - A) \quad \ldots(ii)$$

and $$P(B) = P(A \cap B) + P(B - A) \quad \ldots(iii)$$

Subtracting Eq.(iii) from Eq. (ii), we get

$$P(A \cup B) - P(B) = P(A) - P(A \cap B)$$

or $$P(A \cup B) = P(A) + P(B) - P(A \cap B)$$

The above result can further be verified by observing the Venn diagram

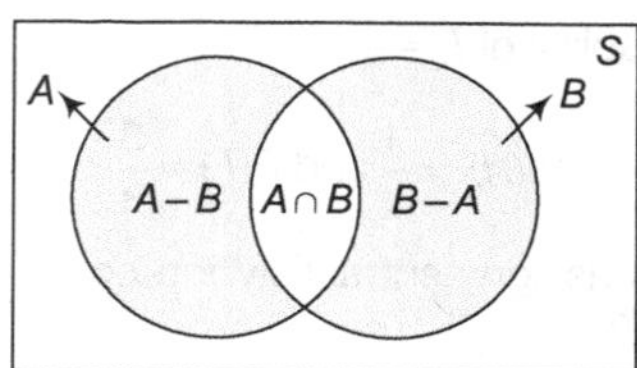

If A and B are disjoint sets, *i.e.*, they are mutually exclusive events, then $A \cap B = \phi$

Therefore, $P(A \cap B) = P(\phi) = 0$

Thus, for mutually exclusive events A and B, we have

$$P(A \cup B) = P(A) + P(B)$$

which is axiom (iii) of probability.

73. *(c)* If all the outcomes 1, 2, ..., 10 are considered to be equally likely, then the probability of each outcome is $\frac{1}{10}$

Now, $P(A) = P(2) + P(4) + P(6) + P(8)$

$$= \frac{1}{10} + \frac{1}{10} + \frac{1}{10} + \frac{1}{10} = \frac{4}{10} = \frac{2}{5}$$

Also event 'not A' $= A' = \{1, 3, 5, 7, 9, 10\}$

Now, $P(A') = P(1) + P(3) + P(5) + P(7) + P(9) + P(10)$

$$= \frac{6}{10} = \frac{3}{5}$$

Thus, $P(A') = \frac{3}{5} = 1 - \frac{2}{5} = 1 - P(A)$

Also, we know that A' and A are mutually exclusive and exhaustive events *i.e.*,

$$A \cap A' = \phi \text{ and } A \cup A' = S \text{ or } P(A \cup A') = P(S)$$

Now, $P(A) + P(A') = 1$, by using axioms (ii) and (iii)

or $P(A') = P(\text{not } A) = 1 - P(A)$

74. *(a)* When a card is drawn from a well shuffled deck of 52 cards, the number of possible outcomes is 52.

A. Let A be the event 'the card drawn is a diamond'
Clearly, the number of elements is set A is 13.
Therefore, $P(A) = \frac{13}{52} = \frac{1}{4}$
i.e., Probability of a diamond card $= \frac{1}{4}$

B. We assume that the event 'card drawn is an ace' is B
Therefore, 'card drawn is not an ace' should be B'.
We know that, $P(B') = 1 - P(B) = 1 - \frac{4}{52} = 1 - \frac{1}{13} = \frac{12}{13}$

C. Let C denotes the event 'card drawn is black card'
Therefore, number of elements in the set $C = 26$
i.e., $P(C) = \frac{26}{52} = \frac{1}{2}$
Thus, probability of a black card $= \frac{1}{2}$

D. We assumed in A. above that A is the event 'card drawn is a diamond', so the event 'card drawn is not a diamond' may be denoted as A' or 'not A'
Now $P(\text{not } A) = 1 - P(A) = 1 - \frac{1}{4} = \frac{3}{4}$

E. The event 'card drawn is not a black card' may be denoted as C' or 'not C'.
We know that, $P(\text{not } C) = 1 - P(C) = 1 - \frac{1}{2} = \frac{1}{2}$
Therefore, probability of not a black card $= \frac{1}{2}$

75. *(d)* There are 9 discs in all, so the total number of possible outcomes is 9.

Let the events A, B, C be defined as

A : 'the disc drawn is red'.

B : 'the disc drawn is yellow'.

C : 'the disc drawn is blue'.

I. The number of red discs = 4, *i.e.*, $n(A) = 4$
Hence, $P(A) = \frac{4}{9}$

II. The number of yellow discs = 2, *i.e.*, $n(B) = 2$
Therefore, $P(B) = \frac{2}{9}$

III. The number of blue discs = 3, *i.e.*, $n(C) = 3$
Therefore, $P(C) = \frac{3}{9} = \frac{1}{3}$

IV. Clearly the event 'not blue' is 'not C'.
We know that, $P(\text{not } C) = 1 - P(C)$
Therefore, $P(\text{not } C) = 1 - \frac{1}{3} = \frac{2}{3}$

V. The event 'either red or blue' may be described by the set 'A or C'.
Since, A and C are mutually exclusive events, we have

$$P(A \text{ or } C) = P(A \cup C) = P(A) + P(C)$$
$$= \frac{4}{9} + \frac{1}{3} = \frac{7}{9}$$

76. *(d)* Let E and F denote the events that Anil and Ashima will qualify the examination, respectively.

Given that,

$$P(E) = 0.05, P(F) = 0.10 \text{ and } P(E \cap F) = 0.02.$$

Then,

I. The event 'both Anil and Ashima will not qualify the examination' may be expressed as $E' \cap F'$.
Since, E' is 'not E', *i.e.*, Anil will not qualify the examination and F' is 'not F', *i.e.*,
Ashima will not qualify the examination

Also $E' \cap F' = (E \cup F)'$ (by Demorgan's law)

Now $P(E \cup F) = P(E) + P(F) - P(E \cap F)$

or $P(E \cup F) = 0.05 + 0.10 - 0.02 = 0.13$

Therefore, $P(E' \cap F') = P(E \cup F)' = 1 - P(E \cup F)$
$= 1 - 0.13 = 0.87$

II. P (atleast one of them will not qualify)
$= 1 - P$ (both of them will qualify)
$= 1 - 0.02 = 0.98$

III. The event only one of them will qualify the examination is same as the event either (Anil will qualify and Ashima will not qualify) or (Anil will not qualify and Ashima will qualify) *i.e.*, $E \cap F'$ or $E' \cap F$, where $E \cap F'$ and $E' \cap F$ are mutually exclusive.
Therefore, P(only one of them will qualify)

$$= P(E \cap F' \text{ or } E' \cap F)$$
$$= P(E \cap F') + P(E' \cap F)$$
$$= P(E) - P(E \cap F) + P(F) - P(E \cap F)$$
$$= 0.05 - 0.02 + 0.10 - 0.02 = 0.11$$

77. *(b)* The total number of persons = 2 + 2 = 4. Out of these four persons, two can be selected in $^4C_2$ ways.

I. No men in the committee of two means there will be two women in the committee. Out of two women, two can be selected in $^2C_2 = 1$ way.

Therefore, $P(\text{no man}) = \dfrac{^2C_2}{^4C_2}$

$$= \frac{1 \times 2 \times 1}{4 \times 3} = \frac{1}{6}$$

II. One man in the committee means that there is one woman. One man out of 2 can be selected in $^2C_1$ ways and one woman out of 2 can be selected in $^2C_1$ ways. Together they be selected in $^2C_1 \times {}^2C_1$ ways.

Therefore, $P(\text{one man}) = \dfrac{^2C_1 \times {}^2C_1}{^4C_2} = \dfrac{2 \times 2}{2 \times 3} = \dfrac{2}{3}$

III. Two men can be selected in $^2C_2$ way.

Hence, $P(\text{two men}) = \dfrac{^2C_2}{^4C_2} = \dfrac{1}{^4C_2} = \dfrac{1}{6}$

78. *(a)* (i) Sum of all probabilities
$= 0.1 + 0.01 + 0.05 + 0.03 + 0.01 + 0.2 + 0.6 = 1$

Hence, the assignment is valid as sum is 1.

(ii) Valid, as sum is 1 (same as part (i).

(iii) Sum of all probabilities
$= 0.1 + 0.2 + 0.3 + 0.4 + 0.5 + 0.6 + 0.7 = 2.8$

Hence, assignment is not valid as sum is $2.8 > 1$.

(iv) Not valid, as probabilities of any event can't be negative.

(v) Not valid, as $\dfrac{15}{14} > 1$ which is not possible.

Note *Probability of any event cannot be negative or cannot be greater than unity.*

79. *(d)* The sample space is $S = \{HH, HT, TH, TT\}$

Let E be the event of getting atleast one tail

$E = \{HT, TH, TT\}$

$\therefore$ Required probability P

$$= \frac{\text{Number of favourable outcomes}}{\text{Total number of outcomes}}$$

$$= \frac{n(E)}{n(S)} = \frac{3}{4}$$

80. *(b)* When a die is thrown, then any number from 1 to 6 can be appear, so sample space is $S = \{1, 2, 3, 4, 5, 6\}$

A. Let E be the event, a prime number will appear.

$\therefore \quad E = \{2, 3, 5\}$

$\therefore$ Probability of event getting prime number

$$= \frac{\text{Number of favourable outcomes}}{\text{Total number of outcomes}}$$

$$= \frac{n(E)}{n(S)} = \frac{3}{6} = \frac{1}{2}$$

B. Let E be the event, a number greater than or equal to 3 will appear.

$\therefore \quad E = \{3, 4, 5, 6\}$

$\therefore$ Probability of event getting a number greater than or equal to 3.

$$P = \frac{n(E)}{n(S)} = \frac{4}{6} = \frac{2}{3}$$

C. Let E be the event a number less than or equal to one will appear.

$\therefore \quad E = \{1\}$

$\therefore$ Probability of event getting a number less than or equal to 1.

$$P = \frac{n(E)}{n(S)} = \frac{1}{6}$$

D. Let E be the event, a number more than 6 will appear.

$E = \{\} = \phi$

($\because$ on the die there is no number greater than 6)

$\therefore$ Probability of event getting a number greater than 6.

$$P = \frac{n(E)}{n(S)} = \frac{0}{6} = 0$$

E. Let E be the event, the number less than 6 will appear.

$E = \{1, 2, 3, 4, 5\}$

$\therefore$ Probability of event getting a number less than 5.

$$P = \frac{n(E)}{n(S)} = \frac{5}{6}$$

81. *(d)* A. There are 52 cards.Hence, total elements in sample space = 52

B. There is one ace of spade

So, $\quad n(E) = 1$

$\therefore$ Probability of event getting an ace of spade

$$= \frac{n(E)}{n(S)} = \frac{1}{52}$$

C. There are four aces in a pack of cards.

$\Rightarrow \quad n(E) = 4$

$\therefore$ Probability of event getting an ace

$$P = \frac{n(E)}{n(S)} = \frac{4}{52} = \frac{1}{13}$$

D. There are 26 black cards in a pack of cards.

$\Rightarrow \quad n(E) = 26$

$\therefore$ Probabilities of event getting a black card

$$P = \frac{n(E)}{n(S)} = \frac{26}{52} = \frac{1}{2}$$

82. *(d)* I.

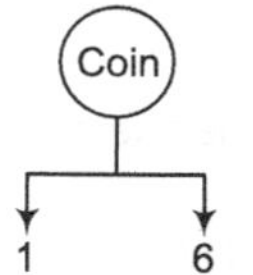

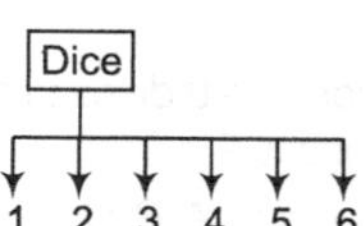

Total number of possible cases in coin = 2

Number of possible cases in dice = 6

(i) The sum is 3 only, if 1 turns up on coin and 2 turns up on dice *i.e.,* only possibility is (1, 2).

∴ Required probability

$$= \frac{\text{Number of favourable outcomes}}{\text{Total number of outcomes}}$$

$$= \frac{1}{2 \times 6} = \frac{1}{12}$$

(ii) The sum is 12 only if 6 turns up on coin and 6 turns up on dice *i.e.,* only possibility is (6, 6).

∴ Required probability $= \frac{1}{2 \times 6} = \frac{1}{12}$

II. Let E be the event that woman is selected.

∴ $n(E) = 6$

Number of elements in sample space $n(S)$ = Total number of persons in the council

$= 4 + 6 = 10$

∴ $P = \frac{n(E)}{n(S)} = \frac{6}{10} = \frac{3}{5}$

83. *(b)* I. If a coin is tossed four times, then total number of possible outcomes

$= 2^4 = 16$

For these cases, sample space can be written as

Sample space	Amount
HHHH	1 + 1 + 1 + 1 = 4
HH HT	1 + 1 + 1 − 1.50 = 3 − 1.50 = 1.50
HH TH	1 + 1 − 1.50 + 1 = 3 − 1.50 = 1.50
HHTT	1 + 1 − 1.50 − 1.50 = 2 − 3 = − 1.00
HTHH	1 − 1.50 + 1 + 1 = 3 − 1.50 = 1.50
HTHT	1 − 1.50 + 1 − 1.50 = 2 − 3 = − 1.00
HTTH	1 − 1.50 − 1.50 + 1 = 2 − 3 = − 1.00
HTTT	1 − 1.50 − 1.50 − 1.50 = 1 − 4.50 = − 3.50
THHH	− 1.50 + 1 + 1 + 1 = − 1.50 + 3 = 1.50
THHT	− 1.50 + 1 + 1 − 1.50 = 2 − 3.00 = − 1.00
THTH	− 1.50 + 1 − 1.50 + 1 = 2 − 3.00 = − 1.00
THTT	− 1.50 + 1 − 1.50 − 1.50 = 1 − 4.50 = − 3.50
TTHH	− 1.50 − 1.50 + 1 + 1 = 2 − 3.00 = − 1.00
TTHT	− 1.50 − 1.50 + 1 − 1.50 = 1 − 4.50 = − 3.50
TTTH	− 1.50 − 1.50 − 1.50 + 1 = − 4.50 + 1 = − 3.50
TTTT	− 1.50 − 1.50 − 1.50 − 1.50 = − 6.00

Hence, from above sample space, we get five types of different amounts

i.e., 4, 1.50, −1.00, −3.50, −6.00 *i.e.,*

Amounts	Number of times occurrence
4.00	1
1.50	4
− 1.00	6
− 3.50	4
− 6.00	1
Total	16

(negative sign indicates the losing value of amount)

⇒ P (winning ₹ 4.00)

$$= \frac{\text{Number of favourable outcomes}}{\text{Total number of outcomes}} = \frac{1}{16}$$

⇒ P (having ₹ 1.50) $= \frac{4}{16} = \frac{1}{4}$

⇒ P (having ₹ − 1.00) $= \frac{6}{16} = \frac{3}{8}$

⇒ P (having ₹ − 3.50) $= \frac{4}{16} = \frac{1}{4}$

⇒ P (having ₹ − 6.00) $= \frac{1}{16}$

II. If A be any event, then given

$$P(A) = \frac{2}{11}$$

∴ P (not A) $= P(A') = 1 - P(A)$

$$= 1 - \frac{2}{11} = \frac{9}{11}$$

[∵ sum of probabilities of an event (A) and its complementary event (A') is always 1]

84. *(c)* If three coins are tossed, then total number of possible outcomes $= 2^3 = 8$ which are

S = {HHH, HHT, HTH, HTT, THH, THT, TTH, TTT}

A. If E be event of getting 3 heads.

⇒ $n(E) = 1$

⇒ Probability of getting 3 heads

$$P = \frac{\text{Number favourable outcomes}}{\text{Total number of outcomes}}$$

$$= \frac{n(E)}{n(S)} = \frac{1}{8}$$

B. If E be event of getting 2 heads.

⇒ $n(E) = 3$

⇒ Probability of getting 2 heads $P = \frac{n(E)}{n(S)} = \frac{3}{8}$

C. If E be the event of getting atleast 2 heads.

(atleast means we can consider the cases having 2 heads and higher than 2)

⇒ $n(E)$ = 2 heads + 3 heads = 3 + 1 = 4

⇒ Probability of getting atleast 2 heads

$$P = \frac{n(E)}{n(S)} = \frac{4}{8} = \frac{1}{2}$$

D. If E be the event of getting atmost 2 heads.

⇒ $n(E)$ = 2 heads + 1 head + 0 head

(atmost means we cannot consider the cases of having heads higher than 2) = 3 + 3 + 1 = 7

⇒ Probability of getting atmost 2 heads

$$P = \frac{n(E)}{n(S)} = \frac{7}{8}$$

E. If E be the event of getting no head.

⇒ $n(E) = 1$

⇒ Probability of getting no head

$$P = \frac{n(E)}{n(S)} = \frac{1}{8}$$

85. *(d)* (i) If E be the event of getting 3 tails.

$\Rightarrow \quad n(E) = 1 \Rightarrow P = \frac{n(E)}{n(S)} = \frac{1}{8}$

(ii) If E be the event of getting exactly 2 tails.

$\Rightarrow \quad n(E) = 3$

$\Rightarrow$ Probability of getting 2 tails $P = \frac{n(E)}{n(S)} = \frac{3}{8}$

(iii) If E be the event of getting no tail.

$\Rightarrow \quad n(E) = 1$

$\Rightarrow$ Probability of getting no tails $P = \frac{n(E)}{n(S)} = \frac{1}{8}$

(iv) If E be the event of getting atmost 2 tails.

$\Rightarrow \quad n(E) = 2$ tails $+ 1$ tail $+ 0$ tail $= 3 + 3 + 1 = 7$

$\Rightarrow$ Probability of getting atmost 2 tails

$$P = \frac{n(E)}{n(S)} = \frac{7}{8}$$

86. *(b)* I. Given word is A S S A S S I N A T I O N

Number of vowels $= 3(A) + 2(I) + 1(O) = 6$

Number of consonant $= 7$

Total number of letters $= 13$

(i) P (vowels) $= \frac{6}{13}$ (ii) P (consonant) $= \frac{7}{13}$

II. 6 numbers out of 20 numbers can be chosen in $^{20}C_6$ ways in which only one combination of numbers is correct.

Let the probability of winning the prize be E.

i.e., $n(E) = 1$, ($\because$ only one prize can be won)

Number of elements in sample space $n(S) = {}^{20}C_6$

$\therefore$ Required probability $= \frac{1}{^{20}C_6}$

$$= \frac{1}{20!/(20-6)!\,6!} \qquad \left(\because {}^nC_r = \frac{n!}{(n-r)!\,r!}\right)$$

$$= \frac{6 \times 5 \times 4 \times 3 \times 2 \times 1}{20 \times 19 \times 18 \times 17 \times 16 \times 15} = \frac{1}{38760}$$

87. *(b)* (i) Here, $P(A \cap B) \not\leq P(A)$

So, given data is not consistent.

(ii) Here, $P(A) = 0.5, P(B) = 0.4, P(A \cup B) = 0.8$

$\Rightarrow \quad P(A \cup B) = P(A) + P(B) - P(A \cap B)$

$\Rightarrow \quad 0.8 = 0.5 + 0.4 - P(A \cap B)$

$\Rightarrow \quad P(A \cap B) = 0.9 - 0.8 = 0.1 < P(A)$ and $P(B)$.

$\Rightarrow \quad P(A)$ and $P(B)$ are consistent.

88. *(a)* I. $\because P(A \cup B) = P(A) + P(B) - P(A \cap B)$

$$P(A \cup B) = \frac{1}{3} + \frac{1}{5} - \frac{1}{15}$$

$$= \frac{5 + 3 - 1}{15} = \frac{8 - 1}{15} = \frac{7}{15}$$

II. $0.6 = 0.35 + P(B) - 0.25$

$(\because P(A \cup B) = P(A) + P(B) - P(A \cap B))$

$\Rightarrow \quad 0.6 = 0.10 + P(B)$

$\Rightarrow \quad P(B) = 0.6 - 0.10 = 0.5$

III. $0.7 = 0.5 + 0.35 - P(A \cap B)$

$(\because P(A \cup B) = P(A) + P(B) - P(A \cap B))$

$0.7 = 0.85 - P(A \cap B)$

$\Rightarrow \quad P(A \cap B) = 0.85 - 0.7 = 0.15$

89. *(b)* I. $\because A$ and B are mutually exclusive events

$\Rightarrow \quad A \cap B = \phi \Rightarrow P(A \cap B) = 0$

$\therefore \quad P(A \text{ or } B) = P(A \cup B) = P(A) + P(B) - P(A \cap B)$

$$= \frac{3}{5} + \frac{1}{5} - 0 = \frac{4}{5}$$

II. Given, $P(E) = \frac{1}{4}, P(F) = \frac{1}{2}$,

$P(E \text{ and } F) = \frac{1}{8} = P(E \cap F)$

(i) $P(E \text{ or } F) = P(E \cup F)$

$= P(E) + P(F) - P(E \cap F)$

($\because A$ or B means $A \cup B$)

$$= \frac{1}{4} + \frac{1}{2} - \frac{1}{8} = \frac{2 + 4 - 1}{8} = \frac{5}{8}$$

(ii) $P(\text{not } E \text{ and not } F) = P(E' \cap F')$

$= P(E \cup F)' = 1 - P(E \cup F)$

[$\because$ Demorgan's law $E' \cap F' = (E \cup F)'$]

$$= 1 - \frac{5}{8} = \frac{3}{8}$$

90. *(b)* I. Given $P(\text{not } E \text{ or not } F) = 0.25$

$\Rightarrow \quad P(E' \cup F') = P(E \cap F)' = 0.25$

($\because$ Demorgan's law $E' \cup F' = (E \cap F)'$)

$\because \quad P(E \cap F)' = 1 - P(E \cap F)$

$\Rightarrow \quad P(E \cap F) = 1 - P(E \cap F)'$

$\therefore \quad P(E \cap F) = 1 - 0.25 = 0.75 \neq 0$

Hence, E and F are not mutually exclusive.

II. Given, $P(A) = 0.42, P(B) = 0.48$

$P(A \text{ and } B) = P(A \cap B) = 0.16$

(i) $P(\text{not } A) = P(A') = 1 - P(A)$

(sum of an event and its complementary event is always 1)

$\Rightarrow \quad P(A') = 1 - 0.42 = 0.58$

(ii) $P(\text{not } B) = P(B') = 1 - P(B)$ $\quad [\because P(B) + P(B') = 1]$

$= 1 - 0.48 = 0.52$

(iii) $P(A \text{ or } B) = P(A \cup B)$

$= P(A) + P(B) - P(A \cap B)$

$= 0.42 + 0.48 - 0.16$

$= 0.90 - 0.16 = 0.74$

91. *(d)* Let M and B denote the students of Mathematics and Biology.

Given, $P(M) = 40\%, P(B) = 30\%$

$$P(M \cap B) = 10\%$$

($\because$ 10% of the class study both subjects *i.e.*, they are common in both subjects)

$\therefore P(M \text{ or } B) = P(M \cup B)$

$$= P(M) + P(B) - P(M \cap B)$$

$$= \frac{40}{100} + \frac{30}{100} - \frac{10}{100} = \frac{60}{100} = 60\% = 0.6$$

92. *(b)* I. Here, $P(\text{I}) = 0.8, P(\text{II}) = 0.7$

$P(\text{atleast one of I and II}) = P(\text{I} \cup \text{II}) = 0.95$

Now, $P(\text{I} \cup \text{II}) = P(\text{I}) + P(\text{II}) - P(\text{I} \cap \text{II})$

$\therefore \quad 0.95 = 0.8 + 0.7 - P(\text{I} \cap \text{II})$

$\Rightarrow \quad 0.95 = 1.5 - P(\text{I} \cap \text{II})$

$\Rightarrow \quad P(\text{I} \cap \text{II}) = 1.5 - 0.95 = 0.55$

II. Here, $P(H \cap E) = 0.5$ (given)

$P(H' \cap E') = P(H \cup E)' = 0.1$

(the probability of passing in both English and Hindi is $P(H \cap E)$, then probability of passing neither will be $P(H \cup E)'$ because it is complementary to 1st event)

$\therefore \quad P(H \cup E) = 1 - P(H \cup E)'$

$$= 1 - 0.1 = 0.9$$

Also, $P(E) = 0.75$

$\Rightarrow \quad P(H \cup E) = P(H) + P(E) - P(H \cap E)$

$\Rightarrow \quad 0.9 = P(H) + 0.75 - 0.5$

$\Rightarrow \quad 0.9 = P(H) + 0.25$

$\Rightarrow \quad P(H) = 0.9 - 0.25 = 0.65$

93. *(d)* Let A and B denote the students in NCC and NSS respectively.

Here, $n(A) = 30, n(B) = 32$

$$n(A \cap B) = 24$$

($\because$ 24 students opted for both NCC and NSS *i.e.*, they are common in both)

$\therefore \quad P(A) = \frac{30}{60}$

$P(B) = \frac{32}{60}$ and $P(A \cap B) = \frac{24}{60}$

A. P (student opted for NCC or NSS)

$P(A \cup B) = P(A) + P(B) - P(A \cap B)$

$[\because P(A \cup B) = P(A \text{ or } B)]$

$$= \frac{30}{60} + \frac{32}{60} - \frac{24}{60}$$

$$= \frac{30 + 32 - 24}{60} = \frac{62 - 24}{60} = \frac{38}{60} = \frac{19}{30}$$

B. P (student has opted neither NCC nor NSS)

$= 1 - P$ (student opted for NCC or NSS)

$$= 1 - \frac{19}{30} = \frac{30 - 19}{30} = \frac{11}{30}$$

C. P (the student has opted NSS but not NCC)

$$= P(B) - P(A \cap B)$$

$$= \frac{32}{60} - \frac{24}{60} = \frac{8}{60} = \frac{2}{15}$$

94. *(b)* 2-digit positive integers are 10, 11, 12, ..., 99. Thus, there are 90 such numbers. Since, out of these, 30 numbers are multiple of 3, therefore, the probability that a randomly chosen positive 2-digit integer is a multiple of 3, is $\frac{30}{90} = \frac{1}{3}$.

95. *(b)* A PIN is a sequence of four symbols selected from 36 (26 letters + 10 digits) symbols.

By the fundamental principle of counting there are $36 \times 36 \times 36 \times 36 = 36^4 = 1679616$ PINs in all. When repetition is not allowed the multiplication rule can be applied to conclude that there are

$36 \times 35 \times 34 \times 33 = 1413720$ different PINs.

The number of PINs that contain atleast one repeated symbol

$$= 1679616 - 1413720 = 265896$$

Thus, the probability that a randomly chosen PIN contains a repeated symbol is

$$\frac{265896}{1679616} = .1583$$

96. *(d)* (i) Since, $P(D) = -0.20$, this is not possible as $0 \le P(A) \le 1$ for any event A.

(ii) $P(S) = P(A \cup B \cup C \cup D)$

$$= \frac{9}{120} + \frac{45}{120} + \frac{27}{120} + \frac{46}{120} = \frac{127}{120} \ne 1.$$

This violates the condition that $P(S) = 1$

97. *(b)* Given that, M and N are two events, then the probability that atleast one of them occurs is

$$P(M \cup N) = P(M) + P(N) - P(M \cap N)$$

98. *(a)* Given that, A and B are two mutually exclusive events Then,

$$P(A \cup B) = P(A) + P(B) \quad [\because (A \cap B) = \phi]$$

$$P(A) + P(B) \le 1$$

$$P(A) + 1 - P(\bar{B}) \le 1$$

$$P(A) \le P(\bar{B})$$

99. *(a)* Given that, $P(A \cup B) = P(A \cap B)$

$\Rightarrow \quad A = B \Rightarrow P(A) = P(B)$

100. *(b)* I. Given word is ALGORITHM

Total number of letters = 9

These 9 letters can be arranged in 9! ways

Taking GOR as one letter, there are $(6 + 1) = 7$ letters

These 7 letters can be arranged in 7! ways

Then, required probability $= \frac{7!}{9!} = \frac{7!}{9 \times 8 \times 7!} = \frac{1}{72}$

II. Let the couple occupies adjacent desks. Taking these two as one, therefore total $(4+1)=5$ to be assigned now, the probability that the couple has adjacent desk is P

$$\therefore \quad P=\frac{5!2!}{6!}=\frac{5!2!}{6\cdot 5!}=\frac{2\times 1}{6}=\frac{1}{3}$$

The probability that the couple has non-adjacent desks $=1-\frac{1}{3}=\frac{2}{3}$

101. *(d)* Given that, probability of each odd number is twice as likely to occur as each even number *i.e.*,

$$P(O)=2P(E)$$

When a die is rolled, then the sample is S

$$S=\{1, 2, 3, 4, 5, 6\}$$

Now, $\quad P(O)=\frac{4}{6}=\frac{2}{3}$

i.e., $\quad P(E)=\frac{1}{2}\left(\frac{2}{3}\right)=\frac{2}{6}=\frac{1}{3}$

Now, G is the event greater than 3

$\therefore$ Number are 4, 5, 6

$$P(G)=2\cdot P(E)\cdot P(O)$$
$$=2\left(\frac{1}{3}\right)\left(\frac{2}{3}\right)=\frac{4}{9}$$

102. *(d)* A. $\quad P(A)=0.35$

$$P(A')=1-0.35=0.65$$
$$P(B)=0.45$$
$$P(B')=1-0.45=0.55$$

Now $\quad P(A')+P(B')=0.65+1.20=1.20$

B. $P(A\cup B)=P(A)+P(B)=0.35+0.45=0.8$

C. $P(A\cap B)=0$

D. $P(A\cap B')=P(A)=0.35$

E. $P(A'\cap B')=1-P(A\cup B)=1-0.8=0.2$

103. *(b)* We know that, in a non-leap year there are 365 days. Which have 52 weeks and 1 day. If this day is a Tuesday or Wednesday, then the year will have 53 Tuesdays or 53 Wednesdays.

$\therefore \quad$ Required probability $=\frac{2}{7}$

104. *(b)* Three numbers chosen from

1 to 20 $={}^{20}C_3=\frac{20\times 19\times 18\times 17!}{17!\times 3\times 2\times 1}=20\times 19\times 3=1140$

Now set of three consecutive numbers = 18

P(they are consecutive) $=\frac{18}{1140}=\frac{3}{190}$

P(they are not consecutive) $=1-\frac{3}{190}=\frac{187}{190}$

105. *(c)* There are 26 red cards and 26 black cards

i.e., total number of cards = 52

P(both cards of different colours)

$$=P(B)\,P(R)+P(R)\,P(B)$$
$$=\frac{26}{52}\times\frac{26}{51}+\frac{26}{52}\times\frac{26}{51}=2\left(\frac{26}{52}\times\frac{26}{51}\right)=\frac{26}{51}$$

106. *(c)* Given, number of persons = 7

Now, treating the two persons as one, we have 6 persons.

Total number of sitting arrangements = 7!

Favourable number of arrangements = 6!

2 persons can be arranged in two ways.

Total number of favourable arrangments = 2(6!)

$\therefore$ Required probability $=\frac{2(6!)}{7!}=\frac{2}{7}$

107. *(a)* A. Probability = 0.95

That means it is very likely to happen.

B. Probability = 0.02

That mean it is very little chance of happening.

C. Probability = − 0.3

We know that, $0\le P(E)\le 1$

So, it is an incorrect assignment.

D. Probability = 0.5

That means as much chance of happening as not

E. Probability = 0

That means no chance of happening.

108. *(c)* Given that, number of nails = 6

Number of nuts = 10

Total number of items = 6 + 10 = 16

and half of them are rusted

$\therefore$ Total rusted items = 3 + 5 = 8

Unrusted nails = 3

P(the items is rusted or nail) $=\frac{3+8}{16}=\frac{11}{16}$

109. *(c)* Total number of tickets = 10000

Now, number of tickets numbered such that it is divisible by 20

$$=\frac{10000}{20}=500$$

Required probability $=\frac{500}{10000}=\frac{1}{20}$

110. *(d)* Given numbers = 0, 2, 3, 5

Th	H	T	U
3	3	2	1

A number is divisible by 5, if the digit at unit place is 0 or 5

Total numbers are $=3\times 3\times 2\times 1=18$

Favourable events are

when 0 is at unit place $=3\times 2\times 1=6$

when 5 is at unit place $=2\times 2\times 1\times 1=4$

Total favourable events are 6 + 4 = 10

$\therefore \quad$ Required probability $=\frac{10}{18}=\frac{5}{9}$

111. *(a)* Total number of possible hands = ${}^{52}C_7$

(i) Number of hands with 4 kings = ${}^{4}C_4 \times {}^{48}C_3$ (other 3 cards must be chosen from the rest 48 cards)

Hence, P(a hand will have 4 kings)

$$= \frac{{}^{4}C_4 \times {}^{48}C_3}{{}^{52}C_7} = \frac{1}{7735}$$

(ii) Number of hands with 3 kings and 4 non-king cards $= {}^{4}C_3 \times {}^{48}C_4$

Therefore, $P(3 \text{ king}) = \frac{{}^{4}C_3 \times {}^{48}C_4}{{}^{52}C_7} = \frac{9}{1547}$

(iii) $P(\text{atleast 3 king}) = P(3 \text{ kings or } 4 \text{ kings})$

$$= P(3 \text{ kings}) + P(4 \text{ kings})$$

$$= \frac{9}{1547} + \frac{1}{7735}$$

$$= \frac{46}{7735}$$

112. *(a)* The number of arrangement (orders) in which Veena can visit four cities A, B, C or D is 4!, *i.e.*, 24. Therefore, $n(S) = 24$.

Since, the number of elements in the sample space of the experiment is 24 all of these outcomes are considered to the equally likely. A sample space for the experiment is

S = {$ABCD$, $ABDC$, $ACBD$, $ACDB$, $ADBC$, $ADCB$, $BACD$, $BADC$, $BDAC$, $BDCA$, $BCAD$, $BCDA$, $CABD$, $CADB$, $CBDA$, $CBAD$, $CDAB$, $CDBA$, $DABC$, $DACB$, $DBCA$, $DBAC$, $DCAB$, $DCBA$}

A. Let the event 'she visits A before B' be denoted by E

Therefore, E = {$ABCD$, $CABD$, $DABC$, $ABDC$, $CADB$, $DACB$, $ACBD$, $ACDB$, $ADBC$, $CDAB$, $DCAB$, $ADCB$}

Thus, $P(E) = \frac{n(F)}{n(S)} = \frac{12}{24} = \frac{1}{2}$

B. Let the event 'Veena visits A before B and B before C' be denoted by F.

Here, F = {$ABCD$, $DABC$, $ABDC$, $ADBC$}

Therefore, $P(F) = \frac{n(F)}{n(S)} = \frac{4}{24} = \frac{1}{6}$

C. Veena can visit city A first and city B last in any one of the following ways

$ACDB$, $ADCB$

$\therefore$ Required probability $= \frac{2}{24} = \frac{1}{12}$

D. Veena can visit city A first in 3! ways or city A second in 3! ways.

$\therefore$ Number of ways in which Veena can visity city A either first or second = 3! + 3! = 12

$\therefore$ Required probability $= \frac{12}{24} = \frac{1}{2}$

E. Taking AB together A, B, C, D can be arranged in 3! ways.

So, required probability $= \frac{3!}{24} = \frac{1}{4}$

113. *(c)* Consider $E = B \cup C$, so that

$$P(A \cup B \cup C) = P(A \cup E)$$
$$= P(A) + P(E) - P(A \cap E) \quad \ldots\text{(i)}$$

Now, $P(E) = P(B \cup C)$
$$= P(B) + P(C) - P(B \cap C) \quad \ldots\text{(ii)}$$

Also, $A \cap E = A \cap (B \cup C) = (A \cap B) \cup (A \cap C)$

[using distribution property of intersection of sets over the union].

Thus,

$$P(A \cap E) = P(A \cap B) + P(A \cap C) - P[(A \cap B) \cap (A \cap C)]$$
$$= P(A \cap B) + P(A \cap C) - P[A \cap B \cap C] \quad \ldots\text{(iii)}$$

Using Eqs. (ii) and (iii) in Eq. (i), we get

$$P[A \cup B \cup C] = P(A) + P(B) + P(C) - P(B \cap C) - (A \cap B) - P(A \cap C) + P(A \cap B \cap C)$$

114. *(a)* If we consider the sample space consisting of all finishing orders in the first three places, we will have ${}^{5}P_3$, *i.e.*, $\frac{5!}{(5-3)!} = 5 \times 4 \times 3 = 60$ sample points, each with a probability of $\frac{1}{60}$.

A. A, B and C finish first, second and third, respectively. There is only one finishing order for this, *i.e.*, ABC.

Thus, P(A, B and C finish first, second and third respectively) $= \frac{1}{60}$

B. A, B and C are the first three finishers. There will be 3! arrangements for A, B and C. Therefore, the sample points corresponding to this event will be 3! in number.

So, P(A, B and C are first three to finish)

$$= \frac{3!}{60} = \frac{6}{60} = \frac{1}{10}$$

115. *(d)*

10 Red marbles
20 Blue marbles
30 Green marbles

Total number of ways selecting 5 marbles out of 60 marbles = ${}^{60}C_5$

I. If E be the event all marbles will be blue, then

$$n(E) = {}^{20}C_5$$

$\therefore$ Required probability $= \frac{{}^{20}C_5}{{}^{60}C_5}$

II. P (atleast one will be green) = 1 − P(no green)

$$= 1 - \frac{{}^{30}C_5}{{}^{60}C_5}$$

116. *(b)*

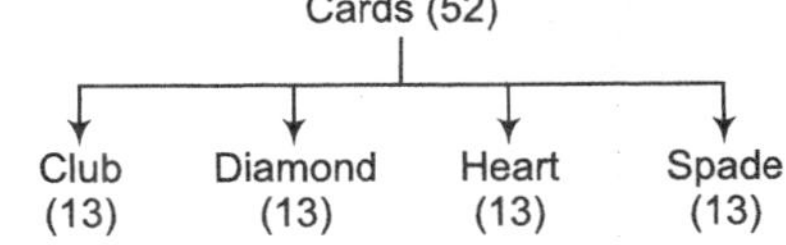

Total number of ways selecting 4 cards out of 52 cards

$= {}^{52}C_4$

If E be the event obtaining 3 diamonds and 1 spade, then

$$n(E) = {}^{13}C_3 \times {}^{13}C_1$$

$\therefore$ Required probability $= \dfrac{{}^{13}C_3 \times {}^{13}C_1}{{}^{52}C_4}$

117. *(b)* Since, repetition is not allowed.

So, first place can be filled in 10 ways

Second place can be filled in 9 ways

Third place can be filled in 8 ways

and fourth place can be filled in 7 ways.

Hence, by fundamental principle of counting, total number of ways

$$= 10 \times 9 \times 8 \times 7 = 5040$$

In which only 1 sequence can open the lock.

$\therefore$ Number of favourable cases = 1

$\therefore$ Required probability $= \dfrac{1}{5040}$

118. *(c)* **Case I** When digits are repeated.

Total four digit numbers which are greater than 5000 from the digits 0, 1, 3, 5,7 *i.e.*, In thousand place, the number will be 5 or 7 and rest of three places any five numbers will come

$\boxed{5}\ \underline{5} \times \underline{5} \times \underline{5} = 125$

$\boxed{7}\ \underline{5} \times \underline{5} \times \underline{5} = 125$

$\therefore$ Total four digit number, $n(S) = 125 + 125 - 1$

$= 250 - 1 = 249$ ($\because$ 5000 is excluded)

Total four digit numbers which are divisible by 5.

i.e., In unit place number will be 0 or 5 and thousand place the number will be 5 to 7 and rest of the two places any five number will be come.

$\boxed{5}\ \underline{5} \times \underline{5}\ \boxed{0} = 25$

$\boxed{7}\ \underline{5} \times \underline{5}\ \boxed{0} = 25$

$\boxed{5}\ \underline{5} \times \underline{5}\ \boxed{5} = 25$

$\boxed{7}\ \underline{5} \times \underline{5}\ \boxed{5} = 25$

Number of favourable cases, $n(E)$

$= 25 + 25 + 25 + 25$

$= 100 - 1 = 99$ ($\because$ 5000 is excluded)

$\therefore$ Required probability $= \dfrac{n(E)}{n(S)} = \dfrac{99}{249} = \dfrac{33}{83}$

Case II When digits are not repeated.

Total four digit numbers which are greater than 5000 from the digits 0, 1, 3, 7, 5 *i.e.*, in thousand place, the number will be 5 or 7. Since, the digit are not repeated, then rest of the place may be filled by 4, 3, 2.

$5\ \ \underline{4} \times \underline{3} \times \underline{2} = 24$

$7\ \ \underline{4} \times \underline{3} \times \underline{2} = 24$

Total four digit number $n(S) = 24 + 24 = 48$

Total four digit numbers which are divisible by 5

$\boxed{5}\ \underline{3} \times \underline{2}\ \boxed{0} = 6$

$\boxed{7}\ \underline{3} \times \underline{2}\ \boxed{0} = 6$

$\boxed{7}\ \underline{3} \times \underline{2}\ \boxed{5} = 6$

Favourable number of cases $n(E) = 6 + 6 + 6 = 18$

$\therefore$ Required probability $= \dfrac{18}{48} = \dfrac{3}{8}$

119. *(b)* A. If E be the event getting a number 2, then $n(E) = 3$

($\because$ there are three faces each with number 2)

$\therefore$ Probability of getting 2

$$= \frac{\text{Number of favourable outcomes}}{\text{Total number of outcomes}}$$

$$= \frac{n(E)}{n(S)} = \frac{3}{6} = \frac{1}{2}$$

(in case of die total number of outcomes is 6.)

B. If E be the event getting a number 1 or 3, then $n(E) = 2 + 1 = 3$

$\therefore$ Probability of getting 1 or 3. $= \dfrac{n(E)}{n(S)} = \dfrac{3}{6} = \dfrac{1}{2}$

($\because$ there are two faces each with number 1 and one face with number 2)

C. If E be the event getting a number not 3, then $n(E) = 2 + 3 = 5$

(not 3 means there will be the number on die other than 3 *i.e.*, 1 and 2. There are two faces with number 1 and three faces with number 2.)

$\therefore$ Probability of getting (not 3) $P = \dfrac{n(E)}{n(S)} = \dfrac{5}{6}$

120. *(a)*

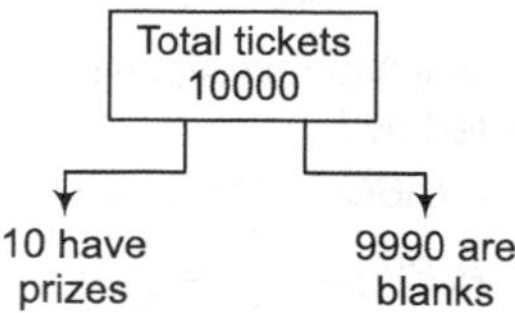

(i) 1 ticket out of 10000 ticket can be selected in $= {}^{10000}C_1$ ways.

Number of ways in which 1 ticket without prize is obtained $= {}^{9990}C_1$

$\therefore$ Required probability P

$$= \frac{\text{Number of favourable outcomes}}{\text{Total number of outcomes}}$$

$$= \frac{{}^{9990}C_1}{{}^{10000}C_1} = \frac{9990}{10000} \qquad (\because {}^{n}C_1 = n)$$

$$= \frac{999}{1000}$$

(ii) Number of ways in which 2 tickets can be selected out of 10000 tickets $= {}^{10000}C_2$

Number of ways in which 2 tickets without prizes can be selected out of 9990 tickets

$$= {}^{9990}C_2$$

$\therefore$ Required probability $= \dfrac{{}^{9990}C_2}{{}^{10000}C_2}$

(iii) Number of ways in which 10 tickets can be selected out of 10000 tickets $= {}^{10000}C_{10}$

Number of ways in which 10 tickets without prizes can be selected out of 9990 tickets $= {}^{9990}C_{10}$

$\therefore$ Required probability $= \dfrac{{}^{9990}C_{10}}{{}^{10000}C_{10}}$

121. *(b)* Let A be the event that selected person is male or B be the event that selected person is over 35 yr.

$$\therefore \quad P(A) = \frac{{}^3C_1}{{}^5C_1} = \frac{3}{5}$$

$$P(B) = \frac{{}^2C_1}{{}^5C_1} = \frac{2}{5}$$

$$\therefore \quad P(A \cup B) = P(A) + P(B) - P(A \cap B)$$
$$= \frac{3}{5} + \frac{2}{5} - \frac{1}{5} = \frac{5}{5} - \frac{1}{5} = \frac{4}{5}$$

122. *(b)* Let the two sections be A and B having 40 and 60 students respectively.

I. When both enter the same section

There are two cases,

Case I Either both are in section A or

Case II Either both are in section B.

I. If both are in section A, then 40 students out of 100 students can be selected $= {}^{100}C_{40}$ ways

and 38 students can be selected out of 98 students

$$= {}^{98}C_{38} \text{ ways}$$

$\therefore$ Probability $P = \dfrac{{}^{98}C_{38}}{{}^{100}C_{40}}$

$$= \frac{98!}{38!\,60!} \times \frac{40!\,60!}{100!} \quad \left(\because {}^nC_r = \frac{n!}{r!(n-r)!}\right)$$

$$= \frac{98! \times 40 \times 39 \times 38!}{38! \times 100 \times 99 \times 98!}$$

$$= \frac{40 \times 39}{100 \times 99} = \frac{2}{5} \times \frac{13}{33} = \frac{26}{165} \quad \ldots\text{(i)}$$

II. If both are in section B. Then, 60 students out of 100 students can be selected $= {}^{100}C_{60}$ ways

58 students out of 98 students can be selected

$$= {}^{98}C_{58}$$

$\therefore$ Probability $= \dfrac{{}^{98}C_{58}}{{}^{100}C_{60}} = \dfrac{98!}{58!\,40!} \times \dfrac{60!\,40!}{100!}$

$$= \frac{98! \times 60 \times 59 \times 58! \times 40!}{58! \times 40! \times 100 \times 99 \times 98!}$$

$$= \frac{60 \times 59}{100 \times 99} = \frac{3}{5} \times \frac{59}{99} = \frac{59}{5 \times 33} = \frac{59}{165} \quad \ldots\text{(ii)}$$

Hence, required probability that students are either in section A or B

$$= \frac{26}{165} + \frac{59}{165} \quad \text{(from Eqs. (i) and (ii))}$$
$$= \frac{85}{165} = \frac{17}{33}$$

II. The probability that both enter the different section

$= 1 -$ Probability that they both enter the same section

$$= 1 - \frac{17}{33} = \frac{16}{33}$$

123. *(c)* Let the envelopes be denoted by E_1, E_2, E_3 and the corresponding letters by L_1, L_2, L_3. 1 letter in correct envelope and 2 in wrong envelope may be put as

$(E_1L_1, E_2L_3, E_3L_2), (E_1L_3, E_2L_2, E_3L_1)$ and (E_1L_2, E_2L_1, E_3L_3)

Two letters or consequently all the letters are in correct envelope may be put in one way

i.e., (E_1L_1, E_2L_2, E_3L_3)

$\therefore$ Number of cases $= 3! = 6$

Number of favourable cases $= 4$

Probability that atleast one letter is in its proper envelope

$$= \frac{4}{6} = \frac{2}{3}$$

124. *(a)* Using the relation,

A. $P(A \cup B) = P(A) + P(B) - P(A \cap B)$
$= 0.54 + 0.69 - 0.35$
$= 1.23 - 0.35 = 0.88$

B. $P(A' \cap B') = P(A \cup B)'$
$= 1 - P(A \cup B)$
$= 1 - 0.88 = 0.12$

C. $P(A \cap B') = P(A \text{ only})$
$= P(A) - P(A \cap B)$
$= 0.54 - 0.35 = 0.19$

D. $P(B \cap A') = P(B \text{ only})$
$= P(B) - P(B \cap A) = 0.69 - 0.35 = 0.34$

125. *(d)* Let B be the event that a truck stopped at the roadblock will have faulty brakes and T be the event that it will have badly worn tires. We have $P(B) = 0.23$, $P(T) = 0.24$ and $P(B \cup T) = 0.38$

and $\quad P(B \cup T) = P(B) + P(T) - P(B \cap T)$

So $\quad 0.38 = 0.23 + 0.24 - P(B \cap T)$

$\Rightarrow \quad P(B \cap T) = 0.23 + 0.24 - 0.38 = 0.09$

126. *(b)* Since, a leap year has 366 days and hence 52 weeks and 2 days. The 2 days can be SM, MT, TW, WTh, ThF, FSt, StS.

Therefore, $P(53 \text{ Sundays or } 53 \text{ Mondays}) = \dfrac{3}{7}$

127. *(d)* Since, a 3-digit number cannot start with digit 0, the hundredth place can have any of the 4 digits. Now, the tens and units place can have all the 5-digits. Therefore, the total possible 3-digit numbers are $4 \times 5 \times 5$ *i.e.*, 100.

The total possible 3-digit numbers having all digits same $= 4$

Hence, $P(\text{3-digit number with same digits}) = \frac{4}{100} = \frac{1}{25}$

128. *(c)* Total number of mappings from a set A having n elements onto itself is n^n.

Now, for one to one mapping the first element in A can have any of the n images in A; the 2nd element in A can have any of the remaining $(n-1)$ images, counting like this, the nth element in A can have only 1 image.

Therefore, the total number of one to one mappings is $n!$.

Hence, the required probability is $\frac{(n)!}{n^n} = \frac{n(n-1)!}{nn^{n-1}} = \frac{(n-1)!}{n^{n-1}}$

129. *(b)* Let $3P(A) = 2P(B) = P(C) = p$ which gives $P(A) = \frac{p}{3}$, $P(B) = \frac{p}{2}$ and $P(C) = p$

Now, since A, B, C are mutually exclusive and exhaustive events, we have

$$P(A) + P(B) + P(C) = 1$$

$$\Rightarrow \quad \frac{p}{3} + \frac{p}{2} + p = 1$$

$$\Rightarrow \quad p = \frac{6}{11}$$

Hence, $P(A) = \frac{p}{3} = \frac{2}{11}$

130. *(d)*

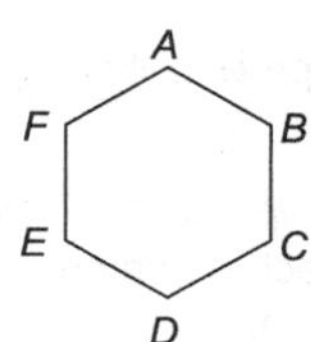

$ABCDEF$ is a regular hexagon. Total number of triangles ${}^6C_3 = 20$. (since, no three points are collinear). Of these only ΔACE; ΔBDF are equilateral triangles.

Therefore, required probability $= \frac{2}{20} = \frac{1}{10}$